Mesoscale Meteorological Modeling

Mesoscale Meteorological Modeling

Third Edition

Roger A. Pielke Sr.

Senior Research Scientist
Cooperative Institute for Research in the Atmosphere and
The Department of Atmospheric and Oceanic Sciences,
University of Colorado, Boulder, CO 80309, USA

And

Professor Emeritus
Department of Atmospheric Science
Colorado State University, Fort Collins
CO 80523, USA

AMSTERDAM • BOSTON • HEIDELBERG • LONDON
NEW YORK • OXFORD • PARIS • SAN DIEGO
SAN FRANCISCO • SINGAPORE • SYDNEY • TOKYO

Academic Press is an imprint of Elsevier

Academic Press is an imprint of Elsevier
225, Wyman Street, Waltham, MA 02451, USA
The Boulevard, Langford Lane, Kidlington, Oxford OX5 1GB, UK
Radarweg 29, PO Box 211, 1000 AE Amsterdam, The Netherlands

Library of Congress Cataloging-in-Publication Data
A catalog record for this book is available from the Library of Congress

British Library Cataloguing-in-Publication Data
A catalogue record for this book is available from the British Library

ISBN: 978-0-12-385237-3

For information on all Academic Press publications
visit our website at http://store.elsevier.com

Contents

Preface to the First Edition

The purpose of this monograph is to provide an overview of mesoscale numerical modeling, beginning with the fundamental physical conservation relations. An overview of the individual chapters is given in the introduction. This book is an outgrowth of my article entitled "Mesoscale Numerical Modeling" which appeared in Volume 23 of *Advances in Geophysics*.

The philosophy of the book is to start from basic principles as much as possible when explaining specific subtopics in mesoscale modeling. Where too much preliminary work is needed, however, references to other published sources are given so that a reader can obtain the complete derivation (including assumptions). Often only an investigator's recent work is listed; however, once that source is found it is straightforward to refer to his or her earlier work, if necessary, by using the published reference list appearing in that paper. An understanding of the assumptions upon which the mathematical relations used in mesoscale modeling are developed is essential for fluency in this subject. To address as wide an audience as possible, basic material is provided for the beginner as well as a more in-depth treatment for the specialist.

The author wishes to acknowledge the contributions of a widely proficient group of people who provided suggestions and comments during the preparation of this book. The reading of all or part of the draft material for this text was required for a course in mesoscale meteorological modeling taught at the University of Virginia and at Colorado State University. Among the students in that course who provided significant suggestions and corrections are Raymond Arritt, David Bader, Charles Cohen, Omar Lucero, Jeffrey McQueen, Charles Martin, Jenn-Luen Song, Craig Tremback, James Toth, and George Young. P. Flatau is acknowledged for acquainting me with several Soviet works of relevance to mesoscale meteorology. Suggestions and aid were also provided by faculty members in the Atmospheric Science Department at Colorado State University, including Duane E. Stevens, Richard H. Johnson, Wayne H. Schubert, and Richard Pearson, Jr.

Several chapters were also sent to a number of acknowledged experts in certain aspects of mesoscale meteorology. These scientists included Andrè Doneaud (Chapters 1–5, 7, and 8), George Young (Chapter 5), Tzvi Gal-Chen (Chapter 6), Raymond Arritt (Chapter 7), Richard McNider (Chapter 7), Steven Ackerman (Chapter 8), Andrew Goorch (Chapter 8), Larry Freeman (Chapter 8), Michael Fritsch (Chapter 9), William Frank (Chapter 9), Jenn-Luen Song (Chapter 9), R.D. Farley (Chapter 9), Harold Orville (Chapter 9), Robert Lee (Chapter 10), Mike McCumber (Chapter 11), Joseph Klemp (Chapter 12), Mordecay Segal (Chapters 2, 3, 10, 11, and 12), and Robert Kessler (Chapter 12). For their help in reviewing the material I am deeply grateful.

I would also like to thank the individuals who contributed to the summary tabulation of models in Appendix B. Although undoubtedly not a comprehensive list (since not every modeling group responded or could be contacted), it should provide a perspective of current mesoscale modeling capabilities.

I would also like to acknowledge the inspiration of William R. Cotton and Joanne Simpson, who facilitated my entry into the field of mesoscale meteorology. In teaching the material in this text and in supervising graduate research. I have sought to adopt their philosophy of providing students with the maximum opportunity to perform independent, innovative investigations. I would also like to give special thanks to Andrè Doneaud and Mordecay Segal, whose patient, conscientious reading of portions of the manuscript has significantly strengthened the text. In addition, I would like to express my sincere appreciation to Thomas H. Vonder Haar who provided me with an effective research environment in which to complete the preparation of this book.

In writing the monograph, I have speculated in topic areas in which there has been no extensive work in mesoscale meteorology. These speculative discussions, most frequent in the sections on radiative effects, particularly in polluted air masses, also occur in a number of places in the chapters on parameterization, methods of solution, boundary and initial conditions, and model evaluation. Such speculation is risky, of course, because the extensive scientific investigation required to validate a particular approach has not yet been accomplished. Nevertheless, I believe such discussions are required to complete the framework of the text and perhaps may be useful in providing some direction to future work. The introduction of this material is successful if it leads to new insight into the field of mesoscale modeling.

Finally, the writing of a monograph or textbook inevitably results in errors, for which I must assume final responsibility. It is hoped that they will not significantly detract from the usefulness of the book and that the reader will benefit positively from ferreting out mistakes. In any case, I would appreciate comments from users about errors of any sort, including the neglect of relevant current work.

The drafts and final manuscripts were typed by the very capable Ann Gaynor, Susan Grimstedt, and Sara Rumley. Their contribution in proofreading the material to achieve a manuscript with a minimal number of errors cannot be overstated. The drafting was completed by Jinte Kelbe, Teresita Arritt, and Judy Sorbie. Portions of the costs of preparing this monograph were provided by the Atmospheric Science Section of the National Science Foundation under Grants ATM 81-00514, ATM 82-42931, and ATM 8304042, and that support is gratefully acknowledged.

Finally and most importantly, I would like to acknowledge the support of my family—Gloria, Tara, and Roger Jr.—in completing this time-consuming and difficult task.

Preface to the Second Edition

Mesoscale meteorological modeling has matured greatly since the first edition was published. From a research tool, mesoscale models are routinely used in operational numerical weather prediction. These models have also been extended into longer-term weather studies, such as seasonal weather prediction and even in climate change studies.

As a result of the proliferation of this atmospheric science modeling tool, the number of published papers has greatly expanded. While I attempted to be reasonably comprehensive in listing this work in the first edition, it is now virtually impossible to be comprehensive today. In fact, with the introduction of the Internet and electronic library searches, the best way to obtain relevant research papers is to access through the World Wide Web!

This new edition has new material but also deletes sections. The section on the finite element solution technique in Chapter 10, for example, has been removed since despite its promise, it remains an approach that is used by only a very small subset of mesoscale modelers. Problems have been added to the new addition, based on work in the course Mesoscale Meteorological Modeling (AT 730) which I have taught almost every two years, both at the Department of Environmental Science at the University of Virginia, and in the Department of Atmospheric Science at Colorado State University.

One new perspective in this text is the introduction of a new perspective in dissecting meteorological modeling capabilities. There are two emphases to this perspective. First, once the models are stripped to their most basic level, what is their accuracy as a function of wavelength? For the fundamental terms in the equations, this involves the numerical approximation of the local temporal derivative, the advection, the pressure gradient force, and the Coriolis term. For derived terms, this involves the numerical approximations of the vertical and horizontal subgrid-scale fluxes, and the source/sink terms in the conservation equations. Secondly, by defining the individual terms in separate levels of detail, it is straightforward to dissect the expressions (parameterizations) and ascertain how uncertainty (error) propagates throughout the parameterizations to the level at which their effects are introduced into the conservation equations.

There are quite a few colleagues who provided me comments, corrections, and suggestions with respect to the First Edition. This includes the extensive and thorough cross-checking of the First Edition by Xingzhen Zhang, Changxin Yang, Linsheng Chen, and Jifan Chou in their translation into Chinese.

These colleagues also include Pinhas Alpert, Ray Arritt, Louis Berkofsky, Bob Bornstein, Chris Castro, Guy Cautenet, Tom Chase, Linsheng Cheng, Mike Flannigan, Zhu Fu-Cheng, Piotr Flatau, Louie Grasso, Mark Hadfield, Hartmut Kapitza, Richard Krasner, Rene Laprise, Alan Lipton, Glen Liston, Guta Mihailovic, Chuck Mollenkamp, Mike Moran, Joseph Mukabana, Peter Olsson, Bill Physick, Andy Pitman, Jim Purdom, Feng Zhi Qiang, Nelson Seaman, Moti Segal, Qingqiu

Shao, Graeme Stephens, Lou Steyaert, Roger Stocker, Gene Takle, Craig Tremback, Sue Van Den Heever, Pier Luigi Vidale, Tomi Vukićević, Roger Wakimoto, Bob Walko, Doug Wesley, Xubin Zeng, and Conrad Ziegler.

I particularly thank Ytzhaq Mahrer, Roni Avissar, Bill Cotton, and Joanne Simpson who have always provided encouragement in the field of mesoscale meteorology and whose counsel and advice I value so much.

I want to acknowledge Dallas Staley who performed an exceptional, outstanding job in typing and editing the text. Her very significant contribution was essential to the completion of the book, and I am very fortunate to have her work with me on this.

Finally, as with the First Edition, my family has been very supportive. I want to dedicate this book to them—Gloria, Roger Jr., Tara, Julie, Richard, Harrison, Megan, and Jacob!

Preface to the Third Edition

The field of mesoscale meteorological modeling has continued to mature since the 2nd Edition of this book was written. Indeed, regional and mesoscale modeling have merged into a continuum of modeling efforts. Mesoscale models, which at the time of the 1st Edition were used almost entirely in research and for selected case studies, are now being run each day in the preparation of routine weather forecasts. The basic foundation of meteorological modeling, however, is, of course, unchanged. Thus the 3rd Edition retains its core discussion of the basics of this engineering tool in Chapters 2–6 (Basic Set of Equations, Simplification of the Basic Equations, Averaging the Conservation Relations, Physical and Analytic Modeling, and Coordinate Transformations).

In Chapter 5, Mel Nicholls presents a discussion of the role of compressibility in mesoscale models. The parameterizations in the first two Editions were merged into a new Chapter 7 titled Traditional Parameterizations where Nelson Seaman's overview of several cumulus parameterizations is presented. The adoption of a new type of parameterization is proposed in Chapter 8 titled New Parameterization Approaches. This methodology has the potential to not only significantly increase computational speed, but also improve the accuracy of the parameterizations. Chapter 9 then presents Methods of Solution with a contribution from Bob Walko on a finite volume solution technique and a section on a fully Lagrangian solution technique by John Lin. Chapters 10 and 11 are on Boundary and Initial Conditions and Model Evaluation and Xubin Zeng provided information on how to represent land- and ocean-atmospheric interactions in Chapters 7 and 10. In Chapter 11, Pinhas Alpert updated his text on factor separation, including examples. Chapter 12 is a new contribution, authored by Toshi Matsui of NASA on a Mesoscale Modeling and Satellite Simulator, which is a new application of models. Chapter 13, as in the first two Editions, has Examples of Mesoscale Models, including a section on modeling of the Martian, Titan, and Venus mesoscale atmosphere by Scot Rafkin, and a section on air quality-meteorological interactions by George Kallos. Another new subsection in that chapter is on dynamic downscaling. Finally, Chapter 14 is a new chapter on Synoptic-Scale Background which adds an important new perspective to the larger-scale environment in which mesoscale systems develop. Appendix B provides a list of a number of mesoscale and regional models including a summary of their components, based on the structure discussed in the text. Toshi Matsui provided Appendix C. We also received assistance from Gail Cordova and Wayne Schubert at Colorado State University in obtaining needed figures. Jason McDonald provided the final design for the cover of the book.

Since the 2nd Edition, we have received comments and corrections from other colleagues and these include Bob Bornstein, Jerry Davis, Giovanni Dolif, Ann Gravier, Erica McGrath-Spangler, and Fedor Mesinger.

I want to thank Dallas Staley, whose work on the book has been outstanding. She has worked with me for nearly 30 years. Throughout this period, her professionalism has been at the very highest level and her work on this book is yet another example. It has been a pleasure to work with her all of these years!

I also want to acknowledge the University of Colorado at Boulder (CIRES, ATOC, and the Vice Chancellor for Research) and the National Science Foundation under Grants AGS 1219833 and AGS 0831331 for the support they have provided.

Finally, as with the first two editions, my family continues to support my research and academic interests. I want to dedicate this book to them—Gloria, Roger Jr., Tara, Julie, Richard, Harrison, Megan, Jacob, Calvin, Emma, and Annabel!

Foreword

The growth of the field of mesoscale meteorology noted in the Foreword to the 2nd edition of this book has only accelerated since its publication. One factor in this acceleration has been the subsequent extraordinary growth in the computational power of computer clusters available for three-dimensional numerical mesoscale atmospheric model simulations. This has allowed for the inclusion of ever more sophisticated parameterizations of the aerosol, radiative, cumulus, cloud microphysics, boundary layer, and urban processes within such models. The increased sophistication of data assimilation techniques has also allowed for the inclusion of a wider range of in situ and remotely sensed meteorological observations in mesoscale simulations, as well for the use of outputs from larger-scale weather models as boundary condition drivers in mesoscale models. Finally, the linkage of numerical mesoscale atmospheric with ocean, hydrological, ecological, and air quality models has allowed for applications of the resulting modeling systems to an ever wider range of weather, air quality, climate, and climate change cases.

Roger A. Pielke Sr. is a seminal figure in the birth of the field of mesoscale meteorology, and has remained in the forefront of the ongoing development of mesoscale models. During his career, Professor Pielke worked as a Research Meteorologist for the NOAA Experimental Meteorology Lab in Miami, and as a faculty member in the Departments of Environmental Sciences at the University of Virginia and Atmospheric Science at Colorado State University. He is currently Senior Researcher Scientist at the Cooperative Institute for Research in the Atmosphere at the University of Colorado.

The three-dimensional mesoscale numerical model he developed during his graduate studies was truly pioneering and an inspiration to many model developers, including myself. This model has continually evolved and remains as a state-of-the-art mesoscale simulation system. His ongoing association with ecologists and hydrologists has given Professor Pielke the wide knowledge and experience to advance the discipline of atmospheric modeling, as he has pioneered the introduction of such processes into mesoscale models. His early accomplishments were recognized by the American Meteorological Society, which awarded him the Leroy Mesinger Award in 1977 for "fundamental contributions to mesoscale meteorology through numerical modeling of the sea breeze and interaction among the mountains, oceans, boundary layer, and the free atmosphere."

Having been at the forefront of research into mesoscale numerical model development and application for the past 40 years, and having served as Chief Editor for both the *Monthly Weather Review* and the *Journal of the Atmospheric Sciences*, Professor Pielke has been in a unique position to write the seminal text on this topic and to update its later editions with important current developments, clearly demonstrated by the more than 2000 references included in the 3rd edition of the book.

The updated completeness and organization of the new edition makes the text even more indispensable for providing students with understandings of the formulation, solution techniques, applications, and limitations of mesoscale models, as it reflects the unique experience, knowledge, and deep insights gained by Professor Pielke in teaching his own Mesoscale Meteorological Modeling courses at the University of Virginia, Colorado State University, and University of Colorado. Students with a solid background in fluid dynamics and numerical methods can cover the material in one semester. The list of supplemental reading sources provided at the end of each chapter provides students with additional background material on chapter topics, while insightful problems will help them retain and reinforce key points made in the text. I am certain that meteorology instructors and students alike will find this book an indispensable resource for their personal libraries.

Robert Bornstein
Professor Emeritus
Department of Meteorology and Climate Science
San Jose State University
San Jose, California, USA

CHAPTER

1 Introduction

To utilize mesoscale dynamical simulations of the atmosphere effectively, it is necessary to understand the basic physical and mathematical foundations of the models and to have an appreciation of how the particular atmospheric system of interest works. This text provides such an overview of the field and should be of use to the practitioner as well as to the researcher of mesoscale phenomena. Because the book starts from fundamental concepts, it should be possible to use the text to evaluate the scientific basis of any simulation model that has been or will be developed.

Mesoscale can be descriptively defined as having a horizontal spatial scale smaller than the conventional rawinsonde network, but significantly larger than individual cumulus clouds. This implies that the horizontal scale is on the order of a few kilometers to several hundred kilometers or so. The vertical scale extends from tens of meters to the depth of the troposphere.

Mesoscale can also be defined as those atmospheric systems that have a horizontal extent large enough for the hydrostatic approximation to the vertical pressure distribution to be valid, yet small enough for the geostrophic and gradient winds to be inappropriate as approximations to the actual wind circulation above the planetary boundary layer. This scale of interest, then, along with computer and cost limitations, defines the domain and grid sizes of mesoscale models. In this text, examples of specific circulations will be presented, illustrating scales of mesoscale circulations.

Mesoscale Meteorological Modeling, Third Edition, Volume 98. http://dx.doi.org/10.1016/B978-0-12-385237-3.00001-3

CHAPTER

Basic Set of Equations

2

CHAPTER OUTLINE HEAD

The foundation for any model is a set of conservation principles.[1] For mesoscale atmospheric models these principles are

1. conservation of mass,
2. conservation of heat,
3. conservation of motion,
4. conservation of water, and
5. conservation of other gaseous and aerosol materials.

These principles form a coupled set of relations that must be satisfied simultaneously and that include sources and sinks in the individual expressions.

The corresponding mathematical representations of these principles for atmospheric applications are developed as follows.[2]

[1] These principles do not mean that the quantity is conserved, but that all processes which influence the quantity are accounted for; i.e., this is the *"conservation principle"*.

[2] A conservation relationship can also be written for electric charge, but in mesoscale modeling, electromagnetic effects are not considered to be dynamically or thermodynamically important on the model-resolved mesoscale. They are certainly important on cloud and precipitation microphysics, and can therefore affect mesoscale and larger processes, but this would need to be included through parameterizations of the microphysics.

Mesoscale Meteorological Modeling, Third Edition, Volume 98. http://dx.doi.org/10.1016/B978-0-12-385237-3.00002-5

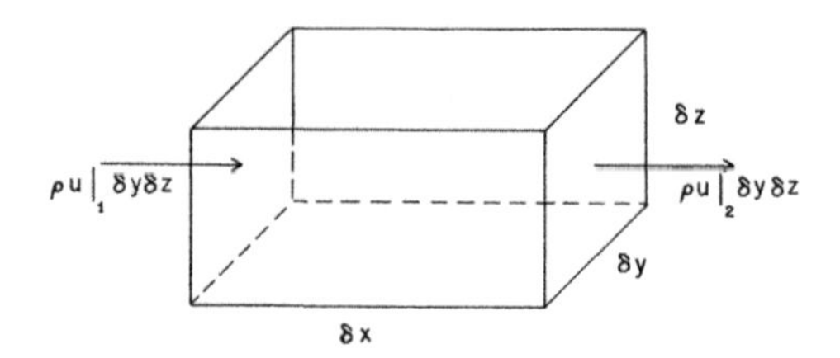

FIGURE 2.1

A schematic of the volume used to derive the conservation of mass relation.

2.1 Conservation of Mass

In the Earth's atmosphere, mass is assumed to have neither sinks nor sources.[3] Stated another way, this concept requires that the mass into and out of an infinitesimal box must be equal to the change of mass in the box. Such a volume is sketched in Fig. 2.1, where $\rho u|_1 \delta y\, \delta z$ is the mass flux into the left side and $\rho u|_2 \delta y\, \delta z$ the mass flux out of the right side. The symbols δx, δy, and δz represent the perpendicular sides of the box, ρ the density, and u the velocity component normal to the $\delta z\, \delta y$ plane.

If the size of the box is sufficiently small, the change in mass flux across the box can be written as

$$\begin{aligned}\left[\rho u|_1 - \rho u|_2\right] \delta y\, \delta z &= \left[\rho u|_1 - \rho u|_1 - \left.\frac{\partial \rho u}{\partial x}\right|_1 \delta x - \frac{1}{2} \left.\frac{\partial^2 \rho u}{\partial x^2}\right|_1 (\delta x)^2 - \cdots \right] \delta y\, \delta z \\ &= \frac{\delta M}{\delta t}\end{aligned}$$

where $\rho u|_2$ has been written in terms of a one-dimensional Taylor series expansion and $\delta M/\delta t$ is the rate of increase or decrease of mass in the box. Neglecting terms in the series of order $(\delta x)^2$ and higher, this expression can be rewritten as

$$-\left.\frac{\partial \rho u}{\partial x}\right|_1 \delta x\, \delta y\, \delta z \simeq \frac{\delta M}{\delta t},$$

and since the mass M is equal to ρV (where $V = \delta x\, \delta y\, \delta z$ is the volume of the box), this expression can be rewritten as

$$-\left.\frac{\partial \rho u}{\partial x}\right|_1 \delta x\, \delta y\, \delta z \simeq V \frac{\delta \rho}{\delta t},$$

assuming the volume is constant with time.

[3]Gases and aerosols that move into or out of the Earth's land and water bodies and those that are lost to space are presumed to have an inconsequential effect on the mass present.

If the mass flux through the sides $\delta x\,\delta y$ and $\delta x\,\delta z$ is considered in a similar fashion, the complete equation for mass flux in the box can be written as

$$-\frac{\partial}{\partial x}\rho u\bigg|_1 \delta x\,\delta y\,\delta z - \frac{\partial}{\partial y}\rho v\bigg|_1 \delta x\,\delta y\,\delta z - \frac{\partial}{\partial z}\rho w\bigg|_1 \delta x\,\delta y\,\delta z \simeq V\frac{\delta\rho}{\delta t},$$

and dividing by volume, the resulting equation is

$$-\frac{\partial \rho u}{\partial x}\bigg|_1 - \frac{\partial \rho v}{\partial y}\bigg|_1 - \frac{\partial \rho w}{\partial z}\bigg|_1 \simeq \frac{\delta\rho}{\delta t}.$$

If the time and spatial increments are taken to zero in the limit, then

$$\lim_{\substack{\delta x\to 0,\ \delta y\to 0\\ \delta z\to 0,\ \delta t\to 0}} \left(-\frac{\partial \rho u}{\partial x}\bigg|_1 - \frac{\partial \rho v}{\partial y}\bigg|_1 - \frac{\partial \rho w}{\partial z}\bigg|_1\right) = \lim_{\substack{\delta x\to 0,\ \delta y\to 0\\ \delta z\to 0,\ \delta t\to 0}} \frac{\delta\rho}{\delta t},$$

since the remainder of the terms in the Taylor series expansion contain δx, δy, or δz. Written in an equivalent fashion,

$$-\left[\frac{\partial}{\partial x}\rho u + \frac{\partial}{\partial y}\rho v + \frac{\partial}{\partial z}\rho w\right] = \frac{\partial\rho}{\partial t}, \tag{2.1}$$

where the subscript 1 has been dropped because the volume of the box has gone to zero in the limit. Equation 2.1 is the mathematical statement of the conservation of mass. It is also called the *continuity equation.* In vector notation, it is written as

$$-\left(\nabla\cdot\rho\vec{V}\right) = \partial\rho/\partial t \tag{2.2}$$

2.2 Conservation of Heat

The atmosphere on the mesoscale behaves very much like an ideal gas and is considered to be in local *thermodynamic equilibrium.*[4] The first law of thermodynamics for the atmosphere states that differential changes in heat content dQ are equal to the sum of differential work performed by an object dW and differential increases in internal energy dI. Expressed more formally, the first law of thermodynamics states that

$$dQ = dW + dI. \tag{2.3}$$

[4]Coulson (1975:10) and Kondratyev (1969:23,24) discuss thermodynamic equilibrium. To be in equilibrium, the intensity of radiation cannot be dependent on direction (i.e., radiation must be *isotropic*), and temperature cannot depend on the frequency and direction of electromagnetic radiation (i.e., the Stefan-Boltzmann law (8-8) must apply). In other words, temperature must be controlled by molecular collisions rather than by interaction of the molecules with the radiation field. At levels less than 50 km or so in the Earth's atmosphere, the density of air is sufficiently great so that over short distances, molecular collisions dominate and a state of local equilibrium occurs.

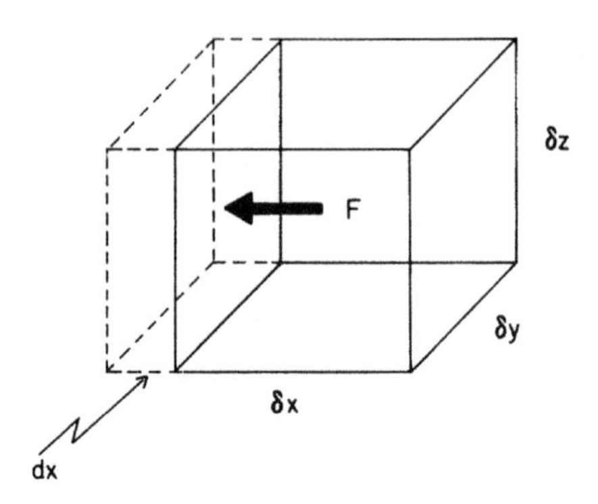

FIGURE 2.2

A schematic of the change in size of a volume of gas resulting from a force F exerted over the surface $\delta z\, \delta y$.

If we represent the region over which Eq. 2.3 applies as a box (Fig. 2.2), with volume $\delta x\, \delta y\, \delta z$, then an incremental increase in the x direction, caused by a force F, can be expressed as

$$dW = F\, dx,$$

and since force can be expressed as a pressure P exerted over an area $\delta y\, \delta z$,

$$dW = p\, \delta y\, \delta z\, \delta x. \tag{2.4}$$

The term $\delta y\, \delta z\, \delta x$ represents a change in volume dV, so that Eq. 2.4 can be rewritten as

$$dW = p\, dV.$$

For a unit mass of material, it is convenient to rewrite the expression as

$$dw = p\, d\alpha, \tag{2.5}$$

where α is the specific volume (i.e., volume per unit mass). In an ideal gas, which the atmosphere closely approximates as discussed subsequently, the pressure in Eq. 2.5 is exerted uniformly on all sides of the gas volume.

The expression for work in Eq. 2.3 could also have included external work performed by such processes as chemical reactions, phase changes, or electromagnetism; however, these effects are not included in this derivation of work.

The ideal gas law, referred to previously, was derived from observations of the behavior of gases at different pressures, temperatures, and volumes. Investigators in the 17th and 18th centuries found that, for a given gas, pressure times volume equals a constant at any fixed temperature (Boyle's law) and that pressure divided by temperature equals a constant at any fixed volume (Charles's law). These two relations can be stated more precisely as

$$p\alpha = F_1(T) \tag{2.6}$$

and

$$p/T = F_2(\alpha), \quad (2.7)$$

where a unit mass of gas is assumed. If Eq. 2.6 is divided by T and Eq. 2.7 multiplied by α, then

$$p\alpha/T = \left(1/T\right) F_1(T) = \alpha F_2(\alpha). \quad (2.8)$$

Since the two right-hand expressions are functions of two different variables, the entire expression must be equal to a constant, conventionally denoted as R. Thus Eq. 2.8 is written as

$$p\alpha/T = R, \quad (2.9)$$

where R has been found to be a function of the chemical composition of the gas. The extent to which actual gases obey Eq. 2.9 specifies how closely they approximate an ideal gas.

The value of the gas constant R for different gases is determined using Avogadro's hypothesis that at a given temperature and pressure gases containing the same number of molecules occupy the same volume. From experimental work, for example, it has been shown that at a pressure of 1 atm ($P_0 = 1014$ mb) and a temperature of $T_0 = 273$ K, 22.4 kliter of a gas (V_0) will have a mass in kilograms equal to the molecular weight of the gas μ. This quantity of gas is defined as 1 kmol.

Using this information, the ideal gas law (Eq. 2.9) and the definition $\alpha_0 = V_0/\mu$, then

$$P_0 V_0/\mu T_0 = R,$$

or by definition

$$P_0 V_0/\mu T_0 = R = R^*/\mu \left[\text{so that } P_0 V_0/T_0 = R^*\right], \quad (2.10)$$

where R^* is called the universal gas constant and μ has units of kilograms per kilomole. From experiments, $R^* = 8.314472 \times 10^3$ J K^{-1}kmol^{-1} (Mohr and Taylor 2000). Since Eq. 2.10 is valid for any combination of pressure, temperature, and volume,

$$p\alpha/T = R = R^*/\mu \quad (2.11)$$

In the atmosphere, the apparent molecular weight of air μ_{atm} is determined by the fractional contribution by mass of each component gas (Table 2.1) from the equation

$$\mu_{atm} = \sum_{i=1}^{N} m_i \Big/ \sum_{i=1}^{N} \left(m_i/\mu_i\right),$$

where m_i is the fractional contribution by mass of the N individual gases in the atmosphere ($\sum_{i=1}^{N} m_i = 1$) and μ_i represents their respective molecular weights.[5]

[5]The form of μ_{atm} is derived using Dalton's law of partial pressures; Dalton's law is $p = \sum_{i=1}^{N} p_i$, when p_i represents the pressure contribution of the individual gases which make up the gas mixture (e.g., see Haltiner and Martin 1957:10).

Table 2.1 Molecular weight and fractional contribution by mass of major gaseous components of the atmosphere (from Wallace and Hobbs 1977).

Gas	Molecular Weight	Fractional Contribution by Mass
N_2	28.016	0.7551
O_2	32.00	0.2314
Ar	39.94	0.0128
H_2O	18.02	variable

For the gaseous components in Table 2.1, excluding water vapor

$$\mu_{\text{dry atm}} = \frac{0.7551 + 0.2314 + 0.0128}{(0.7551/28.016) + (0.2314/32.00) + (0.0128/39.94)} = 28.98,$$

so that the dry gas constant of the atmosphere R_{d} is

$$R_{\text{d}} = R^*/28.98 = 287 \text{ J K}^{-1}\text{kg}^{-1}.$$

When water vapor is included, the apparent molecular weight can be written as

$$\mu_{\text{atm}} = 1 \Bigg/ \left(\frac{1-q}{28.98} + \frac{q}{18.02} \right),$$

where q is the specific humidity or ratio of the mass of water vapor M, to the mass of dry air M_{d}. Expanding this relation

$$\begin{aligned} \mu_{\text{atm}} = \frac{1}{(1/28.98)(1 - q + (28.98/18.02)q)} &= \frac{28.98}{1 - q + (28.98/18.02)q} \\ &= \frac{28.98}{1 + q[(28.98/18.02) - 1]} \\ &= \frac{28.98}{1 + 0.61q}, \end{aligned}$$

and inserting μ_{atm} into Eq. 2.11 gives

$$p\alpha = R_{\text{d}}(1 + .61q)T. \tag{2.12}$$

This form of the ideal gas law includes the contribution of water vapor and is often written as

$$p\alpha = R_{\text{d}}T_{\text{V}}, \tag{2.13}$$

where T_{V} is called the *virtual temperature*, or the temperature required in a dry atmosphere to have the same value of $p\alpha$ as in an atmosphere with a specific humidity q of water vapor. For typical atmospheric conditions (e.g., $q = 0.006$ kg per kg) the difference between the virtual and actual temperatures is about 1°C. Since $T_V \geq T$, air at the same pressure and temperature is less dense when water vapor is present that when it is not. The virtual temperature is generally used by convention in preference to recomputing the gas constant $R = R_d(1 + .61q)$.

The total heat of a parcel of air is defined in terms of moist enthalpy (Pielke et al. 2004)

$$H = C_p T + L_v q$$

where C_p is the specific heat of air at constant pressure, T is the air temperature, and L_v is the latent heat of vaporization and is a weak function of temperature (Henderson-Sellers 1984), although for most mesoscale applications a constant value of $L_v = 2.5 \times 106\ \mathrm{J\ kg^{-1}}$ is assumed. The variable q is the specific humidity.

H is in units of Joule and can be scaled into degree units in order to obtain equivalent temperature for easy comparison to air temperature

$$T_E = H/C_p$$

This equation can be also written as

$$T_E = T + L_v q/C_p$$

where L_v is in units of Joules per kilogram and C_p is in units of Joules per kilogram per degree K. As q is dimensionless (i.e., kg per kg), the ratio has units of degree K.

The importance of using moist enthalpy can be illustrated from Fig. 2.3 where the hottest time of the day does not correspond to the time of the day with the highest heat content. Studies which have used moist enthalpy to determine atmospheric heat content include Pielke et al. (2004), Davey et al. (2006), Fall et al. (2010), and Peterson et al. (2011).

To complete the derivation of the first law of thermodynamics for an ideal gas, it is useful to introduce the concept of exact differentials. If a function F exists such that

$$F = F(x, y),$$

where x and y are two independent variables,[6] then

$$dF = (\partial F/\partial x)dx + (\partial F/\partial y)dy = M\,dx + N\,dy$$

by the chain rule of calculus. If

$$\partial M/\partial y = \partial N/\partial x,$$

then

$$\partial^2 F/\partial x\,\partial y = \partial^2 F/\partial y\,\partial x, \tag{2.14}$$

[6]For an ideal gas, the independent variables are any two of temperature, pressure, and specific volume. Given two of these variables, the gas law determines the third. These variables are also referred to as *state variables*.

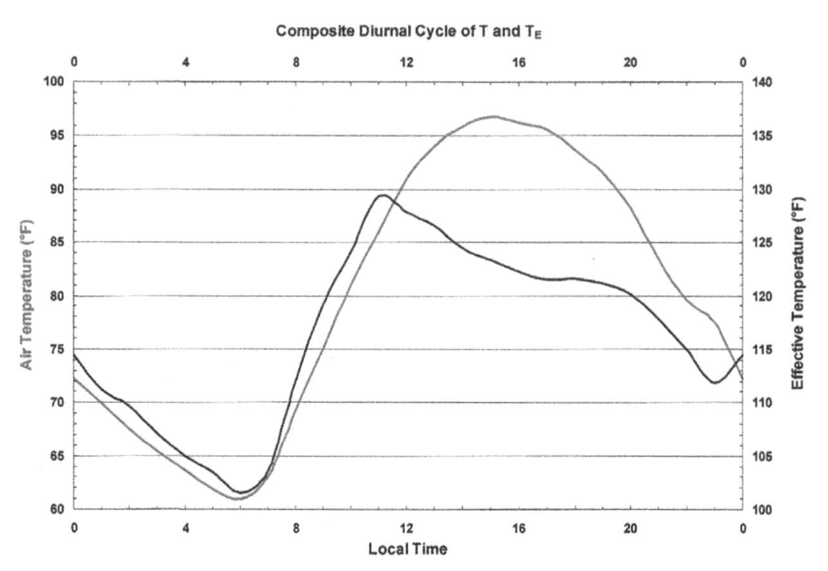

FIGURE 2.3

A daily composite of air temperature (gray line) and effective temperature (dark line) for Fort Collins, Colorado, USA. The composite is created by averaging hourly data during the five days with highest air temperature in each of the three years considered in this section – fifteen days total. This shows the pattern of heating and cooling on the station's extreme hottest days. Note how the effective temperature peaks approximately four hours before the air temperature peaks. Typically, the hottest days are characterized by exceptionally low relative humidity in the late afternoon, which explains the premature drop in effective temperature (from Pielke et al. 2006a). (The color version of this figure is presented in the plate section in the back of the book and the online web version.)

and F is an exact differential. Stated more physically, if Eq. 2.14 is valid, then the path over which the function is evaluated (e.g., $\partial/\partial x$ first, then $\partial/\partial y$ as contrasted with $\partial/\partial y$ first, then $\partial/\partial x$) is unimportant. If the left- and righthand sides of Eq. 2.14 are not equal, however, then dF is an inexact differential and different paths taken to compute it will give different answers.

To ascertain if the change in work given by Eq. 2.5 is an exact differential or not, it is useful to rewrite the expression as

$$dw = d(p\alpha) - \alpha\, dp$$

using the product rule of differentiation $[d(p\alpha) = p\, d\alpha + \alpha\, dp]$. Thus by the gas law Eq. 2.13,

$$dw = R_{\text{d}} dT_{\text{V}} - \alpha\, dp.$$

To check for exactness, let $M = R_{\text{d}}$ and $N = -\alpha$, then

$$\partial M/\partial p = \partial R_{\text{d}}/\partial p = 0 \quad \text{and} \quad \partial N/\partial T_{\text{V}} = \partial(-\alpha)/\partial T_{\text{V}} = -R_{\text{d}}/p,$$

therefore, dw is not an exact differential. The path in which work is performed is important in determining its value.

The internal energy I, in Eq. 2.3, expressed for a unit mass of material, can be written as

$$e = e\left(T_V, \alpha\right), \tag{2.15}$$

where, as a result of the ideal gas law, the virtual temperature T_V and the specific volume α can be used to determine the internal energy of the material. From the chain rule of calculus

$$de = \left(\partial e/\partial T_V\right) dT_V + (\partial e/\partial \alpha) d\alpha,$$

but from experiments with gases that closely follow the ideal gas law (Eq. 2.13), internal energy changes only when temperature changes (i.e., $\partial e/\partial \alpha = 0$). And if we define heat per unit mass from Eq. 2.3 as h, then

$$dh = dw + de = p\,d\alpha + \left(\partial e/\partial T_V\right) dT_V,$$

and

$$\partial h/\partial T_V = \partial e/\partial T_V = C_\alpha,$$

where C_α is defined as the specific heat at constant volume.

Experiments have shown C_α to be only a slowly varying function of temperature. Thus the internal energy relationship for an ideal gas is expressed as

$$de = C_\alpha dT_V$$

Since $M = \partial e/\partial T_V = C_\alpha$ and $N = \partial e/\partial \alpha = 0$, it is obvious that $\partial M/\partial \alpha = \partial N/\partial T = 0$ so that internal energy for an ideal gas is an exact differential.

Our first law of thermodynamics (Eq. 2.3) can now be written as

$$đh = de + đw = C_\alpha dT_V + p\,d\alpha, \tag{2.16}$$

where the diagonal slash through the two terms indicates that they are inexact differentials ($đh$ is inexact because the sum of an exact and an inexact differential must be inexact). It is not convenient to work with this form of the first law, however, because the path taken to go from one set of temperature and pressure, for example, to a different set will affect the amount of heat lost or gained and the amount of work performed.

To eliminate this dependency on path, Eq. 2.16 can be made an exact differential by dividing by temperature T_V, and using the ideal gas law (Eq. 2.13) so that

$$đh/T_V = C_\alpha \left(dT_V/T_V\right) + \left(R_d/\alpha\right) d\alpha. \tag{2.17}$$

Since $M = C_\alpha/T_V$ and $N = R_d/\alpha$ then $\partial M/\partial \alpha = 0$ and $\partial N/\partial T_V = 0$ so that

$$đh/T_V = ds$$

is an exact differential, where s is defined as *entropy*.

Unfortunately, Eq. 2.17 is not in a convenient form for use by meteorologists because temperature and pressure are measured, not specific volume. To generate a more useful form of Eq. 2.17, we differentiate the ideal gas law (Eq. 2.13) logarithmically so that

$$(1/\alpha)d\alpha = (1/T_V)\,dT_V - (1/p)\,dp$$

and substituting into Eq. 2.17 yields

$$ds = \not{d}h/T_V = (C_\alpha dT_V/T_V) + (R_d dT_V/T_V) - (R_d/p)\,dp,$$

or

$$ds = \not{d}h/T_V = (C_\alpha + R_d)\,(dT_V/T_V) - (R_d/p)\,dp. \tag{2.18}$$

Since

$$\not{d}h = (C_\alpha + R_d)\,dT_V - \alpha\,dp$$

then

$$\partial h/\partial T_V = C_\alpha + R_d = C_p,$$

where C_p is defined as the specific heat at constant pressure. Therefore, Eq. 2.18 is written as

$$ds = C_p\,(dT_V/T_V) - (R_d/p)\,dp. \tag{2.19}$$

For an ideal monotomic gas, the ratio of $C_p : C_\alpha : R_d = 5 : 3 : 2$, whereas for a diatomic gas (such as the atmosphere closely approximates) the ratio of $C_p : C_\alpha : R_d$ is $7 : 5 : 2$.

For the situation when no heat is gained or lost (e.g., $ds = 0$),

$$dT_V/T_V = (R_d/C_p)\,dp/p$$

which can be rewritten as

$$d\ln T_V = (R_d/C_p)\,d\ln p. \tag{2.20}$$

If a parcel of air moves between two points with temperatures and pressures given by (T_{V_1}, P_1) and (T_{V_2}, P_2), then integrating Eq. 2.20 gives

$$\int_{T_{V_1}}^{T_{V_2}} d\ln T_V = (R_d/C_p)\int_{P_1}^{P_2} d\ln P = \ln\left(T_{V_2}/T_{V_1}\right) = (R_d/C_p)\ln\left(P_2/P_1\right).$$

Taking antilogs yields

$$T_{V_2}/T_{V_1} = \left(P_2/P_1\right)^{R_d/C_p},$$

which is customarily called *Poisson's equation*. If we set $P_2 = 1000$ mb and T_{V_2} is defined as the *potential temperature* θ then

$$\theta = T_V(1000/p)^{R_d/C_p}, \tag{2.21}$$

where p is in millibars.

To determine the relationship between the potential temperature θ and the entropy s, logarithmically differentiate Eq. 2.21 and multiply by C_p, which yields

$$(C_p/\theta)\,d\theta = (C_p/T_V)\,dT_V - (R_d/p)\,dp,$$

which is identical to Eq. 2.19 so that

$$(C_p/\theta)\,d\theta = ds = đh/T_V. \tag{2.22}$$

Thus a change in potential temperature is equivalent to a change in entropy.

If the change in potential temperature is observed following a parcel, then Eq. 2.22 can be written as

$$\frac{C_p}{\theta}\frac{d\theta}{dt} = \frac{ds}{dt} = \frac{1}{T_V}\frac{đh}{dt} = S_\theta \frac{C_p}{\theta}, \tag{2.23}$$

where S_θ represents the sources and sinks of heat as expressed by changes in potential temperature. The contributors to S_θ include the sum of the following processes:

$$S_\theta = \begin{bmatrix} + \text{freezing} \\ - \text{melting} \end{bmatrix} + \begin{bmatrix} + \text{condensation} \\ - \text{evaporation} \end{bmatrix} + \begin{bmatrix} + \text{deposition} & \text{(vapor to solid)} \\ - \text{sublimation} & \text{(solid to vapor)} \end{bmatrix}$$

$$+ \begin{bmatrix} + \text{exothermic chemical reactions} \\ - \text{endothermic chemical reactions} \end{bmatrix} \tag{2.24}$$

$$+ \begin{bmatrix} + \text{net radiative flux convergence} \\ - \text{net radiative flux divergence} \end{bmatrix} + \begin{bmatrix} \text{dissipation of kinetic energy} \\ \text{by molecular motions} \end{bmatrix}$$

The precise evaluation of these terms can be complicated, and further discussion of them will be deferred until later chapters. In Eq. 2.23, the transfer of heat by molecular processes is not included. The neglect of molecular transfers of heat, or other property of the air, on the mesoscale is justified by the relative contributions to such exchanges through the motion of the fluid as contrasted with molecular diffusion. This neglect is discussed further in Section 2.3.2 of this chapter as well as in Sections 3.3.2 and 5.1.

The term $d\theta/dt$ denotes changes of potential temperature following a parcel, with the operator d/dt often called the *Lagrangian derivative*. Since θ is a function of the three coordinate directions x, y, and z of a parcel at a given time t [i.e., $\theta = \theta(x(t), y(t), z(t), t)$], then by the chain rule of calculus

$$\frac{d\theta}{dt} = \frac{\partial\theta}{\partial t} + \frac{\partial\theta}{\partial x}\frac{dx}{dt} + \frac{\partial\theta}{\partial y}\frac{dy}{dt} + \frac{\partial\theta}{\partial z}\frac{dz}{dt} = S_\theta,$$

or

$$\frac{\partial\theta}{\partial t} = -u\frac{\partial\theta}{\partial x} - v\frac{\partial\theta}{\partial y} - w\frac{\partial\theta}{\partial z} + S_\theta = -\vec{V}\cdot\nabla\theta + S_\theta \tag{2.25}$$

where $\partial\theta/\partial t$ represents local changes in potential temperature and the operation $\partial/\partial t$ is called the *Eulerian derivative*. This equation is a standard form of the conservation of heat relation (often called the potential temperature equation) used in mesoscale models.

It should be noted, however, that since $\not{d}h/dt = \left(C_p T_V/\theta\right) d\theta/dt$, the potential temperature equation is proportional, but not equal to changes in heat content. The proportionality term is given by $C_p T_V/\theta$. The conservation of heat relation is represented by a potential temperature equation rather than by $\not{d}h/dt$ because, as pointed out previously, the latter form is an inexact differential and thus depends on the path taken to accomplish a change. However, $d\theta/dt$ is independent of path.

2.3 Conservation of Motion

The conservation of motion is expressed by Newton's second law, which states that a force exerted on an object causes an acceleration as given by

$$\vec{F} = M\vec{a}$$

where $\vec{F}$ and $\vec{a}$ are the force and acceleration vectors, respectively, and M the mass of the object. It is conventional in atmospheric science to work with force normalized by mass so that this expression can be written as

$$\vec{F}/M = \vec{f} = \vec{a}. \tag{2.26}$$

Since acceleration represents the change of velocity with time following an object, $\vec{a}$ can be written as

$$\vec{a} = d_n\vec{V}_n/dt, \tag{2.27}$$

where the subscript n refers to a non-accelerating coordinate system. However, because atmospheric motions are referenced to a rotating Earth, the acceleration must be expressed in a different form.

If the Earth is rotating with constant angular velocity $\vec{\Omega}$, the velocity $\vec{V}_n$ of an object or parcel of air may be written as the sum of the velocity relative to the Earth and the velocity resulting from rotation. Expressed mathematically,[7]

$$\vec{V}_n = \vec{V} + \vec{\Omega} \times \vec{R}, \tag{2.28}$$

where $\vec{R}$ represents the position vector of the parcel as measured from the origin of the Earth's center, as sketched in Fig. 2.4. The time differential operator can similarly be described by the sum of a derivative relative to the Earth's surface and changes resulting from the rotation rate of the planet, as given by the operation

$$\frac{d_n}{dt} = \frac{d}{dt} + \vec{\Omega}\times. \tag{2.29}$$

[7]The symbol $\times$ is used to indicate a vector cross product.

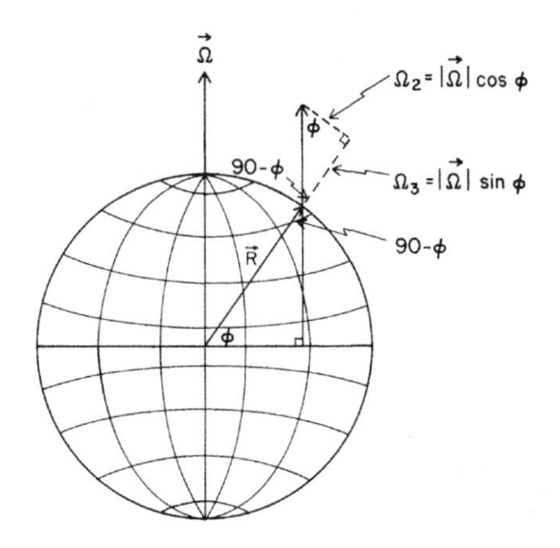

FIGURE 2.4

The components of the angular velocity of the Earth $\vec{\Omega}$ as a function of latitude ϕ. For the Earth $|\vec{\Omega}| = \Omega = 2\pi/24$ h.

Substituting Eqs. 2.29 and 2.28 into 2.27 yields

$$\vec{a} = \left(\frac{d}{dt} + \vec{\Omega}\times\right)\vec{V}_{\text{n}} = \frac{d\vec{V}_{\text{n}}}{dt} + \vec{\Omega} \times \vec{V}_{\text{n}} = \frac{d}{dt}(\vec{V} + \vec{\Omega} \times \vec{R}) + \vec{\Omega} \times (\vec{V} + \vec{\Omega} \times \vec{R}).$$

Simplifying and rearranging results in

$$\vec{a} = (d\vec{V}/dt) + 2(\vec{\Omega} \times \vec{V}) + \vec{\Omega} \times (\vec{\Omega} \times \vec{R}), \qquad (2.30)$$

where the relation $\vec{V} = d\vec{R}/dt$ has been used.

The first term on the right side of Eq. 2.30 is the acceleration as viewed from the rotating Earth. The second term is the *Coriolis acceleration*, which operates only when there is motion, and the last term, called the *centripetal acceleration*, acts on a parcel at all times.

After describing acceleration relative to the Earth, it is necessary to specify the forces that cause changes in motion. In performing this analysis, it is convenient to consider forces as acting *external* and *internal* to a parcel. External forces include those resulting from pressure gradients, gravity, etc., and are independent of motion, and internal forces which are caused by fluid interactions with itself involving frictional dissipation by molecules. This concept of external and internal forces is related to our idea of a parcel that, although assumed to be "infinitesimally" small so that we can apply the concepts of differential calculus, is still presumed large relative to individual molecules. In other words, this parcel must be sufficiently large that only the statistical properties of molecules are important (and are expressed in terms of such so-called macroscopic quantities as pressure and temperature).

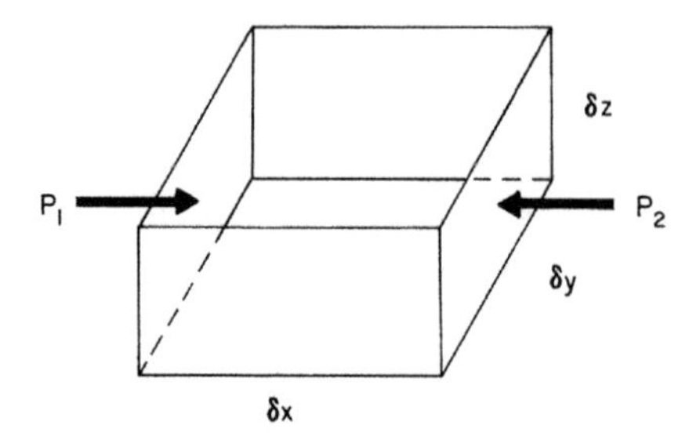

FIGURE 2.5

A schematic of a volume with pressure (P_1 and P_2) on two opposing sides.

2.3.1 External Forces

The pressure gradient force can be derived in a similar fashion to that used for the continuity-of-mass equation (Section 2.1). The pressure difference across a box, depicted in Fig. 2.5, can be expanded in a one-dimensional series and expressed as

$$P_2 - P_1 = \left.\frac{\partial p}{\partial x}\right|_1 \delta x + \frac{1}{2}\left.\frac{\partial^2 p}{\partial x^2}\right|_1 (\delta x)^2 + \cdots + \mathrm{O}\left(\left(\delta x\right)^3\right).$$

Since pressure is force per unit area and is directed toward lower pressure, the force per unit mass in the x direction f_{PGF_x} required in Eq. 2.26 can be written as

$$f_{\mathrm{PGF}_x} = -\frac{(P_2 - P_1)\,A}{M} = -\left.\frac{\partial p}{\partial x}\right|_1 \frac{\delta x A}{M} - \frac{1}{2}\left.\frac{\partial^2 p}{\partial x^2}\right|_1 \frac{(\delta x)^2 A}{M} + \cdots, \tag{2.31}$$

where

$$A = \delta y\,\delta z \text{ and } M = \rho V = \rho\,\delta x\,\delta y\,\delta z.$$

Substituting A and M into Eq. 2.31 yields

$$f_{\mathrm{PGF}_x} = -\frac{1}{\rho}\left.\frac{\partial p}{\partial x}\right|_1 - \frac{1}{2\rho}\left.\frac{\partial^2 p}{\partial x^2}\right|_1 \partial x - \cdots,$$

and if we require δx to become very small,

$$f_{\mathrm{PGF}_x} = \lim_{\delta x \to 0}\left[-\frac{1}{\rho}\left.\frac{\partial p}{\partial x}\right|_1 - \frac{1}{2\rho}\left.\frac{\partial^2 p}{\partial x^2}\right|_1 \partial x - \cdots\right] = -\frac{1}{\rho}\frac{\partial p}{\partial x}.$$

An equivalent derivation in the y and z directions[8] results in a pressure gradient force given by

$$\vec{f}_{\mathrm{PGF}} = -\frac{1}{\rho}\left[\frac{\partial p}{\partial x}\vec{i} + \frac{\partial p}{\partial y}\vec{j} + \frac{\partial p}{\partial z}\vec{k}\right] = -\frac{1}{\rho}\nabla p,$$

where $\vec{i}$, $\vec{j}$, and $\vec{k}$ are the unit vectors in the three spatial directions.

[8]More appropriately, a three-dimensional Taylor series expansion should be applied for each component of the pressure difference (and also in deriving the conservation-of-mass equation in Section 2.1) which results in cross derivative terms. However, in the limit as the horizontal distance approaches zero, the result is the same differential relationship.

Gravity is another external force. If the gravitational force between the Earth and an air parcel is defined as $\vec{G}$, it is customary to include the centripetal acceleration, given in Eq. 2.30, in the definition of a *modified gravitational force*. The force $\vec{G}$ is directed toward the center of the Earth with a magnitude proportional to the mass of the Earth and inversely proportional to the square of the distance of a parcel from the center of the Earth. Subtracting the centripetal acceleration from $\vec{G}$ produces the modified gravity given as

$$-g\vec{k} = \vec{G} - \vec{\Omega} \times (\vec{\Omega} \times \vec{R}).$$

In its application to atmospheric flows, variations of g due to height above the ground or location on the Earth's surface are sometimes considered; however, for mesoscale circulations these small variations in the troposphere are customarily ignored and the modified gravity is treated as a constant ($g = 9.80665\ \mathrm{m\,s^{-2}}$; Mohr and Taylor 2000).

Other external forces, such as electromagnetism, could be included, but for mesoscale circulations within the troposphere only gravity and the pressure gradient are typically included as external forces.

2.3.2 Internal Forces

Internal forces are required to account for the dissipation of momentum by molecular motions. Defined in terms of postulates, the effects of these forces on the momentum are expressed in terms of the viscosity of the gas (or liquid) and the deformation of the momentum field. In the atmosphere, on the mesoscale, the viscosity is sufficiently small and the velocities large enough that the influence of the internal forces is ignored. We demonstrate the reasons for the neglect of these forces more quantitatively in Sections 3.3.2 and 5.1.

The conservation-of-motion relation, Eq. 2.26, can now be written as

$$d\vec{V}/dt = -(1/\rho)\nabla p - g\vec{k} - 2\vec{\Omega} \times \vec{V}, \tag{2.32}$$

where the last term on the right, although only an apparent force arising because of the coordinate frame of reference, is referred to as the *Coriolis force*.

Since

$$\vec{V} = \vec{V}(x(t), y(t), z(t), t)$$

(that is, the velocity is a function of time and the spatial location at a given time), then by the chain rule of calculus

$$\frac{d\vec{V}}{dt} = \frac{\partial \vec{V}}{\partial x}\frac{dx}{dt} + \frac{\partial \vec{V}}{\partial y}\frac{dy}{dt} + \frac{\partial \vec{V}}{\partial z}\frac{dz}{dt} + \frac{\partial \vec{V}}{\partial t},$$

or

$$\frac{d\vec{V}}{dt} = \vec{V} \cdot \nabla\vec{V} + \frac{\partial \vec{V}}{\partial t}.$$

Therefore, Eq. 2.32 can be rewritten as

$$\frac{\partial \vec{V}}{\partial t} = -\vec{V} \cdot \nabla \vec{V} - \frac{1}{\rho}\nabla p - g\vec{k} - 2\vec{\Omega} \times \vec{V}, \tag{2.33}$$

which is a standard form of the conservation of momentum, often called the *equation of motion*.

2.4 Conservation of Water

Water can occur in three forms: solid, liquid, and vapor. To write a conservation law for this substance, it is therefore necessary to keep track of the changes of phase of water as well as to follow its movement through the atmosphere.

The conservation law for water can be written as

$$dq_n/dt = S_{q_n}, \quad n = 1, 2, 3, \tag{2.34}$$

where q_1, q_2, and q_3 are defined as the ratio of the mass of the solid, liquid, and vapor forms of water, respectively, to the mass of air in the same volume. The source-sink term S_{q_n} refers to the processes whereby water undergoes phase changes as well as to water generated or lost in chemical reactions. For most mesoscale applications, chemical changes in water mass can be neglected and the terms can be expressed as contributions owing to the following processes:

$$S_{q1} = \begin{bmatrix} +\text{ freezing} \\ -\text{ melting} \end{bmatrix} + \begin{bmatrix} +\text{ deposition} & \text{(vapor to solid)} \\ -\text{ sublimation} & \text{(solid to vapor)} \end{bmatrix} + \begin{bmatrix} +\text{ fallout from above} \\ -\text{ fallout to below} \end{bmatrix}$$

$$S_{q2} = \begin{bmatrix} +\text{ melting} \\ -\text{ freezing} \end{bmatrix} + \begin{bmatrix} +\text{ condensation} \\ -\text{ evaporation} \end{bmatrix} + \begin{bmatrix} +\text{ fallout from above} \\ -\text{ fallout to below} \end{bmatrix}$$

$$S_{q3} = \begin{bmatrix} +\text{ evaporation} \\ -\text{ condensation} \end{bmatrix} + \begin{bmatrix} +\text{ sublimation} & \text{solid to vapor} \\ -\text{ deposition} & \text{vapor to solid} \end{bmatrix}.$$

The manner in which these terms are expressed mathematically can be very involved. In cumulus cloud models, for example, the condensation of water onto aerosols, and their subsequent development into hydrometeors that fall to the ground, is accounted for by categorizing cloud droplets into a spectrum of interacting size classes. The incorporation of the ice phase creates an even more complex set of interactions.

By contrast, the simplest representation of these sources and sinks of water is to prohibit relative humidities greater than 100%[9] and liquid or solid water below 100%. Excess water vapor over 100% is immediately condensed (or deposited) and falls out as rain or snow. As it falls through an unsaturated environment, water evaporates (or sublimates) to the water vapor phase, thereby elevating the relative humidity.

[9] Relative humidity is defined with respect to water or ice depending on the temperature and the availability of ice nuclei.

Using the chain rule, Eq. 2.34 can be written in terms of the local time rate of change as

$$\partial q_n/\partial t = -\vec{V} \cdot \nabla q_n + S_{q_n}, \quad n = 1, 2, 3. \tag{2.35}$$

Further discussion regarding the source-sink term S_{q_n} is given in Chapter 7.

2.5 Conservation of Other Gaseous and Aerosol Materials

Conservation relations of the form given by Eq. 2.34 can be written for any gaseous or aerosol material in the atmosphere, expressed mathematically as

$$d\chi_m/dt = S_{\chi_m}, \quad m = 1, 2, 3, \ldots, M, \tag{2.36}$$

where χ_m refers to any chemical species except water [which is explained by Eq. 2.35] and is expressed as the mass of the substance to the mass of air in the same volume. Examples of important occasional constituents in the atmosphere include carbon dioxide, methane, sulfur dioxide, sulfates, nitrates, ozone, and the herbicide 2-4-5-T. The source-sink term S_{χ_m} can be written to include changes of state (analogous to that performed for water) as well as chemical transformations, precipitation, and sedimentation.[10] In the atmosphere, for instance, it is well known that SO_2 will convert to sulfate within several days after release. In general, the mathematical representation of this source-sink term can be very complex.

Using the chain rule, Eq. 2.36 can be written as

$$\partial \chi_m/\partial t = -\vec{V} \cdot \nabla \chi_m + S_{\chi_m}, \quad m = 1, 2, \ldots, M. \tag{2.37}$$

As more researchers begin to realize the serious impact of air pollution on our health and economic well being and of trace gases and aerosols within the Earth's climate system, they are including this conservation relation in their mesoscale models.

2.6 Summary

Equations 2.2, 2.25, 2.33, 2.35, and 2.37 are listed together as

$$\partial \rho/\partial t = -\left(\nabla \cdot \rho\vec{V}\right), \tag{2.38}$$

$$\partial \theta/\partial t = -\vec{V} \cdot \nabla\theta + S_\theta, \tag{2.39}$$

$$\partial \vec{V}/\partial t = -\vec{V} \cdot \nabla\vec{V} - 1/\rho\nabla p - g\vec{k} - 2\vec{\Omega} \times \vec{V}. \tag{2.40}$$

$$\partial q_n/\partial t = -\vec{V} \cdot \nabla q_n + S_{q_n}, \quad n = 1, 2, 3, \tag{2.41}$$

$$\partial \chi_m/\partial t = -\vec{V} \cdot \nabla \chi_m + S_{\chi_m}, \quad m = 1, 2, \ldots, M. \tag{2.42}$$

[10] *Sedimentation* refers to the fallout of material that has undergone no phase change nor been produced as the result of a chemical reaction; precipitation, in contrast, is a fallout of material that has been produced as the result of a phase change or chemical reaction.

When we use these equations in the remainder of the text, it is convenient to adopt the formalism of tensor notation. By doing so, the equations are much easier to handle, providing the following simple rules are used:

1. Repeated indices are summed (e.g., in a three-dimensional space, $a_{ii} = a_{11} + a_{22} + a_{33}$).
2. Single indices in a term are called free indexes and refer to the order of a tensor (e.g., a_i is a tensor of order one [a vector]. a_{ij} is a tensor of order two [a matrix], a is a tensor of order zero [a scalar]). The maximum value a free index can attain depends on the spatial dimensions of the system ($i = 3$ for the atmosphere).
3. Only tensors of the same order can be added.
4. Multiplication of tensors can be performed as for scalars (they are commutative with respect to addition and multiplication, a definite advantage as compared with vectors).
5. Parameters are defined to simplify the writing of the gravitational and Coriolis accelerations, i.e.,

$$\delta_{ij} = \begin{bmatrix} 1 & 0 & 0 \\ 0 & 1 & 0 \\ 0 & 0 & 1 \end{bmatrix} = \begin{cases} 1 & \text{for } i = j, \\ 0 & \text{for } i \neq j \end{cases}$$

where i refers to the row and j refers to the column, and

$$\epsilon_{i,j,k} = \underbrace{\overbrace{\begin{bmatrix} 0 & 0 & 0 \\ 0 & 0 & 1 \\ 0 & -1 & 0 \end{bmatrix}}^{j} \begin{bmatrix} 0 & 0 & -1 \\ 0 & 0 & 0 \\ 1 & 0 & 0 \end{bmatrix} \begin{bmatrix} 0 & 1 & 0 \\ -1 & 0 & 0 \\ 0 & 0 & 0 \end{bmatrix}}_{k} \Bigg\} i$$

where the following device has been used: for 0, $i = j, i = k$, or $j = k$; for 1, even permutations of i, j, and k; and for -1, odd permutations of i, j, and k. Using this notational device, along with the requirement that the independent spatial variables $x_1 = (x, y, z)$ are perpendicular to each other at all locations. Equations 2.38-2.42 can be rewritten as

$$\frac{\partial \rho}{\partial t} = -\frac{\partial \rho u_j}{\partial x_j} \tag{2.43}$$

$$\frac{\partial \theta}{\partial t} = -u_j \frac{\partial \theta}{\partial x_j} + S_\theta, \tag{2.44}$$

$$\frac{\partial u_i}{\partial t} = -u_j \frac{\partial u_i}{\partial x_j} - \frac{1}{\rho} \frac{\partial p}{\partial x_i} - g\delta_{i3} - 2\epsilon_{ijk}\Omega_j u_k, \tag{2.45}$$

$$\frac{\partial q_n}{\partial t} = -u_j \frac{\partial q_n}{\partial x_j} + S_{q_n}, \quad n = 1, 2, 3 \tag{2.46}$$

$$\frac{\partial \chi_m}{\partial t} = -u_j \frac{\partial \chi_m}{\partial x_j} + S_{\chi_m}, \quad m = 1, 2, \ldots, M. \tag{2.47}$$

The definition of potential temperature, given by Eq. 2.21, is

$$\theta = T_{\rm V}\left(1000/p(\text{in mb})\right)^{R_{\rm d}/C_{\rm p}}, \tag{2.48}$$

and the ideal gas law (Eq. 2.13) can be written as

$$p = \rho R_{\rm d} T_{\rm V}, \tag{2.49}$$

where density ρ is the inverse of specific volume. The virtual temperature is given by

$$T_{\rm V} = T\left(1 + 0.61q_3\right) \tag{2.50}$$

from Eq. 2.12 and Eq. 2.13.

Equations 2.43-2.50 represent a simultaneous set of $11 + M$ nonlinear partial differential equations in the $11 + M$ dependent variables ($\rho, \theta, T, T_{\rm V}, p, u_i, q_n$, and χ_m) that must be solved if mesoscale circulations are to be studied quantitatively. The independent variables are time t and the three-space coordinates $x_1 = x, x_2 = y$, and $x_3 = z$. The remainder of the text discusses methods of simplification and solution for these fundamental physical relations. In working with mesoscale models and their results, it is imperative that investigators *always determine the extent to which the equations used in specific simulations correspond to these fundamental basic principles.*

Chapter 2 Additional Readings

There are several useful texts which are available to provide additional in-depth information on the material in this chapter. These include:

Curry, J., and P. Webster, 1999: *Thermodynamics of Atmospheres and Oceans*. International Geophysics, Volume 65, Academic Press, 471 pp.

Holton, J.R., 2004: *An Introduction to Dynamic Meteorology*. Fourth Edition, Academic Press, 535 pp.

Lin, Y.-L., 2007: *Mesoscale Dynamics*. Cambridge University Press, New York, 674 pp.

Lindzen, R.S., 2005: *Dynamics in Atmospheric Physics*. Cambridge University Press, New York, 324 pp.

Lovejoy, S., and D. Schertzer, 2013: *The Weather And The Climate: Emergent Laws And Multifractal Cascades*. Cambridge University Press, 496 pp.

Mak, M., 2011: *Atmospheric Dynamics*. Cambridge University Press, 536 pp.

North, G.R., and T. L. Erukhimova, 2009: *Atmospheric Thermodynamics: Elementary Physics and Chemistry*. Cambridge University Press, New York, 280 pp.

Tsonis, A.A., 2007: *An Introduction to Atmospheric Thermodynamics*. Second Edition, Cambridge University Press, 198 pp.

Warner, T., 2011: *Numerical Weather and Climate Prediction*. Cambridge University Press, 548 pp.

Zdunkowsk, W., and A. Bott, 2004: *Thermodynamics of the Atmosphere: A Course in Theoretical Meteorology*. Cambridge University Press, 266 pp.

CHAPTER

3 Simplification of the Basic Equations

CHAPTER OUTLINE HEAD

Equations 2.43-2.47 can be simplified for specific mesoscale meteorological simulations, and by mathematical operations, some of these relations can also be changed in form. In this chapter, commonly made assumptions will be reviewed and the resultant equations presented. In all cases, the Eqs. 2.43-2.47 are altered in form or simplified, or both, to permit their solutions to be made in an easier, more economical fashion.

The method of *scale analysis*[1] is often used to determine the relative importance of the individual terms in the conservation relations. This technique involves the estimation of their order of magnitude through the use of representative values of the dependent variables and constants that make up these terms. Scale analysis can be applied either to individual terms in the fundamental conservation equations, as applied in this chapter, or to analytic solutions of a coupled set of the conservation equations, as discussed in Section 5.2.2. The most rigorous analysis procedure, of course, is to evaluate specific solutions of the conservation equations with and without

[1] Scale analysis is specifically a procedure to use the dimensions of a set of variables and inter-relate then. Buckingham (1914) originally developed this method of dimensional analysis. Overviews are reported in Schmidt and Housen (1995) and Vogel (1998). Hicks (1978) and Meroney (1998) discuss misinterpretation of dimensional analyses, if the nondimensional variables are not properly selected.

Mesoscale Meteorological Modeling, Third Edition, Volume 98. http://dx.doi.org/10.1016/B978-0-12-385237-3.00003-7

particular terms to establish their importance. An example of such an analysis for the hydrostatic assumption in sea-breeze simulations, is described in Section 5.2.3.

In this chapter, the use of scale analysis is summarized. Discussions by investigators such as Thunis and Bornstein (1996) provide additional descriptions of the use of scale analysis to investigate scales of motion on the mesoscale.

3.1 Conservation of Mass

In Chapter 2, the mass-conservation relation was given by Eq. 2.43. To determine appropriate and consistent approximate forms of this equation, portions of the scale analysis of this equation by Dutton and Fichtl (1969) are utilized in the following discussion.

Using the relationship between density and specific volume given by

$$\rho = 1/\alpha,$$

Eq. 2.43 can be rewritten as

$$\frac{\partial \alpha}{\partial t} = -u_j \frac{\partial \alpha}{\partial x_j} + \alpha \frac{\partial u_j}{\partial x_j} \tag{3.1}$$

If it is assumed that

$$\alpha = \alpha_0 + \alpha',$$

where α_o is defined as a synoptic-scale reference specific volume and α' is the mesoscale perturbation from this value,[2] then Eq. 3.1 can be rewritten as

$$\frac{\partial\left(\alpha_0 + \alpha'\right)}{\partial t} = -u_j \frac{\partial}{\partial x_j}\left(\alpha_0 + \alpha'\right) + \left(\alpha_0 + \alpha'\right)\frac{\partial u_j}{\partial x_j}. \tag{3.2}$$

To simplify the scale analysis of this equation, it is assumed that

$$\left|\frac{\partial \alpha_0}{\partial t}\right| \ll \left|\frac{\partial \alpha'}{\partial t}\right|; \quad \left|\frac{\partial \alpha_0}{\partial x}\right| \ll \left|\frac{\partial \alpha'}{\partial x}\right|; \quad \left|\frac{\partial \alpha_0}{\partial y}\right| \ll \left|\frac{\partial \alpha'}{\partial y}\right|, \tag{3.3}$$

so that Eq. 3.2 becomes

$$\frac{\partial \alpha'}{\partial t} = -u\frac{\partial \alpha'}{\partial x} - v\frac{\partial \alpha'}{\partial y} - w\frac{\partial \alpha'}{\partial z} - w\frac{\partial \alpha_0}{\partial z} + \alpha_0\left(1 + \frac{\alpha'}{\alpha_0}\right)\left(\frac{\partial u}{\partial x} + \frac{\partial v}{\partial y} + \frac{\partial w}{\partial z}\right). \tag{3.4}$$

For mesoscale atmospheric circulations, the adoption of the assumptions given in Eq. 3.3 requires that the synoptic state change much more slowly than the mesoscale system and that the horizontal synoptic gradients are much less than the mesoscale gradients.

[2]In Section 4.1, α_0 and α' are defined more formally.

It is also assumed that

$$|\alpha'/\alpha_0| \ll 1,$$

so that Eq. 3.4 reduces to

$$\frac{\partial \alpha'}{\partial t} = -u\frac{\partial \alpha'}{\partial x} - v\frac{\partial \alpha'}{\partial y} - w\frac{\partial \alpha'}{\partial z} - w\frac{\partial \alpha_0}{\partial z} + \alpha_0\left(\frac{\partial u}{\partial x} + \frac{\partial v}{\partial y} + \frac{\partial w}{\partial z}\right). \quad (3.5)$$

This requirement on the ratio of the mesoscale perturbation specific volume to the synoptic-scale reference value is reasonable when realistic values of temperature and pressure are inserted into the ideal gas law (Eq. 2.13).

For example, a representative climatological value of α_0 at sea level is $0.80\,\mathrm{m^3 kg^{-1}}$. Since $\alpha = R_d T_V/p$, upper and lower bounds on specific volume can be estimated for realistic mesoscale situations using the highest temperature and lowest pressure and the lowest temperature and highest pressure likely to occur at a given location over a reasonably short time period (say 12-24 h). If

$$T_V = 40°\mathrm{C}, \quad p = 990 \text{ mb}$$

and

$$T_V = 20°\mathrm{C}, \quad p = 1030 \text{ mb},$$

then $\alpha = 0.91\ \mathrm{m^3kg^{-1}}$ and $\alpha = 0.82\ \mathrm{m^3kg^{-1}}$ for the two cases, so that $|\alpha'|/\alpha_0$ is at most around $\pm 15\%$.

The method of scale analysis is used to estimate the magnitude of the remaining terms in Eq. 3.5, so that

$$\begin{gathered}
\left|\frac{\partial \alpha'}{\partial t}\right| \sim \frac{\alpha'}{t_\alpha}, \quad \left|u\frac{\partial \alpha'}{\partial x}\right| \sim U\frac{\alpha'}{L_x}, \quad \left|v\frac{\partial \alpha'}{\partial y}\right| \sim V\frac{\alpha'}{L_y}, \quad \left|w\frac{\partial \alpha'}{\partial z}\right| \sim W\frac{\alpha'}{L_z}, \\
\left|w\frac{\partial \alpha_0}{\partial z}\right| \sim W\frac{\alpha_0}{H_\alpha}, \quad \left|\alpha_0\frac{\partial u}{\partial x}\right| \sim \alpha_0\frac{U}{L_x}, \quad \left|\alpha_0\frac{\partial v}{\partial y}\right| \sim \alpha_0\frac{V}{L_y}, \quad (3.6) \\
\left|\alpha_0\frac{\partial w}{\partial z}\right| \sim \alpha_0\frac{W}{L_z},
\end{gathered}$$

where t_α^{-1} represents the characteristic frequency of variations in specific volume on the mesoscale; U, V, and W are representative values of the components of velocity; L_x, L_y, and L_z are the spatial scales of the mesoscale disturbance; and H_α, the *density-scaled height* of the atmosphere, is defined as

$$H_\alpha^{-1} = \frac{1}{\alpha_0}\frac{\partial \alpha_0}{\partial z}.$$

In the Earth's troposphere, H_α is approximately 8 km, as illustrated schematically in Fig. 3.1.

3.1.1 Deep Continuity Equation

To examine the necessity for retaining individual terms in Eq. 3.5, it is customary to examine their ratio relative to one of the terms that is expected to remain. In the first

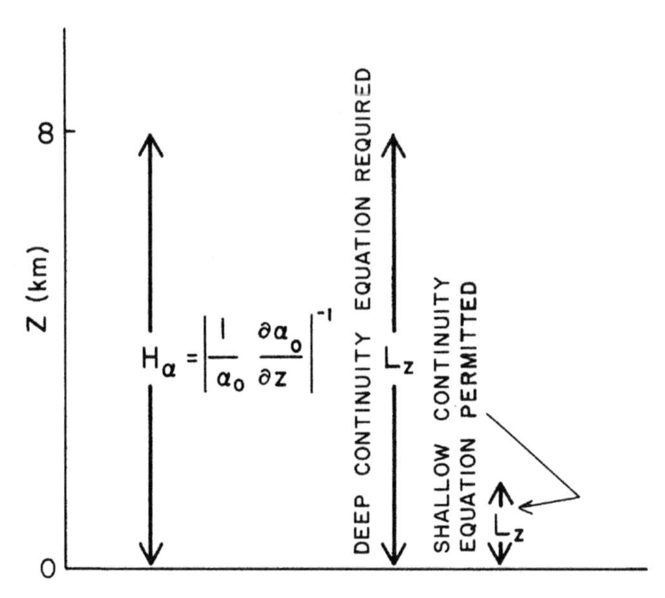

FIGURE 3.1

Schematic illustration contrasting deep and shallow atmospheric circulations. The depth L_z corresponds to the vertical extent of the circulation. In the Earth's atmosphere, $H_\alpha = 8$ km.

case to be examined, the terms are divided by the order of magnitude estimate for $w\,\partial\alpha_0/\partial z$, resulting in

$$\left|\frac{\partial\alpha'}{\partial t}\right| \Big/ \left|w\frac{\partial\alpha_0}{\partial z}\right| \sim \frac{\alpha'}{\alpha_0}\frac{H_\alpha}{t_\alpha W}, \qquad \left|u\frac{\partial\alpha'}{\partial x}\right| \Big/ \left|w\frac{\partial\alpha_0}{\partial z}\right| \sim \frac{\alpha'}{\alpha_0}\frac{U}{W}\frac{H_\alpha}{L_x},$$

$$\left|v\frac{\partial\alpha'}{\partial y}\right| \Big/ \left|w\frac{\partial\alpha_0}{\partial z}\right| \sim \frac{\alpha'}{\alpha_0}\frac{V}{W}\frac{H_\alpha}{L_y},$$

$$\left|w\frac{\partial\alpha'}{\partial z}\right| \Big/ \left|w\frac{\partial\alpha_0}{\partial z}\right| \sim \frac{\alpha'}{\alpha_0}\frac{H_\alpha}{L_z}, \qquad \left|\alpha_0\frac{\partial u}{\partial x}\right| \Big/ \left|w\frac{\partial\alpha_0}{\partial z}\right| \sim \frac{U}{W}\frac{H_\alpha}{L_x}$$

$$\left|\alpha_0\frac{\partial v}{\partial y}\right| \Big/ \left|w\frac{\partial\alpha_0}{\partial z}\right| \sim \frac{V}{W}\frac{H_\alpha}{L_y},$$

and

$$\left|\alpha_0\frac{\partial w}{\partial z}\right| \Big/ \left|w\frac{\partial\alpha_0}{\partial z}\right| \sim \frac{H_\alpha}{L_z}.$$

Since $|\alpha'/\alpha_0| \ll 1$, the terms $u\,\partial\alpha'/\partial x$, $v\,\partial\alpha'/\partial y$, and $w\,\partial\alpha'/\partial z$ are much less than $\alpha_0\,\partial u/\partial x$, $\alpha_0\,\partial v/\partial y$, and $\alpha_0\,\partial w/\partial z$ and can be neglected in Eq. 3.5, provided

$$\text{(i) } L_z \sim H_\alpha, \quad \text{(ii) } \frac{U}{W}\frac{H_\alpha}{L_x} \sim 1, \quad \text{(iii) } \frac{V}{W}\frac{H_\alpha}{L_y} \sim 1 \quad \text{(iv) } \frac{H_\alpha}{t_\alpha W} \sim 1. \tag{3.7}$$

Then Eq. 3.5 can be written as

$$w\frac{\partial \alpha_0}{\partial z} - \alpha_0\left(\frac{\partial u}{\partial x} + \frac{\partial v}{\partial y} + \frac{\partial w}{\partial z}\right) = 0 \qquad (3.8)$$

Since L_z is approximately equal to H_α, conditions (ii) and (iii) in Eq. 3.7 require that

$$U/L_x \sim W/L_z \quad \text{and} \quad V/L_y \sim W/L_z. \qquad (3.9)$$

If therefore, L_x is one or two orders of magnitude larger than L_z, it is expected that the vertical velocity W should be one or two orders of magnitude less than U and V (e.g., for $L_x \sim 80$ km $L_z \sim 8$ km, then $W \sim 0.1U$). This relation between velocities and the horizontal and vertical scales of atmospheric circulations results from the condition that $|\alpha'/\alpha_0| \ll 1$. Because of this constraint, velocities in a longer, horizontal leg of an atmospheric circulation must be proportionately stronger to preserve mass continuity without creating large fluctuations in specific volume.

The last requirement that needs to be justified in Eq. 3.7 is condition (iv). The scaled variable t_α represents a time period in which significant variations in specific volume occur on the mesoscale and can be approximated by

$$t_\alpha \sim L/C, \qquad (3.10)$$

where L is the wavelength over which variations occur and C the rate of movement of these variations. If the changes are caused by advection, then U, V, or W is used to represent C, whereas a characteristic group velocity C_g will be utilized if wave propagation is dominant. The wavelength L is estimated as L_x and L_y for horizontal motion and L_z for vertical movement. When changes in specific volume are assumed primarily to be caused by advection or when the wave group velocity has approximately the same speed as the wind velocity,[3] then

$$t_\alpha W \sim LW/C \sim \begin{cases} L_x W/U \sim L_y W/V \sim H_\alpha \\ L_z W/W \sim L_z \sim H_\alpha \end{cases}$$

where conditions (i)-(iii) in Eq. 3.7 have been used. Therefore,

$$\left|\frac{\partial \alpha'}{\partial t}\right| \Big/ \left|w\frac{\partial \alpha_0}{\partial z}\right| \sim \left|\frac{\alpha'}{\alpha_0}\right|,$$

so that local variations in density can be neglected in the conservation of mass relation if $|\alpha'/\alpha_0| \ll 1$.

[3] As will be discussed in Chapter 5, one of the significant types of wave motions on the mesoscale are internal gravity waves. Speeds as high as 30 m s^{-1} or so are typical of such features (e.g., see Gedzelman and Donn 1979). Also, as illustrated by Dutton (1976, Fig. 12-5) the relationship between the group velocity for internal gravity waves and their wavelength is given by an expression similar to Eq. 3.9 (i.e., $C_{gx}/L_x \sim C_{gz}/L_z$, where C_{gx} and C_{gz} are the group velocities in the x and z directions).

Finally, if $L_y \ll L_x$, mesoscale variations in the x direction are expected to be dominant and the y derivatives in Eq. 3.8 can be ignored. In this case the equation reduces to the two-dimensional form given by

$$w \frac{\partial \alpha_0}{\partial z} - \alpha_0 \left(\frac{\partial u}{\partial x} + \frac{\partial w}{\partial z} \right) = 0.$$

Equation 3.8 is customarily written to include the terms $u\, \partial \alpha_0 / \partial x$ and $v\, \partial \alpha_0 / \partial y$ and is given by

$$\frac{\partial}{\partial x_j} \left(\alpha_0^{-1} u_j \right) = 0,$$

or, returning to the use of density instead of specific volume,

$$\frac{\partial}{\partial x_j} \left(\rho_0 u_j \right) = 0, \tag{3.11}$$

where $\rho_0 = 1/\alpha_0$.

Dutton and Fichtl (1969) refer to this relation as the *deep convection continuity equation* because the vertical depth of the circulation is on the same order as the density scale depth. As originally shown by Ogura and Phillips (1962), and discussed by Lipps and Hemler (1982), and as will be shown in Section 5.2.2, the use of this form of the conservation-of-mass relation eliminates sound waves as a possible solution, which led Ogura and Phillips to refer to Eq. 3.11 as the *anelastic*, or *soundproof assumption*. Because such waves are of no direct interest in most applications of mesoscale meteorology, this equation is often employed to represent mass conservation in mesoscale models in lieu of the more complete prognostic conservation-of-mass equation given by Eq. 3.1. Moreover, as discussed in Section 9.3, the elimination of sound waves permits more economical use of certain numerical solution techniques because their computational stability is dependent on having a time step less than or equal to the time it takes a wave to travel between grid points. Sound waves are the fastest non-electromagnetic waveform in the atmosphere.

3.1.2 Shallow Continuity Equation

A more restrictive mass-conservation relation is derived by dividing the scaled terms in Eq. 3.6 by the scaled form of $\alpha_0\, \partial w / \partial z$ resulting in

$$\left| \frac{\partial \alpha'}{\partial t} \right| \Big/ \left| \alpha_0 \frac{\partial w}{\partial z} \right| \sim \frac{\alpha'}{\alpha_0} \frac{L_z}{W t_\alpha}, \quad \left| u \frac{\partial \alpha'}{\partial x} \right| \Big/ \left| \alpha_0 \frac{\partial w}{\partial z} \right| \sim \frac{\alpha'}{\alpha_0} \frac{U}{W} \frac{L_z}{L_x},$$

$$\left| v \frac{\partial \alpha'}{\partial y} \right| \Big/ \left| \alpha_0 \frac{\partial w}{\partial z} \right| \sim \frac{\alpha'}{\alpha_0} \frac{V}{W} \frac{L_z}{L_y}, \quad \left| w \frac{\partial \alpha'}{\partial z} \right| \Big/ \left| \alpha_0 \frac{\partial w}{\partial z} \right| \sim \frac{\alpha'}{\alpha_0},$$

$$\left| w \frac{\partial \alpha_0}{\partial z} \right| \Big/ \left| \alpha_0 \frac{\partial w}{\partial z} \right| \sim \frac{L_z}{H_\alpha}, \quad \left| \alpha_0 \frac{\partial u}{\partial x} \right| \Big/ \left| \alpha_0 \frac{\partial w}{\partial z} \right| \sim \frac{U}{W} \frac{L_z}{L_x},$$

$$\left| \alpha_0 \frac{\partial v}{\partial y} \right| \Big/ \left| \alpha_0 \frac{\partial w}{\partial z} \right| \sim \frac{V}{W} \frac{L_z}{L_y}.$$

As in the previous analysis, since $|\alpha'/\alpha_0| \ll 1$, $u\partial\alpha'/\partial x$, $v\partial\alpha'/\partial y$, and $w\partial\alpha'/\partial z$ can be neglected in Eq. 3.5, provided

$$L_z/Wt_\alpha \sim 1, \quad L_z/H_\alpha \ll 1, \tag{3.12}$$

then Eq. 3.5 can be written as

$$\alpha_0 \left(\frac{\partial u}{\partial x} + \frac{\partial v}{\partial y} + \frac{\partial w}{\partial z} \right) = 0. \tag{3.13}$$

The first condition in Eq. 3.12 is easier to satisfy than the equivalent requirement in Eq. 3.7 because $L_z \ll H_\alpha$. This requirement implies that the neglect of specific volume variations in Eq. 3.13 is even less important than in Eq. 3.11. The second condition in Eq. 3.12 requires that the depth of the circulation be much less than the scale depth of the atmosphere. For this reason Dutton and Fichtl refer to Eq. 3.13 as the *shallow convection continuity equation*. Written in tensor form, Eq. 3.13 is given by

$$\partial u_j/\partial x_j = 0, \tag{3.14}$$

and this relation is often referred to as the *incompressibility assumption*. This expression not only removes sound waves, but ignores spatial variations in density. For the case of a *homogeneous* (constant density) fluid, this would be the exact form of the conservation-of-mass equation.

Mesoscale models have traditionally utilized this expression to represent the conservation of mass. There is a certain irony in its use, of course, because although air closely follows the ideal gas law, it is also accurately approximated by the incompressible form of the mass-conservation equation when the atmospheric circulations have a limited vertical extent. This apparent discrepancy is explained by realizing that air is, generally, not physically constrained in its movement. When air moves into one side of a parcel, for example, the density can either increase by compression or an equivalent mass of air can move out of the other side of the parcel. As long as the atmospheric parcel is not restricted to a fixed volume, the creation of a pressure gradient between the two sides of the parcel as a result of the different velocities will force the air out of one side so that mass conservation is closely approximated by the incompressible relation (Eq. 3.14).

In mesoscale models either the *prognostic equation* (time-dependent equation) for density (Eq. 2.43), or the *diagnostic equation* (no time-tendency term; Eq. 3.11 or 3.14) can be used to represent mass conservation.

3.2 Conservation of Heat

In mesoscale meteorology, an equivalent, rather exhaustive scale analysis of the conservation of heat relation (Eq. 2.44) is not generally made. This is because of the complex mathematical form of the source-sink term S_θ. The conservation-of-mass relation, in contrast, has no such source-sink term so that the analysis is relatively simple.

Equation 2.44 is modified by making simplifying assumptions regarding the form of S_θ. The development of simplified mathematical representations for any of the source-sink terms is one type of *parameterization*. The most stringent assumption for the conservation of potential temperature relation is to require that all motions be adiabatic, so that $S_\theta = 0$ and Eq. 2.44 reduces to

$$\partial\theta/\partial t = -u_j\,\partial\theta/\partial x_j, \tag{3.15}$$

or, equivalently,

$$d\theta/dt = 0.$$

It is valid to use this assumption in representing mesoscale atmospheric systems provided

$$|S_\theta| \ll \left|\frac{\partial\theta}{\partial t}\right|;\quad \left|u\,\frac{\partial\theta}{\partial x}\right|;\quad \left|v\,\frac{\partial\theta}{\partial y}\right|;\quad \left|w\,\frac{\partial\theta}{\partial z}\right|.$$

This condition is most closely fulfilled when the following are met.

1. The atmosphere is dry with no phase changes of water occurring.
2. Comparatively short time periods are involved so that radiational heating or cooling of the air is relatively small.
3. The heating or cooling of the lowest levels of the atmosphere by the bottom surface is of comparatively small magnitude.

More specific examples of when S_θ must be retained in Eq. 2.44 and how this term can be parameterized are discussed in Chapters 7 and 8.

3.3 Conservation of Motion

A wide range of assumptions have been utilized either to simplify or to alter the form of the conservation of motion equation (Eq. 2.45). In performing scale analysis on this equation, it is convenient to decompose the equation into its vertical and horizontal components since the gravitational acceleration is only included in the equation for vertical acceleration.

3.3.1 Vertical Equation of Motion

The vertical component of Eq. 2.45 can be written as

$$\frac{\partial w}{\partial t} + u\,\frac{\partial w}{\partial x} + v\,\frac{\partial w}{\partial y} + w\,\frac{\partial w}{\partial z} = -\frac{1}{\rho}\frac{\partial p}{\partial z} - g + 2\Omega u\cos\phi. \tag{3.16}$$

or equivalently,

$$\frac{dw}{dt} = -\alpha\,\frac{\partial p}{\partial z} - g + 2\Omega u\cos\phi. \tag{3.17}$$

As in Section 3.1, these terms can be estimated by the *method-of-scale analysis*. The terms on the left side of Eq. 3.16 can be estimated as

$$\left|\frac{\partial w}{\partial t}\right| \sim \frac{W}{t_w}; \quad \left|u\frac{\partial w}{\partial x}\right|; \quad \left|v\frac{\partial w}{\partial y}\right|; \quad \left|w\frac{\partial w}{\partial z}\right| \sim \frac{U}{L_x}W,$$

where t_w is a time scale related to the period required for significant changes in vertical velocity to occur. A similar definition was given in Section 3.1 for t_α. As before, W and U are the characteristic vertical and horizontal velocities, which if $|\alpha'/\alpha_0| \ll 1$, permit Eq. 3.9 to be used to relate the magnitudes of the two velocities (V and L_y are assumed here to have the same magnitudes as U and L_x). Using the same justification for the vertical velocity time scale t_w as used for t_α, then

$$t_w \sim L_x/U,$$

so that the four terms on the left of Eq. 3.16, and, therefore the total vertical acceleration given on the left of Eq. 3.17, have the scale of

$$\left|\frac{\partial w}{\partial t}\right|; \quad \left|u\frac{\partial w}{\partial x}\right|; \quad \left|v\frac{\partial w}{\partial y}\right|; \quad \left|w\frac{\partial w}{\partial z}\right|; \quad \left|\frac{dw}{dt}\right| \sim \frac{UW}{L_x} = \frac{U^2L_z}{L_x^2}.$$

To estimate the vertical pressure gradient term in Eq. 3.17, it is convenient to utilize the ideal gas law (Eq. 2.13), yielding

$$\alpha\frac{\partial p}{\partial z} = R_\mathrm{d}\frac{\partial T}{\partial z} - \frac{RT}{\alpha}\frac{\partial \alpha}{\partial z},$$

where, for convenience, the subscripts d and V have been deleted from R and T. To simplify the scale analysis, it is assumed that the atmosphere is isothermal ($\partial T/\partial z = 0$) and that

$$\frac{1}{\alpha}\frac{\partial \alpha}{\partial z} \simeq \frac{1}{\alpha_0}\frac{\partial \alpha_0}{\partial z} = H_\alpha^{-1}$$

so that

$$\left|\alpha\frac{\partial p}{\partial z}\right| \sim R_\mathrm{d}T/H_\alpha.$$

The two remaining terms are given as $|g|$ and $|2\Omega u\cos\phi| \sim 2\Omega U$.

The significance of the individual terms in Eq. 3.17 in relation to the vertical pressure gradient term, for example, can then be estimated using these scale estimates. The ratio of the orders of magnitude of the vertical acceleration to the vertical pressure gradient is given by

$$\left|\frac{dw}{dt}\right| \bigg/ \left|\alpha\frac{\partial p}{\partial z}\right| \sim \frac{H_\alpha L_z}{L_x^2}\frac{U^2}{RT} = R_w. \tag{3.18}$$

For the case when Eq. 3.8 is used as the continuity equation, $L_z \sim H_\alpha$ so that

$$\left|\frac{dw}{dt}\right| \bigg/ \left|\alpha\frac{\partial p}{\partial z}\right| \sim \frac{H_\alpha^2}{L_x^2}\frac{U^2}{RT} = R_w \tag{3.19}$$

The scale analysis ratio given by Eq. 3.19 implies that the vertical accelerations become more important as the horizontal velocity increases, whereas higher temperatures and longer horizontal wavelength decrease its importance.

According to this scale analysis, therefore, to neglect the vertical acceleration term in Eq. 3.17, R_w must be much less than unity. Unfortunately, however, this analysis is not complete because we have shown only that the magnitude of the gravitational and pressure gradient accelerations are separately much larger than the vertical acceleration. Of more significance is the magnitude of the difference between these two terms. To better examine this relationship, it is convenient to define a large-scale averaged atmosphere that has an exact balance between the gravitational and pressure gradient terms. This can be defined as

$$\partial p_0/\partial z = -\rho_0 g,$$

where the zero subscript is used to indicate a large-scale average (such an average could be defined as given by Eq. 4.12). If any dependent variable is defined to be equal to such an average value plus a deviation from that average (i.e., $\phi = \phi_0 + \widehat{\phi}$, where ϕ is any one of the dependent variables), then Eq. 3.17 can also be written as

$$\frac{dw}{dt} = -\alpha_0 \frac{\partial \hat{p}}{\partial z} + g \frac{\widehat{\alpha}}{\alpha_0} + 2\Omega u \cos\phi, \tag{3.20}$$

where $|\widehat{\alpha}|/\alpha_0 \ll 1$ has been assumed. In Eq. 3.20, the first two terms on the righthand side are much less in magnitude than the first two terms on the righthand side of Eq. 3.17.

The magnitude of the perturbation vertical pressure gradient can be estimated from

$$\left|\frac{1}{\rho_0}\frac{\partial \hat{p}}{\partial z}\right| \simeq \frac{1}{\rho_0}\left|T_0 R \frac{\partial \hat{\rho}}{\partial z} + R\widehat{T}\frac{\partial \rho_0}{\partial z} + R\rho_0 \frac{\partial \widehat{T}}{\partial z}\right|$$
$$\sim \left|-R\frac{\delta T}{L_z}\right| + \left|R\frac{\delta T}{H_\alpha}\right| + \left|R\frac{\delta T}{L_z}\right| \sim R\frac{|\delta T|}{L_z},$$

where for the purposes of this scale analysis, the large-scale atmosphere is assumed isothermal and the linearized ideal gas law of the form $\delta\rho/\rho \simeq -(\delta T/T) + (\delta p/p) \simeq -\delta T/T$ has been used. For this analysis, the term with the vertical gradient of the large-scale density ρ_0 can be neglected if $L_z \ll H_\alpha$.

The ratio of the vertical acceleration to this perturbation pressure gradient term is thus given as

$$\widehat{R}_w = \frac{L_z^2}{L_x^2}\frac{UC_x}{R\delta T} = \frac{H_\alpha^2}{L_x^2}\frac{UC_x}{R\delta T} \text{ (when } L_z \sim H_\alpha\text{)}. \tag{3.21}$$

This relationship is more restrictive than R_w since δT is in the denominator rather than T [note that $\widehat{R}_w$ does not increase without bound because UC_x results from horizontal gradients in temperature (e.g., see Section 3.3.2), so as δT goes to zero in Eq. 3.21, so must UC_x].[4] If $\widehat{R}_w \ll 1$, $g\widehat{\alpha}/\alpha_0$ must be of the same order of magnitude as the perturbation vertical pressure gradient (since $2\Omega u \cos\phi$ is small relative to the first two terms in Eq. 3.20 under all expected atmospheric conditions as discussed shortly).

To illustrate the magnitude of $\widehat{R}_w$ for representative values on the mesoscale, δT and U are set equal to 10°C and 10 m s^{-1}, respectively (based on observed values), C_x is set equal to U, and R is equal to 287 J K^{-1}kg^{-1}. This yields

$$\widehat{R}_w = 0.03 H_\alpha^2/L_x^2,$$

where $H_\alpha \simeq 8$ km (Wallace and Hobbs 1977). If the depth of the circulation is less than the scale height ($L_z < H_\alpha$), $\widehat{R}_w$ is proportionately smaller (e.g., if $L_z = 0.1H_\alpha$, $\widehat{R}_w = 0.0003\ H_\alpha^2/L_x^2$). Thus from this analysis a conservative estimate for neglecting vertical accelerations relative to the vertical pressure gradient term in Eq. 3.17 is

$$H_\alpha/L_x \precsim 1. \tag{3.22}$$

The remaining two terms are evaluated by

$$g/|\alpha\ \partial p/\partial z| \sim gH_\alpha/RT$$

and

$$|2\Omega u \cos\phi|/\ |\alpha\ \partial p/\partial z| \sim 2\Omega U H_\alpha/RT,$$

and if $g \sim 10$ m s^{-2}, $T = 270$ K, $\Omega \sim 7 \times 10^{-5}$s^{-1} $= 2\pi$/day, $U \sim 20$ m s^{-1}, $H_\alpha \sim 8$ km, and $R = 287$ J K^{-1}kg^{-1}, for example, then

$$g/|\alpha\ \partial p/\partial z| \sim 1$$

and

$$|2\Omega \cos\phi|/\ |\alpha\ \partial p/\partial z| \sim 3 \times 10^{-4}.$$

Thus the influence of the rotation of the Earth on the vertical acceleration is inconsequential for any reasonable velocity and can be neglected in Eq. 3.17. The gravitational acceleration cannot be neglected, however, relative to the vertical pressure gradient.

If Eq. 3.22 is valid, Eq. 3.17 reduces to

$$\partial p/\partial z = -g/\alpha = -\rho g \tag{3.23}$$

[4]In addition, if the horizontal pressure gradient force and the horizontal acceleration could be assumed to be of equal magnitude, then using the analysis discussed in Section 3.3.2, Eq. 3.21 can also be written as $\widehat{R}_w = (L_z^2 \delta_x T)/(L_x^2 \delta T)$, where the subscript x on δ in the numerator is used to indicate that the difference in T is in the horizontal direction (in contrast to past analyses by other investigators, δT and $\delta_x T$ need not be of the same magnitude). Thus for this situation, a larger difference in temperature over the same horizontal scale L_x results in a larger vertical acceleration, whereas a greater change in the perturbation temperature for a specific vertical scale L_z reduces the importance of the vertical acceleration relative to the vertical pressure gradient. This latter result implies that the more rapidly a temperature disturbance decreases with height (perhaps due to a more stable large-scale atmosphere), the greater is the reduction in the importance of the vertical acceleration.

and is called the *hydrostatic equation*.[5] In utilizing this relationship, it must be emphasized that the results of the scale analysis imply only that the magnitude of the vertical acceleration is much less than the magnitude of the pressure gradient force *not* that the magnitude of the vertical acceleration is identically zero (e.g., $|dw/dt| \ll |\alpha\, \partial p/\partial z|$ *not* that $|dw/dt| = 0$).

3.3.2 Horizontal Equation of Motion

The horizontal component of Eq. 2.45 can be written as

$$\frac{\partial u}{\partial t} = -u\,\frac{\partial u}{\partial x} - v\,\frac{\partial u}{\partial y} - w\,\frac{\partial u}{\partial z} - \frac{1}{\rho}\frac{\partial p}{\partial x} + 2v\Omega\sin\phi - 2w\Omega\cos\phi \tag{3.24}$$

and

$$\frac{\partial v}{\partial t} = -u\,\frac{\partial v}{\partial x} - v\,\frac{\partial v}{\partial y} - w\,\frac{\partial v}{\partial z} - \frac{1}{\rho}\frac{\partial p}{\partial y} - 2u\Omega\sin\phi. \tag{3.25}$$

To estimate the approximate magnitude of the pressure gradient term in Eqs. 3.24 and 3.25, two methods will be used. First, it is useful to apply the vertically integrated form of Eq. 3.23, assuming the density ρ is a constant ρ^* so that

$$p_{z=0} = \int_{z=0}^{z=D} \rho^* g\,dz = \rho^* g \int_{z=0}^{z=D} dz = \rho^* g D,$$

where the pressure at height D is assumed equal to zero. Thus if the magnitude of the pressure gradient at $z = 0$ is representative of the pressure gradient at any level in the troposphere, then

$$\left|\frac{1}{\rho}\frac{\partial p}{\partial x}\right| \sim g\,\frac{\delta D}{L_x}, \quad \left|\frac{1}{\rho}\frac{\partial p}{\partial y}\right| \sim g\,\frac{\delta D}{L_y},$$

where L_x and L_y are the representative horizontal scales of the mesoscale system and δD is the representative change of D over L_x and L_y. The parameter δD represents the relation between horizontal pressure gradient and horizontal gradients in the depth of a homogeneous atmosphere (i.e., $\delta p/L_x \sim \rho^* g\,\delta D/L_x$).

The second technique involves replacing pressure in Eq. 3.24 with the ideal gas law (Eq. 2.13) yielding for the x derivative (for convenience subscripts d and V have been dropped)

$$\frac{\partial p}{\partial x} = \rho R\,\frac{\partial T}{\partial x} + TR\,\frac{\partial \rho}{\partial x} \simeq \rho_0 R\,\frac{\partial T}{\partial x}, \tag{3.26}$$

where to estimate the magnitude of the pressure gradient, the term $T\,\partial\rho/\partial x$ is neglected and ρ_0 is used to approximate ρ.

[5] This form of the hydrostatic equation is actually valid only for a flat world, since the $x-y-z$ Cartesian system was used in its derivation. However, as discussed by Bannon et al. (1997), since the Earth is a sphere, the actual pressure change with altitude is slightly modified from that in Eq. 3.23. The actual surface pressure, for example, would be about 0.25% less than calculated using Eq. 3.23.

Thus utilizing these two analyses,

$$\left|\frac{1}{\rho}\frac{\partial p}{\partial x}\right| \sim \frac{R\,\delta T}{L_x} \sim g\,\frac{\delta D}{L_x}, \quad \left|\frac{1}{\rho}\frac{\partial p}{\partial y}\right| \sim \frac{R\,\delta T}{L_y} \sim \frac{g\,\delta D}{L_y},$$

where δT is the representative magnitude of the horizontal temperature variations across the mesoscale system. From these relations, $\delta D \sim (R/g)\delta T$.

The advection terms are given by

$$\begin{aligned} |u\,\partial u/\partial x| &\sim U^2/L_x, \quad |v\,\partial u/\partial y| \sim UV/L_y, \quad |w\,\partial u/\partial z| \sim WU/L_z, \\ |u\,\partial v/\partial x| &\sim UV/L_x, \quad |v\,\partial v/\partial y| \sim V^2/L_y, \quad |w\,\partial v/\partial z| \sim WV/L_z. \end{aligned} \tag{3.27}$$

If $U \sim V$ and $L_x \sim L_y$ (as would be expected in general) and since $W/L_z \sim U/L_x$ from relation 3.9, then these terms are of the same order. If $L_y \gg L_x$, the y derivative terms in Eq. 3.27 can be neglected, and Eqs. 3.24 and 3.25 reduce to their two-dimensional forms. Henceforth, in this section U^2/L_x will be used to represent the advective terms and V and L_y will be replaced by U and L_x whenever they appear.

The local tendency terms in Eqs. 3.24 and 3.25 are estimated by

$$|\partial u/\partial t| \sim U/t_{\mathrm{u}} \sim UC/L_x, \quad |\partial v/\partial t| \sim V/t_{\mathrm{v}} \sim VC/L_y \sim UC/L_x,$$

where, as for the vertical component of the conservation-of-motion equation, $C \simeq U$ for both advective and internal gravity wave changes in mesoscale systems can often be assumed.

The remaining three terms in Eqs. 3.24 and 3.25 can be represented as

$$|2u\Omega\sin\phi| \sim |fu| \sim |f|U, \quad |2v\Omega\sin\phi| \sim |fv| \sim |f|V \sim |f|U,$$

$$|2w\Omega\cos\phi| \sim |\hat{f}w| \sim |\hat{f}|W \sim |\hat{f}|\left(L_z/L_x\right)U,$$

where $f = 2\Omega\sin\phi$ and is called the *Coriolis parameter* and $\hat{f} = 2\Omega\cos\phi$. $R_0 = U/|f|L_x$ is the *Rossby number.*

Some of the interpretations that can be made from this analysis include the following:

1. The local time tendency terms $\partial u/\partial t$ and $\partial v/\partial t$ can be neglected if the movement of the mesoscale system is much less than the advecting wind speed (e.g., $C \sim C_{\mathrm{g}} \ll U$). Such a system is said to be *steady state* if $C_{\mathrm{g}} = 0$ or *quasi-steady* if $C_{\mathrm{g}} \neq 0$ but $C_{\mathrm{g}} \ll U$.
2. The terms associated with the rotation of the Earth (fu, fv, and $\hat{f}w$) can be neglected if $R_0 \gg 1$. Since R_0 is inversely proportional to L_x, the larger the horizontal scale of the mesoscale system, the more inappropriate it becomes to neglect those terms. The magnitude of the $\hat{f}w$ term is dependent on the ratio of L_z/L_x (e.g., if $L_z = 0.1L_x$, it is expected to be about 10% of fu and fv).

3. The ratio of the pressure gradient force and advective terms is inversely proportional to the square of the wind speed and proportional to the horizontal temperature gradient. Since the horizontal pressure gradient is approximately a linear term

$$\left(\text{e.g.,}\quad \frac{1}{\rho}\frac{\partial p}{\partial x} \simeq \frac{1}{\rho_0}\frac{\partial p}{\partial x} \quad \text{since} \left|\frac{\rho'}{\rho_0}\right| \sim \left|\frac{\alpha'}{\alpha_0}\right| \ll 1\right),$$

whereas the advective terms are nonlinear, solutions to the conservation relations are greatly simplified if the ratio is large and the nonlinear contribution of advection to the equations can be ignored. Equations 3.24 and 3.25 are linearized if $\rho = \rho_0$ is used in the pressure gradient term, and the advective terms are deleted.

Representative values of R_0, D, and δD can be estimated for mesoscale systems in the Earth's troposphere. With $|f| \sim 10^{-4}\text{s}^{-1}$, $U \sim 20 \text{ m s}^{-1}$, $p_{z=0} \sim 1000$ mb, $g = 10 \text{ m s}^{-2}$, $\rho_0 = 1.25 \text{ kg m}^{-3}$, $\delta T \sim 5°\text{C}$, $L_x = 100$ km, and $L_z = D$, for example,

$$D = 8 \text{ km} \sim H_\alpha, \quad R_0 = 2.0, \quad \delta D = 143 \text{ m}, \quad L_z/L_x = 0.08.$$

Thus for this case it would be appropriate to neglect the term $\hat{f}w$ but not fu and fv. Moreover, because the ratio of the magnitudes of the pressure gradient force term to the advective terms is on the order of four, neither of these terms can be neglected in Eqs. 3.24 and 3.25. The local tendency terms $\partial u/\partial t$ and $\partial v/\partial t$ can also not be ignored if $C \sim U$, since for this case their ratio to the pressure gradient force term is approximately 0.3.

These scaled terms need to be reevaluated for each study of different mesoscale systems to determine the appropriate form of Eqs. 3.24 and 3.25 to use. If in doubt as to whether a term should be excluded, it is, of course, consistent to include it.[6] In the limiting case of $R_0 \ll 1$, $L_z \ll L_x$, $U \sim C$, and $R\delta T \gg U^2$ with $\delta D \sim |f|UL_x/g$ (which is the same as $\delta T \sim |f|UL_x/R$), Eqs. 3.24 and 3.25 reduce to

$$v_g = \frac{1}{\rho f}\frac{\partial p}{\partial x}, \quad u_g = -\frac{1}{\rho f}\frac{\partial p}{\partial y}, \tag{3.28}$$

where v_g and u_g are called the *geostrophic wind components*.[7] At the other extreme, if $R_0 \gg 1$, the Coriolis terms can be neglected relative to the advective terms. This is the assumption used in cumulus and other smaller-scale models.

A final analysis of the vertical and horizontal equations of motion is to examine the magnitude of the molecular viscous forces to the terms in Eq. 2.45. As stated in the discussion of internal forces in Section 2.3.2, molecular forces are assumed

[6] In addition, quantitative linear (e.g., see Section 5.2.3) and nonlinear (e.g., see Section 11.5) models should be integrated with and without selected terms in the conservation equations to determine their importance for specific atmospheric circulations.

[7] When curvature of the synoptic horizontal pressure field is included, the resultant wind is called the *gradient wind*.

insignificant on the mesoscale and are ignored in Eq. 2.45. To justify this assumption, the method of scale analysis is used in which the molecular dissipation of motion D is approximated by

$$D_i = v\partial^2 u_i/\partial x_j \partial x_j, \quad |D| \sim vS/L^2, \tag{3.29}$$

where L represents L_x, L_y, or L_z (for $j = 1, 2$, or 3) and S represents U, V, or W (for $i = 1, 2$, or 3). The kinematic viscosity v is about 1.5×10^{-5} m^{2}s^{-1} for air. To examine the significance of the viscous force, it is customary to compare it with the advective terms, which have a magnitude of U^2/L_x in the horizontal equation and W^2/L_z in the vertical equation of motion. For this analysis, if it is assumed that $L_x \sim L_z \sim L$, so that $W \sim U \sim S$; then the ratio of the magnitude of the advective terms to the viscous force is given by

$$\frac{S^2/L}{vS/L^2} = \frac{SL}{v} = Re,$$

where Re is called the *Reynolds number*. When $Re \gg 1$, changes in motion by advection are much more important than the dissipation of velocity by molecular interactions. Under this condition, the flow is said to be *turbulent* and transfers of all properties of the air (e.g., heat, water vapor) are performed through the movement of air from one point to another. By contrast, when $Re \ll 1$, the molecular dissipation of velocity dominates and the flow is said to be *laminar*. Under this condition properties of the air are transferred on the molecular scale. When $Re \sim 1$, both effects are important and initial and boundary conditions imposed on the flow will determine whether laminar or turbulent mixing predominates.[8]

On the mesoscale, the Reynolds number is very large and the flow is highly turbulent. With $S = W = 1.5$ cm s^{-1} and $L = L_z = 1$ km, for example, $Re = 10^6$. Only near the ground do molecular transfers become important. For this situation, air flow with $L_z \sim 0.1$ mm and $W \sim 1.5$ cm s^{-1}, for example, result in a Reynolds number of 0.1, so that the movement of air is laminar. However, since atmospheric flow above a centimeter or so off the ground have Reynolds numbers much greater than unity, the viscous dissipation term is neglected in Eq. 2.45.

Using the scale analysis in this section, a more formal definition of mesoscale can be given than presented in the Introduction. The criteria are that

1. *the horizontal scale be sufficiently large so that the hydrostatic equation can be applied, and*
2. *the horizontal scale be sufficiently small so that the Coriolis term is small (although it can still be significant) relative to the advective and pressure gradient forces, resulting in a flow field that is substantially different from the gradient wind relation even in the absence of friction effects.*

[8] Tennekes and Lumley (1972) reported that boundary layer flows in a zero pressure gradient become turbulent at $Re \sim 600$ whereas a fluid far from a boundary attains this condition at values of the Reynolds number much closer to unity.

This definition of mesoscale is similar to that proposed by Emanuel (1982a). He defined this scale to occur when both ageostrophic advection and Coriolis accelerations are important, which is essentially the same as given in the second criteria just listed. In addition, since in this text the second criteria forms the upper bound in terms of horizontal scale of mesoscale circulations, this meteorological definition is a function of latitude. At the equator, for example, much larger atmospheric features would be expected to be mesoscale since the winds are not constrained by the gradient wind relation. By contrast, at high latitudes, the upper bound on mesoscale is more limited since the Coriolis term is larger and the wind closely approximates gradient balance above the boundary layer even for relatively small horizontal scales. Thus at low latitudes, mesoscale mass adjustment dominates, even for relatively large circulation features (e.g., several thousand kilometers), whereas at higher latitudes for the same sized features, this restructuring of the mass field is performed by near gradient wind synoptic motions.[9] On other planets with their different rotation rates and diameters, the maximum horizontal scale as function of latitude at which the systems are mesoscale will differ from that of the Earth's.

Scales of motion in which vertical accelerations become important can be referred to as the microscale and correspond to the meso$-\gamma$ scale as defined by Orlanski (1975). This scale of motion, somewhat smaller than mesoscale, has also been referred to as the cumulus scale, with the smallest sizes referenced as the turbulence scale. With respect to turbulence, Pinel and Lovejoy (2013) summarize its behavior as:

> *"It is now clear that a prime characteristic of fully developed turbulence is that most of the important fluxes are concentrated in highly sparse (fractal) sets so that much of the flow appears relatively calm. The modern understanding is that by its very nature, turbulence is highly intermittent so that on any realization of a turbulent process, there will be violent regions in proximity to ones of relative calm. However, examination of the apparently calm regions shows that they also have embedded regions of high activity and as we zoom into smaller and smaller regions this strong heterogeneity continues in a scaling manner until we reach the dissipation scale."*

Scales larger than the mesoscale where the Coriolis effect becomes of the same magnitude as the pressure gradient are termed the regional (synoptic) scale and correspond to Orlanski's meso$-\alpha$ and larger. Therefore, mesoscale as defined in this text corresponds closely to Orlanski's meso$-\beta$ scale.

If the atmospheric feature is on the mesoscale, as defined here, Eq. 3.23 could replace the vertical equation of motion ($i = 3$) from Eq. 2.45, and Eqs. 3.14 or 3.11 could be used in lieu of Eq. 2.43 to represent the conservation of mass. The vertical equation of motion can be retained, of course, even though the atmospheric feature being modeled is on the mesoscale. Indeed, this practice has been generally adopted as discussed in Sections 9.3 and 9.4 of this text. However, knowledge that a system

[9] The *omega-equation* (Carlson 1991; Bluestein 1992, 1993; Pielke 2002c) is an effective mathematical framework to assess the resulting synoptic-scale vertical motion, as discussed in Section 14.3.5.3.

is mesoscale or larger permits us to directly obtain the pressure field from knowledge of the vertical temperature profile as shown in Chapter 4. As a result of these simplifications, mesoscale circulations are defined to be *anelastic, hydrostatic*, and *significantly nongradient wind* meteorological systems. In Sections 5.2.2 and 5.2.3, the scales of motion in which the hydrostatic assumption applies will be examined more quantitatively since scale analysis provides only very qualitative guidance.

3.4 Conservation of Water and Other Gaseous and Aerosol Contaminants

As with the conservation relation for heat, the source-sink expressions for the conservation of water and other gaseous and aerosol contaminants S_{q_n} and S_{χ_m} are complex. The most restrictive relation is to assume that q_n and χ_m are conserved so that

$$\partial q_n/\partial t = -u_j \partial q_n/\partial x_j, \quad n = 1, 2, 3,$$

and

$$\partial \chi_m/\partial t = -u_j \partial \chi_m/\partial x_j, \quad m = 1, 2, \ldots, M,$$

or equivalently,

$$dq_n/dt = 0, \quad d\chi_m/dt = 0.$$

It is valid to use these equations to represent atmospheric circulation if

$$\left|S_{q_n}\right| \ll \left|\frac{\partial q_n}{\partial t}\right|; \quad \left|u\frac{\partial q_n}{\partial x}\right|; \quad \left|v\frac{\partial q_n}{\partial y}\right|; \quad \left|w\frac{\partial q_n}{\partial z}\right|,$$

$$\left|S_{\chi_m}\right| \ll \left|\frac{\partial \chi_m}{\partial t}\right|; \quad \left|u\frac{\partial \chi_m}{\partial x}\right|; \quad \left|v\frac{\partial \chi_m}{\partial y}\right|; \quad \left|w\frac{\partial \chi_m}{\partial z}\right|.$$

The first of these conditions is most closely fulfilled when the amount of water undergoing phase changes between solid, liquid, and gas, and when that created by chemical reactions is much less than changes caused by the advection of water. Similarly, the second condition is satisfactory if the aerosol and gaseous constituents undergo phase and chemical changes that are much smaller in magnitude than the advective changes. Scale analysis of S_{q_n} and S_{χ_m} could be performed, however the complexity and high degree of parameterization of these terms would make such a qualitative evaluation of dubious value. Therefore, a discussion of the scale analysis of these source-sink terms will not be presented here. In Chapter 7, S_{q_n} is discussed in more detail.

Chapter 3 Additional Readings

Cotton, W.R., G. Bryan, and S. van den Heever, 2011: *Storm and Cloud Dynamics*, 2nd edition. Academic Press, Elsevier Publishers, The Netherlands, 820 pp.

Lilly, D.K., 1996: A comparison of incompressible, anelastic and Boussinesq dynamics. *Atmos. Res.*, **40**, 143-151.

Thunis, P., and R. Bornstein, 1996: Hierarchy of mesoscale flow assumptions and equations. *J. Atmos. Sci.*, **53**, 380-397.

Warner, T.T., 2011: *Numerical Weather and Climate Prediction*. Cambridge University Press, 548 pp.

CHAPTER

Averaging the Conservation Relations

4

CHAPTER OUTLINE HEAD

4.1 Definition of Averages

Equations 2.43-2.47 and their simplified forms introduced in Chapter 3 are defined in terms of the differential operators ($\partial/\partial t$, $\partial/\partial x_i$), and thus in terms of mathematical formalism, are valid only in the limit when δt, δx, δy, and δz approach zero. In terms of practical application, however, they are valid only when the spatial increments δx, δy, and δz are much larger than the spacing between molecules (so that only the statistical characteristics of molecular motion, rather than the movement of individual molecules themselves are important), but are small enough so that the differential terms over these distances and over the time interval δt can be represented accurately by a constant. If these terms vary significantly within the intervals, however, Eqs. 2.43-2.47 must be integrated over the distance and time intervals over which they are being applied.

Stated more formally, if

$$l_{\mathrm{m}} \ll \delta x, \delta y, \text{ and } \delta z, \tag{4.1}$$

where l_{m} is the representative spacing between molecules, and if

$$\frac{\partial}{\partial x}\rho u \gg \frac{(\delta x)}{2}\frac{\partial^2 \rho u}{\partial x^2}; \quad \frac{\partial}{\partial y}\rho v \gg \frac{(\delta y)}{2}\frac{\partial^2 \rho v}{\partial y^2}; \quad \frac{\partial \rho}{\partial t} \gg \frac{(\delta t)}{2}\frac{\partial^2 \rho}{\partial t^2}, \ldots, \text{ etc.} \tag{4.2}$$

then using Eqs. 2.43-2.47 (or a simplified form of this system of equations) is justified.

In the atmosphere, the criteria given by Eqs. 4.1 and 4.2 limit the direct application of Eqs. 2.43-2.47 to space scales on the order of about a centimeter and to time scales

Mesoscale Meteorological Modeling, Third Edition, Volume 98. http://dx.doi.org/10.1016/B978-0-12-385237-3.00004-9

of a second or so. Therefore, to use Eqs. 2.43-2.47 to represent the atmosphere accurately, they must be evaluated over those space and time intervals. Because mesoscale circulations have horizontal scales on the order of 10 to 100 km or more and a vertical size of up to approximately 10 km, these equations would have to be solved at 10^{18} to 10^{20} locations. This amount of information, unfortunately, far exceeds the capacity of any existing or foreseeable computer system.

Therefore, to circumvent this problem, it is necessary to integrate the conservation equations over specified spatial and temporal scales, whose sizes are determined by the available computer capacity, including its speed of operation. For a specific mesoscale system, the smaller these scales, the better the *resolution* of the circulation.

In performing this integration, it is convenient to perform the following decomposition:

$$\phi = \bar{\phi} + \phi'',$$

where ϕ represents any one of the dependent variables and

$$\bar{\phi} = \int_t^{t+\Delta t} \int_x^{x+\Delta x} \int_y^{y+\Delta y} \int_z^{z+\Delta z} \phi \, dz \, dy \, dx \, dt / (\Delta t)(\Delta x)(\Delta y)(\Delta z). \quad (4.3)$$

Thus $\bar{\phi}$ represents the average of ϕ over the finite time increment Δt and space intervals Δx, Δy, and Δz. The variable ϕ'' is the deviation of ϕ from this average and is often called the *subgrid-scale perturbation*. In a numerical model, Δt is called the *time step* and Δx, Δy, and Δz represent the model *grid intervals*, as illustrated schematically in Fig. 4.1.

One convenient decomposition is to define the averaging volume such that the subgrid-scale perturbation includes the nonhydrostatic part of the solution and the resolvable scale is accurately represented by the hydrostatic assumption. Thus the influence of the nonhydrostatic component of a model would be parameterized and the hydrostatic portion would be explicitly resolved.

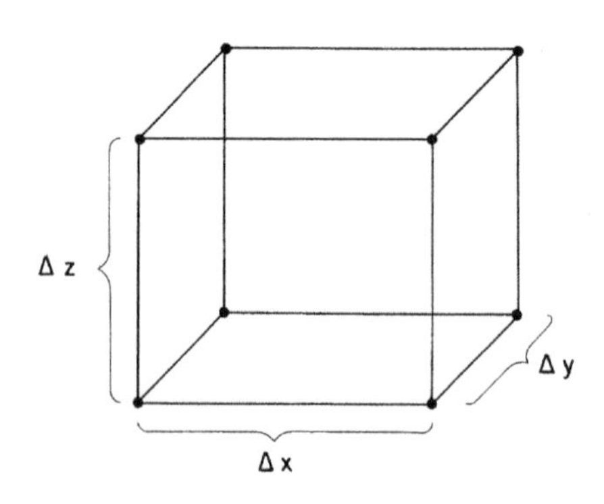

FIGURE 4.1

A schematic of a grid volume. Dependent variables are defined at the corners of the rectangular solid.

Using this definition Eq. 2.45 can be rewritten as

$$\frac{\partial\left(\bar{u}_i+u_i''\right)}{\partial t}=-\left(\bar{u}_j+u_j''\right)\frac{\partial}{\partial x_j}\left(\bar{u}_i+u_i''\right)-\left(\bar{\alpha}+\alpha''\right)\frac{\partial\left(\bar{p}+p''\right)}{\partial x_i} \\ -g\delta_{i3}-2\epsilon_{ijk}\Omega_j\left(\bar{u}_k+u_k''\right). \tag{4.4}$$

Performing the integration of Eq. 4.4 over the intervals Δx, Δy, Δz, and Δt yields

$$\overline{\frac{\partial}{\partial t}\left(\bar{u}_i+u_i''\right)}=-\overline{\left(\bar{u}_j+u_j''\right)\frac{\partial}{\partial x_j}\left(\bar{u}_i+u_i''\right)}-\overline{\left(\overline{\alpha}+\alpha''\right)\frac{\partial\left(\bar{p}+p''\right)}{\partial x_i}} \\ -g\delta_{i3}-2\epsilon_{ijk}\overline{\Omega_j\left(\bar{u}_k+u_k''\right)}, \tag{4.5}$$

where the overbar represents the integral operation

$$\overline{(\ )}=\int_t^{t+\Delta t}\int_x^{x+\Delta x}\int_y^{y+\Delta y}\int_z^{z+\Delta z}(\ )\,dz\,dy\,dx\,dt\,/(\Delta t)(\Delta x)(\Delta y)(\Delta z) \tag{4.6}$$

as performed on ϕ in Eq. 4.3. This operation is often called *grid-volume averaging* since it is performed over the spatial increments Δx, Δy, and Δz.

Equation 4.6 can be generalized to an *ensemble average* as discussed in Cotton and Anthes (1989:Chapter 3). In that approach, an ensemble average can be defined as $<(\ )>=\lim_{N\to\infty}\sum_{i=1}^{N}\overline{(\ )}$ where Eq. 4.6 is an average for a particular case (e.g., a model *realization*). The number of realizations is N. When the results are sensitive, for example, to initial conditions there will be a spread of realization results which is called the ensemble. The grid-volume average has been stochastically chosen for some applications from an ensemble as shown, for example, in Pielke (1984:Section 7.5) and Uliasz et al. (1996) for estimating atmospheric dispersion. More generally, however, as shown in Chapters 7-9 parameterizations exclusively use ensemble average representations.

To simplify Eq. 4.5, it is convenient to assume that the averaged dependent variables change much more slowly in time and space than do the deviations from the average. This *scale separation* between the average and the perturbation implies that $\bar{\phi}$ is approximately constant and ϕ'' highly fluctuating across the distance, Δx, Δy, and Δz and time interval Δt. Examples illustrating when this scale separation is valid and when it is not are given in Fig. 4.2. In addition, the grid intervals and time increments are also presumed not to be a function of location or time, so that the derivatives (i.e., $\partial/\partial t$, $\partial/\partial x_j$) can be straightforwardly removed from inside the integrals. Equation 4.5 then reduces to

$$\frac{\partial\bar{u}_i}{\partial t}=-\bar{u}_j\frac{\partial\bar{u}_i}{\partial x_j}-\overline{u_j''\frac{\partial u_i''}{\partial x_j}}-\bar{\alpha}\frac{\partial\bar{p}}{\partial x_i}-\overline{\alpha''\frac{\partial p''}{\partial x_i}}-g\delta_{i3}-2\epsilon_{ijk}\Omega_j\bar{u}_k, \tag{4.7}$$

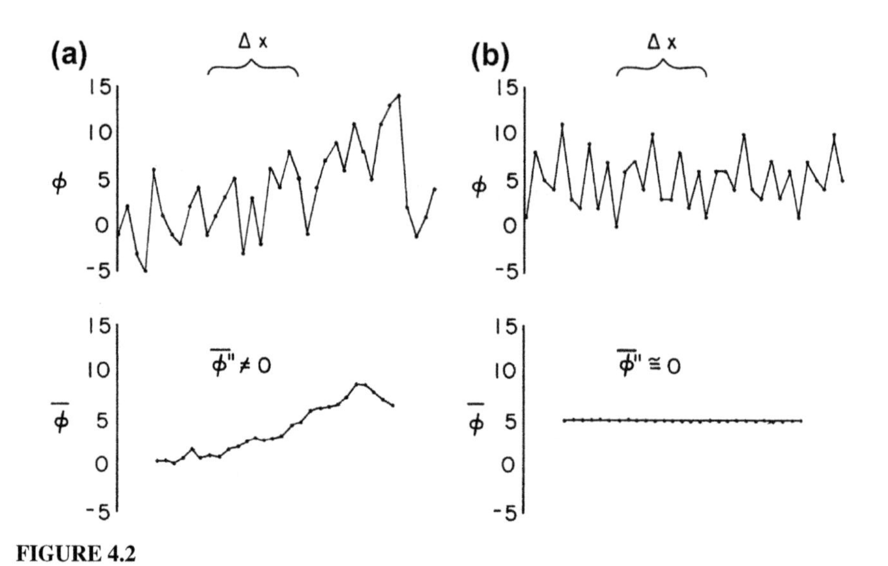

FIGURE 4.2

A one-dimensional schematic illustration of a situation where (a) $\overline{\phi''} \neq 0$ and where (b) $\overline{\phi''} \simeq 0$. The averaging length is illustrated by the interval Δx drawn on the figure ($\bar{\phi} = \frac{1}{10}\int_{x_i}^{x_i+10} \phi\, dx$). Since $\phi = \bar{\phi} + \phi''$, $\overline{\phi''} = 0$ only if $\bar{\phi} = \bar{\bar{\phi}}$.

where the conditions prescribed previously permitted the following type of simplifications to be made:[1]

$$\bar{\bar{u}}_i = \bar{u}_i, \overline{u_i''} = 0, \quad \overline{\frac{\partial u_i}{\partial t}} = \frac{\partial \bar{u}_i}{\partial t}, \quad \overline{\frac{\partial u_i}{\partial x_j}} = \frac{\partial \bar{u}_i}{\partial x_j}, \quad \text{etc.} \tag{4.8}$$

The stipulation that the average of the deviations is zero ($\overline{\phi''} = 0$) is commonly called the *Reynolds assumption.*

Even with the simplifications, Eq. 4.7 contains two additional terms not found in Eq. 2.45, which involve the correlation between the subgrid-scale variables. The second of these terms, $\overline{\alpha'' \partial p'' / \partial x_i}$, could be eliminated, if the assumption is made in Eq. 4.4 that $|\alpha''|/\bar{\alpha} \simeq |\alpha''|/\alpha_0 \ll 1$ (α_0 was defined as a synoptic-scale specific volume during the derivation of the approximate forms of the conservation-of-mass relation in Section 3.1. The mathematical definition of this synoptic scale is given by Eq. 4.12).

With this requirement on specific volume and the assumption that Eq. 3.11 can be written as

$$\frac{\partial}{\partial x_j}\rho_0 u_j \simeq \frac{\partial}{\partial x_j}\bar{\rho} u_j \quad \left(\text{since } \frac{|\alpha''|}{\bar{\alpha}} \simeq \frac{|\alpha''|}{\alpha_0} \ll 1\right), \tag{4.9}$$

[1] Raupach and Shaw (1982:80-82) discuss the situation in which these assumptions fail at interfaces with rigid objects. Galmarini and Thunis (1999), and Galmarini et al. (2000) discuss the errors that are introduced when the Reynold's assumption is invalid.

using the simplifying assumptions given by Eq. 4.8, Eq. 4.5 can be written as

$$\bar{\rho}\frac{\partial \bar{u}_i}{\partial t} = -\frac{\partial}{\partial x_j}\bar{\rho}\,\bar{u}_j\,\bar{u}_i - \frac{\partial}{\partial x_j}\overline{\bar{\rho}u''_j u''_i} - \frac{\partial \bar{p}}{\partial x_i} - \bar{\rho}g\delta_{i3} - 2\epsilon_{ijk}\Omega_j\bar{u}_k\bar{\rho}, \qquad (4.10)$$

where, since $|\alpha''|/\bar{\alpha} \ll 1$, the pressure gradient term is represented by $\bar{\alpha}\,\partial\bar{p}/\partial x$. The remaining *subgrid-scale correlation term* $\overline{\bar{\rho}u''_j u''_i}$ represents the contributions of the smaller scales on the resolvable grid scale resulting from fluctuating velocity components and is, in general, very important in all aspects of dynamic meteorology. This term, also called the *turbulent velocity flux*, must be *parameterized* in terms of averaged quantities to assure that the number of unknowns is equal to the number of equations. Terms of this sort, which arise in the averaged conservation relations, are often of the same order or even larger than the terms that involve only resolved dependent variables. The proper specification of this, and similar subgrid-scale correlation terms, as a function of resolvable, averaged quantities, is referred to as the *closure* problem and is discussed further in Chapter 7.

Equation 4.10 can be written in a second way by assuming

$$\bar{\phi} = \phi_0 + \phi', \qquad (4.11)$$

where (as before ϕ represents any one of the dependent variables)

$$\phi_0 = \int_x^{x+D_x}\int_y^{y+D_y}\bar{\phi}\,dx\,dy/D_x D_y \qquad (4.12)$$

and is called the *layer domain-averaged variable*. The symbols D_x and D_y represent distances that are large compared with the mesoscale system of interest (perhaps the horizontal size [*domain*] of the mesoscale model representation), so that ϕ_0 is assumed to represent the synoptic-scale atmospheric conditions, as referred to previously. The variable ϕ', then, represents the mesoscale deviations from this larger scale.

Substituting Eq. 4.11 into Eq. 4.10 and rearranging results in the equation

$$\begin{aligned}\frac{\partial(u_{i_0}+u'_i)}{\partial t} &= -\left(u_{j_0}+u'_j\right)\frac{\partial}{\partial x_j}\left(u_{i_0}+u'_i\right)\\ &\quad -\frac{1}{\rho_0[1+(\rho'/\rho_0)]}\frac{\partial}{\partial x_j}\rho_0\left(1+\frac{\rho'}{\rho_0}\right)\overline{u''_j u''_i} \qquad (4.13)\\ &\quad -\alpha_0\left(1+\frac{\alpha'}{\alpha_0}\right)\frac{\partial\left(p_0+p'\right)}{\partial x_i} - g\delta_{i3} - 2\epsilon_{ijk}\Omega_j\left(u_{k_0}+u'_k\right).\end{aligned}$$

If $|\alpha'|/\alpha_0 \ll 1$ is required, then the horizontal and vertical components of Eq. 4.13 can be rewritten as

$$\frac{\partial \bar{u}_i}{\partial t} = -\bar{u}_j\frac{\partial \bar{u}_i}{\partial x_j} - \frac{1}{\rho_0}\frac{\partial}{\partial x_j}\rho_0\overline{u''_j u''_i} - \alpha_0\frac{\partial \bar{p}}{\partial x_i} - 2\epsilon_{ijk}\Omega_j\bar{u}_k, \quad i = 1, 2 \qquad (4.14)$$

and

$$\frac{\partial \bar{w}}{\partial t} = -\bar{u}_j \frac{\partial \bar{w}}{\partial x_j} - \frac{1}{\rho_0} \frac{\partial}{\partial x_j} \rho_0 \overline{u''_j w''} - \alpha_0 \frac{\partial p'}{\partial z} + \frac{\alpha'}{\alpha_0} g + 2\bar{u}\Omega \cos\phi, \quad i = 3 \quad (4.15)$$

where it has been assumed that the synoptic-scale pressure field is hydrostatic $(\partial p_0/\partial z = -g/\alpha_0)$ so that

$$\left(\alpha_0 + \alpha'\right) \frac{\partial}{\partial z} \left(p_0 + p'\right) + g = \alpha_0 \left(1 + \frac{\alpha'}{\alpha_0}\right) \frac{\partial p'}{\partial z} - g + g - \frac{\alpha'}{\alpha_0} g \simeq \alpha_0 \frac{\partial p'}{\partial z} - \frac{\alpha'}{\alpha_0} g.$$

The term $\left(\alpha'/\alpha_0\right) g$ is the only expression retained in Eqs. 4.14 and 4.15 that contains temporal variations in specific volume when $|\alpha'|/\alpha_0 \ll 1$. Neglecting temporal variations of density (as given by α') except for $\left(\alpha'/\alpha_0\right) g$ is called the *Boussinesq approximation*. This term can also be rewritten by logarithmically differentiating the ideal gas law so that

$$\frac{d\alpha}{\alpha} = \frac{dT}{T} - \frac{dp}{p}. \quad (4.16)$$

When the changes of α, T, and p are assumed to be much less than their absolute magnitudes (e.g., $|d\alpha| \ll \alpha$, $|dT| \ll T$, and $|dp| \ll p$), and if $\alpha \simeq \alpha_0$, $T \simeq T_0$, and $p \simeq p_0$, then Eq. 4.16 can be approximated by

$$\frac{\alpha'}{\alpha_0} \simeq \frac{T'}{T_0} - \frac{p'}{p_0}. \quad (4.17)$$

Since $|\alpha'|/\alpha_0$ has already been assumed much less than unity, the requirement that $|T'|/T_0 \ll 1$ and $|p'|p_0 \ll 1$ is implied from Eq. 4.17. The vertical component of acceleration can, therefore, also be written as

$$\frac{\partial \bar{w}}{\partial t} = -\bar{u}_j \frac{\partial \bar{w}}{\partial x_j} - \frac{1}{\rho_0} \frac{\partial}{\partial x_j} \rho_0 \overline{u''_j w''} - \alpha_0 \frac{\partial p'}{\partial z} + \left(\frac{T'}{T_0} - \frac{p'}{p_0}\right) g + 2\bar{u}\Omega \cos\phi. \quad (4.18)$$

Finally, since $\theta = T\left(1000/p\right)^{R_d/C_p}$, then α'/α_0 can also be approximated by

$$\frac{\alpha'}{\alpha_0} = \frac{\theta'}{\theta_0} - \frac{C_v}{C_p} \frac{p'}{p_0}, \quad (4.19)$$

which represents another form of the ideal gas law when $|\alpha'|/\alpha_0 \ll 1$. Here the relation $C_p = C_v + R_d$ has been used to obtain the given form. This approximate form for α'/α_0 can be simplified when the vertical scale of the circulation L_z is much smaller than the scale depth of the atmosphere H_α. Using the scale analysis procedure introduced in Chapter 3,

$$|\alpha_0 \, \partial p'/\partial z| \sim \alpha_0 |p'|/L_z,$$

and assuming that the vertical mesoscale pressure perturbation and the density perturbation terms are of the same order of magnitude, then

$$|\alpha_0 \, \partial p'/\partial z| \sim |\alpha'/\alpha_0| g,$$

so that

$$\frac{RT_0|p'|}{L_z p_0} \sim \frac{|\alpha'|}{\alpha_0} g,$$

or

$$\frac{|p'|}{p_0} \sim \frac{L_z g}{RT_0} \frac{|\alpha'|}{\alpha_0} = \frac{L_z g}{p_0 \alpha_0} \frac{|\alpha'|}{\alpha_0} = \frac{L_z}{D} \frac{|\alpha'|}{\alpha_0} \sim \frac{L_z}{H_\alpha} \frac{|\alpha'|}{\alpha_0},$$

where the results from Chapter 3 that $p_0 \sim \rho_0 g D$ and $D \sim H_\alpha$ have been used. Thus if $L_z \ll H_\alpha$,

$$|p'|/p_0 \ll |\alpha'|/\alpha_0 \quad \text{and} \quad |\alpha'|/\alpha_0 \sim |\theta'|/\theta_0,$$

so that for this situation, Eq. 4.18 can be written as

$$\frac{\partial \bar{w}}{\partial t} = -u_j \frac{\partial \bar{w}}{\partial x_j} - \frac{1}{\rho_0} \frac{\partial}{\partial x_j} \rho_0 \overline{u_j'' w''} - \alpha_0 \frac{\partial p'}{\partial z} + \frac{\theta'}{\theta_0} g + 2\bar{u}\Omega \cos\phi. \tag{4.20}$$

Thus when the shallow form of the conservation-of-mass equation (Eq. 3.14) can be used, it is appropriate to use Eq. 4.20 as the vertical equation of motion.

Equations 4.14 and 4.15 can, therefore, be written in tensor form as

$$\begin{aligned} \frac{\partial \bar{u}_i}{\partial t} = -\bar{u}_j \frac{\partial \bar{u}_i}{\partial x_j} - \frac{1}{\rho_0} \frac{\partial}{\partial x_j} \rho_0 \overline{u_j'' u_i''} - \alpha_0 \frac{\partial p'}{\partial x_i} - \alpha_0 \left(\frac{\partial p_0}{\partial x} \delta_{i1} + \frac{\partial p_0}{\partial y} \delta_{i2} \right) \\ + \frac{\alpha'}{\alpha_0} g \delta_{i3} - 2\epsilon_{ijk} \Omega_j \bar{u}_k, \end{aligned} \tag{4.21}$$

where

$$\alpha'/\alpha_0 = (\theta'/\theta_0) - (C_v p'/C_p p_0)$$

for deep atmospheric circulations and

$$\alpha'/\alpha_0 = \theta'/\theta_0$$

for shallow systems. Either Eq. 4.21 or 4.10 can be used to predict velocity fluctuations.

The remainder of the prognostic equations (Eqs. 2.43, 2.44, 2.46, and 2.47), can be averaged in the same manner as performed for the conservation-of-motion relations.

The complete conservation-of-mass equation (2.43), after using the averaging operation given by Eqs. 4.3 and 4.6 along with the assumptions listed in Eq. 4.8, would be given as

$$\frac{\partial \bar{\rho}}{\partial t} = -\frac{\partial}{\partial x_j} \bar{\rho}\, \bar{u}_j - \frac{\partial}{\partial x_j} \overline{\rho'' u_j''}, \tag{4.22}$$

whereas averaging the approximate forms of this relation given by Eqs. 3.11 and 3.14 yields

$$\frac{\partial}{\partial x_j} \rho_0 \bar{u}_j \simeq \frac{\partial}{\partial x_j} \bar{\rho}\, \bar{u}_j = 0 \quad \text{and} \quad \frac{\partial \bar{u}_j}{\partial x_j} = 0. \tag{4.23}$$

The remainder of the prognostic equations (Eqs. 2.44, 2.46, and 2.47), are similar in form, and the equivalent averaged forms can be written as

$$\frac{\partial \bar{\theta}}{\partial t} = -\bar{u}_j \frac{\partial \bar{\theta}}{\partial x_j} - \frac{1}{\rho_0} \frac{\partial}{\partial x_j} \rho_0 \overline{u_j''\theta''} + \bar{S}_\theta, \tag{4.24}$$

$$\frac{\partial \bar{q}_n}{\partial t} = -\bar{u}_j \frac{\partial \bar{q}_n}{\partial x_j} - \frac{1}{\rho_0} \frac{\partial}{\partial x_j} \rho_0 \overline{u_j''q_n''} + \bar{S}_{q_n}, \quad n = 1, 2, 3, \tag{4.25}$$

and

$$\frac{\partial \bar{\chi}_m}{\partial t} = -\bar{u}_j \frac{\partial \bar{\chi}_m}{\partial x_j} - \frac{1}{\rho_0} \frac{\partial}{\partial x_j} \rho_0 \overline{u_j''\chi_m''} + \bar{S}_{\chi_m}, \quad m = 1, 2, ..., M, \tag{4.26}$$

where $\bar{S}_\theta$, $\bar{S}_{q_n}$, and $\bar{S}_{\chi_m}$ represent the integrated contributions of the source-sink terms over the intervals defined by Eq. 4.6.

Using the conservation-of-mass relation, given by Eq. 4.23, the advection terms in Eqs. 4.24-4.26 can be given as

$$-\bar{u}_j \frac{\partial \bar{\theta}}{\partial x_j} = -\frac{1}{\rho_0} \frac{\partial}{\partial x_j} \rho_0 \bar{u}_j \bar{\theta}, \tag{4.27}$$

$$-\bar{u}_j \frac{\partial \bar{q}_n}{\partial x_j} = -\frac{1}{\rho_0} \frac{\partial}{\partial x_j} \rho_0 \bar{u}_j \bar{q}_n, \tag{4.28}$$

and

$$-\bar{u}_j \frac{\partial \bar{\chi}_m}{\partial x_j} = -\frac{1}{\rho_0} \frac{\partial}{\partial x_j} \rho_0 \bar{u}_j \bar{\chi}_m, \tag{4.29}$$

where the righthand side is often referred to as the *flux form* of the advection terms.

The averaged conservation-of-motion equations given by 4.21 (or by 4.10) along with Eqs. 4.23-4.26 are often called the *primitive equations* because they are derived straightforwardly from the original conservation principles presented in Chapter 2. As evident from the assumptions required to obtain them, however, they are not the most fundamental form of the conservation laws as implied by the word "primitive".

4.2 Diagnostic Equation for Nonhydrostatic Pressure

The use of Eq. 4.21 or 4.10, however, requires an evaluation of pressure. If $H_\alpha / L_x \lesssim 1$ such that the motion can be assumed hydrostatic, it is straightforward to integrate Eq. 3.23 to obtain pressure at any level. Using the averaging procedures discussed in this chapter, the vertical equation of motion in Eq. 4.10, in its hydrostatic form, can be represented by

$$\partial \bar{p} / \partial z = -\bar{\rho} g, \tag{4.30}$$

and the equivalent expression for use in the third component of Eq. 4.21 is

$$\partial p' / \partial z = g\alpha' / \alpha_0^2, \tag{4.31}$$

which can be replaced by

$$\partial p'/\partial z = g\rho_0\theta'/\theta_0 \tag{4.32}$$

if $L_z \ll H_\alpha$.

If $|\alpha''|/\alpha_0$ is not assumed to be much less than unity and if the hydrostatic assumption is not expected to be valid or is otherwise not used, then pressure $\bar{p}$ can be computed from the ideal gas law and from Eq. 4.22 (assuming $\overline{\rho'' u_j''}$ can be parameterized in terms of known quantities, or ignored). If either of the approximate forms of the conservation-of-mass relation given by Eq. 4.23 are used, however, pressure cannot be computed in this fashion. In this case, the divergence $\partial/\partial x_i$ (equivalent to $\nabla\cdot$ in vector notation) of Eq. 4.21 yields

$$\begin{aligned}\frac{\partial}{\partial t}\frac{\partial}{\partial x_i}\rho_0\bar{u}_i =& -\frac{\partial^2}{\partial x_i\partial x_j}\left(\rho_0\bar{u}_j\bar{u}_i\right) - \frac{\partial^2}{\partial x_i\partial x_j}\rho_0\overline{u_j'' u_i''} - \frac{\partial^2 p'}{\partial x_i^2}\\ &-\frac{\partial}{\partial x_i}\left[\frac{\partial p_0}{\partial x}\delta_{i1} + \frac{\partial p_0}{\partial y}\delta_{i2}\right] + \delta_{i3}g\frac{\partial}{\partial x_i}\frac{\alpha'}{\alpha_0^2} - 2\epsilon_{ijk}\Omega_j\frac{\partial}{\partial x_i}\rho_0\bar{u}_k.\end{aligned}$$

Using the lefthand side of Eq. 4.23 along with the approximation for α'/α_0 given by Eq. 4.19 gives the *diagnostic* second-order differential equation for the mesoscale pressure perturbation

$$\begin{aligned}&\frac{\partial^2 p'}{\partial x_i^2} + \frac{gC_v}{C_p}\frac{\partial}{\partial z}\left(\frac{\rho_0 p'}{p_0}\right)\\ &= -\frac{\partial^2}{\partial x_i\partial x_j}\left(\rho_0\bar{u}_j\bar{u}_i\right) - \frac{\partial^2}{\partial x_i\partial x_j}\rho_0\overline{u_j'' u_i''} - \frac{\partial}{\partial x_i}\left[\frac{\partial p_0}{\partial x}\delta_{i1} + \frac{\partial p_0}{\partial y}\delta_{i2}\right]\\ &+ g\frac{\partial}{\partial z}\left(\frac{\rho_0\theta'}{\theta_0}\right) - 2\epsilon_{ijk}\Omega_j\frac{\partial}{\partial x_i}\rho_0\bar{u}_k.\end{aligned} \tag{4.33}$$

When $L_z \ll H_\alpha$, the second term on the lefthand side of Eq. 4.33 does not appear, in which case Eq. 4.33 is referred to as a *Poisson partial differential equation* (e.g., see Hildebrand 1962).

When Eq. 4.33 is used to diagnose pressure, the vertical component of Eq. 4.21 must be dropped, otherwise the system of equations that include 4.19, 4.21, 4.23-4.26, and 4.33 would be *overspecified* (i.e., one more equation than the number of unknowns). Vertical velocity can be diagnosed directly from the appropriate form of Eq. 4.23 (the right side of 4.23 can be used if $L_z \ll H_\alpha$).

The advantages of using Eq. 4.33 to compute pressure include the following.

1. The horizontal wavelength of the mesoscale system L_x can be of any size without concern for when the hydrostatic assumption is valid.
2. By using Eq. 4.23, sound waves, which in general are not expected to be meteorologically important on the mesoscale, are excluded (this is demonstrated in Section 5.2.2).

The disadvantages include the following.

1. The required computation time is increased since pressure must be evaluated from the involved formulation given by Eq. 4.33.
2. The mathematical operation of differentiation magnifies errors in the evaluation of Eq. 4.33.

Models that use Eq. 4.33 to determine pressure, or the more general method of using the complete conservation-of-mass equation (i.e., Eq. 4.22), along with the ideal gas law and definition of potential temperature to obtain pressure, are referred to as *nonhydrostatic models*. Of course, it is actually not appropriate to refer to such models as nonhydrostatic since, as shown in Song et al. (1985) and discussed in detail in Section 5.2.3.2, the pressure perturbation typically remains close to hydrostatic balance even when nonhydrostatic pressure effects become important. The nonhydrostatic pressure is also referred to as the *dynamic pressure*. The relative importance of the hydrostatic and dynamic pressure are discussed in Chapter 5.

Investigators who have used an equation of the form given by Eq. 4.33 include Ogura and Charney (1961) and Neumann and Mahrer (1971). Models such as those reported by Tapp and White (1976), Cotton and Tripoli (1978), and Pielke et al. (1992) preferred to retain the more complete conservation-of-mass relation of the form given by Eq. 4.22, and this movement to a compressible model framework has continued (e.g., Cotton et al. 2003), since as shown in Section 7.4.3, the computational problems associated with retaining sound waves as a solution to the model equations have been eliminated.

4.3 Scaled Pressure Form

Finally, in mesoscale modeling, it has often been convenient to replace the pressure gradient term in Eq. 2.45 with

$$\frac{1}{\rho}\frac{\partial p}{\partial x_i} = \theta\frac{\partial \pi}{\partial x_i} \tag{4.34}$$

where the ideal gas law (Eq. 2.49) and the definition of potential temperature (Eq. 2.48) have been used, so that

$$\pi = C_{\mathrm{p}}\left(p/p_{00}\right)^{R_{\mathrm{d}}/C_{\mathrm{p}}} = C_{\mathrm{p}}T_{\mathrm{v}}/\theta.$$

The variable π is often referred to as the *Exner function*.[2] Using the definitions of averaging presented in this chapter,

$$\overline{\theta\frac{\partial \pi}{\partial x_i}} = \bar{\theta}\frac{\partial \bar{\pi}}{\partial x_i} + \overline{\theta''\frac{\partial \pi''}{\partial x_i}}.$$

[2]The term "Exner function" is also often used for $\pi = (p/p_{00})^{R_{\mathrm{d}}/C_{\mathrm{p}}}$. Such a definition yields a nondimensional pressure variable. However, C_{p} is the specific heat of air and, therefore, can be treated as nearly a constant. Consequently, it is reasonable to incorporate C_{p} into the definition of π.

The subgrid-scale correlation term on the righthand side of this expression is of the same form as the $\overline{\alpha'' \partial p'' / \partial x_i}$ term in Eq. 4.7, however, it cannot be eliminated by using the approximate form of the conservation-of-mass relation (Eq. 4.9) as was done in creating Eq. 4.10. To remove this term, it is necessary to utilize results from measurements (e.g., Lumley and Panofsky 1964) that show that

$$\overline{u''_j \partial u''_i / \partial x_j} \gg \overline{\theta'' \partial \pi'' / \partial x_i}.$$

In mesoscale systems, this inequality is reasonable since u''_j often has variations in magnitude over short distances equal to or greater than $\bar{u}_j$ (e.g., if $\bar{u}_. = 5$ m s^{-1}, it is common to have wind gusts to 10 m s^{-1} ($u'' = +5$ m s^{-1}), whereas if $\bar{\theta}$ and $\bar{\pi}$ have magnitudes around 300 K and 10^3 J kg^{-1}K^{-1}, respectively, θ'' and π'' are observed to vary at most 10 K and 3 J kg^{-1}K^{-1}, respectively, (assuming a p'' of 10 mb over the same distances).

Therefore,

$$\overline{\theta \frac{\partial \pi}{\partial x_i}} \simeq \bar{\theta} \frac{\partial \bar{\pi}}{\partial x_i} \tag{4.35}$$

is a reasonable approximation for the averaged pressure gradient term.

Decomposing $\bar{\theta}$ and $\bar{\pi}$ into synoptic and mesoscale components yields

$$\begin{aligned} \theta_0 \left(1 + \frac{\theta'}{\theta_0}\right) \frac{\partial \bar{\pi}}{\partial x_i} &\simeq \theta_0 \frac{\partial \bar{\pi}}{\partial x_i}, \quad i = 1, 2, \\ (\theta_0 + \theta') \frac{\partial}{\partial z} (\pi_0 + \pi') + g &= \theta_0 \left(1 + \frac{\theta'}{\theta_0}\right) \frac{\partial \pi'}{\partial z} - g + g - \frac{\theta'}{\theta_0} g \\ &\simeq \theta_0 \frac{\partial \pi'}{\partial z} - \frac{\theta'}{\theta_0} g. \end{aligned} \tag{4.36}$$

where the synoptic scale is presumed hydrostatic (i.e., $(1/\rho_0)(\partial p_0/\partial z) = \theta_0 \, \partial \pi_0 / \partial z = -g$).

The total resolvable form of the conservation-of-motion relation, equivalent to Eq. 4.10, using $\bar{\pi}$ as the scale pressure can then be written as

$$\frac{\partial \bar{u}_i}{\partial t} = -\bar{u}_j \frac{\partial \bar{u}_i}{\partial x_j} - \frac{1}{\rho_0} \frac{\partial}{\partial x_j} \rho_0 \overline{u''_j u''_i} - \bar{\theta} \frac{\partial \bar{\pi}}{\partial x_i} - g \delta_{i3} - 2 \epsilon_{ijk} \Omega_j \bar{u}_k, \tag{4.37}$$

and the form equivalent to Eq. 4.21 is given by

$$\begin{aligned} \frac{\partial \bar{u}_i}{\partial t} &= -\bar{u}_j \frac{\partial \bar{u}_i}{\partial x_j} - \frac{1}{\rho_0} \frac{\partial}{\partial x_j} \rho_0 \overline{u''_j u''_i} - \theta_0 \frac{\partial \pi'}{\partial x_i} \\ &+ \theta_0 \left[\frac{\partial \pi_0}{\partial x} \delta_{i1} + \frac{\partial \pi_0}{\partial y} \delta_{i2} \right] + \frac{\theta'}{\theta_0} g \delta_{i3} - 2 \epsilon_{ijk} \Omega_j \bar{u}_k. \end{aligned} \tag{4.38}$$

If the hydrostatic assumption is used, the vertical equation of motion ($i = 3$) from Eq. 4.37 is replaced with

$$\partial\bar{\pi}/\partial z = -g/\bar{\theta} \tag{4.39}$$

and Eq. 4.38 yields

$$\partial\pi'/\partial z = g\theta'/\theta_0^2 \tag{4.40}$$

for the same situation.

The major advantages of writing the equations in this form are as follows.

1. The pressure perturbation term p', or its equivalent π', does not occur in the gravity term even if the depth of the circulation is on the same order as the scale depth of the atmosphere ($L_z \sim H_\alpha$).
2. The need to compute the density perturbation ($\rho' = 1/\alpha'$) is removed, thus one less equation needs to be evaluated.
3. The vertical gradient of π is much less than that of p, hence approximating that term using finite difference techniques introduces less error (i.e., $[\partial\bar{\pi}/\partial z/(\partial\bar{p}/\partial z)] = 1/(\bar{\rho}\bar{\theta}) \ll 1$).

The disadvantages include that in deriving an anelastic equation for π', equivalent to Eq. 4.33, a first-derivative term in π' arises in all three spatial directions;

$$\begin{aligned}\rho_0\theta_0\frac{\partial^2\pi'}{\partial x_i^2} + \frac{\partial\pi'}{\partial x_i}\frac{\partial\rho_0\theta_0}{\partial x_i} &= -\frac{\partial}{\partial x_i}\rho_0\bar{u}_j\frac{\partial\bar{u}_i}{\partial x_j} - \frac{\partial^2}{\partial x_i\partial x_j}\rho_0\overline{u_j''u_i''}\\ &\quad -\frac{\partial}{\partial x_i}\rho_0\theta_0\left[\frac{\partial\pi_0}{\partial x}\delta_{i1} - \frac{\partial\pi_0}{\partial y}\delta_{i2}\right] + \frac{\partial}{\partial x_i}\rho_0\frac{\theta'}{\theta_0}g\delta_{i3}\\ &\quad -2\epsilon_{ijk}\Omega_j\frac{\partial}{\partial x_i}\bar{u}_k\rho_0, \quad (\text{assumes } |\bar{u}_i\partial\rho_0/\partial t| \ll |\rho_0\partial\bar{u}_i/\partial t|)\end{aligned}$$

The inequality given by Eq. 3.22 suggests when either of the hydrostatic equations given by Eqs. 4.39 and 4.40 can be used in lieu of the third equation of motion. Since, as shown in Chapter 9, the smallest-sized horizontal feature that can be resolved with reasonable accuracy in a mesoscale model has a size corresponding to $4\Delta x$, then

$$H_\alpha/L_x = H_\alpha/4\Delta x \lesssim 1 \tag{4.41}$$

defines a restriction for the use of the hydrostatic equation. If $\Delta x = 2$ km, for example, the adequacy of the hydrostatic approximation seems assured, at least, based on the analysis given in Chapter 3. Additional and more quantitative evaluation of the accuracy of the hydrostatic assumption are given in Chapter 5, Sections 5.2.2 and 5.2.3.

4.4 Summary

Using the averaging technique presented in this chapter, consistent sets of the conservation relations can be derived. To develop a consistent set of equations, the number of equations must be the same as the number of dependent variables. These averaged equations can be presented in several forms.

In determining the specific grid-volume averaged equations, it is essential to identify the form used for:

1. the conservation-of-mass equation (a scalar equation);
2. the conservation-of-heat equation (a scalar equation);
3. the conservation-of-motion equation (a vector equation);
4. the conservation-of-water equation (three scalar equations);
5. the conservation-of-other gaseous and aerosol materials equations (M scalar equations); and
6. the equation of state.

If simplified forms of any of these equations are used, the assumptions need to be stated.

The terms that constitute these equations also needs to be identified. The terms are in one of the following forms.

1. Externally prescribed variables (examples include u_{i_0}, θ_0).
2. Dependent variables (examples include $u'_i, \theta', \overline{u''_j u''_i}, \overline{u''_j \theta''}, \bar{S}_\theta$).
3. Prescribed constants or functions (examples include g, f).

The dependent variables can be further categorized into two types:

1. those that are predicted from the conservation equations (for example, u'_i, θ'), and
2. those for which parameterizations are often introduced (for example, $\overline{u''_j u''_i}$, $\overline{u''_j \theta''}$, $\bar{S}_\theta$). Parameterizations are based on the variables which are predicted. As shown in Chapter 7, terms such as $\overline{u''_j u''_i}$ can be predicted or parameterized. For each model, the approach adopted needs to be identified. It is important to realize that these parameterizations are *engineering code* which, while often having a fundamental physics basis, involve tunable parameters and functions that are based on only a limited subset of real world conditions.

In Chapters 7-8, examples of methods to parameterize grid-volume average values variables are described. Chapter 9 describes how the predicted variables are represented in the conservation equations.

Problems for Chapter 4

Select an atmospheric model.

1. List the specific form of the following grid-volume conservation equations used. Assume flat terrain so that the equations appear in a simpler form. State each of the assumptions used to obtain the equations.

(a) mass
(b) heat
(c) motion
(d) water
(e) other gases and aerosols

2. List the following. Of the dependent variables used in the selected model, which are predicted from the conservation equations and which are parameterized. Does the number of conservation equations agree with the number of dependent variables which are predicted? To be a consistent model, they must be the same number!

(a) externally prescribed variables
(b) the dependent variables
(c) the prescribed constants

3. To select a model to investigate, refer to Appendix B for examples and references.

Chapter 4 Additional Readings

Cotton, W.R., G. Bryan, and S. van den Heever, 2011: *Storm and Cloud Dynamics*, 2nd edition. Academic Press, Elsevier Publishers, The Netherlands, 820 pp.

Pielke, R.A., J. Eastman, L.D. Grasso, J. Knowles, M. Nicholls, R.L. Walko, and X. Zeng, 1995a: Atmospheric vortices. In: *Fluid Vortices*, S. Green, Editor, Kluwer Academic Publishers, The Netherlands, 617-650.

CHAPTER

Physical and Analytic Modeling

5

CHAPTER OUTLINE HEAD

There are two fundamental methods of simulating mesoscale atmospheric flows – *physical models* and *mathematical models*. With the first technique, scale model replicas of observed ground surface characteristics (e.g., topographic relief, buildings) are constructed and inserted into a chamber such as a wind tunnel (water tanks are also used). The flow of air or other gases or liquids in this chamber is adjusted so as to best represent the larger-scale observed atmospheric conditions. Mathematical modeling, by contrast, utilizes such basic analysis techniques as algebra and calculus to solve directly all or a subset of Eqs. 2.43 to 2.50. As will be discussed subsequently in this chapter, certain subsets of Eqs. 2.43 to 2.50 can be solved exactly, whereas other subsets, including the complete system of equations, requires the approximate solution technique called *numerical modeling*. Numerical modeling methods are described in Chapters 7-11.

Models are used for three purposes: diagnostic evaluations, process studies, and predictions. Diagnostic models, for example, use the conservation equations

Mesoscale Meteorological Modeling, Third Edition, Volume 98. http://dx.doi.org/10.1016/B978-0-12-385237-3.00005-0

combined with whatever observations are available to interpolate data throughout a region. Process models utilize the conservation equations in order to improve physical understanding of atmospheric dynamics and thermodynamics. Predictive models are designed to provide forecasts. Predictive models which use the conservation equations need to improve upon the forecast skill of statistical prediction models, in order to demonstrate that they have forecast skill (Landsea and Knaff 2000). A series of papers on the topic of prediction as a link between science and decision making, are published in Sarewitz et al. (2000).

Bankes (1993) separates models into consolidative and exploratory models. Consolidative models are intended to be a surrogate for an actual system. Exploratory models are used to explore the implications of various assumptions and hypotheses, but recognizing that there is incomplete knowledge to actually represent the real system. In the context of mesoscale models, if, for example, the parameterizations of the physical system are not universally applicable, the model is an exploratory model.

5.1 Physical Models

As shown in Chapter 3, the ratio of the individual terms in Eqs. 2.43-2.47 have representative orders of magnitudes that are dependent on the time and space scales of the phenomena being studied. In that chapter, nondimensional parameters such as the Rossby number and Reynolds number were introduced to examine the relative importance of the individual terms.

In constructing a physical-scale model of a real tropospheric circulation, it is desirable that these dimensionless parameters have the same order of magnitude in the model as in the actual atmosphere. Indeed, neglecting to observe proper scaling in such endeavors as the filming of scale models for motion pictures is easily evident to viewers, such as when explosions destroy replica buildings much too easily or a fire flickers too quickly.

Using the order of magnitude estimates for the dependent variables introduced in Chapter 3, and assuming that L and S are the representative length and velocity scale of the circulation of interest (i.e., no distinction is made here between U, V, W nor L_x, L_y, and L_z), then the scaled version of Eq. 4.37 can be written as

$$\begin{aligned}\left[\frac{S^2}{L}\right]\frac{\partial\widehat{\bar{u}}_i}{\partial\hat{t}} = &-\left[\frac{S^2}{L}\right]\widehat{\bar{u}}_j\,\frac{\partial\widehat{\bar{u}}_i}{\partial\hat{x}_j} - \left[\frac{e_{u_i}^2}{L}\right]\frac{\partial}{\partial\hat{x}_j}\widehat{\overline{u_j''u_i''}} - \left[\frac{R\,\delta\theta}{L}\right]\widehat{\theta}_0\,\frac{\partial\widehat{\pi}'}{\partial\hat{x}_i}\\ &-\left[\frac{R\,\delta\theta}{L}\right]\widehat{\theta}_0\left\{\frac{\partial\widehat{\pi}_0}{\partial\hat{x}}\,\delta_{i1} + \frac{\partial\widehat{\pi}_0}{\partial\widehat{y}}\,\delta_{i2}\right\}\\ &+\left[\frac{\delta\theta}{\theta_0}\,g\right]\widehat{\theta}'\delta_{i3} - [\Omega S]\,2\epsilon_{ijk}\widehat{\Omega}_j\widehat{\bar{u}}_k,\end{aligned} \tag{5.1}$$

where a circumflex ($\widehat{\ }$) over a dependent or independent variable indicates that it is nondimensional. The scaled parameter e_{u_i} is a measure of the subgrid-scale velocity correlations that can be estimated from the mean subgrid-scale kinetic energy given by

$$e_{u_i} = \left[\left(\overline{u_i''^2}/2\right)\right]^{1/2}.$$

Including the estimate for the molecular viscous dissipation given by Eq. 3.29, and multiplying Eq. 5.1 by L/S^2 (to obtain a nondimensional equation for the local acceleration) results in

$$\begin{aligned}\frac{\partial \widehat{\bar{u}}_i}{\partial \hat{t}} &= -\widehat{\bar{u}}_j \frac{\partial \widehat{\bar{u}}}{\partial \hat{x}_j} - \left[\frac{e_{u_i}^2}{S^2}\right] \frac{\partial}{\partial \hat{x}_j} \widehat{\overline{u_j'' u_i''}} - \left[\frac{R\,\delta\theta}{S^2}\right] \widehat{\theta}_0 \frac{\partial \widehat{\pi}}{\partial \hat{x}_i} \\ &- \left[\frac{R\,\delta\theta}{S^2}\right] \widehat{\theta}_0 \left\{\frac{\partial \widehat{\pi}_0}{\partial \hat{x}} \delta_{i1} + \frac{\partial \widehat{\pi}_0}{\partial \hat{y}} \delta_{i2}\right\} + \left[\frac{gL\delta\theta}{\theta_o S^2}\right] \widehat{\theta} \delta_{i3} \\ &- \left[\frac{\Omega L}{S}\right] 2\epsilon_{ijk} \widehat{\Omega}_j \widehat{\bar{u}}_k - \left[\frac{\upsilon}{LS}\right] \frac{\partial^2 \widehat{\bar{u}}_i}{\partial \hat{x}_j \partial \hat{x}_j}.\end{aligned} \tag{5.2}$$

To use a scaled physical model to represent accurately the conservation-of-motion relation in the atmosphere, it is essential that

1. the individual bracketed terms be equal in the model and in the atmosphere, or
2. the bracketed terms that are not equal must be much less in magnitude than the other bracketed terms in Eq. 5.2.

When these conditions are met, the actual and modeled atmospheres are said to have *dynamical similarity*.

Two of these bracketed terms are of the same form as defined in Section 3.3.2 and are given as

$$\Omega L/S = 1/R_0$$

and

$$\upsilon/LS = 1/\mathrm{Re}$$

where R_0 is the Rossby number and Re the Reynolds number, and

$$gL\left(\delta\theta/\theta_0\right)/S^2 = \mathrm{Ri}_{\mathrm{bulk}}$$

is called the *bulk Richardson number*. ($\delta\theta$ represents the potential temperature perturbation and is the same order as δT used in Section 3.3.)

From Eq. 5.2, to maintain dynamic similarity, it is implied that to represent all of the terms in the equation properly:

1. the ratio of the subgrid-scale kinetic energy to the grid-volume average kinetic energy must be kept constant;
2. reducing the length scale L in the physical model requires:

(a) an increase in the magnitude of the horizontal temperature perturbation $\delta\theta$ or a reduction in the simulated wind flow speed S or both,
(b) an increase in the rotation rate Ω or a reduction in S or both,
(c) a decrease in the viscosity v or an increase in S or both;

3. an increase in $\delta\theta$ in the pressure gradient term necessitates that S also increase.

Unfortunately, it is impossible to satisfy all of these requirements simultaneously in existing physical models of mesoscale atmospheric circulations. Such physical models are constructed inside of buildings, which limits the dimensions of the simulated circulations to the size of meters, whereas actual mesoscale circulations extend over kilometers.

To illustrate the difficulty of obtaining dynamic similarity in a mesoscale physical model for all terms in Eq. 5.2, let the horizontal scale of a mountain ridge be 10 km, whereas the physical model of this geographic feature utilizes a 1 m representation. The scale reduction is, therefore, 10^4. Thus if $S = 10$ m s^{-1} in the real situation and air is used in the scaled model atmosphere, then the simulated wind speed would have to be 10^5 m s^{-1} to maintain identical Reynolds number similarity! In addition, to have the same Rossby number for this example, the physical model must rotate 10,000 times more rapidly than the Earth or the wind speed must be reduced by 10,000. Reducing the speed, of course, is contradictory to what is required to obtain Reynolds number similarity! Only if the results are relatively insensitive to changes in these nondimensional quantities, as suggested, for example, by Cermak (1975) for large values of the Reynolds number in simulations of the atmospheric boundary layer, can one ignore large differences in the non-dimensional parameters.

From the example just given, however, it should be clear that it is impossible to obtain *exact* dynamic similarity between mesoscale atmospheric features and the physical model when all of the terms in Eq. 5.2 are included.

Nonetheless, investigators who use physical models have proposed a type of similarity between actual mesoscale atmospheric circulations in which $e^2_{u_i}/S^2 \gg v/LS$ and physical model representations in which $e^2_{u_i}/S^2 \ll v/LS$. With this type of simulation, it is assumed that the mixing by molecular motions, as expressed by Eq. 3.29, acts in the same manner as the mixing by air motions as given by the turbulence flux divergence term $(\partial/\partial\hat{x}_j)\,\widehat{\overline{u''_j u''_i}}$. In its dimensional form, this latter term can be approximated by

$$\frac{\partial}{\partial x_j}\overline{u''_j u''_i} \simeq \frac{\partial}{\partial x_j}\left(-K\frac{\partial \bar{u}_i}{\partial x_j}\right) \tag{5.3}$$

(this approximation is discussed in Chapter 7), where K is called the *turbulent exchange coefficient* and is analogous to the kinematic viscosity v. If a *turbulent Reynolds number* $\mathrm{Re_{turb}}$ is then defined as the ratio of the advective terms to the subgrid-scale correlation terms, then by scale analysis

$$\mathrm{Re_{turb}} = L_{\mathrm{meso}}\, S_{\mathrm{meso}}/K,$$

where the subscript “meso” refers to the mesoscale. Similarity of flow between the real atmosphere and the physical model are then assumed to occur when

$$\mathrm{Re}_{\mathrm{turb}} = L_{\mathrm{meso}}\, S_{\mathrm{meso}}/K = \mathrm{Re} = L_{\mathrm{model}}\, S_{\mathrm{model}}/v$$

where the subscript “model” refers to the scaled physical representation.

If both the actual and simulated wind speeds are equal $S_{\mathrm{meso}} = S_{\mathrm{model}}$, then if $L_{\mathrm{meso}} = 10^4 L_{\mathrm{model}}$, for example, $K = 10^4 v$ is used to justify similarity between the mesoscale and the physical model, *when subgrid-scale mixing is the dominant forcing term* in Eq. 5.2. If air is used in the physical model, then $v = 1.5 \times 10^{-5}\ \mathrm{m^2 s^{-1}}$, so that K must be equal to $1.5 \times 10^{-1}\ \mathrm{m^2 s^{-1}}$ – a condition that may be fulfilled near the ground when the air is very stably stratified.

Using this analysis, physical modelers assume that turbulent mesoscale atmospheric circulations are accurately simulated by laminar laboratory models, provided that the appropriate ratio between the eddy exchange coefficient and kinematic viscosity is obtained.

Using the same assumptions applied to produce Eq. 5.1, the conservation-of-heat relation, represented by the potential temperature equation (4.24), can be written as

$$\left[\frac{\delta\theta\ S}{L}\right]\frac{\partial\widehat{\bar{\theta}}}{\partial\hat{t}} = -\left[\frac{\delta\theta\ S}{L}\right]\widehat{\bar{u}}_j\frac{\partial\widehat{\bar{\theta}}}{\partial\hat{x}_j} - \left[\frac{e_\theta e_{u_i}}{L}\right]\frac{\partial}{\partial\hat{x}_j}\widehat{\overline{u_j''\theta''}} + \left[\frac{\delta\theta\ S}{L}\right]\widehat{\bar{S}}_\theta, \tag{5.4}$$

where $e_\theta e_{u_i}$ is a measure of the subgrid-scale correlation between the fluctuating velocities and temperatures, with e_θ perhaps represented by

$$e_\theta = \left[\left(\overline{\theta''^2}/2\right)\right]^{1/2}.$$

If the molecular conduction of potential temperature C_θ is included in Eq. 5.4 and represented in analogy with the viscous dissipation term as

$$C_\theta = \frac{k_\theta}{\rho C_\mathrm{p}}\frac{\partial^2\bar{\theta}}{\partial x_j \partial x_j};\quad |C_\theta| \sim \frac{k_\theta}{\rho_0 C_\mathrm{p}}\frac{\delta\theta}{L^2},$$

where k_θ is the potential temperature molecular conduction coefficient, then multiplying Eq. 5.4 by $L/\delta\theta\ S$ and including the order of magnitude estimate of C_θ yields

$$\frac{\partial\widehat{\bar{\theta}}}{\partial\hat{t}} = -\widehat{\bar{u}}_j\frac{\partial\widehat{\bar{\theta}}}{\partial\hat{x}_j} - \left[\frac{e_\theta e_{u_j}}{\delta\theta\ S}\right]\frac{\partial}{\partial\hat{x}_j}\widehat{\overline{u_j''\theta''}} + \left[\frac{k_\theta}{\rho_0 C_\mathrm{p} v}\right]\left[\frac{v}{LS}\right]\frac{\partial^2\widehat{\bar{\theta}}}{\partial\hat{x}_j^2} + \widehat{\bar{S}}_\theta. \tag{5.5}$$

The ratio

$$k_\theta/\rho_0\, C_\mathrm{p}\, v = \mathrm{Pr}^{-1}$$

where Pr is called the *Prandtl number* and is of order unity for air.

Thus to obtain *thermal similarity* between the mesoscale circulation and its laboratory representation, the Reynolds number must also be very large and the partitioning of heat transport between the subgrid-scale and resolvable fluxes must be the same. If in Eq. 5.2, for example, the temperature perturbation $\delta\theta$ must be increased in the bulk Richardson number $\mathrm{Ri}_{\mathrm{bulk}}$ to compensate for a decrease of L in the laboratory model, then in Eq. 5.5 the turbulent fluctuations in the simulated atmosphere must also be increased.

The nondimensional source-sink term for potential temperature $\hat{S}_\theta$ is included in the analysis. However, the mathematical procedure of representing it as a single variable masks its physical complexity. As discussed in Chapters 7 and 8, this term includes such effects as radiative flux divergence, phase changes of water, etc. and is an involved function of the dependent variables. Thus it is extremely difficult to evaluate this term using scale analysis, and in practice, physical modelers exclude it in their representation of mesoscale atmospheric flows. An equivalent similarity analysis can be performed for water substance and other aerosol and gaseous contaminants. Because of the inability to accurately represent the sources and sinks of these variables (i.e., $\widehat{\bar{S}}_{q_n}$ and $\widehat{\bar{S}}_{\chi_m}$), however, physical modelers have only studied the movement of nonreactive, conservative pollutants around terrain and building obstacles.

In utilizing physical models, the conservation-of-mass relation given by Eq. 2.43 must also be satisfied, and as long as the ratio of the variations of specific volume to the average specific volume in the physical model are much less than unity, Eqs. 3.11 and 3.14 are satisfactory approximations. The scaled version of the incompressible conservation-of-mass equation (3.14) shows that

$$W \sim L_z U / L_x, \quad W \sim L_z V / L_y$$

so that if the ratio of the vertical to the horizontal scales of the circulation is kept constant between the physical model and the atmosphere, *kinematic similarity* is obtained. This requirement could be satisfied providing the horizontal to vertical representation of the terrain and other physical features of the ground surface in the physical model are not exaggerated. This latter condition is called *geometric similarity*.

The final similarity conditions needed in physical models include the requirement that air flowing into the simulated mesoscale region have velocity and temperature profiles scaled according to the nondimensional relations given by Eqs. 5.2 and 5.5 and that the flow be close to equilibrium (i.e., $\partial\widehat{\bar{u}}_i/\partial\hat{t}$ and $\partial\widehat{\bar{\theta}}/\partial\hat{t}$ are small, relative to the remaining terms in Eqs. 5.2 and 5.5). In addition, such bottom conditions as surface temperature and aerodynamic roughness must be scaled so as to produce kinematic, dynamic, and thermal similarity in the lowest levels of the physical model. These requirements are referred to as *boundary similarity* and their creation necessitates a comparatively long fetch from the input region of the laboratory apparatus to the region of simulation, as well as obstacles such as a lattice placed upwind in the flow to generate specific velocity profiles and turbulence characteristics.

FIGURE 5.1

Visualization of advective dispersion over Port Arguello, California in a stably-stratified laminar flow in a wind tunnel (from Cermak 1971).

With all of these requirements, physical modeling of the mesoscale has been primarily limited to stably-stratified flows over irregular terrain. Even for this case, however, such observed features of the real atmosphere as the veering of the winds with height, radiational cooling, and condensation cannot be reproduced.

The main advantage of physical models of the mesoscale, therefore, has been to provide qualitative estimates of airflow over terrain obstacles during dense, overcast, or nighttime situations. Figure 5.1, reproduced from Cermak (1971), illustrates such a simulation for a scale model of Port Arguello, California, where helium is released as a tracer to represent pollution dispersal. The influence of the model topography on the flow was very marked and corresponded well with the observed trajectories and concentrations. A number of other physical model simulations have been performed of relevance for the mesoscale including those of Cermak (1970), Chaudhry and Cermak (1971), Yamada and Meroney (1971), SethuRaman and Cermak (1973), Hunt et al. (1978), Meroney et al. (1978), Baines (1979), Baines and Davis (1980), Lee et al. (1981), Mitsumoto et al. (1983), Noto (1996), Poreh (1996), and Chen et al. (1999a). Egan (1975) also discuss physical modeling over complex terrain. Avissar et al. (1990) provides a review as to the ability of meteorological wind tunnels to simulate sea and land breezes. Cermak (1996) provides a review of physical modeling. Figure 5.2 from Avissar et al. (1990) illustrates overlap where both physical and numerical mesoscale models realistically simulate sea and land breezes. Nicholls et al. (1993, 1995) discuss how numerical and physical models can be used together to improve the understanding of airflow over buildings. The relationship between grid increments in the numerical model and resolution are discussed in detail in Chapter 9. The spatial resolution required for the hydrostatic assumption to be accurate is discussed later in this chapter.

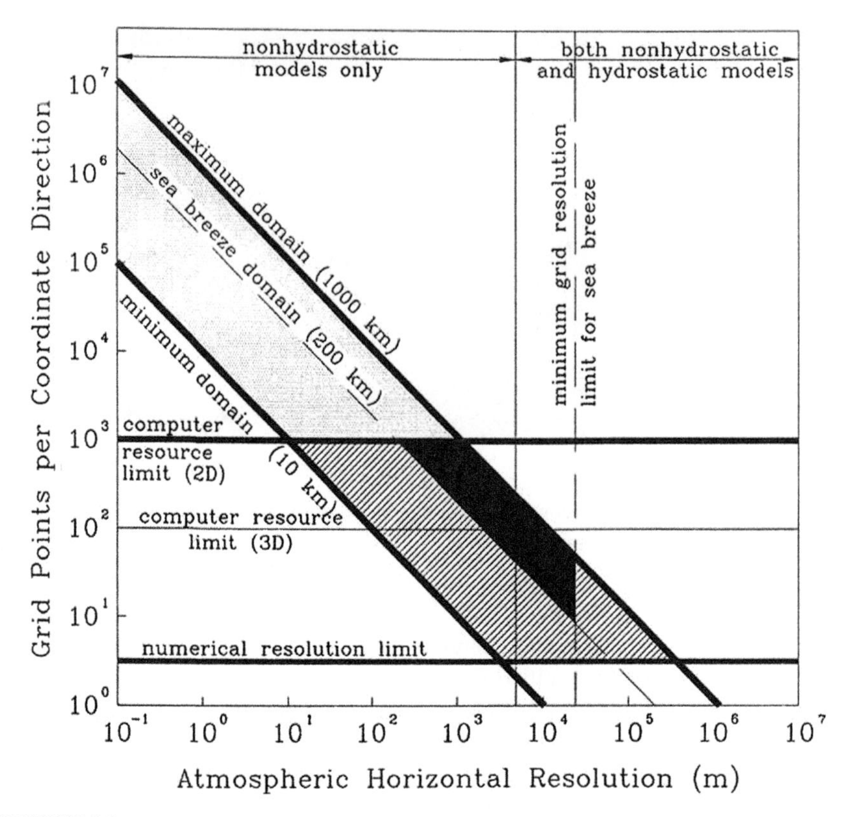

FIGURE 5.2

The operating range (OR) of mesoscale numerical models for sea- and land-breeze simulations: number of model grid points per horizontal coordinate axis vs. atmospheric horizontal resolution. The stippled area denotes the potential OR for mesoscale numerical models in a homogeneous synoptic environment given more powerful computers. The hatched and blackened areas indicate the mesoscale model OR as of 1990. The blackened area denotes the current OR for SLB simulations (from Avissar et al. 1990).

Because physical models are severely limited in their applicability to the mesoscale, however, it is necessary to utilize the techniques of mathematical modeling. The remainder of the text will be devoted to this methodology.

5.2 Linear Models

As discussed previously in the test, the system of equations given by Eqs. 2.43-2.50, represent a *simultaneous set of nonlinear partial differential equations.* They are termed *simultaneous* because each conservation relation must be satisfied at any given time, and they involve *partial derivatives* because four independent variables, x, y, z, and t are involved. The *nonlinear* character of the equations occurs because products of the dependent variables (e.g., $\bar{u}\, \partial \bar{u}/\partial x$, $\bar{w}\, \partial \bar{\theta}/\partial z$ etc.) are included in the relationships.

Over the last several hundred years, mathematical techniques have evolved that permit exact solutions of a range of algebraic and differential equations. However, except for a few highly simplified and idealized situations, no method exists to solve exactly general sets of nonlinear equations. To solve these nonlinear equations, the differential operators must be *approximated* for use in a *numerical model*, as discussed in Chapter 9, so that the results obtained are *not exact*.

To obtain exact[1] solutions to the conservation relationships, it is necessary to remove the nonlinearities in the equations. Results from such simplified, linear equations are useful for the following reasons.

1. The exact solutions of simplified, linear differential equations gives some idea as to the physical mechanisms involved in specific atmospheric circulations. Because precise solutions are obtained, an investigator can be certain the results are not caused by computational errors, such as can be true with numerical models.
2. The use of results from these linear equations can be contrasted with those obtained from a numerical model in which the magnitude of the nonlinear terms are determined to be small, relative to the linear terms. An accurate nonlinear numerical model must be able to reproduce the linear results closely when the products of the dependent variables are small.

Linear representations of the conservation relations have been used to investigate wave motions in the atmosphere, as well as to represent actual mesoscale circulations. Kurihara (1976), for example, applied a linear analysis to investigate spiral bands in a tropical storm. Klemp and Lilly (1975) used such an approach to study wave dynamics in downslope wind storms to the lee of large mountain barriers. Other linear models of airflow over mountain barriers include Wang and Lin (1999).

Simultaneous sets of linear, partial differential equations relevant to atmospheric circulations are derived from Eqs. 2.43-2.50, or from their approximate forms. For example, with shallow, adiabatic atmospheric circulations, Eqs. 4.21, 4.23, and 4.24 can be written in linear form as

$$\frac{\partial \bar{u}_i}{\partial t} = -u_{j0}\frac{\partial \bar{u}_i}{\partial x_j} + K\frac{\partial^2 \bar{u}_i}{\partial x_j^2} - \alpha_0 \frac{\partial p'}{\partial x_i} - \alpha_0\left[\frac{\partial p_0}{\partial x}\delta_{i1} + \frac{\partial p_0}{\partial y}\delta_{i2}\right]$$
$$+ \frac{\theta'}{\theta_0} g\delta_{i3} - 2\epsilon_{ijk}\Omega_j \bar{u}_k, \tag{5.6}$$

$$\frac{\partial \bar{u}_j}{\partial x_j} = 0, \tag{5.7}$$

[1]Exact solutions are also referred to as *analytic* solutions.

and

$$\frac{\partial\bar{\theta}}{\partial t} = -u_{j0}\frac{\partial\bar{\theta}}{\partial x_j} + K\frac{\partial^2\bar{\theta}}{\partial x_j^2}, \tag{5.8}$$

where the subgrid-scale correlation terms have been replaced by

$$\frac{1}{\rho_0}\frac{\partial}{\partial x_j}\rho_0\overline{u_j''u_i''} \simeq -K\frac{\partial^2\bar{u}_i}{\partial x_j^2}, \tag{5.9}$$

and

$$\frac{1}{\rho_0}\frac{\partial}{\partial x_j}\rho_0\,\overline{u_j''\theta''} \simeq -K\frac{\partial^2\bar{\theta}}{\partial x_j^2}, \tag{5.10}$$

with K equal either to a constant or to a function of the independent variables. Besides these subgrid-scale terms, only the advective terms $\overline{u_j}\,\partial\bar{u}_i\partial x_j$ and $\bar{u}_j\,\partial\bar{\theta}/\partial x_j$ are directly affected by the linearization for this particular atmospheric system. Since $|\alpha'|/\alpha_0 \ll 1$, the pressure gradient force and the conservation-of-mass relation were already assumed linear in Chapter 4, and the term involving the rotation of the Earth $2\epsilon_{ijk}\Omega_j\bar{u}_k$ was a linear contribution in the original derivation of the conservation-of-motion given in Chapter 2.

The linearizing assumptions made to obtain Eqs. 5.6 and 5.8 from Eqs. 4.21 and 4.24 are major oversimplifications of the atmosphere and are made only so that exact solutions are obtainable. Observations, nonlinear numerical model results (e.g., see Chapter 13), and even linear solutions (as will be shown in this chapter) demonstrate that the mesoscale velocity perturbations u_j' are not, in general, much less than the synoptic components u_{j0}. In addition, as discussed in Chapter 7, subgrid-scale mixing is not represented accurately for most situations using such a simple form of the exchange coefficient as assumed in Eqs. 5.9 and 5.10.

In the remainder of this chapter, examples of simplified linear models are discussed.

5.2.1 Tank Model

5.2.1.1 Single Homogeneous Fluid

Among the simplest of these models is one that represents a homogeneous fluid ($\bar{\rho} = \rho_0 =$ a constant) situated in a two-dimensional rectangular tank. This model is frequently used to test numerical computational schemes since the physical response of this system is straightforward and easily understood.

Figure 5.3 illustrates this tank where P_h is the pressure at the free surface h, ρ_0 the density of the fluid, and P_B the pressure at the bottom. A set of equations to describe the flow in this tank can be obtained from Eqs. 4.10, 4.30, and 4.23 and written in component form to yield

$$\frac{\partial u'}{\partial t} = -\frac{\partial}{\partial x}u'^2 - \frac{\partial}{\partial z}w'u' - \frac{1}{\rho_0}\frac{\partial\bar{p}}{\partial x}, \tag{5.11}$$

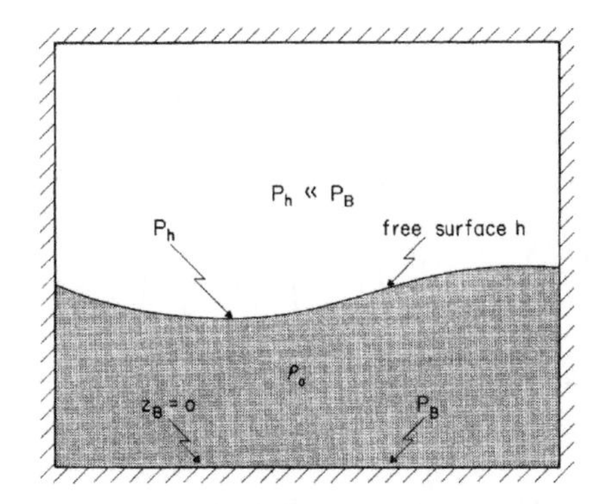

FIGURE 5.3

A two-dimensional tank containing one fluid with constant density ρ_0. The pressure at the free surface h is P_h. On the flat bottom, pressure is indicated by P_B.

$$\frac{\partial \bar{p}}{\partial z} = -\rho_0\, g, \tag{5.12}$$

$$\frac{\partial u'}{\partial x} + \frac{\partial w'}{\partial z} = 0, \tag{5.13}$$

where the hydrostatic relation has been presumed valid, the thermodynamic equation is not required since α' is identically zero, the subgrid-scale mixing and the rotation of the Earth are ignored, and the large-scale velocities (u_0, w_0) and their gradients are set identically to zero. Further, the flow is assumed to be two-dimensional. Since the fluid is homogeneous, Eq. 5.13 is the exact form of the conservation of mass relation. In this system of equations, only the advective terms in Eq. 5.11 are nonlinear.

Using the homogeneous nature of the fluid in the tank, Eqs. 5.11-5.13 can be simplified further. The vertical derivative of the horizontal pressure gradient force can be given as

$$\frac{\partial}{\partial z}\frac{1}{\rho_0}\frac{\partial \bar{p}}{\partial x} = \frac{1}{\rho_0}\frac{\partial}{\partial x}\frac{\partial \bar{p}}{\partial z} = \frac{1}{\rho_0}\frac{\partial}{\partial x}(-\rho_0 g) = 0$$

since ρ_0 and g are constants. Thus the horizontal pressure gradient does not change with height.

By integrating Eq. 5.12 between the bottom $z = 0$ and the surface of the fluid h yields

$$\int_0^h \frac{\partial \bar{p}}{\partial z}\, dz = P_h - P_B = -\rho_0\, gh \tag{5.14}$$

and if $P_h \ll P_B$ (if a vacuum exists above h, P_h is identically equal to zero), then

$$P_B = \rho_0\, gh. \tag{5.15}$$

Differentiating Eq. 5.15 with respect to x, and rearranging, results in

$$\frac{1}{\rho_0}\frac{\partial P_B}{\partial x} = g\,\frac{\partial h}{\partial x}, \tag{5.16}$$

which, since the horizontal pressure gradient is invariant with height (i.e., $(1\rho_0)(\partial P_B/\partial x) = 1(1/\rho_0)(\partial \bar{p}/\partial x)$), permits the horizontal pressure gradient in Eq. 5.11 to be replaced by the righthand side of Eq. 5.16.

Next, differentiating Eq. 5.11 with respect to height and rearranging, yields

$$\frac{\partial}{\partial t}\left(\frac{\partial u'}{\partial z}\right) + u'\frac{\partial}{\partial x}\left(\frac{\partial u'}{\partial z}\right) + w'\frac{\partial}{\partial z}\left(\frac{\partial u'}{\partial z}\right) = 0.$$

Where $(\partial u'/\partial x + \partial w'/\partial z) = 0$ was applied. Thus if $\partial u'/\partial z = 0$ initially, then it can never be generated, since the terms in this expression require existing velocity shear to be nonzero.

With this result, Eq. 5.11 can now be written as

$$\frac{\partial u'}{\partial t} = -u'\frac{\partial u'}{\partial x} - g\frac{\partial h}{\partial x} \tag{5.17}$$

Equation 5.13 can also be given in a different form by integrating between $z = 0$ and $z = h$, so that

$$\int_0^h \frac{\partial w'}{\partial z}dz = w'_h - w'_0 = w'_h = -\int_0^h \frac{\partial u'}{\partial x}dz, \tag{5.18}$$

where $w_0 = 0$ since the bottom is flat. Since

$$\frac{\partial}{\partial z}\frac{\partial u'}{\partial x} = \frac{\partial}{\partial x}\frac{\partial u'}{\partial z} = 0,$$

because $\partial u'/\partial z$ is identically equal to zero, then the horizontal gradient of u' is not a function of height, and Eq. 5.18 can be written as

$$w'_h = -\frac{\partial u'}{\partial x}h. \tag{5.19}$$

If h is defined to be a *material surface*, which moves up and down with the vertical velocity, then

$$w'_h = \frac{dh}{dt} = \frac{\partial h}{\partial t} + u'\frac{\partial h}{\partial x},$$

using the chain rule of calculus, where it is assumed that $h = h(t, x(t))$ and $u_0 = 0$.

Equation 5.19 can, therefore, be written as

$$\frac{\partial h}{\partial t} = -u'\frac{\partial h}{\partial x} - h\frac{\partial u'}{\partial x} \tag{5.20}$$

Along with Eq. 5.17, these two relations are in the form as most commonly applied to the single-fluid tank model. Even though this situation of a homogeneous fluid in a tank is conceptually simple, Eqs. 5.17 and 5.20 are, of course, still nonlinear and no general analytic solution is obtainable.

One way to linearize this system of equations is to set $h = h_0 + h'$, where h_0 is a constant defined to be equal to the average depth of the fluid, so that Eq. 5.20 is written as

$$\frac{\partial h}{\partial t} = -u'\frac{\partial h'}{\partial x} - h_0\frac{\partial u'}{\partial x} - h'\frac{\partial u'}{\partial x}. \tag{5.21}$$

Then, by neglecting products of the dependent variables in Eqs. 5.17 and 5.21, a set of two simultaneous linear partial differential equations is obtained, given by

$$\frac{\partial u'}{\partial t} = -g\frac{\partial h'}{\partial x}, \tag{5.22}$$

$$\frac{\partial h'}{\partial t} = -h_0\frac{\partial u'}{\partial x}. \tag{5.23}$$

This system of equations is rigorously fulfilled when $h' \ll h_0$ so that, for example, in a tank 1 m deep, a perturbation height of 1 cm (a 1% deviation) might be said to satisfy this inequality.

The method of solving Eqs. 5.22 and 5.23 involves representing h' and u' as a function of wavenumber k and frequency ω (i.e., a Fourier transform). This relationship can be expressed mathematically as

$$u'(x,t) = \int_{-\infty}^{\infty}\int_{-\infty}^{\infty} \tilde{u}(k,\omega)e^{i(\omega t + kx)}dkd\omega \tag{5.24}$$

$$h'(x,t) = \int_{-\infty}^{\infty}\int_{-\infty}^{\infty} \tilde{h}(k,\omega)e^{i(\omega t + kx)}dkd\omega \tag{5.25}$$

where $\tilde{u}(k,\omega)$ $\tilde{h}(k,\omega)$, and $e^{i(\omega t+kx)}$ are complex variables. The advantage of performing this transformation is that the *linear partial differential* equations given by Eqs. 5.22 and 5.23 are replaced by two *algebraic* equations. The exponential term can also be written as

$$e^{i(\omega t+kx)} = \cos(\omega t + kx) + i\sin(\omega t + kx),$$

which corresponds to a unit vector of components $\cos(\omega t + kx)$ on the real axis and $\sin(\omega t + kx)$ on the imaginary axis. The frequency ω and wavenumber k can also be expressed as a complex number. As an example, if

$$\omega = \omega_{\rm r} + i\omega_{\rm i},$$

then

$$e^{i\omega t} = e^{i(\omega_{\rm r}+i\omega_{\rm i})t} = e^{-\omega_{\rm i}t}e^{i\omega_{\rm r}t} = e^{-\omega_{\rm i}t}\left(\cos\,\omega_{\rm r}t + i\,\sin\,\omega_{\rm r}t\right), \tag{5.26}$$

so that $e^{-\omega_{\rm i}t}$ can indicate whether $u'(x,t)$ and $h'(x,t)$ damps ($\omega_{\rm i} > 0$) or amplifies ($\omega_{\rm i} < 0$) with time, and the term $\cos\omega_{\rm r}t + i\,\sin\omega_{\rm r}t$ is used to determine changes in

$u'(x, t)$ and $h(x, t)$ owing to propagation. A similar decomposition can be applied to the wavenumber k, where $e^{-k_i x}$ denotes the damping or amplification of the dependent variables in the x direction as a function of wavelength, and $\cos k_r x + i \sin k_r x$ refers to the periodic portion of the spatial distribution of the dependent variables.

When the complex form is used to solve a system of differential equations, only the real part of the solution gives information concerning the magnitude of the dependent variables. The use of complex variables is a valuable tool in the solution of equations that are expected to have periodic solutions.

One crucial advantage of linearizing a system of differential equations is that any single term inside of the integral of Eqs. 5.24 and 5.25 is separately a solution to Eqs. 5.22 and 5.23. In a nonlinear system, products of integrals arise (e.g., from the multiplication of the Fourier representation of u' by that for $\partial h'/\partial x$), reflecting the interactions between different scales of motion. No such interaction is possible, however, with a linear system.

Since Eqs. 5.22 and 5.23 are linear equations,

$$u'(x, t) = \tilde{u}(k, \omega)e^{i(\omega t + kx)} \tag{5.27}$$

and

$$h'(x, t) = \tilde{h}(k, \omega)e^{i(\omega t + kx)} \tag{5.28}$$

can therefore, be used to represent the two dependent variables in those equations. The complete solution can be obtained by adding together the solutions for all possible wavenumbers in a particular problem. Since no sources or sinks of velocity are permitted in Eqs. 5.22 and 5.23, we shall also assume that $\tilde{u}$, $\tilde{h}$, ω, and k are all real so that no terms of the form given by $e^{-w_i t}$, for example, are produced. It is also assumed that the fluid extends indefinitely in the horizontal direction (i.e., there are no lateral walls). Substituting Eqs. 5.27 and 5.28 into the two differential equations 5.22 and 5.23, yield the two algebraic equations

$$\begin{aligned} \omega\tilde{u}e^{i(kx+\omega t)} + gk\tilde{h}e^{i(kx+\omega t)} &= 0, \\ \omega\tilde{h}e^{i(kx+\omega t)} + h_0 k\tilde{u}e^{i(kx+\omega t)} &= 0, \end{aligned}$$

Since $e^{i(kx+\omega t)}$ cannot equal zero, these equations can be reduced to

$$\begin{aligned} \omega\tilde{u} + gk\tilde{h} &= 0, \\ h_0 k\tilde{u} + \omega\tilde{h} &= 0, \end{aligned} \tag{5.29}$$

which in matrix form is given as

$$\begin{bmatrix} w & gk \\ h_0 k & \omega \end{bmatrix} \begin{bmatrix} \tilde{u} \\ \tilde{h} \end{bmatrix} = \begin{bmatrix} 0 \\ 0 \end{bmatrix} \tag{5.30}$$

The solution to this set of two algebraic equations can be determined either by algebraic rearrangement of Eq. 5.29, or by applying concepts of linear algebra to Eq. 5.30. In the first case, solving the top equation for $\tilde{u}$ and substituting it into the bottom relation yields

$$\tilde{h}\left[-\frac{h_0 k^2 g}{\omega} + \omega\right] = 0.$$

Since $\tilde{h}$ does not equal zero in general, the bracketed quantity must, so that the equation

$$\omega/k = \pm\sqrt{g h_0}$$

expresses the relationship between ω and k in Eqs. 5.22 and 5.23.

The determination of the solution using the matrix form is not as straightforward, but is introduced here because it plays an important role in the evaluation of the computational stability of numerical solution techniques (Chapter 9) and in the solution of Eq. 5.44. Matrix equation 5.30 represents a system of two *linear, homogeneous algebraic* equations in two unknowns. It is linear because the coefficients of the 2×2 matrix on the lefthand side of Eq. 5.30 are not functions of $\tilde{u}$ and $\tilde{h}$, and it is homogeneous since there is no nonzero function in the equation that is not a function of $\tilde{u}$ and $\tilde{h}$ (this is the reason that the righthand side of Eq. 5.30 is zero).

As shown, for example, by Murdoch (1957), a system of homogeneous algebraic equations has a *nontrivial solution* (a solution other than when the dependent variables are identically equal to zero) only if the determinant of the coefficients is zero. Therefore,[2]

$$\begin{vmatrix} \omega & gk \\ h_0 k & \omega \end{vmatrix} = 0 = \omega^2 - g h_o k^2$$

or

$$\omega/k = \mp\sqrt{g h_0}. \tag{5.31}$$

Finally, since $\omega = 2\pi/P$ and $k = 2\pi/L$, where P is the period and L a wavelength, then setting the exponents of Eqs. 5.27 and 5.28 equal to zero (i.e., $\omega t + kx = 0$; $\omega/k = -x/t$), yields $\omega/k = -c$, where c is the phase velocity. Therefore,

$$c = \pm\sqrt{g h_o}, \tag{5.32}$$

and the movement in the tank model corresponds to two waves that can propagate in two opposite directions with a speed given by Eq. 5.32. This wave is called an *external gravity wave*, since it is found only at the top of the fluid and requires a gravitational acceleration for it to occur. For $g = 9.8$ m s^{-2} and $h_0 = 10$ km, for example, $c \simeq \pm 313$ m s^{-1}.

[2]A two-by-two determinant, $\begin{vmatrix} a_{11} & a_{12} \\ a_{21} & a_{22} \end{vmatrix}$ is equal to $a_{11}a_{22} - a_{21}a_{12}$.

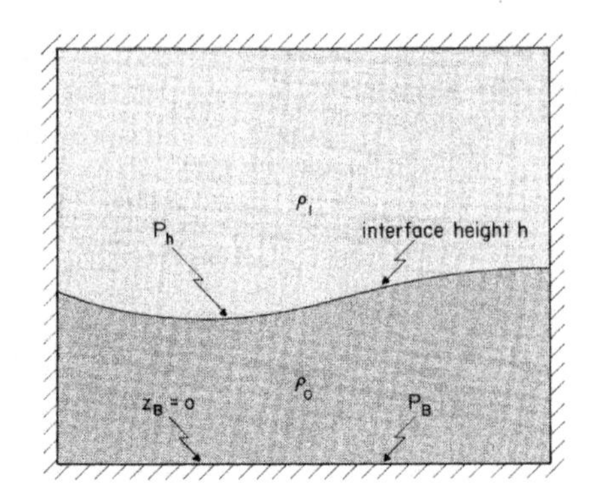

FIGURE 5.4

A two-dimensional tank containing two fluids with densities ρ_0 and ρ_1 ($\rho_0 > \rho_1$). The pressure at the interface h is P_h. On the flat bottom, pressure is indicated by P_B. The depth of the upper fluid is assumed to be much greater than the displacement height, δh, of the interface.

5.2.1.2 *Two-Layered Fluids*

A somewhat more complicated tank model can be derived if it is assumed that P_h is not much less than P_B in Eq. 5.14.[3] Such a situation is depicted in Fig. 5.4, where a fluid of one uniform density ρ_1 overlies a second homogeneous fluid with a larger density ρ_0. The pressure at the rigid top of the second fluid is assumed much less than the pressure at the interface P_h. In the lower fluid, pressure is determined by Eq. 5.12, whereas the hydrostatic relation

$$\partial \bar{p}/\partial z = -\rho_1 g$$

is presumed valid in the upper fluid.

Integrating Eq. 5.12 between the bottom and the initial interface height h_0 yields

$$\int_0^{h_0} \frac{\partial \bar{p}}{\partial z} dz = P_{h_0} - P_B = -\rho_0 g h_0. \tag{5.33}$$

The pressure variation at h_0 is given by

$$\delta P_{h_0} = \rho_0 g\, \delta h - \rho_1 g\, \delta h,$$

since, when $\delta h > 0$, fluid with density ρ_0 moves above h_0, displacing fluid of density ρ_1, whereas the reverse is true if $\delta h < 0$. Therefore,

$$\frac{\delta P_{h_0}}{\delta x} = (\rho_0 - \rho_1)\, g \frac{\delta h}{\delta x},$$

or

$$\lim_{\delta x \to 0} \frac{\delta P_{h_0}}{\delta x} = \frac{\partial P_{h0}}{\partial x} = \frac{\partial \bar{p}}{\partial x} = (\rho_0 - \rho_1)\, g \frac{\partial h}{\partial x}, \tag{5.34}$$

[3]The following material is based on the analysis given in Holton (1972:169-171).

where the invariance of the pressure gradient with height in both fluids has been applied. The horizontal pressure gradient for this case can, therefore, be written as

$$-\frac{1}{\rho_0}\frac{\partial \bar{p}}{\partial x} = -\frac{\left(\rho_0 - \rho_1\right)}{\rho_0} g \frac{\partial h}{\partial x},$$

so that the equation equivalent to 5.17 for two fluids becomes

$$\frac{\partial u'}{\partial t} = -u'\frac{\partial u'}{\partial x} - \frac{\left(\rho_0 - \rho_1\right)}{\rho_0} g \frac{\partial h}{\partial x}. \tag{5.35}$$

Equation 5.20 has the same form for this situation, except h now refers to the interface between the two fluids.

The linear forms of Eqs. 5.35 and 5.20 are written as

$$\frac{\partial u'}{\partial t} = -\frac{\rho_0 - \rho_1}{\rho_0} g \frac{\partial h}{\partial x},$$

$$\frac{\partial h}{\partial t} = -h_0 \frac{\partial u'}{\partial x}.$$

Applying the same solution technique to this model and assuming the fluid extends indefinitely in the horizontal direction, as for the single fluid tank, results in the phase velocity equation

$$c = \pm\sqrt{\Delta\rho g h_0/\rho_0}, \tag{5.36}$$

where $\Delta\rho = \rho_0 - \rho_1$.

Waves that form on such discontinuous interfaces between fluids are one type of *internal gravity waves*. In Section 5.2.2, it is shown that internal waves can also occur even without a density discontinuity. Equation 5.36 reduces to the external gravity wave when the overlying layer has a much smaller density than the bottom fluid (i.e., $\rho_0 \gg \rho_1$), such as is the case for air ($\rho_1 \sim 1.25$ kg m^{-3}) and water ($\rho_0 \sim 1000$ kg m^{-3}). From Eq. 5.36, it is evident that internal gravity waves of this type always travel more slowly than external waves. For $g = 9.8$ m s^{-2}, $h_0 = 1$ km, and $\Delta\rho/\rho_0 = 0.1$, for example, $c \simeq 31$ m s^{-1}. This speed of motion is close to that observed on frontal interfaces in the atmosphere when cold, dense air is being overrun by warmer, less dense air aloft (e.g., Gedzelman and Donn 1979).

These tank models are, of course, gross oversimplifications to any type of mesoscale circulation. Even in terms of real tanks with one or two fluids, however, these models also have serious shortcomings.

1. Although propagation speeds of waves are specified once motion is initiated, there are no mechanisms in these models to generate waves.
2. As is well known, waves generated in tanks can attain a sufficient size such that breaking and overturning occurs. These models are unable to represent such an event.

3. The influence of bottom and lateral friction, and of side walls have been excluded. Thus waves in these models will persist indefinitely in space and time.

Although some of these shortcomings can be minimized by adding linear terms to the original tank model equations, 5.11-5.13, the basic problem with such models is the neglect of nonlinear effects. Breaking of waves, for example, can occur when h' is a significant fraction of h_0 so that $h' \ll h_0$ is no longer fulfilled. Nevertheless, because exact, analytic solutions are obtainable, these simple models are often used to evaluate computational schemes as well as to demonstrate procedures used in numerical models. The tank model is used in this context in Section 9.1.4.

5.2.2 Generalized Linear Equations

The tank models discussed in the previous section represent a situation in which the conservation laws are greatly reduced in form, thereby permitting a straightforward, comparatively simple solution to the possible forms of motion. Unfortunately, by simplifying, only one wave solution was obtained.

To obtain a more complete representation of the types of wave motion in the mesoscale atmosphere, however, it is desirable to retain as many terms as possible in the conservation relations, yet still linearize the equations so that exact solutions are possible. These solutions will not only aid in the evaluation of the more complete nonlinear numerical models, but should also provide insight into the physical mechanisms involved in atmospheric circulations.

In the following derivation, the main focus is to obtain the characteristic phase velocity of different classes of atmospheric wave motion. This information will be essential in successfully applying computational techniques to numerical mesoscale models. To illustrate how the more general equations can be linearized, Eqs. 4.19, 4.21, 4.22, and 4.24 can be reduced to the linear form as follows:

$$\rho_0 \frac{\partial u'}{\partial t} = -\frac{\partial p'}{\partial x} + \rho_0 f v', \tag{5.37}$$

$$\frac{\partial v'}{\partial t} = -f u', \tag{5.38}$$

$$\lambda_1 \rho_0 \frac{\partial w'}{\partial t} = -\frac{\partial p'}{\partial z} - \rho' g, \tag{5.39}$$

$$\frac{\partial \theta'}{\partial t} = -w' \frac{\partial \theta_0}{\partial z}, \tag{5.40}$$

$$\lambda_2 \frac{\partial \rho'}{\partial t} = -\rho_0 \left[\frac{\partial u'}{\partial x} + \frac{\partial w'}{\partial z} \right] - w' \frac{\partial \rho_0}{\partial z}, \tag{5.41}$$

$$\rho' = \rho_0 \frac{C_v}{C_p} \frac{p'}{p_0} - \rho_0 \frac{\theta'}{\theta_0}, \tag{5.42}$$

where

1. the layer-domain averaged fields ($u_0, v_0, w_0, \theta_0, p_0, \rho_0$) are hydrostatic, horizontally homogeneous, and unchanging in time;
2. all subgrid-scale correlation terms are ignored;
3. gradients in the y direction ($i = 2$) are neglected;
4. the Coriolis term is neglected in the vertical equation of motion as is the term involving w' and the Coriolis term $\hat{f}$;
5. all motion is adiabatic ($\bar{S}_\theta = 0$);
6. u_0, v_0, and w_0 are assumed identically equal to zero;
7. the vertical gradients of θ_0 and ρ_0 are constant throughout the atmosphere;
8. variations of ρ', p', and θ' are assumed much less than the magnitudes of ρ_0, p_0, and θ_0, so that $\rho_0 \simeq \bar{\rho}$;
9. products of the mesoscale dependent variables are removed (i.e., $u'_j\, \partial u'_i/\partial x_j$, $u'_j\, \partial\theta'/\partial x_j$, etc.); and
10. moisture and effects of other gaseous and aerosol atmospheric materials are ignored.

The parameters λ_1 and λ_2 are defined as

$$\lambda_1 = \begin{cases} 1 & \text{for the nonhydrostatic representation} \\ 0 & \text{for the hydrostatic representation} \end{cases}$$

$$\lambda_2 = \begin{cases} 1 & \text{for the compressible representation} \\ 0 & \text{for the anelastic representation} \end{cases}$$

and are used to keep track of the terms that are neglected when either the hydrostatic or the anelastic assumptions are made.

This simplified set of 6 linear algebraic and differential equations in the 6 unknowns, u', v', w', p', θ', and ρ' can be solved using the Fourier representation method such as that applied to the linear tank model equations. In the application of this method to Eqs. 5.37-5.42, the following forms are used where the dependent variables are assumed to have a periodic wave form in each of the independent variables, so that

$$\begin{aligned}
u'(x,z,t) &= \tilde{u}\left(k_x, k_z, \omega\right) e^{i(k_x x + k_z z + \omega t)}, \\
v'(x,z,t) &= \tilde{v}\left(k_x, k_z, \omega\right) e^{i(k_x x + k_z z + \omega t)}, \\
w'(x,z,t) &= \tilde{w}\left(k_x, k_z, \omega\right) e^{i(k_x x + k_z z + \omega t)}, \\
\theta'(x,z,t) &= \tilde{\theta}\left(k_x, k_z, \omega\right) e^{i(k_x x + k_z z + \omega t)}, \\
p'(x,z,t) &= \tilde{p}\left(k_x, k_z, \omega\right) e^{i(k_x x + k_z z + \omega t)}, \\
\rho'(x,z,t) &= \tilde{\rho}\left(k_x, k_z, \omega\right) e^{i(k_x x + k_z z + \omega t)}.
\end{aligned} \tag{5.43}$$

As discussed during the solution of the linearized tank model equations, any individual term (also called a *harmonic*) in a Fourier representation is a solution. Any linear combination is also a solution with the complete representation given by adding together all of the harmonics. Thus Eq. 5.43 is used to represent the dependent variables in Eqs. 5.37-5.42.

Rearranging and substituting Eq. 5.43 into these equations yields the simultaneous set of 6 linear, homogeneous algebraic equations in 6 unknowns:

$$\begin{aligned}
\rho_0 i\omega\tilde{u} + ik_x\tilde{p} - \rho_0 f\tilde{v} &= 0,\\
i\omega\tilde{v} + f\tilde{u} &= 0,\\
\lambda_1\rho_0 i\omega\tilde{w} + ik_z\tilde{p} + g\tilde{\rho} &= 0,\\
i\omega\tilde{\theta} + \tilde{w}\frac{\partial\theta_0}{\partial z} &= 0,\\
\lambda_2 i\omega\tilde{\rho} + \rho_0 ik_x\tilde{u} + \rho_0 ik_z\tilde{w} + \tilde{w}\frac{\partial\rho_0}{\partial z} &= 0,\\
\tilde{\rho} + \frac{\rho_0}{\theta_0}\tilde{\theta} - \frac{\rho_0}{p_0}\frac{C_v}{C_p} &= 0.
\end{aligned}$$

Rewriting this system of equations in matrix form gives

$$\begin{bmatrix}
\rho_0 i\omega & -f\rho_0 & 0 & 0 & 0 & ik_x\\
f & i\omega & 0 & 0 & 0 & 0\\
0 & 0 & \lambda_1\rho_0 i\omega & 0 & g & ik_z\\
0 & 0 & \dfrac{\partial\theta_0}{\partial z} & i\omega & 0 & 0\\
\rho_0 ik_x & 0 & \left[\rho_0 ik_z + \dfrac{\partial\rho_0}{\partial z}\right] & 0 & \lambda_2 i\omega & 0\\
0 & 0 & 0 & \dfrac{\rho_0}{\theta_0} & 1 & -\dfrac{\rho_0 C_v}{p_0 C_p}
\end{bmatrix}
\begin{bmatrix}\tilde{u}\\ \tilde{v}\\ \tilde{w}\\ \tilde{\theta}\\ \tilde{\rho}\\ \tilde{p}\end{bmatrix}
=
\begin{bmatrix}0\\0\\0\\0\\0\\0\end{bmatrix} \qquad (5.44)$$

As briefly discussed for the tank model, a nontrivial solution exists for a homogeneous system of equations, only if the determinant of the coefficients of the matrix equation is zero. Large determinants such as obtained from Eq. 5.44 are, unfortunately, not as simple to interpret as was the two-by-two determinant of the tank model, nor is it as simple to evaluate. Programs such as Mathematica (Wolfram 1999) fortunately provide a computationally efficient procedure to solve Eq. 5.44.[4]

[4]The introduction of symbolic algebra (e.g., Hearn 1973) onto computers has removed much of the drudgery of expanding such large determinants. Derickson and Pielke (2000) have used this technique to investigate the influence of nonlinear interactions associated with advection.

Expanding the determinants of Eq. 5.44, and rearranging yields the fifth-order equation for ω

$$
\begin{aligned}
|A| = (\omega^2 - f^2)\Bigg\{ & \frac{\rho_0}{\theta_0}\frac{\partial \theta_0}{\partial z}\lambda_2 \omega k_z - \frac{\rho_0^2}{p_0} i\omega^3 \frac{C_v}{C_p}\lambda_2\lambda_1 \\
& + \omega k_z \left[\rho_0 i k_z + \frac{\partial \rho_0}{\partial z}\right] - \frac{\rho_0}{p_0} i\omega \frac{C_v}{C_p} g \left[\rho_0 i k_z + \frac{\partial \rho_0}{\partial z}\right]\Bigg\} \\
& + \omega^3 k_x^2 \lambda_1 \rho_0 i - g\omega k_x^2 \frac{\rho_0}{\theta_0} i \frac{\partial \theta_0}{\partial z} = 0,
\end{aligned} \tag{5.45}
$$

which can also be written as

$$
\begin{aligned}
|A| = & \left[\frac{\rho_0^2}{p_0} i \frac{C_v}{C_p}\lambda_2\lambda_1\right]\omega^4 + \Bigg[-k_x^2\lambda_1\rho_0 i + \frac{\rho_0}{p_0} i \frac{C_v}{C_p} g \left(\rho_0 i k_z + \frac{\partial \rho_0}{\partial z}\right) \\
& - k_z\left(\rho_0 i k_z + \frac{\partial \rho_0}{\partial z}\right) - f^2 \frac{\rho_0^2}{p_0} i \frac{C_v}{C_p}\lambda_2\lambda_1 - \frac{\rho_0}{\theta_0}\frac{\partial \theta_0}{\partial z}\lambda_2 k_z\Bigg]\omega^2 \\
& + f^2 \frac{\rho_0}{\theta_0}\frac{\partial \theta_0}{\partial z}\lambda_2 k_z + f^2 k_z \left(\rho_0 i k_z + \frac{\partial \rho_0}{\partial z}\right) \\
& - f^2 \frac{\rho_0}{p_0} i \frac{C_v}{C_p} g \left(\rho_0 i k_z + \frac{\partial \rho_0}{\partial z}\right) + g k_x^2 \frac{\rho_0}{\theta_0} i \frac{\partial \theta_0}{\partial z} = 0,
\end{aligned} \tag{5.46}
$$

where the trivial solution $\omega = 0$ has been removed. Ogura and Charney (1961) obtained a similar fourth-order algebraic equation in frequency in the analysis of a similar set of the conservation relations.

Despite our linearization of the conservation equations, along with our simplifying assumptions (e.g., $\bar{S}_\theta = 0$, $\bar{u}_0 = 0$, etc.), Eqs. 5.45 and 5.46 are not simple to evaluate. The tedious effort to get to this point, however, is not in vain since it is possible to make further simplifying assumptions to aid in the interpretation of these relations.

If it is assumed that $k_x = 0$ (no variations of the dependent variables are permitted in the x directions), that $f = 0$ (the Earth's rotation is neglected), and k_z and ω are assumed real (so that there is no amplification or decay of waves in the vertical direction or in time), then, after rearranging, Eq. 5.46 reduces to

$$
\begin{aligned}
& -\lambda_1\lambda_2\omega^2 - igk_z - \frac{g}{\rho_0}\frac{\partial \rho_0}{\partial z} + \frac{p_0}{\rho_0}\frac{C_p}{C_v}k_z^2 \\
& - \frac{p_0}{\rho_0^2}\frac{C_p}{C_v}k_z i \frac{\partial \rho_0}{\partial z} - \frac{C_p}{C_v} i \frac{p_0}{\rho_0\theta_0}\frac{\partial \theta_0}{\partial z}\lambda_2 k_z = 0.
\end{aligned} \tag{5.47}
$$

Using the ideal gas law and definition of potential temperature for the synoptic scale ($p_0 = \rho_0 R T_0$ and $\theta_0 = T_0(1000/p_0 \text{ (in mb)})^{R_d/C_p}$), then

$$\frac{1}{\theta_0}\frac{\partial \theta_0}{\partial z} = \left(1 - \frac{R}{C_p}\right)\frac{1}{p_0}\frac{\partial p_0}{\partial z} - \frac{1}{\rho_0}\frac{\partial \rho_0}{\partial z} = \frac{C_v}{C_p}\frac{1}{p_0}\frac{\partial p_0}{\partial z} - \frac{1}{\rho_0}\frac{\partial \rho_0}{\partial z}$$
$$= -\frac{g\rho_0}{p_0}\frac{C_v}{C_p} - \frac{1}{\rho_0}\frac{\partial \rho_0}{\partial z},$$

where the synoptic scale is assumed hydrostatic ($\partial p_0/\partial z = -\rho_0 g$).

Substituting this expression into Eq. 5.47 results in

$$-\lambda_2\lambda_1\omega^2 + (\lambda_2 - 1)(igk_z) + (\lambda_2 - 1)\, i\, \frac{C_p p_0}{C_v \rho_0^2} k_z \frac{\partial \rho_0}{\partial z} - \frac{g}{\rho_0}\frac{\partial \rho_0}{\partial z} + \frac{p_0}{\rho_0}\frac{C_p}{C_v} k_z^2 = 0$$

and if the compressible continuity of mass relation is used ($\lambda_2 = 1$), then

$$\lambda_2\lambda_1\omega^2 = \frac{p_0}{\rho_0}\frac{C_p}{C_v} k_z^2 - \frac{g}{\rho_0}\frac{\partial \rho_0}{\partial z}. \tag{5.48}$$

The ratio of the two terms on the right is

$$\left|\frac{g}{\rho_0}\frac{\partial \rho_0}{\partial z}\right| \Big/ \left|\frac{p_0}{\rho_0}\frac{C_p}{C_v} k_z^2\right| \simeq \frac{1}{H_\alpha D k_z^2} = \frac{L_z^2}{(2\pi)^2 H_\alpha^2}$$

(since $(1/\rho_0)\partial \rho_0/\partial z = H_\alpha^{-1}$, $p_0/\rho_0 g = D \sim H_\alpha$ and $k_z = 2\pi/L_z$). Even if $L_z \sim H_\alpha$, this ratio is still less than unity ($\left(\frac{1}{2\pi}\right)^2 = 0.03$), so that the solution to Eq. 5.48 can be reasonably estimated by dropping the second term on the right in Eq. 5.48. Therefore, when $\lambda_2 = \lambda_1 = 1$

$$\frac{\omega^2}{k_z^2} \simeq \frac{p_0}{\rho_0}\frac{C_p}{C_v} = R T_0 \frac{C_p}{C_v} \simeq c^2.$$

The wave propagation speed (*phase speed*) c for this situation

$$c \simeq \pm\left(R T_0 \frac{C_p}{C_v}\right)^{1/2} \tag{5.49}$$

is that of a *vertically-propagating sound (acoustic) wave*, which moves upward and downward at the same speed. When either $\lambda_1 = 0$ (the hydrostatic assumption is applied) *or* $\lambda_2 = 0$ (the local time tendency of density is neglected) a wave solution to Eq. 5.48 *does not exist*. This is the reason that the conservation-of-mass relation given by Eq. 3.11 is referred to as the *anelastic* assumption since vertically-propagating sound waves (and as will be shown shortly, sound waves with horizontal components) are eliminated as a possible solution. For reasonable values of temperature ($T_0 = 300$ K, for example), $c \simeq 350$ m s^{-1} in the Earth's lower troposphere.

Another type of wave motion can be isolated from Eq. 5.46 by prescribing $\lambda_2 = 0$ and neglecting the terms associated with the rotation rate of the Earth (e.g., terms multiplied by f). Equation 5.46 then reduces to

$$\left[-k_x^2\lambda_1\rho_0 i + \frac{\rho_0}{p_0} i \frac{C_v}{C_p} g\left(\rho_0 i k_z + \frac{\partial\rho_0}{\partial z}\right) - k_z\left(\rho_0 i k_z + \frac{\partial\rho_0}{\partial z}\right)\right]\omega^2$$
$$+ g k_x^2 \frac{\rho_0}{\theta_0} i \frac{\partial\theta_0}{\partial z} = 0,$$

which can be rewritten as

$$\left[-\rho_0 i\left(\lambda_1 k_x^2 + k_z^2\right) + \frac{\rho_0}{p_0} i \frac{C_v}{C_p} g \frac{\partial\rho_0}{\partial z} - \frac{\rho_0^2}{p_0}\frac{C_v}{C_p} g k_z - k_z \frac{\partial\rho_0}{\partial z}\right]\omega^2 + g k_x^2 \frac{\rho_0}{\theta_0} i \frac{\partial\theta_0}{\partial z} = 0. \tag{5.50}$$

In this equation, it is specified that k_x and ω are real (waves are assumed not to amplify or decay in the horizontal direction or in time). The vertical wave number k_z will be prescribed as complex, however, to account for the imaginary terms that appear in this relation.

Rewriting Eq. 5.50 and rearranging with $k_z = k_{z_r} + i k_{z_i}$ yields the equation

$$\left[\rho_0\left[\lambda_1 k_x^2 + k_{z_r}^2 - k_{z_i}^2\right] - g \frac{\rho_0}{p_0}\frac{C_v}{C_p}\frac{\partial\rho_0}{\partial z} + \frac{\rho_0^2}{p_0}\frac{C_v}{C_p} g k_{z_i} + k_{z_i}\frac{\partial\rho_0}{\partial z}\right]\omega^2$$
$$- g k_x^2 \frac{\rho_0}{\theta_0}\frac{\partial\theta_0}{\partial z} + i\left[2\rho_0 k_{z_i} k_{z_r} - \frac{\rho_0^2}{p_0}\frac{C_v}{C_p} g k_{z_r} - k_{z_r}\frac{\partial\rho_0}{\partial z}\right]\omega^2 = 0. \tag{5.51}$$

Thus to assure that Eq. 5.51 is a real equation

$$k_{z_i} = \frac{\rho_0 C_v}{2 p_0 C_p} g + \frac{1}{2\rho_0}\frac{\partial\rho_0}{\partial z}$$

is required so that Eq. 5.51 reduces to

$$\left\{\rho_0\left(\lambda_1 k_x^2 + k_z^2\right) + \frac{1}{4}\frac{\rho_0^3}{p_0^2}\left(\frac{C_v}{C_p}\right)^2 g^2 + \frac{1}{4\rho_0}\left(\frac{\partial\rho_0}{\partial z}\right)^2 - \frac{1}{2}\frac{\rho_0}{p_0}\frac{C_v}{C_p} g \frac{\partial\rho_0}{\partial z}\right\}\omega^2$$
$$- g k_x^2 \frac{\rho_0}{\theta_0}\frac{\partial\theta_0}{\partial z} = 0, \tag{5.52}$$

where k_{z_r} is written as k_z to simplify the notation. In permitting the vertical wave number to be complex, the assumed solutions given by Eq. 5.43 include the exponential term $e^{-\beta z}$, where

$$\beta = \frac{\rho_0}{2 p_0}\frac{C_v}{C_p} + \frac{1}{2\rho_0}\frac{\partial\rho_0}{\partial z}.$$

Thus for example,

$$\phi'(x, z, t) = \tilde{\phi}\left(k_x, k_z, \omega\right) e^{-\beta z} e^{1(k_x x + k_{z_r} z + \omega t)},$$

where ϕ represents any one of the dependent variables in Eq. 5.43.

Using scale analysis, the ratios of the magnitudes of the three righthand terms within the braces of the first term can be examined.

$$\left|\frac{1}{4}\frac{\rho_0^3}{p_0^2}\left(\frac{C_v}{C_p}\right)^2 g^2\right| \Big/ \left|\rho_0\left(\lambda_1 k_x^2 + k_z^2\right)\right| \sim \frac{1}{4}\left(\frac{C_v}{C_p}\right)^2 \frac{g^2}{R^2 T_0^2 \bar{k}^2}$$

$$\approx 4.2 \times 10^{-11} L^2 (\text{in meters}),$$

$$\left|\frac{1}{4\rho_0}\left(\frac{\partial\rho_0}{\partial z}\right)^2\right| \Big/ \left|\rho_0\left(\lambda_1 k_x^2 + k_z^2\right)\right| \sim \frac{1}{4H_\alpha^2\bar{k}^2} \approx 9.9 \times 10^{-11} L^2 (\text{in meters}),$$

$$\left|\frac{1}{2}\frac{\rho_0}{p_0}\frac{C_v}{C_p} g \frac{\partial\rho_0}{\partial z}\right| \Big/ \left|\rho_0\left(\lambda_1 k_x^2 + k_z^2\right)\right| \sim \frac{1}{2}\frac{C_v}{C_p}\frac{g}{R\, T_0 H_\alpha \bar{k}^2}$$

$$\approx 13.1 \times 10^{-11} L^2 (\text{in meters}),$$

where $\bar{k}$ is used to represent the wavenumber ($\bar{k} \sim (k_z^2 + k_x^2)^{1/2}$) and $\bar{k} = 2\pi/L$ where L is the representative wavelength of the atmospheric circulation. A temperature of 300 K and $H_\alpha = 8$ km were used in this scale estimate along with the values of R, C_p, and C_v for dry air (287, 1004, and 717 J $\text{deg}^{-1}\text{kg}^{-1}$, respectively). Thus if we assume that a ratio of 0.01 is sufficient justification to neglect the largest of these three terms, then

$$L \lesssim 10 \text{ km}$$

is required. For longer wavelengths, and for quantitative analyses, these terms must be retained. The terms are about equal for wavelengths on the order of 100 km.

If it is appropriate to neglect the three terms, Eq. 5.52 reduces to[5]

$$\omega^2 \simeq \frac{k_x^2}{\lambda_1 k_x^2 + k_z^2}\frac{g}{\theta_0}\frac{\partial\theta_0}{\partial z}. \tag{5.53}$$

As shown for the tank model, the negative of the phase speed is equal to the frequency ω divided by the wavenumber in the direction of wave propagation k. The phase speeds in component form are given by

$$c_x = \frac{-\omega k_x}{k_x^2 + k_z^2}; \quad c_z = \frac{-\omega k_z}{k_x^2 + k_z^2}.$$

Equation 5.53 can be written as

$$c^2 \simeq \frac{k_x^2}{\left(k_x^2 + k_z^2\right)\left(\lambda_1 k_x^2 + k_z^2\right)}\frac{g}{\theta_0}\frac{\partial\theta_0}{\partial z} \tag{5.54}$$

or when $\lambda_1 = 1$,

$$c \simeq \pm\frac{k_x}{k_x^2 + k_z^2}\left(\frac{g}{\theta_0}\frac{\partial\theta_0}{\partial z}\right)^{1/2}. \tag{5.55}$$

[5]The quantity $[(g/\theta_0)\partial\theta_0/\partial z]^{1/2}$ is called the *Brunt-Väisälä frequency*. Durran and Klemp (1982a) discuss the proper form of the Brunt-Väisälä frequency in a saturated atmosphere.

On the one hand, if the wave motion is primarily in the horizontal, $c_z \ll c_x$ and, therefore, $k_z \ll k_x$, so that Eq. 5.55 is written as

$$c \simeq \pm \frac{1}{k_x}\left(\frac{g}{\theta_0}\frac{\partial \theta_0}{\partial z}\right)^{1/2}. \tag{5.56}$$

On the other hand, if the wave motion is predominantly in the vertical so that $c_z \gg c_x$ and, therefore, $k_z \gg k_x$,

$$c \sim \pm \frac{k_x}{k_z^2}\left(\frac{g}{\theta_0}\frac{\partial \theta_0}{\partial z}\right)^{1/2}. \tag{5.57}$$

This wave motion is a type of internal gravity wave that can occur in a *continuously- and uniformly-stratified* fluid. In the two-layer tank model discussed previously, an internal gravity wave that can occur at a *density discontinuity* was presented. Gravitational acceleration is required for both of these types of waves to occur.

If Eq. 3.22 (i.e., $H_\alpha/L_x \precsim 1$) is applicable, then the hydrostatic assumption can be applied ($\lambda_1 = 0$) and Eq. 5.53 reduces to

$$\omega^2 = \frac{k_x^2}{k_z^2}\frac{g}{\theta_0}\frac{\partial \theta_0}{\partial z}. \tag{5.58}$$

Comparing Eqs. 5.53 and 5.58, an alternate justification for applying the hydrostatic assumption is valid, if the atmospheric circulation of interest is primarily influenced by this kind of internal gravity wave. In this case, if

$$k_x^2 \ll k_z^2,$$

or equivalently,

$$L_x^2 \gg L_z^2, \tag{5.59}$$

then the use of the hydrostatic assumption is valid.

Using Eq. 5.55, speeds of propagation of internal gravity waves can be estimated, and as discussed in Section 9.1.4, this information is essential in successfully utilizing numerical simulation methods. Using representative values of the following parameters in Eqs. 5.56 and 5.57 ($\theta_0 = 300$ K, $\partial\theta_0/\partial z = 1°/100$ m, $g = 9.8$ m s^{-2}), for example, $c \simeq 15$ m s^{-1} for $L_x = 5$ km and $L_z = 25$ km, and $c \simeq 3$ m s^{-1} for $L_x = 25$ km and $L_z = 5$ km. Thus in a hydrostatic system with a constant temperature lapse rate, the phase speed of the internal gravity wave is primarily upward with a relatively slow propagation speed. When the hydrostatic assumption is not valid, wave propagation tends to be more horizontal and somewhat faster, but still over an order of magnitude slower than the phase speed of vertically-propagating sound waves.

From Eq. 5.55, it is seen that a nonzero value of $\partial\theta_0/\partial z$ is needed to have an internal gravity wave. Thus if we require that the vertical gradient of potential temperature is zero and neglect terms that include the gravitational acceleration, then it is possible to determine wave solutions from Eq. 5.46 that are not gravity waves. To simplify the analysis, we also assume that $\partial\rho_0/\partial z$ is zero[6] and that the rotation of the Earth is negligible. With these conditions, Eq. 5.46 reduces to

$$\lambda_2\lambda_1\omega^2 = \frac{p_0}{\rho_0}\frac{C_p}{C_v}\left(\lambda_1 k_x^2 + k_z^2\right) = R\,T_0\frac{C_p}{C_v}\left(\lambda_1 k_x^2 + k_z^2\right), \tag{5.60}$$

and since $\omega^2/\left(k_x^2 + k_z^2\right) = c^2$,

$$c = \pm\left(R\,T_0 C_p/C_v\right)^{1/2} \tag{5.61}$$

when $\lambda_1 = \lambda_2 = 1$. The phase speed given by Eq. 5.61 is the same as that given by Eq. 5.49 and corresponds to the speed of sound propagation. From Eq. 5.60, it is evident that either the hydrostatic or the anelastic assumptions eliminate this form of wave propagation.[7] In numerical models, as shown in Chapter 9, the ability to resolve a wave form in a model correctly (and often even to produce stable results), requires that the time step used is less than or equal to the time it takes for a wave to travel between grid points. With sound waves removed by applying the hydrostatic or the anelastic assumption, the fastest waves are gravity waves.

A final wave form that we shall look at includes the influence of the rotation of the Earth. It is easiest to illustrate this relation by assuming the lapse rate is adiabatic ($\partial\theta_0/\partial z = 0$), and either the hydrostatic assumption is used ($\lambda_1 = 0$) or no variations are permitted in the x-direction. In this case, Eq. 5.45 reduces to the simple frequency equation for an *inertial wave*[8]

$$\omega = \pm f. \tag{5.62}$$

Our final discussion regarding types of wave motions is concerned with the effect of two or more waves traveling with different propagation speeds. As evident from Eq. 5.54, for example, the phase speed is dependent on the wavenumber. Thus the complete solution to the system of linear equations (5.37-5.42) is a linear superposition of the solutions given by all of the harmonics of the form given by Eq. 5.43. When the speeds of the different waves are in phase, *constructive reinforcement* occurs and the amplitude of the solution is a maximum, whereas if the waves are out of phase, *destructive reinforcement* results and the amplitudes of the individual waves can sum

[6] In the atmosphere, if $\partial\theta_0/\partial z = 0$, using the ideal gas law and definition of potential temperature, it can be shown that ρ_0 will decrease with height. In the analysis used to obtain Eq. 5.61, this change of density with height is ignored.

[7] If only the hydrostatic assumption is made, periodic motion called the *Lamb wave* can occur that has a speed on the same order as sound waves (e.g., see Haltiner and Williams 1980:35). Pielke et al. (1993a) and Nicholls and Pielke (1994a,b; 2000) show how the magnitude of the Lamb wave can be directly associated with the magnitude of diabatic heating in cumulus convection. Bannon (1995a) discusses how the Lamb wave is related to hydrostatic adjustment. Tijm and Van Delden (1998) discuss the role of sound waves in sea breezes. Rõõm and Männik (1999) conclude that a linear nonhydrostatic, compressible model provides the most realistic simulation at all spatial scales, but that the hydrostatic version is equally as accurate at space scales from 10 km to 500-700 km.

[8] Egger (1999) provides the solution for inertial motion when a more complete version of the equations are retained.

to zero. The speed of propagation of the locations of constructive and destructive reinforcement is called the *group velocity*.

To illustrate this more mathematically, let two waves of the same amplitude but different wavelengths coexist such that

$$\phi_1(x,t) = \tilde{\phi}\cos\left(k_1 x + \omega_1 t\right) = \tilde{\phi}\cos\left(2\pi/l_1\right)\left(x - c_1 t\right) \tag{5.63}$$

$$\phi_2(x,t) = \tilde{\phi}\cos\left(k_2 x + \omega_2 t\right) = \tilde{\phi}\cos\left(2\pi/l_2\right)\left(x - c_2 t\right) \tag{5.64}$$

where $\tilde{\phi} = \tilde{\phi}(k_1,\omega_1) = \tilde{\phi}(k_2,\omega_2), \omega_1/k_1 = -c_1$, and $\omega_2/k_2 = -c_2$. Defining $c_1 = c - \delta c, c_2 = c + \delta c, l_1 = l - \delta l$, and $l_2 = l + \delta l$, where δc and δl are small incremental changes in phase speed and wavelength, then for ϕ_1,

$$\begin{aligned}
\frac{2\pi}{l_1}\left(x - c_1 t\right) &= \frac{2\pi}{l - \delta l}\left(x - ct + t\,\delta c\right) \simeq \frac{2\pi}{l}\left(1 + \frac{\delta l}{l}\right)\left(x - ct + t\,\delta c\right) \\
&+ \frac{2\pi}{l}\left(x - ct\right) + \frac{2\pi}{l}\left(t\,\delta c + \frac{\delta l}{l}x - \frac{\delta l}{l}ct + \frac{t\,\delta l\,\delta c}{l}\right) \\
&\simeq \frac{2\pi}{l}\left(x - ct\right) + \frac{2\pi}{l}\left(t\,\delta c + \frac{\delta l}{l}x - \frac{\delta l}{l}ct\right) \\
&= \frac{2\pi}{l}\left(x - ct\right) + \frac{2\pi}{l^2}\,\delta l\left[x - \left(c - l\frac{\delta c}{\delta l}\right)t\right],
\end{aligned}$$

where the binomial expansion (i.e., $(1-\epsilon)^{-n} \simeq 1+\epsilon$ for $\epsilon \ll 1$) has been used, and δc and δl are assumed much less than c and l. A similar expression can be derived for the second wave

$$\frac{2\pi}{l_2}\left(x + c_2 t\right) \simeq \frac{2\pi}{l}(x + ct) + \frac{2\pi}{l^2}\,\delta l\left[-x + \left(c - l\frac{\delta c}{\delta l}\right)t\right].$$

Defining $\alpha = (2\pi/l)(x - ct)$ and $\beta = (2\pi/l^2)\delta l\{x - [c - l(\delta c/\delta l)]t\}$, to simplify the notation, the linear superposition of the two waves of similar but not equal phase speeds and wavelengths given by Eqs. 5.63 and 5.64 yields

$$\begin{aligned}
\phi_1(x,t) + \phi_2(x,t) &= \tilde{\phi}\left[\cos\left(\alpha + \beta\right) + \cos\left(\alpha - \beta\right)\right] \\
&= \tilde{\phi}\left[\cos\alpha\cos\beta - \sin\alpha\sin\beta + \cos\alpha\cos\beta + \sin\alpha\sin\beta\right] \\
&= 2\tilde{\phi}\left[\cos\alpha\cos\beta\right]0. \\
&= 2\tilde{\phi}\left[\cos\frac{2\pi}{l}(x - ct)\cos\frac{2\pi\,\delta l}{l^2}\left(x - \left(c - l\frac{\delta c}{\delta l}\right)t\right)\right].
\end{aligned}$$

The second term in the brackets represents the linear interaction between the two waves leading to constructive reinforcement if $x = (c - l(\delta c/\delta l))t$. The quantity $c - l(\delta c/\delta l)$ is the *group velocity* and in the limit when δc and δl approach zero can be written as

$$c_{\rm g} = c - l\frac{dc}{dl} = \frac{d}{dk}\left(kc\left(k\right)\right) = -\frac{d\omega}{dk}$$

since $c = -\omega/k$.

The group velocity c_g in the coordinate directions can be written as

$$c_{gi} = -\partial\omega/\partial k_i \tag{5.65}$$

by the change rule, since in general, ω is a function of the three components of the wavenumber (e.g., $\omega = \omega(k_x, k_y, k_z) = \omega(k_i);\ i = 1, 2, 3$).

Thus using Eq. 5.65, it is possible to determine the group velocity for any of the wave forms that we have derived. The components of the internal gravity wave group velocity when $L_x^2 \gg L_z^2$, for example, can be calculated from Eq. 5.58 as

$$c_{g_x} = \pm\left(\frac{g}{\theta_0}\frac{\partial\theta_0}{\partial z}\right)^{1/2}\frac{L_z}{2\pi},$$

and

$$c_{g_z} = \pm\left(\frac{g}{\theta_0}\frac{\partial\theta_0}{\partial z}\right)^{1/2}\frac{L_z}{2\pi}\left(\frac{L_z}{L_x}\right),$$

so that when the hydrostatic assumption is valid, the group velocity is predominantly in the horizontal direction, in contrast with the phase velocity, which is primarily vertical.

With the definition of group velocity, our discussion of specialized wave forms in the atmosphere is concluded. On the mesoscale, the internal gravity and inertial waves are the most important classes of oscillatory motion generated. Both internal waves in a continuously stratified atmosphere and on density inversions are important. Sound waves, by contrast, are considered insignificant on the mesoscale, and it is desirable to eliminate them from the solutions if computational problems arise because of their presence. All of these wave forms, of course, were derived after linearizing the conservation relations, as well as making additional simplifying assumptions that permitted us to remove certain terms from the equations. In the atmosphere, of course, such limitations on the modes of interaction are not present. Hence, it is seldom possible to observe the idealized wave forms that have been derived here. Such features as velocity shear and multiple temperature inversions, for example, even if the response of the atmosphere were linear, would produce different phase and group velocities of internal gravity waves from what we have derived. Nonetheless, an understanding of these idealized wave forms is essential if numerical solution techniques are to be applied effectively. Numerical models, however, are the preferred tool, since they can include nonlinear interactions and therefore, provide a more physically complete representation.

From the analysis of internal waves, the ratio of the vertical-to-horizontal scale of the atmospheric circulation (Eq. 5.59) suggests whether the hydrostatic assumption should be used or not in their simulation. If the smallest horizontal feature that can be resolved with reasonable accuracy in a numerical model is $4\ \Delta x$, as used to obtain Eq. 4.41, and $H_\alpha \simeq 8$ km is used to estimate the largest expected vertical wavelength, then $\Delta x \geq 6$ km appears to be needed to assure that the hydrostatic assumption is valid within about a 10% error for all internal waves formed in a continuously-stratified medium in that model representation. If the predominant horizontal wavelength of such waves is assumed to be determined by bottom surface variations (e.g., a mountain) then such forcings must have a horizontal scale of 25 km or more. This criterion for use of the hydrostatic assumption is more restrictive than that given by Eq. 4.41, but it must be emphasized that it only applies in mesoscale models in which internal gravity waves propagating in a continuously-stratified medium are an important part of the physical solution.

5.2.3 Mesoscale Linearized Equations

5.2.3.1 *Defant Model*

Although the linear analysis presented in Section 5.2.2 illustrates characteristic wave motions expected in the atmosphere, it does not represent any actual mesoscale atmospheric system. This inability to represent such features results from the assumed periodic solution in time and space given by Eq. 5.43. To relax this constraint, at least part of the solution to the linearized equations must include nonperiodic spatial structure.

Linear models have been developed from the basic conservation relations for a number of different mesoscale features (e.g., Walsh 1974, sea breeze; Klemp and Lilly 1975, forced airflow over a mountain) to improve the fundamental understanding of mesoscale systems. During the years before computers, linear models provided the only means to represent atmospheric circulations mathematically.

To illustrate a method of solving a linear mesoscale model, a modified version of Defant's (1950) sea- and land-breeze formulation is used in this section. The analysis presented here was extracted and slightly modified from that presented in Martin (1981) and Martin and Pielke (1983).

Equations 5.37-5.42 can be written as

$$\frac{\partial u'}{\partial t} = -\alpha_0 \frac{\partial p'}{\partial x} + f v' - \sigma_x u', \tag{5.66}$$

$$\frac{\partial v'}{\partial t} = -f u' - \sigma_y v', \tag{5.67}$$

$$\lambda_1 \frac{\partial w'}{\partial t} = \gamma \theta' - \alpha_0 \frac{\partial p'}{\partial z} - \lambda_1 \sigma_z w', \tag{5.68}$$

$$\frac{\partial u'}{\partial x} + \frac{\partial w'}{\partial z} = 0, \tag{5.69}$$

$$\frac{\partial \theta'}{\partial t} + w'\beta = K \left\{ \frac{\partial^2 \theta'}{\partial x^2} + \frac{\partial^2 \theta'}{\partial z^2} \right\}, \tag{5.70}$$

where $\beta = \partial \theta_0 / \partial z$ and $\gamma = g/\theta_0$. The assumptions used here are the same as given following Eq. 5.42 except the incompressible form of Eq. 5.41 is used and several of the subgrid flux terms are retained from Eqs. 4.21 and 4.24. These flux terms are represented by

$$\frac{1}{\bar{\rho}}\frac{\partial}{\partial x}\bar{\rho}\overline{u''^2}+\frac{1}{\bar{\rho}}\frac{\partial}{\partial z}\bar{\rho}\,\overline{u''\,w''}=\sigma_x u'; \quad \frac{1}{\bar{\rho}}\frac{\partial}{\partial x}\bar{\rho}\,\overline{v''u''}+\frac{1}{\rho}\frac{\partial}{\partial z}\bar{\rho}\,\overline{v''w''}=\sigma_y v'$$

$$\frac{1}{\bar{\rho}}\frac{\partial}{\partial x}\bar{\rho}\overline{u''w''}+\frac{1}{\bar{\rho}}\frac{\partial}{\partial z}\bar{\rho}\,\overline{w''^2}=\sigma_z w'; \quad \frac{1}{\bar{\rho}}\left\{\frac{\partial}{\partial z}\bar{\rho}\,\overline{w''\theta''}+\frac{\partial}{\partial x}\bar{\rho}\overline{u''\theta''}\right\}$$
$$=-K\left\{\frac{\partial^2\theta'}{\partial x^2}+\frac{\partial^2\theta'}{\partial z^2}\right\}.$$

Where we assume that $\sigma_x=\sigma_y$. The terms that involve $(\partial/\partial y)\,\bar{\rho}\overline{v''^2}$, $(\partial/\partial y)\,\bar{\rho}\overline{u''v''}$, and $(\partial/\partial y)\,\bar{\rho}\overline{w''v''}$ are neglected (or they could be considered included in the parameterizations involving σ_x and σ_z). When $\lambda_1=0$, the equations are hydrostatic.

To solve Eqs. 5.66-5.70 in five unknowns (i.e., u', v', w', p', and θ'), Defant recognized that u' and v' must be 90° out of phase with w', p', and θ' since the first two dependent variables are expressed in terms of derivatives of the others. Moreover, the solutions should be a function of height above the ground surface, rather than simply a periodic function since the sea and land breeze does not extend upward indefinitely. Defant, therefore, assumed solutions of the form

$$\begin{aligned} w'(x,z,t)&=\tilde{w}(z)e^{i\omega t}\sin k_x x, \quad u'(x,z,t)=\tilde{u}(z)e^{i\omega t}\cos k_x x,\\ p'(x,z,t)&=\tilde{p}(z)e^{i\omega t}\sin k_x x, \quad v'(x,z,t)=\tilde{v}(z)e^{i\omega t}\cos k_x x,\\ \theta'(x,z,t)&=\tilde{\theta}(z)e^{i\omega t}\sin k_x x, \end{aligned} \tag{5.71}$$

with the boundary conditions

$$w'(z=0)=w'(z\to\infty)=\theta'(z\to\infty)=0; \quad \theta'(z=0)=Me^{i\omega t}\sin k_x x,$$

where M is the amplitude of the maximum mesoscale perturbation surface potential temperature. In general, $\tilde{w}$, $\tilde{p}$, $\tilde{\theta}$, $\tilde{u}$, and $\tilde{v}$ are complex valued variables. The wave number k_x is equal to 2π divided by the wavelength L_x of the assumed periodic function. In this model, $0.5L_x$ corresponds to the size of land in which the maximum heating occurs $0.25L_x$ inland from the coast. The frequency ω represents the temporal periodic variation in the system, which for a sea-land breeze simulation, corresponds to the diurnal period.

The assumed solutions given by Eq. 5.71 are substituted into Eqs. 5.66-5.70, which after simplification, yields

$$i\omega\tilde{u}=-k_x\alpha_0\tilde{p}+f\tilde{v}-\sigma_x\tilde{u}, \tag{5.72}$$

$$i\omega\tilde{v}=-f\tilde{u}-\sigma_x\tilde{v}, \tag{5.73}$$

$$i\omega\lambda_1\tilde{w}=-\lambda_1\sigma_z\tilde{w}-\alpha_0\frac{d\tilde{p}}{dz}+\gamma\tilde{\theta}, \tag{5.74}$$

$$-k_x\tilde{u}+\frac{d\tilde{w}}{dz}=0, \tag{5.75}$$

$$i\omega\tilde{\theta} = -\tilde{w}\beta - K\tilde{\theta}k_x^2 + K\frac{d^2\tilde{\theta}}{dz^2}. \tag{5.76}$$

These equations, with boundary conditions, are solved simultaneously for the dependent variables $\tilde{u}$, $\tilde{v}$, $\tilde{w}$, $\tilde{\theta}$, and $\tilde{p}$ which are now only functions of z.

Rearranging Eqs. 5.72-5.76 yields

$$\tilde{u} = \frac{1}{k_x}\frac{d\tilde{w}}{dz}, \tag{5.77}$$

$$\tilde{v} = \frac{-f}{i\omega + \sigma_x}\tilde{u}, \tag{5.78}$$

$$\tilde{p} = -\frac{1}{\alpha_0 k_x}\left[\frac{(i\omega + \sigma_x)^2 + f^2}{(i\omega + \sigma_x)}\right]\tilde{u}, \tag{5.79}$$

$$\frac{d^2\tilde{w}}{dz^2} = \eta^2\tilde{w} + r\tilde{\theta}, \tag{5.80}$$

$$\frac{d^2\tilde{\theta}}{dz^2} = \epsilon\tilde{w} + s\tilde{\theta}. \tag{5.81}$$

where

$$\eta^2 = k_x^2\frac{(i\omega + \sigma_x)\lambda_1(i\omega + \sigma_z)}{(i\omega + \sigma_x)^2 + f^2}; \quad r = -\frac{\gamma k_x^2(i\omega + \sigma_x)}{(i\omega + \sigma_x)^2 + f^2}; \quad \epsilon = \frac{\beta}{K};$$

$$s = \frac{i\omega}{K} + k_x^2.$$

The complex-valued variables $\tilde{w}$ and $\tilde{\theta}$ are assumed to have a solution of the form

$$\tilde{w}(z) = Ae^{az} + Be^{-bz} \tag{5.82}$$

and

$$\tilde{\theta}(z) = Ce^{az} + De^{-bz}. \tag{5.83}$$

Applying the lower boundary condition $\tilde{w}(z = 0) = 0$ to Eq. 5.82 results in

$$A = -B,$$

so that

$$\tilde{w}(z) = A(e^{az} - e^{-bz}). \tag{5.84}$$

Substituting this expression for $\tilde{w}$ into Eq. 5.80 gives

$$A(a^2e^{az} - b^2e^{-bz}) = \eta^2 A(e^{az} - e^{-bz}) + r\tilde{\theta} \tag{5.85}$$

The lower boundary condition on θ' requires that

$$M\, e^{i\omega t}\sin k_x x = \left[C + D\right]e^{i\omega t}\sin k_x x, \tag{5.86}$$

which reduces to $D = M - C$, as long as $e^{i\omega t}$ or $\sin k_x x$ does not equal to zero. Equation 5.83 then becomes

$$\tilde{\theta} = M\, e^{-bz} + C(e^{az} - e^{-bz}). \tag{5.87}$$

Substituting this expression for $\tilde{\theta}$ into Eq. 5.85 and setting z to zero yields

$$A = -rM(b^2 - a^2)^{-1}. \tag{5.88}$$

To obtain C, Eqs. 5.81, 5.87, and 5.88 are combined with z set equal to zero, resulting in

$$b^2 M\, e^{-bz} + C(a^2 e^{az} - b^2 e^{-bz}) = \epsilon \left(\frac{-rM}{b^2 - a^2}\right)(e^{az} - e^{-bz}) + s\left[M\, e^{-bz} + C(e^{az} - e^{-bz})\right],$$

which reduces to

$$C = \frac{b^2 - s}{b^2 - a^2} M. \tag{5.89}$$

All the coefficients in the assumed solutions are now expressed in terms of the physical parameters, with the exception of a and b. Using the equations for A, B, C, and D and Eqs. 5.87, 5.84, and 5.77-5.80, the analytic forms for $\tilde{\theta}$, $\tilde{w}$, $\tilde{u}$, $\tilde{v}$, and $\tilde{p}$ are given by

$$\tilde{\theta}(z) = M\, e^{-bz} + \frac{b^2 - s}{b^2 - a^2} M\{e^{az} - e^{-bz}\}, \tag{5.90}$$

$$\tilde{w}(z) = \frac{-rM}{b^2 - a^2}(e^{az} - e^{-bz}), \tag{5.91}$$

$$\tilde{u}(z) = -\frac{1}{k_x}\frac{rM}{b^2 - a^2}(ae^{az} + be^{-bz}), \tag{5.92}$$

$$\tilde{v}(z) = \frac{f}{i\omega + \sigma_x}\frac{1}{k_x}\frac{rM}{b^2 - a^2}(ae^{az} + be^{-bz}), \tag{5.93}$$

$$\tilde{p}(z) = \frac{1}{\alpha_0 k_x^2}\frac{(i\omega + \sigma_x)^2 + f^2}{(i\omega + \sigma_x)}\frac{rM}{b^2 - a^2}(ae^{az} + be^{-bz}). \tag{5.94}$$

To solve for the parameters a and b, Eq. 5.83 is substituted into Eq. 5.85 (using $D = M - C$, as derived previously), yielding, after rearranging and simplifying,

$$(Aa^2 - A\eta^2 - rC)e^{az} - (Ab^2 - A\eta^2 + rM - rC)e^{-bz} = 0. \tag{5.95}$$

A similar relation is found by differentiating Eq. 5.83 twice with respect to z, resulting in

$$d^2\tilde{\theta}/dz^2 = Ca^2 e^{az} + (M - C)b^2 e^{-bz},$$

and then combining this with Eqs. 5.81, 5.84, and 5.87 to give

$$(Ca^2 - \epsilon A - sC)e^{az} + (Mb^2 - Cb^2 + \epsilon A - sM + sC)e^{-bz} = 0, \tag{5.96}$$

The quantities e^{az} and e^{-bz} are independent functions, and $e^{az} - e^{-bz} \neq 0$ in general, if $a \neq b$. If this is true, each of the exponential coefficients in Eqs. 5.95 and 5.96 are equal to zero, leading to the following two systems of equation:

$$\begin{aligned} A(a^2 - \eta^2) - rC &= 0, \\ -A\epsilon + C(a^2 - s) &= 0, \\ A(b^2 - \eta^2) + r(M - C) &= 0, \\ A\epsilon + (M - C)(b^2 - s) &= 0. \end{aligned}$$

Since the determinant of the coefficients of these two systems of algebraic equations must equal zero, the quadratic equations in a^2 and b^2 given by

$$\begin{aligned} (a^2 - \eta^2)(a^2 - s) - \epsilon r = 0, \text{ or } a^4 - (\eta^2 + s)a^2 + (\eta^2 s - \epsilon r) = 0, \\ (b^2 - \eta^2)(b^2 - s) - \epsilon r = 0, \text{ or } b^4 - (\eta^2 + s)b^2 + (\eta^2 s - \epsilon r) = 0, \end{aligned}$$

are produced.

The quadratic formula is used to compute the complex valued parameters a^2 and b^2, yielding

$$a^2; b^2 \left\{ = \frac{\eta^2 + s}{2} \pm \frac{1}{2}\sqrt{\left(\eta^2 + s\right)^2 - 4\left(\eta^2 s - \epsilon r\right)}. \right. \tag{5.97}$$

It follows that a and b can have the values

$$a = \pm\sqrt{a^2} \text{ and } b = \pm\sqrt{b^2}. \tag{5.98}$$

The choice remains of which roots to choose in Eqs. 5.97 and 5.98. To avoid division by zero, a^2 and b^2 must be opposite roots of Eq. 5.97. Solutions of the model equations, however, showed that identical results were obtained whether a^2 was the first root and b^2 the second root of Eq. 5.97, or vice versa. Furthermore, to satisfy the boundary condition $w(z \to \infty) = 0$, in conjunction with Eq. 5.91, b must have a positive real part and a must have a negative real part. This is no restriction since the square roots of a complex number will yield one with a positive real part and one with a negative real part.

With this information, the analytic solution to Defant's linear model is obtained. Values of the dependent variables w', p', θ', u', and v' as a function of x, z, and t are determined by calculating the real parts of Eqs. 5.71 and 5.90-5.94, using Eqs. 5.97, 5.98, and the formulas following Eq. 5.81 to determine the values of $\eta^2, \gamma, s, \epsilon, a^2, b^2, a$, and b.

Figure 5.5 illustrates the $\tilde{u}$, $\tilde{w}$, $\tilde{\theta}$, and $\tilde{p}$ fields at 6 h after simulated sunrise obtained using this linear model. The values of the parameters used in the model were

$$\left.\begin{aligned} \partial\theta_0/\partial z = \beta = 1°\text{C/1 km}; \quad K = 10\ \text{m}^2\text{s}^{-1}; \quad \alpha_0 = 0.758\ \text{m}^3\text{kg}^{-1} \\ \sigma_x = \sigma_z = 10^{-3}\text{s}^{-1}; \quad f = 1.031 \times 10^{-4}\text{s}^{-1}; \quad g = 9.8\ \text{m s}^{-2}; \\ \theta_0 = 273\ \text{K}; \quad M = 10°\text{C}; \quad k_x = 2\pi/100\ \text{km}. \end{aligned}\right\} \tag{5.99}$$

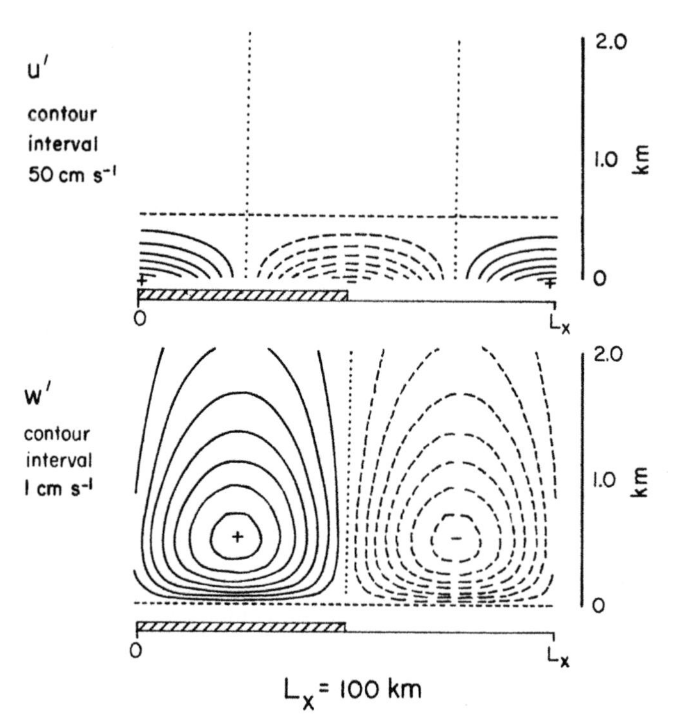

FIGURE 5.5

The horizontal and vertical velocity fields predicted from Defant's (1950) model 6 h after sunrise using the input parameters given by Eq. (5.99) (from Martin 1981).

The symmetric circulation evident in Fig. 5.5 is a result of the horizontal periodicity assumed in the solutions. Land and water are differentiated in the model only by the magnitude of k_x (the same dependent variables over water and land are always of opposite sign because of the form of the assumed solution, Eq. 5.71).

This solution illustrates the interrelation between the dependent variables. Because of the prescribed heating-cooling pattern in the model, pressure falls develop in the region of heating, whereas rises occur where cooling is specified. This pressure pattern causes horizontal accelerations toward regions of lower pressure, as evident from Eq. 5.66. Since mass conservation is required from Eq. 5.69, upward motion necessarily results in the region of heating, whereas subsidence occurs in the areas of cooling.

Varying the parameters in the model, as was done by Martin (1981), also provides insight into the physics of this specific mesoscale circulation. To examine the importance of the hydrostatic assumption, for example, it is possible to calculate nonhydrostatic and hydrostatic results from this model by letting $\lambda_1 = 1$ and $\lambda_1 = 0$. Since the scale analysis presented previously in this book (i.e., Eqs. 3.22 and 5.59) suggest that the horizontal-scale length of a mesoscale circulation L_x is the most important indicator of whether the hydrostatic assumption is valid or not, it is desirable to determine, as a function of L_x, the influence of several of the parameters in Eq. 5.99 on the hydrostatic assumption.

To perform this analysis, following Martin (1981), the relative error between the nonhydrostatic and hydrostatic amplitudes of a given dependent variable is given by

$$E = 2\frac{|\phi_{\rm h}| - |\phi_{\rm nh}|}{|\phi_{\rm h}| + |\phi_{\rm nh}|},$$

where $|\phi_{\rm n}|$ and $|\phi_{\rm nh}|$ are the maximum absolute amplitudes over time and space for a given set of parameters such as listed in Eq. 5.99. The subscripts h and nh correspond to hydrostatic and nonhydrostatic versions, respectively.

Figure 5.6 illustrates one such comparison where E is evaluated as a function of the domain-averaged lapse rate, β, and L_x. As the atmosphere becomes more stably stratified, according to Defant's model, the hydrostatic relation is a more accurate assumption for a given horizontal scale of the circulation.

With a value of $\beta = 1°\text{C}/100$ m, for instance, the maximum error is less than 2% even with $L_x = 1$ km, whereas an equivalent level of accuracy is not attained for $\beta = 0.01°\text{C}/100$ m until L_x is about 10 km. Figure 5.7 shows a similar analysis for the magnitude of the exchange coefficient for heat K, which is assumed a constant in a given solution of Defant's model. In the model, as the rate in which heat is mixed up increases, the hydrostatic relation becomes a poorer assumption for the pressure distribution in the model. With values of $K = 10^2$ m^{2}s^{-1} and $L_x = 1$ km, for example, the maximum error is about 4%, whereas it increases to over 14% for $K = 10^3$ m^{2}s^{-1}.

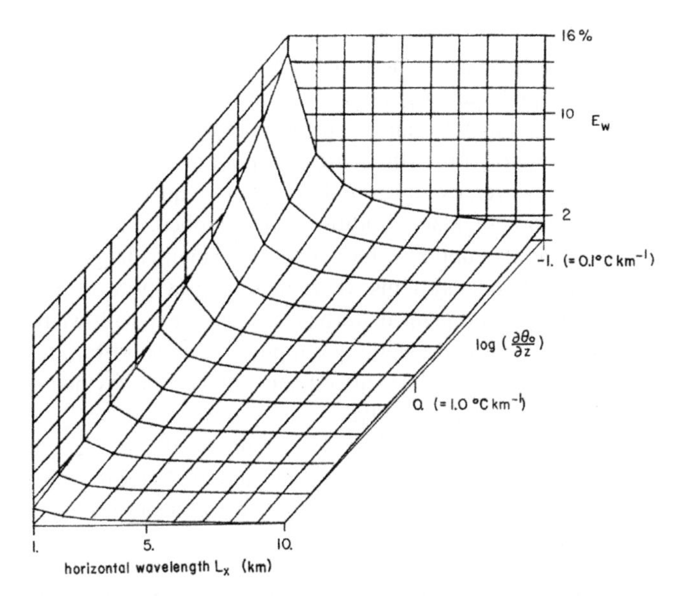

FIGURE 5.6

Relative error in vertical velocity E_w between nonhydrostatic and hydrostatic models. The units for $\partial\theta_0/\partial z = \beta$ are in degrees Celsius per kilometer (from Martin 1981).

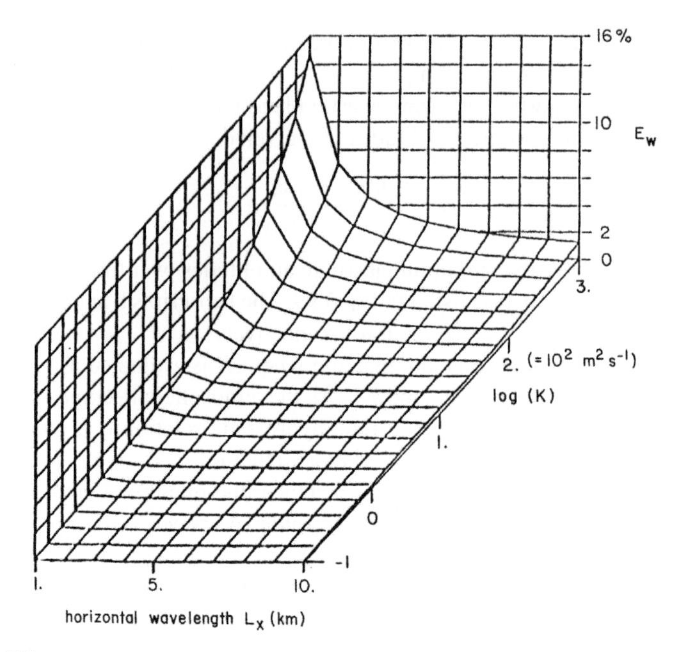

FIGURE 5.7

Relative error in vertical velocity E_w between nonhydrostatic and hydrostatic models. The units for K are in meters squared per second (from Martin 1981).

These results are at variance to the conclusion reached by Orlanski (1981) and Wipperman (1981). Wipperman suggested that the hydrostatic assumption is only valid for horizontal scales larger than 10 km or so, whereas Orlanski claimed that $H_\alpha/\Delta x \ll 1$ is needed before the hydrostatic assumption can be accurately applied. Both of these studies, however, examined only frequency equations of the form given in Section 5.2.2, where no boundary conditions were applied. Thus their conclusion regarding the hydrostatic assumption, which is consistent with that given by Eq. 5.59, is only valid for meteorological systems where internal gravity wave propagation in the free atmosphere is the dominant disturbance. Consistent with Martin's (1981) result, Wipperman found the hydrostatic assumption to be valid for smaller scales when the atmosphere is more stable. He also stated that increased wind speed has the same effect as decreased thermal stability.

As illustrated here, such solutions provide insight into the physical mechanisms that generate and influence the strength of the sea and land breezes. Unfortunately, in Defant's model, physical interactions such as the following are inappropriately represented, or not even included:

1. The subgrid-scale parameterizations for σ and K are assumed independent of time and space so that the intensity of the land breeze is equal to that of the sea breeze. Because vertical mixing is known to be reduced at night over land, however, the land breeze is usually observed to be more shallow and weaker than

the sea breeze (e.g., Mahrer and Pielke 1977a; see also Section 13.2.1). Moreover, realistic parameterizations of the subgrid-scale mixing are nonlinear functions of the dependent variables as discussed in Chapter 7.

2. Advection of temperature and velocity are ignored. Even if the large-scale prevailing flow is zero, the marine air is known to move the region of maximum upward motion inland when a sea breeze occurs (e.g., Estoque 1961).
3. The vertical profile of the large-scale potential temperature is assumed linear. In general, such a condition does not exist and subgrid-scale mixing causes changes in potential temperature owing to curvature in the large scale as well as in the perturbation field.
4. The surface temperature perturbation is prescribed, whereas, in reality, it is a function of the mesoscale circulation (e.g., Physick 1976; see also Fig. 10.23).
5. No interactions are permitted among the dependent variables. Although a necessary condition to obtain analytic results, if $u'\partial' u'/\partial x$, for example, attains a magnitude on the order of $(1/\rho_0)(\partial p'/\partial x)$, nonlinear effects need to be considered.

Although some of these shortcomings were eliminated in later linear models (e.g., Smith 1957 included a linear advection term), it is still impossible to solve the conservation equations analytically when one or more of the terms involve products of dependent variables. Such nonlinear terms arise in the representation of the subgrid-scale processes, the source-sink terms as well as in the expression for advection. The effect of the nonlinear advection (e.g., $u'\ \partial u'/\partial x$, $u'\ \partial\theta'/\partial x$, etc.) on numerical model results for two values of surface heating in a sea-breeze model, as described by Martin (1981), is discussed in Section 10.2.1.1 and illustrated by Fig. 10.3. Those calculations show that the sea-breeze circulation becomes asymmetric owing to nonlinear advection for larger values of surface heating, which results in intensified low-level convergence and weakened low-level divergence. Such asymmetry develops because the advection enhances the convergence in the region of heating, thereby causing a larger horizontal pressure gradient. This increased pressure gradient generates additional convergence because of horizontal advection and this *positive feedback* continues until surface frictional retardation, horizontal turbulent mixing, or the cooling of the surface limits the horizontal velocity acceleration.

With regard to the hydrostatic assumption, however, the reduction in horizontal scale caused by nonlinear advection, dictates that a complete nonlinear model must be used to more completely examine the importance of nonhydrostatic pressure forces when such advection is present. This question is examined in Section 11.5.

5.2.3.2 Further Exploration of the Nonhydrostatic Pressure Perturbations Using the Defant Model

The Defant model can be used to investigate additional aspects of the nonhydrostatic and hydrostatic pressure components (Song et al. 1985). As shown in Section 5.2.3.1, when the hydrostatic assumption is made, $\lambda_1 = 0$, thus $\eta^2 = 0$, and a and b are obtained from

$$a_H^2 - b_H^2 = \frac{s}{2} \pm \frac{1}{2}\left(s^2 + 4\epsilon r\right)^{1/2} \tag{5.100}$$

$$a_H = \pm\sqrt{a_H^2}, \quad b_H = \pm\sqrt{b_H^2} \tag{5.101}$$

where the subscript H of any quantity denotes hydrostatic, and a_H and b_H are obtained in the same manner as a and b.

Since all the prognostic variables are functions of a and b, we know that they will have different solutions if a_H and b_H replace a and b. Thus, to obtain the exact solution for the differences between the hydrostatic and nonhydrostatic quantities, we need to consider this difference in all the prognostic variables in the governing equations.

The residual (i.e., nonhydrostatic) pressure perturbation (denoted as R, where $R \equiv p - p_H$) is obtained as follows.

$$\text{Take } \frac{\partial}{\partial x}(1): \quad \frac{\partial^2 p}{\partial x^2} = -\frac{1}{\alpha_0}\frac{\partial}{\partial x}\frac{\partial u}{\partial t} + \frac{f}{\alpha_0}\frac{\partial v}{\partial x} - \frac{\sigma_x}{\alpha_0}\frac{\partial u}{\partial x} \tag{5.102}$$

$$\text{Take } \frac{\partial}{\partial z}(3): \quad \frac{\partial^2 p}{\partial z^2} = -\frac{1}{\alpha_0}\frac{\partial}{\partial z}\frac{\partial w}{\partial t} + \frac{g}{\theta_0\alpha_0}\frac{\partial \theta}{\partial z} - \frac{\sigma_z \partial w}{\alpha_0 \partial z}. \tag{5.103}$$

Similarly for the hydrostatic system, we obtain

$$\frac{\partial^2 p_H}{\partial x^2} = -\frac{1}{\alpha_0}\frac{\partial}{\partial x}\frac{\partial u_H}{\partial t} + \frac{f}{\alpha_0}\frac{\partial v_H}{\partial x} - \frac{\sigma_x}{\sigma_0}\frac{\partial u_H}{\partial x} \tag{5.104}$$

$$\frac{\partial^2 p_H}{\partial z^2} = \frac{g}{\theta_0\alpha_0}\frac{\partial \theta_H}{\partial z}. \tag{5.105}$$

Subtracting Eq. 5.104 from 5.102, and 5.105 from 5.103, and adding the results yields the Poisson equation for the pressure residual term.

$$\begin{aligned} \nabla^2 R = &-\frac{1}{\alpha_0}\frac{\partial}{\partial x}\frac{\partial}{\partial t}\left(u - u_H\right) + \frac{f}{\alpha_0}\frac{\partial}{\partial x}\left(v - v_H\right) \\ &- \frac{\sigma_x}{\alpha_0}\frac{\partial}{\partial x}\left(u - u_H\right) + \frac{g}{\theta_o\alpha_0}\frac{\partial}{\partial z}\left(\theta - \theta_H\right) \\ &- \frac{1}{\alpha_0}\frac{\partial}{\partial z}\frac{\partial w}{\partial t} - \frac{\sigma_z}{\alpha_0}\frac{\partial w}{\partial z}. \end{aligned} \tag{5.106}$$

Using Defant's analytic solutions, listed in Eqs. 5.90-5.94, we have, for example, the first term on the right side of Eq. 5.106.

$$\begin{aligned} -\frac{1}{\alpha_0}\frac{\partial}{\partial x}\frac{\partial}{\partial t}\left(u - u_H\right) = &\frac{i\omega r M}{\alpha_0} e^{i\omega t} \\ &\times \sin k_x x \left[\frac{a_H e^{a_H z} + b_H e^{-b_H z}}{b_H^2 - a_H^2} - \frac{a e^{az} + b e^{-bz}}{b^2 - a^2}\right]. \end{aligned}$$

Similar expressions can be obtained for all other terms on the right side of Eq. 5.106. After rearrangements, the Poisson equation for the pressure residual can be rewritten as:

$$\nabla^2 R = \frac{M}{\alpha_0} e^{i\omega t} \sin k_x x \left[A e^{az} + B e^{-bz} + C e^{a_H z} + D e^{-b_H z} \right] \tag{5.107}$$

where

$$\left.\begin{aligned} A &= \frac{a}{b^2 - a^2} \left[\frac{-rf^2}{i\omega + \sigma_x} - r\sigma_x + r\sigma_z + \frac{g}{\theta_0} \left(b^2 - s \right) \right] \\ B &= \frac{b}{b^2 - a^2} \left[\frac{rf^2}{i\omega + \sigma_x} - r\sigma_x + r\sigma_z + \frac{g}{\theta_0} \left(a^2 - s \right) \right] \\ C &= \frac{a_H}{b_H^2 - a_H^2} \times \left[i\omega r + \frac{rf^2}{i\omega + \sigma_x} + r\sigma_x - \frac{g}{\theta_0} \left(b_H^2 - s \right) \right] \\ D &= \frac{b_H}{b_H^2 - a_H^2} \times \left[i\omega r + \frac{rf^2}{i\omega + \sigma_x} + r\sigma_x - \frac{g}{\theta_0} \left(a_H^2 - s \right) \right] \end{aligned}\right\} . \tag{5.108}$$

Equations 5.107 and 5.108 provide the exact solution for the pressure residual term which represents the analytic difference of pressure perturbation between hydrostatic and nonhydrostatic states in Defant's model. The formulation, however, has quantities belonging to both states (the a, b, and a_H, b_H) which need to be evaluated simultaneously. This means that the complete residual can only be used for diagnostic purposes if applied in a nonlinear numerical model since it would be just as easy to use the complete anelastic equation for p.

In order to obtain a practical method which calculates nonhydrostatic effects using only information available from a hydrostatic model, following Pielke (1972) the equation for the quasi-nonhydrostatic residual (denoted as R_H) is obtained in the same manner as in Eqs. 5.102-5.106, except that the difference between hydrostatic and nonhydrostatic appears only in the time derivative term and the vertical friction term. Thus we have

$$\nabla^2 R_H = -\frac{1}{\alpha_0} \frac{\partial}{\partial x} \frac{\partial u}{\partial t} + \frac{1}{\alpha_0} \frac{\partial}{\partial x} \frac{\partial u_H}{\partial t} - \frac{1}{\alpha_0} \frac{\partial}{\partial z} \frac{\partial w}{\partial t} - \frac{\sigma_z}{\alpha_0} \frac{\partial w}{\partial z}.$$

Using the incompressible continuity, the above equation is reduced to

$$\nabla^2 R_H = \frac{1}{\alpha_0} \frac{\partial}{\partial x} \frac{\partial u_H}{\partial t} - \frac{\sigma_z}{\alpha_0} \frac{\partial w}{\partial z}. \tag{5.109}$$

This is analogous to the form as applied in Pielke (1972) and Martin and Pielke (1983) in their numerical model evaluations.

Comparing Eq. 5.109 with 5.106, we see that Eq. 5.109 can be obtained directly from Eq. 5.106 by neglecting the differences between u, v, θ, and u_H, v_H,

θ_H, respectively, and assuming incompressibility. Using the assumption of incompressibility, Eq. 5.109 becomes

$$\nabla^2 R_H = \frac{1}{\alpha_0}\frac{\partial}{\partial x}\frac{\partial u_H}{\partial t} + \frac{\sigma_z}{\alpha_0}\frac{\partial u_H}{\partial x}. \tag{5.110}$$

It is then clear that Eq. 5.110 is of practical value because only hydrostatic quantities are involved in the estimation of the nonhydrostatic effects, all of which can be obtained in a hydrostatic model. Also, comparing Eq. 5.110 with 5.106, we see that the neglected terms in deriving Eq. 5.110 are the first four terms in Eq. 5.106, which involve the immediate feedbacks associated with the buoyancy, horizontal friction, and the Coriolis terms.

In terms of Defant's analytic solutions, Eq. 5.110 is written as

$$\nabla^2 R_H = \frac{M}{\alpha_0} e^{i\omega t} \sin kx \left[C' e^{a_H z} + D' e^{-b_H z} \right] \tag{5.111}$$

where

$$C' = \frac{a_H}{b_H^2 - a_H^2}\left(i\omega r + r\sigma_z\right)$$

$$D' = \frac{b_H}{b_H^2 - a_H^2}\left(i\omega r + r\sigma_z\right).$$

The solutions for Eqs. 5.107 and 5.111 are obtained using the method of separation of variables. However, it is straightforward to show that the solution for R_H is exactly the solution for R except that a_H and b_H replace a and b. Written formally, the solution for R is

$$R = \frac{M}{\alpha_0} e^{i\omega t} \sin k_x x \left[\frac{A}{a^2 - k_x^2} e^{az} + \frac{B}{b^2 - k_x^2} e^{-bz} + \frac{C}{a_H^2 - k_x^2} e^{a_H z} + \frac{D}{b_H^2 - k_x^2} e^{-b_H z} \right] \tag{5.112}$$

with A, B, C, D defined in Eq. 5.108.

The pressure terms (p, p_H) and the residual terms (R, R_H) can be analyzed as functions of the horizontal length scale, large-scale stability, subgrid-scale heat diffusion, heating amplitude, and surface friction. The purpose of these analyses is to determine how the nonhydrostatic pressure residual varies with changing physical conditions within the framework of Defant's model. Furthermore, as stated before, since the nonhydrostatic effects are evaluated using Eqs. 5.107 and 5.108, rather than using $\lambda = 1$ in Defant's model, it seems necessary to show the consistency of the results obtained from the two independent procedures. Unless otherwise mentioned, the values of the parameters used for the examples are those listed in Table 5.1.

Figure 5.8 shows the maximum amplitude for the pressure terms (p, p_H) and the residual terms (R, R_H) as functions of horizontal length scale (L) plotted on a logarithmic scale. (All the dependent variables in the following section and figures are presented at their maximum.) The range of scales is chosen from 200 m to

Table 5.1 Control values for the parameters.	
$\beta \left(= \frac{d\theta_0}{dz}\right)$	0.01°C km^{-1}
K	10 m^{2}s^{-1}
α_0	0.758 (m^{3}kg^{-1})
σ_x, σ_z	10^{-3}(s^{-1})
P	1 h
θ_0	273 K
M	10°C
L	1 km
z	15 m
f	10^{-4}(s^{-1})

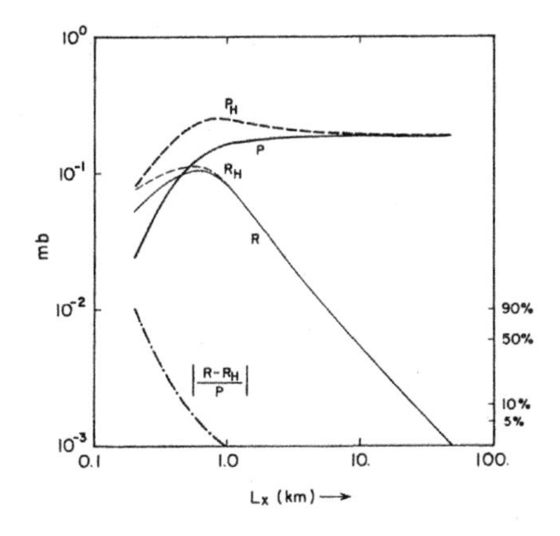

FIGURE 5.8

Variations of the pressure perturbation and the residual terms (p, p_H, and R, R_H; mb) and the absolute error term ($|R - R_H|/|P|$, %) as a function of the horizontal length scale (L_x). Other physical parameters are given in Table 5.1 (from Song et al. 1985).

50 km, which should cover most of the spatial scales in which there is concern regarding the adequacy of the hydrostatic assumption in a model. Since the pressure perturbations are caused by surface heating in this study, and since all perturbation quantities decrease exponentially with height, the perturbations are evaluated near the surface ($z = 15$ m) . All the pressure and the residual terms are given in units of millibars.

As can be seen, when the length scale becomes large, p becomes nearly constant. The difference between p and p_H (i.e., R) becomes negligible for larger scales. This feature can be explained using the definitions of the parameters following Eq. 5.81: as $L \to \infty$, $k_x \to 0$, and $\eta^2 \to 0$, $r \to 0$; also $s = s(\omega, k)$. Thus there is a decreasing dependence on the length scale as it becomes large, resulting in a nearly constant p. Also, it can be seen that there are virtually no differences between a, b and a_H, b_H, respectively, when the scale is large.

For smaller scales, k^2 becomes large and the situation is more complicated. From Fig. 5.8, we see that for scales less than about 1 km, the residuals are of the same order of magnitude as the pressure terms. This indicates that for such small scales, the nonhydrostatic effect is significant, and that p_H is significantly over-estimating the true pressure perturbation. Pielke (1972, Fig. 19) schematically illustrated how the hydrostatic pressure over-estimates the real pressure.

What is also of interest here is how R_H behaves as compared with R. From Fig. 5.8 we see that for length scales larger than about 1 km, there is essentially no difference between R and R_H, while for the smaller scales this difference becomes significant. The quantity $|(R - R_H)/P|$, hereafter called the absolute error and expressed as a percentage, is also plotted. This measure illustrates how much error is introduced as a fraction of the true pressure perturbation, when R_H is used instead of the complete nonhydrostatic pressure residual R. Figure 5.8 shows that this absolute error drops to essentially zero for $L > 1$ km, but increases sharply when $L < 1$ km. It is clear that under this near-neutral condition ($\beta = 0.01°\text{C km}^{-1}$) with a surface heating of 10°C effective for one hour, the quasi-nonhydrostatic residual method gives an accurate measure of the nonhydrostatic effect for horizontal length scales as small as about 1 km.

The dependence of p, p_H, R, and R_H on the large-scale stability is illustrated for $L = 10$ km (Fig. 5.9) and $L = 1$ km (Fig. 5.10). The strength of the surface heating, the heating period, and the strength of the eddy heat diffusion are all the same as used to create Fig. 5.8. In Fig. 5.9, we see that for $L = 10$ km, the difference between p and p_H is negligible for all the chosen stabilities (β; from 0.01 to $20°\text{C km}^{-1}$). R_H and R, although they differ somewhat relative to one another, are both negligibly small compared with the pressure terms. Thus, the absolute error term is very small for all the chosen stabilities (the largest error is 1.5%, occurring at $\beta = 0.4°\text{C km}^{-1}$). The result shown here indicates that for the scales normally considered in mesoscale analyses ($L = 10$ km or larger), in which the driving mechanism is surface heating and the upward transport of this heating is primarily through the associated turbulent eddy processes, the situation is approximately hydrostatic and the residual method can be used to accurately calculate the small nonhydrostatic effects.

When the length scale is reduced to 1 km, however, the residuals become relatively larger than for the previous case, and as seen in Fig. 5.10, R_H departs significantly from R for a wide range of stabilities. From Fig. 5.10, we see that for the less stable situations ($\beta \leq 0.1°\text{C km}^{-1}$) R_H differs more significantly from R.

The dependence of the pressure and the residual terms upon the strength of the eddy heat diffusion is shown in Fig. 5.11. The horizontal length scale is 1 km, and the other parameters are the same as in the above cases. The range of the (constant) diffusion coefficient K is from 0.5 to 50 m^2s^{-1}. It is seen in Fig. 5.11 that as the strength of eddy heat diffusion increases, the pressure perturbations increase. The hydrostatic pressure perturbation consistently exceeds the real pressure perturbation.

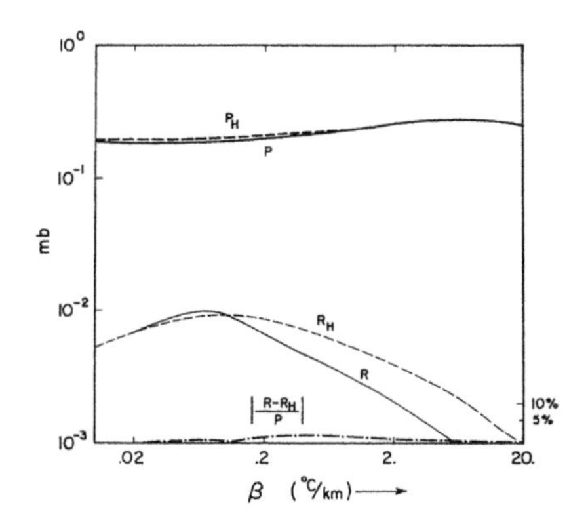

FIGURE 5.9

As in Fig. 5.8 except as a function of the stability parameter (β). $L_x = 10$ km (from Song et al. 1985).

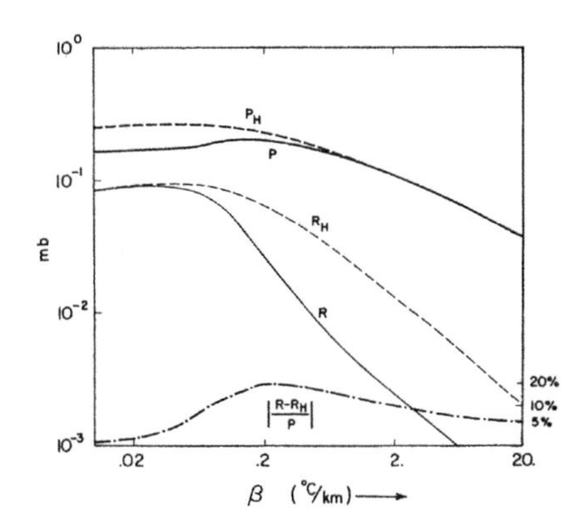

FIGURE 5.10

As in Fig. 5.9 with $L_x = 1$ km (from Song et al. 1985).

For this small horizontal scale (1 km), the residual is the same order of magnitude as the pressure terms. Here R_H gives a very accurate measure of the nonhydrostatic effect associated with the vertical turbulent mixing of heat, except when the diffusion coefficient becomes very large.

Figures 5.12 and 5.13 show the dependence of P, P_H, R, and R_H on the horizontal scale of heating for small K ($K = 1$ m^{2}s^{-1}) and large K ($K = 50$ m^{2}s^{-1}), respectively. In Fig. 5.12, we see that when the eddy heat diffusion is sufficiently small, R_H gives an accurate measure of the nonhydrostatic effects for scales as small as about 300 m. For very large diffusion (Fig. 5.13), the residual is the same order of magnitude as the pressure for scales of a few kilometers or less. The absolute error is rather large for the small scales, and drops to essentially zero at scales greater than

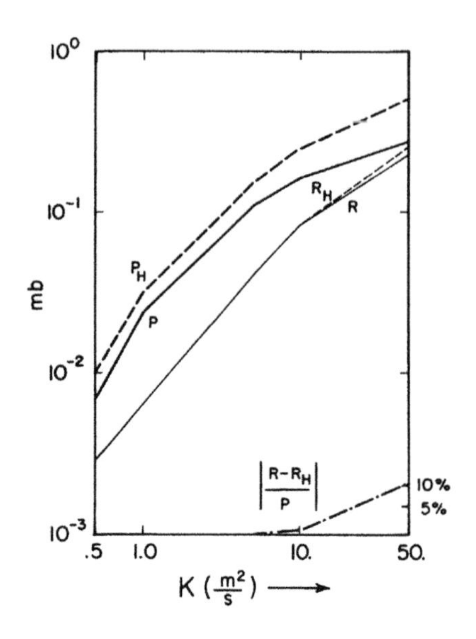

FIGURE 5.11

As in Fig. 5.8 except as a function of the heat diffusion coefficient (K) (from Song et al. 1985).

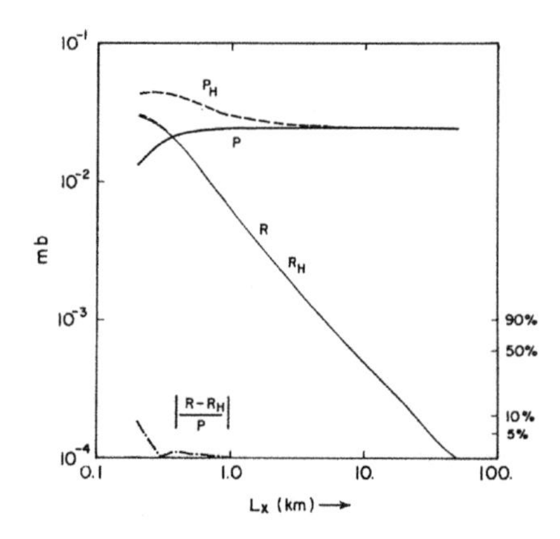

FIGURE 5.12

As in Fig. 5.8 except $K = 1.0\ \mathrm{m^2s^{-1}}$ (from Song et al. 1985).

about 3 km. The increase of the nonhydrostatic effect with increasing strength of the eddy heat diffusion was also illustrated in Martin and Pielke (1983).

The discrepancy between R and R_H for very large K (Fig. 5.11) is found only for small horizontal scales. Figure 5.13 shows that the same strength of heat diffusion ($K = 50\ \mathrm{m^2s^{-1}}$), the difference between R and R_H is essentially zero at scales

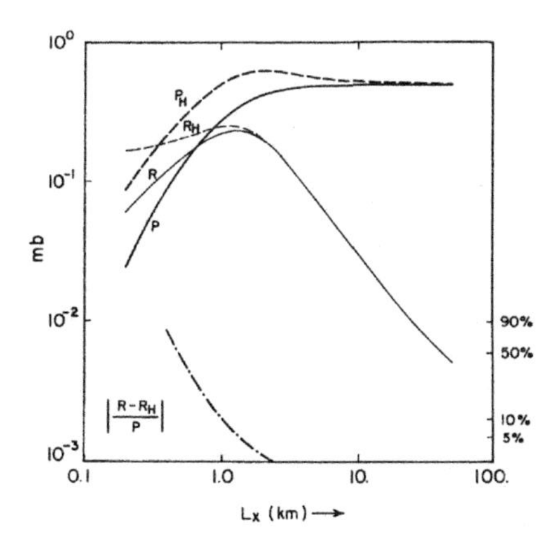

FIGURE 5.13

As in Fig. 5.8 except $K = 50\ \mathrm{m^2s^{-1}}$ (from Song et al. 1985).

larger than about 3 km. This implies that when a strong energy input is coupled with a small horizontal scale, there may be buoyancy oscillations excited which cause departures of R_H from R. From Figs. 5.12 and 5.13, we see that either increasing the horizontal scale or decreasing the strength of heat diffusion will minimize the discrepancy between R_H and R.

The strength of the surface heating is obviously important in producing nonhydrostatic effects. However, this forcing appears only as a constant in Defant's linear model (i.e., see Eq. 5.90 for $\tilde{\theta}(z)$), thus not allowing the interactions between surface heating and mesoscale circulations to take place.

Finally, the effect of the frictional term on pressure and the residual terms is shown in Fig. 5.14. The physical parameters are the same as in Fig. 5.8 except that the (constant) frictional coefficient is reduced by one order of magnitude. Comparing Fig. 5.14 with Fig. 5.8, we see that reducing the friction produces negligible effects upon the pressure perturbations. The absolute error is within 2% for scales larger than about 2 km.

Two other different approximate residual formations (based on the Defant model) can be derived. Using Defant's linear model, the Orlanski (1981) pressure correction term is derived from a vertical integration of the local time derivative of the hydrostatically-obtained vertical velocity. Written in an appropriate form for the comparisons here, the equation for the Orlanski residual (hereafter denoted as R_Q) is

$$\frac{\partial^2}{\partial z^2} R_Q = -\frac{1}{\alpha_0} \frac{\partial}{\partial z} \frac{\partial w_H}{\partial t}. \tag{5.113}$$

It is seen from Eqs. 5.106, 5.109, and 5.113 that R_Q can be derived from Eq. 5.106 by making, in addition to the simplifications made for obtaining R_H, two simplifications concerning the horizontal second derivative of the residual and the

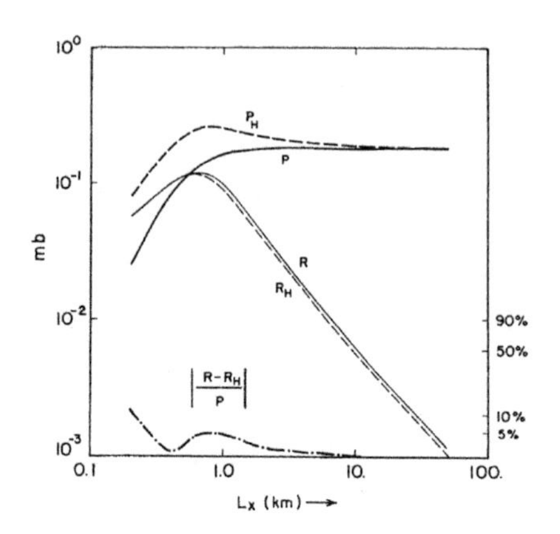

FIGURE 5.14

As in Fig. 5.8 except with a smaller Rayleigh friction coefficient (10^{-4}s^{-1}) (from Song et al. 1985).

vertical friction term. This can be clearly seen if we compare the formal solutions for R_H and R_Q;

$$R_H = \frac{1}{\alpha_0} e^{i\omega t} \sin k_x x \left(\frac{rM}{b_H^2 - a_H^2} \right) \times \left[(i\omega) \left(\frac{a_H}{a_H^2 - k_x^2} e^{a_H z} + \frac{b_H}{b_H^2 - k_x^2} e^{-b_H z} \right) + \sigma_z \left(\frac{a_H}{a_H^2 - k_x^2} e^{a_H z} + \frac{b_H}{b_H^2 - k_x^2} e^{-b_H z} \right) \right]. \tag{5.114}$$

$$R_Q = \frac{1}{\alpha_0} e^{i\omega t} \sin k_x x \left(\frac{rM}{b_H^2 - a_H^2} \right) \times \left[(i\omega) \left(\frac{1}{a_H} e^{a_H z} + \frac{1}{b_H} e^{-b_H z} \right) \right]. \tag{5.115}$$

Neglecting the friction term, we see that R_Q can be obtained directly from R_H by setting k_x to zero. Thus when the horizontal scale of heating becomes large, in the absence of friction (which as discussed previously is a relatively small term), R_Q and R_H are asymptotic to the same value.

Mathematically, the quantity k_x^2 is associated with the x-direction second derivative of the residual, which is derived from the horizontal equation of motion in which the nonhydrostatic effect is explicitly included. That is, from Eq. 5.66 we have

$$\frac{\partial^2}{\partial x^2} (p - p_H) = -\frac{1}{\alpha_0} \frac{\partial}{\partial t} \frac{\partial}{\partial x} (u - u_H) + \frac{f}{\alpha_0} \frac{\partial}{\partial x} (u - u_H) - \frac{\sigma_x}{\alpha_0} \frac{\partial}{\partial x} (u - u_H). \tag{5.116}$$

Therefore, setting k_x^2 to zero is also equivalent to neglecting the nonhydrostatic horizontal momentum residual (i.e., the horizontal velocity residual).

Since in an incompressible system the horizontal velocity gradient is directly related to the generation of vertical acceleration (Eq. 5.69), it is thought necessary to further examine the effect of neglecting k_x^2 (but retaining other important terms). For this purpose, a new residual (hereafter denoted as R_z) is considered which is obtained from the complete vertical equation of motion; i.e.,

$$\frac{\partial^2}{\partial z^2} R_z = -\frac{1}{\alpha_0}\frac{\partial}{\partial z}\left(\frac{\partial w_H}{\partial t}\right) - \frac{\sigma_z}{\alpha_0}\frac{\partial w}{\partial z} + \frac{g}{\theta_0 \alpha_0}\frac{\partial}{\partial z}\left(\theta - \theta_H\right). \tag{5.117}$$

Thus, there are four different residuals to be compared: R, R_H, R_Q, and R_z obtained from Eqs. 5.106, 5.109, 5.113, and 5.117, respectively. Aside from the friction and the Coriolis terms (which are not essential to the main conclusion), R_H and R_Q differ from R and R_z in that the former do not include the nonhydrostatic buoyancy residual (i.e., the potential temperature residual term, as in Eqs. 5.106 and 5.117). On the other hand, R_z and R_Q differ from R and R_H in that the former neglect the nonhydrostatic horizontal momentum residual, or equivalently, *they are based on the assumption of an infinite horizontal length scale.*

In order to have consistent numerical experiments with those in the previous section, the following computations are performed using, unless otherwise mentioned, the physical parameters listed in Table 5.1. Another set of experiments was also performed using the heating period of 12 hours. Since the general patterns of the residuals are similar using either 1 or 12 h as the heating period, only the 1 h results are analyzed here.

In Fig. 5.15, the four residuals are plotted as functions of the horizontal length scale for three selected stabilities. The magnitude of the total pressure perturbations are, for the scale range between about 1 and 10 km, on the order of 10^{-1} mb. We first see that for a given length scale, the nonhydrostatic residuals increase with decreasing stability. For the stable situation ($\beta = 10°\text{C km}^{-1}$), all the residuals are about two orders of magnitude smaller than the total pressure perturbation; while for the near-neutral situation ($\beta = 0.1°\text{C km}^{-1}$), the residuals are comparable to the total pressure perturbation for the smaller length scales. For a given stability, the residuals are generally decreasing with increasing length scale. The closer to the neutral state, the larger the rate of decrease of the residuals with increasing length scale. An exception to this is the non-monotonic variation of R and R_z in the more stable categories which indicates that an optimal horizontal scale exists in which vertical acceleration is maximized as a result of contributions to convergence from opposite coasts (e.g., Abe and Yoshida 1982). This relative maximum is not as significant in the 12-hour period experiments.

With regard to the comparison among the residuals, we see from Fig. 5.15 that in the stable situation, R_z matches with R, while R_H and R_Q deviate from R. On the other hand, in the near-neutral situation, R_H matches with R, while R_Q and R_z deviate from R. In order to more clearly analyze the relative magnitudes of the residuals, vertical profiles of the residuals are plotted for a selected length scale (1 km) and for two stabilities: $\beta = 0.001°\text{C km}^{-1}$ (Fig. 5.16), and $\beta = 10°\text{C km}^{-1}$ (Fig. 5.17). The magnitudes of the corresponding total pressure perturbations are also

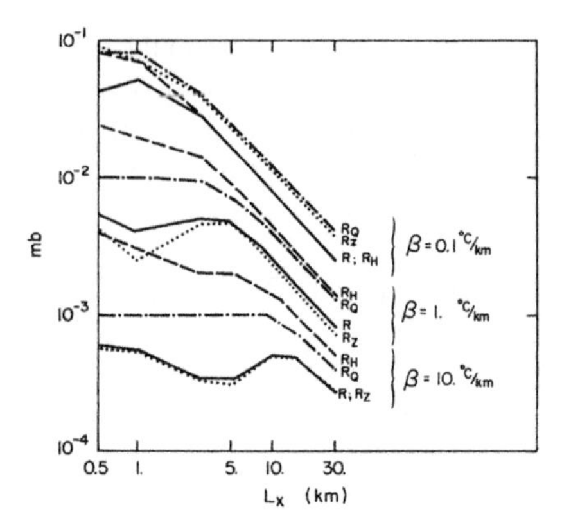

FIGURE 5.15

The magnitudes (in mb) of the four residuals at $z = 15$ m: R (solid), R_H (dashed), R_z (dotted), and R_Q (dash-dot), as functions of the length scale and of three selected stabilities ($\beta = 0.1$, 1, 10°C km^{-1}, as shown) (from Song et al. 1985).

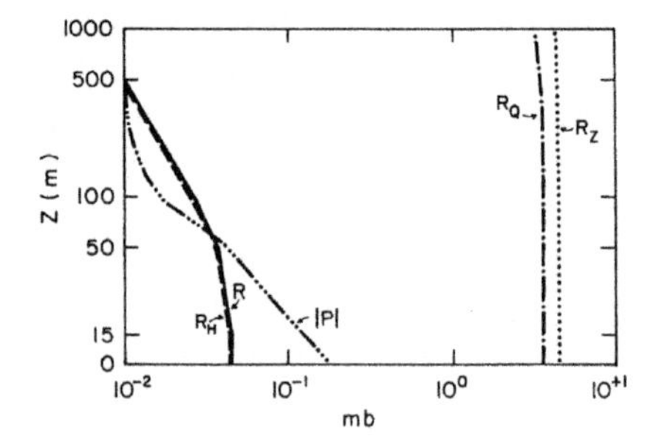

FIGURE 5.16

Vertical profiles of the four residuals (notations are the same as those in Fig. 5.15) and $|p|$ for the near-neutral stability case ($\beta = 0.001$°C km^{-1}). The vertical levels are at $z = 0$, 15, 50, 100, 500, and 1000 m (from Song et al. 1985).

shown to indicate the possible absolute errors which are introduced when a certain residual is used.

From Fig. 5.16 we see that in the near-neutral situation, R_H matches with R everywhere, while R_z and R_Q are both about two orders of magnitude larger than R. Furthermore, both R_z and R_Q are around more than one order of magnitude larger than the total pressure perturbation. It seems clear that the R_Q approach should not be considered for the situations where the environmental stability is near neutral. On the other hand, R_H provides an accurate measure of the nonhydrostatic effect under the near-neutral condition.

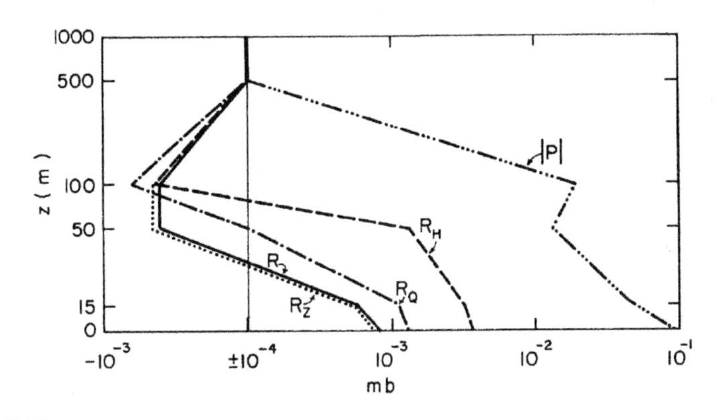

FIGURE 5.17

As in Fig. 5.16 except for $\beta = 10°\text{C km}^{-1}$ (from Song et al. 1985).

From Fig. 5.17, we see that in the stable situation, R_z matches with R while R_Q and R_H deviate somewhat from R, with R_Q slightly better than R_H. In this case, however, all the residuals are almost more than two orders of magnitude smaller than the total pressure perturbation. Clearly this result indicates that nonhydrostatic effects are negligible in the stable situation, and therefore the discrepancies are of little practical importance. As discussed previously, the difference between the exact residual and other residuals is related to the nonhydrostatic buoyancy and horizontal momentum residuals. That is, the horizontal momentum change and the buoyancy associated with the surface heating are the two most important physical mechanisms which contribute to the generation of nonhydrostatic effects for the situations considered in this study. The vertical profiles of θ and θ_H (Fig. 5.18), and u and u_H (Fig. 5.19), are plotted for the same length scale (1 km) and the same stabilities ($\beta = 0.001$; $10°\text{C km}^{-1}$).

From Fig. 5.18, we see that θ is slightly larger than $\theta_H (\theta - \theta_H \leq 0.1°\text{C})$ for the stable situation, while they match with each other everywhere in the near-neutral situation. This explains why R_Q and R_H (in which the $(\theta - \theta_H)$ term is neglected) deviate from R for the stable situation. Physically, this implies that for a system being heated from below, the more thermodynamically stable the system, the larger the fractional contribution of the nonhydrostatic buoyancy to the residual that is generated within the system. In the absolute sense, however, the nonhydrostatic effect is negligible in this case as computed with the total pressure perturbation.

From Fig. 5.19, we see that for the stable situation u and u_H are almost equal, while for the near-neutral situation they differ significantly from each other. Again, this explains why R_Q and R_z (in which the $(u - u_H)$ term is neglected) deviate significantly from R in the near-neutral case, and R_H (which contains the $(u - u_H)$ term) matches with R. Physically, this implies that for an incompressible system, the closer the system's stability is toward neutral stratification, the stronger the wind velocity generated within the system, and therefore the horizontal momentum change plays a more important role in generating nonhydrostatic effects (i.e., vertical

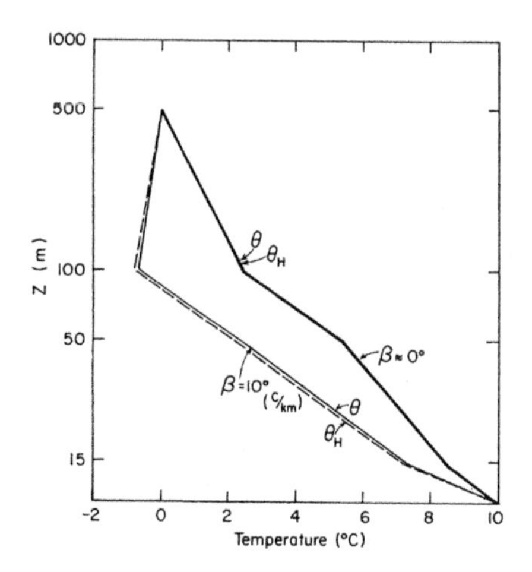

FIGURE 5.18

Vertical profiles of θ (solid) and θ_H (dashed) for the near-neutral case (thick line) and the stable case (thin line) (from Song et al. 1985).

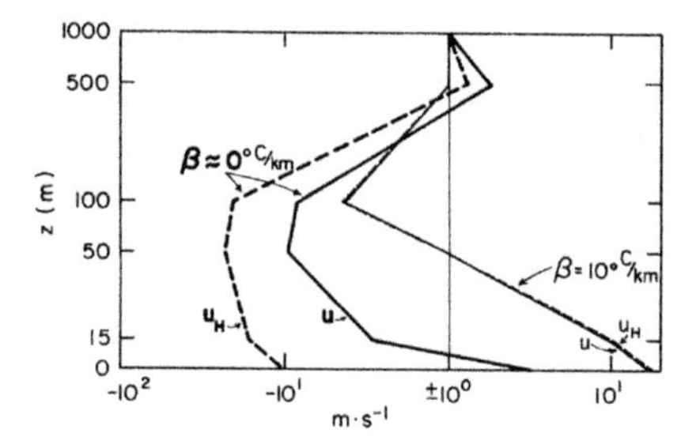

FIGURE 5.19

Vertical profiles of u (solid) and u_H (dashed) for the near-neutral case (thick line) and the stable case (thin line) (from Song et al. 1985).

acceleration) as compared with the situation where there are only weak horizontal velocity perturbations.

Finally, computations presented in Figs. 5.15-5.19 were repeated for various β values between those of the very stable case ($\beta = 10°\text{C km}^{-1}$) and the almost neutral case ($\beta = 0.001°\text{C km}^{-1}$). These results, which are not shown, reflected intermediate features to those presented in these figures.

Defant's linear model is used to derive a mathematically exact solution for the non-hydrostatic pressure residual (total pressure perturbation minus hydrostatic pressure perturbation). From the complete form of this exact residual, we can see that the thermally-induced nonhydrostatic effects are caused, within the linear framework, by physical processes such as horizontal momentum variations, buoyancy effects, frictional effects, and Coriolis effects. Since the complete residual requires both hydrostatic and nonhydrostatic quantities to be evaluated simultaneously, this residual can only be used in a diagnostic analysis for numerical modeling purposes. For the purposes of deriving a prognostic approach to incorporate nonhydrostatic effects into one or more subdomains of a mesoscale model, the complete residual must be simplified so as to neglect those terms which cannot be evaluated without a complete nonhydrostatic model.

One type of simplification made to the exact residual for the purpose of deriving a prognostic approach is to neglect the nonhydrostatic buoyancy residual term. Together with the incompressible continuity, this results in the residual approach introduced in Pielke (1972). Aside from the horizontal friction and the Coriolis terms (which are found not to be critical to the discussions of this study), the Pielke (1972) method differs from the exact residual only in the buoyancy residual term which, in the experiments performed in this study, is relatively important only in the very thermally stable environments. For such stable situations, the nonhydrostatic pressure perturbations are generally about two orders of magnitude smaller than the total pressure perturbation. Thus the discrepancy between the approximate residual and the exact residual is of little practical importance.

For near-neutral stabilities, the Pielke (1972) residual approach has been found to be able to provide accurate approximations to the true pressure perturbation, indicating that it is of practical value for evaluating nonhydrostatic effects within a subdomain of a mesoscale model, when the environment is in a near-neutral state.

Another type of simplification is to neglect the nonhydrostatic horizontal momentum residual term. Within Defant's linear framework, this simplification results in the residual approach introduced in Orlanski (1981). It is found that this residual can be obtained from the exact residual by merely making an assumption that the involved horizontal length scale is very large (i.e., a wavenumber approaching zero). This simplification is equivalent to neglecting the nonhydrostatic velocity perturbation. In a near-neutral environment, it is found that the nonhydrostatic velocity (momentum) residual is relatively much more important than in a stable environment. Neglecting this velocity residual caused the Orlanski residual to overestimate the nonhydrostatic pressure perturbation by about two orders of magnitude.

Physically, the above results imply that for an incompressible system being heated from below, the actual pressure perturbation tends to depart from the hydrostatic pressure perturbation by an amount which depends primarily on the system's environmental thermal stability and horizontal scale of heating. For a sufficiently stable system, there are negligible nonhydrostatic effects. On the other hand, when the stability is near-neutral, relatively stronger perturbations will develop which tend to more closely connect the vertical acceleration with the horizontal momentum variations. In such situations, a residual approach must include the nonhydrostatic momentum residual term, such as the approach of Pielke (1972), in order to accurately evaluate the nonhydrostatic effects. The residual approach presented here has the utility that it can be applied for subregions within a mesoscale model where vertical accelerations are large, while in the remainder of the model, the hydrostatic assumption can be applied.

At the boundaries of the model subdomain, the boundary condition $R_H = 0$ would be applied in the solution of the nonlinear form of Eq. 5.109.

Other studies which have explored the differences in results when hydrostatic and nonhydrostatic versions of a model are used include Sun (1984a), Rõõm and Männik (1999), Cassano and Parish (2000), and Crook and Klemp (2000).

5.3 Role of Compressibility in Mesoscale Models[9]

While Section 5.2.2 showed that we can exclude sound waves from the possible set of solutions by using the anelastic form of the conservation of mass equation, there are reasons to retain the full conservation of mass equation such as Eq. 4.22. The value of this was introduced in a set of papers (Nicholls and Pielke 1994a,1994b; 2000; Schechter et al. 2008). A thermally-induced compression wave is generated when an air parcel is heated. The pressure of the air parcel increases in response to the heating causing it to expand and adjacent air to be compressed. This generates a compression wave that propagates away at the speed of sound. The reduction of the air density within the expanded air parcel causes it to become buoyant.

While an anelastic model can simulate the production of buoyancy for this scenario, it will not generate the compression wave. Nicholls and Pielke (1994b), simulated two-dimensional sea breeze development over a large land surface with a uniform surface heat flux using a fully compressible model, and compared results with a version of the model with the terms allowing the thermal generation of compression waves removed. Compression waves formed in response to the surface heating. Hydrostatic adjustment occurred in the vertical and the compression waves traveling horizontally away from the coasts took the form of Lamb waves (Lamb 1908, 1932).

At the surface these were associated with very broad regions of high pressure perturbations adjacent to the coasts, which rapidly increased with size as the wave fronts propagated away at the speed of sound. These pressure perturbations, whose magnitude was very small, decreased with height as $\exp(-z/\gamma H_s)$, where $H_s = g/RT$ is the scale height and $\gamma = C_p/C_v$, where C_p and C_v, are respectively, the specific heats at constant pressure and volume. R is the gas constant, T is temperature and g the gravitational constant. The surface pressure did not initially decrease uniformly over the land surface, but as the horizontal expansion of the air removed mass off the coasts, low surface pressure worked its way inland at the speed of sound. Eventually a uniform low pressure occurred over land. Sea breezes driven by buoyancy developed at the coasts. It was at these locations that significant winds occurred.

The winds associated with compression waves were relatively insignificant, being two orders of magnitude smaller. In contrast the simulation without the terms allowing the generation of thermal compression waves showed an immediate uniform lowering of surface pressure over the land surface and did not generate Lamb waves. Lowering

[9]This section was written by Dr. Mel Nicholls who pioneered the assessment of the role of compressibility in atmospheric modeling.

of surface pressure over the land surface essentially means there has been a removal of mass since the air is close to a hydrostatic balance and the surface pressure therefore reflects the weight of the air above.

In the fully compressible model the horizontal expansion of air above the land and compression in the Lamb wave regions are properly modeled and mass is correctly conserved. Some of the mass of air that was over land has been transferred horizontally offshore. For the version of the model with the terms allowing thermal compression waves removed, mass is not conserved correctly, and while the surface pressure decreases over the land there is no corresponding increase of surface pressure elsewhere. Nevertheless, the significant meteorological fields were accurately produced by this latter simulation.

Similar considerations apply to total energy conservation. Total energy per unit volume is given to a good accuracy by the sum of internal energy $C_v \rho T$, and gravitational energy $\rho g z$, since kinetic energy is typically negligible in comparison. If internal energy per unit volume is rewritten using the ideal gas law $p = \rho R T$, as $C_v p/R$, it can be seen that the positive pressure perturbation of the Lamb wave produced in the fully compressible sea breeze simulation represents an increase in internal energy. Since density is increased in the compressed air of the Lamb wave there is also an increase in gravitational potential energy. Nicholls and Pielke (1994a), showed that the transfer of total energy perturbations as compression waves propagate away form the source region can be substantial. This gives a new perspective on how total energy can be transferred, since it is not simply a quantity that is advected by the atmospheric winds, but can propagate from one region to another at the speed of sound, without significant meteorological changes taking place.

Nicholls and Pielke (1994b) and Bannon (1995a, 1996) examined hydrostatic adjustment in response to heating in a compressible atmosphere Acoustic waves propagate vertically and vertical displacement of fluid particles occurs as the pressure, density and thermal fields mutually adjust over a finite period. In contrast, an anelastic model produces an instantaneous adjustment since acoustic modes have been filtered from the equations. Bannon (1995b), shows that the steady state solution for the pressure, density and temperature are in agreement with the exact compressible solution. Therefore the total energy in the final state in the anelastic and compressible cases for a heating of finite duration is the same. Since for an anelastic system the pressure is determined within an arbitrary constant this solution has used the condition that the surface pressure remains fixed, which is reasonable for a horizontally infinite atmosphere that is uniformly heated. For the sea breeze case with a heating of finite horizontal extent an anelastic model would use a different condition, such as fixed pressure at a large distance from the land surface, and the solution would show a surface pressure decrease over the heated land surface as in the previous example.

Nicholls and Pielke (2000) further modified the mass conservation equation of the fully compressible version of RAMS to include gaseous sources and sinks of mass, caused for instance by evaporation and condensation, respectively. The fully compressible cloud model was used to simulate a three-dimensional convective storm with both liquid and ice microphysics. The storm developed and decayed over a period of approximately an hour. A Lamb wave emerged from the storm with a leading positive lobe and a trailing negative lobe of perturbation pressure. The reason for the negative trailing lobe of the Lamb wave was not purely a result of cooling due to evaporation and sublimation in the decaying stage of the storms lifetime, but was also shown to be a geometric consequence of the two dimensional nature of the Lamb wave that propagates in the horizontal plane, but not vertically. Another consequence is that as it propagates away from the source region its amplitude decays inversely with the square root of distance traveled. Analysis of the internal and gravitational potential energy fields revealed that the transfer of total energy by the Lamb was approximately equal to the net increase of total energy in the atmosphere brought about by the convective storm. This net increase of total energy was primarily due to latent heat release, but was modified due to the net loss of gaseous mass brought about by phase changes from vapor to liquid and ice water. Furthermore, the effectiveness of condensational heating and evaporative cooling in generating compression waves was countered to some degree by the gaseous mass sink as condensation occurs and gaseous mass source as evaporation occurs, respectively. Fanelli and Bannon (2005) also analyzed the total energy perturbation within a Lamb wave packet that was generated by an idealized thermal forcing. For this dry case they found that the Lamb wave packet contained a total energy anomaly that was more than that input to the atmosphere by heating.

Fully compressible atmospheric models also have potential for investigating sources and propagation characteristics of higher frequency acoustic signals known as infrasound. These are acoustic waves with frequencies less than the lower limit of unimpaired human hearing usually considered to be 20 Hz. Early studies found infrasound emanating from severe storms with periods of a few tens of seconds (e.g., Bowman and Bedard 1971; Georges and Greene 1975). More recently there has been interest in higher frequency signals in the range 0.5-5 Hz that may have a connection with the occurrence of tornadoes (Bedard 2005; Passner and Noble 2006). Theories that have been proposed to explain emission from tornadic storms include axisymmetric oscillations of a vortex (Abdullah 1966; Bedard 2005), vortex Rossby waves (Schechter et al. 2008), and latent heating fluctuations associated with strong turbulence (Akhalkatsi and Gogoeridze 2009, 2011). Tests of the ability of a fully compressible model to simulate infrasound in this frequency range have been conducted and show good agreement with analytical solutions for infrasound produced by vortex Rossby waves and evaporating hydrometeors (Schechter et al. 2008; Schechter and Nicholls 2010). Fully compressible models are likely to play an important role in discerning between these and other theories of infrasonic generation mechanisms from tornadic storms. Increased understanding of infrasound generation mechanisms and propagation may eventually enable infrasonic detection to become a useful diagnostic tool providing important information on the state of a severe weather system.

Problems for Chapter 5

1. Program the Defant model using the analytic solutions given by Eqs. 5.90-5.94. Calculate the solution given for the parameters specified by Eq. 5.99. Then assess the changes of the results as each parameter is altered by 10% of its value. (In Chapter 9, you are asked to write a finite difference version of the Defant model, to ascertain if you can recreate the analytic solutions calculated here.)
2. Using Eq. 5.46, calculate phase speed two different ways. First set $f = 0$, $\lambda_1 = 0$, and $\lambda_2 = 0$. For the second approach, let $\partial \rho_o / \partial z = 0$ and assume the horizontal length scale is less than 10 kilometers (so that you can use scale analysis to remove terms). Show that both approaches converge to Eq. 5.57.

Chapter 5 Additional Readings

There are numerous books and texts which can be used to provide additional quantitative understanding of the material presented in this Chapter. Among those are:

Dutton, J.A., 2002: *The Ceaseless Wind: An Introduction to the Theory of Atmospheric Motion*. Courier Dover Publications, 640 pp.

As already mentioned in Chapter 2, this is a valuable source text. His discussion on wave motions provides considerable, clearly written material on wave dynamics.

Dalu, G.A. and R.A. Pielke, 1989: An analytical study of the sea breeze. *J. Atmos. Sci.*, **46**, 1815-1825.

Dalu, G.A. and R.A. Pielke, 1993: Vertical heat fluxes generated by mesoscale atmospheric flow induced by thermal inhomogeneities in the PBL. *J. Atmos. Sci.*, **50**, 919-926.

Dalu, G.A., R.A. Pielke, M. Baldi, and X. Zeng 1996: Heat and momentum fluxes induced by thermal inhomogeneities with and without large-scale flow. *J. Atmos. Sci.*, **53**, 3286-3302.

Using a linear model, these three Journal of Atmospheric Science research papers describe how spatial variation in land-surface heating influence the magnitude and structure of mesoscale flow. In this paper, the influence of linear advective effects are included. Among the major conclusions is that horizontal turbulent diffusion and horizontal advection both work to horizontally homogenize the atmosphere above small-scale patches.

Lin, Y.-L., 2007: *Mesoscale Dynamics*. Cambridge University Press, 630 pp.

Nappo, C.J., 2002: *An Introduction to Atmospheric Gravity Waves*. Academic Press, 276 pp.

Pedlosky, J., 2003: *Waves in the Ocean and Atmosphere: Introduction to Wave Dynamics*. Springer, 260 pp.

Sutherland, B., 2010: *Internal Gravity Waves*. Cambridge University Press, 394 pp.

Thunis, P., and A. Clappier, 2000: Formulation and evaluation of a nonhydrostatic Mesoscale Vorticity Model (TVM). *Mon. Wea. Rev.*, **128**, 3236-3251.

This paper assesses the adequacy of the hydrostatic and anelastic assumptions in simulating thermally-induced circulations as well as describes their vorticity-based model.

Wolfram, S., 1999: *Mathematica Book, Version 4*. 4th Edition, Cambridge University Press. 1496 pp.

CHAPTER

Coordinate Transformations

6

CHAPTER OUTLINE HEAD

6.1 Tensor Analysis

Thus far in this text we have utilized the independent spatial variables x, y, and z in the derivation of the conservation relationships. These spatial coordinates have been defined to be perpendicular to each other at all locations.

In the application of the conservation relations, however, it is not always desirable to utilize this coordinate representation. In synoptic meteorology, for example, since it is the quantity measured by the radiosonde, pressure p is usually used to replace height z as the vertical coordinate. When a different coordinate form is used, however, the conservation relations which are developed from fundamental physical principles, must be unchanged despite the different mathematical representation. Thus in transforming the conservation relations from one coordinate system to another, the equations must be written so that the physical representation is *invariant* in either

Mesoscale Meteorological Modeling, Third Edition, Volume 98. http://dx.doi.org/10.1016/B978-0-12-385237-3.00006-2

system. The mathematical operation developed to preserve this invariance requires some knowledge of the methods of *tensor analysis*.[1]

If, for example, in the rectangular coordinate system x^i, which has been used in this text up until the present ($x^i = x^1, x^2, x^3 = x, y, z$),

$$r_i = f_i, \tag{6.1}$$

where r_i and f_i represent functions and derivatives of functions (e.g., the lefthand and righthand sides of Eq. 2.45), then in another coordinate system $\tilde{x}^i$, which is related to the x^i coordinate system by a functional transformation, the same physical relation must be

$$\tilde{r}_i = \tilde{f}_i$$

to preserve physical invariance. The components of $\tilde{x}^i$ are $\tilde{x}^1$, $\tilde{x}^2$, and $\tilde{x}^3$. The transformation between coordinate representations is defined in terms of the functional relation between the independent variables in the two coordinate systems. Transformations between coordinate systems are of two types.

The first-order tensor $\tilde{f}_i$ is defined to be *covariant* if the transformation between the x^i and $\tilde{x}^i$ coordinate systems is given by

$$\tilde{f}_i = \frac{\partial x^j}{\partial \tilde{x}^i} f_j, \tag{6.2}$$

where $\partial x^j / \partial \tilde{x}^i$ is the operation that transforms f_j into its proper representation in the $\tilde{x}^i$ coordinate system. The *Jacobian* of the transformation is $\partial x^j / \partial \tilde{x}^i$. The use of a subscript denotes that $\tilde{f}_i$ is a *covariant vector (a tensor of order 1)* since it transforms according to Eq. 6.2. By convention, a superscript in the *denominator* of a derivative quantity (e.g., $\partial / \partial x^j$) is defined as a covariant quantity.

The relation given by Eq. 6.1 can also be written as

$$r^i = f^i,$$

so that in the transformed coordinate system $\tilde{x}^i$ this physical relation must be written as

$$\tilde{r}^i = \tilde{f}^i,$$

where

$$\tilde{f}^i = \frac{\partial \tilde{x}^i}{\partial x^j} f^j. \tag{6.3}$$

When the transformation operation is given by $\partial \tilde{x}^i / \partial x^j$, $\tilde{f}^i$ is called a *contravariant vector (or tensor of order 1)* and is indicated by using a superscript.

[1]The following discussion of tensor analysis utilizes Dutton's (1976) excellent in-depth description of this mathematical tool. Readers who require a more in-depth discussion of tensor analysis should refer to that source.

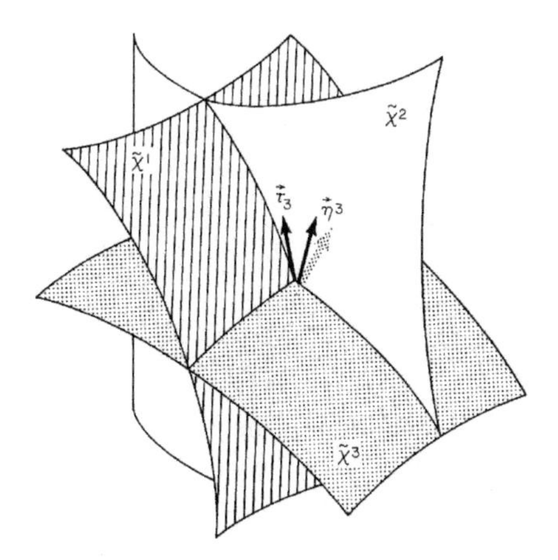

FIGURE 6.1

Illustration of the two types of basis vectors in a nonorthogonal coordinate representation. The vector $\vec{\eta}^3$ is perpendicular to the plane $\tilde{x}^3$ = constant, whereas $\vec{\tau}_3$ is tangent to the curve along which each coordinate except $\tilde{x}^3$ is a constant (from Dutton 1976).

Higher-order tensors are defined in the same way so that

$$\widetilde{B}_{mn} = \frac{\partial x^r}{\partial \tilde{x}^m}\frac{\partial x^s}{\partial \tilde{x}^n} B_{rs}\,; \quad \widetilde{B}^{mn} = \frac{\partial \tilde{x}^m}{\partial x^r}\frac{\partial \tilde{x}^n}{\partial x^s} B^{rs}\,; \quad B^m_n = \frac{\partial \tilde{x}^m}{\partial x^r}\frac{\partial x^s}{\partial \tilde{x}^n} B^r_s,$$

respectively, refer to covariant, contravariant, and mixed tensors of order two.

In the rectangular coordinate system that has been used up to now in the text, the covariant and contravariant forms are identical so that, for example, $u_i = u^i$. In nonorthogonal coordinate systems, however, $\tilde{u}_i \neq \tilde{u}^i$ in general, because $\tilde{u}_i$ is defined in terms of base vectors $\vec{\eta}^i$ that are *perpendicular to the surface* $\tilde{x}^i$ *= constant*, whereas $\tilde{u}^i$ is defined in terms of base vectors $\vec{\tau}_i$ that are *tangent to the curve along which each coordinate except* $\tilde{x}^i$ *is a constant* as illustrated in Fig. 6.1. In the coordinate system we have utilized up to this point, the two sets of basis vectors are coincident and there is no need to differentiate between the covariant and contravariant forms.

In terms of the original rectangular coordinate system, these basis vectors in the transformed coordinate system are defined as

$$\vec{\tau}_j = \frac{\partial}{\partial \tilde{x}^j}\left(x^1\vec{i} + x^2\vec{j} + x^3\vec{k}\right) = \frac{\partial \vec{x}}{\partial \tilde{x}^j}, \tag{6.4}$$

$$\vec{\eta}^i = \vec{i}\frac{\partial}{\partial x^1}\tilde{x}^i + \vec{j}\frac{\partial}{\partial x^2}\tilde{x}^i + \vec{k}\frac{\partial}{\partial x^3}\tilde{x}^i + \vec{\nabla}\tilde{x}^i, \tag{6.5}$$

where $\vec{i}$, $\vec{j}$, and $\vec{k}$ are the orthogonal unit basis vectors in the rectangular coordinate representation. A coordinate system is *orthogonal* when the vector dot product of the basis vectors $\vec{\tau}_j \cdot \vec{\tau}_i$ and $\vec{\eta}^i \cdot \vec{\eta}^j$ is zero at all points except when $i = j$ and *nonorthogonal* when they are not zero. Also in an orthogonal system, $\vec{\tau}_i = \vec{\eta}^i$.

Scalar products[2] involving dependent and independent variables require that a covariant component be multiplied by a contravariant component. This is evident from the scalar product of the basis functions since

$$\vec{\tau}_j \cdot \vec{\eta}^i = \frac{\partial \vec{x}}{\partial \tilde{x}^j} \cdot \vec{\nabla} \tilde{x}^i = \frac{\partial x^l}{\partial \tilde{x}^j} \frac{\partial \tilde{x}^i}{\partial x^l} = \frac{\partial \tilde{x}^i}{\partial \tilde{x}^j} = \delta^i_j,$$

where the chain rule

$$\frac{\partial \tilde{x}^i}{\partial \tilde{x}^j} = \frac{\partial x^l}{\partial \tilde{x}^j} \frac{\partial \tilde{x}^i}{\partial x^l}$$

has been used along with the definition of the Kronecker delta given in Chapter 2, except δ^i_j is now represented as a mixed tensor with one covariant and one contravariant component. Since a vector in the transformed coordinate system can be given in terms of either set of basis vectors, then vectors $\vec{f}$ and $\vec{h}$, which represent physical quantities and are, therefore, invariant between coordinate systems, can be represented by

$$\vec{f} = \tilde{f}_i \vec{\eta}^i = \tilde{f}^j \vec{\tau}_j$$

and

$$\vec{h} = \tilde{h}_i \vec{\eta}^i = \tilde{h}^j \vec{\tau}_j,$$

so that the scalar products $\vec{f} \cdot \vec{f}$ and $\vec{f} \cdot \vec{h}$. For example, are given by

$$\vec{f} \cdot \vec{f} = \vec{\eta}^i \cdot \vec{\tau}_j \tilde{f}_i \tilde{f}^j = \delta^i_j \tilde{f}_i \tilde{f}^j = \tilde{f}_i \tilde{f}^i,$$
$$\vec{f} \cdot \vec{h} = \vec{\eta}^i \cdot \vec{\tau}_j \tilde{f}_i \tilde{h}^j = \vec{\eta}^i \cdot \vec{\tau}_j \tilde{f}^j \tilde{h}_i = \tilde{f}^i \tilde{h}_i = \tilde{f}_i \tilde{h}^i.$$

Thus scalar products required the multiplication of covariant and contravariant components of the same index.

The contravariant and covariant components of a vector are, therefore, found by taking the scalar product of the contravariant and covariant basis functions yielding

$$f^i = \vec{\eta}^i \cdot \vec{f} \quad \text{and} \quad f_i = \vec{\tau}_i \cdot \vec{f}.$$

In our original orthogonal coordinate system, x^i, the square of the length of a differential line segment is expressed by

$$(ds)^2 = dx^i \, dx^i.$$

To express $(ds)^2$ in the transformed coordinates, note that $dx^i = (\partial x^i / \partial \tilde{x}^j) d\tilde{x}^j$, hence

$$(ds)^2 = \left(\frac{\partial x^i}{\partial \tilde{x}^j} d\tilde{x}^j \right) \left(\frac{\partial x^i}{\partial \tilde{x}^m} d\tilde{x}^m \right) \equiv \widetilde{G}_{jm} d\tilde{x}^j \, d\tilde{x}^m, \tag{6.6}$$

where $\widetilde{G}_{jm}$ is the *metric tensor* defined by $(\partial x^i / \partial \tilde{x}^j)(\partial x^i / \partial \tilde{x}^m)$. This metric tensor is fundamental in the requirement that the conservation laws are invariant regardless

[2] In vector terminology, this operation is also called the dot product.

of the functional form of the coordinate transformation. In the rectangular coordinate system used up until now, $G_{jm} = \delta_{jm}$ (thus the individual coordinate axes (x^1, x^2, and x^3) are independent of and orthogonal to one another at all points).

The inverse of the metric tensor is defined by the relation

$$\widetilde{G}^{jl} = \frac{\partial \tilde{x}^j}{\partial x^n}\frac{\partial \tilde{x}^l}{\partial x^n}. \tag{6.7}$$

To verify Eq. 6.7 note that,

$$\begin{aligned}\widetilde{G}_{jm}\widetilde{G}^{jl} &= \frac{\partial x^r}{\partial \tilde{x}^j}\frac{\partial x^r}{\partial \tilde{x}^m}\frac{\partial \tilde{x}^j}{\partial x^n}\frac{\partial \tilde{x}^l}{\partial x^n} = \frac{\partial x^r}{\partial \tilde{x}^j}\frac{\partial \tilde{x}^j}{\partial x^n}\frac{\partial x^r}{\partial \tilde{x}^m}\frac{\partial \tilde{x}^l}{\partial x^n}\\ &= \delta^r_n\frac{\partial x^r}{\partial \tilde{x}^m}\frac{\partial \tilde{x}^l}{\partial x^n} = \frac{\partial x^r}{\partial \tilde{x}^m}\frac{\partial \tilde{x}^l}{\partial x^r} = \delta^l_m.\end{aligned}$$

These forms of the metric tensor can also be expressed as

$$\widetilde{G}_{jm} = \vec{\tau}_j \cdot \vec{\tau}_m \quad \text{and} \quad \widetilde{G}^{jl} = \vec{\eta}^j \cdot \vec{\eta}^l.$$

One advantage of the metric tensor and its inverse is their ability to change a covariant tensor to a contravariant tensor and vice versa. This ability is needed because *only tensors of the same type* (e.g., covariant *or* contravariant) can be added. The reason for this is that covariant tensors are defined in terms of different basis vectors than contravariant tensors, and thus adding them together would be somewhat similar to adding the $\vec{i}$ unit vector to the $\vec{k}$ unit vector in our original x^i coordinate system.

To illustrate this capability of the metric tensor, let

$$\tilde{f}_l = \frac{\partial x^i}{\partial \tilde{x}^l} f_i,$$

then

$$\widetilde{G}^{lj}\tilde{f}_l = \left(\frac{\partial \tilde{x}^l}{\partial x^m}\frac{\partial \tilde{x}^j}{\partial x^m}\right)\frac{\partial x^i}{\partial \tilde{x}^l}f_i = \frac{\partial \tilde{x}^l}{\partial x^m}\frac{\partial x^i}{\partial \tilde{x}^l}\frac{\partial \tilde{x}^j}{\partial x^m}f_i = \delta^i_m\frac{\partial \tilde{x}^j}{\partial x^m}f_i = \frac{\partial \tilde{x}^j}{\partial x^i}f_i = \tilde{f}^j. \tag{6.8}$$

Similarly, the covariant component $\tilde{f}_l$ can be created by multiplying $\tilde{f}^j$ by $\widetilde{G}_{lj}$.

To determine if a quantity is a tensor or not (i.e., transforms according to Eq. 6.2 or 6.3 for *all coordinate systems*), it is necessary to define a third coordinate representation given, for example, by $\bar{x}^i$, which is related to the $\tilde{x}^i$ coordinate system by a functional transformation.

If ϕ is defined here to be a scalar, then using the chain rule

$$\frac{\partial \phi}{\partial \bar{x}^i} = \frac{\partial \tilde{x}^j}{\partial \bar{x}^i}\frac{\partial \phi}{\partial \tilde{x}^j} = \frac{\partial \tilde{x}^j}{\partial \bar{x}^i}\frac{\partial x^l}{\partial \tilde{x}^j}\frac{\partial \phi}{\partial x^l},$$

hence derivatives of scalar quantities transform according to Eq. 6.2, and ϕ is a covariant tensor of order zero.

If $\bar{\phi}_m$ is defined to be a vector representation in the $\bar{x}^i$ coordinate system, however,

$$\frac{\partial \bar{\phi}_m}{\partial \bar{x}^i} = \frac{\partial}{\partial \bar{x}^i}\left(\frac{\partial \tilde{x}^l}{\partial \bar{x}^m}\tilde{\phi}_l\right) = \frac{\partial \tilde{x}^l}{\partial \bar{x}^m}\frac{\partial \tilde{\phi}_l}{\partial \bar{x}^i} + \tilde{\phi}_l \frac{\partial^2 \tilde{x}^l}{\partial \bar{x}^i \partial \bar{x}^m},$$

so that derivatives of a vector are *not* tensors since they do not transform between coordinate systems according to Eq. 6.2 or 6.3.

To circumvent this problem, let

$$\tilde{\phi}_l = \frac{\partial x^j}{\partial \tilde{x}^l}\phi_j,$$

so that $\tilde{\phi}_l$ is a covariant tensor. But

$$\frac{\partial \tilde{\phi}_l}{\partial \tilde{x}^m} = \frac{\partial^2 x^j}{\partial \tilde{x}^m \partial \tilde{x}^l}\phi_j + \frac{\partial \phi_j}{\partial \tilde{x}^m}\frac{\partial x^j}{\partial \tilde{x}^l}, \tag{6.9}$$

is not a tensor, as has already been shown. Since

$$\frac{\partial \phi_j}{\partial \tilde{x}^m} = \frac{\partial x^r}{\partial \tilde{x}^m}\frac{\partial \phi_j}{\partial x^r}$$

by the chain rule, and since

$$\phi_j = \frac{\partial \tilde{x}^s}{\partial x^j}\tilde{\phi}_s,$$

substituting these two relations into Eq. 6.9 and rearranging yields

$$\frac{\partial \tilde{\phi}_l}{\partial \tilde{x}^m} = \frac{\partial^2 x^j}{\partial \tilde{x}^m \partial \tilde{x}^l}\frac{\partial \tilde{x}^s}{\partial x^j}\tilde{\phi}_s = \frac{\partial x^r}{\partial \tilde{x}^m}\frac{\partial x^j}{\partial \tilde{x}^l}\frac{\partial \phi_j}{\partial x^r}. \tag{6.10}$$

Thus a quantity has been created that transforms as a covariant tensor. By convention, Eq. 6.10 is written as

$$\tilde{\phi}_{l;m} = \frac{\partial \tilde{\phi}_l}{\partial \tilde{x}^m} - \tilde{\Gamma}^s_{ml}\tilde{\phi}_s = \frac{\partial x^r}{\partial \tilde{x}^m}\frac{\partial x^j}{\partial \tilde{x}^l}\frac{\partial \phi_j}{\partial x^r}, \tag{6.11}$$

and is called the *covariant derivative*, where

$$\tilde{\Gamma}^s_{ml} = \frac{\partial^2 x^j}{\partial \tilde{x}^m \partial \tilde{x}^l}\frac{\partial \tilde{x}^s}{\partial x^j} \tag{6.12}$$

and is called the *Christoffel symbol*. In the $\bar{x}^i$ coordinate system

$$\bar{\phi}_{t;u} = \frac{\partial \bar{\phi}_t}{\partial \bar{x}^u} - \bar{\Gamma}^i_{ut}\bar{\phi}_i = \frac{\partial \tilde{x}^v}{\partial \bar{x}^u}\frac{\partial \tilde{x}^w}{\partial \bar{x}^t}\frac{\partial \tilde{\phi}_w}{\partial \tilde{x}^v},$$

so that the proper tensorial transformation properties are maintained between coordinate systems. Using an analogous derivation, it is also true that the covariant derivative of a contravariant vector is given by

$$\tilde{\phi}^t_{;u} = \frac{\partial \tilde{\phi}^t}{\partial \tilde{x}^u} + \tilde{\Gamma}^t_{us}\tilde{\phi}^s. \tag{6.13}$$

Other important tensor relations are listed as follows. The Christoffel symbol and the metric tensor are related by

$$\widetilde{\Gamma}^{s}_{ml} = \frac{1}{2}\widetilde{G}^{sj}\left(\frac{\partial \widetilde{G}_{ij}}{\partial \tilde{x}^{m}} + \frac{\partial \widetilde{G}_{mj}}{\partial \tilde{x}^{l}} - \frac{\partial \widetilde{G}_{lm}}{\partial \tilde{x}^{j}}\right), \tag{6.14}$$

as can be shown by substituting for the metric tensor on the right side. In addition, since in the original Cartesian coordinate system $\delta^{ij} = G_{ij} = G^{ij}$, it must also be true in any coordinate system that the covariant derivative of the metric tensor

$$\widetilde{G}_{ij;k} = \widetilde{G}^{ij}_{;k} = 0. \tag{6.15}$$

The covariant derivative of a second-order tensor can also be shown to have a form similar to that given by Eq. 6.11 and 6.13, except two Christoffel symbols appear. If for example, $\tilde{\phi}^{j}_{k}$ is a mixed tensor of order two in the $\tilde{x}_i$ coordinate system, then the covariant derivative is given by

$$\tilde{\phi}^{j}_{k;i} = \frac{\partial \tilde{\phi}^{j}_{k}}{\partial x^{i}} - \widetilde{\Gamma}^{s}_{ki}\tilde{\phi}^{j}_{s} + \widetilde{\Gamma}^{j}_{iu}\tilde{\phi}^{u}_{k}.$$

Using this relation, the product rule of differentiation can be shown to be valid since if $\tilde{\phi}^{j}_{k} = \tilde{a}_k\tilde{b}^j$, then

$$\begin{aligned}\left(\tilde{a}_k\tilde{b}^j\right)_{;i} &= \frac{\partial \tilde{a}_k\tilde{b}^j}{\partial x^{i}} - \widetilde{\Gamma}^{s}_{ki}a_s b^j + \widetilde{\Gamma}^{j}_{iu}a_k b^u \\ &= \tilde{b}^j\frac{\partial \tilde{a}_k}{\partial x^{i}} - \widetilde{\Gamma}^{s}_{ki}a_s b^j + \tilde{a}_k\frac{\partial \tilde{b}^j}{\partial x^{i}} + \widetilde{\Gamma}^{j}_{iu}a_k b^u = \tilde{b}^j\tilde{a}_{k;i} + \tilde{a}_k\tilde{b}^j_{;i}\end{aligned}$$

Moreover, utilizing this rule along with Eq. 6.15

$$\tilde{\phi}_{j;i} = \left(\widetilde{G}_{jl}\tilde{\phi}^{l}\right)_{;i} = \widetilde{G}_{jl}\tilde{\phi}^{l}_{;i},$$

so that the covariant derivative of contravariant and covariant components can be interchanged using the metric tensor.

The determinant of the metric tensor is another important quantity that can be used in specifying the conservation relations in any coordinate system. The determinant is related to the Jacobian of the transformation and its inverse by

$$\widetilde{G}^{1/2} = \left|\frac{\partial x^{i}}{\partial \tilde{x}^{m}}\right| = \left|\frac{\partial \tilde{x}^{m}}{\partial x^{i}}\right|^{-1}, \tag{6.16}$$

where $\widetilde{G}$ is the determinant of the metric tensor $\widetilde{G}_{jm}$. This quantity is very valuable in representing the Christoffel symbol when its contravariant and one of its covariant components are the same such as when u is set equal to t in Eq. 6.13, so that

$$\tilde{\phi}^{t}_{;t} = \frac{\partial \tilde{\phi}^{t}}{\partial \tilde{x}^{t}} + \widetilde{\Gamma}^{t}_{ts}\tilde{\phi}^{s} = \frac{1}{\sqrt{\widetilde{G}}}\frac{\partial}{\partial \tilde{x}^{s}}\left(\sqrt{\widetilde{G}}\tilde{\phi}^{s}\right). \tag{6.17}$$

In obtaining the right side of the expression, the relation

$$\tilde{\Gamma}^t_{ts} = \frac{1}{\sqrt{\tilde{G}}} \frac{\partial}{\partial \tilde{x}^s} \sqrt{\tilde{G}}, \tag{6.18}$$

has been used, where Eq. 6.18 is obtained from Eq. 6.14 using the definition of matrix inverses and derivatives of matrices in terms of cofactors and determinants (e.g., see Dutton 1976:142).

To transform the conservation relations into a separate coordinate system, it is also necessary to require the proper tensorial transformation of the term ϵ_{ijk}, which is used to represent the curl operation in vector notation. This is achieved by the operation

$$\tilde{\epsilon}_{ijk} = \sqrt{\tilde{G}} \epsilon_{ijk} \tag{6.19}$$

since the determinant of the Jacobian of a three-dimensional transformation can be expanded into

$$\sqrt{\tilde{G}} = \frac{\partial x^r}{\partial \tilde{x}^1} \frac{\partial x^s}{\partial \tilde{x}^2} \frac{\partial x^t}{\partial \tilde{x}^3} \epsilon_{rst} \tag{6.20}$$

and therefore,

$$\frac{\partial x^r}{\partial \tilde{x}^i} \frac{\partial x^s}{\partial \tilde{x}^j} \frac{\partial x^t}{\partial \tilde{x}^k} \epsilon_{rst} = \sqrt{\tilde{G}} \epsilon_{ijk} \tag{6.21}$$

The left side of Eq. 6.21 is equal to Eq. 6.20 since the term is zero if any of the indices are equal. The expression given by Eq. 6.20 is only valid when transforming from our original rectangular coordinate system, where $\sqrt{G} = 1$. In the general case,

$$\frac{\partial \tilde{x}^r}{\partial \bar{x}^i} \frac{\partial \tilde{x}^s}{\partial \bar{x}^j} \frac{\partial \tilde{x}^t}{\partial \bar{x}^k} \sqrt{\tilde{G}} \epsilon_{rst} = \bar{\epsilon}_{ijk} \tag{6.22}$$

is used to transform this term properly.

The contravariant transformation form of the term ϵ_{ijk} is obtained by multiplying the covariant form by the inverse of the metric tensor, using the method given in Eq. 6.8, so that

$$\sqrt{\tilde{G}} \tilde{G}^{ir} \tilde{G}^{js} \tilde{G}^{kt} \epsilon_{ijk} = \epsilon_{rst} \Big/ \sqrt{\tilde{G}} = \tilde{\epsilon}^{rst}, \tag{6.23}$$

where $\tilde{G}^{ir} \tilde{G}^{js} \tilde{G}^{kt}$ is equal to the determinant of the inverse of the metric tensor using the same procedures as followed in obtaining Eq. 6.21. This determinant is then equal to $1/\tilde{G}$ using Eq. 6.16.

Using these properties of tensor transformations, it is possible to rewrite the conservation equations in any coordinate system of our choosing with the certainty that the physical representations, which are represented by these conservation relations, remain unchanged. By convention, the equations are written in the contravariant form using the covariant differentiation operation given by Eq. 6.13.

Therefore, the original prognostic conservation relations given by Eqs. 2.43-2.47 in Chapter 2 can, therefore, be written as

$$\frac{\partial \rho}{\partial t} = -\left(\rho \tilde{u}^i\right)_{;i} = -\frac{1}{\sqrt{\tilde{G}}} \frac{\partial}{\partial \tilde{x}^i} \left(\rho \sqrt{\tilde{G}} \tilde{u}^i\right), \tag{6.24}$$

$$\frac{\partial \theta}{\partial t} = \tilde{u}^j \frac{\partial \theta}{\partial \tilde{x}^j} + \widetilde{S}_\theta, \tag{6.25}$$

$$\frac{\partial \tilde{u}^i}{\partial t} = -\tilde{u}^j \tilde{u}^i_{;j} - \widetilde{G}^{ij}\theta \frac{\partial \pi}{\partial \tilde{x}^j} - \frac{\partial \tilde{x}^i}{\partial x^3} g - 2\tilde{\epsilon}^{ijl}\widetilde{\Omega}_j \tilde{u}_l, \tag{6.26}$$

$$\frac{\partial q_n}{\partial t} = -\tilde{u}^j \frac{\partial q_n}{\partial \tilde{x}^j} + \widetilde{S}_{q_n}, \quad n = 1, 2, 3, \tag{6.27}$$

$$\frac{\partial \chi_m}{\partial t} = -\tilde{u}^j \frac{\partial \chi_m}{\partial \tilde{x}^j} + \widetilde{S}_{\chi_m}, \quad m = 1, 2, \ldots, M, \tag{6.28}$$

where the pressure gradient term is represented in terms of the scaled pressure π defined by Eq. 4.34. The definitions of the tensor transformation parameters needed to preserve the physical invariance of these operations (e.g., $\sqrt{\widetilde{G}}$, $\widetilde{G}^{ij}$, $\tilde{\epsilon}^{ijl}$, and $\tilde{u}^i_{;j}$) are given by Eqs. 6.20, 6.7, 6.23, and 6.13. These equations are then valid for *any* functional coordinate representation.

The major rules to follow in obtaining a consistent representation of the conservation laws in the generalized coordinate system are as follows:

1. Require that the individual terms have the same number of contravariant and covariant indices. If not, use the metric tensor to change between covariant and contravariant forms. (Remember that the presence of a superscript on independent variables in the denominator of a derivative indicates, by convention, that it is a covariant form.)
2. Use the definition of covariant differentiation to assure that the derivatives retain physical invariance.
3. Use the Jacobian to transform dependent variables in a covariant representation from one coordinate to the next. Use the inverse of the Jacobian when a contravariant representation is desired.
4. The square root of the determinant of the metric tensor must be used to transfer the parameter ϵ_{ijk} between coordinate systems properly.

6.2 Generalized Vertical Coordinate

In the application of these equations to simulate mesoscale atmospheric flows, only the vertical coordinate in the rectangular system is customarily transformed, and this procedure will be adopted in the discussion that follows. In addition, it is necessary to average the transformed equations since, of course, Eqs. 6.24-6.28 are valid only over spatial and temporal intervals which are much smaller than the mesoscale space and time scales.

The functional form of this generalized vertical coordinate transformation, in terms of the original Cartesian system, can be written as

$$\begin{aligned} \tilde{x}^1 &= x, & x &= \tilde{x}^1, \\ \tilde{x}^2 &= y, & y &= \tilde{x}^2, \\ \tilde{x}^3 &= \sigma(x, y, z, t), & z &= h\left(\tilde{x}^1, \tilde{x}^2, \tilde{x}^3, t\right), \end{aligned}$$

The functional form of σ has been specified in a number of forms including

$$\sigma = \theta, \qquad \sigma = s\left(z - z_G\right) / \left(s - z_G\right),$$
$$\sigma = p, \qquad \sigma = \left(p_G - p\right) / \left(p_G - p_T\right),$$
$$\sigma = p / p_G \qquad \sigma = \left(\theta - \theta_T\right) / \left(\theta_G - \theta_T\right),$$
$$\sigma = \left[\frac{p - p_T}{p_G - p_T}\right]\left[\frac{p_{\text{ref}}(0) - p_T}{p_{\text{ref}(z_G)} - p_T}\right]^{-1}.$$

In these expressions, p_G, θ_G, p_T, and θ_T refer to the pressures and potential temperatures at the bottom and top of the coordinate representation, respectively, and z_G and s specify the terrain height and height of the top, respectively. $p_{\text{ref}}(0)$ and $p_{\text{ref}}(z_G)$ are the pressure at sea level and at z_G using a standard reference atmosphere which is the same across the model (Black 1994). The first two forms of σ on the left are referred to as *isentropic* and *isobaric* representations, respectively, and the remaining six are *terrain-following* coordinate systems, usually called *sigma* representations. The bottom formulation in the right column for σ is a normalized isentropic representation introduced by Branković (1981).

The innovative form of σ at the bottom of the lefthand column is called the "Eta-coordinate system" (Janjić 1990; Mesinger and Black 1992; Black 1994; Mesinger 1996, 1997, 1998; Mesinger et al. 1988, 1997) and has been used by the United States National Centers for Environmental Prediction (NCEP) for one of their regional models. The Eta system has the advantage that this form of the sigma system is nearly horizontal, yet the requirement that the system does not intersect the terrain is retained. Gallus and Klemp (2000) provide a comparison of model simulations of airflow over mountains using the Eta-coordinate and another form of a terrain-following coordinate system which has been adopted by the Weather Research and Forecasting Model (WRF) model (Skamarock et al. 2008), as discussed later in this section. In ocean models, a coordinate system that uses density as a vertical coordinate is often used (e.g., see Bleck and Boudra 1981), while Adcroft et al. (1997) and Marshall et al. (1997) use a partial grid volume coordinate system at their ocean bottom–ocean interface (called "shaved cells"). Laprise (1992a) suggests using hydrostatic pressure as the vertical coordinate.

Phillips (1957) originated the concept where the lowest coordinate surface was coincident with the ground. Deaven (1974, 1996), Friend et al. (1977), Bleck (1978), Uccellini et al. (1979), and Johnson and Uccellini (1983), have utilized an isentropic system well above the ground and a type of sigma representation near the ground. Kasahara (1974) and Sundquist (1979) have discussed various types of vertical coordinates, including the sigma system.

In the expressions for σ, p_T, and s are, generally prescribed as constant in time and space, although several investigators (e.g., Mahrer and Pielke 1975, 1978a) permitted temporal and spatial variations of their top coordinate. Examples of two-dimensional cross sections in several of these representations are given in Fig. 6.2, where a mountain is situated in the center of the region.

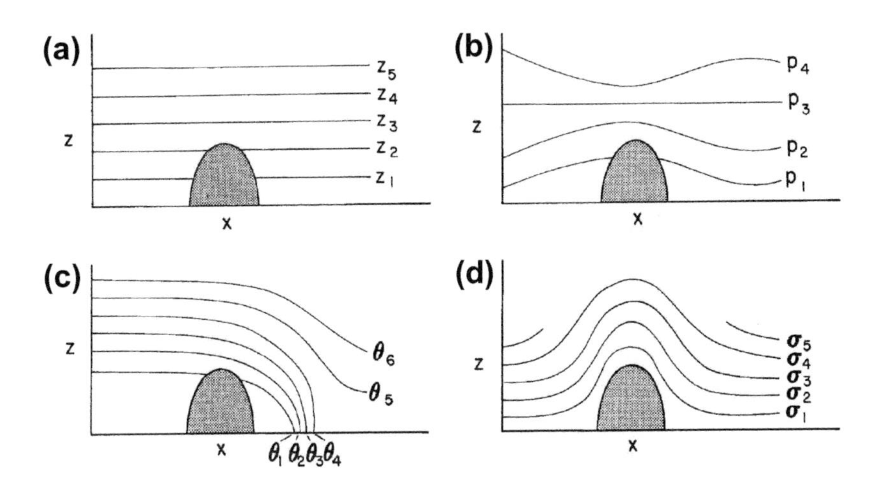

FIGURE 6.2

Schematic illustrations of (a) rectangular, (b) isobaric, (c) isentropic, and (d) sigma coordinate representations as viewed in a rectangular coordinate framework.

This concept of defining a coordinate surface coincident with the bottom topography permits more efficient use of computer resources, and it simplifies the application of lower boundary conditions. In Phillip's original form, adopted by many models (e.g., the U.S. Weather Service forecast models, Rieck 1979), pressure is used to define the independent vertical coordinate σ, where surface pressure is used as the lower boundary. Haltiner (1971), for example, defines $\sigma = p/p_G$, where p_G is the surface pressure whereas p is the pressure at any level. For this example, $\sigma = 1$ corresponds to the ground surface. In mesoscale models, however, σ is often defined using one of its forms which is a function of height rather than pressure. This is advantageous because p_G is a function of time, whereas terrain height is not.

An innovative pressure-based coordinate system was introduced by Laprise (1992a) which uses the hydrostatic pressure as in independent variable. As discussed in that paper, this has the advantage that the conservation equations take a form that closely parallels the form of the hydrostatic equations when written in isobaric coordinates. Laprise suggested that this hydrostatic pressure vertical coordinate could be applied in terrain-following, fully compressible models even when the hydrostatic assumption does not apply.

The WRF model adopted this vertical coordinate system, illustrated in Fig. 6.3 from Skamarock et al. (2008). The σ coordinate system is defined by

$$\sigma = (p_h - p_{ht})/\mu \text{ where } \mu = p_{hs} - p_{ht}.$$

Where p_h is the hydrostatic component of pressure, and p_{hs} and p_{ht} refer to values along the terrain surface and top. The value of this vertical coordinate varies from 1 at the surface to 0 at the top. This vertical coordinate is referred to as a mass vertical

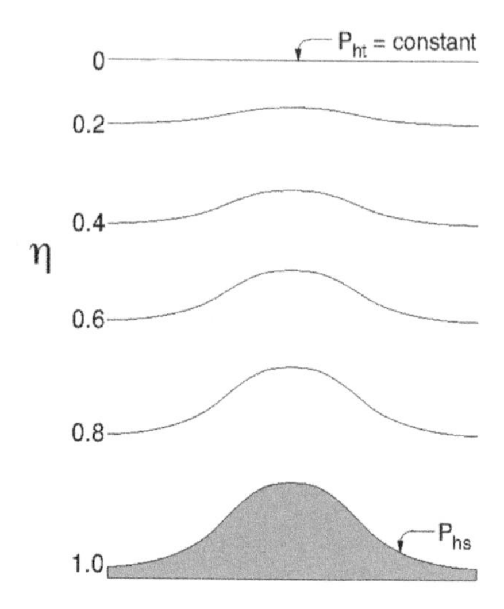

FIGURE 6.3

The form of the σ generalized mass vertical coordinate as used in WRF (from Skamarock et al. 2008).

coordinate since the change of hydrostatic pressure with height is a direct measure of the change of mass of the atmosphere with height.

Using any one of the definitions of the generalized vertical coordinate, the contravariant and covariant forms of the metric tensor $\widetilde{G}^{ij}$ and $\widetilde{G}_{ij}$ for the generalized vertical coordinate are given as

$$\widetilde{G}^{ij} = \frac{\partial \tilde{x}^i}{\partial x^l}\frac{\partial \tilde{x}^j}{\partial x^l} = \begin{bmatrix} 1 & 0 & \frac{\partial \sigma}{\partial x} \\ 0 & 1 & \frac{\partial \sigma}{\partial y} \\ \frac{\partial \sigma}{\partial x} & \frac{\partial \sigma}{\partial y} & \left\{\left(\frac{\partial \sigma}{\partial x}\right)^2 + \left(\frac{\partial \sigma}{\partial y}\right)^2 + \left(\frac{\partial \sigma}{\partial z}\right)^2\right\} \end{bmatrix}$$

$$\widetilde{G}_{ij} = \frac{\partial x^l}{\partial \tilde{x}^i}\frac{\partial x^l}{\partial \tilde{x}^j} = \begin{bmatrix} 1+\left(\frac{\partial h}{\partial \tilde{x}^1}\right)^2 & \frac{\partial h}{\partial \tilde{x}^1}\frac{\partial h}{\partial \tilde{x}^2} & \frac{\partial h}{\partial \tilde{x}^1}\frac{\partial h}{\partial \tilde{x}^3} \\ \frac{\partial h}{\partial \tilde{x}^1}\frac{\partial h}{\partial \tilde{x}^2} & 1+\left(\frac{\partial h}{\partial \tilde{x}^2}\right)^2 & \frac{\partial h}{\partial \tilde{x}^2}\frac{\partial h}{\partial \tilde{x}^3} \\ \frac{\partial h}{\partial \tilde{x}^1}\frac{\partial h}{\partial \tilde{x}^3} & \frac{\partial h}{\partial \tilde{x}^2}\frac{\partial h}{\partial \tilde{x}^3} & \left(\frac{\partial h}{\partial \tilde{x}^3}\right)^2 \end{bmatrix} \tag{6.29}$$

(using Eqs. 6.6 and 6.7), and the only nonzero Christoffel symbol is

$$\widetilde{\Gamma}^3_{jl} = \frac{\partial \sigma}{\partial z}\frac{\partial^2 h}{\partial \tilde{x}^j \partial \tilde{x}^l} \tag{6.30}$$

(from Eq. 6.12), so that the covariant derivative of velocity is given by

$$u^i_{;j} = \begin{cases} \dfrac{\partial \tilde{u}^i}{\partial \tilde{x}^j}, & i = 1, 2. \\ \dfrac{\partial \tilde{u}^3}{\partial \tilde{x}^j} + \widetilde{\Gamma}^3_{jl}\tilde{u}^l, & i = 3, \end{cases} \tag{6.31}$$

(from Eq. 6.13). The determinant of the Jacobian of the transformation,

$$\left(\left| \frac{\partial x^i}{\partial \tilde{x}^j} \right| = \sqrt{\widetilde{G}} \right)$$

(from Eq. 6.16), is given by

$$\left| \frac{\partial x^i}{\partial \tilde{x}^j} \right| = \begin{vmatrix} 1 & 0 & 0 \\ 0 & 1 & 0 \\ \dfrac{\partial h}{\partial \tilde{x}^1} & \dfrac{\partial h}{\partial \tilde{x}^2} & \dfrac{\partial h}{\partial \tilde{x}^3} \end{vmatrix} = \sqrt{\widetilde{G}} = \frac{\partial h}{\partial \tilde{x}^3} \equiv \frac{\partial h}{\partial \sigma}. \tag{6.32}$$

The tangent and normal basis vectors for the generalized vertical coordinate system in terms of the rectangular representation are given by

$$\begin{aligned} \vec{\tau}_1 &= \vec{i} + \vec{k}\frac{\partial h}{\partial \tilde{x}^1}, & \vec{\eta}^1 &= \vec{i}, \\ \vec{\tau}_2 &= \vec{j} + \vec{k}\frac{\partial h}{\partial \tilde{x}^2}, & \vec{\eta}^2 &= \vec{j}, \\ \vec{\tau}_3 &= \vec{k}\frac{\partial h}{\partial \tilde{x}^3}, & \vec{\eta}^3 &= \vec{i}\frac{\partial \sigma}{\partial x} + \vec{j}\frac{\partial \sigma}{\partial y} + \vec{k}\frac{\partial \sigma}{\partial z} \end{aligned} \tag{6.33}$$

(using Eqs. 6.4 and 6.5), where since $\vec{\tau}_i \cdot \vec{\tau}_j$ does not equal zero when $i \neq j$, this coordinate system in general is *nonorthogonal*. In the original rectangular coordinate system, the normal and tangent basis functions are the same (i.e., $\vec{i}$, $\vec{j}$, and $\vec{k}$) and are orthogonal to one another. Since $\vec{\tau}_3$ is tangent to the curve in which only $\tilde{x}^3$ varies, the σ coordinate is vertical at all points (i.e., $\vec{\tau}_3$ is in the direction of $\vec{k}$, from Eq. 6.33).

The individual contravariant and covariant velocity components are found from $\tilde{u}^i = \vec{\eta}^i \cdot \vec{u}$ and $\tilde{u}_i = \vec{\tau}_i \cdot \vec{u}$, respectively, where $\vec{u} = u\vec{i} + v\vec{j} + w\vec{k}$, so that

$$\begin{aligned} \tilde{u}^1 &= u, & \tilde{u}_1 &= u + \frac{\partial h}{\partial \tilde{x}^1}w, \\ \tilde{u}^2 &= v, & \tilde{u}_2 &= v + \frac{\partial h}{\partial \tilde{x}^2}w, \\ \tilde{u}^3 &= u\frac{\partial \sigma}{\partial x} + v\frac{\partial \sigma}{\partial y} + w\frac{\partial \sigma}{\partial z}, & \tilde{u}_3 &= w\frac{\partial h}{\partial \tilde{x}^3}. \end{aligned} \tag{6.34}$$

Kinetic energy is computed from these expressions by

$$e^2 = \frac{1}{2}\left(\tilde{u}^1\tilde{u}_1 + \tilde{u}^2\tilde{u}_2 + \tilde{u}^3\tilde{u}_3\right). \tag{6.35}$$

The Coriolis term in the transformed coordinate system is expressed in terms of the rectangular representation as

$$2\tilde{\epsilon}^{ijl}\tilde{\Omega}_j\tilde{u}_l = 2\epsilon_{ijl}\frac{\partial\sigma}{\partial z}\tilde{\Omega}_j\tilde{u}_l \tag{6.36}$$

(using Eq. 6.23), where

$$2\tilde{\Omega}_1 = 2\left(\Omega_1 + \frac{\partial h}{\partial \tilde{x}^1}\Omega_3\right) = 2\frac{\partial h}{\partial \tilde{x}^1}\Omega_3 = 2\frac{\partial h}{\partial \tilde{x}^1}\Omega\sin\phi = \frac{\partial h}{\partial \tilde{x}^1}f\;(\Omega_1 = 0)$$
$$2\tilde{\Omega}_2 = 2\left(\Omega_2 + \frac{\partial h}{\partial \tilde{x}^2}\Omega_3\right) = 2\Omega\cos\phi + 2\frac{\partial h}{\partial \tilde{x}^2}\Omega\sin\phi = \hat{f} + \frac{\partial h}{\partial \tilde{x}^2}f$$
$$2\tilde{\Omega}_3 = 2\Omega_3\frac{\partial h}{\partial \tilde{x}^3} = 2\frac{\partial h}{\partial \tilde{x}^3}\Omega\sin\phi = \frac{\partial h}{\partial \tilde{x}^3}f$$

with $f = 2\Omega\sin\phi$ and $\hat{f} = 2\Omega\cos\phi$ (Ω is the rotation rate of the Earth and ϕ is the latitude.)

Averaging of Eqs. 6.24-6.28, of course, is required if these equations are to be used in meteorological numerical models with finite grid and time intervals. The averaging operator given by Eq. 4.6 in Chapter 4 is not the correct one, however, because $\Delta x\Delta y\Delta z\Delta t$ is no longer the appropriate averaging volume. In the transformed coordinate system, the appropriate grid-volume averaging operator is defined as[3]

$$\overline{(\;)} = \frac{\int_t^{t+\Delta t}\int_{\tilde{x}^1}^{\tilde{x}^1+\Delta\tilde{x}^1}\int_{\tilde{x}^2}^{\tilde{x}^2+\Delta\tilde{x}^2}\int_{\sigma}^{\sigma+\Delta\sigma}(\;)\,d\sigma\,d\tilde{x}^2\,d\tilde{x}^1\,dt}{(\Delta\tilde{x}^1)(\Delta\tilde{x}^2)(\Delta\sigma)(\Delta t)}. \tag{6.37}$$

The dependent variables can be decomposed into an average and a subgrid-scale perturbation expressed as

$$\phi = \bar{\phi} + \phi'',$$

where ϕ'' is a deviation from the grid-volume average given by Eq. 6.37. The symbol ϕ represents any one of the dependent variables.

Eq. 6.26, for example, can be rewritten using Eq. 6.37 as

$$\frac{\partial\bar{\tilde{u}}^i}{\partial t} = -\bar{\tilde{u}}^j\bar{\tilde{u}}^i_{;j} - \overline{\tilde{u}^{j''}\tilde{u}^{i''}_{;j}} - \tilde{G}^{ij}\bar{\theta}\frac{\partial\bar{\pi}}{\partial\tilde{x}^j} - \frac{\partial\tilde{x}^i}{\partial z}g - 2\tilde{\epsilon}^{ijl}\tilde{\Omega}_j\bar{\tilde{u}}_l. \tag{6.38}$$

In deriving this form it has been assumed that $\theta = \bar{\theta}[1 + (\theta''/\bar{\theta})] \cong \bar{\theta}$ and that

$$\bar{\bar{\tilde{u}}}^i = \bar{\tilde{u}}^i, \quad \overline{\partial u_i/\partial t} = \partial\bar{u}_i/\partial t;\text{ etc. (therefore }\overline{\tilde{u}^{i''}} = 0,\text{ etc.}), \tag{6.39}$$

[3]It must be stressed that Eq. 6.37 does *not* represent the same volume as does Eq. 4.6. To do that the integrand must include the determinant of the Jacobian of the transformation. It is not desirable to do this here because $\tilde{x}^3$, $\tilde{x}^2$, $\tilde{x}^1$, and t are the coordinates of a grid that will be used in a numerical model and are, therefore, the appropriate averaging volume.

as was required in Section 4.1. To make this assumption in the transformed coordinate system, however, it is necessary *to require that changes of the metric tensor over the four-dimensional grid-volume* $\Delta\tilde{x}^1\ \Delta\tilde{x}^2\ \Delta\sigma\ \Delta t$ *are small*, since this tensor appears in Eq. 6.38. Expressed mathematically, this requirement can be written as

$$\overline{\tilde{G}^{ij}} = \frac{\int_t^{t+\Delta t}\int_{\tilde{x}^1}^{\tilde{x}+\Delta\tilde{x}^1}\int_{\tilde{x}^2}^{\tilde{x}^2+\Delta\tilde{x}^2}\int_{\sigma}^{\sigma+\Delta\sigma}\tilde{G}^{ij}\,d\sigma\,d\tilde{x}^2\,d\tilde{x}^1\,dt}{(\Delta t)\,(\Delta\tilde{x}^1)\,(\Delta\tilde{x}^2)\,(\Delta\sigma)} \simeq \tilde{G}^{ij}.$$

This requirement has significant implications on the choice of the vertical generalized coordinate since *it must be selected such that variations of the gradient of the transformed coordinate within the grid volume are small* compared with the grid-volume averaged gradient.

The advection term in Eq. 6.38 is derived from

$$\begin{aligned}\overline{\tilde{u}^j\tilde{u}^i_{;j}} &= \overline{\tilde{u}^j\frac{\partial\tilde{u}^i}{\partial\tilde{x}^j}} + \overline{\tilde{\Gamma}^i_{jl}\tilde{u}^j\tilde{u}^l} \simeq \overline{\tilde{u}^j\frac{\partial\tilde{u}^i}{\partial\tilde{x}^j}} + \bar{\tilde{\Gamma}}^i_{jl}\overline{\tilde{u}^j\tilde{u}^l}\\ &\simeq \bar{\tilde{u}}^j\frac{\partial\bar{\tilde{u}}^i}{\partial\tilde{x}^j} + \overline{\tilde{u}^{j''}\frac{\partial\tilde{u}^{i''}}{\partial\tilde{x}^j}} + \bar{\tilde{\Gamma}}^i_{jl}\left[\bar{\tilde{u}}^j\bar{\tilde{u}}^l + \overline{\tilde{u}^{j''}\tilde{u}^{l''}}\right] = \bar{\tilde{u}}^j\bar{\tilde{u}}^i_{;j} + \overline{\tilde{u}^{j''}\tilde{u}^{i''}_{;j}},\end{aligned}$$

where the assumption that changes of the metric tensor and its derivatives are small permits the removal of the Christoffel symbol from the integrand. This assumption can also be written as

$$\tilde{\Gamma}^i_{jl} = \bar{\tilde{\Gamma}}^i_{jl} + \tilde{\Gamma}''^i_{jl} = \bar{\tilde{\Gamma}}^3_{jl} + \tilde{\Gamma}''^3_{jl} \cong \bar{\tilde{\Gamma}}^3_{jl}, \quad \text{where } \left|\tilde{\Gamma}''^3_{jl}\right| \ll \left|\bar{\tilde{\Gamma}}^3_{jl}\right|.$$

The Coriolis term can be expanded as

$$2\tilde{\epsilon}^{ijl}\tilde{\Omega}_j\bar{\tilde{u}}_l = \frac{2\tilde{\Omega}_j\tilde{G}_{lm}\bar{\tilde{u}}^m\epsilon_{ijl}}{\sqrt{\tilde{G}}} = 2\epsilon_{ijl}\frac{\partial x^r}{\partial\tilde{x}^j}\Omega_r\tilde{G}_{lm}\bar{\tilde{u}}^m\frac{\partial\sigma}{\partial z},$$

with $\Omega_r = (0, \Omega\cos\phi, \Omega\sin\phi) = (0, \hat{f}/2, f/2)$.

In addition

$$\frac{\partial h}{\partial\tilde{x}^3}\frac{\partial\sigma}{\partial x^3} = 1,$$

and

$$\frac{\partial\tilde{x}^1}{\partial z} = \frac{\partial\tilde{x}^2}{\partial z} = 0.$$

Thus with the decomposition of the variables into resolvable and subgrid-scale terms, Eq. 6.38 can be written for the generalized vertical coordinate representation in its component form as

$$\begin{aligned}\frac{\partial\bar{\tilde{u}}^1}{\partial t} &= -\bar{\tilde{u}}^j\frac{\partial\bar{\tilde{u}}^1}{\partial\tilde{x}^j} - \overline{\tilde{u}^{j''}\frac{\partial\tilde{u}^{1''}}{\partial\tilde{x}^j}} - \bar{\theta}\frac{\partial\bar{\pi}}{\partial\tilde{x}^1} - \bar{\theta}\frac{\partial\sigma}{\partial x}\frac{\partial\bar{\pi}}{\partial\tilde{x}^3}\\ &\quad - \hat{f}\left(\frac{\partial h}{\partial\tilde{x}^1}\bar{\tilde{u}}^1 + \frac{\partial h}{\partial\tilde{x}^2}\bar{\tilde{u}}^2 + \frac{\partial h}{\partial\tilde{x}^3}\bar{\tilde{u}}^3\right) + f\bar{\tilde{u}}^2\end{aligned} \tag{6.40}$$

$$\frac{\partial \bar{\tilde{u}}^2}{\partial t} = -\bar{\tilde{u}}^j \frac{\partial \bar{\tilde{u}}^2}{\partial \tilde{x}^j} - \overline{\tilde{u}^{j\prime\prime} \frac{\partial \tilde{u}^{2\prime\prime}}{\partial \tilde{x}^j}} - \bar{\theta}\frac{\partial \bar{\pi}}{\partial \tilde{x}^2} - \bar{\theta}\frac{\partial \sigma}{\partial y}\frac{\partial \bar{\pi}}{\partial \tilde{x}^3} - f\bar{\tilde{u}}^1 \tag{6.41}$$

and

$$\begin{aligned}\frac{\partial \bar{\tilde{u}}^3}{\partial t} &= -\bar{\tilde{u}}^j \frac{\partial \bar{\tilde{u}}^3}{\partial \tilde{x}^j} - \overline{\tilde{u}^{j\prime\prime} \frac{\partial \tilde{u}^{3\prime\prime}}{\partial \tilde{x}^j}} - \bar{\tilde{\Gamma}}^3_{jl}\bar{\tilde{u}}^j\bar{\tilde{u}}^l - \bar{\tilde{\Gamma}}^3_{jl}\overline{\tilde{u}^{j\prime\prime}\tilde{u}^{l\prime\prime}} \\ &- \bar{\theta}\left\{\frac{\partial \sigma}{\partial x}\frac{\partial \bar{\pi}}{\partial \tilde{x}^1} + \frac{\partial \sigma}{\partial y}\frac{\partial \bar{\pi}}{\partial \tilde{x}^2} + \left[\left(\frac{\partial \sigma}{\partial x}\right)^2 + \left(\frac{\partial \sigma}{\partial y}\right)^2 + \left(\frac{\partial \sigma}{\partial z}\right)^2\right]\frac{\partial \bar{\pi}}{\partial \tilde{x}^3}\right\} \\ &+ \left(\hat{f} + \frac{\partial h}{\partial \tilde{x}^2} f\right)\frac{\partial \sigma}{\partial z}\left[\left(1 + \left(\frac{\partial h}{\partial \tilde{x}^1}\right)^2\right)\bar{\tilde{u}}^1 + \frac{\partial h}{\partial \tilde{x}^1}\frac{\partial h}{\partial \tilde{x}^2}\bar{\tilde{u}}^2 + \frac{\partial h}{\partial \tilde{x}^1}\frac{\partial h}{\partial \tilde{x}^3}\bar{\tilde{u}}^3\right] \\ &- \frac{\partial h}{\partial \tilde{x}^1} f \frac{\partial \sigma}{\partial z}\left[\frac{\partial h}{\partial \tilde{x}^2}\frac{\partial h}{\partial \tilde{x}^1}\bar{\tilde{u}}^1 + \left(1 + \left(\frac{\partial h}{\partial \tilde{x}^2}\right)^2\right)\bar{\tilde{u}}^2 + \frac{\partial h}{\partial \tilde{x}^2}\frac{\partial h}{\partial \tilde{x}^3}\bar{\tilde{u}}^3\right] - \frac{\partial \sigma}{\partial z} g.\end{aligned} \tag{6.42}$$

Since from Section 6.1 the vector velocity $\vec{v} = \bar{\tilde{u}}^j \vec{\tau}_j$, $\bar{\tilde{u}}^3$ is in the same direction as the Cartesian velocity $\bar{w}$, whereas $\bar{\tilde{u}}^1$ and $\bar{\tilde{u}}^2$ are, in general, at some angle to $\bar{u}$ and $\bar{v}$ in the original rectangular coordinate system as shown by Eqs. 6.33 and 6.34.

The transformed grid-volume-averaged conservation-of-mass relation from Eq. 6.24, can be written as

$$\frac{\partial \bar{\rho}}{\partial t} = -\frac{\partial \sigma}{\partial z}\frac{\partial}{\partial \tilde{x}^j}\left(\frac{\partial h}{\partial \tilde{x}^3}\overline{\rho \tilde{u}^j}\right), \tag{6.43}$$

which, since $\partial \rho/\partial t = -(\partial/\partial x_j)\rho u_j$ in the rectangular coordinate system (i.e., Eq. 2.43) can be approximated by

$$0 = -\frac{\partial \sigma}{\partial z}\frac{\partial}{\partial \tilde{x}^j}\left(\rho_0 \frac{\partial h}{\partial \tilde{x}^3}\overline{\tilde{u}^j}\right) = -\frac{\partial}{\partial \tilde{x}^j}\left(\rho_0 \frac{\partial h}{\partial \tilde{x}^3}\overline{\tilde{u}^j}\right) \simeq -\frac{\partial}{\partial \tilde{x}^j}\left(\bar{\rho}\frac{\partial h}{\partial \tilde{x}^3}\overline{\tilde{u}^j}\right) \tag{6.44}$$

if it is assumed that $|\alpha''|/\bar{\alpha} \sim |\alpha''|/\alpha_0 \ll 1$, as was done in obtaining Eqs. 3.11 and 4.9. In invoking the assumption that $|\alpha''|$ is much less than $\bar{\alpha}$ and α_0, so that Eq. 6.44 can be written in the form that it is, it must be remembered that the averaging volume is in the transformed coordinate system.

The conservation of heat as represented by potential temperature and the conservation of water substance and other gaseous and aerosol atmospheric material given by Eqs. 6.25, 6.27, and 6.28 can be written in grid-volume-averaged form as

$$\frac{\partial \bar{\theta}}{\partial t} = -\bar{\tilde{u}}^j \frac{\partial \bar{\theta}}{\partial \tilde{x}^j} - \overline{\tilde{u}^j \frac{\partial \theta''}{\partial \tilde{x}^j}} + \bar{\tilde{S}}_\theta, \tag{6.45}$$

$$\frac{\partial \bar{q}_n}{\partial t} = -\bar{\tilde{u}}^j \frac{\partial \bar{q}_n}{\partial \tilde{x}^j} - \overline{\tilde{u}^{j\prime\prime} \frac{\partial q_n''}{\partial \tilde{x}^j}} + \bar{\tilde{S}}_{q_n}, \quad n = 1, 2, 3 \tag{6.46}$$

$$\frac{\partial \bar{\chi}_m}{\partial t} = -\bar{\tilde{u}}^j \frac{\partial \bar{\chi}_m}{\partial \tilde{x}^j} - \overline{\tilde{u}^{j\prime\prime} \frac{\partial \chi_m^{\prime\prime}}{\partial \tilde{x}^j}} + \tilde{\bar{S}}_{\chi_m}, \quad m = 1, 2, ..., M \tag{6.47}$$

In summary, the $8+M$ prognostic equations (6.40, 6.41, 6.42, 6.44, 6.46, and 6.47) in the $8+M$ unknowns $\bar{\tilde{u}}, \bar{\theta}, \bar{\pi}, \bar{q}_n$, and $\bar{\chi}_m$ can be used to represent the conservation relations in any generalized vertical coordinate system, as long as the assumptions such as given by Eq. 6.39 are valid. Because tensor transformation rules were used, one can be certain that the physical representation of the conservation relations is unaffected.

These equations can also be manipulated to obtain equivalent expressions for other forms of the conservation equations. The diagnostic equation for pressure in the transformed system, for example, can be obtained by taking the divergence of Eqs. 6.40-6.42. This operation can be performed by applying the covariant differentiation operation with respect to the free index i (i.e., $(\ )_{;i}$ is applied) to the grid-volume-averaged form of Eq. 6.26.

6.3 The Sigma-z Coordinate System

Terrain-following coordinate systems which are a function of z have been used extensively in regional and mesoscale models (e.g., Mahrer and Pielke 1975; Colton 1976; Blondin 1978; Yamada 1978, and others) in which the hydrostatic assumption has been applied and in mesoscale models in which the hydrostatic assumption has not been made (e.g., Gal-Chen and Somerville 1975a,b; Clark 1977; Pielke et al. 1992; Shi et al. 2000; Saito 2012).

6.3.1 The Hydrostatic Assumption Derivation

In developing hydrostatic model equations, investigators have generally applied the chain rule *separately* in the vertical and horizontal dimensions (utilizing the hydrostatic relation). Using the terrain-following coordinate system defined by

$$\sigma = s \frac{z - z_G}{s - z_G} \tag{6.48}$$

for example, where s is a constant and z_G is a function of x and y, the application of the chain rule to the hydrostatic relation given by Eq. 4.39 yields

$$\frac{\partial \bar{\pi}}{\partial \sigma} = -\frac{s - z_G}{s} \frac{g}{\bar{\theta}}. \tag{6.49}$$

Using Eq. 4.39 as a replacement for the vertical velocity equation is appropriate if the hydrostatic assumption is *exactly* satisfied. However, the invariance of the physical representation is lost if the assumption is not exact, as discussed by Dutton (1976:242), since a correct tensor transformation of the three-dimensional vector velocity equation is required. When the horizontal scales are much larger than the vertical scales of motion, the hydrostatic relation is very closely satisfied and such a separation of the vertical and horizontal equation may be justified. By making the hydrostatic assumption before the coordinate transformation, however, significant insight into the effect of the change of coordinates on the form of the physical invariance of the conservation relations in the transformed system cannot be evaluated. To provide such insight, it is necessary to use the methods of tensor analysis to transform coordinate systems, and then to invoke a more general form of the hydrostatic assumption. A more in-depth understanding of the coordinate transformation is then obtained.

To examine the effect of utilizing the hydrostatic assumption in Eqs. 6.40, 6.41, and 6.42, Eq. 6.48 is defined as the generalized vertical coordinate. The relation between the spatial coordinates in the two representations is given by

$$\begin{aligned} \tilde{x}^1 &= x, & x &= \tilde{x}^1, \\ \tilde{x}^2 &= y, & y &= \tilde{x}^2, \\ \tilde{x}^3 &= \sigma = s\left[z - z_G(x, y)\right] / \left[s - z_G(x, y)\right], & z &= h = (\sigma/s)\left[s - z_G\left(\tilde{x}^1, \tilde{x}^2\right)\right] \\ & & &\quad + z_G\left(\tilde{x}^1, \tilde{x}^2\right), \end{aligned} \tag{6.50}$$

so that the nonzero quantities needed to evaluate the Jacobian and its determinant (Eq. 6.32), metric tensor (Eq. 6.29), and Christoffel symbol (Eq. 6.30) are given as

$$\begin{aligned} &\frac{\partial \sigma}{\partial x} = \frac{\partial z_G}{\partial x}\left(\frac{\sigma - s}{s - z_G}\right); \quad \frac{\partial h}{\partial \tilde{x}^1} = \frac{\partial z_G}{\partial \tilde{x}^1}\left(\frac{s - \sigma}{s}\right); \quad \frac{\partial \sigma}{\partial y} = \frac{\partial z_G}{\partial y}\left(\frac{\sigma - s}{s - z_G}\right); \\ &\frac{\partial h}{\partial \tilde{x}^2} = \frac{\partial z_G}{\partial \tilde{x}^2}\left(\frac{s - \sigma}{s}\right); \quad \frac{\partial \sigma}{\partial z} = \frac{s}{s - z_G}; \quad \frac{\partial h}{\partial \sigma} = \frac{s - z_G}{s} = \sqrt{G} \end{aligned} \tag{6.51}$$

and

$$\begin{aligned} &\tilde{\Gamma}^3_{11} = \frac{s - \sigma}{s - z_G}\frac{\partial^2 z_G}{\partial \tilde{x}^{1^2}}; \quad \tilde{\Gamma}^3_{22} = \frac{s - \sigma}{s - z_G}\frac{\partial^2 z_G}{\partial \tilde{x}^{2^2}}; \quad \tilde{\Gamma}^3_{21} = \frac{s - \sigma}{s - z_G}\frac{\partial^2 z_G}{\partial \tilde{x}^1 \partial \tilde{x}^2}; \\ &\tilde{\Gamma}^3_{23} = -\frac{1}{s - z_G}\frac{\partial z_G}{\partial \tilde{x}^2}; \quad \tilde{\Gamma}^3_{13} = -\frac{1}{s - z_G}\frac{\partial z_G}{\partial \tilde{x}^1}; \end{aligned} \tag{6.52}$$

with $\tilde{\Gamma}^3_{21} = \tilde{\Gamma}^3_{12}$; $\tilde{\Gamma}^3_{23} = \tilde{\Gamma}^3_{32}$; $\tilde{\Gamma}^3_{13} = \tilde{\Gamma}^3_{31}$.

Figure 6.4 schematically illustrates this coordinate transformation as viewed in the Cartesian coordinate framework.

The velocity vector $\vec{V}$, can be expressed as (Pielke and Cram 1989)

$$\begin{aligned} \vec{V} &= \tilde{u}_i \vec{\eta}^i = \tilde{u}^i \vec{\tau}_i = u\vec{i} + v\vec{j} + w\vec{k} \\ \vec{V} &= \tilde{u}_1 \vec{i} + \tilde{u}_2 \vec{j} + \tilde{u}_3\left[\vec{i}\left(\frac{\sigma - s}{s - z_G}\right)\frac{\partial z_G}{\partial x} + \vec{j}\left(\frac{\sigma - s}{s - z_G}\right)\frac{\partial z_G}{\partial y} + \vec{k}\left(\frac{s}{s - z_G}\right)\right] \\ \vec{V} &= \tilde{u}^1\left[\vec{i} + \vec{k}\left(\frac{s - \sigma}{s}\right)\frac{\partial z_G}{\partial \tilde{x}^1}\right] + \tilde{u}^2\left[\vec{j} + \vec{k}\left(\frac{s - \sigma}{s}\right)\frac{\partial z_G}{\partial \tilde{x}^2}\right] + \tilde{u}^3 \vec{k}\left(\frac{s - z_G}{s}\right). \end{aligned} \tag{6.53}$$

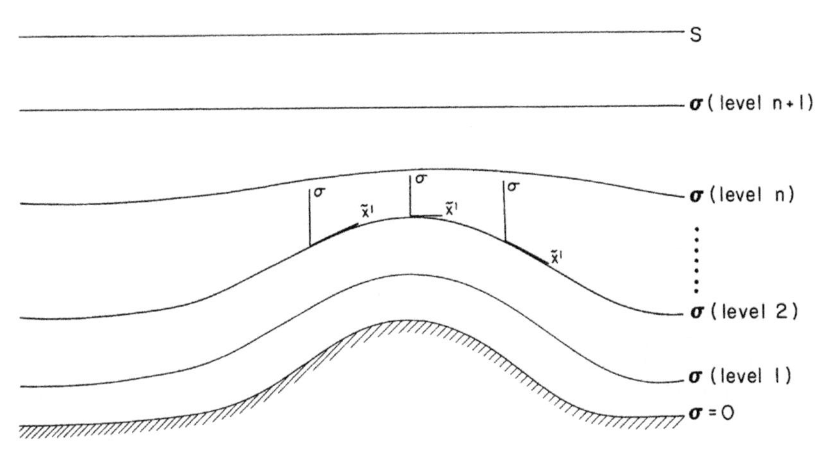

FIGURE 6.4

A schematic representation of σ-coordinate surfaces as they would appear in a rectangular representation, as defined by Eq. 6.48.

The velocities $\tilde{u}_i$ and $\vec{u}^i$ are the covariant and contravariant components, respectively, and are given by

$$\begin{aligned}
\tilde{u}^1 &= u \\
\tilde{u}^2 &= v \\
\tilde{u}^3 &= u\left(\frac{\sigma - s}{s - z_G}\right)\frac{\partial z_G}{\partial x} + v\left(\frac{\sigma - s}{s - z_G}\right)\frac{\partial z_G}{\partial y} + w\left(\frac{s}{s - z_G}\right) \\
\tilde{u}_1 &= u + \left(\frac{s - \sigma}{s}\right)\frac{\partial z_G}{\partial \tilde{x}^1} w \\
\tilde{u}_2 &= v + \left(\frac{s - \sigma}{s}\right)\frac{\partial z_G}{\partial \tilde{x}^2} w \\
\tilde{u}_3 &= w\left(\frac{s - z_G}{s}\right).
\end{aligned} \tag{6.54}$$

Therefore, the vectors in Eq. 6.53 can be rewritten in terms of the Cartesian quantities as

$$\begin{aligned}
\vec{V} &= \left[u + \left(\frac{s - \sigma}{s}\right)\frac{\partial z_G}{\partial \tilde{x}^1} w\right]\vec{i} + \left[v + \left(\frac{s - \sigma}{s}\right)\frac{\partial z_G}{\partial \tilde{x}^2} w\right]\vec{j} \\
&\quad + w\left(\frac{1}{s}\right)\left[\vec{i}(\sigma - s)\frac{\partial z_G}{\partial x} + \vec{j}(\sigma - s)\frac{\partial z_G}{\partial y} + \vec{k}\right] \\
\vec{V} &= u\left[\vec{i} + \vec{k}\left(\frac{s - \sigma}{s}\right)\frac{\partial z_G}{\partial \tilde{x}^1}\right] + v\left[\vec{j} + \vec{k}\left(\frac{s - \sigma}{s}\right)\frac{\partial z_G}{\partial \tilde{x}^2}\right] \\
&\quad + \left[u(\sigma - s)\frac{\partial z_G}{\partial x} + v(\sigma - s)\frac{\partial z_G}{\partial y} + w(s)\right]\vec{k}\left(\frac{1}{s}\right)
\end{aligned} \tag{6.55}$$

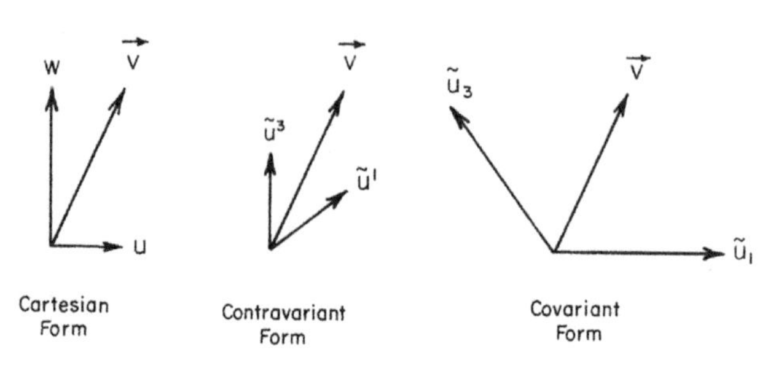

FIGURE 6.5

Vector $\vec{V}$ as expressed in Cartesian, covariant, and contravariant components in a terrain-following coordinate system. The σ-surface is parallel to $\tilde{u}^1$ and perpendicular to $\tilde{u}^3$ (from Pielke and Cram 1989).

Fig. 6.5 shows the vector $\vec{V}$ presented in the Cartesian, covariant, and contravariant forms for a two-dimensional case.

Note that the contravariant form has a component in the vertical direction and a component parallel to the σ-surface which at $\sigma = 0$ is the ground. The covariant form, in contrast, has a component perpendicular to the σ-surface and a horizontal component. The consistent form of kinetic energy is $\frac{1}{2}\vec{V}\cdot\vec{V}$, and can be calculated from

$$\vec{V}\cdot\vec{V} = \tilde{u}_i \vec{\eta}^i \tilde{u}^j \vec{\tau}_j = \delta^i_j \tilde{u}_i \tilde{u}^j = \tilde{u}_i \tilde{u}^i .$$

The covariant components must be multiplied by the corresponding contravariant components, i.e.,

$$\tilde{u}_i \tilde{u}^i = \tilde{u}_1 \tilde{u}^1 + \tilde{u}_2 \tilde{u}^2 + \tilde{u}_3 \tilde{u}^3, \quad \text{or}$$

$$\begin{aligned} \tilde{u}_i \tilde{u}^i &= u\left[u + \left(\frac{s-\sigma}{s}\right)\frac{\partial z_G}{\partial \tilde{x}^1} w\right] + v\left[v + \left(\frac{s-\sigma}{s}\right)\frac{\partial z_G}{\partial \tilde{x}^2} w\right] \\ &+ w\left[u\left(\frac{\sigma - s}{s}\right)\frac{\partial z_G}{\partial x} + v\left(\frac{\sigma - s}{s}\right)\frac{\partial z_G}{\partial y} + w\right] = u^2 + v^2 + w^2 \end{aligned} \tag{6.56}$$

The complete conservation-of-motion equation in the contravariant form can be written in general form for the coordinate transformation as

$$\frac{\partial \tilde{u}^1}{\partial t} = -\tilde{u}^j \frac{\partial \tilde{u}^1}{\partial \tilde{x}^j} - \theta \frac{\partial \pi}{\partial \tilde{x}^1} + \theta \frac{\sigma - s}{s - z_G}\frac{\partial z_G}{\partial x}\frac{\partial \pi}{\partial \tilde{x}^3} - \hat{f}\tilde{u}^3 + f\tilde{u}^2 \tag{6.57}$$

$$\frac{\partial \tilde{u}^2}{\partial t} = -\tilde{u}^j \frac{\partial \tilde{u}^2}{\partial \tilde{x}^j} - \theta \frac{\partial \pi}{\partial \tilde{x}^2} + \theta \frac{\sigma - s}{s - z_G}\frac{\partial z_G}{\partial y}\frac{\partial \pi}{\partial \tilde{x}^3} - f\tilde{u}^1 \tag{6.58}$$

and

$$
\begin{aligned}
\frac{\partial \tilde{u}^3}{\partial t} = & -\tilde{u}^j \frac{\partial \tilde{u}^3}{\partial \tilde{x}^j} - \frac{1}{(s - z_G)} \left[(s-\sigma)\frac{\partial^2 z_G}{\partial \tilde{x}^{1^2}} (\tilde{u}^1)^2 + (s-\sigma)\frac{\partial^2 z_G}{\partial \tilde{x}^{2^2}} (\tilde{u}^2)^2 \right. \\
& \left. + 2(s-\sigma)\frac{\partial^2 z_G}{\partial \tilde{x}^1 \partial \tilde{x}^2} \tilde{u}^1 \tilde{u}^2 - 2\frac{\partial z_G}{\partial \tilde{x}^1} \tilde{u}^1 \tilde{u}^3 - 2\frac{\partial z_G}{\partial \tilde{x}^2} \tilde{u}^2 \tilde{u}^3 \right] \\
& - \theta \left\{ \frac{\partial z_G}{\partial x} \left(\frac{\sigma - s}{s - z_G} \right) \frac{\partial \pi}{\partial \tilde{x}^1} + \frac{\partial z_G}{\partial y} \left(\frac{\sigma - s}{s - z_G} \right) \frac{\partial \pi}{\partial \tilde{x}^2} \right. \\
& + \left[\left(\left(\frac{\partial z_G}{\partial x} \right) \left(\frac{\sigma - s}{s - z_G} \right) \right)^2 + \left(\left(\frac{\partial z_G}{\partial y} \right) \left(\frac{\sigma - s}{s - z_G} \right) \right)^2 \right. \\
& \left. \left. + \left(\frac{s}{s - z_G} \right)^2 \right] \frac{\partial \pi}{\partial \tilde{x}^3} \right\} - \frac{s}{s - z_G} g
\end{aligned}
\tag{6.59}
$$

where, to reduce the notational complexity, the Coriolis term was left out of the $\tilde{u}^3$ equation.[4] Equations 6.57 and 6.58 are applied parallel to σ-surfaces while Eq. 6.59 is applied in the vertical axis along a σ-coordinate (see Eq. 6.53).

6.3.2 Generalized Hydrostatic Equation

A generalized hydrostatic form of Eq. 6.59 can be derived if it is assumed that vertical accelerations are small compared to the remaining terms, which yields

$$
\begin{aligned}
0 = & -\theta \left\{ \frac{\partial z_G}{\partial x} \left(\frac{\sigma - s}{s - z_G} \right) \frac{\partial \pi}{\partial \tilde{x}^1} + \frac{\partial z_G}{\partial y} \left(\frac{\sigma - s}{s - z_G} \right) \frac{\partial \pi}{\partial \tilde{x}^2} \right. \\
& + \left[\left(\left(\frac{\partial z_G}{\partial x} \right) \left(\frac{\sigma - s}{s - z_G} \right) \right)^2 + \left(\left(\frac{\partial z_G}{\partial y} \right) \left(\frac{\sigma - s}{s - z_G} \right) \right)^2 \right. \\
& \left. \left. + \left(\frac{s}{s - z_G} \right)^2 \right] \frac{\partial \pi}{\partial \tilde{x}^3} \right\} - \frac{s}{s - z_G} g.
\end{aligned}
\tag{6.60}
$$

[4]One can manipulate Eqs. 6.57-6.59 as performed by Clark (1977) and Clark (1988; personal communication), such that explicit prognostic equations for $\tilde{u}^1$, $\tilde{u}^2$, and w (the Cartesian vertical velocity) are obtained. This rearrangement makes use of the contravariant velocity component definitions in Eq. 6.54. These equations can be written in a flux form, which is advantageous for computational accuracy, as pointed out by Clark. The mathematical equivalence of the Christoffel symbols still occur in his equations, however, appearing in his diagnostic pressure equation which is derived using Eqs. 6.57 through 6.59.

Rearranging Eq. 6.60 to solve for $\partial\pi/\partial\tilde{x}^3$ produces

$$\frac{\partial\pi}{\partial\tilde{x}^3} = -\left(\frac{s}{s-z_G}\frac{g}{\theta} + \frac{\partial z_G}{\partial x}\left(\frac{\sigma - s}{s-z_G}\right)\frac{\partial\pi}{\partial\tilde{x}^1} + \frac{\partial z_G}{\partial y}\left(\frac{\sigma - s}{s-z_G}\right)\frac{\partial\pi}{\partial\tilde{x}^2}\right) \Big/ \left[\left(\frac{\partial z_G}{\partial x}\left(\frac{\sigma - s}{s-z_G}\right)\right)^2 + \left(\frac{\partial z_G}{\partial y}\left(\frac{\sigma - s}{s-z_G}\right)\right)^2 + \left(\frac{s}{s-z_G}\right)^2\right]. \tag{6.61}$$

Equation 6.61 is a generalized hydrostatic equation since accelerations are neglected in the σ-direction but permitted in the σ-parallel orientation, thereby retaining some nonhydrostatic motion when referred back to the Cartesian hydrostatic equation.[5]

When slope angles are small, Eq. 6.61 reduces to

$$\frac{\partial\pi}{\partial\tilde{x}^3} = -\left(\frac{s-z_G}{s}\right)\frac{g}{\theta} \tag{6.62}$$

which is the form generally applied to represent the hydrostatic assumption in atmospheric models with a terrain-following coordinate system (Eq. 6.48). Equation 6.62 is, therefore, the shallow slope generalized hydrostatic approximation. Physick (1986), however, did apply the complete form given by Eq. 6.61 in his simulation of the flow in the Grand Canyon.

The main results of this analysis is that a generalized hydrostatic equation can be derived which retains some nonhydrostatic motions when referred back to the Cartesian coordinate system. Section 6.4 discusses this approach as applied to drainage flow models.

6.4 Derivation of Drainage Flow Equations Using Two Different Coordinate Representations

In this section, two transformations for drainage flow modeling are examined, and for simplicity, are illustrated for two-dimensional formulations (described in Berkofsky 1993 and Pielke et al. 1985, 1993b). The extension to three dimensions is straightforward. The formulations are:

Transformation I.
$$\tilde{x}^1 = x$$
$$\tilde{x}^3 = z - z_G(x)$$
$$x = \tilde{x}^1$$
$$z = \tilde{x}^3 + z_G\left(\tilde{x}^1\right)$$

Transformation II.
$$\tilde{x}^1 = x\cos\gamma + z\sin\gamma$$
$$\tilde{x}^3 = z\cos\gamma - x\sin\gamma$$
$$x = \tilde{x}^1\cos\gamma - \tilde{x}^3\sin\gamma$$
$$z = \tilde{x}^1\sin\gamma + \tilde{x}^3\cos\gamma$$

where z_G is terrain height and $\gamma = \tan^{-1}[\partial z_G/\partial x]$ is the slope angle of the terrain.

[5]It is important to recognize that while the Cartesian hydrostatic equation represents a balance of forces between the vertical pressure gradient and gravitational forces, the generalized hydrostatic equation as defined here retains those vertical accelerations which are parallel to σ-surfaces.

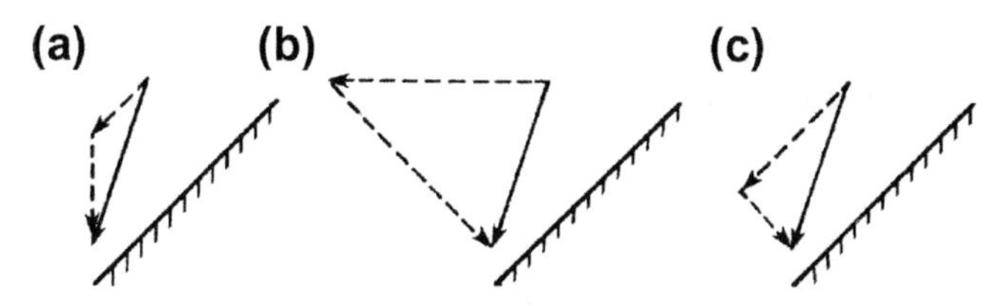

FIGURE 6.6

Representations of a drainage flow wind using the (a) contravariant and (b) covariant forms of the velocity components derived from Transformation I, and (c) using the velocity components obtained by the orthogonal rotation Transformation II. The magnitude and direction of the components of the vector in the different representations have been given in terms of the Cartesian velocity components u and w, and basis vectors $\vec{i}$ and $\vec{k}$. The slope $\partial z_G/\partial x = \alpha$ and γ is the slope angle (from Pielke et al. 1985).

Transformation I represents one form of the nonorthogonal generalized vertical coordinate transformations given at the beginning of Section 6.2. As shown in Fig. 6.4, the $\tilde{x}^3$ coordinate is parallel to the gravity vector and $\tilde{x}^1$ is along the terrain slope.

Transformation II represents an orthogonal rotation in which $\tilde{x}^1$ is parallel and $\tilde{x}^3$ perpendicular to the terrain. Both transformations are illustrated in Fig. 6.6.

The orthogonal rotation is of the form commonly used to develop idealized analytic models of slope flow as summarized, for example, by Mahrt (1982), and applied by McNider (1982). In Mahrt's paper, he suggests that one advantage of such a coordinate transformation is that the

> *...gravitational force perpendicular to the ground is approximately balanced by the pressure gradient force, while the component of the gravitational force parallel to the slope is not balanced and leads to downslope acceleration.*

This type of separation into a hydrostatic part and a nonhydrostatic component is of substantial usefulness in developing analytic (and numerical) slope flow models and its application and generalization are explored using Transformations I and II.

6.4.1 Transformation I

Using the definition of $\widetilde{G}^{ik}$, $\partial\tilde{x}^i/\partial z$, and covariant differentiation and Transformation I, we obtain

$$\widetilde{G}^{ik} = \begin{bmatrix} 1 & -\dfrac{\partial z_G}{\partial x} \\ -\dfrac{\partial z_G}{\partial t} & \left(\dfrac{\partial z_G}{\partial x}\right)^2 + 1 \end{bmatrix}; \quad \frac{\partial \tilde{x}^i}{\partial z} = (0, 1).$$

For two dimensions when subgrid-scale flux[6] terms with $\hat{f}$ and f are ignored, and the pressure gradient force is represented by p' instead of $\tilde{\pi}$, then Eqs. 6.40 and 6.42 can be written as

$$\frac{\partial \tilde{u}^1}{\partial t} + \tilde{u}^j \frac{\partial \tilde{u}^1}{\partial \tilde{x}^j} = -\frac{1}{\rho_0}\frac{\partial p'}{\partial \tilde{x}^1} + \frac{1}{\rho_0}\frac{\partial z_G}{\partial x}\frac{\partial p'}{\partial \tilde{x}^3} \tag{6.63}$$

$$\frac{\partial \tilde{u}^3}{\partial t} + \tilde{u}^j \frac{\partial \tilde{u}^3}{\partial \tilde{x}^j} = +\frac{1}{\rho_0}\left\{\frac{\partial z_G}{\partial x}\frac{\partial p'}{\partial \tilde{x}^1} - \left[\left(\frac{\partial z_G}{\partial x}\right)^2 + 1\right]\frac{\partial p'}{\partial \tilde{x}^3}\right\} + g\frac{\theta'}{\theta_0} \tag{6.64}$$

This type of coordinate transformation has considerable utility because a type of hydrostatic assumption can be assumed valid in the $\tilde{x}^3$ direction but accelerations are still explicitly resolved in the terrain-parallel direction, which for nonzero slope, has a component in the vertical direction. This form of hydrostatic representation is different from that suggested by Mahrt (1982), referenced previously.

If the assumption is made in the $\tilde{x}^3$ direction that $d\tilde{u}^3/dt$ is small relative to the other terms (i.e., a generalized hydrostatic assumption as discussed in Section 6.3.2), Eqs. 6.63 and 6.64 reduce to

$$\frac{\partial \tilde{u}^1}{\partial t} + \tilde{u}^1 \frac{\partial \tilde{u}^1}{\partial \tilde{x}^1} + \tilde{u}^3 \frac{\partial \tilde{u}^1}{\partial \tilde{x}^3} = -\frac{1}{\rho_0}\frac{\partial p'}{\partial \tilde{x}^1} + \frac{1}{\rho_0}\frac{\partial z_G}{\partial x}\frac{\partial p'}{\partial \tilde{x}^3} \tag{6.65}$$

$$\frac{\partial p'}{\partial \tilde{x}^3} = \frac{1}{\left[\left(\frac{\partial z_G}{\partial x}\right)^2 + 1\right]}\left(\frac{g\theta'}{\theta_0}\rho_0 + \frac{\partial z_G}{\partial x}\frac{\partial p'}{\partial \tilde{x}^1}\right) \tag{6.66}$$

Inserting Eq. 6.66 into 6.63 and rearranging yields

$$\frac{\partial \tilde{u}^1}{\partial t} + \tilde{u}^1 \frac{\partial \tilde{u}^1}{\partial \tilde{x}^1} + \tilde{u}^3 \frac{\partial \tilde{u}^1}{\partial \tilde{x}^3} = -\frac{1}{\rho_0}\frac{\partial p'}{\partial \tilde{x}^1} \times \left[1 - \frac{(\partial z_G/\partial x)^2}{[(\partial z_G/\partial x)^2 + 1]}\right] + \frac{\partial z_G/\partial x}{[(\partial z_G/\partial x)^2 + 1]} g\frac{\theta'}{\theta_0},$$

or

$$\frac{\partial \tilde{u}^1}{\partial t} = -\frac{1}{\rho_0}\frac{\partial p'}{\partial \tilde{x}^1}\left(1 - \frac{\alpha^2}{1+\alpha^2}\right) + g\frac{\alpha}{(\alpha^2+1)}\frac{\theta'}{\theta_0} - \tilde{u}^1\frac{\partial \tilde{u}^1}{\partial \tilde{x}^1} - \tilde{u}^3\frac{\partial \tilde{u}^1}{\partial \tilde{x}^3}, \tag{6.67}$$

[6]A straightforward assumption for the subgrid-scale fluxes in the vertical direction in Eq. 6.65 for use in an analytic model could be written as $-\overline{\tilde{u}^{3''}\frac{\partial \tilde{u}^{1''}}{\partial \tilde{x}^3}} = \frac{\partial}{\partial \tilde{x}^3}\left(K\frac{\partial \tilde{u}^1}{\partial \tilde{x}^3}\right)$. Integrating this expression vertically between the surface and the top of a drainage flow, h, yields $\int_{\tilde{x}^3=0}^{\tilde{x}^3=h} \overline{\tilde{u}^{3''}\frac{\partial \tilde{u}^{1''}}{\partial \tilde{x}^3}}\, d\tilde{x}^3 = K\left.\frac{\partial \tilde{u}^1}{\partial \tilde{x}^3}\right|_{\text{at } \tilde{x}^3=0}^{\text{at } \tilde{x}^3=h}$ which could be used in a layered model of drainage flow; $K[\partial \tilde{u}^1/\partial \tilde{x}^3]$ at $\tilde{x}^3 = 0$ could be approximated as $C_D(\tilde{u}^1)^2$, for example, while $K[\partial \tilde{u}^1/\partial \tilde{x}^3]$ at h could be used to represent entrainment at the top of the drainage flow. An additional advantage of using Transformation I is that the integration is in the vertical direction rather than in the direction perpendicular to the $\tilde{x}^3$ surface.

where $\alpha = \partial z_G/\partial x$ has been defined for convenience. For flat terrain, this relation reduces to the original Cartesian horizontal equation-of-motion.

6.4.2 Transformation II

Using the definition of $\widetilde{G}^{ik}$, $\partial\tilde{x}^i/\partial z$ and covariant differentiation, and Transformation II, we obtain

$$G^{ik} = \begin{bmatrix} 1 & 0 \\ 0 & 1 \end{bmatrix}; \quad \frac{\partial\tilde{x}^i}{\partial z} = (\sin\gamma, \cos\gamma).$$

Using Eq. 6.38 in which the same assumptions used to derive Eqs. 6.63 and 6.64 are applied, the simplified equations can be written as

$$\frac{\partial\tilde{u}^1}{\partial t} + \tilde{u}^k\frac{\partial\tilde{u}^1}{\partial\tilde{x}^k} = -\frac{1}{\rho_0}\frac{\partial p'}{\partial\tilde{x}^1} + g\frac{\theta'}{\theta_0}\sin\gamma \tag{6.68}$$

$$\frac{\partial\tilde{u}^3}{\partial t} + \tilde{u}^k\frac{\partial\tilde{u}^3}{\partial\tilde{x}^k} = -\frac{1}{\rho_0}\frac{\partial p'}{\partial\tilde{x}^3} + g\frac{\theta'}{\theta_0}\cos\gamma \tag{6.69}$$

If a hydrostatic-type assumption is made in the $\tilde{x}^3$ direction, as suggested by Mahrt (1982) for sufficiently small slopes, Eqs. 6.68 and 6.69 reduce to

$$\frac{\partial\tilde{u}^1}{\partial t} + \tilde{u}^k\frac{\partial\tilde{u}^1}{\partial\tilde{x}^k} = -\frac{1}{\rho_0}\frac{\partial p'}{\partial\tilde{x}^1} + g\frac{\theta'}{\theta_0}\sin\gamma, \tag{6.70}$$

$$\frac{\partial p'}{\partial\tilde{x}^3} = \rho_0 g\frac{\theta'}{\theta_0}\cos\gamma. \tag{6.71}$$

To quantitatively compare the result of differences in the two transformations, we can develop a layer-integrated momentum and thermodynamic system for a uniform slope configuration and solve it analytically. The layer-integrated thermodynamic equation can be expressed in the general system by

$$\frac{\partial\bar{\theta}}{\partial t} + \tilde{u}^k\frac{\partial\bar{\theta}}{\partial\tilde{x}^k} = \bar{L}_c, \tag{6.72}$$

where $\bar{L}_c$ is the local warming rate for the integrated layer including both radiative and turbulent processes. Using $\bar{\theta} = \theta_0 + \theta'$ where θ_0 is a function of z only, and assuming $\partial\theta_0/\partial t = 0$, then

$$\frac{\partial\theta'}{\partial t} = -\left(\tilde{u}^1\frac{\partial\theta_0}{\partial\tilde{x}^1} + \tilde{u}^1\frac{\partial\theta'}{\partial\tilde{x}^1} + \tilde{u}^3\frac{\partial\theta'}{\partial\tilde{x}^3}\right) + \bar{L}_c. \tag{6.73}$$

For an infinite uniform slope, $\partial\theta'/\partial\tilde{x}^1 = 0$ and mass continuity requires that $\tilde{u}^3 = 0$. Thus using the chain rule

$$\frac{\partial\theta_0}{\partial\tilde{x}^1} = \frac{\partial\theta_0}{\partial z}\frac{\partial x^3}{\partial\tilde{x}^1}, \tag{6.74}$$

so that Eq. 6.73 can be written as

$$\frac{\partial \theta'}{\partial t} = -\tilde{u}^1 \frac{\partial \theta_0}{\partial x^3} \frac{\partial x^3}{\partial \tilde{x}^1} + \bar{L}_c. \tag{6.75}$$

For Transformation I, the generalized hydrostatic system for the uniform slope is

$$\frac{\partial \tilde{u}^1}{\partial t} = g \frac{\tan \gamma}{\left(\tan^2 \gamma + 1\right)} \frac{\theta'}{\theta_0} \tag{6.76}$$

$$\frac{\partial \theta'}{\partial t} = -\tilde{u}^1 \beta \tan \gamma + \bar{L}_c, \tag{6.77}$$

where $\tilde{u}^1$ and θ' represent layer quantities, $\tan \gamma = \partial z_G/\partial x$, $\tan \gamma = \partial x^3/\partial \tilde{x}^1 = \partial z/\partial \tilde{x}^1$ and $\beta = \partial \theta_0/\partial z$. Since the slope is uniform, $\partial p'/\partial \tilde{x}^1 = 0$ is assumed. Following McNider (1982) take $\partial/\partial t$ of Eq. 6.76 and substitute for $\partial \theta'/\partial t$ giving

$$\frac{\partial^2 \tilde{u}^1}{\partial t^2} = \frac{g}{\theta_0} \frac{\tan \gamma}{\left(\tan^2 \gamma + 1\right)} \left(-\tilde{u}^1 \beta \tan \gamma + \bar{L}_c\right) \tag{6.78}$$

or

$$\frac{\partial^2 \tilde{u}^1}{\partial t^2} + \frac{g}{\theta_0} \beta \sin^2 \gamma \tilde{u}^1 - \frac{g}{\theta_0} \cos \gamma \sin \gamma \bar{L}_c = 0. \tag{6.79}$$

Since the slope is uniform and represents a single layer, the equation is an ordinary differential equation which becomes

$$\frac{d^2 \tilde{u}^1}{dt^2} + \frac{g}{\theta_0} \beta \sin^2 \gamma \tilde{u}^1 - \frac{g}{\theta_0} \cos \gamma \sin \gamma \bar{L}_c = 0. \tag{6.80}$$

In a similar manner, the differential equation for Transformation II is

$$\frac{d^2 \hat{u}^1}{dt^2} + \frac{g}{\theta_0} \beta \sin^2 \gamma \hat{u}^1 - \frac{g}{\theta_0} \sin \gamma \bar{L}_c = 0, \tag{6.81}$$

where a caret rather than a tilde is placed over u^1 to indicate that $\tilde{u}^1$ and $\hat{u}^1$ are different velocities. Note that the difference in Eqs. 6.80 and 6.81 is a $\cos \gamma$ coefficient in the last term in Eq. 6.80 so that the variation in the two formulations increases for increasing slope angles. For initial conditions

$$\hat{u}^1 = \tilde{u}^1 = 0; \quad \frac{d\hat{u}^1}{dt} = \frac{d\tilde{u}^1}{dt} = 0$$

the solution for Eq. 6.80 (Transformation I) becomes

$$\tilde{u}^1 = \frac{\bar{L}_c}{\beta \tan \gamma} (1 - \cos \tau t) \tag{6.82}$$

Table 6.1 Comparison of values of $\tan\gamma$ and $\sin\gamma$ for several different slope angles, γ.

γ	$\sin\gamma$	$\tan\gamma$
5°	0.09	0.09
10°	0.17	0.18
15°	0.26	0.27
20°	0.34	0.36
25°	0.42	0.47
30°	0.50	0.58
35°	0.57	0.70
40°	0.64	0.84
45°	0.71	1.00

where

$$\tau^2 = \frac{g\beta}{\theta_0}\sin^2\gamma$$

Likewise, the solution for Eq. 6.81 (Transformation II) is

$$\hat{u}^1 = \frac{\bar{L}_c}{\beta\sin\gamma}(1 - \cos\hat{\tau}t) \tag{6.83}$$

with

$$\hat{\tau}^2 = \frac{g\beta}{\theta_0}\sin^2\gamma,$$

where τ and $\hat{\tau}$ are the period of oscillation in the frictionless results.

Relative values of $\hat{u}^1$ and $\tilde{u}^1$, for both transformations, are tabulated for several slope angles in Table 6.1. The terrain-following formulation (Transformation I) should be more realistic since nonhydrostatic accelerations along the terrain slope can be represented. Nonetheless, slopes greater than about 20° would be required before the difference in the two velocities would become greater than 6%. The oscillation periods, τ and $\hat{\tau}$ are identical with the two coordinate transformations, however.

6.5 Summary

Based on the derivations in this chapter, the conservation equations can be written for any vertical coordinate system, σ

$$\begin{aligned}\frac{\partial\bar{\tilde{u}}^1}{\partial t} &= -\bar{\tilde{u}}^j\frac{\partial\bar{\tilde{u}}^1}{\partial\tilde{x}^j} - \overline{\tilde{u}^{j''}\frac{\partial\tilde{u}^{1''}}{\partial\tilde{x}^j}} - \bar{\theta}\frac{\partial\bar{\pi}}{\partial\tilde{x}^1} - \bar{\theta}\frac{\partial\sigma}{\partial x}\frac{\partial\bar{\pi}}{\partial\tilde{x}^3} \\ &\quad - \hat{f}\left(\frac{\partial h}{\partial\tilde{x}^1}\bar{\tilde{u}}^1 + \frac{\partial h}{\partial\tilde{x}^2}\bar{\tilde{u}}^2 + \frac{\partial h}{\partial\tilde{x}^3}\bar{\tilde{u}}^3\right) + f\bar{\tilde{u}}^2\end{aligned} \tag{6.84}$$

$$\frac{\partial \bar{\tilde{u}}^2}{\partial t} = -\bar{\tilde{u}}^j \frac{\partial \bar{\tilde{u}}^2}{\partial \tilde{x}^j} - \overline{\tilde{u}^{j''} \frac{\partial \tilde{u}^{2''}}{\partial \tilde{x}^j}} - \bar{\theta} \frac{\partial \bar{\pi}}{\partial \tilde{x}^2} - \bar{\theta} \frac{\partial \sigma}{\partial y} \frac{\partial \bar{\pi}}{\partial \tilde{x}^3} - f \bar{\tilde{u}}^1 \tag{6.85}$$

$$\begin{aligned} \frac{\partial \bar{\tilde{u}}^3}{\partial t} = {} & -\bar{\tilde{u}}^j \frac{\partial \bar{\tilde{u}}^3}{\partial \tilde{x}^j} - \overline{\tilde{u}^{j''} \frac{\partial \tilde{u}^{3''}}{\partial \tilde{x}^j}} - \overline{\tilde{\Gamma}}^3_{jl} \bar{\tilde{u}}^j \bar{\tilde{u}}^l - \overline{\tilde{\Gamma}}^3_{jl} \overline{\tilde{u}^{j''} \tilde{u}^{l''}} \\ & - \bar{\theta} \left\{ \frac{\partial \sigma}{\partial x} \frac{\partial \bar{\pi}}{\partial \tilde{x}^1} + \frac{\partial \sigma}{\partial y} \frac{\partial \bar{\pi}}{\partial \tilde{x}^2} + \left[\left(\frac{\partial \sigma}{\partial x} \right)^2 + \left(\frac{\partial \sigma}{\partial y} \right)^2 + \left(\frac{\partial \sigma}{\partial z} \right)^2 \right] \frac{\partial \bar{\pi}}{\partial \tilde{x}^3} \right\} \\ & + \left(\hat{f} + \frac{\partial h}{\partial \tilde{x}^2} f \right) \frac{\partial \sigma}{\partial z} \left[\left(1 + \left(\frac{\partial h}{\partial \tilde{x}^1} \right)^2 \right) \bar{\tilde{u}}^1 + \frac{\partial h}{\partial \tilde{x}^1} \frac{\partial h}{\partial \tilde{x}^2} \bar{\tilde{u}}^2 + \frac{\partial h}{\partial \tilde{x}^1} \frac{\partial h}{\partial \tilde{x}^3} \bar{\tilde{u}}^3 \right] \\ & - \frac{\partial h}{\partial \tilde{x}^1} f \frac{\partial \sigma}{\partial z} \left[\frac{\partial h}{\partial \tilde{x}^2} \frac{\partial h}{\partial \tilde{x}^1} \bar{\tilde{u}}^1 + \left(1 + \left(\frac{\partial h}{\partial \tilde{x}^2} \right)^2 \right) \bar{\tilde{u}}^2 + \frac{\partial h}{\partial \tilde{x}^2} \frac{\partial h}{\partial \tilde{x}^3} \bar{\tilde{u}}^3 \right] - \frac{\partial \sigma}{\partial z} g. \end{aligned} \tag{6.86}$$

$$\frac{\partial \bar{\theta}}{\partial t} = -\bar{\tilde{u}}^j \frac{\partial \bar{\theta}}{\partial \tilde{x}^j} - \overline{\tilde{u}^j \frac{\partial \theta''}{\partial \tilde{x}^j}} + \overline{\tilde{S}}_\theta \tag{6.87}$$

$$\frac{\partial \bar{q}_n}{\partial t} = -\bar{\tilde{u}}^j \frac{\partial \bar{q}_n}{\partial \tilde{x}^j} - \overline{\tilde{u}^{j''} \frac{\partial q''_n}{\partial \tilde{x}^j}} + \overline{\tilde{S}}_{q_n}, \quad n = 1, 2, 3 \tag{6.88}$$

$$\frac{\partial \bar{\chi}_m}{\partial t} = -\bar{\tilde{u}}^j \frac{\partial \bar{\chi}_m}{\partial \tilde{x}^j} - \overline{\tilde{u}^{j''} \frac{\partial \chi''_m}{\partial \tilde{x}^j}} + \overline{\tilde{S}}_{\chi_m}, \quad m = 1, 2, ..., M \tag{6.89}$$

$$\frac{\partial \bar{\rho}}{\partial t} = -\frac{\partial \sigma}{\partial z} \frac{\partial}{\partial \tilde{x}^j} \left(\frac{\partial h}{\partial \tilde{x}^3} \overline{\rho \tilde{u}^j} \right) \tag{6.90}$$

$$\bar{p} = \bar{\rho} R_{\mathrm{d}} \overline{T}_{\mathrm{v}} \tag{6.91}$$

$$\bar{\theta} = \overline{T}_{\mathrm{v}} (1000 \text{ mb}/p)^{R_{\mathrm{d}}/C_{\mathrm{p}}} \tag{6.92}$$

Equations 6.84-6.86 are the conservation-of-velocity equations. Equation 6.87 is the conservation-of-heat equation, Eq. 6.88 is the conservation-of-water substance equation, Eq. 6.89 is the conservation-of-other atmospheric gases and aerosols equation, and Eq. 6.90 is the conservation-of-mass of the air equation. Equation 6.91 is the ideal gas law, which is the equation-of-state for the air, while Eq. 6.92 is the definition of potential temperature. The overbar in these equations is, of course, the generalized form of the grid-volume average, as defined by Eq. 6.37. In the derivation of these equations, the assumptions are made that $(|\theta''|/\bar{\theta}) \ll 1$, $|\overline{\tilde{u}^{i''}}| = 0$, and $\overline{\tilde{G}^{ij}} = \tilde{G}^{ij}$. To apply these equations in a model, they can be manipulated in various ways. However, it must always be possible to work backwards to these fundamental conservation relationships.

Among the standard approximations used in these equations is the replacement of Eq. 6.86 with

$$\frac{\partial \bar{\pi}}{\partial \tilde{x}^3} = -\frac{g}{\bar{\theta}}\left(\frac{\partial \sigma}{\partial z}\right)^{-1} \tag{6.93}$$

where all vertical acceleration terms and terms involving f and $\hat{f}$ are ignored, and $|\partial\sigma/\partial z| \gg |\partial\sigma/\partial x| \sim |\partial\sigma/\partial y|$ is assumed. This makes Eqs. 6.84-6.85, 6.87-6.92, and 6.93 a hydrostatic set of nonlinear partial differential equations. Another common assumption (the "anelastic approximation" is introduced in Section 3.1.1) to replace Eq. 6.90 with Eq. 6.44.

6.6 Application of Terrain-Following Coordinate Systems

Unfortunately, the use of terrain-following coordinate systems introduces several new computational problems to accurate numerical modeling. These can be summarized as follows:

1. The two terms that represent the pressure gradient force along a σ surface, for example (e.g., see Eqs. 6.84 and 6.85) represent differences between large terms. Slight errors in their definition (see Mahrer and Pielke 1977a; Fortunato and Baptista 1996) can introduce significant errors in the model. Sun (1995) suggests using a reference local vertical pressure gradient from which a perturbation pressure gradient in the σ-system is computed; however, two terms (albeit smaller) still appear.
2. When the vertical grid increments ($\Delta\sigma$) are much smaller than the grid increments Δx and Δy, the interpolation needed to define gradients in the x and y directions introduce significant errors, as shown by Mahrer (1984). There is no known solution for this problem, except to limit how small the ratio of $\Delta\sigma/\Delta x$ and $\Delta\sigma/\Delta y$ can be. Mahrer (1984) concludes that $\Delta\sigma \geq \frac{\sigma \Delta z_G}{(s - z_G)}$ is needed where Δz_G is the change in terrain height across one horizontal grid interval).
3. The subgrid-scale flux terms along a σ-surface produce fluxes in the vertical ($\vec{k}$) direction. This is not desirable since turbulence fluxes (Chapter 7) are represented separately in the vertical and horizontal directions.

Mesoscale models, of course, have been developed which successfully utilize terrain-following coordinate system, as summarized in Appendix B. To take advantage of the terrain-following coordinate representation and still retain some of the benefits of a Cartesian coordinate system, models such as RAMS (Pielke et al. 1992; Cotton et al. 2003) interpolate the dependent variables to a Cartesian gradient so that horizontal gradients, rather than σ-parallel gradients are calculated for use in the numerical form of the conservation equations. However, problems such as 2 and 3 listed above still appear as a result of the interpolation.

As another approach, the eta-coordinate system, mass vertical coordinate, or the shaved-coordinate system[7] can be retained for the dynamics of the model (the pressure gradient force, advection, the Coriolis term), while the model parameterizations (Chapters 7 and 8) are evaluated on a Cartesian grid. Since all of the parameterizations are expressed in a one-dimensional framework, there is no need to utilize a terrain-following coordinate system. There is only the need to define the bottom of the parameterization as the Earth surface (e.g., $z_{\text{surface}} = z_G$; with $\tilde{z} = z - z_G$ being the Cartesian height above the surface).

The insertion of at least some subgrid-scale topographic effects into the models (i.e., $z''_G(x)$) has, as reported by Marty Bell (2000, ASTeR - Mission Research Corporation, personal communication), and based in part on the ideas of Roger Ridley, used averaging, silhouette topography, envelope topography (Wallace et al. 1983), or reflected envelope topography to represent these smaller-scale effects. With the averaging approach, the average grid area terrain height is used. The silhouette method fills in the valleys (i.e., $z''_G(x) \leq 0$ within the grid) in computing the grid area-averaged $\bar{z}_G$. The envelope terrain representation uses the grid area-averaged terrain height plus a user-defined variation (such as σ_{z_G}) within the grid area to compute the z_G value applied for that grid point. The reflected envelope topography, which has been applied in RAMS, is the envelope scheme mirrored about the local mean topography heights. If it is a local maximum or minimum, insert the user-defined variation. None of these approaches are completely adequate forms to represent the appropriate grid volume-averaging effect (Eq. 6.37), but is an improvement over not considering subgrid-scale terrain effects at all.

Problems for Chapter 6

1. Show that for flat terrain, the equations listed in Section 6.5 reduce to the grid-volume averaged form of the conservation equations presented in Chapter 4.
2. Substitute one of the specific terrain-following coordinate systems listed at the beginning of Section 6.2 and write the form of each of the conservation equations listed in Chapter 4.
3. Select an atmospheric model which uses a generalized vertical coordinate system (e.g., WRF, RAMS) and use their form of σ to derive their specific form of the equations in Section 6.5.
4. For the model in Problem 3, write the conservation equations that are used for flat terrain. Identify (i) the advective terms, (ii) the vertical subgrid-scale flux terms, (iii) the horizontal subgrid-scale flux terms, (iv) the pressure gradient force term, and (v) the Coriolis terms.

[7] Adcroft et al. (1997) and Marshall et al. (1997) use partial grid volumes at the bottom of their model (called "shaved cells") in an ocean model. Shaved models permit sloping terrain, yet retain a Cartesian framework.

Chapter 6 Additional Readings

Byun, D.W., 1999a: Dynamically consistent formulations in meteorological and air quality models for multiscale atmospheric studies. Part I: Governing equations in a generalized coordinate system. *J. Atmos. Sci.*, **56**, 3789-3807.

Byun, D.W., 1999b: Dynamically consistent formulations in meteorological and air quality models for multiscale atmospheric studies. Part II: Mass conservation issues. *J. Atmos. Sci.*, **56**, 3808-3820.

These papers provide an effective overview of terrain-following coordinate transformations, including excellent figures which illustrate different forms of the transformations.

Dutton, J.A., 1976: *The Ceaseless Wind: An Introduction to the Theory of Atmospheric Motion.* McGraw Hill, New York, 579 pp.

Pages 129-144 and 248-251 were particularly useful in preparing this chapter.

Dutton, J.A., 2002: *The Ceaseless Wind: An Introduction to the Theory of Atmospheric Motion.* Courier Dover Publications, 640 pp.

Dutton provides an outstanding review of tensor analysis and generalized vertical coordinate representations.

Gal-Chen, T., and R.C.J. Somerville, 1975a: On the use of a coordinate transformation for the solution of the Navier-Stokes equations. *J. Comput. Phys.* **17**, 209-228.

This paper provides an excellent overview of the application of a specific terrain-following coordinate system for use in a nonhydrostatic meteorological model. The material on pages 215-219 was referred to frequently during the writing of this chapter.

Several other references which provide supplemental information on the material presented in this chapter are listed below.

Clark, T.L., 1977: A small-scale dynamic model using a terrain-following coordinate transformation. *J. Comput. Phys.* **24**, 186-215.

Kasahara, A., 1974: Various vertical coordinate systems used for numerical weather prediction. *Mon. Wea. Rev.* **102**, 509-522.

Lapidus, A., 1967: A detached shock calculation by second-order finite difference. *J. Comput. Phys.* **2**, 154-177.

Laprise, 1992a: The Euler equations of motion with hydrostatic pressure as independent variable. *Mon. Wea. Rev.*, **120**, 197-207.

Phillips, N.A., 1957: A coordinate system having some special advantages for numerical forecasting. *J. Meteor.* **14**, 184-185.

Skamarock, W.C., J.B. Klemp, J. Dudhia, D.O. Gill, D.M. Barker, W. Wang, and J.G. Powers, 2008: A description of the Advanced Research WRF Version 3. NCAR Technical Note TN-468+STR. 113 pp.

CHAPTER

Traditional Parameterizations 7

CHAPTER OUTLINE HEAD

Mesoscale Meteorological Modeling, Third Edition, Volume 98. http://dx.doi.org/10.1016/B978-0-12-385237-3.00007-4

7.1 Introduction

The grid-volume averaging of the conservation relations as described in Chapters 4 and 6 results in averaged subgrid-scale correlation terms (e.g., $\rho_0 \overline{u''_j u''_i}$ from Eq. 4.21) and averaged source/sink terms (e.g., $\bar{S}_\theta$ from Eq. 4.24).

In this chapter, the representation in mesoscale models of three types of physical processes are introduced. This specification of subgrid-scale and source/sink processes using experimental data and simplified fundamental concepts is called *parameterization.* Usually the parameterizations are not defined in terms of basic conservation principles. A parameterization does not necessarily have to actually simulate the physical processes it is representing, in order to be a realistic representation of these terms.

Indeed, if the quantitative accuracy of a parameterization is not sacrificed, it is desirable to make the parameterization as computationally simple as possible. The three processes to be parameterized are:

1. averaged subgrid-scale fluxes (i.e., $\rho_0 \overline{u''_j u''_i}$, $\rho_0 \overline{u''_j \theta''}$, etc. in Eqs. 4.21 and 4.24-4.26),
2. averaged radiation flux divergence (i.e., part of $\bar{S}_\theta$ in Eq. 4.24), and
3. averaged effects of the change-of-phase of water, including precipitation (i.e., $\bar{S}_{q_n}$ in Eq. 4.25, part of $\bar{S}_\theta$ in Eq. 4.24).

The averaged effects of change-of-phase, precipitation, and/or change into other chemical species of atmospheric gases and aerosols other than water (i.e., S_{χ_m} in Eq. 4.26) is not covered in this text and the reader is referred to texts such as Seinfeld (1975) and Seinfeld and Pandis (1997) for a report on the status of parameterizing these complex effects.

In this section, the parameterization of the averaged vertical subgrid fluxes is described. As discussed in Section 9.4, horizontal subgrid-scale fluxes are utilized only for computational reasons since little is known of horizontal subgrid-scale mixing on the mesoscale (or in other models whenever $\Delta x; \Delta y \gg \Delta z$), although work for ocean mixing such as that of Young et al. (1982) offers an avenue for research.

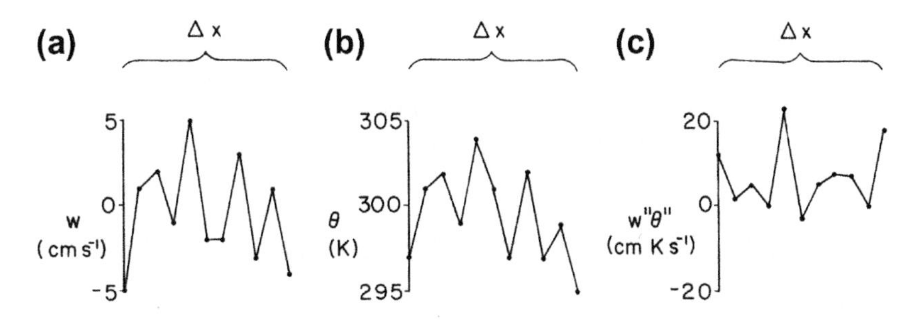

FIGURE 7.1

Schematic illustration of subgrid-scale values of vertical velocity w, potential temperature θ, and the subgrid-scale correlation $w''\theta''$. In this example, the grid-averaged value of vertical motion is required to be approximately zero (i.e., $\bar{w} \simeq 0$), and $\bar{\theta} = 299.5$ K is used. Both grid value averages are assumed constant over Δx. The grid-averaged subgrid-scale correlation $\overline{w''\theta''}$ is equal to 6.9 cm K s^{-1}.

As discussed in Chapter 4, the magnitude of the subgrid-scale variables and fluxes can often be the same or even larger that the resolvable dependent variables. A wind gust of 5 m s^{-1} (representing u''), for example, is not uncommon with an average wind speed of 5 m s^{-1} (representing $\bar{u}$). Figure 7.1 schematically illustrates a subgrid-scale correlation between vertical velocity and potential temperature. In this example, assumed close to flat ground so that the grid-volume averaged vertical velocity is approximately zero (i.e., $\bar{w} \simeq 0$), the ground surface is assumed warmer than the air above so that an upward perturbation vertical velocity tends to transport warm air up, whereas descending motion tends to advect cooler air down.[1] Averaging over the grid interval in this example yields an upward flux of heat ($\overline{w''\theta''} > 0$ with a magnitude of 6.9 cm K s^{-1}). Thus despite the insignificant vertical flux of heat associated with the resolvable dependent variables (i.e., $\bar{w}\bar{\theta} \simeq 0$ since $\bar{w} \simeq 0$), a substantial transport of heat will occur because of the positive correlation between the subgrid-scale vertical velocity and potential temperature perturbation.

In developing subgrid-scale averaged quantities, however, it needs to be recognized that the preferred representation is an ensemble average over the grid volume rather than simply the grid-volume average as defined by Eq. 4.6. The ensemble average represents the most likely value of the subgrid-scale quantity, whereas the grid-volume average represents just one realization. Unless the subgrid-scale quantity is completely deterministic (i.e., without a statistical component), the two averages will not, in general, be the same. In the parameterizations discussed in this chapter, therefore, it is assumed that they are the most likely (i.e., ensemble) estimates. Wyngaard (1982, 1983) and Cotton and Anthes (1989) discuss ensemble averaging in

[1] In this example, and in actual measurements, upward motion does not always transport warmer air aloft, even if the ground is warmer, because cooler air mixed downward at an earlier time or different location may be entrained in an upward moving region.

more depth. Defining a parameterization in terms of a realization from a probability distribution is an opportunity area for future research. Preliminary work in this area has been completed, as reported in Garratt and Pielke (1989), Garratt et al. (1990), and Avissar (1991, 1992), but much more could be done in this research area.

7.2 Parameterization of Subgrid-Scale Averaged Flux Divergence

7.2.1 Basic Terms

To develop parameterization for these subgrid-scale correlations, it is necessary to introduce several basic definitions. In this analysis, to simplify the interpretation, a Cartesian coordinate framework will be applied. The modifications when a generalized vertical coordinate system is used is discussed in Chapter 6.

Neglecting the Coriolis effect, Eq. 4.4 can be rewritten as

$$\frac{\partial}{\partial t}\left(\bar{u}_i + u_i''\right) = -\left(\bar{u}_j + u_j''\right)\frac{\partial}{\partial x_j}\left(\bar{u}_i + u_i''\right) - \left(\theta_0 + \theta' + \theta''\right)\frac{\partial\left(\bar{\pi} + \pi''\right)}{\partial x_i} - g\delta_{i3}, \tag{7.1}$$

where Eq. 4.34 is used to represent the pressure gradient force with θ and π decomposed using the definitions given by Eqs. 4.3 and 4.12. Assuming the synoptic-scale variables are in hydrostatic equilibrium, and fluctuations in potential temperature (i.e., θ' and θ'') are neglected relative to θ_0 except when multiplied by gravity,[2] Eq. 7.1 can be rewritten as

$$\begin{aligned}\frac{\partial}{\partial t}\left(\bar{u}_i + u_i''\right) = &-\left(\bar{u}_j + u_j''\right)\frac{\partial}{\partial x_j}\left(\bar{u}_i + u_i''\right) - \theta_0\frac{\partial\left(\pi' + \pi''\right)}{\partial x_i}\\ &- \theta_0\left[\frac{\partial \pi_0}{\partial x}\delta_{i1} + \frac{\partial \pi_0}{\partial y}\delta_{i2}\right] + \frac{g\theta'}{\theta_0}\delta_{i3} + \frac{g\theta''}{\theta_0}\delta_{i3}.\end{aligned} \tag{7.2}$$

Averaging this equation over a grid volume using Eq. 4.6 and applying the assumptions given by Eq. 4.8 yields

$$\frac{\partial \bar{u}_i}{\partial t} = -\bar{u}_j\frac{\partial}{\partial x_j}\bar{u}_i - \overline{u_j''\frac{\partial}{\partial x_j}u_i''} - \theta_0\frac{\partial \pi'}{\partial x_i} - \theta_0\left[\frac{\partial \pi_0}{\partial x}\delta_{i1} + \frac{\partial \pi_0}{\partial y}\delta_{i2}\right] + g\frac{\theta'}{\theta_0}\delta_{i3}. \tag{7.3}$$

Subtracting Eq. 7.3 from 7.2 gives

$$\frac{\partial u_i''}{\partial t} = -\bar{u}_j\frac{\partial}{\partial x_j}u_i'' - u_j''\frac{\partial u_i''}{\partial x_j} - u_j''\frac{\partial \bar{u}_i}{\partial x_j} + \overline{u_j''\frac{\partial u_i''}{\partial x_j}} - \theta_0\frac{\partial \pi''}{\partial x_i} + g\frac{\theta''}{\theta_0}\delta_{i3}, \tag{7.4}$$

which is a prognostic equation for the subgrid-scale velocity perturbation.

[2] This is essentially the Boussinesq approximation. See Chapter 4, following Eq. 4.15 for a description of this assumption.

Multiplying Eq. 7.4 by u_i'', averaging using Eq. 4.6, and applying the assumption[3] $\overline{u_i''} = 0$ results in

$$\frac{\partial \bar{e}}{\partial t} = -\bar{u}_j \frac{\partial \bar{e}}{\partial x_j} - \overline{u_j'' \frac{\partial e}{\partial x_j}} - \overline{u_j'' u_i''} \frac{\partial \bar{u}_i}{\partial x_j} - \theta_0 \overline{u_i'' \frac{\partial \pi''}{\partial x_i}} + g \frac{\overline{u_i'' \theta''}}{\theta_0} \delta_{i3}, \tag{7.5}$$

where $\bar{e} = \frac{1}{2}\overline{u_i''^2}$ and $e = \frac{1}{2}u_i''^2$. In the context of a numerical model, Eq. 7.5 is the grid-volume averaged, subgrid-scale perturbation kinetic energy equation. This equation is usually referred to as the *turbulent kinetic energy equation*,[4] with $\bar{e}$ called the average turbulent kinetic energy.[5] The individual terms in Eq. 7.5 have the interpretation given in Table 7.1.

If Eq. 7.5 were used to simulate the details of turbulence in a model, the molecular dissipation of average turbulent kinetic energy would be included to guarantee a sink for this energy. In a mesoscale model, however, as discussed in Section 9.4, computational devices such as horizontal filters are applied to prevent the artificial accumulation of kinetic energy at short wavelengths. Such mechanisms are necessitated by the inability to resolve both the mesoscale and the small spatial scales in which the molecular dissipation of kinetic energy becomes significant.[6]

To contrast the relative contribution of the two source/sink terms of $\bar{e}$ in Eq. 7.5, it is useful to define the ratio

$$R_f = \frac{g}{\theta_0}\overline{w''\theta''} \Bigg/ \left[\overline{w''u''}\frac{\partial \bar{u}}{\partial z} + \overline{w''v''}\frac{\partial \bar{v}}{\partial z}\right], \tag{7.6}$$

where R_f is called the *flux Richardson number*. In this expression, horizontal contributions to the shear production of $\bar{e}$ are neglected and $|\partial\bar{u}/\partial z| \simeq |\partial\bar{v}/\partial z| \gg |\partial\bar{w}/\partial z|$. The flux Richardson number is a measure of the relative contribution of the buoyant production or dissipation of averaged, subgrid-scale kinetic energy relative to its generation or extraction by the vertical shear of the averaged horizontal wind.

---

[3] The Reynolds assumption, see following Eq. 4.8.

[4] Although the reference to Eq. 7.5 as a turbulent kinetic energy equation is relatively standard, it is imprecise to do so since molecular viscosity (e.g., Eq. 3.29) was ignored in the original equations (e.g., Eq. 2.45). Therefore, molecular dissipation of turbulent energy is excluded in Eq. 7.5.

[5] It is important to note that the averaging operation given by Eq. 4.6 is not the same as used in turbulence theory. Standard turbulence observations involve measurements at specific points on a tower or along an aircraft track. In the former case, averaging is in time, whereas a one-dimensional space average is used for the second mode of observation. The parameterization of subgrid-scale fluxes, however, utilizes the results from these observational studies as is discussed in this chapter. Such an equivalence is justified only if the measured turbulence characteristics are essentially the same as occur throughout the averaging grid volume. Porch (1982) presents an example of a comparison of volume averaging (using an optical anemometer) and a point measurement (a cup anemometer) for a drainage flow observational study in a California valley. He concluded for that study that point measurements should be averaged for a relatively long time (~2 h) to represent more accurately the volume-averaged values. Unfortunately, of course, if the spatial variations are too large across the volume, no amount of time averaging at a point can provide the appropriate average.

[6] A discussion of the number of grid points required to resolve such a range of scales is given at the beginning of Chapter 4.

Table 7.1 An interpretation of the individual terms in Eq. 7.5.[a]

Term	Interpretation
$\frac{\partial \bar{e}}{\partial t}$	Local grid-volume change of averaged subgrid-scale perturbation kinetic energy $\bar{e}$.
$\bar{u}_j \frac{\partial \bar{e}}{\partial x_j}$	Advection of $\bar{e}$ by the grid-volume averaged velocity.
$\overline{u''_j \frac{\partial e}{\partial x_j}}$	Grid-volume averaged advection of e by the subgrid-scale perturbation velocity.
$\overline{u''_j u''_i} \frac{\partial \bar{u}_i}{\partial x_j}$	Extraction from or input to $\bar{e}$ due to the existence of both an average velocity shear and subgrid-scale velocity fluxes; also referred to as the *shear production* of turbulent kinetic energy.
$\theta_0 \overline{u''_i \frac{\partial \pi''}{\partial x_i}}$	Multiplying this term by ρ_0 and assuming that the anelastic conservation-of-mass equation is valid for the subgrid scale (i.e., $(\partial/\partial x_i)\rho_0 u''_i = 0$), yields $\theta_0(\partial/\partial x_i)\rho_0 \overline{u''_i \pi''}$. Therefore, when the anelastic assumption is valid, this term causes changes in $\bar{e}$ only by advection through the boundaries of the grid volume. As discussed by Lumley and Panofsky (1964), the influence of the correlation between the turbulent velocity and pressure variables is to transfer kinetic energy between the three velocity components.
$g \frac{\overline{u''_i \theta''}}{\theta_0} \delta_{i3} = g \frac{\overline{w'' \theta''}}{\theta_0}$	Extraction or production of $\bar{e}$ by buoyancy. Referred to as the *buoyant production* of turbulent kinetic energy.

[a] *Lumley and Panofsky (1964), from which the derivation of Eq. 7.5 was based, discuss the turbulent kinetic energy equation and its derivation in detail.*

In analogy with the molecular fluxes of heat and momentum (e.g., Eq. 3.29), the vertical subgrid-scale flux terms, $\overline{w''\theta''}$, $\overline{w''u''}$, and $\overline{w''v''}$ are often represented by

$$\overline{w''\theta''} = -K_\theta \frac{\partial \bar{\theta}}{\partial z}; \ \overline{w''q''_k} = -K_\theta \frac{\partial \bar{q}_k}{\partial z}; \quad \overline{w'\chi''_m} - K_\theta \frac{\partial \bar{\chi}_m}{\partial z};$$
$$\overline{w''u''} = -K_m \frac{\partial \bar{u}}{\partial z}; \quad \overline{w''v''} = -K_m \frac{\partial \bar{v}}{\partial z}, \tag{7.7}$$

(as assumed, for example, by Eq. 5.3) where K_θ and K_m are referred to as *exchange coefficients*. This form to represent the grid-volume subgrid-scale fluxes is referred to as *first-order closure*. It is important to note that although molecular mixing is a function of the type of fluid involved, turbulent mixing, as represented by Eq. 7.7, is a function of the flow. Therefore, the turbulent exchange coefficients K_θ and K_m, given in Eq. 7.7 are not constant in time or space. Moreover, the expressions given by Eq. 7.7 require that the subgrid-scale fluxes be *downgradient* as long as the exchange coefficients are positive. Moreover, *countergradient* turbulent fluxes are often observed (e.g., Deardorff 1966), such as on sunny days over land, or when cold air advects over warm lakes and oceans. Nonetheless, Eq. 7.7 has been shown to be a useful representation of subgrid-scale fluxes.

Substituting Eq. 7.7 into Eq. 7.6 yields

$$\begin{aligned} R_f &= K_\theta \frac{g}{\theta_0} \frac{\partial \bar{\theta}}{\partial z} \bigg/ K_m \left[\left(\frac{\partial \bar{u}}{\partial z} \right)^2 + \left(\frac{\partial \bar{v}}{\partial z} \right)^2 \right] \\ &= \frac{K_\theta}{K_m} \frac{g}{\theta_0} \frac{\partial \bar{\theta}}{\partial z} \bigg/ \left[\left(\frac{\partial \bar{u}}{\partial z} \right)^2 + \left(\frac{\partial \bar{v}}{\partial z} \right)^2 \right] = \frac{K_\theta}{K_m} \mathrm{Ri}, \end{aligned} \tag{7.8}$$

where Ri is called the *gradient Richardson number*. The sign of Ri is determined by the sign of the lapse rate of potential temperature. Therefore,

1. $\mathrm{Ri} > 0$ corresponds to $\partial \bar{\theta}/\partial z > 0$, which indicates a stably-stratified layer;
2. $\mathrm{Ri} = 0$ corresponds to $\partial \bar{\theta}/\partial z = 0$, which corresponds to neutral stratification; and
3. $\mathrm{Ri} < 0$ corresponds to $\partial \bar{\theta}/\partial z < 0$, which indicates an unstably-stratified layer.

Theory (e.g., Dutton 1976:79) indicates that when Ri is greater than 0.25, the stable stratification suppresses turbulence sufficiently so that the flow becomes laminar, even in the presence of mean wind shear. This value of Ri is called the *critical Richardson number*.

The unstable-stratified layer itself is broken down into two regions:

1. $|\mathrm{Ri}| \leq 1$, where the shear production of subgrid-scale kinetic energy is important (this regime is referred to as *forced convection*); and
2. $|\mathrm{Ri}| > 1$, where the shear production becomes unimportant relative to the buoyant product of subgrid-scale kinetic energy (this regime is called *free convection*).

The characteristic size of turbulent eddies in the atmosphere are larger during free convection than under forced convection.

Equation 7.4 can also be used to obtain prognostic conservation equations for the subgrid-scale fluxes. Multiplying Eq. 7.4 by u_k'' yields

$$\begin{aligned} u_k'' \frac{\partial u_i''}{\partial t} &= -u_k'' \bar{u}_j \frac{\partial}{\partial x_j} u_i'' - u_k'' u_j'' \frac{\partial u_i''}{\partial x_j} - u_k'' u_j'' \frac{\partial \bar{u}_i}{\partial x_j} \\ &\quad + u_k'' \overline{u_j'' \frac{\partial u_i''}{\partial x_j}} - u_k'' \theta_0 \frac{\partial \pi''}{\partial x_i} + g u_k'' \frac{\theta''}{\theta_0} \delta_{i3} \end{aligned} \tag{7.9}$$

Writing Eq. 7.4 with k as the free index, and then multiplying by u_i'' yields an equation for $u_i'' \partial u_k''/\partial t$. Adding that equation to Eq. 7.9 results in

$$\frac{\partial}{\partial t} \left(u_k'' u_i'' \right) = -\bar{u}_j \frac{\partial}{\partial x_j} u_k'' u_i'' - u_j'' \frac{\partial}{\partial x_j} u_k'' u_i''$$

$$- u_i'' u_j'' \frac{\partial \bar{u}_k}{\partial x_j} - u_k'' u_j'' \frac{\partial \bar{u}_i}{\partial x_j} + \overline{u_i'' u_j'' \frac{\partial}{\partial x_j} u_k''} + \overline{u_k'' u_j'' \frac{\partial u_i''}{\partial x_j}} \tag{7.10}$$
$$- u_i'' \theta_0 \frac{\partial \pi''}{\partial x_k} - u_k'' \theta_0 \frac{\partial \pi''}{\partial x_i} + g u_i'' \frac{\theta''}{\theta_0} \delta_{k3} + g u_k'' \frac{\theta''}{\theta_0} \delta_{i3}$$

Grid-volume averaging Eq. 7.10 using the assumptions given by Eq. 4.8 yields

$$\frac{\partial}{\partial t} \overline{u_k'' u_i''} = -\bar{u}_j \frac{\partial}{\partial x_j} \overline{u_k'' u_i''} - \overline{u_j'' \frac{\partial}{\partial x_j} u_k'' u_i''}$$
$$- \overline{u_i'' u_j''} \frac{\partial \bar{u}_k}{\partial x_j} - \overline{u_k'' u_j''} \frac{\partial \bar{u}_i}{\partial x_j} - \theta_0 \overline{u_i'' \frac{\partial \pi''}{\partial x_k}} \tag{7.11}$$
$$- \theta_0 \overline{u_k'' \frac{\partial \pi''}{\partial x_i}} + \frac{g}{\theta_0} \delta_{k3} \overline{u_i'' \theta''} + \frac{g}{\theta_0} \delta_{i3} \overline{u_k'' \theta''}$$

Prognostic subgrid-scale equations can also be obtained for $\overline{u_i'' \theta''}$, $\overline{u_i'' q_k''}$, and $\overline{u_i'' \chi_m''}$. To illustrate how these equations are derived, multiply Eq. 7.4 by θ'', which results in

$$\theta'' \frac{\partial u_i''}{\partial t} = -\theta'' \bar{u}_j \frac{\partial}{\partial x_j} u_i'' - \theta'' u_j'' \frac{\partial u_i''}{\partial x_j}$$
$$- \theta'' u_j'' \frac{\partial \bar{u}_i}{\partial x_j} + \theta'' \overline{u_j'' \frac{\partial u_i''}{\partial x_j}} - \theta'' \theta_0 \frac{\partial \pi''}{\partial x_i} + g \frac{\theta''^2}{\theta_0} \delta_{i3} \tag{7.12}$$

A prognostic equation for θ'' can be obtained in an analogous manner as used to obtain Eq. 7.4. Equation 2.44 for θ can be written as

$$\frac{\partial}{\partial t} \left(\bar{\theta} + \theta''\right) = -\left(\bar{u}_j + u_j''\right) \frac{\partial}{\partial x_j} \left(\bar{\theta} + \theta''\right) + S_\theta .$$

The source/sink term, S_θ, can be decomposed into a portion which is only dependent on grid-resolved quantities, which are defined here as $\bar{S}_\theta$ and a remainder which is also a function of subgrid-scale effects, S_θ''. Averaging this relation over a grid volume using Eq. 4.6 and applying the assumptions given by Eq. 4.8 yields

$$\frac{\partial \bar{\theta}}{\partial t} = -\bar{u}_j \frac{\partial}{\partial x_j} \bar{\theta} - \overline{u_j'' \frac{\partial \theta''}{\partial x_j}} + \bar{S}_\theta$$

Note that requiring $\overline{S_\theta''} \equiv 0$ assumes that the source/sink term has subgrid-scale effects which average to zero across the grid volume. This is an assumption that is likely to be often unrealistic, but will be assumed here in the derivation of θ''.

Subtracting the equation for $\frac{\partial \bar{\theta}}{\partial t}$ from $\frac{\partial \left(\bar{\theta} + \theta''\right)}{\partial t}$ yields

$$\frac{\partial \theta''}{\partial t} = -\bar{u}_j \frac{\partial \theta''}{\partial x_j} - u_j'' \frac{\partial}{\partial x_j} \left(\bar{\theta} + \theta''\right) + \overline{u_j'' \frac{\partial \theta''}{\partial x_j}} + S_\theta - S_\theta'' \tag{7.13}$$

Equation 7.13 can be multiplied by u_i'' and added to Eq. 7.12. Performing the grid-volume average to this equation, using the assumptions given by Eq. 4.8 results in

$$\begin{aligned}\frac{\partial}{\partial t}\overline{u_i''\theta''} &= -\bar{u}_j\frac{\partial}{\partial x_j}\overline{u_i''\theta''} - \overline{u_j''u_i''}\frac{\partial\bar{\theta}}{\partial x_j} - \overline{\theta''u_j''}\frac{\partial\bar{u}_i}{\partial x_j} \\ &- \overline{u_j''u_i''\frac{\partial\theta''}{\partial x_j}} - \overline{\theta''u_j''\frac{\partial u_i''}{\partial x_j}} - \overline{\theta_0\,\theta''\frac{\partial\pi''}{\partial x_i}} + \frac{g}{\theta_0}\delta_{i3}\overline{\theta''^2} + \overline{u_i''S_\theta''}\end{aligned} \tag{7.14}$$

The prognostic equation for $\overline{\theta''^2}$ can be determined from Eq. 7.13 by multiplying that expression by θ'', and applying the assumption given by Eq. 4.8, which results in

$$\frac{\partial\,\overline{\theta''^2}}{\partial t} = -\bar{u}_j\frac{\partial}{\partial x_j}\overline{\theta''^2} - \overline{u_j''\theta''}\frac{\partial}{\partial x_j}\bar{\theta} - \overline{u_j''\frac{\partial}{\partial x_j}\theta''^2} - \overline{\theta''S_\theta''}.$$

These prognostic equations for the subgrid-scale fluxes are referred to as *second-order closure equations*, since they provide explicit conservation equations which are part of:

(1) the conservation-of-velocity $\left[\partial\left(\overline{u_k''u_i''}\right)/\partial t\right]$
(2) the conservation-of-heat $\left[\partial\left(\overline{u_i''\theta''}\right)/\partial t\right]$
(3) the conservation-of-water $\left[\partial\left(\overline{u_i''q''}\right)/\partial t\right]$
(4) the conservation-of-other atmospheric gases and aerosols $\left[\partial\left(\overline{u_i''\chi_m''}\right)/\partial t\right]$ and
(5) the conservation-of-mass of air $\left[\partial\left(\overline{\rho''u_i''}\right)/\partial t\right]$.

(7.15)

Of these equations, when the assumption that $|\rho''|/\rho_0 \ll 1$ is made, an equation for $\overline{\rho''u_i''}$ is ignored.

The full equations represented by Eq. 7.15 are computationally expensive to solve, and in all cases, either prognostic equations must be developed for the third-order correlation terms (i.e., $\overline{u_j''\frac{\partial}{\partial x_j}\theta''^2}$; $\overline{u_j''\frac{\partial}{\partial x_j}u_k''u_i''}$, etc.) or an assumption regarding their functional form must be made. The development of prognostic equations will introduce fourth-order correlation terms with an even more expensive computational cost. The necessity of truncating the derivation of successively higher-order subgrid-scale prognostic equations is called *closure*. Second-order closure, for example, means that functional forms are assumed for the third-order correlation terms. Third-order closure means that functional terms are specified for the fourth-order closure terms which appear in the prognostic equations for the third-order terms.

A much simpler parameterization of the fluxes is used near the surface. The vertical subgrid-scale fluxes near the ground (the "surface layer") can be obtained using Eqs. 7.11 and 7.14 with the assumption that the only term that matters is the vertical gradient of the flux. Than this term can be represented by the definitions given in Eq. 7.7.

Using dimensional reasoning, the exchange coefficient is then expressed as

$$K_M = kzu_*,$$

where k is called the von Karman constant (which from observations in the atmosphere (e.g., Högström 1996), is estimated to have a value of $k = 0.40 \pm 0.01$ (although Bergmann 1998 reports on a value of $k = 0.3678$). Discussions of this value of the von Karman constant are given in Andreas and Treviño (2000) and Bergmann (2000). The variable u_* is called the friction velocity. The square of the friction velocity can be shown to be equal to the surface shearing stress divided by the density of the air.

The values of the velocity flux in Eq. 7.7 near the surface are then given by the square of the friction velocity. A similar dimensional reasoning is used to define values of θ_*, q_{n_*}, and χ_{m_*}. In Eq. 7.7, the other surface layer fluxes for temperature are $\theta_* u_*$, $q_{n_*} u_*$, and $\chi_{m_*} u_*$.

These fluxes are inserted as the values at the first grid level in the model atmosphere for the computation of the subgrid-scale flux divergences in equations such as 4.21 and 4.24-4.26, as shown by

$$\overline{w''\theta''} = -K_\theta \frac{\partial \bar{\theta}}{\partial z} = -u_* \theta_*,$$

$$\overline{w''q_n''} = -K_q \frac{\partial \bar{q}_n}{\partial z} = -u_* q_{n_*},$$

$$\overline{w''\chi_m''} = -K_\chi \frac{\partial \bar{\chi}_m}{\partial z} = -u_* \chi_{m_*}.$$

Above the surface layer is the transition layer. The top of the boundary layer can be defined (and assessed in model output) as the lowest level in the atmosphere at which the ground surface no longer influences the dependent variables through the turbulent transfer of mass. The transition layer extends from the surface layer to the top of the boundary layer which ranges from 100 m or so to several kilometers or more (and even above the tropopause when surface-forced thunderstorms occur). This top often corresponds to a potential temperature inversion. It is typically shallower at night.

These inversions, for example, are caused by

1. radiational cooling at night, above stratiform clouds and smog layers, or evaporative cooling over moist ground;
2. warming: synoptic subsidence or cumulus-induced subsidence, and
3. advection due to frontal inversions; warm air over cold land, water, or snow, or vertical differences in the horizontal advection of temperature.

Above the surface layer, the mean wind changes direction with height and approaches the free-stream velocity at the top of the boundary layer as the surface-forced subgrid-scale fluxes decrease in magnitude. When the bottom surface is heated, the planetary boundary layer tends to be well mixed, particularly in potential temperature. Specific humidity is somewhat less well mixed because the entrainment of dry air into a growing boundary layer permits a gradient in q_3 to exist between the top of the planetary boundary layer and the (usually) more moist surface (Mahrt 1976). Because of horizontal pressure gradients, winds are the least well mixed. When the surface is cool, relative to the overlying air (i.e., such that the surface layer is stably stratified), vertical gradients in all of the dependent variables exist within the planetary boundary layer. An example of the parameterizations of subgrid-scale fluxes above the surface layer are presented in Section 7.2.3.

For a more in-depth discussion of the surface and transition layers, see Sections 7.2, 7.3.2, and 7.3.3 in Pielke (2002b).

Mellor and Yamada (1974) present a classic overview of subgrid-scale flux closure schemes. They start with the complete subgrid flux equations, such as shown, for example, by Eqs. 7.11 and 7.14. They then define different levels of complexity in which they discriminate into four levels of detail in the parameterizations. Their level-4, for example, retains the complete subgrid flux equations in their prognostic form, while level-1 is of the form given by Eq. 7.7. Shafran et al. (2000) discuss level 1.5 parameterizations where the only prognostic subgrid flux equation used is the equation for subgrid-scale kinetic energy. In the words of Mellor and Yamada, the goal of developing a hierarchical representation is to obtain a parameterization which is "intuitively attractive and which optimizes computational speed and convenience without unduly sacrificing accuracy." Petersen and Holtslag (1999) discuss, for example, a first-order closure for the fluxes and covariances of χ. Sharan et al. (1999) use a level-2 Mellor-Yamada framework to represent σ_w in a stable boundary layer. Glendening (2000) discusses the turbulent kinetic energy budgets for strong shear conditions.

McNider et al. (2012) present a model analysis that shows the complexity of the stable boundary layer to even small perturbations of forcing. Using WRF, Mölders and Kramm (2010) studied wintertime inversions in interior Alaska. Mölders et al. (2011) assess the ability of the WRF model to simulate the boundary layer under conditions of low solar radiation. Holtslag et al. (2012) reviews the latest understanding of the boundary layer with a focus on stable boundary layers.

7.2.2 Viscous Sublayer

The viscous sublayer[7] is defined as the level near the ground $\left(z < z_0;\ \text{with } D \simeq 0\right)$, where the transfer of the dependent variables by molecular motions become important.

Zilitinkevich (1970) and Deardorff (1974) suggest relating temperature and specific humidity at the top of the layer $\bar{\theta}_{z0}$ and $\bar{q}_{z0}$ to the surface values of the variables $\bar{\theta}_G$ and $\bar{q}_G$ using expressions of the form

$$\bar{\theta}_{z0} = \theta_G + 0.0962\left(\theta_*/k\right)\left(u_* z_0/v\right)^{0.45}$$

and

$$\bar{q}_{z0} = q_G + 0.0962\left(q_*/k\right)\left(u_* z_0/v\right)^{0.45} \tag{7.16}$$

By analogy

$$\bar{\chi}_{z0} = \chi_G + 0.0962\left(\chi_*/k\right)\left(u_* z_0/v\right)^{0.45}$$

[7] Traditionally, this layer has been referred to as a "laminar sublayer," although as evident in Eq. 7.16, turbulent fluxes still occur within $z < z_0$ since u_*, q_*, and χ_* still appear.

In these expressions, v is the kinematic viscosity of air ($\sim 1.5 \times 10^{-5}$ m^{2}s^{-1}), k is von Karman's constant with θ_*, u_*, q_*, and χ_*, as one example, defined by Eq. 8.1. Between $z = z_0$ and $z = z_G$, $\bar{u} = \bar{v} = \bar{w} = 0$, whereas variations of $\bar{p}$ and $\bar{\pi}$ across this depth are ignored.

As discussed by Businger (1973), $u_* z_0 / v$ may be considered the Reynolds number of the smallest turbulent eddy in the flow. In addition, he reports on a study by Nikuradse (1933) in which laminar flow occurs with $u_* z_0 / v < 0.13$, whereas turbulent motion dominates with $u_* z_0 / v > 2.5$. In between these two limits, a transition regime exists. The laminar situation is referred to as being *aerodynamically smooth* and the fully turbulent flow is called *aerodynamically rough*.

[8]In general, the term $0.0962\ (u_* z_0 / v)^{0.45}$ in Eq. 7.16 can be replaced by (e.g., Zeng and Dickinson 1998)

$$\ln \frac{z_0}{z_{0t}} = \ln \frac{z_0}{z_{0q}} = \ln \frac{z_0}{z_{0\chi}} = a Re_*^b$$

where $Re_* = u_* z_0 / v$ is the roughness Reynolds number, and z_{0t}, z_{0q}, and $z_{0\chi}$ refer to the roughness lengths for heat, humidity, and tracer gases, respectively. Furthermore, Eq. 7.16 can be derived between $(D+z_0)$ and $(D+z_{0t})$ $[(D+z_{0q})$, or $(D+z_{0\chi})]$ with the assumption that terms involving z_0/L, z_{0t}/L, z_{0q}/L, and $z_{0\chi}/L$ can be omitted, θ (q, or χ) at height $(D+z_{0t})$ $[(D+z_{0q})$, or $(D+z_{0\chi})]$ can be used to represent the surface θ (q, or χ) [i.e., θ_g (q_g, or χ_g)], and β is taken as 1.0.

In general, z_0 is greater than z_{0t} because the momentum transfer in the surface viscous sublayer is affected by both the molecular diffusion and pressure fluctuations (particularly around bluff elements) while the heat transfer is affected by the molecular diffusion only (Zeng and Dickinson 1998). In the original work of Zilitinkevich (1970) and Deardorff (1974), $a = 0.0962$ and $b = 0.45$. In the more recent work based on the comparison of two land models with observations, Zeng et al. (2012) and Zheng et al. (2012) found that $a = 0.8$-0.9 and $b = 0.5$ are more appropriate over bare soil. They also demonstrated that these values are crucial for the realistic simulation of the observed very large diurnal cycle of surface skin temperature (e.g., used for computing upward longwave radiation) over arid regions.

As discussed in Zeng and Dickinson (1998), different values of z_0 and z_{0t} are needed over bare soil only, as energy balance is explicitly considered over canopy which is equivalent to the consideration of z_0/z_{0t}. Further discussion of the formulations for z_{0t} and z_{0q} over ocean and sea ice is presented in Chapter 10.

7.2.3 Parameterization Complexity

It is useful to dissect a parameterization algorithm to determine the number of dependent variables and the adjustable and universal parameters that are introduced. This

[8]The following text in this subsection was written by Professor Xubin Zeng, Department of Atmospheric Sciences, University of Arizona, Tucson.

dissection can be illustrated with the following simple example. Holtslag and Boville (1993) and Tijm et al. (1999a) propose the following form for K_θ above the boundary layer;

$$K_\theta = l_\theta^2 S\, F_\theta(\mathrm{Ri}) \tag{7.17}$$

where

$$\frac{1}{l_\theta} = \frac{1}{kz} + \frac{1}{\lambda_\theta} \tag{7.18}$$

$$S = \left| \frac{\partial \vec{U}_i}{\partial z} \right| \tag{7.19}$$

$$F_\theta(\mathrm{Ri}) = \begin{matrix} (1 - 18\,\mathrm{Ri})^{1/2} & \mathrm{Ri} \le 0 \\ 1/\left(1 + 10\,\mathrm{Ri} + 80\,\mathrm{Ri}^2\right) & \mathrm{Ri} > 0 \end{matrix} \tag{7.20}$$

with

$$\lambda_\theta = \begin{cases} 300\text{ m} & z \le 1\text{ km} \\ 30\text{ m} + 270\exp\left(1 - (z/1000\text{ m})\right) \end{cases} \tag{7.21}$$

In this formulation for K_θ, there are the following dependent variables, parameters, and prescribed constants:

1. In Eq. 7.17, the dependent variables l_θ, S, and F_θ define K_θ.
2. In Eq. 7.18, l_θ is defined with the independent variable z, the dependent variable λ_θ, and the parameter k.
3. In Eq. 7.19, S is defined by the vertical gradient of $\vec{U}_i$.
4. In Eq. 7.20, F_θ (Ri) is defined by the dependent variable Ri (which is defined by Eq. 7.8), and the constants 18, 10, and 80, and the exponent 1/2.
5. In Eq. 7.21, λ_θ is defined by the independent variable z, and the constants 300, 30, 270, and 1000.

Therefore, to represent the term K_θ, in addition to the fundamental variables $\bar{u}_i$ and $\bar{\theta}$, one parameter (k) and 8 constants (18, 10, 80, 1/2, 300, 30, 270, 1000) need to be provided.

A sensitivity analysis can be applied to show how K_θ responds to slight changes in the dependent variables and constants. For example, in Eq. 7.21, if 100 m were used instead of 300 m when λ_θ dominates in Eq. 7.18, K_θ would be 1/9 as large, since K_θ is proportional to l_θ^2. Clearly, the form of Eq. 7.21 will exert a major effect on the parameterized turbulent mixing in a model.

Niyogi et al. (1999) discuss such a sensitivity analysis (in their case for surface fluxes) in terms of the question "what scenarios make a particular parameter significant?" They also appropriately conclude that parameter uncertainty is not only related to its deviation, but is dependent on the values of the other parameters used in a parameterization.

7.3 Parameterization of Radiative Flux Divergence

7.3.1 Basic Terms

The radiative flux divergence source/sink term $\bar{S}_\theta$ in Eq. 4.24 can be written in part as

$$\bar{S}_\theta = \left.\frac{\partial \bar{T}}{\partial t}\right|_{\text{rad}} = -\frac{1}{\bar{\rho} C_{\text{p}}} \frac{\partial \bar{R}}{\partial z}, \tag{7.22}$$

where $\partial \bar{R}/\partial z$ is the grid-volume averaged vertical gradient of absorbed *irradiance* (i.e., radiative energy per area per time) due to all wavelengths of electromagnetic energy. In writing Eq. 7.22, the divergence of $\bar{R}$ in the horizontal direction is neglected, since on the mesoscale, variations of $\bar{R}$ are much larger in the vertical.[9] In addition, changes in pressure resulting from the divergence of $\bar{R}$ are also neglected in Eq. 7.22. It is important to note that $\bar{R}$ contains subgrid-scale as well as resolvable effects. This section discusses methodologies for parameterizing the divergence of irradiance.[10]

The unit of differential area on the surface of a hemisphere can be written as

$$dS = \cos Z \sin Z \, dA \, dZ \, d\psi, \tag{7.23}$$

where as shown in Fig. 7.2, dA is a differential area on a plane through the equator of the hemisphere, Z the angle between the axis of the hemisphere and a line to dS (i.e., the *zenith angle*),[11] and ψ the longitude of dS on the hemisphere (i.e., the *azimuth angle*). The quantity, $\sin Z \, dZ \, d\psi$ is called a *differential solid angle* and has units called steradians.

Using Eq. 7.23, and following Liou (1980), the monochromatic[12] intensity (radiance) of electromagnetic radiation through dS per time per wavelength can be written as

$$I_\lambda(Z, \psi) = de_\lambda / (\cos Z \sin Z \, dA \, dZ \, d\psi \, d\lambda \, dt), \tag{7.24}$$

where de_λ is the differential amount of radiant energy at a given frequency passing through dS in the time interval dt.

[9]This variation occurs because the vertical gradient of such atmospheric properties as density, CO_2, and water vapor are generally much larger than their horizontal gradients. The only major exception to this characteristic of the atmosphere occurs when clouds are present. In this latter situation, however, horizontal gradients of irradiance have been neglected in mesoscale models with the justification that the horizontal resolution on the mesoscale is significantly less, in general, than the vertical resolution. Heterogeneities on the subgrid-scale of clouds, however, has been shown to be important in radiative flux calculations (Tiedtke 1996; O'Hirok and Gautier 1998a,b). Over tropical ocean regions (30°S to 30°N), Tiedtke found that in a global weather forecast model, the net downward shortwave radiative fluxes are increased about 10 W m^{-2} when subgrid-scale effects were included.

[10]The discussion of basic concepts presented in this section makes extensive use of Liou's (1980) excellent treatise on atmospheric radiation.

[11]The zenith angle is also equal to 90° − E, where E is called the *elevation angle.*

[12]Single wavelength.

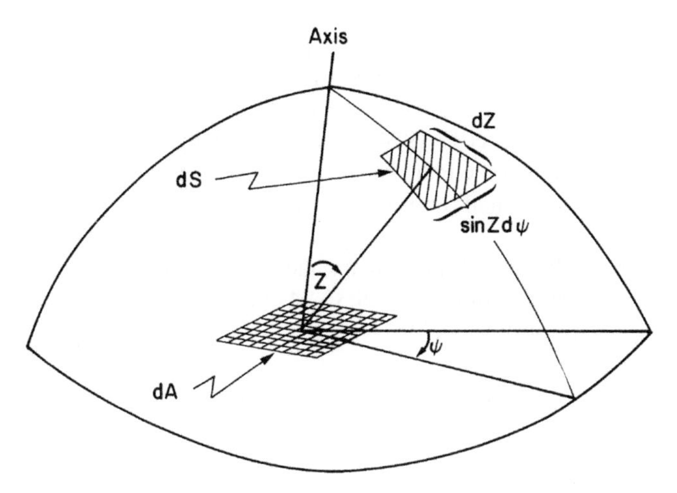

FIGURE 7.2

An illustration of the relation between a differential area dA on a flat surface and the projection of this area dS onto the surface of a hemisphere with a radius of unity. The angles ψ and Z are the longitude and zenith (adapted from Liou 1980:Fig. 1.3).

Integrating Eq. 7.24 over the entire hemisphere yields the monochromatic *irradiance* on dA from all points above that differential surface

$$R_\lambda = \int_0^{2\pi} \int_0^{\pi/2} I_\lambda(Z, \psi) \cos Z \sin Z \, dZ \, d\psi, \tag{7.25}$$

which can be written as

$$R_\lambda = \pi I_\lambda \tag{7.26}$$

when $I_\lambda(z, \psi)$ is independent of direction.[13] The total isotropic irradiance is obtained by integrating Eq. 7.26 over all wavelengths

$$R = \pi \int_0^\infty I_\lambda \, d\lambda. \tag{7.27}$$

The concept of a *blackbody* is essential in the economical parameterization of Eq. 7.22. A blackbody is defined as an object that absorbs all radiation that impinges upon it. As shown, for example, by Coulson (1975), the intensity of electromagnetic radiation emitted by such a blackbody as a function of wavelength λ is given by

$$B_\lambda(T) = C_1 \Big/ \left[\lambda^5 \left(e^{C_2/\lambda T} - 1\right)\right]. \tag{7.28}$$

The monochromatic radiance given by Eq. 7.28, called the *Planck function*, represents the maximum intensity of emitted radiative energy that can occur for particular values of temperature and wavelength.

[13]Radiation that is not dependent on direction is referred to as *isotropic* radiation.

The derivation of Eq. 7.28, using the concepts of quantum mechanics, is described by Liou (1980:9-11, and Appendix C) and Kondratyev (1969:30-32). In Eq. 7.24, $B_\lambda(T)$ has the same units as $I_\lambda(z, \psi)$. The values $C_1 = 1.191 \times 10^{-16}$ W m^{2}st^{-1} and $C_2 = 1.4388 \times 10^{-2}$ m K (NBS 1974) are fundamental physical constants derived from the speed of light and the Planck and Boltzmann constants (see Liou 1980:10-11; also values of fundamental (universal) physical constants are given in Mohr and Taylor 2000).

Integrating Eq. 7.28 over all wavelengths and all directions within the hemisphere depicted in Fig. 7.2 (see Kondratyev 1969:33-34 or Liou 1980:11), where the emitted radiation is assumed isotropic, yields

$$R^* = \pi \int_0^\infty B_\lambda(T)d\lambda = \sigma T^4 \left[\sigma = 5.67400 \times 10^{-8} \text{ W m}^{-2}\text{K}^{-4}\right]$$

Mohr and Taylor (2000) (7.29)

which is called the *Stefan-Boltzmann law*, and σ is the Stefan-Boltzmann constant. The quantity R^* is the blackbody irradiance over all wavelengths and is equal to the maximum amount of radiative energy per unit area an object can emit at a given temperature.

Figure 7.3, obtained using List (1971:412) illustrates the blackbody irradiance computed as a function of wavelength using Eq. 7.28 with $T = 6000$ K (e.g., corresponding to the surface of the Sun) at the distance of the Earth from the Sun, and with $T = 290$ K (e.g., corresponding to the surface of the Earth. As evident in the figure, the two distributions (also called *spectra*) of electromagnetic radiation have almost no overlap. For this reason, irradiance from the Sun is often referred to as *shortwave*

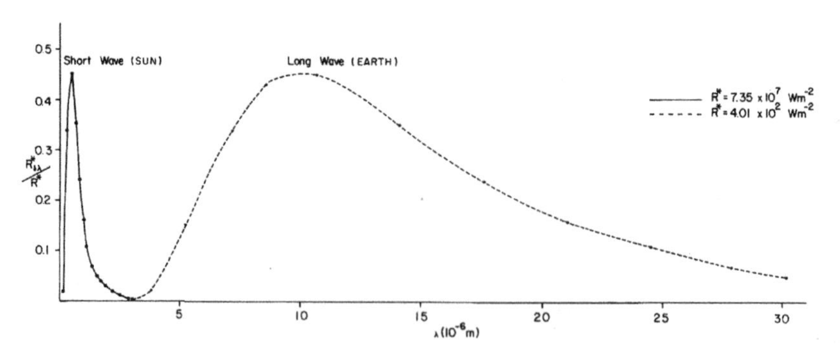

FIGURE 7.3

Radiative flux defined by $R^*_{\delta\lambda} = \int_\lambda^{\lambda+\delta\lambda} E^*_\lambda(T)d\lambda$ divided by the total blackbody radiation R^* (from Eq. 7.29) as a function of wavelength λ. Note that the values of R^* are different for $T = 6000$ K and $T = 290$ K. However, at the distance of the Earth from the Sun, the two values of R^* must be almost the same, since the Earth's total irradiance is always essentially equal to the annual irradiance received from the Sun at the Earth's orbital distance.

radiation,[14] whereas radiation emitted from the Earth is called *longwave (infrared) radiation*. The wavelength separation of these two electromagnetic spectra simplifies the parameterization of Eq. 7.22, as will be shown shortly. Using Eq. 7.28, the peak emission from the Sun's surface has a wavelength of 0.475 μm (blue light), whereas the Earth, using $T = 290$ K, has a peak around 10 μm.

As electromagnetic radiation traverses a layer in the atmosphere, it can be absorbed, reflected, or transmitted. This relation can be written quantitatively as

$$\frac{I_\lambda(\text{absorbed})}{B_\lambda} + \frac{I_\lambda(\text{reflected})}{B_\lambda} + \frac{I_\lambda(\text{transmitted})}{B_\lambda} = 1$$

or

$$a_\lambda + r_\lambda + t_\lambda = 1, \tag{7.30}$$

where a_λ, r_λ, and t_λ are called the *monochromatic absorptivity, reflectivity*, and *transmissivity*, respectively. In addition, the reflectivity can be decomposed into that, which through multiple reflections (*scattering*) is transmitted in the forward direction, and that, through scattering, propagates at an angle that is different from that of the incoming electromagnetic energy. The sum of the absorption and scattering out of the incident direction is called *attenuation* or *extinction* of the electromagnetic radiation (Paltridge and Platt 1976:38). As shown by Kondratyev (1969:22 ff), when an element of volume is in local thermodynamic equilibrium,[15] $a_\lambda = \epsilon_\lambda$ for that volume, where ϵ_λ is the *monochromatic emissivity*. This relationship between absorptivity and emissivity is called *Kirchhoff's law*. Levels below about 50 km or so in the Earth's atmosphere are in local thermodynamic equilibrium. With a blackbody, $a_\lambda = 1$ so that $\epsilon_\lambda = 1$ also. When a_λ is independent of wavelength, but $\epsilon_\lambda = a_\lambda < 1$, the object is called a *graybody*.

Following Liou (1980), the change of radiance as the electromagnetic energy travels a distance in the atmosphere ds can be written as

$$dI_\lambda \big/ \left(k_\lambda \rho\, ds\right) = -I_\lambda + B_\lambda(T) + J_\lambda, \tag{7.31}$$

where ρ is the density of air, and k_λ is called the mass extinction cross section (in S.I. units of meters squared per kilogram of radiatively active material). The first term on the right of Eq. 7.31 is the loss of radiance from attenuation, and $B_\lambda(T)$ is the emission and J_λ the source of radiance from scattering into the line segment ds.

Because of the separation of wavelength in the spectra plotted in Fig. 7.3, it is convenient to develop separate parameterizations for long and short wavelengths.

7.3.2 Longwave Radiative Flux Divergence

7.3.2.1 Clear Atmosphere

In a clear atmosphere, scattering of longwave radiation is neglected relative to the absorption and emission of electromagnetic energy (Liou 1980). Equation 7.31 can,

[14] Shortwave radiation consists of ultraviolet, visible, and near-infrared wavelengths.

[15] Local thermodynamic equilibrium is defined in Footnote 4 in Chapter 2.

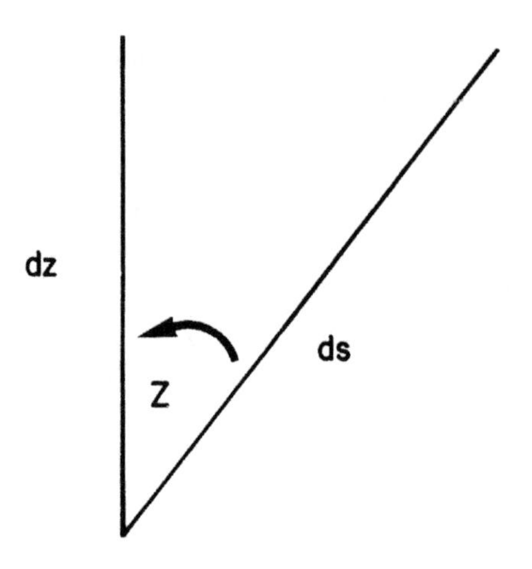

FIGURE 7.4

A schematic illustration of the relation between the zenith angle Z and ds and dz.

therefore, be written as

$$dI_\lambda / (k_\lambda \rho \, ds) = -I_\lambda + B_\lambda(T). \tag{7.32}$$

To apply the solution of Eq. 7.32 to 7.22, it is necessary to define dI_λ with respect to the vertical direction. Since $\cos Z = dz/ds$ (see Fig. 7.4), and attenuation is only owing to absorption, Eq. 7.32 can be written as

$$\cos Z \frac{dI_\lambda}{k_{a\lambda} \rho \, dz} = -\cos Z \frac{dI_\lambda}{d\tau_\lambda} = -I_\lambda + B_\lambda(T), \tag{7.33}$$

where

$$\tau_\lambda = \int_z^\infty k_{a\lambda} \rho \, dz \tag{7.34}$$

is called the *normal optical thickness*.[16]

When solving Eq. 7.33, it is useful to evaluate this differential equation separately for upward and downward radiances. The procedure of solution is to multiply Eq. 7.33 by the integrating factor[17] $e^{\tau_\lambda / \cos Z}$, integrate from τ_λ to τ_{G_λ} and multiply the result by $e^{\tau_\lambda / \cos Z}$. The optical thickness τ_{G_λ} is defined by Eq. 7.34 with z equal to the ground elevation (i.e., $z = z_G$). The result of this evaluation is

$$I_\lambda \uparrow = I_\lambda \uparrow|_G \, e^{-(\tau_{G_\lambda} - \tau_\lambda)/\cos Z} + \int_{\tau_\lambda}^{\tau_{G_\lambda}} B_\lambda(T)\Big|_{\hat{\tau}_\lambda} e^{-[(\hat{\tau}_\lambda - \tau_\lambda)/\cos Z]} \frac{d\hat{\tau}_\lambda}{\cos Z}$$
$$(1 \geq \cos Z > 0), \tag{7.35}$$

[16]Using Leibnitz's rule (e.g., see Hildebrand 1962:360), $d\tau_\lambda = -k_{a\lambda} \rho \, dz$. Note that since k_λ has units of area per mass, $\tau_{a\lambda}$ is called the *mass absorption coefficient*.

[17]See Spiegel (1967), or other introductory texts on ordinary differential equations for a discussion of integrating factors.

where $I_\lambda\uparrow|_G$ is the upward radiance from the ground. The downward radiance is obtained in a similar fashion (i.e., see Liou 1980:24) where Eq. 7.33 is integrated downward from the top τ_{T_λ} to τ_λ.[18]

$$I_\lambda\downarrow = I_\lambda\downarrow|_T\, e^{-(\tau_\lambda-\tau_{T_\lambda})/\cos Z} + \int_{\tau_{T_\lambda}}^{\tau_\lambda} B_\lambda(T)\Big|_{\hat{\tau}_\lambda} e^{-[(\tau_\lambda-\hat{\tau}_\lambda)/\cos Z]}\frac{d\hat{\tau}_\lambda}{\cos Z}$$

$$(1 \geq \cos Z > 0) \tag{7.36}$$

$I_\lambda\downarrow|_T$ represents downward radiance from the model top. Using the definition of irradiance given by Eq. 7.25, assuming that the radiative transfer is independent of azimuth, and that $I_\lambda\uparrow|_G$ and $I_\lambda\downarrow|_T$ emit as blackbodies,[19] Eqs. 7.35 and 7.36 can be written as

$$R_\lambda\uparrow = 2\pi B_\lambda\left(T_G\right) E_3\left(\tau_{G_\lambda} - \tau_\lambda\right) + 2\pi \int_{\tau_\lambda}^{\tau_{G_\lambda}} B_\lambda(T)|_{\hat{\tau}_\lambda}\, E_2\left(\hat{\tau}_\lambda - \tau_\lambda\right) d\hat{\tau}_\lambda, \tag{7.37}$$

$$R_\lambda\downarrow = 2\pi \tilde{B}_\lambda\left(T_T^*\right) E_3\left(\tau_\lambda - \tau_{T_\lambda}\right) + 2\pi \int_{\tau_{T_\lambda}}^{\tau_\lambda} B_\lambda(T)|_{\hat{\tau}_\lambda}\, E_2\left(\tau_\lambda - \hat{\tau}_\lambda\right) d\hat{\tau}_\lambda, \tag{7.38}$$

where

$$E_3(y) = \int_0^1 e^{-y/\cos Z} \cos Z\, d(\cos Z),$$
$$E_2(y) = \int_0^1 e^{-y/\cos Z} d(\cos Z) = -dE_3(y)/dy \tag{7.39}$$

and $\tilde{B}_\lambda\left(T_T^*\right)$ is evaluated including any attenuation above level z_T for wavelength λ. The temperatures T_G and T_T^* correspond to the ground surface temperature and the effective temperature at the model top (if $z, \to \infty$, $T_T^* \to 0$). In Eq. 7.39, y, of course, can correspond to $\tau_{G_\lambda} - \tau_\lambda$, $\tau_\lambda - \tau_{T_\lambda}$, $\hat{\tau}_\lambda - \tau_\lambda$, or $\tau_\lambda - \hat{\tau}_\lambda$.

The total upward and downward radiative flux at level z is obtained by integrating Eqs. 7.37 and 7.38 over all wavelengths giving

$$R\uparrow = \int_0^\infty R_\lambda\uparrow d\lambda; \quad R\downarrow = \int_0^\infty R_\lambda\downarrow d\lambda. \tag{7.40}$$

In principle, as pointed out by Liou (1980), Eq. 7.40 can be used to determine the total irradiance from the integrated longwave spectrum. However, the optical thickness, τ_λ, is a complicated and rapidly varying function of wavelength, as illustrated in Fig. 7.5. In the clear atmosphere, this complex distribution of τ_λ is caused by the absorption spectra of certain gases in the air. The triatomic molecules, carbon dioxide, water vapor, and ozone, in particular, have numerous significant absorption lines in this

[18] In the real atmosphere, τ_{T_λ} would be set equal to zero using Eq. 7.34 with $z \to \infty$. The optical thickness τ_{T_λ} is not set equal to zero here, however, since numerical model tops are generally within the atmosphere.

[19] The Earth's surface radiates in the infrared close to its blackbody value (see Tables 10.3 and 10.8). To make the same assumption for $I_\lambda\downarrow|_{z_T}$, one can specify an effective temperature at the model top to represent the infrared emission from above z_T as if it were a blackbody. More appropriately, of course, the actual values of $I_\lambda\downarrow|_{z_T}$ should be used.

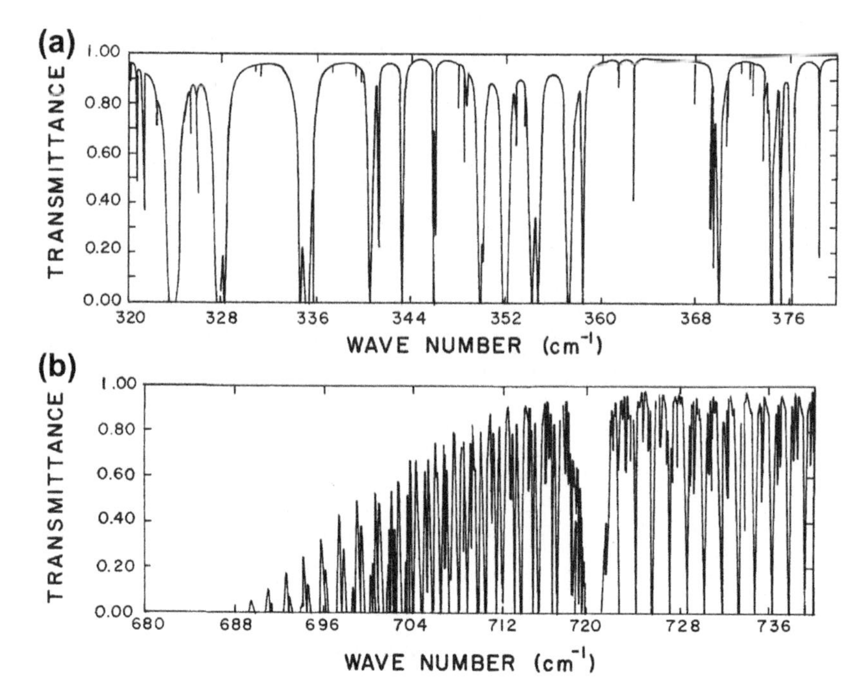

FIGURE 7.5

(a) Monochromatic transmissivity t_λ of water vapor between wavelengths 31 (corresponding to 320 cm^{-1}) and 26 μm (corresponding to 376 cm^{-1}); and (b) carbon dioxide between wavelengths 14.7 (corresponding to 680 cm^{-1}) and 13.6 μm (corresponding to 736 cm^{-1}) from Liou (1980) as adapted from McClatckey and Selby (1972).

portion of the electromagnetic spectrum. Thus, as pointed out by Liou (1980:95), the evaluation of Eq. 7.40 would require the double integration using very small increments of λ and τ_λ to represent the thousands of absorption lines properly within the infrared region.

To circumvent this problem, *a transmission function*, defined as

$$\Gamma_{\bar{\lambda}}(\tau) = \frac{1}{\Delta\lambda}\int_{\lambda_1}^{\lambda_2} e^{-\tau} d\lambda; \quad \Delta\lambda = \lambda_2 - \lambda_1, \tag{7.41}$$

is used where the interval is large enough such that several absorption lines are included, but small enough such that $B_\lambda(T)$ is approximately constant across the interval (i.e., $B_{\bar{\lambda}}(T) \simeq B_\lambda(T)$). Equations 7.37 and 7.38 are then integrated over $\Delta\lambda$ yielding

$$\begin{aligned} R_{\bar{\lambda}}\uparrow &= \int_{\lambda_1}^{\lambda_2} R_\lambda \uparrow \frac{d\lambda}{\Delta\lambda} = 2\pi\, B_{\bar{\lambda}}\left(T_{\mathrm{G}}\right) \int_{\lambda_1}^{\lambda_2} E_2\left(\tau_{\mathrm{G}_\lambda} - \tau_\lambda\right) \frac{d\lambda}{\Delta\lambda} \\ &+ 2\pi \int_{\tau_{\bar{\lambda}}}^{\tau_{\mathrm{G}_{\bar{\lambda}}}} B_{\bar{\lambda}}(T)\Bigg|_{\hat{\tau}_{\bar{\lambda}}} \int_{\lambda_1}^{\lambda_2} E_3\left(\hat{\tau}_\lambda - \tau_\lambda\right) \frac{d\hat{\tau}_\lambda}{\Delta\lambda}\, d\lambda, \end{aligned} \tag{7.42}$$

$$R_{\bar{\lambda}}\downarrow = \int_{\lambda_1}^{\lambda_2} R_\lambda\downarrow \frac{d\lambda}{\Delta\lambda} = 2\pi\, \tilde{B}_{\bar{\lambda}}\left(T_{\mathrm{T}}^*\right) \int_{\lambda_1}^{\lambda_2} E_3\left(\tau_\lambda - \tau_{\mathrm{T}_\lambda}\right) \frac{d\lambda}{\Delta\lambda}$$
$$+\, 2\pi \int_{\tau_{\bar{\lambda}_{\mathrm{T}}}}^{\tau_{\bar{\lambda}}} B_{\bar{\lambda}}(T)\Big|_{\hat{\tau}_{\bar{\lambda}}} \int_{\gamma_1}^{\gamma_2} E_2\left(\tau_\lambda - \hat{\tau}_\lambda\right) \frac{d\hat{\tau}_\lambda}{\Delta\lambda}\, d\lambda. \quad (7.43)$$

The *slab transmission function* is then defined by

$$\Gamma_{\bar{\lambda}}^s(y) = 2 \int_{\lambda_1}^{\lambda_2} E_3(y) \frac{d\lambda}{\Delta\lambda}, \quad (7.44)$$

so that from Eq. 7.39,

$$\frac{d\Gamma_{\bar{\lambda}}^s(y)}{dy} = -2 \int_{\lambda_1}^{\lambda_2} E_2(y) \frac{d\lambda}{\Delta\lambda}. \quad (7.45)$$

Therefore, Eqs. 7.42 and 7.43 can be written as

$$R_{\bar{\lambda}}\uparrow = \pi B_{\bar{\lambda}}\left(T_{\mathrm{G}}\right) \Gamma_{\bar{\lambda}}^s\left(\tau_{\mathrm{G}_\lambda} - \tau_{\bar{\lambda}}\right) - \int_{\tau_{\bar{\lambda}}}^{\tau_{\mathrm{G}_{\bar{\lambda}}}} \pi B_{\bar{\lambda}}(T)\Big|_{\hat{\tau}_{\bar{\lambda}}} \frac{d\Gamma_{\bar{\lambda}}^s}{d\hat{\tau}_{\bar{\lambda}}} \left(\hat{\tau}_{\bar{\lambda}} - \tau_{\bar{\lambda}}\right) d\hat{\tau}_{\bar{\lambda}}, \quad (7.46)$$

$$R_{\bar{\lambda}}\downarrow = \pi \tilde{B}_{\bar{\lambda}}\left(T_{\mathrm{T}}^*\right) \Gamma_{\bar{\lambda}}^s\left(\tau_{\bar{\lambda}} - \tau_{\mathrm{T}_{\bar{\lambda}}}\right) + \int_{\tau_{\bar{\lambda}_{\mathrm{T}}}}^{\tau_{\bar{\lambda}}} \pi B_{\bar{\lambda}}(T)\Big|_{\hat{\tau}_{\bar{\lambda}}} \frac{d\Gamma_{\bar{\lambda}}^s}{d\hat{\tau}_{\bar{\lambda}}} \left(\tau_{\bar{\lambda}} - \hat{\tau}_{\bar{\lambda}}\right) d\hat{\tau}_{\bar{\lambda}}. \quad (7.47)$$

(Note that by averaging over λ, $d\hat{\tau}_\lambda$ becomes $d\hat{\tau}_{\bar{\lambda}}$.)

If a *normal path length*[20] is defined as

$$u = \int_0^z \rho\, dz \quad u_\infty = \int_0^\infty \rho\, dz, \quad (7.48)$$

then Eq. 7.34 can be written as $\tau_\lambda = \int_u^{u_\infty} k_{a\lambda}\, du$, by the change of variable given by Eq. 7.48. In making this coordinate transformation, changes of k_λ with u (e.g., because of temperature changes with height) are neglected.

Using this change of variable, Eqs. 7.46 and 7.47 can be written as

$$R_{\bar{\lambda}}\uparrow = \pi B_{\bar{\lambda}}\left(T_{\mathrm{G}}\right) \Gamma_{\bar{\lambda}}^s(u) + \int_0^u \pi B_{\bar{\lambda}}(T)\Big|_{\hat{u}} \frac{d\Gamma_{\bar{\lambda}}^s}{d\hat{u}} \left(u - \hat{u}\right) d\hat{u}, \quad (7.49)$$

and

$$R_{\bar{\lambda}}\downarrow = \pi \tilde{B}_{\bar{\lambda}}\left(T_{\mathrm{T}}^*\right) \Gamma_{\bar{\lambda}}^s\left(u_{\mathrm{T}} - u\right) + \int_{u_{\mathrm{T}}}^u \pi B_{\bar{\lambda}}(T)\Big|_{\hat{u}} \frac{d\Gamma_{\bar{\lambda}}^s}{d\hat{u}} \left(\hat{u} - u\right) d\hat{u}. \quad (7.50)$$

[20]Note that despite being referred to as a length, u has dimensions of mass per area. The value of u increases monotonically with height z since $\rho > 0$, until $z \rightarrow \infty$.

Integrating over all wavelengths, and using Eq. 7.29,

$$R\uparrow = \sigma T_G^4 t^s(u, T_G) = \int_0^u \sigma T^4 \bigg|_{\hat{u}} \frac{dt^s(u-\hat{u}, T)}{d\hat{u}} d\hat{u}, \quad (7.51)$$

$$R\downarrow = \tilde{B} + \int_{u_T}^u \sigma T^4 \bigg|_{\hat{u}} \frac{dt^s(\hat{u}-u, T)}{d\hat{u}} d\hat{u}, \quad (7.52)$$

where

$$t^s(u, T) = \int_0^\infty \pi B_\lambda(T)\Gamma_\lambda^s(u) d\lambda / \sigma T^4 \quad (7.53)$$

is called the *broadband flux transmissivity*, defined in terms of the transmission function Eq. 7.44, when $\lambda_2 = \lambda_1$. $\tilde{B}$ is the longwave radiative flux reaching the model top which has energy in the radiatively active wavelengths. In deriving Eqs. 7.51 and 7.52, T is assumed constant within each differential path length du so that $B_\lambda(T)$ can be incorporated within the derivative term in the integrals of Eqs. 7.51 and 7.53.

The *broadband emissivity* is then defined as

$$\epsilon^s(u, T) = 1 - t^s(u, T), \quad (7.54)$$

from Eq. 7.30, since $a_\lambda = \epsilon_\lambda$ for all wavelengths, and $r_\lambda = 0$, as assumed by Eq. 7.32.

To apply Eqs. 7.51 and 7.52 to a model, the integral given by Eq. 7.53 is represented by one or more intervals, so that using Eq. 7.29, Eq. 7.54 can be written as

$$\epsilon^s(u, T) = \sum_{i=1}^{I} \pi B_{\bar{\lambda}_i}(T)\left[1 - \Gamma_{\bar{\lambda}_i}^s(u)\right] \Delta_i\lambda / \sigma T^4, \quad (7.55)$$

where $\bar{\lambda}_i$ is the average wavelength within the interval $\lambda_{i+1} - \lambda_i = \Delta_i\lambda$.

The emissivities given by Eq. 7.55 are also often written separately for the major absorbers of infrared radiation in the atmosphere. Atwater (1974), using data from Kuhn (1963), for example, suggests values of broadband emissivity for water vapor, which can be written as

$$\epsilon_{q3}(u, T) \simeq \epsilon_{q3}(\delta P) = \begin{cases} 0.104 \log_{10} \delta P + 0.440, & -4 < \log_{10} \delta P \le -3 \\ 0.121 \log_{10} \delta P + 0.491, & -3 < \log_{10} \delta P \le -1.5 \\ 0.146 \log_{10} \delta P + 0.527, & -1.5 < \log_{10} \delta P \le -1.0 \\ 0.161 \log_{10} \delta P + 0.542, & -1.0 < \log_{10} \delta P \le 0 \\ 0.136 \log_{10} \delta P + 0.542, & \log_{10} \delta P > 0, \end{cases} \quad (7.56)$$

where δP is in grams per centimeter squared and

$$\delta P = \int_z^{z+\delta z} \rho q_3 \, dz$$

is the optical path length for water vapor between z and $z + \delta z$. When $\delta z \to \infty$ and $z \to z_G$, $\delta P \to P$ where P is called *precipitable water*. To use Eq. 7.56, δP must be expressed in units of grams per centimeters squared. Since 1 g of water is 1 cm deep (i.e., $\rho_{q_2} = 1 \text{ g cm}^{-3}$), δP is also equal to the precipitable water in centimeters within the path length. For the application of ϵ_{q_3} in the parameterization of the heating associated with longwave radiation in models, the dependence of ϵ_{q_3} on temperature is inconsequential, compared with that of water vapor. Figure 7.6a

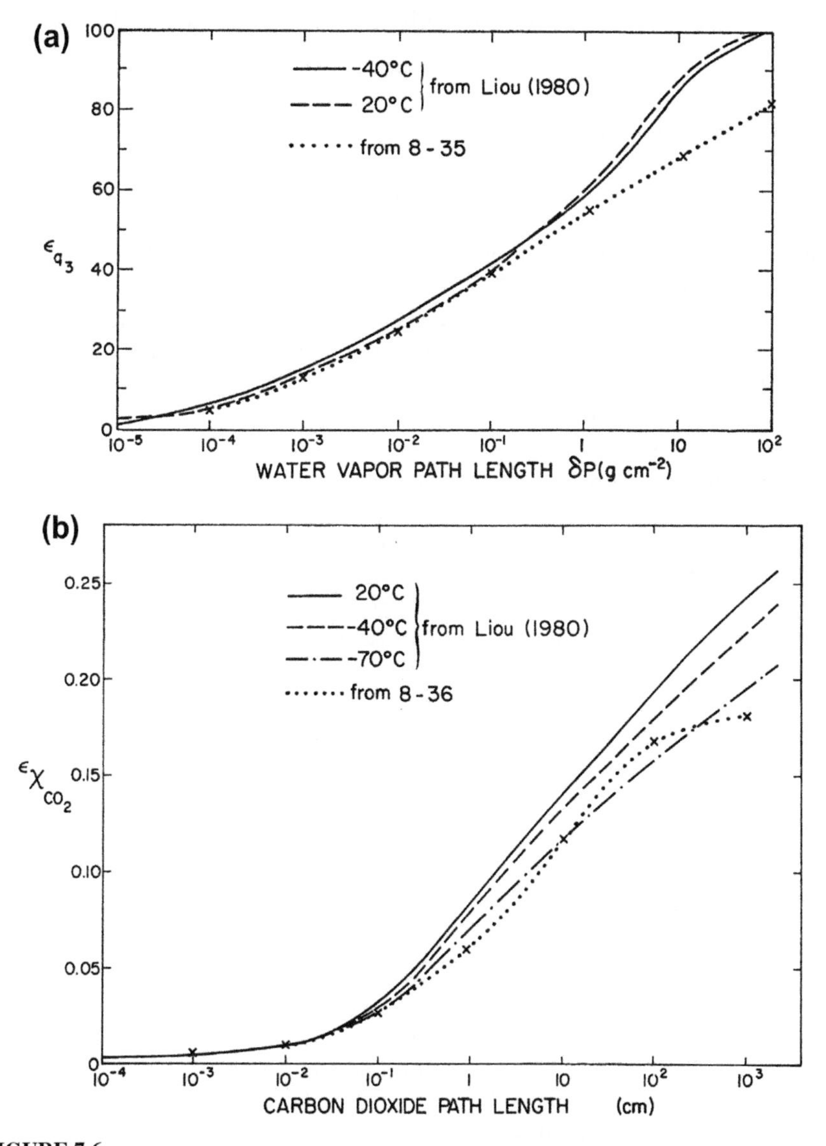

FIGURE 7.6

(a) Broadband emissivity for water vapor $\epsilon_{q_3}(\delta P)$ in percent as a function of δP. (b) Broadband emissivity for carbon dioxide $\epsilon_{\chi_{CO_2}}$ (δH_c) as a function of δH_c (adapted from Liou 1980).

illustrates this relation between ϵ_{q3} and δP as reported by Liou (1980:Fig. 4.8), along with the algorithm given by Eq. 7.56.

For the broadband emission for carbon dioxide, Kondratyev (1969) proposed the formulation

$$\epsilon_{\chi_{CO_2}}(u, T) \simeq \epsilon_{\chi_{CO_2}}(\delta H_c) = 0.185\left[1 - \exp\left(-0.39\,\delta H_c^{0.4}\right)\right], \quad (\delta H_c \text{ is in centimeters}) \tag{7.57}$$

where δH_c in centimeters is given by

$$\delta H_c = 0.252\left(p_0 - p\right) \tag{7.58}$$

with pressure p in millibars at height z, and $p_0 = p$ (at $z = 0$) = 1014 mb (i.e., sea-level pressure) used to obtain Eq. 7.58. Since carbon dioxide is well-mixed vertically and horizontally above the planetary boundary layer in the troposphere and in the stratosphere, the density of CO_2 in the air (prescribed as 320 parts of CO_2 to one million parts of air to obtain Eq. 7.58) is assumed to depend only on pressure in Eq. 7.57. This assumption permits an empirical formulation for δH_c in terms of pressure less than sea-level pressure.

Figure 7.6b illustrates $\epsilon_{\chi_{CO_2}}$ as a function of δH_c. In contrast with the broadband emissivity for water vapor, the emissivity of carbon dioxide depends on temperature for larger values of δH_c. The empirical representation given Eq. 7.58, however, does not consider this effect.

The relative radiative effect of changes in the atmospheric concentration of carbon dioxide and water vapor are presented from an analysis by Norman Woods published in Section 8.2.8 in Cotton and Pielke (2007).

Finally, before using ϵ_{q3} and $\epsilon_{\chi_{CO_2}}$ to represent $\epsilon^s(u, T)$ in Eq. 7.55, it is necessary to assure that the emissivities are not double counted in those portions of the infrared spectrum where absorption by water vapor and carbon dioxide overlaps. To assure that this does not occur, Eq. 7.55 can be written as

$$\begin{aligned} \epsilon^s(u, T) &= \epsilon^s\left(u_{q3}, T\right) + \epsilon^s\left(u_{CO_2}, T\right) - \Delta\epsilon^f\left(u_{q3}, u_{CO_2}, T\right) \\ &\simeq \epsilon_{q3}(\delta P) + \epsilon_{\chi_{CO_2}}\left(\delta H_c\right) - \Delta\epsilon^f\left(u_{q3}, u_{CO_2}, T\right). \end{aligned} \tag{7.59}$$

Work by Staley and Jurica (1970), however, shows that $\Delta\epsilon^f$ is a small correction, being on the order of 0.05 to 0.10 for a pathlength through an entire typical midlatitude atmosphere. Shorter pathlengths and lower temperatures have smaller values of the overlap term. Tables 3-6 in Staley and Jurica provide values of $\Delta\epsilon^f$ as a function of pathlength and temperature, although for most mesoscale model applications, it seems appropriate to neglect this term.

Ozone, methane, and other trace gases could also be included in Eq. 7.55, using an empirical representation for its broadband emissivity, such as performed by Sasamori (1968).

For use in a model, Eqs. 7.51 and 7.52 are often written in a different form. Integrating the integrals in these two equations by parts yields

$$\begin{aligned} R\uparrow &= \sigma T_G^4 + \int_0^u \epsilon\left(u-\hat{u},T\right)\frac{d}{d\hat{u}}\sigma T^4\,d\hat{u}, \\ R\downarrow &= \sigma T_T^4 + \int_{u_T}^u \epsilon\left(\hat{u}-u,T\right)\frac{d}{d\hat{u}}\sigma T^4\,d\hat{u}, \end{aligned} \qquad (7.60)$$

where $T(u=0)$ corresponds to T_G and Eq. 7.54 was used to replace the broadband transmissivity t^s with the broadband emissivity ϵ^s.

Up to this point in the representation of longwave radiative fluxes, grid-volume averaging, as described in Chapter 4, has not been introduced. As indicated by Eq. 7.22, though, such averaged quantities are needed to represent the source/sink term S_θ. Therefore, to use Eq. 7.60, the net upward irradiance over all longwave wavelengths must be written as

$$\begin{aligned} \bar{R}\uparrow - \bar{R}\downarrow &= \overline{\sigma\left(\overline{T}_G + T_G''\right)^4} - \overline{\sigma\left(\overline{T}_T + T_T''\right)^4} \\ &+ \overline{\int_0^u \epsilon\left(u-\hat{u},\overline{T}+T''\right)\frac{d}{d\hat{u}}\sigma\left(\overline{T}+T''\right)^4 d\hat{u}} \\ &- \overline{\int_T^u \epsilon\left(\hat{u}-u,\overline{T}+T''\right)\frac{d}{d\hat{u}}\sigma\left(\overline{T}+T''\right)^4 d\hat{u}} \end{aligned}$$

where the assumption is made that $\rho \simeq \bar{\rho}$ can be used (e.g., see Eq. 4.9). Subgrid-scale correlations of temperature with itself, and with the broadband emissivity, are usually neglected in the parameterization of $R\uparrow$ and $R\downarrow$. Since $|T''| \ll \overline{T}$, in general, such an approximation seems reasonable, although under what circumstances the correlation terms can be neglected needs to be quantitatively assessed. Thus T and ϵ in Eq. 7.60 are replaced by the grid-volume resolvable quantities $\overline{T}$ and $\overline{\epsilon}$.[21]

The vertical gradients of $R\uparrow$ and $R\downarrow$ from Eq. 7.60, needed in Eq. 7.22, can therefore be written as

$$\begin{aligned} \frac{\partial R\uparrow}{\partial u} &= \int_0^u \frac{d\epsilon\left(u-\hat{u},\overline{T}\right)}{du}\frac{d}{d\hat{u}}\sigma\overline{T}^4\,d\hat{u} \\ \frac{\partial R\downarrow}{\partial u} &= \int_{u_T}^u \frac{d\epsilon\left(\hat{u}-u,\overline{T}\right)}{du}\frac{d}{d\hat{u}}\sigma\overline{T}^4\,d\hat{u} \end{aligned} \qquad (7.61)$$

where Leibnitz's rule (see footnote for Eq. 7.34) and the requirement that $\epsilon(0,T)=0$ have been used.

Sasamori (1972) has suggested a simpler (and therefore, more economical) form for Eq. 7.61. If the atmosphere between the top and the ground is assumed isothermal

[21] André et al. (1978) did include such fluctuations in their higher-order closure planetary boundary layer model, although such sophistication is prohibitively expensive for mesoscale applications. For notational convenience, $\overline{\epsilon}$ is replaced by ϵ in the subsequent text.

at the height of interest, then the integrals in Eq. 7.61 reduce to

$$\lim_{u\to 0} \int_0^u \frac{d\epsilon(u-\hat{u},\overline{T})}{du} \frac{d}{d\hat{u}} \sigma \overline{T}^4 \, d\hat{u} \to \sigma \frac{d\epsilon(u,\overline{T})}{du} \left[\overline{T}^4(u) - \overline{T}_G^4\right]$$

$$\lim_{u\to u_T} \int_{u_T}^u \frac{d\epsilon(\hat{u}-u,\overline{T})}{du} \frac{d}{d\hat{u}} \sigma \overline{T}^4 \, d\hat{u} \to \sigma \frac{d\epsilon(u_T-u,\overline{T})}{du} \left[\overline{T}_T^4 - \overline{T}^4(u)\right]$$

using Riemann-Stieltzes integration (e.g., see Lumley and Panofsky 1964:220), where $(d/du)\sigma\overline{T}^4$ is zero everywhere, except infinitesimally close to the top and ground where its value is infinite (i.e., a discontinuous function). The equations in 7.61 using Sasamori's (1972) isothermal approximation, neglecting subgrid-scale quantities, and assuming $\overline{\rho}$ is constant over the interval du (so that u can be replaced with z), can therefore be written as

$$\frac{\partial \bar{R}\uparrow}{\partial z} \simeq \sigma \frac{d\epsilon\left(u,\overline{T}\right)}{dz} \left[\overline{T}^4(z) - \overline{T}_G^4\right].$$
$$\frac{\partial \bar{R}\downarrow}{\partial z} \simeq \sigma \frac{d\epsilon\left(u_T - u,\overline{T}\right)}{dz} \left[\overline{T}_T^4 - \overline{T}^4(z)\right]. \tag{7.62}$$

The vertical derivatives of ϵ are evaluated for $\partial\bar{R}\uparrow/\partial z$ and $\partial\bar{R}\downarrow/\partial z$ at a distance z above the ground. To use Eqs. 7.56 and 7.57 in 7.62, $\delta P = \int_z^{z_T} \rho q_3 \, dz$ and $\delta H_c = 0.252(p - p_T)$ cm must be used when the argument of ϵ is $u_T - u$.

Equation 7.22 can also be written as

$$\bar{S}_\theta = \left.\frac{\partial \overline{T}}{\partial t}\right|_{\text{rad}} = -\frac{1}{\bar{\rho} C_p} \left[\left.\frac{\partial \bar{R}}{\partial z}\right|_{\text{lw}} + \left.\frac{\partial \bar{R}}{\partial z}\right|_{\text{sw}}\right], \tag{7.63}$$

where the subscripts "lw" and "sw" refer to the absorption of longwave and shortwave irradiance, respectively. Equation 7.62 or the more complete form (Eq. 7.61) can be used to represent $\partial\bar{R}/\partial z\,|_{\text{lw}}$, where

$$\left.\frac{\partial \bar{R}}{\partial z}\right|_{\text{lw}} = \frac{\partial \bar{R}\uparrow}{\partial z} - \frac{\partial \bar{R}\downarrow}{\partial z}.$$

Figure 7.7, calculated by R.T. McNider (1981, personal communication), illustrates the accuracy of using Eq. 7.62, in lieu of Eq. 7.61, to calculate temperature change in Eq. 7.63 for the night of DAY 33 of the Wangara experiment. As evident from Fig. 7.7, at least for this particular simulation, Sasamori's simplified form (Eq. 7.62) yields very realistic results, although detailed small-scale variations of temperature changes due to longwave flux divergence are not reproduced.

In a clear atmosphere without pronounced temperature and water vapor discontinuities, Paltridge and Platt (1976:187) suggested a particularly simple formulation to estimate longwave radiational cooling given by

$$-\frac{1}{\bar{\rho} C_p} \left.\frac{\partial \bar{R}}{\partial z}\right|_{\text{lw}} = -\left[0.017\overline{T} + 1.8\right], \tag{7.64}$$

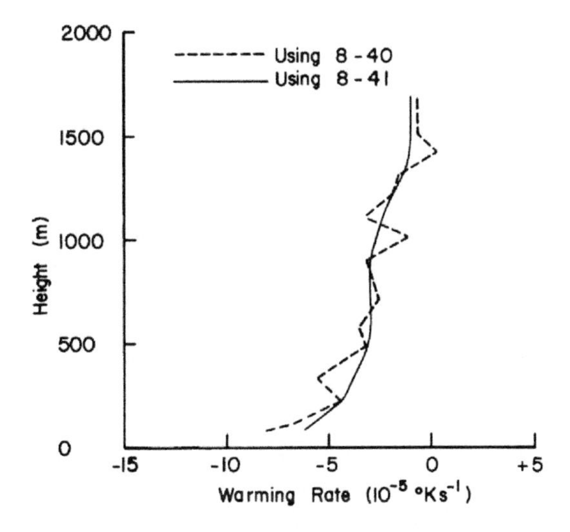

FIGURE 7.7

A comparison of a longwave radiational heating-cooling computation obtained for the night of DAY 33 during the Wangara Experiment using Sasamori's simplified parameterization (Eq. 7.62) and a more complete parameterized form Eq. (7.61). Computed by R.T. McNider (1981, personal communication).

(in units of degrees Celsius per day; $\overline{T}$ in degrees Celsius; and $-90°\text{C} \leq \overline{T} \leq 30°\text{C}$) with a standard deviation of 0.33°C. With this relation, a temperature of 0°C yields a cooling rate of 1.8°C day^{-1}.

Kuo (1979), using a more complete parameterization of longwave radiation, reported that for clear air with a temperature and moisture profile approximating that of the U.S. standard atmosphere, the infrared cooling rate is about 1.2°C day^{-1} with a maximum at the surface and 7.5 km and a minimum at 2 km and the tropopause. As emphasized by Paltridge and Platt, however, representations such as given by Eq. 7.64 are inappropriate when significant vertical gradients of temperature or water vapor occur, or when clouds exist. In general, for clear air, a longwave radiation parameterization given by Eqs. 7.61 or 7.62 appears to be appropriate for use in a mesoscale model.

The importance of longwave radiative fluxes relative to turbulent fluxes in stable nocturnal boundary layers has been investigated by Gopalakrishnan et al. (1998). With weak winds, radiative flux divergence can dominate turbulent fluxes for this situation. Jiang et al. (2000) use large eddy simulations of shallow cumulus convection to assess the relative importance of microphysics and radiative fluxes.

7.3.2.2 *Cloudy Air*

When clouds are present, their water content strongly influences the optical path length for infrared radiation. As reported by Stephens (1978a), a cumulonimbus cloud with a liquid water content of 2.5 g m^{-3} has such a small path length that it radiates as a blackbody beyond a depth of 12 m into the cloud. In contrast, a thin stratus cloud with a liquid water content of 0.05 g m^{-3} requires a depth of at least 600 m before it radiates as a blackbody. Liou and Wittman (1979) have reported that cirrus, because of its relatively low amount of water content (in the ice phase), is too shallow to be treated as a blackbody at any depth. Charlock (1982) has suggested that changes of the liquid water content of the thinnest clouds have the most significant effect on climate. Wyser et al. (1999) discuss the importance of the size of cloud droplets and ice crystals.

The radiational cooling at the top of such cloud layers can be very substantial. Roach and Slingo (1979), for example, found a cooling rate of 8.7°C h^{-1} from the 1 mb layer at the top of nocturnal stratocumulus over England. Such cooling can have a substantial impact on the entrainment rate of higher-level air into the stratocumulus layer (e.g., see Deardorff 1981) and on cold downward moving plumes within the cloud (e.g., see Caughey et al. 1982).

Stephens and Webster (1981) have shown that vertical temperature structure (and, therefore, mesoscale dynamics) is highly sensitive to cloud height, although it is only sensitive to water path for clouds that are shallow relative to the infrared optical path length. They contend that high, thin clouds at low and middle latitudes and all clouds at high latitudes tend to warm the surface compared with a clear sky, whereas all other clouds cool. Platt's (1981) results suggest that cirrus clouds having an optical depth greater than 12 in the tropics and about 5 in the midlatitudes will result in a cooling tendency below the clouds. Mesoscale models (e.g., McCumber 1980; Thompson 1993) have often treated clouds as blackbodies in the longwave portion of the spectrum, where no infrared radiation is transmitted through the cloud.

Clouds consist of liquid and/or ice crystals in a range of distribution sizes, and the details of their radiative properties are very complex, particularly for ice crystals. In the parameterization of infrared fluxes within clouds, therefore, ice crystals are normally considered in terms of a particle with a radius defined in terms of the surface area of the crystal (Paltridge and Platt 1976), since no analytic theory exists to describe the absorption and scattering of irregular particles such as ice crystals. Moreover, scattering by both ice and water droplets is generally ignored because infrared scattering occurs predominantly in the forward direction, so that the direction of the incoming radiation is essentially unaffected (Paltridge and Platt 1976).

Stephens (1978a) suggested a useful parameterization for longwave radiation within a water cloud given by

$$\begin{aligned} \bar{R}\uparrow &= \bar{R}_{\mathrm{CB}}\uparrow\left[1-\epsilon\uparrow\right]+\epsilon\uparrow\sigma\overline{T}^{4}, \\ \bar{R}\downarrow &= \bar{R}_{\mathrm{CT}}\downarrow\left[1-\epsilon\downarrow\right]+\epsilon\downarrow\sigma\overline{T}^{4}, \end{aligned} \qquad (7.65)$$

where $\bar{R}_{\mathrm{CT}}\downarrow$ and $\bar{R}_{\mathrm{CB}}\uparrow$ are the clear-air irradiance at the cloud top z_{CT} and cloud base z_{CB}, respectively, determined from a formulation such as Eq. 7.60.[22] The temperature $\overline{T}$, and $\epsilon\downarrow$ and $\epsilon\uparrow$, called the effective emissivity, are evaluated at desired levels within a cloud.

[22]In the absence of other cloud layers, the top part of Eq. 7.60 would be used to determine $\bar{R}_{\mathrm{CB}}\uparrow$, and the bottom of Eq. 7.60 is applied to calculate $\bar{R}_{\mathrm{CT}}\downarrow$.

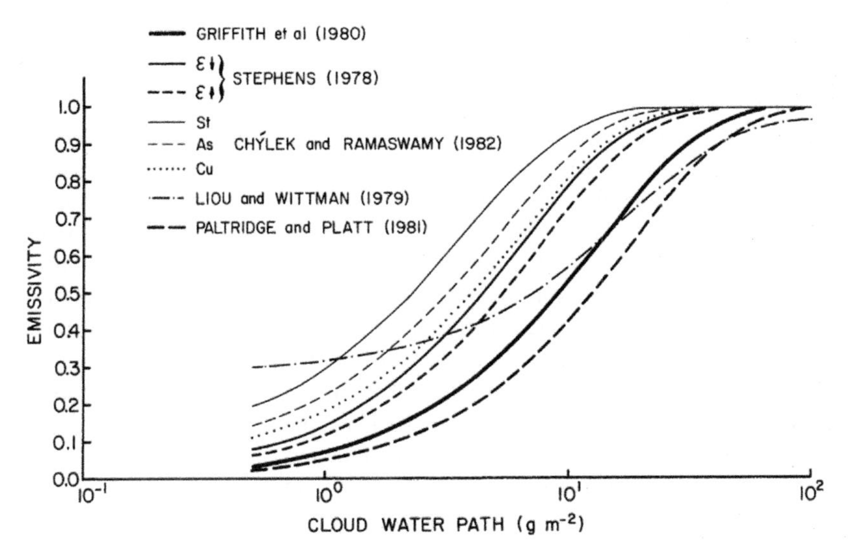

FIGURE 7.8

Cloud emittance as a function of cloud water path for various parameterizations. The parameterization of Liou and Wittman (1979), Griffith et al. (1980), and Paltridge and Platt (1981) included only ice, and the others were for liquid water clouds (from S. Ackerman, CSU, 1983, personal communication).

As reported by Stephens (1978b), these effective emissivities were obtained by solving Eq. 7.65 for $\epsilon\downarrow$ and $\epsilon\uparrow$, using a detailed radiational model with eight cloud types in a U.S. standard atmosphere to obtain $\bar{R}\downarrow$, $\bar{R}\uparrow$, and $\overline{T}$. The values of $\epsilon\downarrow$ and $\epsilon\uparrow$ were then determined as a function of integrated liquid water content (see Fig. 7.8). The resultant empirical formulation is given by[23]

$$\epsilon\uparrow = 1 - e^{-a\uparrow W\uparrow}, \quad \epsilon\downarrow = 1 - e^{-a\downarrow W\downarrow}, \tag{7.66}$$

where $a\uparrow = 0.130\,\mathrm{m^2g^{-1}}$ and $a\downarrow = 0.158\,\mathrm{m^2g^{-1}}$ were found to give the best fit to the data derived from Stephen's (1978b) detailed theoretical model. The integrated water content in the cloud above and below the level of interest ($W\downarrow$ and $W\uparrow$), was found to have the most pronounced influence on the effective emissivity; a conclusion that is further substantiated from the observational study of nocturnal stratocumulus in Great Britain by Slingo et al. (1982). A validation and extension of Stephen's (1978a) parameterized model is also given by Chýlek and Ramaswamy (1982). In that study, they concluded that emissivity is only a function of integrated liquid water for wavelengths of 8 to 11.5 μm, however, they maintain it is also a function of the droplet size distribution for wavelengths greater than 11.5 μm. For typical drop size distributions

[23] $W\uparrow = \int_{z_{CB}}^{z} \rho_i \bar{q}_1\, dz + \int_{z_{CB}}^{z} \rho_l \bar{q}_2\, dz$; $W\downarrow = \int_{z}^{z_{CT}} \rho_i \bar{q}_1\, dz + \int_{z}^{z_{CT}} \rho_l \bar{q}_2\, dz$, where the contribution from the presence of ice has been added to Stephens representation and ρ_i and ρ_l are the densities of ice and water. The appropriateness of including ice in this fashion needs to be determined.

for stratus, altocumulus, and cumulus clouds, they stated that the effect on the flux emissivity in the 8- to 14-μm band is about ±35%. Liou and Ou (1981) also present a parameterization of infrared radiative transfer in the presence of a semitransparent cloud layer and compare their results against observations and a more detailed theoretical model. In the Liou and Ou study, a model with five broadband emissivity values was used to represent the five major absorption regions in the infrared spectrum.

Stephens (1978a), however, concluded that cloud drop distribution, ambient temperature, and water vapor distributions within the cloud are not important in the estimation of this emissivity for water clouds, and this result will be used in the following discussion. It is implied from Paltridge and Platt (1976) that ice clouds can similarly be represented by expressions such as Eqs. 7.66 and 7.65, using the definition of integrated water content given in the footnote to Eq. 7.66. Figure 7.8 illustrates the emissivity as a function of cloud water path for several parameterization schemes to illustrate the uncertainty remaining in the representation of emissivity in clouds. Those of Liou and Wittman (1979), Griffith et al. (1980), and Paltridge and Platt (1981) were for ice clouds and the remainder were for liquid water clouds.

For application to a mesoscale model, Eq. 7.65 can be used when a grid volume is saturated with cloud material, and formulations such as Eq. 7.60 can be used in a clear atmosphere. The vertical gradient of Eq. 7.65, needed in Eq. 7.22, can be written as

$$\begin{aligned}\frac{\partial \bar{R}\uparrow}{\partial z} &= a\uparrow \frac{\partial W\uparrow}{\partial z} e^{-a\uparrow W\uparrow}\left[\sigma \overline{T}^4 - \bar{R}_{\mathrm{CB}}^{\uparrow}\right] + \epsilon\uparrow \frac{\partial}{\partial z}\sigma \overline{T}^4 \\ \frac{\partial \bar{R}\downarrow}{\partial z} &= a\downarrow \frac{\partial W\downarrow}{\partial z} e^{-a\downarrow W\downarrow}\left[\sigma \overline{T}^4 - \bar{R}_{\mathrm{CT}}^{\downarrow}\right] + \epsilon\downarrow \frac{\partial}{\partial z}\sigma \overline{T}^4\end{aligned} \tag{7.67}$$

If the cloud layer is assumed isothermal at the height of interest (analogous to the assumption used by Sasamori 1972 to derive Eq. 7.62), the righthand terms in Eq. 7.67 can be neglected.

When a grid volume is only partially saturated by clouds (usually denoted as a fractional coverage σ_{c}), $\partial \bar{R}/\partial z|_{\mathrm{lw}}$ in Eq. 7.63 can be written using

$$\left.\frac{\partial \bar{R}}{\partial z}\right|_{\mathrm{lw}} = \sigma_{\mathrm{c}}\left[\frac{\partial \bar{R}\uparrow}{\partial z} - \frac{\partial \bar{R}\downarrow}{\partial z}\right] + \left(1 - \sigma_{\mathrm{c}}\right)\left[\frac{\partial \bar{R}\uparrow}{\partial z} - \frac{\partial \bar{R}\downarrow}{\partial z}\right], \tag{7.68}$$

where, for example, the first term on the righthand side could be from Eq. 7.67 and the second from Eq. 7.62.

If multiple cloud layers are present, each clear and each cloudy region can be treated separately, using the adjacent regions as the vertical boundary conditions, as illustrated schematically in Fig. 7.9. For this example, the layer-averaged up and downward irradiances needed to calculate the longwave heating-cooling in Eq. 7.63 are computed from:

$$\begin{aligned}\mathrm{G}-1: \quad \bar{R}_{\mathrm{lw}} &= \left[1 - \left(\sigma_{\mathrm{c}_1} + \sigma_{\mathrm{c}_2} + \sigma_{\mathrm{c}_3}\right)\right] R_A + \sigma_{\mathrm{c}_2}\left(1 - \sigma_{\mathrm{c}_1}\right) R_J \\ &+ \sigma_{\mathrm{c}_1} R_H + \left(1 - \sigma_{\mathrm{c}_1}\right)\left(1 - \sigma_{\mathrm{c}_2}\right)\sigma_{\mathrm{c}_3} R_F\end{aligned}$$

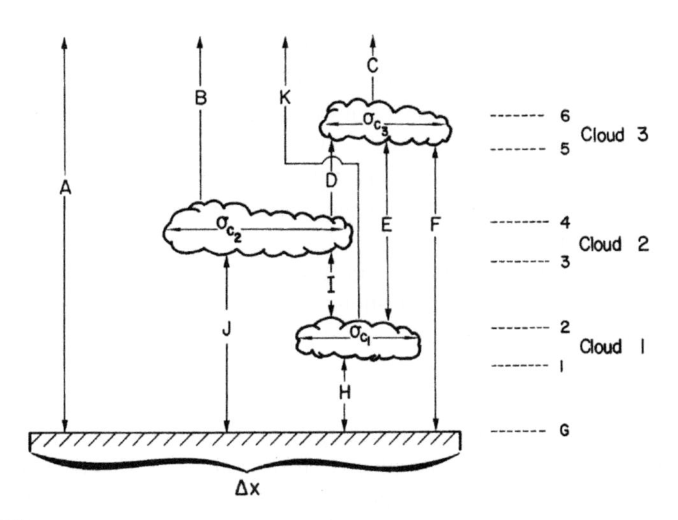

FIGURE 7.9

A schematic representation of a procedure to compute longwave irradiance in a model grid mesh when multiple cloud levels, covering various fractions of the grid increment exist. The letters indicate the subregions outside of the clouds where a formulation such as Eq. 7.60 is used to compute irradiance with the temperature at the ground replaced with $\overline{T}$ at the cloud top height in B, D, C, I, K, and E, and the temperature at model top replaced with T at the cloud base height in J, D, E, F, I, and H. Inside of the clouds, irradiance is computed using a form such as given by Eq. 7.65.

$$
\begin{aligned}
1-2: \quad \bar{R}_{\text{lw}} &= \left[1-\left(\sigma_{c_1}+\sigma_{c_2}+\sigma_{c_3}\right)\right] R_A + \left(1-\sigma_{c_1}\right)\sigma_{c_2} R_J \\
&\quad + \sigma_{c_1} R_{\sigma_{c_1}} + \left(1-\sigma_{c_1}\right)\left(1-\sigma_{c_2}\right)\sigma_{c_3} R_F \\
2-3: \quad \bar{R}_{\text{lw}} &= \left[1-\left(\sigma_{c_1}+\sigma_{c_2}+\sigma_{c_3}\right)\right] R_A + \left(1-\sigma_{c_1}\right)\sigma_{c_2} R_J \\
&\quad + \sigma_{c_2}\sigma_{c_1} R_I + \left(1-\sigma_{c_2}\right)\left(1-\sigma_{c_3}\right)\sigma_{c_1} R_K \\
&\quad + \left(1-\sigma_{c_2}\right)\sigma_{c_1}\sigma_{c_3} R_E + \left(1-\sigma_{c_1}\right)\left(1-\sigma_{c_2}\right)\sigma_{c_3} R_F, \\
3-4: \quad \bar{R}_{\text{lw}} &= \left[1-\left(\sigma_{c_1}+\sigma_{c_2}+\sigma_{c_3}\right)\right] R_A + \sigma_{c_2} R_{\sigma_{c_2}} \\
&\quad + \left(1-\sigma_{c_2}\right)\left(1-\sigma_{c_3}\right)\sigma_{c_1} R_K + \left(1-\sigma_{c_2}\right)\sigma_{c_1}\sigma_{c_3} R_E \\
&\quad + \left(1-\sigma_{c_1}\right)\left(1-\sigma_{c_2}\right)\sigma_{c_3} R_F, \\
4-5: \quad \bar{R}_{\text{lw}} &= \left[1-\left(\sigma_{c_1}+\sigma_{c_2}+\sigma_{c_3}\right)\right] R_A + \sigma_{c_2}\left(1-\sigma_{c_3}\right) R_B + \sigma_{c_2}\sigma_{c_3} R_D \\
&\quad + \sigma_{c_1}\left(1-\sigma_{c_2}\right)\left(1-\sigma_{c_3}\right) R_K + \sigma_{c_1}\left(1-\sigma_{c_2}\right)\sigma_{c_3} R_E \\
&\quad + \left(1-\sigma_{c_1}\right)\left(1-\sigma_{c_2}\right)\sigma_{c_3} R_F, \\
5-6: \quad \bar{R}_{\text{lw}} &= \left[1-\left(\sigma_{c_1}+\sigma_{c_2}+\sigma_{c_3}\right)\right] R_A + \sigma_{c_2}\left(1-\sigma_{c_3}\right) R_B \\
&\quad + \sigma_{c_1}\left(1-\sigma_{c_2}\right)\left(1-\sigma_{c_3}\right) R_K + \sigma_{c_3} R_{\sigma_{c_3}} \\
\text{above } 6: \quad \bar{R}_{\text{lw}} &= \left[1-\left(\sigma_{c_1}+\sigma_{c_2}+\sigma_{c_3}\right)\right] R_A + \sigma_{c_2}\left(1-\sigma_{c_3}\right) R_B \\
&\quad + \sigma_{c_1}\left(1-\sigma_{c_2}\right)\left(1-\sigma_{c_3}\right) R_K + \sigma_{c_3} R_C,
\end{aligned}
$$

where the subscript indicates over what layer the irradiances are evaluated (σ_{c_1}, σ_{c_2}, and σ_{c_3} refer to the three clouds in Fig. 7.9). Atwater (1974) used a similar form to proportionally weight the upward and downward longwave irradiance by the fractional coverage of a grid increment by clouds. Stephens et al. (2010) shows how poor the parameterization of clouds in models still is. They concluded, for example, that "models produce precipitation approximately twice as often as that observed and make rainfall far too lightly."

Clearly, cloud parameterization remains a source for model skill uncertainty, and this includes the ability of parameterizations to accurately mimic both longwave and shortwave radiative flux divergence.

7.3.2.3 Polluted Air

Observations have shown that even naturally occurring aerosols significantly affect the net radiation balance in the atmosphere (Alpert et al. 1998; Cautenet et al. 2000). Carlson and Benjamin (1980), for example, found typical heating rates from the combined short and longwave spectrum to be in excess of 1°C for most of the atmosphere below 500 mb in a region of suspended Saharan Desert dust. Ackerman and Cox (1982) determined that aerosols in the desert air over Saudi Arabia approximately doubled the clear sky shortwave absorption and may play an important role in the maintenance of the heat low over the peninsula.

Activities of man, such as manufacturing, agriculture, and transportation, input large quantities of aerosols and gases into the atmosphere. As reported, for example, by Wallace and Hobbs (1977), a typical urban air mass may have 10^5 or more aerosols per cm^3, whereas an air mass located over land, far from developed areas typically has a concentration of aerosols on the order of 10^4 per cm^3. Such changes of concentration can have a significant effect on the infrared irradiance. Saito (1981), for instance, has found that as visibility over central Tokyo decreases from 20 to 10 to 2 km, the downward longwave radiation increases by 1.3, 2.8, and 6.0%, respectively.

As with cloud droplets and ice crystals (as reported in Section 7.3.2.2), scattering of infrared radiation by suspended pollution particles is generally neglected since scattering occurs predominantly in the forward direction (Paltridge and Platt 1976:227). Moreover, since aerosols are usually smaller than water droplets, they are less effective at absorbing and scattering electromagnetic radiation than an equal number of cloud water particles. Ackerman et al. (1976) have included scattering in their treatment of longwave irradiance in a polluted atmosphere, although Viskanta et al. (1977a) conclude that scattering only slightly increases the importance of absorption by increasing the path length of the electromagnetic radiation.

To perform detailed calculations of the infrared extinction (i.e., absorption plus backscatter) requires knowledge of the size distribution, composition, and refractive indices of the aerosols. The refractive index $\eta = \eta_r + i\eta_i$ (discussed in detail by Liou 1980:78) is a complex number where the real and imaginary components correspond to the scattering and absorption properties, respectively, of a particle. When no absorption occurs, the refractive index has no imaginary component.

The representation of the scattering and absorption of radiation by aerosols is expected to be more complicated than that by water droplets and ice crystals. This results from the diversity of chemical species in an aerosol layer, as contrasted with one chemical substance in a clean water cloud.

In parameterizing the influence of aerosols on the infrared irradiance, Paltridge and Platt (1976:227) have suggested that only a rough estimate of the absorption by these particles is required. To determine this simple representation, a *volume absorption coefficient* (in the S.I. system the units are meter^{-1})

$$\beta_{a\lambda} = k_{a\lambda}\rho_\chi \tag{7.69}$$

is defined, where $k_{a\lambda}$ is the mass absorption coefficient specified by Eq. 7.34 with ρ_χ the density of the contaminant. This definition of an absorption coefficient of course, can also be applied to a gaseous contaminant in the atmosphere. For aerosol materials (or for water), Eq. 7.69 is also written as

$$\beta_{a\lambda} = \pi \int_0^\infty \frac{dn(r)}{dr} r^2 E_{a\lambda}\, dr, \tag{7.70}$$

where $dn(r)/dr$ is the number of particles per radius interval per unit volume at radius r, and $E_{a\lambda}$ is defined as the ratio of the absorption cross section to the geometric cross section of a single spherical particle.[24] As discussed by Paltridge and Platt (1976:78), in using a spherical representation for the aerosols, it is assumed that in a population of aerosols, their orientation is random so that a spherical shape can be used to represent their integrated influence on the flux of radiation (such an assumption, of course, is invalid when irregular-shaped aerosols have a preferential orientation, e.g., thin disks will tend to fall with their long axis more or less parallel to the ground).

From observations (e.g., Paltridge and Platt 1976:Fig. A.4; and Wallace and Hobbs 1977), the distribution of aerosols are well approximated by the relation

$$dn(r)/dr = (b_1/2.3)\, r^{-(1+b_2)} \tag{7.71}$$

in the size range greater than 0.1 μm or so, where b_1 and b_2 are constants (b_1 has dimensions whereas b_2 is dimensionless). This function, called the *Junge distribution*, is recommended by Paltridge and Platt (1976:281) as being particularly useful in radiation problems, at least in moderately polluted continental atmospheres (in clean, maritime air masses, however, Paltridge and Platt state that the Junge distribution is invalid, although for infrared radiation calculations, the low numbers of aerosols in that environment imply that $\beta_{a\lambda}$ will be very small relative to the absorption from water vapor and carbon dioxide). Modified gamma, standard gamma, and lognormal distributions of aerosols, as summarized by Paltridge and Platt (1976:Table A.1), have

[24] An absorption efficiency for the electromagnetic radiation of wavelength λ is represented by $E_{a\lambda}$. In addition, the function $n(r)$, for mathematical clarity, could also be written as $n(r, \delta r)$ to indicate that the magnitude of n is also a function of the chosen radius interval, δr. In this chapter, however, as is standard in publications on atmospheric radiation, this relation to δr will be assumed when the function form $n(r)$ is used.

also been used in lieu of Eq. 7.71. In addition, Abele and Clement (1980) discuss several functional forms, including the use of Chebyshev polynomials, to represent these distributions.

Equation 7.70 can also be written as

$$\begin{aligned}\beta_{a\lambda}\left(r_1, r_2\right) &= \pi \int_{r_1}^{r_2} r^2\, E_{a\lambda}(r) \frac{dn(r)}{dr}\, dr \\ &= \frac{\pi b_1}{2.3} \int_{r_1}^{r_2} E_{a\lambda}(r) r^{1-b_2}\, dr,\end{aligned} \tag{7.72}$$

where $\beta_{a\lambda}\left(r_1, r_2\right)$ is the volume absorption coefficient resulting from aerosols between radius r_1 and r_2, and Eq. 7.71 is used to represent $dn(r)/dr$. As suggested by Paltridge and Platt (1976:227), based on observations the absorption efficiency $E_{a\lambda}$ can be decomposed into the two parts

$$E_{a\lambda} = \begin{cases} b_3 r, & r_{\min} < r < r_m \\ 1.0, & r_m \leq r < r_{\max}, \end{cases}$$

so that

$$\beta_{a\lambda} = \frac{\pi b_1 b_3}{2.3} \left[\int_{r_{\min}}^{r_m} r^{2-b_2}\, dr \right] + \frac{\pi b_1}{2.3} \int_{r_m}^{r_{\max}} r^{1-b_2}\, dr. \tag{7.73}$$

For a Junge distribution with $\eta = 1.55 - 0.1i$, Paltridge and Platt give values of $r_{\min} = 0.01\ \mu\text{m}$, $r_m = 7\ \mu\text{m}$, and $r_{\max} = 70\ \mu\text{m}$. The constant b_1 is proportional to the total number of aerosols in the distribution, whereas b_3 is a function of wavelength λ and the refractive index of the aerosol particle. Obviously, the greater the number of the same type of aerosols, the greater the absorption. The exponent b_2 has been found observationally in polluted air masses to have values between 2 and 4 (Liou 1980:238) with $b_2 = 3$ being the most common (Paltridge and Platt 1976:280). The value for b_3 is determined from[25]

$$b_3 = \frac{dE_{a\lambda}}{dr} = \left(\frac{2\pi}{\lambda}\right) \frac{24\eta_i \eta_r}{\left[\left(\eta_r^2 - \eta_i^2 + 2\right)\right]^2 + \left(2\eta_i \eta_r\right)^2} \tag{7.74}$$

(from Paltridge and Platt 1976:227), where the righthand side of this relation is derived using the concepts of refraction (discussed by Paltridge and Platt 1976:222, 227; and Liou 1980:Appendix D). Using representative values for polluted air of $b_2 = 3$ and $b_1 = 2.3 \times 10^{-11}$ (since $b_2 = 3$, b_1 is dimensionless for this form of Eq. 7.72) and a value of b_3 of $1.34 \times 10^3\ \text{cm}^{-1}$ as given by Paltridge and Platt (1976:228) yields a value for Eq. 7.73 of

$$\beta_{q\lambda} = 3.24 \times 10^{-2}\ \text{km}^{-1} \tag{7.75}$$

Other values of the complex index of refraction, of course, yield different values of $\beta_{a\lambda}$. Paltridge and Platt, for example, given a value of $\beta_{a\lambda} = 1.9 \times 10^{-1}\ \text{km}^{-1}$ for a carbonaceous aerosol with a value of $\eta_i = 0.6$.

[25]In deriving this expression, any variation of η_i and η_r with r is neglected.

When condensation occurs on dry hydroscopic aerosols, the refractive index is assumed to approach a pure water droplet as it grows larger. Hänel (1971) proposed the empirical formulation for shortwave irradiance, which is also equally applicable in the infrared region, given as

$$\begin{aligned}\eta_{\mathrm{r}} &= \eta_{\mathrm{r}_{q2}} + \left(\eta_{\mathrm{r}} - \eta_{\mathrm{r}_{q2}}\right) / \left(\mathrm{r}_{q2}/\mathrm{r}_0\right)^3, \\ \eta_{\mathrm{i}} &= \eta_{\mathrm{i}_{q2}} + \left(\eta_{\mathrm{i}} - \eta_{\mathrm{i}_{q2}}\right) / \left(\mathrm{r}_{q2}/r_0\right)^3,\end{aligned} \tag{7.76}$$

where $\eta_{\mathrm{r}_{q2}}$ and $\eta_{\mathrm{i}_{q2}}$ refer to the real and imaginary components of the refractive index of liquid water, and r_{q2}/r_0 is the ratio of the radius of the liquid droplet to that of the dry aerosol (see Eq. 7.101). Equation 7.76 is also given by Nilsson (1979) for aerosol extinction of longwave radiation.

In applying Eq. 7.73, or other analogous formulations to the atmosphere, it is generally recognized that the absorption of infrared radiation by aerosols will only be important within the wavelength band from 8 to 14 μm. Elsewhere in the longwave region, the absorption by carbon dioxide and water vapor is assumed dominant[26] (e.g., see Ackerman et al. 1976:33, Zdunkowski et al. 1976:2403, and Welch et al. 1978:140).

It is, therefore, necessary to integrate Eq. 7.73 over the wavelengths 8 to 14 μm. As shown by Paltridge and Platt (1976: their Appendix A), however, η_{r} and η_{i} can vary substantially as a function of wavelength within even this small interval. Therefore, one procedure is to evaluate $\beta_{a\lambda}$ over smaller intervals of wavelength in which η_{r} and η_{i} can be assumed constant.

In a polluted atmosphere without clouds, the emissivities could then be computed from[27]

$$\epsilon_{\chi_m}\left(\Delta z_{\mathrm{a}}\right) = \sum_{m=1}^{M} \alpha \sum_{\Delta\lambda} \left[1 - b_{s_m} \exp\left(-\beta_{a\lambda}^{\chi_m} \Delta z_{\mathrm{a}}\right)\right] \quad (8\ \mu\mathrm{m} \leq \lambda \leq 14\ \mu\mathrm{m}), \tag{7.77}$$

where the sum $\sum_{\Delta\lambda}$ (using a constant $\Delta\lambda$) is over as many wavelength intervals needed to represent η_{r} and η_{i} as a constant within that interval. The coefficient α is inserted to represent the average value of $\frac{\pi B_{\bar{\lambda}}(T)}{\sigma T^4}$ within the wavelength interval. Within each interval $\Delta\lambda$ for each chemical species, $\beta_{a\bar{\lambda}}^{\chi_m}$ is the average or gaseous contaminant for each wavelength interval (see the footnote to Eq. 7.77). The quantity Δz_{a} is a depth within an aerosol layer. It may also be necessary to define two separate emissivities, $\epsilon_{\chi_m}\downarrow$ and $\epsilon_{\chi_m}\uparrow$ using the form given by Eq. 7.77, analogous to that required for a

[26]The region of the electromagnetic spectrum between 8 and 14 μm is called the *atmospheric window*, since radiation is transmitted through the air relatively unattenuated at these wavelengths. Carbon dioxide and water vapor do not have substantial absorption lines in this portion of the spectrum.

[27]Equation 7.77 can be used with $b_{s_m} = 1$ only if the absorption of each different chemical species is over different intervals. If not, the regions of overlap in absorption for these materials must be corrected for (e.g., see Eq. 7.59 for CO_2 and H_2O) in order not to include erroneous excessive absorption. To the author's knowledge, there has not been any work using a simple parameterization such as Eq. 7.77 for use in a mesoscale model although it appears to be a reasonable representation.

water cloud (i.e., Eq. 7.66) since the spectral composition of the irradiance at the top and bottom of the aerosol clouds could be different.

To estimate the divergence of infrared irradiance caused by pollution needed in Eq. 7.63, Eq. 7.77 can be used in an expression such as

$$\begin{aligned}\frac{\partial R\uparrow}{\partial z} &= \int_{z_{AB}}^{z} \frac{d\epsilon_{\chi_m}\left(z-\hat{z}\right)}{dz} \frac{d}{d\hat{z}} \sigma \overline{T}^4 \, d\hat{z} \\ \frac{\partial R\downarrow}{\partial z} &= \int_{z_{AT}}^{z} \frac{d\epsilon_{\chi_m}\left(\hat{z}-z\right)}{dz} \frac{d}{d\hat{z}} \sigma \overline{T}^4 \, d\hat{z}\end{aligned} \quad (7.78)$$

where z_{AB} and z_{AT} are the bottom and top of the aerosol layer, respectively. This formulation is analogous to Eq. 7.61. A simplified representation such as Eq. 7.62 could also possibly be used with T_{G} and T_{T} replaced by the temperatures at the base and top of the aerosols, and u defined in terms of distance from the base of the aerosols.

In actual parameterizations of polluted air masses, several methods to represent the absorption of infrared radiation by pollutants have been used. Viskanta et al. (1977a,b) for example, used the gas ethylene, which has a strong absorption in the 8 to 12 μm interval to represent the net effect of all pollutants. Atwater (1971a) assumed a modified gamma distribution given by

$$n(r) = 5.7 \times 10^{19} r^3 \exp\left(-45 r^{0.25}\right)$$

to represent the size distribution of aerosols, and an imaginary part of the refractive index of 0.25 for infrared radiation. His distribution of aerosols assumed a concentration of 10^6 cm^{-3} with a modal radius of 0.005 μm. The value of the volume absorption coefficient that he used to represent the entire infrared spectrum varied from 0 to 0.8 km^{-1} for his summer simulations, and from 0 to 0.5 km^{-1} for his winter simulations over an urban area. The higher values of the absorption coefficient were adopted to represent severe pollution conditions. In Atwater (1971b) he stated that volume absorption coefficients greater than 0.1 km^{-1} cause changes in temperature that exceed those caused by water vapor. Andreyev and Ivlev (1980) found that large aerosols (i.e., ≥ 0.5 μm) tend to be minerals and to absorb primarily between 2 to 15 μm wavelengths.

Welch and Zdunkowski (1976), Zdunkowski et al. (1976), and Welch et al. (1978) used a parameterization developed by Korb et al. (1975) to represent both the scattering and absorption of infrared radiation by aerosols (and water vapor) in the atmospheric window. They used measured values for dry aerosol parameters from the industrialized area of Mainz, Germany and determined that, in general, η_{i} increases with wavelength. For $\lambda = 10$ μm, they found $\eta_{\mathrm{r}} = 1.7$ and $\eta_{\mathrm{i}} = 0.34$ for the dry aerosols with $\beta_{a\lambda} = 4.04 \times 10^{-3}$ km^{-1} when the total particle concentration was 2262 cm^{-3}. These investigators also included the effect of relative humidity on $\beta_{a\lambda}$. They estimated the ratio of $\beta_{a\lambda}$ for a hydroscopic particle in a moist atmosphere to $\beta_{a\lambda}$ of a completely dry aerosol as 1.01, 1.54, 3.12, and 11.03 for relative humidities of 20, 75, 95, and 99%, respectively, with $\lambda = 10$ μm.

Ackerman et al. (1976) also included scattering, as well as absorption, in their simulation of infrared irradiance within a polluted atmosphere, although they neglected the dependence of aerosol properties on relative humidity. Using aerosol data from 342 distributions measured in Los Angeles, they found with an observed average concentration of 10^5 cm^{-3}, excess cooling caused by aerosols in the lowest 1 km of almost 1°C day^{-1}. They also concluded that increasing the concentration of aerosols tends to produce more isotropic scattering and to increase the fraction of infrared energy absorbed.

As with clouds, the skillful parameterization of the effects of aerosols remain a major challenge (and this is true for both longwave and short wave radiative flux divergence). See, for example, the summary table 2-2 on page 40 in the National Research Council (2005) report. There is also a detailed discussion of the radiative flux effects of a wide spectrum of aerosols (and gases including CO_2) in Chapter 2 of Solomon et al. (2007). Bond et al. (2013) documented the substantive effect of black carbon (soot) on the radiation budget and shown it has a major effect on the radiation budget. Additional discussion of the effect of pollution on radiation is provided in Sections 7.3.5 and 13.2.5. Saleeby and van den Heever (2013) also discuss the 2012 improvements to the representation of the aerosol effect (including the representation of radiative flux divergence) in the RAMS model.

7.3.3 Shortwave Radiative Flux Divergence

Shortwave irradiance is composed of the two components

1. direct irradiance, and
2. diffuse irradiance.

Direct shortwave irradiance is that which reaches a point without being absorbed or scattered from its line of propagation by the intervening atmosphere. The image of the Sun's disk as a sharp and distinct object represents that portion of the shortwave radiation that reaches the viewer directly. Diffuse irradiance, by contrast, reaches the observer after first being scattered from its line of propagation. On an overcast day, for example, the Sun's disk is not visible and all of the shortwave irradiance is diffuse. On such days, this diffuse solar radiation may be nearly isotropic (Zdunkowski et al. 1980).

The direct downward solar irradiance reaching a horizontal surface of unit area at the top of the atmosphere $R^{\downarrow}_{\mathrm{sw}0}$, can be written as

$$R^{\downarrow}_{\mathrm{sw}0} = \bar{R}^{\downarrow}_{\mathrm{sw}0} = \begin{cases} S_0\left(a^2/r^2\right)\cos Z, & |Z < 90°| \\ 0, & |Z \geq 90°|. \end{cases} \tag{7.79}$$

The ratio of the average distance of the Earth from the Sun to its location at any time of the year can be calculated, from Paltridge and Platt (1976:57, 63) as

$$\begin{aligned}\frac{a^2}{r^2} &= 1.000110 + 0.034221\cos d_0 + 0.001280\sin d_0 \\ &\quad + 0.000719\cos 2d_0 + 0.000077\sin 2d_0\end{aligned} \tag{7.80}$$

with

$$d_0 = 2\pi m/365. \tag{7.81}$$

The variable m is the day number starting with 0 on 1 January and ending on 31 December. The quantity $S_0 = 1376\ \mathrm{W\,m^{-2}}$ (Hickey et al. 1980) is the irradiance from the Sun on a surface of unit area perpendicular to the direction of propagation of the Sun's electromagnetic energy at the semi-major axis distance of the Earth's elliptical orbit from the Sun a; S_0 is called the *solar constant*. The distance of the Earth from the Sun at any given time varies from $r = 0.98324a$ in early January to $r = 1.01671a$ in early July (List 1971, Table 170 and from Eq. 7.80). The variable Z is the zenith angle (e.g., see Fig. 7.2), defined as 90° when the Sun's disk bisects the horizon and as 0° when it is overhead.

The zenith angle is defined by

$$\cos Z = \cos\phi\ \cos\delta_{\mathrm{sun}}\ \cos h_r + \sin\delta_{\mathrm{sun}}\ \sin\phi, \tag{7.82}$$

where ϕ is latitude, δ_{sun} the declination of the Sun (which ranges between $+23.5°$ on 21 June to $-23.5°$ on 22 December),[28] and h_r the hour angle ($0° \equiv$ noon). Using Eq. 7.82, sunrise and sunset occur when $Z = \pm 90°$ and can be obtained from

$$h_r = \arccos\{-\tan\delta_{\mathrm{sun}}\ \tan\phi\},$$

(when $\tan\delta_{\mathrm{sun}}\ \tan\phi < -1$, night occurs the entire time, whereas for $\tan\delta_{\mathrm{sun}}\tan\phi > 1$, the Sun is up for the entire 24-h period).

7.3.3.1 Clear Air

Direct Irradiance In a clear, clean atmosphere, ozone, water vapor, and the constant gases (see Table 2.1, particularly diatomic oxygen, Kondratyev 1969:261) are the principal absorbers of shortwave irradiance. Of these atmospheric constituents, water vapor is the major source of heating by shortwave absorption within the troposphere (e.g., List 1971:420; Paltridge and Platt 1976:94). Within the boundary layer, heating as a result of shortwave absorption has been shown to be substantial (e.g., Moores 1982). This water vapor absorption occurs at the near-infrared portion of the solar spectrum.

Paltridge and Platt (1976:95) have suggested a formulation for fractional absorption over the entire solar spectrum, based on Yamamoto's (1962) study, which is given by

$$a_{q3} = 2.9\,\delta P / \left[\left(1 + 141.5\,\delta P\right)^{0.635} + 5.925\,\delta P \right],$$

where δP in units of grams per centimeter square is defined following Eq. 7.56 using $\delta z \to \infty$. The slight effect of CO_2 and O_2 absorption is also included in this empirical formulation, which is accurate to within 1% over values of δP from 10^{-2} to $10\ \mathrm{g\,cm^{-2}}$. With $\delta P = 10\ \mathrm{g\,cm^{-2}}$, for instance, $a_{q3} = 0.18$. Note that the δP terms which appear

[28] $\delta_{\mathrm{sun}} = 0.006918 - 0.399912\cos d_0 + 0.070257\sin d_0 - 0.006758\cos 2d_0 + 0.000907\sin 2d_0 - 0.002697\cos 3d_0 + 0.001480\sin 3d_0$; δ_{sun} is in radians (Paltridge and Platt 1976:63).

in this expression would need to be multiplied by $(\cos Z)^{-1}$ to account for slanted pathlengths of sunlight through the atmosphere.

Atwater (1974) also uses a similar representation to represent the absorption of shortwave irradiance. Obtained from McDonald (1960), this empirical relation has been expressed by McCumber (1980) as[29]

$$a_{q3} = 0.077(\delta P/\cos Z)^{0.3}$$

With $\delta P = 10\,\mathrm{g\,cm^{-2}}$, a_{q3} in this relation with $Z=0$ is equal to 0.15. These formulations for absorption are used in Eq. 7.63 as

$$\frac{\partial \bar{R}\downarrow_{\mathrm{sw}}}{\partial z} \simeq -\frac{a^2}{r^2}\frac{\partial S_0 \cos Z a_{q3}}{\partial z} = -S_0\frac{a^2}{r^2}\cos Z\frac{\partial a_{q3}}{\partial z}. \tag{7.83}$$

Upward shortwave reflections from such surfaces as clouds, ground, and water bodies can also influence radiative heating. Unfortunately, because of the small-scale irregularities of these surfaces, the impinging solar radiation is generally not reflected as by a mirror[30] but is reoriented to a wide range of vertical directions. Such reorientation, for example, explains why, except over water, ground reflections looking away from the Sun are about equal in brightness as when looking in the direction of the Sun. Heating from these reflections has not been included in mesoscale models, although its incorporation would be straightforward, assuming isotropic reflection from the surface.

The shortwave irradiance absorbed by the ground can be written as

$$\bar{R}_{\mathrm{SWG}} = \left(\bar{R}^{\downarrow}_{\mathrm{SWG}} + \bar{R}^{\mathrm{D}}_{\mathrm{SWG}}\right)(1 - A), \tag{7.84}$$

where $\bar{R}^{\mathrm{D}}_{\mathrm{SWG}}$ is the diffuse irradiance at the ground (see Section 7.3.3.1), A the *albedo*[31] (reflectance) of the ground, and $\bar{R}^{\downarrow}_{\mathrm{SWG}}$ determined from a formulation such as Eq. 7.83 by integrating through the depth of the atmosphere. The upward reflected irradiance from the ground is, therefore, given by

$$\bar{R}^{\mathrm{up}}_{\mathrm{SWG}} = A\left(\bar{R}^{\downarrow}_{\mathrm{SWG}} + \bar{R}^{D}_{\mathrm{SWG}}\right).$$

Albedo is discussed in more detail in Chapter 9. For use in Eq. 7.63, the divergence of upward irradiance could be written as

$$\partial \bar{R}^{\mathrm{up}}_{\mathrm{sw}}/\partial z = -\bar{R}^{\mathrm{up}}_{\mathrm{SWG}}\,\partial a_{q3}/\partial z. \tag{7.85}$$

[29] For large zenith angles, the influence of the curvature of the Earth should also be considered. For values of $Z \lesssim 80°$, this expression for a_{q3} is very accurate. For larger values of Z, the ratio $1/\cos Z = \sec Z$ gives a result that is too large because of the curvature of the Earth and refraction effects (see List 1971:Table 137).

[30] The exception to this observation is over calm water bodies. Visible satellite imagery often reveals the disk image of the Sun over ocean areas in which the winds are very light.

[31] Representative values of A are given in Table 10.5.

Diffuse Irradiance In a clear, clean atmosphere, scattering of shortwave irradiance occurs as this electromagnetic energy propagates through the gases in the atmosphere. Discussed in detail in a number of texts (e.g., Liou 1980:Section 3.7), this type of multiple reflection is called *Rayleigh scattering* and is roughly inversely proportional to the fourth power of wavelength. Rayleigh scattering, which accounts for the blue color of the sky, occurs when the wavelength of the radiation is much larger than the objects causing the scatter (e.g., visible light has a wavelength much greater than the size of the molecules of gas in the air). The scattering of shortwave irradiance increases its path length and thereby enhances the heating of the atmosphere.

Atwater and Brown (1974) used an expression for fractional transmissivity of shortwave irradiance at the ground that accounted for downward Rayleigh scattering caused by O_3, O_2, and CO_2, as well as the absorption from these gases. Originally presented by Kondratyev (1969), this relation is given by

$$t = 1.03 - 0.08\sqrt{\left[9.49 \times 10^{-4} p(\text{in mb}) + 0.051\right] / \cos Z}.$$

Atwater and Ball (1981) present a similar formulation for t. For $p = 1000$ mb and $Z = 0$, this gives a value of $t = 0.95$; $Z = 45°$ yields $t = 0.93$. Equation 7.84 can then be written as[32]

$$\bar{R}_{\text{sw}_\text{G}} = \left(t - a_{q3}\right)\left(1 - A\right) R^{\downarrow}_{\text{sw}_0} \tag{7.86}$$

where the absorption loss from water vapor is included. Above the surface

$$\bar{R}_{\text{sw}} = \bar{R}^{\downarrow}_{\text{sw}} + \bar{R}^{\text{D}}_{\text{sw}} = \left(t - a_{q3}\right) R^{\downarrow}_{\text{sw}_0}$$

can be used with t and a_{q3} evaluated at the pressure height p. The middle expression in $\bar{R}_{\text{sw}}$ is the direct plus diffuse radiation, and the righthand side includes the astronomical effect (in $R^{\downarrow}_{\text{sw}_0}$, see Eq. 7.79), the extinction due to water vapor absorption (in a_{q3}), and the Rayleigh scattering effect (in t) as the shortwave radiation is transmitted through the atmosphere. The derivative of $\bar{R}_{\text{sw}}$ with height yields the heating from direct plus diffuse solar radiation.

As reported by List (1971:420), as long as the scattering particles are small in comparison with the wavelength of the incident light (i.e., Rayleigh scattering), half of the scattering is downward and half is upward. This simplifies the incorporation of diffuse irradiance into t, such as described by List.

7.3.4 Cloudy Air

The influence of clouds on solar irradiance is significant. Gannon (1978), for example, found that the sea-breeze circulation over south Florida was terminated as the shading from cirrus over the land markedly reduced the solar flux reaching the ground.

[32]Note that a_{q3} can be subtracted from t in this fashion because there is little spectral overlap between absorption and scattering by the standard atmospheric gases and by water vapor. The standard gases scatter predominantly in the shorter visible wavelength, whereas water vapor absorbs mostly in the near infrared.

Carpenter (1982a) demonstrated, using a numerical model, that differential ground heating as a result of cloud shadowing by a bank of altocumulus clouds can generate significant mesoscale ascent. This region of upward motion apparently resulted in substantial thunderstorm activity over England as the cloud bank moved eastward on the case study day that he examined. In a regional model ($\Delta x = \Delta y \simeq 58$ km), Wong et al. (1983a) have illustrated that large errors in precipitation and pressure fields can occur if cloud influences on both longwave and shortwave radiative flux divergence are not included. Sasamori (1972) concluded that when clouds are present, their effect in the solar wavelengths is primarily to reduce shortwave radiation transmission below clouds rather than cause heating within them. Transmission is reduced as the radiation is reflected into space because of the relatively high albedo of the top surfaces of water and ice clouds. Gu et al. (2001) found that a cumulus cloud field can focus solar reflection from clouds enough that the surface insolation can occasionally exceed the solar input at the top of the atmosphere. Harrington et al. (2000) investigated with a model the radiative impacts on the growth of cloud droplets within Arctic stratus.

In cloudy air (and in layers of aerosols), the radiative transfer of shortwave electromagnetic energy is more complicated than in clear air because scattering becomes much more important and involves a complex pattern of multiple reflections when the wavelength of the radiant energy and particle size is about the same. For longwave irradiance, scattering is generally of less importance as discussed in Sections 7.3.2.2 and 7.3.2.3.

The mathematical procedure to represent this scatter is often referred to as *Mie scattering*, after the first individual who solved the equations for radiative transfer when the wavelength and particle size are about equal. Described in detail by Kondratyev (1969:Section 4.4) and Liou (1980:Chapter 5), the formulation for this type of scattering, even for spherical particles is much more involved than for Rayleigh scattering. The expression "Mie scattering" is used when the incident wavelength is equal to or smaller than the particles that cause the scatter. Qualitatively, as discussed by Liou (1980:7), particles tend to scatter radiative energy preferentially in the forward direction, but with complicated side lobes,[33] when the particle size and wavelength of radiation are of the same order. As particles become larger, their forward scatter becomes greater for the same wavelength of irradiance.

As discussed by Paltridge and Platt (1976:103), multiple scattering of shortwave irradiance within a cloud increases its absorption since its path length increased. Although the imaginary component of the refractive index of ice and water (which is proportional to the absorption) is very small at wavelengths less than 2 μm and the wavelengths of absorption of water vapor, and of ice and liquid water, are similar, Paltridge and Platt (1976:105) have suggested that cloud absorption by droplets and ice crystals may be a subtle but important component in cloud dynamics. Such an effect, however, has not yet been included in mesoscale models, although Stephens (1978a) presents one such parameterization.

[33]Side lobes refer to local maxima in scattered radiation at various angles off of the original line of propagation of the electromagnetic energy.

Stephens divides the solar spectrum into intervals:

1. 0.3 μm to 0.75 μm where absorption is neglected,
2. 0.75 μm to 4.0 μm where absorption is included.

Using the definition of optical thickness given by Eq. 7.34, the volume absorption coefficient given by Eqs. 7.69 and 7.70 with $n(r)$ corresponding to a cloud of liquid droplets yields

$$\tau_\lambda = \pi \int_z^{z+\Delta z_c} \int_0^\infty n(r) r^2 E_{a\lambda}\, dr\, dz. \tag{7.87}$$

In this expression for optical thickness, Δz_c is a portion or all of the cloud thickness (in contrast with Eq. 7.34, this expressions for τ_λ is not integrated to the top of the atmosphere). Since cloud droplets are large relative to shortwave irradiance, Eq. 7.87 can be written as

$$\tau \simeq 2\pi \int_z^{z+\Delta z_c} \int_0^\infty n(r) r^2\, dr\, dz, \tag{7.88}$$

where the ratio of the scattering cross section to the geometric cross section (i.e., $E_{a\lambda}$) is 2, as determined from Mie theory. (This condition that $E_{a\lambda} = 2$ is called the *large drop assumption* since the wavelength of the electromagnetic energy is presumed much less than the size of the cloud droplets.) In addition, because $E_{a\lambda}$ is identically equal to 2, the wavelength dependence in τ is removed. Stephens then defined an effective radius of the cloud droplet distribution as

$$r_e = \int_0^\infty n(r) r^3\, dr \Big/ \int_0^\infty n(r) r^2\, dr, \tag{7.89}$$

so that, since volume $= \frac{4}{3}\pi r^3$, Eq. 7.88 can be written as

$$\tau \simeq \frac{3}{2}\, \delta p_{q2} \big/ \left(r_e \rho_{q2}\right). \tag{7.90}$$

The importance of the effective radius in mesoscale models is explored by Wyser et al. (1999). They found, for example, that cloud droplet size information has its largest effect on the shortwave radiative fluxes of thin clouds.

The quantity δp_{q_2}, analogous to δP_{q_3} defined following Eq. 7.56, is given by

$$\delta p_{q_2} = \frac{4}{3}\pi \int_z^{z+\Delta z_c} \int_0^\infty \rho_{q_2} n(r) r^3\, dr\, dz.$$

The integral $\int_0^\infty \frac{4}{3}\pi \rho_{q_2} n(r) r^3\, dr$ is the *liquid water content* of the cloud at a given level; δp_{q_2} is in units of mass per unit area.

Grid-volume averages of Eq. 7.90 (see Eq. 4.6), or a more general form, for τ or τ_λ should be made, of course. However, essentially no information is known regarding subgrid-scale fluctuations on the mesoscale within clouds.

Using Mie theory, Stephens developed empirical formulations for τ as a function of integrated liquid water content given as

$$\log_{10}\tau = \begin{cases} 0.2633 + 1.7095\ \ln\left[\log_{10} W\right], & 0.30\ \mu\text{m} \le \lambda \le 0.75\ \mu\text{m}, \\ 0.3492 + 1.6518\ \ln\left[\log_{10} W\right], & 0.75\ \mu\text{m} \le \lambda \le 4.0\ \mu\text{m}, \end{cases} \quad (7.91)$$

where $W = \delta p_{q_2}$ is used in units of grams per meter squared. For a value of $W = 100\ \text{g m}^{-2}$ (typical of an altostratus cloud), $\tau = 28$ for the shorter wavelengths, and $\tau = 31$ for the longer wavelengths. Twomey (1978) noted that when $\tau \ge 10$, all solar irradiance exiting from the bottom of the cloud is diffuse.

Assuming that the ground surface is nonreflective (i.e., $A = 0$), or any underlying cloud layer is nonreflective, Stephens writes

$$r_\text{c} = \frac{\beta_1\tau/\cos Z}{1+\beta_1\tau/\cos Z}, \quad t_\text{c} = 1 - r_\text{c}, \quad \left(0.3\ \mu\text{m} \le \lambda \le 0.75\ \mu\text{m}\right),$$

$$\left.\begin{aligned} r_\text{c} &= \left(u^2 - 1\right)\left[\exp\left(\tau_\text{eff}\right) - \exp\left(-\tau_\text{eff}\right)\right]/R_\text{c}, \\ t_\text{c} &= 4u/R_\text{c}, \quad a_\text{c} = 1 - r_\text{c} - t_\text{c}, \end{aligned}\right\} \left(0.75\ \mu\text{m} < \lambda \le 4.0\ \mu\text{m}\right) \quad (7.92)$$

where

$$\begin{aligned} u^2 &= \left(1 - \omega + 2\beta_2\omega\right)/(1-\omega), \\ \tau_\text{eff} &= \left\{(1-\omega)\left[1 - \omega + 2\beta_2\omega\right]\right\}^{1/2}\tau/\cos Z, \\ R_\text{c} &= (u+1)^2\exp\left(\tau_\text{eff}\right) - (u-1)^2\exp\left(-\tau_\text{eff}\right). \end{aligned}$$

The β terms represent the fraction of incident radiation backscattered. Using his theoretical model, Stephens (1978b) estimated values of β_1, β_2, and ω are given in Table 7.2.

With these two spectral region parameterizations, radiative heating in model layers within and transmission of radiation through a cloud can be determined using Eq. 7.92, Table 7.2, and Eq. 7.91. Shortwave heating within the cloud, for use in Eq. 7.63, is obtained from

$$\partial\bar{R}_{\text{sw}_\text{c}}/\partial z = -\left(\bar{R}_\text{sw}^{\downarrow} + \bar{R}_\text{sw}^\text{D}\right)_\text{CT}\partial a_\text{c}/\partial z, \quad (7.93)$$

where $(\bar{R}_\text{sw}^{\downarrow} + \bar{R}_\text{sw}^\text{D})_\text{CT}$ is the incident direct and diffuse shortwave irradiance at the top of the cloud. The amount of irradiance transmitted through the cloud is given by Eq. 7.92. For a single cloud layer, the shortwave radiation at the ground given by Eq. 7.86 is modified to

$$\bar{R}_{\text{sw}_\text{G}} = t_\text{c}\left(1 - A\right)\left(\bar{R}_\text{sw}^{\downarrow} + \bar{R}_\text{SW}^\text{D}\right)_\text{CT}, \quad (7.94)$$

where the scattering and absorption below the cloud has been neglected.[34]

[34]For thick, low clouds the neglect of the extinction of shortwave radiation by water vapor and the other gases below clouds is reasonable. When the clouds are optically thin or at high levels, however the reduction of solar radiation as it propagates through the air below the clouds should be included.

Table 7.2 The broadband values of ω,β_1, and β_2 used to determine Eq. 7.92 (from G. Stephens 1982, personal communication).

	cos Z								
τ	**1.0**	**0.8**	**0.7**	**0.6**	**0.5**	**0.4**	**0.3**	**0.2**	**0.1**
(a) Average values of $1-\omega$									
1	.0225	.0222	.0218	.0208	.0199	.0155	.0109	.0059	.0017
2	.0213	.0200	.0179	.0176	.0156	.0118	.0078	.0038	.0010
5	.0195	.0166	.0146	.0125	.0096	.0069	.0043	.0021	.0005
10	.0173	.0138	.0114	.0093	.0070	.0049	.0026	.0013	.0003
16	.0156	.0111	.0090	.0073	.0052	.0035	.0019	.0009	.0002
25	.0115	.0088	.0069	.0052	.0038	.0026	.0014	.0007	.0001
40	.0104	.0055	.0042	.0032	.0023	.0014	.0008	.0003	.0001
60	.0083	.0050	.0038	.0028	.0020	.0013	.0007	.0003	.0001
80	.0069	.0043	.0035	.0022	.0018	.0011	.0006	.0003	.0000
100	.0060	.0043	.0035	.0022	.0018	.0011	.0006	.0003	.0000
200	.0044	.0031	.0025	.0016	.0011	.0007	.0004	.0002	.0000
500	.0026	.0018	.0014	.0010	.0007	.0005	.0003	.0001	.0000
(b) Average values of β_1									
1	.0421	.0557	.0657	.0769	.0932	.1111	.1295	.1407	.1196
2	.0472	.0615	.0708	.0803	.0924	.1017	.1077	.1034	.0794
5	.0582	.0692	.0744	.0782	.0815	.0812	.0776	.0680	.0483
10	.0682	.0726	.0737	.0733	.0723	.0685	.0626	.0527	.0359
16	.0734	.0738	.0728	.0707	.0680	.0631	.0564	.0465	.0310
25	.0768	.0744	.0723	.0691	.0653	.0598	.0526	.0427	.0281
40	.0791	.0749	.0719	.0680	.0636	.0575	.0501	.0402	.0261
60	.0805	.0752	.0717	.0674	.0627	.0563	.0488	.0389	.0251
80	.0812	.0754	.0717	.0672	.0622	.0558	.0481	.0382	.0246
100	.0820	.0757	.0717	.0670	.0619	.0553	.0475	.0376	.0241
200	.0831	.0763	.0721	.0672	.0619	.0552	.0473	.0374	.0241
500	.0874	.0800	.0755	.0703	.0647	.0576	.0494	.0392	.0262
(c) Average values of β_2									
1	.0477	.0627	.0734	.0855	.1022	.1200	.1379	.1465	.1207
2	.0537	.0690	.0788	.0886	.1003	.1090	.1133	.1065	.0794
5	.0660	.0769	.0817	.0850	.0871	.0864	.0801	.0688	.0474
10	.0759	.0793	.0795	.0781	.0757	.0705	.0629	.0516	.0339
16	.0801	.0787	.0766	.0732	.0689	.0626	.0543	.0434	.0277
25	.0807	.0759	.0724	.0678	.0625	.0555	.0471	.0368	.0229
40	.0770	.0700	.0656	.0603	.0545	.0476	.0396	.0302	.0184
60	.0699	.0621	.0575	.0522	.0466	.0401	.0329	.0248	.0148
80	.0634	.0556	.0510	.0460	.0408	.0348	.0283	.0211	.0125
100	.0534	.0461	.0420	.0376	.0330	.0279	.0225	.0166	.0097
200	.0415	.0353	.0319	.0283	.0246	.0206	.0165	.0120	.0068
500	.0251	.0208	.0186	.0163	.0140	.0115	.0090	.0064	.0032

For multiple cloud layers, the irradiance reaching each cloud top can be estimated using a value corresponding to the irradiance which exited downward from the next highest cloud base minus the extinction in the free atmosphere below that cloud base. Since each cloud will absorb and backscatter shortwave radiative energy, the free atmosphere irradiance reaching lower clouds will be progressively less. When clouds are assumed to cover only a portion of a grid mesh (such as sketched in Fig. 7.9), a fractional weighting of the clouds' contribution to absorption and scattering can be performed, although cloud shape (Welch et al. 1980; Welch and Zdunkowski 1981a) and shading of adjacent clouds will influence the flux of shortwave radiation. The second effect will be particularly important when the clouds are in close proximity to one another (Gube et al. 1980). If backscatter is neglected from the ground and overlying clouds, and shading and variations in cloud shape are ignored, however, the shortwave irradiance at the ground can be parameterized for use in mesoscale models as

$$\bar{R}_{\mathrm{SWG}} = \prod_{i=1}^{d} \left[1 - \sigma_{\mathrm{c}_i} \left(1 - t_{\mathrm{c}_i}\right)\right] (1 - A) \left(\bar{R}^{\downarrow}_{\mathrm{SW}} + \bar{R}^{\mathrm{D}}_{\mathrm{SW}}\right)_{\mathrm{CT}} \tag{7.95}$$

from Atwater and Ball 1981), where, as with Eq. 7.94, the contributions to a_{q3} from within the cloud layers should be excluded.

Newiger and Bähnke (1981) reported that aerosol particles, if incorporated into clouds, may be the main absorber of solar radiation in clouds in the visible wavelengths, rather than that from the pure water in the cloud droplets. Knowledge of the liquid water content alone is not sufficient to determine the absorption of solar radiation within a cloud. The aerosol content (and type) within the cloud must also be known. Feingold and Kreidenweis (2000), for example, discuss whether the heterogeneous processing of aerosols increase the number of cloud droplets.

7.3.5 Polluted Air

In contrast with infrared radiation, scattering of shortwave electromagnetic radiation by aerosols is recognized by all investigators as being an important component in the transfer of this energy. This scattering is what causes the white or yellow sky color usually associated with a polluted atmosphere. As discussed by Cerni (1982), and Weber and Baker (1982) unless the air is exceptionally clean, the ratio of diffuse to direct solar radiation is significantly affected by the quantity of pollution as well as being a strong function of zenith angle, particularly when that angle is larger than 70° or so and the optical depth is large.

The absorption of shortwave energy by aerosols is also recognized as being important, although its magnitude, as represented by the imaginary index of refraction, is not known except for a few specific substances. Hänel et al. (1982), for example, have reported heating rates from absorption of solar radiation by aerosols as large as about 0.5°C h^{-1} during the middle of the day under clear sky conditions over Frankfurt, Germany. They concluded that such absorption "is of high climatological importance, especially in industrial areas."

The extinction (absorption plus net backscattering) of solar energy by aerosols can be evaluated using a similar equation as applied to estimate the absorption of longwave energy by aerosols (Section 7.3.2.3). In general, the optical path length for aerosols in the visible wavelengths is about ten times that in the infrared (Paltridge and Platt 1976:215). An analogous equation to Eq. 7.73, but for extinction, is given by

$$\beta_{e\lambda}\left(r_1, r_2\right) = \frac{\pi b_1}{2.3} \int_{r_1}^{r_2} E_{e\lambda}(r) r^{1-b_2}\, dr, \tag{7.96}$$

where $\beta_{e\lambda}$ is a *volume extinction coefficient* and $E_{e\lambda}$ defined as the ratio of the extinction cross section to the geometric cross section of a single spherical particle. As in Section 7.3.2.3, the Junge distribution of aerosols (Eq. 7.71) is used over the size range r_1 to r_2. If the change of variable, given as $\xi = 2\pi r/\lambda$, is used in Eq. 7.96, then

$$\beta_{e\lambda}\left(r_1, r_2\right) = \frac{\pi b_1}{2\pi}\left(\frac{\lambda}{2\pi}\right)^{2-b_2} \int_{\xi_1}^{\xi_2} E_{e\lambda}\left(\xi\right) \xi^{1-b_2}\, d\xi,$$

or

$$\beta_{e\lambda}\left(r_1, r_2\right) \simeq \beta_{e\lambda}\left(0, \infty\right) \simeq \frac{\pi b_1}{2\pi}\left(\frac{\lambda}{2\pi}\right)^{2-b_2} \int_{0}^{\infty} E_{e\lambda}\left(\xi\right) \xi^{1-b_2}\, d\xi, \tag{7.97}$$

if, as assumed by Junge (1963), particles smaller than r_1 are too small and greater than r_2 are too few to contribute significantly to extinction. As discussed following Eq. 7.71, the Junge distribution is a realistic representation of the distribution of aerosols in polluted air masses. As discussed by Paltridge and Platt (1976:217, 222) the refractive index of aerosols is essentially independent of wavelength in the shortwave intervals, and since $E_{e\lambda}(\xi)$ is a function only of the refractive index, then the integral in Eq. 7.97 is a function only of refractive index. Therefore, Eq. 7.97 can be written as[35]

$$\beta_{e\lambda} = b_4 \lambda^{2-b_2}, \tag{7.98}$$

where b_4 is a function of particle refractive index, which itself is a function of radius and chemical characteristic of the aerosol. Using $b_2 = 3$ (see following Eq. 7.74), shortwave irradiance reaching the ground, as given by an expression such as Eq. 7.86 is thus given by

$$\bar{R}_{\mathrm{SW_G}} = \left(t - a_{q3}\right) t_{\mathrm{a}}(1 - A) R^{\downarrow}_{\mathrm{SW_0}}, \tag{7.99}$$

where

$$t_{\mathrm{a}} = \sum_{m=1}^{M} \sum_{\Delta\lambda} \left[b_{5_m} \exp\left(-b_{4_m} \lambda^{-1}\, \Delta z_{\mathrm{a}}\right)\right] \tag{7.100}$$

with b_{4_m} being defined for each aerosol and gaseous contaminant. The quantity Δz_{a} is the depth of a layer within the aerosol cloud. The parameter b_{5_m} is included to

[35]The total aerosol optical depth due to extinction $\tau_{\mathrm{a}} = \int_0^{\infty} \int_{z_{\mathrm{G}}}^{\infty} \beta_{e\lambda}\, dz\, d\lambda$ is called the *turbidity* (Liou 1980:238).

account for the overlap in attenuation when two or more contaminants are present (see following Eq. 7.77). In using Eq. 7.100 in this form, however, changes of the distribution of aerosol sizes and composition with height are ignored.

The effect of humidity on the radius size (which will also, in general, change b_{4_m}) for a moderately polluted air mass can be estimated for relative humidities less than 90% or so, following Kasten (1969), as

$$r = r_0 \left[1 - \left(q_3/q_{3_s}\right)\right]^{-0.23}, \tag{7.101}$$

where q_{3_s} is the saturation specific humidity ($100 q_3/q_{3_s}$, is the relative humidity). According to Eq. 7.101, for a relative humidity of 50%, $r \simeq 1.17\, r_0$, whereas at 90%, $r \simeq 1.7\, r_0$. Zdunkowski and Liou (1976) examined the effects of pollution on the absorption of shortwave irradiance and found it inconsequential for relative humidities from 30 to 70%, although, as implied by Eq. 7.101, a significant effect might be expected only for relative humidities nearer to 100%, where r and r_0 become significantly different.

Although Eq. 7.100 provides the information to evaluate the extinction of solar radiation caused by atmospheric pollutants, it does not permit the determination of the effect of these materials on the flux divergence term in Eq. 7.63. To perform this evaluation, the amount of absorption of solar irradiance by the aerosols is required. Absorption occurs when the imaginary component of the particle refractive index is nonzero. A volume absorption coefficient, analogous to Eq. 7.98, can be written as

$$\beta_{a\lambda} = b_6 \lambda^{2-b_2} = b_6 \lambda^{-1}, \tag{7.102}$$

where b_6 is a function of particle refractive index, and the number of aerosols, but in general, is different than b_4. The shortwave heating within the aerosol layer, for use in Eq. 7.63, can then be obtained from

$$\frac{\partial \bar{R}_{\mathrm{sw_a}}}{\partial z} = -\left(\bar{R}_{\mathrm{sw}}^{\downarrow} + \bar{R}_{\mathrm{sw}}^{\mathrm{D}}\right) \frac{\partial a_{\mathrm{a}}}{\partial z}, \tag{7.103}$$

where $\bar{R}_{\mathrm{sw}}^{\downarrow} + \bar{R}_{\mathrm{sw}}^{\mathrm{D}}$ in Eq. 7.103 is evaluated at the top of the aerosol layer and

$$a_{\mathrm{a}} = \sum_{m=1}^{M} \sum_{\Delta\lambda} \left[1 - b_{7_m} \exp\left(-b_{6_m} \lambda^{-1}\, \Delta z_a\right)\right]. \tag{7.104}$$

The absorptivity a_{a} is defined analogously to Eq. 7.100 with b_{7_m} included to account for the overlap in absorptivities by different aerosols within the same spectral interval $\left(\sum_{m=1}^{M} b_{7_m} = 1\right)$ and b_{6_m} a function of the refractive index for each type of aerosol.

Paltridge and Platt (1976:222), however, question the use of Eq. 7.102 to represent the absorption of solar irradiance by very small aerosol particles ($r < 0.1\ \mu$m – which are referred to as *Aitken particles*). From Mie theory, they present a value of single particle absorption efficiency given by

$$E_{a\lambda} = -4\left(\frac{2\pi r}{\lambda}\right) \operatorname{Im}\left[\left(n_{\mathrm{c}}^2 - 1\right) / \left(\eta_{\mathrm{c}}^2 + 2\right)\right], \tag{7.105}$$

where η_c is the refractive index ($\eta_c = \eta_r - i\eta_i$). Thus

$$\beta_{a\lambda}(r_*) = (3/4r_*) E_{a\lambda}. \tag{7.106}$$

Values of η_c in the visible range are estimated as 1.8–0.5i for graphites, soots, and coals (Twitty and Weinman 1971), 1.55–0.044i for fly ash (Grams et al. 1972), and Paltridge and Platt (1976:287) give a typical value of a dry atmospheric aerosol of about 1.5–0.005i. Liquid water ranges from a value of $1.34–1.86 \times 10^{-9}i$ at $\lambda = 0.4$ μm to $1.33–1.25 \times 10^{-7}i$ at $\lambda = 0.8$ μm (Paltridge and Platt 1976:Table 1.4). To obtain Eq. 7.106 for $\beta_{a\lambda}$ the aerosol cloud is assumed to have only one radius size r_* with the same value of $E_{a\lambda}$. To use this expression, r_* could be defined as the effective radius (see Eq. 7.89) for aerosols less than 0.1 μm. The radiative heating from these Aitken aerosols could then be estimated by Eq. 7.103, where

$$a_a = \sum_{m=1}^{M} \sum_{\Delta\lambda} \left[1 - b_{7_m} \exp\left(-\frac{3}{4r_{*_m}} E^m_{a\lambda} \Delta z_a \right) \right] r_* < 0.1\ \mu\text{m}, \tag{7.107}$$

which is analogous to Eq. 7.104. The quantities $E^m_{a\lambda}$ and r_{*_m} could be defined for each absorbing Aitken aerosol component, (with b_{7_m} suitably defined) or a representative value of r_* and $E_{a\lambda}$, with $b_{7_m} = 1$, used to represent the averaged absorption characteristics.[36] Since $E_{a\lambda}$ is proportional to radius, $E^m_{a\lambda}/r_{*_m}$ is independent of radius in Eq. 7.107 as long as the refractive index is a constant for all sizes of each Aitken aerosol type.

As discussed by Paltridge and Platt (1976:224), radiative heating by aerosols, as given by Eq. 7.107, is most important for highly absorbent materials such as carbonaceous aerosols ($\eta_i = 0.66$). As η_i decreases, the heating by sizes larger than 0.1 μm becomes dominant, although because of their large number in polluted air, Aitken aerosols can still exert a significant effect on the radiative heating.

Parameterized versions of solar irradiance in a polluted atmosphere include that reported by Welch and Zdunkowski (1976), Zdunkowski et al. (1976), and Welch et al. (1978). Using a spherical harmonic representation for the Mie equations of radiative transfer, as described by Zdunkowski and Korb (1974), they represent multiple scattering and absorption separately for three intervals in the solar spectrum. They include the water vapor region, where all absorption bands of water vapor are included; the NO_2 region where absorption and scattering by this gas can occur; and the remainder of the solar spectrum. In all three intervals, absorption and scattering by aerosols and the constant gases are included. Their method, too detailed to present here, is discussed in Welch and Zdunkowski (1976).

Among their results they found that solar irradiance can cause heating in a polluted boundary layer in excess of 4°C h^{-1} with a zenith angle of 45°. Welch et al. (1978) in their two-dimensional simulation of the effects of polluted air on an urban-rural area found temperatures at the ground in the urban area during stagnant synoptic

[36]Paltridge and Platt (1976:225), however, caution against using mean absorption characteristics of a range of different aerosol types to compute absorption.

conditions to be reduced by 2°C because of low-level pollution sources and up to 7°C when upper-level sources occur; a result that was partially caused by the enhanced reflection and absorption of solar radiation by suspended aerosols. (Changes in albedo and roughness over the urban area also influenced these temperature changes.)

Viskanta et al. (1977a,b) and Viskanta and Daniel (1980) also used the method of spherical harmonics (Bergstrom and Viskanta 1973a) to solve the equations of radiative transfer. Using complex indices of refraction, as reported by Bergstrom (1972), they divided the solar spectrum into 12 intervals. They also concluded that the absorption of shortwave irradiance was an important component of the heat budget in polluted air masses, with a magnitude on the same order as for water vapor.

Atwater (1971a,b) used a simpler form to represent shortwave absorption and scattering. To represent $\beta_{a\lambda}$, he used a modified gamma distribution and values of $E_{a\lambda}$ for $\lambda = 0.485$ μm, with height dependence ignored. In Atwater (1971b) he found heating rates of 30°C day^{-1} for 1 ppm of NO_2, while 1 ppm of SO_2, 0.1 ppm ethyl nitrate, 0.1 ppm biacetyl, 0.1 ppm ozone, and 0.1 ppm methyl propenyl ketone resulted in values of around 0.1 to 0.2°C day^{-1}. Nitric acid vapor (1 ppm), hydrogen peroxide (0.1 ppm), and ethyl nitrate (0.1 ppm) had values less than 0.005°C day^{-1}.

Reck and Hummel (1981), using a model with both short and longwave radiative effects, concluded that aerosols can lead to heating or cooling at the surface depending on the surface albedo, and on the imaginary part of the index of refraction η_i. For surface albedos above about 0.38, they found that the presence of absorbing aerosols ($\eta_i \neq 0$) always resulted in heating at the surface. They also concluded that the surface temperature is insensitive to its size distribution, although the chemical composition (and hence η_i) depend on the size spectra. Similarly, Porch and MacCracken (1982) determined that the aerosol size distribution was relatively unimportant in determining the radiative heating and cooling resulting from shortwave radiation in the arctic. From an observational study over a specific region (St. Louis). Method and Carlson (1982) concluded that the total radiative effects of aerosols over the city was negligible, apparently because the aerosols had low absorptance (i.e., η_i small). Saleeby and van den Heever (2013) also discuss improvements to the representation of the aerosol effect (including the representation of radiative flux divergence) in the RAMS model.

It needs to be determined if detailed treatments of radiative transfer such as proposed by Welch and Zdunkowski (1976), and Viskanta et al. (1977a,b) are required for accurate simulations of radiative heating and cooling resulting from aerosols in a mesoscale model. Although a complete treatment is, of course, desirable, *the accuracy of any one parameterization in a model need not be any more precise than the least accurate parameterization of significant physical processes for the atmospheric system of interest*. Additional work is required to test detailed, as opposed to simplified, parameterizations of radiative transfer in mesoscale models.

7.3.6 Parameterization Complexity

In three-dimensional mesoscale models, radiation physics has not yet received the attention given to planetary boundary layer dynamics. The importance of longwave radiative flux divergence to such atmospheric features as nocturnal mesoscale drainage flows and of both short and longwave fluxes in polluted atmospheres over urban areas during the day, however, suggests that the accurate parameterization of radiation in mesoscale models is a fertile area for future research. The detailed analyses of the radiative fluxes in the stable boundary layer by Steeneveld et al. (2011) and McNider et al. (2012) illustrate the type of assessments that are needed.

7.3.6.1 *Equation 7.64*

This algorithm is a particularly simple representation of longwave radiative flux divergence. It requires only $\overline{T}$ which is one of the variables in the conservation-of-heat equation. There are two adjustable (i.e., tunable) coefficients. This formula ignores CO_2 and water vapor effects, so the coefficients would need to be altered for different humidities, and other trace gas concentrations.

7.3.6.2 *Equation 7.62*

A more sophisticated algorithm for longwave radiative flux divergence is given by Eq. 7.62. In this representation, the dependent variables are $\overline{T}$, $\bar{\rho}$, $\bar{q}_3$, and $\bar{p}$. The coefficient, σ, is the Stefan-Boltzmann constant which is assumed to be a fundamental quantity based on quantum mechanics. The emissivities, ϵ_{q_3} and $\epsilon_{\chi_{CO_2}}$, involve a set of observationally determined coefficients (Eqs. 7.56, 7.57, and 7.58). The independent variable is z.

7.3.6.3 *Equation 7.79*

Equation 7.79 for direct solar radiation requires the specification of the solar constant, S_0, the day of the year (using Eq. 7.81), and the latitude, declination of the sun, and the time of day from Eq. 7.82. This equation is based on astronomical measurements to determine the coefficients in Eq. 7.80. Spherical geometry is used to obtain Eq. 7.82.

7.4 Parameterization of Moist Thermodynamic Processes

7.4.1 Basic Concepts[37]

In many mesoscale systems such as the sea breeze and squall line, phase changes of water occur as mesoscale and/or subgrid-scale circulations lift air above its condensation level and as water falls back out of or detrains from clouds and begins to evaporate. The presence of water as solid, liquid, and gas necessitates that the complete form of the conservation equations for water substance (e.g., Eq. 4.25) be included in a mesoscale model. In addition, the proper representation of the source/sink term for diabatic heating (i.e., S_θ in Eq. 4.24) is required. This chapter discusses procedures with which the effects of the phase change of water can be included in grid-volume averaged conservation equations, such as given by Eqs. 4.24 and 4.25.

[37] See also Chapter 14.

To parameterize the effects of phase changes in a mesoscale model, it is helpful to catalog the grid-volume averaged atmosphere in a vertical column as

1. *convectively stable* if $\partial\bar{\theta}_E/\partial z > 0$ everywhere above the condensation-sublimation level of z_{cl}, or
2. *convectively unstable* if for at least one level above z_{cl}, $\partial\bar{\theta}_E/\partial z \leq 0$.

When a layer is convectively stable, forced lifting of the layer must continue to sustain the conversion of water vapor to liquid or solid, once the specific humidity equals the saturation specific humidity. If the layer is convectively unstable, however, clouds continue to grow without further forced lifting of the layer once saturation is reached. Convective instability is also called *potential instability* or *layer instability*.

The variable $\bar{\theta}_E$, the grid-volume averaged *equivalent potential temperature*, is used to determine grid-volume averaged convective instability. This temperature is derived as follows from the conservation-of-heat relation expressed by Eq. 2.23. Let the contribution resulting from the first three terms in Eq. 2.24 for the source/sink term S_θ be written as

$$\frac{C_p}{\theta} S_\theta^* = -\left(\delta_f L_f + \delta_c L_c\right) T_v^{-1} \frac{dq_s}{dt} = \frac{C_p}{\theta} \frac{d\theta}{dt}, \tag{7.108}$$

where q_s is the saturation specific humidity[38] and L_c and L_f are the latent heats of condensation and freezing, respectively ($L_c = 2.5 \times 10^6$ J kg^{-1}; $L_f = 0.33 \times 10^6$ J kg^{-1} at 0°C).[39] The parameters $\delta_f = 1$ if freezing or melting occurs, $\delta_c = 1$ if condensation or evaporation occurs, $\delta_c = \delta_f = 1$ if deposition or sublimation occurs, and zero otherwise. Using Eq. 2.22, Eq. 7.108 can also be written as

$$C_p \frac{d\theta}{\theta} = -L T_v^{-1}\, dq_s \simeq -L\, d\left(q_s/T_v\right), \tag{7.109}$$

where the approximation $T_v^{-1}\, |dq_s| \gg q_s T_v^{-2}\, |dT_v|$ has been used,[40] and L is equal to either L_c or $L_f + L_c$.

If at low temperatures $q_s/T_v \rightarrow 0$ (i.e., the saturation specific humidity goes to zero faster than temperature approaches absolute zero (e.g., see Eq. 7.115), then Eq. 7.109 can be integrated to yield

$$C_p \int_\theta^{\theta_{ES}} d\, \ln\hat{\theta} = -L \int_{q_s/T_v}^{0} d\left(\hat{q}_s/\widehat{T}_v\right),$$

[38] Specific humidity is defined as the ratio of the density of the gas or aerosol to the density of the air including this gas or aerosol. *Mixing ratio* is defined as the ratio of the gas or aerosol to the rest of the air excluding this gas or aerosol. At low concentrations (e.g., less than 5 parts per hundred) the two are almost equal, so that the two terms can be used interchangeably for that situation.

[39] From Wallace and Hobbs (1977).

[40] $q_s T_v^{-2}|dT_v|/T_v^{-1}|dq_s| = (q_s/T_v)|dT_v/dq_s|$ is much less than unity for most reasonable atmospheric conditions. For example, the ratio is about 0.07 at warm temperatures (i.e., $T_v = 300$ K, $q_s = 0.02$) and approximately 0.06 at cold temperatures (i.e., $T_v = 250$ K, $q_s = 0.0002$), where $dT_v|/|dq_s|$ is evaluated over a 5° interval; q_{si} and q_{sw} as a function of T_v are given by Eq. 7.115.

where C_p and L are treated as constants. The upper limit of integration θ_{ES} is called the *saturation equivalent potential temperature* since specific humidity is given by its saturated value q_s. The integrated form of this relation can then be written as

$$\theta_{ES} = \theta \exp\left(\frac{L}{C_p}\frac{q_s}{T_v}\right) = \theta \exp\left(\frac{Lq_s}{\pi\theta}\right), \tag{7.110}$$

where the definition of π, given following Eq. 4.34, has been used. With $L = L_c$, θ_{ES} represents the saturation equivalent potential temperature with respect to liquid water, and with $L = L_c + L_f$, the temperature is defined with respect to ice. This formulation for θ_{ES} is a measure of the change in potential temperature if all the moisture is condensed ($L = L_c$), or deposited, or condensed and frozen ($L = L_c + L_f$), with the heat released used to warm a parcel of air. Because of the approximations made (e.g., Eq. 7.109), the expression is not exact (see Simpson 1978 and Bolton 1980 for a precise derivation of θ_{ES}), however, it is in a suitable form to use in most mesoscale model calculations.[41]

The grid-volume averaged form of Eq. 7.110 is defined by replacing the instantaneous values of the dependent variables in Eq. 7.110 with their grid-volume averaged counterparts. Expressed formally,

$$\bar{\theta}_{ES} = \bar{\theta} \exp\left(L\bar{q}_s/C_p\bar{T}_v\right) = \bar{\theta} \exp\left(L\bar{q}_s/\bar{\pi}\bar{\theta}\right). \tag{7.111}$$

When an air panel is not saturated, $\bar{q}_s$ in Eq. 7.111 is replaced by the specific humidity of the parcel $\bar{q}$ yielding

$$\bar{\theta}_E = \bar{\theta} \exp\left(L\bar{q}/\bar{\pi}\bar{\theta}\right), \tag{7.112}$$

where $\bar{\theta}_E$ is the *equivalent potential temperature*. A layer with $\partial\bar{\theta}_E/\partial z \leq 0$ will become less stable as it is lifted (as shown graphically, for example, by Byers 1959:191), whereas if $\partial\bar{\theta}_E/\partial z > 0$, it will become more stable. It is the vertical distribution of $\bar{\theta}_E$ that is used to assess convective stability.

Betts (1974) has demonstrated that vertical profiles of the difference $\bar{\theta}_{ES} - \bar{\theta}_E$ is a measure of convective regimes. Over Venezuela, he found that in the lowest levels (i.e., the 10 mb layer nearest the ground) $\bar{\theta}_{ES} - \bar{\theta}_E = 40°C$, or so, on dry days while this difference was reduced by about half on disturbed days with extensive cumulus connection.

7.4.2 Parameterization of the Influences of Phase Changes of Water in a Convectively Stable Atmosphere ($\partial\bar{\theta}_E/\partial z > 0$)

If the atmosphere is convectively stable on the resolvable scale, (i.e., $\partial\bar{\theta}_E/\partial z > 0$) everywhere in a column above the lowest saturation level[42] and if a layer is lifted

[41]Betts (1982) introduced a concept called *saturation point* to represent the thermodynamic properties of clear and nonprecipitation-sized cloudy air.

[42]The saturation level is the height at which lifting of a parcel of air causes sufficient cooling to generate condensation or deposition. The saturation level can, of course, vary spatially within a grid volume.

until saturation occurs, then

1. only stratiform clouds will develop if $\partial\theta_E/\partial z$ is also greater than zero everywhere within all of the grid volumes which are saturated,[43], or
2. some cumuliform clouds can develop if $\partial\theta_E/\partial z \leq 0$ locally within one or more grid volumes which are saturated. The number, height, and vigor of these clouds is expected to depend on the magnitude and distribution of these regions of subgrid-scale convective instability.

Figure 7.10 illustrates examples, as seen from below cloud base, of when only stratiform clouds develop (Fig. 7.10a) and of when some cumulus convection develops in an otherwise layered stratiform cloud layer (Fig. 7.10b). Sommeria and Deardorff (1977) discuss the use of a statistical model when only a portion of a grid volume becomes saturated in a planetary boundary-layer model. Vali et al. (1998) illustrate the observed variable spatial structure of stratus clouds.

The mechanics of formation of precipitation that develops within these two categories of equivalent potential temperature stratification are very different. As summarized by Houze (1981), stratiform precipitation occurs with weak ascending motion with precipitation particles forming near the top of the clouds and growing as they fall. Convective precipitation is associated with strong updrafts, where cloud droplets are initiated near cloud base and grow as they are transported upward. Precipitation from convective systems falls to the ground when gravitational sedimentation exceeds the upward velocity within the cloud or when the precipitation is advected to regions within or outside of the cloud where the ascent is weak or negative.

7.4.2.1 Convectively Stable Everywhere within a Column above the Saturation Level

Detailed representation of the Microphysics

In an atmosphere that is convectively stable at all points within a grid volume, the conversion of water between its phases can be represented straightforwardly using formulations such as those developed for cloud models. The degree of sophistication can involve a detailed simulation of the microphysics, including nuclei activation and the growth to precipitation-sized liquid and ice particles. Taylor and Ackerman (1999), for example, found that the microphysical structure and cloud top of stratus clouds were very significantly affected by aerosol emissions from a ship into otherwise clean maritime clouds. Mölders et al. (2010) show how such ship emissions can affect weather even in otherwise pristine locations. Yarker et al. (2010) documented the effect of volcanic emissions on clouds and precipitation.

To study these types of effects, Clark (1973) incorporated a detailed representation of *warm cloud* (i.e., $T > 0°C$ everywhere in the cloud) *microphysics* in his

This level is also called the lifting condensation level. The height of this level will vary depending on the height of origin of the air parcel. These thermodynamic concepts are discussed further in Chapter 14.

[43] θ_E is defined using the decomposition given preceding Eq. 4.3 (i.e., $\theta_E = \bar{\theta}_E + \theta''_E$), where θ''_E is the subgrid-scale equivalent potential temperature.

FIGURE 7.10

(a) A stratus cloud over northern Illinois, 1 December 1974, and (b) a stratocumulus cloud over Gogebic, in the upper peninsula of Michigan, June 1972 (photographed by Ron Holle).

cumulus model. As pointed out by Orville (1980), the conservation-of-water equation (e.g., Eq. 4.25) can be broken into as many as 50 to 100 equations to represent the growth of cloud droplets to precipitation-sized particles. One equation for each size category of liquid or ice particle is used. However, with present computer limitations, such a sophistication is not practical within a mesoscale model. Interested readers who desire a thorough discussion of the procedure used to represent the microphysics in detail are referred to Orville (1980:Section 3.1), Pruppacher (1982), and to the Pruppacher and Klett (1978) extensive review of this subject.

Bulk Representation of the Microphysics

An alternative to the detailed microphysical representation is referred to as the *parameterized microphysical* or *bulk* representation. With this procedure, liquid water and ice can be categorized into the four classes:

1. cloud liquid water,
2. cloud ice,
3. rain, and
4. snow,

so that the equations for $\partial\bar{q}_1/\partial t$ and $\partial\bar{q}_2/\partial t$ in Eq. 4.25 need be composed into only two equations each. The reason for this particular form of categorization is that rain and snow are assumed to have a size such that gravitational sedimentation is appreciable (i.e., they have a significant *fall velocity*),[44] whereas cloud liquid water and cloud ice do not.

The Conservation Equations for Water and Potential Temperature With these decompositions and following Orville (1980), Eq. 4.25 can be written as

$$\begin{aligned}
\frac{\partial\bar{q}_1^{\text{ci}}}{\partial t} &= -\bar{u}_j\frac{\partial}{\partial x_j}\bar{q}_1^{\text{ci}} - \frac{1}{\rho_0}\frac{\partial}{\partial x_j}\rho_0\,\overline{u_j''q_1''}^{\,\text{ci}} + S_{\text{freezing}} + S_{\text{deposition}} - P_{\text{S1}} - P_{\text{S2}} \\
\frac{\partial\bar{q}_1^{\text{s}}}{\partial t} &= -\bar{u}_j\frac{\partial}{\partial x_j}\bar{q}_1^{\text{s}} - \frac{1}{\rho_0}\frac{\partial}{\partial x_j}\rho_0\,\overline{u_j''q_1''^{\text{s}}} - V_T^{\text{s}}\frac{\partial}{\partial z}\bar{q}_1^{\text{s}} + P_{\text{S1}} + P_{\text{S2}} + P_{\text{S3}} + P_{\text{S4}} - P_{\text{S5}} \\
\frac{\partial\bar{q}_2^{\text{cw}}}{\partial t} &= -\bar{u}_j\frac{\partial}{\partial x_j}\bar{q}_2^{\text{cw}} - \frac{1}{\rho_0}\frac{\partial}{\partial x_j}\rho_0\,\overline{u_j''q_2''}^{\,\text{cw}} + S_{\text{condensation}} + S_{\text{freezing}} \\
&\quad - P_{\text{R1}} - P_{\text{R2}} - P_{\text{S3}} \\
\frac{\partial\bar{q}_2^{\text{R}}}{\partial t} &= -\bar{u}_j\frac{\partial}{\partial x_j}\bar{q}_2^{\text{R}} - \frac{1}{\rho_0}\frac{\partial}{\partial x_j}\rho_0\,\overline{u_j''q_2''}^{\,\text{R}} - V_T^{\text{R}}\frac{\partial}{\partial z}\bar{q}_2^{\text{R}} + P_{\text{R1}} + P_{\text{R2}} + P_{\text{S5}} - P_{\text{R3}} \\
\frac{\partial\bar{q}_3}{\partial t} &= -\bar{u}_j\frac{\partial}{\partial x_j}\bar{q}_3 - \frac{1}{\rho_0}\frac{\partial}{\partial x_j}\rho_0\,\overline{u_j''q_3''} - S_{\text{condensation}} - S_{\text{deposition}} - P_{\text{S4}} + P_{\text{R3}}
\end{aligned} \tag{7.113}$$

where $\bar{q}_1^{\text{ci}}, \bar{q}_1^{\text{s}}, \bar{q}_2^{\text{cw}}$, and $\bar{q}_2^{\text{R}}$ are the grid-volume averaged values of specific humidity for cloud ice, snow, cloud water, and rain, respectively. The symbols S_{freezing}, $S_{\text{deposition}}$, and $S_{\text{condensation}}$ represent changes in $\bar{q}_1^{\text{ci}}$, $\bar{q}_2^{\text{cw}}$, and $\bar{q}_3$ resulting from freezing, deposition, and condensation (see following Eq. 2.34), respectively, and P_{R1}, P_{R2}, P_{R3}, P_{S1}, P_{S2}, P_{S3}, P_{S4}, and P_{S5} represent different mechanisms for the conversion of cloud ice and cloud water to snow and rain that are defined, for example, in Section 7.4.4.1. The fall velocities of snow and ice are, respectively, V_T^{s} and V_T^{R}.

[44] Fall velocity is also referred to as terminal velocity.

The vertical subgrid-scale flux terms in Eq. 7.113 (e.g., $\overline{u_j''q_1''^{\mathrm{ci}}}$) can be represented by

$$\overline{w''q_1''^{\,\mathrm{ci}}} = -K_\theta \frac{\partial \bar{q}_1^{\mathrm{ci}}}{\partial z}; \quad \overline{w''q_1''^{\,\mathrm{s}}} = -K_\theta \frac{\partial \bar{q}_1^{\mathrm{s}}}{\partial z}$$

$$\overline{w''q_2''^{\,\mathrm{cw}}} = -K_\theta \frac{\partial \bar{q}_2^{\mathrm{cw}}}{\partial z}; \quad \overline{w''q_2''^{\,\mathrm{R}}} = -K_\theta \frac{\partial \bar{q}^{\mathrm{R}}}{\partial z} \quad \overline{w''q_3''} = -K_\theta \frac{\partial \bar{q}_3}{\partial z};$$

where K_θ could be evaluated using Eq. 7.17 except, with saturated air, the gradient Richardson number should be redefined as

$$R_\mathrm{i} = \frac{g}{\theta_0}\frac{\partial \bar{\theta}_{\mathrm{ES}}}{\partial z} \Big/ \left[\left(\frac{\partial \bar{u}}{\partial z}\right)^2 + \left(\frac{\partial \bar{v}}{\partial z}\right)^2\right].$$

The source/sink terms in Eq. 7.113 can be written as

$$\begin{aligned}
S_{\mathrm{deposition}} &= -\delta_\mathrm{s}\, \bar{w}\, \frac{\partial \bar{q}_{\mathrm{si}}}{\partial z}, & \overline{T}_\mathrm{v} &\le T_0^\mathrm{s}; & \bar{q}_3 &= \bar{q}_{\mathrm{si}} & &\text{then } \delta_\mathrm{s} = 1 \\
S_{\mathrm{condensation}} &= -\delta_\mathrm{c}\, \bar{w}\, \frac{\partial \bar{q}_{\mathrm{sw}}}{\partial z}, & \overline{T}_\mathrm{v} &> T_0^\mathrm{s}; & \bar{q}_3 &= \bar{q}_{\mathrm{sw}} & &\text{then } \delta_\mathrm{c} = 1 \quad (7.114) \\
S_{\mathrm{freezing}} &= \delta_\mathrm{f}\, \frac{\delta \bar{q}^{\mathrm{ci}}}{\delta t}, & \overline{T}_\mathrm{v} &\le T_0^\mathrm{f}; & \bar{q}_2^{\mathrm{cw}} &> 0 & &\text{then } \delta_\mathrm{f} = 1
\end{aligned}$$

where $\bar{q}_{\mathrm{si}}$ and $\bar{q}_{\mathrm{sw}}$ are the saturation specific humidities with respect to ice and water, respectively, and $\delta_\mathrm{s} = 1$ when deposition or sublimation occurs,[45] but zero otherwise. The threshold temperatures[46] in Eq. 7.114 are defined as occurring when the air is cold enough such that direct vapor to ice conversion occurs (T_0^s) and when it is cold enough such that the liquid water freezes (T_0^f). Only vertical gradients are considered in Eq. 7.114 because they are usually much larger, in general, than the horizontal gradients of saturation specific humidity. A similar assumption was made in defining radiative flux divergence for use in mesoscale models (i.e., Eq. 7.22). The quantity $\delta \bar{q}_1^{\mathrm{ci}}/\delta t$, which represents the freezing of cloud water, could be defined such that complete glaciation from cloud water occurs within one time step once $\overline{T}$ becomes less than or equal to T_0^f.

The saturation specific humidity of water vapor with respect to liquid water and to ice is determined using the Clausius-Clapeyron equation (e.g., see Wallace and Hobbs 1977:95). This equation, for liquid water and ice, can be written as

$$de_{\mathrm{sw}}/e_{\mathrm{sw}} = L_\mathrm{c}\, dT \Big/ \left(R_\mathrm{v} T_\mathrm{v}^2\right); \quad de_{\mathrm{si}}/e_{\mathrm{si}} = L_\mathrm{s}\, dT \Big/ \left(R_\mathrm{v} T_\mathrm{v}^2\right);$$

where e_{sw} and e_{si} are the saturation vapor pressures of water vapor with respect to liquid water and ice, respectively, (see List 1971:351-364) for specific vales of e_{sw}

[45] In other words, $\delta_\mathrm{s} = 1$ when $\delta_\mathrm{c} = \delta_\mathrm{f} = 1$ as defined following Eq. 7.108.

[46] The specific values of T_0^s and T_0^f depend on the activation temperature of the ice nuclei that are assumed present.

and e_{si}). The gas constant for water vapor is $R_v (R_v = 461 \text{ J K kg}^{-1}$; Wallace and Hobbs 1977); with T_v the virtual temperature.

Since saturation specific humidity and vapor pressure are related by

$$q_s = 0.622 e_s / \left(p - 0.378 e_s\right) \simeq 0.622 e_s / p, \quad e_s \ll p,$$

then

$$\frac{dq_{si}}{q_{si}} = \frac{L_s}{R_v}\frac{dT}{T_v^2}; \quad \frac{dq_{sw}}{q_{sw}} = \frac{L_c dT}{R_v T_v^2};$$

if the change in saturation vapor pressure is assumed to occur isobarically (i.e., $dp \equiv 0$). As T approaches 0 K, q_{sw} and q_{si} approach 0, since e_{sw} and e_{si} approach 0 at that temperature.

The saturation specific humidity of water vapor with respect to liquid water and ice for reasonable values of temperature and pressure within the troposphere are then given by

$$\begin{aligned} q_{si} &\simeq \frac{3.8}{\bar{p}} \exp\left[\frac{21.9\left(\overline{T}_v - 273.2\right)}{\overline{T}_v - 7.7}\right], \\ q_{sw} &\simeq \frac{3.8}{\bar{p}} \exp\left[\frac{17.3\left(\overline{T}_v - 273.2\right)}{\overline{T}_v - 35.9}\right], \end{aligned} \tag{7.115}$$

(where T_v is in degrees K) using the empirical formulas for e_{sw} and e_{si} given by Murray (1967). A similar formulation for q_{sw} can be derived from Bolton's (1980) representation of e_{sw}. At $\bar{p} = 1000$ mb, the maximum difference between q_{sw} and q_{si} occurs at $\overline{T} \simeq 12°\text{C}$ and is equal to approximately 0.2 g kg^{-1}. At all temperatures less than 0°C, $q_{sw} > q_{si}$.

The influence of these phase changes on the potential temperature in Eq. 4.24 can be written as

$$\begin{aligned} \frac{\partial \theta}{\partial t} = &-\bar{u}_j \frac{\partial \bar{\theta}}{\partial x_j} - \frac{1}{\rho_0}\frac{\partial}{\partial x_j}\rho_0 \overline{u_j''\theta''} - \frac{\bar{\theta}}{\overline{T}_v C_p}\left[\delta_s L_s \left(\bar{w}\frac{\partial \bar{q}_{si}}{\partial z} - P_{S4}\right)\right. \\ &\left. + \delta_c L_c \left(\bar{w}\frac{\partial \bar{q}_{sw}}{\partial z} + P_{R3}\right) - \delta_f L_f \left(\frac{\delta \bar{q}_1^{ci}}{\delta t} - P_{S5} + P_{S3}\right)\right] \end{aligned} \tag{7.116}$$

using Eq. 7.108 with $L_f + L_c = L_s$. In Eq. 7.116, $\bar{\theta}/(\overline{T}_v C_p)$ can be replaced with $1/\bar{\pi}$ using the definition of π given following Eq. 4.34.

The use of Eqs. 7.113 and 7.114, however, requires that changes of heat content caused by the phase changes of water be considered since from Eq. 7.115, $\bar{q}_{si}$ and $\bar{q}_{sw}$ are functions of temperature. One procedure to account for this effect in the absence of precipitation-sized particles is to use an algorithm given by

$$\left.\frac{\partial \bar{\theta}}{\partial t}\right|_{pc}^{n} = \frac{1}{\bar{\pi}}\left[-\delta_s L_s \bar{w}\left.\frac{\partial \bar{q}_{si}}{\partial z}\right|^{n} - \delta_c L_c \bar{w}\left.\frac{\partial \bar{q}_{sw}}{\partial z}\right|^{n} + \delta_f L_f \left.\frac{\delta \bar{q}_1^{ci}}{\delta t}\right|^{n}\right];$$

$$\left.\frac{\partial \bar{q}_1^{\text{ci}}}{\partial t}\right|_{pc}^{n} = -\delta_{\text{s}}\bar{w}\left.\frac{\partial \bar{q}_{\text{si}}}{\partial z}\right|^{n} + \delta_{\text{f}}\left.\frac{\delta \bar{q}_1^{\text{ci}}}{\delta t}\right|^{n} \tag{7.117}$$

$$\left.\frac{\partial \bar{q}_2^{\text{cw}}}{\partial t}\right|_{pc}^{n} = -\delta_{\text{c}}\bar{w}\left.\frac{\partial \bar{q}_{\text{sw}}}{\partial z}\right|^{n} - \delta_{\text{f}}\left.\frac{\delta \bar{q}_1^{\text{ci}}}{\delta t}\right|^{n}; \quad \left.\frac{\partial \bar{q}_3}{\partial t}\right|_{pc}^{n} = +\delta_{\text{c}}\bar{w}\left.\frac{\partial \bar{q}_{\text{sw}}}{\partial z}\right|^{n} + \delta_{\text{s}}\bar{w}\left.\frac{\partial \bar{q}_{\text{si}}}{\partial z}\right|^{n};$$

where $n = 1, 2, 3, \ldots, N$ represents the number of iterations required before the local changes in $\bar{\theta}, \bar{q}_1^{\text{ci}}, \bar{q}_2^{\text{cw}}$, and $\bar{q}_3$ resulting from phase changes approach zero.[47] (Conversions to and from precipitation-sized particles are ignored in Eq. 7.117, although it would be straightforward to add these effects if equations for $\bar{q}_1^{\text{S}}$ and $\bar{q}_2^{\text{R}}$ from Eq. 7.113 were also included.)

During each iteration, Eq. 7.115 is used to determine a new saturation specific humidity and to evaluate the relation of the actual temperature to the threshold temperatures in Eq. 7.114. The initial values of $\bar{\theta}, \bar{q}_1^{\text{ci}}, \bar{q}_2^{\text{cw}}$, and $\bar{q}_3$ in the iteration are evaluated from the first two terms on the right side of Eqs. 7.113 and 7.116. The vertical velocity $\bar{w}$ remains constant during the iteration. McCumber (1980), in his determination of the appropriate values for potential temperature and specific humidity for water vapor in a water cloud, used convergence criteria for $\partial\bar{\theta}/\partial t|_{pc}^{n}$ and $\partial\bar{q}_3/\partial t|_{pc}^{n}$ of 0.05 K/120 s and 0.005 g kg^{-1}/120 s. In his three-dimensional mesoscale simulation of rainfall over south Florida, no more than 17 iterations were ever required.

Simplified forms for the conservation of Water

In the past, most mesoscale models have used a simpler form to represent phase changes than given by Eq. 7.113. McCumber (1980), using a procedure introduced by Asai (1965), for example, determines if a grid-volume has been supersaturated with respect to liquid water. If it has, temperature is adjusted using a formulation similar to Eq. 7.117, with condensate formed such as to reduce the supersaturation to zero.[48] All of the remaining condensate, assumed to be precipitation size, falls to the next grid level. If this layer is subsaturated, some or all of the precipitation will evaporate causing cooling and moistening. If condensate remains and the layer becomes saturated, the precipitation falls to the next level. This process continues until either all of the precipitation evaporates or it reaches the ground.

Nickerson (1979), by contrast, permitted only cloud water and excluded precipitation from occurring. Although not detailed in his paper, cloud water appears to be

[47] Since $S_{\text{deposition}}$ and $S_{\text{condensation}}$ are discontinuous functions at T_0^{s} and S_{freezing} is discontinuous at T_0^{f}, there may be difficulty obtaining convergence when $\overline{T}$ is near these values. The practical solution to this problem is to not permit the threshold criteria in Eq. 7.114 to occur more than once during an iteration.

[48] In some mesoscale applications, supersaturation has been arbitrarily defined to occur at less than 100% relative humidity. McCumber (1980) used a value of 90%. Such a reduction in the saturation value, used in synoptic models has been justified by assuming that although the grid-volume averaged specific humidity is unsaturated, a significant number of subgrid-scale values will become saturated once the relative humidity exceeds a certain value. In a nonturbulent, stably-stratified atmosphere, such an approximation would be inappropriate, but is reasonable with q_3'' is not identically equal to zero.

created using a formulation such as given by $S_{\text{condensation}}$ in Eq. 7.114 and advected with an equation similar to $\partial\bar{q}_2^{\text{cw}}/\partial t$ in Eq. 7.113 except the last three terms are ignored.

Colton (1976) used a somewhat more sophisticated parameterization for clouds and precipitation. Although he only had one equation for liquid water (i.e., he combined $\partial\bar{q}_2^{\text{cw}}/\partial t$ and $\partial\bar{q}_2^{\text{R}}\partial t$ in Eq. 7.113, with none for ice); terminal velocities, developed by Ogura and Takahashi (1971), which are representative of rain and snow, were included. When the temperature was at or below zero, he used a terminal velocity representative of snow, and above freezing, a rainwater value was used. The expressions he used are given as

$$V_{\text{T}}^{\text{R}} = \begin{cases} 31.2\left(\bar{\rho}\bar{q}_2^{\text{cw}}\right)^{0.125}, & \overline{T} > 0°\text{C}, \\ 5.9\left(\bar{\rho}\bar{q}_2^{\text{cw}}\right)^{0.11}, & \overline{T} \leq 0°\text{C} \end{cases}$$

where to use these formulas as given, $\bar{\rho}$ must be in units of grams per centimeters cubed. V_{T}^{R} then is in units of meters per second.

Colton permitted no supersaturation and used a direct method to compute the amount of condensate given by

$$\begin{aligned} \overline{T}_{\text{v}} &= \overline{T}_{\text{v}}^{*} + L_{\text{c}}\delta M_1/C_{\text{p}} \\ \bar{q}_2 &= \bar{q}_2^{*} + \delta M_1 \quad \left(\bar{q}_2 = \bar{q}_2^{\text{R}} + \bar{q}_2^{\text{cw}}\right), \\ \bar{q}_3 &= \bar{q}_3^{*} - \delta M_1, \end{aligned} \tag{7.118}$$

where $\bar{q}_2^{*}$ and $\bar{q}_3^{*}$ are changes in specific humidity for liquid water and water vapor, resulting from advection and subgrid-scale fluxes (i.e., the first two terms in each of the last three equations in Eq. 7.113). If $\delta M = \bar{q}_3^{*} - \bar{q}_{\text{sw}}$, then

$$\delta M_1 = \delta M\left[1 + L_{\text{c}}^2\bar{q}_{\text{sw}}/\left(C_{\text{p}}R_{\text{v}}\overline{T}_{\text{v}}^{*2}\right)\right]^{-1}.$$

The temperature $\overline{T}_{\text{v}}^{*}$ is computed from $\bar{\theta}^{*}$, which is calculated from the first two terms on the right of Eq. 7.116. Using the definition of θ given by Eq. 2.48, and of π given following Eq. 4.34, $\overline{T}^{*} = \bar{\theta}^{*}\bar{\pi}/C_{\text{p}}$. As shown by Asai (1965), and used by Ogura and Takahashi (1973), this expression for δM_1 provides an exact evaluation for the changes in temperature and water content caused by moistening and warming caused by condensation. Evaporation can also be determined using Eq. 7.118 since $\delta M_1 < 0$ when $\bar{q}_3^{*} < \bar{q}_{\text{sw}}$ (i.e., the air is subsaturated). Colton apparently calculated precipitation rates at the surface from the term $V_{\text{T}}^{\text{R}}\bar{q}_2$, which is evaluated from values of these terms at the first model level above the ground. As illustrated in Fig. 7.11, Colton was successful in predicting precipitation rates using his scheme.

7.4.2.2 Subgrid-Scale Regions of Convective Instability above the Saturation Level but with $\partial\bar{\theta}_{\text{E}}/\partial z > 0$

Up to this point in Section 7.4.2, only the situation $\partial\theta_{\text{E}}/\partial z > 0$ everywhere has been considered in the parameterization. When the atmosphere has subgrid regions

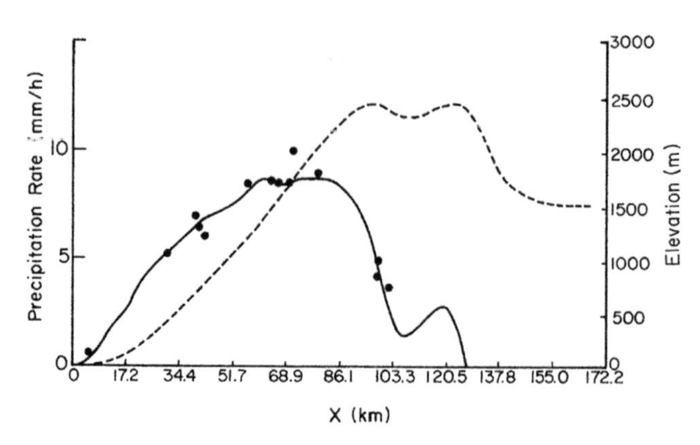

FIGURE 7.11

Model-predicted precipitation rates (solid line) and observed precipitation rates (black dots) along a cross section of the Sierra Nevada (dashed line) for 21-22 December 1964 (adapted from Colton 1976).

that are convectively unstable *and* in which condensation or sublimation occur, the representations for phase change previously mentioned may be unsatisfactory. This is particularly true if these regions of $\partial\theta_E/\partial z < 0$ extend through a significant depth of the atmosphere and cover a substantial portion of the grid domain. To represent these regions in a mesoscale model, when the grid-volume averaged vertical gradient of equivalent potential temperature is stable (i.e., $\partial\bar{\theta}_E/\partial z > 0$), will require innovative parameterization techniques. Indeed, to represent moist thermodynamics accurately for such a situation, it may be necessary to reduce the grid volume.

7.4.3 Parameterization of the Influences of Phase Changes of Water in a Convectively Unstable Atmosphere ($\partial\bar{\theta}_E/\partial z \leq 0$)

If the atmosphere is convectively unstable on the resolvable scale (i.e., $\partial\bar{\theta}_E/\partial z \leq 0$) somewhere in a column above the lowest saturation level and saturation occurs where $\partial\bar{\theta}_E/\partial z$ is less than or equal to zero, then

1. only cumuliform clouds will form if $\partial\theta_E/\partial z \leq 0$ everywhere within saturated grid volumes in that column, or
2. some layer-form clouds can form if $\partial\bar{\theta}_E/\partial z > 0$ locally within a grid volume, which is saturated. The extent and levels of such clouds depend on the distribution of regions that are convectively stable.

Figure 7.12 illustrates examples, as seen from below base, when only cumuliform clouds develop (Fig. 7.12a), and when some layered clouds form in a predominantly cumuliform cloud mass (Fig. 7.12b). Johnson et al. (1999) suggests there are three distinct types of tropical convective clouds: shallow cumulus, cumulus congestus,

FIGURE 7.12

(a) A group of cumulus clouds over southern Florida at 0930 LST, 15 August 1978, and (b) a cumulus congestus complex with layered clouds on its periphery located in southern Arizona at 1615 LST, 23 October 1974 (photographed by Ron Holle).

and cumulonimbus. Nair et al. (1998) used satellite imagery to determine the spatial patterning of cumulus cloud fields.

Cumuliform clouds also form when the top of the planetary boundary exceeds the saturation level and the surface layer is superadiabatic. For this situation, cumulus clouds are the visible manifestation of the turbulent eddies within the planetary boundary layer. When such clouds occur with $\partial\theta_E/\partial z > 0$ everywhere above the boundary layer, their growth into deeper cumulus clouds (e.g., cumulus congestus) will not occur. Sommeria (1976) investigated turbulent processes in a trade wind boundary layer over water when such shallow cumulus clouds formed, using an extension of Deardorff's (1973) sophisticated planetary boundary-layer model.

In contrast with stratiform cloud systems, cumulus clouds generally have smaller spatial dimensions and more irregular patterns of updrafts and downdrafts. Except for cumulonimbus-size systems, individual cumulus clouds have horizontal dimensions which are smaller than can be resolved by a mesoscale model grid. Moreover, the depth that a cumulus cloud attains is more dependent on the magnitude and vertical distribution of convective instability than on the intensity of the mesoscale ascent, once saturation is attained. Along with such effects as precipitation, downdrafts, ground shadowing, and cumulus-induced subsidence, the accurate representation of the influence of cumulus clouds on the mesoscale has been and will remain one of the more difficult problems in mesoscale meteorology.

The ability to represent cumulus clouds accurately in a mesoscale model requires that the mesoscale dynamic and thermodynamic structure control the regions of initiation and development of this moist convective activity. There is evidence that this situation occurs. As illustrated in Fig. 7.13b, for example, Ulanski and Garstang (1978) found significant correlation between boundary-layer convergence patterns and subsequent cumulus-produced rainfall over land. This low-level convergence was found to precede cumulus rainfall by as much as 90 min. They also found that the amount of rainfall was significantly correlated with the duration of the precedent boundary-layer convergence, as shown in Fig. 7.13a. Holle and Watson (1983) have found from their data set that (defining an event period as the time between initial surface convergence and complete dissipation) the first visible cloud occurs at an average of about 1/6 into this period with most rapid cloud growth at around 1/3 of the event lifetime. Doneaud et al. (1983) confirmed the relationship between antecedent low-level convergence and subsequent cumulus convective activity over southeastern Montana, although Achtemeir (1983) has found the relationship between convergence and rainfall to be more complex over St. Louis based on summer METROMEX 1975 data.

On a larger spatial scale, Pielke (1974a) also found qualitative agreement between predicted sea-breeze convergence and the subsequent actual development of cumulonimbus activity, and Simpson et al. (1980) obtained a large positive correlation between merged thunderstorm complexes and sea-breeze convergence for three case study days over south Florida. For a particular summer day over south Florida, Pielke and Mahrer (1978) obtained a 4-h lag between this predicted mesoscale convergence and thunderstorm activity. Pielke et al. (1991) and Pielke (2001a) summarized how mesoscale convergence preconditions the environment for thunderstorm development. Most of the rainfall in these sea-breeze events occurs in large cumulonimbus complexes (Simpson et al. 1980).

In the 1974 GARP Atlantic Tropical Experiment (GATE), Ogura et al. (1979) found low-level convergence to be present or enhanced prior to the development of organized convective systems in all cases considered. Krishnamurti et al. (1983) concluded from the GATE data that the incorporation of the influence of mesoscale convergence is essential for successful cumulus parameterization. Over Oklahoma, Sun and Ogura (1979) observed a well-defined band of low-level convergence, apparently generated by a horizontal temperature gradient, to precede the development of showers and thunderstorms for a particular day.

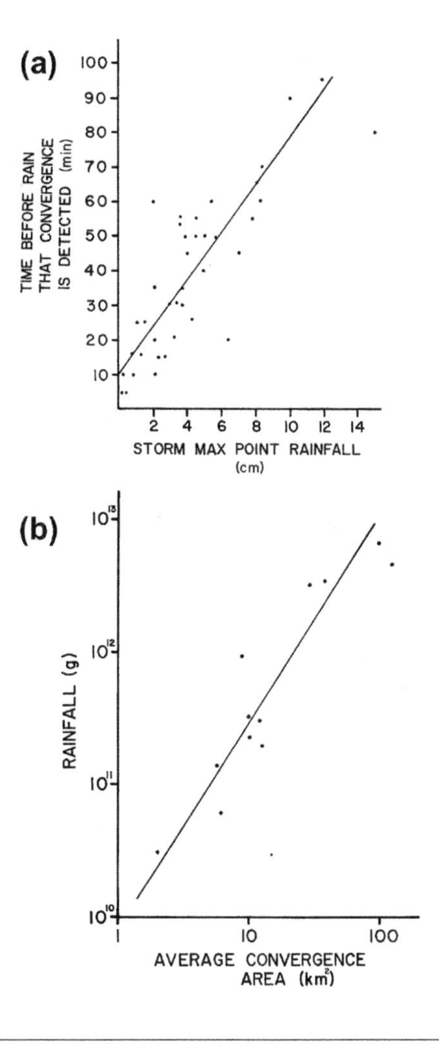

FIGURE 7.13

(a) The relationship between the duration of boundary-layer convergence and subsequent rainfall during a summer season over south Florida, and (b) the relation between convergence area and rainfall amount (from Ulanski and Garstang 1978).

In considering methodologies to represent cumulus clouds in a mesoscale model, it is useful to group them into the four classes:

1. convective adjustment,
2. use of one-dimensional cloud models,
3. use of a cumulus field model, or set of equivalent observations, and
4. explicit representation of moist thermodynamics.

7.4.3.1 Convective Adjustment

With the first method, as discussed by Kurihara (1973), Krishnamurti et al. (1980), and others, the lapse rate is forced to be moist adiabatic over all or part of the model grid when saturation occurs. This is the simplest and cheapest form of cumulus parameterization, although unfortunately with this approach, regions of potential instability are removed too quickly on the mesoscale. In addition, this scheme provides poor vertical profiles of the averaged subgrid-scale heating and moistening. For these reasons, more sophisticated parameterization schemes have been developed (Hong and Pan 1998). Haltiner and Williams (1980) provide a review of convective adjustment schemes for use in larger-scale models.

7.4.3.2 Use of One-Dimensional Cloud Models

In the second group, Kuo (1965, 1974) Krishnamurti and Moxim (1971), Ooyama (1971), Arakawa and Schubert (1974), Yenai (1975), Kreitzberg and Perkey (1976, 1977), Anthes (1977), Johnson (1977), Fritsch and Chappell (1980a,b), Yamazaki and Ninomiya (1981), Molinari (1982), Hong and Pan (1998), Gallus (1999) and others discuss or use one-dimensional cloud models to represent the feedback of cumulus scales to the larger scale. In using one-dimensional models, it has often been assumed that in deep cumulonimbus systems, the vertical distribution of heating on the cloud scale is essentially the same as the vertical distribution of heating on the model grid scale (e.g., see Anthes 1977), although such as assumption is certainly not true for shallow cumulus or when substantial downdrafts exist. Using this approach, Kuo and Raymond (1980) concluded that the main heating from cumulus activity was from latent heat release although subsidence warming is also important, particularly close to cloud top. Yenai (1975) has presented a review of these types of cumulus representations, as does Hsu (1979), who also used an extension of Kuo's (1974) parameterization. An early, but useful summary of cumulus parameterization is given by Ogura (1972). Cotton (1975) and Simpson (1976, 1983) provide reviews of one-dimensional cumulus models. One-dimensional models of clouds are also referred to as *single-column models* (e.g., see Das et al. 1999; Wu et al. 2000a).

The above types of schemes achieve closure by assuming either that moisture convergence in the lower troposphere supplied by the mesoscale or larger scale is necessary for cumulus development, or that cumulus clouds develop when sufficient thermodynamic instability is achieved on the mesoscale or larger scale. Effects of wind shear on precipitation efficiency are included using data such as shown in Fig. 7.14. The relation of cumulus-caused downdrafts to previous cumulus updrafts has been developed using observational data, such as shown in Fig. 7.15.

One of the most commonly used cumulus convection schemes is called the Kuo scheme, originally developed by Kuo (1974). A form of this scheme, as modified by Molinari (1985), is described as follows as reported in Tremback (1990). The source terms in the conservation-of-heat and conservation-of-water equations due to deep

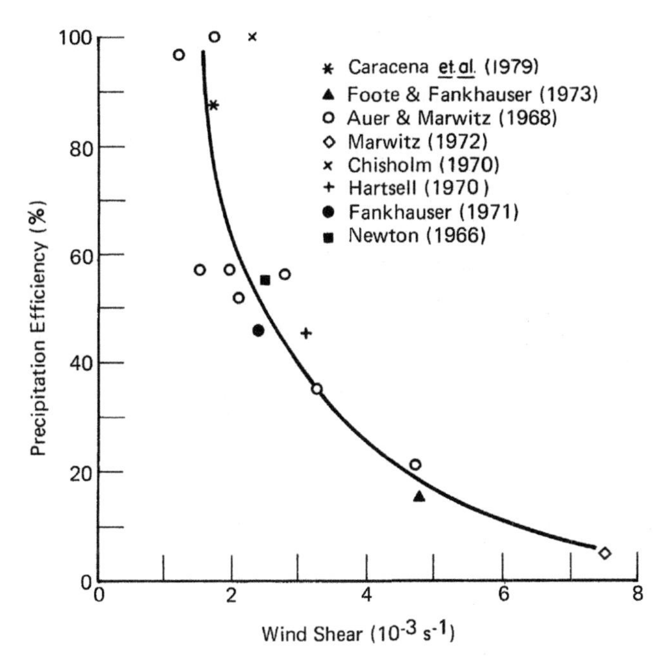

FIGURE 7.14

Precipitation efficiency ϵ_p defined as the ratio of rainout to water vapor inflow as a function of the vertical shear of the horizontal wind in the layer between cloud base and cloud top (reproduced from Fritsch and Chappell 1980a). The legend indicates data sources.

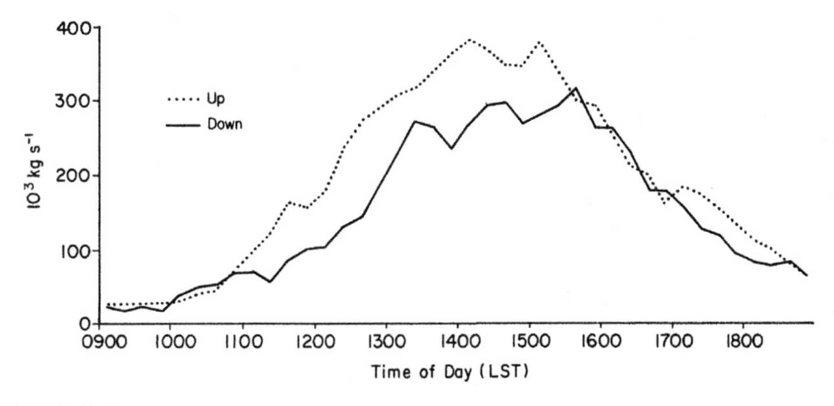

FIGURE 7.15

Average cumulus convective transports over a portion of south Florida on days with substantial convective activity (adapted from Cooper et al. 1982), where the dotted line indicates upward and the solid line downward transport.

cumulus convection are written as

$$\bar{S}_{\theta}\Big|_{cu} = L(1-b)\pi^{-1} I\, Q_1 \Big/ \int Q_1\, dz$$

$$\bar{S}_{q_3}\Big|_{cu} = bI\, Q_2 \Big/ \int Q_2\, dz.$$

I is the rate at which the resolvable scale supplies moisture to a model grid column. Molinari and Corsetti (1985) suggest representing I as the resolvable vertical water vapor flux (i.e., $\bar{w}\bar{q}_3$) through the lifting condensation level. Kuo (1974) defined b as the fraction of I which increases the moisture of the column, while $1-b$ precipitates.

The quantities Q_1 and Q_2 represent the vertical profiles of heating and moistening due to the deep cumulus convection. Q_1 is the difference between the potential temperature outside the cloud, $\bar{\theta}_e$, and the potential temperature within the cloud $\bar{\theta}_c$. This latter temperature is computed as a weighted average between updraft and downdraft profiles. The updraft potential temperature corresponds to the moist adiabat through the lifting condensation level. The downdraft potential temperature is evaluated to begin at the altitude in the outside air where $\bar{\theta}_E$ is a minimum, with the downdraft air temperature equal to the temperature outside the cloud at that height. At the lifting condensation level, the downdraft air is assumed to be 2°C cooler than the outside air. At the surface, the downdraft air is assumed to be 5°C cooler. Other levels are linearly interpolated.

The fractional area covered by downdraft is assumed to be 1% where it is initiated, 10% at the lifting condensation level, 20% at the height of maximum downdraft mass flux and 100% at the surface. The outside cloud temperature is used instead of the updraft below the lifting condensation level. Knupp (1987) discusses the observed kinematic structure of downdrafts in more detail.

For Q_2, the layer below the lifting condensation level is dried at the rate I (since this is the flux of water vapor through the lifting condensation level). The anvil region of the cloud is defined as 2/3 of the height between the level with the highest value of $\bar{\theta}_E$ within 3 km of the ground and the cloud top where moistening is uniform at the rate bI.

The scheme is implemented so that the moist adiabat is never exceeded. If it is, I is arbitrarily reduced in magnitude. The Kuo scheme is activated only if the lapse rate is convectively unstable and $\bar{w} > w_{\text{threshold}}$ at the lifting condensation level where $w_{\text{threshold}}$ is arbitrarily selected. Cloud top is defined as the height where the moist adiabat through the lifting condensation level intersects the temperature outside the cloud (the *equilibrium level*). Pielke (1995; and Chapter 14) summarizes the concepts of lifting condensation level and equilibrium level and shows how they are computed.

The downdraft mass flux over land on days with extensive cumulus activity tends to lag the equivalent updraft values by about one-half hour or so, at least until mid afternoon. Cooper et al. (1982) concluded that downdraft-induced convergence sustains cumulus convection until the available buoyant energy is used up. This available buoyant energy accumulates due to mesoscale horizontal wind convergence (e.g., see Pielke et al. 1991) during the earlier, pre-cumulus portion of the day. This conclusion is consistent with the observation of Fritsch et al. (1976) that the large scale typically requires many hours to generate potential buoyant energy, but once cumulus convection develops, this energy is much more quickly removed.

7.4.3.3 Summary of Several Cumulus Cloud Schemes

Nelson Seaman has provided a summary of these types of vertical (one-dimensional) parameterizations for deep cumulus convection. These parameterizations are summarized below.

The Anthes-Kuo Parameterization

Key References: Kuo (1965, 1974), Anthes (1977)

Key Assumptions:

1. Closure based on the assumption that the intensity of subgrid deep convection is proportional to the vertically-integrated convergence of water-mass in a grid column (the net resolved-scale moisture convergence, M_t, (including surface evaporation) must exceed a critical threshold value, M_c).
2. In order for the moisture convergence to trigger convection, the cloud depth and available buoyant energy (ABE) in the column simultaneously must exceed minimum threshold values.
3. The moisture-convergence closure assumes that the area of a grid box is large (by $\gg 10^2$), compared to the area of convective updrafts.
4. The water convergence can be used to produce rain, or to moisten the column. The fraction rained out, b, is a function of mean relative humidity of the column.

Strong Points:

1. The moisture-convergence closure is well designed for tropics and coarse-grid applications.
2. Tends to be robust for a wide variety of coarse-grid applications (e.g., NCEP's NGM and many global models).
3. Anthes added an easily scaled, empirically-based profile for net heating and moistening due to convection that allowed efficient calculation of feedbacks to the environment.

Weak Points:

1. At mesh sizes of 30 km or less, can produce extreme rainfall similar to Molinar and Dudek's (1992) "grid-point storms" for explicit-microphysics-only applications.
2. Does not include convective downdrafts, so not well suited for simulating mesoscale convective systems influenced strongly by outflow boundaries.

The Arakawa-Schubert Parameterization

Key References: Arakawa and Schubert (1974)

Key Assumptions:

1. A cloud field exists as an ensemble of many smaller clouds with decreasing numbers of successively larger clouds.
2. The closure is based on the assumption that convection intensity is controlled by a cloud-work function, which is a measure of the generation of integrated buoyancy force in the environment, which then is related to kinetic energy generation inside the cloud. Thus the convection is tied closely to the rate of buoyancy production at the grid scale.
3. The cloud model includes effects of entrainment, but detrains only at the cloud top (detrainment has been added by some investigators), and defines a steady-state plume.
4. The rain rate is a fraction of the liquid water in the updraft, which may depend on cloud size and wind shear.

Strong Points:

1. Inclusion of an ensemble of clouds is physically more reasonable than other parameterizations that assume all clouds in a grid box are identical.
2. The scheme is well designed for convection over tropical oceans where the rate of buoyancy generation is gradual.

Weak Points:

1. The cloud-work function closure is not well related to non-steady-state situations, such as explosive convection over midlatitude continents.
2. Can be comparatively expensive due to calculations for cloud-size ensembles.
3. Does not include treatment of convective-scale downdrafts (although Grell 1993 added a downdraft scheme).

The Fritsch-Chappell Parameterization

Key References: Fritsch and Chappell (1980a), Fritsch and Kain (1993)

Key Assumptions:

1. Designed for grid lengths between 10-30 km.
2. The amount of convective activity originates with the concept of Potential Buoyant Energy (PBE, or positive area on a thermodynamic diagram between the Level of Free Convection, or LFC, and the Equilibrium Level). This energy becomes "available" if the negative area below the LFC can be overcome, so that a sub-cloud parcel reaches its LFC with positive vertical motion. Thus the Convective Available Potential Energy (CAPE) is the PBE – the negative area below the LFC.
3. The time scale of the convection t_c, is defined as the advective time, which is the grid length divided by the horizontal wind speed $(DX/|\vec{V}|)$.
4. Closure based on the assumption that convective tendencies are such that all CAPE in the column is removed within one convective time period, t_c.

5. Separate updrafts and downdrafts are calculated. The cloud-model allows for parcel entrainment into the updrafts (an entraining plume cloud-model). Clouds detrain only at their top through the anvil, or at their base due to downdrafts.
6. The area of the updraft initially is assumed to be 1%, and the submodel iterates until the calculated updraft/downdraft removes all CAPE during t_c.
7. The trigger mechanism is based on whether a parcel with T_v and q_v (average values for a subcloud mixed layer), and with a perturbation temperature, DT can reach the LFC with positive buoyancy. The perturbation $DT = C_1 w^{1/3}$, where C_1 is a constant and w is the resolved-scale vertical velocity at the LFC.

Strong Points:

1. Recognized CAPE is a suitable closure for Great Plains storms.
2. Possibly the first convective parameterization specifically designed for mesobeta-scale applications.

Weak Points:

1. Does not conserve water and air mass.

The Betts-Miller Parameterization

Key References: Betts (1986), Betts and Miller (1986)

Key Assumptions:

1. There is a quasi-equilibrium thermodynamic structure toward which the environment is moved due to convection. This structure can be defined in terms of a "mixing line" determined from observational data.
2. For the purpose of representing convection in global models, it is unnecessary to explicitly represent heating and moistening due to the subgrid processes of updrafts, downdrafts, entrainment, and detrainment. On the assumption that simplicity of design is more efficient and less prone to errors, all of these are treated implicitly.
3. The closure assumes that the rate at which convective instability is generated in the environment determines how rapidly the environment profile is changed toward the mixing line. The relaxation time scale for the convective is roughly 2 h.

Strong Points:

1. Mixing-line closure is well designed for tropical oceans, coarse grids, and cases in which the response of the environment evolves slowly.
2. Quite robust for a wide variety of applications and can be adapted for the mesoscale by adjustment of several parameters. It is used operationally in NCEP's Eta model.

Weak Points:

1. Does not include a convective-scale downdraft parameterization (although later versions by some investigators have attempted to add their effects).
2. The mixing-line closure appears to be less appropriate in cases of explosive deep convection and does not directly generate mesobeta-scale highs and lows.

The Grell Parameterization

Key References: Grell et al. (1991), Grell (1993)

Key Assumptions:

1. Assumes deep convective clouds are all of one size.
2. The original Grell scheme used the Arakawa-Schubert cloud work function for its closure, but this was later changed to use a CAPE closure, as in Kain-Fritsch.
3. No direct mixing laterally with the environment (no entrainment or detrainment), except at the levels of origin or termination of updrafts and downdrafts. Thus mass flux is constant with height.
4. Since there is no lateral mixing (Reynolds averaging), it is not necessary to assume that the fractional area coverage of updrafts and downdrafts in the grid column is small. This allows the scheme to operate easily at finer scales, although some degree of scale separation is still important.

Strong Points:

1. Very robust scheme, which has been modified to look more like Kain-Fritsch.
2. Includes effects of downdrafts.
3. Well adapted for grids as fine as 10-12 km.

Weak Points:

1. Original Arakawa-Schubert characteristics of the closure have been mostly replaced (but this has also improved performance for explosive convection).
2. Ignores entrainment-detrainment effects.

The Kain-Fritsch Parameterization

Key References: Kain and Fritsch (1990, 1993)

Key Assumptions:

1. Most of the assumptions of the Fritsch-Chappell parameterization are retained, including the critical CAPE-removal closure assumption.
2. The cloud model is reformulated into an entraining-detraining model, with parcel buoyancy calculated as a function of parcels mixed laterally between the environment and the updrafts.
3. The differencing is reformulated to conserve mass, thermal energy, mass and momentum.
4. Designed for grid sizes of ~20-25 km.

Strong Points:

1. Contains the most complete treatment of in-cloud physical processes of currently available convective parameterizations.

2. Downdrafts parameterization allows better simulation of mesoscale responses than is possible with most schemes.

Weak Points:

1. CAPE closure is not well suited to tropical environments and can result in too vigorous convection.

The PENN State Shallow Convection Parameterization

Key References: Seaman et al. (1996), Deng (1999), Deng et al. (1999, 2000)

Key Assumptions:

1. Closure is based on the assumption that convective intensity, in terms of cloud-base mass flux, is controlled by a hybrid of the boundary-layer turbulent kinetic energy (TKE) and the CAPE in the column.
2. The cloud radius is a function of planetary boundary-layer (PBL) depth and cloud depth.
3. Cloud top height grows at a fraction of the maximum updraft speed because of resistance in the environment above the cloud.
4. Cloud mass detrained from the shallow convective updrafts is not mixed immediately with the environment, but rather becomes part of a nearly-neutrally buoyant cloud (NBC) at its level of detrainment.
5. The area and liquid water content of NBCs are predicted, based on source terms from the cumulus updrafts, advection of cloud properties, and dissipation due to mixing, settling, instability, and precipitation processes.
6. Updraft-initiating parcels are released at the top of the PBL, they have thermal and moisture characteristics that are defined from the air in the lowest 20% of the PBL and a vertical velocity that is based on the maximum TKE in the PBL.
7. Radiative effects of shallow clouds include effects of partial vertical randomness of NBCs.

Strong Points:

1. The hybrid mass-flux closure is consistent with size-dependent cumulus forcing in the atmosphere.
2. The inclusion of an NBC class provides flexibility between stratocumulus and cumulus environments.
3. Smoothly transitions from shallow convection to deep convection (Kain and Fritsch 1990) to solid stratiform cloud (Dudhia 1989).
4. Suitable for marine and land environments and for use in mesoscale models.

Weak Points:

1. A number of parameters and subgrid processes need to be studied further and modeled based on LES and additional observations.
2. As a new scheme, this parameterization needs further testing and evaluation in a variety of 3-D environments.

7.4.3.4 Use of a Cumulus Field Model or Set of Equivalent Observations

With this approach, as described in Golden and Sartor (1978), two- or three-dimensional cumulus field model simulations or sets of observations are evaluated to determine the temporal and spatial response of cumulus clouds to a particular set of mesoscale dependent variables, as well as their subsequent feedback to the mesoscale. Examples of possible models for such use include Hill (1974), Miller and Pearce (1974), Pastushkov (1975), Cotton and Tripoli (1978), Klemp and Wilhelmson (1978a,b), Clark (1979), Schlesinger (1980), and Simpson et al. (1982). Yau and Michaud (1982) have simulated the evolution of a cumulus cloud field in three-dimensions using Hill's method of random surface heating as an initiation mechanism for convection. Schlesinger (1982a) gives a summary of three-dimensional cumulus convection models. Cumulus field models are also referred to as *cloud resolving models* or *cloud ensemble models* (Tao et al. 1999, 2001). McNider and Kopp (1990) have introduced an imaginative procedure to initiate cumulus convection in such models where the boundary-layer parameterization in the model is used to specify the spatial scale and intensity of the initial thermal perturbation that produces the cumulus cloud. Cloud-resolving models have been used in both two and three dimensions (Tompkins 2000). Other examples of cloud-resolving simulations include Grabowski (2000) and Jiang et al. (2000).

Chang and Orville (1973), Cotton et al. (1976), Chen and Orville (1980), Soong and Tao (1980), Tripoli and Cotton (1980), and Schlesinger (1982b), for instance, have examined the response of cloud models to mesoscale convergence and changes in the thermodynamic and wind structure by mesoscale circulations. Fritsch and Chappell (1980a) justified their form for small-scale temperature perturbations by using Chen and Orvilles' result that thermals are stronger and larger when low-level convergence is present, a result that has been replicated by Soong and Tao (1980).

Beniston and Sommeria (1981) have used Deardorff's (1973) fine mesh (50 m horizontal grid increment) planetary boundary-layer model, as modified by Sommeria (1976), to test two cumulus parameterization schemes (e.g., those of Betts 1975, 1976 and Fraedrich 1976) using model generated data, as well as to develop specific empirical relations for the response of shallow, nonprecipitating trade wind cumulus to larger-scale forcing. Among their results are that these shallow clouds (less than 1 km in depth) have a cloud depth to radius ratio of 0.4 ± 0.1 and that cloud base mass flux is related to cloud volume with a correlation coefficient of the order of 0.8 by

$$M_{\mathrm{u}}\left(z_{\mathrm{LCL}}\right) = 3.204 \times 10^{-9}\ \mathrm{s}^{-1}\ \mathrm{kg\ m}^{-5}\ V_{\mathrm{c}} + 2.18\ \mathrm{kg\ m}^{-2}\ \mathrm{s}^{-1},$$

as illustrated in Fig. 7.16a. The variable $M_{\mathrm{u}}(z_{\mathrm{LCL}})$ is the mass flux at cloud base and V_{c} the cloud volume, where $V_{\mathrm{c}} = 10^8\ \mathrm{m}^3$ would correspond to a spherical cloud with a radius of 288 m. They also found that the growth and decay times for individual modeled clouds were about equal and positively related to the maximum area a_{c} of the cloud at its peak of activity. Beniston and Sommeria estimated the lifetime of shallow trade wind cumulus as

$$\tau_{\mathrm{c}} = 3.17 \times 10^{-2}\ \mathrm{s\ m}^{-2}\ a_{\mathrm{c}} + 173\ \mathrm{s},$$

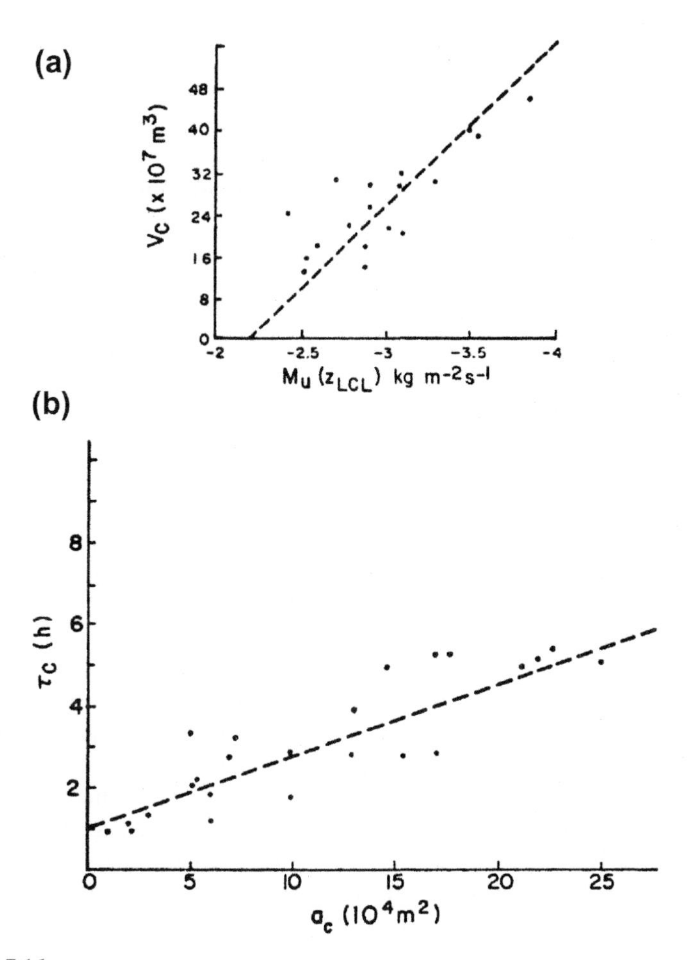

FIGURE 7.16

(a) Cloud volume V_c as a function of the mass flux at cloud base $M_u(z_{LCL})$, and (b) cloud lifetime τ_c as a function of its maximum areal coverage a_c (from Beniston and Sommeria 1981).

as illustrated in Fig. 7.16b. Such a formulation for cloud lifetime, if it can be extended to larger cumulus clouds, could provide an improvement for the value of cloud lifetime used in cumulus parameterization schemes. Beniston and Sommeria also found that thermodynamic fluxes in trade wind cumulus could be represented by a single cloud mass-flux profile rather than by a complex cloud distribution (e.g., that of Arakawa and Schubert 1974). French et al. (1999) observed the actual evolution and pulsation of small cumulus clouds in Florida.

The use of such cumulus models to parameterize the response of cumulus clouds to the larger-scale environment is innovative and offers great hope for improved parameterization schemes. It captures the approach proposed by Randall et al. (2003) except there is no need to actually run such cumulus cloud models within the parent model (whether it be a global model or a smaller-scale model). The development of the parameterization can be completed offline as discussed in Chapter 8.

7.4.3.5 *Explicit Representation of Moist Thermodynamics*

Rosenthal (1979a) has suggested that in tropical mesoscale models, cumulus clouds should be represented explicitly in the same fashion as performed for stratiform clouds. He maintains that the successful implementation of cumulus parameterization schemes, requires a strong coupling between the cloud and larger scale. Thus although tropical cyclogenesis can be well represented with such an approach, squall lines such as reported by Zipser (1971), cannot. As discussed by Rosenthal (1978), these squall lines have a distribution of moist and dry downdrafts such that convection is diminished near the center of the larger-scale system with subsequent deep cumulus clouds forced to develop away from the region of larger-scale vertical ascent. Zipser (1977) gave an example of such destructive interference between the cumulus scale and the larger scale. Weisman et al. (1997) examined the degradation of the simulation of squall lines as the horizontal grid interval was changed from 1 km to 12 km.

Because of this limitation, Rosenthal (1978, 1979a) has replaced one-dimensional cumulus parameterizations with an explicit treatment of moist thermodynamics on a grid with a 20 km horizontal grid mesh interval. With this approach, he was able to simulate tropical storms that represent the constructive reinforcement between the cumulus and mesoscales, as well as tropical squall lines in which the larger-scale fields of dependent variables have little effect after the initiation of the system. Studies by Yamasaki (1977), and Jones (1980) have also shown that realistic hurricane simulations can be obtained when latent heat is released on the resolvable grid scale in such convectively unstable atmospheres. The type of system that develops (i.e., the tropical storm or squall line) depends on the magnitude of the vertical shear of the horizontal wind and the dryness of the middle atmosphere. Rosenthal (1979b) concluded that the further use of cumulus parameterization schemes in hurricane simulations "seems to be of dubious value." He contends that an "experienced numerical experimenter can pick and choose closures that will provide almost any desired result."

Rosenthal (1978, 1980) included an explicit representation for moist thermodynamics using equations for water vapor, rain, and cloud water similar to those given by Eq. 7.113. Ice processes were neglected. His representation closely followed the work of Kessler (1969), although a similar formulation such as proposed by Orville (1980) could presumably also be used. Bhumralkar (1972, 1973) also obtained realistic results using an explicit representation of moist thermodynamics in his two-dimensional model simulation of the airflow over Grand Bahama Island during the day.

A justification for the neglect of cumulus parameterization may occur when, above the lifting condensation level

$$\left|\bar{w}\frac{\partial\bar{\theta}}{\partial z}\right| \gg \left|w''\frac{\partial\theta''}{\partial z}\right|; \quad \left|\bar{w}\frac{\partial\bar{q}_n}{\partial z}\right| \gg \left|w''\frac{\partial q_n''}{\partial z}\right|,$$

so that the vertical transport of heat and water is predominantly on the resolvable rather than on the subgrid scale. Such a condition may be true in the saturated hurricane environment even for relatively coarse horizontal resolution.

Black and Holland (1995), for instance, in Hurricane Kerry off the Australian coast, have documented a peak in the spectrum of $w\theta$ around a wavelength of 20 km with values dropping by a factor of about 7 for wavelengths of 10 km. The use of a 20 km horizontal grid with a bulk parameterization of the microphysics, as applied by Rosenthal (1978), although perhaps somewhat too large (at least for a simulation of a storm such as Kerry), may nevertheless, be a reasonable approach to the realistic simulation of tropical storms. In contrast, for mesoscale convective cluster development over land, Fritsch and Chappell (1980a,b) introduced a subgrid-scale fluctuation temperature term to account for the significant contributions by such subgrid-scale flux terms. Fritsch and Chappell (1980b) also use a 20 km horizontal grid. Weisman et al. (1997) investigated the effect of model resolution on the simulation of cumulus convective systems. Warner and Hsu (2000) provide a valuable analysis of the effects of applying parameterized cumulus convection on the outer coarse grid of a model but of grid-volume resolved cumulus convection on the inner fine grid of a nested model.

The need to satisfy this inequality may determine the appropriate grid sizes needed in a mesoscale model when the atmosphere is convectively unstable in regions of saturation. Since at least four grid intervals or more in each spatial direction are required to represent variables in a numerical model properly (as discussed in Chapter 9), *moist processes such as condensation and sublimation should be predominantly realized on spatial scales at least this large.*

As computational capability continues to increase, the need for cumulus parameterization in mesoscale models is quickly becoming moot, as the important spatial scales can be directly resolved. Grid spacings of 100 m and less in each direction likely would achieve this goal, as smaller scales can be parameterized by subgrid-scale flux representations, such as presented earlier in this chapter. Larger-scale (e.g., global) models and the outer domain of mesoscale and regional models, of course, will still need to parameterize cumulus cloud effects.

7.4.4 Examples of Parameterizations of Diabatic Heating, Precipitation and Fluxes from the Phase Changes of Water

7.4.4.1 Conversion Terms to Use in Eq. 7.113

There are a number of parameterizations for the conversion terms in Eq. 7.113 (e.g., Rutledge and Hobbs 1983, 1984). In this section, the suggested forms proposed by Orville (1980) and Lin et al. (1983) are used to illustrate one possible technique of parameterization. A schematic of such conversion terms is reported in Rotstayn (1999, his Fig. 1). Following the work of these investigators, the conversion terms in Eq 7.113 can be evaluated from

P_S = conversion from cloud ice to snowflakes + accretion of cloud ice crystals by snow + accretion of cloud droplets by snow + depositional growth of snow (or − sublimational loss of snow) − melting of snow to form rain $= P_{S1} + P_{S2} + P_{S3} + P_{S4} - P_{S5}$.

$P_R =$ conversion of cloud droplets to rain + accretion of cloud droplets by raindrops − evaporation of raindrops + melting of snow to form rain $= P_{R1} + P_{R2} - P_{R3} + P_{S5}$.

The dimensional unit of P_S and P_R is s^{-1}. The formulations for the components of P_S and P_R are defined by Orville and Lin et al. as follows for rain and for graupel-like snow of hexagonal type. Other types of snow will require somewhat different values of the constants in these expressions.

$$(1)\ P_{S1} = \alpha_1 \left(\bar{q}_1^{ci} - \bar{q}_*^{ci}\right); \quad \alpha_1 = 10^{-3}\ s^{-1} \exp\left(0.025\left(\overline{T} - 273\ K\right)\right); \quad \overline{T} < 0°C$$

where $\bar{q}_*^{ci}$ is a threshold value of specific humidity for cloud ice. In Lin et al. (1983), this value is set as $\bar{q}_*^{ci} = 0.001$. This formulation for P_{S1}, based on Kessler's (1969) original work, as modified by Chang (1977), is called *autoconversion* since no physical mechanism is included to explain why precipitation-sized ice crystals form only after cloud ice concentrations exceed a certain amount.

This algorithm requires the dependent variables $\bar{q}_1^{ci}$ and $\overline{T}$, and tunable coefficients α_1, and $\bar{q}_*^{ci}$.[49]

$$(2)\ P_{S2} = \frac{0.79\, E_s^{ci} n_{0s} c \bar{q}_1^{ci} \Gamma(3+d)}{b_s^{3+d}} \left(\frac{\bar{\rho}_{ZG}}{\bar{\rho}}\right)^{1/2}$$

when $\overline{T} < 0°C$, which represents the accretion of cloud ice by snow. In this expression

$$E_s^{ci} = \exp\left[0.025\left(\overline{T} - 273\ K\right)\right]$$

is the collection efficiency of snow for cloud ice. The term $(\bar{\rho}_{ZG}/\bar{\rho})^{1/2}$ was included to account for changes of terminal velocity with height, where $\bar{\rho}_{ZG}$ is the density at ground level. The constants d and c were set equal to $d = 0.25$ and $c = 152.93\ cm^{1-d}\ s^{-1} = 152.93\ cm^{0.75}\ s^{-1}$, which were suggested as being representative for the graupel-like snow of hexagonal type.[50]

The values of n_{0s} and b_s are determined from

$$n_s = n_{0s} \exp\left(-b_s D_s\right),$$

where D_s is the diameter of the snow particles and n_s their number per unit volume per increment of diameter size. As listed by Orville, $n_{0s} = 3 \times 10^{-2}\ cm^{-4} = 3 \times 10^6\ m^{-4}$ from measurements by Gunn and Marshall (1958), and

$$b_s = \left(3.14 \rho_s n_{0s} / \bar{\rho} \bar{q}_1^{s}\right)^{1/4}$$

[49]The determination of the dependent variables, tunable coefficients, and universal constants (if any) for each of these conversion terms are left as an exercise in the problem section.

[50]The gamma function $\Gamma(3+d)$ can be evaluated using standard mathematical tables (e.g., Selby 1967:461). Using those tables $\Gamma(3.25) \simeq 5.44$.

with ρ_s the density of the snow crystal, given as 0.1 g cm^{-3} for graupel-like snow of hexagonal type. Other investigators use different values. Scott (1982), for example, used a value of $n_{0s} = 5 \times 10^7$ m^{-4}, although he reports on observational values that range from $n_{0\,s} = 10^8$ m^{-4} at the top of tropical cumulus (Simpson and Wiggert 1969) to $n_{0s} = 3 \times 10^6$ m^{-4} for precipitation water concentrations of less than 1.4 g m^{-3} and temperatures between −2 and −32°C (Houze et al. 1979).

The equation for n_s is called the *Marshall-Palmer* distribution since these investigators originated the concept for rain (Marshall and Palmer 1948).

$$\text{(3)}\quad P_{S3} = \frac{0.79\, E_s^{cw} n_{0s} c \bar{q}_2^{cw} \Gamma(3+d)}{b_s^{3+d}} \left(\frac{\bar{\rho}_{ZG}}{\bar{\rho}}\right)^{1/2}$$

which represents the accretion of cloud water by snow. The collection efficiency of snow for cloud water, E_s^{cw}, is assumed as unity. When the air temperature is less than 0°C, these cloud droplets will freeze and increase the amount of snow.

$$\text{(4)}\quad P_{S4} = \frac{6.28\,(S_i - 1)}{\bar{\rho}\,(a_1 + a_2)} n_{0s} \left[0.78 b_s^{-2} + 0.31 s_c^{1/3} \Gamma\left(\frac{d+5}{2}\right) c^{0.5} \times \left(\frac{\bar{\rho}_{ZG}}{\bar{\rho}}\right)^{1/4} v^{-0.5} b_s^{-(d+5)/2}\right]$$

when $\overline{T} < 0°\text{C}$, $\bar{q}_2^{cw} + \bar{q}_1^{ci} + \bar{q}_1^{s} > 0$ and $\bar{q}_1^{ci} + \bar{q}_1^{s} \neq 0$. This expression represents the depositional growth or sublimation loss of snow. In this equation, S_i is the supersaturation of water vapor over ice, defined as $\bar{q}_3/\bar{q}_{si}$ using Eq. 9.8. $\Gamma\{(d+5)/2\} \simeq 1.79$ for $d = 0.25$. The variables v and s_c are the kinematic viscosity of air ($v \simeq 1.5 \times 10^{-5}$ m^{2}s^{-1}) and the Schmidt number[51] ($s_c \simeq 0.8$). Also

$$a_1 = L_s^2 / \left(k_a R_v \overline{T}^2\right), \quad a_2 = 1/\bar{\rho} q_{si} \psi,$$

where ψ is the molecular diffusivity of water vapor in air and k_a the thermal conductivity of air. The formulation for P_{S4} is based on the work of Byers (1965), with a modification for wind ventilation given by Beard and Pruppacher (1971). Depending on the sign of S_i, P_{S4} is positive or negative.

$$\text{(5)}\quad P_{S5} = -\frac{6.28}{\bar{\rho} L_f}\left[k_a \overline{T}(°\text{C}) - L_c \psi \bar{\rho}\,(q_{si} - \bar{q}_3)\right] n_{0s} \times \left[0.78 b_s^{-2} + 0.31 s_c^{1/3} \Gamma\left(\frac{d+5}{2}\right) c^{0.5} \times \left(\frac{\rho_{ZG}}{\bar{\rho}}\right)^{1/4} v^{-0.5} b_s^{-(d+5)/2}\right] - \frac{c_w T(°\text{C})}{L_f} P_{S3}$$

[51]The Schmidt number is defined as v/ψ. At 0°C and $\overline{p} = 1000$ mb, $\psi \simeq 1.875 \times 10^{-5}$ m^{2}s^{-1} (Beard and Prupaccher 1971).

when $\overline{T} > 0°\text{C}$. This expression, based on Mason (1956), represents the melting of snow to form rain. The specific heat of liquid water c_{w} is 4.187×10^3 J kg^{-1}K^{-1}.

$$(6)\quad P_{\text{R1}} = \bar{\rho}\left(\bar{q}_2^{\text{cw}} - q_*^{\text{cw}}\right)^2 \Big[1.2 \times 10^{-4} + \left\{1.569 \times 10^{-12} n_{\text{R}} / \left(d_0\left[\bar{q}_2^{\text{cw}} - \bar{q}_*^{\text{cw}}\right]\right)\right\}\Big]^{-1},$$

which represents the conversion of cloud droplets by collision and collection (i.e., *coalescence*) to form raindrops. Based on Berry (1967), and modified by Orville and Lin et al., n_{R} is the number concentration of droplets, d_0 the dispersion of the droplets (i.e., their standard deviation around their mean size), and $\bar{q}_*^{\text{cw}}$ is a threshold value of specific humidity for cloud water required before there is a significant probability for the formation of raindrops. Orville and Lin et al. reported a value of $\bar{q}_*^{\text{cw}} = 0.002$.

$$(7)\quad P_{\text{R2}} = \frac{0.79 E_{\text{R}}^{\text{cw}} n_{0\text{R}} \bar{q}_2^{\text{cw}} \Gamma(3+f)}{b_{\text{R}}^{3+f}}.$$

This expression represents the accretion of cloud water by raindrops, with the collection efficiency given by $E_{\text{R}}^{\text{cw}} = 1$. The values of $n_{0\text{R}}$ and b_{R} are determined from

$$n_{\text{R}} = n_{0\text{R}} \exp\left(-b_{\text{R}} D_{\text{R}}\right),$$

where D_{R} is the diameter of the raindrops and n_{R} the number per unit volume per increment of diameter size. As given by Marshall and Palmer (1948), $n_{0\text{R}} = 8 \times 10^{-2}$ cm^{-4}. Scott (1982), using Merceret (1975), suggests a value of $n_{0\text{R}} = 10^7$ m^{-4}. The value of b_{R} is given by

$$b_{\text{R}} = \left(3.14 \rho_2 n_{0\text{R}} / \bar{\rho} \bar{q}_2^{\text{R}}\right)^{1/4},$$

where ρ_2 is the density of liquid water ($\rho_2 = 1$ g cm^{-3} $= 10^3$ kg m^{-3}). Condensational growth of raindrops is insignificant as compared with growth by collision–collection (e.g., see Wallace and Hobbs 1977) and is neglected in P_{R2}.

$$(8)\quad P_{\text{R3}} = -6.28\left(s_{\text{w}} - 1\right) n_{0\text{R}} \left[0.78 b_{\text{R}}^{-2} + 0.31 s_{\text{c}}^{1/3} \Gamma\left[\frac{f+5}{2}\right] a_3^{0.5} \upsilon^{-0.5} \times \left(\frac{\bar{\rho}_{\text{ZG}}}{\bar{\rho}}\right)^{1/4} b_{\text{R}} - \left[\frac{f+5}{2}\right]\right] \frac{1}{\bar{\rho}} \left(\frac{L_{\text{c}}^2}{k_a R_{\text{v}} \overline{T}^2} + \frac{1}{\bar{\rho} q_{\text{sw}} \psi}\right)^{-1}; \quad s_{\text{w}} < 1,$$

where $s_{\text{w}} = \bar{q}_3 / q_{\text{sw}}$. This term represented the evaporation of rainwater in a subsaturated environment. The values of f and a_3 are 0.8 and 2115 cm^{1-f}s^{-1} = 2115 cm$^{0.2}$s^{-1}, respectively. With this value of $f, \Gamma(3+f) \simeq 5.59$ and $\Gamma[(f+5)/2] \simeq 1.92$.

Orville (1980) and Lin et al. (1983) also present conversion terms for such effects as the accretion of rain by cloud ice, the accretion of cloud ice by rain, the accretion of snow by rain, raindrop freezing, and hail generation and growth. Except for raindrop

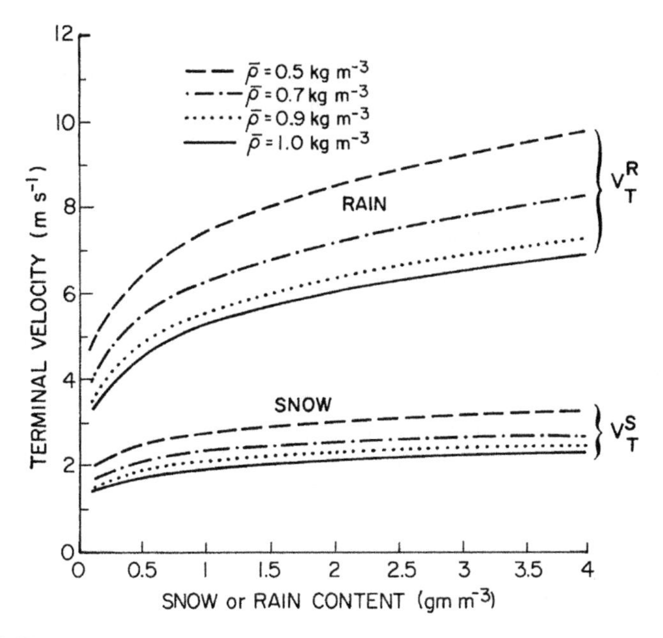

FIGURE 7.17

Mass-weighted mean terminal velocities for rain and snow as a function of rain and snow water content (adapted from Orville 1980).

freezing (i.e., the creation of sleet), these other processes are expected to be significant only when the atmosphere is convectively unstable.

Finally, the terminal velocities of snow and rain required in Eq. 7.113, given by Orville and Lin et al. are [52]

$$V_{\mathrm{T}}^{\mathrm{R}} = \frac{a_3 \Gamma(4+f)}{6\,(b_{\mathrm{R}})^{\mathrm{f}}} \left(\frac{\bar{\rho}_{\mathrm{ZG}}}{\bar{\rho}}\right)^{1/2},$$

$$V_{\mathrm{T}}^{\mathrm{s}} = \frac{c\Gamma(4+d)}{6\,(b_{\mathrm{s}})^{\mathrm{d}}} \left(\frac{\bar{\rho}_{\mathrm{ZG}}}{\bar{\rho}}\right)^{1/2}.$$

These velocities are the mass-weighted mean velocities, defined originally by Srivastava (1971) as

$$V_{\mathrm{T}}^{\mathrm{R}} = \frac{1}{\bar{q}_2^{\mathrm{R}}} \int_0^{\infty} V_{\mathrm{T}}^{\mathrm{R}}(D)\bar{q}_2^{\mathrm{R}}(D)dD, \quad V_{\mathrm{T}}^{\mathrm{s}} = \frac{1}{\bar{q}_1^{\mathrm{s}}} \int_0^{\infty} V_{\mathrm{T}}^{\mathrm{s}}(D)\bar{q}_1^{\mathrm{s}}(D)dD,$$

where $V_{\mathrm{T}}^{\mathrm{R}}(D)$ and $V_{\mathrm{T}}^{\mathrm{s}}(D)$ are terminal velocities or rain and snow for particular diameter particles. The quantities $\bar{q}_2^{\mathrm{R}}(D)$ and $\bar{q}_1^{\mathrm{s}}(D)$ represent the specific humidity of raindrops and snow per unit increment of diameter sizes. Figure 7.17 illustrates the

[52]For $d = 0.25$ and $f = 0.8$, $\Gamma(4+d) \simeq 21.8$ and $\Gamma(4+f) \simeq 22.4$.

relationship between mass-weighted mean velocities and total rain or snow content for several different values of atmospheric density. Values of the terminal fall velocity in meters per second for specific diameters in meters can be estimated by

$$V_{\mathrm{T}}^{\mathrm{R}}(D) = 130D^{0.5} \text{ Kessler (1969)}$$

$$V_{\mathrm{T}}^{\mathrm{S}}(D) = \begin{cases} 2.71D^{0.206} & \text{aggregates of dendrites and plates} \\ & \text{from (Jiusto and Bosworth 1971)} \\ 3.95D^{0.206} & \text{aggregates of columns (Jiusto and Bosworth 1971)} \end{cases}$$

as summarized by Scott (1982). The values of $\bar{q}_2^{\mathrm{R}}(D)$ and $\bar{q}_1^{\mathrm{S}}(D)$ can be estimated from $\bar{q}_2^{\mathrm{R}}(D) = 0.52n_{0\mathrm{R}}\left(\rho_2/\bar{\rho}\right)D^3e^{-b_{\mathrm{R}}D}$, $\bar{q}_1{}^{\mathrm{s}}(D) = 0.52n_{0\mathrm{s}}\left(\rho_1/\bar{\rho}\right)D^3e^{-b_{\mathrm{s}}D}$ using the Marshall-Palmer distribution for rain and snow and assuming the volume of a raindrop and snow crystal can be represented as $0.52D^3$.

Other forms of these conversion terms can be found, for example, in Meyers et al. (1997), Zhao et al. (1997), Grabowski (1998), and Rotstayn et al. (2000).

Problems for Chapter 7

1. Select a parameterization for the subgrid-scale heat fluxes from an atmospheric model of your choice. Dissect the parameterization using the technique outlined in Section 7.2.3. List the additional new dependent variables, and adjustable and universal parameters. Assess the sensitivity in the calculated value of the flux for uncertainties of $\pm 10\%$ as a function of one or more of the adjustable constants and universal parameters.
2. Perform No. 1, except for the subgrid-scale parameterization used for the velocity fluxes. What are the differences between the two parameterizations?
3. Derive an equation in which $\overline{S_\theta''} = 0$ is not assumed.
4. Obtain a clear-sky longwave radiative flux divergence parameterization which is used in a model of your choice. Write the dependent variables, universal constants, and tunable coefficients that are used. Assess the differences that result with a $\pm 10\%$ and $\pm 25\%$ change in the values of the tunable coefficients and the dependent variables.
5. Do the same as No. 4, but for direct and diffuse solar radiation.
6. Repeat No. 4 and No. 5 for a cloudy atmosphere.
7. Repeat No. 4 and No. 5 for a polluted atmosphere.
8. Obtain a copy of the NCAR Community Climate Radiation Model (CRM) and decompose the specific algorithms that are used for the radiative fluxes, using the same techniques as in Problems 4–8. The model version 2.0 is available at http://www.cgd/ucar.edu/cms/crm.
9. Using conversion relations for P_{S} and P_{R} in Section 7.4.4.1, determine the dependent variables, turnable coefficients, and universal constants (if any). Calculate how the individual conversion terms change with a $\pm 10\%$ and $\pm 25\%$ change in the tunable coefficients and the dependent variables.

10. Complete the exercise in No. 9, but use conversion equations from another model.

11. Program the Kuo model, and determine the sensitivity of the parameterization to $\pm 10\%$ and $\pm 25\%$ changes in the tunable coefficients and relationships.

Chapter 7 Additional Readings

To understand the parameterization techniques to represent the subgrid-scale fluxes used in mesoscale models, it is necessary to understand atmospheric turbulence. Among the valuable texts in this area are:

Lumley, J.L. and H.A. Panofsky, 1964: *The Structure of Atmospheric Turbulence.* Interscience Monographs and Texts in Physics and Astronomy, Vol. 12, Interscience, New York, NY, 239 pp.

In this classic book, John Lumley wrote the first half, and the latter portion was written by Hans Panofsky. Lumley's sections provide the mathematical basis for turbulence theory, and Panofsky's portion emphasizes specific applications of this theory to an improved understanding of mixing in the atmosphere.

Stensrud, D.J., 2007: *Parameterization Schemes: Keys to Understanding Numerical Weather Prediction Models.* Cambridge University Press, 478 pp.

This text provides a very effective discussion of parameterizations, and how they are used.

Tennekes, H. and J.L. Lumley, 1972: *A First Course in Turbulence.* MIT Press, Cambridge, MA, 300 pp.

The authors introduce turbulence theory using effective physical examples of such mixing.

This text is a valuable reference source for nomenclature and clear explanations of turbulence theory.

The next contributions provide excellent in-depth discussions on how to parameterize the atmospheric boundary layer.

Bélair, S., J. Mailhot, J.W. Strapp, and J.I. MacPherson, 1999: An examination of local versus nonlocal aspects of a TKE-based boundary layer scheme in clear convective conditions. *J. Appl. Meteor.,* **38**, 1499-1518.

Beljaars, A.C.M., and P. Viterbo, 1998: Role of the boundary layer in a numerical weather prediction model. In: *Clear and Cloudy Boundary Layers*, A.M. Holstlag ad P.G. Duynkerke, Eds., Royal Netherlands Academy of Arts and Sciences, Amsterdam, 287-304.

Blackadar, A.K., 1979: High resolution models of the planetary boundary layer. *Adv. Environ. Sci. Eng.* **I**, J.R. Pfafflin and E.N. Ziegler, Eds., Gordon and Breach Science Publishers, 50-85.

Cuijpers, J.W.M., and A.A.M. Holtslag, 1998: Impact of skewness and nonlocal effects on scalar and buoyancy fluxes in convective boundary layers. *J. Atmos. Sci.*, **55**, 151-162.

Holtslag, A.A.M., 1998: Fluxes and gradients in atmospheric boundary layers. In *Clear and Cloudy Boundary Layers*, A.A.M. Holtslag and P.G. Duynkerke, Eds., Royal Netherlands Academy of Arts and Sciences, PO Box 19121, 1000 G Amsterdam, The Netherlands.

Uliasz, M., 1994: Subgrid scale parameterizations. In: *Mesoscale Modeling of the Atmosphere.* R. Pearce and R.A. Pielke, Eds., American Meteorological Society, 13-19.

There are a series of excellent books on boundary layer theory. The following texts are among the best.

Arya, S.P., 1988: *Introduction to Micrometeorology*. Academic Press, Harcourt Brace Jovanovich Publishers, San Diego, CA, 307 pp.

Garratt, J.R., 1992: *The Atmospheric Boundary Layer*. Cambridge University Press, Cambridge, 316 pp.

Garratt, J.R., and Hess, G.D. 2003. Neutrally stratified boundary layer. In: *Encyclopedia of Atmospheric Sciences*, J.R. Holton, J.A. Curry, and J.A. Pyle, Eds., Academic Press, UK, 262-71.

Garstang, M., and D. Fitzjarrald, 1999: *Observations of Surface to Atmospheric Interactions in the Tropics*. Oxford University Press, New York, 405 pp.

Holtslag, A.A.M., and P.G. Duynkerke, 1998: *Clear and Cloudy Boundary Layers*. Royal Netherlands Academy of Arts and Sciences, 372 pp.

Sorbjan, Z., 1989: *Structure of the Atmospheric Boundary Layer*. Prentice Hall, Englewood Cliffs, NJ, 317 pp.

Stull, R.B., 1988: *An Introduction to Boundary Layer Meteorology*. Kluwer Academic Publishers, The Netherlands, 666 pp.

Stull, R.B., 2000: *Meteorology for Scientists and Engineers*. 2nd Edition, Brooks/Cole Thomson Learning, 502 pp.

A very useful summary of field campaigns and long-term observational facilities to monitor the boundary layer is reviewed in Tunick (1999). Review papers include Avissar (1995) and Brutsaert (1998).

CHAPTER

New Parameterization Approaches

8

CHAPTER OUTLINE HEAD

8.1 Introduction

In Chapter 7, the traditional approach to parameterization of subgrid-scale flux divergence, long- and shortwave radiative flux divergence and heat and moisture changes from stable and convective cloud systems was presented. These traditional parameterizations use the framework of a model to obtain the values of the grid-volume resolved values of dependent variables such as velocity, potential temperature, and absolute humidity (see Fig. 8.1).

However, since as demonstrated in that chapter, these traditional parameterizations are tuned with interpolation formula and adjustable constants and functions obtained from real-world observations (such as from a field program) and/or a higher resolution model. They are not basic physics models.

This is an important distinction as it means that if we can obtain the same result as obtained from the traditional parameterization, but at a fraction of the cost, we should replace the traditional parameterization with the new method (see Fig. 8.2). In addition, we can use the new method to go directly from the real world observations without an intervening layer of another model. Since the traditional parameterizations are often developed from field experiment data, generally using just a small subset of

Mesoscale Meteorological Modeling, Third Edition, Volume 98. http://dx.doi.org/10.1016/B978-0-12-385237-3.00008-6

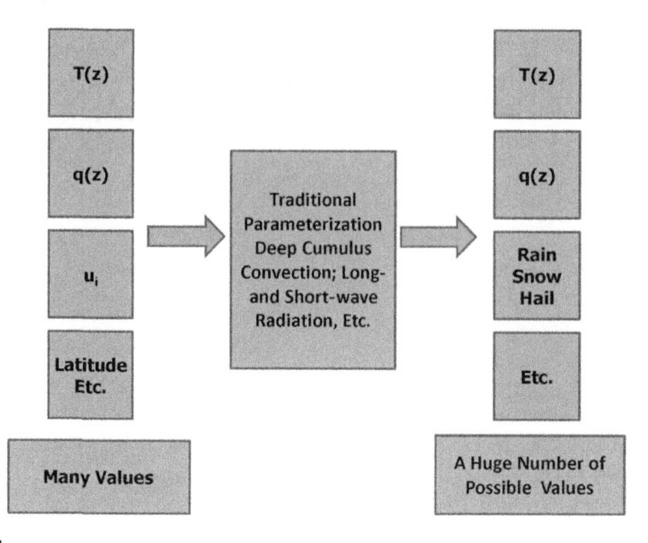

FIGURE 8.1

Schematic illustration of the application of traditional parameterizations.

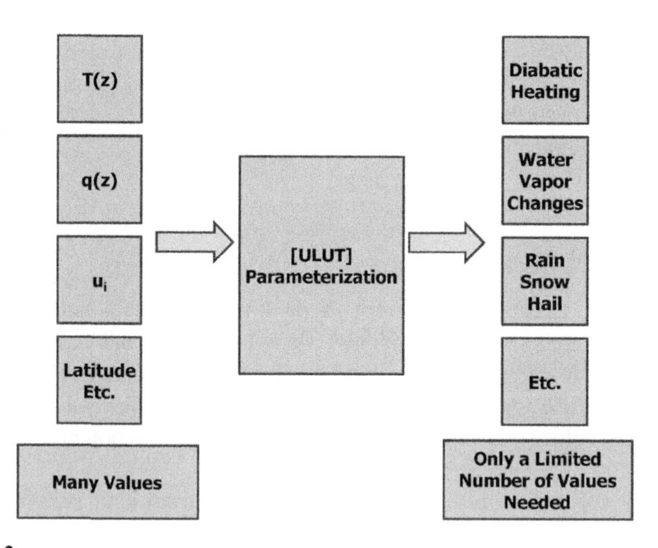

FIGURE 8.2

Schematic illustration of the application of a proposed new parameterization approach.

real world conditions, the direct use of all of the real world observations would make the parameterizations more general.

Moreover, the computational costs of the traditional parameterizations are often much greater than for the dynamical core of a model, particularly as parameterizations introduce greater complexity. Majewski et al. (2002) report the computational cost of different components of high-resolution global operational numerical weather prediction models. They show that parameterizations occupy 46.8% of the total when the radiation code was updated every 2 hours. If the radiation code (which represents 57.5% of the total parameterization cost) were updated on a more realistic time scale, e.g., 10 minutes, this parameterization by itself would cost about 85% of the total computational time!

Chevallier et al. (1998) report that the longwave radiation scheme accounts for 10% and 18%, respectively, of the total computing time required by the general circulation model at the European Centre for Medium-Range Weather Forecasts and the climate model of the Laboratoire de Meteorologie Dynamic. Tests conducted by Leoncini et al. (2008) show that the Regional Atmospheric Modeling System (RAMS) has similar levels of performance and the Harrington radiation parameterization occupies 13% of the total CPU time.

Thus there is a new approach to parameterization. There are three ways to implement this new method, which we can call the "*look-up table (LUT)*" parameterization methodology. A LUT could be an actual look-up table, or interpolation formula that is derived from such a discrete tabular data set. The concept is very straightforward. The LUT parameterization approach can be written as

$$\text{Output}(\Delta\mathbf{f}) = \mathbf{T}[\text{Input}(\mathbf{f}), \mathbf{c}]$$

where the dependent variable response that need to be computed (the Output $\Delta\mathbf{f}$), are obtained from the Input values $\mathbf{f}$ and the prescribed constants, $\mathbf{c}$ of the parameterization, through the transfer function, $\mathbf{T}$. The quantities $\mathbf{f}$ and $\mathbf{c}$ are vectors. The vector $\mathbf{f}$ includes, for example, the grid volume dependent variables, while $\mathbf{c}$ includes the time of year and day, the latitude etc. $\mathbf{T}$ is the LUT parameterization.

$\mathbf{T}$ is obtained from observations and/or a higher resolution model (i.e., through observed data values of $\Delta\mathbf{f}$ as a function of observed (real world) values of $\mathbf{f}$ and $\mathbf{c}$). $\mathbf{T}$ can provide an instantaneous change (i.e., over a time step) or be inserted over a period of time, such as performed with the Fritsch-Chappell (1980a,b) deep cumulus parameterization. This approach is also common in the remote-sensing community (e.g., Jin et al. 2004). Pielke (1984; pages 263-265) proposed this approach to parameterize the response of cumulus clouds to the larger-scale environment.

The traditional paradigm is to exercise the parameterization within the atmospheric model for each grid point for separate physical processes during the period of model integration. However, the LUT approach can significantly reduce the computational cost. The concept is to integrate the parameterization offline for the universe of $\mathbf{f}$ and $\mathbf{c}$, where the number of values of $\mathbf{f}$ that is needed depends on the graining that is chosen. The LUT, expressed as a multidimensional array or fitting function, provides the needed value of $\mathbf{T}$.

There are several issues that permit the feasibility of this approach:

- The physical fidelity of the parameterization is not important as long as the Output ($\Delta\mathbf{f}$) is at least as accurate as the traditional parameterization.
- Existing parameterizations are generally exercised in 1-D vertical columns with the input values of $\mathbf{f}$ obtained from just one x-y grid point. This simplifies significantly the number of calculations that must be performed in creating the LUT.

- Existing parameterizations include mathematical complexity which is not justified by the skill that it has in defining **T**. In other words, the dimensionality (i.e., as represented by its degrees of freedom) of the parameterization is much greater than warranted. Such a large number of combinations results in a large number of physically meaningless inputs that result from the mathematical formulation used to construct a parameterization, rather than based on the data resolution used to construct the parameterization. No parameterization can justify a dimensionality in the millions. This means the number of separate values of **T** can be much less than provided by the parent parameterization. The term graining can be used to describe the number of separate values of **T**.
- The success of LUTs highly depends on the availability of a large repository of the precomputed values, and more critically, on fast, targeted retrieval of this information. Fortunately, those commercial search engines, such as Google and Yahoo, have already demonstrated the feasibility of such an approach. For example, Google can search an index database of billions of web pages in under a second for most user queries. The LUT approach has the added advantage that the total information stored is much more compact, well structured, and much easier to index.
- In an atmospheric model, several different parameterizations usually are used to reproduce the various physical processes. However, it is generally unrealistic to separate the processes in this way since the observations and physics make no such artificial separation. These processes are, in fact, 3-D and interact with each other. Thus the most effective way to implement a LUT is to combine all of the relevant physics that result in diabatic heating and cooling, atmospheric moistening and drying, etc.

With a LUT parameterization, there is therefore no need for millions of data points in a LUT in order to realistically reproduce a parameterization of a specific process with the accuracy needed for use in an atmospheric model.

To illustrate this point, the hyperspace of a transfer function **T**, and how slices through it can be applied to establish the needed resolution of a parameterization, is presented using the Louis surface flux parameterization (Louis 1979) as discussed here (Lu 2004). The Louis surface flux scheme, although a simple parameterization, still requires storage if used as a LUT. The surface heat flux, as calculated from the Louis surface flux parameterization is a function of the wind (u) and the potential temperature (θ) at a height (z), the surface potential temperature (θ_0), and the roughness length (z_0). Figure 8.3 shows one slice through the hyperspace of **T** where the surface heat flux varies with u and θ, while the other variables are fixed ($z = 1.0$ m, $\theta_0 = 300$ K, $z_0 = 0.1$ m). The domain of u is set from 0.05 to 2.05 m s^{-1} with an interval (the graining) of 0.02 m s^{-1} and the domain of θ is from 290 to 310 K with an interval of 0.2 K. This graining of the parameterization (with 100 by 100 data points) indicates that this resolution is sufficient to capture the physically important variations that are represented by the parameterization for the variable space used in this example. Even extending the domains to more realistic ranges (e.g., 250-320 K, and 0.05-50 m s^{-1}) with the same graining, the total number of calculations or data

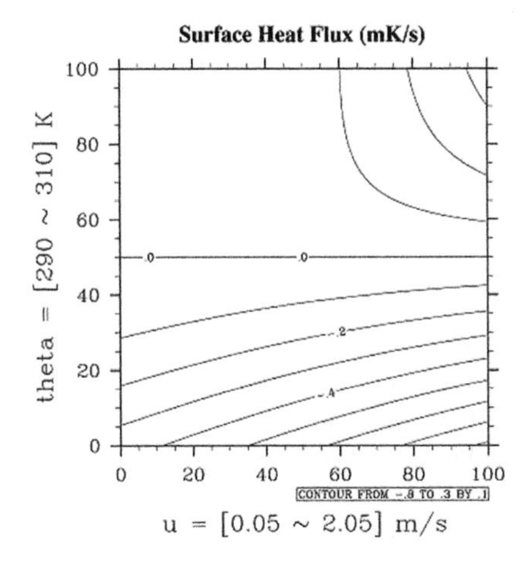

FIGURE 8.3

Surface heat flux calculated from wind (u, m s^{-1}) and potential temperature (θ, K) at $z = 1.0$ m. The u is in domain of 0.05 to 2.05 m s^{-1} with an interval of 0.02. The domain of θ is 290-310 K and the interval is 0.2 K. The surface potential temperature $\theta = 300$ K and the roughness length $z_0 = 0.1$ m (from Lu 2004).

points would still be manageable (875,000 for the above ranges). In the context of a general parameterization, we do not need billions of data points in an LUT in order to realistically parameterize a process for use in an atmospheric model.

One major advantage of the computationally efficient LUT approach includes the ability to create more realizations in the creation of ensemble forecasts. As shown by Kalnay (2002), Mullen and Buizza (2002), and Bright and Mullen (2002), such ensemble forecasts provide a particularly valuable way to improve forecasting skill.

There are three ways a LUT can be applied. A LUT can be used for a subsection of a parameterization (e.g., the surface wind profile or precipitation efficiency as a function of wind shear). This is already done by modelers.

The second approach is to use a LUT for the aggregate effect of a particular physical process, such as longwave radiative flux divergence. This has generally not been done by modelers. The third approach is to use a LUT for the aggregate effect of a set of physical processes such as diabatic heating, and moistening/drying.

Examples of each are discussed next.

8.2 The Look-Up Table Method in Traditional Parameterizations

As previously mentioned, one way to apply this approach is within the traditional parameterizations. This actually is already done in all parameterizations. For example,

LUTs have been used for years to store precomputed rates of hydrometeor collisions, melting, and nucleation in the RAMS microphysics parameterization (Walko et al. 2000a), while the overall parameterization is computed in the conventional way. Schultz (1995) developed an explicit formula-based cloud physics parameterization for use in operational models which encompass the LUT concept.

8.2.1 Subgrid-Scale Fluxes

The function $\psi_{\mathrm{M}}(z/L)$ is referred to as the correction to the logarithmic wind profile resulting from the deviation from neutral stratification. This formula (a type of LUT) is used as part of the calculation of subgrid-scale flux divergence at the lowest level in the model atmosphere. For a neutral stratification, $\psi_{\mathrm{M}} = 0$. Figure 8.4 schematically illustrates the form of $\overline{V}$ when plotted as a function of ln z for stable, unstable, and neutral stratification. Note that z_0 is presumed independent of stability so that each profile is extrapolated to the same value. This is required, of course, since ϕ_{M} approaches unity as z decreases (i.e., $z/L = z_0/L \simeq 0$ if $z_0 \ll L$). Specific observational estimates of ϕ_{M} are discussed in Högström (1996) with one suggested formula presented in Eq. 8.1 (see Fig. 8.5). Expressions analogous to ϕ_{M} and ψ_{M} can also be

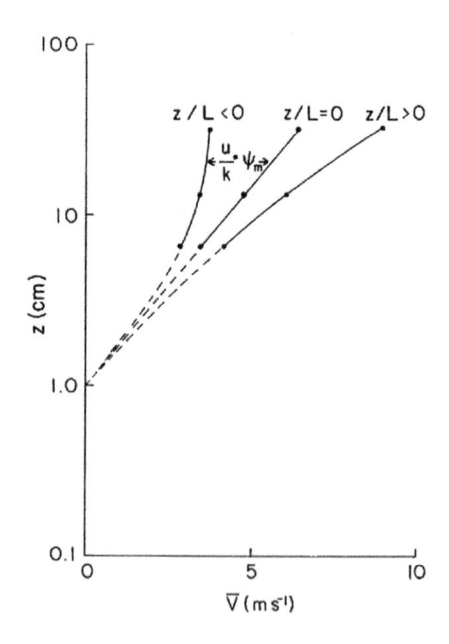

FIGURE 8.4

Schematic illustration of the procedure used to compute the wind profile near the ground from observations of mean wind speed at three levels, along with the knowledge of the stability as measured by z/L. The difference between the logarithmic wind profile and the actual wind profile at any level is given by $(u_*/k)\ \phi_M$ ($\phi_M < 0$ when $z/L > 0$, $\phi_M > 0$ when $z/L < 0$); for $z/L < 0$, see, for example, Eq. 8.1.

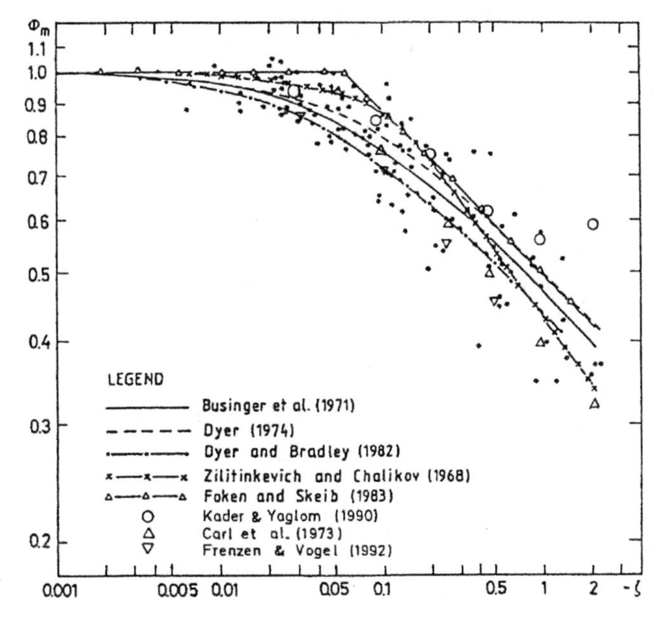

FIGURE 8.5

Plot of ϕ_M against $(z - D)/L$ in log-log representation for unstable stratification. The small dots are data from Högström (1988). The other symbols have been obtained from information from the sources (from Högström 1996).

derived for the vertical subgrid-scale fluxes of potential temperature, water, and other gaseous and aerosol atmospheric materials.

$$\left.\begin{aligned}
\phi_{\mathrm{M}} &= \frac{kz}{u_{*0}}\frac{\partial \overline{V}}{\partial z} \simeq \begin{cases} (1 - 19z/L)^{-1/4} & z/L \leq 0 \\ 1 + 5.3z/L & 0.5 > z/L > 0 \end{cases} \\
\phi_{\mathrm{H}} &= \frac{kz}{\theta_{*0}}\frac{\partial \bar{\theta}}{\partial z} = \frac{kz}{q_{*0}}\frac{\partial \bar{q}_3}{\partial z} = \frac{kz}{\chi_{*m_0}}\frac{\partial \bar{\chi}_m}{\partial z} \simeq \begin{cases} (1 - 11.6z/L)^{-1/2} & z/L \leq 0 \\ 1 + 8.0z/L & 0 \leq z/L \leq 0.5 \end{cases}
\end{aligned}\right\} \tag{8.1}$$

Other examples of LUTs, as used for a traditional parameterization, are illustrated in Figs. 7.14 and 7.15.

8.3 The LUT Approach for the Total Net Effect of Each Separate Physical Process

An example of the second type of LUT application is reported in Matsui et al. (2004). The Fu-Liou radiation parameterization is presented to illustrate this technique. When the LUT-based approach is implemented, the Fu-Liou radiation code becomes 443 times faster than the original code.

Leoncini et al. (2008) develop a LUT for the Harrington radiation parameterization where empirical orthogonal functions (EOFs) of all of the input variables of the parent scheme are computed, run the scheme on the EOFs, and express the output of a generic input sounding exploiting the input-output pairs associated with the EOFs. The weights are based on the difference between the input and EOFs water vapor mixing ratios. Results show very good agreement ($r > 0.91$) between the different transfer schemes and the Harrington radiation parameterization with a very significant reduction in computational cost (at least 95%).

8.4 The Generalized LUT for the Integrated Effect on Diabatic Heating and Other Source/Sink Terms

8.4.1 Methodology

To illustrate the methodology (the third type of LUT application), we focus first here on subgrid-scale diabatic effects. As given by Eq. 2.44, the conservation equation for potential temperature can be written as,

$$\partial\theta/\partial t = -u\partial\theta/\partial x - v\partial\theta/\partial y - w\partial\theta/\partial z + S_\theta \qquad (8.2)$$

The source/sink term S_θ includes all of the diabatic physics, which, in a model, is decomposed into separate parameterizations and a resolvable term for phase changes of water.

The methodology is that instead of using separate physics to compute the terms that comprise S_θ, observed data are used to construct this term in the format of a transfer function (e.g., a "look-up table"), as proposed by Matsui et al. (2004) and Pielke et al. (2006b) for the individual parameterizations that comprise S_θ (Pielke et al. 2007a).

The remote sensing community uses such an approach routinely in its algorithms to convert satellite radiances into variables, for example (e.g., Kidder and Vonder Haar 1995). There is also a direct analogy to the approach several investigators have made in land surface modeling. Like the convective parameterization problem, the land surface parameterizations are fraught with highly complex interactions between vegetation, soil, and moisture. Given this complexity some modelers have resorted to simpler models constrained by satellite observations to recover fluxes as a residual. In particular, McNider et al. (1994), Jones et al. (1998), Alapaty et al. (2001) and others have proposed and used morning satellite surface tendencies to infer the moisture availability and evening tendencies to infer heat capacity (McNider et al. 2005). The triangle method of Gillies et al. (1997) is another example where a look-up table approach is used to derive surface energy fluxes from satellite observed values of vegetation fraction and surface radiant temperature.

We propose a similar methodology for parameterizations in atmospheric models. In the following, the method is illustrated for the physics of diabatic heating, but it can be applied to any quantities that are parameterized within weather and climate models. The procedure is as follows:

1. Satellite and other complementary observations are used to obtain potential temperature, the horizontal winds, water vapor, liquid water, and ice for each of the footprints that are viewed by the satellite over a period of time. These data need to be transferred to a grid point format.
2. The individual terms in Eq. 8.2 are directly computed for the sampling time period of the observations: (a) $\partial\theta/\partial t$ and (b) $u\partial\theta/\partial x$ and $v\partial\theta/\partial y$ while $w\partial\theta/\partial z$ is diagnosed using the spatial gradient of the horizontal wind field (w is diagnosed from the conservation of mass and/or using quasigeostrophic theory when applicable). The value of S_θ is then computed as a residual: $S_\theta = u\partial\theta/\partial x + v\partial\theta/\partial y + w\partial\theta/\partial z + \partial\theta/\partial t$.
3. The model-resolved portion of S_θ can be subtracted out also; $< S_\theta >= S_\theta - Lw\partial q_s/\partial z$, where $L\partial q_s/\partial z$ is the latent heat of model-resolved phase change when q is equal to q_s and $w > 0$ (q_s is the saturation specific humidity). This calculation can be generalized to include phase changes of liquid and ice also, and w is the grid volume averaged vertical velocity. The quantity $< S_\theta >$ is then the subgrid-scale diabatic contribution.

The unified parameterization is the transfer function **T**,

$$\mathbf{T} = [f(\text{observation input of } u\partial\theta/\partial x + v\partial\theta/\partial y, \partial\theta/\partial t, \partial\theta q/\partial x + v\partial q/\partial y, \partial q/\partial t, \text{time of year, latitude})] \rightarrow < S_\theta >,$$

which can be expressed as a look-up table. Two necessary conditions for this approach to provide an accurate unified parameterization are (1) that the satellite observations, complemented with other observations, are sufficiently accurate with the needed spatial and temporal resolution and (2) that the satellite observations encompass a broad and global range of meteorological conditions. The latter condition is satisfied given the global observing systems. The validity of the first condition must be investigated specifically for the available observations, including the cloud library from multiscale modeling framework (Tao et al. 2009), as well as for promising new satellite instrumentation (Geosynchronous Imaging Fourier Transform Spectrometer: GIFTS; see Fig. 2 at http://ams.confex.com/ams/pdfpapers/70174.pdf (Li et al. 2004), and platforms, e.g., Geostationary Operational Environmental Satellite-R; GOES-R). Those massive data sets can be efficiently hardwired through the unified LUT or neural network approach as proposed by Matsui et al. (2004) and Pielke et al. (2006b).

8.4.2 Proof of Concept – An Approach to Test

The proof of concept of the new methodology is to use regional model simulations to construct **T** since each of the input values can be obtained from the model fields and the value of $< S_\theta >$ can be computed by summing each of the diabatic terms that are calculated in the model using the traditional 1-D, separate parameterizations.

The goal is to recreate $< S_\theta >$ from the sum of the turbulent flux divergence, the shortwave radiative flux divergence, the longwave radiative flux divergence, the phase changes of water on the subgrid scale, and cumulus cloud flux divergence of θ. If we refer to this value of $< S_\theta >$ as $<< S_\theta >>$, and the transfer function calculated version as $< S_\theta >$, then the methodology is successful if the diabatic heating calculated by the traditional way using separate parameterizations, $<< S_\theta >>$, produces essentially the same result as using the transfer function approach, $< S_\theta >$; i.e., $<< S_\theta >> \approx < S_\theta >$, and the calculation of $< S_\theta >$ is computationally much more rapid. After we prove the concept, $< S_\theta >$ can then replace the separate, more computationally expensive individual calculations of the subgrid diabatic terms. The transfer function **T** can replace the traditional approach of parameterization.

To demonstrate skill of the method when remotely sensed data are used to construct **T**, the proof of concept experiments will also include computation of the source/sink term from simulated satellite data, denoted by $< S_\theta^* >$. The simulated satellite data are computed from the model fields so as to have spatial and temporal resolution and error characteristics of the actual satellite data, including future sensors. Coarser spatial and temporal resolution of the simulated data relative to the model produces representativeness errors in $< S_\theta^* >$. Amplitude errors assigned to the individual fields (i.e., temperature, humidity, wind) would reflect the data accuracy resulting from the measurement and retrieval errors. The quantity is by definition stochastic because it contains information about the errors in the data. This property implies that an ensemble average of model simulations using from a range within the error margins should be compared with the control model simulation. A small difference between the two model results would imply high skill of the method. This difference represents a global measure of the impact of the data errors. In the proof of concept experiments, the simulated data errors could be varied to determine desired resolution and accuracy in the data to result in a satisfactory estimate of $< S_\theta >$. The results of this analysis will assist satellite developers in decisions with respect to the needed accuracy of future instrumentation and will assist the modeling community in developing improved assimilation and computationally efficient simulation models.

8.5 The "Superparameterization" Approach

It has been proposed (Grabowski 2001, Randall et al. 2003) to embed a cloud-resolving model within a larger-scale model in order to improve the accuracy of simulating cloud interactions with the larger-scale model. This has been called a "superparameterization." Superparameterization (also called Multi-Modeling Frameworks or MMF) refers to using a 2-D or 3-D cloud-resolving model to simulate a process in place of the traditional parameterizations that has been commonly used in weather and climate models in order to keep the computational cost low.

With respect to convective cloud parameterizations, a cumulus parameterization workshop (Tao et al. 2003) concluded that there are three major approaches to cumulus parameterization: traditional, statistical, and super parameterization (the multiscale MMF approach). Most traditional column-based convective parameterization schemes presently use a mass flux (e.g., Grell, 1993; Kain 2004) or quasi-equilibrium approach (Arakawa and Schubert 1974). The statistical approach is a statistical parameterization based on the analysis of cloud resolving model output (Tao et al. 2003) and is of the type discussed in Section 8.2.

The MMF approach uses data from many cloud resolving model (CRM) simulations to diagnose the cloud system response to large-scale parameters. In MMF a full 2-D CRM is embedded within each grid cell of a large-scale model (Randall et al. 2003). However, the MMF is presently very computationally expensive and is, as yet, impractical for operational weather forecasting, ensemble simulations, or climate simulations.

There was a consensus from a cumulus parameterization workshop (Tao et al. 2003) that a consistent, comprehensive cloud database (associated with clouds and cloud systems that developed in different geographic locations) should be generated from the ensemble of CRMs for use in the development and improvement of cumulus parameterization schemes. These model-generated cloud data could be generated in close collaboration with parameterization developers and then used to replace the computationally very expensive MMF with the LUT approach.

There is also the direct use of real world observations of cloud processes as discussed in Section 8.4. This bypasses the need to use CRMs entirely, except perhaps as a way to assimilate the observed data. Satellite and other available observations are used to construct the LUTs, which necessarily include the combined effect of each of the atmospheric physics processes. Since the observations are sampling reality, this assures that 3D interactions are implicitly included. Moreover, as the models (even those with a global domain) achieve increasingly higher spatial resolution, eventually mesoscale and clouds will be explicitly resolved, and LUTs (or other parameterization approaches) would only be needed for the smallest spatial scales.

Chapter 8 Additional Readings

Leoncini, G., R.A. Pielke Sr., and P. Gabriel, 2008: From model based parameterizations to Lookup Tables: An EOF approach. *Wea. Forecasting,* **23**, 1127-1145.

Matsui, T., G. Leoncini, R.A. Pielke Sr., and U.S. Nair, 2004: A new paradigm for parameterization in atmospheric models: Application to the new Fu-Liou radiation code. Atmospheric Science Paper No. 747, Colorado State University, Fort Collins, CO 80523, 32 pp.

Pielke Sr., R.A., T. Matsui, G. Leoncini, T. Nobis, U. Nair, E. Lu, J. Eastman, S. Kumar, C. Peters-Lidard, Y. Tian, and R. Walko, 2006b: A new paradigm for parameterizations in numerical weather prediction and other atmospheric models. *National Wea. Digest,* **30**, 93-99.

Pielke Sr., R.A., D. Stokowski, J.-W. Wang, T. Vukicevic, G. Leoncini, T. Matsui, C. Castro, D. Niyogi, C.M. Kishtawal, A. Biazar, K. Doty, R.T. McNider, U. Nair, and W.K. Tao, 2007a: Satellite-based model parameterization of diabatic heating. *EOS,* **Vol. 88, No. 8**, 20 February, 96-97.

CHAPTER

Methods of Solution

9

CHAPTER OUTLINE HEAD

As explained in Chapter 5, sets of simultaneous, nonlinear, partial differential equations cannot be solved using known analytic methods, but require numerical methods of computation where the equations are discretized and solved on a lattice. This lattice corresponds to the grid volume defined by Eq. 4.6.

There are several broad classes of solution techniques available to represent terms involving the derivatives in the spatial coordinates of these differential equations including:

Mesoscale Meteorological Modeling, Third Edition, Volume 98. http://dx.doi.org/10.1016/B978-0-12-385237-3.00009-8

1. *finite difference* schemes, which utilize a form of truncated Taylor series expansion;
2. *spectral* techniques in which dependent variables are transformed to wavenumber space using a global basis function (e.g., a Fourier transform);
3. the *pseudospectral* method which uses a truncated spectral series to approximate derivatives;
4. *finite element* schemes, which seek to minimize the error between the actual and approximate solutions using a local basis function; and
5. *interpolation* schemes in which polynomials are used to approximate the dependent variables in one or more spatial directions.
6. *finite volume techniques*[1] are based on integrating over the volume of each grid cell.

In mesoscale models, only the finite difference and interpolation schemes have generally been used. The finite element techniques have been applied in mesoscale models by only a few groups and none to my knowledge in recent years. Interested readers can refer to the first edition of this book for the derivation of finite element algorithms. An introduction to finite difference and interpolation schemes will be discussed in this chapter.

The spectral technique has been shown to be highly accurate (e.g., Fox and Deardorff 1972; Machenhauer 1979; Orszag 1971) and eliminates the fictitious feedback of energy to the larger scale called aliasing (discussed in Section 9.4). However, the mathematical expressions that result from the Fourier transformation are cumbersome to handle and have required periodic boundary conditions to make it work effectively. Thus this scheme has not found acceptance among mesoscale models.

The pseudospectral technique has been introduced by Fox and Orszag (1973) and has been contrasted with the spectral technique and conventional finite difference methods by Christensen and Prahm (1976). A brief review of the pseudospectral method is given in Merilees and Orszag (1979). Eliassen (1980) has summarized the uses of these and other representations in air pollution transport modeling. Although the pseudospectral technique appears to be a viable tool for use in mesoscale models, it also has not been adopted (although several researchers, e.g., Lee 1981; Patrinos and Leach 1982, investigate its utility). The major reason for the neglect of these spectral schemes up to the present may be the interest in the interpolation approach, as well as difficulty in handling nonperiodic boundary conditions using either of the spectral techniques.

9.1 Finite Difference Schemes – An Introduction

9.1.1 Advection

Most mesoscale models utilize the finite difference method, because of its comparative ease of coding onto a computer as well as its conceptual simplicity. This technique

[1]The finite volume approach is discussed by Bob Walko in Section 9.6.

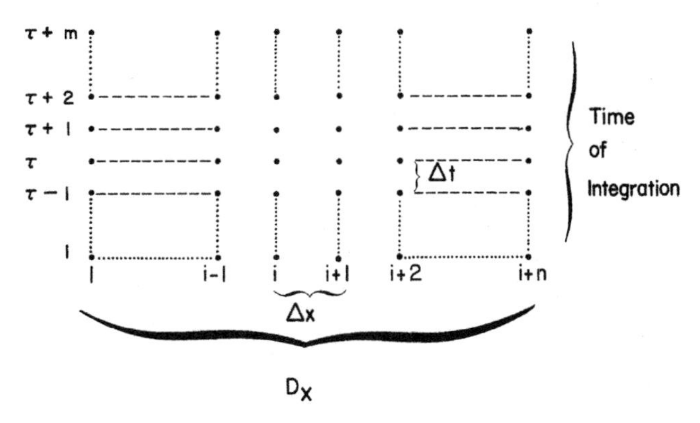

FIGURE 9.1

A schematic illustration of an $x - t$ grid representation where $\Delta x = x(i) - x(i-1) = x(i+1) - x(i)$, etc., and $\Delta t = t(\tau) - t(\tau - 1) = t(\tau + 1) - t(\tau)$, etc. ($n$ and m are integers greater than 2).

simply involves approximating the differential terms, including time, by one or more terms in a Taylor series expansion. For example, the local tendency and advective terms in the prognostic Eqs. 4.21 and 4.24-4.26 can be approximated by

$$\frac{\partial \bar{\phi}}{\partial t} = -\bar{u}\frac{\partial \bar{\phi}}{\partial x} \simeq \frac{\phi_i^{\tau+1} - \phi_i^{\tau}}{\Delta t} = -u_i^{\tau}\frac{\phi_{i+1}^{\tau} - \phi_{i-1}^{\tau}}{2\Delta x}, \tag{9.1}$$

where the overbar has been dropped to simplify the notation, τ is used to indicate the number of time steps taken, i indicates the grid point location in the x direction (as illustrated in Fig. 9.1), $\Delta t = t(\tau + 1) - t(\tau)$, and $\Delta x = x(i+1) - x(i)$. The dependent variable ($\bar{\phi}$) refers to any one of the dependent variables. Note that the tensor subscript notation is not used here, rather a new mathematical shorthand is introduced to represent the time and space locations of the dependent variables. The equation on the righthand side of Eq. 9.1 is called a *difference equation*.

In making such an approximation to a differential equation, several questions are asked concerning its ability to represent the actual differential equation accurately. These include the following.

1. When Δt and Δx approach zero, does the approximate form converge to the differential equation?
2. Is the numerical representation linearly stable to small perturbations?
3. If the scheme is linearly stable, how well are the amplitudes and phases represented for waves of different wavelengths relative to the exact solution?

These three questions must be considered for all computational approximation techniques.

The difference equation given by Eq. 9.1, appears to be a straightforward and reasonable form to apply to represent the corresponding differential equation. However, as will be shown shortly, although the approximation does converge to the correct representation when Δt and Δx approach zero in the limit, the scheme is *linearly unstable* to small perturbations and, therefore, cannot be used.

To illustrate the first criteria, let the dependent variable ϕ have the form

$$\phi = \hat{\phi} \cos kx,$$

(where $\hat{\phi}$ has constant amplitude) so that

$$\partial\phi/\partial x = -\hat{\phi} k \sin kx.$$

One finite difference approximation to this term (as used in Eq. 9.1) is

$$\frac{\phi_{i+1} - \phi_{i-1}}{2\Delta x} = \frac{\hat{\phi}}{2\Delta x}[\cos k(x + \Delta x) - \cos k(x - \Delta x)], \tag{9.2}$$

which, by expanding $\cos k(x + \Delta x)$ and $\cos k(x - \Delta x)$, can be written as

$$\frac{\phi_{i+1} - \phi_{i-1}}{2\Delta x} = -\frac{\hat{\phi}}{\Delta x} \sin kx \sin k\Delta x. \tag{9.3}$$

The ratio of the approximate to actual forms of this differential quantity is thus given by

$$\frac{\phi_{i+1} - \phi_{i-1}}{2\Delta x} \bigg/ \frac{\partial\phi}{\partial x} = \frac{\sin k\Delta x}{k\Delta x}. \tag{9.4}$$

By a Taylor series expansion

$$\sin k\Delta x = k\Delta x - \frac{(k\Delta x)^3}{3!} + \frac{(k\Delta x)^5}{5!} - \cdots$$

thus, when $k\Delta x \ll 1$,[2] Eq. 9.4 can be written as

$$\frac{\phi_{i+1} - \phi_{i-1}}{2\Delta x} \bigg/ \frac{\partial\phi}{\partial x} \sim \frac{k\Delta x}{k\Delta x} = 1.$$

Since $k = 2\pi/L$, then writing L in terms of the grid spacing $L = n\Delta x$, where n is the number of grid points in one cycle of the cosine function, then $k\Delta x \ll 1$ requires that $2\pi/n \ll 1$ or $n \gg 1$. In other words, the cosine wave must have a very long wavelength for its derivative to be represented accurately by Eq. 9.2.

By contrast, if $L = 2\Delta x$,

$$\frac{\phi_{i+1} - \phi_{i-1}}{2\Delta x} \bigg/ \frac{\partial\phi}{\partial x} = \frac{\sin \pi}{\pi} = 0,$$

so that the representation given by Eq. 9.2 fails to resolve a feature that has a wavelength of two grid increments. Examples of a longwave and a shortwave are given in Fig. 9.2.

[2] If Δx approaches zero in the limit, $k\Delta x$ also approaches zero so that Eq. 9.3 is a *consistent* representation of the derivative quantity. If $\Delta x \neq 0$, however, it is clear that the accuracy of the approximation is a function of wavenumber.

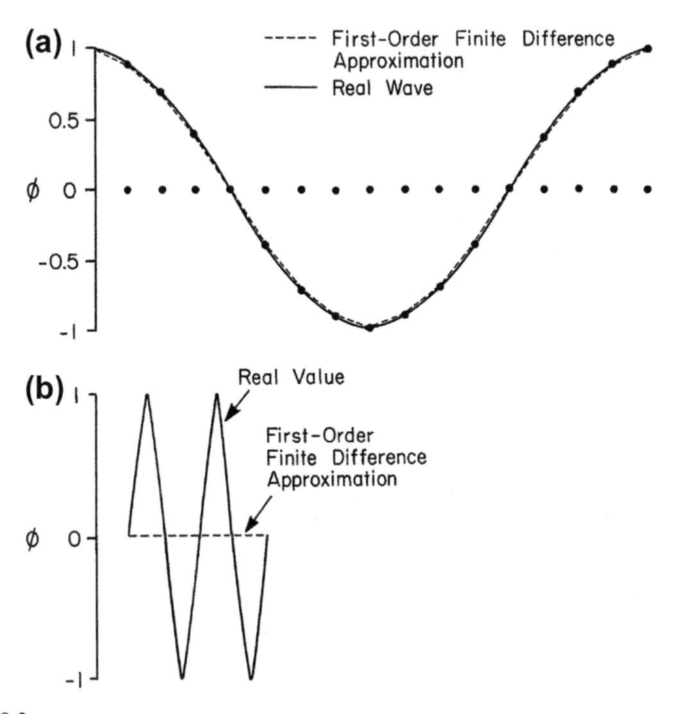

FIGURE 9.2

Centered finite difference representation using Eq. 9.2 for (a) a $16\Delta x$ wave, and (b) a $2\Delta x$ wave.

Thus the representation of the derivative of a function using values at neighboring grid points provides very poor representations of short waves relative to the grid mesh Δx, whereas longer waves are reasonably well resolved. The ability, or lack of it, of a numerical scheme to resolve features of different wavelengths properly is a crucial consideration in the use of a numerical approximation scheme.

The linear stability of Eq. 9.1 can be evaluated using the techniques for representing waves in terms of complex variables introduced in Chapter 5.[3] As discussed there, a dependent variable ϕ, for example, can be represented as

$$\phi(x, t) = \hat{\phi}(k, \omega)e^{i(kx+\omega t)}, \tag{9.5}$$

where $\hat{\phi}$, k, and ω can be complex. In a numerical model, the spatial and temporal independent variables can be written as

$$x = n\Delta x; \quad t = \tau \Delta t,$$

[3]This technique is often referred to as the *Von Neumann Method*. A second method exists called the *Matrix Method of Stability* in which boundary conditions are included in the analysis. This latter methodology, however, is more difficult to apply. It is generally preferable to show the stability, or lack of it, of a numerical scheme using the Von Neumann Method and this approach is adopted in this chapter.

so that Eq. 9.5 can also be written as

$$\phi(x,t) = \phi(n\Delta x, \tau\Delta t) = \hat{\phi}(k,\omega)e^{i(kn\Delta x + \omega\tau\Delta t)}. \qquad (9.6)$$

As discussed in Chapter 5, to use the formulation given by Eq. 9.5 in a differential equation, it is necessary to linearize the equation. As written, Eq. 9.1 is not in a linear form since the righthand side involves products of dependent variables. The procedure, therefore, is to replace the advecting velocity u_i^τ with a constant value U, so that the finite difference approximation becomes

$$\frac{\phi_i^{\tau+1} - \phi_i^\tau}{\Delta t} = -U\frac{\phi_{i+1}^\tau - \phi_{i-1}^\tau}{2\Delta x} \qquad (9.7)$$

This difference equation has one unknown, $\phi_i^{\tau+1}$, since it is assumed that values of ϕ at time τ are known; therefore, the equation is well posed and solvable. The finite difference representation given by Eq. 9.7 is called *forward-in-time, centered-in-space*.

At this point it is useful to comment on the standard procedure for examining the computational linear stability of numerical schemes. As shown in Chapter 5, even the linearized form of the conservation relations can produce complicated solutions. Therefore, investigators generally examine subsets of the linearized relations. Simple numerical approximations to advection, for example, such as given by Eq. 9.7, are examined separately from the remaining terms in the conservation relations. Numerical approximations to the subgrid-scale fluxes and the Coriolis terms are other portions of the conservation relations that we shall investigate later in this chapter. The basic assumption is that if the computational approximation to the individual linearized subsets of the equations is accurate, they will also be accurate representations when they are used in the nonlinear framework. However, as will be discussed in Section 9.4, accuracy of the linear differential equations is a *necessary but not a sufficient condition* to guarantee satisfactory nonlinear solutions.

Using Eq. 9.6, Eq. 9.7 can be rewritten as

$$\hat{\phi}(k,\omega)e^{i(kn\Delta x+\omega\tau\Delta t)}[e^{i\omega\Delta t} - 1] = -\frac{U\Delta t}{2\Delta x}\hat{\phi}(k,\omega)e^{i(kn\Delta x+\omega\tau\Delta t)}[e^{ik\Delta x} - e^{-ik\Delta x}]$$

or

$$[e^{i\omega\Delta t} - 1] = -\frac{U\Delta t}{2\Delta x}[e^{ik\Delta x} - e^{-ik\Delta x}]. \qquad (9.8)$$

At this point, it is convenient to introduce a notational representation suggested by Arthur Mizzi (1979, personal communication) where

$$\psi^1 = e^{i\omega\Delta t}, \quad \psi_1 = e^{ik\Delta x}, \quad \psi_{-1} = e^{-ik\Delta x},$$

so that $\psi_1 - \psi_{-1} = 2i\sin k\Delta x$ (this is shown by expanding $e^{ik\Delta x}$ and $e^{-ik\Delta x}$ in terms of $\cos a + i\sin a$, where $a = k\Delta x$ in the first exponential and $a = -k\Delta x$ in the second).

With these definitions, Eq. 9.8 can be written as

$$(\psi^1 - 1) = -\frac{U\Delta t}{\Delta x} i \sin k\Delta x. \tag{9.9}$$

This mathematical shorthand is a versatile tool to examine linear computational stability and will be used throughout this chapter.

Equation 9.9 can then be rearranged, yielding

$$\psi^1 = 1 - \frac{U\Delta t}{\Delta x} i \sin k\Delta x = 1 - Ci \sin k\Delta x, \tag{9.10}$$

where C is called the *Courant number*. If C is greater than unity, then the distance traveled in one time step due to advection is greater than the grid separation, whereas the opposite is true if C is less than one.

In the original differential (Eq. 9.1), no sources or sinks of velocity appear; however, in the difference representation, Eq. 9.7, damping or amplifying in time of the result is possible because of the imprecise approximation for the derivatives. To evaluate this effect, the frequency ω is decomposed into real and imaginary parts given by

$$\omega = \omega_{\mathrm{r}} + i\omega_{\mathrm{i}}.$$

Using this definition for ω, Eq. 9.10 can be rewritten as

$$\lambda e^{i\omega_{\mathrm{r}}\Delta t} = 1 - Ci \sin k\Delta x, \tag{9.11}$$

where $\lambda = \pm e^{-\omega_{\mathrm{i}}\Delta t}$ and is called the amplitude change of the solution per time step (λ^τ is then the change in amplitude after τ time steps). Since Eq. 9.1 has no sources or sinks of velocity, λ should be identically equal to $+1$.

To solve for λ, $e^{i\omega_{\mathrm{r}}\Delta t}$ is expanded into $\cos\omega_{\mathrm{r}}\Delta t + i \sin\omega_{\mathrm{r}}\Delta t$. The real and imaginary parts of Eq. 9.11 must separately be equal[4] so that Eq. 9.11 can be written as

$$\lambda \cos\omega_{\mathrm{r}}\,\Delta t = 1,$$
$$\lambda \sin\omega_{\mathrm{r}}\,\Delta t = -C \sin k\Delta x.$$

Squaring both sides of these two expressions and adding yields

$$\lambda^2(\cos^2\omega_{\mathrm{r}}\,\Delta t + \sin^2\omega_{\mathrm{r}}\,\Delta t) = \lambda^2 = 1 + C^2 \sin^2 k\Delta x,$$

or

$$\lambda = \pm\sqrt{1 + C^2 \sin k\Delta x}. \tag{9.12}$$

Since, except when the wavenumber becomes very small (e.g., $L \to \infty$), or as the grid increment Δx approaches zero, or for a $2\Delta x$ wave, $C^2 \sin^2 k\Delta x$ is greater than zero, and $|\lambda|$ is greater than one.

The numerical representation given by Eq. 9.7 is thus *linearly unstable* since the solution amplifies each time step. No such amplification will occur in the linear

[4]The real and imaginary parts of a complex number can be considered a vector with two perpendicular components. In an equation the real and imaginary components must separately be equal.

form of the differential equation given by the lefthand side of Eq. 9.1 (i.e., $\partial\bar{\phi}/\partial t = -U\partial\bar{\phi}/\partial x$). If a numerical scheme is linearly unstable, its use to approximate the nonlinear differential equation is rejected since small perturbations with such a representation will grow fictitiously.

The failure of the righthand side equation in Eq. 9.1 to accurately approximate the corresponding differential equation is surprising because it appeared to be the most natural choice. In examining solution techniques, therefore, intuition is not sufficient and stability analyses must be undertaken to determine the accuracy of numerical approximations. The methodology for examining linear computational stability introduced here will be used in the remainder of the chapter to examine the fidelity of different numerical representation techniques.

9.1.1.1 *Forward-Upstream Differencing*

Another approximation to the advection equation, given on the lefthand side of Eq. 9.1, is

$$\frac{\partial\bar{\phi}}{\partial t} = -\bar{u}\frac{\partial\bar{\phi}}{\partial x} \simeq \frac{\phi_i^{\tau+1} - \phi_i^{\tau}}{\Delta t} = \begin{cases} -u_i^{\tau}\dfrac{\phi_{i+1}^{\tau} - \phi_i^{\tau}}{\Delta x}, & u_i \leq 0 \\ -u_i^{\tau}\dfrac{\phi_i^{\tau} - \phi_{i-1}^{\tau}}{\Delta x}, & u_i > 0, \end{cases} \tag{9.13}$$

which is referred to as the *forward, upstream* scheme since the space derivative is evaluated upwind from the grid point. Linearizing this approximation by setting u_i^{τ} equal to a constant advecting velocity U, then for $U > 0$, the representation of this relation in terms of wave number and frequency is[5]

$$\psi^1 = 1 - C(1 - \psi_{-1}),$$

or, equivalently,

$$\lambda e^{i\omega_{\mathrm{r}}\Delta t} = 1 - C(1 - \cos k\Delta x + i \sin k\Delta x).$$

Equating real and imaginary components

$$\begin{aligned} \lambda \cos \omega_{\mathrm{r}}\, \Delta t &= 1 - C(1 - \cos k\Delta x), \\ \lambda \sin \omega_{\mathrm{r}}\, \Delta t &= -C \sin k\Delta x, \end{aligned} \tag{9.14}$$

squaring and adding yields

$$\lambda^2 = [1 - C(1 - \cos k\Delta x)]^2 + C^2 \sin^2 k\Delta x.$$

This expression, after expanding and rearranging can be written as

$$\lambda = \pm\sqrt{1 + 2C(\cos k\Delta x - 1)(1 - C)}.$$

[5] A similar analysis can be performed for $U < 0$, and the result will be of the same form.

In contrast with the forward-in-time, centered-in-space scheme, the upstream representation is linearly stable ($|\lambda| \leq 1$) as long as

$$-1 \leq 2C(\cos k\Delta x - 1)(1 - C) \leq 0.$$

Since $\cos k\Delta x \leq 1$, the quantity inside the lefthand parentheses is always less than or equal to zero, and the inequality holds as long as $C \leq 1$. Numerical approximation techniques that must satisfy certain criteria to have linearly stable results are referred to as *conditionally stable schemes*. When $C = 1$ or $k \simeq 0$ (i.e., $L \to \infty$), $\lambda = 1$ and the solutions neither damp nor amplify. Other values of λ, displayed in Table 9.1 as a function of wavelength $L = n\Delta x$, and C, the Courant number, show that except for the longest waves and $C = 0$ or 1, the scheme damps the solution with the most error at $C = 0.5$. With this latter value of C, wavelengths of $2\Delta x$ are completely eliminated – a result that can be seen most easily by rewriting the linear form of Eq. 9.13 as

$$\phi_i^{\tau+1} = (1 - C)\phi_i^{\tau} + C\phi_{i-1}^{\tau} \quad (\text{for } U > 0).$$

Since a $2\Delta x$ wave can be represented as $\phi_i^{\tau} = (-1)^i a_1 + a_0$, where a_1 and a_0 are constants and i is the spatial index counter, then for $C = 0.5$, $\phi_i^{\tau+1} = 0$.

The predicted phase speed as a function of wavenumber can also be obtained from Eq. 9.14 by dividing the imaginary by the real components yielding

$$\sin \omega_{\mathrm{r}} \Delta t / \cos \omega_{\mathrm{r}} \Delta t = \tan \omega_{\mathrm{r}} \Delta t = -C \sin k\Delta x / [1 + C(\cos k\Delta x - 1)],$$

or since the phase speed c is equal to the negative of the frequency divided by the wavenumber (e.g., see text preceding Eq. 5.32),

$$\tilde{c}_{\phi} = \frac{-1}{k\Delta t} \tan^{-1}\left[\frac{-C \sin k\Delta x}{1 + C(\cos k\Delta x - 1)}\right].$$

Since the actual solution[6] to the differential equation

$$\frac{\partial \bar{\phi}}{\partial t} = -\frac{U \partial \bar{\phi}}{\partial x} \quad \text{is} \quad \frac{\omega}{k} = -\tilde{c}_{\phi} = -U,$$

the ratio of the computational to true solution of the phase speeds is

$$\frac{\tilde{c}_{\phi}}{U} = \frac{1}{kU\Delta t} \tan^{-1}\left[\frac{C \sin k\Delta x}{1 + C(\cos k\Delta x - 1)}\right].$$

The accuracy of a linear numerical solution depends on how well the calculated values of λ and $\tilde{c}_{\phi}$ approximate the exact solutions of the differential equation; λ_{exact} and $c_{\phi\text{exact}}$ where, in this case, $\lambda_{\text{exact}} = 1$ and $\tilde{c}_{\phi\text{exact}} = U$. If $|\lambda| > 1$ for any possible wavelength, the solution technique is *linearly unstable*. If it is not linearly unstable,

[6]To obtain this result, simply replace $\bar{\phi}$ with $\bar{\phi} = \hat{\phi} + e^{i(kx+\omega t)}$ in the differential equation and solve for ω/k.

Table 9.1 Values of the amplitude λ and phase error $\tilde{c}_\phi/U$, per time step as a function of wavelength for different computational approximations to the advection equation $\partial\phi/\partial t = -U\partial\phi/\partial x$.

Scheme		Wavelength	C												
			0.001	0.01	0.1	0.2	0.3	0.4	0.5	0.6	0.7	0.8	0.9	1.0	1.1
I. Forward-in-time linear interpolation upstream	λ	$2\Delta x$	0.998	0.980	0.800	0.600	0.400	0.200	0.000	0.200	0.400	0.600	0.800	1.000	$\|\lambda\| > 1$
		$4\Delta x$	0.999	0.990	0.906	0.825	0.762	0.721	0.707	0.721	0.762	0.825	0.906	1.000	
		$10\Delta x$	1.000	0.998	0.983	0.969	0.959	0.953	0.951	0.953	0.959	0.969	0.983	1.000	
		$20\Delta x$	1.000	1.000	0.996	0.992	0.990	0.988	0.988	0.988	0.990	0.992	0.996	1.000	
	$\tilde{c}_\phi/U$	$2\Delta x$	0.000	0.000	0.000	0.000	0.000	0.000	1.000	1.667	1.429	1.250	1.111	1.000	
		$4\Delta x$	0.637	0.643	0.704	0.780	0.859	0.936	1.000	1.043	1.060	1.055	1.033	1.000	
		$10\Delta x$	0.936	0.937	0.953	0.968	0.981	0.992	1.000	1.005	1.008	1.008	1.005	1.000	
		$20\Delta x$	0.984	0.984	0.988	0.992	0.995	0.998	1.000	1.001	1.002	1.002	1.001	1.000	
II. Centered-in-time, centered-in-space (leapfrog)	λ	$2\Delta x$	1.000	1.000	1.000	1.000	1.000	1.000	1.000	1.000	1.000	1.000	1.000	1.000	$\|\lambda\| > 1$
		$4\Delta x$	1.000	1.000	1.000	1.000	1.000	1.000	1.000	1.000	1.000	1.000	1.000	1.000	
		$10\Delta x$	1.000	1.000	1.000	1.000	1.000	1.000	1.000	1.000	1.000	1.000	1.000	1.000	
		$20\Delta x$	1.000	1.000	1.000	1.000	1.000	1.000	1.000	1.000	1.000	1.000	1.000	1.000	
	$\tilde{c}_\phi/U_{\text{physical mode}}$	$2\Delta x$	0.000	0.000	0.000	0.000	0.000	0.000	0.000	0.000	0.000	0.000	0.000	0.000	
		$4\Delta x$	0.637	0.637	0.638	0.641	0.647	0.655	0.667	0.683	0.705	0.738	0.792	1.000	
		$10\Delta x$	0.935	0.935	0.936	0.938	0.940	0.944	0.950	0.956	0.964	0.974	0.986	1.000	
		$20\Delta x$	0.984	0.984	0.984	0.984	0.985	0.986	0.988	0.989	0.991	0.994	0.997	1.000	
III. Forward-in-time, upstream spline interpolation	λ	$2\Delta x$	1.000	0.999	0.944	0.792	0.568	0.296	0.000	0.296	0.568	0.792	0.944	1.000	1.000
		$4\Delta x$	1.000	1.000	0.997	0.989	0.981	0.975	0.972	0.975	0.981	0.989	0.997	1.000	0.888
		$10\Delta x$	1.000	1.000	1.000	1.000	1.000	1.000	1.000	1.000	1.000	1.000	1.000	1.000	0.996
		$20\Delta x$	1.000	1.000	1.000	1.000	1.000	1.000	1.000	1.000	1.000	1.000	1.000	1.000	1.000
	$\tilde{c}_\phi/U$	$2\Delta x$	0.000	0.000	0.000	0.000	0.000	0.000	0.000	0.000	0.000	0.000	0.000	0.000	0.000
		$4\Delta x$	0.955	0.955	0.958	0.967	0.979	0.980	1.000	1.007	1.009	1.008	1.005	1.000	1.042
		$10\Delta x$	0.999	0.999	0.999	0.999	1.000	1.000	1.000	1.000	1.000	1.000	1.000	1.000	1.001
		$20\Delta x$	1.000	1.000	1.000	1.000	1.000	1.000	1.000	1.000	1.000	1.000	1.000	1.000	1.000

(continued)

Table 9.1 (*Continued*)

Scheme		Wavelength	*C*												
			0.001	**0.01**	**0.1**	**0.2**	**0.3**	**0.4**	**0.5**	**0.6**	**0.7**	**0.8**	**0.9**	**1.0**	**1.1**
IV. Adam-Bashford λ centered-in-space[b]		$2\Delta x$	1.000	1.000	1.000	1.000	1.000	1.000	1.000	1.000	1.000	Computational mode is unstable for at least one of the wavelengths			
		$4\Delta x$	1.000	1.000	1.000	0.999	0.997	0.991	0.977	0.950	0.893				
		$10\Delta x$	1.000	1.000	1.000	1.000	1.000	0.999	0.997	0.994	0.990				
		$20\Delta x$	1.000	1.000	1.000	1.000	1.000	1.000	1.000	1.000	0.999				
	$\tilde{c}_\phi / U_{\text{physical mode}}$	$2\Delta x$	0.000	0.000	0.000	0.000	0.000	0.000	0.000	0.000	0.000	Computational mode is unstable for at least one of the wavelengths			
		$4\Delta x$	0.637	0.637	0.637	0.637	0.638	0.642	0.642	0.661	0.674				
		$10\Delta x$	0.935	0.935	0.935	0.936	0.936	0.937	0.938	0.941	0.945				
		$20\Delta x$	0.984	0.984	0.984	0.984	0.984	0.984	0.984	0.984	0.984				

[a] *It should be noted that in an approximate scheme which is damping (i.e., $|\lambda| < 1$), reducing Δt for the same Δx does not necessarily result in less total damping after a period of time. This results because the solution technique is used more frequently during that time because of the smaller Δt. Therefore, for improved accuracy and computational efficiency, as large a Δt as permitted by the linear stability criteria, should be chosen when an approximation scheme has computational damping (computed by Charlie Martin and Jeff McQueen).*

[b] *The values for the Adams-Bashford scheme were computed by Alex Costa and Sue van den Heever.*

but the absolute value of $\lambda/\lambda_{\text{exact}}$ is less than unity for any wavelength, the scheme is *damping*, and if λ is identically equal to λ_{exact}, the technique yields the correct amplitude (λ_{exact} can be less than one, such as for the diffusion equation; see Section 9.1.2). When $\tilde{c}_\phi \neq c_{\phi\text{exact}}$, the approximation representation is *erroneously dispersive* (the exact solution, of course, is *dispersive* if $c_{\phi\text{exact}}$ is a function of k).

It is also important to determine the damping over a specified time period (such as the time it takes the wave to travel one grid increment). This means a scheme could be slightly damping per time step, but if the time step is small, the accumulated damping over time can be quite large. For example, from Table 9.1 for Scheme I, for $4\Delta x$ and $C = 0.1$, after 10 time steps (the time required for the exact solution to travel Δx) the wave would be 37% of its correct amplitude (i.e., $\lambda^{10} = 0.34$). With $C = 0.5$, while the amplitude change per time step is greater than with $C = 0.1$, the amplitude change after the wave travels one $\Delta x (\lambda^2)$ is 0.5.

Values of $\tilde{c}_\phi/U$ are given in Table 9.1 for various combinations of C and k. As with the amplitude, a wavelength of $2\Delta x$ has, generally, the poorest representation of the proper phase speed. Only at $C = 1$ and at $C = 0.5$ (where the amplitude of a $2\Delta x$ wave is eliminated in one time step) is the phase accurately represented for all wavelengths. When $0.5 < C < 1.0$, waves travel faster in the finite difference representation than the true solution, whereas they are slower when $0 < C < 0.5$.

The damping characteristics of upstream differencing can be examined in a different fashion using a truncated Taylor series approximation to Eq. 9.13, where

$$\phi_i^{\tau+1} \simeq \phi_i^\tau + \tfrac{\partial\phi}{\partial t}\Delta t + \tfrac{1}{2}\tfrac{\partial^2\phi}{\partial t^2}(\Delta t)^2,$$

$$\phi_{i-1}^\tau \simeq \phi_i^\tau - \tfrac{\partial\phi}{\partial t}\Delta x + \tfrac{1}{2}\tfrac{\partial^2\phi}{\partial x^2}(\Delta x)^2,$$

where it is understood that the derivative terms are evaluated at τ and i. These two forms are substituted into the righthand equation in Eq. 9.13 (with the advecting velocity u_i^τ set equal to the constant velocity U; $U > 0$), yielding

$$\phi_i^\tau + \frac{\partial\phi}{\partial t}\Delta t + \frac{1}{2}\frac{\partial^2\phi}{\partial t^2}(\Delta t)^2 - \phi_i^\tau = -C\left[\phi_i^\tau - \phi_i^\tau + \frac{\partial\phi}{\partial x}\Delta x - \frac{1}{2}\frac{\partial^2\phi}{\partial x^2}(\Delta x)^2\right].$$

or, after subtracting identical terms and rearranging,

$$\frac{\partial\phi}{\partial t} + U\frac{\partial\phi}{\partial x} + \frac{1}{2}\frac{\partial^2\phi}{\partial t^2}\Delta t - \frac{1}{2}U\frac{\partial^2\phi}{\partial x^2}\Delta x = 0. \tag{9.15}$$

The first two terms on the left are in the same form as the original linear differential equation (i.e., the lefthand equation in Eq. 9.13, with $\bar{u} = U$), and the righthand side represents the *computational diffusion*, which results from using forward, upstream differencing. As Δt and Δx approach zero in the limit, Eq. 9.15 reduces to the proper differential equation.

This diffusion can be written in another form by differentiating in time the lefthand equation in Eq. 9.13 with $\bar{u} = U$ resulting in

$$\frac{\partial^2 \phi}{\partial t^2} = -U\frac{\partial}{\partial t}\frac{\partial \phi}{\partial x} = -U\frac{\partial}{\partial x}\frac{\partial \phi}{\partial t} = -U\frac{\partial}{\partial x}\left(-U\frac{\partial \phi}{\partial x}\right) = U^2\frac{\partial^2 \phi}{\partial x^2},$$

so the two terms on the right of Eq. 9.15 can also be written as

$$\frac{1}{2}\frac{\partial^2 \phi}{\partial t^2}\Delta t - \frac{1}{2}U\frac{\partial^2 \phi}{\partial x^2}\Delta x = \frac{1}{2}U\Delta x\left(U\frac{\Delta t}{\Delta x} - 1\right)\frac{\partial^2 \phi}{\partial x^2} = \upsilon_c\frac{\partial^2 \phi}{\partial x^2},$$

where υ_c is called the *computational diffusion coefficient*. This coefficient has no physical significance and is simply an artifact of the computational scheme.

Forward-in-time, upstream differencing has been used extensively in mesoscale numerical modeling. Its characteristic damping and failure to preserve the proper phase, however, have generated serious criticisms of this technique. It is now generally believed that it is only appropriate to use this method if advection and wave propagation are not dominant in the conservation relations for a particular mesoscale feature. Furthermore, if subgrid-scale mixing is important, υ_c must be less than the corresponding physically relevant turbulent exchange coefficient. It may be possible, however, to modify this numerical approach to improve its accuracy. Smolarkiewicz (1983), for instance, presents a scheme to reduce the implicit diffusion of upstream differencing by adding a corrective step to the calculation. Brown and Pandolfo (1980) have provided a discussion of upstream differencing. Wang (1996) describes an extension of forward-in-time, upstream differencing for non-uniform and time-dependent advection which achieves better accuracy.

9.1.1.2 Leapfrog Centered-in-Space Differencing Scheme

A third finite difference representation to the advection equation is

$$\frac{\phi^{\tau+1} - \phi^{\tau-1}}{2\Delta t} = -u_i^\tau\frac{\phi_{i+1}^\tau - \phi_{i-1}^\tau}{2\Delta x}, \tag{9.16}$$

where the righthand side is in the same form as the difference representation given in Eq. 9.1 and the lefthand side is *centered-in-time*. This scheme is often called *leapfrog* because ϕ_i^τ does not appear in Eq. 9.16. As will be shown shortly, the use of a centered-in-time representation permits Eq. 9.16 to be linearly stable under certain conditions.

Assuming $u_j = U$, Eq. 9.16 can be rewritten as a function of wavenumber and frequency as

$$\psi^1 = \psi^{-1} - i\alpha, \tag{9.17}$$

where $\alpha = 2C\sin k\Delta x$. From the definitions of ψ^1 and ψ^{-1} (i.e., $\psi^1 = \lambda e^{i\omega_r\Delta t}$, $\psi^{-1} = \lambda^{-1}e^{-i\omega_r\Delta t}$) Eq. 9.17 can be rewritten as

$$\psi^2 + i\alpha\psi^1 - 1 = 0, \tag{9.18}$$

where the identity $\psi^2 = \left(\psi^1\right)\left(\psi^1\right) = \left(\psi^1\right)^2$ has been used. Since the solution of a quadratic equation $ax^2 + bx + c = 0$ is $x = (-b \pm \sqrt{b^2 - 4ac})/2a$, then

$$\psi^1 = \left(-i\alpha \pm \sqrt{(i\alpha)^2 + 4}\right)/2 = \left(-i\alpha \pm \sqrt{4 - \alpha^2}\right)/2.$$

Rewriting ψ^1 in terms of its real and imaginary components yields

$$\left.\begin{aligned} \lambda \cos \omega_r \Delta t &= \pm\sqrt{4-\alpha^2}/2 \\ \lambda \sin \omega_r \Delta t &= -\alpha/2 \end{aligned}\right\} \quad \text{if } \alpha^2 \leq 4$$

and

$$\left.\begin{aligned} \lambda \cos \omega_r \Delta t &= 0 \\ \lambda \sin \omega_r \Delta t &= (-\alpha \pm \sqrt{\alpha^2 - 4})/2 \end{aligned}\right\} \quad \text{if } \alpha^2 > 4.$$

When $\alpha^2 \leq 4$, squaring the real and imaginary components gives

$$\lambda^2 = \frac{1}{4}\alpha^2 + 1 - \frac{1}{4}\alpha^2 = 1, \quad \text{or} \quad \lambda = \pm 1,$$

so that the *amplitude is preserved for all wavelengths* and the scheme is said to be *neutrally stable*.

When $\alpha^2 > 4$, $\cos \omega_r \Delta t = 0$, so that the imaginary component can be written as

$$\lambda = -\frac{1}{2}\alpha \pm \frac{1}{2}\sqrt{\alpha^2 - 4}.$$

To ascertain whether this quantity is less than, or greater than unity, let

$$\alpha = 2 + \epsilon,$$

where $\epsilon > 0$, so that

$$\lambda = -1 - \frac{1}{2}\epsilon \pm \frac{1}{2}\sqrt{4\epsilon + \epsilon^2}.$$

Since either root is possible,

$$\lambda = -1 - \frac{1}{2}\epsilon - \frac{1}{2}\sqrt{4\epsilon + \epsilon^2} < -1,$$

($|\lambda| > 1$) so that when $\alpha^2 > 4$, the leapfrog scheme is linearly unstable. Since $\alpha^2 = 4C^2 \sin^2 k\Delta x$, stability is retained only when

$$C^2 \sin^2 k\Delta x \leq 1,$$

or since the maximum value of $\sin^2 k\Delta x$ is unity for a $4\Delta x$ wave ($k\Delta x = \pi/2$), then

$$|C| \leq 1$$

is a necessary and sufficient condition for the linear stability of the scheme.

The ratio of the predicted phase speed to the advecting velocity for this technique can be obtained by dividing the imaginary by the real component for $\alpha^2 \leq 4$ and solving for the phase speed. Since $|\lambda| = 1$, however, it is also possible to use either the imaginary or real components separately to obtain the phase speed. Using the imaginary part, therefore, gives the ratio of the calculated to analytic phase speeds as

$$\frac{\tilde{c}_\phi}{U} = \frac{1}{Uk\Delta t} \sin^{-1}\left(\pm\frac{\alpha}{2}\right).$$

Because of the quadratic form of Eq. 9.18, two wave solutions occur. One moves downstream ($\tilde{c}_\phi > 0$, when $U > 0$) and is related to the real solution of the advection equation, and the other travels upstream and is called the *computational mode*.[7] The computational mode occurs because the leapfrog is a second-order difference equation. Such separation of solutions by the centered-in-time, leapfrog scheme can be controlled by occasionally averaging in time to assure that the even and odd time steps remain consistent with one another. As long as the time steps are consistent, the amplitude of the computational mode is small.

Values of λ and $\tilde{c}_\phi$ for the physical solution for different values of C and wavelength are displayed in Table 9.1. There is also the computational solution which is shown for the phase of the leapfrog scheme in the first edition of this book. Although the leapfrog scheme preserves amplitudes exactly as long as $|C| \leq 1$, the accuracy of the phase representation deteriorates markedly for the shorter wavelengths. Because the numerical representation of these waves travels more slowly than the true solution, the scheme is said to be *dispersive* since when waves of different wavelengths are linearly superimposed, they will travel with different speeds relative to one another even if the advecting velocity is a constant. The retention of these dispersive shorter waves in the solution can cause computational problems through nonlinear instability, as discussed in Section 9.4.[8] The important conclusion obtained from the analysis of the leapfrog scheme is that *the exact representation of the amplitude does not by itself guarantee successful simulations since the fictitious dispersion of waves of different lengths can generate errors.* Baer and Simons (1970), for example, have reported that in approximating nonlinear advection terms, individual energy components may have large errors when the total energy has essentially none. They further conclude that neither conservation of integral properties nor satisfactory prediction of amplitude is sufficient to justify confidence in the results – one must also assure the accurate calculation of phase speed.

Smolarkiewicz and Margolin (1998) summarize how using the error characteristics of a finite difference scheme can be used to improve the accuracy of the solution. They

[7] To determine which quadrant the $\sin^{-1}\left(\pm\frac{\alpha}{2}\right)$ requires assessing which solution propagates in the same direction as the physical solution. For the leapfrog scheme, the physical mode is quadrant 1 of the sine function, while the computational mode is in quadrant 3 (π to $3\pi/2$).

[8] The retention of $2\Delta x$ features in a model is undesirable since they are poorly resolved and can create nonlinear instability. The challenge is to eliminate this wavelength but leave the longer waves relatively unaffected.

show, for example, how the successive iterative application of the positive definite[9] properties of the upstream difference scheme is used to compensate for the residual truncation error. More iteration results in smaller truncation errors. This numerical technique is also discussed in Wortmann-Vierthaler and Moussiopoulos (1995).

9.1.1.3 Adams-Bashford Differencing Scheme

A fourth representation of advection that is used is the Adams-Bashford algorithm (Durran 1991). It is used, for example, in the Song and Haidvogel (1994) ocean model. The analyses of this scheme was completed by Alex Costa and Sue van den Heever as part of a class at Colorado State University on mesoscale modeling.

The scheme can be written as

$$\frac{\phi^{\tau+1} - \phi^{\tau}}{\Delta t} = \frac{1}{12}[23F(\phi^{\tau}) - 16F(\phi^{\tau-1}) + 5F(\phi^{\tau-2})] \tag{9.19}$$

where in one dimension,

$$F(\phi^{\tau}) = -U\frac{\partial \phi}{\partial x} = -U\frac{\phi_{i+1}^{\tau} - \phi_{i-1}^{\tau}}{2\Delta x} \tag{9.20}$$

is used.

Substituting Eq. 9.20 into Eq. 9.19 yields

$$\phi_i^{\tau+1} = \phi_i^{\tau} - U\frac{\Delta t}{\Delta x}\frac{1}{24}\left[23\left(\phi_{i+1}^{\tau} - \phi_{i-1}^{\tau}\right) - 16\left(\phi_{i+1}^{\tau-1} - \phi_{i-1}^{\tau-1}\right) + 5\left(\phi_{i+1}^{\tau-2} - \phi_{i-1}^{\tau-2}\right)\right]. \tag{9.21}$$

Substituting ψ^1, ψ^0, ψ^{-1}, and ψ^{-2} into Eq. 9.21 yields

$$\psi^3 - \left(1 - \frac{23}{12}i\alpha\right)\psi^2 - \frac{4}{3}i\alpha\psi^1 + \frac{5}{12}i\alpha = 0 \tag{9.22}$$

where $\alpha = U\frac{\Delta t}{\Delta x}\sin k\Delta x$.

The amplitude and phase for the physical solution of ψ^1 as computed by Alex Costa and Sue van den Heever, are given in Table 9.1. The solutions for the two computational amplitudes and phases can be obtained from the solution of Eq. 9.22. A surprising result of the analysis for the computational modes is that for one of the modes, the computational values of λ is greater than unity for values of C larger than about 0.72 (for a $4\Delta x$ wavelength). For this reason, all values of the physical solution are left off of Table 9.1 for these values of C, even though the physical solution itself is not linearly unstable. As soon as a computational mode develops, if $C > 0.72$ and $4\Delta x$ waves occur, the computational mode will quickly swamp the solution of the advection equation.

[9]"Positive definite" means that the finite difference solution never produces negative values.

In both the forward-upstream and leapfrog schemes that we have examined, the time step must be less than or equal to the time it takes for changes at one grid point to be translated by advection to the next grid point downstream. With the Adams-Bashford scheme, the time step must be even less. When we generalize this result to all types of wave propagation, the need to filter rapidly moving waves, which are not considered important on the mesoscale, is apparent. This is the reason that scale analysis is used to derive simplified conservation relations (e.g., the anelastic conservation-of-mass equation 3.11) so that sound waves can be eliminated as a possible solution, as shown in Section 5.2.2.

9.1.2 Subgrid-Scale Flux

As shown by Eq. 7.7, the subgrid-scale correlation terms can be represented as the product of an exchange coefficient and the gradient of the appropriate dependent variable. This relation can be written, for example, as

$$\frac{\partial \bar{\phi}}{\partial t} = \frac{\partial}{\partial z} K \frac{\partial \bar{\phi}}{\partial z} \simeq \frac{\phi_i^{\tau+1} - \phi_i^{\tau}}{\Delta t} = K_{i+\frac{1}{2}} \frac{\phi_{i+1}^{\tau} - \phi_i^{\tau}}{(\Delta z)^2} - K_{i-\frac{1}{2}} \frac{\phi_i^{\tau} - \phi_{i-1}^{\tau}}{(\Delta z)^2}, \quad (9.23)$$

where $\Delta z = z(i+1) - z(i) = z(i) - z(i-1)$ and ϕ represents any one of the dependent variables. This equation is often referred to as the *diffusion equation*. To study the linear stability of this scheme, the exchange coefficient is assumed a constant ($K_{i+\frac{1}{2}} = K_{i-\frac{1}{2}} = K$), and Eq. 9.23 is written as

$$\phi_i^{\tau+1} = \phi_i^{\tau} + K \frac{\Delta t}{(\Delta z)^2} \left(\phi_{i+1}^{\tau} - 2\phi_i^{\tau} + \phi_{i-1}^{\tau} \right). \quad (9.24)$$

The exact solution to the diffusion equation (the lefthand side of Eq. 9.23 with K equal to a constant, i.e., $\partial\bar{\phi}/\partial t = K\partial^2\bar{\phi}/\partial z^2$) can be determined by assuming

$$\bar{\phi} = \phi_0 e^{i(kz+\omega t)} = \phi_0 e^{-\omega_i t} e^{i(k_r z + \omega_r t)}$$

where no damping in the z direction is permitted (i.e., $k_i \equiv 0$). Substituting this expression into the linearized diffusion equation and simplifying, yields

$$i\omega_r - \omega_i = -Kk^2$$

where the subscript r on k has been eliminated to simplify the notation. Equating real and imaginary components shows that $\omega_r \equiv 0$ so that the exact solution can be written as

$$\bar{\phi} = \phi_0 e^{-Kk^2 t} e^{ikz}.$$

Expressing the dependent variables as a function of frequency and wavenumber, Eq. 9.24 can be rewritten as

$$\psi^1 = 1 + \gamma \left(\psi_1 - 2 + \psi_{-1} \right) = 1 + 2\gamma(\cos k\Delta z - 1),$$

where $\gamma = K\Delta t/(\Delta z)^2$ and $\psi_1 + \psi_{-1} = 2\cos k\Delta z$. The nondimensional parameter γ is called the *Fourier number*. Equating real and imaginary components yields

$$\lambda \cos \omega_r \Delta t = 1 + 2\gamma(\cos k\Delta z - 1),$$
$$\lambda \sin \omega_r \Delta t = 0.$$

Since $\sin \omega_r \Delta t$ must be identically equal to zero, $\omega_r \Delta t$ and, therefore, the phase speed are also equal to zero. Thus the solution to Eq. 9.24 does not propagate as a wave but amplifies or decays in place. Since $\cos \omega_r \Delta t = 1$, the real part can be divided by the analytic solution,[10] $\lambda_a = e^{-Kk^2\Delta t} = e^{-\gamma(2\pi)^2/n^2}$ and rewritten as

$$\frac{\lambda}{\lambda_a} = \frac{1 + 2\gamma(\cos k\Delta z - 1)}{e^{-\gamma(2\pi)^2/n^2}}$$

where n is the number of grid points per wavelength. For very long waves ($n \to \infty$) $\lambda_a = 1$ and $\lambda = 1$ since $\cos k\Delta z = \cos(2\pi/L)\Delta z = 1$, and, therefore, no damping or amplification occurs. For the shortest waves that can be resolved ($L = 2\Delta z$; $n = 2$),

$$\lambda = 1 - 4\gamma.$$

To assure that the magnitude of λ is less than unity and, therefore, computationally stable, 4γ must be less than or equal to 2 or

$$\gamma \leq \frac{1}{2}.$$

The condition $\gamma = \frac{1}{2}$, however, causes γ to switch between $+1$ and -1 each application of Eq. 9.24, but the analytic solution is $\lambda_a = e^{-4.95} = 0.00719$. This unrealistic response of $2\Delta z$ wavelength features can cause computational problems in a nonlinear model as is discussed in Section 9.4. To eliminate $2\Delta z$ waves at each application of Eq. 9.24, λ can be set to zero for a $2\Delta z$ wave resulting in $\gamma = \frac{1}{4}$. Thus the standard requirement specified in using this scheme is that

$$\gamma = K\Delta t/(\Delta z)^2 \leq \frac{1}{4},$$

with the expectation that γ is close to $\frac{1}{4}$ so that the presence of $2\Delta z$ waves is minimized.

Up to this point, the approximation to the advective and subgrid-scale flux terms have always been defined at the current time step (i.e., ϕ_i^τ). The predicted dependent variable $\phi_i^{\tau+1}$ only enters through the time tendency term. Such schemes are referred to as *explicit* and can be written in general as

$$\phi^{\tau+1} = f\left(\phi^\tau\right),$$

[10] The exact damping per time step is λ_a.

where the function f, can include spatial derivatives of ϕ^{τ} as well as the variable itself.

In contrast, an *implicit* scheme uses information from the future time step, as well as present values. For this case

$$\phi^{\tau+1} = f(\phi^{\tau+1}, \phi^{\tau}).$$

In general, the use of an implicit representation permits longer time steps than the explicit form without causing linear instability. An implicit form of the lefthand equation in Eq. 9.23 for variable Δz can be written (e.g., Paegle et al. 1976) as

$$\frac{\phi^{\tau+1}-\phi^{\tau}}{\Delta t} = \frac{1}{\Delta z_j}\left[K_{j+\frac{1}{2}}\frac{\beta_\tau(\phi_{j+1}^{\tau}-\phi_j^{\tau})+\beta_{\tau+1}(\phi_{j+1}^{\tau+1}-\phi_j^{\tau+1})}{\Delta z_{j+\frac{1}{2}}} - K_{j-\frac{1}{2}}\frac{\beta_\tau(\phi_j^{\tau}-\phi_{j-1}^{\tau})+\beta_{\tau+1}(\phi_j^{\tau+1}-\phi_{j-1}^{\tau+1})}{\Delta z_{j-\frac{1}{2}}}\right] \tag{9.25}$$

where $\beta_\tau + \beta_{\tau+1} = 1$, $\Delta z_j = z_{j+\frac{1}{2}} - z_{j-\frac{1}{2}}$, $\Delta z_{j+1} = z_{j+1} - z_j$, and $\Delta z_{j-1} = z_j - z_{j-1}$. The use of β_τ and $\beta_{\tau+1}$ weights the current and future contributions to the numerical approximation of the lefthand side of Eq. 9.23. Note that when $\beta_{\tau+1} = 0$ and $\Delta z_j = \Delta z_{j+1} = \Delta z_{j-1} = \Delta z$, the scheme reverts back to the explicit scheme given by the right side of Eq. 9.23. Linearizing Eq. 9.25 by setting $K_{j+\frac{1}{2}}$ and $K_{j-\frac{1}{2}}$ equal to a constant, using a constant grid interval Δz, and representing the dependent variable in terms of wavenumber and frequency results in

$$\psi^1 = 1 + \gamma\left[\beta_\tau(\psi_1 - 2 + \psi_{-1}) + \beta_{\tau+1}(\psi_1^1 - 2\psi^1 + \psi_{-1}^1)\right],$$

where, as with the explicit scheme, $\gamma = K\Delta t/\left(\Delta z\right)^2$. Since

$$\psi_1^1 = \psi^1\psi_1, \text{ and } \psi_{-1}^1 = \psi^1\psi_{-1},$$

$$\psi^1 = 1 + \gamma\beta_\tau\left(\psi_1 - 2 + \psi_{-1}\right) + \gamma\beta_{\tau+1}\psi^1\left(\psi_1 - 2 + \psi_{-1}\right),$$

or

$$\psi^1 = \frac{\left[1+\gamma\beta_\tau\left(\psi_1+\psi_{-1}-2\right)\right]}{\left[1-\gamma\beta_{\tau+1}\left(\psi_1+\psi_{-1}-2\right)\right]} = \frac{1+2\gamma\beta_\tau\left(\cos k\Delta z-1\right)}{1-2\gamma\beta_{\tau+1}\left(\cos k\Delta z-1\right)} = \lambda,$$

where as with the analysis of the explicit representation, the imaginary part is zero so that $\gamma = \psi^1$.

Values of the ratio of the computational approximation of the damping to the analytic damping λ/λ_a are presented in Table 9.2 as a function of wavelength and β_τ. For a given value of γ, the $2\Delta z$ wave is most poorly represented. In addition, the $2\Delta z$ wave is always insufficiently damped and often the value of λ is negative yielding a wave

Table 9.2 Values of the ratio of the computational to analytic damping as a function of wavelength for different forms of the forward-in-time, centered-in-space approximation to the linearized diffusion equation ($\partial\phi/\partial t = K = \partial^2\phi/\partial z^2$). Computed by C. Martin.

		γ									
Scheme	**Wavelength**	**0.1**	**0.2**	**0.3**	**0.4**	**0.5**	**0.6**	**0.7**	**0.8**	**0.9**	**1.0**
Forward-in-time centered-in-space diffusion											
Explicit	$2\Delta x$	1.610	1.440	−3.863	−31.094	<−100					
	$4\Delta x$	1.024	0.983	0.839	0.537	0.0					
$\beta_\tau = 1$	$10\Delta x$	1.001	0.999	0.997	0.992	0.986		$\lambda > 1$ for a $2\Delta z$ wave			
	$20\Delta x$	1.000	1.000	1.000	1.000	0.999					
	$2\Delta x$	1.725	2.554	2.272	−4.202	−34.761	<−100	<−100	<−100	<−100	<−100
	$4\Delta x$	1.038	1.053	1.030	0.952	0.792	0.517	0.079	0.584	−1.555	−2.948
$\beta_\tau = 0.7$	$10\Delta x$	1.001	1.001	1.001	1.000	0.998	0.996	0.992	0.988	0.982	0.975
	$20\Delta x$	1.000	1.000	1.000	1.000	1.000	1.000	1.000	0.999	0.999	0.999
Implicit	$2\Delta x$	1.789	3.085	4.829	5.758	0.00	−33.91	<−100	<−100	<−100	<−100
	$4\Delta x$	1.047	1.092	1.129	1.150	1.145	1.099	0.993	0.800	0.485	0.00
$\beta_\tau = 0.5$	$10\Delta x$	1.001	1.003	1.004	1.005	1.006	1.007	1.007	1.008	1.008	1.008
	$20\Delta x$	1.000	1.000	1.000	1.000	1.000	1.000	1.000	1.001	1.001	1.001
	$2\Delta x$	1.845	3.507	6.718	12.711	23.17	38.97	54.11	33.16	<−100	<−100
	$4\Delta x$	1.055	1.126	1.211	1.307	1.414	1.529	1.648	1.766	1.875	1.965
$\beta_\tau = 0.3$	$10\Delta x$	1.002	1.004	1.006	1.009	1.013	1.017	1.021	1.026	1.031	1.037
	$20\Delta x$	1.000	1.000	1.000	1.001	1.001	1.001	1.001	1.002	1.002	1.003
	$2\Delta x$	1.916	3.999	8.779	19.93	46.35	>100	>100	>100	>100	>100
	$4\Delta x$	1.067	1.170	1.310	1.491	1.717	1.998	2.344	2.769	3.290	3.931
$\beta_\tau = 0.1$	$10\Delta x$	1.002	1.005	1.010	1.016	1.023	1.031	1.040	1.050	1.062	1.074
	$20\Delta x$	1.000	1.000	1.001	1.001	1.002	1.002	1.003	1.004	1.004	1.005

whose amplitude reverses (flip flops) each time step. The solutions become more accurate as γ becomes smaller, and the implicit representation gives reasonable results for large wavelengths even when the explicit form is linearly unstable for all spatial scales.

Equation 9.25 can be written in the following form

$$-\frac{\Delta t K_{j-\frac{1}{2}}\beta_{\tau+1}}{\Delta z_j \Delta z_{j-\frac{1}{2}}}\phi_{j-1}^{\tau+1} + \left[1 + \frac{\Delta t K_{j+\frac{1}{2}}\beta_{\tau+1}}{\Delta z_j \Delta z_{j+\frac{1}{2}}} + \frac{\Delta t K_{j-\frac{1}{2}}\beta_{\tau+1}}{\Delta z_j \Delta z_{j-\frac{1}{2}}}\right]\phi_j^{\tau+1}$$
$$-\frac{\Delta t K_{j+\frac{1}{2}}\beta_{\tau+1}}{\Delta z_j \Delta z_{j+\frac{1}{2}}}\phi_{j+1}^{\tau+1}$$
$$= \phi_j^\tau + \frac{\Delta t}{\Delta z_j}\left[\frac{K_{j+\frac{1}{2}}\beta_\tau\left(\phi_{j+1}^\tau - \phi_j^\tau\right)}{\Delta z_{j+\frac{1}{2}}} - \frac{K_{j-\frac{1}{2}}\beta_\tau\left(\phi_j^\tau - \phi_{j-1}^\tau\right)}{\Delta z_{j-\frac{1}{2}}}\right] \quad (9.26)$$

and solved for nonperiodic boundary conditions using a procedure described in Section 9.2. Its solution for periodic boundary conditions is given in Appendix A.

When $\beta_\tau = \beta_{\tau+1}$, this representation is called the Crank-Nicholson scheme. Paegle et al. (1976) have presented results that show that $\beta_{\tau+1} = 0.75$ provides a representation as accurate as the explicit scheme but with a much longer permissible time step. Figure 9.3, reproduced from Mahrer and Pielke (1978a), illustrates predictions of the growth of a heated boundary layer using both the explicit representation of diffusion given by Eq. 9.23 and the implicit form (Eq. 9.25) with $\beta_{\tau+1} = 0.75$. As

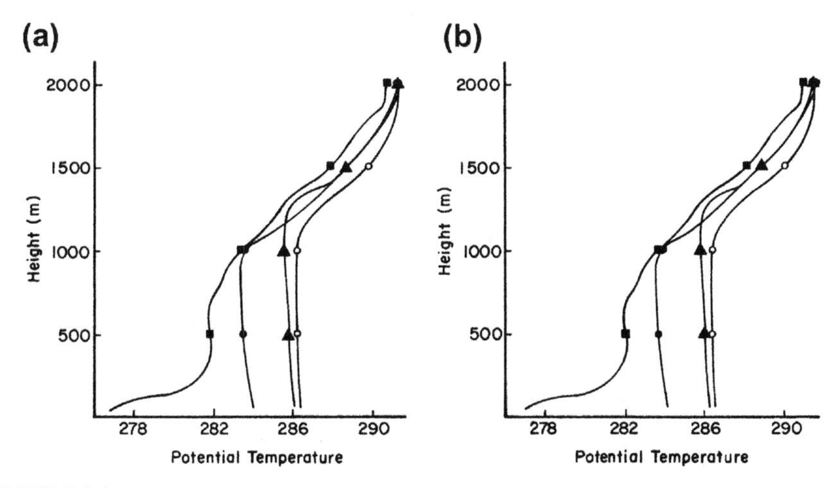

FIGURE 9.3

Vertical profiles of the potential temperature for Wangara Day 33 with (a) the implicit scheme ($\beta_{\tau+1} = 0.75$), and (b) the explicit scheme ($\beta_{\tau+1} = 0$) where the darkened squares are at 0900, the darkened circles at 1200, the darkened triangles at 1500, and the open circles at 1700 (from Mahrer and Pielke 1978a).

reported in that paper, use of the implicit form permitted a much longer time step so that the calculation ran 17 times faster than when the explicit form was used.

9.1.3 Coriolis Terms

The implicit scheme can also be shown to be a necessity for the Coriolis terms. The terms dealing with the rotation of the Earth (see Eq. 4.21) are already in linear form and with $L_z \ll L_x$ can be written as

$$\partial \bar{u}/\partial t = f\bar{v}; \quad \partial \bar{v}/\partial t = -f\bar{u}. \tag{9.27}$$

If these relations are approximated using an explicit representation, they are written as

$$(u_i^{\tau+1} - u_i^{\tau})/\Delta t = f v_i^{\tau}; \quad (v_i^{\tau+1} - v_i^{\tau})/\Delta t = f u_i^{\tau}. \tag{9.28}$$

Rewriting the dependent variables in terms of frequency and wavenumber, and rearranging yields

$$\hat{u}(\psi^1 - 1) - \hat{v}\Delta t f = 0,$$
$$\hat{u} f \Delta t + \hat{v}(\psi^1 - 1) = 0,$$

where $\hat{u}$ and $\hat{v}$ are functions of ω and k. In matrix form, these equations can be written as

$$\begin{bmatrix} \psi^1 - 1 & -\Delta t f \\ \Delta t f & \psi^1 - 1 \end{bmatrix} \begin{bmatrix} \hat{u} \\ \hat{v} \end{bmatrix} = \begin{bmatrix} 0 \\ 0 \end{bmatrix}.$$

As shown preceding Eq. 5.31, this homogeneous set of algebraic equations has a solution only if the determinant of the coefficients is equal to zero, thus

$$(\psi^1 - 1)^2 + (\Delta t)^2 f^2 = \psi^2 - 2\psi^1 + 1 + (\Delta t)^2 f^2 = 0.$$

Using the formula for the solution of a quadratic equation

$$\psi^1 = \left[2 \pm \sqrt{4 - 4\left(1 + (\Delta t)^2 f^2\right)}\right]/2 = 1 \pm i\Delta t f.$$

Equating real and imaginary components

$$\lambda \cos \omega_{\mathrm{r}} \Delta t = 1,$$
$$\lambda \sin \omega_{\mathrm{r}} \Delta t = \pm \Delta t f,$$

and then summing and squaring yields

$$\lambda^2 = 1 + (\Delta t)^2 f^2 \quad \text{so } \lambda = \sqrt{1 + (\Delta t)^2 f^2} \geq 1.$$

Thus except when $\Delta t = 0$ or $f = 0$ (at the equator) the explicit representation of the Coriolis terms given by Eq. 9.28 is *linearly unstable*.[11] The analytic solution, of course, requires that $\lambda = 1$.

A second representation of Eq. 9.27 is to use an implicit form given by

$$\left(u_i^{\tau+1} - u_i^{\tau}\right)/\Delta t = f v_i^{\tau}; \quad \left(v_i^{\tau+1} - v_i^{\tau}\right)/\Delta t = -f u_i^{\tau+1}, \tag{9.29}$$

so that the updated value of u is used in the v equation. Rewriting Eq. 9.29 in terms of frequency and wavenumber, rearranging, and writing in matrix form results in

$$\begin{bmatrix} \psi^1 - 1 & -\Delta t f \\ \psi^1 f \Delta t & \psi^1 - 1 \end{bmatrix} \begin{bmatrix} \hat{u} \\ \hat{v} \end{bmatrix} = \begin{bmatrix} 0 \\ 0 \end{bmatrix}.$$

Setting the determinant of the matrix of coefficients to zero gives the equation

$$\psi^2 + \psi^1\left[(\Delta t)^2 f^2 - 2\right] + 1 = 0,$$

which, using the quadratic formula and rearranging, has the solution

$$\psi^1 = 1 - \frac{(\Delta t)^2 f^2}{2} \pm \frac{(\Delta t) f \sqrt{(\Delta t)^2 f^2 - 4}}{2}. \tag{9.30}$$

Two forms of this relation need to be examined. If $(\Delta t)^2 f^2 > 4$, the entire expression is real and $\omega_r \Delta t = 0$. Thus if $(\Delta t)^2 f^2 = 4 + \epsilon^2$, where $\epsilon > 0$ but $\epsilon^2 \ll 4$, then

$$\psi^1 = \lambda \simeq 1 - \frac{4+\epsilon^2}{2} \pm \epsilon = 1 - 2 - \frac{\epsilon^2}{2} \pm \epsilon.$$

Since both roots are possible

$$\lambda \simeq -1 - \frac{\epsilon^2}{2} - \epsilon < -1 \quad \text{(i.e., } |\lambda| > 1\text{)},$$

so that when $(\Delta t)^2 f^2 > 4$, the representation is linearly unstable.

If $(\Delta t)^2 f^2 \leq 4$, then Eq. 9.30 can be rewritten as

$$\psi^1 = 1 - \frac{(\Delta t)^2 f^2}{2} \pm \frac{i(\Delta t) f \sqrt{4 - (\Delta t)^2 f^2}}{2}.$$

Equating real and imaginary parts yields

$$\lambda \cos \omega_r \Delta t = 1 - \left[(\Delta t)^2 f^2/2\right],$$

$$\lambda \sin \omega_r \Delta t = \left[\pm \Delta t f \sqrt{4 - (\Delta t)^2 f^2}\right]/2.$$

[11] From a practical viewpoint, however, $\lambda \simeq 1$ since $(\Delta t)^2 f^2$ is very small for commonly found values of Δt and f in a mesoscale model (e.g., for $\Delta t = 100$ s and $f = 10^{-4}\text{s}^{-1}$, $\lambda = \sqrt{1 + 10^{-6}}$).

Squaring the two expressions and adding results in

$$\lambda^2 = 1 - (\Delta t)^2 f^2 + \frac{1}{4}\left[(\Delta t)^2 f^2\right]^2 + (\Delta t)^2 f^2 - \frac{1}{4}\left[(\Delta t)^2 f^2\right]^2 = 1.$$

The scheme is, therefore, neutrally stable as long as $(\Delta t)^2 f^2 \leq 4$. Since the maximum value of f on the Earth is $1.45 \times 10^{-4}\text{s}^{-1}$, time steps shorter than 13,793 s assure linear stability of this term.

9.1.4 Pressure Gradient and Velocity Divergence

To investigate different approximations to the pressure gradient force and velocity divergence terms it is useful to use the linear form of the one-fluid tank model introduced in Section 5.2.1.1. The linear equations 5.22 and 5.23 developed in that chapter are given by

$$\frac{\partial u'}{\partial t} = -g\frac{\partial h}{\partial x}; \quad \frac{\partial h}{\partial t} = -h_0\frac{\partial u'}{\partial x}. \tag{9.31}$$

As expressed by Eq. 5.32, the analytic solution to Eq. 9.31 corresponds to a wave traveling with a speed of $\sqrt{gh_0}$. Since the amplitude of this wave is unchanging, a computationally accurate scheme will have $\lambda = 1$ in the stability analysis.

In a forward-in-time, centered-in-space explicit representation, these equations can be approximated by

$$\frac{u_i^{\tau+1} - u_i^{\tau}}{\Delta t} = -g\frac{h_{i+1}^{\tau} - h_{i-1}^{\tau}}{2\Delta x},$$

$$\frac{h_i^{\tau+1} - h_i^{\tau}}{\Delta t} = -h_0\frac{u_{i+1}^{\tau} - u_{i-1}^{\tau}}{2\Delta x}.$$

Representing these two finite difference equations in wavenumber and frequency space yields the algebraic equations

$$(\psi^1 - 1)\hat{u} + \frac{g\Delta t}{2\Delta x}\left(\psi_1 - \psi_{-1}\right)\hat{h} = 0,$$

$$(\psi^1 - 1)\hat{h} + \frac{h_0\Delta t}{2\Delta x}\left(\psi_1 - \psi_{-1}\right)\hat{u} = 0,$$

which after substituting for $\psi_1 - \psi_{-1}$ and rearranging, can be written in matrix form as

$$\begin{bmatrix} (\psi^1 - 1) & \frac{g\Delta t}{\Delta x} i \sin k\Delta x \\ h_0\frac{\Delta t}{\Delta x} i \sin k\Delta x & (\psi^1 - 1) \end{bmatrix} \begin{bmatrix} \hat{u} \\ \hat{h} \end{bmatrix} = \begin{bmatrix} 0 \\ 0 \end{bmatrix}.$$

For a nontrivial solution, the determinant of the coefficients must be zero so that

$$\psi^2 - 2\psi^1 + 1 + gh_0\left(\frac{\Delta t}{\Delta x}\right)^2 \sin^2 k\Delta x = 0,$$

and using the quadratic formula

$$\psi^1 = \frac{2 \pm \sqrt{4 - 4\left(1 + gh_0(\Delta t/\Delta x)^2 \sin^2 k\Delta x\right)}}{2},$$

or

$$\psi^1 = 1 \pm \sqrt{-gh_0(\Delta t/\Delta x)^2 \sin^2 k\Delta x} = 1 \pm i\sqrt{gh_0}(\Delta t/\Delta x)\sin k\Delta x.$$

The real and imaginary components of this expression are

$$\lambda \cos \omega_r \Delta t = 1,$$
$$\lambda \sin \omega_r \Delta t = \pm\sqrt{gh_0}(\Delta t/\Delta x)\sin k\Delta x,$$

Squaring and summing the two equations results in

$$\lambda^2 = 1 + gh_0(\Delta t/\Delta x)^2 \sin^2 k\Delta x,$$

so that $|\lambda| > 1$ and the *explicit representation* is *linearly unstable.*

Since the Coriolis terms are linearly unstable for an explicit scheme, but stable when an implicit representation is used, it seems reasonable to examine a similar form here, where dependent variables are updated to their $\tau + 1$ values in the second equation of the simultaneous set. Equation 9.31 can thus be approximated[12] by

$$\begin{aligned} \frac{u_i^{\tau+1} - u_i^{\tau}}{\Delta t} + g\frac{h_{i+1}^{\tau} - h_{i-1}^{\tau}}{2\Delta x} &= 0, \\ \frac{h_i^{\tau+1} - h_i^{\tau}}{\Delta t} + h_0\frac{u_{i+1}^{\tau+1} - u_{i-1}^{\tau+1}}{2\Delta x} &= 0. \end{aligned} \tag{9.32}$$

If one programs this system of equations on a computer, the choice of initial conditions will determine the initial amplitude and phase. Using $u^{\tau=0} = (g/h)^{1/2}h^{\tau=0}$, for example, with $h^{\tau=0} = \cos kj\Delta k$ will result in an amplitude change for the first time step of $\lambda = 1 + [\frac{1}{2}\tan(kj\Delta x)\sin k\Delta x]$ defining $g/h = 1$ for simplicity. The following stability analysis is only valid after this time.

Rewriting this expression in terms of frequency and wavenumber yields

$$(\psi^1 - 1)\hat{u} + \frac{g\Delta t}{2\Delta x}(\psi_1 - \psi_{-1})\hat{h} = 0,$$

$$(\psi^1 - 1)\hat{h} + \frac{h_0}{2}\frac{\Delta t}{\Delta x}(\psi_1^1 - \psi_{-1}^1)\hat{u} = 0,$$

or in the equivalent matrix form, substituting for $\psi_i - \psi_{-1}$, $(\psi_1^1 - \psi_{-1}^1 = \psi^1(\psi_1 - \psi_{-1}))$,

$$\begin{bmatrix} (\psi^1 - 1) & \frac{g\Delta t}{\Delta x} i \sin k\Delta x \\ \left(\frac{h_0\Delta t}{\Delta x} i \sin k\Delta x\right)\psi^1 & (\psi^1 - 1) \end{bmatrix} \begin{bmatrix} \hat{u} \\ \hat{h} \end{bmatrix} = \begin{bmatrix} 0 \\ 0 \end{bmatrix}.$$

[12] Note that since the second difference equation requires $u_{i+1}^{\tau+1}$ and $u_{i-1}^{\tau+1}$, it is necessary to compute all the values of $u^{\tau+1}$ before beginning the computations for $h^{\tau+1}$.

The determinant of the coefficients must equal zero for this system of equations to have a nontrivial solution so that

$$\psi^2 + \psi^1\left(\gamma^2 - 2\right) + 1 = 0,$$

where $\gamma^2 = gh_0(\Delta t/\Delta x)^2 \sin^2 k\Delta x$ is the resultant quadratic equation. Solving for ψ^1 gives

$$\psi^1 = \left[(2-\gamma^2) \pm \gamma\sqrt{\gamma^2 - 4}\right]/2. \tag{9.33}$$

Two possible situations arise: $\gamma^2 \leq 4$ and $\gamma^2 > 4$. For the first situation ψ^1 can be rewritten as $\psi^1 = \left[(2-\gamma^2) \pm i\gamma\sqrt{4-\gamma^2}\right]/2$, so that equating real and imaginary components yields

$$\lambda \cos\omega_r\,\Delta t = \left(2-\gamma^2\right)/2,$$

$$\lambda \sin\omega_r\,\Delta t = \left(\pm\gamma\sqrt{4-\gamma^2}\right)/2.$$

Squaring and summing these two expressions results in

$$\lambda^2 = \frac{1}{4}\left(4 - 4\gamma^2 + \gamma^4 + 4\gamma^2 - \gamma^4\right) = 1.$$

Since $|\lambda| = 1$ the scheme is neutrally stable and the phase speed as a function of wavelength c can be obtained from either the imaginary or real components. Using the imaginary component the ratio of the calculated to analytic phase speeds is

$$\frac{c}{c_a} = \frac{c}{\pm\sqrt{gh_0}} = \frac{1}{\pm k\Delta t\sqrt{gh_0}}\sin^{-1}\left(\frac{\gamma\sqrt{4-\gamma^2}}{2}\right).$$

From this expression, if $\gamma^2 = 4$, the calculated phase speed c is zero, which is undesirable, of course, because the analytic solution is $\sqrt{gh_0}$. Table 9.3 gives values of c/c_a for selected values of k and γ.

When $\gamma^2 > 4$, the real and imaginary components of Eq. 9.33 are given by

$$\lambda \cos\omega_r\,\Delta t = \frac{1}{2}\left[2 - \gamma^2 \pm \gamma\sqrt{\gamma^2-4}\right],$$

$$\lambda \sin\omega_r\,\Delta t = 0.$$

Since there is no nonzero imaginary contribution, $\cos\omega_r\Delta t = 1(\omega_r = 0$ so the wave has no phase speed) and

$$\lambda = 1 - \frac{1}{2}\left[\gamma^2 \pm \gamma\sqrt{\gamma^2-4}\right].$$

Table 9.3 Values of the amplitude λ and phase error per time step as a function of wavelength and $\sqrt{gh_0}\,\Delta t/\Delta x$ for the centered-in-space, implicit, forward-in-time approximation to the linearized tank model equations. Calculations performed by S. Weidman.

		$\sqrt{gh_0}\,\Delta t/\Delta x$													
	Wavelength	**0.001**	**0.01**	**0.1**	**0.2**	**0.3**	**0.4**	**0.5**	**0.6**	**0.7**	**0.8**	**0.9**	**1.0**	**1.5**	**2.0**
λ	$2\Delta x$	1.0	1.0	1.0	1.0	1.0	1.0	1.0	1.0	1.0	1.0	1.0	1.0	1.0	1.0
	$4\Delta x$	1.0	1.0	1.0	1.0	1.0	1.0	1.0	1.0	1.0	1.0	1.0	1.0	1.0	1.0
	$10\Delta x$	1.0	1.0	1.0	1.0	1.0	1.0	1.0	1.0	1.0	1.0	1.0	1.0	1.0	1.0
	$20\Delta x$	1.0	1.0	1.0	1.0	1.0	1.0	1.0	1.0	1.0	1.0	1.0	1.0	1.0	1.0
c/c_a	$2\Delta x$	0.0	0.0	0.0	0.0	0.0	0.0	0.0	0.0	0.0	0.0	0.0	0.0	0.0	0.0
	$4\Delta x$	0.637	0.637	0.637	0.638	0.639	0.641	0.643	0.647	0.650	0.655	0.660	0.667	0.613	0.0
	$10\Delta x$	0.935	0.935	0.936	0.936	0.937	0.938	0.939	0.940	0.942	0.944	0.947	0.950	0.969	1.0
	$20\Delta x$	0.984	0.984	0.984	0.984	0.984	0.984	0.985	0.985	0.986	0.986	0.987	0.988	0.993	1.0

If $\gamma^2 = 4 + \epsilon^2$, where $\epsilon > 0$, then with $\gamma > 0$,

$$\lambda = -1 - \frac{1}{2}\epsilon^2 - \epsilon < -1,$$

so the scheme is linearly unstable.

Thus to obtain stable results with the implicit finite difference representation to the tank model equations, it is required that

$$\gamma^2 = gh_0(\Delta t/\Delta x)^2 \sin^2 k\Delta x < 4.$$

Since $\sin^2 k\Delta x = 1$ when the wavelength is $4\Delta x (k = 2\pi/4\Delta x)$, the stability condition is

$$\left|\pm\sqrt{gh_0}\Delta t/\Delta x\right| < 2.$$

Sun (1980) has investigated the linear stability of finite difference approximations to equations of the form given by Eqs. 5.37 and 5.39-5.41 with $f = 0$, $\rho'/\rho_0 = \theta'/\theta_0$, and $\lambda_2 = 0$. As shown in Section 5.2.2, this system of equations has internal gravity waves as the solution. Using approximate solution techniques, Sun showed that when the hydrostatic assumption is used ($\lambda_1 = 0$), the dependent variables are staggered in space, updated velocities are used in the potential temperature equation, and the pressure gradient approximated by a centered-in-space scheme, a computationally stable solution results. This sequence of calculations discussed by Sun is called the forward-backward time integration scheme and has been adopted in a number of mesoscale models (e.g., Bhumralkar 1972, 1973; Jones 1973; Pielke 1974a, b. Bhumralkar (1972) similarly found that unless updated values of velocity were used in the computation of potential temperature the results would be linearly unstable. Also, as he and others have concluded, the time step must be less than or equal to the time it takes a disturbance to propagate between grid points, otherwise the solutions will be unstable.

It is also interesting to note that the approximation to the pressure gradient force (e.g., $g\partial h/\partial x$ in Eq. 9.31) utilizes what closely corresponds to a first-order Taylor series approximation to this gradient. As shown by Eq. 2.31, however, such a representation to the pressure gradient force is only valid in the limit as the spatial distance over which this force is evaluated approaches zero. Since this distance is not zero in a numerical model, it should be investigated as to whether the inclusion of higher-order terms in the series expansion (i.e., Eq. 2.31) and their representation by approximate solution techniques would produce improved representations of phase speed.

Finally, other finite difference representations for these and other terms can be examined in the same fashion as presented in this section. At this point, however, we shall investigate other forms of representation for these terms.

9.2 Upstream Interpolation Schemes – An Introduction

A second category of approximation schemes is the interpolation technique.[13] With this method, dependent variables at grid points are used to derive interpolation formula for the space between as well as at the grid points. Such schemes have been used in mesoscale models to represent advection. The finite difference method is one form of interpolation, of course, since upstream differencing, for example, assumes a piecewise continuous function, as sketched in Fig. 9.4a which is linear between grid points.

In the general category of interpolation schemes, however, all the grid points in the domain, or at least all of the grid points in one coordinate direction, are used to approximate the dependent variables. Figure 9.4b and c illustrate two such functions, where one requires that the function equal the dependent variables at the grid point (Fig. 9.4b) and the other to minimize overshoot does not make such a requirement. Texts such as Nielsen (1964) discuss the various types of interpolation formulas, and Goodin et al. (1981) provide an effective, concise discussion of the use of weighted interpolation. In using these formula to represent advection, the change in a dependent variable at time $\tau+1$ caused by advection is determined by going upstream a distance $u_i^\tau \Delta t$ and using the resultant interpolated value at that point to represent the change.

In this section, we shall examine one particular interpolation function (reported by Ahlberg et al. 1967) that has been used effectively in mesoscale models. The technique outlined, of course, can be utilized for any desired interpolation scheme. Rüshøjgaard et al. (1998), for example, describe a different procedure to represent the second derivation of the interpolation scheme.

Let $S(x)$ be the interpolation function and require that

1. $S(x)$, $S'(x)$, and $S''(x)$ are continuous;
2. $S(x)$ is a cubic polynomial over the interval $x_{i-1} \leq x \leq x_i$ (i.e., $S(x) = ax^3 + bx^2 + cx + d$; $S''(x) = 6ax + 2b$); and
3. $S(x_i) = \phi_i$;

where ϕ_i is the value of the dependent variable ϕ at the grid point x_i. If we define

$$M_{i-1} = S''\left(x_{i-1}\right) \quad M_i = S''\left(x_i\right),$$

then

$$S''(x) = M_{i-1}\frac{x_i - x}{h_i} + M_i\frac{x - x_{i-1}}{h_i}, \tag{9.34}$$

where $h_i = x_i - x_{i-1}$ (Eq. 9.34 is of the form $S''(x) = 6ax + 2b$, where $6a = \left(M_i - M_{i-1}\right)/h_i$ and $2b = \left(M_{i-1}x_i - M_i x_{i-1}\right)/h_i$).

Integrating Eq. 9.34 with respect to x gives

$$S'(x) = -M_{i-1}\frac{\left(x_i - x\right)^2}{2h_i} + \frac{M_i\left(x - x_{i-1}\right)^2}{2h_i} + E,$$

[13] The term "semi-Lagrangian scheme" is also used for this technique.

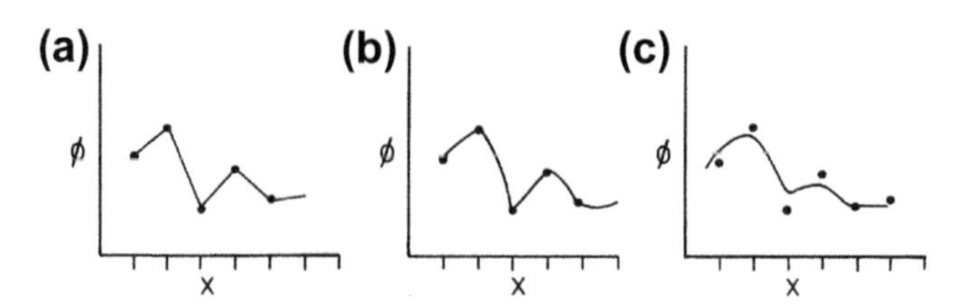

FIGURE 9.4

Schematic examples of interpolation formula that can be used to represent dependent variables on a one-dimensional grid; (a) a piecewise, continuous, linear fit, (b) a polynomial fit with grid point values defined exactly by the interpolation formula, and (c) a polynomial fit that limits the curvature of the interpolation formula.

and integrating again yields

$$S(x) = M_{i-1}\frac{(x_i - x)^3}{6h_i} + \frac{M_i\,(x - x_{i-1})^3}{6h_i} + Ex + F. \tag{9.35}$$

Since $S(x_{i-1}) = \phi_{i-1}$ and $S(x_i) = \phi_i$, then

$$S(x_{i-1}) = M_{i-1}\frac{(x_i - x_{i-1})^2}{6} + Ex_{i-1} + F = \phi_{i-1}, \tag{9.36}$$

and

$$S(x_i) = M_i\frac{(x_i - x_{i-1})^2}{6} + Ex_i + F = \phi_i. \tag{9.37}$$

Subtracting Eq. 9.36 from Eq. 9.37 gives

$$\phi_i - \phi_{i-1} = M_i\frac{h_i^2}{6} - M_{i-1}\frac{h_i^2}{6} + E\,(x_i - x_{i-1}),$$

so that

$$E = \frac{\phi_i - \phi_{i-1}}{h_i} - \frac{h_i}{6}\,(M_i - M_{i-1}).$$

To obtain the constant F, multiply Eq. 9.36 by x_i and Eq. 9.37 by x_{i-1} and subtract Eq. 9.37 from Eq. 9.36 resulting in

$$\frac{\phi_{i-1}x_i - \phi_i x_{i-1}}{h_i} = \frac{M_{i-1}x_i h_i}{6} - \frac{M_i x_{i-1} h_i}{6} + F,$$

so that

$$F = \frac{\phi_{i-1}x_i - \phi_i x_{i-1}}{h_i} + \frac{h_i}{6}\,(M_i x_{i-1} - M_{i-1}x_i).$$

Equation 9.35, after rearranging, can therefore, be written as

$$S(x) = M_{i-1}\frac{(x_i - x)^3}{6h_i} + M_i\frac{(x - x_{i-1})^3}{6h_i} + \left(\phi_{i-1} - \frac{h_i^2}{6}M_{i-1}\right)\frac{(x_i - x)}{h_i}$$
$$+ \left(\phi_i - \frac{h_i^2}{6}M_i\right)\frac{(x - x_{i-1})}{h_i},$$

with the first derivative given by

$$S'(x) = -M_{i-1}\frac{(x_i - x)^2}{2h_i} + \frac{M_i(x - x_{i-1})^2}{2h_i} + \frac{\phi_i - \phi_{i-1}}{h_i} - \frac{M_i - M_{i-1}}{6}h_i. \quad (9.38)$$

In utilizing the spline to approximate advective terms, use is made of the slope $S'(x)$ rather than the second derivative term $S''(x)$.

To do this, let

$$N_i = S'(x_i) = \frac{M_i h_i}{2} + \frac{\phi_i - \phi_{i-1}}{h_i} - \frac{M_i - M_{i-1}}{6}h_i, \quad (9.39)$$

and

$$N_{i-1} = S'(x_{i-1}) = -M_{i-1}\frac{h_i}{2} + \frac{\phi_i - \phi_{i-1}}{h_i} - \frac{M_i - M_{i-1}}{6}h_i. \quad (9.40)$$

Adding the equations for N_i and N_{i-1} and rearranging yields

$$\frac{M_i - M_{i-1}}{6}h_i = N_i + N_{i-1} - \frac{2(\phi_i - \phi_{i-1})}{h_i},$$

so that Eqs. 9.39 and 9.40 can be rewritten as

$$N_i = \frac{M_i h_i}{2} + \frac{3(\phi_i - \phi_{i-1})}{h_i} - N_i - N_{i-1},$$
$$N_{i-1} = -\frac{M_{i-1}h_i}{2} + \frac{3(\phi_i - \phi_{i-1})}{h_i} - N_i - N_{i-1}.$$

Solving these equations for M_i and M_{i-1}

$$M_i = \frac{1}{h_i}\left[4N_i + 2N_{i-1} - \frac{6(\phi_i - \phi_{i-1})}{h_i}\right],$$
$$M_{i-1} = \frac{1}{h_i}\left[-4N_{i-1} - 2N_i + \frac{6(\phi_i - \phi_{i-1})}{h_i}\right],$$

gives the relation between the first and second derivatives at the grid points.

Substituting these expressions for M_i and M_{i-1} back into Eq. 9.38, and after prodigious rearranging, yields

$$S'(x) = \frac{N_i}{h_i^2}\left(x_{i-1} - x\right)\left(2x_i + x_{i-1} - 3x\right) + \frac{N_{i-1}}{h_i^2}\left(x_i - x\right)\left(2x_{i-1} + x_i - 3x\right) + \frac{6\left(\phi_i - \phi_{i-1}\right)}{h_i^3}\left(x_i - x\right)\left(x - x_{i-1}\right). \tag{9.41}$$

Integration of $S'(x)$ yields $S(x)$ in terms of the first derivatives,

$$S(x) = -\frac{N_i}{h_i^2}\left(x - x_{i-1}\right)^2\left(x_i - x\right) + \frac{N_{i-1}}{h_i^2}\left(x_i - x\right)^2\left(x - x_{i-1}\right) + \frac{\phi_{i-1}}{h_i^3}\left(x_i - x\right)^2\left[2\left(x - x_{i-1}\right) + h_i\right] + \frac{\phi_i}{h_i^3}\left(x - x_{i-1}\right)^2\left[2\left(x_i - x\right) + h_i\right], \tag{9.42}$$

and differentiating Eq. 9.41 and rearranging yields

$$S''(x) = -\frac{2N_i}{h_i^2}\left(x_i + 2x_{i-1} - 3x\right) - \frac{2N_{i-1}}{h_i^2}\left(x_{i-1} + 2x_i - 3x\right) + \frac{6\left(\phi_i - \phi_{i-1}\right)}{h_i^3}\left(x_i + x_{i-1} - 2x\right). \tag{9.43}$$

This last equation for $S''(x)$ is used to obtain the values of N_i and N_{i-1} required in the spline interpolation Eq. 9.42.

Since the second derivatives' spline interpolation are required to be continuous at grid points, the value of $S''(x)$ approaching x_i from the $x \geq x_i$ side [denoted by $S''\left(x_i^+\right)$] must equal $S''(x)$ approaching x_i from the $x \leq x_i$ side [denoted by $S''\left(x_i^-\right)$]. From Eq. 9.43 after setting $x = x_i$ and rearranging

$$S''\left(x_i^-\right) = \frac{4N_i}{h_i} + \frac{2N_{i-1}}{h_i} - \frac{6\left(\phi_i - \phi_{i-1}\right)}{h_i^2}.$$

By replacing $i-1$ by i and i by $i+1$, $S''\left(x_i^+\right)$ can be obtained from Eq. 9.43, giving, after rearranging,

$$S''\left(x_i^+\right) = -\frac{2N_{i+1}}{h_{i+1}} - \frac{4N_i}{h_{i+1}} + \frac{6\left(\phi_{i+1} - \phi_i\right)}{h_{i+1}^2},$$

where $h_{i+1} = x_{i+1} - x_i$.

Since continuity is required

$$S''\left(x_i^+\right) = S''\left(x_i^-\right),$$

so that

$$\frac{3\left(\phi_{i+1}-\phi_i\right)}{h_{i+1}^2}+\frac{3\left(\phi_i-\phi_{i-1}\right)}{h_i^2}=\frac{N_{i-1}}{h_i}+\frac{2N_i}{h_i}+\frac{2N_i}{h_{i+1}}+\frac{N_{i+1}}{h_{i+1}}. \tag{9.44}$$

To put in a simpler form, let

$$\alpha_i=h_{i+1}\big/\left(h_i+h_{i+1}\right);\quad \mu_i=1-\alpha_i=h_i\big/\left(h_{i+1}+h_i\right).$$

Using these definitions and multiplying Eq. 9.44 by $h_i h_{i+1}\big/\left(h_i+h_{i+1}\right)$ yields

$$\frac{3\mu_i}{h_{i+1}}\left(\phi_{i+1}-\phi_i\right)+\frac{3\alpha_i}{h_i}\left(\phi_i-\phi_{i-1}\right)=\alpha_i N_{i-1}+2N_i+\mu_i N_{i+1}. \tag{9.45}$$

In matrix form, for nonperiodic boundary conditions, this equation can be written as

$$\begin{bmatrix} b_1 & b_2 & 0 & 0 & 0 & \cdots & 0 & 0 & 0 \\ \alpha_2 & 2 & \mu_2 & 0 & 0 & \cdots & 0 & 0 & 0 \\ 0 & \alpha_3 & 2 & \mu_3 & 0 & \cdots & 0 & 0 & 0 \\ 0 & 0 & \alpha_4 & 2 & \mu_4 & \cdots & 0 & 0 & 0 \\ \vdots & \vdots & \vdots & \vdots & \vdots & \cdots & \vdots & \vdots & \vdots \\ 0 & 0 & 0 & 0 & 0 & \cdots & \alpha_{D-1} & 2 & \mu_{D-1} \\ 0 & 0 & 0 & 0 & 0 & \cdots & 0 & b_{D-1} & b_D \end{bmatrix} \begin{bmatrix} N_1 \\ N_2 \\ N_3 \\ N_4 \\ \vdots \\ N_{D-1} \\ N_D \end{bmatrix} = \begin{bmatrix} d_1 \\ d_2 \\ d_3 \\ d_4 \\ \vdots \\ d_{D-1} \\ d_D \end{bmatrix} \tag{9.46}$$

where D is the number of grid points in the x direction. For i not equal to 1 or D,

$$d_i=\frac{3\mu_i}{h_{i+1}}\left(\phi_{i+1}-\phi_i\right)+\frac{3\alpha_i}{h_i}\left(\phi_i-\phi_{i-1}\right),\quad \text{for } i=2,3,\ldots,D-2,D-1.$$

The matrix of coefficients in this system of equations is called *tridiagonal* because only elements along the three central diagonals are nonzero.

The values of b_1, b_2, b_{D-1}, and b_D are determined from the form of the assumed boundary conditions.[14] One type of boundary condition is obtained from $S''(x_1^+)$ and $S''(x_D^-)$, so that

$$\phi_1''=S''\left(x_1^+\right)=-\frac{2N_2}{h_2}-\frac{4N_1}{h_2}+\frac{6\left(\phi_2-\phi_1\right)}{h_2^2},$$

and

$$\phi_D''=S''\left(x_D^-\right)=\frac{4N_D}{h_D}+\frac{2N_{D-1}}{h_D}-\frac{6\left(\phi_D-\phi_{D-1}\right)}{h_D^2}.$$

If $\phi_1'', \phi_D'', \phi_1$, and ϕ_D are specified, then

$$2N_1+N_2=\frac{3\left(\phi_2-\phi_1\right)}{h_2}-\frac{\phi_1''}{2}h_2,$$

[14]The solution of Eq. 9.45 with periodic boundary conditions is described in Appendix A.

and

$$N_{D-1} + 2N_D = \frac{3\left(\phi_D - \phi_{D-1}\right)}{h_D} - \frac{\phi''_D h_D}{2},$$

so that in the matrix equation

$$d_1 = \frac{3\left(\phi_2 - \phi_1\right)}{h_2} - \frac{\phi''_1}{2} h_2; \quad d_D = \frac{3\left(\phi_D - \phi_{D-1}\right)}{h_D} - \frac{\phi''_D h_D}{2};$$

and

$$b_1 = 2, \quad b_2 = 1, \quad b_{D-1} = 1, \quad \text{and} \quad b_D = 2.$$

In specifying ϕ'' and ϕ at the boundaries, larger-scale information, if available, could be used, or data from one or more grid points inside the grid domain could be interpolated to the boundaries. Boundary conditions will be discussed in more detail in the next chapter.

The determination of the first derivatives from the matrix Eq. 9.46 is performed by the *method of Gaussian elimination* (e.g., see Hadley 1962:51).[15] With this technique, the first row is multiplied by α_2/b_1 and subtracted from the second, giving the values of

$$\left(0, 2 - \left(b_2\alpha_2/b_1\right), \mu_2, 0, 0, \ldots, 0\right) = \left(0, P_2, \mu_2, 0, 0, \ldots, 0\right)$$

on the left side and

$$d_2 - \left(d_1\alpha_2/b_1\right) = G_2$$

on the right. This row of coefficients is used to replace the second row of Eq. 9.46.

Next this new second row is multiplied by α_3/P_2 and subtracted from the third yielding

$$\left(0, 0, 2 - \left(\mu_2\alpha_3/P_2\right), \mu_3, 0, 0, \ldots, 0\right) = \left(0, 0, P_3, \mu_3, 0, \ldots, 0\right)$$

on the left side and

$$d_3 - \left(G_2\alpha_3/P_2\right) = G_3$$

on the right. This row of coefficients is then used to replace the third row of Eq. 9.46. This same type of operation continues for the remainder of the rows.

The object of these algebraic operations on the matrix equation is to create all zeros on the lefthand side of the diagonal of the matrix. As will be seen shortly, rewriting the matrix equation in this form creates an efficient algorithm for solving for the first derivative of the spline.

[15] Modern computers also have standard algorithms to efficiently solve such tridiagonal matrix equations (e.g., Adams et al. 1975).

With these operations of the matrix equation and using $b_2 = b_{D-1} = 1$; $b_1 = b_D = 2$, Eq. 9.46 can be rewritten as

$$\begin{bmatrix} 2 & 1 & 0 & & & \cdots & & 0 \\ 0 & P_2 & \mu_2 & 0 & & \cdots & & 0 \\ 0 & 0 & P_3 & \mu_3 & 0 & \cdots & & 0 \\ 0 & 0 & 0 & P_4 & \mu_4 & \cdots & & 0 \\ \vdots & \vdots & \vdots & \vdots & \vdots & \cdots & & \vdots \\ & & & & & & P_{D-1} & \mu_{D-1} \\ 0 & 0 & 0 & 0 & 0 & \cdots & 0 & P_D \end{bmatrix} \begin{bmatrix} N_1 \\ N_2 \\ N_3 \\ N_4 \\ \vdots \\ N_{D-1} \\ N_D \end{bmatrix} = \begin{bmatrix} d_1 \\ G_2 \\ G_3 \\ G_4 \\ \vdots \\ G_{D-1} \\ G_D \end{bmatrix},$$

where

$$P_i = 2 - \left[\mu_{i-1}\alpha_i/P_{i-1}\right]$$
$$G_i = d_i - \left[G_{i-1}\alpha_i/P_{i-1}\right]$$

with $i = (2, \ldots, D)$; $G_{i-1} = \mu_{i-1} = 0$; $P_{i-1} \neq 0$ for $i = 1$. In the method of Gaussian elimination, the algebraic operations required to create zeros to the left of the diagonal is called the *forward sweep*.

With the matrix in this form, the slopes of the spline at the grid points are determined by a *backward sweep* starting with N_D and continuing until N is determined. Expressed algebraically

$$N_D = G_D/P_D \qquad (i = D),$$
$$N_i = \left(G_i - \mu_i N_{i+1}\right)/P_i \quad (i = 1, 2, \ldots, D-1).$$

To eliminate the need to compute G_i, these expressions are often written in an equivalent fashion as

$$N_D = R_D \qquad (i = D),$$
$$N_i = R_i - \left(\mu_i N_{i+1}/P_i\right) \quad (i = 1, 2, \ldots, D-1), \tag{9.47}$$

where

$$R_i = \left(d_i - R_{i-1}\alpha_i\right)/P_i \left(\text{with } R_{i-1} = 0 \text{ for } i = 1\right).$$

With these expressions for N_i, it is possible to compute $S(x)$ from Eq. 9.42 for any point in the domain. In the interpolation scheme, estimating changes caused by advection at time level $\tau + 1$ is performed by evaluating the value of the spline (or other approximation formula) a distance of $|u_i^\tau|\Delta t$ *upstream* from the grid point of interest. The assumption is that the value of the dependent variable at $|u_i^\tau|\Delta t$ upstream of grid point i at time level τ is equal to its value at $\tau + 1$ at the grid point. This distance can also be written as $|u_i^\tau|\Delta t = Ch_i$, where C is the Courant number defined using the absolute value of the velocity at a grid point ($C = |u_i^\tau|\Delta t/h_i$).

One could also be tempted to use the values of N_i directly in the conservation relations to approximate the spatial derivatives. Experience has shown (e.g.,

Price and MacPherson 1973), however, that this method can cause undesirable growth of short wavelength noise for a nonlinear problem if used without smoothing.

To determine the value of the spline at $x_i - Ch_i$ for the case of $u_i^\tau \geq 0$, the value of x in Eq. 9.42 is replace with $x_i - Ch_i$ yielding

$$\begin{aligned} S\left(x_i - Ch_i\right) = & N_{i-1}\left(x_i - x_i + Ch_i\right)^2\left(x_i - Ch_i - x_{i-1}\right)/h_i^2 \\ & - N_i\left(x_i - CH_i - x_{i-1}\right)^2\left(x_i - x_i + Ch_i\right)/h_i^2 \\ & + \phi_{-i-1}\left(x_i - x_i - Ch_i\right)^2\left[2\left(x_i - Ch_i - x_{i-1}\right) + h_i\right]/h_i^3 \\ & + \phi_i\left(x_i - Ch_i - x_{i-1}\right)^2\left[2\left(x_i - x_i + Ch_i\right) + h_i\right]/h_i^3. \end{aligned}$$

Expanding this relation and rearranging yields

$$\begin{aligned} S\left(x_i - Ch_i\right) = & \phi_i^\tau - CN_ih_i + C^2\left[N_{i-1}h_i + 2N_ih_i + 3\left(\phi_{i-1}^\tau - \phi_i^\tau\right)\right] \\ & - C^3\left[h_iN_{i-1} + h_iN_i + 2\left(\phi_{i-1}^\tau - \phi_i^\tau\right)\right] \quad \left(u_i^\tau \geq 0\right). \end{aligned} \quad (9.48)$$

For the case of $u_i^\tau < 0$, the spline must be evaluated between grid points x_i and x_{i+1}. An expression for the spline, analogous to Eq. 9.42 (Eq. 9.42 is for application between grid points x_{i-1} and x_i), is obtained by replacing $i-1$ by i, and i by $i+1$ in Eq. 9.42 yielding

$$\begin{aligned} S(x) = & -\frac{N_{i+1}}{h_{i+1}^2}\left(x - x_i\right)^2\left(x_{i+1} - x\right) + \frac{N_i}{h_{i+1}^2}\left(x_{i+1} - x\right)^2\left(x - x_i\right) \\ & + \frac{\phi_i}{h_{i+1}^3}\left(x_{i+1} - x\right)^2\left[2\left(x - x_i\right) + h_{i+1}\right] \\ & + \frac{\phi_{i+1}}{h_{i+1}^3}\left(x - x_i\right)^2\left[2\left(x_{i+1} - x\right) + h_{i+1}\right]. \end{aligned}$$

Substituting $x_i + Ch_{i+1}$ for x in this expression and rearranging gives

$$\begin{aligned} S\left(x_i + Ch_{i+1}\right) = & \phi_i^\tau + CN_ih_{i+1} - C^2\left[N_{i+1}h_{i+1} + 2N_ih_{i+1} + 3\left(\phi_i^\tau - \phi_{i+1}^\tau\right)\right] \\ & + C^3\left[h_{i+1}N_i + h_{i+1}N_{i+1} + 2\left(\phi_i^\tau - \phi_{i+1}^\tau\right)\right] \quad \left(u_i^\tau < 0\right). \end{aligned} \quad (9.49)$$

Equations 9.48 and 9.49 are thus used to determine the changes in the dependent variable due to advection from

$$\left.\phi_i^{\tau+1}\right|_{\text{advective changes}} = S\left(x_i - Ch_i\right)\left(u_i^\tau \geq 0\right). \quad (9.50)$$

$$\left.\phi_i^{\tau+1}\right|_{\text{advective changes}} = S\left(x_i + Ch_{i+1}\right)\left(u_i^\tau < 0\right). \quad (9.51)$$

To determine the linear computational characteristics of this scheme, let $u_i^\tau = U \geq 0$ and $h_i = h_{i+1} = \Delta x$ (so that $\alpha_i = \mu_i = 1/2$). The equation for the slope of the spline (Eq. 9.45) interior to the boundaries for this situation is given by

$$(3/\Delta x)\left(\phi_{i+1}^\tau - \phi_{i-1}^\tau\right) = N_{i-1} + 4N_i + N_{i+1}.$$

Expressing the values of ϕ and N in terms of frequency and wavenumber yields

$$\left(3\hat{\phi}/\Delta x\right)\left(\psi_1 - \psi_{-1}\right) = \widehat{N}\left(\psi_{-1} + 4 + \psi_1\right).$$

Solving for $\widehat{N}$ in terms of $\hat{\phi}$ gives

$$\widehat{N} = \frac{3\hat{\phi}\left(\psi_1 - \psi_{-1}\right)}{\Delta x\left(\psi_{-1} + 4 + \psi_1\right)} = \frac{3i\hat{\phi}}{\Delta x}\frac{\sin k\Delta x}{(\cos k\Delta x + 2)} = \frac{Gi\hat{\phi}}{\Delta x}. \tag{9.52}$$

For notational convenience, G is defined equal to $3\sin k\Delta x/(\cos k\Delta x + 2)$.

Equation 9.50 (where the righthand side is determined from Eq. 9.48) can similarly be rewritten in terms of frequency and wavenumber as

$$\begin{aligned}\hat{\phi}\psi^1 &= \hat{\phi} - C\Delta x\widehat{N} + C^2\left[\Delta x\psi_{-1}\widehat{N} + 2\widehat{N}\Delta x + 3\hat{\phi}\left(\psi_{-1} - 1\right)\right] \\ &- C^3\left[\Delta x\widehat{N}\psi_{-1} + \Delta x\widehat{N} + 2\hat{\phi}\left(\psi_{-1} - 1\right)\right].\end{aligned}$$

Substituting for $\widehat{N}$ from Eq. 9.52 and rearranging results in

$$\begin{aligned}\psi^1 &= 1 + C^2(G\sin k\Delta x + 3\cos k\Delta x - 3) + C^3(2 - G\sin k\Delta x - 2\cos k\Delta x) \\ &+ i\Big[-GC + C^2(2G + G\cos k\Delta x - 3\sin k\Delta x) \\ &+ C^3(2\sin k\Delta x - G\cos k\Delta x - G)\Big].\end{aligned} \tag{9.53}$$

Clearly when the Courant number goes to zero, $\psi^1 = 1$ so that the spline upstream interpolation scheme is a consistent representation of the advective equation.

However, the evaluation of amplitude and phase characteristics from Eq. 9.53 when $C \neq 0$ is not as straightforwardly performed as was possible with the finite difference representations. Equating real and imaginary components of Eq. 9.53.

$$\begin{aligned}\lambda\cos\omega_r\Delta t &= 1 + C^2(G\sin k\Delta x + 3\cos k\Delta x - 3) \\ &+ C^3(2 - G\sin k\Delta x - 2\cos k\Delta x)\end{aligned}$$

and

$$\begin{aligned}\lambda\sin\omega_r\Delta t &= -GC + C^2(2G + G\cos k\Delta x - 3\sin k\Delta x) \\ &+ C^3(2\sin k\Delta x - G\cos k\Delta x - G).\end{aligned}$$

To compute λ and $\tilde{c}_\phi/U (\tilde{c}_\phi = -\omega_r/k; k$ is real) from these equations it is most convenient to substitute particular values of C and k into these relations and solve for λ and ω_r on the computer. This is done either by using complex arithmetic, which is routinely available on most computers, or simply performing the squaring, summing, and division of the imaginary component by the real component numerically.

Table 9.1 presents values of λ and $\tilde{c}_\phi/U$ for this scheme. As contrasted with the leapfrog technique, the amplitude is not preserved for all wavelengths and $2\Delta x$ waves are eliminated completely when $C = 0.5$. The phase for the $2\Delta x$ wave is also incorrectly represented because it remains stationary. This latter problem with $2\Delta x$ waves is not particularly severe, however, especially with values of C between 0.2 and 0.8, because features with this wavelength are rapidly damped with time (e.g., at $C = 0.2$, the amplitude is reduced by 90% after 10 time steps; $\lambda^{10} = 0.097$). At $4\Delta x$, the spline provides a much more accurate representation of the advection. At $C = 0.5$, for example, which is the worst case[16] in terms of amplitude, $\lambda = 0.972$ and $\tilde{c}_\phi = U$. The leapfrog scheme, in contrast, has a value of $\lambda = 1.0$ at $C = 0.5$ for a $4\Delta x$ wave, but $\tilde{c}_\phi/U = 0.67$ or a 33% error in phase.

Thus although the leapfrog preserves amplitude exactly, it does a poor job of representing the phase characteristics of the shorter waves. The amplitude conservation of leapfrog, therefore, becomes a liability since the presence of $2\Delta x$ waves in the wrong location can create erroneous results through nonlinear interactions when physical forcings such as, for example, condensation occurs. The spline technique, by contrast, has the desirable feature of damping wavelengths in which the phase characteristics are poorly represented. Therefore, in choosing a computational scheme it appears desirable to select one in which the accuracy of λ and $\tilde{c}_\phi/U$ are closely correlated (i.e., $|\lambda|$ much less than unity when the numerical phase accuracy is poor).

The most substantial problem with the spline is its tendency to overshoot. If, for example, a field has a string of zeros followed by nonzero values, the spline can create small values less than zero near the interface with the zeros because of its cubic form. Although not a serious source of error, it is nonphysical to have negative values in such fields as specific humidity. Requiring values to exceed or equal zero is one correction for this inconsistency.

In using schemes such as the spline in numerical models, the interpolation formula could be derived in terms of more than one spatial direction (e.g., as a two-dimensional spline in x and y). Krishnamurti et al. (1973) used a bilinear interpolation scheme (Krishnamurti 1962; Mathur 1970) defined in two space dimensions in his synoptic tropical prediction model.

Unfortunately, however, most interpolation schemes become very complicated in form and expensive to use if more than one spatial coordinate is used. An alternate approach is to use the spline separately in each spatial direction. This approach of evaluating each coordinate direction separately is called *splitting*[17] (e.g., Long and

[16]The spline has its worst representation of phase speed in Table 9.1 for a $4\Delta x$ wave ($\tilde{c}_\phi/U = 0.955$) at $C = 0.001$.

[17]Splitting is also referred to as the *Marchuk method* (Mesinger and Arakawa 1976).

Hicks 1975; Mesinger and Arakawa 1976) and is used to represent the spline in the mesoscale model reported by Mahrer and Pielke (1978a). Using this technique with the spline in a two-dimensional model, for instance, where i and k are the grid points in x and z, respectively, and $u_i^\tau \geq 0$ and $w_k^\tau \geq 0$,

$$\phi_{i,k}^{\tau+1^*} \mid \text{changes caused by advection in } x = S_u\left(x_i - C_u h_i\right) \tag{9.54}$$

(where $C_u = |u_i^\tau|\Delta t/h_i$, with values of ϕ at time τ used in S_u) is determined for all i, then

$$\phi_{i,k}^{\tau+1} \mid \text{changes caused by advection in } z = S_w\left(z_k - C_w h_k\right) \tag{9.55}$$

(where $C_w = |w_i^\tau|\Delta t/h_k$, with values of ϕ at $\tau + 1^*$ determined from Eq. 9.54 used in S_w) is calculated for all k. Pepper et al. (1979) have shown that splitting does not degrade the accuracy of the solutions.

Other interpolation schemes besides the spline could also be used to represent advection. The methodology would be developed in an analogous fashion to that of the spline. Bates and McDonald (1982) provide a useful study of the application of such upstream interpolation techniques to atmospheric models. Smolarkiewicz and Margolin (1997) contrast the use of interpolation, and centered-in-time and space finite difference schemes. Benoit et al. (1997a, b, c) describe the use of an interpolation scheme in the Canadian MC2 model. Other examples of the use of interpolation schemes to represent advection include Janjić (1995) Pinty et al. (1995), Böttcher (1996), Héreil and Laprise (1996), Li and Bates (1996), Makar and Karpik (1996), Ritchie and Tanguay (1996), Sun et al. (1996), Lin and Rood (1997), Sun and Yeh (1997), Caya et al. (1998), Finkele (1998), Qian et al. (1998), McDonald (1999), and Xiao (2000). Laprise and Plante (1995) discuss the use of the interpolation solution technique both upstream and downstream.

Tremback et al. (1987) provides a detailed analysis of the extension of the forward-in-time upstream advection approach to a higher order accuracy. A sixth-order scheme was found to have the best balance between efficiency and accuracy. Finkele (1998) describes the accuracy of a third-order interpolation advection scheme for use in her simulation of sea breezes. Another paper which discusses this solution technique is that of Pellerin et al. (1995).

9.3 Time Splitting

In Section 5.2.2, it was shown that the neglect of local variations in density in the conservation-of-mass relation eliminate rapidly propagating sound waves as a possible solution. Use of the resultant anelastic (soundproof) conservation-of-mass equation in atmospheric flows, therefore, limit the fastest wave speeds to gravity waves. As implied from the results presented in this chapter, time steps must be selected such that the distance traveled by a wave in one time step is less than or equal to the appropriate grid spacing (i.e., $c\Delta t \leq \Delta$, where c is the propagation speed of the fastest traveling wave where, for instance, $c = U$ and $\Delta = \Delta x$ in the advection equation; $c = \sqrt{gh_0}$ and $\Delta = 2\Delta x$ in the tank model; and by analogy, $c = \sqrt{RT_0(C_p/c_v)}$ and $\Delta = 2\Delta x$ when sound waves are present). If this criteria is not attained, either the approximate solutions are linearly unstable or the representation is very inaccurate.

If the anelastic equation is used to represent mass conservation and, therefore, to eliminate sound waves, however, the solution of a complex elliptic partial differential equation for pressure (e.g., Eq. 4.33) is required. To eliminate the need to compute pressure from such an equation, but still retain stable and accurate solutions in a nonhydrostatic model, it is possible to compute the time tendency of the terms that generate the sound waves separately from the remainder of the dynamic equations. Thus a very short time step is used for the relatively few terms that generate sound, and a much more economical time interval can be selected for the features that are considered more important on the mesoscale. Klemp and Wilhelmson (1978a, b), and Tripoli and Cotton (1982) used this approach in three-dimensional cloud models, and Tapp and White (1976) utilized it in their mesoscale model. Daley (1980) discussed the splitting of fast gravity wave computations from slower speed gravity and Rossby waves using a procedure called model normal mode expansion for the fast waves.

Using the analysis of the forms of wave motion discussed in Section 5.2.2, the equations and terms that give rise to sound waves can be written in the form

$$\begin{aligned}
\frac{\partial \bar{u}_i}{\partial t} + \frac{1}{\bar{\rho}}\frac{\partial \bar{p}}{\partial x_i} &= f_{u_i}, \\
\frac{\partial \bar{\rho}}{\partial t} + \frac{\partial}{\partial x_j}\bar{\rho}\bar{u}_j &= f_{\rho}, \\
\frac{\partial \bar{\theta}}{\partial t} &= f_{\theta} \\
\bar{p}(\text{in mb}) = C_p^{C_p/C_v} R_d^{C_p/C_v} (1000\text{ mb})^{-R_d/C_v} \left(\bar{\rho}\bar{\theta}\right)^{C_p/C_v} &= f_p, \quad (9.56)
\end{aligned}$$

where f_{u_i}, f_{ρ}, f_{θ}, and f_p contain the terms in these conservation relations that are not written explicitly here. These 6 equations in the 6 unknowns u_i, $\bar{\rho}$, $\bar{p}$, and $\bar{\theta}$ can then be written in an appropriate numerical approximation form and evaluated quickly and efficiently for short time steps where f_{u_i}, f_{ρ}, f_{θ}, and f_p are kept fixed for a specified period of time (e.g., say equal to 10 times the time step used to represent the sound waves). With this approach, sound waves are simulated in an accurate fashion, and a longer time step can be used to update values of f_{u_i}, f_{ρ}, f_{θ}, and f_p. Alternatively, the sound waves can be evaluated using an implicit differencing scheme such that longer time steps are used although the accuracy of the representation of the sound waves themselves are degraded by the implicit scheme. Discussions of the time-splitting schemes are reported in Saito (1997), and Wicker and Skamarock (1998). The first reference to this very effective time-splitting approach appeared in Derickson (1974).

9.4 Nonlinear Effects – Aliasing

There has been little discussion in this text regarding the influence of products of dependent variables on the approximated solutions of the conservation relations. This neglect, of course, results from the inability of existing mathematical techniques to provide analytic solutions to nonlinear differential equations, except for idealized specific cases. Thus, even though the conservation relations are nonlinear, most of Chapter 5, for example, concentrated on solutions to linear equations, with previous sections of this chapter emphasizing linear analysis tools in the investigation of solution techniques that are to be used in nonlinear models. In this section, the actual and computational results of nonlinear interactions are discussed.

In the atmosphere on the mesoscale there are spatial scales in which kinetic energy is being produced (e.g., the scale of the horizontal temperature gradient associated with a seacoast such as illustrated in Fig. 13.3) and scales in which this kinetic energy is being dissipated into heat by molecular interactions. In the first case, scales of motion are on the order of a hundred meters to a hundred kilometers, for example, whereas the sizes of motion significantly affected by molecular interactions are a centimeter or less. Somewhere in between exists scales of motion that are not directly influenced by either molecular dissipation or the forces generating the kinetic energy in the first place.

Thus it is expected that in this region the kinetic energy per unit wavenumber per unit mass as a function of wavenumber $E(k)$ is proportional to only the spatial scale of the motion (as specified by its wavenumber) and by the rate at which energy is being removed at the much smaller scales. Moreover, since kinetic energy does not accumulate once the larger-scale forcing is terminated, the energy must be transformed with time into smaller and smaller sizes until they are of a centimeter or less in size and can be removed by molecular interactions. From dimensional arguments, if $E(k)$ is dependent only on wavenumber and dissipation rate

$$E(k) = a\epsilon^{2/3}k^{-5/3}, \tag{9.57}$$

where a is a proportionality constant and ϵ has units of energy per unit time per unit mass. This region, where kinetic energy is independent of the original forcings of the motion and of its dissipation by molecular viscosity, is called the *inertial subrange*. This name arises because the advective terms (in this context these terms are also called the inertial terms) transfer kinetic energy among the three components of velocity as well as generate smaller and smaller sizes of circulations.

Lumley and Panofsky (1964) have provided a detailed discussion of the transfer of kinetic energy by turbulence. Tennekes (1978), Gage (1979), Lilly (1983), and Moran (1992) discussed the observed occurrence of such a $k^{-5/3}$ relation in mesoscale and larger atmospheric features.

In a numerical mesoscale model, however, this cascade of energy to smaller scales cannot occur because the smallest feature that can be resolved has a wavelength of two times the grid spacing.[18] If, for example,

$$\phi_1 = \phi_0 \cos k_1 \Delta x, \quad \text{and} \quad \phi_2 = \phi_0 \cos k_2 \Delta x$$

represent two waves in a model with equal amplitudes ϕ_0, then a nonlinear interaction between the two can be represented by

$$\phi_1 \phi_2 = \phi_0^2 \cos k_1 \Delta x \cos k_2 \Delta x$$

or, using trigonometric identities

$$\phi_1 \phi_2 = \frac{1}{2} \phi_0^2 \left[\cos \left(k_1 + k_2 \right) \Delta x + \cos \left(k_1 - k_2 \right) \Delta x \right]. \tag{9.58}$$

Thus two waves result from this interaction with wavenumbers $k_1 + k_2$ and $k_1 - k_2$.

Suppose, as an example, a $2\Delta x$ and a $4\Delta x$ wave interact ($k_1 = 2\pi/2\Delta x$, $k_2 = 2\pi/4\Delta x$), the resultant waves are given by

$$\phi_1 \phi_2 = \frac{1}{2} \phi_0^2 \left(\cos 2\pi \left(\frac{6}{8} \right) \Delta x + \cos 2\pi \left(\frac{1}{4} \right) \Delta x \right),$$

which corresponds to a $1.33\Delta x$ and a $4\Delta x$ wave. The latter size wave, of course, can be resolved but the $1.33\Delta x$ wave cannot! Instead it will be fictitiously seen as a $4\Delta x$ since that size is the first integer multiple of $\frac{4}{3}\Delta x$ equal to $n\Delta x$, where n is also an integer and $n \geq 2$. Waves that appear erroneously in this fashion are said to have *aliased* or *folded* to longer wavelengths. As seen from Eq. 9.58, to have aliasing, one of the waves must be less than $4\Delta x$ in length in order to generate physical solutions less than $2\Delta x$.

Listed in Table 9.4 are examples of wave-wave interactions that will produce aliased waves. Figure 9.5 illustrates how a $1.33\Delta x$ wave would be misinterpreted as a $4\Delta x$ wave on a computational grid. Even if no $2\Delta x$ waves are initially present, they will be created since the interaction of two $4\Delta x$ waves will generate a $2\Delta x$ wave, whereas longer wave-wave interactions will produce $4\Delta x$ waves. Two interactive waves, both with $2\Delta x$ wavelengths will not produce a wave of $1\Delta x$ because identical values will result at each grid point. In this case, the energy in the $1\Delta x$ wave will be seen fictitiously as the addition of a constant value of energy to the model.

Thus when wave interactions occur in the real world, smaller and larger wavelengths result. Eventually the smaller waves attain a size in which molecular dissipation can eliminate motion. In a numerical model, however, which has a discrete grid, waves smaller than $2\Delta x$ are erroneously seen as larger-scale waves. These erroneous larger-scale waves interact and again transfer their energy to larger and smaller scales. Because the proper cascade of energy to smaller and smaller scales is interrupted, a

[18]As discussed in Section 4.1, due to computer resource limitations, the grid intervals in meteorological models cannot be made small enough to represent molecular processes, and yet still simulate mesoscale atmospheric phenomena.

Table 9.4 Examples of wave-wave interactions that will produce aliased waves.

Interactive Wavelengths	Should Produce	Will Produce Due to Aliasing
$2\Delta x$ and $2\Delta x$	$1\Delta x$	*Add a constant to the entire model*
$2\Delta x$ and $4\Delta x$	$1.33\Delta x$	$4\Delta x$
$2\Delta x$ and $6\Delta x$	$1.5\Delta x$	$3\Delta x$
$2\Delta x$ and $8\Delta x$	$1.6\Delta x$	$8\Delta x$
$2\Delta x$ and $10\Delta x$	$1.67\Delta x$	$5\Delta x$

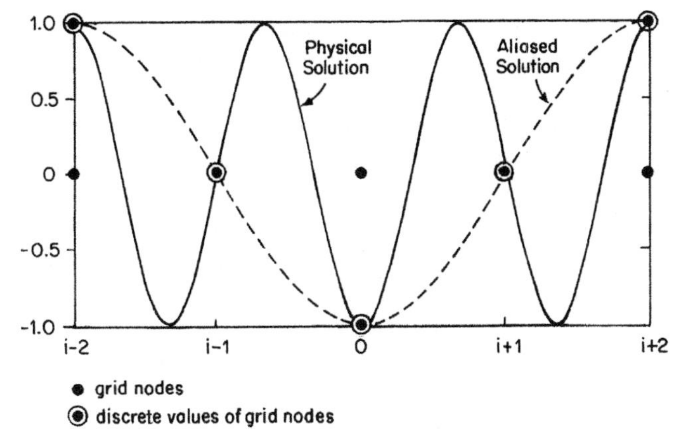

FIGURE 9.5

Schematic illustration of how a physical solution with a wavelength of $1.33\Delta x$, caused by the nonlinear interaction of waves of $2\Delta x$ and $4\Delta x$ in length, is seen as a computational $4\Delta x$ wave in the computational grid.

fictitious energy buildup occurs as energy continues to be added to the model through the forcing terms, but with its dissipation improperly represented.

Therefore, even if a computational solution technique is linearly stable, the results can degrade into physically meaningless computational noise. Indeed, with many computational solution schemes, this erroneous accumulation of energy can cause the model dependent variables to increase in magnitude without bound – an error that is referred to as *nonlinear instability*.[19]

Wavelengths less than $4\Delta x$ are required for aliasing to occur. As shown previously in this chapter, such short waves are inadequately resolved on a computational grid and even in the linearized equations are poorly represented in terms of amplitude

[19]Unbounded growth of nonlinear instability can be controlled by requiring conservation of energy in the selected differencing scheme. However, even with these schemes the solution can degrade into small-scale noise. Initial conditions which contain considerable shortwave features can result in nonlinear instability developing more rapidly.

and/or phase. For these reasons, and because they are expected to cascade to even smaller scales anyway, it is desirable to remove these waves.

There are two methods to accomplish this task:

1. by the proper parameterization of the subgrid-scale correlation terms (i.e., $\overline{u_i'' u_j''}$, $\overline{\theta'' u_j''}$, etc.) so that energy is extracted from the averaged equations in a manner consistent with reality; or
2. by the use of a spatial smoother (also referred to as a filter) which removes the shortest waves, but leaves the longer ones relatively unaffected.

The first method is the most attractive, of course, because it is based on fundamental physical concepts. Unfortunately, however, as shown in Chapter 7, only the vertical subgrid-scale correlation terms are reasonably well known and can be parameterized accurately in terms of the dependent variables. Horizontal subgrid mixing in mesoscale models, in contrast, can be estimated only crudely since the horizontal averaging scale in such models is typically larger than the vertical scale (i.e., $\Delta x \simeq \Delta y \gg \Delta z$ in Eq. 4.6). Moreover, there has been little theoretical or observational studies on the structure of horizontal mixing over heterogeneous ground surfaces.

The forms used in the representation of the horizontal subgrid correlation terms, therefore, have been chosen to *control nonlinear aliasing, not to represent actual physical processes accurately*. One form of horizontal mixing used (which has the appearance of a physical parameterization but, in a mesoscale model, is not) is

$$-\overline{u_i'' u_j''} = K_H \partial \bar{u}_i / \partial x_j \quad (i = 1, 2), \tag{9.59}$$

where

$$K_H = \alpha (\Delta x)^2 \left[\frac{1}{2} \left(\frac{\partial \bar{v}}{\partial x} + \frac{\partial \bar{u}}{\partial y} \right)^2 + \left(\frac{\partial \bar{u}}{\partial x} \right)^2 + \left(\frac{\partial \bar{v}}{\partial y} \right)^2 \right]^{1/2}$$

with the coefficient α arbitrarily adjusted until $2\Delta x$ wavelengths do not appear to degrade the solutions significantly. This procedure, which is completely ad hoc, permits solutions to be changed in magnitude simply by changing the value of α. Tag et al. (1979) provided a useful discussion of several forms of these variable eddy coefficient formulations. Laprise et al. (1998) provides a very useful analysis of different forms of K_H.

The alternative to an explicit diffusion equation is to formally apply a filter, such as discussed by Shapiro (1970). Cullen (1976) compared solutions in a simplified synoptic-scale model using several types of filters and explicit diffusion representations and showed that selective filters can be used more effectively in finite element representations than with finite difference techniques because of the greater accuracy of the first method. The use of optimal filters is also discussed by von Storch (1978), and Jones (1977a) outlines a smoothing technique to control computational noise as information is transferred between coarse and fine grids in a hurricane model.

When a terrain-following coordinate system is used (e.g., see Fig. 6.4 and associated text), a modeler must assure that the explicit horizontal diffusion or the smoother be applied on the $x - y$ plane since $\tilde{x}^1$ and $\tilde{x}^2$ are not, in general, horizontal. If an intended horizontal filter is mistakenly applied to the $\tilde{x}^1 - \tilde{x}^2$ surface, diffusion will be inadvertently input into the z direction. This smoothing is not desirable if a physically realistic representation of vertical diffusion is included in the model and if the vertical component of the computational diffusion, erroneously evaluated on the $\tilde{x}^1 - \tilde{x}^2$ surface, is the same order as the physical vertical mixing.

Smoothers can be explicit or implicit. With implicit smoothing, computational techniques are chosen, such as upstream differencing or the cubic spline interpolation, which include inherent damping (e.g., see the values of λ in Table 9.1). Explicit smoothers, by contrast, require the addition of an operation to the prognostic equations that generates smoothed dependent variables from the original predicted dependent variables. The ideal smoother is one that eliminates wavelengths smaller than $4\Delta x$ each time step but leaves the larger sizes unaffected. Such smoothers are referred to as *low-pass filters* and are said to be *selective*.

Pepper et al. (1979) reported on a highly selective filter of the form

$$(1-\delta)\phi_{i+1}^* + 2(1+\delta)\phi_i^* + (1-\delta)\phi_{i-1}^* = \phi_{i+1} + 2\phi_i + \phi_{i-1}, \tag{9.60}$$

where ϕ and $\phi*$ are the dependent variables to be smoothed and the smoothed value, and δ an arbitrarily chosen weighting parameter for the smoothed values. As shown subsequently, this filter eliminates $2\Delta x$ waves at each application, and its smoothing of longer waves is dependent on the value of δ.

Let

$$\phi_{i+1}^* = \psi_1^1\phi; \quad \phi_i^* = \psi^1\phi; \quad \phi_{i-1}^* = \psi_{-1}^1\phi; \tag{9.61}$$

which is similar to the form used in the section on linear stability analysis, except here

$$\psi^1 = \lambda$$

corresponds to the change in magnitude of the solution per application of the filter. One application of the filter, performed per time step, can also be written as

$$\lambda = e^{-Kk^2\Delta t} = e^{-\gamma(2\pi)^2/n^2},$$

where $\gamma = K\Delta t(\Delta x)^2$. Thus when λ is known, it is possible to compute a value of K that will give the same smoothing when the linear diffusion equation[20] is used as when Eq. 9.60 is applied.

Using the decomposition of the dependent variable given by Eq. 9.61, 9.60 can be rewritten as

$$(1-\delta)\psi_1^1 + 2(1+\delta)\psi^1 + (1-\delta)\psi_{-1}^1 = \psi_1 + 2 + \psi_{-1}.$$

Solving for $\psi^1 = \lambda$ gives

$$\lambda = \frac{\cos k\Delta x + 1}{(1-\delta)\cos k\Delta x + 1 + \delta}, \tag{9.62}$$

[20] $\partial\bar{\phi}/\partial t = K\partial^2\bar{\phi}/\partial x^2$.

and the equivalent value of K (Long 1979, personal communication) can be determined from

$$K(k, \delta) = -\left(1/k^2 \Delta t\right) \ln \lambda. \qquad (9.63)$$

As evident from these expressions, $\lambda = 0\ (K = \infty)$ for a $2\Delta x$ wave $[\cos(2\pi/2\Delta x)\ \Delta x = \cos \pi = -1]$. For a very long wave, $\cos k\Delta x$ approaches 1 and $\lambda \simeq 1\ (K \simeq 0)$, as it should. For wavelengths intermediate between these two values, the amount of damping is dependent on the value of δ. Table 9.5 illustrates the damping by Eq. 9.60 for several selected values of wavelength and δ. As evident from the Table, for small values of δ, the formulation given by Eq. 9.60 is highly selective, and its influence on longer wavelengths is very small. It should be noted, however, that in separately applying the filter to more than one spatial direction (i.e., splitting), the application of this filter to the second direction can reintroduce $2\Delta x$ wavelengths into the first direction of application. This could be controlled to some extent (as suggested by Xubin Zeng, personal communication) by alternating the order of the direction of smoothing between time steps (i.e., at the n time, filter x then in y; at the $n+1$ time, filter y then x).

All mesoscale models employ some sort of horizontal filtering to control nonlinear aliasing. Either explicit smoothers, such as Eq. 9.60 are used, or implicit computational diffusion inherent to the numerical approximation, such as with upstream differencing or the cubic spline, provide the necessary removal of the shortest wavelengths.

The use of a smoother can also prevent linear instability as long as the magnitude of λ for the smoother times λ of the linearly unstable numerical scheme is less than or equal to unity. For example, if

$$\left|\left(\frac{\cos k\Delta x + 1}{(1-\delta)\cos k\Delta x + 1 + \delta}\right)\left(\sqrt{1 + C^2 \sin^2 k\Delta x}\right)\right| \leq 1,$$

(where Eq. 9.62 gives λ for the smoother, and the positive root of Eq. 9.12 is used for λ resulting from the forward-in-time, centered-in-space representation to the advection equation) then the solutions can be made stable. Unfortunately, however, in a mesoscale model, without performing a Fourier decomposition, it is impossible to

Table 9.5 Values of λ per application of Eq. 9.60 as a function of wavelength and δ.

	δ			
Wavelength	**0.001**	**0.005**	**0.10**	**0.50**
$2\Delta x$	0	0	0	0
$4\Delta x$	0.999	0.995	0.900	0.667
$6\Delta x$	1.000	0.998	0.968	0.857
$8\Delta x$	1.000	0.999	0.988	0.945
$10\Delta x$	1.000	0.999	0.990	0.950
$12\Delta x$	1.000	1.000	0.993	0.966
$14\Delta x$	1.000	1.000	0.995	0.974

determine the wavelengths of all the features at each time step in order to select the proper value of the smoother to counteract the linear instability of the difference scheme. Thus, in general, the use of Eq. 9.60 to control the linear instability of a scheme will create a damped system, and even more importantly, the phase characteristics of such a scheme will be poor. This approach, therefore, should not be used in numerical models.

9.5 A Fully-Lagrangian Approach to Solving Atmospheric Dynamics[21]

9.5.1 Introduction

The schemes examined earlier in this chapter, with the exception of Section 9.2, are based on the Eulerian perspective which uses grid points fixed in space. An alternative is to adopt the Lagrangian framework, in which movement along air parcels is examined, and to construct an advection scheme based on this different perspective (Fig. 9.6).

Traditionally, the term "Lagrangian atmospheric model" has usually been used to refer to offline trajectory models (Stohl 1998). In these models, meteorological fields outputted from Eulerian atmospheric models (e.g., NWP, mesoscale, or climate models) have been used to advect computational particles (representing air parcels) to trace out their trajectories (Lin et al. 2011). Some widely used examples of such models include HYSPLIT (Draxler and Hess 1998), STILT (Lin et al. 2003), FLEXPART (Stohl et al. 2005), NAME (Ryall and Maryon 1998), and HYPACT (Uliasz et al. 1996).

Differing from erstwhile uses of the term, the Lagrangian atmospheric modeling approach discussed in this section solves the atmospheric dynamical equations of motion in the Lagrangian frame instead of relying on output from other Eulerian models. The Lagrangian perspective is potentially advantageous in several ways:

1. Realism: The Lagrangian approach is more physically realistic. At its essence, the atmosphere is Lagrangian, when we consider an air parcel and imagine how its constituent molecules are transported around. This means that the Lagrangian approach more closely approximates physical phenomena and has the potential, in theory, to better capture phenomena that exist in atmospheric flows (e.g., turbulent eddies, entrainment at the top of the mixed layer).
2. Stability: Lagrangian advection is numerically stable (Smolarkiewicz and Pudykiewicz 1992), permitting larger timesteps and reducing computational cost. Fundamentally, this stability can be traced to its mathematical origins, which is the fact that in the Lagrangian framework, conservation equations are linear when advective terms are absorbed within the Lagrangian derivative d/dt (Eq. 2.25).

[21] This section was written by Professor John Lin of the University of Utah.

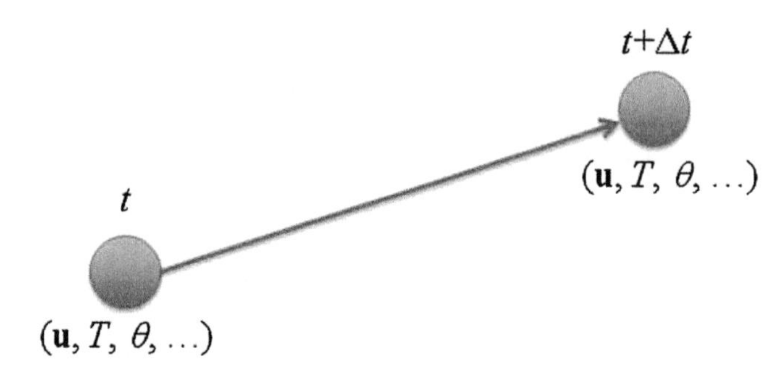

FIGURE 9.6

Advection of an air parcel in a fully-Lagrangian approach, in which the state variables carried by particles (representing air parcels of equal mass) are retained during a timestep Δt.

3. Minimal Numerical Diffusion: Lagrangian advection does not smear out scalar gradients like its Eulerian counterpart and preserves such gradients when they are found in the atmosphere (McKenna et al. 2002; Smolarkiewicz and Pudykiewicz 1992; Shin and Reich 2009).
4. Mass Conservation: Conservation of mass is the most important conservation property to be adhered to (or approximated) by an atmospheric model's dynamical core (Thuburn 2006). By treating each Lagrangian air parcel as one of equal mass, global mass conservation can be preserved trivially by retaining all parcels.
5. Availability of Trajectory Information: By simulating motions of air parcels, information regarding the trajectories of the parcels can be saved and retrieved (Haertel and Straub 2010). Trajectories have provided invaluable information for a variety of scientific purposes, from quantifying greenhouse gas emissions to tracing the fate of pollutants and volcanic ash (Lin et al. 2011).

The "semi-Lagrangian" approach to advection, introduced in Section 9.2, is related but somewhat different. In the semi-Lagrangian approach, interpolation is carried out between Eulerian grid points to determine the upstream value of a particular state variable; this is the value which will arrive at the grid point after a single timestep, due to advection. In contrast, the 'fully Lagrangian' approach considered here allows the Lagrangian parcels to retain their identity, and their movements during each timestep naturally provide the advection of atmospheric state variables (Fig. 9.6).

Despite its promise, the fully-Lagrangian advection scheme has seldom been adopted or discussed in the literature for modeling geophysical fluid dynamics from the mesoscale to the global scale. A few papers in recent years, however, have identified the potential for a fully-Lagrangian approach. For instance, Haertel and Randall (2002) and Haertel et al. (2004) have adopted the fully-Lagrangian approach for oceanic modeling. As one of the first atmospheric applications of the fully-Lagrangian approach at the mesoscale Alam and Lin (2008) introduced and tested the fully-Lagrangian approach in a two-dimensional setting using a sea breeze model that yields analytical solutions.

The lack of popularity of atmospheric models adopting the fully-Lagrangian approach for modeling atmospheric dynamics can likely be traced to two reasons: 1) irregularity of the particle mesh (Welander 1955; Staniforth and Cote 1991) and 2) concern about the large number of particles necessary and the associated computational cost. In Alam and Lin (2008), the irregularity of the Lagrangian particle mesh is removed by the adoption of an Eulerian grid at one step of the algorithm. They also showed that the fully-Lagrangian approach is actually comparable or even more computationally efficient than the Eulerian or semi-Lagrangian methods.

The following summarizes the methodology and results of Alam and Lin (2008).

9.5.2 Methodology

The Alam and Lin (2008) Lagrangian computational algorithm consists of the following:

1. Computational particles representing air parcels are initialized by distributing them according to air density. In this way, the particles represent air parcels of equal mass. Therefore, particle density would follow the air density profile, decaying exponentially with altitude. And then, during each timestep:
2. Advect each particle while allowing each particle to carry the values of state variables as they are advected (Fig. 9.6).
3. Update values of state variables while fixing the particles' locations. The state variables are updated by using an 11×20 staggered Eulerian grid system. As a simple initial approach, the values retained by particles within a grid cell are averaged to get the value at a grid point.

The methodology here was tested to satisfy a key physical criterion in all Lagrangian models - the "well-mixed criterion" (Thomson 1987), which states that particles which are initially distributed according to air density must remain so, a reflection of the Second Law of Thermodynamics.

In 2), an Eulerian grid system is introduced into the algorithm, so it might be argued that the method is not fully-Lagrangian, in a strict sense. However, the method preserves the essence of being fully-Lagrangian, because the advection portion in 1) retains the identity of individual particles and the values of state variables they advect (Fig. 9.6).

9.5.3 Sea Breeze System

The 2-D sea-breeze system adopted for comparison between Lagrangian simulations and analytical solutions is the one described by Defant's equations (Section 5.2.3.1). The non-linear advection terms are neglected in order to derive Defant's analytical solutions. Therefore, the analytical solutions are expected to more closely approximate the sea breeze system subject to smaller forcing, represented by the amplitude of the maximum mesoscale perturbation surface potential temperature M (a measure of the land-sea temperature contrast). $M = 1°C$ is adopted in the results shown here.

9.5.4 Results

A comparison between the Defant analytical solutions and the Lagrangian simulations for $M = 1°C$ and after $t = 24$ hours is shown in Fig. 9.7. Overall, the Lagrangian simulations show close correspondence with the analytical solutions. Some small differences can be seen toward the model top, set at 4 km. This is expected because of the differing boundary conditions. In the Defant solutions, the state variables are expected to decay to 0 as $z \rightarrow \infty$, while a rigid lid boundary condition is imposed in the Lagrangian model due to its finite vertical extent.

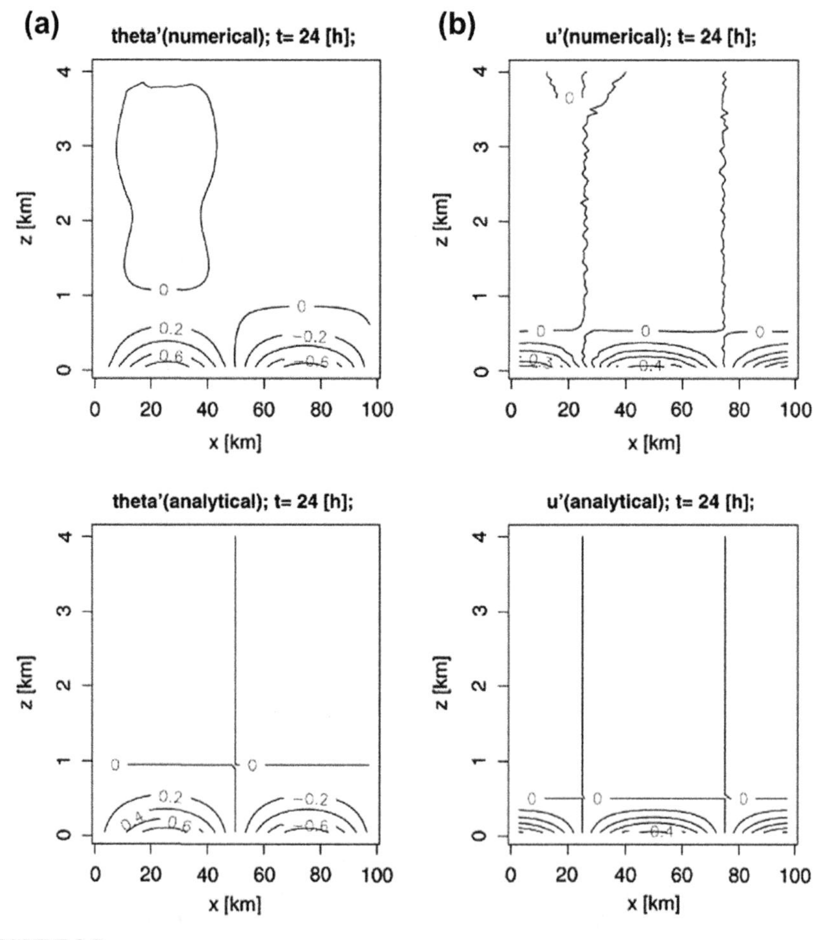

FIGURE 9.7

Comparisons between the fully-Lagrangian approach (top panels) and the Defant analytical solutions (bottom panels) for a) potential temperature θ' and b) horizontal velocity u'. M was set to 1°C. From Alam and Lin (2008).

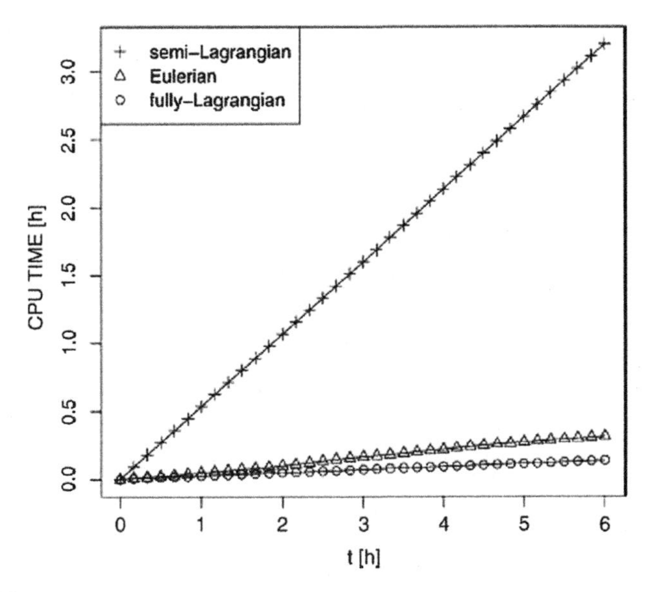

FIGURE 9.8

Comparisons of computational cost (in terms of CPU time) between the semi-Lagrangian, Eulerian, and fully-Lagrangian approaches to simulating the sea breeze system. From Alam and Lin (2008).

Alam and Lin (2008) examined the sensitivity of model results to the number of particles. As expected, the error decreased as the particle number increased. For the sea breeze system, the error leveled off at ~20 particles within each cell. It is worth noting that this number is expected to be system-dependent, with more particles necessary for more complicated flows.

The fully Lagrangian method was compared against the Eulerian advection method as well as the semi-Lagrangian method for both accuracy and computational cost. Semi-Lagrangian and fully Lagrangian methods were shown to have comparable accuracy when evaluated against the Defant solutions, while the Eulerian method exhibited larger errors. Finally, the CPU time for the three advection methods were tallied (Fig. 9.8). The results indicated that the fully-Lagrangian method is the most computationally efficient: running faster than the Eulerian method and significantly faster than the semi-Lagrangian method. The higher cost associated with the semi-Lagrangian method is due to the cubic spline interpolation.

9.5.5 Summary/Conclusions

Although the system examined is a simple 2-D sea breeze case, these results suggest that great potential exists for the fully-Lagrangian method to simulate atmospheric advection, as well as atmospheric dynamics in general.

Given the availability of computational power using modern high-performance computing clusters, the concern in the past regarding irregularity of the particle mesh (Welander 1955; Staniforth and Cote 1991) can be overcome by filling the entire model domain with Lagrangian particles. By using a large number of particles, irregularities in the particle mesh due to advection of the particles can be suppressed. Even as the particles move, the particle mesh remains regular, in the sense that particle density would follow the atmospheric density's exponential profile.

More recently, Shin and Reich (2009) extended the Lagrangian particle methods to simulate non-hydrostatic atmospheric flow regimes and demonstrated that conserved quantities such as potential temperature under adiabatic conditions can be simulated accurately. Haertel and Straub (2010) implemented a fully Lagrangian parameterization of moist convection and demonstrated its efficacy in both the single column mode and over an entire aquaplanet. This work shows that the fully Lagrangian approach can properly simulate moist convective processes, as well as generating realistic equatorial waves.

Ultimately, one can envision a fully-Lagrangian atmospheric modeling system at the mesoscale or even the global-scale, in which the entire atmosphere is simulated using the air parcels moving in the Lagrangian framework.

9.6 Finite Volume and Cut-Cell Solution Technique[22]

Drs. Robert Walko and Professor Roni Avissar are the developers of the Ocean Land Atmosphere Model (OLAM) and have developed a finite volume approach for the representation of a coordinate system. A description of this approach follows. We have retained their use of the vector form of the conservation equations since subscripts are used in the finite volume methodology, which if the tensor subscripts were also used, would make the equations harder to follow.

OLAM is built on the conservation laws for mass, momentum, and internal energy discussed in Chapter 2; e.g., see Eqs. 2.43 to 2.50. In differential form, the OLAM equations are written as follows: Conservation of momentum component in direction

$$\frac{\partial V_i}{\partial t} = -\nabla \cdot \left(v_i \vec{V}\right) - (\nabla p)_i - \left(2\rho\vec{\Omega} \times \vec{v}\right)_i + \rho g_i + F_i \tag{9.64}$$

Mass continuity:

$$\frac{\partial \rho}{\partial t} = -\nabla \cdot \vec{V} + M \tag{9.65}$$

Energy conservation:

$$\frac{\partial (\rho\Theta)}{\partial t} = -\nabla \cdot (\Theta\vec{V}) + H \tag{9.66}$$

Conservation of general scalar quantities:

$$\frac{\partial (\rho s)}{\partial t} = -\nabla \cdot (s\vec{V}) + Q \tag{9.67}$$

[22]This section was written by Dr. Robert Walko of the University of Miami, Coral Gables, FL.

Equation of state:

$$p = \left[\left(\rho_d R_d + \rho_v R_v\right)\theta\right]^{\frac{C_p}{C_v}} \left(\frac{1}{p_0}\right)^{\frac{R_d}{C_v}} \tag{9.68}$$

In these equations, $\vec{v}$ and $\vec{V} \equiv \rho\vec{v}$ are velocity and momentum vectors, $\vec{g}$ and $\vec{\Omega}$ are the Earth's gravity and angular velocity vectors, subscript i represents a vector component in the x_i direction, t is time, p is pressure, θ is potential temperature, Θ is ice-liquid potential temperature (defined below), C_p and C_v are the specific heats of dry air at constant pressure and constant volume, R_d and R_v are gas constants for dry air and water vapor, p_0 is a reference pressure equal to 10^5 Pa, and ρ is total density, given by the sum of densities of dry air, water vapor, and liquid plus ice condensate:

$$\rho = \rho_d + \rho_v + \rho_c = \rho\left(s_d + s_v + s_c\right). \tag{9.69}$$

Scalar variable s represents the specific density or concentration (relative to ρ) of individual scalar quantities, such as various classes of ice and liquid hydrometeors, aerosols, and water vapor. F_i, M, H, and Q are source terms for momentum, mass, internal energy, and scalar fields that represent radiative transfer, surface fluxes, and/or microphysical processes. In these continuous equations, turbulent transport is not distinguished from resolved advective transport; both are represented by point correlation terms, such as $v_i\vec{V}$ and $\Theta\vec{V}$.

The ice-liquid potential temperature, described in Tripoli and Cotton (1981) and Walko et al. (2000a), is used in OLAM and RAMS as the prognostic internal energy variable, and has the highly desirable property of being constant in a parcel for processes of transport and internal phase change. It is empirically related to potential temperature by:

$$\theta = \Theta\left[1 + \frac{q_{lat}}{C_p \max(T, 253)}\right], \tag{9.70}$$

where q_{lat} is the latent heat required to vaporize any liquid and ice water present and T is air temperature. Application of 9.69 and 9.70 to 9.68 gives the following equation for pressure,

$$p = \left[(\rho\Theta)\beta\right]^{\frac{C_p}{C_v}} \left[\frac{1}{p_0}\right]^{\frac{R}{C_v}}, \tag{9.71}$$

where

$$\beta = \left[s_d R_d + s_v R_v\right]\frac{\theta}{\Theta}, \tag{9.72}$$

The coefficient defined in 9.72 depends primarily on the specific densities of water in all three phases, although a weak dependence on air temperature arises from 9.70.

9.6.1 Finite Volume Discretization

Following Adcroft et al. (1997), 9.64-9.67 are discretized by integrating over control volumes Ψ bounded by surface areas σ that represent individual grid cells. Control volume integrals of divergence quantities are transformed to surface integrals by application of the Gauss divergence theorem,

$$\int_{\Psi} \nabla \cdot \vec{\Phi} d\Psi = \oint \vec{\Phi} \cdot d\vec{\sigma}, \tag{9.73}$$

where $d\vec{\sigma} = \hat{n} d\sigma$, $\hat{n}$ is the outward normal unit vector to the surface element, and $d\vec{\sigma}$ is a differential element of surface area. The resulting integral conservation laws for momentum, mass, internal energy, and conservative scalar fields on model grid cells are:

$$\frac{\partial}{\partial t} \int V_i \, d\Psi = -\oint \left(v_i \vec{V}\right) \cdot d\vec{\sigma} - \int \frac{\partial p}{\partial x_i} d\Psi - \int \left(2\rho \vec{\Omega} \times \vec{v}\right)_i d\Psi$$
$$+ \int \rho g_i d\Psi + \int F_i \, d\Psi \tag{9.74}$$

$$\frac{\partial}{\partial t} \int \rho \, d\Psi = -\oint \vec{V} \cdot d\vec{\sigma} + \int M \, d\Psi \tag{9.75}$$

$$\frac{\partial}{\partial t} \int \rho\Theta \, d\Psi = -\oint \left(\Theta \vec{V}\right) \cdot d\vec{\sigma} + \int H \, d\Psi \tag{9.76}$$

$$\frac{\partial}{\partial t} \int \rho s \, d\Psi = -\oint \left(s \vec{V}\right) \cdot d\vec{\sigma} + \int Q \, d\Psi. \tag{9.77}$$

Following the C-grid stagger, the normal momentum component is defined and prognosed on each face j, $V_j = \vec{V}_j \cdot \hat{n}_j$, of a control volume for scalar variables. Volume integrals in Eqs. 9.74-9.77 are represented as products of the mean value in the grid cell (for which we use an overbar symbol) and the volume of the cell:

$$\int \phi \, d\Psi \equiv \bar{\phi} \Psi. \tag{9.78}$$

Surface integrals of fluxes in Eqs. 9.74-9.77 are discretized as a summation over individual faces of the control volumes, and over each face area, σ_j, the flux is partitioned into mean and subgrid-scale (SGS) contributions:

$$\oint \left(\phi \vec{V}\right) \cdot d\vec{\sigma} = \sum_j \left[\int \left(\phi_j \vec{V}_j\right) \cdot d\vec{\sigma}_j\right] \equiv \sum_j \left[\left(\bar{\phi}_j \bar{V}_j + SGS\left\{\phi_j, V_j\right\}\right) \sigma_j\right]. \tag{9.79}$$

With application of Eqs. 9.78-9.79, Eqs. 9.74-9.77 may be written

$$\frac{\partial \bar{V}_i}{\partial t} \Psi = -\sum_j \left[\left(\bar{v}_{ij} \bar{V}_j + SGS\left\{v_{ij}, V_j\right\}\right) \sigma_j\right] - \frac{\Delta \bar{p}}{\Delta x_i}$$
$$- \left(2\bar{\rho} \vec{\Omega} \times \vec{\bar{v}}\right)_i \Psi + \bar{\rho} g_i \Psi + \bar{F}_i \Psi \tag{9.80}$$

$$\frac{\partial \bar{\rho}}{\partial t} \Psi = -\sum_j \left[\bar{V}_j \sigma_j\right] \tag{9.81}$$

$$\frac{\partial \overline{\rho\Theta}}{\partial t}\Psi = -\sum_j \left[\left(\bar{\Theta}_j \bar{V}_j + SGS\left\{ \Theta_j, V_j \right\} \right) \sigma_j \right] + \bar{H}\Psi \tag{9.82}$$

$$\frac{\partial \overline{\rho s}}{\partial t}\Psi = -\sum_j \left[\left(\bar{s}_j \bar{V}_j + SGS\left\{ s_j, V_j \right\} \right) \sigma_j \right] + \bar{Q}\Psi \tag{9.83}$$

Advective fluxes $\bar{v}_{ij}\bar{V}_j$, $\bar{\Theta}_j\bar{V}_j$, and $\bar{s}_j\bar{V}_j$ are evaluated at each face j of a scalar control volume, which is where $\bar{V}_j$ is defined. (The one exception is the control volume for vertical momentum, to whose top and bottom faces $\bar{V}_j$ must be vertically interpolated.) Advected scalar quantities $\bar{v}_i$, $\bar{\Theta}$, and $\bar{s}$ are defined as means over their respective control volumes and must be interpolated to face j according to the advection algorithm used; an added subscript denotes their interpolated values. Note that the direction of $\bar{v}_i$ is in general different from $\bar{V}_j$ at a control volume surface.

As pointed out by Adcroft et al. (1997), equations in this form are exact representations of the conservation laws, even though they apply to discrete volumes (here, we have defined the SGS terms to exactly compensate for truncation error). Adcroft et al. also point out that although in practice, numerical evaluations of averaged and SGS terms are not exact, global conservation is still followed provided that each surface flux removes as much substance from one cell as it adds to the adjacent cell. This principle is followed explicitly in OLAM. For each flux computation at a control volume face, the donor cell and receptor cell are identified, and the appropriate amount of substance is subtracted from one cell and added to the other.

All areas σ and volumes appearing in Eqs. 9.80-9.83 are pre-computed and stored in arrays in OLAM. Wherever terrain partially or completely obstructs a grid cell face, σ_j is reduced accordingly, leading to proportionate reduction in any flux across that face. Such reduction is considered to be uniform over the face as if regulated by a Venetian blind, not (as suggested by Fig. 9.9) only over the submerged portion of the face. Similarly, volume reduction of a partially submerged cell is considered to be uniform, and any mean scalar quantity in the cell remains centered within the full cell, not within the unsubmerged portion. This rule maintains the physical location of each cell and face center and preserves the separability of horizontal and vertical gradient computations.

9.7 Distinction Between Grid Increment and Resolution

As reported in Pielke (2001b) the issue of the definition of the term *resolution*, as we use it in the physical sciences, is more than a semantic issue, as also suggested by Durran (2000). This subject was originally discussed in Pielke (1991) and Laprise (1992b), and has been added to by the articles of Walters (2000), and Grasso (2000). Walters, for example, concludes that

"Due only to the geometric relationship between the numerical grid and the true solution, as many as 10 grid points may be required to assure a reasonable (e.g., greater than 95%) representation of the true solution's amplitude and its first

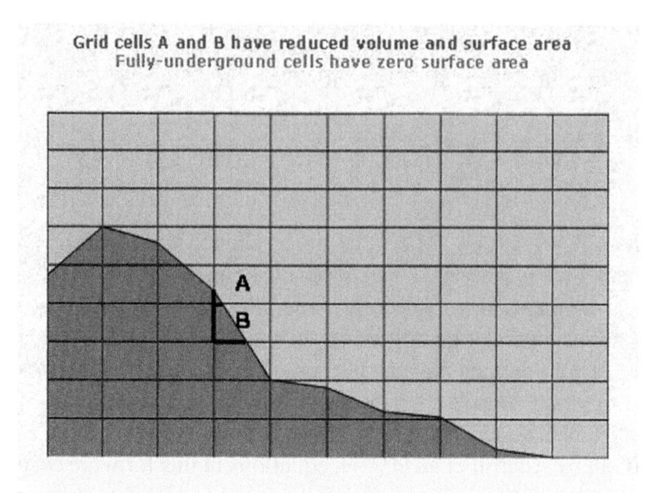

Grid cells A and B have reduced volume and surface area
Fully-underground cells have zero surface area
A
B

FIGURE 9.9

Schematic of cut-cell representation of topography. Bold lines along the faces of cells A and B represent the fractional closures of those face areas, while the dark gray area represents the fractional reduction in cell volumes. (The color version of this figure is presented in the plate section in the back of the book and the online web version.) Courtesy of Robert Walko.

horizontal derivative...the shortest wavelength that can be effectively resolved by a grid point numerical model may be considerably longer than $4\delta x$."

Users of numerical model output, for example, such as ecologists, social scientists, and hydrologists, generally interpret resolution as the scale at which a physical phenomenon is actually resolved. However, when we actually mean grid increment but use the term resolution, we are misleading the users to assume the data (observed or modeled) have a finer spatial scale than is actually true. Figure 9.2 illustrates this relationship between grid increment and resolution.

The definition of resolution, as applied to physics in Webster's New World Dictionary of American English (1988) is "*the capability of an optical system, or other imaging system, of making clear and distinguishable the separate parts or components of an object.*" In the context of models, the 'image' is the physical feature of interest, such as a cumulus convective cloud system.

As Durran (2000) correctly states, the number of grid intervals (i.e., sampling points) needed to adequately resolve a feature has a subjective component. Ten grid intervals in each spatial direction to resolve one wavelength provides better resolution than four grid intervals. One grid interval obviously cannot resolve a feature with that wavelength, however, so that it is clear that the term resolution is not the same as grid separation or grid spacing. The use of resolution and grid increment interchangeably is clearly inappropriate and misleading to the users of model output data.

To avoid confusion, the terms *grid increment, grid separation, and/or grid interval* should always be used (instead of *resolution*) when we refer to δx, δy, and δz values.

9.8 Summary

In this chapter the concepts of linear and nonlinear stability have been introduced along with examples of specific computational solution techniques. Among the conclusions of relevance to mesoscale modeling that are implied from the chapter are the following.

1. When advection is considered an important component of the mesoscale circulation, the advection terms should be approximated with formulations that provide accurate predictions of phase *and* amplitude for wavelengths of $4\Delta x$ and longer. The interpolated schemes are examples of solution techniques that have this attribute.
2. When vertical turbulent diffusion is considered an important component of a mesoscale circulation, it is desirable to use a scheme such as the implicit formation that provides accurate solutions yet is economical to apply.
3. When the pressure gradient force is an important component of a mesoscale circulation, a forward-in-time, centered-in-space representation is an accurate representation. This result is implied from the linear tank model solutions. The gradient terms that appear in the conservation-of-mass equation can similarly be accurately represented by a centered-in-space approximation. To obtain these accurate solutions, however, an implicit formulation between the pressure gradient and conservation-of-mass relation must be used.
4. In a nonlinear representation, aliasing causes the fictitious accumulation of energy on wavelengths shorter than $4\Delta x$. To eliminate this problem, it is necessary to apply horizontal diffusion either through an explicit representation, such as a smoother, or implicitly as part of the computation scheme.
5. The approximation of advection, the pressure gradient force, and subgrid-scale diffusion can only be reasonably well represented for spatial scales of at least four grid increments in each spatial direction. The term *model resolution*, therefore, should be used only for spatial scales that are four grid increments or larger (Pielke 1991; Grasso 2000). This conclusion also applies to spectral models (Laprise 1992b).
6. The separation of a nonlinear model into its linear and nonlinear components may provide improved solution accuracy. The linear portion could be computed analytically. Only the remaining nonlinear portion would need to be solved numerically.
7. Parameterizations should appropriately be inserted in a model so that their spatial scale is at least four grid increments in each direction. Otherwise, the approximate forms of the advection, subgrid-scale mixing, and the pressure gradient force will produce significant errors in the computational solution.
8. Fully Lagrangian solution and finite volume techniques offer promising new directions to improve the accuracy of approximating the differential conservation relationships for motion, heat, mass, moisture, and other constituents.

Problems for Chapter 9

For Problems 1-4, calculate the analytic solution using the solution techniques for difference equations given in Section 9.1. Since the exact solution of the advection equation ($\partial\phi/\partial t = U\phi/\partial\phi/\partial x$) is $\lambda_{act} = 1$ and $c_\phi = U$, tabulate the ratios of the analytic to exact solutions for Courant numbers of 0.001, 0.01, 0.1, 0.2, 0.3, 0.4, 0.5, 0.6, 0.7, 0.8, 0.9, 1.0, 1.5, and for wavelengths of $2\Delta x, 4\Delta x, 10\Delta x$, and $20\Delta x$ for the following finite difference equations.

1. $$\frac{\phi_i^{\tau+1}-\phi_i^\tau}{\Delta t} = -\frac{U}{6\Delta x}\left[\phi_{i+2}^\tau - 2\phi_{i+1}^\tau + 9\phi_i^\tau - 10\phi_{i-1}^\tau + 2\phi_{i-2}^\tau\right] \quad U>0$$

Reference: Kawamura, T., H. Takami, and K. Kuwahara, 1986: Computation of high Reynolds number around a circular cylinder with surface roughness. *Fluid Dyn. Res.*, **1**, 145-162.

2. $$\frac{\phi_i^{\tau+1}-\phi_i^\tau}{\Delta t} = -\frac{U}{6\Delta x}\left[2\phi_{i+1}^\tau + 3\phi_i^\tau - 6\phi_{i-1}^\tau + \phi_{i-2}^\tau\right] \quad U>0$$

Reference: Agarwal, R.K., 1981: A third-order-accurate upwind scheme for Navier-Stokes solutions at high Reynolds numbers. *AIAA 19th Aerospace Sciences Meeting*, January 12-15, 1981, St. Louis, Missouri, Paper No. AIAA-81-0112, 14 pp.

3. $$\frac{\phi_i^{\tau+1}-\phi_i^\tau}{\Delta t} = -\frac{U}{\Delta x}\left[b_1(\phi_{i+1}^\tau - \phi_{i-1}^\tau) + b_2(\phi_{i+2}^\tau - \phi_{i-2}^\tau)\right] \quad U>0$$

$b_1 = 0.785398 \quad b_2 = -0.155076$

Reference: Derickson, R.G., 1992: Finite difference methods in geophysical flow simulations. Colorado State University, Ph.D. Dissertation, Department of Civil Engineering, 320 pp.

4. $$\frac{\phi_i^{\tau+1}-\phi_i^{\tau-1}}{\Delta t} = -\frac{U}{\Delta x}\left[\frac{4}{3}\left(\phi_{j+1}^\tau - \phi_{j-1}^\tau\right) - \frac{1}{6}\left(\phi_{j+2}^\tau - \phi_{j-2}^\tau\right)\right]$$ which is a 4th order in space, leapfrog differencing scheme.

5. Program the difference equation 9.7 using cyclic boundary conditions. Compare your amplifying solution with that of Eq. 9.10. Why do the results differ after only a few time steps?

6. Program the difference Eq. 9.13 for $U > 0$. Using the cyclic boundary conditions, integrate the model forward for 100 time steps. Evaluate the change of amplitude with time and the phase speed of your results for the values of Courant number and wavelength given in Table 9.1. Compare the numerically computed values with the analytic results. They should be identical.

7. Perform Problem #6 but with Eq. 9.16. Note that leap-frog has two solutions; only one of them is physical.

8. Program Eq. 9.25 with cyclic boundary conditions and constant K, and calculate the change of amplitude after integrating for 100 time steps with $\beta_\tau = 1$ and $\beta_\tau = 0.5$. Refer to Appendix A with respect to how to solve an implicit equation.

Use the same values of γ and wavelength as in Table 9.2, and compare your numerical results with the solutions tabulated in Table 9.2.

9. Program Eq. 9.32 with cyclic boundary conditions and calculate the change of amplitude and phase after 100 time steps. Use the same values of $\sqrt{gh_0}\Delta t/\Delta x$ and wavelength as shown in Table 9.3. Compare your numerical results with the analytic results.

10. From Problem #4 in Chapter 6, write the approximate form for each term (for flat terrain). Describe the phase and amplitude change per time step for the linear form of each term.

11. Obtain the reference Porte-Agel et al. (2000; their Appendix) and confirm the error that is introduced when a finite difference algorithm is used to compute the vertical gradient of the horizontal velocity. Then assess the error when a non-neutral surface layer is presented (using for the wind profile in the surface layer such as given by 8.1).

Chapter 9 Additional Readings

Coiffier, J., 2012: Fundamentals of Numerical Weather Prediction. Cambridge University Press. ISBN 978-1-107-00103-9.

Derickson, R.G., 1992: Finite difference methods in geophysical flow simulations. Colorado State University, Ph.D. Dissertation, Department of Civil Engineering, 320 pp.

This Ph.D. Dissertation provides a thorough detailed analysis of the minimum resolution possible as a function of the number of grid points. Derickson used the computational error characteristics of the solution techniques to produce more accurate solutions.

Foufoula-Georgiou, E., and P. Kumar, Eds., 1994: *Wavelets in Geophysics*. Academic Press, New York, 373 pp.

This book discusses a procedure called "wavelet analysis" which uses a localized transform in space and time with which to analyze geophysical signals including atmospheric features.

Haltiner, G.J., and R.T. Williams, 1980: *Numerical Prediction and Dynamic Meteorology*. 2nd Edition, John Wiley & Sons, New York, 477 pp.

Although this text is oriented toward synoptic-scale numerical weather prediction, the discussion in their Chapters 5 and 6 on numerical solution technique complements the material discussed in this chapter.

Kalnay, E., 2002: *Atmospheric Modeling, Data Assimilation and Predictability*. Cambridge University Press, 364 pp.

This is an excellent text on atmospheric modeling including how to apply this tool to ensemble prediction.

Krishnamurti, T.N., and L. Bounoua, 1995: *An Introduction to Numerical Weather Prediction Techniques*. CRC Press, Washington DC, 304 pp.

This book provides an introduction to finite difference techniques as well as information on parameterizations in larger-scale models.

Mesinger, F., 1997: Dynamics of limited-area models: Formulation and numerical methods. *Meteor. Atmos. Phys.*, **63**, 3-14.

This paper provides a valuable perspective on regional and mesoscale modeling. As one of the pioneers in this field, and an innovator in the use of these models in operational weather forecasting, this is a must read.

Mesinger, F., and A. Arakawa, 1976: Numerical methods used in atmospheric models. *GARP Publ. Ser.* **17**, 1-64.

This publication provides a useful discussion of finite difference techniques. Volume 2 of this GARP series (published in 1979) presents summaries of the spectral, pseudospectral, and finite element techniques.

Pielke, R.A., and R.W. Arritt, 1984: A proposal to standardize models. *Bull. Amer. Meteor. Soc.*, **65**, 1082.

A recommendation to develop plug-compatible components of models is proposed.

Pielke, R.A., L.R. Bernardet, P.J. Fitzpatrick, S.C. Gillies, R.F. Hertenstein, A.S. Jones, X. Lin, J.E. Nachamkin, U.S. Nair, J.M. Papineau, G.S. Poulos, M.H. Savoie, and P.L. Vidale, 1995b: Standardized test to evaluate numerical weather prediction algorithms. *Bull. Amer. Meteor. Soc.*, **76**, 46-48.

Students who completed a mesoscale modeling class using the first edition of this book summarized recommended procedures to evaluate modeling algorithms.

Richtmyer, R.D., and K.W. Morton, 1967: *Difference Methods for Initial-Value Problems.* Interscience Publishers, New York, 405 pp.

Although somewhat dated, this text provides a valuable fundamental discussion of numerical solution techniques.

Sun, W.-Y., 1993: Numerical experiments for advection equation. *J. Comput. Phys.*, **108**, 264-271.

This article provides a concise, effective summary of two additional computational representations of advection.

Warner, T., 2011: *Numerical Weather and Climate Prediction.* Cambridge University Press, 548 pp.

This text offers an effective summary of atmospheric modeling including computational solution techniques.

CHAPTER

Boundary and Initial Conditions

10

CHAPTER OUTLINE HEAD

10.1 Introduction

Chapter 9 provided an introduction to methodologies for obtaining solutions to the conservation relations were introduced. In those discussions, it was shown, for example, that certain approximate representations of the differential equations produce more accurate solutions than others. Linear forms of the conservation equations (e.g., one-dimensional advection, one-dimensional diffusion, Coriolis terms) were examined independently of one another with the assumption that a *necessary requirement*

Mesoscale Meteorological Modeling, Third Edition, Volume 98. http://dx.doi.org/10.1016/B978-0-12-385237-3.00010-4

for satisfactory solutions to the nonlinear partial differential equations is an approximation that is accurate for the linearized version. Moreover, these components of the conservation equations were studied independently from one another, even in the linear version, with the presumption that each portion of the relations must separately be accurate.

Once optimal approximate forms of the equations are selected, however, it is still necessary to define the domain and grid structure over which the equations will be evaluated. In addition, boundary and initial conditions are required to provide unique solutions for any set of differential equations. The modeler has to also ask whether the boundary is independent of what occurs in the atmosphere, or is interactive with the atmosphere. When the latter condition occurs, the boundary is more appropriately called a *flux interface*.

10.2 Grid and Domain Structure

The selection of the domain size and grid increments in a mesoscale model are dictated by the following constraints:

1. What is the dimensionality of the forcing?
2. What are the spatial scales of the physical response to this forcing?
3. What are the available computer resources?

Mesoscale models can have either two or three independent spatial coordinates. In addition, models have been developed where the dependent variables are evaluated either at specific grid points or averages of the resolvable dependent variables are obtained over a layer. With the second methodology, the averaging operation can be defined as

$$\langle \phi \rangle = \int_z^{z+\delta h} \bar{\phi}\, d\hat{z}/\delta h,$$

where $\bar{\phi}$ is defined by Eq. 4.6 and δh is the depth of the layer average.

The following are examples of mesoscale models and their spatial representation.

1. *x-y layered representation* (e.g., Lavoie 1972, 1974, Lee 1973). In these formulations, the dependent variables are averaged over discrete layers, such as the planetary boundary layer, so that the explicit vertical dependency in the conservation relation is removed. Such a layered representation is also often used in oceanographic models (e.g., O'Brien and Hurlburt 1972).
2. *x-σ representation*[1] (e.g., Estoque 1961, 1962). This formulation has been used to provide horizontal and vertical resolution of mesoscale structure, but without the added cost of a second horizontal coordinate. The x-σ coordinate form is only appropriate for those mesoscale features that are predominantly forced by

[1] σ is used here to indicate the generalized vertical coordinate discussed in Chapter 6. In many model applications, the rectangular vertical coordinate z is used.

two-dimensional features (e.g., a uniform mountain ridge) and for the theoretical analysis of the conservation relations. In addition, an axisymmetric formulation, where radius r replaces x as the horizontal coordinate, has also been used to better simulate circular atmosphere features. Examples of such axisymmetric models include hurricane models (e.g., Rosenthal 1970), circular lake and island circulations (e.g., Neumann and Mahrer 1974, 1975), and cumulus clouds (Murray 1970).

3. *x-y-σ representation* (e.g., McPherson 1970 and Pielke 1974a). This is the most general form of spatial coordinates used and should provide the best representation of actual mesoscale features.

The spatial scales of the forcings and of the resultant perturbation fields determine the necessary domain size of the model as well as its grid spacing. To represent mesoscale systems properly, it is required that:

1. the meteorologically significant variations in the dependent variables caused by the mesoscale forcing be contained within the model, and
2. the averaging volume used to define the model grid spacings must be sufficiently small so that the mesoscale forcings and responses are accurately represented.

Figure 10.1 schematically illustrates one problem that can arise if a domain is inappropriately selected. In this example, the actual drainage flow on a clear night with synoptic winds is expected to originate along both slopes of a mountain ridge as discussed in more detail in Section 13.2.3. To save money, however, a modeler elects

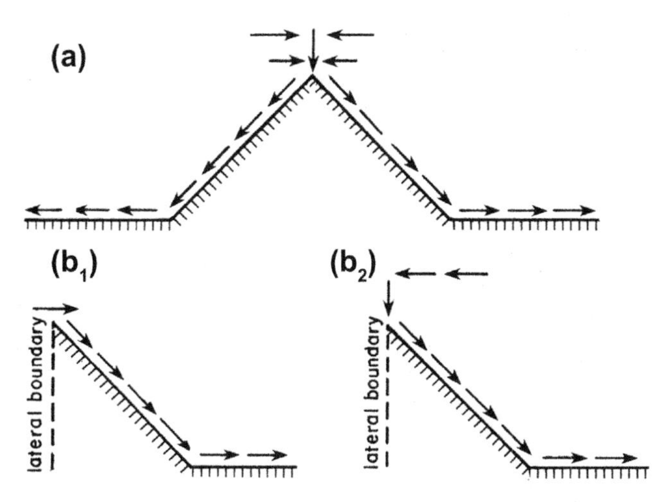

FIGURE 10.1

(a) Schematic of a nocturnal drainage flow in the absence of synoptic winds where the replacement air at ridge crest comes from both aloft and from the same level. (b_1) A simulation with an open lateral boundary at ridge top, and (b_2) a simulation with a closed lateral boundary at ridge top; $\partial\theta_0/\partial z > 0$ is assumed.

to truncate the domain at the ridge crest so that only one slope is represented. Because of this constraint, however, downslope air to the east of the ridge line in a model with $\partial\theta_0/\partial z > 0$ will be predominantly replaced by air from the lefthand lateral boundary (if it is an open boundary[2]) or from aloft (if it is a closed boundary). In other words, *the lateral boundary will determine the solutions*. In the real atmosphere, by contrast, downslope winds can develop on both eastern and western slopes so that the origin of the replacement air needed to preserved mass will depend only on such physical factors as the magnitude of the overlying thermodynamic stability. To represent this fundamental physical interaction correctly, it is therefore necessary to include the complete mesoscale variations within the model.

10.2.1 Horizontal Grid

10.2.1.1 Grid Size

The grid size used in a numerical model depends on the anticipated, or if available, observed, spatial sizes of the mesoscale feature of interest. If surface topography is considered to be the dominant forcing, the ragged landscape of Grand Teton National Park will obviously require a smaller grid interval than the undulating Flint Hill region of central Kansas, for example.

The representation of surface topography as a function of wavelength can be used to determine the characteristic scales of the terrain. Figure 10.2 illustrates the contribution of topographic features of different horizontal scales to the total variations in ground surface elevation for a northwest-southeast cross section of the Blue Ridge Mountains of central Virginia. A one-dimensional Fourier transformation, as shown by Panofsky and Brier (1968), given by

$$z_G(x_j) = \bar{z}_G + \sum_{n=1}^{I_x/2} \left[a_n \sin\left(2\pi nj\Delta x/D_x\right) + b_n \cos\left(2\pi nj\Delta x/D_x\right)\right]$$

$$(j = 1, 2, \ldots, I_x),$$

was used where

$$a_n = \frac{2}{I_x}\sum_{n=1}^{(I_x/2)-1} z'_G(x_j)\sin(2\pi nj\Delta x/D_x); \quad a_{I_x/2} = 0$$

$$b_n = \frac{2}{I_x}\sum_{n=1}^{(I_x/2)-1} z'_G(x_j)\cos(2\pi nj\Delta x/D_x); \quad b_{I_x/2} = -z'_G(x_j)/I_x.$$

In this expression, I_x is an even integer number of grid points of separation Δx used to represent the terrain height within the interval D_x. The variable n is referred to as the number of the harmonic. The quantity $\left(a_n^2 + b_n^2\right)^{1/2}$ represents the contribution of each wavelength of size $2\Delta x, 3\Delta x, 4\Delta x, \ldots,$ up to $I_x\Delta x/2$ to the function $z'_G(x_j)$. The domain-averaged terrain height is given by $\bar{z}_G$ with z'_G the variations from this value.

[2]Open and closed boundaries are defined in Section 10.4.1.1.

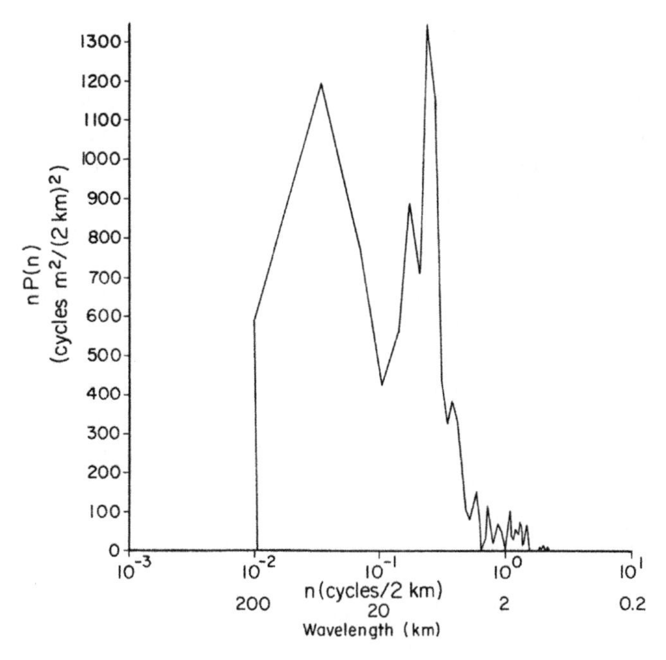

FIGURE 10.2

The variance of topography plotted as a function of horizontal wavelength for a cross section across a portion of the Blue Ridge Mountains in central Virginia (from Pielke and Kennedy 1980).

Ideally one would prefer to use a two-dimensional Fourier transformation, but such programs are not routinely available for large amounts of data. The two-dimensional terrain data from which the information used to compute Fig. 10.2 was derived, had 61 m intervals over a 200 by 200 km area. Using this number of grid points (about 10 million) in a two-dimensional transform is expensive to process. If only one-dimensional transforms are calculated, it is necessary to perform a series of cross sections through the regions of most rapidly varying terrain.

The cross section shown in Fig. 10.2 demonstrates that most of the terrain features vary significantly over scales substantially larger than 2 km. The predominant horizontal scales for this cross section are at 40 and 10 km with 95% of the variance[3] of topography having horizontal wavelengths greater than 6 km. Therefore, for this example, a horizontal grid increment of 1.5 km or less is a necessary condition to resolve 95% of the terrain irregularities with a resolution of $4\Delta x$ or larger. Pielke and Kennedy (1980), Young and Pielke (1983), Young et al. (1984), Steyn and Ayotte (1985),

[3]The integral over a given set of wavelengths in Fig. 10.2 is equal to the variance of the topography over this interval. Integrating over all wavelengths yields the total variance of the terrain, assuming that significant terrain variations do not occur on scales less than twice the minimum resolution in the topographic data.

and Salvador et al. (1999) have discussed the spatial analysis of terrain scales in more detail. McQueen et al. (1995) investigated the influence of grid increment size with respect to the resolution of terrain in an operational mesoscale model. Gollvik (1999) explored the ability of different horizontal grid increments (22 km, 11 km, and 5.5 km) and resolution of topography in model simulations of precipitation in a regional model. Salvador et al. (1999) used two-dimensional spectral analysis to determine the needed spatial resolution of mesoscale model simulations for a portion of the east coast of Spain.

Salmon et al. (1981) have discussed the implications of high wavenumber terrain features in a model and concluded that such features cause noisy solutions that tend to overemphasize the real impact of these small-scale terrain variations. Unfortunately, since the horizontal gradient of a pressure perturbation is proportional to horizontal wavenumber (e.g., see following Eq. 5.43 where $\partial p'/\partial x = ik_x\tilde{p}$ after the Fourier transform) as well as the spatial scale of terrain variations (e.g., see Eq. 6.57 where the pressure gradient force contains a term with $\partial z_G/\partial x$); the relative contribution of short wavelength terrain features to velocity accelerations would be expected to be larger than implied from a decomposition of terrain variations alone (Walmsley et al. 1982).

Observational studies such as Lenschow et al. (1979) and Mahrt and Heald (1981) show that terrain slope and small-scale three-dimensional features exert a substantial influence on surface temperature distribution and boundary layer structure over even mildly irregular terrain. From observations of drainage wind fluctuations in the Geysers area of California, Porch (1982) found that a correspondence may exist between prominent spectral peaks of wind velocity and characteristic variations in complex terrain as represented by a two-dimensional Fourier transform of the terrain relief. The approach of analyzing surface inhomogeneities (whether they are terrain features, land-water contrasts, or whatever) is therefore, a necessary (although not a sufficient) tool in establishing the horizontal grid size required in a mesoscale model.

If the conservation relations were linear, the spatial scale of the forcing would equal the spatial scale of the resultant atmospheric circulation, as illustrated for the sea and land breeze in Section 5.2.3.1, and the use of Fourier transforms of terrain, by itself, could yield the minimum spatial grid resolution required. As shown in Fig. 10.3, however, for two separate nonlinear sea-breeze simulations (with maximum surface temperature amplitudes $T_{G_{\max}}$ of 1°C and 10°C, respectively) when the terrain forcing becomes sufficiently strong (i.e., for $T_{G_{\max}} = 10$°C), the nonlinearity of the conservation relations act to decrease the horizontal spatial scale of the circulation L_x from that of the forcing. In practice, the only way to assure that the correct spatial scales are simulated in a nonlinear model is to perform integrations with progressively finer resolution. When the results do not significantly change for a given scale of forcing with further reduction of the grid mesh, the model has achieved sufficient spatial resolution.

10.2.1.2 Grid Mesh

In setting up a horizontal grid, grid increments can be kept constant or allowed to stretch. The advantages of a constant grid include the relative ease of coding such a framework onto the computer, as well as the comparative simplicity of inputting

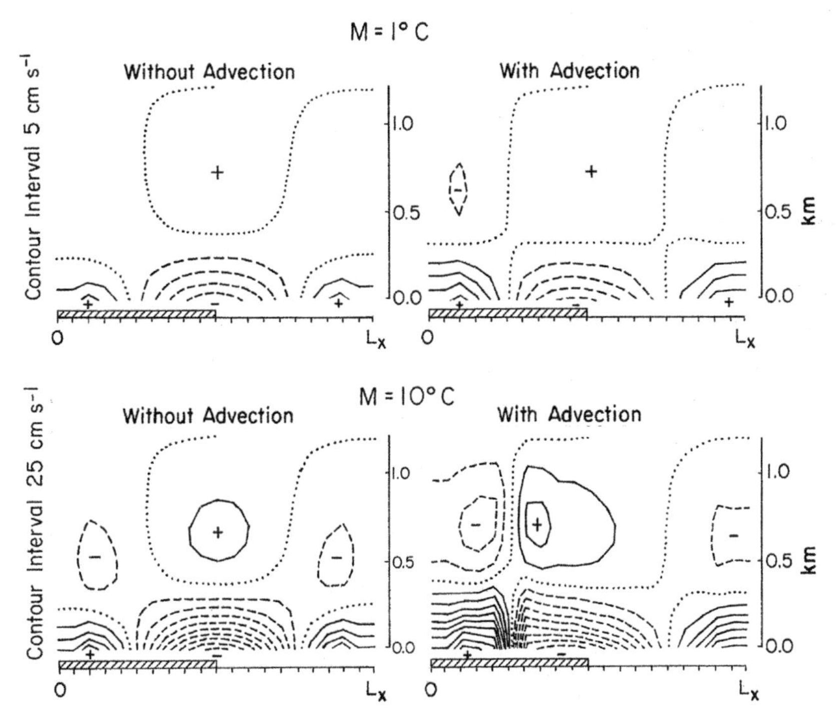

FIGURE 10.3

The horizontal velocity field 6 h after sunrise (at the time of maximum heating) calculated with the nonlinear analog to Defant's sea-breeze model introduced in Section 5.2.3.1. The large-scale vertical gradient of potential temperature was 1 K km^{-1}, the horizontal scale of heating was $L_x = 100$ km with a maximum surface temperature perturbation M of (a) 1°C, and (b) 10°C. Defant's analytic results discussed in Section 5.2.3.1 were used to initialize the model runs (from Martin 1981).

geographic features into the model. Disadvantages occur using an economically feasible number of grid points, however, including the close proximity of the sides of the model to the region of interest as well as the difficulty of properly incorporating large- and small-scale features within the same model domain.

Two different horizontal grid representations have been developed that are designed to reduce the problems associated with the constant grid. These are the *stretched grid* and the technique of *grid meshing*.

With a stretched grid it is possible to remove the boundaries of the model as far from the area of interest as one would like, so that a simulation of a much larger domain than with the constant grid can be made with the same number of grid points. In practice, stretched grids are either simply specified by assigning values to the grid locations (e.g., $x_1 = 0$; $x_2 = 50$ km; $x_3 = x_2 + 40$ km; $x_4 = x_3 + 30$ km;

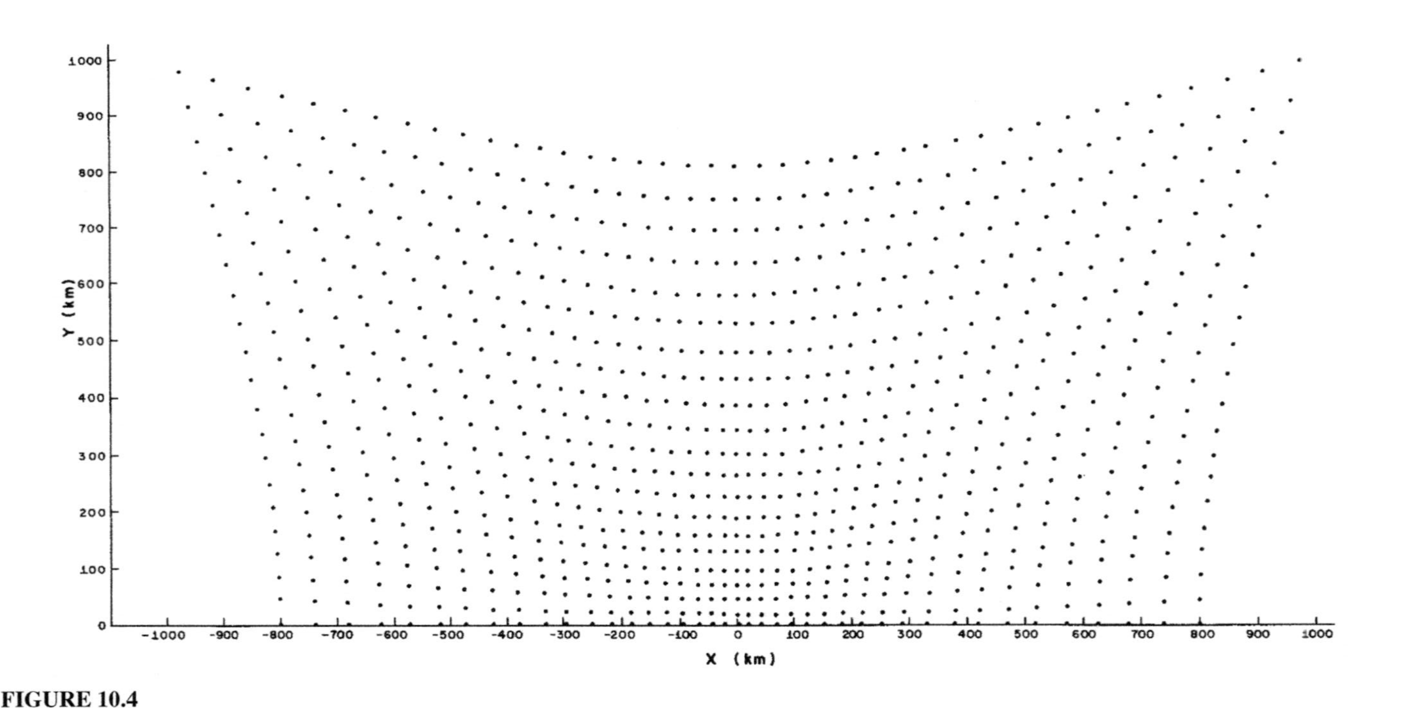

FIGURE 10.4

The northern half of a two-dimensional variable horizontal stretched grid with a minimum grid separation of 20 km (from Anthes 1970).

$x_5 = x_4 + 20$ km; ...; $x_{I_x-1} = x_{I_x-2} + 40$ km; $x_{I_x} = x_{I_x-1} + 50$ km) or the independent spatial coordinates are transformed by a mathematical relation.

Lee (1973), for example, in his simulation of the airflow over the island of Barbados, used a normalized transformation[4] adapted from Schulman (1970) given by

$$s(x) = c\{ax + \tanh[(x - x_0)/\eta] + b\};$$
$$b = \tanh x_0/\sigma, \ x_0 = 0.5, a = 0.5, \eta = 0.044;$$
$$c = [a + b + \tanh(1 - x_0)/\eta]^{-1},$$

and Anthes (1970) offers an example of a stretched horizontal grid in two dimensions (Fig. 10.4). Fox-Rabinovitz et al. (1997) discuss the stretched grid approach, including the problems it introduces and recommended solutions including the introduction of diffusion-type filters and uniformly-stretched grids.

In Lee's formulation, η is a stretching scale factor and $s(x = 0) = 0$; $s(x = 1) = 1$. With this type of representation, the chain rule of calculus is used to replace the spatial derivatives in the conservation relations with terms such as

$$\frac{ds}{dx}\frac{\partial}{\partial s} = \frac{\partial}{\partial x}; \quad \left(\frac{ds}{dx}\right)^2 \frac{\partial^2}{\partial s^2} + \frac{d^2 s}{dx^2}\frac{\partial}{\partial s} = \frac{\partial^2}{\partial x^2}. \tag{10.1}$$

In the stretched horizontal coordinate system $\Delta s = s_{i+1} - s_i$ is a constant. This constant grid interval, however, is not found to be superior to simply specifying the stretched grid in terms of the original independent variable x since *the advantage of a constant grid* Δs *is nullified by the need to compute extra terms*, as given in Eq. 10.1. The term ds/dx, for example, must be computed as the average value over the particular interval Δx to which Δs corresponds. Since ds/dx itself is a function of x and therefore, varies over Δs, no advantage is gained by using this mathematical transformation. Simply specifying the grid locations of x_i is adequate.

In addition, the number of grid points between an interior location and the boundary is of equal importance to the size of the grid spacing. *Irrespective of the size of* Δs *(or* Δx*), for example, a location one grid length inside a boundary will be influenced by the boundary after only one time step.*

Using a fine-mesh grid inserted inside of a coarse grid is an alternative to the stretched grid approach.[5] In this case, a constant grid representation with grid increments Δ is surrounded by a grid with separation $n\Delta = \delta$, where $n > 1$. In contrast with the stretched grid where changes in grid size are defined by a continuous function, the meshed grid approach requires a discontinuity between the fine and coarse grid. Figure 10.5, reproduced from Mathur (1974), illustrates such a grid mesh representation in a hurricane model. Using this approach, Mathur was able to simulate the fine structure of the hurricane near the eye wall accurately, as well as the larger-scale environment that was influencing the storm intensity and movement. More applications of the nested grid representation are described in Clark and Farley (1984), Pielke et al. (1992), and Walko et al. (1995).

[4] $x = 0$ and $x = 1$ correspond to the model domain sides so that $0 \leq x \leq 1$.

[5] The use of two or more distinct grid meshes is referred to as a *nested grid*. Elsberry (1978) provides a short review of nested grid procedures.

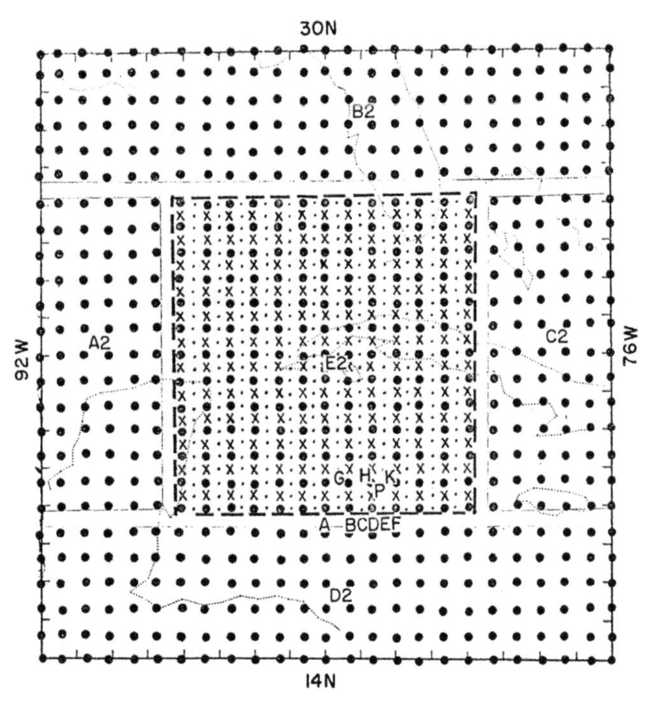

FIGURE 10.5

A fine mesh with a grid increment of approximately 37 km, inserted inside of a coarse mesh with grid lengths of approximately 74 km. Variables within the fine mesh are defined as the small dots and x's, as well as the darkened circles (from Mathur 1974).

The meshed grids (as well as the stretched and constant horizontal grids) can also be defined to move relative to the Earth's surface. Schlesinger (1973) has utilized a *movable grid* to prevent a simulated thunderstorm from exiting his domain, and Jones (1977b) used three meshed grids with lengths of 10, 30, and 90 km to simulate the dynamics of a moving hurricane, with the smaller two grids moving with the storm.

Problems arise, however, in using stretched or grid mesh representations. As shown in Chapter 9 (e.g., Table 9.1), waves that have lengths that are short relative to the model grid size, propagate erroneously relative to the exact analytic solution. Thus a wave in a fine mesh with a wavelength of 8Δ would have a representation of 2δ in the coarser grid if the grid separation was four times larger in the coarse grid ($\delta = 4\Delta$). The wave, therefore, would be poorly represented in the coarse grid. Moreover, as this wave travels from one grid region to another, the change in grid resolution can cause reflection and refraction of the wave (Morse 1973) in much the same fashion as occurs when electromagnetic waves travel from one physical medium to another (e.g., as light travels from air into water).

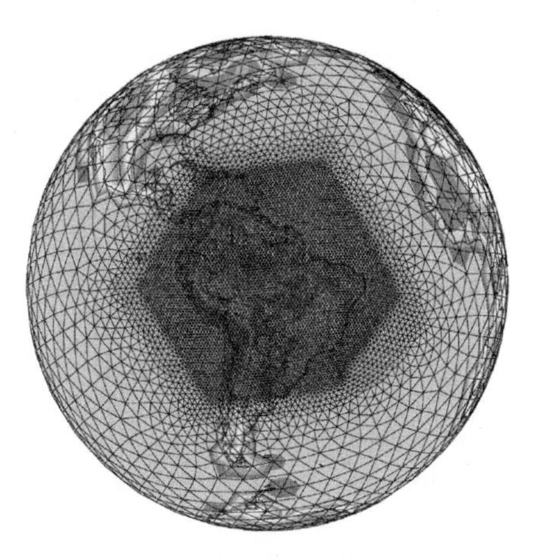

FIGURE 10.6

A global grid coverage with mesoscale grid spacing in selected regions using triangulation based on an icosahedron in the OLAM model. Figure provided courtesy of Robert Walko. (The color version of this figure is presented in the plate section in the back of the book and the online web version.)

Thus although stretched and nested grids increase the domain size, another source of computational error is introduced. For meshed grids, the minimization of these errors is best accomplished by judicious use of filtering near the boundary between the coarse and fine mesh, and the reader is referred to papers such as Perkey and Kreitzberg (1976) and Jones (1977a) for specifics concerning their schemes. Another discussion of procedures for nested grids is provided in Panin et al. (1999).

In applying boundary conditions between coarse and fine meshes, modelers can choose to permit perturbations to enter and leave the fine mesh (*two-way interaction*) or to prevent waves in the fine-grid mesh from exiting the fine mesh, but permitting coarse waves to enter (*parasitic grid representation*, Perkey and Kreitzberg 1976; Baumhefner and Perkey 1982). The use of nested grids to obtain higher spatial resolution in a region is referred to as *downscaling*. Ron Cionco was among the first to introduce this concept, as summarized in Cionco (1994).

Walko and Avissar (2008a,b) have introduced a horizontal coordinate system which has global coverage with much higher resolution in selected regions of interest (e.g., see Fig. 10.6). In this sense it is similar to Fig. 10.4 but has global coverage and thus no lateral boundary conditions are required.

10.2.2 Vertical Grid

As with the horizontal grid, the vertical grid of a mesoscale model is selected to have the most resolution in and near the region of interest. Uniform grid spacing at all levels is not feasible, in general, because of limitations of computer storage and of cost. The concept of a representative vertical scale length of the circulation can be used to estimate the required resolution, since as discussed in Chapter 9, as many as ten or more grid increments may be needed to resolve the atmospheric system adequately (e.g., see Tables 9.1, 9.2, and 9.3).

If vertical turbulent mixing is an important component of the circulation, its characteristic length scale provides a measure of the needed grid resolution. As discussed in Section 7.3, for example, the turbulent length scale when the surface layer is stable is a function of height above the ground near the surface, and a prescribed length scale, or function of local shear and temperature gradients above that level. The definition of l_θ by Eq. 7.18 gives one form of this length scale. In contrast, when the surface layer is neutral or unstable, the representation for the length scale remains a function of distance above the ground even well removed from the surface. For this reason, mesoscale models, in general, have the smallest grid increments near the ground, with the grid mesh expanding upward.

Modelers who use such an expanding grid generally attempt to make the transition from fine to coarse resolution as smooth as possible. Levels of the vertical grid are either arbitrarily selected or output from a functional form. Orlanski et al. (1974), for example, tested two forms

$$s = \ln\left(\frac{z + 30.5}{30.5}\right) \bigg/ \ln\left(\frac{S + 30.5}{30.5}\right) \quad \text{and } s = \frac{z}{1600} + \frac{1}{17.9} \ln \frac{z + 15}{15}$$

to represent his vertical grid. In the first formulation, S is the top of the model (they used $S = 20$ km). With the lefthand expression for s, using 70 levels, the grid resolution was 3 m near the ground and 1700 m near the top, whereas using 80 levels, the righthand form varied more slowly from 3.7 m near the surface to approximately a constant grid of 175 m from 4000 m to the top (which was set at 10.16 km in that representation). Orlanski et al. (1974) rejected the lefthand form for s because the extremely coarse resolution in the upper levels produced a significant distortion of gravity waves, which propagated upward from the active lower layers. This problem was much less apparent using the second coordinate stretching, despite the lower top.

Orlanski et al. (1974) utilized a large number of levels in their two-dimensional model simulation. In a three-dimensional simulation, the use of 70 or 80 vertical levels is computationally expensive. Pielke (1974b) examined the amount of vertical resolution necessary in a two-dimensional model to resolve a sea-breeze circulation properly. Such an evaluation appears necessary if the investigator is to establish whether the numerical grid separation rather than the physics determines the form of the solution.

In the study reported in Pielke (1974b), a sea-breeze simulation was performed as a control. Two separate experiments were performed – one in which the depth of the model was doubled but the same resolution below 4.22 km was retained, and the other where the vertical grid spacings were halved but the initial model depth was the same as the control experiment. The predicted sea-breeze patterns 8 h after simulated sunrise, shown in Fig. 10.7, illustrate that for this particular situation, the results were not forced by the grid spacing. In those experiments, the sea-breeze convergence zones remained below 3 km with about the same vertical length scale, regardless of the grid resolution. Sensitivity experiments such as this are required to establish the needed vertical resolution.

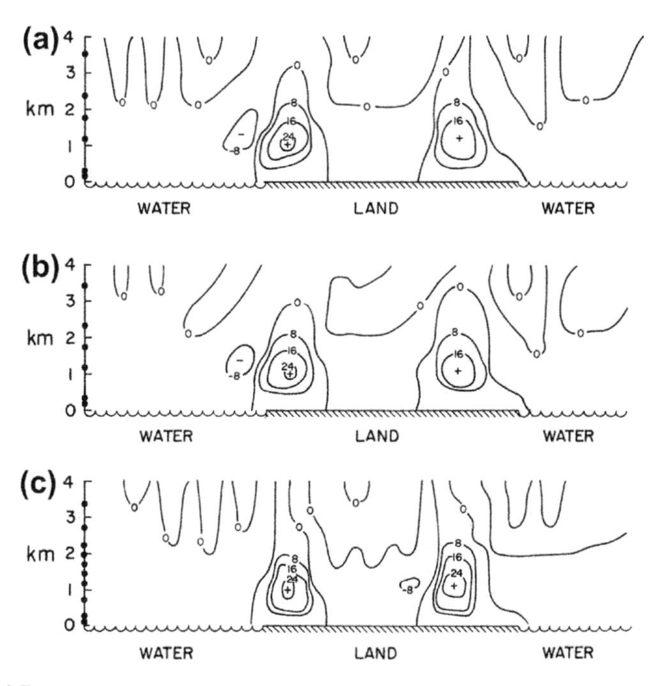

FIGURE 10.7

(a) The vertical motion field in a two-dimensional sea-breeze model with 7 vertical levels and an initial top of 4.22 km, (b) with 13 vertical levels and an initial top of 12.02 km, and (c) 13 vertical levels with an initial top of 4.22 km. Grid point levels 4 km and below are indicated by darkened circles (Pielke 1974b).

Two examples of observationally determined vertical and horizontal scales of mesoscale motion are shown in Figs. 10.8 and 10.9. In Fig. 10.8 the sea and land breezes were observed to have relatively shallow vertical depth (consistent with the results given in Fig. 10.7). Forced airflow over rough terrain creates a deep tropospheric perturbation as illustrated in Fig. 10.9.

10.2.3 Definition of Grid Points

In setting up the model grid, the locations at which the dependent variables are defined must be specified. In the differential representation, no such problem arises, of course, since all variables are defined at every point. Although the dependent variables could be defined at the same grid point, in general, the variables are *staggered* with respect to one another. Lilly (1961), for example, presented a staggered grid representation that helps preserve such properties as total kinetic energy[6] within a model domain. Batteen and Han (1981) examine the use of different spatial distributions of dependent variables on a rectangular grid for use in ocean models.

[6]See Section 11.6.2 for a discussion of kinetic energy conservation.

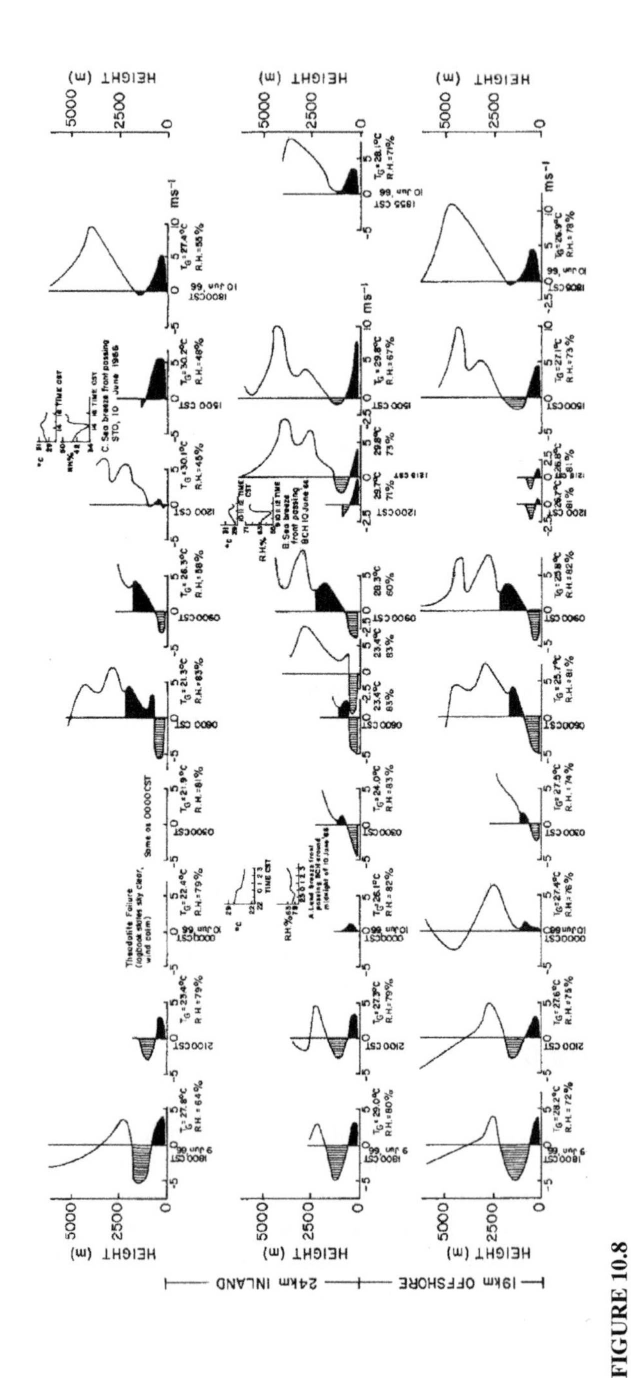

FIGURE 10.8

The $z - t$ variations of land and sea breezes normal to the coast from 1800 CST on 9 June to 1855 CST on 10 June 1966, at three stations perpendicular to the Texas coast. The average top of the sea breeze during the day was 570 m and the average height of the return flow was 1800 m (from Hsu 1970).

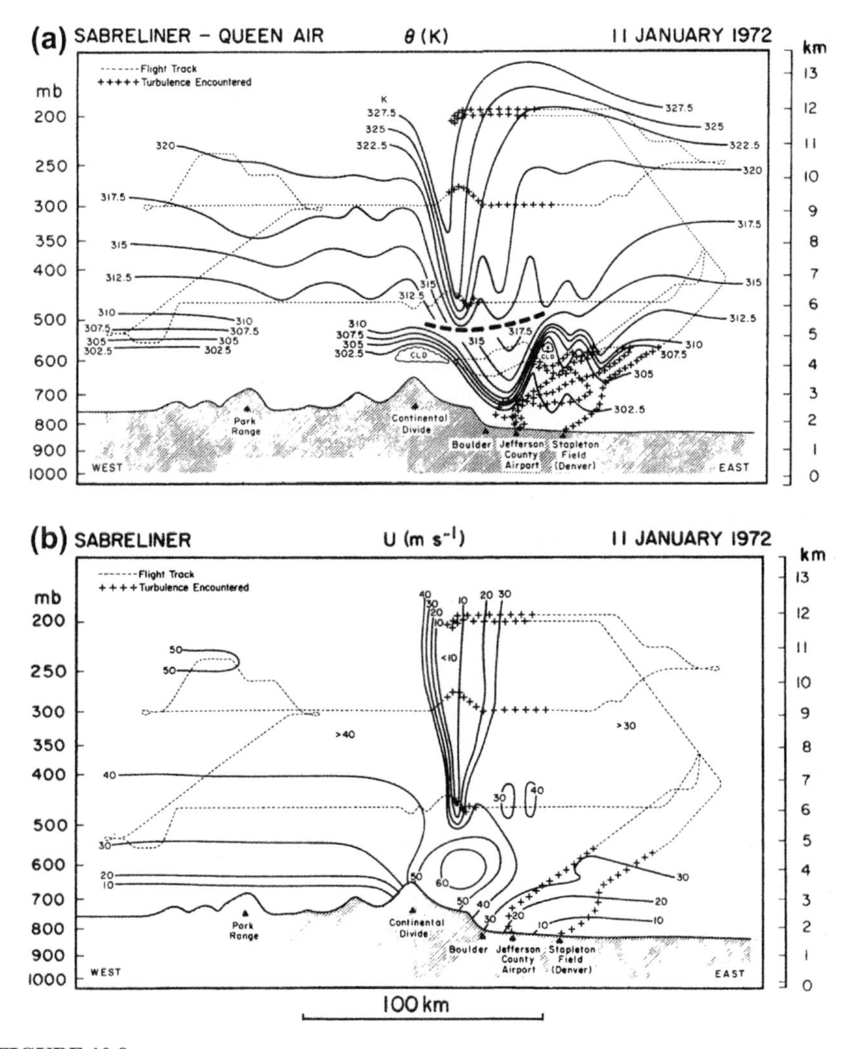

FIGURE 10.9

The $x - z$ structure of (a) isentropes and (b) winds observed across a portion of central Colorado during a downslope wind storm (from Lilly and Zipser 1972).

The need for a staggered grid is motivated by the differential nature of the conservation relations. In the two-dimensional form of the incompressible continuity equation (e.g., Eq. 4.23 with $\partial \bar{v}/\partial y = 0$), for example, it is convenient to stagger the vertical and horizontal velocities $\bar{w}$ and $\bar{u}$. The numerical finite difference approximation for a constant horizontal and vertical grid can then be written as

$$\bar{w}_{i,k} = \bar{w}_{i,k-1} - \frac{\bar{u}_{i+\frac{1}{2},k-\frac{1}{2}} - \bar{u}_{i-\frac{1}{2},k-\frac{1}{2}}}{\Delta x} \Delta z, \tag{10.2}$$

(where $\bar{u}$ is defined to be located halfway between the grid points at which $\bar{w}$ is defined) instead of a form such as

$$\bar{w}_{i,k} = \bar{w}_{i,k-1} - \frac{\hat{u}_{i+\frac{1}{2},k-\frac{1}{2}} - \hat{u}_{i-\frac{1}{2},k-\frac{1}{2}}}{\Delta x} \Delta z, \tag{10.3}$$

where

$$\hat{u}_{i+\frac{1}{2},k-\frac{1}{2}} = \left(\bar{u}_{i+1,k} + \bar{u}_{i,k} + \bar{u}_{i+1,k-1} + \bar{u}_{i,k-1}\right)/4,$$

and

$$\hat{u}_{i-\frac{1}{2},k-\frac{1}{2}} = \left(\bar{u}_{i,k} + \bar{u}_{i-1,k} + \bar{u}_{i,k-1} + \bar{u}_{i-1,k-1}\right)/4,$$

It is found that staggering the dependent variables as given by Eq. 10.2 increases the effective resolution by a factor of two, since derivatives are defined over an increment Δx, for instance, rather than $2\Delta x$, yet without requiring averaging as in Eq. 10.3. The horizontal and vertical pressure gradient term (e.g., in Eq. 4.21), is also effectively represented by such staggering.

Anthes (1971, personal communication) has also reported that staggering the horizontal and vertical velocities in the mass-continuity equation, such as done in Eq. 10.2, can minimize the direct influence of the lateral boundaries on the computation of vertical velocity. As illustrated in Fig. 10.10, for example, using Eq. 10.2 $\bar{w}$ need be computed using only interior values of $\bar{u}$. As suggested by Anthes and replicated by others (e.g., Pielke 1974a), the elimination of the lateral boundary values of $\bar{u}$ (and $\bar{v}$) from the computation of vertical velocity in the continuity equation provides, in general, much better behaved solutions.

Winninghoff (1968), and Arakawa and Lamb (1977) introduced classes of grid staggering which is referred to extensively today (Fig. 10.11). The mesoscale models presented in Appendix B usually refer to the types of grid stagger which are presented in the Figure.

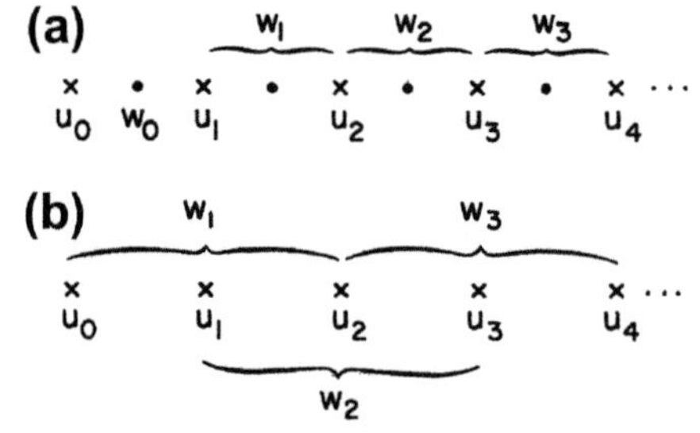

FIGURE 10.10

A schematic of (a) a staggered grid, and (b) a nonstaggered grid for the computation of $\bar{u}$ and $\bar{w}$. The symbols u_0 and w_0 represent the boundary values. The braces indicate which values are used in the computation of w from the continuity equation; (10.2) for (a), and (11.3) for (b). In (a), the horizontal velocity is defined at $\times$ points, the vertical velocity at points with blackened circles, whereas in (b) both velocities are defined at $\times$ points.

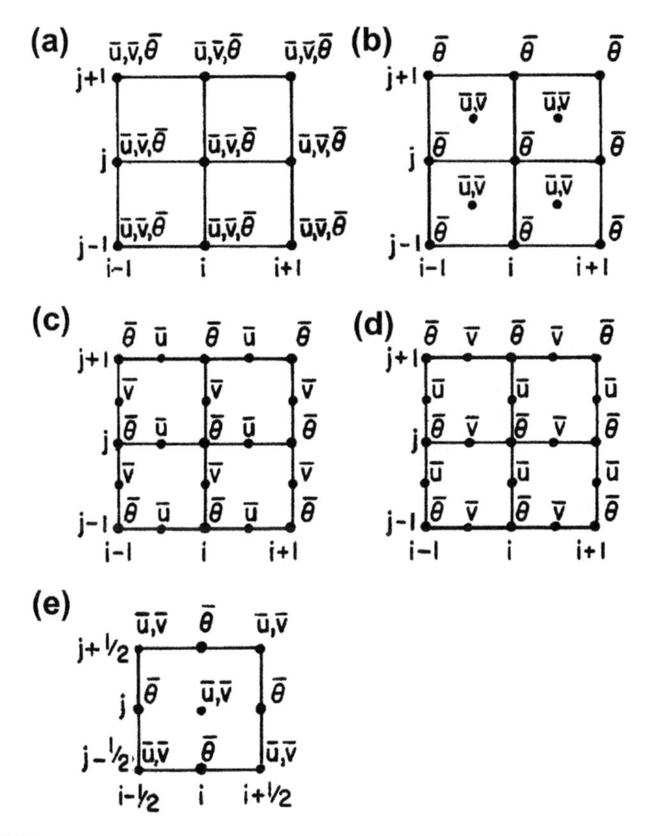

FIGURE 10.11

Illustration of the Arakawa and Lamb grid stagger for a two dimensional grid for the variables $\bar{u}$, $\bar{v}$, and $\bar{\theta}$ for a two-dimensional grid (adapted from Arakawa and Lamb 1977).

Once the domain size, vertical and horizontal grid increments, and locations at which the variables are defined are established, it is necessary to specify the temporal and spatial boundary conditions for the conservation relations. Temporal boundary values are required because the differential conservation equations represent an *initial value* problem, and spatial boundary information is needed because the domain size is finite yielding a *boundary-value* problem. The procedure to determine the values of the dependent variables required to commence the integration of the model equations is called *initialization*, and the values assigned to the perimeter of the model domain are referred to as *boundary conditions*.[7]

[7]The prognostic form of the conservation equations are referred to as hyperbolic differential equations and require both initial and spatial boundary conditions for their solution. Certain subsets of these conservation equations (e.g., Eq. 4.33), however, are elliptic differential equations and only spatial boundary conditions are required. Most texts on partial differential equations discuss the determination of the particular types of partial differential equations in more detail. Haltiner and Williams (1980) gave a brief summary of types of partial differential equations.

10.3 Initialization

The dependent variables that appear in a model representation require initial values for the integration of the equations to begin. For instance, values of $\bar{u}$, $\bar{v}$, $\bar{w}$, $\bar{\pi}$, $\bar{\theta}$, $\bar{q}_n$ and $\bar{\chi}_m$ are required at the startup time of the simulation.[8]

In terms of initialization of the wind and temperature fields, mesoscale and synoptic models are very different. Recall from Section 3.3.2 that for the synoptic and larger scales, the Rossby number R_0 is much less than unity. Hence, the wind is seldom far from gradient wind balance indicating that the mass field dominates the response of the wind field. Thus for these scales it is more important to measure the temperature distribution horizontally and with height than the wind field. The temperature field is used to obtain the mass field through the hydrostatic equation, with the distribution of mass represented by the pressure field (i.e., $\partial\bar{p}/\partial z = -\bar{\rho}g$, so that $\partial \ln \bar{p}/\partial z = -g/R\bar{T}$). A useful approximation to the winds can then be diagnosed from the mass field.

As the horizontal scales of the circulation are reduced, however, the relation between the wind and temperature (i.e., mass) fields becomes more complex. The ratio of the advective to the horizontal pressure gradient terms can be used to estimate whether the velocity or the mass field dominates the structure of the mesoscale circulation.

Using the scale analysis for the advection and horizontal pressure gradient force, their ratio can be defined as

$$I_0 = U^2/R\delta T. \tag{10.4}$$

Thus when $I_0 \gg 1$, the velocity field is expected to be dominant, whereas the temperature distribution should be more important for $I_0 \ll 1$. When $I_0 \simeq 1$, both fields are equally significant in determining the form of the mesoscale system.

Over heated land under light synoptic flow, for example, I_0 would be much less than unity and the temperature distribution will dominate the development and evolution of the sea breeze. The model simulation depicted in Fig. 10.3 illustrates such a circulation, where the pattern of the sea breeze was predominantly determined by the horizontal temperature gradient, even though advection was important in determining its detailed structure. Further demonstration of the importance of a proper temperature initialization, as opposed to an accurate wind initialization, in a mesoscale sea-breeze model is given by Carpenter and Lowther (1982). For atmospheric systems near the ground, wind speeds are smaller than aloft because of ground friction (e.g., see Section 7.2), so that the mass field would be expected to exert a substantial control on the wind field at low levels (D. Keyser 1980, personal communication). Conversely, if the winds are strong, I_0 can be much larger than unity and the wind field will dominate the temperature pattern.

[8]Guiraud and Zeytounian (1982) have briefly discussed the influence on initialization of the reduced number of initial conditions needed when a model is made hydrostatic.

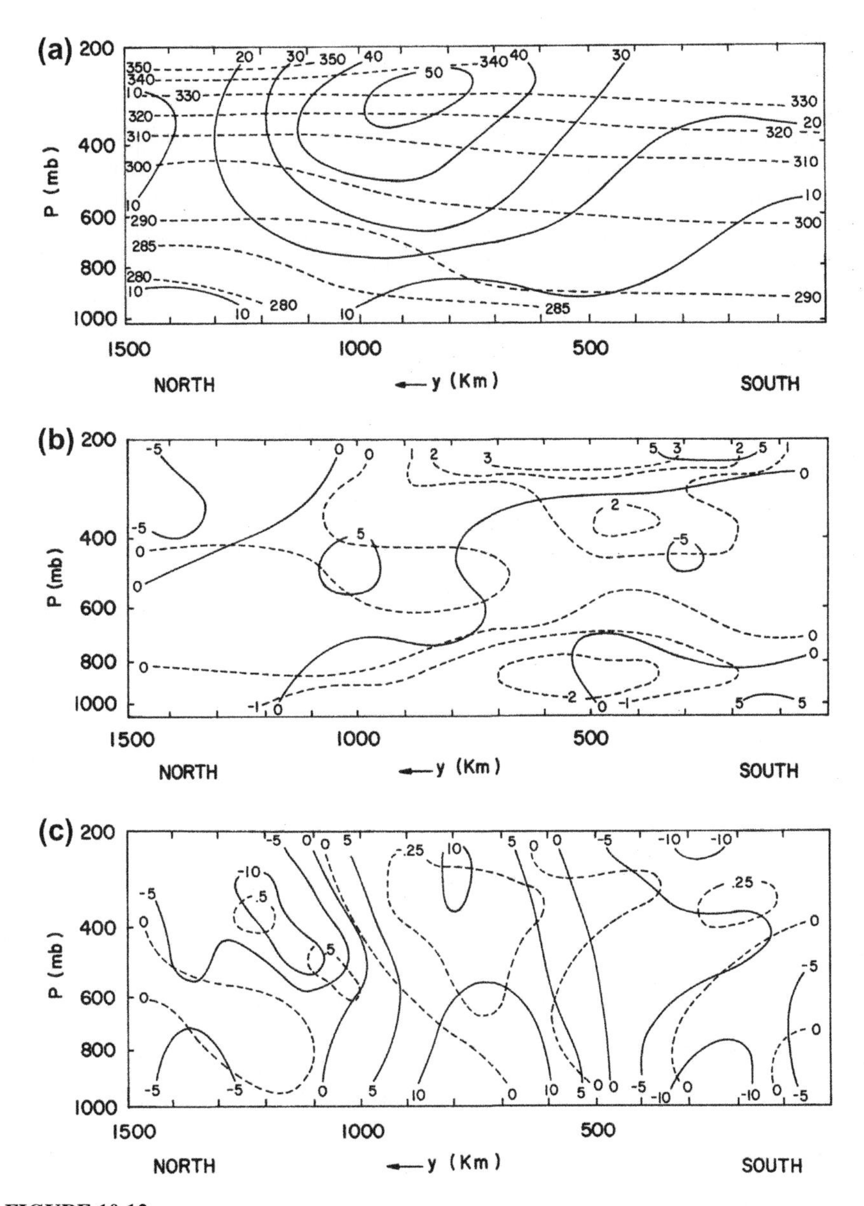

FIGURE 10.12

(a) North-south cross section through Manawski, Ontario of the observed winds (solid lines; m s^{-1}) and geostrophically-balanced potential temperature field (dashed line, K) for 0700 EST 16 October 1973. (b) Same as (a) except for the difference field between actual and computed winds and temperature where assumed fields had perfect initial winds (i.e., as given in (a), but erroneous initial temperatures). (c) Same as (b) except with perfect initial temperature [i.e., from (a)] but erroneous initial winds (from Hoke and Anthes 1976).

Hoke and Anthes (1976) performed experiments to evaluate the relative need to determine the winds as opposed to the temperatures in a two-dimensional simulation of a jet stream with a north-south extent of approximately 700 km (Fig. 10.12a). The first experiment attempted to generate the jet utilizing the observed winds, but ignored the observed temperatures, and instead linearly interpolated the temperatures inside the domain from values on the side boundary. This assumed temperature distribution gave a horizontally uniform geostrophic speed with a maximum value of 34 m s^{-1}. Figure 10.12b presents the errors produced for the two fields. A second experiment was then performed where the temperature field was forced to agree with the observations at each stage of the initialization process,[9] but the first guess of the wind field used the geostrophic wind at each grid point. The error field for this experiment is given in Fig. 10.12c.

Comparing Fig. 10.12b and c, it is evident that the wind velocity errors are smaller, and therefore the jet core is better represented when the wind field is known (i.e., the maximum error in the wind field is greater than 10 m s^{-1} in Fig. 10.12c, but is substantially less in Fig. 10.12b).

The calculation of the *root mean square error* (RMSE) is a convenient tool to quantitatively compare these two results. These are calculated for the velocity and temperature from

$$\begin{aligned} \mathrm{RMSE}_u &= \sum_{i=1}^{I_x}\sum_{k=1}^{K_z}(u_{\mathrm{obs}} - u_{\mathrm{pred}})^2 \Big/ I_x K_z, \\ \mathrm{RMSE}_T &= \sum_{i=1}^{I_x}\sum_{k=1}^{K_z}(T_{\mathrm{obs}} - T_{\mathrm{pred}})^2 \Big/ I_x K_z, \end{aligned} \tag{10.5}$$

where I_x and K_z are the number of grid points in the horizontal and vertical directions, respectively, and the subscripts "obs" and "pred" refer to the observed and predicted values of east-west velocity and temperature. For the experiment in which the observed winds were used (Fig. 10.12b) $\mathrm{RMSE}_u = 2.8$ m s^{-1} and $\mathrm{RMSE}_T = 1.1$ K, with $\mathrm{RMSE}_u = 6.9$ m s^{-1} and $\mathrm{RMSE}_T = 0.2$ K when the observed temperatures were used alone (Fig. 10.12c).[10] Thus in mesoscale circulations where $I_0 \gg 1$, it appears to be much more important to measure winds than temperature to initialize a model simulation. The initialization of the other dependent variables in a mesoscale model, in contrast, is similar to a synoptic model.

10.3.1 Initialization Procedures

The methodology for initializing a mesoscale model can be grouped into four categories:

[9] This initialization process, called dynamic initialization, is defined in Section 10.3.1.

[10] The small value of RMSE_T from Fig. 10.12c resulted because temperature was continually forced back to the observed value during the initialization. No such adjustment, however, was used on the wind field in obtaining Fig. 10.12b.

1. objective analysis,
2. dynamic initialization,
3. normal mode initialization, and
4. the adjoint method.

An excellent presentation of model initialization and data assimilation techniques is presented in Kalnay (2002). Initialization of the model involves dependent variables in the atmosphere and at the surface interface (Liston et al. 1999; Pielke et al. 1999a). Model solutions are often very sensitive to initial conditions. Park and Droegemeier (2000), for example, illustrated the sensitivity of the simulation of deep cumulus convection to errors in the water vapor field.

With *objective analysis*, available observational data are extrapolated to grid points either by using simple weighting functions, where the initial dependent variables are a function of the distance from the observation or by utilizing a *variational analysis* routine (O'Brien 1970; Sasaki 1970a,b,c; Sasaki and Lewis 1970; Sinha et al. 1998), where one or more of the conservation relations are applied to minimize the variance of the difference between the observations and the analyzed fields. This technique utilizes concepts of variational calculus, where the fundamental equations along with specified constraints as to the amount of agreement between the observations and analysis are given (e.g., the local time derivative must be very small to remove high-frequency motions). An explanation of this approach is beyond the scope of the text and the reader is referred to Sasaki (1970a) for a detailed discussion of the technique.

The *dynamic initialization* technique offers an alternative to objective analysis. With this approach, the model equations are integrated over a period of time so that values in the observed fields that are not representative of mesoscale resolution data are minimized.[11] By using the model equations, an approximate dynamic balance is achieved among the dependent variables since computational features generated by data inconsistency will be removed through damping or outward propagation through the side boundaries of the model as it adjusts toward equilibrium. As shown by Hoke and Anthes (1976), this propagation will appear as gravity, inertial, or inertial-gravitational waves but will have no physical significance. If large inconsistencies in the measurements occur, however, large computational adjustments will result in completely erroneous solutions.

One form of dynamic initialization suggested by Hoke and Anthes (1976) involves performing an *initialization integration*, where terms are added to the conservation relations that nudge the solutions toward the observed conditions. The conservation-of-motion equation in the x direction, for example, can be rewritten to include this term as

$$\frac{\partial \bar{u}}{\partial t} = -\bar{u}_j \frac{\partial \bar{u}}{\partial x_j} - \frac{1}{\rho_0} \frac{\partial}{\partial x_j} \rho_0 \overline{u''_j u''} - \theta_0 \frac{\partial \bar{\pi}}{\partial x} + f v + \hat{f} \bar{w} + G_u \left(u_{\text{obs}} - \bar{u}\right),$$

[11] The mesoscale model requires values of the dependent variables averaged over a grid volume as defined in Chapter 4 (e.g., $\bar{u}, \bar{\theta}$, etc.), and the observations provide values over the sampling volume of the instrumentation. This difference gives rise to observational data that are not consistent with the mesoscale averaged variables needed for input to a model.

where G_u is referred to as the *nudging coefficient*. The other prognostic equations can be written in a similar fashion. By integrating the model equations for a period of time (say 12 h), where the geostrophic wind may be the first guess, the imbalances in the solutions are reduced and large unrealistic accelerations will not occur when the simulation experiment actually commences.

Nudging coefficients added to each prognostic equation can be assumed to be a function of observation accuracy, of the distance between an observation and the grid point, of the variable nudged, and of the typical magnitudes of the other terms in the prognostic equations. By using dynamic initialization, the conservation relations themselves are used to distribute initial values of the dependent variables throughout the model in a physically consistent fashion. The major disadvantage of this approach is the cost in computer resources of an extended initialization integration, which could be 12 h or so in a mesoscale model simulation, whereas the model experiment may be only 24 h in length. Examples of dynamic initialization include those of Temperton (1973), Anthes (1974a), Hoke and Anthes (1977), Kurihara and Tuleya (1978), Kurihara and Bender (1979), and Kuo and Guo (1989). Douville et al. (1999) discuss the use of nudging and an optimal interpolation scheme to insert soil moisture into models. Liston et al. (1999) describes a procedure to more effectively insert snow cover into regional and mesoscale models.

Nudging applied during model integration has become referred to as four-dimensional assimilation when it is applied throughout the model integration using observed data that is available during this time period. This type of nudging can be applied just near the lateral boundaries and/or in the interior. When only low frequency information is assimilated, this has been called spectral nudging (e.g., see Rockel et al. 2008). This type of nudging is routinely used in dynamic downscaling from global models (Pielke and Wilby 2012). Dynamic downscaling is discussed further in Chapter 13.

Nonlinear normal mode initialization provides an alternative to dynamic initialization. As summarized by Daley (1981), this approach eliminates the integration time period needed by the dynamic initialization method to remove inconsistencies in the input data. This approach, which is discussed in Daley's review paper, has been adopted by the weather service of the Canadian government and by the European Centre for Medium Range Forecasts for use in their synoptic models.

The nonlinear normal mode initialization scheme involves segregating high-frequency and low-frequency components of the initial input data using horizontal and vertical structure functions[12] for the initial atmosphere. By removing the high frequencies, which are assumed to have no meteorological significance, only relevant low-frequency information is assumed to remain. This scheme offers promise as an effective initialization scheme provided that sufficient observation data are available.

[12]Structure functions arise when a dependent variable is decomposed into products of dependent variables that are individually functions of only a subset of the entire range of independent variables. Each separate function represents a different subset of the independent variables. A vertical structure function, for example, can be written as $\phi(x, y, z, t) = \Phi(z)\psi(x, y, t)$. Baer and Katz (1980), and Daley (1981) described the advantages of such a decomposition of the dependent variables.

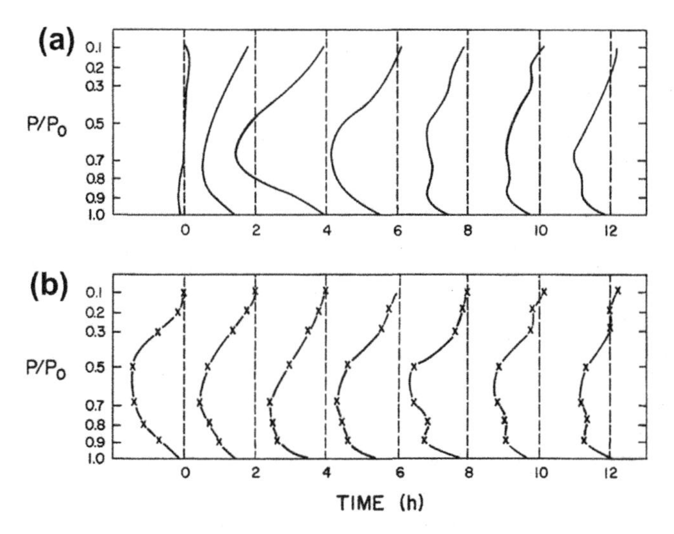

FIGURE 10.13

The variation of vertical motion with time near Labrador, Canada for a synoptic model prediction (a) without, and (b) with nonlinear normal mode initialization (from Daley 1979).

Kasahara (1982) has presented a methodology in which the normal mode initialization procedure could be used in limited area numerical weather prediction models. Figure 10.13, reproduced from Daley (1981), illustrates the reduced period of initialization required using the normal mode method as contrasted with the dynamic initialization scheme.

Intensive interest in the normal mode initialization technique continues. Wergen (1981) discusses nonlinear normal model initialization in the presence of a 2000 m mountain with a horizontal grid interval of 63.5 km and a slope of up to 1 to 63. Lipton and Pielke (1986) evaluated the vertical normal modes for the type of terrain-following coordinate system discussed in Section 6.3.

The *adjoint method* is overviewed in the following discussion provided by Vukićević and Hess (2000), and Vukićević et al. (2001). The time rate of change (i.e., the evolution) of mesoscale flow can be represented by the following general governing equation:

$$\frac{\partial \vec{X}}{\partial t} = M\vec{X} + \vec{F}(t) \tag{10.6}$$

where $\vec{X}$ is a vector of physical quantities that are studied (e.g., surface temperature, wind, CO_2 flux, as a function of time and space, in general), t is time, M is a physical/dynamical model, and $\vec{F}(t)$ is time-dependent forcing (e.g., precipitation and radiation for the land-surface processes). The solution of Eq. 10.6 depends on the initial condition vector ($\vec{X}_0$), the boundary condition vector ($\vec{X}_b$), and a set of free physical parameters denoted by an $\vec{\alpha}$.

We are interested in the system's response measured by the change of a diagnostic function, defined in general form as:

$$J(x) = \int_0^T \int_\Omega g(\vec{X}) d\omega dt \tag{10.7}$$

where $[0, T]$ is time interval, Ω is the spatial domain, $g(\vec{X})$ is a diagnostic operator (e.g., $g(\vec{X}) = \vec{X}$; $g(\vec{X})$ = flux of moisture, etc.). It is obvious that the change of J can occur either when the control parameters or the forcing are varied. The dependence of the change of J on the control parameters is expressed as

$$\Delta J = \frac{\partial J}{\partial \vec{Y}} \delta\vec{Y} + O\left(\delta\vec{Y}^2\right) \tag{10.8}$$

where, for brevity, $\vec{Y}$ is defined as a vector of variations of the control parameters: $\delta\vec{Y} = (\delta\vec{X}_0, \delta\vec{X}_b, \delta\alpha)$.

Assuming that second and higher order terms are small

$$O\left(\delta\vec{Y}^2\right) \ll O(\delta\vec{Y}) \tag{10.9}$$

and using the theory of variations, Eq. 10.8 becomes

$$\Delta J = \int_0^T \int_\Omega \Lambda(t,\omega) M_{\vec{\alpha}} \delta\vec{\alpha} d\omega dt - \int_\Omega [\Lambda(t,\omega)\delta\vec{X}]_0^T - \int_0^T [\Lambda(t,\omega)\delta\vec{X}]_{O(\Omega)} \tag{10.10}$$

where Λ is solution of an adjoint system associated with the system (Eq. 10.6). $M_{\vec{\alpha}}$ is a portion of the model in Eq. 10.6 that depends on the parameters only.

Also, from Green's function theory of solutions for partial differential equations (e.g., Roach 1970), the solution of the system (Eq. 10.6) for the given (i.e., fixed) set of control parameters can be expressed as a function of the adjoint solution and the forcing $\vec{F}$.

$$x(t) = \int_0^t \int_\Omega \Lambda(t,\tau)\vec{F}(\tau) d\tau d\omega \tag{10.11}$$

Substitution of Eq. 10.11 in 10.7 gives relationship between J and the adjoint solution and forcing.

Because the adjoint solution multiplies variations of the control parameters and the forcing function in the expressions 10.10 and 10.11, respectively, these expressions show that to know the sensitivity of the diagnostic function J to the variation of control parameters or the forcing, one can compute the solution of the adjoint system (Λ). The adjoint system is readily derived from the homogeneous part of the original system (Eq. 10.6) (e.g., Roach 1970). The adjoint solution is a function of time and space and represents a map in phase space of the influences of the controlling factors on the system that is studied. From this map we can learn, for example, about the physical mechanisms of interactions between different components of the mesoscale system.

The adjoint sensitivity analysis is exact for the linear systems (i.e., when M in Eq. 10.6 is linear, implying $O(\delta\vec{Y}^2) = 0$). For the nonlinear systems (i.e., M is nonlinear) this analysis is exact under the assumption that variations of the controlling factors are small. In the nonlinear case, therefore, the adjoint analysis produces a "first order" sensitivity. Consequently, the adjoint analysis of a mesoscale system examines the first order interactions between the components of the system. Understanding the first order interactions is beneficial as shown in the studies where the adjoint analysis has been used to examine nonlinear systems (e.g., Zou et al. 1993, Vukićević 1998).

The sensitivity of the system to control parameters can also be studied by perturbation sensitivity experiments whereby the value of a parameter within one of the model's parameterizations is changed by a small amount and the new model solution is computed. This method is, however, inefficient for systems where the number of parameters is large. Moreover, this method can be very inefficient for computation of the influence functions associated with the forcing (i.e., Green's functions). In contrast, only one adjoint model solution is required to evaluate the sensitivity of the given J to all controlling factors.

Detailed examples of the use of the adjoint method can be found in Rabier et al. (1992), Marchuk (1995), Vukićević and Raeder (1995), Kaminski et al. (1997), Vukićević (1998), Vukićević and Hess (2000), Fukumori et al. (2004), Bugnion et al. (2006), Mahfouf and Bilodeau (2007), Moore et al. (2009), and Doyle et al. (2012a). The use of the adjoint method is also applied by Uliasz et al. (1996).

10.4 Spatial Boundary Conditions

Since the mesoscale model domain is artificially enclosed with sides, it is necessary to specify the values of the dependent variables at this perimeter surface of the model. Such values are referred to as *boundary conditions* and are required to integrate in time the approximate forms of the conservation relations.

In discussing boundary conditions in mesoscale models, it is convenient to discuss the top, lateral, and bottom sides separately. Because of the finite domain of these models, the top and lateral perimeters are incorporated only because of computational necessity and have no physical meaning. The bottom, however, is a real boundary, and the transfer of such physical properties as heat and moisture across this interface plays a fundamental role in most mesoscale meteorological circulations.

The number of boundary conditions that can be applied in a model depends on the form of the differential equations that are used. Model equations that have the correct number are said to be *well posed*, and if a greater number are used than required, the model equations are said to be *overspecified*. As maintained by Oliger and Sundström (1976), who discussed the mathematical properties of boundary conditions of initial-spatial boundary value problems in detail, conservation relations that are represented by nondissipative approximate solutions (e.g., leapfrog) and that are overspecified, generate physically erroneous shortwave features that travel across the model grid at the fastest wave speed permitted in the model. Such waves are generated at the boundary. Oliger and Sundström have argued that models that are hydrostatic are ill posed for any choice of local[13] boundary conditions (except for the unlikely case where the exact solution is known on the boundary without error), and thus erroneous wave motions are expected to be created at the boundaries in such a model. Chen (1973) also reports that overspecification of boundary conditions can excite computational modes, however, smoothing at points next to the boundaries, suppresses these erroneous perturbations. Oliger and Sundström (1976) contend that the anelastic nonhydrostatic form of the conservation relations can be written in a well-posed form.

In the three-dimensional nonhydrostatic cloud model of Klemp and Wilhelmson (1978a,b), in which the compressible form of the conservation equations is used, they suggested, based on the work of Oliger and Sundström (1976), that all prognostic variables but one should be specified on the inflow boundary of the model, whereas only one such boundary conditions should be applied at the outflow boundary. Such boundary conditions, they maintain, are required to retain a well-posed set of differential equations.

As a practical problem, erroneous solutions generated at the boundaries are only serious when they propagate from the boundary into the region of significant mesoscale perturbation of the flow from the larger-scale environment. Since they are shortwave features, they can be effectively removed by a selective low-pass filter such as Eq. 9.60. Larger-scale trends in the model variables caused by the boundaries can also introduce very serious errors as discussed in Section 10.4.1.

A serious problem also arises when modelers differentiate the conservation relations so as to permit additional boundary conditions (Neumann and Mahrer 1971). For example, the incompressible conservation-of-mass relation (Eq. 4.23) can be written as

$$\frac{\partial \bar{w}}{\partial z} = -\left(\frac{\partial \bar{u}}{\partial x} + \frac{\partial \bar{v}}{\partial y}\right). \tag{10.12}$$

Integrating Eq. 10.12 with respect to z permits one boundary condition from this relation. To add an additional boundary condition some investigators (e.g., Estoque 1961; Vukovich et al. 1976) have differentiated Eq. 10.12 with respect to z yielding

$$\frac{\partial^2 \bar{w}}{\partial z^2} = -\frac{\partial}{\partial z}\left(\frac{\partial \bar{u}}{\partial x} + \frac{\partial \bar{v}}{\partial y}\right). \tag{10.13}$$

With this form, two boundary conditions are required, and such an equation has been used to specify a rigid boundary at the bottom and top of a model. As shown by Neumann and Mahrer (1971), however, integrating Eq. 10.13 with respect to z yields

$$\int \frac{\partial^2 \bar{w}}{\partial z^2} dz = \frac{\partial \bar{w}}{\partial z} + F(x, y, t), \tag{10.14}$$

where $F(x, y, t)$ is the constant of integration. Unless the limits of the integral on the left side are properly specified, so that $F(x, y \cdot t)$ is identically zero, mass is *not*

[13]A local boundary condition is one that is generated at the boundary and is not a function of interior grid points.

conserved. Equation 10.13 is, therefore, in general not a proper form of the conservation-of-mass relation and should not be used. A similar criticism, of course, can be applied to any differential operation on the original conservation relations if the proper integration constants are not applied.

10.4.1 Lateral Boundary Conditions

As already mentioned, the lateral boundaries of a mesoscale model are required only because the simulated domain must be limited in horizontal extent as a result of constraints in computer resources. However, because it is impossible to specify values on this boundary properly, at least in a hydrostatic model (and even in compressible models physically accurate values are difficult to find, in general), it is desirable to remove this boundary as far from the region of interest as possible. Expanding the grid in the horizontal is one available mechanism to minimize the effect of the lateral boundary as discussed in Section 10.2.1.2.

Anthes and Warner (1978) have demonstrated the serious errors in mesoscale model results that can occur if the lateral boundary conditions are incorrectly specified. Following the procedure introduced by Anthes and Warner, the sensitivity of a mesoscale model to erroneous values on the boundaries can be illustrated using the east-west component of Eq. 4.21, which is written as

$$\frac{\partial \bar{u}}{\partial t} = -\frac{\partial}{\partial x}\left(\frac{\bar{u}^2}{2}\right) - \alpha_0 \frac{\partial p'}{\partial x} + R, \tag{10.15}$$

where R represents the remaining terms. Integrating Eq. 10.15 using the domain-volume average given by Eq. 4.12 (where D_x and D_y correspond exactly to the domain size of the model) and assuming symmetry in the y direction to simplify the analysis, yields

$$\frac{\partial u_0}{\partial t} = \frac{\bar{u}_{\mathrm{W}}^2 - \bar{u}_{\mathrm{E}}^2}{2D_x} - \alpha_0 \frac{p'_{\mathrm{E}} - p'_{\mathrm{W}}}{D_x} + \int_x^{x+D_x} \int_y^{y+D_y} R\, dx\, dy / D_x D_y. \tag{10.16}$$

The quantity $\partial u_0/\partial t$ represents the average acceleration of the entire model domain at level z, and the subscripts "E" and "W" refer to the east and west boundaries, respectively, of the model.

From Eq. 10.16, it is evident that an error in the specification of the values of $\bar{u}$ and p' at either boundary will introduce a fictitious acceleration of the entire model domain at that level. The error is inversely proportional to the size of the domain D_x. Table 10.1 illustrates values of velocity and pressure difference across a model domain that will generate a domain acceleration of 1 m $\mathrm{s}^{-1}\mathrm{h}^{-1}$ for three values of D_x. In preparing Table 10.1, the eastern velocity component $\bar{u}_{\mathrm{E}}$ is written in terms of the western component $\bar{u}_{\mathrm{W}}$ as $\bar{u}_{\mathrm{E}} = \bar{u}_{\mathrm{W}} + \Delta u$, so that

$$\left(\bar{u}_{\mathrm{E}}^2 - \bar{u}_{\mathrm{W}}^2\right) / 2 = \left(\Delta u \bar{u}_{\mathrm{W}}\right) + \left[(\Delta u)^2 / 2\right]. \tag{10.17}$$

Table 10.1 The values needed in Eq. 10.16 to generate a domain-averaged acceleration at level z of 1 m $s^{-1}h^{-1}$. The velocity difference Δu can be determined from Eq. 10.17 using the quadratic equation (i.e., $\Delta\bar{u} = -\bar{u}_W \pm \sqrt{\bar{u}_W^2 + 2F}$, where $F = \bar{u}_E^2 - \bar{u}_W^2/2$). A value of $\alpha_0 = 1$ m^3kg^{-1} was used in computing the pressure gradient force (adapted from Anthes and Warner (1978).

D_x(km)	$(\bar{u}_E^2 - \bar{u}_W^2)/2$ (m^2s^{-2})	$\alpha_0(p'_E - p'_W)$ (m^2s^{-2})	Δu in m s^{-1} for $\bar{u}_W$ in m s^{-1} of			$p'_E - p'_w$ (mb)
			0	**10**	**20**	
1000	278.0	278.0	23.6	15.6	10.9	2.78
100	28.0	28.0	7.5	2.5	1.4	0.28
10	2.8	2.8	2.4	0.3	0.1	0.03

The analysis of Eq. 10.16, as tabulated in Table 10.1, shows that for small domain sizes, small errors in the prescription of wind speed at the lateral boundaries can cause a significant acceleration within the model domain. This effect becomes more serious for larger wind speeds because of the quadratic form of the advective term in Eq. 10.16. The model results are even more sensitive to the specification of the mesoscale pressure perturbation on the lateral boundaries. Even a fraction of a millibar of error in pressure on one of the boundaries can generate substantial accelerations throughout the smaller domain sizes. Pielke et al. (1989) discuss the significance of those errors in the context of mesoscale observational networks.

Unfortunately, there has been little quantitative discussion in the literature of techniques to control such erroneous acceleration. Other work has concentrated on the minimization of the backward reflection into the model domain of outward propagating advective and gravity waves.

10.4.1.1 Types of Lateral Boundary Conditions

Lateral boundary conditions can be *open* (i.e., mesoscale perturbations can pass into and out of the model domain) or *closed* (i.e., such perturbations are not permitted to exit or enter).[14] There are several types of these lateral boundary conditions, some of which are designed to minimize the reflection of erroneous information back into the model domain, yet still permit the input of larger-scale flow into the region. The types of boundary conditions include the following.

1. *Constant Inflow, Gradient Outflow Conditions*
   With this procedure, air entering the model is assumed to be unaffected by the downstream mesoscale perturbation to the flow, so that the dependent variables remain unchanged at inflow boundaries (i.e., a closed boundary). Air exiting the model, however, is assumed to instantaneously have the same value as found one

[14]Closed boundaries could also be defined where no flow occurs at that perimeter of a model. This definition seems too limited, however, so that in this text a closed boundary means that $u', v', w' \equiv 0$ at the boundary, rather than $\bar{u}, \bar{v}, \bar{w} \equiv 0$. Such a boundary is closed to mesoscale flow but not to larger-scale motions.

grid point upstream (thus the term gradient boundary condition since, for example, $\partial\phi/\partial x \simeq \phi(N-1) - \phi(N) = 0$, where ϕ is any one of the dependent variables and N is the outflow boundary). Inflow and outflow are defined in terms of the wind direction at the boundaries. Unfortunately, this procedure cannot properly handle disturbances that propagate upstream (e.g., internal gravity waves) and simultaneously correctly handle changes to the downstream boundary as advection and wave propagation move information at a finite speed from the last interior grid point to the boundary.

Mason and Sykes (1979) use a modification to this scheme that is applied to the velocity components as $u_N^{\tau+1} = 1.5u_{N-1}^{\tau} - 0.5u_{N-3}^{\tau}$. This condition is applied at outflow points. They concluded that this representation, although it results in some reflection at boundaries, is not only extremely simple, but is stable and effective. (In testing this outflow boundary condition, u_{N-2}^{τ} produced improved results compared to the use of u_{N-3}^{τ}.)

2. *Radiative Boundary Conditions*

With this procedure, the variables at the lateral boundaries are changed in value so as to minimize reflection of outward propagating perturbations to the flow, back into the model domain. Several procedures have been introduced to implement radiative boundary conditions using, for the east-west boundary, an equation of the form

$$\partial\bar{u}/\partial t = -c\partial\bar{u}/\partial x.$$

(The other prognostic conservation equations can similarly be evaluated at the boundary from $\partial\bar{\phi}/\partial t = -c\partial\bar{\phi}/\partial x$, where $\bar{\phi}$ is any one of the prognostic dependent variables.) These methods include those of

(a) Orlanski (1976) where

$$c = \frac{-\partial u}{\partial t} \Big/ \frac{\partial u}{\partial x}$$

is evaluated at the last grid point immediately in from the boundary with the requirement that $0 \leq c \leq \Delta x/\Delta t$. Miller and Thorpe (1981) proposed improvements to Orlanski's method, including the use of an upstream formulation rather than a leapfrog scheme. Carpenter (1982b) describes how the Miller and Thorpe scheme can be generalized to include changes introduced at the boundary from a larger-scale model.

(b) Klemp and Lilly (1978), and Klemp and Wilhelmson (1978a) where c is a constant, equal to the model domain height times the Brunt-Väisälä frequency[15] divided by π, chosen to represent the dominant phase velocity of an internal gravity wave.

(c) Hack and Schubert (1981) where c is evaluated separately for individual phase speeds for all of the internal gravity wave modes near the model boundary. The resultant change of the dependent variable at the boundary is evaluated by the summation of the changes resulting from each wave. The advantage

[15]The Brunt-Väisälä frequency is defined by footnote 5 in Chapter 5.

of Hack and Schuberts' technique over Orlanski's is that the evaluated wave speeds are determined from atmospheric structure through the depth of the model, whereas Orlanski's condition is determined separately at each vertical level. Hack and Schubert reported on a comparison of results for a hurricane model using these and other more reflective lateral boundary conditions. Using their results as a control, Orlanski's condition was nevertheless found to be very effective even though it used information at one level.

Lilly (1981) has sought to explain the importance of lateral radiative conditions using an idealized linear model of convection. In the context of a convective storm simulation, he concluded that the effect of this boundary condition on the mass flow into or out of the model produced by the convection is more important than the avoidance of reflection of waves at the boundary. However, he found that a boundary condition designed to minimize reflection is also nearly optimal to control the mass flow. Tripoli and Cotton (1982) offered an approach where both backward reflection and domain-averaged acceleration are controlled using a larger-scale compensation region external to the primary model domain.

3. *Sponge Boundary Conditions*

The use of enhanced filtering near the lateral boundaries can be used to damp advective and wave disturbances as they move toward the periphery of the model domain. These filters are added either by increasing the value of a horizontal exchange coefficient in an explicit diffusion formulation[16] near the boundary (e.g., Deaven 1974) or by applying larger smoothers[17] in that region (e.g., Perkey and Kreitzberg 1976; Jones 1977a). Alternatively, the prognostic equations can be written in the form

$$\frac{\partial \bar{\phi}}{\partial t} = -\bar{u}\frac{\partial \bar{\phi}}{\partial x} - r\left(\bar{\phi} - \phi_0\right),$$

where r is called a *relaxation coefficient* (Davis 1983) and ϕ_0 the desired value of $\bar{\phi}$ at the boundary. The relaxation coefficient is defined to become nonzero within some distance of the boundary, reaching a maximum at the boundary. Durran (1981) used such a formulation to represent an absorbing layer[18] at the top of his model flow simulation.

As implied by the results of Morse (1973), the increased filtering cannot be applied abruptly at some selected distance from the lateral boundary because erroneous reflection back into the center of the model domain will result. As stated in Section 10.2.1.2, such reflections are analogous to those found in the study of optics, when electromagnetic radiation travels from a material of one index of refraction to another.

The sponge boundary condition is a form of radiative boundary condition in which increasingly greater explicit viscosity is applied close to the lateral sides of

[16]See, for example, Eq. 9.59, where α would increase near the lateral boundaries.
[17]See, for example, Eq. 9.60, where δ would increase near the lateral boundaries.
[18]See Section 10.4.2 for a discussion of the absorbing layer.

a model. In contrast to the types of radiative boundary conditions discussed previously, however, the sponge condition requires a number of grid points near the boundary to permit the smoothing to increase gradually. These added grid points contribute to the computational cost of a model simulation.

Sponge boundary conditions are frequently used in meshed models, such as those discussed in Section 10.2.1.2. In the parasitic form of grid meshing reported by Perkey and Kreitzberg (1976), for example, a sponge boundary condition is applied near the exterior of the fine mesh to prevent disturbances generated within that region from propagating to the coarse grid. Since the sponge is a low-pass filter, longer wave features are permitted to move from the coarse grid into the interior of the fine mesh.

4. *Periodic Boundary Conditions*
The values of the dependent variables at one boundary of the model domain are assumed identically equal to the values at the other end [e.g., $\phi(x_D) = \phi(x_0)$]. Although of considerable value in comparing a numerical model with an exact analytic solution, realistic mesoscale simulations generally do not permit the reintroduction of perturbations to the flow into the inflow region of the model after they have exited the boundary. The application of periodic boundary conditions to selected approximations to the advection equation and to the diffusion equation is given in Appendix A.
5. *Larger-Scale Model or Analyzed Boundary Conditions*
Davis (1983) has provided a more complete summary of lateral boundary conditions including an analysis of their advantages and disadvantages. Among his results, he concluded that Oliger and Sundström's (1976) analysis regarding the ill-posed characteristic of any local boundary condition in a hydrostatic model does not result in serious errors in such models when the relaxation form of lateral boundary condition is used. Anthes (1983) has also presented an overview of lateral boundary conditions including a discussion of the use of nested models for regional-scale forecasting. The use of analyzed large-scale fields by the National Center for Environmental Prediction (NCEP), as reported in Kalnay et al. (1996), has been found to be very useful in providing initial and lateral boundary conditions to regional and mesoscale models (e.g., see Liston and Pielke 2000).

As discussed in Pielke and Wilby (2012) and Pielke et al. (2012), with respect to mesoscale and regional models in which the initial conditions are forgotten (such as for year-long and multi-decadal climate predictions), however, the lateral boundary conditions (and interior nudging, if used) are dominate drivers of the mesoscale and regional model predictions (e.g., see also Rockel et al. 2008; Lo et al. 2008). If there are errors in the parent model, there is no way for the limited area model to correct for those errors.

10.4.1.2 Summary of Lateral Boundary Conditions

A necessary test to apply to any selected lateral boundary condition is to enlarge the model domain progressively (i.e., by adding grid points) until successive enlargements have no appreciable changes on the solutions within the region of interest. Smaller domain sizes would result in mesoscale circulations that are significantly altered as the lateral boundaries are moved.

In summary, the main recommendations concerning lateral boundary conditions are the following:

1. Remove the lateral boundaries far enough from the region of interest so that a subsequent further enlargement has no appreciable change to the solutions within the interior. Enlarging the model requires that the number of grid points as well as the grid spacing be increased in the vicinity of the lateral boundary.
2. The radiative lateral boundary condition appears to be the form that permits the least expansion of the model domain. Several forms of the radiative boundary condition exist, and each modeler will probably need to test them individually for their particular applications. Orlanski's (1976) form has been reported by several investigators (e.g., Clark 1979; Hack and Schubert 1981) to, perhaps, have a general utility.
3. The sensitivity of the mesoscale model domain to grid-volume averaged accelerations also indicates that the model domain must have as large a horizontal scale as possible. Although changes in the large-scale field can be input through these boundaries, the sensitivity of the results for mesoscale domain sizes to even small variations in pressure and velocity (Table 10.1) makes the acquisition of observational input with suitable accuracy difficult to obtain (Anthes and Warner 1978).
4. The lateral boundary conditions must be accurate (skillful) values of the grid-volume dependent variables. Otherwise, errors are being fed into the limited area model which will only result in an erroneous result. This issue is a particular problem for Type 2, 3, and 4 downscaling as reported in Pielke and Wilby (2012), and discussed earlier in this Chapter, as well as in Chapter 13.

An early example of a successful incorporation of larger-scale information into a mesoscale model is that of Carpenter (1979), who has succeeded in changing boundary conditions to represent a varying synoptic regime in his simulation of the sea breezes over England for a case study day. He found that his simulated sea-breeze fronts were very sensitive to the position of synoptic-scale features. Ballentine (1980) has been equally successful in applying synoptic tendencies to a simulation of New England coastal frontogenesis. In both of these studies, however, the mesoscale systems were primarily forced by the underlying terrain, rather than by input through the side walls of the models (i.e., terrain-induced mesoscale systems as discussed in Section 13.2).

Seth and Giorgi (1998) discuss the role of lateral domain size with respect to accurate regional atmospheric model simulations. A summary on issues associated with lateral boundary conditions is given in Warner et al. (1997). Another paper that discusses the role of lateral boundary conditions is Fukutome et al. (1999).

10.4.2 Top Boundary Conditions

The top of the mesoscale model, as with the lateral boundaries, should be removed as far as possible from the region of significant mesoscale disturbance. Ideally this would place the top at $\bar{p} = 0$, so that the density of air is zero.

It is not necessary to go this high in mesoscale models, however, because of the deep layer of stable thermodynamic stratification, which always exists throughout the stratosphere and in much of the upper troposphere. Such layers, almost always stable even to the lifting of saturated air, inhibit vertical advection and tend to generate circulations that have larger horizontal than vertical scales. As shown, for example, by Pielke (1972), increased stratification causes shallower circulations to develop and makes the hydrostatic assumption more applicable since the characteristic horizontal length scale L_x, becomes larger and the vertical length scale L_z becomes smaller. Only through vertical propagation by wave motion can information from near the surface be propagated upward through this stable region.

Using this characteristic of the Earth's atmosphere, modelers have placed the tops of their domains:

1. deep within the stratosphere (e.g., Klemp and Lilly 1978; Peltier and Clark 1979).
2. at the tropopause (e.g., Mahrer and Pielke 1975), or
3. within the stable layer of the troposphere (e.g., Estoque 1961; Pielke 1974a).

The selection of these levels is based on numerical experimentation with different depths for a model, as well as from linear models, which suggest the depth in the atmosphere to which the mesoscale circulation will extend. In sea-breeze models over flat terrain, for example (see Fig. 10.7), the simulation results are essentially unaffected by increasing the depth of the model from 4.2 to 12 km. With strong airflow over rough terrain, however, Peltier and Clark (1979) maintain that the lower stratosphere must be resolved to represent turbulence at those levels properly. They concluded that the turbulence can reflect upward propagating internal gravity waves, thereby causing large amplifications in the flow over the mountain near the ground.

The form of the top is also important. Modelers have utilized rigid tops (e.g., Estoque 1961, 1962), impervious material surfaces (e.g., Pielke 1974a), porous lids (e.g., Lavoie 1972), and absorbing layers (e.g., Anthes and Warner 1978; Klemp and Lilly 1978; Mahrer and Pielke 1978b). In the past, rigid tops were utilized since they purportedly eliminated rapidly moving external gravity waves from the solutions thereby permitting longer time steps in the explicit finite difference schemes that were used.

With a rigid top, the vertical velocity is set to zero and pressure is adjusted to account for mesoscale perturbations at that level. As can be expected, however, unless the solutions would naturally approach zero at that level, the solutions are arbitrarily constrained. Estoque (1973) performed a set of equivalent experiments using a sea-breeze model with and without a rigid lid. When the rigid lid was used, pressure was changed on the top boundary to compensate for the restrictive requirement on vertical velocity, whereas in the second experiment, pressure was set equal to a constant at that level and vertical velocity changes were permitted. Solutions from these two experiments, given in Fig. 10.14, are significantly different. It is not certain, of course, if either solution is realistic, yet it is clear that if a rigid lid is to be imposed, it must be located well above the region of significant mesoscale disturbance.

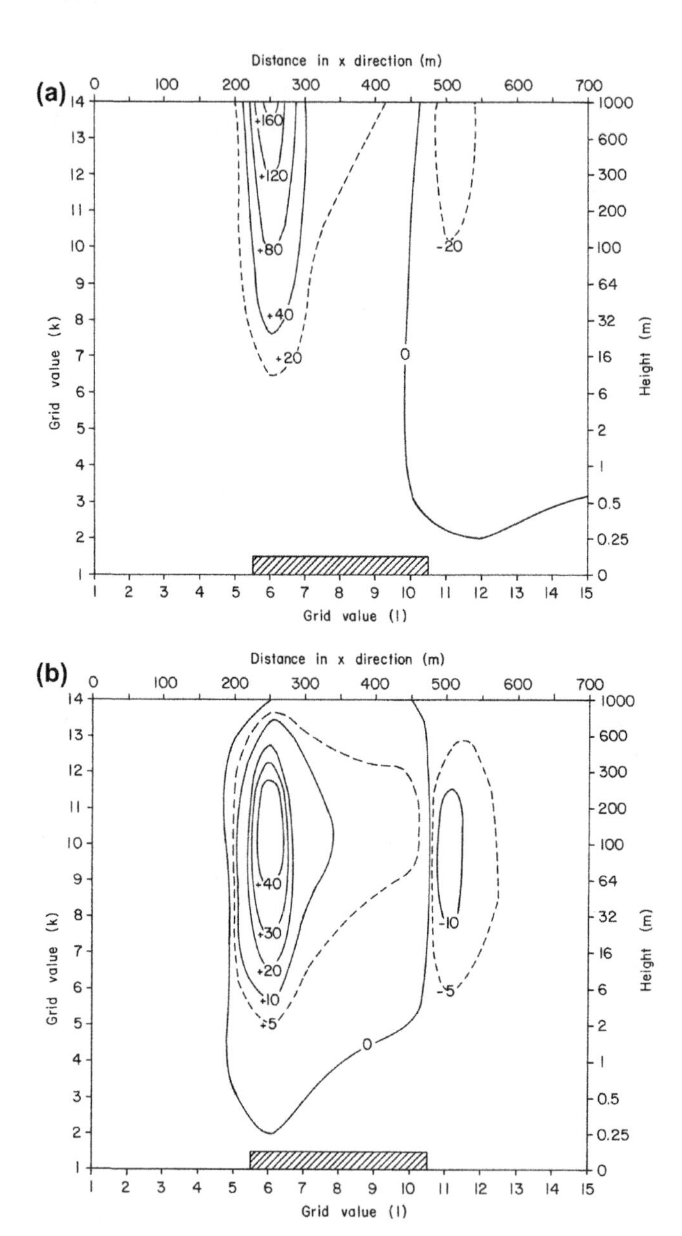

FIGURE 10.14

The vertical motion in centimeters per second owing to airflow over an island having a roughness length of 100 cm (between grid points 6 and 10, inclusive), and initial wind speed of 10 m s^{-1}; $\bar{\theta}$ at the surface and at 1 km is 303 and 309 K, respectively. The horizontal grid spacing is 50 m. In run (a), the pressure of the top is kept fixed at 900 mb while vertical motion at that level varies. In run (b), the top is made rigid and pressure varies (from Estoque 1973).

The impervious material surface lid was introduced to remedy some of the criticisms of the rigid top assumption. The form of this lid used by Pielke (1974a) illustrates this approach. Assuming the incompressible form of the conservation-of-mass relation as valid (i.e., Eq. 4.23), that expression is integrated between the highest fixed grid level in the model z_t, and a material surface s_θ, yielding

$$\bar{w}_s = \bar{w}_{z_t} - \int_{z_t}^{s_\theta} \left(\frac{\partial \bar{u}}{\partial x} + \frac{\partial \bar{v}}{\partial y} \right) dz,$$

where $\bar{w}_{z_t}$ and $\bar{w}_s$ are the vertical velocities at z_t and of the material surface. In the absence of diabatic effects at these levels, the material surface corresponds to a surface of constant potential temperature hence the use of the subscript "θ" on s.

Defining $\bar{w}_s = ds_\theta/dt$ and using the chain rule of calculus with $s_\theta = s_\theta(t, x(t), y(t))$, then

$$\frac{\partial s_\theta}{\partial t} = -\bar{u}\frac{\partial s_\theta}{\partial x} - \bar{v}\frac{\partial s_\theta}{\partial y} + \bar{w}_{z_t} - \int_{z_t}^{s_\theta} \left(\frac{\partial \bar{u}}{\partial x} + \frac{\partial \bar{v}}{\partial y} \right) dz. \tag{10.18}$$

In contrast to the rigid top, the material surface moves in response to divergence below and is considered to be a more realistic representation of conditions at the top of the model. This material surface is usually defined coincident with a surface of constant potential temperature and is placed at the tropopause level. If diabatic changes and vertical subgrid-scale mixing at this level are small relative to changes at a location due to advection, such a representation should closely correspond to the movement of the tropopause.

With the assumption that s_θ is a potential temperature surface, $\bar{\theta}$ is kept constant on it. The remaining dependent variables, however, must be estimated. One procedure is to use the initial values of these variables at whatever height is predicted for s_θ. This method, however, assumes that changes below s_θ have no influence on the variables, other than $\bar{\theta}$, at s_θ and above. Another possibility, as yet untested, is to insert several additional potential temperature layers above s_θ and to integrate an adiabatic form of the conservation relations at those levels in an isentropic coordinate representation in order to permit a dynamic adjustment of the dependent variables on s_θ. Another alternative is to treat s_θ as an interface between fluids of two different densities and thereby include a pressure gradient force on s_θ analogous to that derived for the two layer tank model in Section 5.2.1.2, where h corresponds to s_θ.

The use of a porous model top differs from the impervious lid in that mass transport is permitted across s_θ. Lavoie (1972) used such an approach since the lid of his model was the top of the planetary boundary layer, and he needed to entrain mass into it as it grew. Deardorff's (1974) prognostic equation for the depth of the planetary boundary layer is of the same form. Other than using such a level as a cap to the boundary layer, such an approach has not been used in mesoscale models.

The use of an *absorbing layer* with multiple levels to represent the top of the model was introduced by Klemp and Lilly (1978) in their simulation of airflow over rough terrain. As maintained in that paper, and based in part on results reported in Klemp and Lilly's (1975) linear model results, vertically propagating internal gravity wave energy can be erroneously reflected downward by a single level top. From their linear theory, some downward reflection is expected when discontinuities of temperature and wind occur, but the bulk of the energy is usually expected to propagate into the stratosphere and there to be dissipated by small-scale turbulence. In a linear model, the incorporation of a radiation boundary condition which permits this wave energy to leave is straightforward, but in a nonlinear model, Klemp and Lilly (1978) argued that local boundary conditions were unable to handle this effect properly.

Klemp and Lilly stated than an absorbing layer must be placed above the main portion of the model domain. In this region, horizontal filtering is increased from the base of the absorbing layer to the top of the model to prevent energy from being reflected down erroneously. To prevent reflections caused by the smoothing, the filter must be increased gradually. This approach is analogous to the sponge method used to minimize lateral boundary effects, as discussed in Section 10.4.1.1.

Based on the studies of top boundary conditions, the following conclusions are made.

1. If the vertical propagation of internal gravity wave energy is equal to or more important than advective properties in a mesoscale model, then an absorbing layer (or a boundary condition) is required.
2. Otherwise a material surface top, placed along an isentropic surface well removed from the region where advective changes are significant is sufficient.
3. The use of a rigid top in a mesoscale model is inappropriate unless advective effects are dominant and the depth of the model is much greater than the region of mesoscale disturbance. In this case, the precise form of the top (e.g., material surface, rigid top, or absorbing layer, etc.) is unimportant since perturbations that reach that level will be inconsequential for any of the dependent variables.

10.4.3 Bottom Boundary Condition

In a mesoscale atmospheric model, the bottom is the only boundary that has physical significance. Moreover, it is the differential gradient of the dependent variables along this surface that generates many mesoscale circulations (i.e., surface boundary-forced mesoscale systems, see Section 13.2) and that has a pronounced influence on the remaining mesoscale flows (i.e., initial value/lateral boundary-forced mesoscale systems, see Section 13.3). Changes in this lower boundary over time (e.g., from anthropogenic activity or overgrazing by animals) can cause substantial climatic changes such as desertification (Otterman 1975; Idso 1981). Because of the crucial importance of this boundary for mesoscale atmospheric systems, it is essential that it be represented as accurately as possible.

In discussing this component of mesoscale models, it is convenient to consider the land and water surfaces separately. This is done because water is translucent to solar radiation and overturns much more easily than land.

10.4.3.1 Water Bodies

To represent the surface of water bodies such as lakes, bays, and oceans properly in a mesoscale atmospheric model, it is necessary to permit dynamic and thermodynamic interactions between the air and the water (Pielke 1981). Such a connection can involve small-scale boundary layer interactions such as gaseous interchanges between the water–air interface, for example, or larger-scale transports of heat by wind-driven currents. Since these interactions generally involve complex nonlinear processes, it is necessary to use an oceanographic model to simulate these interactions properly and to provide appropriate bottom boundary conditions over water for the meteorological model.

Clancy et al. (1979) have attempted some simulations using a coupled sea-breeze and upwelling oceanographic model to study the interactions between these two geophysical phenomena. In their model, the sea surface temperature generated by upwelling was used as a bottom condition in the sea-breeze part of the model, and the wind shearing stress at the sea surface was utilized to influence the intensity of upwelling. Mizzi (1982), and Mizzi and Pielke (1984) continued this work using a more complete atmospheric model. Both studies determined that the interaction between upwelling and sea-breeze intensity was weak for this particular region of the Oregon coast. The direction of the *synoptic* flow, however, is important. Hawkins and Stuart (1980), for example, reported that northerly flow along the Oregon coast is associated with upwelling and strong sea breezes, whereas southerly synoptic flow results in a cessation of upwelling and weak sea breezes.

Avissar and Pan (2000) used the Regional Atmospheric Modeling System (RAMS) to simulate summer hydrometeorological processes of Lake Kinneret (Sea of Galilee) in Israel as a first step in developing a coupled atmosphere-lake coupled modeling system. Costa et al. (2001) has coupled RAMS to the Princeton Ocean Model in order to investigate deep cumulus cloud-ocean interactions in the western tropical Pacific Ocean. Xue et al. (2000a) describe results from a coupled mesoscale atmospheric-water body modeling system.

Over the Gulf stream, Sweet et al. (1981) have shown that the large horizontal sea-surface temperature gradient on the north side of that well-defined current generates lines of low-level cloud paralleling the edge of the Gulf stream. In addition, the sea state is significantly affected by the difference in stability across the Gulf stream boundary with rougher seas and stronger low-level winds on the warm side of the boundary and calmer, more humid conditions north of the current. Jacobs and Brown (1974) performed preliminary three-dimensional simulations of the air-lake interactions over Lake Ontario. Zeng et al. (1999a) uses observed data to describe the relation between sea-surface skin temperature, and wind speed and a sea-surface bucket temperature.

Other past work documents a number of the effects of wind on ocean dynamics. These include the following:

1. An increase of wind speed produces a deepening of the ocean-mixed layer (Marchuk et al. 1977; Chang and Anthes 1978; Elsberry 1978; Kondo et al. 1979).

2. An increase in wind speed produces small-scale wave breaking. This change over from an aerodynamically smooth to a rough water surface occurs over a rather narrow velocity range ($u_* \simeq 23$ cm s^{-1}; Melville 1977).
3. Spatial and temporal variations in wind velocity cause currents in coastal waters (Emery and Csanady 1973; Blackford 1978; Sheng et al. 1978; Svendsen and Thompson 1978).
4. Changes in wind speed and direction along a coastline alter the upwelling-downwelling pattern (Csanady 1975; Knowles and Singer 1977; Hamilton and Rattray 1978; Allender 1979).
5. Changes in wind speed alter the circulation in estuaries and harbors through mixing and the resultant creation of horizontal gradients of temperature and salinity in the water (Hachey 1934; Weisberg 1976; Long 1977; Wang 1979).
6. Wind energy absorbed by coastal waters is a function of the spectral energy of the wind (Lazier and Sandstrom 1978).
7. Wind velocity affects the drift of coastal pack ice (McPhee 1979).
8. The wind velocity field influences the movement of pollutants in the water (Pickett and Dossett 1979).
9. The wind causes the formation of helical circulation patterns in the water with the resultant accumulation of surface debris in lines parallel to the surface wind direction (Gross 1977).
10. The orientation of the coastline and ocean bathymetry influence wind-induced upwelling (Hua and Thomasset 1983).
11. The wind profile near the ocean surface is influenced significantly by blowing sea spray and rain during strong winds (Pielke and Lee 1991).
12. Diurnal variations of the sea-surface temperature can result in significant variations of surface turbulent fluxes (Zeng and Dickinson 1998).

The effects of mesoscale circulations on the coastal waters, therefore, often include diurnal changes in the vertical gradients of temperature, of salinity, and of other gaseous and aerosol materials in the upper levels of the water caused by changes in the temporal and spatial fields of the atmospheric dependent variables over the water surface. Pielke (1991) suggested that mesoscale resolution of ocean upwelling is needed to properly simulate the ocean uptake of carbon dioxide. The advection of aerosol may also affect the temperature of the coastal waters (through changes in the turbidity of the water) and hence feedback to the intensity of mesoscale circulations, as well as to local baroclinic circulations in the water. In general, however, the most significant influence of mesoscale atmospheric circulations on ocean dynamics is a result of the low-level wind that, through surface shearing stresses, influences currents and vertical turbulent mixing in the water. The water body, by contrast, primarily influences atmospheric mesoscale circulations through its surface temperature, including its time and space variability. Pielke (1981) discussed the interactions between coastal waters and the atmosphere in more detail. An overview of air-sea interactions is given in Rogers (1995).

10.4.3.2 Computation of Ocean Surface Fluxes[19]

To compute atmospheric turbulence in the surface layer, displacement height D, roughness lengths (z_0, z_{0t}, and z_{0q}), and surface potential temperature θ_g and humidity q_g are needed, as has been discussed in Chapter 7. The surface ocean current and sea ice (which moves with ocean current) is much smaller than the near-surface wind speed, and hence is usually omitted.

Over ocean or large lakes, D can be taken as zero, but z_0 cannot be taken as constant (e.g., see Chapter 8). Zeng et al. (1998) and Brunke et al. (2003) reviewed various formulations of roughness lengths, and the formulations from Zeng et al. are:

$$z_0 = 0.013\frac{u_*^2}{g} + 0.011\frac{\nu}{u_*}$$

$$ln\frac{z_0}{z_{0t}} = ln\frac{z_0}{z_{0q}} = 2.67Re_*^{1/4} - 2.57$$

where the roughness Reynolds number Re_* has been discussed in Section 7.2.2.

In general, ocean waves should be explicitly considered in the computation of z_0 (e.g., in the ECMWF operational model). At present, however, most applications still apply above equation directly.

These formulations are found to be valid for near-surface wind speed $V(z)$ from 0-23 m s^{-1}. For stronger wind conditions (e.g., under hurricane wind conditions), very limited observational data are available. Based on field measurements over open ocean, Black et al. (2007) suggested that z_0 is nearly constant for 10 m wind speed greater than 22-23 m s^{-1}. This means that the above equation can be used to compute z_0 for all wind conditions by taking $V = min(V, 23.0)$ (in unit m s^{-1}).

Based on high-wind wind tunnel experiments, Donelan et al. (2004) suggested that z_0 reaches its maximum value at 10 m wind speed of about 35 m s^{-1}. Then different formulations for z_0 (e.g., from Zweers et al. 2010) could be used. Under strong wind conditions, the determination of z_{0t} and z_{0q} is also uncertain. Black et al. (2007) suggested that $\ln(10/z_0)/\ln(10/z_{0t})$ is nearly constant ($\sim$0.7).

In atmospheric data assimilation, weather forecasting, and atmospheric (mesoscale or global) modeling, the term sea surface temperature (SST) usually refers to the (five-day to monthly) product of blended satellite retrievals and in situ measurements at a depth of a few centimeters to a few meters from buoys and ships (Reynolds and Smith 1994). In oceanic and ocean-atmosphere coupled modeling, the term SST refers to the mean temperature of the top ocean layer of about 10 meters in depth. Numerous studies (e.g., Fairall et al. 1996) have demonstrated that these temperatures are significantly different from the sea-surface skin temperature (T_g).

The skin temperature (T_g) needs to be computed from an ocean model coupled with a skin temperature parameterization (e.g., Zeng and Beljaars 2005) or specified from observed SST coupled with a skin temperature parameterization. In the Zeng

[19]The following text was written by Professor Xubin Zeng, Department of Atmospheric Sciences, University of Arizona, Tucson.

and Beljaars scheme, the temperature difference across the skin layer (with depth δ, typically a few micrometers to a few millimeters), $T_g - T_{-\delta}$, is computed from the net flux (including the ocean surface latent and sensible heat fluxes, net long-wave radiative flux, and the solar radiative flux absorbed in this thin layer), while the temperature difference across the warm layer, $T_{-\delta} - T_{-d}$ (with minimal diurnal temperature variation at depth d), is computed from the net flux and ocean turbulent mixing. This scheme has been implemented into the ECMWF operational model.

Then θ_g can be computed from T_g based on the definition of potential temperature, and q_g can be computed as the saturated specific humidity over saline sea water:

$$q_g = 0.98 q_{sat}(T_g, P_g)$$

where q_{sat} is the saturation specific humidity for pure water at T_g and surface pressure P_g. The factor of 0.98 is an approximation to $(1 - 0.527s)$ for the average oceanic salinity s of 34 parts per thousand. Note that the salinity factor of 0.98 could be important for the computation of ocean surface evaporation. For instance, for $q_{sat} =$ 20 g kg^{-1}, the surface-air humidity difference $(q_{sat} - q_a)$ of 4.0 g kg^{-1} would differ from $(q_g - q_a)$ of 3.6 g kg^{-1} by 10% over tropical ocean. Proportional to this difference, the surface evaporation would also differ by about 10%.

Over sea ice, D can be taken as zero, and Brunke et al. (2006) intercompared various formulations of roughness lengths. Their recommended formulations are

$$z_{0t} = z_{0q} = 5 \times 10^{-4} m$$
$$z_0 = \alpha_1[7 \times 10^{-4} - 6.5 \times 10^{-4} exp(-\alpha_2 max[0, u_* - \alpha_3])]$$

where $\alpha_1 = 1.4 + 0.2\,max(\alpha_4, T_g), \alpha_2 = 10 + 3\,max(\alpha_4, T_g), \alpha_3 = 0.05$ m s^{-1}, $\alpha_4 = -2°$C, and $T_g(\leq 0°\text{C})$ in the unit °C in these expressions.

Sea ice skin temperature (T_g) needs to be computed from a sea ice model or prescribed from observations. Then θ_g can be computed from T_s based on the definition of potential temperature, and q_g can be computed as the saturated humidity over ice at T_g and surface pressure.

10.4.3.3 *Land Surfaces*

The representation of land surfaces as a bottom boundary requires different types of models than required to represent the water interface properly. In contrast to water, the ground is opaque and does not readily overturn. To represent land as a bottom surface, it is convenient to consider bare soil separately from vegetated ground. The former characterization is much easier to simulate and mesoscale models have become increasingly more sophisticated in its representation. Vegetation effects, by contrast, are very complex. A variety of field campaigns have been performed to develop improved understanding of land-atmosphere interactions (Hall 1999 and references therein; LeMone et al. 2000; Nair et al. 2011; Pielke et al. 2011). Overviews of land-atmosphere interactions are given, for example, Pielke and Avissar (1990), Avissar (1995), Entekhabi (1995), Dolman et al. (2003), Cotton and Pielke (2007), and Pielke et al. (2007b).

Bare Soil As discussed in Chapter 7, the mesoscale horizontal velocity is zero at a roughness height z_0. Over bare soil, micrometeorological observations have shown the value of z_0 to be very small, in general, because rocks, stones, and grains of soil are usually small and offer relatively little resistance to the wind. A tabulation of representative values of z_0 is given in Table 10.2 for bare soils as well as for various vegetative surfaces. The mesoscale vertical velocity perpendicular to the ground surface is also equal to zero, so that the velocity and vertical subgrid momentum flux at the ground surface can be approximated by[20]

$$u\left(z_0\right) = v\left(z_0\right) = w\left(z_0\right) = 0; \quad \overline{u''w''} = -u_*^2 \cos\mu; \quad \overline{v''w''} = -u_*^2 \sin\mu. \tag{10.19}$$

The specification of potential temperature $\bar{\theta}$, water substance $\bar{q}_n$, other gaseous and aerosol atmospheric materials $\bar{\chi}_m$, and pressure $\bar{\pi}$ at the land surface are not as simple to estimate as are $\bar{u}$, $\bar{v}$, and $\bar{w}$. The variables $\bar{\theta}$, $\bar{q}_n$, and $\bar{\chi}_m$ generally depend on fluxes of these quantities into and out of the ground. Walko et al. (2000b), based on the work of Garratt (1992), for example, show that $T(z_0) = T_G + 2\theta_*/k$, over bare soil.

The pressure at level z_0 can be diagnosed from Eqs. 4.30 or 4.39 if the hydrostatic representation is used and if the pressure at some arbitrary level above z_0 is known. Equations 4.30 and 4.39 can then be integrated downward from that level. In an anelastic or a compressible model local hydrostatic equilibrium between the first grid point above the ground and the surface has often been assumed to diagnose pressure at the surface.

The need to introduce a boundary condition on Eq. 4.39 (or 4.30), however, introduces a problem in the integration of these relations. If the top of the model s_θ (defined by Eq. 10.18) is the level where the pressure boundary condition is needed, a method to estimate pressure at that level must be devised. Defining it as a constant is not satisfactory since net warming in a column must *initially* result in its expansion so that the pressure at s_θ will rise while remaining a constant at z_0 (Nicholls and Pielke 1994a).

To illustrate the problem, Eqs. 4.30 and 4.39 can be differentiated with time and integrated between z_0 and s_θ yielding[21]

$$\left.\frac{\partial \bar{\pi}}{\partial t}\right|_{s_\theta} - \left.\frac{\partial \bar{\pi}}{\partial t}\right|_{z_0} = g \int_{z_0}^{s_\theta} \frac{1}{\bar{\theta}^2}\frac{\partial \bar{\theta}}{\partial t} dz, \tag{10.20}$$

$$\left.\frac{\partial \bar{p}}{\partial t}\right|_{s_\theta} - \left.\frac{\partial \bar{p}}{\partial t}\right|_{z_0} = -g \int_{z_0}^{s_\theta} \frac{\partial \bar{\rho}}{\partial t} dz. \tag{10.21}$$

[20]If a terrain-following coordinate system is used (see Chapter 6), the boundary conditions given by Eq. 10.19 must be transformed properly to the new coordinate system.

[21]Note that even when the incompressible continuity Eq. 4.23 is used $\partial\bar{\rho}/\partial t \neq 0$. As shown in Section 3.1, the use of the incompressible and anelastic forms of the conservation-of-mass are permitted when temporal changes of density are small relative to the other terms in Eq. 2.38, but not necessarily equal to zero.

Table 10.2 Representative values of aerodynamic roughness for a uniform distribution of these types of ground cover.

	Aerodynamic Roughness z_0	Height of Ground Cover	Displacement Height D
Ice[a]	0.001 cm		
Smooth mud flats[f]	0.001 cm		
Snow[a]	0.005–0.01 cm		
Sand[a]	0.03 cm		
Smooth desert[f]	0.03 cm		
Smooth snow on short grass[c]	0.005 cm		
Snow surface, natural prairie[c]	0.1 cm		
Soils[a]	0.1–1 cm		
Short grass[a]	0.3–1 cm	2–10 cm	
Mown grass[c]	0.2 cm	1.5 cm	
	0.7 cm	3 cm	
	2.4 cm with $\overline{V}$ at 2 m $= 2$ m s^{-1}	4.5 cm	
	1.7 cm with $\overline{V}$ at 2 m $= 6.8$ m s^{-1}		
Long grass[a]	4–10 cm	25 cm to 1 m	
Long grass (60–70 cm)	15 cm[f], 9 cm[c] with $\overline{V}$ at 2 m $= 1.5$ m s^{-1}		
	11 cm[f], 6.1 cm[c] with $\overline{V}$ at 2 m $= 3.5$ m s^{-1}		
	8 cm[f], 3.7 cm[c] with $\overline{V}$ at 2 m $= 6.2$ m s^{-1}		
Agricultural crops[a]	4–20 cm[d]	~40 cm to 2 m[d]	~27 – ~1.3 m[e]
Orchards[a]	50–1 m[d]	~5 m to 10 m[d]	~3.3 – ~6.7 m[e]
Deciduous forests[a]	1–6 m[d]	~10 m to 60 m[d]	~6.7 – ~40 m[e]
Coniferous forests[a]	1–6 m[d]	~10 m to 60 m[d]	~6.7 – ~40 m[e]
Rural Delmarva peninsula[b]	33 cm (for NW flow)		
Pakistan desert[c]	0.03 cm		

[a] *From Oke (1978).*
[b] *From Snow (1981).*
[c] *From Priestly (1959).*
[d] *Using Eq. 7.36 from Pielke (2002b).*
[e] *Using type Eq. 7.35 from Pielke (2002b).*
[f] *From Sellers (1965).*

Using the definition of $\bar{\pi}$ (following Eq. 4.34) of potential temperature (Eq. 2.48), and the ideal gas law (Eq. 2.49), Eqs. 10.20 and 10.21 are related by

$$\frac{\partial \bar{\pi}}{\partial t} = \frac{1}{\bar{\rho}\bar{\theta}}\frac{\partial \bar{p}}{\partial t},$$

so that Eq. 10.20 can be rewritten as

$$\left.\frac{\partial \bar{\pi}}{\partial t}\right|_{s_\theta} - \left.\frac{\partial \bar{\pi}}{\partial t}\right|_{z_0} = g\int_{z_0}^{s_\theta} \frac{1}{\bar{\theta}^2}\frac{\partial \bar{\theta}}{\partial t}dz = \frac{-g}{\bar{\rho}\bar{\theta}}\int_{z_0}^{s_\theta} \frac{\partial \bar{p}}{\partial t}dz. \quad (10.22)$$

As evident in Eq. 10.22, for a fixed s_θ, a change in the heat content of a column is equivalent to a change of the mass in that column for a hydrostatic atmosphere. Such changes in heat could occur due to physical mechanisms such as radiative flux divergence (i.e., see Eq. 7.22). With fixed pressure on s_θ (i.e., $\partial \bar{\pi}/\partial t|_{s_\theta} \equiv 0$), the result of heating, as represented by the middle term in Eq. 10.22, would be a drop in surface pressure.

In the real atmosphere, by contrast, net heating in a column would result in a thickness adjustment such that in the absence of advection or wave propagation in the horizontal

$$-g\int_{z_0}^{s_\theta} \frac{1}{\bar{\theta}}dz = C; \quad (10.23)$$

where C is a negative constant (since $\bar{\pi}_{s_\theta} < \bar{\pi}_{z_0}$). In the atmosphere such a thickness adjustment can be performed by compressibility (Nicholls and Pielke 1994a), which is not permitted in the model if either of the conservation-of-mass relations given by Eq. 4.23 are used. Equation 10.23 is obtained by integrating Eq. 4.39 between z_0 and s_θ. Differentiating Eq. 10.23 with respect to time and using Leibnitz's rule (e.g., see Hildebrand 1962:360) yields

$$\int_{z_0}^{s_\theta} \frac{1}{\bar{\theta}^2}\frac{\partial \bar{\theta}}{\partial t}dz - \frac{1}{\bar{\theta}_{s_\theta}}\frac{\partial s_\theta}{\partial t} = 0,$$

where $\bar{\theta}_{s_\theta}$ is the value of potential temperature on s_θ. Solving for $\partial s_\theta/\partial t$ gives

$$\frac{\partial s_\theta}{\partial t} = \bar{\theta}_{s_\theta}\int_{z_0}^{s_\theta} \frac{1}{\bar{\theta}^2}\frac{\partial \bar{\theta}}{\partial t}dz, \quad (10.24)$$

so that net heating results in an increase in the thickness of the layer (i.e., $\partial s_\theta/\partial t > 0$), whereas cooling generates a decrease in the thickness. Feliks and Huss (1982) discussed the need to include this effect in mesoscale models. The change in thickness caused by expansion or contraction in Eq. 10.24 should be added to the expression for s_θ in Eq. 10.18 (i.e., $\partial s_\theta/\partial t = \partial s_\theta/\partial t|$ from Eq. 10.18 $+\partial s_\theta/\partial t|$ from Eq. 10.24).

Surface Temperature At the surface, early mesoscale models prescribed potential temperature as a periodic heating function and permitted no feedback between the mesoscale circulation and the ground surface temperature. Pielke (1974a) used such a representation given as

$$\bar{\theta}(z_0) = \bar{\theta}_0(z_0) + \Delta\bar{\theta}_{max} \sin \frac{2\pi t}{\text{day}}, \tag{10.25}$$

where day was equal to twice the length of daylight, t is the time after sunrise, $\bar{\theta}_0(z_0)$ the potential temperature at z_0 at sunrise, and $\Delta\bar{\theta}_{max}$ the maximum temperature attained during the day. Downward-looking radiometer measurements from aircraft were used to estimate $\Delta\bar{\theta}_{max}$.

A more general periodic form was used by Neumann and Mahrer (1971), and Mahrer and Pielke (1976) where observed temperature data were fitted to a series of periodic functions of the form suggested by Kuo (1968). Such a periodic form can be written following Panofsky and Brier (1968), as

$$\bar{\theta}(z_0) = \bar{\theta}_T(z_0) + \sum_{n=1}^{N/2} \left(a_n \sin \frac{2\pi nt}{T} + b_n \cos \frac{2\pi nt}{T} \right) \tag{10.26}$$

where

$$\begin{aligned} a_n &= \frac{2}{N} \sum_{n=1}^{(N/2)-1} \bar{\theta}^+ (z_0) \sin \frac{2\pi nt}{T}; \quad a_{N/2} = 0 \\ b_n &= \frac{2}{N} \sum_{n=1}^{(N/2)-1} \bar{\theta}^+ (z_0) \cos \frac{2\pi nt}{T}; \quad b_{N/2} = -\frac{\bar{\theta}^+ (z_0)}{N} \end{aligned} \tag{10.27}$$

with T usually chosen as 1 day (i.e., to correspond to the diurnal cycle) with $t = 0$ corresponding to midnight, and N an even integer number of observations. The value of $\bar{\theta}_T(z_0)$ corresponds to the average temperature over T and is assumed constant within T. The quantity $\bar{\theta}^+(z_0)$ is the deviation of potential temperature at time t from $\bar{\theta}_T(z_0)$. In practice, a different temperature than $\bar{\theta}^+(z_0)$ is often used to obtain a_n and b_n. Neumann and Mahrer (1971), for example, used a series expansion equivalent to Eq. 10.26 with $N = 4$ and values of temperature from the paper of Kuo (1968, Fig. 6) at a soil depth of 0.5 cm. Mahrer and Pielke (1976) applied an expansion with $N = 8$ for use in Eq. 10.26 using radiometer data collected over Barbados in 1969 and analyzed for the diurnal temperature wave by Holley (1972). Estoque et al. (1976) used a representation of the form given by Eq. 10.26 with t defined in terms of hours after 0700 LST using infrared measurements of surface temperature from aircraft flights over Lake Ontario.

Although simple to apply, such formulations imply an infinite reservoir of heat and permit no feedbacks between the ground and the atmosphere.

Surface Heat Energy Budget Physick (1976) developed the first mesoscale model that permitted feedbacks between the temperature of the ground surface and the atmosphere. He used a heat budget technique where the ground surface was assumed to have zero heat storage. Similar approaches were subsequently adopted by Estoque and Gross (1981), Mahrer and Pielke (1977a,b), and others.

With the heat budget method, the conductive Q_G, convective Q_C, and radiative Q_R contributions of heat to the ground surface are balanced resulting in an equilibrium surface temperature. Written more formally,

$$-Q_G + Q_C + Q_R = 0$$

In the soil, in general,

$$|Q_G| \gg |Q_C|; \quad |Q_G| \gg |Q_R|,$$

so that only ground conduction is retained as the principal contribution to the heat balance (when rain falls or snow melts and percolates into the soil, substantial heat can be transferred so that these inequalities may not be satisfied).

In the atmosphere, the inequalities are reversed, i.e.,

$$|Q_C| \gg |Q_G|; \quad |Q_R| \gg |Q_G|,$$

since molecular transfers of heat in turbulent air are ineffective compared with radiation and convection (as discussed in Section 5.1 for convection; i.e., $e_{u_i}^2/S^2 \gg v/LS$) in Eq. 5.2.

Retaining the symbol Q_G to refer to ground heat conduction, the convective and radiative heat transfers in the atmosphere can be written as

$$Q_C = -\bar{\rho} C_p \overline{w''\theta''} - \bar{\rho} L_v \overline{w''q_3''};$$

and

$$Q_R = Q_N = (1 - A)\left(\bar{R}\downarrow_{\text{SWG}} + \bar{R}^{\text{D}}_{\text{SWG}}\right) + \bar{R}\downarrow_{\text{LWG}} - \bar{R}\uparrow_{\text{LWG}},$$

using algorithms for the turbulent sensible and latent heat fluxes Q_C and Q_R is composed of direct and diffuse shortwave radiation (such as from Eq. 7.84) and of upward and downward longwave radiation (such as from Eq. 7.60 evaluated at the ground level). The quantity Q_N is called the *net radiation*. An example of this balance of heat fluxes

$$+Q_G + \bar{\rho} C_p \overline{w''\theta''} + \bar{\rho} L_v \overline{w''q_3''} - (1 - A)\left(\bar{R}\downarrow_{\text{SWG}} + \bar{R}^{\text{D}}_{\text{SWG}}\right) - \bar{R}\downarrow_{\text{lwG}} + \bar{R}\uparrow_{\text{lwG}} = 0 \tag{10.28}$$

is illustrated in Fig. 10.15 for a location in Saudi Arabia. The length of the arrows plotted in the figure represent examples of the relative magnitudes of these fluxes, where the vector sum is zero since the interface is assumed to be infinitesimally thin and has no heat storage.

An additional term can be added to Eq. 10.28 if anthropogenic or natural sources and sinks of heat exist (Grimmond and Oke 1999, Masson 2000). When this is important, the heat budget should be written as

$$+Q_G + \bar{\rho} C_p \overline{w''\theta''} + \bar{\rho} L_v \overline{w''q_3''} - (1 - A)\left(\bar{R}\downarrow_{\text{SWG}} + \bar{R}^{\text{D}}_{\text{SWG}}\right) - \bar{R}\downarrow_{\text{lwG}} + \bar{R}\uparrow_{\text{lwG}} + Q_m = 0, \tag{10.29}$$

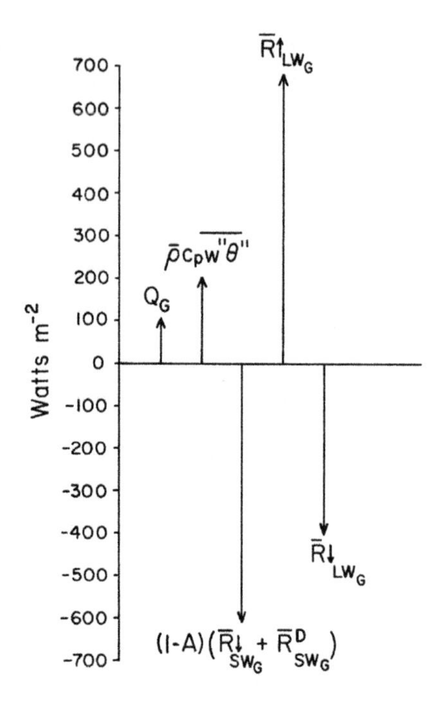

FIGURE 10.15

Average heat fluxes at 1300 LST observed over the Empty Quarter of Saudi Arabia. A positive value represents a loss to the surface and a negative value is a gain. The sum of the fluxes equals zero. Over the Empty Quarter, the latent heat flux is essentially zero (from Eric Smith, CSU 1982, personal communication).

where $Q_m > 0$ could refer to waste heat from heating or air conditioning in a city, heat from a forest fire or volcano, etc. Ichinose et al. (1999), for example, found that Q_m in central Tokyo, Japan exceeded 400 watts per meter squared in the daytime in the winter, with a maximum value of 1590 watts per meter squared.

Since at the ground $u = 0$, in the expression for the grid-volume average for $R\uparrow_{lw}$ in Eq. 7.60, $\bar{R}\uparrow_{lw_G} = \sigma\overline{T}_G^4$, if a ground emissivity of $\epsilon_G = 1$ is assumed. The temperature of the ground surface $\overline{T}_G$ is called the *equilibrium surface temperature*. If $\epsilon_G < 1$, the upward longwave flux from the ground should be written as

$$\bar{R}\uparrow_{lw_G} = \epsilon_G\sigma\overline{T}_G^4 + (1 - \epsilon_G)\,\bar{R}\downarrow_{lw_G}. \tag{10.30}$$

The second term on the right of Eq. 10.30 is the reflection of downward longwave radiation when $\epsilon_G \neq 1$. For this situation, the ground does not radiate as a blackbody (the definition of a blackbody is given following Eq. 7.27). Most ground surfaces, however, radiate close to a blackbody, as seen in Table 10.3. As discussed by Lee (1978:71), a value of $\epsilon_G = 0.95$, $\overline{T}_G = 25°C$, and an effective radiating temperature

Table 10.3 Emissivities of longwave radiation for representative types of ground covers.

Ground Cover	ϵ
Fresh snow	0.99[a]
Old snow	0.82[a]
Dry sand	0.95[b], 0.914[c]
Wet sand	0.98[b], 0.936[c]
Dry peat	0.97[b]
Wet peat	0.98[b]
Soils	0.90-0.98[a]
Asphalt	0.95[a], 0.956[c]
Concrete	0.71-0.90[a], 0.966[c]
Tar and gravel	0.92[a]
Limestone gravel	0.92[b]
Light sandstone rock	0.98[b]
Desert	0.84-0.91[a]
Grass lawn	0.97[b]
Grass	0.90-0.95[a]
Deciduous forests	0.97-0.98[a], 0.95[b]
Coniferous forests	0.97-0.98[a], 0.97[b]
Range over an urban area	0.85-0.95[a]
Pure water	0.993[c]
Water, plus thin film of petroleum oil	0.972[c]

[a] *From Oke (1973:15, 247).*
[b] *From Lee (1978:69).*
[c] *From Paltridge and Platt (1976:135).*

of the air above the surface[22] of 20°C results in a relative error of -1% when Eq. 10.30 is used with $\epsilon_G = 1$.

Soil Heat Flux The rate of conduction of heat within the soil $Q_G = \upsilon \partial T/\partial z|_G$ can be evaluated from a one-dimensional diffusion equation[23] within the soil, i.e.,

$$\frac{\partial T}{\partial t} = \frac{\partial}{\partial z}\frac{\upsilon}{\rho c}\frac{\partial T}{\partial z}, \tag{10.31}$$

where υ, c, and ρ are the thermal conductivity, specific heat capacity and soil density, respectively. The $\upsilon/\rho c = k_s$ is called the *thermal diffusivity*.[24] The quantity k_s determines the speed of penetration of a temperature wave into the soil, and υ indicates the rate of heat transport. The temperature gradient at ground level is $\partial T/\partial z|_G$. Examples of values of these parameters for different types of soils are given in Table 10.4.

[22]The effective radiating temperature T_* is defined from $R{\downarrow}_{\mathrm{lwG}} = \sigma T_*^4$.
[23]On the mesoscale, the horizontal conduction of heat in the soil is neglected since the horizontal grid is much larger than the vertical grid in the soil.
[24]The thermal diffusivity k_s is also referred to as *heat conductivity* and *thermometric conductivity* (Huschke 1959).

Table 10.4 Representative values of thermal conductivity υ, specific heat capacity c, density ρ, and thermal diffusivity k_S, for various types of surfaces.

Concrete	4.60[a]	879[a]	2.3[a] $\times 10^{3}$	2.3[a] $\times 10^{-6}$
Rock	2.93[a]	753[a]	2.7[a] $\times 10^{3}$	1.4[a] $\times 10^{-6}$
Ice	2.51[a]	2093[a]	0.9[a] $\times 10^{3}$	1.3[a] $\times 10^{-6}$
		2100[b]	0.92[b] $\times 10^{3}$	1.16[b] $\times 10^{-6}$
Snow				
New	0.14[a]	2093[a]	0.2[a] $\times 10^{3}$	0.3 $\times 10^{-6}$
	0.08[b]	2090[b]	0.10[b] $\times 10^{3}$	0.1[b] $\times 10^{-6}$
Old	1.67[a]	2093[a]	0.8[a] $\times 10^{3}$	1.0[a] $\times 10^{-6}$
	0.42[b]	2090[b]	0.48[b] $\times 10^{3}$	0.4[b] $\times 10^{-6}$
Nonturbulent air	0.03[a], 0.02[c], 0.025[b]	1005	0.0012[a] $\times 10^{3}$	21[a] $\times 10^{-6}$
***Clay soil* (40% pore space)**				
Dry	0.25[b]	890[b]	1.6[b] $\times 10^{3}$	0.18[b] $\times 10^{-6}$
10% liquid water	0.63[a]	1005[a]	1.7[a] $\times 10^{3}$	0.37[a] $\times 10^{-6}$
20% liquid water	1.12[a]	1172[a]	1.8[a] $\times 10^{3}$	0.53[a] $\times 10^{-6}$
30% liquid water	1.33[a]	1340[a]	1.9[b] $\times 10^{3}$	0.52[b] $\times 10^{-6}$
40% liquid water	1.58[b]	1550[b]	2.0[b] $\times 10^{3}$	0.51[b] $\times 10^{-6}$
***Sand soil* (40% pore space)**				
Dry	0.30[b]	800[b]	1.6[b] $\times 10^{3}$	0.24[b] $\times 10^{-6}$
10% liquid water	1.05[a]	1088[a]	1.7[a] $\times 10^{3}$	0.57[a] $\times 10^{-6}$
20% liquid water	1.95[a]	1256[a]	1.8[a] $\times 10^{3}$	0.85[a] $\times 10^{-6}$
30% liquid water	2.16[a]	1423[a]	1.9[a] $\times 10^{3}$	0.80[a] $\times 10^{-6}$
40% liquid water	2.20[b]	1480[b]	2.0[b] $\times 10^{3}$	0.74[b] $\times 10^{-6}$
***Peat soil* (80% pore space)**				
Dry	0.06[b]	1920[b]	0.3[b] $\times 10^{3}$	0.10[b] $\times 10^{-6}$
10% liquid water	0.10[a]	2302[a]	0.4[a] $\times 10^{3}$	0.12[a] $\times 10^{-6}$
40% liquid water	0.29[a]	3098[a]	0.7[a] $\times 10^{3}$	0.13[a] $\times 10^{-6}$
70% liquid water	0.43[a]	3433[a]	1.0[a] $\times 10^{3}$	0.13[a] $\times 10^{-6}$
80% liquid water	0.50[b]	3650[b]	1.1[a] $\times 10^{3}$	0.12[b] $\times 10^{-6}$
Light soil with roots	0.11[c]	1256[c]	.03[c] $\times 10^{3}$	0.30[c] $\times 10^{-6}$
Liquid water	0.63[a], 0.57[b]	4186[a]	1.0[a] $\times 10^{3}$	0.15 $\times 10^{-6}$

[a] *From Lee (1978:87).*
[b] *From Oke (1973:38).*
[c] *From Rosenberg (1974:66).*

Soil conductivity depends on a number of factors including the conductivity of the individual soil particles, their sizes, the compaction of the soil as measured by porosity, and the soil moisture. Soil particle size, for example, can vary widely from on the order of 1 μm for clay to 100 μm for sand. In addition, although not considered here, heat can be transferred by the percolation of water and change of phase of water. The freezing and thawing of soils, for example, contributes significantly to the heat budget within the soil (e.g., see Viterbo et al. 1999).

When water is present in the soil, its heat capacity, density, and thermal conductivity vary depending on the amount of water present. As reported in McCumber (1980)

$$\rho c = \left(1 - \eta_{\mathrm{s}}\right) \rho_i c_i + \eta \rho_{\mathrm{w}} c_{\mathrm{w}}, \tag{10.32}$$

where η_s is the saturation moisture content (its porosity), η is the volumetric moisture content ($\text{cm}^3\text{cm}^{-3}$), and $\rho_i c_i$ is the product of the density and heat capacity of the dry soil type i (η and η_s are discussed in more detail shortly). The density and specific heat capacity of water $\rho_w c_w$ are given in Table 10.4. Equation 10.32 is simply a weighting of the contributions to the volumetric heat capacity of the dry soil and of the liquid water that is present. The heat capacity of air has been omitted since it is negligibly small compared with the other two terms.

The thermal conductivity varies over several orders of magnitude as soil dries out. McCumber (1980), referring to Al Nakshabandi and Kohnke's (1965) empirical data,

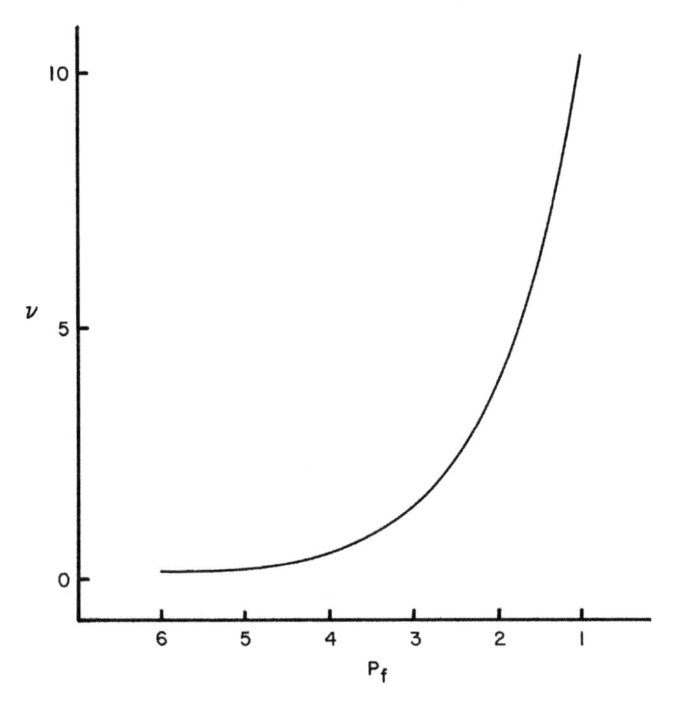

FIGURE 10.16

Dependence of soil thermal conductivity on moisture potential. The magnitude of the base ten logarithm of the moisture potential in centimeters is P_{f} [i.e., $P_{\mathrm{f}} = \log_{10} \Psi$, (cm)]. The thermal conductivity υ is plotted in units of J /(m s°C). This diagram corresponds to Fig. 4 in Al Nakshabandi and Kohnke (1965) (from McCumber 1980, Fig. 2).

expresses v in units of J /(m s°C) as

$$v = \begin{cases} 419 \exp - \left[\left(P_{\text{f}} + 2.7 \right) \right] & P_{\text{f}} \leq 5.1 \\ 0.172 & P_{\text{f}} > 5.1, \end{cases}$$

where P_{f} is the base ten logarithm of the magnitude of the moisture potential Ψ (moisture potential is defined following Eq. 10.39). Figure 10.16 illustrates the strong dependence of heat diffusivity on moisture potential. As shown by Al Nakshabandi and Kohnke (1965) moisture potential is virtually independent of the soil type, so that it is convenient to express v in this form.

Equation 10.31 is of the same form as discussed in Section 9.1.2 and can be solved on a vertical grid lattice by the implicit finite difference solution technique given by Eq. 9.25. Since the largest gradients in the diurnal variation of temperature are observed within 50 cm or so of soil surfaces, very fine resolution is required in such representations. McCumber (1980), and McCumber and Pielke (1981) used a vertical grid of 0, 0.5, 1.5, 3, 5, 8, 12, 18, 26, 36, 48, 62, 79, and 100 cm in his simulation of heat and moisture fluxes within various types of soils.

Soils are generally not homogeneous with depth. Nonetheless, this simple model provides a useful theoretical framework to understand the rate of transfer of heat into the ground.

An integrated form of Eq. 10.31 can also be used to compute the soil heat flux Q_{G}. Assuming ρ and c are constant with depth, Eq. 10.31 can be rewritten as

$$c\rho \frac{\partial T}{\partial t} = \frac{\partial}{\partial z} v \frac{\partial T}{\partial z} = \frac{\partial}{\partial z} Q_{\text{s}}, \tag{10.33}$$

where Q_{s} is the heat flux within the soil. Integrating Eq. 10.33 between the surface and a depth δ_z sufficiently deep so that $Q_{\text{s}} \equiv 0$, yields

$$Q_{\text{G}} = \rho c \int_{-\delta z}^{0} \frac{\partial T}{\partial t} dz. \tag{10.34}$$

Integrating the temporal temperature changes within the soil using this expression could provide a more accurate estimate of the soil heat flux at ground level than computing Q_{G} from $Q_{\text{G}} = v \partial T / \partial z|_{\text{G}}$. The quantity Q_{G} is also referred to as the *heat storage* term.

Albedo The amount of solar radiation impinging upon a horizontal surface is a function of such factors as latitude, cloud cover, time of year and of day (see Section 7.3.3). In this discussion, the only additional information required is the effect of terrain slope $\partial z_{\text{G}} / \partial x$ and $\partial z_{\text{G}} / \partial y$, albedo A, and ground wetness. This latter input is required, for example, to estimate the relative partitioning between latent and sensible turbulent heat fluxes in Eqs. 10.28 and 10.29.

The albedo of a surface is the fractional reflectance of radiation that reaches it. Although albedo is a function of wavelength, in mesoscale meteorology it usually refers to the reflectance of solar radiation, both direct and diffuse. Examples of representative albedos for a range of characteristic soils and other bare surfaces are given in Table 10.5 where it is seen, for instance, that fresh snow can reflect up to 95% of the solar radiation that reaches it, whereas dark soil (e.g., wet peat) reflects only 5%.

The albedo is not a constant at a given location, even with a uniform surface, but varies as a function of Sun zenith angle as well as the wetness of the soil (e.g., McCumber 1980). As reported in McCumber (1980), the variability of albedo with zenith angle Z has been estimated empirically, based on the work of Idso et al. (1975a), and the influence of wetness has been given by Idso et al. (1975a), and Gannon (1978). The effect of zenith angle and wetness on albedo can be represented mathematically as

$$A = A_Z + A_s \tag{10.35}$$

where

$$A_Z = \left[\exp\left(0.003286Z^{1.5}\right) - 1\right]/100,$$

and A_s is a function of the ratio of the volumetric moisture content η, and porosity η_s, of the soil. A plot of the functional form of A_Z is given in Fig. 10.17. The functional form of A_s is not available for all soil types, with Idso et al. (1975a) giving its value for Avondale loam soil as

$$A_s = \begin{cases} 0.31 - 0.34\Delta & \Delta \leq 0.5 \\ 0.14 & \Delta > 0.5, \end{cases}$$

with $\Delta = \eta/\eta_s$. For Florida peat, Gannon gives

$$A_s = \begin{cases} 0.14(1 - \Delta) & \Delta \leq 0.5 \\ 0.07 & \Delta > 0.5. \end{cases} \tag{10.36}$$

Using these expressions for A_Z and A_s in Eq. 10.35 it is evident that albedos are larger when the Sun is lower in the sky and for drier soils. As given by A_s for Avondale loam, for example, the albedo is 0.31 for dry soil (i.e., $\Delta = 0$) but decreases by over one-half to 0.14 for wet soils (i.e., $\Delta > 0.5$). Idso et al. (1975b) have provided additional details concerning the influence of soil moisture on albedo, and other components of the surface energy budget. Otterman (1974), Berkofsky (1977), and others have argued that changes in albedo can have a profound influence on average vertical motion with increased albedo (caused by overgrazing, for example) causing subsidence and a tendency toward *desertification* in arid areas. Viterbo and Betts (1999) demonstrated the improvement in synoptic weather prediction accuracy when a more accurate representation of the albedo of snow in the boreal forest is included.

Determination of Land Surface Albedo[25] Much progress has been made on land surface albedo along several lines. First, the albedos for visible and near-infrared bands (usually separating at wavelength of 0.7 μm) are computed, as the visible band is involved in the vegetation photosynthesis which is linked to the stomatal conductance

[25]The following text was written by Professor Xubin Zeng, Department of Atmospheric Sciences, University of Arizona, Tucson.

Table 10.5 Albedo of shortwave radiation for assorted types of ground covers.[a]

Ground Cover	*A*
Fresh snow	0.75–0.95[b], 0.70–0.95[c], 0.80–0.95[d], 0.95[e]
Fresh snow (low density)	0.85[f]
Fresh snow (high density)	0.65[f]
Fresh dry snow	0.80–0.95[g]
Pure white snow	0.60–0.70[g]
Polluted snow	0.40–0.50[g]
Snow several days old	0.40–0.70[b], 0.70[c], 0.42–0.70[d], 0.40[e]
Clean old snow	0.55[f]
Dirty old snow	0.45[f]
Clean glacier ice	0.35[f]
Dirty glacier ice	0.25[f]
Glacier	0.20–0.40[e]
Dark soil	0.05–0.15[b], 0.05–0.15[g]
Dry clay or gray soil	0.20–0.35[b], 0.20–0.35[g]
Dark organic soils	0.10[f]
Clay	0.20[f]
Moist gray soils	0.10–0.20[g]
Dry clay soils	0.20–0.35[d]
Dry light sand	0.25–0.45[b]
Dry light sandy soils	0.25–0.45[g]
Dry sandy soils	0.25–0.45[d]
Light sandy soils	0.35[f]
Dry sand dune	0.35–0.45[b], 0.37[c]
Wet sand dune	0.20–0.30[b], 0.24[c]
Dry light sand, high sun	0.35[f]
Dry light sand, low sun	0.60[f]
Wet gray sand	0.10[f]
Dry gray sand	0.20[f]
Wet white sand	0.25[f]
Dry white sand	0.35[f]
Peat soils	0.05–0.15[d]
Dry black coal spoil, high sun	0.05[f]
Dry concrete	0.17–0.27[b], 0.10–0.35[e]
Road black top	0.05–0.10[b]
Asphalt	0.05–0.20[e]
Tar and gravel	0.08–0.18[e]
Long grass (1.0 m)	0.16[e]
Short grass (2 cm)	0.26[e]
Wet dead grass	0.20[f]
Dry dead grass	0.30[f]
Typical fields	0.20[f]
Dry steppe	0.25[f], 0.20–0.30[g]
Tundra and heather	0.15[f]
Tundra	0.18–0.25[e], 0.15–0.20[g]

(Continued)

Table 10.5 Continued

Ground Cover	*A*
Meadows	0.15–0.25[g]
Cereal and tobacco crops	0.25[f]
Cotton, potatoes, and tomato crops	0.20[f]
Sugar cane	0.15[f]
Agricultural crops	0.18–0.25[e], 0.20–0.30[d]
Rye and wheat fields	0.10–0.25[g]
Potato plantations	0.15–0.25[g]
Cotton plantations	0.20–0.25[g]
Orchards	0.15–0.20[e]
Deciduous forests, bare of leaves	0.15[e]
Deciduous forests, leaved	0.20[e]
Deciduous forests	0.15–0.20[g]
Deciduous forests, bare with snow on the ground	0.20[d]
Mixed hardwoods in leaf	0.18[f]
Rain forest	0.15[f]
Eucalyptus	0.20[f]
Forest	
pine, fir, oak	0.10–0.18[c]
coniferous forests	0.10–0.15[g], 0.10–0.15[d]
red pine forests	0.10[f]
Urban area	0.10–0.27 with an average of 0.15[e]
Water	$-0.0139 + 0.0467 \tan Z \quad 1 \geq A \geq 0.03$[h]

[a] *The smaller number is for high solar zenith angles, while the larger albedo is more representative for low Sun angles.*
[b] *From Sellers (1965:21).*
[c] *From Munn (1966:15).*
[d] *From Rosenberg (1974:27).*
[e] *From Oke (1973:15, 247).*
[f] *From Lee (1978:58–59).*
[g] *From de Jong (1973).*
[h] *From Atwater and Ball (1981:879).*

for vegetation transpiration (e.g., Oleson et al. 2010). Second, the albedo's dependence on Z has been considered for each vegetation type (including desert) [e.g., Wang et al. 2005, 2007 based on the MODIS bidirectional reflectance distribution functions – BRDF]. Based on observational data, Yang et al. (2008) parameterized the direct-beam albedo as a product of the direct-beam albedo at $Z = 60°$ (or the diffuse albedo), which varies with surface type or geographical location and/or season, and a function that depends only on Z.

When the Z-dependent direct-beam albedo and Z-independent diffuse albedo are considered, the fourth term in Eq. 10.28 for the surface-absorbed solar radiation has to be computed for the direct-beam and diffuse radiation, separately. Third, instead of

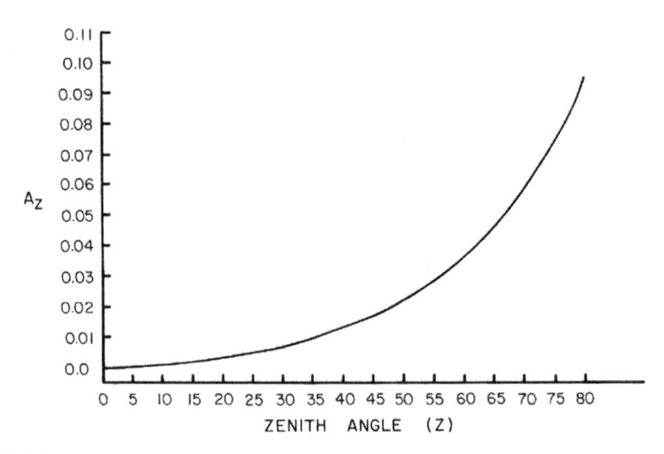

FIGURE 10.17

Albedo as a function of zenith angle Z. The enhancement to the surface albedo is expressed as a fractional increase relative to its value when the Sun is directly overhead (from McCumber 1980).

computing the land (vegetation and ground combined) albedo, the radiative transfer through canopy can be computed (e.g., based on the two-stream approximation, Oleson et al. 2010).

The forest shading of the underlying snow has received much attention (e.g., Viterbo and Betts 1999). Barlage et al. (2005) developed a new global 0.05° maximum albedo for snow covered land from the MODIS BRDF/albedo, reflectance, and land cover data. They found that the global mean maximum snow albedos vary from 0.35 for forests, 0.56 for savanna, 0.70 for grassland, to 0.84 for barren.

Soil Moisture Flux Representation. The determination of η and η_s requires a model of the flux of moisture into and out of the soil. Detailed relationships for computing the temporal fluctuations in soil moisture content are available (e.g., Philips 1957) and have been used in atmospheric models such as Sasamori (1970) and Garrett (1978). Beljaars et al. (1996) and Viterbo and Betts (1999) demonstrated the importance of accurate soil moisture values in the context of numerical weather prediction. In this section, the approach utilized by McCumber will be discussed. Other useful discussion of the parameterization of heat and moisture flows in and at the surface of soils are given in Camillo and Schmugge (1981), Sievers et al. (1983), Lee and Pielke (1992), and Mihailovic et al. (1993, 1995, 1999a). Shao and Irannejad (1999) compare the impact of four soil hydraulic models on land-surface processes.

The derivation of equations for soil moisture flux is presented in most courses in hydrology. For use in mesoscale meteorological models where the determination of the soil moisture at the surface W_G is required, the following form (McCumber 1980) can be used.

In Chapter 2, the conservation-of-motion relation for the atmosphere was derived. In the soil a similar expression can be derived for the conservation-of-motion of water within the soil. For short-term ($\sim$24 hours) mesoscale atmospheric applications, the vertical transport of water in the soil is usually much more significant than horizontal transport so that only the vertical equation-of-water movement need be considered. The neglect of horizontal advection within the ground on these time periods arises because the movement of water through soil is usually relatively slow, relative to atmospheric transports, and the horizontal gradients of water are generally assumed much less than vertical gradients. For longer time periods, however, as shown by Walko et al. (2000b), surface and subsurface flow over water can significantly alter the amount of water available to evaporate and transpire into the atmosphere. In this section, however, we will retain the short-term mesoscale focus of this book.

Therefore, using only the vertical equation-of-motion and assuming horizontal homogeneity

$$\frac{\partial w}{\partial t} + w\frac{\partial w}{\partial z} = -\frac{1}{\rho_W}\frac{\partial p}{\partial z} - g + F_W,$$

where F_W is the dissipation of water motion by viscous forces. In the atmosphere, as discussed following Eq. 3.29, scale arguments were used to show that, away from the ground, molecular dissipation of air motion was much smaller than the advective transport of air. This is not generally the case in soil.

If it is further assumed that local changes in vertical velocity are small relative to the other terms, the Reynolds number is much less than unity, and F_W can be parameterized[26] by

$$F_W = \frac{\mu_W}{\rho_W}\nabla^2 w,$$

(where ρ_W and μ_W are the density and the dynamic viscosity of water, respectively) then the vertical equation-of-water movement is given by

$$0 = -\frac{1}{\rho_W}\frac{\partial p}{\partial z} - g + \frac{\mu_W}{\rho_W}\nabla^2 w. \tag{10.37}$$

As with the atmospheric equations (as discussed in Chapter 4), this differential equation is averaged over a grid volume. The need for averaging in the soil, however, is different than the reason for performing this operation in the atmosphere. In a mesoscale atmospheric model averaging is required because sufficient computer resources are unavailable to solve the conservation relations at intervals of 1 cm or so. In the soil, however, averaging is needed because, otherwise, the conservation relation for vertical water movement would have to be evaluated separately for each grain of soil and interconnecting air space.

The averaging volume is defined as

$$\Delta V = \int_{\Delta v} dv = \int_{\eta_s \Delta V} dv + \int_{(1-\eta_s)\Delta V} dv, \tag{10.38}$$

where $\Delta V = \Delta x \Delta y \Delta z$, and η_s (the *porosity*) is the fraction of the soil volume containing air space (also called void space). Because the transfer of water is much more

[26]The equivalent atmospheric term is given by Eq. 3.29 in Chapter 3.

rapid through the air spaces between soil grains than through the soil material itself, Eq. 10.37 is averaged using the first integral on the righthand side of Eq. 10.38.

The first term on the right of Eq. 10.37, after averaging, becomes

$$\int_{\eta_s \Delta V} \frac{1}{\rho_w}\frac{\partial p}{\partial z} dv = \frac{1}{\rho_w}\overline{\frac{\partial p}{\partial z}} \eta_s \Delta V = \frac{1}{\rho_w}\frac{\partial \bar{p}}{\partial z} \eta_s \Delta V,$$

where $\overline{\partial p/\partial z} = \partial \bar{p}/\partial z$ because the averaging volume is not a function of depth. The gravitational acceleration term is straightforward and becomes

$$\int_{\eta_s \Delta V} g dv = g \eta_s \Delta V.$$

The third term on the right of Eq. 10.37 is rewritten as

$$\frac{\mu_w}{\rho_w} \nabla^2 w = \alpha \frac{\mu_w}{\rho_w} \frac{u_d}{d^2} \tag{10.39}$$

where d is a length scale, u_d a velocity scale (called the Darcian velocity) representing the mean velocity of water flow through the soil, and α is a proportionality factor. This formulation is based on the dimensions of the component terms in $\nabla^2 w$. Assuming that each term in Eq. 10.39 except α is constant in the averaging volume, the integrated form of this expression is given as

$$\int_{\eta_s \Delta V} \frac{\mu_w}{\rho_w} \alpha \frac{u_d}{d^2} dV = \frac{\mu_w}{\rho_w} \bar{\alpha} \frac{u_d}{d^2} \eta_s \Delta V.$$

Equation 10.37 can now be written as

$$0 = -\frac{1}{\rho_w}\frac{\partial \bar{p}}{\partial z} \eta_s \Delta V - g \eta_s \Delta V + \frac{\mu_w}{\rho_w} \bar{\alpha} \frac{u_d}{d^2} \eta_s \Delta V,$$

or, after rearranging,

$$g + \frac{1}{\rho_w}\frac{\partial \bar{p}}{\partial z} = \frac{\mu_w}{\rho_w} \bar{\alpha} \frac{u_d}{d^2}.$$

Solving for u_d yields

$$u_d = g \frac{\rho_w d^2}{\mu_w \bar{\alpha}} \frac{\partial}{\partial z} \left(z + \frac{\bar{p}}{g \rho_w} \right)$$

or

$$u_d = K_\eta \frac{\partial}{\partial z} \left(z + \Psi \right),$$

where $K_\eta = \left(g d^2 / \mu_w \bar{\alpha} \right) \rho_w$ is called the *hydraulic conductivity* and $\Psi = -\bar{p}/g\rho_w$ the *moisture potential.* The exchange coefficient K_η accounts for the influence of gravity drainage in the viscous soil, and Ψ represents the potential energy needed to extract water against capillary and adhesive forces in the soil.

The soil moisture flux is thus given as

$$W_s = \rho_w u_d = \rho_w K_\eta \frac{\partial}{\partial z}(z + \Psi). \tag{10.40}$$

With the specification of the soil moisture flux, the local time rate of change of the volumetric moisture content η is given by

$$\frac{\partial \eta}{\partial t} = \frac{1}{\rho_w} \frac{\partial W_s}{\partial z}, \tag{10.41}$$

where, consistent with the derivation of Eq. 10.37, advection of water is neglected. In addition, sources and sinks of water such as from rainfall are neglected.

Equation 10.40 can also be written as

$$W_s = D_\eta \rho_w \frac{\partial \eta}{\partial z} + K_\eta \rho_w, \tag{10.42}$$

where the chain rule of calculus has been used to write $D_\eta = K_\eta \partial\Psi/\partial\eta$.

The parameters of K_η, D_η, and Ψ are related to η using a set of empirical relations reported in Clapp and Hornberger (1978) and given as

$$\Psi = \Psi_s \left(\frac{\eta_s}{\eta}\right)^b; \quad K_\eta = K_{\eta_s} \left(\frac{\eta}{\eta_s}\right)^{2b+3}; \quad D_\eta = -\frac{b K_{\eta_s} \Psi_s}{\eta} \left(\frac{\eta}{\eta_s}\right)^{b+3},$$

where Ψ_s and K_{η_s} refer to the saturated soil values. McCumber (1980) has listed a table of these parameters (reproduced here as Table 10.6) as a function of 11 U.S. Department of Agriculture (1951) soil textural classes plus peat.

Another option uses the work of van Genuchten (1980). In his work, as reported in Walko et al. (1998, 2000b),

$$K_\eta = K_{\eta_s} \eta_e^{0.5} \left[1 - \left(1 - \eta^{1/m}\right)^m\right]^2$$

$$\Psi = \frac{1}{a_1} \left[(\eta_e)^{-1/m} - 1\right]^{1/n}$$

with

$$\eta_e = \frac{\eta - \eta_r}{\eta_s - \eta_n}.$$

The variable η_n is the minimum possible value of η, and a, m, and n are empirical parameters given by Carsel and Parrish (1988).

In these formulations, K_{η_s} can be assumed to decrease exponentially based on the work of Beven (1982, 1984).

$$K_{\eta_s}(z) = K_0 e^{f\hat{z}}$$

where f^{-1} is the *e*-folding depth of K_{η_s}. The height $\hat{z}$ is the distance below the ground surface. The use of these formulations is discussed further in Walko et al. (2000a).

Table 10.6 Soil parameters as a function of 11 United States Department of Agriculture (USDA 1951) textural classes plus peat. Units for soil porosity (η_s) are centimeters per centimeters cubed, saturated moisture potential (Ψ_s) is in centimeters, and saturated hydraulic conductivity (K_{η_s}) is expressed in centimeters per second. The exponent b is dimensionless. Permanent wilting moisture content (η_{wilt}) is in centimeters cubed per centimeters cubed, and it corresponds to 153 m (15b) suction. Dry volumetric heat capacity ($\rho_i c_i$) is in joules per centimeter cubed per degree Celsius. The first four variables for the USDA textures are reproduced from Clapp and Hornberger (1978). Table adapted from McCumber (1980). Values for the saturated hydraulic conductivity for peat range from 1×10^{-5} centimeters per second for deeply humidified sapric peat to 2.8×10^{-2} centimeters per second in relatively undercomposed fibric peat are presented. Average soil porosity ranges from 0.83 to 0.93 (Letts et al. 2000).

Soil Type	η_s	Ψ_s	K_{η_s}	b	η_{wilt}	$\rho_i c_i$
Sand	.395	−12.1	.01760	4.05	.0677	1.47
Loamy sand	.410	−9.0	.01563	4.38	.0750	1.41
Sandy loam	.435	−21.8	.00341	4.90	.1142	1.34
Silt loam	.485	−78.6	.00072	5.30	.1794	1.27
Loam	.451	−47.8	.00070	5.39	.1547	1.21
Sandy clay loam	.420	−29.9	.00063	7.12	.1749	1.18
Silty clay loam	.477	−35.6	.00017	7.75	.2181	1.32
Clay loam	.476	−63.0	.00025	8.52	.2498	1.23
Sandy clay	.426	−15.3	.00022	10.40	.2193	1.18
Silty clay	.492	−49.0	.00010	10.40	.2832	1.15
Clay	.482	−40.5	.00013	11.40	.2864	1.09
Peat	.863	−35.6	.00080	7.75	.3947	0.84

Solving Eq. 10.41 at the surface permits the determination of A_s for use in Eq. 10.35 which is needed in the surface heat budget calculation. In addition, as shown following Eq. 10.43, η is needed in the determination of the equilibrium moisture value at the ground surface.

Figures 10.18, 10.19, 10.20, and 10.21,[27] reproduced from McCumber (1980), illustrate the influence of different types of bare soil on diurnal variations of surface temperature, specific humidity, sensible heat flux, and latent heat flux. As evident from Fig. 10.18, for example, a sand surface has a much larger variation in temperature than the other soils (as evident by anyone who has walked on a beach in the summer), whereas peat and marsh have the largest excursions in surface moisture (Fig. 10.19). McCumber concluded that the large excursions in the sand temperature were a result of its inability to hold water effectively and to transfer it up from below, whereas peat

[27]The procedure to calculate the equilibrium moisture value that is discussed following Eq. 10.39 was used to make these calculations.

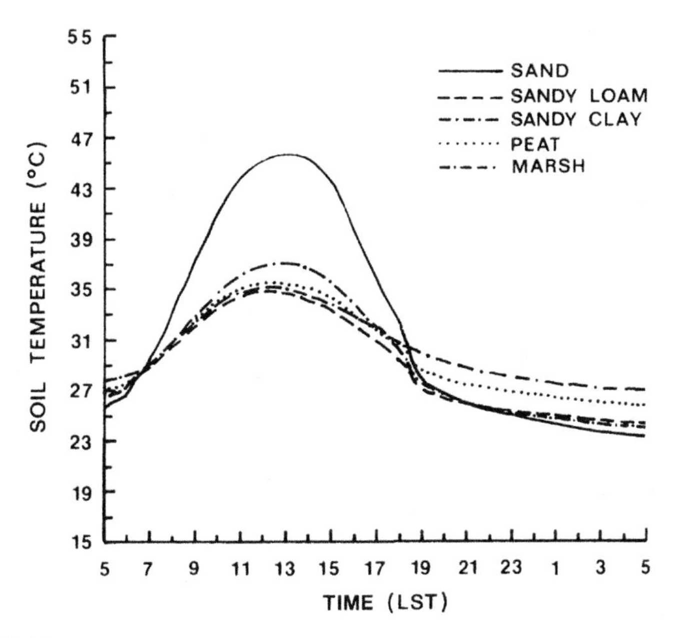

FIGURE 10.18

Predicted soil surface temperature (°C) as a function of soil type for a July summer day in south Florida (from McCumber 1980).

and marsh, even with their much lower albedos than sand, had less variability because of the efficiency with which incoming radiation was utilized to cause evaporation as opposed to an elevation in surface temperature.

From McCumber's results, he concluded that soil moisture and albedo were the two most important controls that regulate feedback to the atmosphere from bare soils. The amount of soil moisture determined the fractional partitioning of sensible and latent subgrid-scale fluxes, and the albedo was a crucial parameter in determining the available radiation reaching the surface. Physick (1980) showed that the relative contributions of the sensible and latent heat fluxes determined the inland penetration rate of sea breezes. Larger values of sensible heat provide more direct heating of the atmosphere, thereby resulting in a larger horizontal pressure gradient and more rapid inland propagation of the sea breeze. Idso et al. (1975b,c) documented observationally major changes in evaporation that can occur from the wetting of a dry soil surface. Wetzel (1978) used a one-dimensional boundary layer model to show that even the irrigation (i.e., wetting) of a relatively small percentage of land (~20%) can have a large effect on an unstably stratified boundary layer even well downstream from the irrigated region. Zhang and Anthes (1982) illustrated, using a one-dimensional model, the strong sensitivity of the planetary boundary layer to moisture availability.

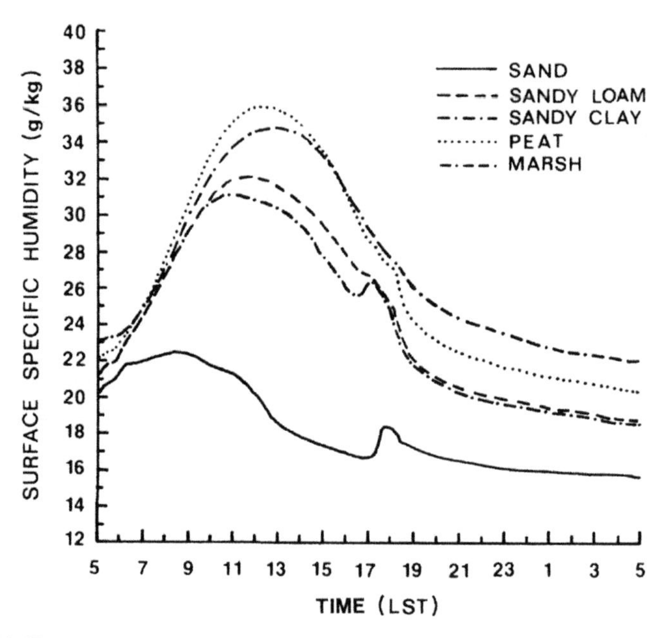

FIGURE 10.19

Same as Fig. 10.18 except for soil surface specific humidity (from McCumber 1980).

Calculation of Soil Moisture[28] The numerical solution of the soil moisture governing equation (10.41) requires special attention, as the hydrostatic subtraction is needed just as in the non-hydrostatic vertical momentum equation in the atmosphere (Zeng and Decker 2009). The vertical momentum equation in the atmosphere is dominated by the approximate hydrostatic balance between the pressure gradient force and the gravitational force, but the hydrostatic pressure is not directly involved in the vertical movement of air parcels (see Chapter 4). If the original form of the vertical momentum equation is used directly for numerical solutions, the typical (and finite) grid spacing in atmospheric models would result in large truncation errors in the computation of vertical velocity even though these errors do converge to zero as the grid spacing and time step approach zero. To solve this problem, the vertical momentum equation with the hydrostatic pressure subtracted, rather than its original form, is used for numerical solutions in atmospheric models (e.g., Eq. 4.31).

Similar to the above atmospheric equation, the soil moisture equation 10.41 contains the hydrostatic equilibrium, as can be seen clearly in the derivation of Eq. 10.40. Zeng and Decker (2009) found that the widely used numerical scheme to solve Eq. 10.41 is deficient when the water table is within the model domain. Furthermore, these deficiencies cannot be reduced by using a smaller grid spacing. The numerical

[28]This section was written by Professor Xubin Zeng, Department of Atmospheric Sciences, University of Arizona, Tucson.

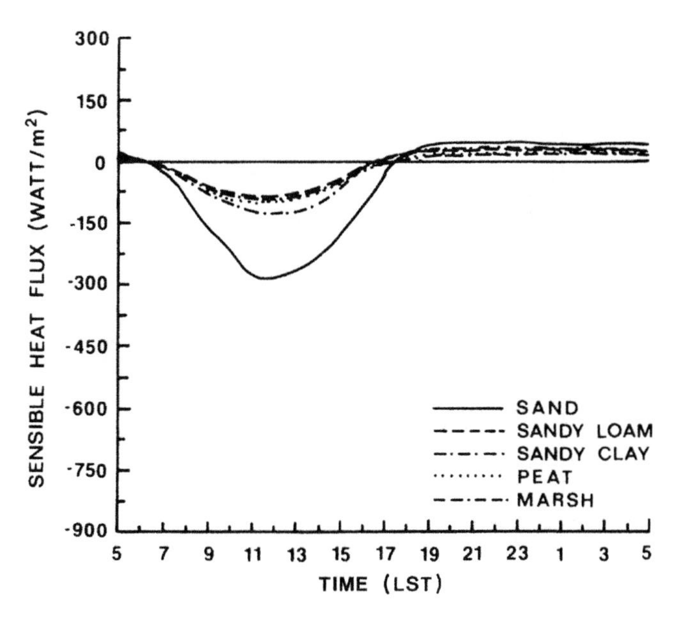

FIGURE 10.20

Same as Fig. 10.18 except for surface layer turbulent sensible heat flux (from McCumber 1980).

errors are much smaller when the water table is below the model domain. These deficiencies were overlooked in the past, most likely because of the more dominant influence of the free drainage bottom boundary condition used by many land models. Zeng and Decker fixed these deficiencies by explicitly subtracting the hydrostatic equilibrium soil moisture distribution from Eq. 10.41, similar to the hydrostatic subtraction in the atmosphere (e.g., Eq. 4.31). However, this equilibrium distribution is time-dependent (rather than time-independent in the atmosphere) and can be derived at each time step from a constant hydraulic (i.e., capillary plus gravitational) potential above the water table, representing a steady-state solution of Eq. 10.41). This has a significant and positive effect on global and regional land surface fluxes (Decker and Zeng 2009) and hence has been implemented in the Community Land Model (CLM4.0; Oleson et al. 2010).

Anthropogenic Sources of Surface Heating The contributions to the surface heat budget by anthropogenic and natural sources, Q_m have been estimated based on such factors as population density, heat content of a forest fire, etc. Such studies include those of Orville et al. (1981) who examined the influence of cooling towers on cumulus convection using a two-dimensional cloud model.

Hanna and Swisher (1971), Hanna and Gifford (1975), and Pielke (1976) have reported estimates of heat released because of some of these activities. Using their values, along with other estimates, the power per unit area and the area of input that

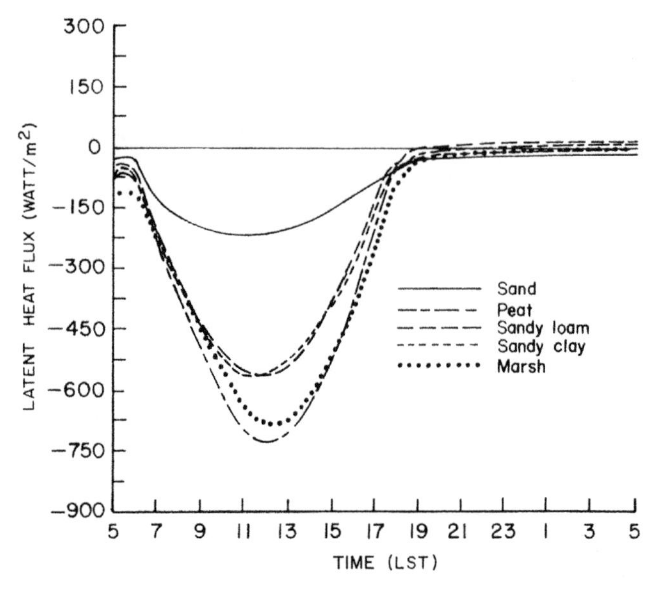

FIGURE 10.21

Same as Fig. 10.18 except for surface layer turbulent latent heat flux (from McCumber 1980).

would be included in Q_m are given in Table 10.7. Harrison and McGoldrick (1981) have provided levels of current artificial heat input on a 10 × 10 km grid over Great Britain. Over this size area they calculated that in February, values of Q_m are nearly 30 W m^{-2} in the London and Birmingham areas, whereas over 1 km^2 areas in Teesside, England, Q_m exceeds 600 W m^{-2}.

Based on previous mesoscale model calculations, Pielke (1976) estimated that heat releases on the mesoscale (e.g., 10^3 km^2) of 10 W m^{-2} would have no detectable effect on local weather, whereas 100 W m^{-2} input uniformly over the same area would result in influences on weather that could be detected statistically. If 1000 W m^{-2} were input, however, the response of the mesoscale system would be significant and immediate. By inputting various values of Q_m into a mesoscale model for different situations, useful information can be obtained regarding inadvertent weather modification effects.

Also evident in Table 10.7 is the influence of area on the observed meteorological response. Even an immense input of heat, such as from a Saturn rocket booster, had only a very localized effect if the area of input is small. Hence substantially smaller rates of heating, if over a mesoscale area, have a much more pronounced effect on the mesoscale meteorological response; a conclusion that is substantiated using the linear model results reported in Dalu and Pielke (1993) and Dalu et al. (1996).

Table 10.7 Representative values of heat input due to anthropogenic and natural sources.

Feature	Heat (W m^{-2})	Area (km^2)	Observational Effect
Suburban area	4	10	Negligible
Urban area	100	1,000	Effects on local climate, local convergence due to a city results in enhanced precipitation downwind
Tropical island	400	600	Influence on rainfall pattern downwind due to enhanced convergence
Australian bushfire	2,000	50	Cumulonimbus cloud generation
Surtsey volcano	100,000	1	Deep cumulus cloud, water spouts
Saturn V booster rocket tests	4,900,000	0.0003	Cumulus cloud

Ground Wetness McCumber (1980) has improved the heat budget to include a more realistic representation of the turbulent latent heat flux term through a better representation of the value of specific humidity at the ground surface q_G (q_G is related to $\bar{q}_{z_0}$ by Eq. 7.16).

McCumber computes the equilibrium surface specific humidity from

$$q_G = hq_s(T_G),$$

where

$$q_s(T_G) = 0.622\left[\frac{e_s(T_G)}{P(\text{in mb}) - 0.378e_s(T_G)}\right] \tag{10.43}$$

(from the definition of specific humidity),

$$e_s(T_G) = 6.1078\exp\left[\left(\frac{T_G - 273.16}{T_G - 35.86}\right)17.269\right]$$

(from Teten's formula; e.g., see Murray 1970, with T_G in degrees Kelvin) and

$$h = \exp\left(\frac{+g\Psi_G}{R_v T_G}\right).$$

The variable Ψ_G is the moisture potential (expressed as a head of water) at the ground surface, and R_v is the gas constant for water vapor ($R_v = 461$ J K^{-1}kg^{-1}). The soil moisture potential at the ground surface is obtained by McCumber (1980) from a finite difference analog to Eq. 10.40 and is given by

$$\Psi_G = \Psi_{G-\Delta z} + \left[\frac{W_s|_G}{\rho_w K_\eta|_G} - 1\right]\Delta z, \tag{10.44}$$

where $W_s|_G$ is the soil moisture flux at the surface. Continuity of moisture flux is required at $z = 0$, so that

$$W_s|_G - \bar{\rho}\overline{w''q''} \simeq W_s|_G + \bar{\rho}u_*q_* \simeq 0$$

must be obtained. Garrett (1978) defines this to occur when

$$\left| \frac{\bar{\rho}u_*q_* - W_s|_G}{\bar{\rho}u_*q_*} \right| < 0.001. \qquad (10.45)$$

In determining Ψ_G, at each time step at each grid point, $W_s|_G$ is initially determined using values of Ψ_G and $\Psi_{G-\Delta z}$ from the last time step. Subsequently

$$W_s|_G^{n+1} = \delta\, W_s|_G^n + (1-\delta)\bar{\rho}u_*q_*,$$

where δ is a weighting factor $(0 \le \delta \le 1)$ used to promote a convergent solution (as defined by Eq. 10.45) and the superscript $n+1$ refers to the next guess in the iteration. The value of $W_s|_G^{n+1}$ is used to recompute a value of Ψ_G from Eq. 10.44, which then provides updated values of η, and then K_η and D_η at the surface. McCumber found the fastest convergence to a solution when $\delta > 0.5$, with dry soils (e.g., $h < 0.70$) requiring values closer to unity. Once the surface moisture characteristics are computed, then Eq. 10.42 can be used to calculate moisture fluxes within the soil.

Influence of Terrain Slope As of yet in our discussion, the influence of sloping terrain on the amount of shortwave radiation that reaches the ground has not been considered in the surface heat budget. In Section 7.3.3, it was shown that in a horizontal layer of the atmosphere the amount of solar radiation on a unit cross-sectional area is dependent on such factors as the time of the year, the opacity of the atmosphere, the latitude, and the time of day. For irregular terrain, the orientation of the ground surface with respect to the Sun must also be considered.

Kondratyev (1969) presented a formula that accounts for the influence of sloping terrain on the incident direct solar radiation per unit area, and his analysis was used by Mahrer and Pielke (1977b) to investigate the effect of irregular terrain heating on mesoscale circulations. Figure 10.22, adapted from Kondratyev (1969), illustrates the angle i at which direct solar radiation impinges on sloping terrain, where $S_H = \bar{R}\downarrow_{\mathrm{sw}_G} / \cos Z$ is the value of direct solar radiation at the ground on a unit cross-sectional area perpendicular to the Sun's rays. The value of $\bar{R}\downarrow_{\mathrm{sw}_G}$ can be obtained for clear skies from the integrated form of the Eq. 7.83 evaluated at the ground.

Expressed mathematically

$$\bar{R}\downarrow^{\mathrm{sl}}_{\mathrm{sw}_G} = S_H \cos i, \qquad (10.46)$$

where

$$\cos i = \cos\alpha \cos Z + \sin\alpha \sin Z \cos(\beta - \gamma).$$

The zenith angle Z was defined by Eq. 7.82, and α the slope of the terrain is given by

$$\alpha = \tan^{-1}\left[\left(\frac{\partial z_G}{\partial x}\right)^2 + \left(\frac{\partial z_G}{\partial y}\right)^2\right]^{1/2},$$

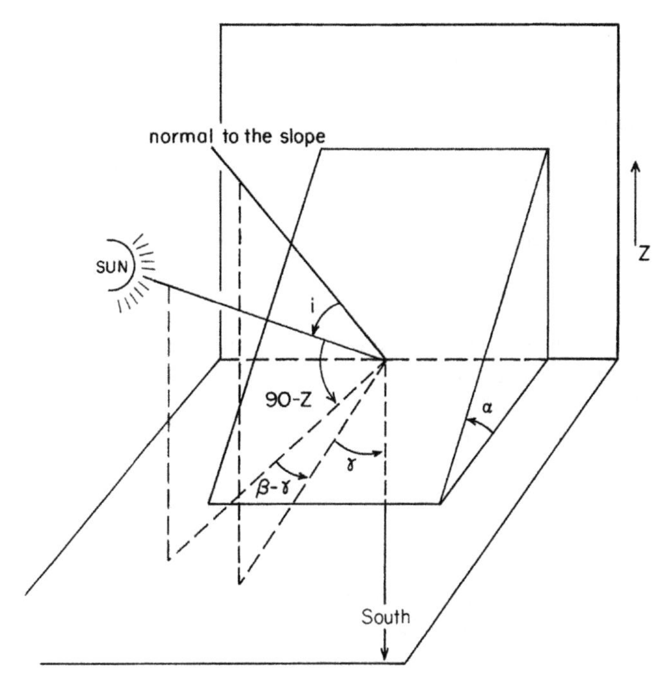

FIGURE 10.22

An illustration of the angles used in the definition of Eq. 10.46 (adapted from Kondratyev 1969, Fig. 5.38).

where z_G is terrain height. The quantity $\beta - \gamma$ is the orientation of the Sun's azimuth β with respect to the azimuth of the terrain slope γ. The slope azimuth is expressed by

$$\gamma = \frac{\pi}{2} - \tan^{-1}\left(\frac{\partial z_G}{\partial y} \bigg/ \frac{\partial z_G}{\partial x}\right)$$

(so that south has zero azimuth), and the azimuth of the Sun is obtained by projecting the location of the Sun onto a horizontal surface and also requiring that south has zero azimuth. The expression for β is given by

$$\beta = \sin^{-1}\left(\frac{\cos\delta_{\text{sun}} \sin h_r}{\sin Z}\right),$$

where, as in Section 7.3.3, δ_{sun} is the declination of the Sun and h_r is the hour angle. Spherical trigonometric identities such as given, for example, in Selby (1967:161-164) are used to derive γ and β.

The following example is presented to illustrate the dependence of direct solar heating on terrain slope and azimuth. Assuming $\phi = 40°$N, clear sky, local noon, $\gamma = 0°$ (i.e., south-facing slope), June 21 ($\delta_{\text{sun}} = 23.5°$), $r = 1.015a$, $S_0 = 1380$ W m^{-2}, $A = 0.2$, $\alpha = 10°$, and $R\downarrow_{\text{sw}_0}$ in Eq. 7.79 is attenuated by 40% between

Table 10.8 Solar radiation at noon on a slope of 10° oriented in four different directions on June 21 and December 21 using the values given in the text. Units are watts per meter squared.

Variable	Date	South	East	North	West
$\bar{R}^{\text{sl}}_{\text{swG}} \downarrow$	21 June	625	594	563	594
$\bar{R}^{\text{sl}}_{\text{swG}} \downarrow$	21 December	215	159	103	159

the top of the atmosphere and the ground, Eq. 10.46 yields $\bar{R}\downarrow^{\text{sl}}_{\text{swG}} = 625$ W m^{-2}. Table 10.8 gives the same calculation except $\gamma = \pi/2; \pi;$ and $3\pi/2$ corresponding to an east-, north-, and west-facing slope, respectively. Also represented in Table 10.8 are calculations for the same situation except for December 21 ($\delta_{\text{sun}} = -23.5°$) with $r = 0.985a$.

These, and even smaller slopes, can have a substantial influence on mesoscale circulations. Mahrer and Pielke (1977b:Fig. 4) found the eastern slope of a 1 km mountain (with a slope of about 2°) to be about 1° to 2°C warmer in the morning and cooler by the same amount in the afternoon than the same location on the western slope. Observed solar radiation on four different slopes in Kansas is discussed in Nie et al. (1992).

The influence of slope on the diffuse shortwave radiation $\bar{R}^{\text{D}}_{\text{swG}}$, however, is considered negligible for slopes of less than about 20°, as explained by Lee (1978:61-62). Although the diffuse radiation is reduced in magnitude because only a portion of the sky is visible, if the diffuse radiation is isotropic (see Footnote 13 of Chapter 7), then $\bar{R}^{\text{D}}_{\text{swG}}$ received on a flat, open surface is only modified by

$$\bar{R}^{\text{D}_{\text{sl}}}_{\text{swG}} = \bar{R}^{\text{D}}_{\text{swG}} \cos^2(\alpha/2).$$

As given by Lee, for $\alpha = 16.7°$ (a slope of 0.3), the difference between $\bar{R}^{\text{D}}_{\text{swG}}$ and $\bar{R}^{\text{D}_{\text{sl}}}_{\text{swG}}$ is only 2%. Lee also showed that the contribution to the total solar radiation by the reflection of total solar radiation from surrounding terrain is only about 3% or less of $\bar{R}^{\text{sl}}_{\text{swG}}$ for slopes of less than 20° $\left(\bar{R}^{\text{sl}}_{\text{swG}} = \bar{R}\downarrow^{\text{sl}}_{\text{swG}} + \bar{R}^{\text{D}}_{\text{swG}}\right)$. Ignoring attenuation in the intervening atmosphere of the reflected light from the surrounding terrain, the increased solar absorption on the slope for a uniform albedo is

$$\Delta R\downarrow_{\text{swG}} = (1 - A)A\left(\bar{R}\downarrow_{\text{swG}} + \bar{R}^{\text{D}}_{\text{swG}}\right)\sin^2(\alpha/2).$$

There is also a small effect on albedo since the effective zenith angle in Eq. 10.35 is the angle between the solar beam and the angle normal to the slope. As seen in Fig. 10.17 for slopes less than 20° the effect of terrain slope on albedo is relatively small even for large zenith angles. With $\alpha = 0.3$ sloping toward the north, for example, a value of Z of 50° at noon, would provide an effective zenith angle along the slope of 66.7° resulting in a change of albedo of about 2.5%.

The influence of slight slopes on the longwave radiation balance is also small, as discussed by Lee (1978:71) using the same type of geometric argument as applied to derive $\bar{R}^{D_{sl}}_{swG}$ and $\Delta R\downarrow_{swG}$; i.e., for an equilateral triangular valley, $\bar{R}\downarrow^{sl}_{lw} = \bar{R}\downarrow \cos^2(\alpha/2) + \bar{R}_{lw}\sin^2(\alpha/2)$, where $\bar{R}_{lw}$ is the longwave radiation reaching one side of the valley from the other and $\bar{R}\downarrow$ comes from the atmosphere above the slope (e.g., using Eq. 7.60).

Zdunkowski et al. (1980), and Welch and Zdunkowski (1981b), have provided a discussion of the influence of slope on the amount of direct and diffuse solar radiation incident on sloped surfaces as a function of optical depth and solar zenith angle. They report, for example, that at 40° of latitude during the equinox, a north-facing slope receives more integrated daily shortwave radiation on a cloudy day than on a clear day.

A consequence of this analysis is that the direct shortwave radiation term $\bar{R}\downarrow_{swG}$ in Eqs. 10.28 and 10.29 must be replaced with $\bar{R}\downarrow^{sl}_{swG}$ (Eq. 10.46) over sloping terrain.

Snow Sturm et al. (1995) describes snow on the ground in seven separate classes: tundra (0.38), taiga (0.26), alpine, maritime (0.35), ephemeral, prairie, and mountain, where their typical bulk density in g cm^{-3} is listed parenthetically, when listed in their paper. Snow can be represented similar to soil with the important additional characteristic that snow can change its phase (melt, refreeze, sublimate). Snow can also saltate (i.e., bounce along the surface), and blow through the air. Liston and Sturm (1998) developed a model to simulate this drifting of snow. Greene et al. (1999) applied this model to a portion of the Continental Divide in Colorado, where it was found that as much as 30% of the snow can sublimate into the air. Although not yet done, this drifting/blowing snow model could also be used to simulate the movement of sand and dust by the wind.

Liston (1999) also developed a snow melt model which uses solar input and wind speed, along with the terrain slope and azimuth, to represent this phase change. Liston illustrates the close interrelationships among the spatial snow distribution, snow melt, and snow cover depletion from sublimation, which in windy areas is often a large fraction of the reduction of snow cover.

Examples of Snow Model Parameterizations[29] As an example, the Community Land model (CLM4.0; Oleson et al. 2010) uses a fairly sophisticated snow submodel. Snow can have up to five layers, and the state variables for snow are the mass of water, mass of ice, layer thickness, and temperature. The water vapor phase is omitted. Also considered are the black and organic carbon and mineral dust in snow that originate from atmospheric aerosol deposition and influence the radiative transfer in snow.

Snow compaction includes three types of processes: destructive metamorphism of new snow (crystal breakdown due to wind or thermodynamic stress); snow load or overburden (pressure), and melting (changes in snow structure due to melt-freeze cycles plus changes in crystals due to liquid water).

[29]This section was written by Professor Xubin Zeng, Department of Atmospheric Sciences, University of Arizona, Tucson.

A particularly challenging issue is the treatment of snow-vegetation interactions. This also depends on how snow is computed in a land model. For instance, CLM considers the soil, vegetation, and snow separately, while the community Noah land model (as used in the NCEP global and regional operational models and the community regional/mesoscale WRF) considers the soil, vegetation, and snow as a single unit over an atmospheric grid cell.

Partly for this reason, a widely known deficiency of Noah is that snowmelt occurs much too early. By explicitly considering the vegetation shading effect on snow sublimation and snowmelt and with other revisions, Wang et al. (2010) are able to significantly improve Noah simulations of all snow processes such as snow water equivalent, snow depth, and sensible and latent heat fluxes over both forest and grass sites.

These modifications maintain the Noah model structure and do not introduce new prognostic variables, allowing easy implementation into NCEP operational models and into WRF. Furthermore, they are found to be as good as, or slightly better than, a much more complicated land model in the snow simulation over the three forest sites.

Other examples of model representations of snow processes include Marshall and Oglesby (1994), Marshall et al. (1994), Horne and Kavvas (1997), and Walko et al. (2000b). Over snow cover during the day, and with a calm wind, for example, Halberstam and Schieldge (1981), have demonstrated the importance of radiative flux divergence and moisture flux from the surface in determining temperature and wind profiles. Because of the high albedo of the snow and the creation of a moist layer of air immediately above the surface, both downward and upward reflected solar radiation are absorbed just above the surface, thereby creating a local region of enhanced warming within the surface layer.

Vegetation When vegetation is introduced at the ground, the proper representation of the bottom boundary conditions becomes more difficult than for bare soil, since the observational and theoretical information concerning the fluxes of heat, moisture, momentum, and other gaseous and aerosol materials into and out of the vegetation remains limited. Since much of the world is vegetated and the vegetation dynamically changes over time (Eidenshink and Haas, 1992; see also Tables 10.9 and 10.10), however, it would be inappropriate to neglect this important component of the ground characteristics in a mesoscale model. Even in arid regions with a sparsity of vegetation, Otterman (1981a,b) has illustrated the importance of protruding plant material on the flux of heat.

In this section of the chapter, one type of parameterization for the influence vegetation on the boundary layer is presented (that of McCumber 1980). McCumbers' soil-vegetation-atmosphere transfer (SVAT) scheme was among the earliest developed. Other early formulations are described for drainage flow simulations by Garrett (1983a) and for boundary layer structure over flat terrain by Yamada (1982). Terjung and O'Rourke (1981) have examined the influence of vegetation on the magnitude of the terms in the surface energy budget in an urban area. Although the accuracy of physically elaborate parameterizations of vegetation, such as that of Deardorff

Table 10.9 Representative values of leaf area index L_A, transmissivity, absorption, and albedo for shortwave radiation $\bar{\tau}$, $\bar{a}$, and A_f, and emissivity for longwave radiation ϵ_f for various types of vegetation. Values were estimated from data given in papers published in Monteith (1975a,b). Additional values of A_f are given in Table 10.5. Similarly, additional values of ϵ_f are listed in Table 10.3.

	L_A	$\bar{\tau}$	$\bar{a}$	A_f	ϵ_f
Maize, Rice					
June 1	0	~0.90	~0.0	~0.10	0.95
Mid July	1.8	~0.55	~0.30	~0.15	0.95
September 1	4	~0.15	~0.65	~0.20	0.95
Mid October	6	~0.10	~0.70	~0.20	0.95
Cotton	2	~0.23	~0.57	~0.20	0.95
Wheat, Barley	4	~0.25	~0.55	~0.20	0.95
Prairie Grasslands					
Green	~1				0.96
Dead	~4				0.96
Meadow	2		~0.48		0.96
	4		~0.72		0.96
	6		~0.82		0.96
Coniferous Forest	2	~0.20	~0.70	~0.10	0.97
	4	~0.08	~0.82	~0.10	0.97
Deciduous Forest					
Aspen (in foliage)	2	~0.45	~0.35	~0.20	0.95[a]
	4	~0.23	~0.65	~0.12	0.95[a]
	6	~0.06	~0.82	~0.12	0.95[a]
Oak					
30-year old stand					
No foliage		~0.63	~0.25	0.12	0.95[a]
Full foliage		~0.12	~0.72	0.16	0.95[a]
160-year old stand					
No foliage		~0.30	~0.58	0.12	0.95[a]
Full foliage		~0.25	~0.59	0.16	0.95[a]

[a] *These values of ϵ_f were obtained from Lee (1978:69).*

(1978), have been questioned (e.g., Monteith 1981), realistic representations of this ground-air interaction must be included in mesoscale models.

There are three temporal scales of interaction between vegetation and the atmosphere: *biophysical, biogeochemical, and biogeographic* (Pielke 1998). Vegetation processes have been shown to be very important in both weather and climate studies (e.g., see Pielke et al. 2011). Biophysical influences include controls on the transpiration of water vapor through the stoma of plants. Biogeochemical effects include above and below ground vegetation growth such as represented by the CENTURY model (Lu et al. 2001). Biogeographic models include changes in the mixture of vegetation

Table 10.10 Characteristic values of surface, biological resistance of a canopy to loss of water. Values are estimated from published values given in papers presented in Monteith (1975a,b).

Type of Vegetation	r_c (s m^{-1})
Cotton field	
0600 LST	~130
(noon)	~17
1800 LST	~330
Sunflower field	
$L_A = 1.8$	110
$L_A = 3.6$	80
Coniferous forest	
0600 LST	
Spruce	~20
Hemlock	~240
Pine	~50
Noon	
Spruce	~100
Hemlock	~150
Pine	~130
1800 LST	
Spruce	~120
Hemlock	~200
Pine	~310
Prairie grasslands	
0800 LST	
late July	~100
mid September	~150
1200 LST	
late July	~100
mid September	~500
1800 LST	
late July	~150
mid September	~550

species and the spatial movement of biomes. The General Energy and Mass Transport model (GEMTM, see Chen and Coughenour 1994, Eastman et al. 2001a) represents both biophysics and biogeochemistry. Benoit et al. (2000) discuss the use of coupled atmospheric-hydrologic modeling. In this chapter, since the focus is on the use of mesoscale models in weather forecasting, only biophysics models are discussed.

Incorporating vegetation into a mesoscale model can be very important. Figure 10.23, for example, (reproduced from McCumber 1980 in one of the very first atmospheric model simulations which included such a more realistic representation of vegetation), using input synoptic meteorological data for 17 July 1973 applicable

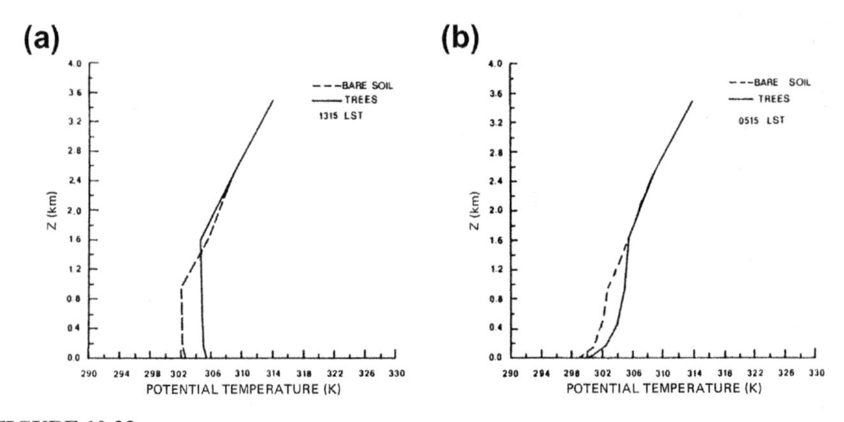

FIGURE 10.23

Vertical profile of (a) potential temperature for early afternoon and (b) at sunrise the following day for bare sandy loam ($\sigma = 0.0$) soil and for a forested area, where $\sigma_f = 0.90$. Initial synoptic data were for 17 July 1973 over south Florida (from McCumber 1980).

to south Florida shows substantially different profiles of potential temperature in the afternoon and morning over a forested area (with $\sigma_f = 0.90$) for a sandy loam soil and over a bare soil of the same soil type ($\sigma_f = 0.0$). The differences in temperature are as large as 3°C, with the depth of the mixed layer during the day differing by over 300 meters. Figure 10.24, also adopted from McCumber (1980), illustrates large variations of foliage, canopy, and ground temperatures over four different types of vegetation soil combinations. Because the surface temperature dominates the forcing for many types of mesoscale systems, *different vegetation and soils*, therefore, *play an important role in such atmospheric phenomena* and must be included in model simulations.

Evaluations using later forms of this SVAT model (now referred to as the Land Atmosphere Ecosystem Feedback – LEAF model) are referred to in Lee et al. (1995), and Shaw et al. (1997). A 1999 version, referred to as LEAF-2, is described in Walko et al. (2000a), and is used by Pielke et al. (1999b). LEAF-2 is also able to represent runoff, an important immediate loss of water vapor return to the atmosphere. Famiglietti and Wood (1991) provide a discussion of the role of runoff.

Figures 10.25-10.27 illustrate how SVATS represents atmosphere-surface interactions for three different models (BATS, LAPS, LEAF). A figure of this type for three other SVATS is also shown by Schultz et al. (1998, their Fig. 1). As evident in the figure, the transfer of moisture and heat between different components of the vegetation and soil system is represented by the electrical analog of a resistor.

Basic equations used in three SVATS were presented in Appendix D of Pielke (2002b). The original sources for these models include Mihailovic and Ruml (1996), Mihailovic and Kallos (1997), Mihailovic et al. (1993, 1998, 1999a), and Walko et al. (2000b). Mihailovic et al. (1999b) discuss how to represent deep soil in such models. Vegetation data sets have been described in Zeng et al. (2000a). Jiang et al. (2011)

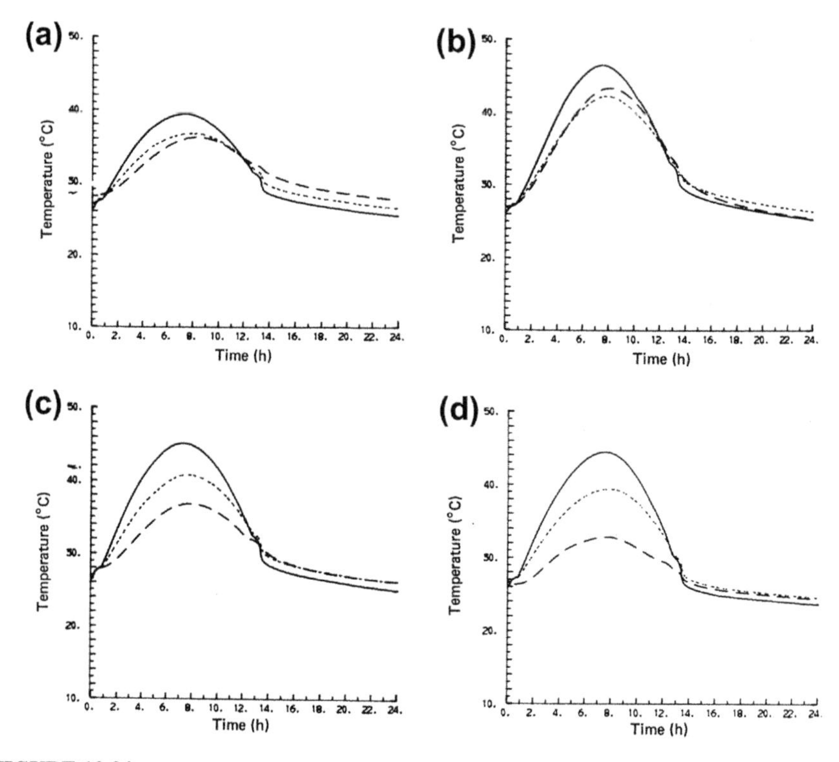

FIGURE 10.24

Diurnal variation of predicted foliage temperature T_f (solid line), canopy air temperature T_{af} (short dashed line), and ground temperature T_G (long dashed line) for (a) grass on top of peat soil ($\sigma_f = 0.85$); (b) grass on top of sandy soil ($\sigma_f = 0.75$); (c) trees on top of sandy clay ($\sigma_f = 0.90$); and (d) trees on top of sandy loam ($\sigma_f = 0.90$). The abscissa is time in hours after sunrise. Simulation for 17 July 1973 conditions over south Florida (from McCumber 1980).

present an evaluation of the SVAT SiB3. A summary of land-surface data sets is given in Hof (1999). Reviews of land-atmosphere interactions are given in Monteith (1981), Avissar (1995), Betts et al. (1996), Chen et al. (2001), Pielke (2001c), Pitman (2003), and Pielke et al. (2011). The Project for Intercomparison of Land-Surface Parameterization Schemes (PILPS) was introduced to assess the ability of the SVATS to represent actual observed surface data (Henderson-Sellers et al. 1993, 1995; Shao and Henderson-Sellers 1996).[30]

[30]Other papers that discuss SVATS and their use include Dickinson (1984), Sellers et al. (1986), Noilhan and Planton (1989), Pinty et al. (1989), Avissar and Pielke (1991), Ye and Pielke (1993), Lakhtakia and Warner (1994), Lee et al. (1995), Pleim and Xiu (1995), Viterbo and Beljaars (1995), Chen et al. (1996, 1997), Gao et al. (1996), DeRidder and Schayes (1997), Dai and Zeng (1997), Mölders and Raabe (1997), Niyogi and SethuRaman (1997), Betts et al. (1998), Bosilovich and Sun (1998), Dai et al. (1998), Qu et al. (1998), Qingcun et al. (1998), Schultz et al. (1998), Yang et al. (1998, 1999a,b), Ashby (1999), Bastidas et al. (1999), Bonan et al. (1999), Boone et al. (1999), Chang et al. (1999),

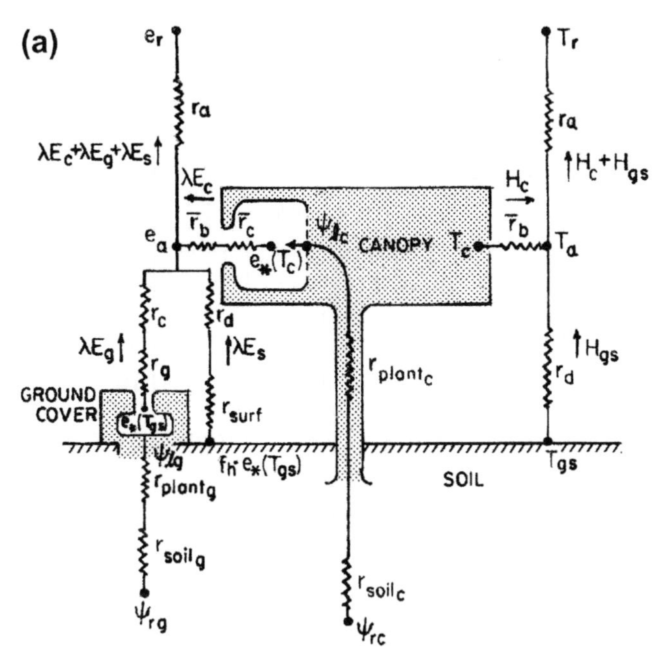

FIGURE 10.25

(a) Framework of the Simple Biosphere (SiB) model. The transfer pathways for latent and sensible heat flux are shown on the left and righthand sides of the diagram, respectively (from Sellers et al. 1986). Symbols are described in the text. (b) Same as (a) except for the Biosphere-Atmosphere Transfer Scheme (BATS). (c) Same as (a) except for the Land Ecosystem Atmosphere Feedback (LEAF) model (from Lee et al. 1993).

Problems for Chapter 10

1. Using Eq. 10.46, compute the lowest latitude at which the Sun does not set on 21 June. Compute the latitudes at which the Sun does not rise on 21 December. (Hint: you are computing the latitude of the Arctic Circle.)
2. Using the simple energy budget $\sigma T^4 = S(1 - A)/4$ where S is the solar constant, A is the albedo, σ is the Stefan-Boltzmann constant, and T is the temperature,

Chapin et al. (1999), Entin et al. (1999), Gupta et al. (1999), Liu et al. (1999), Lynch et al. (1999a), Oki et al. (1999), Pauwels and Wood (1999), Zeng et al. (1999b), Xu et al. (1999), Boone et al. (2000), Ding et al. (2000), Dirmeyer et al. (2000), Mihailovic et al. (2000), Mohr et al. (2000), Schlosser et al. (2000), Sen et al. (2000), van den Hurk et al. (2000), Verseghy (2000), Walko et al. (2000b), Wu et al. (2000b), Zeller and Nikolov (2000), Zeng et al. (2000b), and Marshall et al. (2003). As evident from this extensive list of cites, SVAT research has become an active area of research.

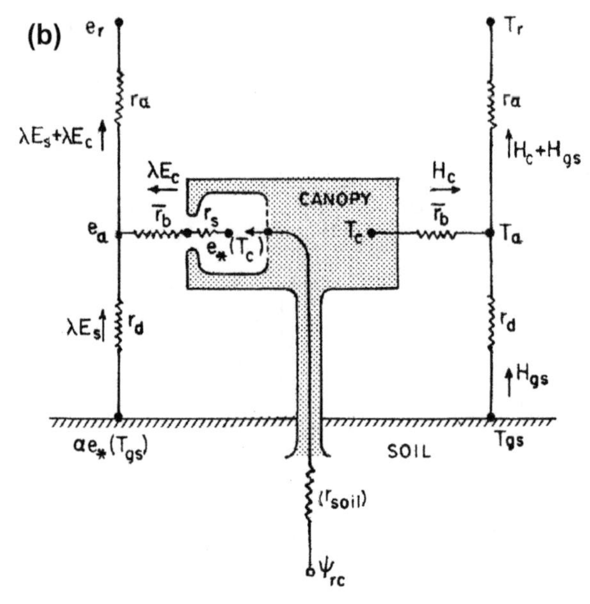

FIGURE 10.26

Same as 10.25 except for the Biosphere-Atmosphere Transfer Scheme (BATS). Symbols are described in the text (from Lee et al. 1993).

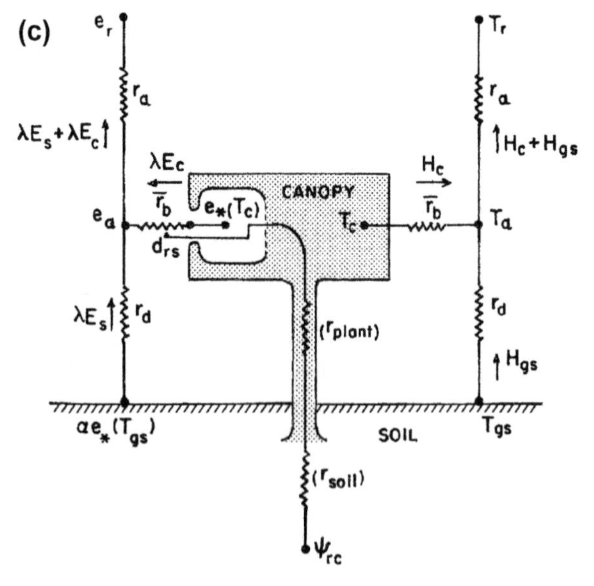

FIGURE 10.27

Same as 10.25 except for the Land Ecosystem Atmosphere Feedback (LEAF) model (from Lee et al. 1993).

derive the quantity $\partial T / \partial A$. Calculate the change of albedo that is required in order to obtain a 1°C change in temperature.

3. Program the tank model in Section 9.1.4. Run with cyclic boundary conditions in order to show that the values in Table 9.3 can be reproduced. Then run the tank model with (i) constant inflow-gradient outflow, and (ii) constant inflow-radiative outflow (see Section 10.4.1.1). Discuss the resultant solutions and how they differ from the results with cyclic boundary conditions.
4. Select a land-atmospheric model from the peer-reviewed literature and sketch its processes using the framework given in Fig. 10.25.

Chapter 10 Additional Readings

There are a variety of books and review papers that go into greater depth on the subjects in this Chapter. They include:

Brutsaert, W., 1982a: *Evaporation into the Atmosphere: Theory, History and Applications,* D. Reidel, Norwell, MA, 299 pp.

Halldin, S., and S.-E. Gryning, 1999: Boreal forests and climate. *Agric. Forest Meteor.,* **98-99**, 1-4.

Hayden, B.P., 1998: Ecosystem feedbacks on climate at the landscape scale. *Phil. Trans. R. Soc. Lond. B.,* **353**, 5-18.

Kabat, P., Claussen, M., Dirmeyer, P.A., J.H.C. Gash, L. Bravo de Guenni, M. Meybeck, R.A. Pielke Sr., C.J. Vorosmarty, R.W.A. Hutjes, and S. Lutkemeier, Editors, 2004: Vegetation, water, humans and the climate: A new perspective on an interactive system. Springer, Berlin, Global Change - The IGBP Series, 566 pp.

National Research Council, 2005: Radiative forcing of climate change: Expanding the concept and addressing uncertainties. Committee on Radiative Forcing Effects on Climate Change, Climate Research Committee, Board on Atmospheric Sciences and Climate, Division on Earth and Life Studies, The National Academies Press, Washington, D.C., 208 pp.

Parlange, M.B., A.T. Cahill, D.R. Nielsen, J.W. Hopmans, and O. Wendroth, 1998: Review of heat and water movement in field soils. *Soil and Tillage Research,* **47**, 5-10.

Pielke Jr., R.A., 2010: *The Climate Fix: What Scientists and Politicians Won't Tell You About Global Warming*. Basic Books, 276 pp.

Tenhunen, J.D., and P. Kabat, Editors, 1999: Integrating hydrology, ecosystem dynamics, and biogeochemistry in complex landscapes. Report of the Dahlem Workshop on Integrating Hydrology, Ecosystem Dynamics, and Biogeochemistry in Complex Landscapes, January 18-23, 1998, Wiley, New York, 367 pp.

There are several special journal issues of early land-surface field campaigns. These include (listed by Editors) Murphy (1992), Sellers et al. (1997), Avissar and Lawford (1999), and Hall (1999). Reviews of field campaigns are presented in Kabat (1999) and LeMone et al. (2000).

CHAPTER

Model Evaluation

11

CHAPTER OUTLINE HEAD

11.1 Evaluation Criteria

In utilizing a mesoscale numerical model, there are six suggested basic requirements that must be met before the credibility of simulations performed with that model can be established by the scientific community. In reading papers in the published literature, the same criteria must be considered when evaluating the results and conclusions of those papers.

These requirements are as follows.

Mesoscale Meteorological Modeling, Third Edition, Volume 98. http://dx.doi.org/10.1016/B978-0-12-385237-3.00011-6

1. The model must be compared with known analytic solutions. To perform these experiments, the mesoscale model is forced by very small perturbations so that essentially linearized results are produced, or the initial and boundary conditions are idealized so that exact solutions to the nonlinear equations are possible.
2. Nonlinear simulations with the model must be compared with the results from other models, which have been developed independently.
3. The mass, moisture, and energy budgets of the model must be computed to determine the conservation of these important physical quantities.
4. The model predictions must be quantitatively compared with observations.
5. The computer logic of the model must be available on request, so that the flow structure of the code can be examined.
6. The published version of the model must have been subjected to peer review. For this reason, model results presented in recognized professional journals (e.g., *Monthly Weather Review, The Journal of Atmospheric Science, The Quarterly Journal of the Royal Meteorological Society, Tellus, The Journal of the Meteorological Society of Japan, The Chinese Journal of Atmospheric Sciences, Atmosfera, Atmosphere-Ocean, Boundary-Layer Meteorology, The Journal of Geophysical Research, Meteorology and Atmospheric Physics, Russian Meteorology and Hydrology*, etc.) should carry more weight than those distributed in report formats. A particularly appealing new journal approach is illustrated by the journal *Atmospheric Chemistry and Physics* which is open access with public peer-review and interactive public discussion.

Hanna (1994) provides a similar list of evaluation criteria. In this chapter, several of these criteria will be examined in more detail.

11.2 Types of Models

There are three types of applications of these models: for process studies, for diagnosis, and for forecasting.

Process studies: The application of models to improve our understanding of how the system works is a valuable application of these tools. The term "sensitivity study" can be used to characterize a process study. In a sensitivity study, a subset of the forcings and/or feedback of the system may be perturbed to examine its response. The model of the system might be incomplete and not include each of the important feedbacks and forcings. Marshall et al. (2004) is one example of this type of model application.

This is also the type of modeling being used to make climate projections (predictions) in climate assessments such as the IPCC reports (e.g., see Pielke 2002a). This application of the models is made despite their inability to show multi-decadal regional and mesoscale skill in forecasting changes in climate statistics when run in a hindcast mode (e.g., see Pielke 2013, and also Section 13.5). These IPCC model runs are actually process model simulations.

Diagnosis: The application of models, in which observed data is assimilated into the model, to produce an observational analysis that is consistent with our best understanding of the system as represented by the manner in which the fundamental concepts and parameterizations are represented. An example of this approach is the North American Regional Reanalysis [http://www.emc.ncep.noaa.gov/mmb/rreanl/].

Forecasting: The application of models to predict the future state of the system. Forecasts can be made from a single realization, or from an ensemble of forecasts which are produced by slightly perturbing the initial conditions and/or other aspects of the model. An example of such predictions can be viewed at http://www.mmm.ucar.edu/wrf/users/forecasts.html.

11.3 Comparison with Analytic Theory

To compare a numerical model with its analytic analog, the equations in the computational model must be used in the same form as utilized to develop the solution for the analytic version. Except for special cases, the development of an analogous system of equations in a numerical model usually requires that the equations be linearized. In addition, in order to minimize computational errors, the grid resolution of the model must be sufficiently small such that the spatial scale of the forcing (e.g., L_x and L_z) are adequately resolved, as summarized in Section 9.8.

Figure 11.1, reproduced from Martin (1981), illustrates a numerical simulation performed to validate the model against Defant's (1950) exact linear solution. Defant's analytic model was derived in Section 5.2.3.1. Although there are minor differences between the fields in Fig. 11.1 and those evaluated from Defant's solution (e.g., Fig. 5.5), the solutions are almost identical.

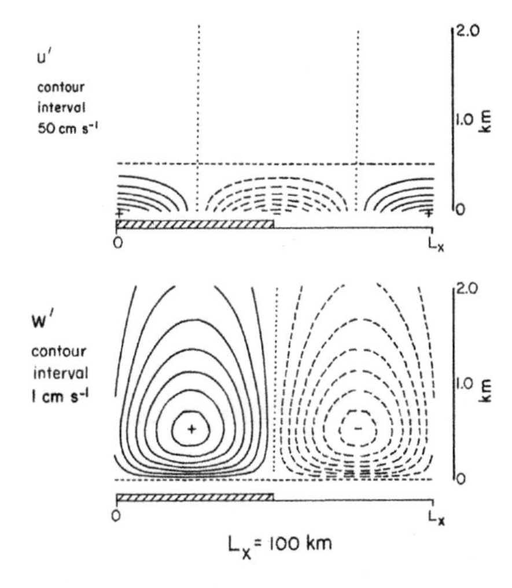

FIGURE 11.1

The horizontal and vertical velocity fields 6 h after sunrise predicted by a numerical model analog to Defant's (1950) analytic model. The input parameters are given by Eq. 5.99; the results correspond to the exact solution given in Fig. 5.5 (from Martin 1981).

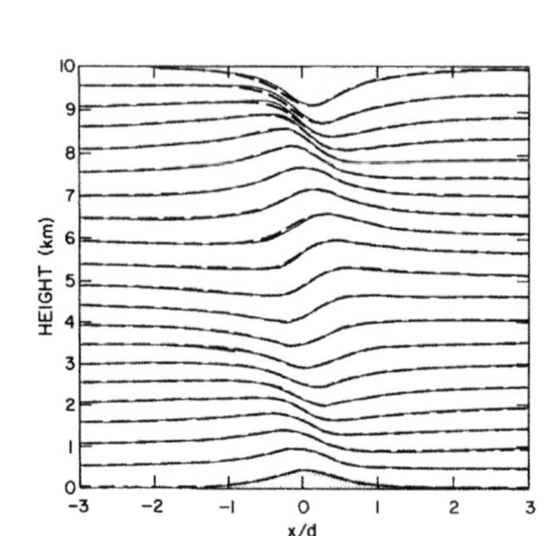

FIGURE 11.2

Comparison of predicted contours of potential temperature $\bar{\theta}$ for an analytic solution (dashed line) and the equivalent numerical solution (solid line) for a bell-shaped mountain of 10 m. Results have been amplified by 50 for illustration purposes. The normalizing factor d is the characteristic half-width of the mountain. The atmosphere in this simulation was prescribed as isothermal initially with a large-scale wind flow of 20 m s^{-1}, constant with height (from Klemp and Lilly 1978).

Klemp and Lilly (1978) have performed similar comparisons between analytic and numerical solutions for airflow over rough terrain. One example is reproduced in Fig. 11.2. In addition to validations against linear theory, Klemp and Lilly (1978), and Lilly and Klemp (1979) also performed comparisons against analytic solutions of a subset of the nonlinear conservation equations, developed by Long (1953,1955; see Section 5.3 of Pielke 2002b for a derivation of the Long model). Although Long's solutions are valid only for the special case when the flow is steady state and the density multiplied by the domain-averaged horizontal velocity squared are independent of height, such comparisons offer some evidence of the accuracy of the numerical computations. The limitations of Long's solution to actual stratified flows over an obstacle is discussed by Baines (1977). Durran (1981) has referenced studies by other investigators who obtained exact solutions for specialized sets of the nonlinear conservation equations.

11.4 Comparison with Other Numerical Models

To evaluate a numerical model, it is useful to compare its results for a particular simulation with those from a model from a different set of investigators such as reported by Cox et al. (1998). Although all models start with the conservation equations discussed in Chapter 2, such model components including the computational schemes, parameterizations, and the particular simplified form of the conservation equations result in different model formulations. Although similar model results do not necessarily indicate a realistic reproduction of the actual atmospheric system, they are useful experiments to ascertain if independent researchers using different model structures can replicate each others results.

Mahrer and Pielke (1977b) performed a qualitative evaluation of their three-dimensional simulation of the airflow over the White Sands Missile Range in New Mexico against that of Anthes and Warner (1974), however, no quantitative measures of degree of agreement were used. Tapp and White (1976), Hsu (1979), and MacDonald et al. (2000) performed a similar qualitative comparison of their results against the sea-breeze simulation reported in Pielke (1974a). An example of an intercomparison between Tapp and White, and Pielke's (1974a) results are illustrated in Fig. 11.3. Carpenter and Lowther (1981) have shown that these Florida sea-breeze results are relatively insensitive to changes in the vertical grid mesh used. This result is consistent with the two-dimensional vertical grid resolution sensitivity test illustrated in Fig. 10.7.

Using two-dimensional models, Kessler and Pielke (1982), Mahrer and Pielke (1978b), and Peltier and Clark (1979) have simulated the airflow over rough terrain for the Colorado Front Range windstorm of 11 January 1972 (described by Lilly and Zipser, 1972), and compared their results against those of Klemp and Lilly (1978). A model intercomparison of the simulations of this wind storm (using 11 different models) is reported in Doyle et al. (2000).

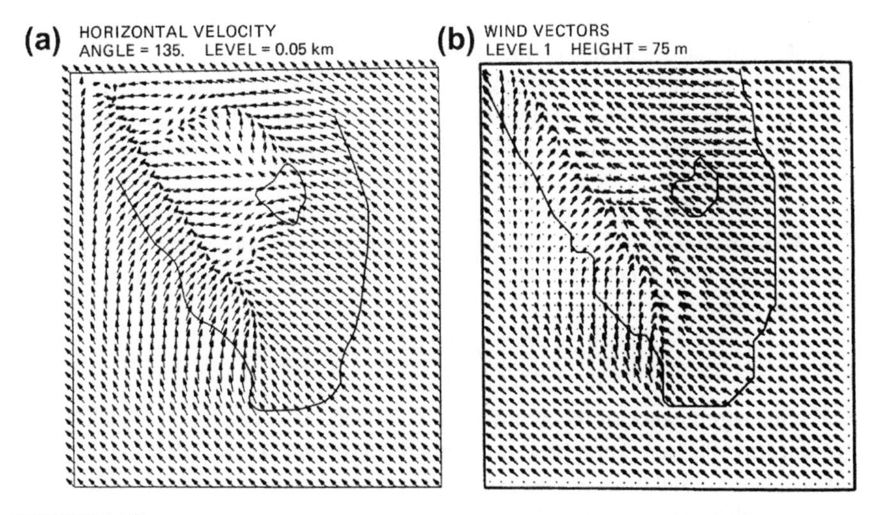

FIGURE 11.3

The predicted horizontal winds (a) at 50 m, 10 h after sunrise (from Pielke 1974a); and (b) at 75 m, 12 h after sunrise (from Tapp and White 1976). The synoptic geostrophic wind for both simulations was from the southeast to 6 m s^{-1}, and the maximum land-surface temperature during the day was 10°C warmer than the surrounding ocean. The distance between one grid point (indicated by the origin of the arrows) corresponds to 6 m s^{-1}.

11.5 Comparison Against Different Model Formulations

Instead of comparing results with thc model of a different investigator, alternative forms of the same model can be examined. One form of *sensitivity* study[1] involves changes to the model including different computational schemes, other approximations to the conservation equations, etc. Tapp and White (1976), for example, contrasted the use of forward-in-time, upstream differencing (Scheme I in Table 9.1) of the advective terms in their sea-breeze model with the use of a second-order, leapfrog representation (Scheme II in Table 9.1). Although the results were similar, the use of upstream differencing resulted in smoother vertical and horizontal velocity fields. The noisier fields resulting from the leapfrog representation may have occurred due to the poor handling of phase speed with that scheme. Mahrer and Pielke (1978b) performed a test of the upstream spline interpolation (Scheme III in Table 9.1) and the upstream differencing in a two-dimensional sea-breeze simulation and found no significant differences in the results. (In the same paper, however, Mahrer and Pielke found that the upstream differencing scheme produced mountain wave solutions with excessive damping, as contrasted with the more realistic appearing solutions obtained with the spline. Sea-breeze simulations can produce reasonable solutions with upstream differencing because such a mesoscale feature is strongly controlled by vertical subgrid-scale mixing, whereas simulations of forced airflow over rough terrain require a much more accurate representation of advection and gravity wave propagation.)

The evaluation of nonlinear model results with and without the hydrostatic assumption is of particular interest. For the sea-breeze circulation, Pielke (1972), Martin (1981), and Martin and Pielke (1983) examined the relative magnitude of the nonhydrostatic pressure in a nonlinear model in considerable depth. One procedure used to evaluate this pressure is to derive a Poisson equation for the hydrostatic component of the pressure, $\bar{p}_{\mathrm{H}} = p'_{\mathrm{H}} + p_0$, as was performed for the Defant sea breeze model in Section 5.2.3.2. The difference between the total and hydrostatic pressure, defined here as R', represents a grid-volume averaged nonhydrostatic pressure residual. In an anelastic formulation, since $\partial p_0/\partial x_i$ is already required to be in hydrostatic balance, the ratio given by

$$\left|\frac{\partial p'_{\mathrm{H}}}{\partial x_i} + \frac{\partial R'}{\partial x_i}\right| \Bigg/ \left|\frac{\partial p'_{\mathrm{H}}}{\partial x_i}\right|, \quad i = 1, 2, \tag{11.1}$$

indicates the significance of the nonhydrostatic effect.

To illustrate the derivation of R' for a nonlinear model, assume that the depth of the atmospheric circulation of interest is much smaller than the density scale depth of the atmosphere (i.e., $L_z \ll H_\alpha$) so that the shallow continuity equation in 4.23 can be used, the second term on the left of Eq. 4.33 can be ignored, and α'/α_0 can be approximated by θ'/θ_0. In addition, to simplify the analysis (without losing the generality of the result since p_0 is assumed hydrostatically determined) assume

[1]The other type of sensitivity study involves changes in the physical parameters within the model (e.g., the initial wind speed, Coriolis value, ground roughness, etc.)

$(\partial/\partial x_i)\rho_0 = 0 (i = 1, 2)$. For this situation, differentiating Eq. 4.32 with respect to z and Eq. 4.14 with respect to x and y (i.e., $\partial/\partial x_i$ with $i = 1, 2$), where $\bar{p}$ is replaced with $\bar{p}_{\mathrm{H}}$, and adding the two equations yields

$$\left(\frac{\partial^2}{\partial x^2}+\frac{\partial^2}{\partial y^2}+\frac{\partial^2}{\partial z^2}\right) p'_{\mathrm{H}} = \nabla^2 p'_{\mathrm{H}} = \frac{\partial^2}{\partial x_i \partial x_j}\rho_0 \bar{u}_i \bar{u}_j - \frac{\partial^2}{\partial x_i \partial x_j}\rho_0 \overline{u''_j u''_i} - 2\rho_0 \epsilon_{ijk}\Omega_j \frac{\partial}{\partial x_i}\bar{u}_k - \frac{\partial}{\partial x_i}\rho_0 \frac{\partial \bar{u}^*_i}{\partial t} + g\frac{\partial}{\partial z}\rho_0 \frac{\theta'}{\theta_0} \quad (i = 1, 2), \tag{11.2}$$

where $\partial \bar{u}^*_i/\partial t$ is evaluated from Eq. 4.14 using $\bar{p}_{\mathrm{H}}$ in place of the total pressure $\bar{p}$. Subtracting Eq. 11.2 from the form of Eq. 4.33 for a shallow atmospheric system and with $\partial \rho_0/\partial x_i (i = 1, 2) = 0$, and assuming that the velocities that occur in Eq. 11.2 not involving a time tendency term are the same as the equivalent velocities in Eq. 4.33, results in the equation.[2]

$$\frac{\partial^2 R'}{\partial x_i^2} = -\frac{\partial^2}{\partial z \partial x_j}\rho_0 \bar{u}_j \bar{w} - \frac{\partial^2}{\partial z \partial x_j}\rho_0 \overline{u''_j w''} + \frac{\partial}{\partial x}\rho_0 \frac{\partial \bar{u}^*}{\partial t} + \frac{\partial}{\partial y}\rho_0 \frac{\partial \bar{v}^*}{\partial t}, \tag{11.3}$$

where $R' = \bar{p} - \bar{p}_{\mathrm{H}} = p' - p'_{\mathrm{H}}$. Since the magnitude of $\partial/\partial x_i$ in Eq. 11.1 is over the same distance for each term, an examination of

$$\epsilon = \left|R'/p'_{\mathrm{H}}\right|$$

at each grid point over a model simulation is an adequate test of the adequacy of the hydrostatic assumption.

Pielke (1972) found for sea-breeze simulations that for the same scale of horizontal heating, ϵ became larger as the heating was increased and as the thermodynamic stratification was made less stable. This result agrees with that found by Martin for the linear model results discussed in Section 5.2.3 and illustrated in Figs. 5.6 and 5.7. The variation of the maximum nonhydrostatic pressure residual as a function of heat input and overlying stratification from Pielke's (1972) results are illustrated in Fig. 11.4 as a function of the horizontal grid scale used. For each experiment, heat was input within the lowest 300 m over a horizontal distance of $9\Delta x$ and over a time scale such that the maximum heating was reached at the time indicated at the top of the Figure. Even for relatively small scales of horizontal heating over short time periods

[2]The solution to this diagnostic differential equation can be obtained using the procedure of sequential relaxation discussed in presented in Section 10.3 in Pielke (2002b). Haltiner and Williams (1980, Chapter 5) also present procedures to solve Eq. 11.3 using direct matrix procedures. Mason and Sykes (1978) discuss the use of a direct method to solve for pressure in a Cartesian coordinate framework even when topography exists in the model domain.

Also, in deriving Eq. 11.2, the velocities on the right of Eq. 4.14, were assumed to be the velocities obtained when the complete nonhydrostatic pressure, $\bar{p}_{\mathrm{H}} + R'$, was used (e.g., as available at the beginning of a time step). This assumption does not have to be made, however (it simply results in more terms in Eq. 11.3 if it is not made). As long as changes of R' at a grid point between time steps are small relative to the magnitude of R'; it is a reasonable assumption.

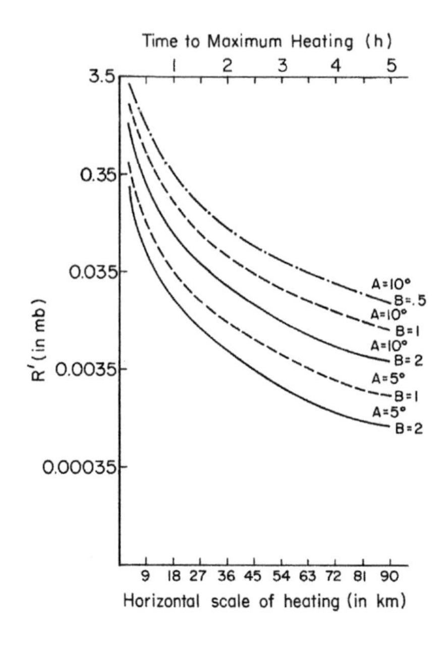

FIGURE 11.4

The maximum absolute value of the nonhydrostatic pressure residual R' as a function of horizontal scale of heating and time to maximum heating (from Pielke 1972). To determine R' from Pielke (1972:26) a large-scale pressure of 1000 mb was used. The magnitude of maximum heating is A (i.e., using $A = \Delta\bar{\theta}_{\text{max}}$ in Eq. 10.25) and B is the value of $\partial\theta_0/\partial z$ in the middle and lower levels of the model in terms of $B°\text{C}/300$ m.

(e.g., with $L_x = 9$ km , the time to maximum heating was 30 min), the hydrostatic relation appears to be a valid assumption for the pressure distribution.

Figures 11.5 and 11.6 illustrate results from Pielke (1972) for a hydrostatic model run, where p'_{H} is used to represent the horizontal pressure gradient, and for a nonhydrostatic simulation $p' = p'_{\text{H}} + R'$ is used for that horizontal gradient. The scales of horizontal heating in the calculation are 2.7 and 9 km, with a maximum temperature amplitude, $\Delta\bar{\theta}_{\text{max}}$, in Eq. 10.25 of 5°, and a potential temperature gradient in the lowest 2.7 km of the model of $1°\text{C}/300$ m. In Figs. 11.5 and 11.6, day in Eq. 10.25 was defined as 2160 s and 7200 s, respectively. Despite the short time period of heat input, however, the differences between the hydrostatic and nonhydrostatic simulations for $L_x = 9$ km were small. With $L_x = 2.7$ km , the hydrostatic solution had substantially larger amplitude although the locations of the convergence zones were similar. Figure 11.7 illustrates the contribution of the nonhydrostatic pressure residual, R' to the total pressure for Pielke's (1972) sea-breeze calculations. In a nonhydrostatic model, the vertical accelerations act to diminish the magnitude of the hydrostatic horizontal pressure gradients.

Martin's (1981) study substantiated Pielke's (1972) investigation of the relative influence of the nonhydrostatic pressure residual. In Martin's thesis, the nonlinear

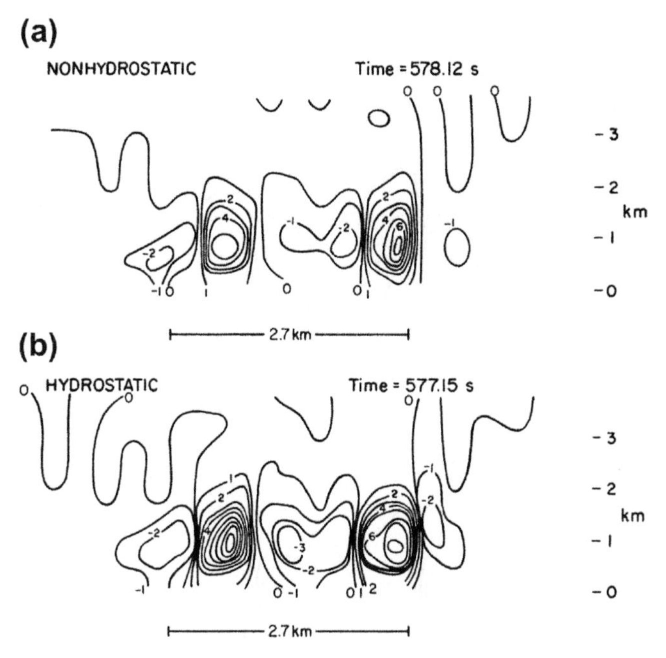

FIGURE 11.5

The vertical velocity in centimeters per second in (a) an anelastic, nonhydrostatic, and (b) a hydrostatic model; where $L_x = 2.7$ km, $\Delta x = 0.3$ km, $\Delta\bar{\theta}_{\max}$ and day in Eq. 10.25 of 5°C and 2160 s, and $\partial\theta_0/\partial z = 1°\text{C}/300$ m in the lowest 2.7 km. The horizontal scale of heating is indicated at the bottom (from Pielke 1972).

advection terms are added to Defant's (1950) analytic equations given by Eqs. 5.66-5.70 (i.e., $u'\,\partial u'/\partial x$ and $w'\,\partial u'/\partial z$ to Eq. 5.66; $u'\,\partial v'/\partial x$ and $w'\,\partial v'/\partial z$ to Eq. 5.67; $u'\,\partial w'/\partial x$ and $w'\,\partial w'/\partial z$ to Eq. 5.68; and $u'\,\partial\theta'/\partial x$ and $w'\,\partial\theta'/\partial z$ to Eq. 5.70). A hydrostatic model is formed from these equations using $\partial p'_{\mathrm{H}}/\partial z = \rho_0\theta'/\theta_0$ in place of Eq. 5.68 (i.e., $\lambda_1 = 0$) and a nonhydrostatic version is derived of the form given by Eq. 11.3 where $p' = p'_{\mathrm{H}} + R'$ is used in Eqs. 5.66 and 5.68 with $\lambda_1 = 1$. Figure 11.8 illustrates predicted results for horizontal velocity at the time of maximum heating where $L_x = 6.25$ km and the largest temperature perturbation is 2.5°C. As in Pielke's (1972) earlier study, the nonhydrostatic and hydrostatic results are similar even for this relatively small spatial scale of heating.

Tag and Rosmond (1980) extended the hydrostatic-nonhydrostatic comparison to a three-dimensional cloud simulation. Among their findings was that moist processes magnified the nonhydrostatic effect, although by increasing the stability from 1 to 2°C km^{-1}, the differences caused by the nonhydrostatic effect were almost eliminated.

One of the advantages of using Eq. 11.3 to compute the nonhydrostatic pressure residual is that it need be computed only in a region where significant vertical

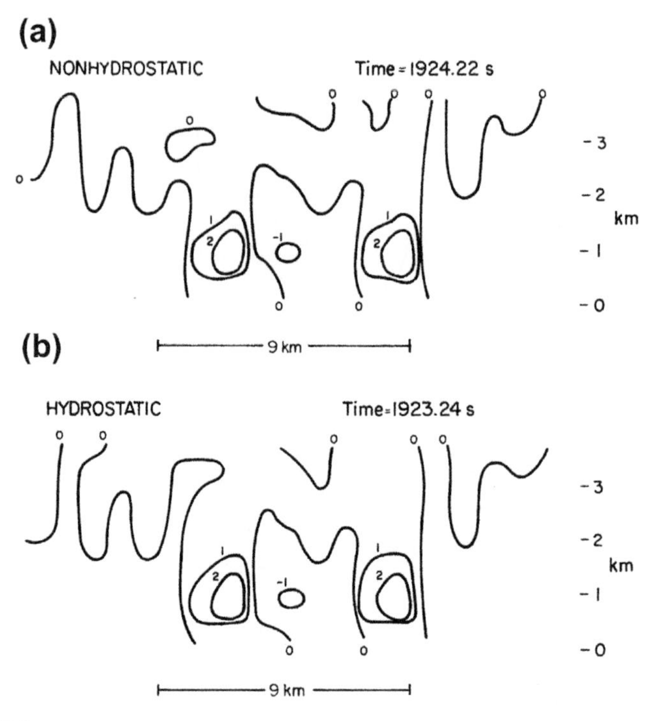

FIGURE 11.6

Same as Fig. 11.5 except $L_x = 9$ km and day $= 7200$ s (from Pielke 1972).

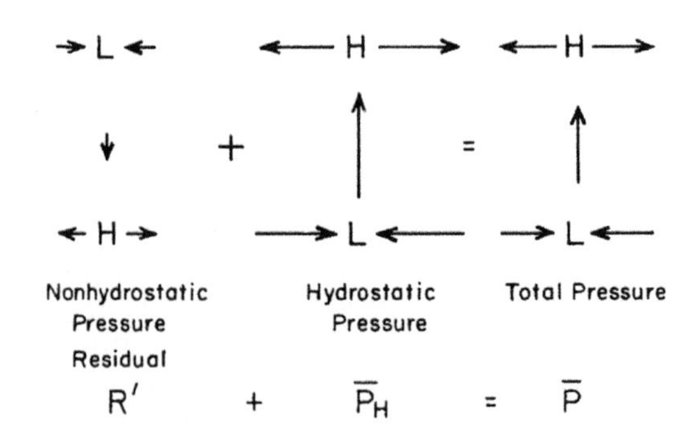

FIGURE 11.7

A schematic of the relative contributions of the nonhydrostatic pressure residual and the hydrostatic pressure to the total pressure at a location over land in the center of the lowest pressure in the sea-breeze convergence zone. The arrows illustrate the instantaneous horizontal winds that would be expected from these pressure distributions (adapted from Pielke 1972).

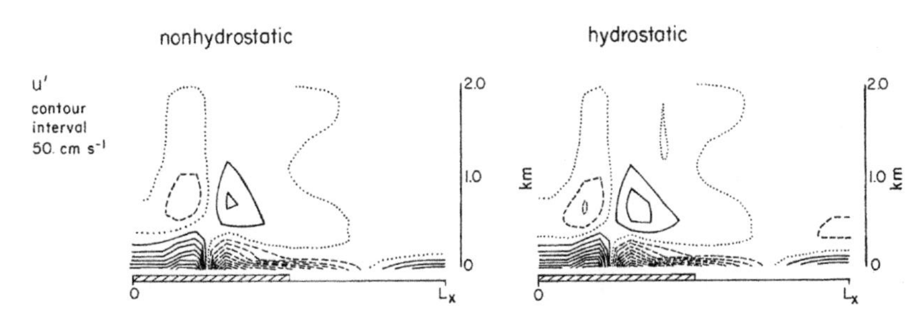

FIGURE 11.8

Nonhydrostatic and hydrostatic simulations for $L_x = 6.25$ km after 2700 s with a surface heating function of $\bar{\theta} = \theta_0(t = 0) + |\Delta\theta|_{\max} \sin(2\pi x/L_x) \sin(\pi t/T)$, where $|\Delta\theta|_{\max} = 2.5$°C and $T = 3$ h. The horizontal and vertical grid spacings were $\Delta x = 0.306$ km and $\Delta z = 100$ m. Periodic lateral boundary conditions were used. Other prescribed values include $\partial\theta_0/\partial x = 10$°C km^{-1}, $K_\theta = 10$ m^{2}s^{-1}, $f = 1.301 \times 10^{-4}$s^{-1}, and $\sigma_{\rm H}$ and $\sigma_{\rm v} = 10^{-3}$s^{-1} (the meaning of these symbols are given in Section 5.2.3.1). Positive values are given by the solid line and negative values by dashed lines with zero indicated by the dotted line (from Martin 1981).

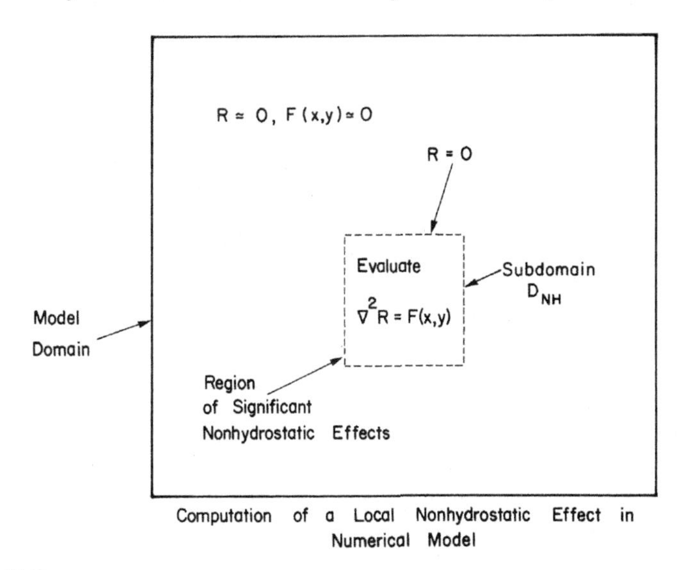

FIGURE 11.9

The definition of a subdomain $D_{\rm NH}$ in a model where nonhydrostatic effects are significant. Such a domain is defined to enclose those regions where $F(x, y)$ is significantly different from zero.

accelerations exist. As illustrated in Fig. 11.9, the boundary condition for R on the subdomain is straightforward to apply since $R = 0$ where the motions are hydrostatic.

The importance of the nonhydrostatic residual has also been examined, to some extent, for forced air over rough terrain. Reproduced from Durran (1981), Fig. 11.10

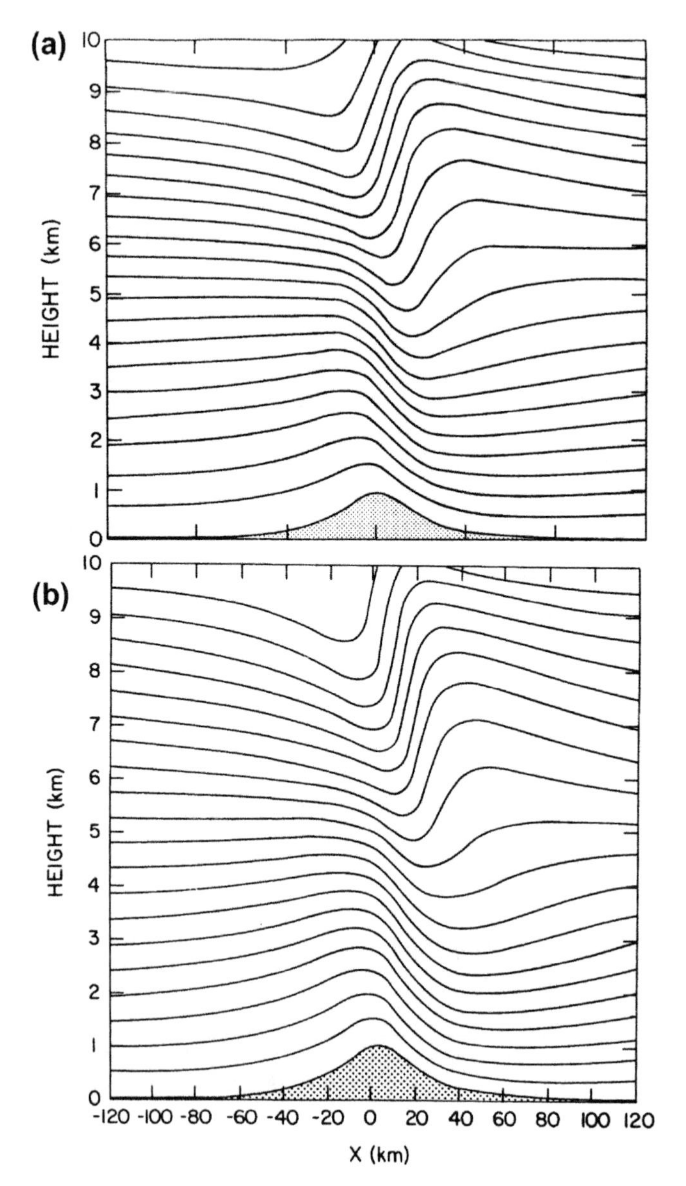

FIGURE 11.10

Contours of potential temperature from (a) a nonlinear hydrostatic simulation by Klemp and Lilly (1978), and (b) a nonlinear nonhydrostatic simulation by Durran (1981). The upstream winds were 20 m s^{-1} and the upstream stratification within the lowest 10 km was 4°C km^{-1}. Note that because of the scale differences between the top and bottom figures, the height scale in (b) is about 8% larger than that in (a). To more quantitatively compare the two results, measure the trough to crest difference for equivalent initial inflow potential temperature heights.

shows potential temperature surfaces from one of Klemp and Lilly's (1978:Fig. 10) hydrostatic simulations and an equivalent result performed by Durran for a nonhydrostatic simulation. In both model runs the upstream wind was 20 m s^{-1}, constant with height, the mountain height reached 1 km, $\partial\theta_0/\partial z = 4°\text{C km}^{-1}$ within the lowest 10 km, and an isothermal absorbing layer (see Section 10.4.2) was prescribed between 10 and 20 km in height. The half-width of the mountain was 20 km, where the terrain was defined as

$$z_G = b^2 z_{G_{max}} \Big/ \left(x^2 + b^2\right), \tag{11.4}$$

where b is the half-width and $z_{G_{max}}$ is the maximum height of the terrain. Although the nonhydrostatic simulation produced a slightly steeper wave in the upper stratosphere, the hydrostatic and nonhydrostatic results are almost identical. Klemp and Lilly (1980) concluded that for realistic atmospheres with simple, uniform structure (i.e., constant large-scale velocity and static stability with height), ratios of $(b/|u_0|)[(g/\theta_0)(\partial\theta_0/\partial z)]^{1/2} \geq 10$ or so yield nearly identical hydrostatic and nonhydrostatic results. With $\partial\theta_0/\partial z = 1\text{ K }100\text{ m}^{-1}$, $\theta_0 = 300$ K, and $u_0 = 20$ m s^{-1}, for example, $b \geq 10$ km or so satisfies this requirement. For more general atmospheric structure, however, it is desirable to check the hydrostatic assumption for each situation.

11.6 Calculation of Model Budgets

The evaluation of the budgets of such physical quantities as kinetic energy and mass are useful not only to improve our understanding of mesoscale physical processes, but also as a diagnostic tool to examine the fidelity of the computer program logic. Before discussing such budgets of mesoscale models, however, it is useful to examine this approach with the single fluid tank model introduced in Section 5.2.1.1.

11.6.1 Mass and Energy Equations for the Homogeneous Tank Model

The mass budget for this simplified model given by Eqs. 5.17 and 5.20 is particularly easy to compute since the fluid is assumed homogeneous. Using the product rule of differentiation and integrating Eq. 5.20 over the tank model domain D_x gives

$$\int_0^{D_x} \frac{\partial h}{\partial t}\, dx + \int_0^{D_x} \frac{\partial}{\partial x} u' h\, dx = \int_0^{D_x} \frac{\partial h}{\partial t}\, dx + u' h\Big|_0^{D_x} = 0$$

If the sides of the tank ($x = 0, D_x$) are rigid (i.e., $u'_0 = u'_{D_x} = 0$),[3] or if periodic boundary conditions for h and u' are used (i.e., $u'h$ at $x = 0$ is equal to $u'h$ at

[3]If $u' \equiv 0$ at the boundaries, then $\partial u'/\partial x$ and $\partial u'/\partial t$ must also be identically zero at these locations. Therefore, from Eq. 5.17 $g\,\partial h/\partial x = 0$ at the boundaries so that no slope to the fluid is permitted at the walls. Numerical approximations to the tank model equations with rigid walls must utilize this boundary condition on h, recognizing that $\partial h/\partial x = 0$ at the boundary does not mean that $\Delta h/\Delta x = 0$ between the boundary and one grid point inside when Δx is finite.

$x = D_x$), then

$$\int_0^{D_x} \frac{\partial h}{\partial t} dx = \frac{\partial}{\partial t} \int_0^{D_x} h\, dx = \frac{\partial \bar{h}}{\partial t} = 0,$$

so that the average height of the fluid must be constant in time. In a numerical model, the conservation-of-mass, as represented by the depth of the fluid, can be checked each time step by

$$\bar{h} = \frac{1}{I_D} \sum_{i=1}^{I_D} h_i,$$

where I_D is the number of grid points. If $\Delta\bar{h}/\Delta t \neq 0$, mass is not conserved.

The kinetic energy of the tank model can be computed by multiplying Eq. 5.17 by hu' and Eq. 5.20 by $u'^2/2$ yielding the two simultaneous partial differential equations

$$hu' \frac{\partial u'}{\partial t} + hu'^2 \frac{\partial u'}{\partial x} + ghu' \frac{\partial h'}{\partial x} = h \frac{\partial u'^2/2}{\partial t} + hu' \frac{\partial}{\partial x}\left(\frac{u'^2}{2}\right) + ghu' \frac{\partial h}{\partial x} = 0$$

$$\frac{u'^2}{2} \frac{\partial h}{\partial t} + \frac{u'^2}{2} u' \frac{\partial h}{\partial x} + \frac{u'^2}{2} h \frac{\partial u'}{\partial x} = \frac{u'^2}{2} \frac{\partial h}{\partial t} + \frac{u'^2}{2} \frac{\partial}{\partial x}(hu') = 0.$$

Adding these two equations utilizing the product rule of differentiation and multiplying through by the constant density ρ_0 yields

$$\frac{\partial}{\partial t}\left(\rho_0 \frac{u'^2}{2} h\right) + \frac{\partial}{\partial x}\left(\rho_0 \frac{u'^2}{2} hu'\right) + \rho_0 ghu' \frac{\partial h}{\partial x} = 0, \tag{11.5}$$

where $\rho_0(u'^2/2h)$ as units of kinetic energy per unit area.

The potential energy equation is obtained by multiplying Eq. 5.20 by $\rho_0 gh$ resulting in

$$\rho_0 gh \frac{\partial h}{\partial t} + \rho_0 gh \frac{\partial}{\partial x} hu' = \rho_0 g \frac{\partial h^2/2}{\partial t} + \rho_0 gh \frac{\partial u'h}{\partial x} = 0, \tag{11.6}$$

where $\rho_0 gh^2/2$ has units of potential energy per unit area.

To obtain the total energy, add Eqs. 11.5 and 11.6 and rearrange giving

$$\frac{\partial}{\partial t}\left[\rho_0 \frac{u'^2}{2} h + \rho_0 g \frac{h^2}{2}\right] + \frac{\partial}{\partial x}\left[hu'\left(\rho_0 \frac{u'^2}{2} + \rho_0 gh\right)\right] = 0. \tag{11.7}$$

The first term on the left is the local change of total energy per unit area E and the second term is proportional to the horizontal flux divergence of this energy per unit area.

Integrating this expression over the model domain (i.e., the size of the tank) gives

$$\int_0^{D_x} \frac{\partial E}{\partial t} dx = -hu'\left(\rho_0 \frac{u'^2}{2} + \rho_0 gh\right)\Bigg|_0^{D_x} = \frac{\partial}{\partial t} \int_0^{D_x} E\, dx = \frac{\partial E_0}{\partial t},$$

which is equal to zero if the walls are rigid or if cyclic boundary conditions are applied. If $\partial E_0/\partial t = 0$, then total energy is conserved in the tank model. Numerical approximations of the tank model equations, therefore, should also conserve total energy (i.e., evaluating the approximate form for the first bracketed term on the left of Eq. 11.7 at each grid point and summing across the domain should yield a number that is identical at each time step). Simulations that differ significantly from such mass- and energy-conservation criteria, therefore, are suspect and results from them should be used cautiously, if at all.

11.6.2 Mass and Energy Equations for a Mesoscale Model

In mesoscale models, much more involved conservation relations are used, however, it is similarly desirable to conserve mass and energy. To illustrate the procedure, the hydrostatic, anelastic form of the equations given by Eqs. 6.84-6.87 and 6.90 with σ defined by Eq. 6.48 are used to derive the kinetic energy and mass-conservation relations.

The shallow slope, hydrostatic form of Eq. 6.87 (Eq. 6.62) can be differentiated with time and the order of operation reversed to yield

$$\frac{\partial}{\partial \tilde{x}^3}\frac{\partial \bar{\pi}}{\partial t} = \frac{g}{\bar{\theta}^2}\frac{s - z_G}{s}\frac{\partial \bar{\theta}}{\partial t}.$$

Integrating between $\tilde{x}^3 = 0$ and s_θ and rearranging yields[4]

$$\left.\frac{\partial \bar{\pi}}{\partial t}\right|_{z_G} = \left.\frac{\partial \bar{\pi}}{\partial t}\right|_{s_\theta} - g\,\frac{s - z_G}{s}\int_0^{s_\theta}\frac{1}{\bar{\theta}^2}\frac{\partial \bar{\theta}}{\partial t}\,d\tilde{x}^3, \tag{11.8}$$

which is the same form as Eq. 10.20 except $z_G \neq 0$ in Eq. 11.8. The pressure tendency $\partial\bar{\pi}/\partial t|_{s_\theta}$ must be specified as a boundary condition (i.e., see following Eq. 10.18), and the integrated term in Eq. 11.8 is evaluated using Eq. 6.87. Integrating the righthand side of Eq. 11.8 over the model domain gives[5]

$$\begin{aligned}\frac{1}{D_{\tilde{x}^1}D_{\tilde{x}^2}}\int_0^{D_{\tilde{x}^1}}\int_0^{D_{\tilde{x}^2}}\left.\frac{\partial \bar{\pi}}{\partial t}\right|_{z_G}\left(\frac{s - z_G}{s}\right)\left(\frac{s}{s - z_G}\right)d\tilde{x}^2\,d\tilde{x}^1 \\ = \frac{1}{D_{\tilde{x}^1}D_{\tilde{x}^2}}\frac{\partial}{\partial t}\int_0^{D_{\tilde{x}^1}}\int_0^{D_{\tilde{x}^2}}\bar{\pi}_*\,d\tilde{x}^2\,d\tilde{x}^1 = \frac{\partial \Pi_*}{\partial t}\end{aligned} \tag{11.9}$$

Values of $\partial\bar{\theta}/\partial t$ needed at $\tilde{x}^1 = 0, D_{\tilde{x}^1}$ and at $\tilde{x}^2 = 0, D_{\tilde{x}^2}$ are obtained from the assumed lateral boundary condition on $\bar{\theta}$ (see Section 10.4.1).

---

[4]As used here and in Section 6.3, s is a constant, usually defined to correspond to the initial value of s_θ as defined by Eq. 10.18 and following material. The variable s_θ corresponds to a movable potential temperature surface.

[5]As shown by Dutton (1976:144), differential area on a constant $\tilde{x}^3$ surface can be written as $dS = |\vec{n}^3|\sqrt{\tilde{G}}\,d\tilde{x}^2 d\tilde{x}^1$. For the terrain-following coordinate system defined by 6.48, $\sqrt{\tilde{G}} = (s - z_G)/s$ (from Eqs. 6.32 and 6.51) and for small slopes $|\vec{n}^3| \simeq s/(s - z_G)$ (from Eqs. 6.33 and 6.51).

Since from Eq. 10.22 a change of π at the surface is equivalent to a change of mass above that level, Π_* in Eq. 11.9 provides the value of the average mass change per unit area over the model domain. This value of Π_* can be compared against the integrated value of $\overline{\Pi}_{zG}$ computed directly from Eq. 6.62, i.e.,

$$\frac{\overline{\Pi}_{zG}}{\partial t} = \frac{1}{D_{\tilde{x}^1} D_{\tilde{x}^2}} \frac{\partial}{\partial t} \int_0^{D_{\tilde{x}^1}} \int_0^{D_{\tilde{x}^2}} \bar{\pi}_{zG}\, d\tilde{x}^2\, d\tilde{x}^1,$$

where

$$\bar{\pi}_{zG} = +g\,\frac{s - z_G}{s} \int_0^{s_\theta} \frac{d\tilde{x}^3}{\bar{\theta}} + \overline{\Pi}|_{s_\theta}.$$

The difference $(\partial/\partial t)(\Pi_{zG} - \Pi_*)$ is proportional to the mass loss.

A kinetic energy equation for the flow parallel to the terrain can also be derived from the set of equations. Since from Eq. 6.34, $\tilde{u}^1 = u$ and $\tilde{u}^2 = v$, multiplying Eq. 6.57 by $\rho_0 \bar{\tilde{u}}^1 (s - z_G)/s$, and Eq. 6.58 by $\rho_0 \bar{\tilde{u}}^2 (s - z_G)/s$, adding the two equations, and using the anelastic conservation-of-mass equation, after multiplying by $\bar{k} = \frac{1}{2}\left(\bar{\tilde{u}}^1 \bar{\tilde{u}}^1 + \bar{\tilde{u}}^2 \bar{\tilde{u}}^2\right)$ results in the terrain-following kinetic energy equation

$$\begin{aligned}
\rho_0 \frac{s - z_G}{s} \frac{\partial \bar{k}}{\partial t} = &- \frac{\partial}{\partial \tilde{x}^j} \rho_0 \frac{s - z_G}{s} \bar{\tilde{u}}^j \bar{k} - \bar{\tilde{u}}^1 \rho_0 \frac{s - z_G}{s} \overline{\tilde{u}^{j''} \frac{\partial \tilde{u}^{1''}}{\partial \tilde{x}^j}} \\
&- \bar{\tilde{u}}^2 \rho_0 \frac{s - z_G}{s} \overline{\tilde{u}^{j''} \frac{\partial \tilde{u}^{2''}}{\partial \tilde{x}^j}} - \bar{\tilde{u}}^1 \rho_0 \frac{s - z_G}{s} \\
&\times \left\{ \theta \frac{\partial \bar{\pi}}{\partial \tilde{x}^1} - g \frac{\sigma - s}{s} \frac{\partial z_G}{\partial x} \right\} - \bar{\tilde{u}}^2 \rho_0 \frac{s - z_G}{s} \\
&\times \left\{ \theta \frac{\partial \bar{\pi}}{\partial \tilde{x}^2} - g \frac{\sigma - s}{s} \frac{\partial z_G}{\partial y} \right\} - \bar{\tilde{u}}^1 \rho_0 \frac{s - z_G}{s} \hat{f}\, \bar{\tilde{u}}^3.
\end{aligned} \tag{11.10}$$

Equation 11.10 can be integrated over the model domain yielding

$$\begin{aligned}
\frac{\partial K_*}{\partial t} = &- \int_0^{s_\theta} \left\{ \int_0^{D_{\tilde{x}^2}} \rho_0 \frac{s - z_G}{s} \bar{\tilde{u}}^1 \bar{k} \Big|_0^{D_{\tilde{x}^1}} d\tilde{x}^2 + \int_0^{D_{\tilde{x}^1}} \rho_0 \frac{s - z_G}{s} \bar{\tilde{u}}^2 \bar{k} \Big|_0^{D_{\tilde{x}^2}} d\tilde{x}^1 \right. \\
&+ \int_0^{D_{\tilde{x}^2}} \int_0^{D_{\tilde{x}^1}} \rho_0 \frac{s - z_G}{s} \left[\bar{\tilde{u}}^1 \overline{\tilde{u}^{j''} \frac{\partial \tilde{u}^{1''}}{\partial \tilde{x}^j}} + \bar{\tilde{u}}^2 \overline{\tilde{u}^{j''} \frac{\partial \tilde{u}^{2''}}{\partial \tilde{x}^j}} \right] d\tilde{x}^1\, d\tilde{x}^2 \\
&+ \int_0^{D_{\tilde{x}^2}} \int_0^{D_{\tilde{x}^1}} \rho_0 \frac{s - z_G}{s} \left[\bar{\tilde{u}}^1 \bar{\theta} \frac{\partial \bar{\pi}}{\partial \tilde{x}^1} + \bar{\tilde{u}}^2 \bar{\theta} \frac{\partial \bar{\pi}}{\partial \tilde{x}^2} \right] d\tilde{x}^1\, d\tilde{x}^2 \\
&- \int_0^{D_{\tilde{x}^2}} \int_0^{D_{\tilde{x}^1}} \rho_0 \frac{s - z_G}{s} \left[\bar{\tilde{u}}^1 g \frac{\sigma - s}{s} \frac{\partial z_G}{\partial x} + \bar{\tilde{u}}^2 g \frac{\sigma - s}{s} \frac{\partial z_G}{\partial y} \right] d\tilde{x}^1\, d\tilde{x}^2 \\
&- \left. \int_0^{D_{\tilde{x}^2}} \int_0^{D_{\tilde{x}^1}} \rho_0 \frac{s - z_G}{s} \bar{\tilde{u}}^1 \hat{f}\, \bar{\tilde{u}}^3\, d\tilde{x}^1\, d\tilde{x}^2 \right\} d\tilde{x}^3 \\
&+ \int_0^{D_{\tilde{x}^2}} \int_0^{D_{\tilde{x}^1}} \rho_0 \frac{s - z_G}{s} \bar{k}_{s_\theta} \frac{\partial s_\theta}{\partial t}\, d\tilde{x}^1\, d\tilde{x}^2,
\end{aligned} \tag{11.11}$$

where

$$K_* = \int_0^{s_\theta} \int_0^{D_{\tilde{x}^1}} \int_0^{D_{\tilde{x}^2}} \left(\rho_0 \frac{s - z_G}{s} \bar{k} \right) d\tilde{x}^2 \, d\tilde{x}^1 \, d\tilde{x}^3.$$

In deriving Eq. 11.11, the condition that $\bar{\tilde{u}}^3 = 0$ at z_G and s_θ has been used. The last term in Eq. 11.11 arises from Leibnitz's rule[6] since s_θ is a function of time. Each of the variables in the last term are evaluated at s_θ.

The first two terms on the righthand side of Eq. 11.11 are proportional to the net flow of terrain-following kinetic energy through the sides of the model domain, and the next term represents the change in kinetic energy from subgrid-scale effects. The terms involving $\partial\bar{\pi}/\partial\tilde{x}^1$ and $\partial\bar{\pi}/\partial\tilde{x}^2$ are proportional to the conversion of potential to kinetic energy by cross-isobaric flow, and the expressions containing the gradients of terrain represent the conversion of potential to kinetic energy through upslope and downslope flow. The next to last term in Eq. 11.11 (with $\hat{f}$) would not appear in a three-dimensional kinetic energy equation using the complete conservation-of-motion equation (i.e., without the hydrostatic assumption), since the Coriolis force arises solely because of a coordinate transformation (i.e., see Section 2.3) and, therefore, cannot do work. Therefore, to have a physically consistent terrain-following energy equation, it is necessary to remove this term in Eq. 11.11 and in Eq. 6.84.

In using Eq. 11.11 to determine the total terrain-following kinetic energy changes, it is imperative that the approximation technique used to evaluate the individual terms in that expression be the same as that used in the original approximate form of the conservation relation.

The time rate of change of terrain-following kinetic energy can also be evaluated directly at each individual grid point and then summed, i.e.,

$$\begin{aligned} \frac{\partial K}{\partial t} &= \int_0^{s_\theta} \int_0^{D_{\tilde{x}^2}} \int_0^{D_{\tilde{x}^2}} \rho_0 \frac{s - z_G}{s} \frac{\partial \bar{k}}{\partial t} d\tilde{x}^2 \, d\tilde{x}^1 \, d\tilde{x}^3 \\ &+ \int_0^{D_{\tilde{x}^2}} \int_0^{D_{\tilde{x}^1}} \rho_0 \frac{s - z_G}{s} \bar{k}_{s_\theta} \frac{\partial s_\theta}{\partial t} d\tilde{x}^1 \, d\tilde{x}^2 \end{aligned} \tag{11.12}$$

is used to obtain an estimate of the total terrain-following kinetic energy change instead of Eq. 11.11. If the kinetic energy changes computed by the numerical approximation to this expression and the approximated form of Eq. 11.11 closely agree, the modeler can be certain that mistakes, such as coding errors, are not causing significant sources of unexplained changes of kinetic energy. Note that since the last term in Eqs. 11.11 and 11.12 is the same, there is no need to compute it for a comparison of K and K_*.

Anthes and Warner (1978) discuss the use of kinetic energy budgets in mesoscale models in further detail as a tool to check the model code, as well as to seek additional insight into the energetics of mesoscale systems. Among their results they have shown that the flux of kinetic energy through the side walls of a mesoscale model crucially affect the solutions in the interior. They conclude that because of the

[6] Leibnitz's rule is given in such sources as Hildebrand (1962:360) and Dutton (1976:115).

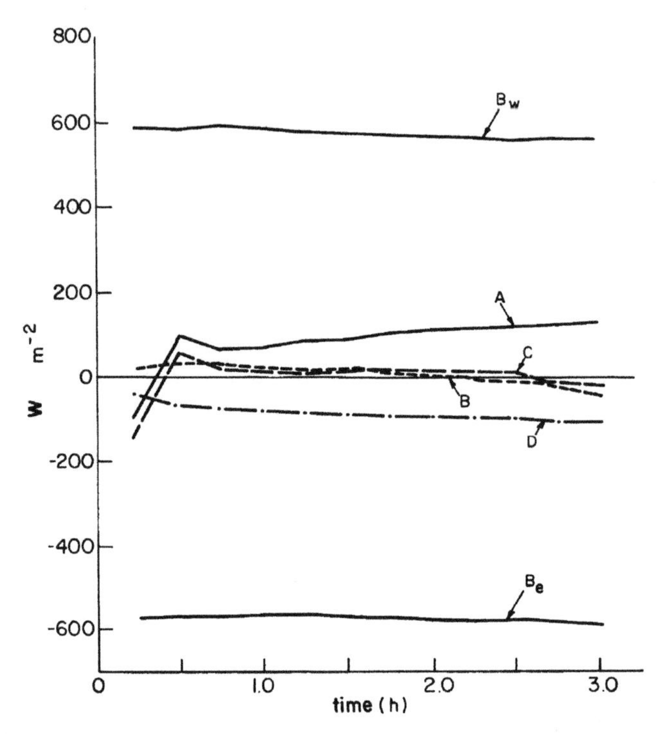

FIGURE 11.11

Individual components of a two-dimensional form of the domain-averaged kinetic energy equation, which is equivalent to Eq. 11.11. A is the generation of kinetic energy by cross-isobaric flow (from the terms with $\bar{\tilde{u}}^1[\bar{\theta}(\partial\tilde{\pi}/\partial\tilde{x}^1) - g\{(\sigma - s)/s\}(\partial z_G/\partial x)]$); B_w and B_e are the flux of kinetic energy across the west and east boundaries from the two terms evaluated from $\bar{\tilde{u}}^1\bar{k}\big|_0^{D\tilde{x}^1}$); B is the net flux across the west and east boundaries (from $B_e - B_w$); C is the domain-averaged change of kinetic energy; and D the dissipation of kinetic energy by horizontal diffusion (from the term with $\bar{\tilde{u}}^1\overline{\tilde{u}^{1''}\partial\tilde{u}^{1''}/\partial\tilde{x}^1}$.). Analogs to the last two terms in Eq. 11.11 were not evaluated by Anthes and Warner (reproduced from Anthes and Warner 1978).

extreme sensitivity of mesoscale model results to domain size and the form of lateral boundary conditions, studies of the energetics of real-world mesoscale systems will be very difficult to perform and very sensitive to errors and small-scale variations of wind, potential temperature, and pressure at the model boundaries. Figure 11.11 illustrates the magnitude of individual terms as a function of time in a two-dimensional analog of Eq. 11.11, computed by Anthes and Warner for strong airflow over rough terrain. Of particular importance is the large magnitude of the boundary fluxes of kinetic energy through the west and east boundaries. Even small percentage errors in these terms can cause serious errors in the results, a conclusion that was illustrated by Table 10.1. In a different study, Tag and Rosmond (1980) discuss energy conservation in a three-dimensional small-scale (nonhydrostatic) model in considerable detail. Pearson (1975), Dalu and Green (1980), and Green and Dalu (1980), provide additional studies of the energetics of mesoscale systems.

Avissar and Chen (1993) use a mesoscale kinetic energy equation similar to Eq. 11.11 in order to develop a parameterization of mesoscale fluxes for use in larger-scale models.

11.6.3 Momentum Flux

Another useful diagnostic tool for model evaluation involves the calculation of the momentum flux. Used more extensively for the study of the dynamics of forced air over rough terrain (e.g., Klemp and Lilly 1978) it is straightforward to calculate.

To illustrate its evaluation, in the absence of the Coriolis term for two-dimensional flow, Eq. 4.14 for $i = 1$ can be written as

$$\frac{\partial \rho_0 \bar{u}}{\partial t} = -\frac{\partial}{\partial x}\rho_0 \bar{u}^2 - \frac{\partial}{\partial z}\rho_0 \bar{w}\bar{u} - \frac{\partial}{\partial x}\rho_0 \overline{u''^2} - \frac{\partial}{\partial z}\rho_0 \overline{u''w''} - \frac{\partial \bar{p}}{\partial x}, \tag{11.13}$$

where the conservation-of-mass Eq. 4.23 has been used. Assuming a steady state and that $\bar{u}$, $\bar{p}$, and $\overline{u''^2}$, far enough upstream and downstream of a two-dimensional barrier to the flow, are constant, Eq. 11.13 can be integrated to yield

$$-\frac{\partial}{\partial z}\int_{-\infty}^{\infty}\left(\rho_0 \bar{w}\bar{u} + \rho_0 \overline{w''u''}\right) dx = 0,$$

or

$$\int_{-\infty}^{\infty}\rho_0\left(\bar{w}\bar{u} + \overline{w''u''}\right) dx = m_1, \tag{11.14}$$

where m_1 is a constant with dimensions of kilogram per seconds squared. Equation 11.14 can be written as

$$\int_{-\infty}^{\infty}\rho_0\left[w_0 u_0 + w' u_0 + w_0 u' + w' u' + \overline{w''u''}\right] dx = m_1 \tag{11.15}$$

using the definition of a mesoscale perturbation from the domain-averaged (i.e., synoptic) value given by Eq. 4.11. For the case $w_0 = 0$ and u_0 equal to a positive constant, Eq. 11.15 reduces to

$$\int_{-\infty}^{\infty}\rho_0\left[w' u' + \overline{w''u''}\right] dx = m_2;$$

since $(\partial/\partial z)\rho_0 \bar{w}\bar{u}$ can be written as $(\partial/\partial z)\rho_0 w' u'$ in Eq. 11.13. Assuming nonturbulent flow, the equation can be further reduced to

$$\int_{-\infty}^{\infty}\rho_0 w' u'\, dx = m_3. \tag{11.16}$$

The constant m_3 is less than zero if the source of the mesoscale motion is the ground surface and there is no downward reflection or generation of perturbed flow above the ground. For this situation the movement of a parcel upward (i.e., $w' > 0$), which is toward a level of higher potential energy, results in a reduction of kinetic energy (i.e., $u' < 0$). The converse is true for the downward movement of a parcel. Hence $w'u' < 0$ is required to satisfy the conservation of total energy.

Equation 11.16 is of the form most commonly applied in the diagnosis of a mesoscale simulation of airflow over rough terrain. In a numerical model, u_0 equal to a constant, $w_0 = 0$, and nonturbulent flow can be assumed for an atmosphere of constant large-scale velocity and static stability, and $w'u'$ calculated to ascertain if it satisfies Eq. 11.16. To prevent aliasing problems (described in Section 9.4), however, long-term inviscid calculations (i.e., without any explicit or computational smoothing) can only be performed for small mountain perturbations where the nonlinear effects are minimal. Klemp and Lilly (1978) show for inviscid, isothermal analytic solutions over a bell-shaped mountain given by Eq. 11.4 with u_0 equal to a constant and $w_0 = 0$, that the momentum flux is of the form

$$m_{3a} \simeq \frac{-\pi}{4} \rho_0 u_0 z_{G_{\max}}^2 \left(\frac{g}{\theta_0} \frac{\partial \theta_0}{\partial z} \right)^{1/2} \simeq -\frac{\pi}{4} \rho_0 z_{G_{\max}}^2 \sigma u_0^2. \tag{11.17}$$

In Eq. 11.17, the definition of σ where

$$\sigma^2 = -\frac{g}{\bar{\rho} u_0^2} \frac{d\bar{\rho}}{dh_0}$$

with $\bar{\rho}$ linear in h_0. The relation between density and potential temperature vertical gradients is given in Klemp and Lilly (1978) and Pielke (2002b). Figure 11.12,

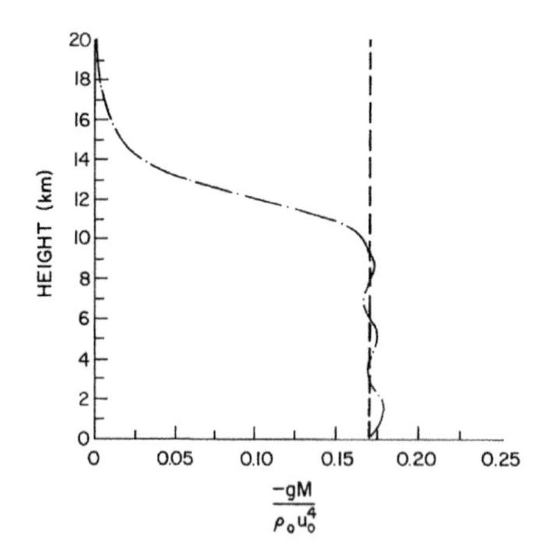

FIGURE 11.12

A plot of m_3 calculated from Eq. 11.16 from a numerical model result for a bell-shaped mountain (dash-dotted line) and m_{3_a} evaluated using Eq. 11.17 for the same mountain shape (dashed line), as a function of height for $u_0 = 20$ m s^{-1}, $z_{G_{\max}}$ equivalent to 100 m and $\rho_0 = 1$ kg m^{-3}. The atmosphere is isothermal and M corresponds to m_3 and m_{3_a} (adapted from Klemp and Lilly 1978).

reproduced from Klemp and Lilly, illustrates a comparison of a numerically computed horizontally integrated momentum flux (from Eq. 11.16) for a simulation of airflow with $u_0 = 20$ m s^{-1} over a mountain of the form given by Eq. 11.4, along with the analytic result given by Eq. 11.17. The numerical model was integrated with $z_{G_{max}} = 10$ m and m_3 multiplied by 10^2 to compare against the linear solution m_{3_a}. The results are almost coincident up to 10 km, thereby providing proof to the fidelity of the numerical model. Above 10 km, the numerical model utilizes an absorbing layer (see Section 10.4.2) to mimic the radiation boundary condition of the analytic model.

Another useful parameter that can be calculated from a model of forced air over rough terrain is the surface drag. This drag occurs because the sloping terrain is a partial barrier and impedes the large-scale wind flow. The wave drag for a two-dimensional mountain can be written as

$$D = \int_{-\infty}^{\infty} \bar{p}\left(z_G\right) \frac{\partial z_G}{\partial x} dx, \tag{11.18}$$

where $\bar{p}$ is evaluated on z_G. The momentum flux (Eq. 11.16) evaluated at z_G is identical to the surface drag given by Eq. 11.18. The integrand of Eq. 11.18 arises from the force per unit area exerted on a two-dimensional mountain in the x direction.[7] For a mountain that is symmetric around its crest, for example, an asymmetric pressure field will result in a value of D that is not equal to zero. Moreover, since total energy must be conserved, the generation of internal gravity waves by a mountain must result in the extraction of energy from the ground.

From the Lilly and Klemp (1979) solution to Long's model presented in Section 5.3 of Pielke (2002b):

$$D_{LK} = -\rho_0 u_0^2 \int_{-\infty}^{\infty} \sigma \left(z_{G_i} - \left(\frac{\sigma z_{G_i}^2}{2} \right) \right) \frac{\partial z_G}{\partial x} dx, \tag{11.19}$$

where the subscript LK is used to indicate it is from the Lilly-Klemp solution to the Long model. Lilly and Klemp (1979) contrast the drag (Eq. 11.19) from their solution to Long's equation for an isothermal atmosphere with constant velocity using a nonlinear bottom boundary condition, with the drag (Eq. 11.17) computed for a linear lower boundary condition. Among their results they found that the drag was enhanced compared with linear theory for mountains with a gentle upslope and steep downslope terrain.

11.7 Standardizing Model Code

As originally proposed by Pielke and Arritt (1984), although the mathematical and physical framework of many of the subroutines used to construct mesoscale models

[7]The change in force per unit area in the x direction can be written as $\bar{p}(z_G)$ cos α $\Delta n/\Delta x$, where α is the terrain slope and Δn is distance along the slope. Since Δn cos $\alpha = \Delta z$, at $z = z_G$ the change in force per unit area in the x direction becomes $\bar{p}(z_G)\Delta z_G/\Delta x$. In the limit as Δz_G and Δx approach zero, and integrating from + to $-\infty$ yields Eq. 11.18.

are similar, there has been little initiative to standardize the computer logic which links these subroutines with the mainframe of a model between modeling groups. These subroutines (i.e., modules) include such components of a program as the cumulus parameterization scheme, the planetary boundary layer representation, the lateral boundary conditions, etc. that were discussed earlier in the book. When only one investigating group uses a model and does not intend to share their modules, there is no need to standardize these modules. If, on the other hand, a researcher would like to utilize a module from someone else's code (e.g., their boundary layer parameterization), scientific productivity would be greatly facilitated if the borrowed module could be simply "plugged into" an existing code.

In addition, researchers who specialize in one specific area of relevance to modeling could develop a module (i.e., a parameterization) that has a standardized interface so it could be used with relatively little effort by other investigators in their models. This standardization of software, of course, is common practice in the microcomputer arena, but it has not been widely adopted in our field except within each modeling group. This approach would be most easily implemented in those codes which have a skeletal mainframe driver to the model which calls the modules as subroutines. Pielke et al. (1995b) further discusses how to intercompare codes between models in order to decide which modules to adopt.

11.8 Comparison with Observations

The validation of a model using observations can be cataloged into two general classes:

1. subjective evaluation, or
2. point and pattern quantitative validation.

Using subjective verification, one or more of the predicted fields are qualitatively compared against observations of a related phenomena. Pielke (1974a), for example, compared the simulated vertical motion at an elevation of 1.22 km over south Florida with the observed locations of rain showers as seen via a 10 cm radar located at Miami. The justification for the comparison is that the primary control for rain shower development over south Florida during a synoptically undisturbed summer day is the location and intensity of the low-level convergence (Pielke et al. 1991). Since 1.22 km is approximately at the top of the planetary boundary layer, the predicted vertical velocity at that level yields an appropriate estimate for low-level convergence. Figure 11.13 illustrates one such comparison for 29 June 1971, 9-1/2 hours after sunrise. As evident in the Figure, the wishbone pattern of the rain showers is closely correlated with the distribution of vertical motion at 1.22 km.

Point-to-point correspondence between model prediction and observation of the same meteorological parameter provides a quantitative test of model skill. Keyser and Anthes (1977) utilize a useful technique where if (i) ϕ_i and $\phi_{i_{\mathrm{obs}}}$ are individual predictions and observations at the same grid point, respectively; (ii) ϕ_0 and $\phi_{0_{\mathrm{obs}}}$ are

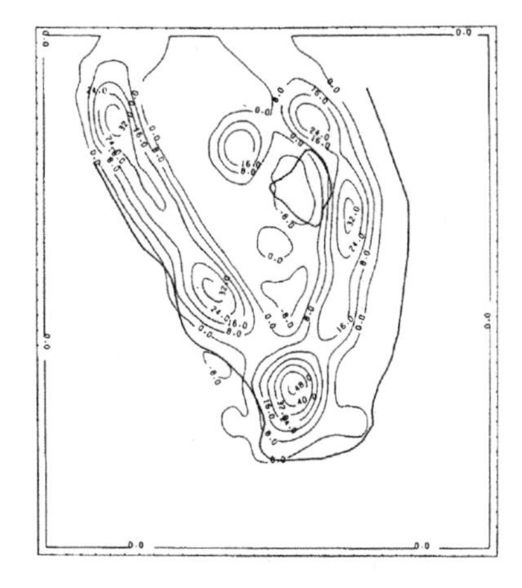

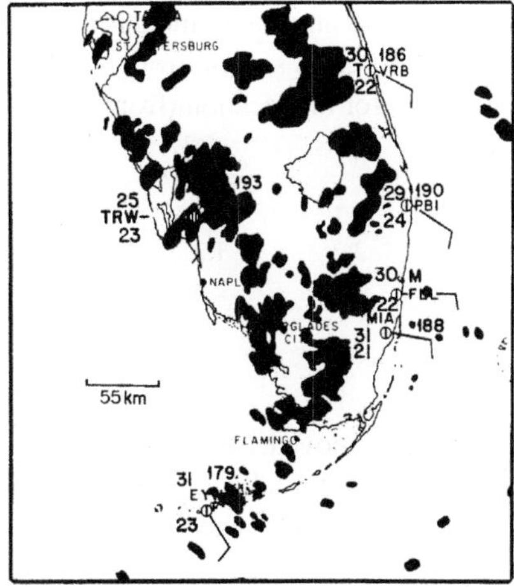

FIGURE 11.13

A model-predicted vertical motion field at 1.22 km and the radar echo map at about 9.5 h after sunrise for 29 June 1971. The large-scale horizontal velocity used in the model simulation above the initial height of the planetary boundary layer was 2.5 m s^{-1} from the east-southeast. The model used in this simulation is reported in Pielke (1974a).

the average values of ϕ_i and $\phi_{i_{\text{obs}}}$ at a level, respectively; and (iii) $\#N$ is the number of observations, then

$$E = \left\{ \sum_{i=1}^{\#N} \left(\phi_i - \phi_{i_{\text{obs}}}\right)^2 \Big/ \#N \right\}^{1/2},$$

$$E_{UB} = \left\{ \sum_{i=1}^{\#N} \left[\left(\phi_i - \phi_0\right) - \left(\phi_{i_{\text{obs}}} - \phi_{0_{\text{obs}}}\right)\right]^2 \Big/ \#N \right\}^{1/2}, \tag{11.20}$$

$$\sigma_{\text{obs}} = \left\{ \sum_{i=1}^{\#N} \left(\phi_{i_{\text{obs}}} - \phi_{0_{\text{obs}}}\right)^2 \Big/ \#N \right\}^{1/2},$$

and

$$\sigma = \left\{ \sum_{i=1}^{\#N} \left(\phi_i - \phi_0\right)^2 \Big/ \#N \right\}^{1/2}$$

can be used to determine the skill of the model results. The parameter E is the *root mean square error* (RMSE), E_{UB} the RMSE after a constant bias is removed, and σ and σ_{obs} the standard deviations of the predictions and the observations, respectively.[8] Keyser and Anthes found that the RMSE can be reduced significantly when a

[8]The use of root mean square analysis to examine the skill of model results for different sets of initial conditions was also discussed in Section 10.3 associated with Eq. 10.5.

Table 11.1 Error analysis of model-predicted winds and temperature using Eq. 11.20 for an east-west cross section from Naples to just north of Fort Lauderdale, Florida. See Fig. 11.13 for the location (from Pielke and Mahrer 1978).

Variable	E	E_{UB}	σ	σ_{obs}	E_{UB}/E	E_{UB}/σ_{obs}
$\bar{u}$ (m s^{-1})	3.1	3.1	1.2	2.2	1.0	1.4
$\bar{v}$ (m s^{-1})	2.2	1.2	0.8	1.2	0.5	1.0
$\overline{T}$ (°C)	5.1	2.8	3.9	4.6	0.5	0.6

constant bias is removed. Such a bias, they suggested, could be a result of an incorrect specification of the initial and/or bottom and lateral boundary conditions.

Skill is demonstrated when

1. $\sigma \simeq \sigma_{obs}$,
2. $E < \sigma_{obs}$, and
3. $E_{UB} < \sigma_{obs}$.

Pielke and Mahrer (1978) applied these criteria to their simulation of the sea breezes over south Florida to show that the model had skill in predicting wind velocity and temperature at 3 m. Temperature predictions over the entire daylight period, as given in Table 11.1, for example, had a ratio of $E_{UB}/\sigma_{obs} = 0.6$. Segal and Pielke (1981) have applied this analysis tool over the Chesapeake Bay region to evaluate the skill of a mesoscale model prediction of biometeorological heat load during the daylight hours. For temperature, for example, Segal and Pielke found $E/\sigma_{obs} = 0.53$ with $\sigma_{obs} = 2.12$°C and $\sigma = 2.24$°C. This evaluation technique has also been applied by Shaw et al. (1997) in the modeling of a Great Plains dryline.

One problem with point-to-point validation, however, is that spatial and temporal displacement of the predicted from the observed fields could yield a poor verification according to Eq. 11.20, even though the shape and magnitude of the simulated pattern could be almost exact. Although not yet attempted in a mesoscale model, rigid translation of the predicted results on the model grid (e.g., in one grid interval increments) relative to the observations, and re-computation of E and E_{UB} in Eq. 11.20 offer one possibility to consider the effect of displacement on the accuracy of the results.

A quantitative measure of model skill to predict observed meteorological fields such as displayed in Fig. 11.13 is also possible using concepts of set theory. Pielke and Mahrer (1978) applied this technique to determine the degree of correspondence between predicted low-level convergence zones (as estimated by the vertical velocity $\bar{w}$, at 1.22 km) and the locations of radar echoes over south Florida. The two major questions answered using this technique were:

1. what fraction of the predicted convergence zones are covered by showers, and
2. what fraction of the showers that occur lie inside of the predicted convergence zones?

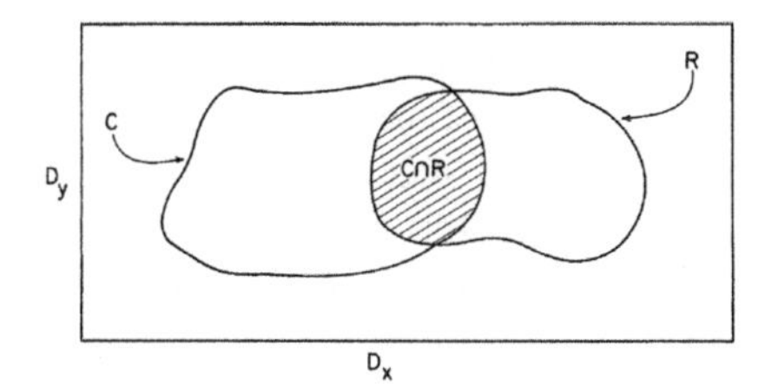

FIGURE 11.14

A schematic illustration of the juxtaposition of a field of radar echoes R and low-level convergence of a given magnitude and larger C. The two fields are coincident at $C \bigcap R$. The quantity, $F_c = (C \bigcap R)/C$, indicates the fraction of a convergence zone covered by radar echoes, and $F_E/F_m = [(C \bigcap R)/R]/(C/D_x D_y)$ measures the ratio of echoes within the convergence zone to the fraction of the model domain covered by that magnitude of convergence and larger.

To illustrate the procedure of analysis, let $D_x D_y$ be the model domain area, C the area of the model domain covered by predicted convergence of a given magnitude or larger, and R the area of the model domain covered by radar echoes of a specified intensity and greater. With these definitions:

1. $F_E = (C \bigcap R)/R$ is the fraction of echoes in convergence zones with values equal to or greater than a certain value of convergence (where the symbol $\bigcap$ is an intersection in set theory symbolism),
2. $F_m = C/D_x D_y$ is the fraction of the model domain covered by a specified value of convergence and larger,
3. $F_c = (C \bigcap R)/C$ is a measure of the fraction of convergence zones, of a given magnitude and larger, covered by echoes.

Skill is demonstrated if $F_E/F_m > 1$ since the ratio would be expected to be unity by random chance. A necessary condition for perfect skill is $F_c = 1$ since the entire convergence zone would be covered with echoes in that case.[9]

This methodology is illustrated schematically in Fig. 11.14 for an idealized distribution of radar echoes and convergence. Results for an actual model simulation of a sea breeze over south Florida (for 1 July 1973) from Pielke and Mahrer (1978) are given in Table 11.2, where the ratio of F_E/F_m was greater than unity in 26 out of 30 categories. The ratio was larger than 2.0 for 20 of the categories. By contrast, F_c was much less than unity indicating that most of the convergence zones were not covered by rain showers – a result indicating that sea-breeze convergence alone does not completely explain the spatial variability of radar echoes over south Florida during the summer.

The application of this analysis procedure to other meteorological variables such as cloud cover, rainfall, etc., is straightforward. Simpson et al. (1980), for example, quantitatively examined the skill of the mesoscale model predictions over south

[9]Since R can be larger than C, $F_c = 1$ is not a sufficient measure of perfect skill.

Table 11.2 The fraction of convergence zones F_c of a given magnitude and larger, covered by radar echoes, and the ratio of the fraction of echoes in convergence zones of a given magnitude and larger to the fraction of the model domain covered by that magnitude of convergence and larger; this ratio is given by F_E/F_m. Convergence is defined by vertical velocity $\bar{w}$ at 1.22 km. For 1200-1800 EST, (i) $\bar{w} > 0$ cm s^{-1}, (ii) $\bar{w} > 8$ cm s^{-1}, (iii) $\bar{w} > 16$ cm s^{-1}, (iv) $\bar{w} > 24$ cm s^{-1}, (v) $\bar{w} > 32$ cm s^{-1}. For 1900 EST (i) $\bar{w} > 0$ cm s^{-1}, (ii) $\bar{w} > 8$ cm s^{-1}, (iii) $\bar{w} > 24$ cm s^{-1}, (iv) $\bar{w} > 40$ cm s^{-1}, (v) $\bar{w} > 56$ cm s^{-1} (from Pielke and Mahrer 1978).

Time	F_c					F_E/F_m				
(EST)	(i)	(ii)	(iii)	(iv)	(v)	(i)	(ii)	(iii)	(iv)	(v)
1200	0.0049	0.0118	0.0	0.0	0.0	1.31	2.67	0.0	–	–
1300	0.0571	0.1237	0.0072	0.0	0.0	2.16	3.17	0.20	–	–
1400	0.0945	0.1770	0.3111	0.0	0.0	1.98	2.78	3.50	0.0	–
1500	0.1396	0.1337	0.1942	0.3040	0.0	2.25	2.56	3.29	6.00	–
1600	0.0889	0.1765	0.1156	0.00	0.0	2.44	2.95	2.00	0.0	–
1700	0.1609	0.2829	0.2950	0.2379	0.0	2.49	3.16	3.00	2.33	–
1800	0.1451	0.1976	0.3403	0.3866	0.0909	2.19	3.07	3.75	3.75	0.10
1900	0.0759	0.1205	0.1313	0.0211	0.0	1.06	1.43	1.67	1.67	–

Florida on several days during the summer to predict locations of shower mergers as seen by radar. The technique, of course, can be applied to other geographic areas and to different mesoscale systems.

Anthes (1983) provides an effective summary of additional evaluations of model skill. These include

$$TS = \frac{CFA}{FA + OA - CFA}$$

where TS is called the "threat score". CFA is the correctly forecast area, FA is the forecast area, and OA is the observed area. These quantities are equivalent to $CFA = C \bigcap R$, $FA = C$, and $OA = R$ used to obtain Table 11.2.

The threat score can also be defined as

$$TS = C/(F + O - C)$$

where C is the number of locations in which a forecast is defined to be correct, F is the number of locations for which a forecast is made, and O is the number of locations which observed the forecast quantity.

A bias score, B, can be defined as

$$B = FA/OA$$

and by

$$B = F/O.$$

Colle et al. (1999) discuss the changes in bias scores as the spatial grid increment in the MM5 model is made smaller.

To assess model skill, Mielke (1984, 1991) introduced a new statistical evaluation scheme referred to as the Multivariate Randomized Block Permutation (MRBP) procedure. His approach has the advantage that regression relations and comparisons between model and observed data is based on the absolute value of the differences, rather than the square of the distances.

A summary of the MRBP technique is provided in Lee et al. (1995), and is reproduced here. As described by Sheynin (1973), the initial known use of regression by D. Bernoulli (circa 1734) for astronomical problems involved the least sum of absolute deviations (LADs) regression. The distance function associated with LAD regression is the common Euclidean distance between observed and predicted response values. Further work in developing LAD regression was accomplished by R.J. Boscovich (circa 1755), P.S. Laplace (circa 1789), and C.F. Gauss (circa 1809). Sheynin (1973) points out that Gauss developed linear programming for the sole purpose of estimating the parameters associated with LAD regression. Gauss consequently had to introduce the least sum of squared deviations (LSDs) regression (also termed least squares regression) simply because calculus provided an efficient way to estimate the parameters associated with LSD regression. Thus LSD regression is a default procedure which was introduced only because Gauss lacked appropriate computational equipment to solve linear programming problems. The American mathematician and astronomer N. Bowditch (circa 1809) immediately attacked LSD regression because squared deviations unduly overemphasize questionable observations in comparison to the absolute deviations associated with LAD regression (Sheynin 1973).

The MRBP procedure developed by Mielke (1984, 1991) is based on the LAD regression. Specifically, MRBP randomly permutes the observed vector of values ($\vec{X}$) relative to the model-predicted vector of values ($\tilde{\vec{X}}$) with the agreement measure, ρ, defined by

$$\rho = \frac{\mu_\delta - \delta}{\mu_\delta} \tag{11.21}$$

where $\delta = (1/n)\sum_{i=1}^{n} |\vec{X}_i - \tilde{\vec{X}}_i|$ is the average distance between n-observed and model-predicted data pairs and μ_δ is the average value of δ over all $n!$ permutations. Note that the Euclidean distance between vector value pairs is used to evaluate the agreement measure and that good predictions are associated with relatively small values of δ. The LAD regression used here is both multivariate (n vectors of two or more dependent variables may be involved) and nonlinear. The remaining problem is to determine if a realized value of δ for observed and model-predicted values is merely due to chance. The standard measurement for this purpose is the P value, i.e., the probability of obtaining a value of δ which is not larger than a realized value of δ given that each of the $n!$ values of δ occurs with equal probability. Although the exact calculation of all $n!$ values of δ is seldom computationally feasible, an approximate P value is based on the standardized test statistic given by

$$T = \left(\delta - \mu_\delta\right)/\sigma_\delta \tag{11.22}$$

where σ_δ is the exact standard deviation of δ, and T is approximately distributed as the Pearson type III distribution (Mielke 1984, 1991). Examples of the use of the MRBP evaluation technique in mesoscale modeling are reported in Cotton et al. (1994), Lee et al. (1995), and Mielke and Berry (2000).

11.9 Model Sensitivity Analyses[10]

Numerical models provide a powerful tool for atmospheric research. One of the most common ways of utilizing a model is by performing sensitivity experiments. Their purpose is to isolate the effects of different factors on certain atmospheric fields, in one or more case studies. Factors that have been tested in sensitivity studies include, for example, surface sensible and latent heat fluxes, latent heat release, horizontal and vertical resolution, sea surface temperatures, horizontal diffusion, surface stress, initial and boundary conditions, topography, surface moisture, atmospheric stability, and radiation. Sensitivity studies are performed either with real data case studies, or with idealized atmospheric situations.

The interactions between factors are usually neglected; their significant role in some cases has been pointed out (e.g., Uccellini et al. 1987; Mailhot and Chouinard 1989). No sensitivity studies have been proposed in atmospheric studies to isolate these interactions before the factor separation methodology was introduced by Stein and Alpert (1993).

The evolution history of the Factor Separation (in brief, FS) method in atmospheric science starts with P. Alpert and U. Stein's publication employing Numerical Simulations (Stein and Alpert 1993). This study is the basis for many investigations which have followed in the field. The FS method itself and its mathematical platform were first demonstrated with two factors, topography and surface heat fluxes, and with their effect on rainfall distribution for cyclonic evolution in the Mediterranean. One of the most important conclusions from the study was that if synergic contributions are not calculated and separated, then the comparison of factors may be misleading, particularly when the synergic effect is not insignificant.

The applications of the FS method can assist in studying the effects of local factors such as terrain, land-sea differences, land use, snow cover, air-sea interactions, sea surface temperature, urban areas, and air pollution. The net effects of the aforementioned factors on rainfall, cloud cover, and air temperature, for instance, are critical in many studies and over the whole range of atmospheric scales. In addition, the atmospheric circulations induced by each factor separately or by an interaction among two or more factors are essential for better understanding of the primary atmospheric processes. This feature of the FS method was later shown to play a vital role in understanding ongoing processes in various atmospheric models.

The Alpert and Sholokhman (2011) book entitled "Factor Separation in the Atmosphere, Applications and Future Prospects" presents the method with many applications, showing a consistent and simple approach for isolating the resulting fields due

[10]This section was written by Professor Pinhas Alpert of the Department of Dynamic Meteorology and Planetary Sciences at Tel-Aviv University, Israel.

to any interactions among factors, as well as that due to the pure factors, using linear combinations of a number of simulations. One of the most attractive features of the factor separation approach is its ability to quantify synergies or interaction processes, which were found to play a central role in many atmospheric applications discussed in this book. There seems to be some basic psychological tendency in human thinking towards linearization, resulting in the simplified assumption that synergies are small or can be ignored. Nonlinearities in the atmosphere, however, are often quite significant and therefore need to be calculated and separated from the pure contributions of each factor. A number of scientists have applied FS methodology to a range of modeling problems, including paleoclimatology, limnology, regional climate change, rainfall analysis, cloud modeling, pollution, crop growth, and other forecasting applications. Alpert and Sholokhman (2011) describe the fundamentals of the method, and collect a number of applications in the atmosphere that cover a range of temporal and spatial modeling scales spanning a time scale of a few minutes such as in plant-canopy interaction up to paleoclimate studies of millions of years.

The FS standard method works on the principle of "on-off". In other words, the factors are switched off (zeroed) one-by-one, and the intermediate result is investigated independently for each case. The case where all the chosen factors are switched "off" is the basic case that, obviously, does not depend on any of the factors and/or their combinations. The opposite case, with all of the factors switched "on", is the "control" or "full" result that includes all the factors and their synergistic contributions. The FS method provides the methodology to distinguish between the pure influence of each and every factor as well as their mutual influence i.e. the synergies that come into play when two or more factors are "switched on" together.

Following this analysis procedure, if f_0 is the control and f_1, f_2 and f_3 represent three alteration experiments, then

$$\hat{f}_0 = f_0$$
$$\hat{f}_1 = f_1 - f_0$$
$$\hat{f}_2 = f_2 - f_0$$
$$\hat{f}_3 = f_3 - f_0$$

where $\hat{f}_1$, $\hat{f}_2$, $\hat{f}_3$ represents the individual effects of making just one alteration to the control. In the past, this is where most sensitivity experiments ended. However, as shown by Stein and Alpert (1993), the following interaction terms are important,

$$\hat{f}_{12} = f_{12} - (f_1 + f_2) + f_0$$
$$\hat{f}_{13} = f_{13} - (f_1 + f_3) + f_0$$
$$\hat{f}_{23} = f_{23} - (f_2 + f_3) + f_0$$
$$\hat{f}_{123} = f_{123} - (f_{12} + f_{13} + f_{23}) + (f_1 + f_2 + f_3) - f_0$$

Hence, the result of the FS is not only the separation of the factors for $\hat{f}_1$, $\hat{f}_2$, $\hat{f}_3$, but also all of the possible combinations of these factors, i.e., $\hat{f}_{12}$, $\hat{f}_{23}$, $\hat{f}_{13}$ and $\hat{f}_{123}$.

The factor $\hat{f}_{123}$, for instance, is the contribution due to the triple interaction among the three factors under evaluation.

Hence, eight simulations are necessary with 3 factors. Here, we present one example of Factor Separation on a medium scale in the atmosphere, often referred to as mesoscale or in general, mesometeorology.

The following example is that of a deep Genoa cyclogenesis (Alpert et al. 1996a), that has been observed during the Alpine Experiment (ALPEX) in March 1982, and then studied intensively by several research groups.

11.9.1 The 3-6 March 1982 Lee Cyclone Development

A relatively large number of studies have been devoted to cyclogenesis, with particular attention given to the processes responsible for the lee cyclone generation. Early studies of Lee Cyclogenesis (henceforth LC) focused on observations, and indicated the regions with the highest frequencies (Petterssen 1956). Several theories have been advanced to explain the LC features, and they are frequently separated for convenience into two groups, as follows: those modified (by the lower boundary layer) baroclinic instability approach, as reviewed by Tibaldi et al. (1990) and Pierrehumbert (1985), and the directional wind shear suggested by Smith (1984).

One of the most studied lee cyclogenesis both observationally (Buzzi et al. 1985, 1987) and by numerical simulations (Tafferner and Egger 1990; Tibaldi and Buzzi 1983; Bleck and Mattocks 1984; Dell'Osso 1984) is the 3-6 March ALPEX case. It was the most intense (Buzzi et al. 1985) lee cyclogenesis deepening during the ALPEX international special observational period, that was aimed at a better understanding of mountain role in weather. A 8.7 hPa (mb) deepening in the lee of the Alps occurred within 24 hours, with most of the pressure fall (6.1 hPa) occurring within the last 12 hr period.

Following earlier studies, four most relevant factors were chosen: topography (t), surface latent heat flux (l), surface sensible heat flux (s), and the latent heat release (r) in the deep clouds developing within the cyclonic system. In order to calculate all the possible sixteen (24) contributions, simulations with the PSU/NCAR mesoscale model version MM4 were performed. Horizontal resolution was 80 km with a 46×34 mesh and 16 levels up to 16 km. It should be pointed out that in the early 1990s, 80 km was still referred to as mesoscale. Further details on model and simulations can be found in Alpert et al. (1996a), and in SA. The initial time for the simulations was 4 March 1200 UTC. The sea-level pressure change at the center of the control run cyclone was then partitioned into contributions by each process or combination of processes. These included four "pure" contributions (t, l, s, and r), six double interactions, four triple interactions (tsl, tsr, tl, and slr), and even one quadruple interaction ($tslr$). The residual was referred to as the "large-scale contribution." This term is excluded from the following discussion because only the local processes were investigated.

The mechanical effect of the Alpine topography (t) is the first to generate the rapid deepening within six hours, while all other contributions are still much smaller. This corresponds well to the aforementioned first deepening phase. In the second phase, the latent heat release (r) or "convection only" becomes the dominant contributor (42-54 h), while the "pure" topographic contribution quickly diminishes to become later a major cyclolytic (destruction of cyclone) factor. This is probably due to the cyclone's motion beyond the favorable lee region. Of interest here are the large contributions of the synergistic double and triple interactions. For example, the mountain-induced convection (tr) is the second contributor at the first stage (27-36 hr). Each of the synergistic terms can be associated with a specific meaning, thus shedding light on the complex physical mechanisms under investigation. The triple interaction *tlr*, found to be quite important at the 45-60 hr phase, for example, represents the contribution to the deepening by the terrain-induced convection with local moisture which is the triple synergism of terrain, convection, and surface latent heat flux. Obviously, these results, as well as in the other examples, are valid only to the extent that the model simulations simulate well the real atmosphere.

11.9.2 Absolute Comparisons for Effects of Several Processes

However, Alpert et al. (1995) link the sensitivity of the FS results to the number and type of factors to be selected. They show that the increase in the number of factors diminishes the individual contribution of a particular factor. If, for example, topography, which is a crucial factor, is included in each of the eight potential sets (t), (t, s), (t, l), (t, r), (t, s, l), (t, s, r), (t, l, r) and (t, s, l, r), its contribution varies significantly from case to case as shown in Fig. 11.15. Notice that the notation with ( )

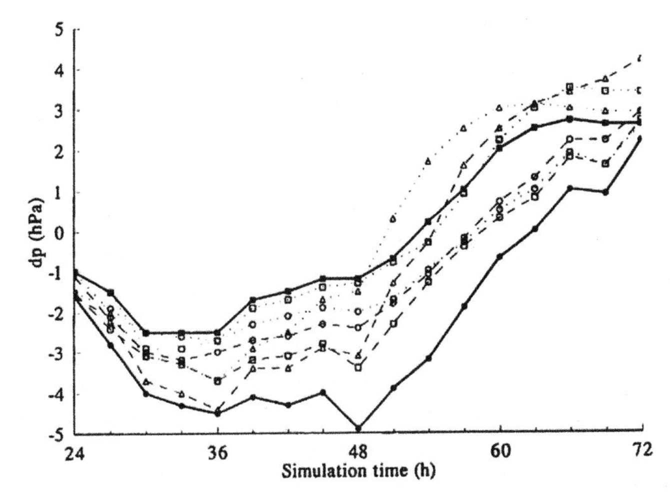

FIGURE 11.15

The pure topographic contribution (t) in h Pa, to the pressure fall at the center of the control cyclone as a function of the simulation time. Each curve represents one of the 8 possible sets of factors (see text). Heavy lines indicate the maximum choice of the one factor only (t) and the maximum choice of the 4 factors, i.e., (t, s, l, r).

refers to a potential numerical experiment for which the chosen factors or processes to be tested are listed within the parentheses as a group. This notation should not be confused with that of a specific contribution isolated for a particular experiment. For example, the triple interaction tlr in the preceding section was isolated in an experiment for which four factors were chosen, i.e., the set (t, s, l, r). Of course, each set is a legitimate choice for a modeler who investigates the role of topography (t) in the lee cyclogenesis, and such examples are found in the literature (see SA). Early, we have chosen the specific set with all four factors included, i.e., (t, s, l, r). Figure 11.15 presents the "pure" topographic contribution to the cyclone deepening for each of the eight possible sets of factors to be investigated. It is not unexpected that the topographic contribution is largest when only one factor (t) was considered, and the smallest, at least until the 48th hour of simulation was reached, when the largest set of factors (t, s, l, r) was chosen, since, as the number of factors increases, synergistic contributions between the new factors and topography are extracted out from the original t contribution. In the extreme case where only topography was chosen as a factor, all synergistic contributions with the topography are tacitly assumed to be part of the topographic contribution, making it the largest – see the bottom curve in Fig. 11.15. For example, the cyclogenetic contribution of the latent heat release synergism with topography (tr) is associated with the topography for the case when the factor r is not an investigated factor.

11.9.3 Spread of the Factor Separation Model Simulations – another By-Product Study Tool

Earlier, an experiment with a set of four factors was described. The FS method therefore required 16 = 24 different simulations with all possible combinations of factors switched on/off. Obviously, in each simulation, the cyclone center may have a different location, which reflects the specific dynamical evolution due to the contribution of factors switched on. Figure 11.16 presents the sixteen different locations of the cyclone center, following 45 hours of simulation time. In this notation, the factors switched on are listed, so TLR, for instance, indicates the simulation where only the factors t, l, and r were operating, while s was switched off.

A remarkable feature in the spread presented in Fig. 11.16 is the tendency of the factors to attract the cyclone center toward different regions. For instance, the convection (r) drifts the cyclones to the east-northeast while topography does similarly but to the north-northwest. As may be expected, the latent and sensible heat fluxes (l and s) tend to move the cyclone further south, towards the sea. This feature is even more pronounced as the spread of the solutions with time is analyzed. As time evolves, and the cyclone spread region moves away from the lee cyclogenetic area, at 60 hours, a clear change of the aforementioned tendencies takes place (not shown). The topographic factor delays the cyclone at the west over Sardinia, while sea fluxes move the cyclone to the east, where a larger and warmer body of water dominates. Alpert et al. (1996b) point out that as the cyclone centers spread region increases

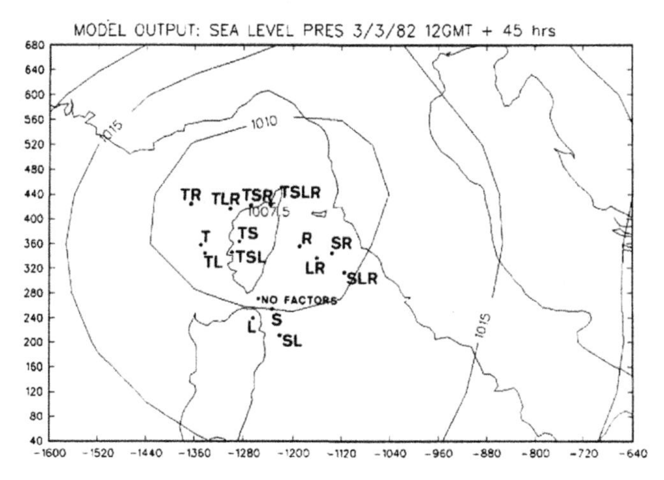

FIGURE 11.16

The 16 locations of the cyclone center following 45 hr of simulation for the 16 simulations with all possible combinations of factors switched on/off. In this notation, the factors switched on are listed: hence, TLR, for instance, indicates the simulation where only the factors t, l, and r were operating while s switched off.

further with time the FS for a very specific point, like the center of the control run, becomes less and less meaningful.

A list of examples for applying this methodology for sensitivity analysis is reported by Alpert and Sholokhman (2011).

Problems for Chapter 11

1. Using the one-layer tank model derived in Chapter 9, program the equations in Section 11.6.1 to compute the time rate of change of the mass and energy budget of the model by summing over the grid points of the model and then compare to what they should be from the requirement that h is a constant and Eqs. 11.5 and 11.7. Use cyclic lateral boundary conditions.
2. Repeat #1 with constant inflow and gradient outflow lateral boundary conditions.
3. Repeat #1 with constant inflow and radiative outflow lateral boundary conditions.
4. Select a mesoscale model and describe which of the model evaluations reported in this Chapter have been used.

Chapter 11 Additional Readings

Several publications that provide additional examples of mesoscale model evaluations include:

Alpert, P., and T. Sholokhman, Eds., 2011: *Factor Separation Method in the Atmosphere-Applications and Future Prospects*. Cambridge University Press, 292 pp.

Cox, R., B.L. Bauer, and T. Smith, 1998: A mesoscale model intercomparison. *Bull. Amer. Meteor. Soc.,* **79**, 265-283.

Hanna, S.R., and R. Yang, 2001: Evaluations of mesoscale models' predictions of near-surface winds, temperature gradients, and mixing depths. *J. Appl. Meteor.,* **40**, 1095-1104.

Snook, J.S., P.A. Stamus, J. Edwards, Z. Christidis, and J.A. McGinley, 1998: Local-domain mesoscale analysis and forecast model support for the 1996 Centennial Olympic Games. *Wea. Forecasting,* **13**, 138-150.

CHAPTER

Mesoscale Modeling and Satellite Simulator*

12

CHAPTER OUTLINE HEAD

Satellite remote sensing is a technique that estimates geophysical parameters from the electromagnetic energy reflected or emitted from the Earth. With a-priori information of the unique scattering/absorbing characteristics of molecular, atmospheric particles and surface properties, scientists can estimate ocean/land surface characteristics, atmospheric profiles (temperature and humidity), surface wind vectors, aerosol, cloud, and precipitation processes. Because footprint sizes of satellite observations are typically $1 \sim 30$ km, they have been widely used for applications to mesoscale meteorological modeling. Especially, in the past decade, a large number of the Earth Observing Satellites (EOS) have been launched, and are providing greater opportunity and potential for the meteorological community.

Satellite-derived sea surface temperature, vegetation index, and land-cover maps are important bottom boundary conditions in mesoscale modeling (Section 10.4.3). Vegetation index and land-cover maps have also been used to investigate the sensitivities of mesoscale meteorology to surface boundary conditions (e.g., Eastman et al. 2001a,2001b; Matsui et al. 2004) or to improve forecast skills (e.g., Lu et al. 2001; Case et al. 2011; Wen et al. 2012). Satellite observations of surface wind, atmospheric chemistry, aerosols, clouds, and precipitation have been often used for evaluating the

*This chapter was written by Dr. Toshi Matsui, NASA Goddard Space Flight Center, Greenbelt, MD.

Mesoscale Meteorological Modeling, Third Edition, Volume 98. http://dx.doi.org/10.1016/B978-0-12-385237-3.00012-8

performance of mesoscale modeling. Clouds are visible mesoscale phenomena so that many studies utilized satellite images and products to evaluate different aspects of cloud parameterization and microphysics in mesoscale models (Mocko and Cotton 1995; Wang et al. 1998; Mechem and Kogan 2003; Sato et al. 2012). Precipitation is one of the most important parameters in the evaluation of regional climate models, and satellite-based rainfall data plays important roles for evaluating precipitation variability (e.g., Wu et al. 2005; Monaghan et al. 2012). Synthetic Aperture Radar (SAR) provides estimation of high-resolution wind speed and vectors, which evaluated surface wind dynamics in the mesoscale model (Koch and Feser 2006).

This small set of examples suggests that satellite-observed geophysical parameters are grant assets for the mesoscale modeling community. However, scientists must pay extra attention to the uncertainties in satellite-derived products. This is because satellite retrievals always face ill-posed problems: i.e., the number of uncertainties is larger than the number of satellite-received signals. This requires that a satellite retrieval has to make unique assumptions to estimate geophysical parameters. For example, a visible imager can estimate cloud optical thickness, effective radius, and liquid water path through visible and near-infrared channels (Nakajima and King 1990). Depending on the choice in the wavelength, this two-channel retrieval results in quite different values in the cloud effective radius, because different near-infrared radiances are sensitive to the cloud properties at different cloud depths, becoming sensitive to the presence of drizzle droplets within the clouds (Nakajima et al. 2010). Visible imagers, microwave sensors, and cloud radar often estimate different quantities of rainfall, cloud ice amounts over the same area (Berg et al. 2008; Waliser et al. 2009; Eliasson et al. 2010). Simultaneously, different sets of channels, samplings, footprints, sensors wavelengths, and retrieval assumptions result in different aerosol optical depths among different satellite products (Myhre et al. 2004). These discrepancies are more commonly observed after the initial completion of the A-Train Satellite Constellations (Stephens et al. 2004). As a result, modeling communities have been frustrated by precise evaluation of meteorological model parameters using satellite-derived datasets. A more solid bypass must be established between mesoscale modeling and satellite observations.

One solution, discussed in this chapter, is to utilize the satellite raw signals, technically called Level-1B (L1B) signals, in the unit of radiance in the visible imager, brightness temperature in infrared and microwave sensors, equivalent radar reflectivity in radar instruments, or backscattering coefficient in LIDAR instruments. As long as sensors are well calibrated, these are the most direct signals from the satellite without any retrieval assumptions. A question is how to utilize satellite L1B signals and compare them with mesoscale modeling. To this end, various satellite simulators have emerged and been coupled to various mesoscale models (Donovan et al. 2004; Han et al. 2006; Haynes et al. 2007; Masunaga et al. 2011; Matsui et al. 2013). A satellite simulator is the tool that predicts satellite L1B signals from a set of geophysical parameters simulated in mesoscale modeling.[1] In this way, mesoscale modeling can

[1] Discussion of the Level 2 simulator for the general circulation model has been omitted in this chapter.

harness calibrated L1B observations from various satellite instruments, and provide diverse opportunities such as model evaluation, data assimilation, and satellite mission supports. Concurrently, this requires specific knowledge related to the radiative transfer and satellite sensor characteristics in addition to knowledge of mesoscale modeling. The following sections address more details in satellite instrumental simulators (12.1), and their applications to mesoscale meteorological modeling (12.2).

12.1 Satellite Instrumental Simulator

A satellite simulator is a tool to convert model-simulated geophysical parameters to satellite-instrument observable signals. This section briefly describes various functions in a *satellite simulator for mesoscale modeling*. These include i) prediction of satellite orbit (satellite orbital prediction and sensor scanning geometry), ii) estimation of single-scattering properties and radiative transfer, and iii) antenna gain patterns and scanning geometry of satellite sensors. Physics and mathematical foundations are described in the following paragraphs.

12.1.1 Satellite Orbit

A satellite orbit follows the Kepler orbit, which is explained by the six Keplerian elements. Orientation of the orbital plane can be explained by inclination (i: the tilt angle of the ellipse plane with respect to the reference plane), argument of perigee (ω: the angle of the ellipse perigee with respect to the reference plane), and right ascension of the ascending node (Ω: the angle of the ascending node with respect to the reference plane). Satellites undergo an elliptical orbit around the Earth. The shape and size of ellipse can be defined by i) eccentricity (e): degree of perturbation from the perfect circle, and ii) the semi-major angle (a): the half distance between the perigee and apogee in the ellipse. Finally, the angle between the perigee and the satellite with respect to the Earth's center can be defined by true anomaly (f). These Keplerian elements (i, ω, Ω, e, a, and f) determine the satellite position in a three-dimensional Earth-Center Earth-Fixed (ECEF) coordinate system[2] (see Fig. 12.1 and 12.2). i, e, and a are fixed parameters, while ω, Ω, and f are prognostic parameters derived from the following equations.

$$\omega = \omega_0 + \dot{\omega}\, dt \tag{12.1}$$

$$\Omega = \Omega_0 + \Omega_{rot} + \dot{\Omega}\, dt \tag{12.2}$$

$$f = a \tan 2\left(\sqrt{1 - e^2} \sin E, \cos E - e\right) \tag{12.3}$$

where ω_0 and Ω_0 are initial (epoch time, $dt = 0$) values of ω and Ω respectively, dt is satellite progress time since epoch time, Ω_{rot} is the Earth rotational contribution

[2]ECEF is a three-dimensional Cartesian coordinate system defined as $(x, y, z) = (0, 0, 0)$ at the center of mass of the Earth. The x-axis is pointed toward 0° of latitude and 0° of longitude; the y-axis is pointed toward 0° of latitude and 90° of longitude; and the z-axis is pointed toward the North Pole.

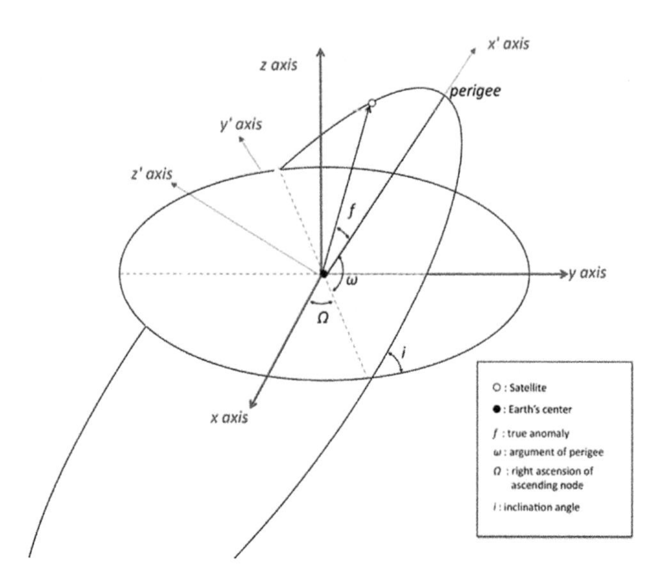

FIGURE 12.1

Keplerian elements for satellite motion.

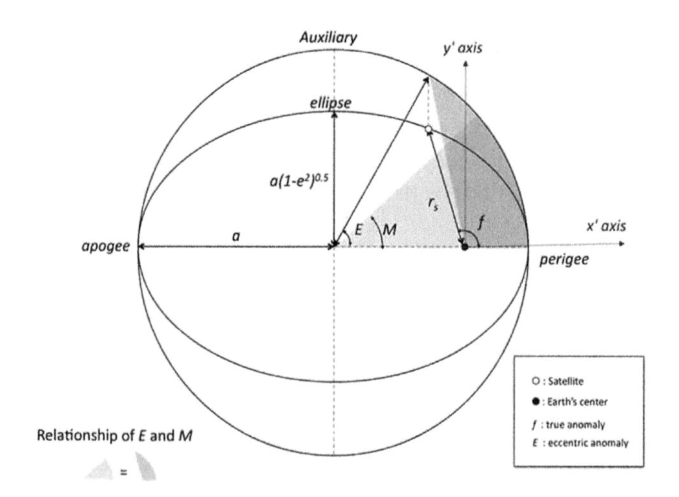

FIGURE 12.2

Relationship between eccentric anomalies (E) and mean anomalies (M). (The color version of this figure is presented in the plate section in the back of the book and the online web version.)

to Ω, and E are eccentric anomalies that link to the mean anomalies (M) (Fig. 12.3).

$$M = E - e \sin E \tag{12.4}$$

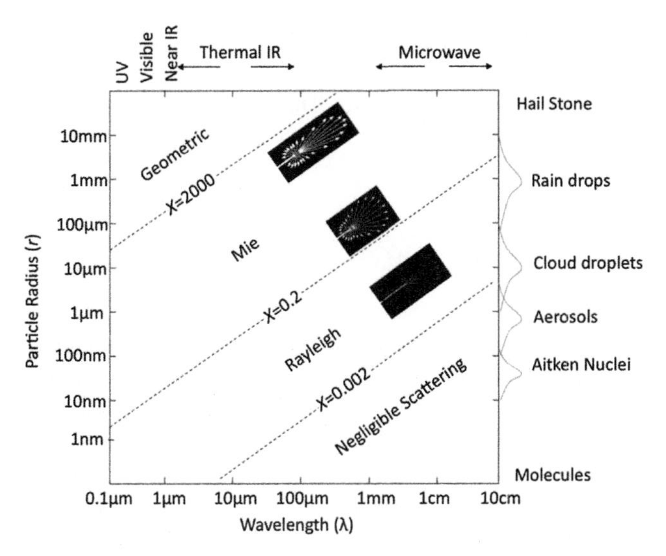

FIGURE 12.3

Non-dimensional shape parameter (X) as a function of particle radius (r) and wavelength (λ) adapted from Petty (2006). The Rayleigh regime corresponds to X ranging from 0.002 to 0.2, where scattering is relatively small and isotropic. The Mie regime corresponds to X ranging from 0.2 to 2000, where scattering is stronger and peaked more in the forward direction. The geometric regime corresponds to X greater than 2000, where most of the scattering is directed forward. (The color version of this figure is presented in the plate section in the back of the book and the online web version.)

and M is derived from the initial mean anomalies (M_0) and perturbation anomalies $\left(\dot{M}\right)$.

$$M = M_0 + \dot{M}\,dt \tag{12.5}$$

The mean perturbation constants ($\dot{\omega}$, $\dot{\Omega}$, and $\dot{M}$) of Eqs. 12.1, 12.2, and 12.5 are derived via Kozai's first-order perturbation theory that accounts for gravitational anomalies due to Earth's oblate shape.

$$\dot{\omega} = \frac{3}{2} J_2 \frac{r_{eq}^2}{\left(a(1-e^2)\right)^2} \dot{M} \left(2 - \frac{5}{2}\sin^2 i\right) \tag{12.6}$$

$$\dot{\Omega} = -\frac{3}{2} J_2 \frac{r_{eq}^2}{\left(a(1-e^2)\right)^2} \dot{M} \cos^2 i \tag{12.7}$$

$$\dot{M} = \dot{M}_0 \left[1 + \frac{3}{2} J_2 \frac{r_{eq}^2}{a\left(1-e^2\right)^2} \left(1 - \frac{3}{2}\sin^2 i\right)\right] \tag{12.8}$$

where $J_2(=1.082549 \times 10^{-3})$ is the unitless perturbation term in gravitational representation, $r_{eq}(=6378.145$ km) is Earth's mean equatorial radius, and other terms are identical to those in the previous equations. The unperturbed mean-motion constant ($\dot{M}_0$) of the satellite is explained by the balance between angular moment and the satellite-Earth gravitational force:

$$\dot{M}_0 = \sqrt{\frac{G}{a^3}} \tag{12.9}$$

where $G(=398601.2$ km^{3}sec^{-2}) is the Earth's gravitational constant, and a is the semi-major axis. Considering the near-circular orbit ($e \approx 0$), it is obvious that the semi-major axis (a) is the most important parameter that determines satellite orbit speed and orbital period. In general, polar-orbiting satellites have ~7000 km of the semi-major axis so that satellites have approximately 100 min of orbital period. In this period, meteorological satellite instruments with a wide swath width (~2000 km) overpass particular locations on Earth twice a day (AM and PM pass). Geostational satellites are placed in orbit with a large semi-major axis (42,164 km) with a zero inclination angle in order to synchronize satellite orbit and Earth rotation, allowing more frequent observations over a dish of the Earth.

Finally, the satellite position vector can be transformed in the ECEF Cartesian coordinate for given Keplerian elements via the matrix rotation method (Kidder and Vonder Haar 1995). Assuming that the orbital plane lies in the pseudo $x' - y' - z'$ Cartesian coordinate for a given true anomaly (f) (Fig. 12.3), the pseudo position vector, $\vec{P}_{x'-y'-y'}$, can be described as

$$\vec{P}_{x'-y'-z'} = \begin{pmatrix} x' \\ y' \\ z' \end{pmatrix} = \begin{pmatrix} r_s \cos f \\ r_s \sin f \\ 0 \end{pmatrix} \vec{P}_{x'-y'-z'} = \begin{pmatrix} x' \\ y' \\ z' \end{pmatrix} = \begin{pmatrix} r_s \cos f \\ r_s \sin f \\ 0 \end{pmatrix} \tag{12.10}$$

where r_s is the distance between Earth's center and the satellite:

$$r_s = a\frac{1 - e^2}{1 + e \cos f} \tag{12.11}$$

Then, the pseudo position vector, $\vec{P}_{x'-y'}$, is i) rotated for argument of perigee, ω, about the z axis, ii) rotated for inclination angle, i, about the x axis, and iii) rotated for the right ascension of the ascending node, Ω, about the z axis. This will lead to the satellite position vector in the ECEF Cartesian coordinate (x, y, z), $\vec{P}$.

$$\vec{P} = \begin{pmatrix} x_p \\ y_p \\ z_p \end{pmatrix} = \left(\vec{R}_z(\Omega) \times \left(\vec{R}_x(i) \times \left(\vec{R}_z(\omega) \times \vec{P}_{x'-y'-z'}\right)\right)\right) \tag{12.12}$$

where

$$\vec{R}_z(\omega) = \begin{pmatrix} \cos\omega & -\sin\omega & 0 \\ \sin\omega & \cos\omega & 0 \\ 0 & 0 & 1 \end{pmatrix}, \quad \vec{R}_x(i) = \begin{pmatrix} 1 & 0 & 0 \\ 0 & \cos i & -\sin i \\ 0 & \sin i & \cos i \end{pmatrix}, \quad \text{and}$$

$$\vec{R}_z(\Omega) = \begin{pmatrix} \cos\Omega & -\sin\Omega & 0 \\ \sin\Omega & \cos\Omega & 0 \\ 0 & 0 & 1 \end{pmatrix}.$$

12.1.2 Radiative Transfer and Single-Scattering Theory

Satellite instruments receive emerging electromagnetic energy through atmospheric media. This process can be numerically and theoretically predicted through the principles of radiative transfer. Co-polarized radiative transfer at specific frequency (v) for infinitesimal homogeneous plane-parallel media is described as

$$\mu = \frac{dI_v(\mu, \phi)}{d\tau_v} = I_v(\mu, \phi) - J_v(\mu, \phi) \tag{12.13}$$

where I_v is radiation intensity, μ is the cosine of emergent angle, τ_v is optical depth, ϕ is azimuth angle, and J_v is the source term expressed as

$$J_v(\mu, \phi) = (1 - \tilde{\omega}_v)B_v(T) + \frac{\tilde{\omega}_v}{4\pi} \int_0^{2\pi} \int_{-1}^{+1} P_v\left(\mu, \phi, u', \phi'\right) I_v\left(u', \phi'\right) d\mu' d\phi' \tag{12.14}$$

where $\tilde{\omega}_v$ is single-scattering albedo, B_v is the Plank function of temperature (T) in local thermodynamic equilibrium, P_v is the scattering phase function, which is the probability of scattering from radiation intensity Iv from the direction (μ', ϕ') to the direction (μ, ϕ) defined as

$$\frac{1}{4\pi} \int_0^{2\pi} \int_{-1}^{+1} P_v\left(\mu, \phi, u', \phi'\right) d\mu' d\phi' = 1 \tag{12.15}$$

The 1st and 2nd term of the righthand side of Eq. 12.14 express the emission and scattering term, respectively. For clear-sky atmosphere, the emission term dominates ($\tilde{\omega}_v \approx 0$) from the infrared to the microwave spectrum, while the scattering term dominates ($\tilde{\omega}_v \approx 1$) from the ultraviolet to the visible spectrum. The emission and scattering terms become equally important in the near-infrared spectrum. For all sky situations, the strength of the scattering term depends on the type of particles in the atmosphere. It is convenient to define the non-dimensional relative scale of particle size to radiation wavelength, called size parameter (X).

$$X \equiv \frac{2\pi r}{\lambda} \tag{12.16}$$

where r is the equi-volume spherical radius of particles, and λ is the wavelength of interest. For example, for most of the microwave wavelength, the presence of cloud droplets, aerosols, and molecules in the atmosphere do not contribute to radiation scattering, although they strongly enhance scattering contributions in visible, near-infrared, and thermal IR wavelength. Even in the microwave regime, raindrops, or hailstones contribute to scattering signals, which are commonly used in microwave rainfall algorithms over land (see Fig. 12.3). Thus, understanding X is a practical way to understand the relationship between particles and radiation signals at different wavelengths in remote sensing. In further detail, particle scattering properties also depend on the refractive index and particle shapes (discussed later).

Molecular density, properties (composites, sizes, and shapes) of aerosols, cloud, and/or precipitation in a mesoscale model should be represented consistently in a satellite simulator to calculate atmospheric single-scattering properties (τ_v, $\tilde{\omega}_v$, P_v) in Eqs. 12.13 and 12.14. At each grid element, single-scattering properties should be integrated by the sum of molecular, aerosols, and cloud-precipitation species in a mesoscale model as follows.

$$\tau_v = l\left(\sum_{i=1}^{S_{gas}} k_{gas,v} + \sum_{i=1}^{S_{aero}} k_{aero,v} + \sum_{i=1}^{S_{cloud}} k_{cloud,v}\right) \tag{12.17}$$

$$\tilde{\omega}_v = \frac{l}{\tau_v}\left(\sum_{i=1}^{S_{gas}} k_{gas,v}\tilde{\omega}_{gas,v} + \sum_{i=1}^{S_{aero}} k_{aero,v}\tilde{\omega}_{aero,v} + \sum_{i=1}^{S_{cloud}} k_{cloud,v}\tilde{\omega}_{cloud,v}\right) \tag{12.18}$$

$$P_v = \frac{l}{\tau_v\tilde{\omega}_v}\left(\sum_{i=1}^{S_{gas}} k_{gas,v}\tilde{\omega}_{gas,v}P_{gas,v} + \sum_{i=1}^{S_{aero}} k_{aero,v}\tilde{\omega}_{aero,v}P_{aero,v} + \sum_{i=1}^{S_{cloud}} k_{cloud,v}\tilde{\omega}_{cloud,v}P_{cloud,v}\right) \tag{12.19}$$

where k is the extinction coefficient, gas represents gaseous species, $aero$ represents aerosol species, $cloud$ represents cloud-precipitation species, and their number of species in gases (S_{gas}), aerosols (S_{aero}),[3] and clouds (S_{cloud}) depend on the complexity in a mesoscale model. Scattering of gaseous species are estimated through the Rayleigh approximation (only at ultraviolet and visible spectrum), while gaseous extinction is estimated at various spectrum resolution of satellite sensors. There are a number of techniques, including the correlated-K, band, and Elaser method, all of which have a basis on the HITRAN spectroscopic databases (Rothman et al. 1998). Furthermore, aerosols and clouds are often unique size distributions ($N(r)$) so that single-scattering properties of a single particle must be integrated over the size spectra, consistent to the assumption of a mesoscale model. For examples of cloud species,

$$k_{cloud} = \int \pi r^2 Q_{ext}(r)N(r)dr \tag{12.20}$$

[3]Aerosol terms become negligible at the microwave spectrum.

$$\tilde{\omega}_{cloud} = \frac{1}{k_{cloud}} \int \pi r^2 Q_{scat}(r) N(r) dr \tag{12.21}$$

$$P_{cloud,v} = \frac{\int \pi r^2 p_{scat}(r) Q_{scat}(r) N(r) dr}{\int \pi r^2 Q_{scat}(r) N(r) dr} \tag{12.22}$$

where $Q_{ext}(r)$, $Q_{scat}(r)$, $p_{scat}(r)$ are extinction efficiency, scattering efficiency, and phase function of a single particle with equi-volume radius (r). The single-scattering properties of a single particle can be estimated through various methods. One of the widely accepted methods is the Lorenz-Mie solution, which is the exact analytical solution of the Maxwell theory for a spherical particle. Cloud droplets and water-coated hygroscopic aerosols form a sphere shape so that the Lorenz-Mie solution is commonly used. Their characteristics for different refractive index and size parameters are described in various publications (e.g., Hansen and Travis 1974; Petty 2006; Liou 2002). Another widely used method is the T-Matrix method for rotationally symmetric non-spherical particles (e.g., Mishchenko et al. 1996). The T-matrix is often used for the first-order approximation of single-scattering properties for non-spherical particles, because of the computational efficiency in comparison to complex numerical solutions. Other well-known numerical methods are the Discrete Dipole Approximation (Draine and Flatau 1994) and the Finite-Difference Time-Domain method (Yang et al. 2005). These two methods are computationally far more intensive than the Lorenz-Mie or the T-matrix solutions, while they can apply to complex-shape particles, such as mineral dust or ice crystals in the Mie and Geometric scattering regimes (Fig. 12.3).

For passive instruments (e.g., radiometer), once single-scattering properties are computed following the assumptions of a mesoscale model, emerging radiance at the top of atmospheric layers is computed through various numerical radiative transfer schemes that resolve Eqs. 12.10 and 12.11 with surface boundary conditions. Ideally, it must be solved in a three-dimensional atmosphere, however due to computational time, it is commonly solved in a one-dimensional plane-parallel atmosphere. One of the widely utilized numerical solutions is the discrete-ordinate method, which expands radiation into the finite number of discrete stream by expanding the phase function through Legendre polynomial function with Gauss quadrate function (Stamnes et al. 1988). This method is a generalized solution for a scattering and non-scattering atmosphere so that the number of scattering angles (stream) can be adjusted to the degree of scattering contributions in the radiative transfer (Fig. 12.3). Vector versions of the discrete ordinate method have been developed to fully resolve polarized radiative transfer at the ultraviolet or microwave spectrum, in which polarization is important (Weng 1992; Spurr 2001; Spurr et al., 2001). If scattering terms are relatively weak, faster radiative transfer code can be formed. The Delta-Eddington two-stream model is one of the most practical radiative transfers at the microwave spectrum widely used in the microwave community (Kummerow 1993). Another solution is the doubling and adding (DA) method (Hansen 1971; Evans and Stephens 1991). The DA method is computationally demanding, while it gives generalized solutions for fully polarized radiation with high-order expansion of phase function. Various new techniques have been developed over the DA solution for faster numerical solution (Liu and Weng 2006; Heidinger et al. 2006). Thus, choice of radiative transfer schemes depends on the degree of scattering terms and treatment of polarization; a preferred choice is always a numerically fast solution with enough accuracy. Computational time is especially important for the application to data assimilation.

Satellite active instruments transmit electromagnetic energy as a signal. When the signal hits an object in the atmosphere, a fraction of the energy returns to the receiver. Since the speed and power of transmitted electromagnetic energy are known, received power is applied to profile the distance and characteristics of the object in the atmosphere. For the microwave spectrum, an instrument of using this technology is called radar, while it is called Light Detection And Ranging (LIDAR) for the visible spectrum in general. For this technique, an instrument transmits and receives backscattered electromagnetic energy in the same path, and the speed of electromagnetic energy is well known. Therefore, the radiative transfer equation for active sensors can be greatly simplified to form radar or LIDAR equations.

For radar, the quantity of received power is linked to the equivalent radar reflectivity factor at frequency $v(Z_v)$, which is derived through two-way path-integrated optical depth and total backscattering coefficient at the range of l.[4]

$$Z_v = \frac{c^4}{v^4 \pi^5 \mid K_v \mid^2} \sigma_{back,y} l \exp\left[-2 \int_0^l \tau_v(l) dl\right] \tag{12.23}$$

where c is the speed of light, K is the radar constant derived from dielectric constants of a particle, and l is the path length from an instrument to the object. Exponential terms of Eq. 12.23 represent two-way attenuation of transmitted and backscattered energy, which represent how much transmitted energy can be reduced during the travel from the instrument to the object, and back to the instrument. The total backscattering coefficient ($\sigma_{back,y}$) is computed by a sum of cloud-precipitation species in a mesoscale model.[5]

$$\sigma_{back,y} = \sum_{i=1}^{S_{cloud}} \sigma_{cloud,y} \tag{12.24}$$

Backscattering coefficients are integrated over the particle size spectra identical to the assumption in a mesoscale model.

$$\sigma_{cloud,y} = \frac{\int \pi r^2 \sigma_{back}(r) N(r) dr}{\int \pi r^2 Q_{scat}(r) N(r) dr} \tag{12.25}$$

where backscattering efficiency is a product of scattering efficiency and phase function at π (180°).

$$\sigma_{back}(r) = \frac{Q_{scat}(r) p(r, \Theta = \pi)}{4\pi} \tag{12.26}$$

[4] The multiple-scattering effect has not been accounted for in this equation. For more information, please read Battaglia et al. (2010).

[5] Scattering of aerosols and gaseous species are negligible at the microwave spectrum (Fig. 12.3).

Single-scattering properties of a single particle (Q_{ext}, Q_{scat}, p) between a passive microwave simulator and a radar simulator should be derived through identical single-scattering assumptions (size, shape, refractive index, and methods) for the multi-instrumental approach described in Section 12.2.

For LIDAR, the quantity of received power is linked to the total backscattering coefficient at frequency $v(\beta_v)$, which is derived through a two-way path-integrated optical depth and total backscattering coefficient at the range of l.

$$\beta_v = \sigma_{back,v}(l) \exp\left[-2\int_0^l \tau_v(l)dl\right] \tag{12.27}$$

Multiple scattering terms were omitted from the equation for simplicity. For further information, please review Hogan (2008). Because of the shorter wavelength, LIDAR can detect finer particles such as molecules, aerosols, and clouds particles at a higher range resolution (Fig. 12.3). Thus, the total backscattering cross section is a sum of properties from molecular, aerosols, and clouds.

$$\sigma_{back,v} = \sum_{i=1}^{S_{gas}} \sigma_{gas,v} + \sum_{i=1}^{S_{aero}} \sigma_{aero,v} + \sum_{i=1}^{S_{cloud}} \sigma_{cloud,v} \tag{12.28}$$

Backscattering coefficients are integrated by size distributions and each single-scattering property. Single-scattering properties (Q_{ext}, Q_{scat}, p) of single particles between the passive visible-IR simulator and the LIDAR simulator should be derived through identical single-scattering assumptions (Platt 1973).

12.1.3 Sensor Scanning and Antenna Gain Function

Horizontal grid spacing of a mesoscale model ranges from a few hundred meters to a few tens of kilometers, which is equivalent to a footprint of satellite observations. Often, horizontal grid spacing of mesoscale models becomes significantly finer than a satellite footprint size, especially common in passive microwave sensors.[6] In this case, simulated L1B signals at grid spacing of a mesoscale model must be convolved to the satellite instrument footprint size. First of all, a satellite instrumental scanning system must be considered. There are a few major scanning mechanics. A cross-track "wiskbroom" scanner has a sensor or mirror rotating along the axis of the satellite propagating direction, which is common in a visible-IR imager, microwave radiometer/sounder. A fixed-angle linear array "pushbroom" sensor receives radiance at all arrays simultaneously, which is commonly used in spectrometers. A conically rotating sensor has the tilted conical receiver rotating normal to the Earth's surface, and are commonly used in microwave imagers. If one assumes the curvature of the

[6]Unlike visible or infrared remote sensing, TOA upwelling radiation of microwaves is less intense than the visible and IR spectrum and as a result it requires a larger antenna to gain more electromagnetic energy.

Earth's ellipsoidal surface as a local sphere, a combination of satellite altitude, Earth radius, and scanning angle can determine the location of the field of view (FOV) on the reference ellipsoid (see Appendix C).

Generally, satellite instruments receive electromagnetic energy around the center of the FOV. Footprint size is defined by the half-power beam width (HPBW), which is the angular threshold of received electromagnetic energy reduction 50% from the FOV center. Antenna gain function (G) is the angular pattern of the electromagnetic energy received within the antenna, which is often mimicked by the Gaussian distribution with HPBW (θ_{HP}).

$$G(\theta) = \frac{1}{s\sqrt{2\pi}} \exp\left[-\left(\frac{\theta}{s\sqrt{2}}\right)^2\right], \tag{12.29}$$

and

$$s = \frac{\theta_{HP}}{2\sqrt{2\ln(2)}} \tag{12.30}$$

where θ is the beam-width angle between the center of the FOV and the surrounding points toward the satellite location.

Gain function can be computed at each grid element of a mesoscale model using the satellite and IFOV center positions, then grid-scale radiance ($I_{i,j}$) can be integrated with the weight of gain function that yield sensor-level radiance of IFOV (I_{IFOV}).

$$I_{IFOV} = \frac{\sum_i \sum_j G_{i,j} I_{i,j}}{\sum_i \sum_j G_{i,j}}. \tag{12.31}$$

Further, for a passive microwave sensor, I_{IFOV} is temporally integrated during the sampling period to compute radiance at the effective FOV (I_{EFOV}). The example for a passive microwave sensor is illustrated in Fig. 12.4. Eventually, I_{EFOV} or I_{IFOV} becomes the satellite-observable L1B signals.

We have reviewed mathematics and physics foundations of each function in the satellite simulator. The complexity and maturity of these functions depends on the package of the available satellite simulators (Donovan et al. 2004; Han et al. 2006; Haynes et al. 2007; Masunaga et al. 2011; Matsui et al. 2013). Some satellite simulators will not have the complex antenna gain function or scanning geometry. Some do not employ detailed satellite orbit simulation. Radiative transfer schemes also greatly vary across different platforms depending on the supporting satellite platform (https://sites.google.com/site/satellitesimulators/). Some simulators are much more complex and able to support detailed instrumentation to enable the concept of future satellite development (Tanelli et al. 2005; Battaglia and Tanelli 2011). These different methods in radiative transfer and single-scattering calculation are remaining uncertainties in satellite simulators. However, it should be noted that satellite retrieval contains inversion uncertainties in addition to these forward-model uncertainties.

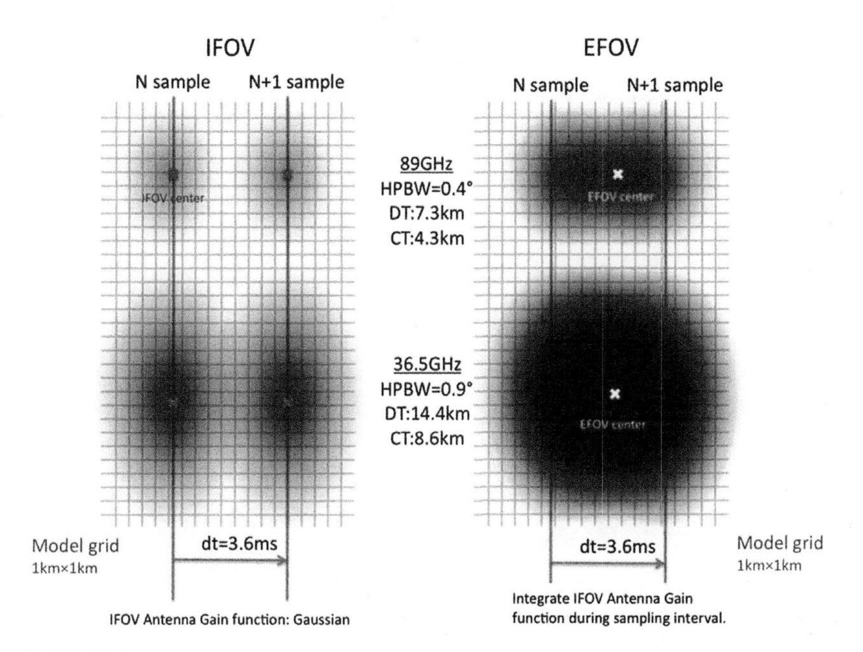

FIGURE 12.4

Scheme of satellite instrument antenna gain patterns over the mesoscale model grid (1 km of grid spacing). Example I is set for half-power beam width (HPBW) = 0.4° and 0.9°, which is approximately 7.3×4.3 km^2 and 14.4×4.3 km^2 of the footprint size for a given satellite altitude (6776.14 km) and off-nadir angles (48.5°) in a conically rotating sensor. (The color version of this figure is presented in the plate section in the back of the book and the online web version.)

12.2 Application of Satellite Simulators to Mesoscale Meteorological Modeling

This section briefly describes the application of *satellite simulators to mesoscale modeling*. The satellite simulator is often called a "forward model" in comparison with an inverse model (retrieval) in the remote sensing community. Satellite simulators are also referred to as an "operator" in the data assimilation community. As discussed in the previous section, satellite simulators can predict satellite-observable L1B signals through a set of radiative transfer, electromagnetic scattering and absorption, and satellite orbit and instrumental scanning patterns. Figure 12.5 (left column) shows an example of multi-instrumental L1B observations from the A-Train Constellation (Stephens et al. 2002). The Advanced Microwave Scanning Radiometer for EOS (AMSR-E) in the Aqua satellite observes TOA upwelling microwave brightness temperature (Tb).[7] Very cold Tb at 89 GHz indicates the presence of convective rain

[7]Tb is the black-body temperature equivalent to the radiance intensity.

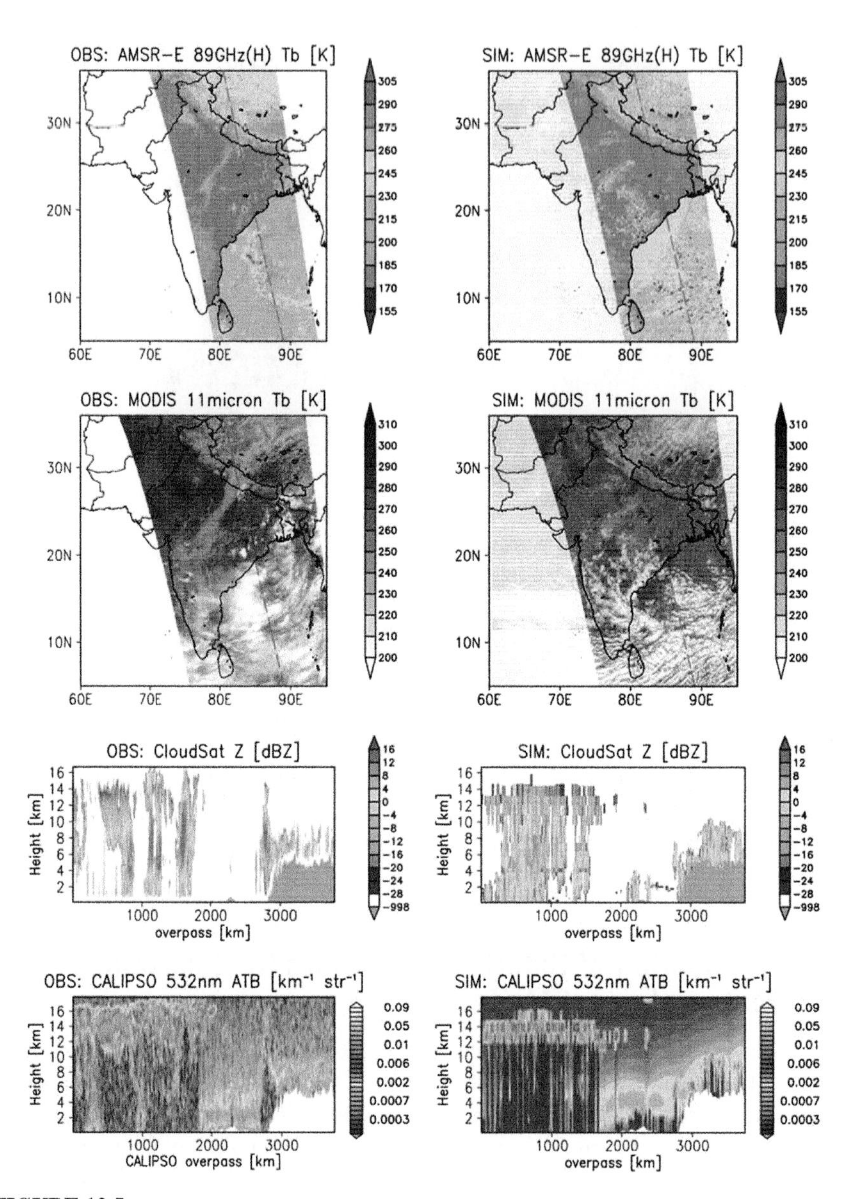

FIGURE 12.5

Observed (left panel) and simulated (right panel) satellite Level1B signals over India. (The color version of this figure is presented in the plate section in the back of the book and the online web version.)

associated with large ice aloft, because large-size solid raining particles strongly scatter surface-emitting upwelling microwave radiance back to the surface (Fig. 12.3). The Moderate Resolution Imaging Spectroradiometer (MODIS) in the Aqua satellite observes spectrum radiance at the visible to infrared spectrum. At the infrared window channel (11 μm), molecular absorption is minimal so that surface or cloud emission becomes the dominant source of TOA upwelling Tb. Thus, MODIS 11 μm Tb represents distributions of cloud-top or surface temperature. Active sensors can capture the curtain image of the atmosphere in detail. CloudSat Cloud Profiling Radar (CPR) observes equivalent radar reflectivities at 94 GHz, which represents a profile of cloud layers. Cloud-Aerosol Lidar and Infrared Pathfinder Satellite Observations (CALIPSO) Cloud-Aerosol Lidar with Orthogonal Polarisation (CALIOP) measures the LIDAR backscattering coefficient at 532 nm, which is sensitive to aerosols and smaller cloud layers. In contrast to CloudSat CPR images, LIDAR profiles are strongly attenuated below thick clouds due to attenuation.

The right column of Fig. 12.5 shows the corresponding L1B fields predicted from a coupled aerosol-cloud-land surface mesoscale model[8] through multi-instrumental satellite simulators described in the previous section (Matsui et al. 2009). This figure depicts the similarity and difference between observations and simulations, and performance is now being evaluated of the mesoscale model with satellite L1B data instead of comparing geophysical parameters. In this manner, there are three major applications of the satellite simulator to mesoscale meteorological modeling: i) satellite radiance-based model evaluation, ii) satellite radiance-based data assimilation, and iii) support of future/current satellite missions with mesoscale modeling. The following sections review such applications from these categories.

Cloud-precipitation processes are visible mesoscale phenomena. They are critical pieces in understanding weather and climate systems related to the Earth's radiation and water budgets/cycles. Thus, these processes and physics are a priority metric for the assessment of mesoscale model simulations. Development of microphysics schemes have been attempted during past decades but were not successful due to lack of model evaluation datasets especially for deep convective clouds. Aircraft sampling provides useful cloud-precipitation parameters, but its spatial and temporal coverage are limited. Satellite observations have a great advantage of spatial coverage, but retrieval uncertainties and assumption hamper their application to cloud model evaluations. Thus, various satellite radiance-based model evaluation studies have been conducted. The simplest approach is to use satellite near-infrared or infrared Tb to evaluate and improve cloudiness in mesoscale models using geostational satellites (Chaboureau et al. 2002; Grasso and Greenwald 2004; Otkin et al. 2009; Grasso and Lindsey 2011; Jankov et al. 2011). Another method is to use satellite microwave Tb to evaluate rainfall intensity and distributions (Wu and Petty 2010). Radar reflectivity profiles from space-born satellites helped to evaluate the profile of rain-sensitive particles in a mesoscale model (Zhou et al. 2007; Zeng et al. 2011).

[8]This model is the NASA-Unified Weather Research and Forecast (NU-WRF) model that couples with NASA Goddard radiation, microphysics, aerosol, and land-surface models.

Simultaneous uses of multi-instrumental L1B data are even better approaches to evaluate various aspects of cloud and precipitation structure and microphysics: e.g., combinations of microwave and infrared Tb for evaluating cloud cover, top temperature, and precipitation simultaneously (Meirold-Mautner et al. 2007; Chaboureau et al. 2002), microwave Tb and radar reflectivity profiles for evaluating column precipitation and microphysics profile (Han et al. 2010; Li et al. 2010; Shi et al. 2010), infrared Tb and radar reflectivity for evaluating cloud- and rain-top height to classify cloud-precipitation systems (Stephens et al. 2004; Masunaga et al. 2008), and multi-spectrum visible-infrared radiance for evaluating various aspects of cloud-top information (Greenwald et al. 2002; Grasso et al. 2010).

Matsui et al. (2009) developed a comprehensive and statistical multi-sensor radiance-based evaluation framework for cloud-precipitation processes in mesoscale models (12.6). This method uses three instruments of the Tropical Rainfall Measuring Mission (TRMM) satellite and three evaluation steps, hence denoted as the TRMM Triple-sensor Three-step Evaluation Framework (T3EF). T3EF uses the collocated TRMM triple-sensor L1B measurements, including Precipitation Radar (PR) 13.8 GHz attenuation-corrected reflectivity profiles, Visible and Infrared Scanner

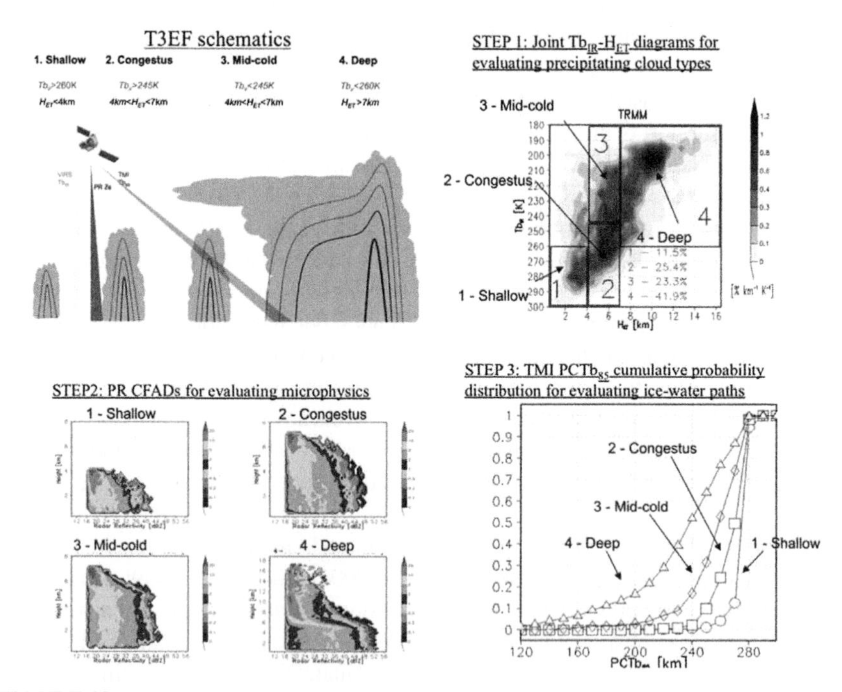

FIGURE 12.6

Schematic of the T3EF step 1, 2, and 3 from the TRMM L1B observation over the South China Sea during May–June in 1998. (The color version of this figure is presented in the plate section in the back of the book and the online web version.)

(VIRS) 12 μm infrared brightness temperature (Tb_{IR}), and TRMM Microwave Imager (TMI) 85.5 GHz polarization-corrected microwave brightness temperature ($PCTb_{85}$). Using the multi-instrument satellite simulators, TRMM-consistent L1B field can be constructed from the mesoscale model simulations. The first step of the T3EF is to evaluate macrostructure of cloud-precipitation processes by creating joint diagrams from Tb_{IR} and H_{ET} (radar echo-top height). Tb_{IR} represents cloud top temperature and H_{ET} represents rain-top height so that using the artificial thresholds of Tb_{IR} and H_{ET}, one can categorize cloud type into shallow, cumulus congestus, mid-cold, and deep classes, and determine their population. The second step of the T3EF is to evaluate the profile of rain microphysics by constructing contoured frequency with altitude diagrams (CFADs) of PR reflectivity for each cloud class. Precipitation microphysics (because it's falling) can be classified vertically, because of their microphysics process involved in terminal velocity, breakup, collection, and sedimentation. The third step is to evaluate the column ice water path by constructing the cumulative relative histogram of $PCTb_{85}$. The presence of more dense ice particles, such as graupel, tends to depress $PCTb_{85}$ so that, since the wavelength of PR reflectivity (14 GHz) and TMI Tb (85 GHz) is about 2.1 cm and 3.5 mm, TMI Tb is more sensitive to finer cloud particles than PR reflectivity. For example, PR CFADs between congestus and mid-cold are very similar, whereas cumulative histograms of mid-cold is flatter than congestus. Overall, this evaluation method evaluates the population of different types of observed and simulated precipitating cloud systems, and columns and profiles of microphysics information is evaluated for different cloud types. In this way, microphysics performance can be statistically evaluated by multi-instruments, while being isolated from forecast skill. Thus, cloud microphysics improvement can be guided to a better direction by constraining multi-sensor observations.

With these approaches, spatial comparison between the observed and simulated radiance evaluates mesoscale forecast skill related to the initial and boundary conditions, while statistical comparison can reveal biases in model physics. These methods, again noted, are essentially unaffected from the retrieval uncertainties in satellite remote sensing. If these spatial/statistical errors in a mesoscale model and uncertainties in a satellite simulator are formed in the background and observational error covariance, satellite radiance can be effectively blended with mesoscale models to optimally constrain meteorological modeling fields. This is radiance-based satellite data assimilation. While data assimilation techniques has been commonly developed upon dynamics fields (temperature, humidity, and wind fields) in the global model (Parrish and Derber 1992), the mesoscale model (especially grid spacing less than 5 km) is more suitable for assimilating cloud-precipitation processes. This is because i) initial and lateral boundary conditions from a global model have already assimilated large-scale dynamic fields, ii) a meso-γ mesoscale model can explicitly resolve cloud-precipitation processes without a subgrid convection scheme so that cloud-precipitation-affected radiance can be simulated more straightforward, and iii) assimilation of explicit cloud-precipitation processes improve geolocation of latent heat, radiative heating, and mesoscale dynamics. Infrared Tb from the geostational satellites has been assimilated in the cloudy atmosphere through 4-dimensional variational data assimilation (Vukicevic et al. 2006; Seaman et al. 2010; Polkinghorne and Vukicevic 2011) and the ensemble-based method (Zupanski et al. 2011a). Microwave Tb from passive microwave radiometers has been assimilated in the rainy atmosphere through the ensemble-based method (Zupanski et al. 2011b).

In comparison with assimilating balanced synoptic fields (e.g., temperature, humidity, balanced wind fields), cloud-precipitation assimilation is involved in many issues: i) assimilating nonlinear ageostrophic systems, ii) high temporal and spatial variability of cloud-precipitation processes, iii) uncertainties in cloud microphysics profiles from the radiances, iv) biases in the cloud microphysics, v) larger computational demands on simulating cloud-resolving mesoscale models (Vukicevic et al. 2006). Although a larger effort is required to conquer various issues (Errico et al. 2007), various numerical weather prediction (NWP) centers start assimilating rain- or cloud-contaminated radiance as their NWP models become operating at higher resolutions (Bauer et al. 2011).

These satellite radiance-based evaluation and assimilation techniques can be potentially applied to aerosol-transport models. Similar to cloud-precipitation retrievals, any satellite retrieval using a passive imager assume aerosol types, sizes, and optical properties in order to obtain column aerosol optical depths (AODs) at certain ultraviolet-visible spectrum. However, these background assumptions of aerosol species, size distributions, and chemical composition may be inconsistent with those in aerosol transport models, and this might overshadow concrete evaluation of aerosol microphysics using satellite-retrieved products. Aerosol microphysics in most of the existing aerosol transport models is a one-moment scheme, i.e., predicting mass-mixing ratio only. This simply represents that particle size distributions (PSDs) and, hence, optical properties are diagnostic. Therefore, we should first establish the radiance-based aerosol-transport model evaluation framework similar to the radiance-based cloud-precipitation evaluation framework. Aerosol PSDs may be constrained by joint statistical composites of multi-sensor multi-wavelength radiances such as spectrometer, imager, and LIDAR instruments. Establishment of a multi-sensor aerosol-transport model evaluation framework can be eventually extended to an aerosol data assimilation framework (Weaver et al. 2007).

If mesoscale model simulations are accurate and detailed enough relative to the complexity and uncertainties of retrieval algorithms, mesoscale models and satellite simulators can, alternatively, support satellite algorithm development. Of course, in this approach, the level of complexity in mesoscale models and satellite simulators must be equal to or better than the complexity in satellite retrieval assumptions. For example, if at present, simulators and retrieval algorithms rely on the Mie calculation for scattering parameters, the algorithm errors related to the non-spherical parameter cannot be studied. If the optical depth of the target geophysical parameter is very small, realistic surface boundary conditions (emissivity or albedo field) in satellite simulators are required for simulating realistic satellite signals.

This approach began two decades ago for the application of precipitation-observing satellite sensors. At that time, there were no field observations available to construct profiles of cloud-precipitation microphysics within tropical convection. Cloud-resolving models with one-moment bulk microphysics provided realistic three-dimensional fields of cloud-precipitation particles (Tao and Simpson 1993). Adler et al. (1991) and Haferman et al. (1993) simulated satellite-observable microwave Tb for studying details in the precipitation-Tb relationship. Meneghini (1996) simulated realistic spaceborne precipitation radar with the same cloud-resolving model results, and exercised a radar precipitation algorithm to prepare the launch of the TRMM satellite. The impact of three-dimensional radiation effects has also been studied (Battaglia et al. 2007).

Grasso et al. (2008) utilized the mesoscale model with two-moment microphysics for generating virtual L1B data of the Geostationary Operational Environmental Satellite R (GOES-R, ~2015 launch) Advanced Baseline Imager (ABI) and the *Joint Polar Satellite System* (*JPSS*,[9] ~2016 launch) Visible/Infrared Imager/Radiometer Suite (VIIRS). Their approach is to validate and improve the performance of the mesoscale model and the coupled multi-spectrum satellite simulators (Greenwald et al. 2002; Grasso and Greenwald 2004), and by using the well-validated mesoscale model database to generate future satellite imagery. They proposed the simulated future satellite L1B data could be utilized for testing cloud-retrieval, upper-tropospheric wind, and/or fire-detection algorithms before these satellites were placed in their orbits.

The Global Precipitation Measurement (GPM) is a next-generation multi-satellite global precipitation-measuring mission (Hou et al. 2013). The GPM Core Observatory plays a central role in the GPM mission by calibrating the precipitation algorithms of the GPM Constellation satellites. Before the launch (February 2014), the GPM Core precipitation algorithms must be established upon the existing database, especially targeting on mid- to high-latitude precipitation systems beyond the ranges of the TRMM satellite. To this end, *the Synthetic GPM Simulator* has been established on the GPM Ground Validation (GV) sites (Matsui et al. 2013). The Synthetic GPM Simulator is the unified framework that integrates GPM Ground Validation (GV) observations, the cloud-resolving mesoscale model with spectral-bin microphysics, and a unified GPM instrument simulator. Essentially, GV in situ observations were used to constrain the cloud-precipitation simulation focusing on the vertical profiles of cloud-precipitation microphysics (Iguchi et al. 2012a,b). GV-constrained mesoscale model simulations are then fed to the GPM simulator to generate GPM-observable signals, including satellite footprint L1B signals (microwave Tb and radar reflectivity) and retrieved L2 geophysical parameters (e.g., rainfall rate). This GPM simulator consists of detailed satellite orbital and sensor-scanning modules, unified radiative transfer modules for microwave radiometer and precipitation radars, and realistic antenna gain modules upon the fields of the mesoscale model. The generated GPM orbital testbed is a critical database for pre-launch precipitation algorithms for the upcoming GPM Core satellite launch (Matsui et al. 2013).

Beside uncertainties in the performance of mesoscale models, the coupled high-resolution model and simulators system proves to be a great benefit for remote sensing

[9] Previously called the National Polar-Orbiting Operational Environmental Satellite System (NPOESS).

algorithm development, because it is a physically-based approach so that scientists explicitly know the physical parameters and radiative properties as a priori. Even within the field campaign framework, it is impossible to know the details of atmospheric parameters in the entire space and time. But the high-resolution mesoscale model and simulators system allow scientists to explicitly know the physical parameters in the entire four-dimensional space.

To conclude, satellite simulators can bridge mesoscale models and satellite observations tighter than ever, and will be more often used in the mesoscale model community in future. This approach encourages the mesoscale modeling community to consider details in remote sensing techniques and application for model evaluation. Alternatively, the remote sensing community can be appreciated for the benefit of detailed mesoscale modeling simulations in support of developing the remote sensing algorithm.

Problems for Chapter 12

1. Search online or in the literature and find how modern mesoscale models are coupled with satellite data and simulators.
2. Review satellite algorithm ATBD, and determine what channels of specific satellite instruments are useful in detecting different geophysical parameters.

Chapter 12 Additional Readings

Kidder, S.Q., and T.H. Vonder Haar, 1995: *Satellite Meteorology*. Academic Press, 466 pp.

Liou, K.-N., 2002: *An Introduction to Atmospheric Radiation*. (2nd Ed.), Academic Press, 583 pp.

Petty, G.W., 2006: *A First Course in Atmospheric Radiation*. (2nd Ed.), Sundog Publishing, 460 pp.

CHAPTER

13 Examples of Mesoscale Models

CHAPTER OUTLINE HEAD

Mesoscale atmospheric systems can be divided into two groups:

1. those forced primarily by surface inhomogeneities (terrain- and physiographically-induced mesoscale systems), or

Mesoscale Meteorological Modeling, Third Edition, Volume 98. http://dx.doi.org/10.1016/B978-0-12-385237-3.00013-X

2. those forced primarily the lateral boundaries or from internal atmospheric instabilities (synoptically-induced mesoscale systems).

The synoptically-induced systems that occur in a region when they propagate inward (i.e., through the lateral boundaries of the mesoscale model) can be called propagating synoptically-induced mesoscale systems. Those that develop within the regional model domain as a result of small-scale atmospheric features within the mesoscale model domain can be called internally generated synoptically-induced mesoscale systems. These atmospheric developments within the mesoscale domain are often referred to as "dynamic instabilities."

Mesoscale features forced primarily by surface inhomogeneities (terrain- and physiographically-induced mesoscale systems) includes sea and land breezes, mountain-valley winds, urban circulations, and forced airflow over rough terrain. A necessary condition for their development is an accurate specification of the surface landscape including land-water boundaries, vegetation distribution, urban area, sea surface temperatures, topography, etc. These features are the least difficult mesoscale features to simulate, since the sources of these mesoscale circulations are geographically fixed and they recur frequently. These mesoscale systems do not generally move far from their point of origin, and in general do not require a detailed spatial representation of the initial and the lateral and top boundary conditions for the dependent variables.

Propagating synoptically-induced mesoscale systems include squall lines, hurricanes, and traveling mesoscale cloud clusters. When these features propagate into the mesoscale model domain, they must be accurately resolved through the lateral boundaries of the model. Since, as discussed in Chapter 9, at least four grid increments in each spatial direction are required to accurately resolve these features, the propagating system must be at least this large at the lateral boundaries. This is a challenge, however, as the source of that information (i.e., a larger-scale model or reanalysis) must have fine enough spatial and temporal resolution to resolve the feature. Features that are smaller than this, even if they occur in the real world situation, will be missed at the lateral boundaries.

The challenge is even greater for internally generated synoptically-induced mesoscale systems. While propagating synoptically-induced mesoscale systems can spawn other synoptically-induced mesoscale systems (e.g., a secondary squall line) as well as be modulated by surface inhomogeneities, any systems that develop on their own within the atmosphere simulated by the mesoscale model domain must result, at least in part, from high spatial information in the initialization of the model. This requires high spatial resolution observed mesoscale information. However, as we concluded in the discussion of Table 10.1, such a high spatial requirement requires very small tolerances of errors. An example of an internally generated synoptically-induced mesoscale systems generated by atmospheric instability includes those produced by conditional instability of the second kind (CISK; see Holton 1972; Wallace and Hobbs 1977:442; Mak 1981; Ooyama 1982). A summary study by Clark et al. (2012) shows the large progress that has been made in simulating mesoscale features associated with synoptically-induced mesoscale systems.

In the real world, of course, the actual weather is a combination of terrain- and physiographically-induced mesoscale systems and propagating and internally generated synoptically-induced mesoscale systems. A large percentage of the rainfall over the Earth results from the tendency for synoptically-driven weather disturbances to organize into such synoptically-induced mesoscale-sized precipitating cloud systems (Houze and Hobbs 1982). Browning (1980) provides a similar categorization of mesoscale systems as described in this chapter.

In the following two sections, examples of mesoscale model results are presented. These examples are broken into specific types in order to assist in understanding why they occur. However, in the real world, weather is made up, of course, of a combination of several types. Since the first and second editions of this book were written, mesoscale modeling has matured into an integral part of operational weather forecasting and research. These include, as reported in Hong and Dudhia (2012) the North American Mesoscale model (NAM; United States), the Application of Research to Operations at Mesoscale model (AROME; France), the Unified Model (UM; United Kingdom and also used in Korea), the High-Resolution Limited-Area Model (HIRLAM)-Aire Limitée Adaptation Dynamique Développement International (ALADIN) Regional/Mesoscale Operational NWP in Europe (HARMONIE; Europe), the nonhydrostatic model (NHM; Japan), and the Weather Research and Forecasting model (WRF). In the Hong and Dudhia (2012) article, they report that the models at grid sizes of 1-5 km provide reliable information for weather forecasts, in particular, precipitation. Krishnamurthy et al. (2013) illustrate how mesoscale models, combined with Doppler lidar measurements, are being used to site wind turbines. A summary of a set of models that resolve the mesoscale are presented in Appendix B. At the end of each subsection in this Chapter, I have alerted readers, with examples, to books and articles where further references can be found with respect to each type of mesoscale system.

13.1 Spatial Scales at which Mesoscale Circulations are Important

The minimum spatial scale of mesoscale circulations is determined by the sizes below which turbulent mixing diffuses horizontal spatial gradients of temperature such that they very quickly become too small to create a horizontal gradient of hydrostatic pressure that is large enough to create a coherent circulation. Circulations, of course, can occur below this scale (Pielke et al. 1995a) such as tornadoes, but these are not hydrostatically balanced systems. In the surface heated planetary boundary layer, strong vertical mixing can rapidly mix heat horizontally, in addition to the vertical mixing, such that significant horizontal gradients in temperature throughout the boundary layer seldom persist for spatial scales smaller than a few kilometers. It is for this reason that parameterizations of surface landscape heterogeneities have adopted the so-called *mosaic* or *tile* approach in which surface fluxes are calculated using a

1-D vertical model and then they are fractionally weighted to insert into the lower boundary of a mesoscale (or other) model. This approach is discussed in Avissar and Pielke (1989) where they write

> *"In each surface grid element of the numerical model similar homogeneous land patches located at different places within the element are regrouped into subgrid classes. Then, for each one of the subgrid classes, a sophisticated micrometeorological model of the soil-plant-atmosphere system is applied to assess the surface temperature, humidity, and fluxes to the atmosphere. The global fluxes of energy between the grid and the atmosphere are obtained by averaging according to the distribution of the subgrid classes. In addition to the surface forcing, detailed micrometeorological conditions of the patches are assessed for the domain simulated by the atmospheric model."*

Investigators have also applied an average value of surface parameters (e.g., albedo, aerodynamic roughness, etc.) to represent landscape heterogeneities that are smaller than the grid cell of the mesoscale (or other) model. This approach, uses, for example, a concept called "effective roughness" to represent the cumulative effect of the heterogeneities on surface roughness (Schmid and Bünzli 2007). While an effective roughness is a useful concept for terrain variations on scales smaller than the model grid area (e.g., see Menenti and Ritchie 1994), the mosaic approach is concluded to be better for landscape heterogeneities.

From experiences with modeling and field data, a horizontal grid spacing of 1 km (which means features larger than 4 km are resolved, as discussed in Chapter 9) is sufficient to capture the scale at which mesoscale features occur. For features that are 4 km and larger, the hydrostatic assumption is quite accurate (e.g., as discussed in Chapters 4 and 5), even if a model is used which includes the ability to simulate non-hydrostatic pressure effects. For the upper spatial scale at which mesoscale features occur, while they of course, blend into synoptic-scale features, they can be defined as those circulations that deviate significantly from gradient wind balance even above the boundary layer. The determination of the spatial scale of these circulations has been examined analytically in a series of studies (Dalu et al. 1991, 1996, 2000; Dalu and Pielke 1993; Baldi et al. 2005, 2008). The role of horizontal grid interval on mesoscale simulation was investigated by Salvador et al. (1999). For diurnally-forced surface systems such as sea breezes and with no large-scale flow, for example, the intensity of mesoscale cells increases for increasing values of the wave number, reaching a maximum value when the wavelength of the forcing is of the order of a local Rossby radius, and then decreases as the wavelength of the forcing decreases.

This upper spatial scale is determined by the time and space required a parcel of air placed in motion above the atmospheric boundary layer to achieve a near balance between the horizontal pressure gradient force and the Coriolis force (i.e., often referred to as *geostrophic adjustment* but more accurately should be called *gradient wind adjustment*). The sea breeze can be used to illustrate this adjustment time. Because of its diurnal forcing, it is a particularly ideal mesoscale circulation to examine in order to address this issue. This is even more true where the sea breeze occurs on most days during a season (e.g., Neumann and Mahrer 1971). The inland penetration of the sea breeze is an effective way to assess the maximum spatial scale that these non-gradient wind atmospheric circulations can achieve. Simpson (2006), for example, in an observational study over 12 years, out of 76 sea breezes that they studied only 12 made it as far inland at 85 km. Dalu and Pielke (1989) concluded that the sea breeze has a characteristic time scale which is a combination of the inertial period and an e-folding time due to friction. For a time larger than this characteristic time scale, the inland penetration of the sea breeze is confined by a Rossby deformation radius, which includes a frictional effect. Friction and inertia reduce not only the intensity, but also the horizontal scale of motion. At the equator, the controlling parameter for the intensity and penetration of the sea breeze is friction. Rotunno (1983) concluded that

> *"Given that the earth's atmosphere may be idealized as a rotating, stratified fluid characterized by the Coriolis parameter f and the Brunt–Vaisala frequency N, and that the diurnal cycle of heating and cooling of the land relative to the sea acts as a stationary, oscillatory source of energy of frequency $\omega (= 2\pi\ day^{-1})$, it follows from the linear theory of motion that where $f > \omega$ the atmospheric response is confined to within a distance $Nh(f{-}2{-}\omega{-}2)^{-1/2}$ of the coastline, where h is the vertical scale of the heating. When $f < \omega$, the atmospheric response is in the form of internal-inertial waves which extend to "Infinity" along ray paths extending upward and outward from the coast. Near the ground, the horizontal extent of the sea breeze is given by the horizontal wale of the dominant wave mode, $Nh(\omega^2 - f^{-2})^{-1/2}$."*

All atmospheric circulations will necessarily be influenced by such a constraint. Since features at mid- and high-latitudes above a few hundred kilometers in horizontal scale are close to gradient wind balance above the boundary layer, mesoscale features lie between tens and several hundred of kilometers in horizontal scale. In the lower latitudes, where the Coriolis effect is weaker, the spatial scale will be larger.

13.2 Terrain- and Physiographically-Induced Mesoscale Systems

13.2.1 Sea- and Land-Breezes over Flat Terrain

Of all the mesoscale phenomena, sea and land breezes over flat terrain appear to have been the most studied, both observationally and theoretically. This is undoubtedly a result of the geographically fixed nature of the phenomenon (the location of land–water boundaries), as well as the repetitive nature of the event. The sea breeze is defined to occur when the wind is onshore, whereas the land breeze occurs when the opposite flow exists.

During the case of nonexistent large-scale winds, it is comparatively easy to describe the diurnal variations of the coastal wind circulations. Defant (1951) presented an excellent qualitative description for this condition, which is illustrated in Fig. 13.1. For this case, the idealized sequence of events is as follows.

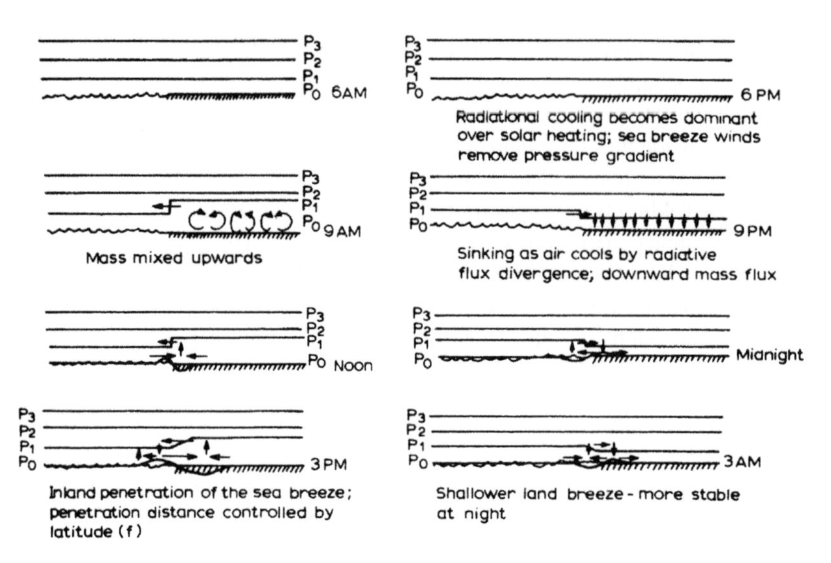

FIGURE 13.1

Schematic of the diurnal evolution of the sea and land breeze in the absence of synoptic flow (from Pielke 1981).

1. At some time in the early morning the pressure surfaces become flat and no winds occur (e.g., at 0800 LST – perhaps an hour after sunrise).
2. Later in the morning, mass is mixed upward over land by turbulent mixing in the unstably stratified boundary layer creating an offshore pressure gradient at some distance above the ground.[1] Over water, its translucent character and ability to mix prevent significant heating of the surface (e.g., at 1100 LST).
3. The resultant offshore flow of air above the ground near the coast creates a low-pressure region at the ground, and onshore winds (*the sea breeze*) develop (e.g., at 1300 LST).
4. The onshore winds transport cooler marine air over the land, thereby advecting the horizontal temperature gradient and, hence, the sea breeze inland. The distance the sea breeze travels inland depends most directly on the intensity of the total heat input to the air (Pearson 1973) and the latitude (e.g., at 1600 LST).
5. As the sun sets, longwave radiational cooling becomes dominant over solar heating, and the local wind field removes the horizontal temperature gradient. The pressure surfaces again become horizontal (e.g., at 1900 LST).
6. As longwave cooling continues, the air near the ground becomes more dense and sinks. The resultant lowering of the pressure surfaces a short distance above the ground creates an onshore wind at that level (e.g., at 2200 LST).

[1]Nicholls and Pielke (1994a,b) demonstrate that the expansion of the volume, as required by the ideal gas law, also elevates the pressure surface at heights above the surface.

7. In response to the loss of mass above the surface over the water a pressure minimum develops at the ocean interface immediately off the coast. The offshore wind that then develops near the surface is called *the land breeze* (e.g., at 0100 LST).
8. The distance of offshore penetration of the land breeze deepens on the amount of cooling over the land. Because the planetary boundary layer over land is stably stratified at night and, therefore, vertical mixing is weaker and closer to the ground, the land breeze is a shallower and weaker phenomenon than the daytime sea breeze.

When the coastline is irregular, local regions of enhanced or weakened low-level convergence develop, as illustrated for the daytime portion of the cycle in Fig. 13.2. (Such zones of preferential convergence help explain the preference for showers in certain locations in south Florida during the summer, as seen, for example, in Fig. 11.13.) This preference for showers results from the enhancement of convective

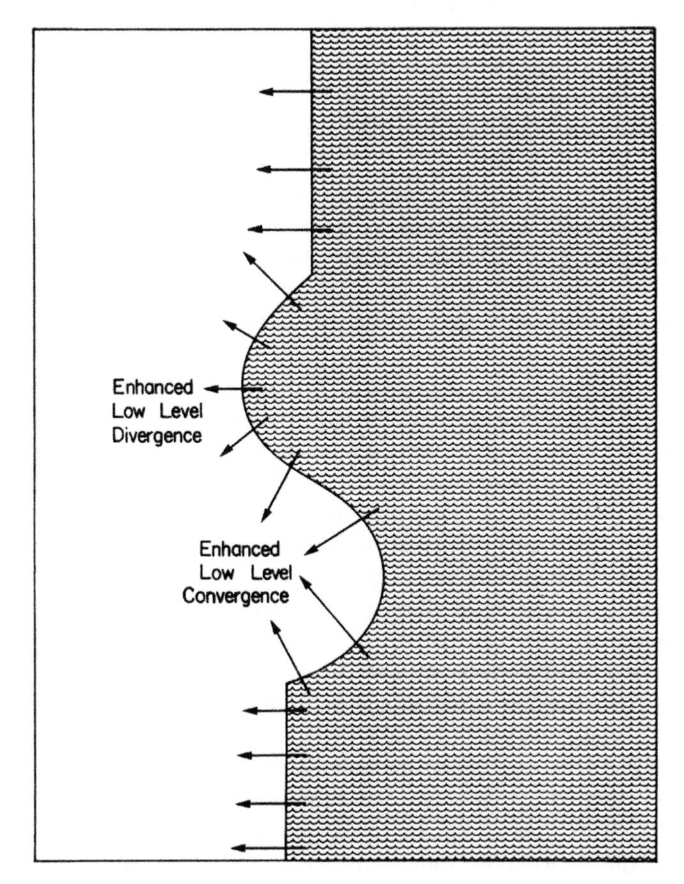

FIGURE 13.2

Schematic of the influence of coastline configuration on the sea breeze in the absence of large-scale flow.

potential energy associated with the horizontal convergence of water vapor (Pielke et al. 1991) and the moistening of the deeper troposphere as repeated cumulus convection in this region detrains water vapor into the same atmospheric column.

The evolution of the sea breeze is somewhat more complicated when a weak or moderate (i.e., $\lesssim 6$ m s^{-1}) prevailing synoptic flow is included. For the two distinct situations of comparatively cold water and comparatively warm water relative to land, a synoptic wind direction from the colder to the warmer surface weakens the intensity of the local wind by diminishing the horizontal temperature gradient. By contrast, when a prevailing larger-scale flow of the same strength is from the warmer to the colder surface, the temperature gradient is strengthened and the subsequent local wind flow is stronger.

Examples of water that is warm relative to the land include the eastern sides of continents in the tropics and midlatitudes at night and over coastal water during a polar outbreak. Situations with water that is cold relative to the adjacent land include the eastern sides of continents in tropical and midlatitudes during sunny days, along the west side of continents in which upwelling is occurring, as well as along polar coastal areas in the summer.

Fog and low stratus often form over the relatively cold water in polar and upwelling ocean areas (e.g., Noonkester 1979; Pilié et al. 1979). Noonkester discussed fog formation caused by the offshore movement of warm, dry air along the coast of southern California and its subsequent movement back toward and over land in the sea breeze. Estoque (1962) performed numerical experiments showing the influence of the prevailing synoptic flow on sea-breeze convergence, and Yoshikado (1981) used observations to illustrate the influence of the geostrophic wind direction on the sea-breeze pattern.

Figure 13.3 illustrates predicted sea-breeze results for weak and moderate onshore synoptic flow. With the weaker winds, the large horizontal gradient of potential temperature (and, therefore, large horizontal gradient of pressure) results in a tight and well-defined sea-breeze circulation as it moves inland. When the prevailing onshore flow is stronger, however, such a large pressure gradient cannot develop because of the rapid inland movement and greater warming of the marine air. In this and subsequent figures, the ends of the dumbbells are spaced 100 km apart to illustrate the approximate horizontal grid spacing of a large-scale model, as contrasted with mesoscale grid resolution.

The magnitude of the effect of a particular horizontal temperature gradient can be estimated from existing observational and numerical studies (e.g., Hanna and Swisher 1971, Hanna and Gifford 1975). From these and other related works it has been found that, in the tropics and midlatitudes, a horizontal gradient of less than about 10 W m^{-2} per 30 km has only a minor influence on local wind patterns. With a gradient of 100 W m^{-2} per 30 km, however, significant effects are discernible from the statistical evaluation of observational data, whereas at 1000 W m^{-2} per 30 km the influence on local wind patterns is very pronounced in case-by-case studies.

Observational studies of significance to this phenomenon are numerous – a sampling includes those of Byers and Rodebush (1948), Day (1953), Carson (1954),

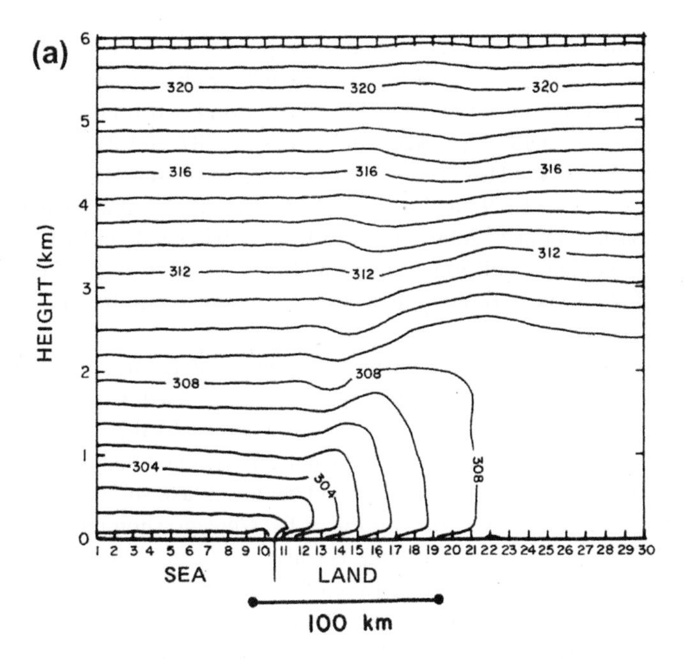

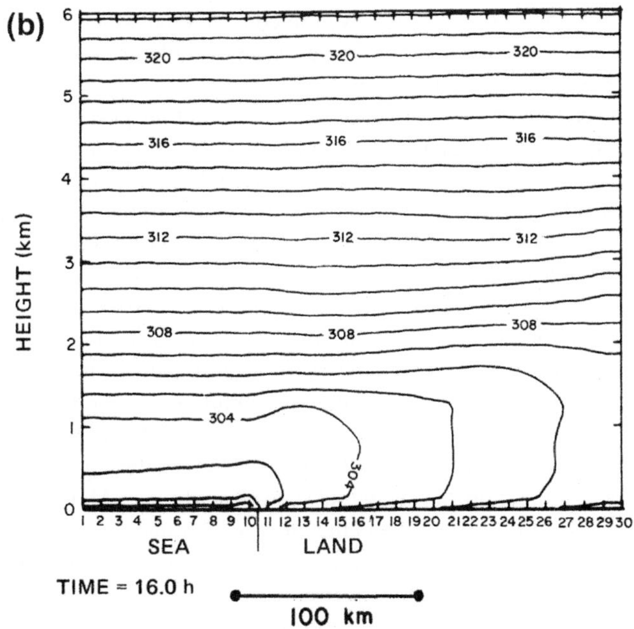

FIGURE 13.3

The vertical cross section of potential temperature along a coastline at 1600 LST for (a) a 1 m s^{-1} onshore synoptic wind, and (b) a 6 m s^{-1} onshore synoptic wind. Initial input was for a typical summer day over south Florida (from Mahrer and Pielke 1978b).

Gentry and Moore (1954), Randerson and Thompson (1964), Plank (1966), Frank et al. (1967), Pielke and Cotton (1977), Burpee (1979), Schwartz and Bosart (1979), Cunning et al. (1982), and Blanchard (1983) for Florida; Kozo (1982a) for part of the Alaskan coast; Lyons (1975), Keen and Lyons (1978), and Ryznar and Touma (1981) for Lake Michigan; Hsu (1969) for the Texas coast; Neumann (1951), Doran (1979), Skibin and Hod (1979), and Bitan (1981) for Israel; Druilhet et al. (1982) for the southern coast of France; Johnson and O'Brien (1973) and Darby (2001) along the coasts of Oregon and California; Simpson (1996) for southern England; Skinner and Tapper (1994) for islands along the coast of northern Australia, and Lalas et al. (1983) for Athens, Greece.

Using observational data, Biggs and Graves (1962), and Lyons (1972) have developed indices to estimate whether or not a sea breeze will occur. Lyons, for example, has shown that when $V_g^2/\Delta T$ is greater than 10 (where V_g is the 0600 CST surface geostrophic wind speed in meters per second and ΔT the maximum temperature difference between the inland air temperature and the mean lake surface temperature in degrees Celsius), a sea breeze will not occur at the Chicago shoreline. A sea breeze does not develop when this ratio is large because the horizontal pressure gradient generated by the differential heating between the land and the lake is insufficient to overcome the kinetic energy of the large-scale flow.

These studies have demonstrated that land and sea breezes (and other similar mesoscale circulations) are poorly resolved in conventional weather-observing network systems. Such a lack of resolution creates serious problems in developing routine operational forecasts of these phenomena.

Lyons and Keen (1976) also concluded that studies of transport and diffusion over land are generally invalid when applied to the coastal environment. Lyons and Cole (1976), and Eastman et al. (1995), for example, have discussed the accumulation of pollutants that results from the recirculation associated with the Lake Michigan sea breeze – an effect that is not considered in commonly used dispersion models. Keen et al. (1979) also concluded that size sorting of aerosols occurs within this lake breeze system. In a different geographic area, Carroll and Baskett (1979) concluded that the most serious degradation of air quality in Yosemite National Park occurs because of the transport of material from several hundred kilometers away by the sea breeze from the Pacific coast, as well as by the valley-mountain circulation generated by the Sierras.

In southern California, Sackinger et al. (1982) have used a tracer study to document that essentially all of the released material transported to sea during a land breeze was advected back across the shore during the subsequent sea breeze. McRae et al. (1982a) has similarly described for this situation that material emitted into an elevated stable layer at night is transported offshore, fumigated to the surface, and returned onshore during the daytime. Fumigation occurred over the ocean because it was warmer than the overlying air. Blumenthal et al. (1978) documented the relation between smog in the Los Angeles Basin and the sea breeze, and Lalas et al. (1982) discussed the SO_2 concentrations in Athens, Greece and the need to determine the local sea breeze, heat island, and terrain circulations.

Examples of early analytic studies of direct relevance to the sea-breeze phenomenon include those of Haurwitz (1947), Schmidt (1947), Defant (1950), Malkus and Stern (1953), Stern and Malkus (1953), and Smith (1955, 1957), and other studies of this sort include those of Geisler and Bretherton (1969), Walsh (1974), Neumann (1977), Kimura and Eguchi (1978), Dalu and Green (1980), Sun and Orlanski (1981a), Rotunno (1983), Uedo (1983), Dalu and Pielke (1989), and Dalu et al. (1996, 2000). The first nonlinear numerical modeling studies of this phenomenon, performed using two-dimensional models, include that of Estoque (1961, 1962), followed by Fisher (1961), Moroz (1967), Neumann and Mahrer (1971), and others. Pielke (1974b), Neumann and Mahrer (1974, 1975), Estoque et al. (1976, 1994), Physick (1976, 1980), Patrinos and Kistler (1977), Dalu (1978), Gannon (1978), Sahashi (1981), Sun and Orlanski (1981b), Kozo (1982b), Okeyo (1982), Maddukuri (1982), Alpert and Neumann (1983), Segal et al. (1983a), Xian and Pielke (1991), Nicholls et al. (1991), Tijm et al. (1999a,b), and Baker et al. (2001) also used such two-dimensional models to improve our understanding of the physical processes associated with the sea breeze.

Anthes (1978), for instance, has shown using a two-dimensional model that with a zero large-scale prevailing flow, the return flow of the sea breeze occurs entirely above the boundary layer, whereas the onshore winds are confined below that level. Abe and Yoshida (1982) examined the influence of peninsula width on the intensity of the sea breeze and found that a width of 30 to 50 km produces the strongest upward vertical velocities. Ozoe et al. (1983) used a two-dimensional model to investigate the local pollution patterns and average parcel trajectories in the presence of land and sea breezes. Satomura (2000) used a two-dimensional model to study the diurnal variation of precipitation over the Indo-China region.

Since the early 1970s, computer capability has improved sufficiently to permit three-dimensional simulations, and in recent years, operational mesoscale models are used to prepare our daily weather forecasts. McPherson (1970) was the first to report such calculations of the sea breeze and was followed, for instance, by studies such as those of Pielke (1974a), Mahrer and Pielke (1976, 1978b), Warner et al. (1978), Carpenter (1979), Hsu (1979), Kikuchi et al. (1981), Xu et al. (1996), Kotroni et al. (1997), Pielke et al. (1999b), Cai and Steyn (2000), Sheng et al. (2000), and Yimin and Lyons (2000). In these studies, valuable new insight into the sea breeze was attained, including the conclusion that along coastlines, under undisturbed synoptic conditions during the summer in the tropics and subtropics, the sea breeze exerts a dominant influence on the sites of formation and the movement of thunderstorm complexes (Pielke 1974a). A sea breeze (or lake breeze) also significantly influences the transport and dispersion of pollution (Eastman et al. 1995; Kassomenos et al. 1995; Lyons et al. 1992, 1995; Kotroni et al. 1999a).

Figure 13.4 presents a sea-breeze model calculation during an afternoon over the Chesapeake Bay, illustrating the need for three-dimensional simulations. A quantitative numerical study of the sea breeze in this region is presented in Segal and Pielke (1981), and Segal et al. (1982a). Savijärvi and Järvenoja (2000) modeled the mesoscale atmospheric conditions over Lake Tanganyika in Africa. Marshall et al.

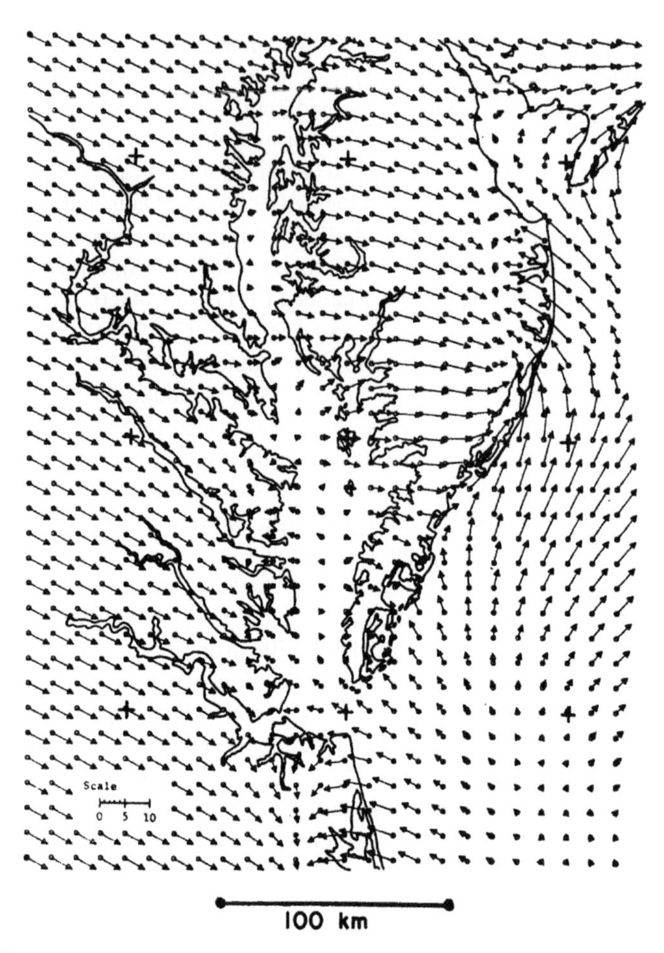

FIGURE 13.4

Predicted winds at 4 m at about 1500 LST over the Chesapeake Bay for 9 August 1975. Scale bar in meters per second (model simulations were performed by W. Snow at the University of Virginia).

(2004) presents two-month averaged 10 m winds and divergence for south Florida for a model simulation with natural and 1993 landscape, as shown in Fig. 13.5.[2]

The sea breeze may also be involved in the generation of severe local weather. Clarke et al. (1981), for example, have reported on the generation of wind squalls accompanied with spectacular roll clouds, which move onshore in northern Australia, whose origin appears frequently to be related to the interaction of a sea-breeze front and a developing nocturnal inversion.

[2]The landscape change that occurred over this time period is shown on the cover of the book.

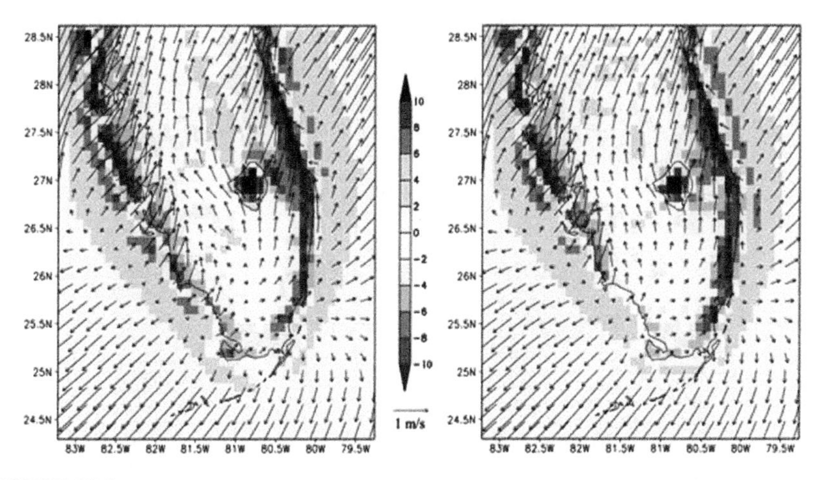

FIGURE 13.5

Two-month average of the 1600 UTC 10 m horizontal wind (vectors) and derived divergence field (color shaded; 10^5 s^{-1}) from the simulations of Jul–Aug 1989 with (left) pre-1900 land cover and (right) 1993 land use (from Marshall et al. 2004). (The color version of this figure is presented in the plate section in the back of the book and the online web version.)

The establishment of a mesoscale horizontal temperature gradient along a coastline can cause or enhance low-level jets as a mesoscale generated wind seeks to adjust to gradient wind balance. This mechanism has been used to explain observed wind maxima along the south Texas (McNider et al. 1982) and Oregon (Mizzi and Pielke 1984) coasts.

Other mesoscale modeling studies of the low-level jet include Zhong et al. (1996). A review of coastal meteorology is given in Rogers (1995). For more recent studies of land and sea breezes see Cotton and Pielke (2007), Simpson (2007), and Grant et al. (2013).

13.2.2 Vegetation, Snow, and Sea Ice Breezes

Since horizontal variations of vegetation coverage and type of snow cover and sea ice can result in differences in surface sensible heat fluxes as large as those between the land and adjacent sea, it should expected that landscape heterogeneities will also generate mesoscale flows (Cotton and Pielke 1995; Pielke 2001a). Figure 13.6 and 13.7 from Avissar and Liu (1996), for example, illustrates the role of different shapes of deforestation in a tropical region, on the generation of mesoscale circulations.

Modeling and observational studies of the effects resulting from vegetation and soil moisture variations and change include those of Lanicci et al. (1987), Segal et al. (1988, 1989a,b), Pielke and Zeng (1989), Bryant et al. (1990), Pielke and Avissar (1990), Chang and Wetzel (1991), Kimura and Takahashi (1991), Pielke et al.

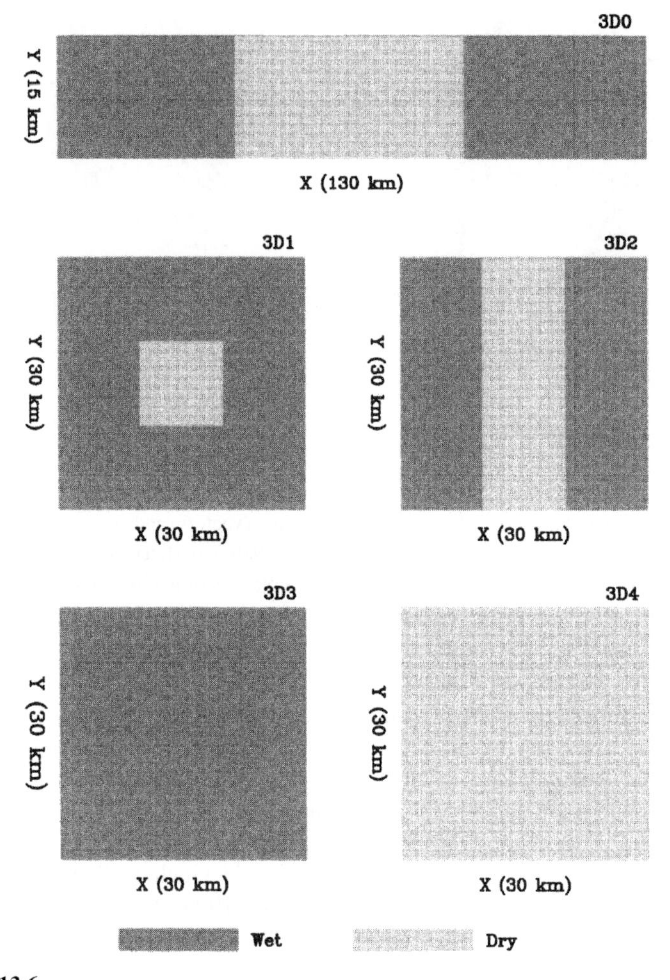

FIGURE 13.6

Schematic representation of the simulated horizontal domains (from Avissar and Liu 1996).

(1991), Avissar (1992), Entekhabi et al. (1992), Chen and Avissar (1994), Dirmeyer (1994), Clark and Arritt (1995), Copeland (1995), Copeland et al. (1996), Cotton and Pielke (1995), Cutrim et al. (1995), Hong et al. (1995), Klink (1995), Crook (1996), Eltahir (1996), Lyons et al. (1996), Nadezhina and Shklyarevich (1996), Schrieber et al. (1996), Taylor et al. (1997), Vidale et al. (1997), Eastman et al. (1998, 2001a), Emori (1998), Stohlgren et al. (1998), Chase et al. (1999), Kalthoff et al. (1999), Mölders (1999a), Zeng and Neelin (1999), Friedrich et al. (2000), Chen et al. (2001), Freedman et al. (2001), Lu et al. (2001), Lu and Shuttleworth (2002), Raddatz (2005, 2007), Hanesiak et al. (2009), Raddatz et al. (2009), Brimelow et al. (2010a,b),

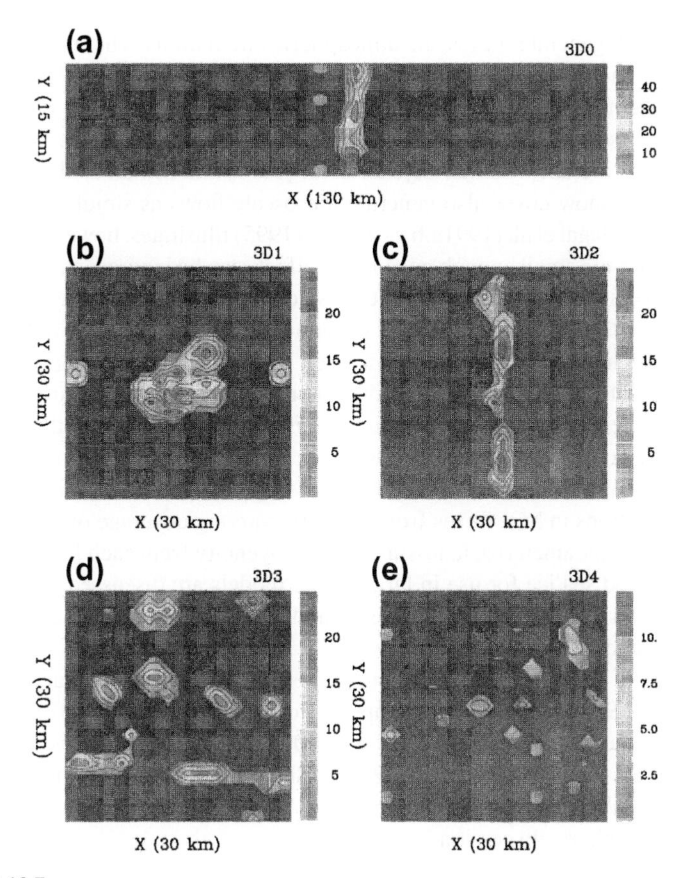

FIGURE 13.7

Accumulated precipitation (millimeters), at 1800 LST in domain (a) 3D0, (b) 3D1, (c) 3D2, (d) 3D3, and (e) 3D4. Contour intervals are 2 mm in 3D0, 1 mm in 3D1, 3D2, and 3D3, and 0.05 mm in 3D4 (from Avissar and Liu 1996).

Mahmood et al. (2010, 2013a,b), Yang et al. (2011), Lu (2011), Pitman et al. (2012), Shrestha et al. (2012), Comarazamy et al. (2013), Harding et al. (2013), Lo and Famiglietti (2013), and Ray (2013). Mölders (1999b) discusses how floods can alter the local mesoscale system. Nykanen et al. (2001) discuss how small-scale precipitation variability influences the larger-scale organization of land-atmosphere fluxes. Sud et al. (1993, 1995) discusses the dependence of rainfall on vegetation. Mölders and Kramm (2007), and Knowles (1993) show how changes in landscape due to fires can affect clouds and precipitation. Lu et al. (2005) investigated mesoscale circulations and atmospheric CO_2 variations over a heterogeneous landscape of forests, pastures, and large rivers in Brazil.

Fires also can generate mesoscale atmospheric circulations. The influence of fire on the surface energy budget is discussed, for example, in Amiro et al. (1999), and Bremer and Ham (1999). The influence of the heat of the fire on the physical and chemical of properties of soil are described in Giovannini et al. (1988). Modeling of forest fire events includes Gómez-Tejedor et al. (1999) and Millán et al. (1998).

Variations of snow cover also generate mesoscale flows as simulated by Taylor et al. (1998), and Segal et al. (1991a,b,c). Liston (1995) illustrates, however, that when the snow patches are small enough, mesoscale effects can be ignored, and the individual surface fluxes simply added using the mosaic approach discussed by Avissar and Pielke (1989). Liston and Sturm (1998) present a model that shows how winds can create snow patches of varying depth as a result of drifting and blowing of the snow. Hartman et al. (1999) show that wind-driven sublimation of snow must be known in order to properly predict moisture fluxes. Liston (1999) discusses this approach further. Greene et al. (1999) assessed how landscape influences snow-cover depletion and regional weather. Raddatz et al. (2013) presents analyses of the significance of mesoscale variations in heat fluxes from spatially-varying coverage of sea ice.

Procedures to parameterize land-surface heterogeneity from such features as vegetation, snow, and sea ice for use in larger-scale models are discussed, for example, in Zeng and Pielke (1993, 1995a,b), Avissar and Chen (1995), Lynn et al. (1995a,b), Arola (1999), and Liu et al. (1999a).

Overviews of the role of the land surface in weather include Avissar (1995) and Pitman et al. (1999). For more recent papers and books also see Kabat et al. (2004), Cotton and Pielke (2005), Georgescu et al. (2009a, 2011a, 2013), Nair et al. (2011), Pielke et al. (2011), Tuinenburg et al. (2011), Shrestha et al. (2012), and Mahmood et al. (2013a,b). Wu et al. (2009) examines how land-surface heterogeneity produces mesoscale atmospheric dispersion.

13.2.3 Mountain–Valley Winds

In a region with irregular terrain, local wind patterns can develop because of the differential heating between the ground surface and the free atmosphere at the same elevation some distance away. A larger diurnal temperature variation usually occurs at the ground, so that during the day the higher terrain becomes an *elevated heat source*, whereas at night it is an *elevated heat sink*.

Two categories of mountain–valley winds are generally recognized:

1. *slope flow*, and
2. *valley winds*.

These types are easiest to recognize when the prevailing large-scale flow is light. Slope flow refers to cool, dense air flowing down elevated terrain at night, with warm, less dense air moving toward higher elevations during the day. Such air movement is often referred to, respectively, as *nocturnal drainage flow* and the *daytime upslope*. The nocturnal drainage flow is also called a *katabatic wind* (e.g., Manins and Sawford

1979a,b).[3] Manins and Sawford (1979b), for example, found that such drainage winds are three-dimensional phenomena and that a critical gradient Richardson number of about 0.25 is required to maintain mixing between the katabatic and ambient winds. Other studies of drainage flows include those of Andersen (1981), Egger (1981), Horst and Doran (1981), Mahrt (1982), Mahrt and Larsen (1982), McNider (1982), Clements and Nappo (1983), Doran and Horst (1983), Arritt and Pielke (1986), Ye et al. (1989, 1990a), Banta and Gannon (1995), and Winstead and Young (2000). Upslope winds have been studied by Ye et al. (1987, 1990b). The Atmospheric Studies in Complex Terrain (ASCOT) program[4] (e.g., Dickerson and Gudiksen 1980, 1981) studied this type of wind field in detail in the Geysers area of California. Yamada (1981) performed a three-dimensional numerical simulation of the drainage winds in the ASCOT study area, and Lange (1981) reports on the use of a diagnostic[5] wind field model (that of Sherman 1978) to model the movement of tracers in the drainage flow of this valley.

Valley winds, the second category of mountain-valley flow, are up- and down-valley circulations that develop from along-valley horizontal pressure gradients in one segment of a valley, which occur because of the input into that part of a valley by the slope flow of air of a different temperature structure than occurs adjacent to that segment. Since both slope flows and these horizontal pressure gradients along the valley floor must be resolved, three-dimensional models are required to simulate valley winds (e.g., see McNider 1981; he models a valley flow that develops at the exit of a valley). Slope flows generally occur when topographic gradients along the slope are steeper than those found along the valley bottom, hence, slope winds tend to develop more quickly than valley flow.

When slope flows occur but valley winds cannot develop as a result of blocking by the terrain configuration, the valley can be referred to as a *trapping valley*, as suggested by T. McKee of Colorado State University. McKee refers to valleys with a substantial valley flow, by contrast, as *flushing valleys*. Magono et al. (1982) concluded that extremely low temperatures can develop in snow-covered trapped valleys.

During sunny days, slope winds tend to be deeper than at night, as with the sea breeze, because the heating of the ground by the sun is mixed upward effectively by turbulent heat fluxes. At night, radiational cooling predominates if the winds are light and the resultant perturbation flow field is more shallow. Figure 13.8 (reproduced from Mahrer and Pielke 1977b) illustrates the differences in depth and strength of

[3]Some authors distinguish between a drainage flow that is caused by a source of cool air high on a slope and a katabatic flow that is continuously cooled from below as the air sinks. For analytic solutions of downhill flows such a division is useful, however, in the real atmosphere this categorization is generally not distinct. A general discussion of gravitationally forced flows for a number of geophysical phenomena is given in Simpson (1982).

[4]The ASCOT program was designed to develop the technology needed to assess atmospheric properties and the impact of new energy sources on air quality in areas of complex terrain as described by M.H. Dickerson in the Foreword of Orgill (1981), and by Knox et al. (1983:15-19). The report by Orgill presents a review of past meteorological and diffusion work in complex terrain.

[5]Diagnostic models are briefly discussed in Section 13.2.4.

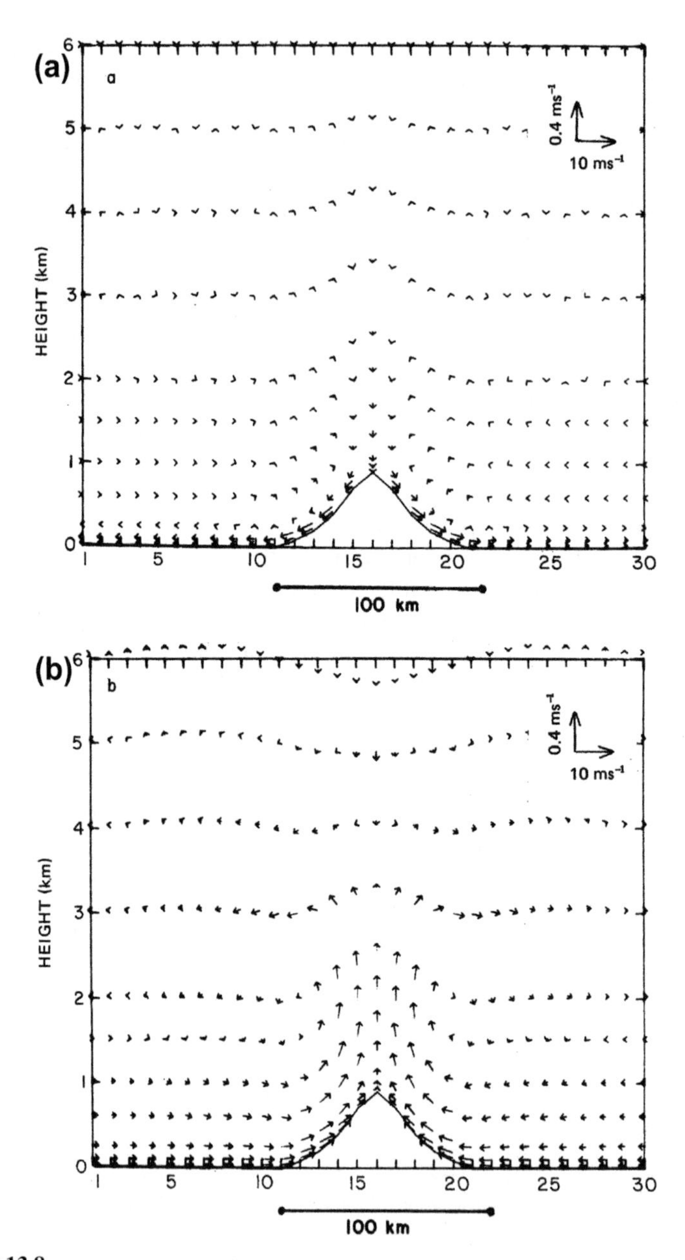

FIGURE 13.8

Two-dimensional simulation of (a) nocturnal drainage flow, and (b) upslope flow with no prevailing synoptic flow. Input condition typical of summer in midlatitudes (from Mahrer and Pielke 1977b).

upslope and downslope winds in a two-dimensional simulation in the absence of a prevailing synoptic flow. These Figures also illustrate that the airflow tends to form a closed circulation, so that if pollutants were continuously released in one segment of the flow, they would tend to accumulate in a region. Such *recirculation* is ignored in the Gaussian plume models commonly used to estimate concentrations of pollutants in irregular terrain. Kossmann et al. (1999) has shown, using tracer data, that discontinuities in the boundary-layer height associated with mountainous terrain and its associated mesoscale flow can enhance the transfer of boundary-layer air into the free atmosphere.

The diurnal evolution of the planetary boundary layer in mountainous topography, however, is more complicated than observed over flat terrain. Whiteman (1982), and Whiteman and McKee (1978, 1982), for example, discussed the breakup of temperature inversions within deep valleys because of upward heat flux after sunrise and the sinking of the inversion layer over the valley center as upslope flows develop along the valley walls. When the inversion height becomes low enough, turbulent mixing resulting from the heating of the ground eliminates the inversion and a relatively deep boundary layer is produced [over Colorado during the summer, McKee (1982, personal communication) reported observed mixed layers as high as 5 km or so above the surface]. Before the inversion is eliminated, however, enhanced air pollution can occur since the vertical depth of mixing of an effluent becomes more limited.

When a large-scale flow (often including a vertical shear of the horizontal wind), variable surface characteristics, and/or three-dimensional topographic features are present, the resultant mesoscale flow field can become even more complex. Without extensive observations or an accurate three-dimensional mesoscale model, it is generally impossible to anticipate the details of the diurnal variations in the wind field. An example of a study of three-dimensional mountain-valley type flow patterns is that of Hughes (1978).

Wipperman and Gross (1981) have used a two-dimensional mesoscale model to construct a wind rose in irregular terrain for stable atmospheric stratification at Mannheim, West Germany, utilizing 12 computations with different geostrophic wind directions and speeds. The large-scale winds were taken from averaged synoptic values over a 5-year period. In their study, they concluded that a nonhydrostatic version provided a somewhat better replication of the wind rose than the hydrostatic version, where their horizontal grid spacing was 2 km. Wipperman and Gross, however, did not explain the reason for the difference between the two models in which the assumed stable stratification in the lowest 1800 m ($\partial\bar{\theta}/\partial z = 0.6°\text{C}/100$ m) would be expected to minimize vertical accelerations (e.g., see Fig. 5.6).

Modeling simulations of mountain-valley winds include those of McNider (1981), Mannouji (1982), Kimura and Kuwagata (1995), Kuwagata and Kimura (1997), and Chase et al. (1999), and observational studies of this atmospheric feature are represented by MacHattie (1968), Wooldridge and Orgill (1978), George (1979), Banta and Cotton (1981), Broder et al. (1981), Ohata et al. (1981), Banta (1982, 1984), Hootman and Blumen (1983), and DeWekker et al. (1998). Whiteman (1980), Bader (1981), and Whiteman et al. (1999) have studied the breakup of temperature inversions in Colorado mountain valleys after sunrise. Also in Colorado, the South Park Area Cumulus Experiment (SPACE, Danielson and Cotton 1977; Cotton et al. 1982a; Knupp and Cotton 1982a,b) was an investigation of the influence of mountain winds on cumulus cloud development. Nair et al. (1997) simulated the development of cumulus cloud convection over the Black Hills of South Dakota. Poulos and Bossart (1995) modeled dispersion within complex terrain. Doran and Zhong (2000) investigated thermally-driven gap winds in the Mexico City area.

As an example of results from these studies, MacHattie (1968) found that the synoptic wind was most strongly coupled to the flow in the direction of the main valley. The diurnal perturbation was, therefore, reduced more in that direction than it was normal to the valley axis. Also, the diurnal variation of the wind was observed to be less well defined on days with intense solar radiation because the strong heating was effective in developing a deep planetary boundary layer, which enhanced mixing of the gradient wind down to the ground.

Whiteman and Doran (1993) illustrates the influence of terrain and the horizontal pressure gradient force orientation on the diurnal variation of the wind in complex terrain. They illustrate very effectively, as shown in Fig. 13.9, the differences in the

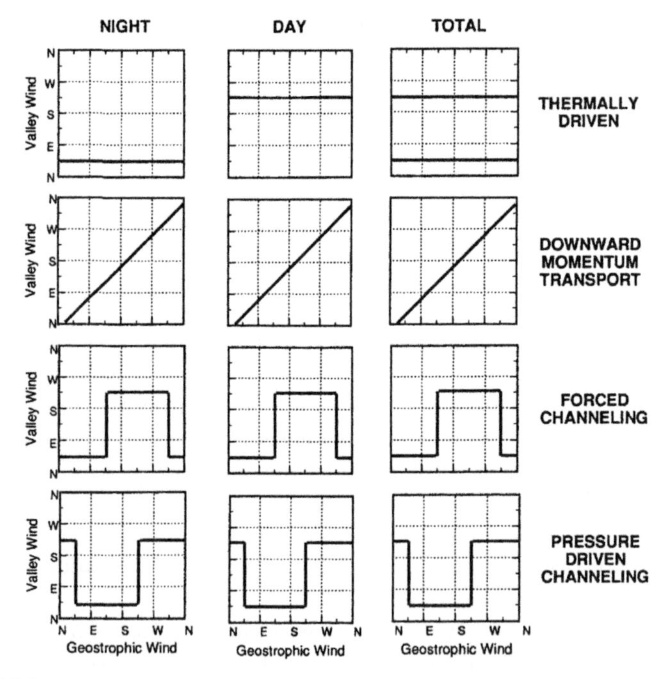

FIGURE 13.9

Relationships between above-valley (geostrophic) and valley wind directions for four possible forcing mechanisms: thermal forcing, downward momentum transport, forced channeling, and pressure-driven channeling. The valley is assumed to run from northeast to southwest (from Whiteman and Doran 1993).

diurnal variation of the valley wind direction as a function of the wind direction above the valley for four distinct mechanisms which can control the wind direction. *Thermally-driven winds* are independent of the above-valley winds and are solely controlled by locally-developed along-valley pressure gradients. *Downward momentum transport* of the above-valley winds (such as associated with a deep convective boundary layer) would produce similar wind directions at all levels. *Forced channeling* occurs when the valley flow aligns itself as to whether the above-valley flow has a net flow down- or up-valley. A sharp transition occurs within the valley when this net flow changes from a down- or up-valley direction. The *pressure-driven channeling* (which is 90° out of phase with the forced channeling) occurs when the winds in the valley respond only to the large-scale horizontal pressure, and not to winds that occur above the valley. The valley winds blow directly toward low pressure synoptic systems in this case. Pressure-driven channeling should be important when there is a lack of downward momentum transport (such as when the valley is stably stratified) and thermal mesoscale wind effects are weak.

A stable surface layer that persists throughout the day can produce a less defined mesoscale system. Ohata et al. (1981), for instance, documented the major influence of snow cover in minimizing the strength of the local mountain-valley wind circulation in Nepal.

Weaver and Kelly (1982) documented that mountain ranges in Colorado are preferred locations for thunderstorm development during the summer, and Johnson and Toth (1982a,b), and Smith and McKee (1983) have shown the influence of topography in northeastern Colorado on controlling the local diurnal wind field. Johnson and Toth found that near the Front Range of the Rockies, upslope winds tend to dominate during most of the day with downslope winds at night and later in the afternoon. The downslope winds before sunset apparently result from the entrainment of synoptic westerly winds aloft down to lower levels, as well as from downdrafts initiated by cumulus convection over the higher terrain to the west. This preference for westerly flow of air throughout the troposphere in the late afternoon causes cumulonimbus convection to propagate toward the east at the time of their maximum development.

Holroyd (1982) summarized climatologically the occurrence and movement of mountain-generated cumulonimbus rainfall over the northern Great Plains during the May-July period. In this study, the maximum rainfall from these systems was found about 400 km northeast of the eastern boundary of the Rockies. Cotton et al. (1983), Wetzel et al. (1983), McAnelly and Cotton (1986), Tripoli and Cotton (1989a,b), Tremback (1990), Nachamkin and Cotton (2000), and Nachamkin et al. (2000) have investigated indepth these mountain-generated thunderstorm complexes.

The observed monthly mean temperature distribution in areas of complex topography is also strongly dependent on terrain as illustrated in Fig. 13.10 for central Virginia.

Sea- and land-breeze circulations interacting with mountain-valley systems have also been studied (e.g., Doran and Neumann 1977; Mahrer and Pielke 1977a,b; Ookouchi et al. 1978; Garrett 1980; Alpert et al. 1982; Segal et al. 1982b; Millán et al. 1987, 1996; Salvador et al. 1997; Alonso et al. 2000; Gangoiti et al. 2001, 2002).

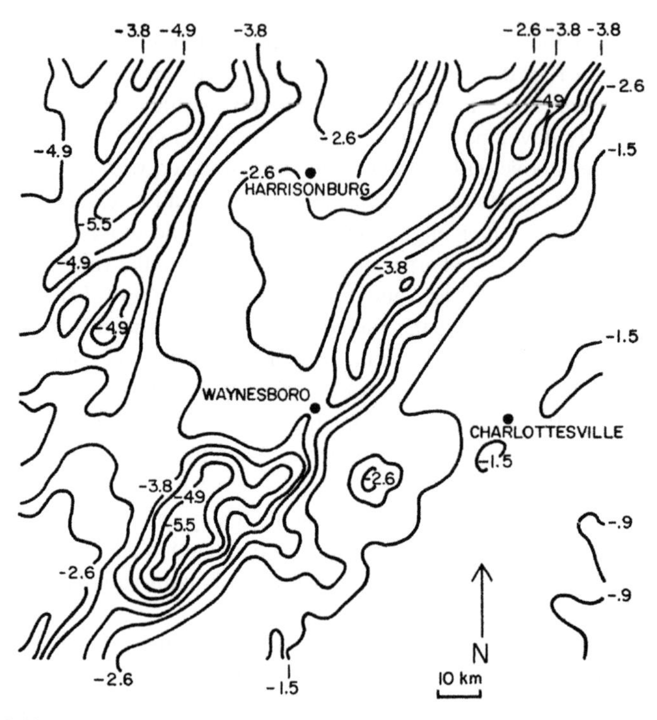

FIGURE 13.10

The estimated mean temperatures in degrees Celsius at Stevensen screen height (~2 m) over central Virginia in January 1961 (from Pielke and Mehring 1977).

Such interactions can be very complex and, as shown by Segal et al. (1983b), are not simply a superposition of the two different phenomena. Rather, mountains along coastal regions, acting as elevated heat sources, create subsidence over and just inland from the coastal waters, thereby influencing the intensity and distribution of the sea breeze. Mass (1982) illustrated how such a terrain configuration can influence the diurnal pattern of rainfall in the Puget Sound region of Washington State.

Other related studies include that of Asai and Mitsumoto (1978) who investigated the influence of slope on sea and land breezes with a linear and nonlinear representation, and of Kikuchi et al. (1981) who used a three-dimensional model to examine the importance of elevated terrain along the coast of Japan around Tokyo in the evolution of the sea breeze. Sahashi (1981) reported on the use of a two-dimensional nonhydrostatic model to study sea breezes over irregular terrain. A schematic of the onshore-offshore diurnal wind pattern over the island of Hawaii is reproduced from Garrett (1980) in Fig. 13.11. In Greenland, Gryning and Lyck (1983) discuss the possible interplay of a drainage wind and a sea breeze along the coast in influencing the transport and diffusion of tracer material. Regional-scale mountain wind circulations are reported by Bossert and Cotton (1994a,b).

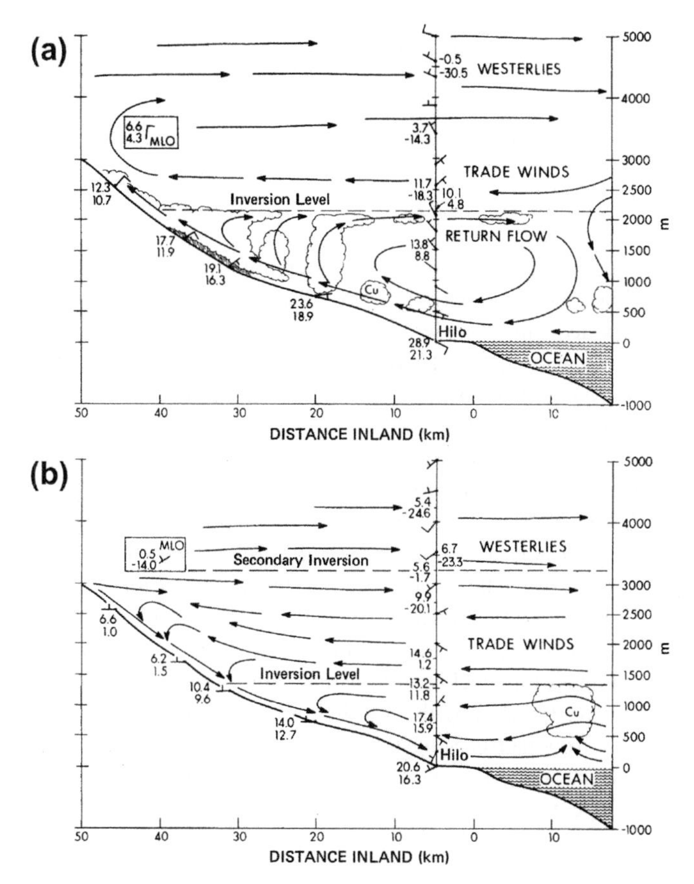

FIGURE 13.11

An east-west cross section of the onshore-offshore flow pattern over the island of Hawaii (a) during midafternoon, and (b) in the early morning. Temperature (°C), dewpoint temperature (°C), and wind velocity (using wind barbs) are from observations plotted together with the schematic flow field (from Garrett 1980).

For other studies of mountain-valley winds see Whiteman (2000), Whiteman et al. (2010), Colle et al. (2012), Lehner and Whiteman (2012; 2013), Chrust et al. (2013), Lareau et al. (2013), and Zardi and Whiteman (2013).

13.2.4 Forced Airflow Over and Around Rough Terrain

When air flows over terrain features that have horizontal scales of 25-100 km or so, another type of mesoscale system develops. This atmospheric feature is different from the sea and land breezes, and mountain-valley winds because forced ascent of air in a prevailing stably-stratified air mass, rather than differential heating of the ground by the sun, generates the mesoscale perturbation. The intensity of this mesoscale system is directly proportional to the pressure gradient generated by this forced movement of air.

Since the pressure gradient force (of the form

$$-\bar{\theta}\,\partial\bar{\pi}/\partial\tilde{x}^1 + g\left(\left(\sigma - s\right)/s\right)\partial z_G/\partial x + f\bar{\tilde{u}}^2,$$

for example, as obtained from Eq. 6.84) is of such importance in the evolution of this type of flow and because it is approximately a linear term, exact analytic wave solutions have been applied with considerable success. Early investigators who used exact solutions include Queney (1947, 1948), Scorer (1949), Eliassen and Palm (1960), Covez (1971), and Vergeiner (1971). Eliassen and Palm (1960) found that, depending on wavelength, 65-100% of the wave energy generated as airflow is forced over mountains could be reflected downward from layers of strong wind in the upper troposphere. An analytic study is presented in Kumar et al. (1998).

Klemp and Lilly (1975) have had some success at applying a linear model to estimate the occurrence or non-occurrence of extreme downslope winds in the lee of the Colorado Rockies, and Sangster (1977) tested a statistical procedure to forecast these winds using observed synoptic information and parameters derived from linear theory. Klemp and Lilly found that the maximum downslope winds occur when an inversion is present near mountaintop level upstream, and if the temperature and wind profiles are such that the wave induced by the terrain approximately reverses phase between the surface and the tropopause. Sangster determined that temperature differences in the vertical and strong westerly winds at 700 mb were important parameters in causing strong downslope winds, although in contrast with Klemp and Lilly's result such information as the vertical wavelength, Scorer parameter (defined subsequently), and the presence of an inversion were not. Extending Klemp and Lilly's work, Hyun and Kim (1979) gave another example of a linear two-dimensional model of this type of mesoscale system.

Three-dimensional linear models (e.g., Blumen and McGregor 1976; Somieski 1981) provide guidance as to what fraction of the airstream goes around topographic barriers and how much advects over it as a function of factors such as the thermodynamic stability. When air can neither go over nor go around because it is too stable, and the terrain feature too elongated, the influence of the mountain propagates rapidly upwind – a process called *blocking*. Baker (1971) and Richwien (1978, 1980) have given examples of such blocking by segments of the Appalachian Mountains in the eastern United States, and Schwerdtfeger (1974) and Kozo (1982c) described the effect of blocking by the Brooks Range in Alaska on the local wind field along the Beaufort sea coast. Schwerdtfeger (1975) discussed the influence of blocking due to a peninsula in Antarctica, while Wesley and Pielke (1990) describe how blocking east of the Front Range of the Colorado Rockies can produce convergence zones which locally enhance precipitation. Reason et al. (1999) studied and simulated with a mesoscale model, the development of propagating mesoscale systems along the coast of southeast Australia that are associated to some extent with topography.

Smith (1982a) suggests this blocking causes a deformation of the potential temperature surfaces such as to create an unstable layer upstream of the mountain. Parrish (1982), using observational and modeling results concluded that low-level, mountain parallel jets can form during the winter in the Sierra Nevadas of California as a result of blocking. Over the Kanto plains of Japan, Harada (1981) also suggested that mountains to the west may play a role in the generation of low-level jets.

Manins and Sawford (1982), using observational data from a small valley in southeast Australia, found that blocking occurred when $F_r \lesssim 1.6$, where F_r is called the *Froude number* and is defined by

$$F_r = \overline{V}_g \left(z^2_{G_{max}} g \frac{\partial \theta_0}{\partial z} \Big/ \theta_0 \right)^{-1/2},$$

$\overline{V}_g$ is the ridge-perpendicular, large-scale wind speed above the maximum terrain $z_{G_{max}}$, and θ_0 the large-scale potential temperature. When F_r was greater than about 1.6, Manins and Sawford found that the air within the valley became coupled with the large-scale flow and was flushed out. It would be expected that this critical value of F_r would vary for different terrain configurations. The critical Froude number, as discussed by Manins and Sawford, represents the relative magnitudes of kinetic energy of the large-scale wind to the potential energy change needed to move an air parcel near the surface over a terrain barrier. Lyons and Steedman (1981) found a critical value of $F_r = 1.5$ for a shallow valley in western Australia.

Linear theory predicts that the vertical wavelength of lee waves induced by a single ridge is given

$$L_z = 2\pi / S_0^{1/2} = 2\pi\, \overline{V}_g / \left[(g/\theta_0)\, (\partial\theta_0/\partial z) \right]^{1/2},$$

where S_0 is called the *Scorer parameter* (e.g., Alaka 1960; Anthes and Warner 1978). According to linear theory for well-developed waves to develop (as listed by Anthes and Warner) the Scorer parameter must be less in the upper troposphere than at lower levels. This requires that if $\partial\theta_0/\partial z$ is constant, $\overline{V}_g$ must increase with height, whereas if $\overline{V}_g$ is a constant, $\partial\theta_0/\partial z$ must be less stable in the higher levels. According to linear theory, in the absence of the Coriolis effect two types of wave motions are induced as air flows over rough terrain – the *forced wave*, which is collocated with the underlying topography, and the *lee wave*, which propagates downstream. Trapped lee waves (i.e., which propagate indefinitely downstream in the absence of friction but which decay in amplitude rapidly with height) are a commonly seen type of air motion to the lee of mountain barriers when S_0 decreases rapidly with height. Only the forced wave is realistically simulated in a hydrostatic model as evident from the reviews of Smith (1979), and Klemp and Lilly (1980).

The use of nonlinear models to simulate the airflow over mountains originated with Hovermale (1965), who felt that the large perturbation velocities observed in actual mountain flows violated the requirements of linear theory in which the products of perturbations must be small. Nonlinear studies have continued with the work of such investigators as Furukawa (1973), Anthes and Warner (1974),

Gal-Chen and Somerville (1975b), Deaven (1976), Clark (1977), Clark and Peltier (1977), Klemp and Lilly (1978), Mahrer and Pielke (1978b), Peltier and Clark (1979, 1983), Seaman and Anthes (1981), Seaman (1982), Arritt et al. (1987), Poulos and Pielke (1994), Sun and Chern (1994), Pinty et al. (1995), Snook and Pielke (1995), Kang et al. (1998), Papineau (1998), and Mayr and Gohm (2000). Peltier and Clark (1979, 1983), for example, disagree with Klemp and Lilly's (1975) explanation for strong downslope wind events and suggest that downward reflection of energy from breaking waves in the stratosphere is the primary mechanism. Poulos et al. (2000) investigated the interaction between large-scale airflow over the Front Range mountains of Colorado and nocturnal drainage flow. Figure 13.12 illustrates a simulation by Klemp and Lilly (1978) for a particular windstorm in Colorado on 11 January 1972, a study day also studied by Mahrer and Pielke (1978b), and Peltier and Clark (1979).

Other geographic areas have also been studied. Clark and Gall (1982) have provided observational comparison of several model predicted and observed parameters at different levels over the Elk Mountain region of Wyoming. Seaman (1982) also performed a model prediction of the airflow over this terrain feature.

Peltier and Clark imply that a nonhydrostatic model is necessary to simulate the windstorm properly on this day, whereas Klemp and Lilly, and Mahrer and Pielke claim a hydrostatic representation is adequate. The scale analysis introduced previously in this book (e.g., Eq. 5.59 and following) indicates that the hydrostatic formulation is adequate in representing this windstorm; however, additional quantitative experiments are needed to settle the issue conclusively. An exchange of correspondences regarding the different mechanisms for the generation of downslope wind storms is given by Lilly and Klemp (1980), and Peltier and Clark (1980). Durran (1986) describes that these wind storms behave as a generalized hydraulic jump, which is apparently the reason that Mahrer and Pielke (1978b) were successful in producing a simulated wind storm even though the diffusive forward upstream differencing scheme was used to represent advection.

When the Coriolis effect and boundary layer dynamics are included, the response of the atmosphere to terrain is more complex. As shown by Kessler and Pielke (1982), air that becomes ageostrophic after passing over one ridge does not adjust again to equilibrium for a long distance downstream. Therefore, if a second ridge is situated a short distance downstream, its upstream wind profile would be markedly different from that obtained if the Coriolis effect were not included. Smith's (1982b) results support this conclusion. Using a linear model, he demonstrated that although the pressure and vertical motion fields over mesoscale sized mountains are unaffected by the Coriolis force, the horizontal trajectories are altered, with a significant ageostrophic component being produced.

Kessler and Pielke also suggested that net boundary layer warming could occur downwind of a second ridge relative to the first, even in the absence of the Coriolis effect, because of the enhanced mixing of potentially warmer air downward as it accelerates over the upstream ridge. This net warming can occur in the absence of precipitation on the upwind side of the mountain barrier.

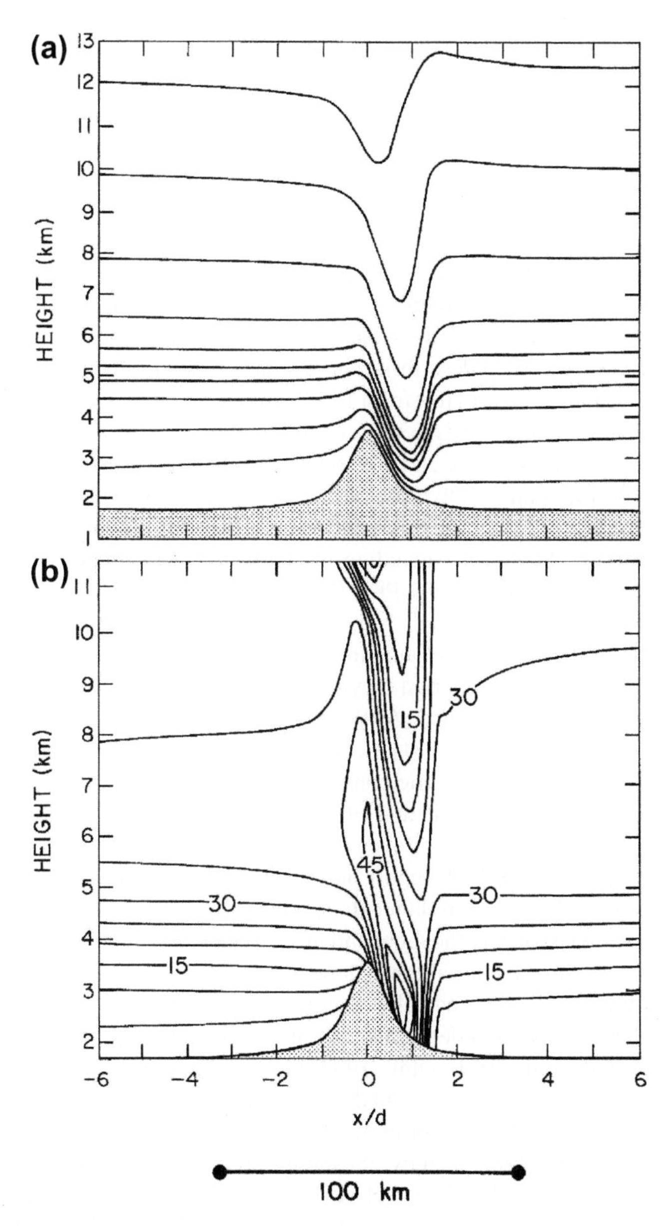

FIGURE 13.12

(a) The predicted potential temperature cross section, and (b) horizontal wind field in m s^{-1} for a simulation of the 11 January 1972 windstorm along the east slopes of the Colorado Rockies (from Klemp and Lilly 1978).

Lee et al. (1989) discuss how a cold pool of air downstream from a mountain barrier inhibits strong downslope wind flow. They find that the development of a large amplitude mountain wave is inhibited. Based on observations and modeling results, they concluded that in the absence of significant surface heating, a large-scale horizontal surface pressure gradient directed away from the mountain must be present to remove the cold pool before the downslope winds can reach the surface.

Observational studies of strong airflow over rough terrain include those of Lilly and Zipser (1972), Lilly and Kennedy (1973), Brinkmann (1974), Hoinka (1980), Lilly et al. (1982), and Zipser and Bedard (1982). Lilly and Zipser, for example, observed wind gusts of 166-200 km h^{-1} associated with a chinook immediately downwind of the Rockies. Reed (1981) described a case study of downslope winds with gusts to around 100 km h^{-1} downwind from low sections of the Cascade Mountains of Washington. Under lighter winds in this geographic area, Mass (1981) reported on a zone of preferential convergence and precipitation in the Puget Sound area associated with the wind flow around the Olympic Mountains. Walter and Overland (1982) have discussed theoretically and observationally several different synoptic flows over and around the Olympic Peninsula, including the damaging winds associated with the Hood Canal Bridge disaster. The ALPEX (Kuettner 1986; Pichler et al. 1995; Alpert et al. 1996a) and PYREX (Bougeault et al. 1990, 1997) field campaigns were designed to study airflow associated with mountain barriers in Europe.

In Europe, Pettré (1982) reported on violent winds associated with forced airflow down the Rhône Valley in France. At the top of Sierra Grande Mountain in New Mexico, Barnett and Reynolds (1981) measured winds for a 6 month period to assess the potential for wind energy electric conversion systems. A report by the Centre for Advanced Studies in Atmospheric and Fluids Science (1983) summarize studies of the influence of strong wind flows over the Himalayas on the local wind fields in India, and Arakawa et al. (1982) report on an observational study of forced airflow over rough terrain in Japan. In this latter study, the wind to the lee of the terrain barrier is found to be strongest during the night and morning at which time the synoptic flow, the land breeze, and the nocturnal downslope winds are superimposed.

Over many mountainous and hilly regions of the world, this forced lifting on the upwind slopes causes condensation and/or sublimation and precipitation (e.g., Marwitz 1983; Passarelli and Boehme 1983) and is an important factor in the local water budget. Blocking and the resultant deformation of the upstream isentropes could also create regions of convective instability even if no such instability were present in the upwind synoptic flow, thereby enhancing precipitation immediately upwind of the mountains (Smith 1982a), as well as increasing the spatial irregularity of the precipitation (Gocho 1982). The merger of cumulus clouds in the blocked flow may also increase precipitation amounts as suggested by the work of Sakakibara (1981). Huge snow packs of over 800 cm, for instance, occur in the San Juan Mountains of southwestern Colorado in large part because of this effect.

Because of the increase in potential temperature that results from the release of latent heat (and entrainment of potentially warmer air from above the planetary boundary layer), comparatively dry, and even arid regions often occur in the lee of

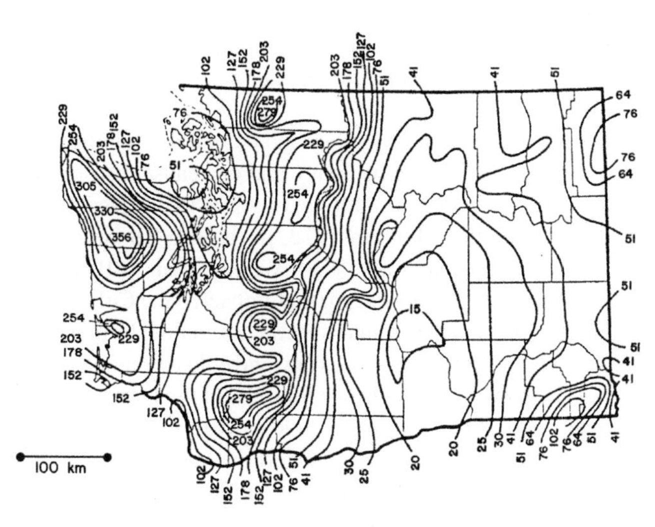

FIGURE 13.13

The annual rainfall (in centimeters) over Washington State (from Climate and Man 1941).

mountains, particularly when the prevailing flow is persistently from one direction.[6] As suggested by Smith and Lin (1982) from a linearized model result, this latent heat release also alters the structure of wave motions as air is forced over rough terrain.

Figure 13.13 illustrates the annual rainfall pattern in Washington State, which is controlled in large measure by the distribution of terrain relative to the prevailing, generally southwest synoptic flow during the wet season. Figure 7.11 (from the two-dimensional results of Colton 1976) illustrates an example of the predicted and observed orographic rainfall pattern over the Sierra Nevadas of California, with precipitation confined to the windward slope. Over south Wales in Great Britain, Hill et al. (1981) documented observationally the average enhancement by a factor of three of frontal rainfall during southwest wind flow by even hills of modest height (maximum elevations of 600 m).

Simulations of clouds or precipitation, or both, over rough terrain using three-dimensional models include those of Chappell et al. (1978) for the San Juan Mountains, Nickerson and Magaziner (1976), and Nickerson (1979) for the island of Hawaii, and Abbs and Pielke (1987), and Snook and Pielke (1995) for Colorado. Lavoie (1974) presented a one-layer simulation of Oahu in Hawaii and Chang (1970) applied Lavoie's model to the Black Hills of South Dakota. Raddatz and Khandekar (1979) also successfully applied Lavoie's model to the western plains of Canada using a 47.6 km horizontal grid. Hutchison (1995, 1998a,b) discussed the coherence of

[6]The warm wind that occurs to the lee of mountains as a result of such forced ascent and descent is called, for example, a *chinook* in North American and a *föhn* in Germany. When cold advection is occurring, strong downslope winds are called a *bora*.

weather that occurs as a result of the presence of topographic features, which permits an interpolation of weather information to smaller scales than the model-simulated resolution.

When the atmosphere is particularly moist and potential instability is released, the resultant rains over rough terrain can be heavy and can cause disastrous flash floods such as occurred in Fort Collins, Colorado in July 1998 (Petersen et al. 1999), and over the Black Hills of South Dakota in 1972, in the Big Thompson watershed in Colorado in 1976 (e.g., Caracena et al. 1979). Wesley et al. (2013) reported on the extraordinary 16-20 March 2003 snowstorm that impacted the Colorado Front Range and some surrounding areas. They found several anomalously high local snowfall gradients associated with upslope flow, blocking, and cold air damming. Snowfall that exceeded 2.5 meters in a few locations in the foothills of the mountains occurred with this storm. Mesoscale models provide an effective tool to explain these extreme events.

Accurate simulations of airflow over rough terrain when precipitation and cloudiness occur must not only properly represent the complex terrain but also the dynamic and thermodynamic changes caused by the phase transformations of water. Hill (1978), for example, found that circulation cells are formed over mountain areas by the precipitation itself, and Reid et al. (1976) determined that cloud shadowing over irregular terrain also affects the intensity of the airflow over mountains. Fraser et al. (1973) and Hobbs et al. (1973) described a diagnostic two-dimensional simulation of the airflow over the Cascades of Washington State in which a detailed description of the cloud and precipitation microphysics is included. Gocho (1978) used a two-dimensional linear steady-state model to investigate the influence of microphysical processes on rainfall over the Suzuka Mountains in Japan. Cotton et al. (1982b), and Meyers and Cotton (1992) applied an ice-phase parameterization to a two-dimensional model simulation of stable wintertime orographic flow.

Durran (1981), Durran and Klemp (1982b), and Kessler and Pielke (1983), in a confirmation of the Smith and Lin (1982) conclusions, have shown that latent heat release can substantially alter the structure of internal waves over mountainous terrain. Kessler and Pielke found that the release of heat of condensation over a ridge top results in a more symmetric wind field over the mountain than when no phase change of water occurs.

Simulations of wind flow over rough terrain using a form of dynamic initialization or objective analysis include those reported in Collier (1975, 1977), Fosberg et al. (1976), Danard (1977), Rhea (1977), Bell (1978), Dickerson (1978), Sherman (1978), and Patnack et al. (1983). Ludwig and Byrd (1980) outlined what they claim to be a particularly efficient procedure to compute mass consistent flow fields from wind observations in rough terrain. Such models are called *diagnostic*, even if the conservation relations are used, because they are not used to forecast forward-in-time through the integration of the conservation relations.

Diagnostic models are very economical and appear to be effective mesoscale analysis tools when

1. the dominate forcing is the terrain,

2. below the highest terrain heights a strong, well-defined inversion exists at the top of the planetary boundary layer, and
3. sufficient observational data are available to input to the analysis.

Whiteman (2000) summarizes studies of forced airflow over and around rough terrain.

Recent papers include Doyle and Durran (2002), Epifanio and Durran (2002), and Chen et al. (2005, 2007, 2007). Chen et al. (2005) show the assumption that the mountain-wave momentum flux will be constant with height (away from critical layers) in steady inviscid flow with perturbations dropping to zero downstream (thereby eliminating the trapped wave cases), does not hold in flows that are almost steady (changing on a 2-day time scale). Doyle and Durran (2007) show that in three dimensions, rotors break down into more chaotic substructures, so one should not expect them to consist of one giant coherent swirling horizontal tube of air. Epifanio and Durran (2002) reported that vertical stretching plays a key role in intensifying the vertical vorticity in lee vortices. This extends the early work of Smolarkiewicz and Rotunno (1989) documenting the development of vertical vorticity in the lee of isolated tall mountains by the baroclinic generation of horizontal vorticity and subsequent tilting of that vorticity into the vertical. According to Epifanio and Durran (2002), the Smolarkiewicz and Rotunno mechanism yields vertical vorticity values that are too weak without the help of vertical stretching.

Recent papers also include Kirshbaum and Durran (2005) and Kirshbaum et al. (2007). Kirshbaum and Durran present observations and modeling showing that relatively modest terrain features can trigger stationary bands aligned with large-scale winds. Kirshbaum et al. (2007) provide an analysis of how lee waves trigger cumulus convective bands associated with small-scale topographic obstacles. Whether a topographic obstacle in this region is able to trigger a strong rainband depends on the phase of its lee wave at cloud entry. Convective growth only occurs downstream of obstacles that give rise to lee-wave-induced displacements that create positive vertical velocity anomalies and nearly zero buoyancy anomalies at the point where air parcels undergo saturation.

13.2.5 Urban Circulations

Urban circulations are similar to sea and land breezes, and mountain-valley winds in that it is the differential heating and cooling between the rural and urban areas that generates and sustains the wind system. Over the cities and suburban areas, such alterations as asphalting, buildings, and the removal of vegetation have markedly altered the surface heat budget, and therefore, the intensity of heat flux to the air.

The influence of these urban areas on the local weather pattern has received increased attention as the areal extent of such regions expand, and as we begin to realize the major influence of industrial and populated areas on climate, and on human health and well being. It is in the study of urban circulations that Eq. 4.26 becomes an important component in the conservation laws relevant to mesoscale atmospheric flows. As reported by Pielke (1978), an estimated 15,000 deaths per year in the United

States are due to air pollution. This number exceeds the annual average number of fatalities of all the remaining weather-related hazards combined.

Health effects from poor air quality occur throughout the worlds. In Italy during July 1976, for example, the accidental venting of the highly toxic organic compound called dioxin (2, 3, 7, 8-tetrachlorodibenzo-*p*-dioxin) from a factory and its transport and dispersal by the local flow field caused death to farm animals and sickness to people and forced the permanent evacuation of individuals from their homes (Science News 1976, Fuller 1978). Seinfeld (1975) summarized the effects of air pollution on human health, as known up to that time.

The impact of anthropogenic gases and aerosol contaminants in the atmosphere demands greater complexity in mesoscale models since the number of interactions is greater. In addition, Gaussian plume simulations, as originally proposed by Pasquill (1961) and used by the Environmental Protection Agency (e.g., Turner 1969), are inaccurate representations of pollutant distributions when such phenomena as the recirculation of the urban air occurs. Calder (1977), for example, in a serious over-simplification of urban meteorology, assumes a single wind speed and direction are representative of an entire city area for 1 h periods in his multiple-source plume model formulation. van Egmond and Onderdelinden (1981), however, found that even with 108 stations monitoring SO_2 over a $15 \times 220\ km^2$ area in Holland, relative errors occur of 20% in the construction of analyzed fields of this pollutant. The errors are due to measurement errors and small-scale influences of local sources (including the local wind and turbulence fields).

Urban models have tended to evolve separately in the areas of air chemistry and meteorology. In the former case, models have treated detailed chemical interaction (e.g., Peterson 1970; Appel et al. 1978; Kowalczyk et al. 1978; Brewer et al. 1981; Seigneur 1994) but have neglected to adequately handle the mesoscale dynamics. Such models are often called *Box models* (e.g., Schere and Demerjian 1978). Surveys of the knowledge of atmospheric chemistry are given by Hales (1975), McEwan and Phillips (1975), and Heicklen (1976). EPA (1980) provides a summary of selected photochemical grid models for use over urban areas.

Mesoscale meteorological models have been used to estimate the transport and dispersion of pollutants. McNider (1981), for example, simulated the movement of effluent under drainage wind flow. McRae et al. (1982b) used the advection-diffusion equation of the form given by Eq. 4.26, including surface removal processes and parameterized photochemistry, to estimate urban air pollution. Other work to formulate transport and dispersion representations of pollutants for use in three-dimensional models includes that of Yamada (1977), and Uliasz et al. (1996). Sheih (1977) and others have considered the influence of thermal coagulation and gravitation on pollution concentrations. Hane (1978) simulated the wet deposition of pollutants over St. Louis, Missouri using a two-dimensional squall line model.

In general, however, the meteorological observations and simulations have concentrated on the effects of an urban area on the wind, temperature, and moisture fields, rather than chemical interactions. Loose and Bornstein (1977), for instance, investigated the influence of New York City on the synoptic flow and observed that when a heat island was well developed, synoptic fronts decelerated over the upwind-half of the city, and accelerated over its downwind-half in response to the higher surface roughness of the urban area. Bornstein, as Adviser, directed a study of the influence of New York City on the sea breeze (Anderson 1979; Fontana 1979; Thompson 1979) and further illustrated the large drag effect of the buildings in this urban area. Bornstein and Thompson (1981) described the influence of this wind retardation on sulfur dioxide concentrations in New York City. Among their results they found decreased concentrations near the coast, but larger concentrations on the downwind side of the city associated with the inland passage of the sea breeze. Such an observation is explained as the contamination of relatively clean, onshore flow as it traverses the city.

The St. Louis area has been studied extensively as part of the METROMEX program (see Project METROMEX 1976 for a summary and the May 1978 issue of the *Journal of Applied Meteorology* for a series of articles with results from this program). Vukovich et al. (1976), for example, found from a mesoscale model that the urban effect of St. Louis depends on wind direction, when the synoptic wind is above a certain threshold. From the observational data, Oochs and Johnson (1980) concluded that both radar echo tops and bases were lower over St. Louis compared to their rural counterparts. They attributed this difference to weaker updrafts in urban clouds. Changnon (1982) reported on the substantial reduction in visibility over this city due to locally generated pollutants. Shreffler (1982) discussed the variability of winds over the St. Louis area and its influence on short-term air quality predictions, while Ching et al. (1983) described the relative importance of vertical heat flux and horizontal advection over this city. Other mesoscale modeling studies of this urban area include those of Vukovich et al. (1979), Vukovich and King (1980), and Hjelmfelt (1980, 1982).

Changnon (1980) found results over Chicago, Illinois to be similar to those observed over St. Louis, although the rainfall increase due to the urban area was less, a result he attributed to the proximity of Lake Michigan. He observed a 15% increase of rainfall over central Chicago compared with the surrounding rural areas. Fujita and Wakimoto (1982) illustrated a reduction of mean wind speeds over Chicago by over one-half as compared with the open terrain to the west of the city.

Additional observational and theoretical urban studies include those of Anderson (1971), Olfe and Lee (1971), Oke and Fuggle (1972), Lee and Olfe (1974), Taylor (1974), Yap and Oke (1974), Fuggle and Oke (1976), Nunez and Oke (1976), Oke (1976), Loose and Bornstein (1977), Sawai (1978), Sisterson and Dirks (1978), Shreffler (1979), Goodin et al. (1980), Goldreich et al. (1981), Leduc et al. (1981), Tapper et al. (1981), Sorbjan and Uliasz (1982), Yonetani (1983), Ulrickson and Mass (1990), Kallos et al. (1993), Pilinis et al. (1993), Sailor (1995), Tso (1995), Banta et al. (1998), Grimmond and Oke (1999), Hafner and Kidder (1999), Ichinose et al. (1999), Kallo and Owen (1999), Philandras et al. (1999), Bornstein and Lin (2000), Jazcilevich et al. (2000), Kanda and Inoue (2001), and Sharan et al. (2000). Air pollution studies in India are reported in Centre for Advanced Studies in Atmospheric and Fluids Science (1983), while Kotroni et al. (1999a) simulate an air pollution episode in Athens, Greece. Air pollution studies using mesoscale models also include Kallos et al. (1998), Cautenet et al. (1999), Varinou et al. (1999), Yamada (1999), Seaman and Michelson (2000), Warner and Sheu (2000), Seigel and van den Heever (2012a), and Saleeby and van den Heever (2013). Also, in Section 13.4, presents examples of the application of the use of mesoscale models for air quality studies.

Arritt et al. (1988) used a mesoscale model to investigate the spatial variations of the deposition velocity of sulfur dioxide resulting from mesoscale flow. Taha (1999) and Taha et al. (1999) discuss the importance of urban heat storage on mesoscale weather.

Among the results of these types of studies, Oke (1973) determined from an observational study that the heat island effect of a city on its surroundings under cloudless skies is inversely proportional to the large-scale wind speed and directly related to the logarithm of the population. The heat island effect is also apparently dependent on the culture and age of settlement. Oke (1982:Fig. 3), for instance, found less heat island intensity in Europe than in North America for the same population size. Over Australia, Manton and Ayers (1982) reported that aerosol production is proportional to the town population and has a rate of input of 8×10^{13} s^{-1}per person.

Nkemdirim (1980) found for Calgary, Alberta, Canada that the urban heat island intensity is directly proportional to the magnitude of the stable lapse rate and inversely proportional to the wind speed at the upwind edge of the city. Large-scale wind speeds greater than 15 m s^{-1} or so essentially eliminated the heat island effect, as did near neutral lapse rates. Palumbo and Mazzarella (1980) ascertained that urbanization of Naples, Italy has resulted in a local increase of rainfall. Yonetani (1982) reported on an increase of precipitation in the urbanized area of Tokyo, as compared with its suburbs. Also in Japan, Harada (1981) suggested the possibility that the increase between 1927 and 1976 of rainfall and thunder occurrences during the warm season at the town of Muroran was due to increased industrial activity associated with the making of steel.

Schultz and Warner (1982), using a two-dimensional numerical model, investigated the importance of the sea-breeze, mountain-valley, and urban circulations in the Los Angeles Basin and concluded that the urban heat island effect was negligible in their simulation. Van der Hoven (1967) reported on an early transport and diffusion study in the same geographic area. Goodin et al. (1980), used a diagnostic model (see Section 13.2.4) to construct a three-dimensional mass-conservative wind field over the Los Angeles Basin. McRae and Seinfeld (1983) reported on an evaluation of an urban photochemical air pollution model that contains chemical reactivity for the Los Angeles Basin, although measured surface wind fields were used to estimate air parcel trajectories rather than a mesoscale model simulation. As shown by Reible et al. (1983) the behavior of plumes cannot be determined from surface winds alone but will also depend on the wind directional shear with height.

Studies of islands have also been performed to estimate the influence of urban areas on climate and weather (as well as to study the effects of the islands themselves, of course). Such investigations are particularly useful because pollution is not generally

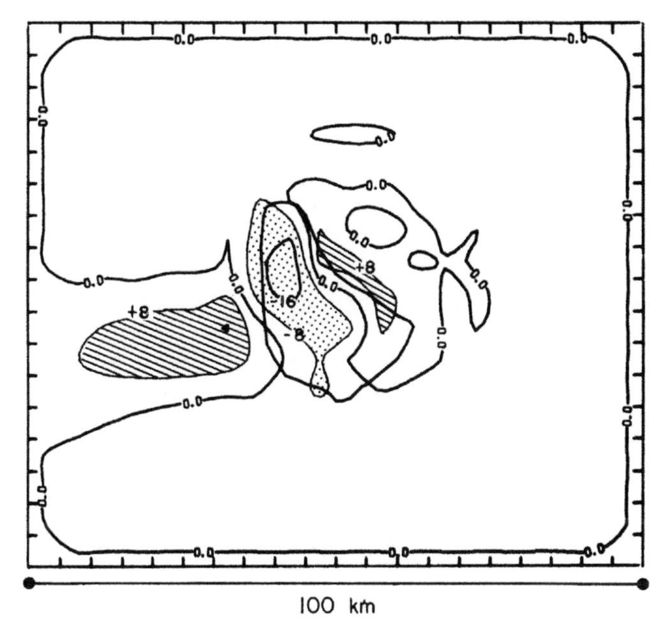

FIGURE 13.14

The vertical velocity field at 1 km above the water and land surface at 1300 LST in the vicinity of Barbados during a typical summer afternoon. The contour interval is 8 cm s^{-1}. The synoptic geostrophic wind was 10 m s^{-1} from the east (from Mahrer and Pielke 1976).

significant over an island, whereas it may be in the city environment. Mahrer and Pielke (1976) performed such a study, using the island of Barbados in the West Indies, and found a downwind pressure minimum that was created by the advection offshore of the heat generated by the island. Figure 13.14 illustrates the resultant low-level convergent zone that was produced downwind of the island. Scofield and Weiss (1977), and the principal investigators of Project METROMEX (1976) reported on preferred regions of thunderstorm development downwind of urban areas, apparently at least partially due to this type of convergent wind field. Figure 13.15 (reproduced from the University of Wisconsin, CIMSS, 2012) illustrates the higher surface temperatures of a metropolitan area (in this case, London, U.K.) relative to the rural area for a calm night in mid-May.

Matson et al. (1978) used satellite imagery to illustrate maximum urban-rural differences ranging from 2.6 to 6.5°C in the midwestern and northeastern United States on a particular summer day. Price (1979), using high-resolution satellite imagery, found peak rural-urban temperature differences as large as 17°C over New York City – a result that is substantially larger than surface-based observed temperatures. He suggests that this difference could be due to satellite sensing of industrial areas, rooftops, as well as the trapping of energy within urban canyons (Nunez and Oke 1977), which

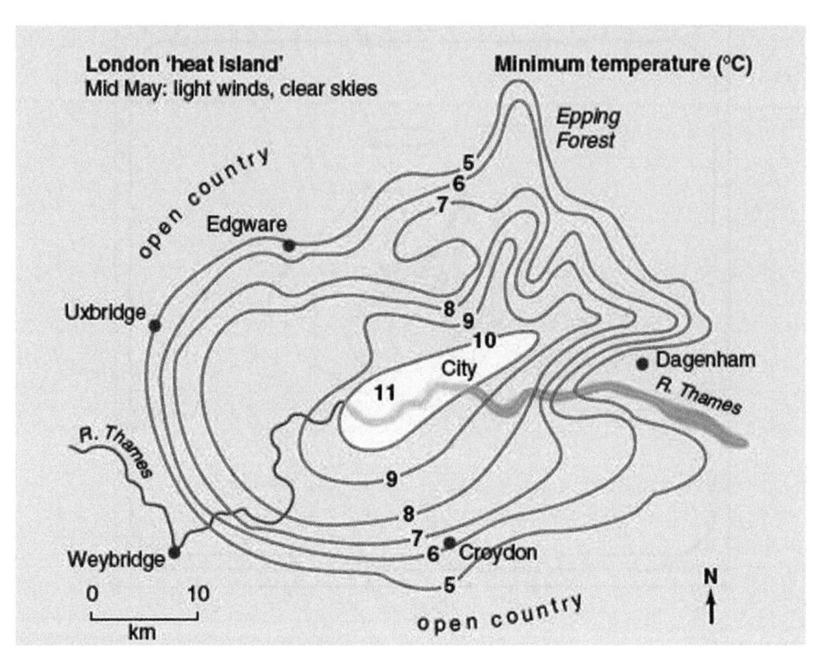

FIGURE 13.15

Illustrates the higher surface temperatures of a metropolitan area (in this case, London, U.K.) relative to the rural area for a calm night in mid-May (from the University of Wisconsin, CIMSS, 2012; available online at: http://cimss.ssec.wisc.edu/climatechange/globalCC/lesson7/UHI2.html). (The color version of this figure is presented in the plate section in the back of the book and the online web version.)

are not sensed by the surface observations. Actual heat fluxes into the atmosphere must, therefore, be proportionally larger than if only surface-based observed data are used to estimate fluxes in mesoscale models of the urban circulation. The cover of the March 1980 issue of the *Bulletin of the American Meteorological Society* (Matson and Legeckis 1980) illustrates another satellite image of urban heat islands (this example is for eastern New England), and Carlson and Augustine (1978) present an image for the Los Angeles area. Using surface station data, Winkler et al. (1981) show that the difference in the January and July mean temperatures between downtown Minneapolis St. Paul and the surrounding countryside is about 2 and 3°C, respectively.

Other studies of heated islands include those of Estoque and Bhumralkar (1969), Delage and Taylor (1970), Bhumralkar (1972), Lee (1973), Lal (1979), and Carbone et al. (2000). Chopra (1973) summarized these and other aspects of the influence of islands on atmospheric flow patterns. Melgarejo (1980) used a two-dimensional boundary layer model to help assess wind energy over the island of Gotland, Sweden. Garstang et al. (1975) presented a summary of heat island studies.

On a somewhat larger scale, Keyser and Anthes (1977) report on predictions of planetary boundary-layer depths over the middle Atlantic States using a mesoscale model. The concentration of pollution, of course, is closely related to the boundary-layer depth. Sheih (1978), Hanna (1979), and McNider et al. (1980), provide more accurate representations of pollution dispersion for use in urban, and other types of mesoscale models. de Wispelaere (1981) and Zannetti (1990) provided summaries of a number of techniques to estimate pollution concentration over mesoscale sized areas. Moran and Pielke (1996a,b) and Moran (1992) compared mesoscale and regional model simulations of dispersion with tracer data.

Efforts to couple air chemistry and meteorology are required. Swan and Lee (1980), for example, urged that because of the highly nonlinear interactions between chemical reactions in the atmosphere and the meteorology, both chemistry and meteorology must be considered simultaneously (e.g., using the same time steps) to make an accurate assessment of the effects of individual sources on air quality.

Gas and aerosol pollutants are known to have large temporal fluctuations and large spatial mesoscale variations. Health standards for such pollutants as carbon monoxide, ozone, nitrogen oxide, sulfur dioxide, lead, and fine aerosol particles have been developed. Many urban areas do not satisfy air quality standards and are said to be in *nonattainment*. Although these air quality standards were chosen primarily for health reasons (and to monitor visibility), the effect on local weather could also be substantially influenced by lesser concentrations than mandated as upper air quality limits.[7] Viskanta and Weirich (1978) used a two-dimensional mesoscale model in which initial pollutant concentrations at the surface in a rural and an urban area were 20 and 100 $\mu g\ m^{-3}$, respectively, for gases and aerosols. They showed that surface temperatures are reduced about 0.3°C during midday downwind of an urban area in both winter and summer in midlatitudes, whereas around sunrise it was 0.8°C warmer in winter and 0.5°C warmer in summer. Even upwind of the city, changes in surface temperature were predicted due to the influence of changes in radiative flux divergence on the urban circulation. These pollutants also affect visibility (e.g., Mumpower et al. 1981).

Other studies of the effect of pollution on urban weather include Atwater (1971a,b, 1974, 1977), Bergstrom and Viskanta (1973a,b), Pandolfo et al. (1976), Zdunkowski et al. (1976), Viskanta et al. (1977b), Welch et al. (1978), and Viskanta and Daniel (1980). This last article reported on a two-dimensional simulation of St. Louis during the summer and found a maximum heat island effect of +3°C. Viskanta et al. found that the interaction of radiation with air pollution acts to decrease stability near the ground at night and to increase it during the day. Welch et al. (1978) found significant changes in the modeled planetary boundary-layer structure over and downwind from a city due to changes in atmospheric turbidity, roughness, heating, and soil types over the urban area. Atwater (1977) investigated urban effects in desert, tropical, midlatitude, and tundra locations and concluded that the largest thermal effects were in the

[7]Carbon dioxide is also emitted from human activities in urban areas (and elsewhere). It is important to realize, however, that other emissions, such as sulphur dioxide, carbon monoxide, lead, mercury, etc., have no positive benefits in the troposphere or at the surface. They are appropriately called "pollutants." Carbon dioxide, however, while a climate forcing (which can have negative effects by altering climate), can also have positive effects on the climate system both in terms of its radiative forcing effects (e.g., resulting in warmer minimum temperatures at night in cold, dry regions), and benefits to vegetation growth.

tundra, whereas the smallest were in the tropics and deserts. In contrast to Viskanta and Daniel (1980) and others, however, he concluded that except in the tundra, pollutants are only minor factors in the formation of heat islands. Pandolfo et al. (1976) also suggested that NO_2 and the particulate aerosols are the only commonly found anthropogenic pollutant constituents with significant radiative effects, and that their concentrations can be represented as a fixed fraction of CO, which is the only pollutant they predicted explicitly.

Robinson (1977), Bornstein and Oke (1980), Landsberg (1981), and Bennett and Saab (1982) reviewed the influence of pollution and urbanization on urban climate. The 1983 AMS/EPA Specialty Conference on Air Quality Modeling of the Nonhomogeneous, Nonstationary Urban Boundary Layer reviewed our current understanding of urban meteorology with papers on such topics as ground-air exchange (Garrett 1983b), parameterization of subgrid-scale fluxes (Lewellen 1983), pollution removal mechanisms (Hales 1983), parameterization of radiation (Kerschgens 1983), modeling techniques (Warner 1983), transport and diffusion (McNider 1983), and synoptic influences in urban circulations (McKee et al. 1983). Seaman (2000) provides a review of the use of meteorological models for air quality assessments. Accurate modeling simulations of urban areas require that both air chemistry and the meteorology must interact. The primary interactions are as follows:

1. The rate of input, transport, diffusion, and fallout of pollutants, as well as the types and rates of chemical reactions depend on the mesoscale meteorological dynamics and thermodynamics.
2. Mesoscale circulations are influenced by alterations in radiative characteristics due to changes in the clarity of the atmosphere because of pollutant gases and aerosols.

More recent studies of urban circulations include that of Jin et al. (2005), Jin and Shepherd (2008), Carrió et al. (2010), Shepherd et al. (2010a,b,c; 2011), Carrió and Cotton (2011), Jin et al. (2010), Mitra et al. (2011), Niyogi et al. (2011), Shepherd and Mote (2011), Carter et al. (2012), Leelasakultum et al. (2012), Mölders et al. (2012), Stone (2012), Tran and Mölders (2012a,b), Andersen and Shepherd (2013), Mölders (2013), Paredes-Miranda et al. (2013), and Shepherd (2013). Nobis (2010) examined the ability of an urban parameterization to improve operational NWP characterization of the sensible weather in Washington, DC.

Review papers on urban mesoscale features include Chow et al. (2012), and NRC (2012). See also Millán et al. (1997, 2000, 2002), Millán (2002), Palau et al. (2005a,b), Georgescu et al. (2009b,c 2011b, 2012b, 2013), and Yang et al. (2012).

13.2.6 Lake Effect

When cold air advects over warmer ocean or lake water, the sensible and latent heat fluxes to the atmosphere can be very large, deepening the planetary boundary layer as the air continues its traverse over water. Over the East China sea during arctic outbreaks, total heat fluxes of more than 450 W m^{-2} are supplied from the sea surface (Nitta 1976), and Bosart (1981) found heat fluxes averaging 600 W m^{-2} in cold arctic air in the advance of an extratropical cyclone. This latter heating was apparently critical in the subsequent explosive deepening of the cyclone. Chou and Atlas (1981, 1982), and Stage (1983) have shown how these ocean-air heat fluxes can be estimated during cold air outbreaks using satellite imagery. In a related study, Atlas and Chou (1982), and Atlas et al. (1983) calculated that the top of the boundary layer rose from about 1 to 1.4 km over a distance of about 250 km as polar air advected south-southeast from the New York City area over relatively warm ocean water on 17 February 1979. Garstang et al. (1980) documented using both mesoscale observations and model simulations that arctic air initially accelerates substantially, as it advects offshore of the Delmarva peninsula during cold outbreaks.

When the body of water is sufficiently broad, marked changes in local weather occur along the windward shore, relative to conditions found on the lee coast as a result of these large heat fluxes. The shorelines of the Great Lakes (e.g., see Hill 1971; Jiusto and Kaplan 1972; Strommen and Harman 1978) and the Sea of Japan (e.g., see Takeda et al. 1982), for example, suffer from major, localized snowstorms due to the over-water advection of arctic and polar air during the winter. Lavoie (1972), and Estoque and Ninomiya (1976) successfully represented this phenomena using a one-layer mesoscale model, and Hjelmfelt and Braham (1983) used a three-dimensional numerical model. In the Hjelmfelt-Braham study, it was found that horizontal grid lengths of 24 and 40 km were unable to simulate lake-effect snow adequately on the windward shore of Lake Michigan. Only with a horizontal grid interval of 8 km was an accurate simulation achieved.

Boudra (1977) with a 50 km and Boudra (1981) with a 40-45 km grid interval, reported on somewhat larger domain simulations of the influence of all the Great Lakes on the flow field. Ellenton and Danard (1979) have applied a model with a 48 km grid interval to Lake Huron and vicinity. Studies of lake effect snowstorms include Schultz (1999), and Sousounis et al. (1999).

Lavoie's (1972) work found that upslope winds over low topographic relief enhance precipitation. This type of phenomenon, highly localized in space, occurs over many midlatitude windward coastal regions of the world during cold air outbreaks. The weather over the entire Great Lakes region, for instance, is substantially influenced by this juxtaposition of land and water (e.g., see Fig. 13.16). Figure 13.17 (reproduced from Jiusto et al. 1970) gives the observed snowfall downwind of Lakes Ontario and Erie after several days of arctic airflow over the area. The large spatial variability, evident in both of these Figures, illustrates the strong influence of lake effect snowstorms on local weather. Indeed, the effect of the lakes extend far inland. Leffler and Foster (1974), for instance, reported on probable annual snowfall in excess of 5 m at elevations above 1.3 km in West Virginia. Such large precipitation amounts are

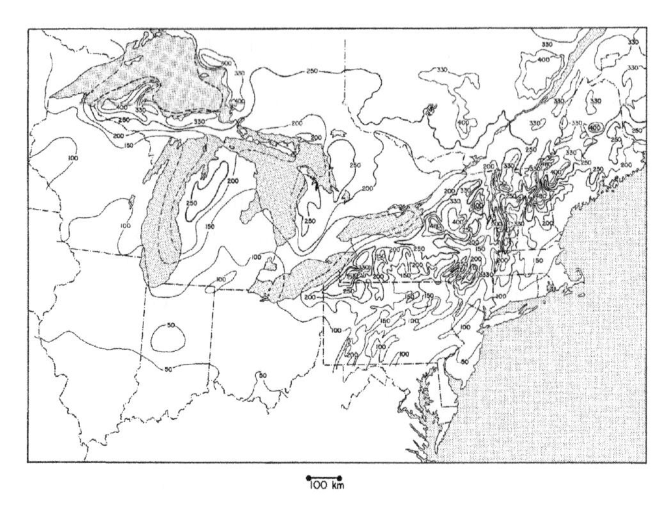

FIGURE 13.16

The mean winter snowfall in centimeters over a portion of eastern North America. U.S. data 1951-1960; Canadian data 1931-1960 (from Muller 1966).

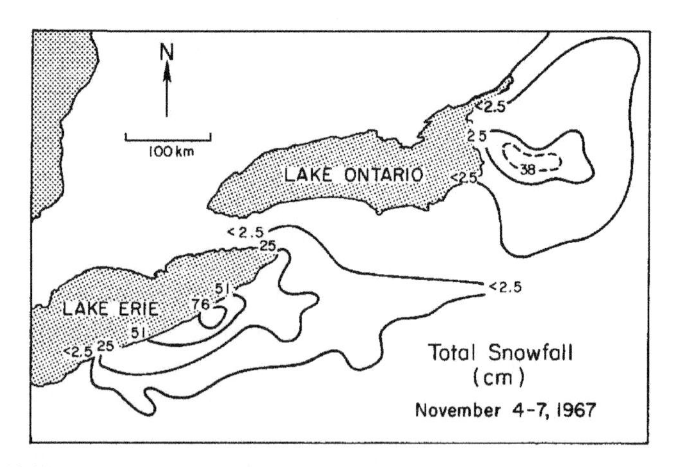

FIGURE 13.17

The observed snowfall resulting from a severe lake effect storm. All of the precipitation was attributed to this terrain-induced mesoscale phenomena. Adapted from Jiusto et al. (1970), and Jiusto and Kaplan (1972).

primarily due to orographic lifting of lake-moistened air during northwesterly, cold low-level synoptic flow.

A review book that includes studies of lake effect mesoscale systems (as well as other aspects of winter storms) is Kocin and Ucccellini (2004).

13.3 Mesoscale Systems Primarily Forced Through Lateral Boundaries or from Internal Atmospheric Instabilities

13.3.1 Convective Bands Embedded in Stratiform Cloud Systems

In extratropical cyclones, along synoptic-scale fronts and associated with tropical weather systems, precipitation is often not uniformly distributed, but occurs in well-organized mesoscale-sized bands of heavier snow or rain (e.g., Akiyama 1978). These smaller-scale systems are generally a part of the synoptic system but are usually not resolvable with conventional meteorological observations, except by satellite and radar. They occur when organized local regions of the atmosphere are convectively unstable, whereas the mean atmosphere is stable to moist adiabatic displacements in the vertical. Such bands can be reinforced by terrain inhomogeneities, such as the development of small-scale baroclinic zones (small-scale fronts) along the coast associated with the passage of extratropical storms as reported by Bosart et al. (1972), Bosart (1975, 1981), and Marks and Austin (1979), or they can be disrupted or destroyed as they descend larger terrain barriers as reported by Hobbs (1978). Over the open ocean they are well defined in satellite imagery and clearly an important component of the extratropical storm system.

In the vicinity of northwest Europe, distinct subsynoptic-scale disturbances of this type (referred to as *polar lows*; Rasmussen 2000), apparently driven to some extent by latent heat release (e.g., Oerlemans 1980; Rasmussen 1979, 1981, 1983) are relatively common features in the polar outbreaks, associated with extratropical storms. Similar appearing features also occur in the wintertime over the north Pacific (e.g., Mullen 1979; Reed 1979). Locatelli et al. (1982), in a case study of several such mesoscale disturbances over and off of the northwest coast of Washington and Oregon, concluded that the systems that they investigated were deep baroclinic disturbances that extended as high as the 400 mb level. In general, however, the relative contributions of latent heat release from cumulus activity and of horizontal thickness gradients to the generation and evolution of these synoptically-induced mesoscale systems is not completely understood. A study contrasting polar lows in the Pacific and Atlantic is that of Sardie and Warner (1983), where they concluded that moist cumulus convection plays a major role in their evolution in the Atlantic, whereas such a mechanism is not necessary in the north Pacific region. Winter mesoscale cyclogenesis to the east of Korea is modeled by Lee et al. (1998).

The interactions between the mesoscale and synoptic scales for this type of mesoscale system are complex. Ballentine's (1980) work represents an early numerical study of the quasi-stationary coastal fronts reported by Bosart, and the Cyclonic Extratropical Storms Project – CYCLES (e.g., Herzegh and Hobbs 1980, 1981, Hobbs et al. 1980, Matejka et al. 1980, Parsons and Hobbs 1983; Rutledge and Hobbs 1983; Wang et al. 1983) represents an extensive observational program to understand these systems along the northwest Washington coast. In the CYCLES study, for example, Houze et al. (1981a) found that the presence of low-level, mesoscale ascent was crucial in generating significant cloud condensate at the lower levels. The accretion of this condensate by hydrometeors falling from higher levels, resulted in substantially larger precipitation rates in the warm frontal region of extratropical cyclones than would have occurred otherwise. From the CYCLES project, mesoscale rainbands have been cataloged into the five basic types (Hobbs 1978); warm-frontal, warm-sector, cold-frontal (wide and narrow), prefrontal cold-surge (wide and wavelike) and post-frontal bands. Mesoscale model simulations of embedded convective systems include Nicosia and Grumm (1999).

For more recent studies of convective bands embedded in stratiform cloud systems, see Novak et al. (2009, 2010), Truong et al. (2009), Han et al. (2010), and Cotton et al. (2011).

13.3.2 Squall Lines

Along with the convective bands embedded in stratiform clouds, the squall line is among the most difficult of mesoscale phenomena to simulate. Although the squall line is undoubtedly influenced (and often generated) by fixed geographic features such as terrain, the squall line is highly variable in space and transient in time so that accurate lateral boundary and initial conditions, essential to satisfactory predictions, are difficult and expensive to obtain. Uccellini (1980), for instance, documented the strong synoptic forcing of upper- and lower-level tropospheric jets, which are frequently associated with squall line formation. Therefore, although it may be possible to build a climatology of mesoscale model forecasts for representative conditions for terrain-induced mesoscale systems, it is necessary to perform squall line predictions for each event which is one of the main goals in the development of the Center for Analysis and Prediction of Storms (CAPS) at the University of Oklahoma. An illustration of the explosive growth of a squall line over Virginia is illustrated in Fig. 13.18a-e.

Squall lines often form in association with synoptic weather features such as cold and warm fronts, dry lines (e.g., Schaeffer 1974, Ogura and Liou 1980, Koch and McCarthy 1982, McCarthy and Koch 1982, Shapiro 1982, Homan and Vincent 1983), and tropical waves (e.g., Fortune 1980), although they typically travel at a greater speed than these larger-scale weather phenomena. Squall lines develop in air masses that are convectively unstable and, as shown by Weiss and Purdom (1974), their appearance is strongly influenced by factors such as the occurrence of early morning cloudiness.

Negri and Vonder Haar (1980) have used 5 min interval satellite imagery to determine the magnitude of moisture convergence in the presquall line environment. They found maximum values of this convergence to be 2.2×10^{-3} g kg^{-1}s^{-1} for a severe storm outbreak in the midwest in April 1975. Purdom and Marcus (1982) provided evidence that the merger and intersection of thunderstorm-produced outflow boundaries are the major source of low-level convergence for subsequent deep cumulus convection over the southeastern United States during the summer, particularly in the mid and late afternoons. Holle and Maier (1980) documented the formation of a tornado caused by the intersection of thunderstorm outflows over south Florida.

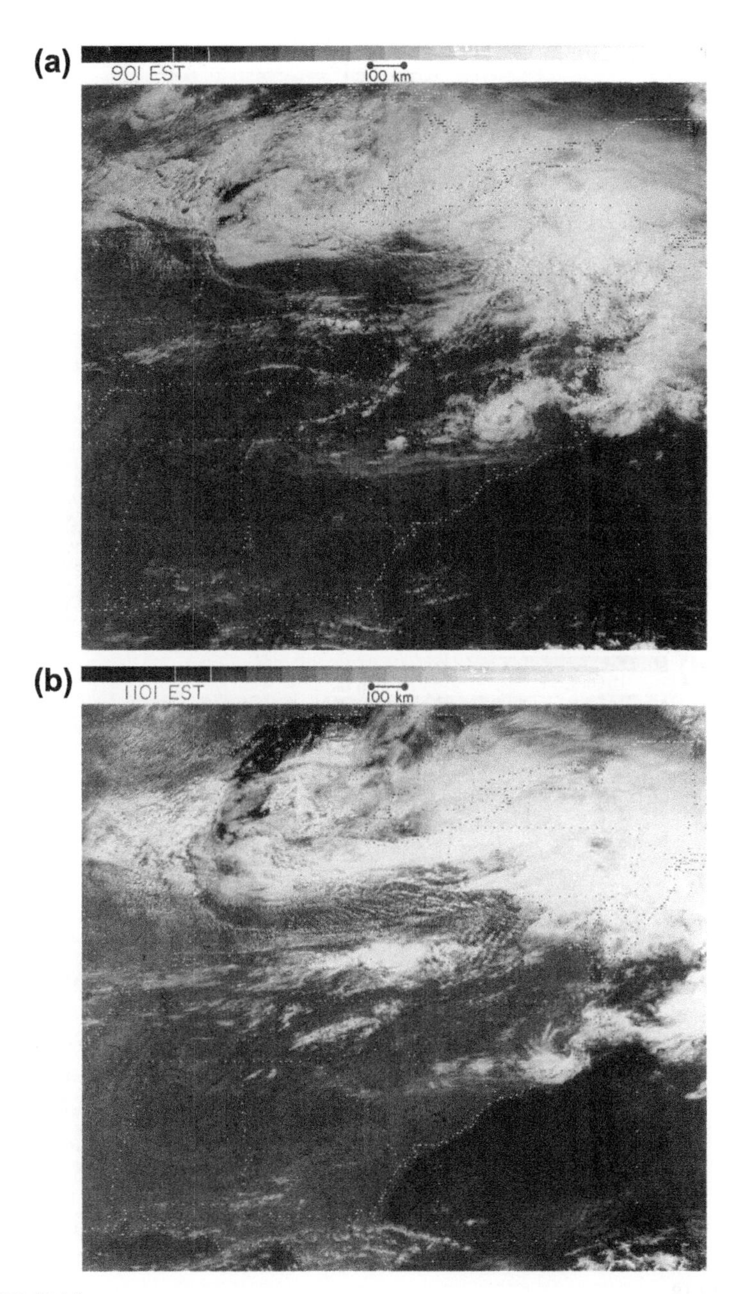

FIGURE 13.18

The development of a squall line over the middle Atlantic states on 6 June 1977 as observed at 2 h intervals by geostationary satellite imagery. The times of observation are given in Eastern Standard Time (EST) in the upper left.

(c)

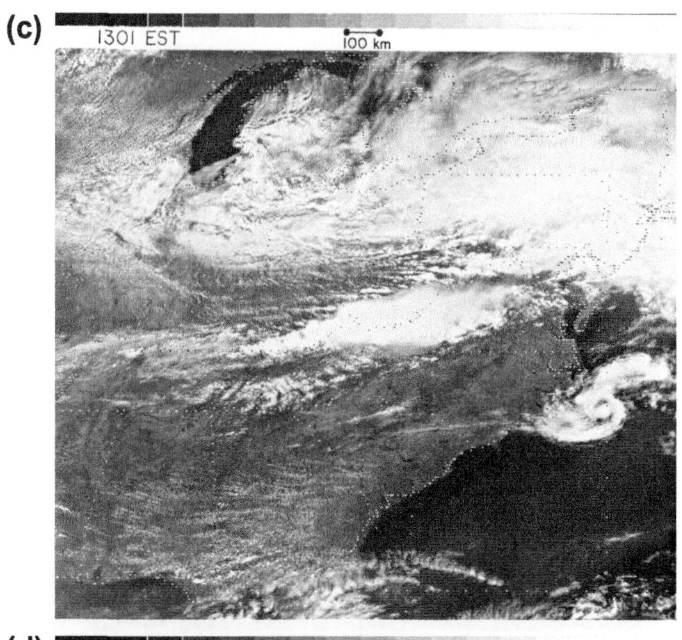

(d)

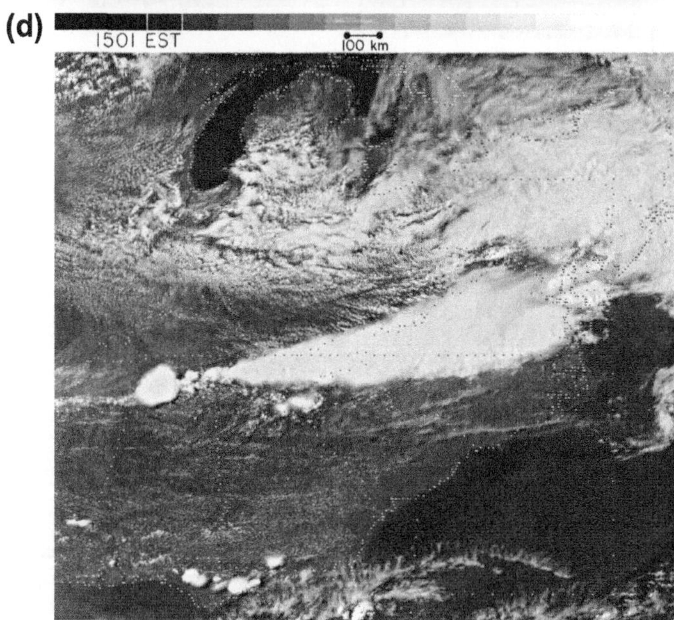

FIGURE 13.18

Continued.

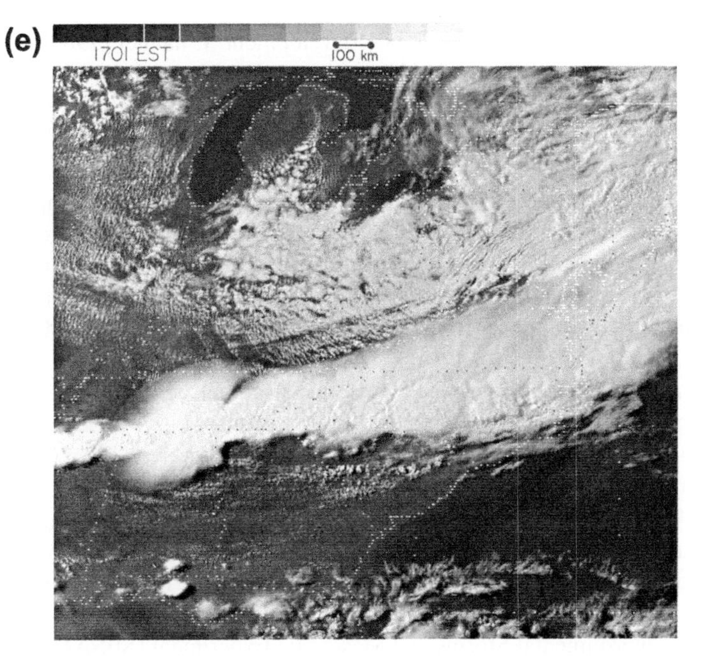

FIGURE 13.18

Continued.

Using linearized forms of the conservation relations, studies (e.g., Raymond 1975) indicate that squall lines apparently propagate as waves with particularly intense convective activity occurring where two or more of these waves constructively reinforce one another and where the atmosphere is convectively unstable. Ley and Peltier (1981) provided a detailed analysis of a propagating wave-cloud formation with a horizontal wavelength of 10 km whose origin was assumed to be due to the transient latent heat released by a severe storm some distance away. Silva Dias (1979) performed a linear analysis of tropical squall lines. Also using a linear model, Raymond (1983) examined the influence of cumulonimbus-induced downdrafts on subsequent convection.

Sun and Ogura (1979) suggested that squall lines may also be initiated through differential temperature gradients in the boundary layer interacting with the synoptic flow in an analogous manner to that causing sea and land breezes. From a case study for 8 June 1966 over Oklahoma, Sun and Ogura observed a well-defined band of horizontal convergence at low levels prior to the appearance of the first radar echoes in a region of large horizontal temperature contrast. Colby (1980) showed that large convective instability was also a necessary prerequisite for the squall line development on this day, which explained why the convection formed in the eastern portion of the region of upward vertical motion. Uccellini and Johnson (1979) examined the role of tropospheric jet streaks in squall line development. Emanuel (1982b) discussed the possibility that squall lines are self exciting and involve a CISK-like cooperative interaction between the cumulus and mesoscale.

Hane (1973), Schaeffer (1974), Perkey (1976), Ross and Orlanski (1978), Chang et al. (1981), Kondo (1981), Kaplen et al. (1982), Thorpe et al. (1982), Wong et al. (1983b), Cram et al. (1992a,b), and Finley et al. (2001), provide examples of non-linear mesoscale simulations of features related to squall lines, and a review of the understanding of squall line dynamics is presented by Lilly (1979). Brown (1979) has modeled mesoscale unsaturated downdrafts driven by rainfall evaporation from precipitating cloud features such as anvils created by squall lines. Bradberry (1981) suggested from an observational study of a squall line over Oklahoma on 26 April 1969 that mesoscale ascent associated with condensational heating about 5 km behind the leading edge of radar echoes, and descent due to evaporative cooling from cumulus cloud tops and from the dissipation of a middle cloud layer 10 km ahead of the echoes may be typical of large mature convective storm systems.

Squall lines that become stagnant over one geographic location (e.g., Johnstown, Pennsylvania in July 1977, Hoxit et al. 1978; London, England in August 1975, Bailey et al. 1981) can produce devastating floods. Squall lines also often produce devastating tornado outbreaks (e.g., see Fritsch 1975; Hoxit and Chappell 1975a,b). The Severe Environmental Storms and Mesoscale Experiment (SESAME; e.g., Lilly 1975, Alberty and Barnes 1979, Alberty et al. 1979) was a mesoscale observational program designed to improve our understanding of the influence of the local environment on the generation of these intense cumulus convective systems over the Great Plains of the United States. The GARP Atlantic Tropical Experiment (GATE: e.g., Frank 1978, 1980, Zipser and Gautier 1978, Warner et al. 1979, Houze and Betts 1981), the Venezuelan International Meteorological and Hydrological Experiment (VIMHEX; e.g., Betts et al. 1976), and the winter monsoon experiment (MONEX; e.g., Warner 1982) represent similar observational programs in the tropical eastern Atlantic, over land in tropical South America, and over water southeast of Vietnam.

Examples of studies of squall lines observed in the tropics include those of Fernandez (1982), Gamache and Houze (1982), and Ishihara and Yanagisawa (1982). Zipser et al. (1981) documents a weaker mesoscale convective system off the coast of west Africa. Johnson and Nicholls (1983) analyzed boundary-layer structure associated with tropical squall lines over the eastern Atlantic. Additional discussions of squall lines and their dynamics are given, for example, in Dudhia and Moncrieff (1989), and Houze (1989).

For more recent studies of squall lines, see Cotton et al. (2011), Seigel et al. (2012), Storer and van den Heever (2012), Adams-Selin et al. (2013a), Bender and Freitas (2013), Seigel and van den Heever (2013).

13.3.3 Mesoscale Convective Clusters

Since geostationary satellite imagery became routinely available over the United States, the frequent occurrence of persistent mesoscale areas of cumulus convection,

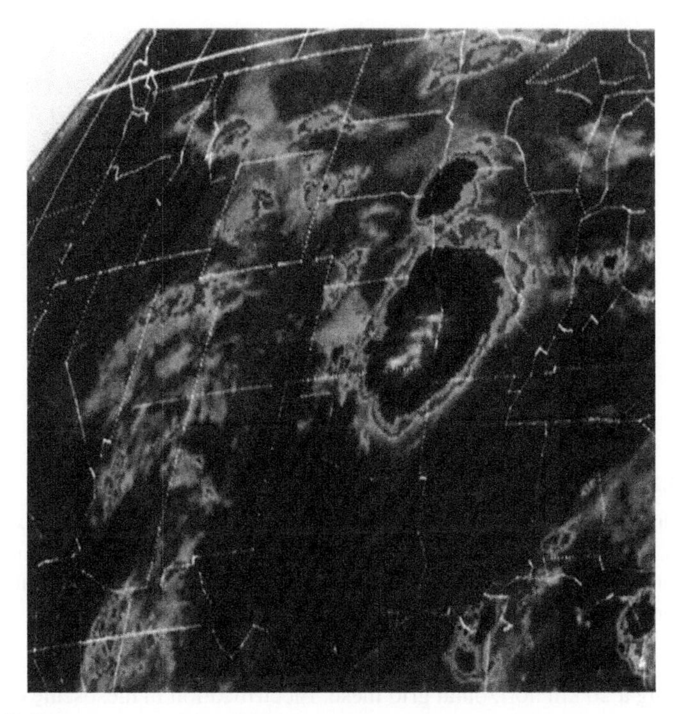

FIGURE 13.19

Enhanced infrared (IR) image of the United States at 0300 MDT 4 August 1977, from the GOES satellite at 70°W longitude. The stepped gray shades of medium gray, light gray, dark gray and black are thresholds for areas with apparent blackbody temperatures colder than −32, −42, −53 and −59°C, respectively. Temperatures progressively lower than −63°C appear as a gradual convective development of 4 August over the Colorado mountains. The intense meso-α-scale convective complex (MCC) centered over eastern Kansas originated in the eastern Rockies and western plains the previous evening (from McAnelly and Cotton 1986).

such as illustrated in Fig. 13.19 and 13.20, has been noted, as first suggested by W.R. Cotton (1977, personal communication). Defined as *mesoscale convective clusters* (MCCs) by Maddox (1980a,c), these systems appear to originate over and downstream of the Rocky Mountains and propagate from west to east (Nachamkin and Cotton 2000; Nachamkin et al. 2000; Cotton et al. 2011).[8] A warm core system, they have been shown (Fritsch and Maddox 1980, 1981a; Maddox 1980b; Maddox et al. 1981, Keyser and Johnson 1982) to cause major alterations in the synoptic flow field. Fritsch et al. (1980) presented a scheme to initialize model simulations of these mesoscale convective clusters, and Fritsch and Maddox (1981b) perform a model simulation of

[8]To include mesoscale convective cloud features that do not fit into the MCC classification given in Table 13.1, the more inclusive term *mesoscale convective system* (MCS) is used.

FIGURE 13.20

Perspective drawing of an idealized, weakly rotating mesoscale convective complex as viewed from the rear. The apex region is located where the heavier rain is drawn. The cumulonimbus towers at right form on the mesoscale cold front. The mesoscale warm front and the forward part of the stratiform region are not shown in this view (from Fortune 1989).

an MCC using a 20 km horizontal grid mesh. Electrification in mesoscale convective systems is discussed in Schuur and Rutledge (2000a,b).

Cotton et al. (1983) and Wetzel et al. (1983) document the initiation of an MCC over the Rockies and its eventual passage into the western Atlantic 2 days later. Park and Sikdar (1982) documented an MCC over Oklahoma and illustrate its complex interaction with the synoptic scale. Ninomiya et al. (1981) illustrated what is apparently a small MCC moving over and off of the coast of China toward Japan at about 32°N. Its origin was just east of the Tibetan Plateau. Other studies of these mesoscale convective systems include Chen et al. (1999b), Tucker and Zentmire (1999), and Bernardet et al. (2000). Table 13.1, adapted from Maddox (1980c) gives his definition of an MCC.

Table 13.1 The definition of a mesoscale convective cluster – MCC (from Maddox 1980c).

Size	Contiguous cloud shield with temperatures less than or equal to −32°C covering an area of more than 100,000 km^2 and an interior cloud region with temperatures less than or equal to −52°C covering an area of more than 50,000 km^2.
Duration	The above two conditions must last for at least 6 h. The MCC is then defined as an entity until the two conditions listed above no longer apply.
Shape	The eccentricity (i.e., ratio of minor to major axis) must be greater or equal to 0.7 at the time of its maximum extent.

Bosart and Sanders (1981) concluded from an observational analysis of the data for the July 1977 flood in Johnstown, Pennsylvania, that an MCC, rather than a stagnant squall line, was responsible for the excessive rainfall. Mesoscale convective clusters also are a major component of the tropical atmosphere and are also referred to as *cloud clusters*. Observational studies of these tropical systems include those of Houze and Cheng (1981), Zipser et al. (1981), and Houze (1982). In the Asia tropics, Houze et al. (1981b), Johnson (1982), and Johnson and Kriete (1982) documented a diurnally-varying mesoscale precipitating cloud system over the ocean that is related to the winter monsoon flow pattern and the daily heating cycle of the adjacent land. Potty and Sethu Raman (2000) simulated the structure and track of monsoon depressions over India. Zipser (1982) briefly summarized our understanding of mesoscale convective clusters.

A review of mesoscale convective systems is given in Smull (1995). A review book that summarizes studies of mesoscale convective clusters by the originator of that concept is Cotton et al. (2011). See also recent papers by Seigel and van den Heever (2012a), and Adams-Selin et al. (2013b). Sheffield et al. (2013) examine the effect of pollution aerosols on the growth of cumulus congestus clouds.

13.3.4 Tropical Cyclones

Tropical cyclones, which are generally smaller than the extratropical cyclones that form along the polar front, represent one of the larger mesoscale phenomena. Their location of formation and subsequent movement are strongly influenced by the synoptic scale. Ooyama (1982) in his review article, for instance, defined the tropical cyclone as "a mesoscale power plant with a synoptic-scale supportive system."

Tropical storms form when the heating due to cumulonimbus activity positively reinforces the low-level convergent wind flow such that an increasingly more intense mesoscale vortex develops. Simulations of this phenomenon include two-dimensional studies such as Rosenthal (1970, 1971), and Kurihara (1975). Anthes et al. (1971), Anthes (1972), Kurihara and Tuleya (1974), and Jones (1980), and others have performed idealized three-dimensional calculations. Kurihara and Bender (1982), and Bender and Kurihara (1983) performed a three-dimensional hurricane simulation with 5 km horizontal grid increments in the finest grid of their quadruply nested model. Anthes et al. (1971) contrasted solutions using two- and three-dimensional simulations.

Ceselski (1974) and Mathur (1974, 1975) have given pioneering examples of simulations for observed tropical storms (an example of the wind field predicted by Mathur (1974) for a Caribbean hurricane is given in Fig. 13.21). Eastman (1995), Eastman et al. (1996), Liu et al. (1997, 1999b), and Zhang et al. (2000) have simulated Hurricane Andrew (1992) as it approached and made landfall in south Florida. Lagouvardos et al. (1999) modeled a system in the Mediterranean Sea which had characteristics of a tropical cyclone.

The effect of landfall on tropical cyclone structure has been studied by Moss and Jones (1978), Tuleya and Kurihara (1978), Tuleya et al. (1984), and Powell et al. (1996).

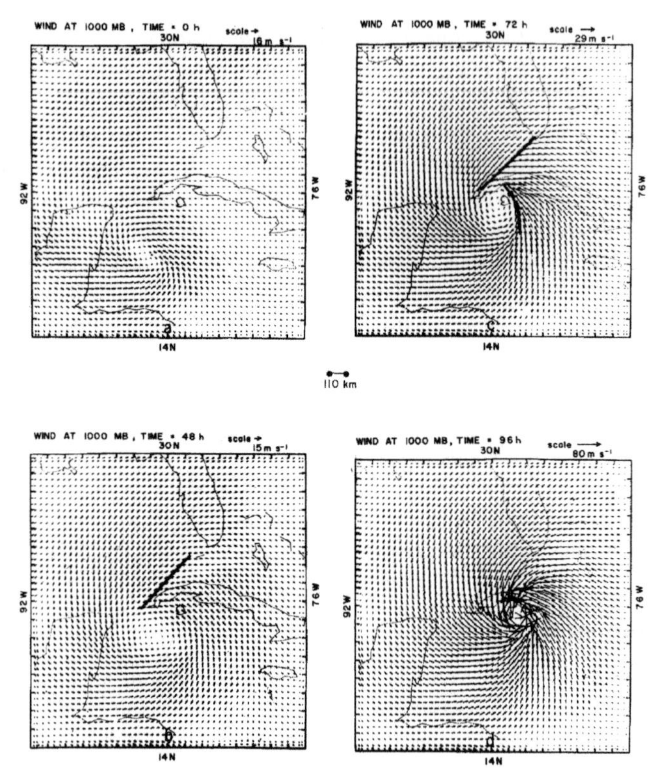

FIGURE 13.21

Three-dimensional model prediction of the winds at 1000 mb associated with the development of Hurricane Isabel in 1964 (from Mathur 1974).

Wind flow in mountainous terrain caused by such storms has been investigated by a diagnostic model and a physical model (Brand et al. 1979), and using numerical models (Chang 1982; Lin et al. 1999). In Chang's study, his model simulation demonstrated that the original low-level center of a tropical storm is blocked by the mountainous terrain of a large island, with a secondary low-level circulation forming in the lee of the island as the upper-level center propagates over the region immediately downwind of the island.

Using observations of damage, Fujita (1980) estimated the wind field pattern of landfalling hurricanes, and Powell (1980) examined the use of several boundary layer models to estimate wind speed near the surface in hurricanes. Surface observations associated with hurricanes are reported in Cione et al. (2000).

An overview of tropical storm modeling is given by Simpson and Pielke (1976), and more extensive reviews are presented by Anthes (1974b, 1982), Pielke (1990), and Pielke Jr. and Pielke Sr. (1997). Krishnamurti and Kanamitsu (1973) simulated the more common tropical disturbance referred to as the non-developing tropical wave.

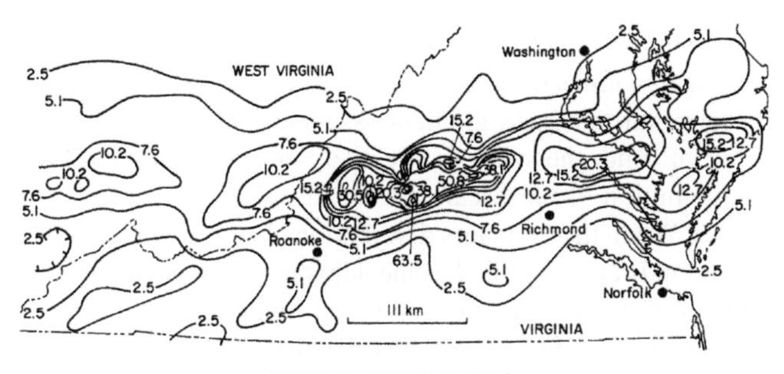

FIGURE 13.22

The observed precipitation (in centimeters) from the remnants of Hurricane Camille from noon 19 August to midnight 20 August 1969 (from the December 1969 issue of the *Virginia Climatological Summary*, NOAA, Asheville, North Carolina).

The latter model is described in Krishnamurti et al. (1973). Another example of a mesoscale tropical cyclone simulation is that of Mathur (1997). Gall et al. (2013) summarize a set of mesoscale model characteristics in his Table 13.3 and associated text, which illustrates the current (2013) state of high spatial resolution hurricane modeling.

Figure 13.22 (reproduced from the December 1969 issue of the *Virginia Climatological Summary*) illustrates the complex precipitation patterns that can occur over land due to the overland track of a tropical storm interacting with a cold front (in this case Hurricane Camille). Zhou (1980) presented a similar study over central China. See also the study of the extreme rain event in Vietnam (Truong et al. 2009).

Books that summarize studies of tropical cyclones include Pielke Jr. and Pielke Sr. (1997), and Emanuel (2005). Recent papers on this subject include Zhang et al. (2007, 2009), Carrió and Cotton (2010), and Cotton et al. (2012).

13.3.5 Frontal Circulations

Synoptic-scale fronts are directly related to a horizontal temperature gradient averaged over a depth in the atmosphere (e.g., between 1000-850 mb, 1000-500 mb). Such a mean temperature gradient is referred to as a *thickness gradient*, as described in detail in most texts on basic meteorology (e.g., Wallace and Hobbs 1977; Pielke 1995). This relation between the mean temperature between two pressure surfaces and the thickness results directly from the hydrostatic assumption. Synoptic-scale cold, warm, and stationary fronts are found on the leading warm edge of these baroclinic zones, and occluded fronts lie in the middle of a polarward extension of warm thickness. An early discussion of fronts and their relation to extratropical cyclones is given in Bjerknes and Solberg (1921). Synoptic fronts are associated with the genesis of extratropical low pressure systems. Gyakum et al. (1996) summarize the ability of regional models to simulate explosive ocean extratropical low development. Kuo et al. (1992) simulate the dynamics of an occluded marine cyclone.

When the horizontal thickness gradient associated with a front has been steady for a long period of time, the wind field closely approximates gradient wind balance. Using the definition of mesoscale presented in Chapter 1, these fronts are not mesoscale features. When the thickness distribution changes rapidly, however, significant nongradient winds are generated as the adjustment toward a new balanced state begins. Williams (1967), for example, suggested that cold frontogenesis is created from the large-scale baroclinic field by the divergent component of the wind. Such frontal circulations are mesoscale and are an integral component in frontogenesis and cyclogenesis.

Studies of frontal circulations include those of Palmén and Newton (1969), Browning and Harrold (1970), Rao (1971), Hoskins and Bretherton (1972), Shapiro (1981), Bennetts and Sharp (1982), Lagouvardos et al. (1996), and Kotroni et al. (1999b,c). Uccellini et al. (1981) examined the importance of circulations associated with jet streaks in frontal zones during intense cyclogenesis along the east coast of the United States, and Uccellini (1980) discussed the role of jet streaks in the formation of low-level jets over the Great Plains. Carbone (1982) documented intense rainfall associated with a cold front in central California, and Hobbs and Persson (1982) reported on precipitation associated with a cold front moving onshore along the Washington coast. As suggested at the end of Section 13.3.1, these mesoscale frontal circulations are closely related to the convective bands embedded in stratiform cloud systems.

Ross and Orlanski (1982) used a three-dimensional numerical model to simulate the 48 h evolution of a cold front in early May 1967 over the southeastern United States, while Sergeev (1983) used a parameterization of precipitation forming processes to examine the influence of precipitation on frontal dynamics. Dickison et al. (1983) documented that flying insects can accumulate in the convergence zones associated with cold fronts. Lagouvardos et al. (1998) simulated an extreme cold frontal surge over Greece, which produced the worst snowfall in 100 years. A cold outbreak in Kenya is discussed by Okoola (2000).

In eastern Asia, a frontal circulation which extends southwest to northeast for several thousand km is established each year associated with the Asian summer monsoon. Mesoscale model simulations of this feature (which is called the "Baiu front" or the "Mei-Yu front") have been performed by Sun (1984b), and Hsu and Sun (1994). Observational studies of this front have been reported, for example, in Ninomiya (1992, 2000), Yamazaki and Chen (1993), and Takahashi et al. (1996).

Williams (1972) contrasted the creation of fronts by a nondivergent horizontal wind field that contains stretching deformation, with the generation of fronts by the divergent component of the wind. He concluded that the first mechanism requires too much time to cause frontogenesis. Apparently, the low-level convergence of the thickness gradient by nonlinear horizontal advection (such as discussed for the sea breeze in Section 10.2.1.1 and Fig. 10.3) is required to create fronts in the observed time period.

Shaw et al. (1997), Ziegler et al. (1995, 1997), Ziegler and Rasmussen (1998), and Grasso (2000) discuss the initiation of deep cumulus convection along another type of frontal zone called the *dryline*. The dryline is an interface between shallow moist and deep dry boundary layers. This feature, for example, is frequently found in the western High Plains of the United States in the summer. Other mesoscale model simulations of the dryline include Sun (1987), and Sun and Wu (1992).

Recent examples of studies of frontal circulations include Grim et al. (2007), Han et al. (2007, 2009), Schultz and Vaughan (2011), Naud et al. (2012), Seigel and van den Heever (2012b), Truong et al. (2012), Igel and van den Heever (2013), and Igel et al. (2013).

13.4 Integrated Applications on Air Quality – Meteorology Interactions[9]

Atmospheric composition is important for determining both the radiative transfer and cloud formation in the atmosphere (Levin et al. 1996; Myhre et al. 2003; Seinfeld et al. 2004; Givati and Rosenfeld 2004; Ramanathan et al. 2007, Solomon et al. 2007; Solomos et al. 2011). The concentration and physicochemical properties of airborne particles are responsible for significant modification of weather and climate (direct and indirect aerosol effects). The concepts of CCN/IN activation of aerosols, radiative transfer implications and chemical weather modeling can be addressed in the framework of online coupled air quality/meteorology models. Such an example is the application of the Integrated Community Limited Area Modeling system (an extended version of the Regional Atmospheric Modeling System) RAMS/ICLAMS (Pielke et al. 1992; Cotton et al. 2003; Solomos et al. 2011) for the simulation of a thunderstorm over Eastern Mediterranean that was accompanied by transportation of Saharan dust as seen in the satellite image of Fig. 13.23.

The model was set up with three nested grids (15 km, 3 km, and 750 m) as seen in Fig. 13.24. In the vertical, 32 levels were used starting from 50 m near the ground and reaching up to 18 km with a stretching ratio of 1.2. Dust and sea salt were the dominant pollutants close to the cloud base as indicated by in situ aircraft measurements (Levin et al. 2005) and model results in Fig. 13.25. Online treatment of the airborne particles as prognostic variables in the model resulted in significant modification of cloud development and precipitation amounts in the area (Fig. 13.26). Assuming different partitioning of soluble/insoluble particle fractions resulted in higher cloud tops for the scenarios with higher numbers of efficient CCN and IN and in a delay of about one hour in the initiation of precipitation. These clouds included more liquid droplets and ice elements and the competition for the available water inhibited these condensates from growing to precipitation sizes. Comparison of modeled precipitation amounts with ground observations indicated a significant variability in modeling results that was related to the difference in aerosol properties. Overall, including the contribution

[9] This section was written by Professor George Kallos of the University of Athens, Greece.

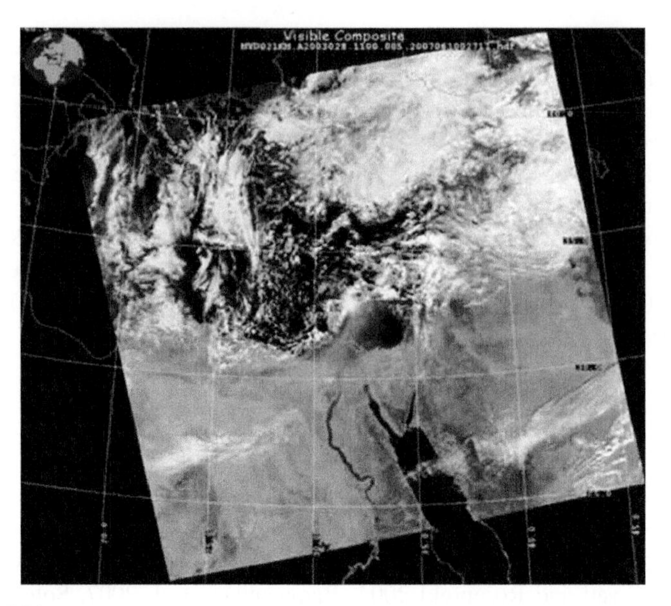

FIGURE 13.23

MODIS-Aqua visible channel on 28 January 2003 1100 UTC (from Solomos et al. 2011). (The color version of this figure is presented in the plate section in the back of the book and the online web version.)

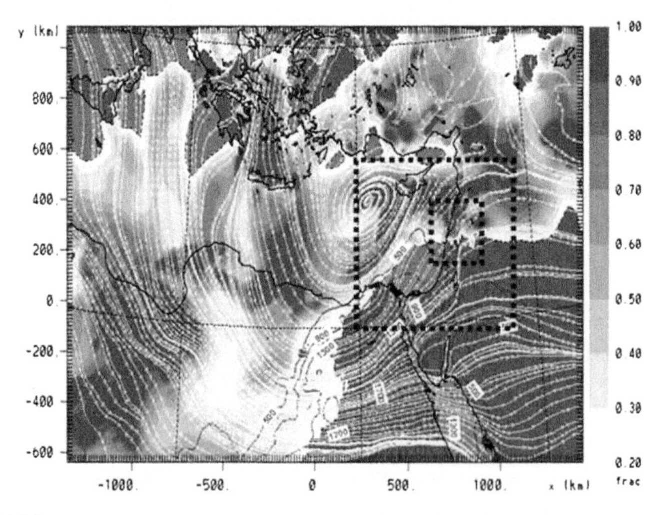

FIGURE 13.24

Cloud cover percentage (grayscale), streamlines at first model layer (green contours), dust – load (red contours in mg m^{-2}). The dashed rectangulars indicate the location of the nested grids (from Solomos et al. 2011). (The color version of this figure is presented in the plate section in the back of the book and the online web version.)

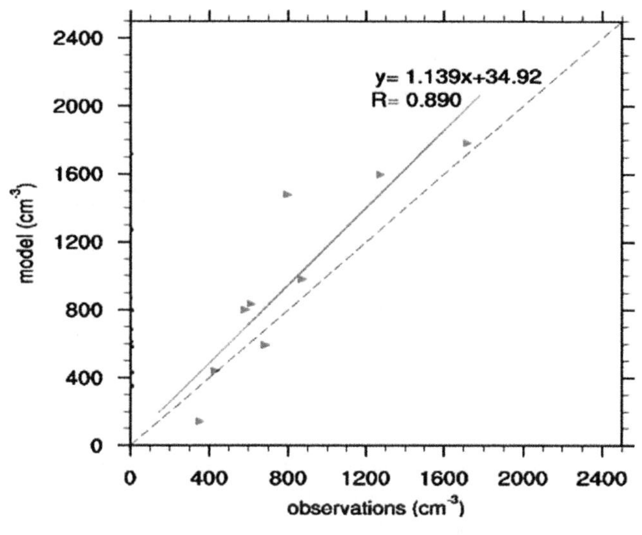

FIGURE 13.25

Comparison of aircraft measurements of natural particles with modeled dust and salt concentrations inside the dust layer (below 2 km). The solid line indicates the linear regression line while the dotted line indicates the $y = x$ line (from Solomos et al. 2011). (The color version of this figure is presented in the plate section in the back of the book and the online web version.)

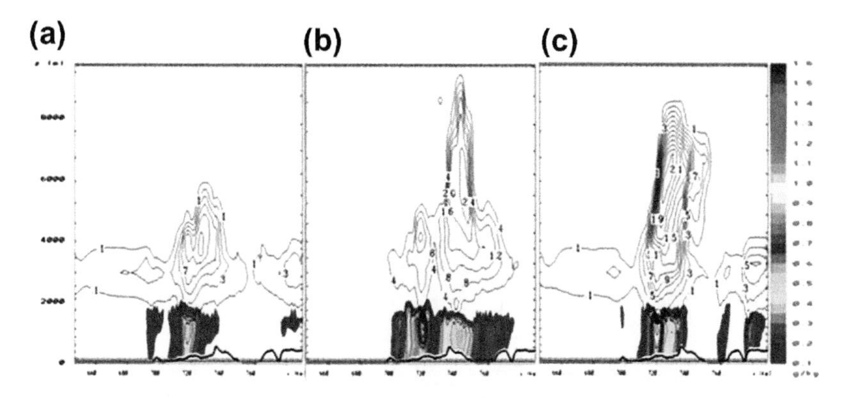

FIGURE 13.26

West to east cross-section of rain mixing ratio (color palette in g kg^{-1}) and ice mixing ratio (black line contours in g kg^{-1}) at the time of highest cloud top over Haifa. a) 0900 UTC 29 January 2003 assuming 5% hygroscopic dust (EXP1). b) 1000 UTC 29 January 2003 assuming 20% hygroscopic dust (EXP2). c) 0900 UTC 29 January 2003 assuming 5% hygroscopic dust and INx10 (EXP3). (The color version of this figure is presented in the plate section in the back of the book and the online web version.)

of airborne particles in meteorology improved the model performance for most cases especially in high precipitation thresholds. This is an indication of the importance of the aerosol processes in the atmosphere and of the need for an integrated atmospheric-air quality modeling approach.

13.4.1 Production of Dust Due to Convective Outflows

Dust modeling is in general based on the concept of saltation and bombardment of desert sand due to wind excess over a specific threshold (e.g., Marticorena and Bergametti 1995; Spyrou et al. 2010). A not so conventional and less studied source of dust in the atmosphere is the so-called "haboobs". These are extended dust fronts that are often found over arid areas in several places over the world (e.g., Sahara, Middle East, Arabian Peninsula, Taklimakan, Gobi, Phoenix, USA, Australia, and others). The formation of these systems is due to convectionally driven density currents that are generated from the evaporative cooling of thunderstorms outflows. An example of "haboob" formation south of the Atlas Mountains on 31 May 2006 has been seen simulated with the use of the limited area model RAMS/ICLAMS (Solomos et al. 2011). The model for this study was set up with four nesting grids: a coarse 12 km, an intermediate of 2.4 km and two grids of 800 m horizontal resolution (Fig. 13.27). In the vertical, the grid increment was 20 m near the ground stretching up to 18 km with a factor of 1.10. The location of the model grids is indicated in Fig. 13.27. The low pressure system

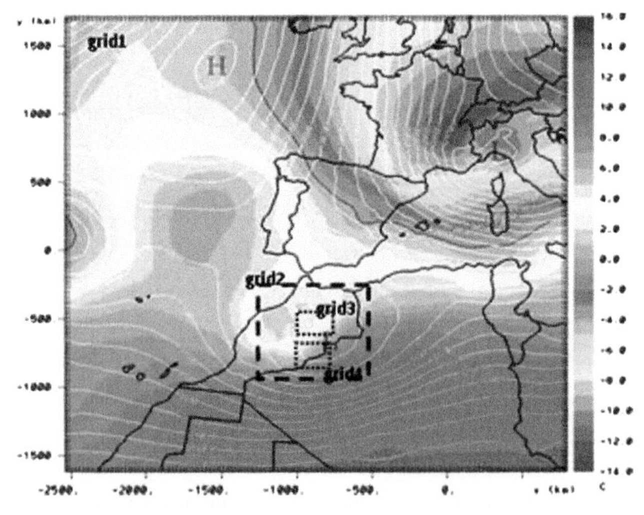

FIGURE 13.27

Geopotential height (white contour lines every 10 gpm) and temperature at 700 hPa (color palette in °C) – 1100 UTC on 31 May 2006. The dashed rectangulars indicate the locations of the nested grids (from Solomos et al. 2012). (The color version of this figure is presented in the plate section in the back of the book and the online web version.)

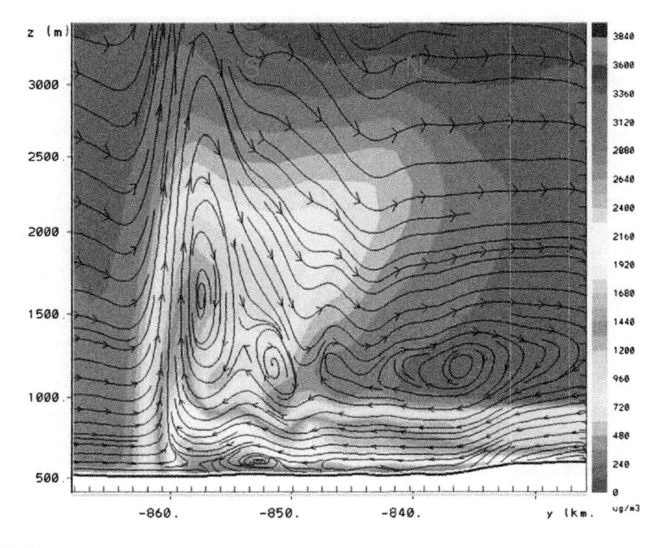

FIGURE 13.28

Dust concentration (color scale in μg m^{-3}) and streamlines at 1850 UTC for a reference frame relative to the propagating speed. The location of the cross section is at 4.4°E and from 31.50°N to 29.80°N. The direction of the motion is from north to south as illustrated with the red arrow on top of the figure (from Solomos et al. 2012). (The color version of this figure is presented in the plate section in the back of the book and the online web version.)

that is evident in Fig. 13.28 over Morocco resulted in significant convective activity and rainfall south of the Atlas Mountains. The evaporation of cloud and rain droplets close to the ground resulted in the formation of a density current that propagated southwards and was accompanied by dust production along the frontal line. As seen in the vertical cross section from the finest (800 × 800 m) model grid (Fig. 13.28), the inner structure of the density current was characterized by an area of reverse flow close to the surface and intense dust production along the propagating system. The highest dust fluxes were collocated with the flow reversal area and the elevated particles were transported up to 2 km inside the density current head. Most of the dust remained trapped inside the density current. Injection of dust in the free troposphere occurred mainly at the top of the propagating head first due to efficient uplifting of preexisting airborne dust along the frontal updrafts and second due to turbulent mixing.

13.4.2 Atmospheric Dispersion Modeling Applications

Another area of mesoscale modeling applications is the coupling of atmospheric with air quality models. Langragian dispersion modeling can benefit from coupling with a prognostic meteorological model especially over complex terrain areas. An example of such an application is the simulation of the fire smoke plume after an industrial accident at the bay of San Francisco on 26 August 2012 (Fig. 13.29). For

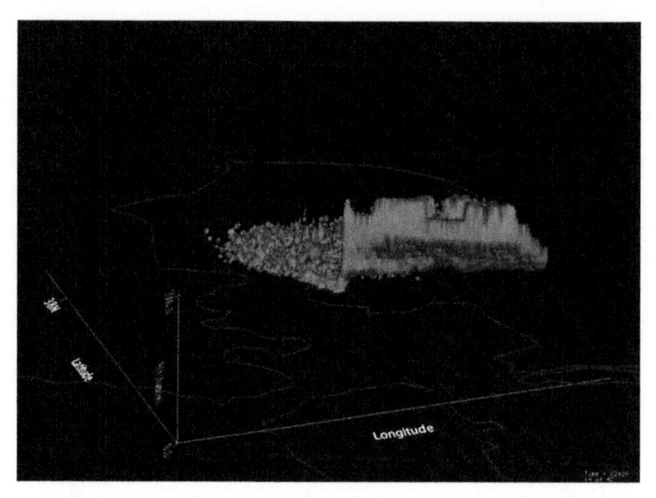

FIGURE 13.29

Simulation of the plume rise due to an industrial accident at the area of San Francisco Bay (26 August 2012). (The color version of this figure is presented in the plate section in the back of the book and the online web version.)

this case study RAMS (Pielke et al. 1992; Cotton et al. 2003) was set up with four nested grids: 9.6 × 9.6 km, 2.4 × 2.4 km, 600 × 600 m, 200 × 200 m. The finest model grid over the area of interest allowed the proper description of the complex sea-land partitioning and the description of wind temporal and spatial variability. The Hybrid Particle and Concentration Transport Model HYPACT (Walko et al. 2001) was driven with RAMS meteorology at 10 minutes intervals and the release was described as Lagrangian source. The shape of the source was assumed to be a 300 m diameter column from the surface up to 1000 m. As seen in Fig. 13.29 a number of particles were forced towards the N close to the surface but the main plume was directed SE due to the clockwise wind veering with height.

13.4.3 Air Quality Applications

Coupling of chemical transport models with advanced meteorological models is a complex task. However, it remains an emerging need in the framework of chemical weather forecasting (Grell and Baklanov 2011). An example of the integrated treatment of the air quality components in an online model is the calculation of the photolysis rates. The direct coupling of the photolytic mechanism with the meteorological parameters gives the ability to take into account any abrupt modification in the radiation budget. This ability is essential over areas with frequent mineral dust events such as the Greater Mediterranean Area (GMA). One such event took place over East Mediterranean during 17-18 April 2004. Significant loads of dust particles were transported from the Libyan coast over South Greece due to a well-established and persistent SW flow over North Africa. Dust particles were mobilized and uplifted over the Mediterranean Sea forming a dust layer that reached 4 km height. The existence of this dust layer modified the radiation budget over the area resulting in modified photolysis rates. Two simulations were performed over the area of interest: a basic run with no impact of dust aerosols on the radiation fluxes (BR) and a second run that takes into account the radiative effect of airborne particles (DR). As seen in Fig. 13.30a the

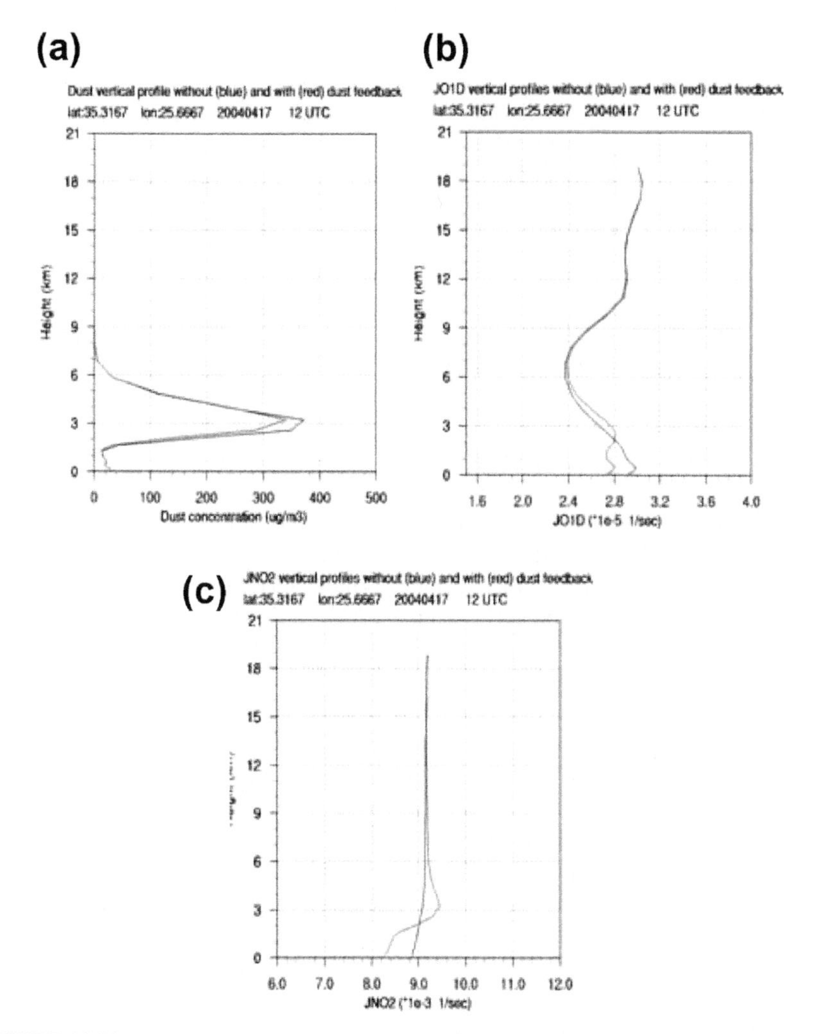

FIGURE 13.30

Vertical profiles of a) dust concentration ($\mu g\ m^{-3}$) and photolysis rates of b) JO1D ($10^{-5}\ sec^{-1}$), and c) JNO2 ($10^{-3}\ sec^{-1}$), with dust impact on radiation (red) and without dust impact on radiation (blue) on 17 July 1200 UTC at Finokalia station, Crete (lat = 35.3°, lon = 25.7°). (The color version of this figure is presented in the plate section in the back of the book and the online web version.)

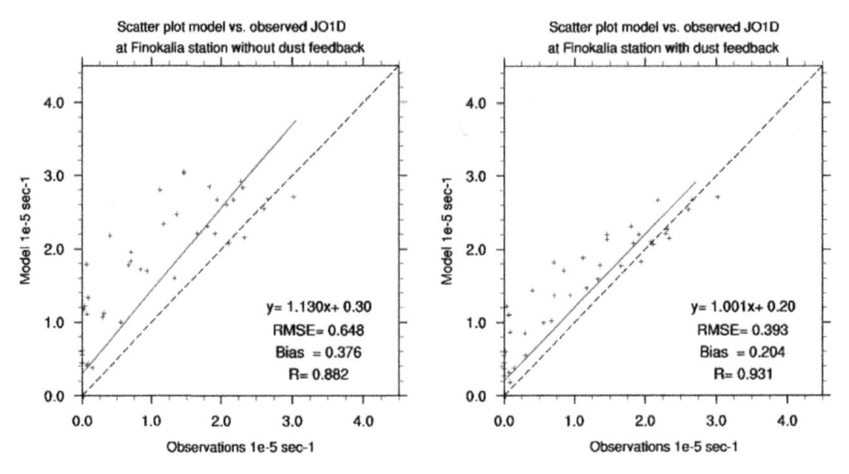

FIGURE 13.31

Scatter plots of modeled versus observed photolysis rates of a) JO1D ($10^{-5}sec^{-1}$) without dust impact on solar radiation, b) JO1D ($10^{-5}sec^{-1}$) with dust impact on solar radiation, at Finokalia station, Crete, on 17 and 18 April 2004. The black dashed line represents the $x = y$ line while the solid line represents the regression line of each plot. (The color version of this figure is presented in the plate section in the back of the book and the online web version.)

dust concentration is affected by the inclusion of the dust impact on radiation due to the perturbance of the wind field. The DR photolysis rates below and at the bottom of the dust layer are lower than the BR rates for both JO1D (Fig. 13.30b) and JNO2 (Fig. 13.30c). This is attributed to the shading effect caused by the dust particles near surface that leads to the decrease in the available actinic flux as discussed in Astitha and Kallos (2009). Within and above the dust layer the situation is reversed due to the scattering of the incoming solar radiation from the dust particles at the top of the dust layer. The differences smooth at about 7 km.

Comparison of the model results with available measurements from the Finokalia station (Crete, 35°19'N, 25°40'E, altitude 250 m above sea level) shows that the BR simulation overestimates the photolysis rates since it does not take into account the decrease in the incoming solar radiation due to the dust layer (Fig. 13.31). The inclusion of this interaction leads to a more accurate representation of the photolytic activity in the area.

13.5 Dynamic Downscaling

A major application of regional and mesoscale models in recent years has been to create high spatial and temporal resolution atmospheric fields for use in hydrologic, ecological, agricultural, and other impact studies.

As reported in Castro et al. (2005) and Pielke and Wilby (2012), the term "downscaling" refers to the use of either fine spatial-scale numerical atmospheric models (dynamical downscaling), or a statistical relationship (statistical downscaling) using a "limited area model" (LAM) in order to achieve detailed regional, mesoscale, and local atmospheric data. The starting point for downscaling is typically a larger-scale atmospheric or coupled oceanic-atmospheric model run globally (GCM). The downscaled high resolution data can then be inserted into other types of numerical simulation tools such as hydrological, agricultural, and ecological models.

Dynamical downscaling can be classified into four distinct types:

1. Type 1: LAM forced by lateral boundary conditions from a numerical weather prediction GCM or global data reanalysis at regular time intervals (typically 6 or 12 h), by bottom boundary conditions (e.g., terrain), and specified initial conditions. A numerical weather prediction GCM is a GCM in which the global initial atmospheric conditions are not yet forgotten.
2. Type 2: LAM initial atmospheric conditions have been forgotten, but results are still dependent on the lateral boundary conditions from an NWP (numerical prediction model) GCM or global data reanalysis and on the bottom boundary conditions.
3. Type 3: Lateral boundary conditions are provided from a GCM which is forced with specified surface boundary conditions.
4. Type 4: Lateral boundary conditions from a completely coupled Earth system global climate model in which the atmosphere-ocean-biosphere and cryosphere are interactive.

Tables 13.2 and 13.3 illustrate examples of each type of downscaled model, including the model dependence on the indicated constraints for each of the four types. Examples of Type 2 dynamic downscaling is reported in Lo et al. (2008), Kanamitsu and Kanamaru (2007), and Rockel et al. (2008), while examples of Type 3 are given in Castro et al. (2007) and Veljovic et al. (2010). A summary of Type 4 dynamical downscaling is reported in Chapter 10 of the 2001 Intergovernmental Panel on Climate Change (IPCC) science report (Houghton et al. 2001).

With short-term numerical weather prediction (Type 1), the observations used in the analysis to initialize a model retain a component of realism even when degraded to the coarser model resolution of a global model (i.e., the data are sampled from a continuous field). This realism persists for a period of time (up to a week or so), when used as lateral boundary conditions for a weather prediction LAM. This is not true with Type 4 simulations, where observed data does not exist to influence the predictions (Pielke 2001c). LAMs cannot significantly increase predictability if the solution is highly dependent on the large-scale forcing supplied by the lateral boundaries. Even when the model solution is strongly influenced by the surface boundary, improved skill still cannot be achieved without accurate lateral boundary conditions.

Figure 13.32 illustrates the degradation of the results as the downscaled model domain (using RAMS) is enlarged and as the spatial resolution is made coarser. Rockel et al. (2008) showed that a spectral nudging technique permits more added variability at smaller scales than a four-dimensional internal grid nudging on large domains, but both of these nudging approaches adjusts the model results towards the larger-scale model. In that sense, the downscaled results are a slave of the parent model.

Table 13.2 Dependence of Regional Model on Indicated Constraints. From Castro et al. (2005)

	TYPE I	TYPE II	TYPE III	TYPE IV
Bottom Boundary Conditions	Terrain;	Terrain;	Terrain;	Terrain; Soils
	LDAS[a];	Climatological Vegetation;	Climatological Vegetation;	
	Observed SSTs	Observed SSTs	Observed SSTs;	
		Deep Soil Moisture	Deep Soil Moisture	
Initial Conditions	ETA Analysis Field	none	none	none
Lateral Boundary Conditions	Global Forecast System Atmospheric Model[b]	NCEP Reanalysis[c]	Global Model Forced by Observed SSTs	IPCC[d]; US National Assessment[e]
Regional	ETA[f], MM5[g], RAMS[h], ARPS[i]	PIRCS[j]	COLA[k] -ETA[l]	RegCM[m]

[a] *http://ldas.gsfc.nasa.gov/.*
[b] *http://wwwt.emc.ncep.noaa.gov/gmb/moorthi/gam.html.*
[c] *Kalnay et al. (1996).*
[d] *Houghton et al. (2001).*
[e] *http://www.gcrio.org/NationalAssessment.*
[f] *Black (1994).*
[g] *Grell et al. (1994).*
[h] *Pielke et al. (1992).*
[i] *Xue et al. (2000b, 2001).*
[j] *Takle et al. (1999).*
[k] *http://www-pcmdi.llnl.gov/modldoc/amip/14cola.html.*
[l] *Mesinger et al. (1997).*
[m] *Giorgi (1993a, b).*

Type 1 dynamic downscaling has been a very successful application of this technology. High spatial and temporal resolution weather forecasts, with savings in lives, has resulted. Type 2 dynamic downscaling is used to assess what is the upper level of skill possible from Type 3 and Type 4 downscaling with the later the most difficult. Type 3 dynamic downscaling (on seasonal time scales) is at the forefront of our current capabilities and even here its value-added is limited to special cases such as when the regional/mesoscale model has surface information (such as saturate soils) at a much higher spatial scale than in the parent model.

Type 4 dynamic downscaling on time scales longer than a season, despite its widespread use (e.g., Solomon et al. 2007) has not provided any value added, and,

Table 13.3 Examples of predictability. From top to bottom of table: more constraints to fewer constraints; from bottom to top of table: less predictive skill to greater predictive skill. From Castro et al. (2005).

	Type	Constraints
Day-to-day weather prediction	1	initial conditions; lateral boundary conditions topography; other bottom land boundary conditions; solar irradiance; well-mixed greenhouse gases
Seasonal weather simulation	2	lateral boundary conditions; topography; other bottom land boundary conditions; solar irradiance; well-mixed greenhouse gases
Season weather prediction	3	topography; other bottom land boundary conditions; sea-surface temperatures; solar irradiance; well-mixed greenhouse gases
Multi-year climate prediction	4	topography; solar irradiance; well-mixed greenhouse gases

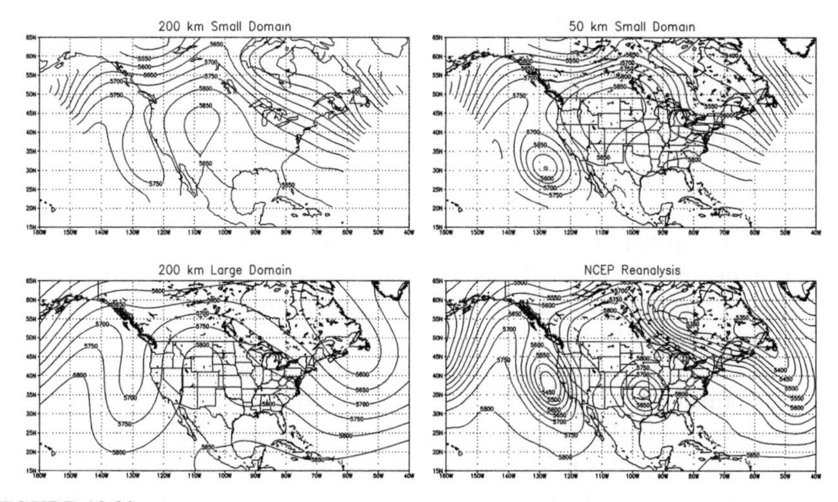

FIGURE 13.32

The 500 mb height (m) on 0Z UTC, 12 May 1993, for the indicated model experiments and NCEP Reanalysis (from Castro et al. 2005).

indeed, is misleading the impacts community regarding its value. As reported in Pielke et al. (2011), there are several reasons for its failure. These can be summarized as

1. As a necessary condition for an accurate Type 4 downscaling prediction, the multi-decadal global climate model simulations must include all first-order climate

forcings and feedbacks. However, they do not (see for example: NRC 2005; Pielke Sr. et al. 2009).

2. The global multi-decadal predictions themselves, are unable to skillfully simulate major atmospheric circulation features such the Pacific Decadal Oscillation (PDO), the North Atlantic Oscillation (NAO), El Niño and La Niña, and the South Asian monsoon (Annamalai et al. 2007; Wyatt 2012).
3. While dynamic regional downscaling yield higher spatial resolution, the regional climate models are strongly dependent on the lateral boundary conditions and interior nudging by their parent global models (e.g., see Rockel et al. 2008). Large-scale climate errors in the global models are retained and could even be amplified by the higher spatial resolution Type 4 models. Examples that document the serious errors in global models include Anagnostopoulos et al. (2010), Kundzewicz and Stakhiv (2010), Stephens et al. (2010), Fyfe et al. (2011), Driscoll et al. (2012), Goddard et al. (2012), Jiang et al. (2012), Sakaguchi et al. (2012), van Haren et al. (2012), van Oldenborgh et al. (2012), Xu and Yang (2012), and Sun et al. (2013).
4. Since the global multi-decadal climate model predictions cannot accurately predict circulation features such as the PDO, NAO, El Niño, and La Niña they cannot provide accurate lateral boundary conditions and interior nudging to the Type 4 models (Compo et al. 2011; Wyatt 2012).
5. The regional models themselves do not have the domain scale (or two-way interaction) to skillfully predict these larger-scale atmospheric features.
6. There is also only one-way interaction between the regional/mesoscale and global models which is not physically consistent. If the regional/mesoscale model significantly alters the atmospheric and/or ocean circulations, there is no way for this information to alter the larger-scale circulation features which are being fed into the regional model through the lateral boundary conditions and nudging.
7. When higher spatial analyses of land use and other forcings are considered in the regional/mesoscale domain, the errors and uncertainty from the larger model still persists thus rendering the added complexity and details ineffective (Mishra et al. 2010; Ray et al. 2010).
8. The lateral boundary conditions for input to Type 4 downscaling require regional-scale information from a global forecast model. However the global model does not have this regional-scale information due to its limited spatial resolution. This is, however, a logical paradox since the Type 4 model needs something that can only be acquired by a regional/mesoscale model. Therefore, the acquisition of lateral boundary conditions with the needed spatial resolution becomes logically impossible.

Finally, there is sometimes an incorrect assumption that although global climate models cannot predict future climate change as an initial value problem, they can predict future climate statistics as a boundary value problem (Palmer et al. 2008). With respect to weather patterns, for the Type 4 downscaling regional/mesoscale (and global) models to add value over and beyond what is available from the historical, recent paleo-record, and worse case sequence of days, however, they must be able to skillfully predict the *changes* in the regional weather statistics. There is no evidence yet that they have this skill.

13.6 Mesoscale Modeling of Extraterrestrial Atmospheres[10]

The reinvigoration of Mars exploration in the early 1990s combined with the maturation of terrestrial mesoscale modeling codes provided fertile ground for the application of suitably modified models to the atmosphere of Mars (Rafkin et al. 2001; Toigo et al. 2002; Tyler et al. 2002). The number of such Mars models has grown steadily in the subsequent decade (Moudden and McConnell 2005; Wing and Austin 2006; Kauhanen et al. 2008; Spiga and Forget 2009). In addition to Mars, limited data return from missions to Venus and Titan have proven sufficient to conduct equally limited convective cloud modeling studies of these atmospheres (Baker et al. 1998; Hueso and Sánchez-Lavega 2006; Barth and Rafkin 2007; McGouldrick and Toon 2007; Barth 2010). While all of these studies benefit from the previous decades of terrestrial work, each planetary body presents unique challenges and often highlights many of the assumptions taken for granted in terrestrial modeling applications. Still, the use of mesoscale models, particular for Mars, has become an integral part of interpreting the data returned from missions. Furthermore, the models are increasingly used to provide constraints and bounds on environmental conditions in support of mission planning and operations (e.g., Rafkin and Michaels 2003; Tyler et al. 2008; Toigo and Richardson 2003; Michaels and Rafkin 2008).

13.6.1 Mars

Mars presents many of the same meteorological phenomena found on Earth: A mean overturning meridional circulation (e.g., Haberle et al. 1993; Wilson 1997; Zalucha et al. 2010) seasons associated with a planetary obliquity of 23° (e.g., Hess et al. 1977; Jakosky 1983; Guo et al. 2009) baroclinic disturbances and surface fronts (e.g., Barnes et al. 1993; Collins et al. 1995; Tyler and Barnes 2005) water and carbon dioxide condensate clouds (e.g., Smith et al. 2001; Hinson and Wilson 2004; Benson et al. 2006; Michaels et al. 2006; Colaprete et al. 2008) katabatic and anabatic diurnal circulations (Tyler and Barnes 2005; Sta. Maria et al. 2006; Spiga 2011) dust storms (e.g., Gierasch 1974; Cantor et al. 2001; Basu et al. 2006; Wang 2007) and dust devils (e.g., Ferri et al. 2003; Fisher et al. 2005; Ellehoj et al. 2010). However, in comparison to Earth, the short radiative timescale, low atmospheric density, extreme topographic relief, and highly variable amounts of radiatively active dust aerosol, all require special consideration when attempting to numerically model the atmosphere of Mars.

[10]This section was written by Dr. Scot Rafkin of the Southwest Research Institute in Boulder, Colorado.

The radiative forcing environment of Mars is a key element in atmospheric modeling. Although the atmosphere is dominantly CO_2, the global mean surface pressure of ~600 Pa provides little in the way of greenhouse warming (Gierasch and Goody 1968) and the radiative time constant is short (less than 1 Mars day = 1 sol = ~24 h 20 min). Further, without a large oceanic and atmospheric thermal mass like Earth, there are large diurnal swings of temperature. A hot, tropical day on Mars might reach 270 K, but drop to 170 K at night. The polar night temperatures are limited by the CO_2 condensation temperature of ~150 K with a slight dependence on pressure. Atmospheric dust also has a strong radiative impact (Pollack et al. 1979). Dust is bright in the visible spectrum, but it is also very absorptive and will result in an overall heating within a dusty layer. Dust also has a strong absorption band near 9 μ m, which provides a greenhouse warming. Overall, atmospheric dust will reduce the near-surface daytime temperatures and increase the overnight temperatures much like clouds on Earth (e.g., Conrath 1975; Haberle et al. 1982; Pollack 1982).

Unlike the atmospheric surface layer on Earth that is heated in large part by turbulent eddy fluxes, Mars is heated most strongly by radiative flux convergence (Haberle et al. 1993; Larsen et al. 2002; Michaels and Rafkin 2004; Savijärvi and Kauhanen 2008). During the day, the hot surface radiates upward where dust and CO_2 absorb the upwelling energy. This creates deep convective boundary layers and very unstable lapse rates (Hinson et al. 2008, Smith et al. 2004). At night, the situation is reversed and very strong nocturnal inversions develop (e.g., Smith et al. 2004, Spiga et al. 2011). The effect of radiative heating is magnified by the low atmospheric mass; a unit of energy input will create almost a 100 times larger increase in temperature than on Earth due solely to the difference in density. For the same reason, the efficiency of turbulent energy transport is greatly reduced on Mars. For a given eddy covariance, the lower density of Mars would result in a flux roughly 100 times smaller than Earth. However, the strong heating on Mars usually means covariances are almost an order of magnitude larger than Earth so that turbulent fluxes may only be one rather than two orders of magnitude larger than on Earth (Haberle et al. 1993; Tillman et al. 1994; Michaels and Rafkin 2004; Sorbjan 2007).

A major consequence of the differences in atmospheric heating between Earth and Mars is that elevation exerts comparatively little influence on the near-surface temperature of Mars (Fig. 13.33). Isotherms tend to follow topography on Mars (i.e., mountain peaks are nearly the same temperature as the surrounding plains) whereas isotherms tend to be quasi-horizontal on Earth (i.e., mountain peaks are colder than the surrounding plains). From hydrostatic considerations, the horizontal pressure gradient driven by thermal differences will be much larger on Mars than on Earth and thermally driven slope flows should be concomitantly large (Gierasch and Sagan 1971; Mahrt 1982; Ye et al. 1990b; Baron et al. 1998). Ye et al. (1990b) estimated that the upslope circulations should be 2.5 times stronger than on Earth. Subsequent mesoscale modeling studies are in general agreement with these findings (Rafkin et al. 2001, 2002; Rafkin and Michaels 2003; Toigo and Richardson 2003; Spiga et al. 2011) and further demonstrated that even relatively minor topographical features such as hills and crater rims can produce substantially thermally-driven circulations

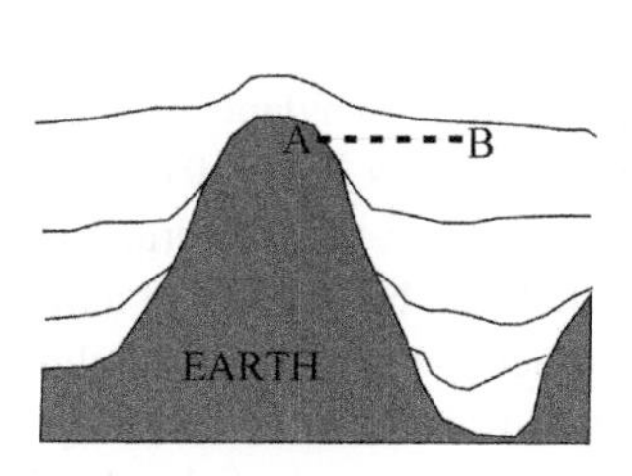

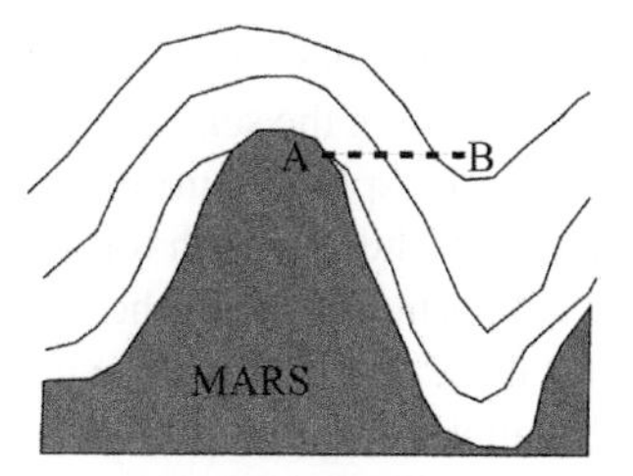

FIGURE 13.33

Schematic of Mars vs. Earth temperature structure with topography.

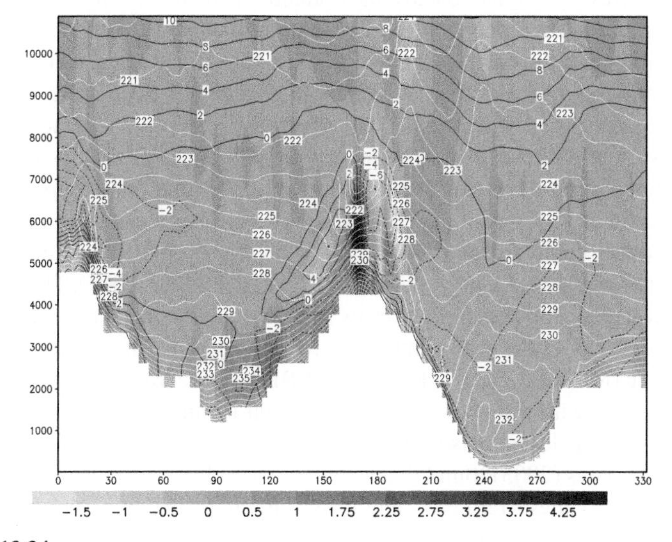

FIGURE 13.34

Example of topo-thermal circulation from mesomodel.

(Fig. 13.34). Another consequence of the topography-following isotherms of Mars is that the techniques developed for initializing Earth mesoscale models from larger-scale models must be largely abandoned at Mars. When mapping coarse atmospheric data to more detailed topography in a mesoscale model the use of lapse rates to extrapolate or extend atmospheric data fails. Instead, methods that preserve the vertical thermal structure near the surface must be employed while simultaneously ensuring that adiabatically unstable lapse rates and spurious horizontal pressure gradients between adjacent columns are not introduced (Rafkin et al. 2001; Tyler et al. 2002).

The presence of dust amplifies the thermal tide compared to that of Earth (e.g., Wilson and Hamilton 1996; Banfield et al. 2000). The daily heating and cooling of the atmospheric column due to solar absorption produces the second largest pressure signal with the first being the seasonal cycle associated with deposition of 25% of the dominantly CO_2 atmosphere onto the winter pole (Hess et al. 1977; Hourdin et al. 1993). Interaction of the tide with topography and wave-wave interaction with itself produces diurnal harmonics with the semidiurnal tide being of high amplitude. The tide has at least three major impacts when it comes to mesoscale modeling. The first is that model domains and lateral boundary conditions must be carefully configured so as to reduce or eliminate the distortions and reflections of the thermal tide as it propagates into and out of the modeling domain. Tyler et al. (2002) found that a superhemispheric mother domain centered on the pole and spanning beyond the subtropics in the antipode hemisphere provided the best treatment of the tide since the bulk of the wave never had to leave the domain. While this configuration provides for optimal numerical treatment it is disadvantageous in that the grid is large and is not optimal for simulations in the tropics or subtropics. The second impact is that boundary conditions must be updated frequently enough so that the boundaries can provide proper forcing associated with the tide. With a period of $\sim$ 12 hours for the semidiurnal tide, updates of at least every three hours are needed to minimize aliasing. One and one-half hours are more typical. The third impact is that thermal compression must be accounted for in the dynamical core. An intercomparison of Martian mesoscale models found that nonhydrostatic dynamical cores overestimated the diurnal surface pressure cycle compared to hydrostatic cores (Tyler and Barnes 2005). This effect was traced to incomplete compressibility terms in the nonhydrostatic cores. The RAMS-based acoustic model core was completely redone adding compressibility into the implicit solver (Michaels et al. 2006). The WRF-based core replaced the pressure tendency equation with a geopotential equation that included diabatic effects (Spiga and Forget 2009). Tides can also interfere with or generate mesoscale circulations. Rafkin and Michaels (2003) found that as the tide traveled from east to west through the 1000 km long Valles Marineris canyon system, also aligned nearly east to west, the tidal pressure gradient and canyon channeling strongly accelerated along canyon winds to near 50 m s^{-1}.

Direct validation of many Mars mesoscale modeling results is not yet possible. There have been only four in situ meteorological experiments on the surface of Mars: Viking Lander 1 and 2 (VL1 and VL2) (Hess et al. 1976, 1977) the atmospheric structure experiment on the Mars Pathfinder Lander (Schofield et al. 1997) and the Phoenix Scout Lander (Tamppari et al. 2008, 2010). VL1 returned data for nearly a Mars year in the subtropical latitudes of the northern hemisphere plain, and VL2 returned data from the northern middle latitudes for over two Mars years, including a period during a planet encircling dust storm. The Pathfinder experiment lasted 83 sols. The Phoenix mission was of similar duration at a location in the higher northern latitudes. All of these missions took measurements in the relatively flat northern plains and thus were not able to fully validate the strong thermal flows simulated by models. Mesoscale simulations of these landing sites have all shown reasonably good agreement with the observations (Tyler et al. 2002, 2008; Rafkin and Michaels 2003; Toigo and Richardson 2003; Siili et al. 2006; Michaels and Rafkin 2008).

More typically, mesoscale model validation has come indirectly from imagery and occasionally by thermal remote sensing. As an example, early morning linear cloud features were observed by the Viking Orbiter (e.g., Briggs et al. 1977) that were suspected to be caused by bore waves (Hunt et al. 1981) or a hydraulic jump (Kahn and Gierasch 1982). The strong nocturnal inversion and steep topography of Mars is conducive to these types of disturbances; extremely cold and dense katabatic flows propagating into the inversion present is a classic scenario. Sta. Maria et al. (2006) found modeled bore waves that closely matched the observed characteristics. Dust streaks and other imaged albedo features provide indications of wind direction (e.g., Kuzmin et al. 2001; Greeley et al. 2003; Fenton et al. 2005; Michaels 2006) while wave clouds can provide indirect clues about wind direction and stability. Spiga (2011) was able to identify the signature of adiabatically warmed air predicted to descend downslope in the surface thermal signal sensed from orbit.

Water ice and dust clouds are fairly common above the peaks of the Tharsis Mountains that rise 20 km above the surrounding plains (Benson et al. 2006). Michaels et al. (2006) simulated water ice clouds with the observed coverage, opacity, and temporal variations consistent with observations. The cloud formation processes was found to result from two interacting and related circulations. First, the strong upslope circulation during the afternoon vented the water vapor from the lower plains up to 20 km above the mountain peak. This water vapor was then entrained into cross-mountain flow and resulted in a mountain wave cloud. Hinson and Wilson (2004) had previously identified the thermal tide as an additional mechanism that cooled the air to saturation. Also imaged above the mountain Arsia Mons was a spectacular spiral dust cloud. Rafkin et al. (2002) were able to successfully reproduce this feature, showing how anabatic winds racing up the slopes spiraled inward and lifted dust to produce the feature. In both the water and dust studies (Rafkin et al. 2002; Michaels et al. 2006) the strong plumes of rising motion above the mountains were found to contribute to a substantial amount of the total vertical flux of the mean meridional circulation. Thus, mesoscale motions have an important contribution to the global mass, energy, dust, and water cycle. Recent observations by the Mars Climate Sounder (McCleese et al. 2007) has revealed the vertical structure of dust as never before, and shows a persistent elevated layer of dust 15-25 km above the surface with a corresponding minimum above the boundary layer (Kleinböhl et al. 2009; McCleese et al. 2010; Heavens et al. 2011). Rafkin (2012) has interpreted this as further evidence to the venting transport of topographically-anchored mesoscale circulations (Fig. 13.35).

Perhaps one of the most remarkable results from all the mesoscale modeling studies performed for Mars to date is that they have all achieved success without the use of actual reanalysis or, with the exception of atmospheric dust loading, observational atmospheric data in initialization or boundary conditions. Instead, the simulations have all used input from Mars general circulation models that lack data assimilation or observational initialization from an observed atmospheric state. The Mars GCMs are started from rest with an isothermal atmosphere; the GCM data are realizations from a climate model. A reasonable representation of the atmospheric dust load and distribution is critical to properly simulating the thermal structure and

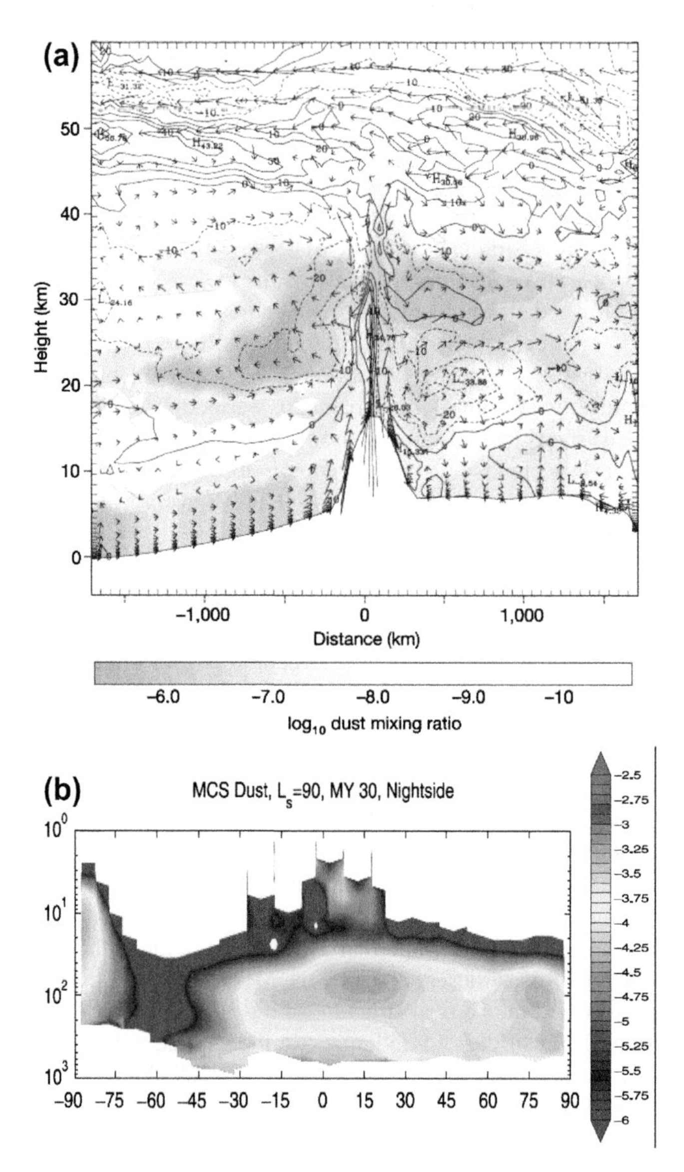

FIGURE 13.35

(a) A vertical, east–west cross-section from grid two simulated by the Mars Regional Atmospheric Modelling System (MRAMS), cutting through the centre of Arsia Mons. (b) MCS image of the dust profile. From Rafkin et al. (2002). (The color version of this figure is presented in the plate section in the back of the book and the online web version.)

therefore the winds. Observed maps of total column dust opacity are available for several Mars years (Smith et al. 2001; McCleese et al. 2010). The vertical distribution of dust is usually assumed to follow a profile with a relatively well mixed boundary layer with decreasing mixing ratio above (Conrath 1975; Wolkenberg et al. 2009). The validity of this profile is questionable given recent observations of the vertical dust distribution (Heavens et al. 2011). On Earth, initialization with observationally unconstrained GCM data would be unlikely to successfully reproduce specifically observed phenomena like spiral dust clouds. On Mars, however, the short radiative time constant means that there is little memory in the atmospheric system; thermal perturbations are rapidly dissipated. Further, the strong, repeatable global thermal tide and regional, thermally-forced slope flows can overwhelm perturbations at the synoptic scale. Weather forecasting and modeling on Mars is very much a boundary condition problem where the circulations are strongly driven by heating of topography and the thermal tide. The surface properties (topography, thermal inertial, albedo) are now well established from orbital measurements (Smith et al. 2001; Putzig et al. 2005). As long as the dust distribution is reasonably correct, the atmosphere quickly responds to the heating and topography. Baroclinic disturbances can produce non-repeatable variations, but the overall circulation quickly adjusts back to the radiatively- and thermally-driven state after passage of the system. Recent work has focused on Mars data assimilation (Lewis and Barker 2005; Montabone et al. 2005; Lewis et al. 2007; Hoffman et al. 2010; Kahre et al. 2010) which will almost assuredly improve the prediction skill of the models. However, the improvement is unlikely to be as large as that for Earth where proper initialization with an observed state is absolutely necessary to demonstrate forecasting skill over any meaningful duration.

The one area where Mars does show significant non-predictability and perhaps demonstrates some climate memory is within the dust cycle. As dust settles from dust storms, it changes the surface thermal inertia and albedo, which can then feedback to slightly alter the circulation (Newman et al. 2002; Basu et al. 2006; Kahre et al. 2006; Fenton et al. 2007). In this way, the system can retain some memory of previous events over seasonal or even annual timescales. The origin of some dust storms is known. For example, when baroclinic circulations are properly phased with the tide, dust lifted along the front can be advected from one hemisphere to another (Wang et al. 2003: whereas inopportune phasing keeps the dust confined locally. Dust storms originating along the polar caps are largely attributable to katabatic flows and baroclinic storms (Cantor et al. 2001; Toigo et al. 2002). There are a very large number of storms, however, which have no obvious generation mechanism. This includes the large, planet encompassing storms that occur every few Mars years (Haberle 1986; Smith et al. 2002; Clancy et al. 2010) as well as numerous almost daily local storms and regional scale storms that occur multiple times in a year (Cantor et al. 2001). For now, these storms are unpredictable. This may change when reanalysis data become available for initialization.

In addition to the currently unpredictable nature of many dust storms, the instability mechanism that allows some to grow rapidly from common daily storms to regional storms that last for days is unknown. The first attempt to explain this growth was through a feedback mechanism similar to the growth of tropical disturbances to tropical storms (Gierasch 1974). Radiative heating of a localized dust disturbance would preferentially heat the atmospheric column, hydrostatically lower the pressure, and accelerate the winds to pick up more dust. This idea was largely abandoned when subsequent observations failed to detect anything resembling rotation tropical disturbances. Recent work (Rafkin 2009, 2012) suggests dismissal of the idea may have been premature. A feedback process may still be operating even if there are no obvious or outward signs of organized rotation. Further, dust does not respond to vertical motions like water vapor making identification of any dynamically coherent structure from orbit very difficult. The descending eye of a hurricane is clear because liquid water evaporates in the adiabatically warming air. Dust will not evaporate, which would preclude the development of a visible eye wall. Likewise, water condenses where air is rising, making the spiral arm band structures visible through clouds. Dust does not provide such a convenient tracer and more often than not forms an optically thick obscuring haze that hides whatever structure may be below. Additional observations, and ideally, in situ measurements will be required to test whether there exists a Wind-Enhanced Interaction of Radiation and Dust (WEIRD) process (Rafkin 2009) on Mars that is analogous to the Wind-Induced Sensible and Latent Heat Exchange (WISHE) mechanism for hurricanes (Emanuel 1991) on Earth. WISHE may be operating in many tropical disturbances that never grow to cyclones due to damping processes (e.g., shear, cooling water, etc.). WEIRD may likewise be operating in many dust disturbances even if most do not grow to fully developed dynamical systems.

13.6.2 Titan and Venus

To date, there have been no true mesoscale modeling studies of any planetary bodies other than Earth and Mars. However, the clouds of Titan and Venus have been numerically simulated with what are mesoscale model cores or with codes sharing heritage with mesoscale models. Even so, these cloud modeling studies are extremely limited owing to the continued paucity of observational data. The Cassini Orbiter (Matson et al. 2002) and the Huygens descent probe (Lebreton and Matson 2002; Zarnecki 2002; Lebreton et al. 2005) have identified cloud activity in the dominantly N_2 Titan atmosphere (Griffith et al. 1998; Porco 2005; Rannou et al. 2006; Schaller et al. 2006). Methane, which is stable as a vapor, liquid and solid in the cryogenic temperature regime of Titan plays the analogous role of water in the hydrological cycle of Earth (Lunine and Atreya 2008). Much less is known about other weather systems, including mesoscale circulations and disturbances, on Titan. The time variations, rapid changes of cloud height, and cloud morphology of the clouds observed near Titan's South Pole (corresponding to the presumed rising branch of the Hadley cell in the austral summer) strongly suggest deep convective clouds (Griffith et al. 1998; Brown et al. 2002; Mitchell et al. 2006). The temperature profile for Titan is fairly well constrained. Further, due to CH_4 radiative properties and strong N_2 pressure broadening, the radiative time constant for Titan is greater than a Titan year; there is only minor seasonal and latitudinal variation of temperature (e.g., Coustenis and Bézard 1995). The methane distribution is more poorly constrained. The one in situ humidity observation is from the Huygens probe in the tropics, which measured a surface humidity of ~45% and a well mixed CH_4 profile up to ~8 km where the atmosphere saturated (Niemann et al. 2005). The temperature and methane profile measured by the probe contains near zero convective available potential energy (CAPE). Given the near globally invariant temperature profile, the methane abundance in the vicinity of the convective clouds must be sufficiently different than the tropics and provide sufficient CAPE so as to generate clouds with characteristics similar to those observed. Hueso and Sánchez-Lavega (2006) were the first to numerically demonstrate that by increasing the boundary layer abundance (and therefore CAPE) that methane thunderstorms could be produced. Barth and Rafkin (2007, 2010) further demonstrated how varying the methane abundance controlled cloud top height, precipitation rates, microphysical characteristics, and cloud optical properties. Methane abundance corresponding to ~65% will produce cloud tops near the tropopause at 40 km.

Venus is also known to have convective clouds in the lowest cloud layer between ~49 to 55 km (Seiff and Kirk 1982; Linkin et al. 1986; Ingersoll et al. 1987; Young et al. 1987). In situ measurements by the Vega balloons measured updrafts and downdrafts of 1 to 3 m s^{-1} at ~54 km altitude. The only study to investigate these clouds with a coupled dynamic-microphysics model is Baker et al. (1998). Somewhat similar to closed-cell convective clouds on Earth, the simulated Venus clouds were found to have strong coherent downdrafts with more gradual and broader upwelling. The strongest vertical velocities in the simulated clouds were in the downdraft in excess of 7 m s^{-1}. The downdrafts penetrated several kilometers into the stable layer below and resulted in compressional heating several factors higher than radiative heating. The rapid transport and penetrative nature of the convection may have implications for the overall Venus chemical cycle which is more typically modeled with 1D diffusive column models (e.g., James et al. 1997; Krasnopolsky 2006; McGouldrick and Toon 2007).

Chapter 13 Additional Readings

Books and articles which provide in-depth discussions of the different types of mesoscale systems include:

Anthes, R.A., 1982: *Tropical Cyclones, Their Evolution, Structure and Effect.* AMS monograph, Boston, Massachusetts.

Atkinson, B.W., 1981: *Mesoscale Atmospheric Circulations.* Academic Press, New York, 495 pp.

Baines, P.G., 1995: *Topographic Effects in Stratified Flows.* Cambridge University Press, New York, 482 pp.

Banta, R., and W. Blumen, 1990: *Atmospheric Processes Over Complex Terrain.* Meteorological Monographs (American Meteorological Society), W. Blumen, Ed., No. 45. 323 pp.

Browning, K.A., (Ed.), 1982: *Nowcasting.* Academic Press, London and New York, 256 pp.

Browning, K.A., 1986: Conceptual models of precipitation systems. *Wea. Forecasting*, **1**, 23-41.

Cotton, W.R., and R.A. Pielke Sr., 2007: *Human Impacts on Weather and Climate*. Cambridge University Press, 330 pp.

Cotton, W.R., G. Bryan, and S. van den Heever, 2011: *Storm and Cloud Dynamics*, 2nd edition. Academic Press, Elsevier Publishers, The Netherlands, 820 pp.

Markowski, P. and Y. Richardson, 2010: Mesoscale Meteorology in Midlatitudes. Wiley and the Royal Meterological Society. ISBN: 978-0-470-74213-6. 430 pages.

Pielke Jr., R.A., and R.A. Pielke, Sr., 1997: *Hurricanes: Their Nature and Impacts on Society*. John Wiley and Sons, England, 279 pp.

Pielke Jr., R.A., and R.A. Pielke, Sr., Eds., 2000: *Storms*. Volumes I and II. Routledge Press, London.

Pielke Sr., R.A., 1990: *The Hurricane*. Routledge Press, London, England, 228 pp.

Pielke, R.A., and R.P. Pearce, Editors, 1994: *Mesoscale Modeling of the Atmosphere*. American Meteorological Society Monograph, Volume 25, 167 pp.

Rotunno, R., J.A. Curry, C.W. Fairall, C.A. Friehe, W.A. Lyons, J.E. Overland, R.A. Pielke, D.P. Rogers, S.A. Stage, G.L. Geernaert, J.W. Nielsen, and W.A. Sprigg, 1992: *Coastal meteorology – A review of the state of the science*. Panel on Coastal Meteorology, Committee on Meteorological Analysis, Prediction, and Research, Board on Atmospheric Sciences and Climate, Commission on Geosciences, Environment, and Resources, National Research Council, National Academy Press, Washington, D.C., 99 pp.

Smith, R.B., 1979: Influence of mountains on the atmosphere. *Adv. Geophys.*, **21**, 87-217.

Trapp, R.J., 2013: *Mesoscale-Convective Processes in the Atmosphere*, Cambridge University Press, 400 pp.

Warner, T., 2011: *Numerical Weather and Climate Prediction*. Cambridge University Press, 548 pp.

Whiteman, C.D., 2000: *Mountain Meteorology: Fundamentals and Applications*, Oxford University Press, 376 pp.

Yih, C.-S., 1980: *Stratified Flows*. Academic Press, New York, 418 pp.

Zannetti, P., 1990: *Air Pollution Modeling: Theories, Computational Methods, and Available Software*. Computational Mechanics Publications, Boston, 444 pp.

The value of predictive models, which includes mesoscale models, is discussed in:

Droegemeier, K.K., J.D. Smith, S. Businger, C. Doswell III, J. Doyle, C. Duffy, E. Foufoula-Georgiou, V. Krajewski, M. Lemone, C. Mass, R. Pielke Sr., P. Ray, S. Rutledge, and E. Zipser, 2000: Hydrological aspects of weather prediction and flood warnings: Report of the Ninth Prospectus Development Team of the U.S. Weather Research Program. *Bull. Amer. Meteor. Sci.*, **81**, 2665-2680.

Mass, C.F., and Y.-H. Kuo, 1998: Regional real-time numerical weather prediction: current status and future potential. *Bull. Amer. Meteor. Soc.*, **79**, 253-263.

Pielke Jr., R.A., D. Sarewitz, R. Byerly Jr., and D. Jamieson, 1999: Prediction in the Earth sciences and environmental policy making. *Eos, Trans. Amer. Geophys. Union*, **80**, 1, 312-313.

Pielke, R.A., 1994: The status of mesoscale meteorological models. *Planning and Managing Regional Air Quality: Modeling and Measurement Studies*, P.A. Solomon and T.A. Silver, Eds., Lewis Publishers, 435-458.

Sarewitz, D., R.A. Pielke Jr., and R. Byerly (Eds.), 2000: *Prediction: Science Decision Making and the Future of Nature*. Island Press, Covelo, CA, 405 pp.

CHAPTER

Synoptic-Scale Background 14

CHAPTER OUTLINE HEAD

Mesoscale Meteorological Modeling, Third Edition, Volume 98. http://dx.doi.org/10.1016/B978-0-12-385237-3.00014-1

14.1 Introduction

This chapter introduces the vertical and horizontal structure and dynamics of synoptic-scale systems, which, as discussed in Chapter 3, are close to a balance with the pressure gradient and Coriolis terms in the conservation of horizontal motion equations. Synoptic systems are also nearly in hydrostatic balance. This permits the derivation of fundamental concepts which provides the background state for mesoscale atmospheric motions. More detailed analyses of synoptic systems can be found in the excellent texts of Bluestein (1992, 1993), Carlson (2012), and Lackmann (2012).

14.2 Quantitative Measures of the Vertical Profile of the Atmosphere

The first law of thermodynamics as shown in Eq. 2.19 can be written as:

$$ds = C_{\rm p}\left(\frac{dT_v}{T_v}\right) - \left(\frac{R_d}{p}\right)dp \tag{14.1}$$

where s is entropy, $C_{\rm p}$ is the specific heat at constant pressure (1005 J K^{-1}kg^{-1} for dry air), T_v is virtual temperature,[1] R_d is the gas constant for dry air (287 J K^{-1}kg^{-1}), and p is pressure.

[1] $T_v = T(1 + 0.61w)$ where w is defined by Eq. 14.6. T_v is always greater than or equal to T and is used in order to account for the observation that water vapor has a smaller atomic weight than the dry air which it replaces in a volume.

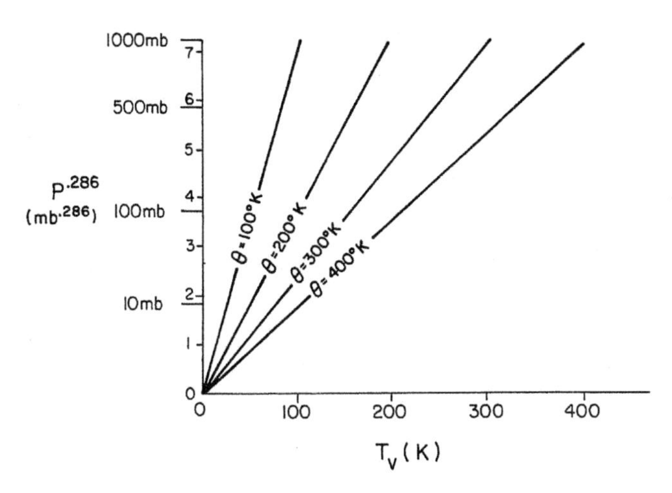

FIGURE 14.1

Relation between T_v, p, and θ in dry air.

When no heat is gained or lost, $ds = 0$, so that:

$$d\left(\ln T_v\right) = \left(R_d/C_p\right) d \ln p \tag{14.2}$$

Integrating Eq. 14.2 between two points with a temperature and pressure combination of T_{v1}, P_1, and T_{v2}, P_2, and taking antilogs yields:

$$T_{v2}/T_{v1} = \left(P_2/P_1\right)^{R_d/C_p} \tag{14.3}$$

Let $P_2 = 1000$ mb and $T_{v2} = \theta$, then Eq. 14.3 becomes:

$$\theta = T_v[1000/p \text{ (in mb)}]^{R_d/C_p} \tag{14.4}$$

or

$$T_v = \theta(1000 \text{ mb})^{C_p/R_d}\, p^{R_d/C_p} = \left(\text{constant}\right) \theta\, p^{0.286} \tag{14.5}$$

A graphical solution of Eq. 14.5 is illustrated in Fig. 14.1. In atmospheric applications, the p scale is inverted since pressure decreases with height. Normally the up direction on a page is related to a positive upward vertical direction.

As an example, for $T_v = 300$ K, $p = 1000$ mb; and for $T_v = 270.5$ K, $p =$ 700 mb, and $\theta = 300$ K in both cases. For $T_v = 248$ K, $p = 700$ mb; $\theta = 275$ K [0°C = 273.16 K].

Other quantities can be plotted on such a thermodynamic diagram.[2] The *mixing ratio* is defined as the ratio of the mass of water vapor, m_v, to the mass of dry air, m_d,

$$w = \frac{m_v}{m_d} = \frac{\rho_v}{\rho_d} \tag{14.6}$$

[2]Henceforth, in this chapter, with just a few exceptions, T will be used to represent virtual temperature.

Since the volume of the air in which the water vapor and dry air is contained is the same, the corresponding densities, ρ, can be inserted in place of the mass.

From the ideal gas law:

$$\rho_v = \frac{p_v}{R_v T} = \frac{e}{R_v T}; \quad \rho_d = \frac{p_d}{R_d T} \tag{14.7}$$

where $R_v = 461\ \mathrm{J\ K^{-1}kg^{-1}}$ and by convention $e = p_v$ is used. The temperature of the dry air and water vapor are assumed the same. The pressure of the dry air can be rewritten as $p - e$, where p is the total pressure using Dalton's law of partial pressures.

Equation 14.6 can then be written as:

$$w = \frac{e}{R_v T} \Big/ \left(\frac{p-e}{R_d T}\right) = 0.622e/\left(p - e\right) \tag{14.8}$$

since $R_d/R_v = 0.622$.

The *maximum* amount of water vapor that a parcel of air can hold at a given temperature is written as:

$$w_s = 0.622e_s/\left(p - e_s\right) \tag{14.9}$$

where e_s is the saturation vapor pressure.

Experimental work (e.g., see Murray 1967) has permitted the specification of e_s as a function of temperature, so that, as one formula:

$$w_s \cong \frac{3.8}{p} \exp\left[\frac{21.9(T - 273.2)}{(T - 7.7)}\right]; \ T \text{ in Kelvin} \tag{14.10}$$

for typical tropospheric values with $p \gg e_s$.

An important distinction in Eq. 14.10 is that w_s is saturated with respect to liquid water. A different formulation (maximum difference of about 0.2 g $\mathrm{kg^{-1}}$) is applicable for saturation with respect to ice. Since w_s is a function of p and T_v, values of w_s could also be drafted on Fig. 14.1 and on whatever other type of thermodynamic representation is chosen.

Relative humidity is defined from w and w_s as:

$$RH = 100w/w_s \tag{14.11}$$

The value of w can be determined by measuring the *dewpoint temperature*, T_D. This is the temperature at which *condensation will occur if the atmosphere is cooled at constant pressure* (i.e., isobarically). Thus, when $T = T_D$, $w = w_s$ (i.e., a value of w at temperature T corresponds to w_s at temperature T_D).

When the atmospheric water content and pressure are constant, w and T_D remain constant. Since w_s is a function of temperature, however, the relative humidity will vary as a function of temperature. An example of this situation is illustrated in Fig. 14.2 (e.g., see Eq. 14.10).

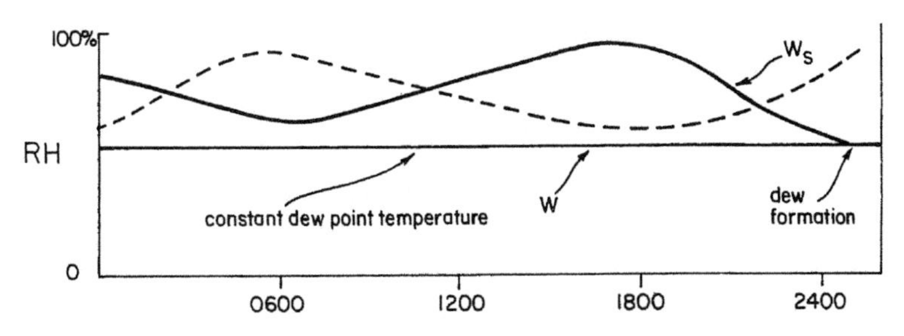

FIGURE 14.2

Schematic illustration of the relation of relative humidity (dashed line) to dewpoint temperature. This pattern is characteristic of a summer day in which diurnal heating causes a variation of air temperature, but the absolute water content of the air does not vary during the day.

A *frost point temperature*, T_F, can similarly be defined as the temperature at which deposition (e.g., frost formation) occurs when air is cooled isobarically at constant pressure until the air becomes saturated with respect to ice.

Since the saturation vapor pressure of liquid water is always greater than or equal to the saturation vapor pressure of ice:

$$T_D \geq T_F.$$

When phase changes of water occur, heat is added or deleted from a parcel of air so that potential temperature, θ, is no longer conserved. For the situation of liquid phase changes, the first law of thermodynamics can be written as:

$$C_\mathrm{p}\frac{dT}{T} - \frac{R_d}{p}dp = \frac{-L}{T}dw_s \tag{14.12}$$

where L is the latent heat of condensation.

The term on the right of Eq. 14.12 can be written as:

$$\frac{1}{T}dw_s \cong d\left(\frac{w_s}{T}\right)$$

as long as $dw_s/T \gg w_s dT/T^2$ which it is for reasonable tropospheric conditions (e.g., see Pielke 1984, pg. 268).

Equation 14.12 can then be written as:

$$C_\mathrm{p}\frac{dT}{T} - \frac{R_d}{P}dp \cong -L\,d\left(\frac{w_s}{T}\right) \tag{14.13}$$

From Eqs. 14.1 and 14.4, Eq. 14.13 can be rewritten as:

$$\frac{C_\mathrm{p}}{\theta}d\theta \cong -L\,d\left(\frac{w_s}{T}\right) \tag{14.14}$$

If it is required (as seems intuitive) that $w_s/T \rightarrow 0$ as $T \rightarrow 0$ K (e.g., see Fig. 2.7 of Wallace and Hobbs 1977; saturation moisture content goes to zero faster than temperature goes to zero) then an *equivalent potential temperature* is defined from:

$$C_{\text{p}} \int_{\theta}^{\theta_E} d \ln \theta \cong -L \int_{w_s/T}^{0} d\left(\frac{w_s}{T}\right),$$

which after integrating and taking antilogs yields:

$$\theta_E \cong \theta \exp\left(\frac{L}{C_{\text{p}}} \frac{w_s}{T}\right). \tag{14.15}$$

The quantity θ_E corresponds to the temperature of a parcel if it were moved to 1000 mb with all of its water vapor condensed and heat released. Obviously, from Eq. 14.15, $\theta_E \geq \theta$, with $\theta_E \cong \theta$ at low water vapor contents (e.g., in cold atmospheres).

Contours of constant θ_E are placed on thermodynamic diagrams since θ_E is *conserved* to saturation – it is a measure of the potential heat increase due to complete saturation of the water vapor present and is independent of the actual amount that has condensed.

There are a wide range of presentations of thermodynamic properties on thermodynamic charts. Four examples, reproduced from AWS (1979), are presented in Figs. 14.3a-d. Byers (1959, pp. 177-180) also discuss different forms of presentation of thermodynamic properties, such as the Stüve diagram which has T by $\ln p$ axes and the tephigram which has T by θ axes.

14.2.1 Dry Adiabatic Lapse Rate

If no heat is added to or removed from a parcel, the potential temperature must be constant as shown previously. Therefore, for this situation,

$$\frac{d\theta}{dz} = 0 \tag{14.16}$$

is a statement that there is no heat changes for a vertically displaced parcel. From the definition[3] of θ $\left[\theta = T \ \left(1000 \text{ mb}/p\right)^{R_d/C_{\text{p}}}\right]$, therefore, after differentiating logarithmically with height:

$$\frac{1}{\theta}\frac{d\theta}{dz} = 0 = \frac{1}{T}\frac{dT}{dz} - \frac{R_d}{C_{\text{p}} p}\frac{dp}{dz} \tag{14.17}$$

is an equivalent statement of Eq. 14.16. Assuming hydrostatic balance (i.e., $dp/dz = -\rho g$), Eq. 14.17 after rearranging, becomes:

$$\frac{dT}{dz} = \frac{-R_d T \rho}{p} \frac{g}{C_{\text{p}}} = -\frac{g}{C_{\text{p}}} = -\gamma_d \tag{14.18}$$

[3]The appropriate temperature to use is the virtual temperature, defined as $T = T_{\text{dry}}(1 + 0.61w)$ where w is the mixing ratio. See Chapter 2 for a derivation of virtual temperature. T_{dry} is the thermometer temperature.

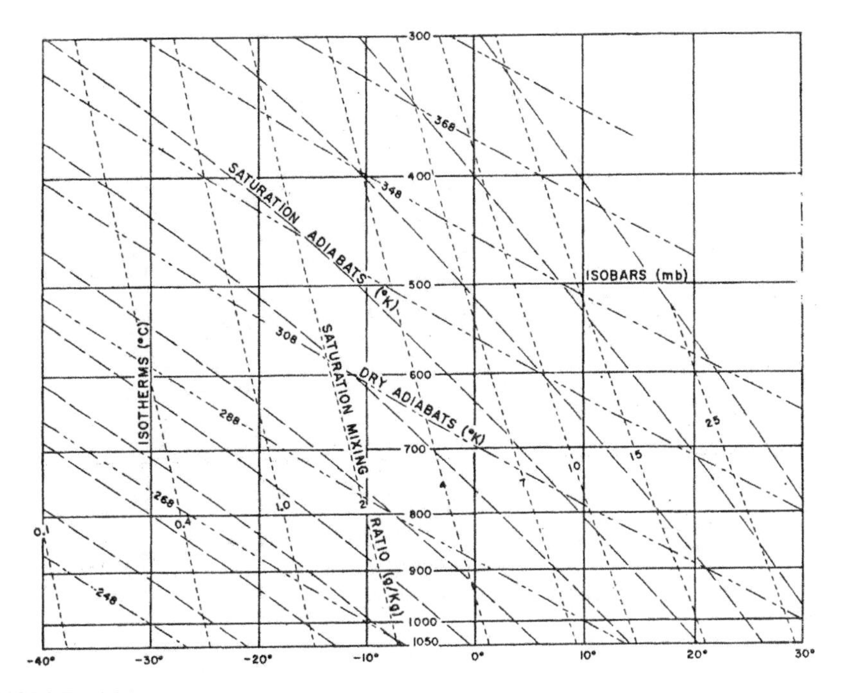

FIGURE 14.3a

Coordinate system of the emagram (from AWS 1979).

where the ideal gas law ($p = \rho RT$) has been applied. For the Earth's troposphere $g/C_p \cong 1°\text{C}/100$ m. The variable γ_d is referred to as the *adiabatic lapse rate.*

On a thermodynamic diagram, lines of constant θ correspond to a temperature lapse rate equal to $-1°\text{C}/100$ m.

14.2.2 Wet Adiabatic Lapse Rate

When a parcel is lifted, temperature decreases as evident from Eq. 14.18. Since air cannot hold as much water vapor at colder temperatures, (e.g., see Wallace and Hobbs 1977, pg. 73) sufficient lifting will result in condensation (deposition) when the vapor pressure of the water vapor, e, becomes equal to the saturation vapor pressure, e_s, with respect to water (e.g., ice). Since $e/e_s = 1$ corresponds to $w/w_s = 1$, where w is the mixing ratio defined in Section 14.1, the height where $w = w_s$ is first calculated on a thermodynamic diagram is referred to as the *lifting condensation level* (LCL).

The value of w for the parcel is determined by measuring the *dewpoint temperature*. The dewpoint temperature is defined as the temperature at which condensation first occurs as a result of cooling *at constant pressure*. An analogous temperature (*the frost point*) is defined for the first occurrence of deposition due to cooling at constant temperature.

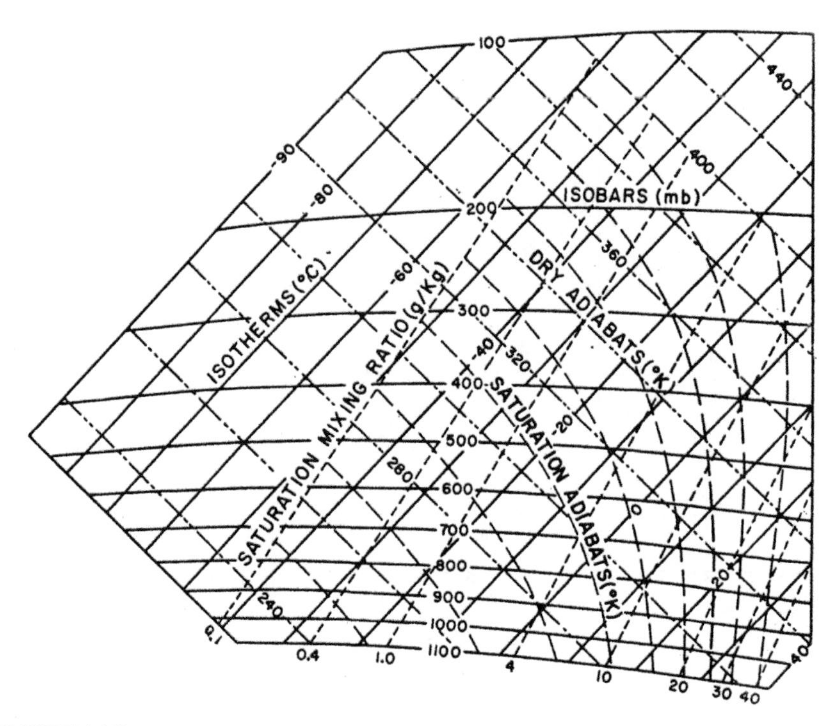

FIGURE 14.3b

Coordinate system of the tephigram (from AWS 1979).

The temperature of the parcel determines the maximum amount of water vapor that can be contained without condensation or sublimation. This relation between saturation mixing ratio and temperature is given for realistic tropospheric conditions as:

$$w_s \cong \frac{3.8}{\overline{p}} \exp\left[\frac{21.9\left(\overline{T} - 273.2\right)}{\overline{T} - 7.7}\right] \quad \text{(for liquid water)} \qquad (14.19)$$

$$w_{s_i} \cong \frac{3.8}{\overline{p}} \exp\left[\frac{17.3\left(\overline{T} - 273.2\right)}{\overline{T} - 35.9}\right] \quad \text{(for ice)} \qquad (14.20)$$

(from Pielke 1984, pg. 234). Lines of constant saturation mixing ratio from a formulation such as Eq. 14.19 are usually drawn on thermodynamic diagrams as dashed or dotted lines.

Since water vapor content up to the LCL is constant, w is a constant as a parcel ascends or descends below the LCL. It is important to recognize, however, that a constant value of w does not indicate that the dewpoint temperature is constant with height. As the parcel ascends, expansion results in a reduction in the vapor pressure, e, with height as seen from the ideal gas law (i.e., $e = \rho_v R_v T$ where ρ_v is the density and

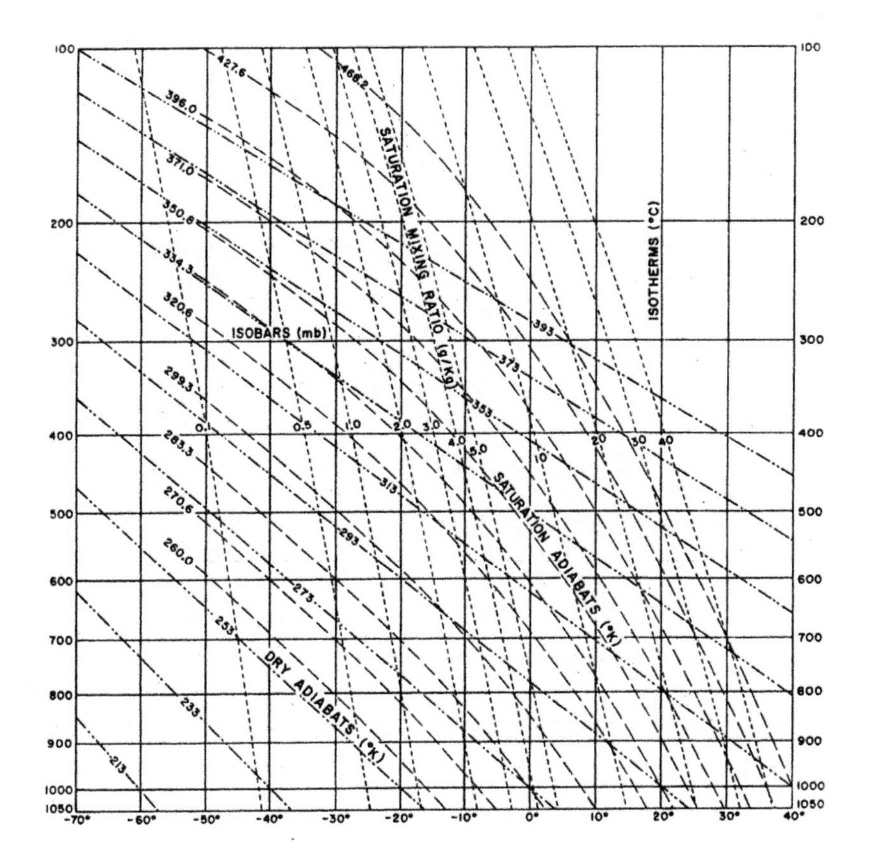

FIGURE 14.3c

Coordinate system of the Stüve ("pseudo-adiabatic") diagram (from AWS 1979).

R_v is the gas constant of the water vapor). Expansion requires that ρ_v become less and Eq. 14.18 indicates that temperature decreases as well. Thus, the temperature to which a parcel must be cooled *isobarically* in order to achieve condensation (sublimation) becomes lower at higher heights (i.e., lower pressures) since e decreases with height. Therefore while $w = \rho_v/\rho$ is constant with height below the LCL, dT_D/dz is less than zero.

The phase change of water at and above the LCL permits a source of a heat. Equation 14.17 can be generalized to represent this source term as:

$$\frac{1}{\theta}\frac{d\theta}{dz} = \frac{1}{T}\frac{dT}{dz} - \frac{R_d}{C_p p}\frac{dp}{dz} = -\frac{L}{C_p T}\frac{dw_s}{dz} \tag{14.21}$$

where dw_s/dz is the change of saturation mixing ratio with height which when negative represents the amount of water vapor converted to another phase. The latent heat of phase change is given by L.

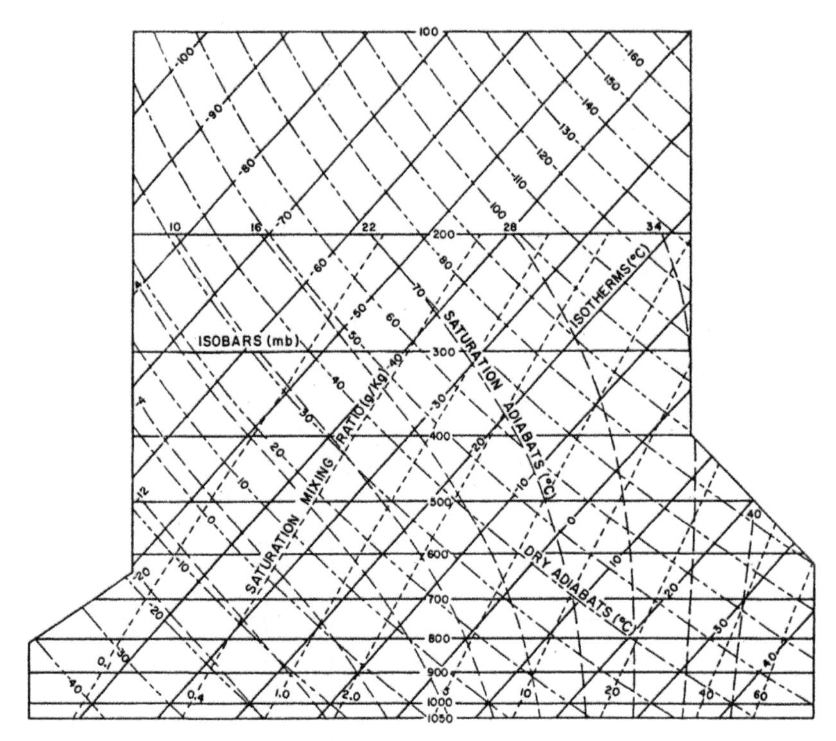

FIGURE 14.3d

Coordinate system of the skew T, log P diagram (from AWS 1979).

Rearranging Eq. 14.21, and substituting the hydrostatic relation yields:

$$\frac{dT}{dz} = -\frac{L}{C_p}\frac{dw_s}{dz} - \frac{R_d T \rho g}{p C_p} = -\frac{L}{C_p}\frac{dw_s}{dz} - \frac{g}{C_p} \tag{14.22}$$

where the gas law has been applied to simplify the last term on the right. By the chain rule of calculus, dw_s/dz can be written as:

$$\frac{dw_s}{dz} = \frac{dw_s}{dT}\frac{dT}{dz}. \tag{14.23}$$

Substituting Eq. 14.23 into Eq. 14.22 and rearranging yields:

$$\frac{dT}{dz} = -\frac{g}{C_p} \Big/ \left[1 + \frac{L}{C_p}\frac{dw_s}{dT}\right] = \frac{-\gamma_d}{\left[1 + \frac{L}{C_p}\frac{dw_s}{dT}\right]} = -\gamma_m \tag{14.24}$$

where γ_m is referred to as the *moist adiabatic lapse* rate. (Note that $\gamma_m \leq \gamma_d$ because of the heat liberated by the phase change of water.) When the phase change is from water vapor to liquid water, L corresponds to the latent heat of condensation ($L \cong 2.5 \times 10^6\,\mathrm{J\,kg^{-1}}$) and w_s is the saturation mixing ratio with respect to liquid water (e.g., Eq. 14.19) and γ_m is often called a *water adiabat*. When the phase change is to ice, L is the latent heat of deposition ($L \cong 2.88 \times 10^6$ J kg^{-1}) and w_s is the saturation mixing ratio with respect to ice (e.g., Eq. 14.20). On most thermodynamic diagrams, except those specifically designed for the upper troposphere where, for instance, the prime forecast consideration is the analysis for jet contrails, water adiabats are the ones most frequently plotted.

From Eq. 14.24, irrespective of which moist adiabat is used, since dw_s/dT is positive:

$$\gamma_m \leq \gamma_d$$

in all circumstances. Also, since dw_s/dT becomes small for colder temperatures, $\gamma_m \cong \gamma_d$ in cold air. Lines of γ_m are often indicated on thermodynamic diagrams as dashed lines in the same color as the solid lines of γ_d.

There are two interpretations of moist ascent along a γ_m lapse rate. If the liquid water or ice is carried along with the parcel, then during subsequent descent this water can convert back to water vapor (i.e., evaporation for liquid water, sublimation for ice). For this situation, the phase change process is completely reversible and the lines of γ_m are referred to as *saturated adiabats*. On the other hand, if the liquid water or ice is interpreted to precipitate out of the parcel, a subsequent descent of the parcel will not permit the attainment of the original water vapor content. With this interpretation, the lines of γ_m are referred to as *pseudo-adiabats* and the process of lifting above the LCL is considered *irreversible*.

Two additional quantities are used to describe the moist thermodynamic stratification of the atmosphere – the *equivalent potential temperature* and the *wet bulb potential temperature*. To illustrate these two quantities, they will be derived in their approximate forms. Iribarne and Godson (1973) provide more precise derivations. Equation 14.21 can be written in the form:

$$\frac{1}{\theta}\frac{d\theta}{dz} = -\frac{L}{C_\text{p}T}\frac{dw_s}{dz} \cong -\frac{L}{C_\text{p}}\frac{d}{dz}\left(\frac{w_s}{T}\right) \tag{14.25}$$

as long as the approximation $T^{-1}\,|dw_s/dz| \gg w_s T^{-2}\,|dT/dz|$ is valid (which it is for reasonable atmospheric conditions within the troposphere). The saturation mixing ratio corresponds to the value of w of a saturated parcel of air. Saturation of the parcel will only occur when it is cooled sufficiently so that the saturation value of w is attained.

Since at low temperatures saturation mixing ratio goes to zero more rapidly than temperature, w_s/T approaches zero at absolute zero. Therefore, treating L and C_p as constants, Eq. 14.25 can be integrated from an observed temperature to absolute zero yielding the approximate formula[4]:

$$\theta_{es} = \theta \exp\left(\frac{Lw_s}{C_\text{p}T}\right). \tag{14.26}$$

[4]A more exact form of θ_{es} can be obtained from $T_{es} = T_v + \frac{L}{C_\text{p}} w_s$ where T_e is the isobaric equivalent temperature and, thus, $\theta_{es} = T_{es}\left(\frac{1000}{p}\right)^{Rd/C_\text{p}}$ can be used to compute θ_{es}.

The equivalent potential temperature, therefore, represents the potential temperature that would occur if *all* of the water was condensed (when L corresponds to the latent heat of condensation) and the resultant heat is used to warm the parcel to a higher potential temperature.

The wet bulb potential temperature is also derived from Eq. 14.25 and can be rewritten in its approximate form as:

$$\frac{d\left(\ln\theta\right)}{dz} = -\frac{L}{C_{\mathrm{p}}}\frac{d}{dz}\left[\frac{w_s}{T}\right]. \qquad (14.27)$$

Equation 14.27 can be integrated between current values of θ and w_s and values of potential temperature, θ_W, and saturation mixing ratio, w_s', it would have if water vapor were added to the air parcel so as to cause saturation. This yields:

$$\int_{\theta}^{\theta_W}\frac{d}{dz}\ln\theta\,dz = -\frac{L}{C_{\mathrm{p}}}\int_{w_s}^{w_s'}\frac{d}{dz}\left[\frac{w_s}{T}\right]dz$$

or

$$\ln\frac{\theta_W}{\theta} = -\frac{L}{C_{\mathrm{p}}}\left[\frac{w_s}{T} - \frac{w_s'}{T_W}\right]$$

or

$$\theta_W = \theta exp\left[-\frac{L}{C_{\mathrm{p}}}\left(\frac{w_s}{T} - \frac{w_s'}{T_W}\right)\right]. \qquad (14.28)$$

The value of the wet bulb temperature in Eq. 14.28 can be obtained for an isobaric process from the first law of thermodynamics in the form:

$$C_{\mathrm{p}}\,dT = -L dw_s,$$

which after integrating over the same limits as applied to obtain Eq. 14.28, yields:

$$C_{\mathrm{p}}\left(T - T_W\right) = -L\left[w_s - w_s'\right]$$

or

$$T_W = T - \frac{L}{C_{\mathrm{p}}}\left(w_s' - w_s\right). \qquad (14.29)$$

Since moistening an air parcel elevates the dewpoint temperature, while the evaporation of water cools the temperature, $T_D \leq T_W \leq T$.

θ_E and θ_W correspond to lines of constant moist adiabatic lapse rate, γ_m. Both θ_E and θ_W are derived so as to account for the decrease in temperature with height of a saturated air parcel, as latent heat is continually released.

14.2.3 Lifting Condensation Level

An air parcel ascends dry adiabatically $\left(\frac{d\theta}{dz} = 0\right)$ until saturation is attained. The moisture content of a parcel is specified by the mixing ratio as discussed previously.

The height at which the ascending parcel first becomes saturated is called the *lifting condensation level* (LCL). Below the LCL θ, θ_E, and θ_W remain constant in the absence of entrainment of air with different thermodynamic properties. All three forms of potential temperature, therefore, are conserved with respect to dry air motions. Above the LCL, however, only θ_E and θ_W of a parcel remain constant in the absence of entrainment. Therefore, only θ_E and θ_W, and not θ, are conserved with respect to saturated air motions.

14.2.4 Concept of Static Stability

Since force is equal to a mass times an acceleration (i.e., Newton's second law), the vertical equation of motion in the atmosphere can be written as:

$$\frac{d^2z}{dt^2} = \frac{dw}{dt} = -\frac{1}{\rho}\frac{\partial p}{\partial z} - g. \tag{14.30}$$

where, in Eq. 14.30, w is vertical motion. The two forces on the right side of Eq. 14.30 are the vertical pressure gradient force and gravitational acceleration. When these two forces are equal and opposite, the atmosphere is said to be in hydrostatic balance. Correspondingly, an imbalance of the two forces results in an acceleration.

In terms of an air parcel, it is convenient to write the hydrostatic version of Eq. 14.30 for the ambient (i.e., surrounding) atmosphere (denoted by a subscript "e") and the complete form of Eq. 14.30 (denoted by a subscript "p") for the parcel.

The relation of the parcel to the surrounding atmosphere is schematically illustrated in Fig. 14.4:

$$\left.\begin{aligned} \left.\frac{\partial p}{\partial z}\right|_e &= -\rho_e g \\ \frac{dw}{dt} &= -\frac{1}{\rho_p}\left.\frac{\partial p}{\partial z}\right|_p - g \end{aligned}\right\}. \tag{14.31}$$

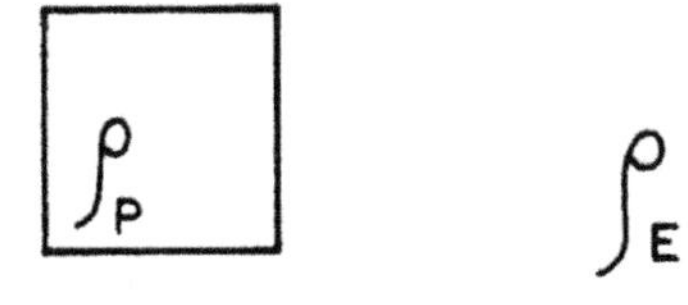

FIGURE 14.4

Schematic of the relation between a parcel and the surrounding atmosphere.

In applying Eq. 14.31, it is assumed that the vertical pressure gradient acting on the parcel is identical to the vertical pressure gradient of the atmosphere at the same level, i.e.,

$$\left.\frac{\partial p}{\partial z}\right|_e = \left.\frac{\partial p}{\partial z}\right|_p = \frac{\partial p}{\partial z}.$$

The bottom expression in Eq. 14.31 can then be written, after rearranging, as:

$$\frac{dw}{dt} = g\left[\frac{\rho_e - \rho_p}{\rho_p}\right]. \tag{14.32}$$

Thus, a parcel starting at rest will accelerate upward if it is less dense than the surrounding air. If $\rho_e = \rho_p$, a parcel at rest will stay at rest, while a parcel in motion will continue to move at a constant speed.

Using the ideal gas law for the parcel and for the ambient atmosphere:

$$\rho_e = p/R_d T_e; \quad \rho_p = p/R_d T_p,$$

Eq. 14.32 can be rewritten as:

$$\frac{dw}{dt} = g\frac{T_p - T_e}{T_e}. \tag{14.33}$$

Therefore, a parcel, starting at rest, will accelerate upward if it is warmer than the surrounding air.

Using a Taylor series expansion, the response of a parcel to forced motion from its height of origin in the atmosphere can be evaluated:

$$T_e = T_o + \left.\frac{dT}{dz}\right|_e \delta z + \left.\frac{1}{2}\frac{d^2T}{dz^2}\right|_e (\delta z)^2 + \cdots$$

$$T_p = T_o + \left.\frac{dT}{dz}\right|_p \delta z + \left.\frac{1}{2}\frac{d^2T}{dz^2}\right|_p (\delta z)^2 + \cdots$$

where T_o is the temperature at the level at which the parcel originated.

If δz is small,

$$T_e \cong T_o - \gamma\,\delta z; \quad T_p \cong T_o - \gamma_d\,\delta z$$

where $\gamma = -dT/dz|_e$ is the lapse rate of the environment and γ_d is the lapse rate of a parcel undergoing dry adiabatic motion (i.e., $\gamma_d = \frac{g}{C_p} = -\frac{dT}{dz}|_p$; see Eq. 14.18. Equation 14.33 can therefore be approximated by:

$$\frac{dw}{dt} \cong g\frac{(\gamma - \gamma_d)}{T_e}\delta z \tag{14.34}$$

In terms of Eq. 14.34, the following definitions are used when referring to a dry atmosphere,

- an unstable equilibrium exists when $\gamma > \gamma_d$
- a neutral equilibrium exists when $\gamma = \gamma_d$
- a stable equilibrium exists when $\gamma < \gamma_d$.

In the atmosphere, $\gamma < \gamma_d$ at almost all locations except,

a) near the ground on sunny days,
b) over water when colder air advects over it, and
c) at the top of clouds, particularly at sunset.

Since values of constant potential temperature θ are equivalent to γ_d,

$$\left.\begin{array}{l} \gamma > \gamma_d \\[2ex] \gamma = \gamma_d \\[2ex] \gamma < \gamma_d \end{array}\right\} \quad \text{is equivalent to} \quad \left\{\begin{array}{l} \dfrac{\partial\theta}{\partial z} < 0 \\[2ex] \dfrac{\partial\theta}{\partial z} = 0 \\[2ex] \dfrac{\partial\theta}{\partial z} > 0 \end{array}\right. \tag{14.35}$$

Partial derivatives are used here to emphasize that the potential temperature lapse rate is referred to rather than the value of θ following a parcel. When $\partial\theta/\partial z < 0$, the lapse rate γ is said to be *superadiabatic*.

Corresponding definitions can be made for a saturated environment except γ_d is replaced by γ_m. If the air is saturated,

- $\gamma > \gamma_m$ is unstable and cumuliform clouds result,
- $\gamma = \gamma_m$ is neutral and cumuliform clouds occur,
- $\gamma < \gamma_m$ is stable and stratiform clouds result.

Since values of constant equivalent and wet bulb potential are equivalent to constant values of γ_m, a general terminology relating lapse rates and the different θ forms of potential temperature can be derived. These relationships are summarized in Table 14.1.

Up to this point, thermodynamic stability has referred to parcel motion. Often, however, entire layers of the atmosphere are lifted as a result of large-scale ascent. As will be illustrated in the following discussion, this lifting can result in significant changes in the atmospheric lapse rates, γ.

Figures 14.5a and b schematically represent two basic types of atmospheric stratification. In Fig. 14.5a, $\partial\theta_E/\partial z > 0$ so the atmosphere is absolutely stable, while $\partial\theta_E/\partial z < 0$ in Fig. 14.5b which is a conditionally unstable atmosphere as long as $\partial\theta/\partial z > 0$. The stable atmosphere is characterized by relatively dry air capped by relatively wetter air aloft. The converse is true for the conditionally unstable atmosphere.

Table 14.1 Relation between lapse rates.

Thermodynamic Stability	Lapse Rate Relationship	Potential Temperature Relationship	Parcel Behavior
Absolutely stable	$\gamma < \gamma_m$	$\frac{\partial \theta_W}{\partial z} = \frac{\partial \theta_E}{\partial z} > 0$	Regardless of whether saturation occurs in the absence of entrainment, the parcel will move back to its original position if displaced
Conditionally stable	$\gamma_m \leq \gamma < \gamma_d$	$\frac{\partial \theta_W}{\partial z} = \frac{\partial \theta_E}{\partial z} < 0$ but $\frac{\partial \theta}{\partial z} > 0$	The parcel accelerates away from its original position if displaced and if saturated. If not saturated it will move back to its original position. In addition, even if the parcel is displaced sufficiently such that saturation occurs, the parcel will still move back towards its original position until further saturated ascent overcomes the temperature deficit that resulted from the earlier dry adiabatic upward motion
Absolutely unstable	$\gamma \geq \gamma_d$	$\frac{\partial \theta}{\partial z} \leq 0$	Parcel will accelerate away from its original position when displaced

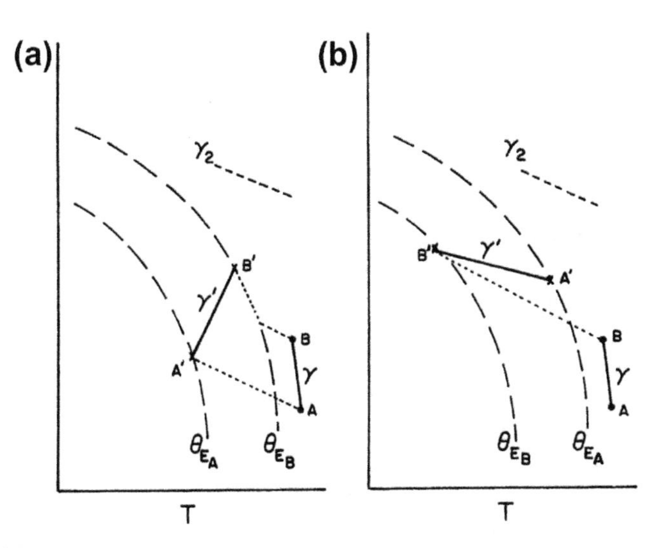

FIGURE 14.5

Schematic of (a) a convectively stable, and (b) convectively unstable air mass which is lifted from heights A-B to A'-B'. The original lapse rate is γ and the new lapse rate after lifting is γ'.

The changes in stratification due to the lifting of the layer A-B to A′-B′ are shown in the figure. The equivalent potential temperatures, θ_{E_A} and θ_{E_B} at the heights A and B, and which are invariant under both dry and moist ascent, are also illustrated. The lifting of the layer proceeds adiabatically (i.e., constant θ) until saturation is achieved, whereupon values of constant θ_E are followed.

As evident in the figure, when $\partial\theta_E/\partial z > 0$, the layer becomes more stable, while when $\partial\theta_E/\partial z < 0$, the stratification becomes less stable.

When $\partial\theta_E/\partial z$ becomes more negative, the atmosphere becomes more conducive to cumuliform convection.

When $\partial\theta_E/\partial z < 0$, the layer is variously referred to as:

i)	Convectively unstable	–	since cumulus convection results when saturation is realized in such an atmosphere.
ii)	Potential instability	–	since organized lifting must occur before saturation is actually realized.
iii)	Layer instability	–	since it is the lifting of a layer of the atmosphere that increases the instability and permits saturation to occur.

As a qualitative guide, dry air above moist air is a fingerprint of a convectively unstable atmosphere, and is one criteria looked for in predicting severe thunderstorm outbreaks.

In using these on thermodynamic definitions, it is important to remember that conditional instability refers to a parcel, while convective instability refers to a layer.

14.2.5 Convective Parameters

There are several derived thermodynamically-related parameters that are valuable in estimating if and when convection will occur, and how intense it will be. These are:

a) Equilibrium Level (EL) – This is the height in the atmosphere at which the temperature of an air parcel, T_p, equals the temperature of the environment at that level, T_e. Below this height for some distance, $T_p > T_e$.
This height closely corresponds to the average heights of cumulus cloud tops. Cumulus clouds which exceed this height are referred to as *overshooting* tops since they exceed their equilibrium level.

b) Convective Temperature (T_c) – This is a temperature near the surface corresponding to a dry adiabatic environmental lapse rate (created as a result of surface heating by solar insolation and resultant mixing) which is high enough so that parcel ascent from the shallow superadiabatic layer near the surface reaches a height at which condensation occurs. It often closely corresponds to the maximum daytime surface temperature.

c) Convective Condensation Level (CCL) – This is the height of condensation associated with T_c. Condensation at this height is manifested initially by shallow cumulus clouds which represent the tops of buoyant turbulent eddies within the boundary layer. Once the CCL is attained, surface temperatures generally do not exceed T_c as a result of the shading of the ground by the clouds and the increased winds near the surface as the cumulus clouds themselves begin to enhance mixing within the layers below the CCL. The CCL is always higher than or equal to the LCL. The most accurate way to compute the CCL is to compute the average w within a height z_i from the surface. The depth z_i corresponds to the height of the layer with a near adiabatic lapse rate. When z_i reaches a height such that the value of the average w over the depth z_i attains its saturated value at z_i, then z_i corresponds to the CCL.

d) Level of Free Convection (LFC) – This is the height at which a parcel mechanically lifted from near the surface will initially attain a temperature warmer than the ambient air. The parcel will subsequently rise from its own buoyancy.

e) Positive Buoyant Energy – This energy is proportional to the temperature excess of a parcel between the LFC and the EL. The mechanical energy required to lift a parcel to the LFC is termed a negative buoyant energy. Positive buoyant energy is also called Convective Available Potential Energy (CAPE).

f) Convective Inhibition (CIN) – This is the heat energy that must be added to the lower levels of the profile in order to make the potential temperature at the LFC equal to the potential temperature near the surface (i.e., $\partial\theta/\partial z = 0$). This energy removes the negative buoyant energy.

g) Lifted Index (LI) – This measure of stability is defined as:

$$\mathrm{LI} = T_{500\ \mathrm{mb}} - T_{p500\ \mathrm{mb}}$$

where $T_{p500\text{ mb}}$ is the temperature of a parcel lifted at a constant θ to the LCL and at a constant θ_E to 500 mb. $T_{500\text{ mb}}$ is the observed temperature at 500 mb. Values of LI > 0 are generally associated with no significant cumulus convection; $0 > \text{LI} > -4$ with showers; $-4 > \text{LI} > -6$ thunderstorms; and $\text{LI} < -6$ with severe thunderstorms.

h) K-Index (K) – This measure of stability is defined as:

$$K = T_{850\text{ mb}} + T_{d850\text{ mb}} - T_{500\text{ mb}} - \left(T_{700\text{ mb}} - T_{d700\text{ mb}}\right);$$

$K > 30$ is generally associated with thunderstorms. The formula needs to be modified for use in higher elevations of the western United States.

i) Mixing Condensation Level (MCL) – This is the height that condensation will occur as a result of strong winds mixing a layer so as to attain uniform with altitude potential temperature and mixing ratio. The MCL is the minimum height at which uniform mixing of θ and w results in saturation. The effect of surface heating is ignored.

j) Precipitable Water (P) – Vertical integral of water depth if all water vapor in a column were condensed out. Defined in terms of $\text{g/cm}^2 \equiv 1$ cm of water depth as:

$$P = \int_{\text{surface}}^{\infty} \rho_v \, dz = \int_{\text{surface}}^{\infty} w \rho dz$$

k) Helicity – A property of a moving fluid which represents the potential for helical flow (i.e., flow which follows the pattern of a corkscrew) to evolve. Helicity is proportional to the strength of the flow, the amount of vertical wind shear, and the amount of turning in the flow (i.e., vorticity). Atmospheric helicity is computed from the vertical wind profile in the lower part of the atmosphere (usually from the surface up to 3 km), and is measured relative to storm motion. Higher values of helicity (generally, around 150 m^2/s^2 or more) favor the development of mid-level rotation (i.e., mesocyclones). Extreme values can exceed 600 m^2/s^2. (NWS, http://forecast.weather.gov/glossary.php?word=HELICITY; Lilly 1986).

14.2.6 Schematic Illustration of the Calculation of Variables on a Thermodynamic Diagram

14.2.6.1 *Potential Temperature, (θ)*

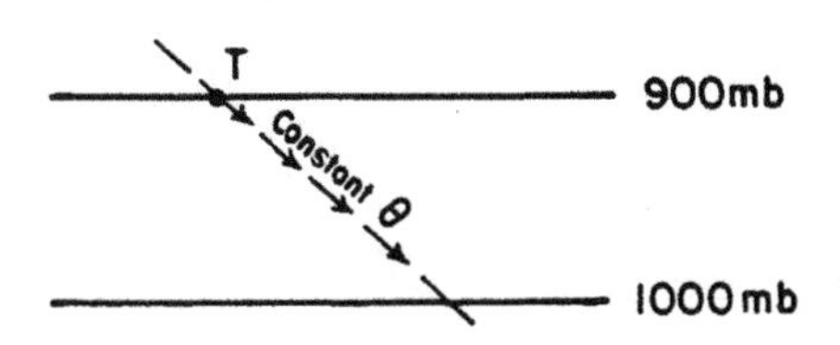

14.2.6.2 *Lifted Condensation Level (LCL)*

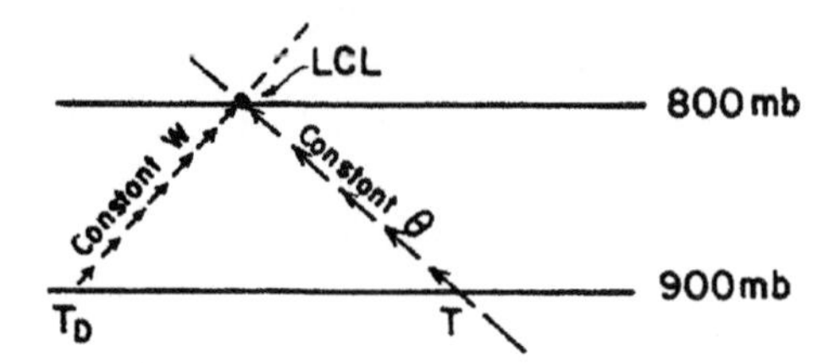

14.2.6.3 *Wet Bulb Potential Temperature (θ_W) and wet bulb temperature (T_W)*

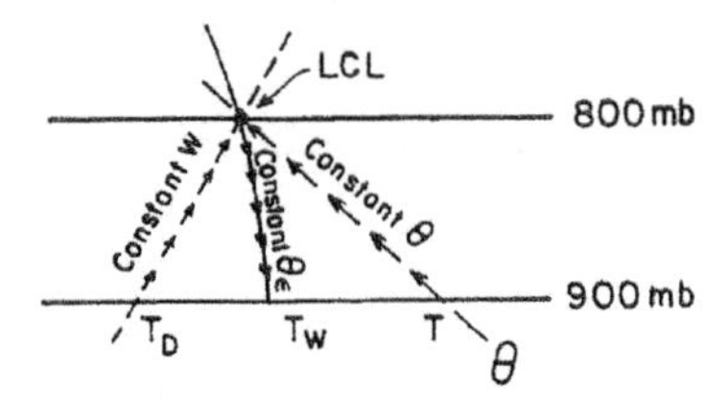

14.2.6.4 *Equivalent Potential Temperature (θ_E)*

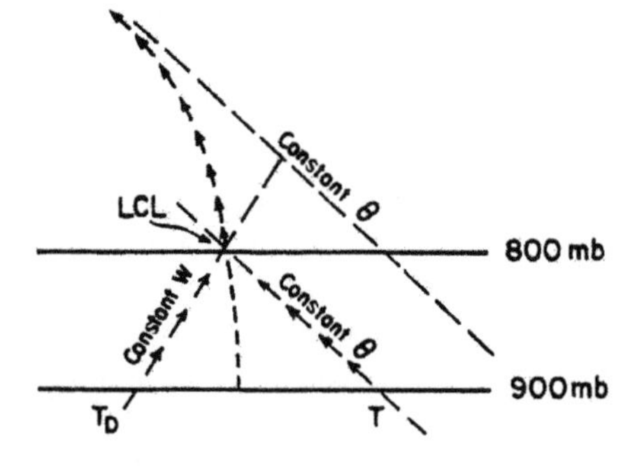

14.2.6.5 *Level of Free Convection (LFC), Equilibrium Level (EL), and Area of Positive Buoyancy*

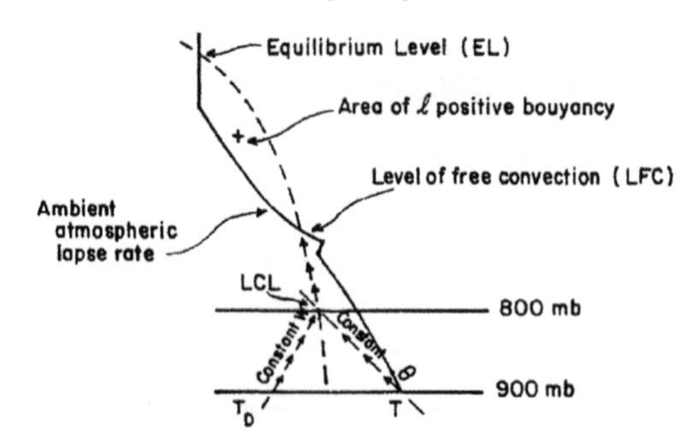

Table 14.2 Uses of common thermodynamic parameters.

Parameter	Use
θ	Used as a tracer for dry motions where $\partial\theta/\partial z > 0$ (therefore, most useful in the mid and upper troposphere).
θ_E, θ_W	Used as a tracer for both dry and wet motions. Major problem with its use as a tracer is that θ_E and θ_W are often multi-valued with height (i.e., $\partial\theta_E/\partial z = \partial\theta_W/\partial z$ often change sign one or more times with height).
LCL	Provides an estimate of stratiform cloud base as a result of forced lifting over mountains or along frontal surfaces.
CCL	Provides an estimate of cumulus cloud base due to surface heating.
MCL	Provides an estimate of cloud base when winds increase so as to mix a layer near the ground.
LFC	Provides an estimate of the depth to which forced ascent is required before cumuliform clouds will spontaneously grow.
EL	Provides an estimate of average cumulus cloud top within a region. Overshooting cumulus tops will be higher than the EL.
Area of positive buoyancy	Provides a quantitative measure of the intensity of cumulus growth if clouds reach the CCL.
Precipitable water	Provides a liquid water equivalent if all the water vapor in a column was condensed. Used to estimate average precipitation potential over an area.
LI	LI > 0: no deep cumulus clouds expected; 0 > LI > − 4: showers expected; −4 > LI >−6: thunderstorms expected; LI < − 6: severe thunderstorms expected.
K	Used to estimate whether or not cumulus convection will occur.
$\frac{\partial\theta_E}{\partial z}; \frac{\partial\theta_W}{\partial z}$	Determines layers in which forced lifting will destabilize the layer resulting in more intense cumulus convection if it develops $\left(\frac{\partial\theta_E}{\partial z} = \frac{\partial\theta_W}{\partial z} < 0\right)$.
T_W	Provides an estimate for the observed temperature when precipitation falls into a layer, evaporating as it falls.
T_D	Provides an estimate of the temperature that must be reached on clear nights before frost or dew will form (T_D with respect to deposition or condensation are close enough for the purposes of this estimate). Except, due to advection, nighttime temperatures seldom fall much below T_D, since heat is released at the surface through the phase change of water.
Helicity	Provides an estimate for the potential for rotating thunderstorm complexes, which thus increases the likelihood for large and strong tornadoes.

14.2.6.6 *Convective Temperature (T_C) and Convective Condensation Level (CCL)*

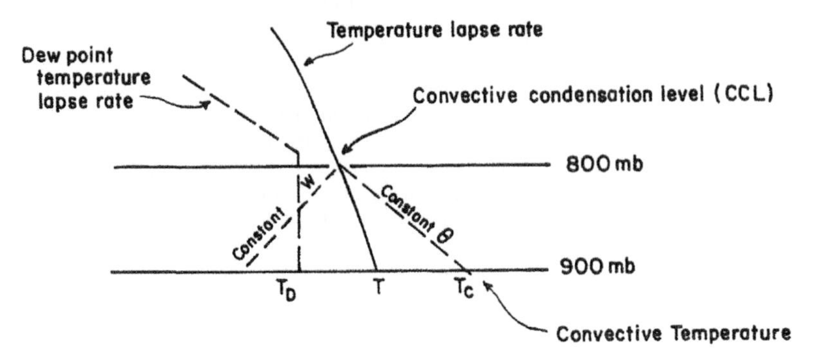

The most accurate way to compute the CCL is to compute the average w within a height z_i from the surface. The depth z_i corresponds to the height of the layer with a near adiabatic lapse rate. When z_i reaches a height such that the value of the average w over the depth z_i attains its saturated value at z_i, then z_i corresponds to the CCL.

14.2.6.7 *Mixing Condensation Level (MCL)*

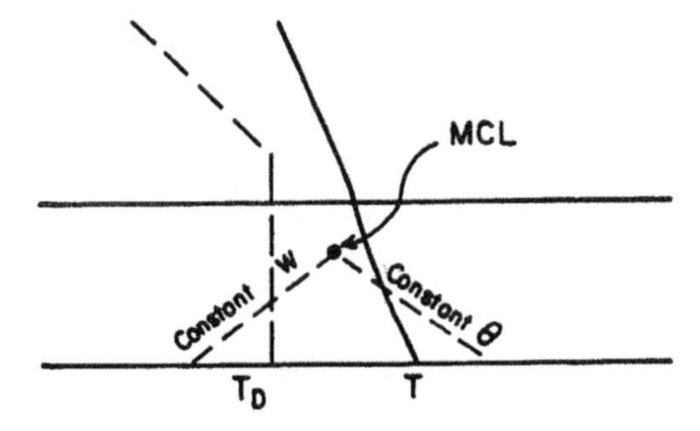

The MCL is the minimum height at which uniform mixing of θ and w results in saturation.

A summary of the uses of the common thermodynamic parameters is presented in Table 14.2.

14.3 Depiction of the Horizontal Structure of the Atmosphere

14.3.1 Balance Winds

The balanced wind which characterizes the synoptic scale is called the "gradient wind." The geostrophic wind (see Eq. 3.28) is a subset of the gradient wind. The relationship between the horizontal pressure gradient and the geostrophic wind can be seen in Fig. 14.6. In terms of which vertical coordinate system to use in deriving

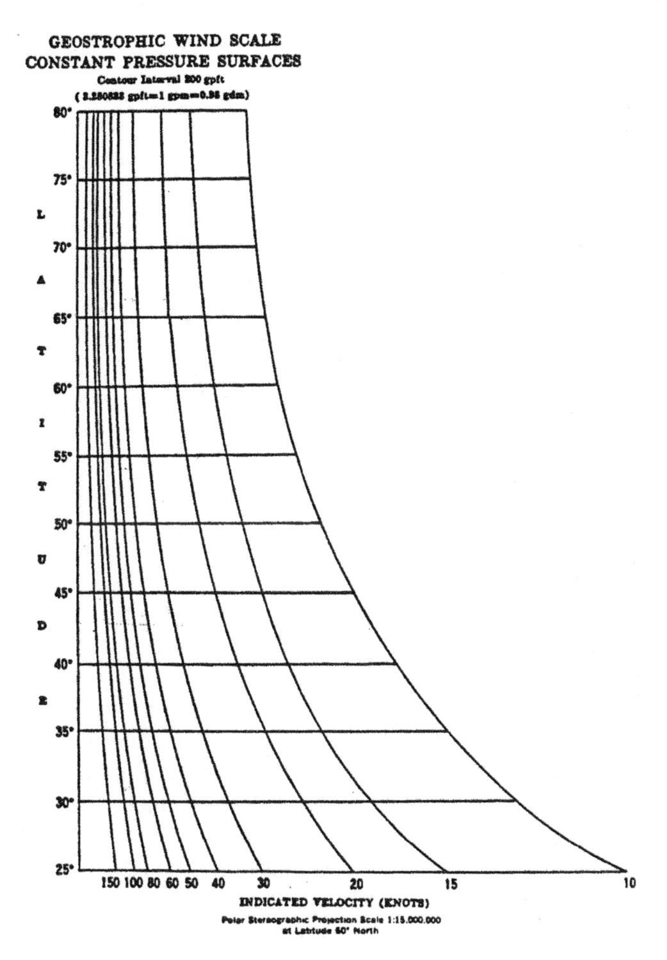

FIGURE 14.6

Relation between geostrophic wind speed and latitude.

these variables (e.g., see Chapter 6), since radiosondes traditionally have measured height about the surface using the pressure measured as the balloon ascends, synoptic meteorology has generally used pressure as its vertical coordinate, although potential temperature has also been applied in research studies (e.g., Uccellini and Johnson 1979).[5] Another characteristic of the synoptic scale is that the magnitude of the vertical motion on the synoptic scale is much smaller than the horizontal motion, as shown in Chapter 3.

[5]In the USDA publication Climate and Man (1941) and Rossby (1941), it was proposed to use potential temperature as the vertical coordinate as this facilitates following air parcels on a coordinate surface when adiabatic motion occurs. However, for synoptic weather forecasts, the pressure framework was selected. Artz et al. (1985) provides an evaluation of differences obtained when an isentropic trajectory analysis is used instead of a constant height evaluation.

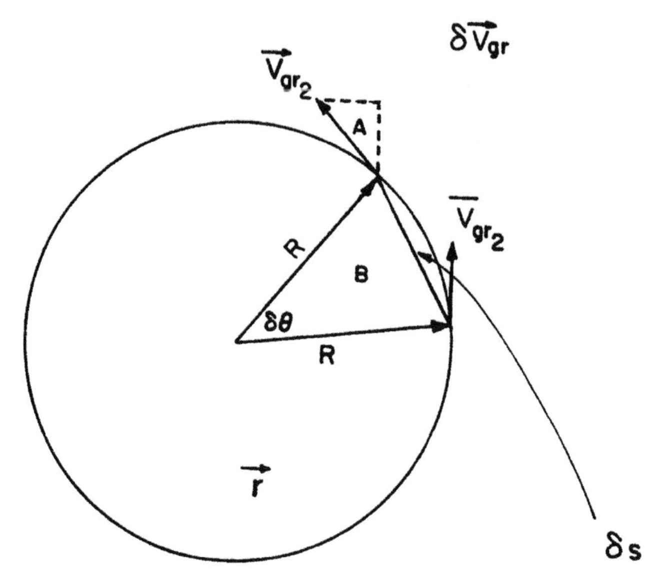

FIGURE 14.7

Illustration of curved flow (based on Holton 1972 and Richards et al. 1962).

The *gradient wind* is derived as follows. Figure 14.7 illustrates an example of curved flow around a circle.[6] The change in the gradient wind over a small time period is given by $\delta \vec{V}_{gr}$ where $\delta\theta$ is the angular displacement. The vector $\vec{r}$, with magnitude, R_T, is the position vector.

Triangles A and B, defined in Fig. 14.7, are similar triangles because A has the two sides with magnitude $|\vec{V}_{gr}|$ and B has the two sides R_T. Therefore, the angle opposite the side with $\delta \vec{V}_{gr}$ is also $\delta\theta$, i.e.,

$$\frac{\left|\delta \vec{V}_{gr}\right|}{\left|\vec{V}_{gr}\right|} = \frac{\delta s}{R_T} = \sin \delta\theta \cong \delta\theta \quad \text{for small } |\delta\theta|$$

Thus, $\left|\delta \vec{V}_{gr}\right| = \left|\vec{V}_{gr}\right| \delta\theta$. Dividing by δt, taking the limit as $\delta t \rightarrow 0$ and noting that $\delta \vec{V}_{gr}$ becomes directed towards the origin, yields:

$$\lim_{\delta t \rightarrow 0} \frac{\delta \vec{V}_{gr}}{\delta t} \longrightarrow \frac{d\vec{V}_{gr}}{dt} = \left|\vec{V}_{gr}\right| \frac{d\theta}{dt} \left(\frac{-\vec{r}}{R_T}\right) \qquad (14.36)$$

where the unit vector $\vec{r}/R_t$ provides the direction to $d\vec{V}_{gr}/dt$. Since,

$$|\vec{V}_{gr}| = \frac{d\theta}{dt} R_T$$

[6]The following discussion is based on material in Holton (1972) and in Richards et al. (1962).

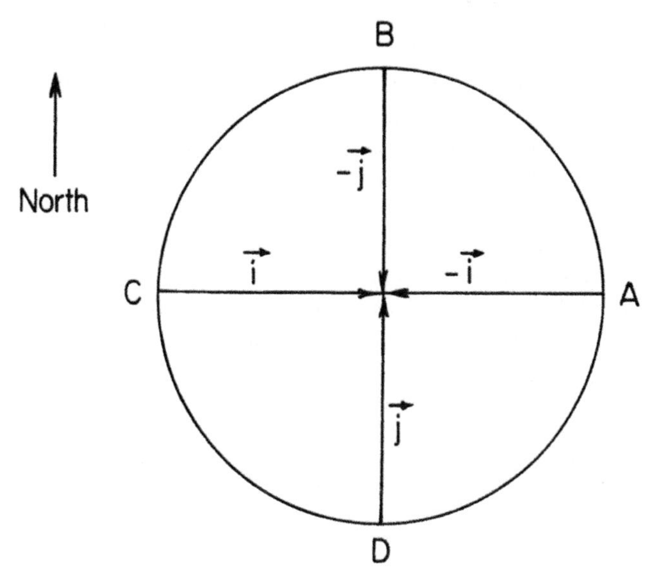

FIGURE 14.8

Circular curved flow used to evaluate Eq. 14.37. The direction of the unit vector $-\vec{r}$ at points A,B,C, and D are denoted by the appropriate Cartesian unit vector. North is towards the top of the page.

(e.g., $2\pi R_T \equiv$ circumference of the circle, so that $2\pi R_T$ / time is speed), then Eq. 14.36 can be rewritten as:

$$\frac{d\vec{V}}{dt} = \frac{|\vec{V}_{gr}|^2}{R_T^2}(-\vec{r}). \tag{14.37}$$

In order to interpret Eq. 14.37, this relation will be evaluated at the four points illustrated in Fig. 14.8.

In component form, Eq. 14.37 at the four points noted in Fig. 14.8 are:

$$\left.\begin{aligned}\frac{du}{dt} &= 0\\ \frac{dv}{dt} &= \frac{-U_{gr}^2}{R_T}\end{aligned}\right\} \text{at } B \qquad \left.\begin{aligned}\frac{du}{dt} &= 0\\ \frac{dv}{dt} &= \frac{U_{gr}^2}{R_T}\end{aligned}\right\} \text{at } D$$

$$\left.\begin{aligned}\frac{dv}{dt} &= 0\\ \frac{du}{dt} &= \frac{-V_{gr}^2}{R_T}\end{aligned}\right\} \text{at } A \qquad \left.\begin{aligned}\frac{dv}{dt} &= 0\\ \frac{du}{dt} &= \frac{V_{gr}^2}{R_T}\end{aligned}\right\} \text{at } C$$

In interpreting these relations it is important to note that it is the radius of the *trajectory*, R_T, not the streamline (which could be denoted as R_s) which determines the curvature of the flow. A streamline and a trajectory will be the same only if the flow field is spatially and temporally fixed. Also, the trajectory motion does not have to be circular as used for convenience to derive these relations. Since $\delta t \to 0$ in the derivation of Eqs. 14.36 and 14.37, only the instantaneous radius of the curved path of the trajectory need be used.

In order to illustrate the balanced wind that develops when the acceleration due to curvature is included along with the pressure gradient force and Coriolis terms, the specification of Eq. 14.37 at point A will be used. There is no loss of generality since a coordinate system could always be rotated so that a point of interest corresponds to location A.

At A, $u_g = 0$, while $v_g > 0$ for a low and $v_g < 0$ for a high in the Northern Hemisphere. For this case, using Eq. 14.37 in Eq. 4.21 with $i = 1$ and 2 with the subgrid-scale flux divergence $\hat{f} w$ Coriolis terms ignored, yields:

$$\frac{-v_{gr}^2}{R_T} = -\frac{1}{\rho}\frac{\partial p}{\partial x} + f v_{gr},$$

and since $(-1/\rho)(\partial p/\partial x) = -f v_g$,

$$\frac{-v_{gr}^2}{R_T} = -f v_g + f v_{gr} = f\left(v_{gr} - v_g\right) \tag{14.38}$$

The velocity, v_{gr}, which solves this relation is referred to as the *gradient wind*. Rearranging of this expression results in:

$$v_{gr}^2 + f R_T v_{gr} - R_T f v_g = 0 \tag{14.39}$$

Using the quadratic equation formula, v_{gr} in Eq. 14.39 is given as:

$$v_{gr} = \left(-f R_T \pm \sqrt{f^2 R_T^2 + 4 R_T f v_g}\right) \Big/ 2 \tag{14.40}$$

For a cyclone in the Northern Hemisphere,[7] $v_g > 0$ at A so that the radical in Eq. 14.40 is always real for this case and there is no limit in this relation to the magnitude of the gradient wind.

In contrast, however, for an anticyclone, $v_g < 0$ at A in the Northern Hemisphere so that $f^2 R_T^2 > 4 R_T f v_g$, i.e.,

$$v_g < f R_T / 4 \tag{14.41}$$

[7] The same result, of course, applies in the Southern Hemisphere because with a cyclone there $v_g < 0$ but $f < 0$, also.

is required for the radical to be real. Therefore, Eq. 14.40 suggests that there is a constraint on the magnitude of the pressure gradient force in anticyclones that does not exist for synoptic lows. This is the reason that lows on the synoptic weather map often have tight gradients while highs do not.

Equation 14.39 can also be rewritten as:

$$\frac{v_{gr}^2}{R_T f} + v_{gr} = v_g \qquad (14.42)$$

Thus, $v_{gr} < v_g$ for a cyclone since $v_g > 0$ at A but $|v_{gr}| > |v_g|$ for an anticyclone since $v_g < 0$ at A. These inequalities indicate that for the same pressure gradient (as represented by the geostrophic wind), the gradient balanced wind is stronger around a high than a low. This stronger wind around a high (and the associated greater curved centripetal acceleration) is the reason that a limit to the strength of the pressure gradient force, as represented by Eq. 14.41, occurs.

The balance of forces that are contained in Eq. 14.38 are illustrated in Fig. 14.9. The gradient winds associated with the low are *subgeostrophic* because the centripetal acceleration term helps balance the acceleration due to the pressure gradient force. Therefore, the Coriolis terms, $f v_{gr}$ and v_{gr}, need not be large.

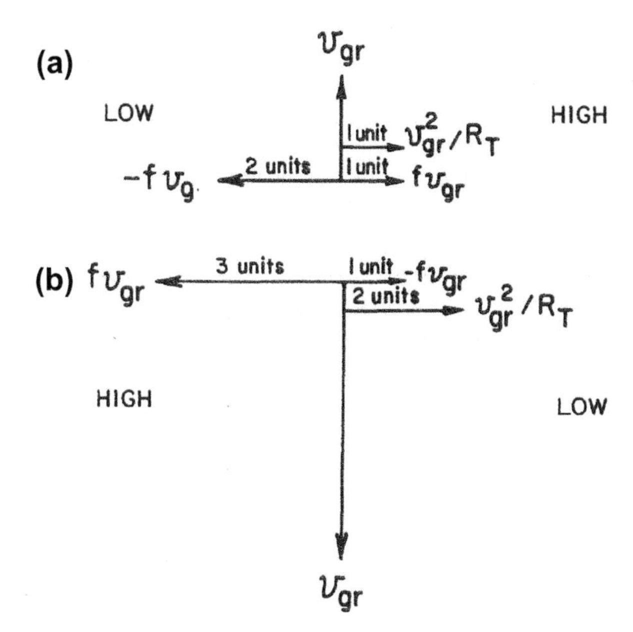

FIGURE 14.9

Schematic of the balance of acceleration in Eq. 14.37 for (a) a cyclone, and (b) an anticyclone.

In contrast, the winds associated with the anticyclone are *supergeostrophic* because a large Coriolis acceleration (and hence a large value of v_{gr}) is needed to balance the sum of the acceleration due to the pressure gradient force and the centripetal acceleration.

When the Coriolis force is neglected in Eq. 14.38, the equation represents a balance between the centripetal acceleration and the pressure gradient force. This balance, referred to as *cyclostrophic wind balance*, is used to estimate wind speeds in small-scale vortices such as tornadoes and dust devils. From Eq. 14.38 the balance can be written as:

$$\frac{-v_d^2}{R_T} = -f v_g = -\frac{1}{\rho}\frac{\partial p}{\partial x} \tag{14.43}$$

where v_g can be used to define the pressure gradient. The cyclostrophic wind is thus obtained from:

$$v_d = \pm\sqrt{\frac{R}{\rho}\frac{\partial p}{\partial x}} \tag{14.44}$$

As an example, a tornado with a radius of 0.5 km, a pressure gradient of 100 mb km^{-1}, and an air density of 1.25 kg m^{-3} would have a cyclostrophic wind balance of 63 m s^{-1}.

While both plus and minus solutions of Eq. 14.44 are possible, the circulation is usually cyclonic since the parent thunderstorm is generally rotating cyclonically due to Coriolis turning as a result of wind flow, on a fairly large scale, into the cumulonimbus system. It is thought that intense updrafts in the thunderstorm vertically stretch some of its circulation, thereby creating the strong vortex of the tornado. Intense horizontal shears between the vortex and surrounding ambient air create even smaller-scale vortices called *suction vortices*, which may be the source of the greatest tornado damage. The most severe tornadoes often occur in families when the atmosphere is favorable for their development over a large area. Over 300 people were killed in the midwest and southeast U.S. in one day in April, 1974 during a tornado family outbreak.

An evaluation of the influence of friction on the resultant balance can also be achieved by retaining the subgrid-scale flux divergence terms in 4.21 represented here by F_u and F_v, in the derivation of Eq. 14.38. By defining, for example,

$$F_u = C_D u^2; \quad F_v = C_D v^2,$$

the more general form of Eq. 14.38 (i.e., with friction) can be written as:

$$\frac{V_F^2}{R_T} + f V_F = f V_g + C_D V_F^2 \tag{14.45}$$

where V_F is, in general, at some angle to the gradient wind and $F_V = C_D V_F^2 = C_D\left(u_F^2 + v_F^2\right)$. The subscript, F, indicates that frictional effects are included.

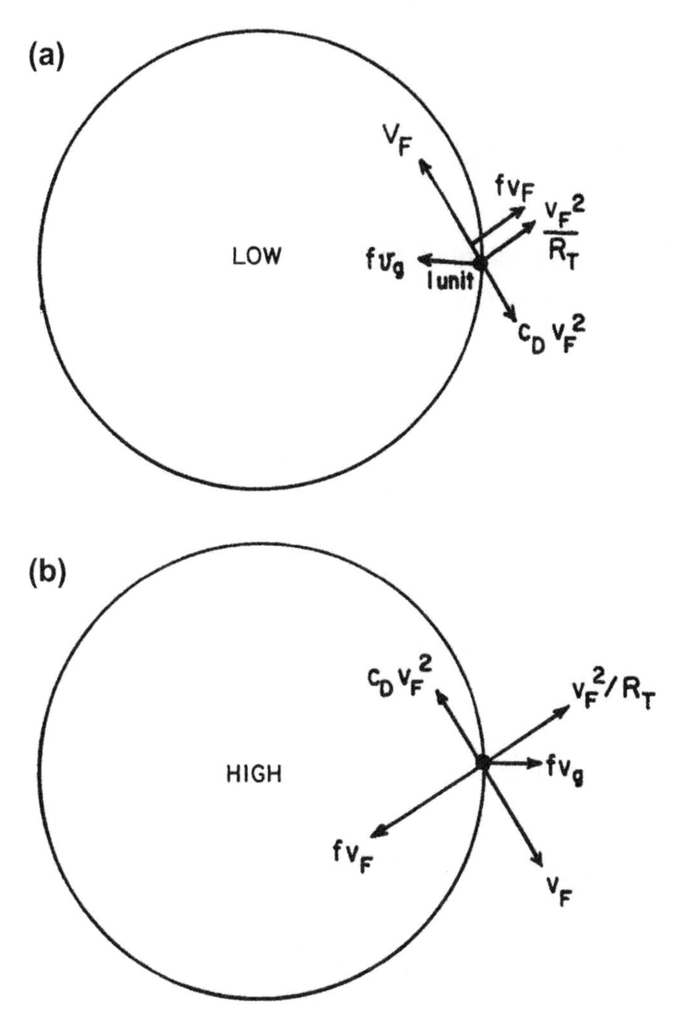

FIGURE 14.10

A schematic of the resultant balance wind when accelerations due to a pressure gradient, centripetal, Coriolis, and frictional accelerations are present for (a) a cyclone, and (b) an anticyclone.

The variable C_D is a drag coefficient which is a function of height above the ground and the thermodynamic stability. The geostrophic wind speed is $V_g = |\vec{V}_g|$, and is defined by $V_g = \frac{g \partial z}{f \partial n}$ where n is perpendicular to the height contours. A schematic illustration of a balance wind that can result from this balance of forces is illustrated in Fig. 14.10. Of particular importance is the deceleration of the flow and the turning of the wind towards low pressure. This results in low-level divergence out of anticyclones and low-level convergence into cyclones. Note that:

- the frictional acceleration acts directly opposite to the direction of the wind;
- the Coriolis acceleration is perpendicular to the wind direction; and
- the centripetal acceleration is also perpendicular to the instantaneous wind direction.

14.3.2 The Thickness Relationship

As will be shown in this section, on the synoptic scale the vertical distance between two pressure surfaces is proportional to the mean temperature in the intervening layer. The hydrostatic relation, Eq. 4.30, is used to obtain the relation, i.e.,

$$\frac{\partial p}{\partial z} = -\rho g = \frac{-pg}{R_d T}$$

using the ideal gas law ($p = \rho RT$). This equation can also be written as:

$$\frac{\partial \ln p}{\partial z} = -\frac{g}{R_d T} \quad (14.46)$$

Integrating Eq. 14.46 between two heights $(z_1, z_2; z_2 > z_1)$ corresponding to two pressure surfaces $(P_1, P_2; P_2 < P_1)$ yields:

$$\int_{z_1}^{z_2} \frac{\partial \ln p}{\partial z} dz = \ln \frac{P_2}{P_1} = -\frac{g}{R_d} \int_{z_1}^{z_2} \frac{dz}{T} = -\frac{g}{R_d} (z_2 - z_1) \left(\overline{\frac{1}{T}} \right) \quad (14.47)$$

where the mean value theorem of calculus has been used to remove $1/T$ from the integrand. If it is assumed that $\left(\overline{1/T} \right) \cong 1/\overline{T}$, then Eq. 14.47 can be written, after rearranging, as:

$$(z_2 - z_1) = \Delta z = \frac{R_d \overline{T_v}}{g} \ln (P_1/P_2) . \quad (14.48)$$

Therefore, as stated earlier in this section, the *thickness* between two pressure surfaces is proportional to the mean temperature, $\overline{T}$, in that layer.

Characteristic thickness distributions for four types of synoptic weather features are schematically illustrated in Fig. 14.11. Note that the subtropical anticyclone (a *warm core* system) and the extratropical cyclone (a *cold core* system) increase in intensity with height. Therefore, these systems would be expected to be even better defined at upper tropospheric levels than at the surface. In contrast, the tropical cyclone (a *warm core* system) and the polar anticyclone (a *cold core* system) become less intense with height.

The magnitude of the thickness is often used to characterize local weather. In the eastern United States near sea level for instance, $\Delta z = 5400$ m for the 1000 to 500 mb layer (i.e., $\overline{T} = -1.6°$C) closely corresponds to the 50% probability between rain and snow if precipitation is occurring, while $\Delta z \geq 5700$ m (i.e., $\overline{T} = 13.5°$C)

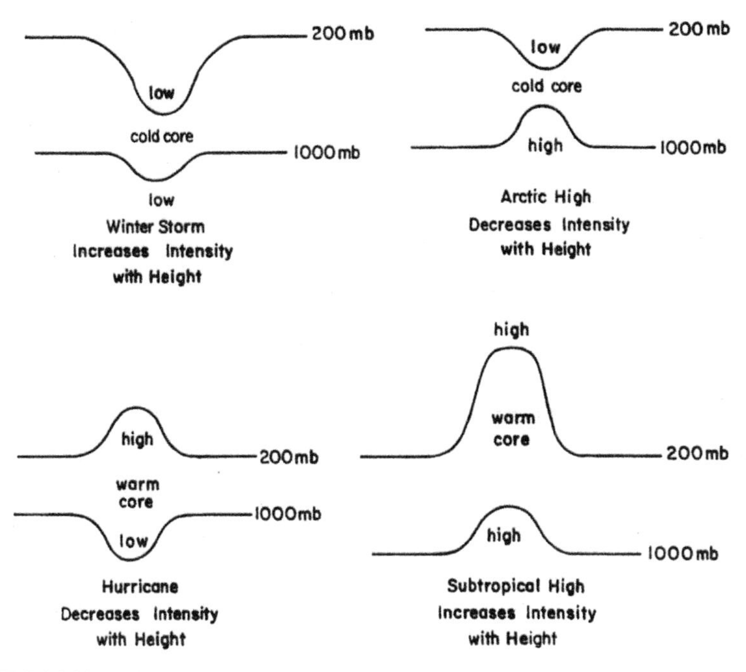

FIGURE 14.11

Schematic illustration of the thickness distribution associated with four types of synoptic weather features.

usually is associated with sultry weather. Specific examples for selected U.S. cities of the thickness values associated with a 50% probability of rain or snow are listed in Table 14.3. Note the substantially higher values of $\Delta z|_{\text{rain/snow}}$ for sites such as Denver and Cheyenne which are located at relatively high altitudes so that only a portion of the 1000-500 mb thickness actually exists. Generally, the warmest part of the layer does not occur. The lowest thicknesses are along the west coast where the distribution of temperature with height tends to be less stable during the precipitation events than in the eastern United States. This is because most precipitation in the winter occurs poleward of warm fronts in the eastern United States where the atmosphere tends to be more stably stratified. In the coastal regions of the western United States the precipitation is frequently post-cold frontal where a much less stable sounding occurs. Thus, a lower value of $\overline{T}$ is required along the west coast than the east coast for snow to reach the surface. Despite relatively cold temperatures aloft along the west coast, the relatively warm temperatures near the surface melt the snow before it reaches sea level. (In the mountains along the west coast, however, the more rapid decrease of temperature with height results in a lower snow level than would generally occur in the eastern United States with the same sea-level temperature.)

Thickness analyses are also used to locate synoptic fronts and to estimate their intensity. By definition, a synoptic front must be associated with a horizontal gradient

Table 14.3 Thickness values associated with a 50% chance of snow given that precipitation is occurring (from TPBS, 1974).

	$\Delta z\|_{\text{rain/snow}}$
Bismark, North Dakota	5413
Bristal, Tennessee	5387
Baltimore, Maryland	5362
Beckley, West Virginia	5426
Chicago, Illinois	5391
Columbia, Missouri	5400
El Paso, Texas	5467
Goodland, Kansas	5461
Denver, Colorado	5501
Cheyenne, Wyoming	5509
Grand Junction, Colorado	5454
Havre, Montana	5410
Lander, Wyoming	5501
Medford, Oregon	5262
Minneapolis, Minnesota	5408
Missoula, Montana	5396
Norfolk, Virginia	5371
Philadelphia, Pennsylvania	5361
Pueblo, Colorado	5489
Reno, Nevada	5427
Seattle-Tacoma, Washington	5205

of thickness. The surface intersections of fronts are located on the warm side of the thickness gradient and they usually tilt poleward with height. The method to determine the specific location of surface fronts is discussed in Section 14.3.5.1.

The procedure to graphically analyze thickness from two constant pressure analyses is illustrated in Fig. 14.12. An alternative is to directly calculate Δz from the sounding data, plot it on a base map, and analyze the plotted fields.

14.3.3 Thermal Wind

From Eq. 3.28, the geostrophic wind can be written (using vector notation) as:

$$\vec{V}_g = \vec{k} \times \frac{g}{f} \vec{\nabla}_p z$$

Differentiating this expression with respect to pressure yields:

$$\frac{\partial \vec{V}_g}{\partial p} = \frac{g}{f} \vec{k} \times \vec{\nabla}_p \frac{\partial z}{\partial p}$$

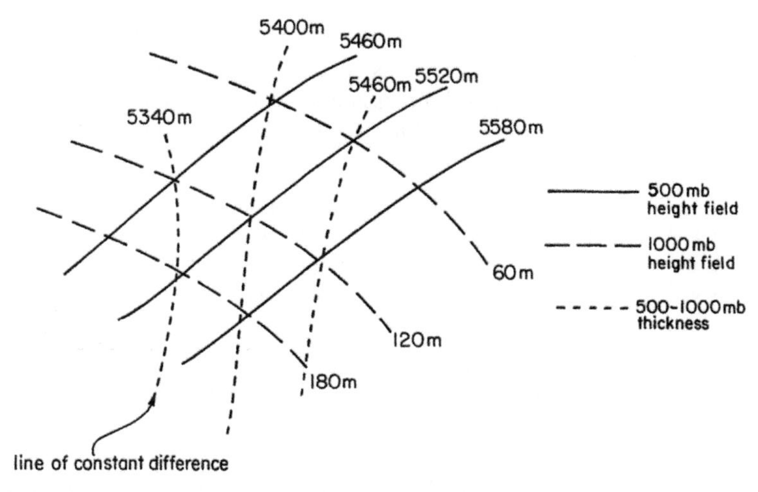

FIGURE 14.12

A schematic illustrating the procedure to graphically determine thickness. Two rules are to *include all intersections* of the two constant pressure height fields and *not to cross contours*.

which, since $\partial z/\partial p = -1/\rho g = -R_d T/gp$ by the hydrostatic relation and the gas law, results after rearranging in:

$$\frac{\partial \vec{V}_g}{\partial \ln p} = -\frac{R_d}{f}\vec{k} \times \vec{\nabla}_p T \tag{14.49}$$

Therefore, the change of geostrophic wind with pressure is proportional to the gradient of temperature on a constant pressure surface.

Integrating Eq. 14.49 between two pressure surfaces and rearranging gives:

$$\int_{\ln p_1}^{\ln p_2} \frac{\partial \vec{V}_g}{\partial \ln p} d\ln p = \vec{V}_{g_{p_2}} - \vec{V}_{g_{p_1}} = \Delta\vec{V}_g = \frac{R_d}{f}\vec{k} \times \vec{\nabla}_p \overline{T}\left[\ln\frac{P_1}{P_2}\right] \tag{14.50}$$

where the mean value theorem of calculus has been used to extract $\nabla_p \overline{T}$ from the integral. The quantity ΔV_g is referred to as the *thermal wind.*

Equation 14.50 can also be written in terms of the thickness gradient. Performing the gradient operation on Eq. 14.48 yields:

$$\nabla_p(\Delta z) = \frac{R_d}{g}\ln(P_1/P_2)\vec{\nabla}_p\overline{T},$$

so that Eq. 14.50 can be rewritten as:

$$\Delta\vec{V}_g = \frac{g\vec{k}}{f} \times \vec{\nabla}_p(\Delta z) \tag{14.51}$$

Since the magnitude of $\Delta\vec{V}_g$ is related to the value of the average horizontal temperature gradient through the thickness equation, $|\Delta\vec{V}_g|$ is used to classify the strength of synoptic fronts. Using the pressure surfaces 1000 mb and 500 mb, the following criteria have been established for use by the U.S. Weather Service;

$$
\begin{array}{rlll}
 & |\Delta\vec{V}_g| & < 12.5\text{ m s}^{-1} & \longrightarrow \text{no front} \\
12.5\text{ m s}^{-1} & < |\Delta\vec{V}_g| & \leq 25.0\text{ m s}^{-1} & \longrightarrow \text{weak front} \\
25.0\text{ m s}^{-1} & < |\Delta\vec{V}_g| & \leq 37.5\text{ m s}^{-1} & \longrightarrow \text{moderate front} \\
37.5\text{ m s}^{-1} & < |\Delta\vec{V}_g| & & \longrightarrow \text{strong front}
\end{array}
$$

Subjectively, the category of frontal strength is raised one level by the Weather Service if the weather along the front is unusually active, or lowered one level if there is very little activity.

Since $\vec{V}_g$ at 1000 mb is usually small, to the extent that the geostrophic wind approximates the actual wind, strong winds at 500 mb are indicative of a strong front. In addition, since the same sign of the synoptic temperature gradient generally exists up to the tropopause, the geostrophic wind continues to increase with height. Above the tropopause, the temperature gradient reverses sign so that the geostrophic wind decreases with height. The region of strongest geostrophic wind near the tropopause height is referred to as the *jet stream*.

The magnitude and direction of the thermal wind can be used to estimate temperature advection. As illustrated in Fig. 14.13, geostrophic winds which rotate counterclockwise (i.e., *back*) with height are associated with *cold advection* in the Northern Hemisphere. Geostrophic winds that turn clockwise (i.e., *veer*) with height are due to *warm advection*. In the Southern Hemisphere the reverse is true (i.e., backing winds with height are associated with warm advection). Cold advection in a layer of the atmosphere simply means that,

$$\frac{\partial \overline{T}}{\partial t} < 0 \text{ because } -\overline{\vec{V}} \cdot \nabla_p \overline{T} < 0$$

while warm advection is:

$$\frac{\partial \overline{T}}{\partial t} > 0 \text{ because } -\overline{\vec{V}} \cdot \nabla_p \overline{T} > 0,$$

where $\overline{\vec{V}}$ is the average wind vector in the layer.

In the case of no temperature advection, as illustrated in Fig. 14.13, only the speed of the geostrophic wind, not the direction, changes with height. With cold air towards the pole, this requires that the westerlies increase in speed with height with a low-level westerly geostrophic wind. With warm air towards the north the westerlies would decrease with height for this situation.

An illustration as to why geostrophic winds back with height in the Northern Hemisphere under cold advection, while geostrophic winds veer with height during warm advection, is illustrated in Fig. 14.14. As sketched in that figure, troughs (often

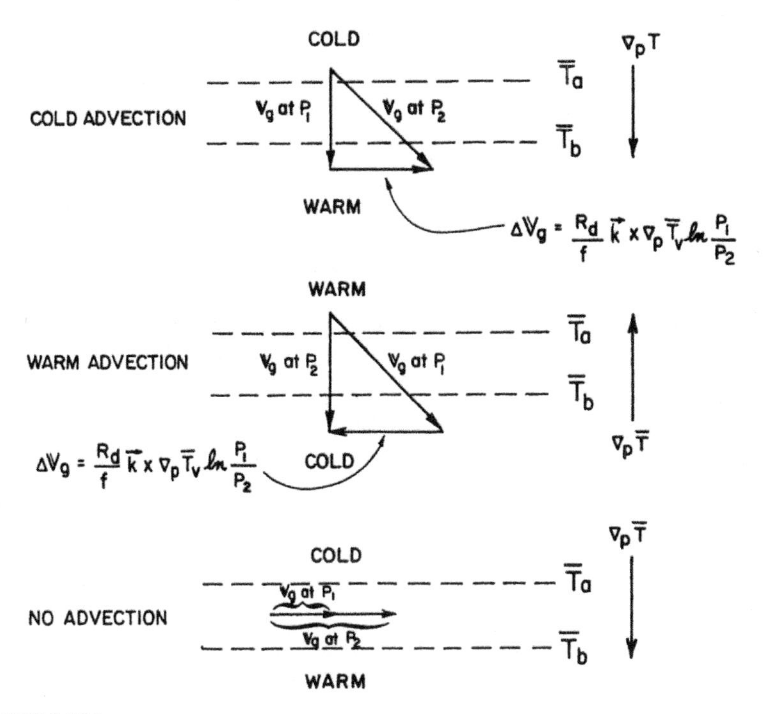

FIGURE 14.13

A schematic illustration of the relation between the thermal wind and temperature advection. In this figure, $\overline{T_a} < \overline{T_b}$ and $P_1 > P_2$.

written as "trofs") tilt towards colder air with height while ridges align towards warmer air with height. This is a direct result of the thickness relation (i.e., Eq. 14.48) which states that the vertical distance between pressure surfaces is directly proportional to the mean temperature of the intervening layer.

14.3.4 Horizontal Vorticity with Gradient Wind Balance

A valuable concept in describing synoptic atmospheric circulations is that of *vertical vorticity* (see also Section 4.2 in Pielke 2002b and Pielke et al. 1995a). Vorticity is defined as,

$$\vec{\nabla} \times \vec{V} \equiv \text{vorticity} = \xi,$$

and represents a measure of circulation. In synoptic meteorology, the vertical component of vorticity is of primary interest. This component can be written as:

$$\xi_z = \vec{k} \cdot \vec{\nabla} \times \vec{V} = \left(\frac{\partial v}{\partial x} - \frac{\partial u}{\partial y} \right) \qquad (14.52)$$

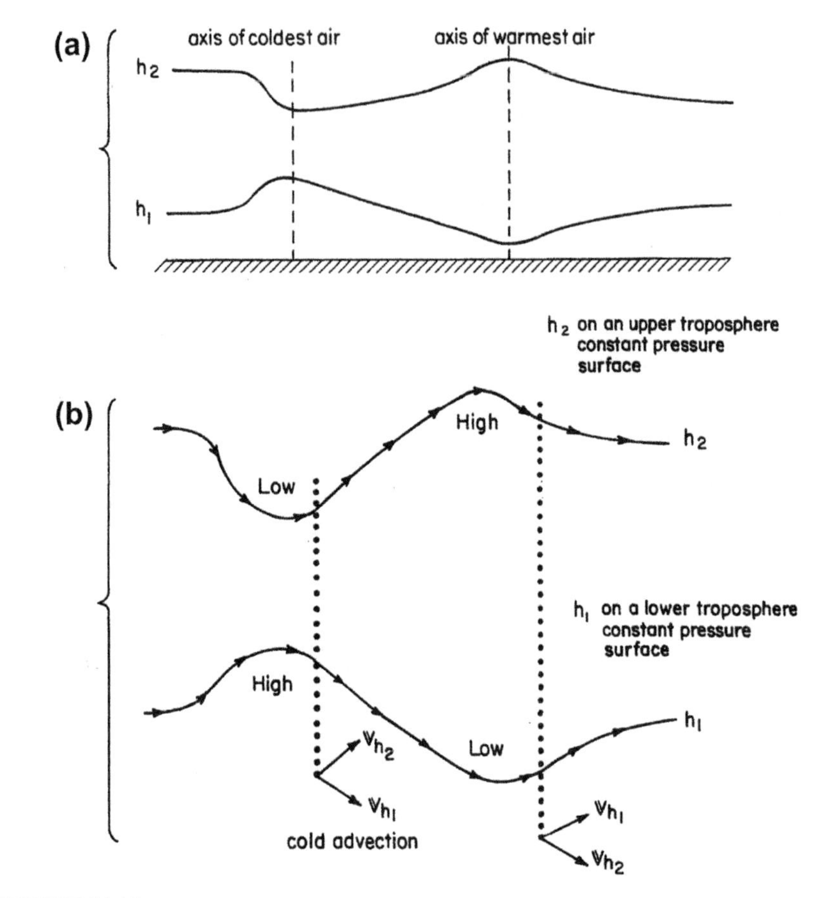

FIGURE 14.14

Schematic illustrating the reason that cold advection is associated with backing winds and warm advection with veering winds with height in the Northern Hemisphere. A vertical cross section is given in (a), a horizontal cross section in (b).

Using the Stokes theorem (e.g., see Kaplan 1952, pg. 275), Eq. 14.52 can be integrated over an area yielding:

$$\iint_S \xi_z \, ds = \iint_S \vec{k} \cdot \vec{\nabla} \times \vec{V} \, ds = \oint V \cdot d\vec{r} = C$$

where S is the area of integration and $\oint$ represents a closed line integral around the perimeter of the area which is perpendicular to $\vec{k}$. The distance differential is represented by $d\vec{r}$, so that the circulation C is defined by the integrated velocity along the line integral. Therefore, circulation represents an area average of the vorticity.

An equation for *vertical vorticity* is obtained from Section 14.3.1. For convenience in the following derivation, the $\hat{f}w$ term is neglected since the magnitude of w is assumed much less than u and v on the synoptic scale and friction is ignored. Performing the partial derivative in y on the first equation and the partial derivative in x on the second equation 4.21 with $i = 1$ and 2 and subtracting the resultant second equation from the first yields, after rearrangement yields:

$$\frac{d}{dt}(\xi_z + f) = -(f + \xi_z)\left[\frac{\partial u}{\partial x} + \frac{\partial v}{\partial y}\right] + \left[\frac{\partial w}{\partial y}\frac{\partial u}{\partial z} - \frac{\partial w}{\partial x}\frac{\partial v}{\partial z}\right] - \frac{1}{\rho^2}\left[\frac{\partial p}{\partial x}\frac{\partial \rho}{\partial y} - \frac{\partial p}{\partial y}\frac{\partial \rho}{\partial x}\right] \quad (14.53)$$

In obtaining Eq. 14.52, the observation that $\frac{df}{dt} = v\frac{\partial f}{\partial y}$ is used. The quantity $\xi_z + f$ is referred to as *absolute vorticity* with ξ_z called the *relative vorticity*. The Coriolis parameter f is the Earth's vorticity which results because of the planet's rotation.

The terms in Eq. 14.52 are defined as follows:

$\dfrac{d(\xi_z + f)}{dt}$: change of absolute vorticity, following a parcel.

$(f + \xi_z)\left[\dfrac{\partial u}{\partial x} + \dfrac{\partial z}{\partial y}\right]$: represents changes of vertical absolute vorticity as a result of horizontal velocity divergence. To the extent that the atmosphere is incompressible, $\dfrac{\partial u}{\partial x} + \dfrac{\partial v}{\partial y} = -\dfrac{\partial w}{\partial z}$ can be substituted into this term.

$\left[\dfrac{\partial w}{\partial y}\dfrac{\partial u}{\partial z} - \dfrac{\partial w}{\partial x}\dfrac{\partial v}{\partial z}\right]$: represents changes of vertical absolute vorticity due to conversion from or to horizontal vorticity. This term is called the *tilting term*.

$\dfrac{1}{\rho^2}\left[\dfrac{\partial \rho}{\partial x}\dfrac{\partial p}{\partial y} - \dfrac{\partial \rho}{\partial y}\dfrac{\partial p}{\partial x}\right]$: represents changes of absolute vorticity as a result of differential heating or cooling. This term is referred to as the *solenoidal term*.

Relative vorticity can also be expressed in so-called natural coordinates. Natural coordinates are defined with respect to the motion of a parcel. Using this approach:

$$\xi_z = \frac{\partial v}{\partial x} - \frac{\partial u}{\partial y} = \frac{V}{R_T} - \frac{\partial V}{\partial n} \quad (14.54)$$

where V/R_T represents the angular velocity of solid rotation of an air parcel about a vertical axis with an instantaneous radius of curvature, R_T. The radius, R_T, is defined positive for counterclockwise rotation. The lateral shear term, $-\partial V/\partial n$, represents the effective angular velocity of an air parcel produced by distortion due to horizontal velocity differences at its boundaries. An illustration of the two terms on the

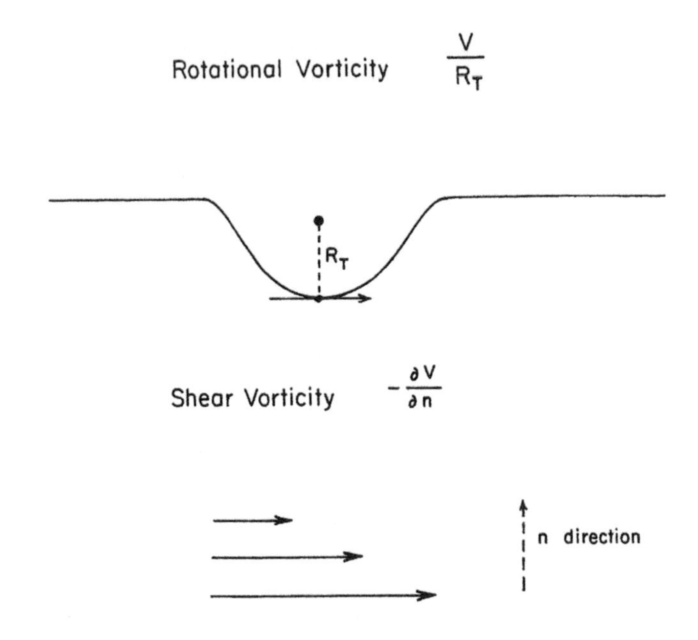

FIGURE 14.15

Examples of rotational and shear cyclonic vorticity illustrated in natural coordinates.

right of Eq. 14.54 are given in Fig. 14.15. A useful interpretation of Eq. 14.54 is obtained by assuming a stick in an airstream represented by Fig. 14.15 with a pivot point midway along the stick. In both cases, for this example, the stick would rotate counterclockwise, indicating a cyclonic circulation which is represented by positive relative vorticity.

In order to illustrate the use of Eq. 14.53, the tilting and solenoidal terms are ignored, so that,

$$\frac{d(\xi_z + f)}{dt} = -\left(f + \xi_z\right)\left(\frac{\partial u}{\partial x} + \frac{\partial v}{\partial y}\right) = -\left(f + \xi_z\right) \text{div}_H \vec{V},$$

where the convention of writing $\frac{\partial u}{\partial x} + \frac{\partial v}{\partial y}$ as $\text{div}_H \vec{V}$ has been adopted. Clearly $\text{div}_H V < 0$ indicates convergent horizontal winds, while $\text{div}_H V > 0$ specifies divergent horizontal winds. Figure 14.16 illustrates a characteristic ridge-trough parcel trajectory in the upper troposphere. Since wind speeds in the westerlies in midlatitudes usually increase monotonically with height within the troposphere, the wind (and, therefore, the vorticity) field at the upper levels exerts a major control on the synoptic vertical motion field as represented by $\text{div}_H \vec{V} \cong -\partial w/\partial z$.

To the extent that the parcel trajectory is in gradient wind balance and, therefore, represented by Eq. 14.45 with $C_D = 0$, the parcel at the upper levels will decelerate as it moves from the ridge crest into the trough. Correspondingly, the parcel will

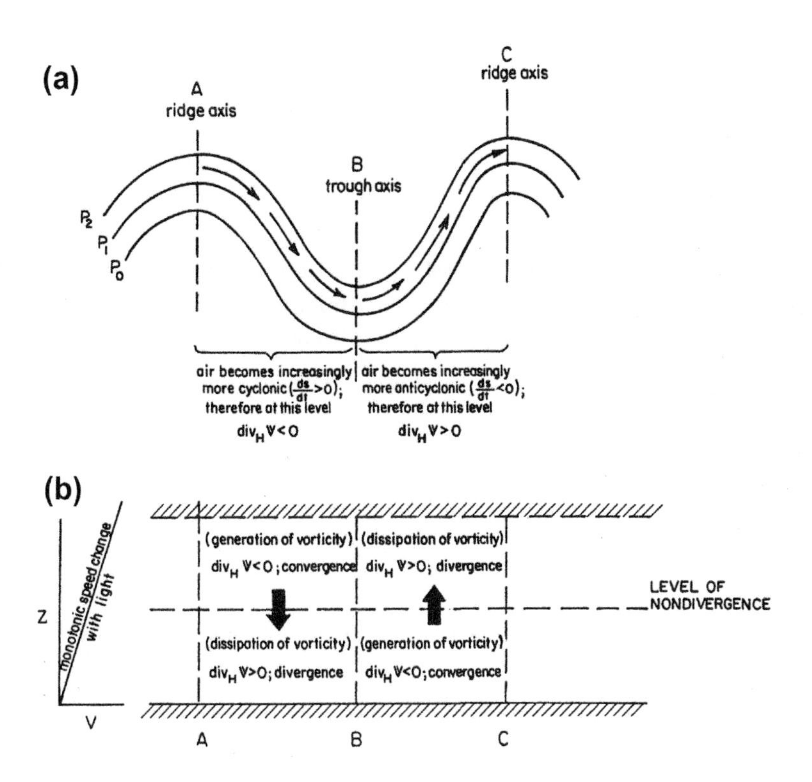

FIGURE 14.16

Schematic illustration of the inferred change of vorticity and resultant vertical motion (b) as an air parcel in gradient wind balance moves through a constant pressure gradient wind field in the upper troposphere given in (a).

accelerate as it approaches the ridge crest from the trough. Since, except for a modification due to the decrease of average air density with height, $\mathrm{div}_H\ \vec{V} \cong -\partial w/\partial z$ is a good approximation in the Earth's troposphere, and the distribution of vertical velocities in Fig. 14.16b would result in order to conserve mass. On flat ground, $w = 0$, of course, while at the tropopause $w \cong 0$ will be assumed in this analysis since the strong thermal stratification in the lower stratosphere strongly inhibits vertical motion.

The distribution of vertical motion in the midtroposphere, associated with gradient wind balance around ridges and troughs, provides one explanation for the observed preference for clouds east of a trough axis and west of a ridge crest (additional explanations are provided in Section 14.3.5). The level of nondivergence indicated in Figure 14.16b occurs around 600 mb.

Values of realistic changes in gradient wind speed as a parcel moves through a ridge and trough are listed in Table 14.4. Assuming a height of 10 km, and a linear change of speed with height (with $|\vec{V}| = 0$ at the surface) yields the values of vertical

Table 14.4 Estimate of induced vertical motion in cm s^{-1} at the level of nondivergence (a height of 5 km is used here) assuming steady flow and a wind speed which increases linearly with height (V at the surface is zero). The distance over which change takes place is on the left of the table. Analysis requires a large enough radius of curvature of parcel trajectory when the flow is anticyclonic such that the inequality given by Eq. 14.41 is satisfied.

Distance (km)	Change in Gradient Wind Speed from Ridge Crest to Trough Axis in the Upper Troposphere (m s^{-1})			Change in Gradient Wind Speed from Trough Axis to Ridge Crest in the Upper Troposphere (m s^{-1})		
	10	50	100	10	50	100
500	−10	−50	−100	10	50	100
1000	−5	−25	−50	5	25	50
2000	−2.5	−12.5	−25	2.5	12.5	25
5000	−1	−5	−10	1	5	10

velocity, w, at the level of nondivergence given in the table. As concluded by Palmén and Newton (1969, pg. 145),

> *"The magnitude of the gradient-wind divergence is greatest if the wave amplitude is large, the wavelength small, and the wind speed is large and significantly different from that at the level of nondivergence."*

As stated further by Palmén and Newton (1969), since the inferred vertical motion is larger when a more substantial vertical gradient wind shear exists, stronger thickness gradients associated with ridges and trough couplets are associated with more intense vertical motion patterns.

14.3.5 Extratropical Cyclones

14.3.5.1 Determination and Definition of Fronts

A front on the synoptic scale is characterized by the following three criteria.

i) A front separates different air mass types.
ii) A thickness gradient defines the approximate location of the front, which is located near the warm side of the gradient.
iii) The type of front depends on the direction of movement of the colder air mass.

At the surface, the following indicators are applied to precisely locate a front, once the criteria listed above are satisfied:

a) a wind shift line
b) a pressure trough

c) a temperature discontinuity
d) a dewpoint temperature discontinuity
e) the pressure tendency pattern
f) horizontal variations in visibility
g) horizontal variations in precipitation type

There are five basic types of fronts, i.e.,

- cold front
- warm front
- stationary front
- cold occlusion
- warm occlusion

The symbols used on a weather map for each are shown (note that both types of occlusions use the same symbols). As an example, the common characteristics of criteria a)–g) for a cold and a warm front in the Northern Hemisphere are listed in Table 14.5.

An illustration of the relation between fronts and the thickness gradient is illustrated in Fig. 14.17.

The development of fronts are associated with low-level synoptic flow which is convergent in the presence of a thickness gradient in the low levels. The development of fronts is termed *frontogenesis*. In contrast, low-level synoptic flow which becomes divergent in the vicinity of an existing front is *frontolytic* (i.e., causing the front to weaken).

The two types of each front are:

- active fronts, and
- inactive fronts.

An active front exists when the warmer air mass is overrunning the cold air mass. This occurs when the warmer air mass has a wind component with respect to the frontal motion through a significant depth of the atmosphere which is blowing towards the front. This overrunning can result in clouds and precipitation if the overrunning is deep and extends high enough over the frontal surface to reach the lifting condensation level.

Figure 14.18 schematically illustrates the differences between active and inactive cold, warm, and stationary fronts. Occlusions can be defined in a similar fashion where the direction of movement of the low-level air on the warmer side determines whether it is an active or inactive occlusion.

Cross sections through cold and warm occlusions are illustrated in Fig. 14.19. The tongue of warm air aloft represented by the bulge in the thickness contours occurs when a cold front catches up with a warm front, thereby lifting the warmer air between the air which was originally between the two fronts. If the air behind the cold front is warmer than the air ahead of the warm front, a warm occlusion results. In the United States, warm occlusions are more common than cold occlusions. The surface

Table 14.5 Characteristics of warm and cold fronts at the surface.

	Cold Front	Warm Front
A wind shift line	a) Ahead of the front in the warmer air, surface winds are southwesterly, often gusty close to the front. Behind the front, winds veer to the west or northwest and increase in strength as the cold advection destabilizes the air, thereby mixing the stronger winds aloft down to the surface.	Ahead of the front in the colder air, winds are southeasterly to northeasterly in direction with little gustiness, generally. Behind the front, winds veer to the southwest or south.
A pressure trough	b) The development of a pressure trough is associated with convergent synoptic-scale low-level winds which concentrate the thickness gradient, thereby strengthening the front. Frictional convergence into the trough enhances the gradient further.	Same as for cold fronts.
A temperature discontinuity	c) The air is generally colder at the surface behind the front in the air mass with lower thicknesses. An exception occurs, particularly in the summer when although the thickness is smaller behind a front so that the average temperature is colder, stronger vertical mixing and/or removal of cloud cover behind the front can result in warmer surface temperatures.	Colder air ahead of the warm front. The gradient of temperature at the surface is usually less ahead of a warm front than is the case for a cold front. This is attributed to the greater frictional retardation of the forward movement of cold air by the ground than is the case for the retreat of cold air ahead of the warm front.
A dewpoint temperature discontinuity	d) Lower dewpoint temperature air is generally found behind cold fronts since cold air can hold less moisture than warmer air.	Lower dewpoint temperature air is generally found ahead of warm fronts although the relative humidity is generally much higher ahead of the front.
The pressure tendency pattern	e) Pressure falls ahead of the front; rises behind the front; the rate of pressure change depends on the strength of the pressure trough associated with the front and its rate of movement. Pressure changes are generally more rapid with a cold front than a warm front of equal space because the temperature gradient (and therefore, horizontal pressure gradient) in low levels is larger for the reason mentioned in c) of these criteria.	Pressure falls comparatively slowly ahead of the warm front unless the trough associated with the front is intensifying rapidly; pressure rises slowly or is steady behind the front.
Horizontal gradient in visibility	f) Visibility is often excellent behind the cold front due to the enhanced vertical mixing.	Often poor visibility ahead of the warm front improving after the warm front passes.
Horizontal variation in precipitation type	g) Deep cumulus convection often occurs ahead of the cold front. Behind the cold front, if they occur, showers are usually more shallow in vertical extent.	Often steady precipitation ahead of the front, replaced by showers or no precipitation behind the front.

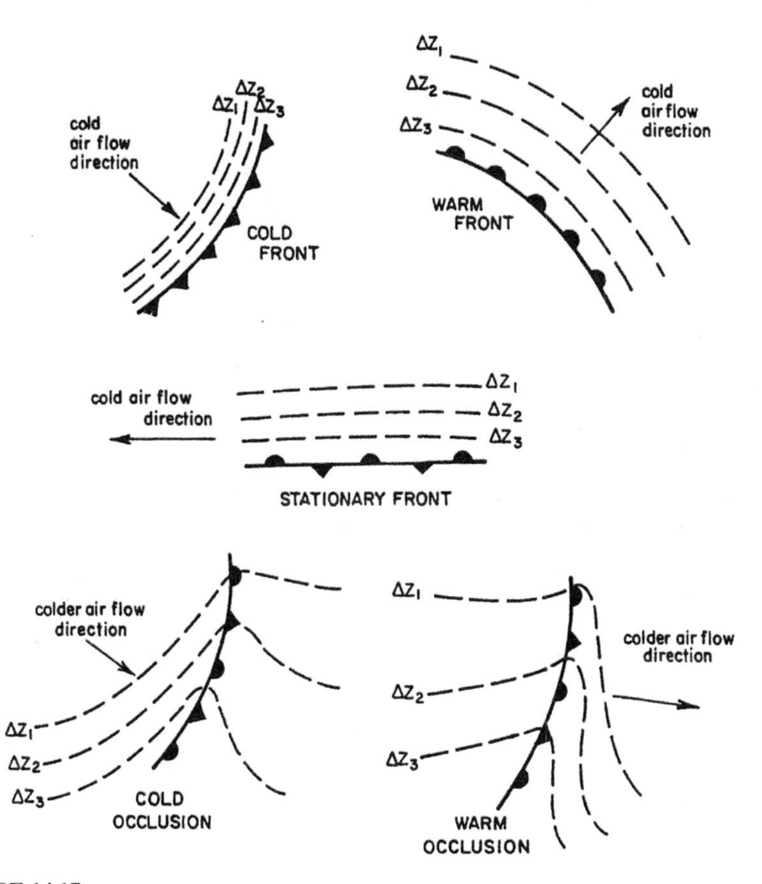

FIGURE 14.17

Schematic illustration of the different types of fronts. Fronts always move in the direction in which the more dense (i.e., colder) air mass is moving, $\nabla_{z3} > \nabla_{z1}$.

frontal position closely corresponds to the position of the warm thickness bulge. This thickness bulge is generally associated with a minimum of pressure (i.e., the pressure trough associated with fronts). Cold occlusions tend to have the characteristics of a cold front, while warm occlusions are more similar to warm fronts.

14.3.5.2 *Surface Depiction of Extratropical Storm Development*

The existence of well-defined cyclonic circulations in mid and high latitudes (i.e., extratropical cyclones) were first defined at the surface where observations were sufficiently dense to resolve the feature. The association of the cyclone with fronts (and the existence of fronts) was originally noted by Norwegian meteorologists shortly after World War I (see Namias 1983 for a historical discussion of the discovery of fronts and the extratropical cyclone). The term front was adopted because of its

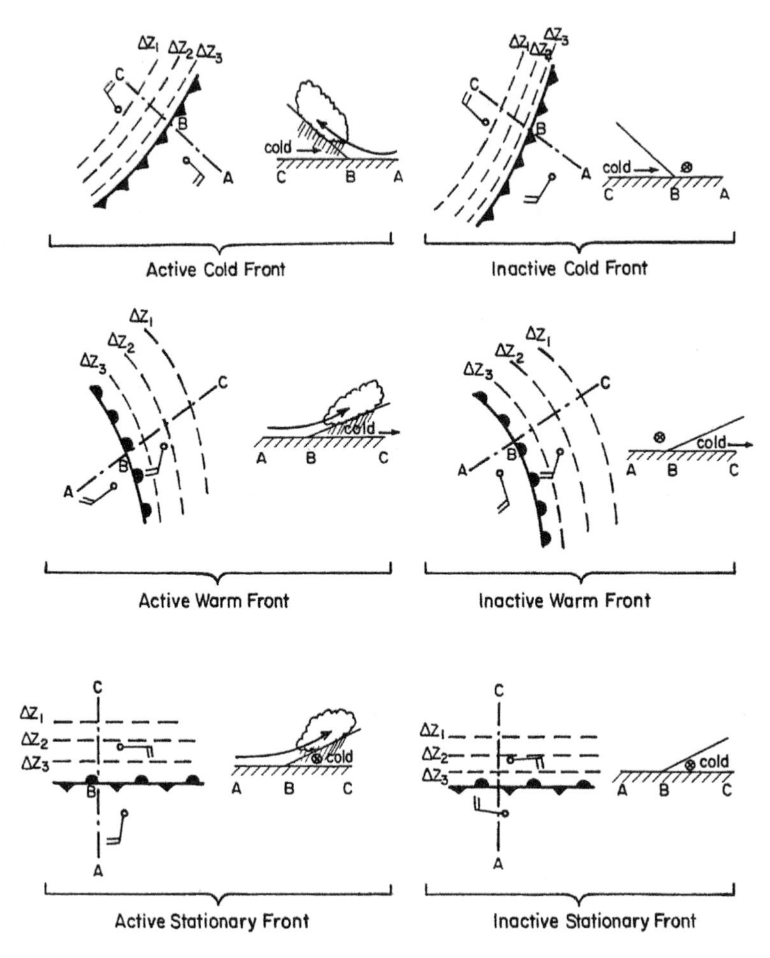

FIGURE 14.18

Schematic illustration of active and inactive cold, warm, and stationary fronts.

association with the frontal and trench warfare of the war, in which well-defined defensive positions separated opposing armies.

Figure 14.20 illustrates three major stages in the evolution of a cyclone from a wave on a stationary front to its mature form. This process is referred to as cyclogenesis. All cyclones do not evolve through each stage, however. Many terminate at the initial stage and are termed *wave cyclones*. The difference in degree of cyclonic development is associated with the concept of *baroclinic instability*. In simple terms, however, a cyclone evolves through all three stages when there is a continuous conversion of potential energy (as represented by the juxtaposition of cold and warm air) to kinetic energy (as represented by the cyclonic wind flow around the low pressure system). As an extratropical cyclone evolves through its different stages, its minimum

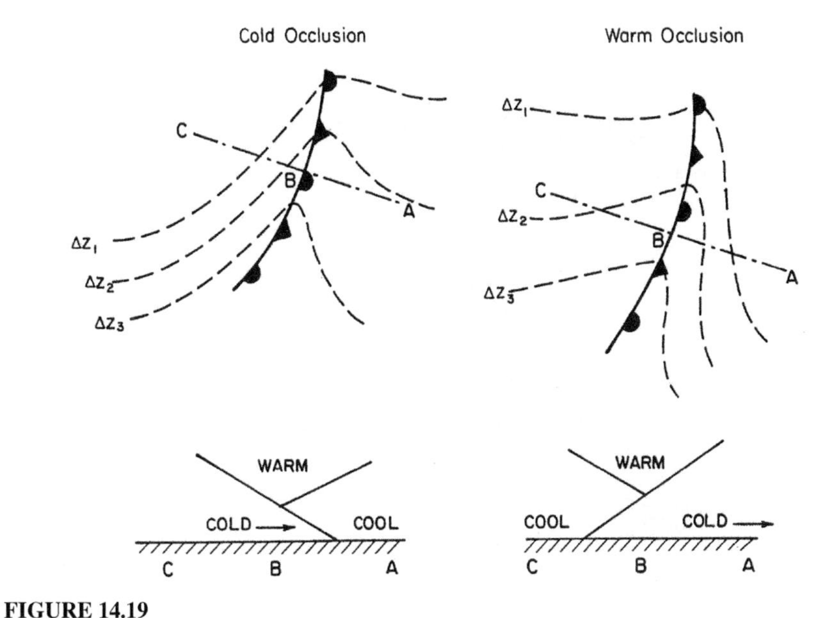

FIGURE 14.19

Schematic illustration of the juxtaposition of air masses in warm and cold occlusions. The movement of the cold air is indicated by the arrow.

pressure decreases and moves toward the colder air. The most rapid falls occur in the development stage – when the falls are excessive the extratropical storm is said to be undergoing *explosive cyclogenesis.*

The juncture of the warm, cold, and occlusion fronts during the mature stage is referred to as the *triple point*. Often a new low pressure center may develop at the triple point. This new development, referred to as *secondary cyclogenesis*, frequently occurs off the east coast of continents in midlatitudes during the winter as this zone of horizontal thickness gradient and associated cold air moves over warmer waters. The development processes associated with cyclogenesis are discussed in the next sections.

14.3.5.3 The Omega Equation

In order to describe the vertical motion pattern above the surface associated with extratropical cyclones and other types of synoptic weather features, it is useful to combine the vorticity equation and first law of thermodynamics into a single relation.

Using the same procedure to derive Eq. 14.52, except on a constant pressure surface, Eq. 4.21 with $i = 1$ and 2 can be used to derive a vertical vorticity equation on a constant pressure surface (i.e., the vertical σ coordinate pressure, p; see Chapter 6). Differentiating Eq. 4.21 with $i = 1$ by $\frac{\partial}{\partial y}$, and Eq. 4.21 with $i = 2$ by $\frac{\partial}{\partial z}$, subtracting the resultant first expression from the second and neglecting the tilting and friction

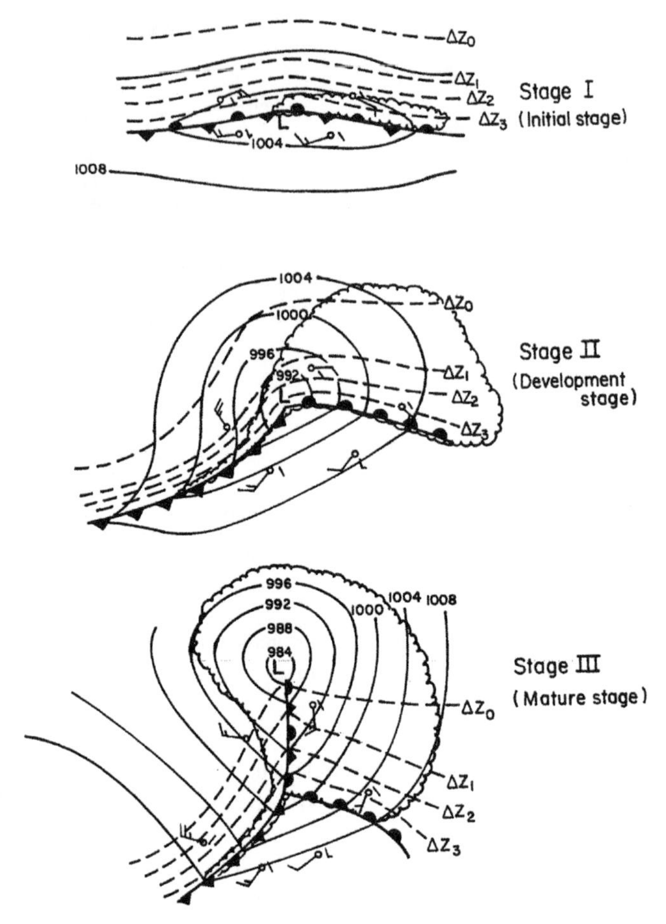

FIGURE 14.20

Schematic illustration of the evolution of an extratropical cyclone from a wave cyclone to a mature cyclone. The scalloped area represents cloud covered areas. Thickness is given by Δz where $\Delta_{z3} > \Delta_{z_o}$. Pressure is in millibars. Examples of station observations as plotted on U.S. surface map analyses are shown.

terms yields:

$$\frac{\partial \xi_p}{\partial t} + \vec{V} \cdot \vec{\nabla}_p \left(\xi_p + f\right) = \left(\xi_p + f\right) \frac{\partial \omega}{\partial p}, \tag{14.55}$$

where ξ_p is the relative vorticity on a constant pressure surface. The quantity $\frac{\partial u}{\partial x} + \frac{\partial v}{\partial y}$ on a constant pressure service is replaced by $\frac{\partial \omega}{\partial p}$ with $\omega = \frac{dp}{dt}$ (e.g., see Haltiner 1971, pp. 6-8).

Since $\vec{V}_g = \frac{g}{f}\vec{k} \times \nabla_p z$, a *geostrophic vorticity*, defined as ξ_g, can be obtained from:

$$\vec{\nabla} \times \vec{V}_g = \frac{g}{f}\nabla_p^2 z = \xi_g \tag{14.56}$$

Since this vorticity is defined in terms of the geostrophic wind, it is a synoptic-scale definition. This vorticity, and the equations derived from it are not appropriate for mesoscale systems, but are appropriate for synoptic-scale weather features. A practical use of Eq. 14.56 is that vorticity can be estimated from the curvature of the height contours on a constant pressure analysis. Substituting Eq. 14.56 into Eq. 14.55, where ξ_p is set equal to ξ_g, yields:

$$\frac{\partial}{\partial t}\frac{g}{f}\nabla_p^2 z + \vec{V} \cdot \vec{\nabla}_p \left(\xi_g + f\right) = \left(\xi_g + f\right)\frac{\partial \omega}{\partial p}, \tag{14.57}$$

where $\omega = dp/dt$.

From Eq. 14.14, the first law of thermodynamics can be written as:

$$C_p \frac{d \ln \theta}{dt} = Q/T \tag{14.58}$$

where Q represents changes of the sensible heat of a parcel (i.e., diabatic effects). Q can include explicit synoptic-scale phase changes of water as represented by $-Ldw_s$ in Eq. 14.12, as well as radiative flux divergence and subsynoptic-scale phase changes of water due to cumulus clouds.

Equation 14.58 can be rewritten as:

$$C_p \left[\frac{\partial \ln \theta}{\partial t} + u\frac{\partial \ln \theta}{\partial x} + v\frac{\partial \ln \theta}{\partial y} + \omega\frac{\partial \ln \theta}{\partial p}\right] = Q/T. \tag{14.59}$$

Since using the gas law:

$$\theta = T\left[1000/p\right]^{R_d/C_p} = \frac{p\alpha}{R_d}\left(\frac{1000}{p}\right)^{R_d/C_p}$$

then

$$\frac{\partial \ln \theta}{\partial x} = \frac{\partial \ln \alpha}{\partial x}; \quad \frac{\partial \ln \theta}{\partial y} = \frac{\partial \ln \alpha}{\partial y}; \quad \frac{\partial \ln \theta}{\partial t} = \frac{\partial \ln \alpha}{\partial t}$$

on a constant pressure surface and Eq. 14.59 can be written as:

$$C_p \left[\frac{\partial \alpha}{\partial t} + u\frac{\partial \alpha}{\partial x} + v\frac{\partial \alpha}{\partial y} + \omega\alpha\frac{\partial \ln \theta}{\partial p}\right] = \frac{\alpha}{T}Q \tag{14.60}$$

From the hydrostatic relation in a pressure coordinate framework $\left(\text{i.e.,} \frac{\partial z}{\partial p} = -\alpha/g\right)$:

$$\alpha = -g\frac{\partial z}{\partial p}$$

so that Eq. 14.60 can also be written as:

$$C_p\left[-\frac{\partial}{\partial t}\left(g\frac{\partial z}{\partial p}\right)-u\frac{\partial}{\partial x}\left(g\frac{\partial z}{\partial p}\right)-v\frac{\partial}{\partial y}\left(g\frac{\partial z}{\partial p}\right)+\omega\alpha\frac{\partial \ln\theta}{\partial p}\right]=\frac{\alpha}{T}Q \quad (14.61)$$

By convention:

$$\sigma=-\alpha\frac{\partial \ln\theta}{\partial p}=g\frac{\partial z}{\partial p}\frac{\partial \ln\theta}{\partial p}$$

is defined so that Eq. 14.61 becomes, after rearranging:

$$\frac{\partial}{\partial t}\left(-g\frac{\partial z}{\partial p}\right)-\vec{V}\cdot\vec{\nabla}_p\left(g\frac{\partial z}{\partial p}\right)-\omega\sigma=\frac{\alpha}{C_pT}Q=\frac{R_d}{pC_p}Q. \quad (14.62)$$

Performing the operation $\partial/\partial p$ on Eq. 14.57 yields:

$$g\nabla_p^2\frac{\partial}{\partial t}\frac{\partial z}{\partial p}+f\frac{\partial}{\partial p}\left[\vec{V}\cdot\nabla_p\left(\xi_g+f\right)\right]=f\left(f+\xi_g\right)\frac{\partial^2\omega}{\partial p^2};$$

performing the operation ∇_p^2 on Eq. 14.62 and assuming that σ is a function of pressure only yields:

$$-g\nabla_p^2\frac{\partial}{\partial t}\left(\frac{\partial z}{\partial p}\right)-\nabla_p^2\left[\vec{V}\cdot\nabla_p\left(g\frac{\partial z}{\partial p}\right)\right]-\sigma\nabla_p^2\omega=\frac{R_d}{pC_p}\nabla_p^2Q.$$

Adding the last two equations produces:

$$f\frac{\partial}{\partial p}\left[\vec{V}\cdot\nabla_p\left(\xi+f\right)\right]-\nabla_p^2\left[\vec{V}\cdot\nabla_p\left(g\frac{\partial z}{\partial p}\right)\right]-\sigma\nabla_p^2\omega$$
$$=\frac{R_d}{C_pp}\nabla_p^2Q+f\left(f+\xi_g\right)\frac{\partial^2\omega}{\partial p^2}.$$

Since $\partial z/\partial p=-\alpha/g=-RT/gp$, this relation can also be written as:

$$\sigma\nabla_p^2\omega+f\left(f+\xi_g\right)\frac{\partial^2\omega}{\partial p^2}$$
$$=f\frac{\partial}{\partial p}\left[\vec{V}\cdot\vec{\nabla}_p\left(\xi_g+f\right)\right]+\frac{R_d}{p}\nabla_p^2\left[\vec{V}\cdot\vec{\nabla}_pT\right]-\frac{R_d}{C_pp}\nabla_p^2Q. \quad (14.63)$$

This equation is called the *Omega equation* and represents a diagnostic second order differential equation for $\frac{dp}{dt}$.

To illustrate the importance of the different terms in Eq. 14.63, note that the left-hand side of Eq. 14.63 is of the form $\nabla^2\omega$ (although it cannot be written so simply in a quantitative solution to Eq. 14.63 because the coefficient of $\nabla_p^2\omega$ is different from $\partial^2\omega/\partial p^2$). If $\nabla^2\omega$ has a wave form, i.e.,

$$\nabla^2\omega=-k^2A\sin kx$$

where A is a constant, and $k = 2\pi/L$ where L is the wavelength, then:

$$\omega \sim A \sin kx$$

Since,

$$\omega = \frac{dp}{dt} = \frac{\partial p}{\partial t} + u\frac{\partial p}{\partial x} + v\frac{\partial p}{\partial y} + w\frac{\partial p}{\partial z} \cong w\frac{\partial p}{\partial z} = -w\rho g;$$

for typical synoptic values of pressure tendency and pressure gradient, then:

$$\omega \sim -w \quad \text{and} \quad w \sim \nabla^2 \omega.$$

In Eq. 14.63, the three terms on the right side represent the following:

$\frac{\partial}{\partial p}\left[\vec{V} \cdot \vec{\nabla}_p \left(\xi_g + f\right)\right]$ → vertical variation of the advection of absolute vorticity on a constant pressure surface.

$\nabla_p^2\left[\vec{V} \cdot \vec{\nabla}_p T\right]$ → the curvature of the advection of temperature on a constant pressure surface.

$\nabla_p^2 Q$ → the curvature of diabatic heating on a constant pressure surface.

These three terms can be interpreted more easily.

Using the relation between $\partial/\partial p$ and $\partial/\partial z$, and our observation that $\nabla^2\omega \sim w$,

$$w \sim \frac{\partial}{\partial p}\left[\vec{V} \cdot \vec{\nabla}_p \left(\xi_g + f\right)\right] \sim -\frac{\partial}{\partial z}\left[\vec{V} \cdot \vec{\nabla}_p \left(\xi_g + f\right)\right] \tag{14.64}$$

In most situations in the atmosphere, the vorticity advection is much smaller in the lower troposphere than in the middle and upper troposphere since $\vec{V}$ and ξ_g are usually smaller near the surface. As shown in Section 14.3.3, on the synoptic scale, cold air towards the poles requires that $\vec{V}$ becomes more positive with height.

Using this observation of the behavior of $\vec{V}$ and ξ_g with height:

$$w \sim -\vec{V} \cdot \vec{\nabla}_p \left(\xi_g + f\right) \tag{14.65}$$

In other words, vertical velocity is proportional to vorticity advection. Since upper-level vorticity patterns are usually geographically the same as at midtropospheric levels (since troughs and ridges are nearly vertical in the upper troposphere; C. Kreitzberg, personal communication, 1970), the 500 mb level is generally chosen to estimate vorticity advection. This level is also close to the level of nondivergence (see Fig. 14.16) in which creation or dissipation of relative vorticity is small, so that the conservation of absolute vorticity is a good approximation.

Thus for the Northern Hemisphere where $\xi_g > 0$ for cyclonic vorticity,

$w > 0$ if $-\vec{V} \cdot \vec{\nabla}_p (\xi_g + f) > 0$	positive vorticity advection (PVA)
$w < 0$ if $-\vec{V} \cdot \vec{\nabla}_p (\xi_g + f) < 0$	negative vorticity advection (NVA)

To generalize this concept to the Southern Hemisphere, PVA should be called cyclonic vorticity advection; NVA should be referred to as anticyclonic vorticity. The curvature of the advection of temperature on a constant pressure term can be represented as:

$$\nabla_p^2 \left[\vec{V} \cdot \vec{\nabla}_p T \right] \sim -k^2 B \sin kx$$

where B is a constant. Therefore,

$$\vec{V} \cdot \vec{\nabla}_p T \sim B \sin kx$$

Since:

$$w \sim \nabla_p^2 \left[\vec{V} \cdot \vec{\nabla}_p T \right]$$

then

$$w \sim -\vec{V} \cdot \vec{\nabla}_p T$$

Thus,

$w > 0$ if $-\vec{V} \cdot \vec{\nabla}_p T > 0$	warm advection
$w < 0$ if $-\vec{V} \cdot \vec{\nabla}_p T < 0$	cold advection

The 700 mb surface is often used to evaluate the temperature advection patterns since the gradients of temperature are often larger at this height than higher up and the winds are significant in speed. The 850 mb height can be used (when the terrain is low enough) although the values of $\vec{V}$ are often substantially smaller.

Finally, since $\nabla_p^2 Q \sim -k^2 C \sin kx$ can be assumed in this form, $w \sim -\nabla_p^2 Q$, and $Q \sim w$ results.

Therefore,

$w > 0$	diabatic heating
$w < 0$	diabatic cooling

An example of diabatic heating on the synoptic scale is deep cumulonimbus activity. An example of diabatic cooling is longwave radiative flux divergence.

In summary, the preceding analysis suggests the following relation between vertical motion, vorticity and temperature advection, and diabatic heating.

$$w > 0 \begin{cases} \text{positive vorticity advection} \\ \text{warm advection} \\ \text{diabatic heating.} \end{cases} \qquad w < 0 \begin{cases} \text{negative vorticity advection} \\ \text{cold advection} \\ \text{diabatic cooling.} \end{cases}$$

When combinations of terms exist which would separately result in different signs of the vertical motion (e.g., positive vorticity advection with cold advection), the resultant vertical motion will depend on the relative magnitudes of the individual contributions. Also, remember that this relation for vertical motion is only accurate as long as the assumptions used to derive Eq. 14.63 are valid.

Using synoptic analyses the following rules of thumb usually apply:

i)	vorticity advection:	evaluate at 500 mb.
ii)	temperature advection:	evaluate at 700 mb; at elevations near sea level, also evaluate at 850 mb.
iii)	diabatic heating:	contribution of major importance in synoptic weather patterns (particularly cyclogenesis) are areas of deep cumulonimbus. Refer to geostationary satellite imagery and radar for determination of locations of deep convection.

An article by Trenberth (1978) shows how the Omega equation, in a form close to that given by Eq. 14.63, can be rearranged in order to combine the vorticity and temperature advection terms. The reason to combine them is that the two terms can be opposite in sign with respect to their contribution to vertical velocity.

By rearranging the Omega equation, Trenberth shows that, as an approximation,

- upward motion occurs where there would be cyclonic vorticity advection by the thermal wind;
- descent occurs where there would be anticyclonic vorticity advection by the thermal wind.

Actual advection by the thermal wind, of course, does not occur since the thermal wind is a velocity vector difference and is not a true wind. Nonetheless, the combination of the vorticity advection and temperature advection terms in Eq. 14.63 permit an interpretation of vertical motion in the context of the advection of vorticity by a vertical shear of the horizontal geostrophic wind. This vertical shear, the thermal wind, is related to the thickness gradient as shown by Eq. 14.51.

14.3.5.4 ***Petterssen's Development Equation***

Equation 14.53 can be written as:

$$\frac{\partial\left(\xi_z+f\right)}{\partial t}+\vec{V}_H\cdot\nabla_p\left(\xi_z+f\right)=0 \qquad (14.66)$$

if vertical advection of absolute vorticity, the tilting term, and the solenoidal term are ignored, and Eq. 14.66 is assumed valid at the level of nondivergence (~500 mb). $\vec{V}_H$ is the wind on the pressure surface. Since, if the wind is in geostrophic balance:

$$\vec{V}_{H500}=\vec{V}_{H_{SFC}}+\Delta\vec{V}_g$$

where $\Delta\vec{V}_g$ is the geostrophic wind shear (see Eq. 14.51). Thus,

$$(\xi_z+f)_{500}=(\xi_z+f)_{SFC}+(\xi_z+f)_T$$

since $\nabla \times \vec{V}_{H500} = \left(\nabla \times \vec{V}_{H_{SFC}}\right) + (\nabla \times \Delta\vec{V})$. Equation 14.66 can be written as:

$$\frac{\partial(\xi_z + f)_{SFC}}{\partial t} = -\vec{V}_{H500} \cdot \nabla_p(\xi_z + f)_{500} - \frac{\partial(\xi_z + f)_T}{\partial t} \tag{14.67}$$

From Eq. 14.51,

$$\nabla \times \Delta\vec{V}_g = \frac{g}{f}\nabla_p^2(\Delta z)$$

where $\Delta z = z_{500} - z_G$ with z_{500} the 500 mb height and z_G the surface elevation so that,

$$\frac{\partial(\xi_z + f)_T}{\partial t} = \frac{g}{f}\nabla_p^2\frac{\partial(\Delta z)}{\partial t} \tag{14.68}$$

Integrating Eq. 14.62 between the surface pressure, p_{SFC}, and 500 mb yields, after rearranging:

$$-g\int_{p_{SFC}}^{500}\frac{\partial}{\partial t}\left(\frac{\partial z}{\partial p}\right)dp = -g\frac{\partial}{\partial t}\int_{z_G}^{z_{500}} dz = -g\frac{\partial(\Delta z)}{\partial t}$$
$$= \int_{p_{SFC}}^{500\text{ mb}}\left(\vec{V}\cdot\nabla_p\left(g\frac{\partial z}{\partial p}\right) + \omega\sigma + \frac{R}{pC_p}Q\right)dp \tag{14.69}$$

Performing ∇_p^2 on Eq. 14.69, substituting into Eq. 14.68 and then Eq. 14.67 yields:

$$\frac{\partial(\xi_z + f)_{SFC}}{\partial t} = -\vec{V}_{H500}\cdot\nabla_p(\xi_z + f)_{500} + \frac{g}{f}\nabla_p^2\int_{p_{SFC}}^{500}\vec{V}_H\cdot\nabla_p\left(\frac{\partial z}{\partial p}\right)dp$$
$$+\frac{\nabla_p^2}{f}\int_{p_{SFC}}^{500}\omega\sigma\,dp + \frac{R\nabla_p^2}{fC_p}\int_{p_{SFC}}^{500}\frac{Q}{p}dp \tag{14.70}$$

This is the Petterssen development equation for the change of surface absolute vorticity due to:

- $-\vec{V}_{H500}\cdot\nabla_p(\xi_z + f)_{500}$: horizontal vorticity advection at 500 mb.
- $\frac{g}{f}\nabla_p^2\cdot\int_{p_{SFC}}^{500}\vec{V}_H\cdot\nabla_p\left(\frac{\partial z}{\partial p}\right)dp = -\frac{R}{f}\nabla_p^2\int_{p_{SFC}}^{500}\frac{\vec{V}_H\cdot\nabla_p}{p}(T)\,dp$: proportional to a pressure-weighted horizontal temperature advection between the surface and 500 mb.
- $\frac{\nabla_p^2}{f}\int_{p_{SFC}}^{500\,mb}\sigma\omega dp$: proportional to vertical motion through the layer.
- $\frac{R\nabla_p^2}{fC_p}\int\frac{Q}{p}dp$: proportional to a pressure-weighted diabatic heating pattern.

14.3.5.5 Upper-Level Flow Associated with Extratropical Storm Development

Throughout the troposphere, an extratropical cyclone is characterized by the following during its development stage:

- surface pressure falls,
- ageostrophic[8] wind convergence acceleration (i.e., isallobaric wind) toward the developing low pressure system, thereby generating kinetic energy,
- convergence generates cyclonic vorticity at low levels,
- mid-level ascent and associated cloudiness as a result of the convergence, and
- upper-level divergent geostrophic winds which generate anticyclonic relative vorticity, which builds the upper-level ridge downstream.

Figure 14.21 schematically illustrates the relation between 500 mb and 700 mb flows, the thickness gradient, and the surface front. Note that there are warm advection and positive vorticity advection ahead of the upper-level trough. In terms of estimating vorticity advection, the 500 mb flow is generally used as discussed in Section 14.3.5.3. Thickness advection is usually evaluated at 700 mb, although in areas near sea level the 850 mb flow is also used. Since surface pressure is falling, the divergence aloft must be removing mass more rapidly than it can be replaced by convergent inflow at low levels. This characteristic of developing low pressure systems implies that they are forced by upper tropospheric airflow patterns. The surface pressure field, to a first approximation at least, is a response to this upper-level forcing.

As an extratropical cyclone evolves into Stage II, the cold and warm advection patterns become very well defined. It is the juxtaposition of this temperature advection pattern that provides the energy associated with the increase in circulation (i.e., vorticity) of the cyclone. Warm air lifted over the cold air of the warm front results in higher heights aloft which results in a divergent component of the flow at that altitude ahead of the cyclone in the upper troposphere. This loss of mass aloft permits the surface cyclone to deepen further, thereby strengthening the warm advection further. To the rear of the surface cyclone, the cold advection and sinking as the air moves equatorward more rapidly than the thickness gradient propagates, generates convergence aloft and a subsequent building of the high pressure in the low troposphere. The higher pressure provides stronger cold advection. The lower heights tilt towards the cold air with decreasing pressure – a direct result of the thickness relation.

As the cyclone occludes in Stage III, the lower heights at all levels within the troposphere tend to become coincident, resulting in little or no temperature or vorticity advection. Near the triple point, however, where the fronts juncture, significant temperature and vorticity advection often remain such that this is a favorable area for secondary cyclogenesis.

Within the cyclone the strongest vertical motion within the low- and midtroposphere usually occurs when there is warm advection and positive vorticity advection combined. As seen in the schematic, for example, in Stage II the largest ascent (and likely greatest rate of stratiform precipitation) would be to the north and northwest of the cyclone center at 700 mb. Actual cyclones, of course, can have different advection patterns.

[8]The term used to refer to flow which is not in gradient balance is "*ageostrophic*." This terminology is, of course, misleading since gradient winds and geostrophic winds, in general, are not equal. A more accurate term could be "*agradient*" flow.

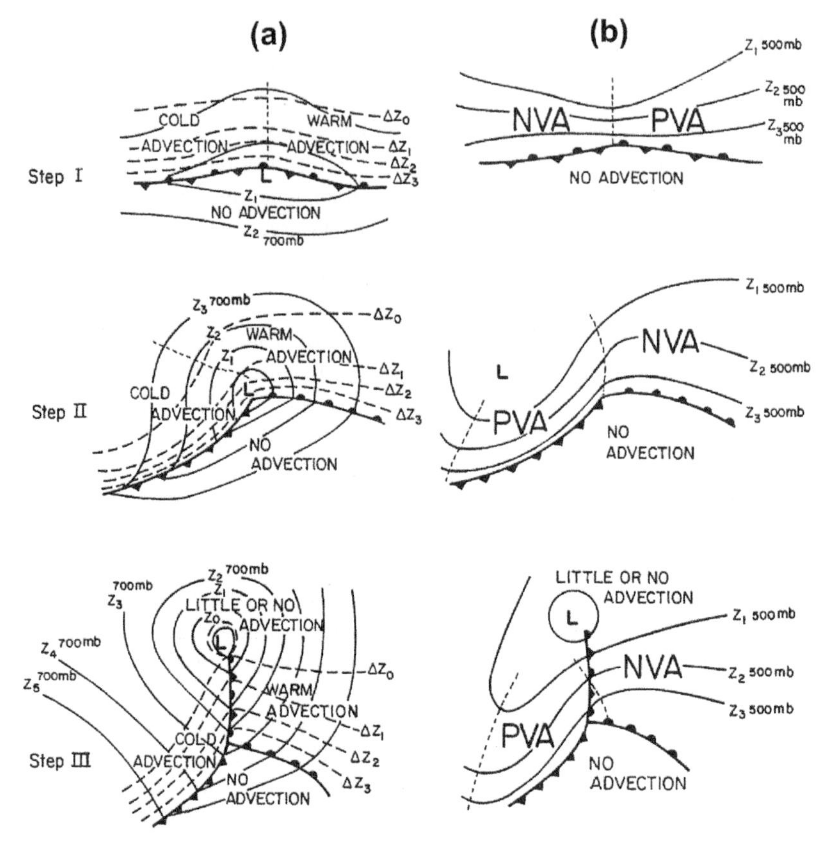

FIGURE 14.21

Schematic illustration of thickness advection and vorticity advection associated with extratropical cyclone development. The 700 mb height contours (solid lines), 1000-500 mb thickness contours (dashed lines), and surface frontal location are given in (a). The 500 mb height contours (solid lines) and surface frontal location are shown in (b).

Subsiding air would occur in the mid and lower troposphere where they are both cold and have negative vorticity advection. The subsidence associated with the negative vorticity advection can often negate the de-stabilization that occurs because of the cold advection. A frequent satellite and surface observation is that the division between NVA and PVA in the area of cold advection is often denoted by a significant break in cloud type – towering cumulus and showers often occur in the air with PVA; the cumulus clouds, however, are frequently flattened at the top when NVA develops.

A summary of the conditions associated with the development of extratropical storms is presented below:

- Favorable conditions:
  - the existence of a thickness gradient in the lower troposphere (i.e., a front); particularly when it is anticyclonically curved;

  - the presence of an upper-level trough with cold advection to its rear and warm advection ahead; and
  - release of latent heat near the center of the surface low by deep cumulonimbus and stratiform precipitation.

- Unfavorable conditions
  - a weakening thickness gradient as a result of low-level divergent flow; and
  - the absence of an upper-level trough or a trough with cold advection ahead of it, and warm advection behind resulting in a trough which will decrease in intensity with time.

Extratropical cyclones are different from hurricanes and tropical storms because their energy is primarily from the juxtaposition of cold and warm air masses (i.e., a horizontal thickness gradient). Tropical cyclones, in contrast, derive their energy through heating around the central core as a result of deep cumulonimbus. In addition, the wind field of extratropical cyclones, although spread over a large area, has weaker maximum speeds since the pressure gradient is not as strong as found in a mature, well-developed hurricane. Oceanic extratropical cyclones are less of a danger to shipping than hurricanes because the seas are not as chaotic since the wind direction does not vary through 360° around a small center as it does for the tropical system.

14.3.5.6 *Synoptic Momentum Transport Patterns*

The north-south transport of planetary and air flow momentum on opposite sides of a trof will influence the subsequent evolution of the trof. As illustrated in Fig. 14.22, a positively tilted trof exports this momentum such that unless cold advection can maintain the region of low height, the trof will weaken.

14.3.6 Examples of the Use of Synoptic Dynamics to Explain Weather Dynamics

14.3.6.1 *Post-Cold Frontal Cloudiness*

From Holton (1972, pg. 69), the conservation of potential vorticity can be written as:

$$\frac{d}{dt}\left[\left(\xi_z + f\right)\frac{\partial \theta}{\partial p}\right] = 0 \qquad (14.71)$$

Since,

$$\frac{df}{dt} = v\frac{\partial f}{\partial y} = v\beta,$$

with southward-moving air, as with a cold arctic outbreak in the Northern Hemisphere:

$$v < 0; \quad \beta v < 0$$

and

$\frac{d\xi_z}{dt} < 0$ for anticyclonic curvature to the air parcel trajectories

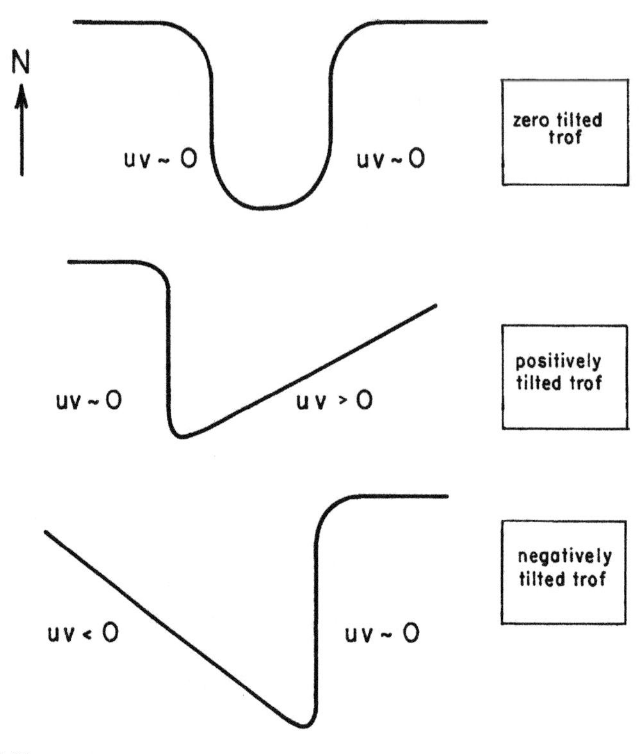

FIGURE 14.22

Influence of trof structure on its evolution over time.

$$\frac{d\xi_z}{dt} > 0 \text{ for cyclonic curvature to the air parcel trajectories}$$

Two cases can be examined with $\beta v < 0$. In both cases the flow will be assumed to be in the lower troposphere:

a) For $\frac{d\xi_z}{d}t < 0$, $\frac{d}{dt}\left(\frac{\partial\theta}{\partial p}\right) > 0$ in order to maintain $\left(\xi_z + f\right)\left(\partial\theta/\partial p\right)$ as a constant. $d\left(\partial\theta/\partial p\right)/dt > 0$ indicates that surfaces of constant θ are being brought closer together. Since $w = 0$ at the ground surface, this requirement on $\partial\theta/\partial p$ implies sinking motion.
Stratocumulus clouds, if any, that form in this environment as a result of the destabilizing effect of cold advection, tend to be flattened at the top as a result of the sinking air. This subsidence also exerts a warming component on the atmosphere by compression, thereby counteracting to some extent the cold advection.

b) For $\frac{d\xi_z}{dt} > 0$ there exists a region where,

$$\left|\frac{d\xi_z}{dt}\right| > |\beta v|$$

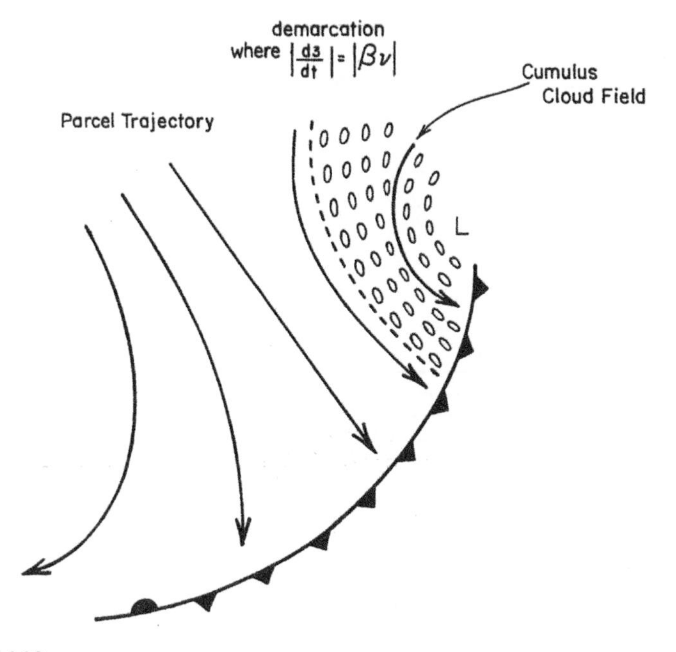

FIGURE 14.23

Schematic of the relation between parcel trajectory and the cloud field behind a cold front.

so that $d\left(\xi_z + f\right)/dt$ is increasing. Therefore, $\frac{\partial\theta}{\partial p}$ must decrease following a parcel. With $w = 0$ at the ground surface, the spreading of potential temperature surfaces with respect to pressure require that ascent occur.

In this region the destabilizing effect of cold advection and the ascent required by the conversion of potential vorticity provide an ambient atmosphere conducive to towering cumulus and showers. The demarcation between the regions where $|d\xi_z/dt|$ is greater than and less than $|\beta v|$ is often characterized by a sharp break in cloud type visible from both satellite and from the ground. Deep cumulus extend from this demarcation toward the surface low center, while flattened cumulus or clear skies occur in the opposite direction. Figure 14.23 schematically illustrates this characteristic of the flow and associated clouds.

14.3.6.2 *Troughing East of Mountains*

It is observed on surface weather maps that troughs tend to form downwind of large-scale elongated mountain barriers when a substantial wind blows over the barrier. Such troughs, referred to as *lee-side troughs*, can be explained in terms of the conservation of potential vorticity.

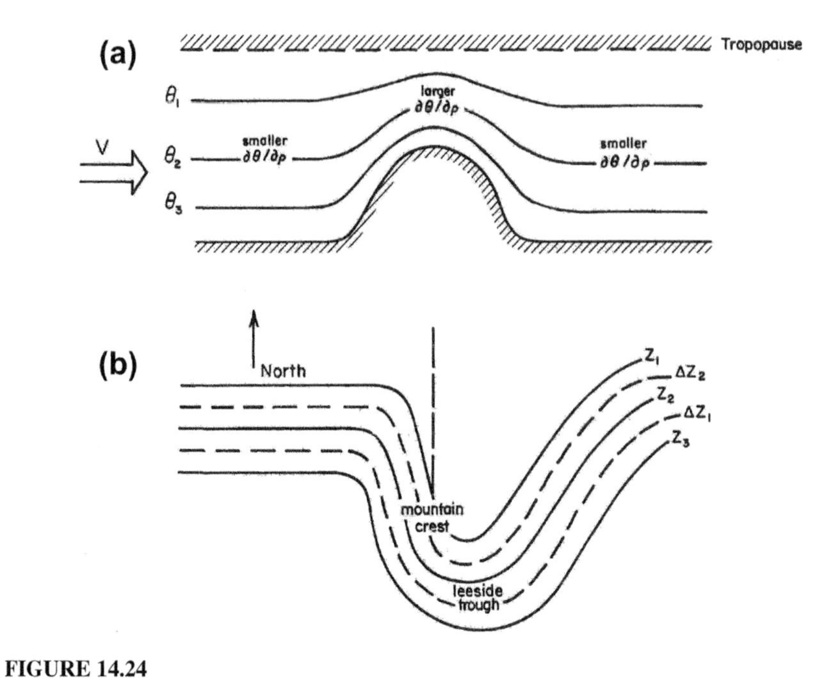

FIGURE 14.24

(a) $x - z$, and (b) $x - y$ cross section of particle trajectories across a large-scale elongated ridge. Stratification, height, and thickness are drawn and steady-state flow is assumed.

As illustrated in Fig. 14.24a for the Northern Hemisphere, since vertical motion is inhibited by the stable stratification above the tropopause as it crosses the barrier, $\partial\theta/\partial p$ must become larger using Eq. 14.71, resulting in a reduction of relative vorticity (i.e., $d\xi_z/dt < 0$). If the initial flow is straight and without horizontal shear, ξ_z therefore becomes anticyclonic and the parcel motion is as depicted in Fig. 14.24b.[9] The turn to anticyclonic flow occurs as soon as the ground begins to elevate.

Upon reaching the lee slope, the parcel direction is from the northwest and moves to the south of its original latitude. When reaching its original latitude at the edge of the lee slope, ξ_z, and therefore $\partial\theta/\partial p$ return to their original values. Moving further south, however, f decreases so ξ_z becomes cyclonic, eventually attaining enough curvature so that the parcel turns northward.

In the absence of friction, a series of troughs and ridges would extend downstream from the barrier indefinitely.

In the real world, the tendency for lee side troughing provides a thickness gradient (due to the resultant displacement of temperature contours by advection) which is conducive to cyclogenesis, as illustrated in Fig. 14.24b.

[9]If the flow were not uniform in the direction perpendicular to the barrier, the parcel could have attained anticyclonic shear vorticity instead of just anticyclonic curvature vorticity.

The path of a parcel crossing the ridge can be discussed using the conservation of potential vorticity, i.e.,

$$\frac{\partial\theta}{\partial p}\left(\xi_z + f\right) = \text{constant.}$$

If we assume all of the relative vorticity is due to curvature of the flow then:

$$\xi_z = V/R_T$$

where R_T is the radius of the air parcel trajectory and V is the wind speed. If a subscript "1" denotes the upstream value and "2" indicates its modification by the ridge then for east-west flow:

$$\left(\xi_1 + f\right)\left.\frac{\partial\theta}{\partial p}\right|_1 = \left(\xi_2 + f\right)\left.\frac{\partial\theta}{\partial p}\right|_2 \tag{14.72}$$

For nonsheared upstream flow $\xi_1 = 0$. As the airflow becomes more stratified, as it passes over the ridge, $\left.\frac{\partial\theta}{\partial p}\right|_2$ can be written as:

$$\left.\frac{\partial\theta}{\partial p}\right|_2 = \alpha\left.\frac{\partial\theta}{\partial p}\right|_1 \qquad \alpha > 1$$

Using these definitions Eq. 14.72 can be written, after rearranging as:

$$\xi_2 = \frac{V}{R_T} = \frac{f\left(1-\alpha\right)}{\alpha}$$

which can be rearranged to solve for R_T yielding:

$$R_T = \frac{\alpha}{\left(1-\alpha\right)f}V$$

Since $\alpha > 1$ and $V > 0$, $R_T < 0$ which requires that the flow be anticyclonic upon rising over the ridge.

As an example let $\alpha = 2$ and $f = 10^{-4}\ \text{s}^{-1}$; thus,

$$R_T = -2\times10^4 V\ \left(-2\times10^4 \text{ has units of seconds}\right)$$

For $V = 10\ \text{m s}^{-1}$, $50\ \text{m s}^{-1}$, and $100\ \text{m s}^{-1}$, $R_T = -200$ km, -1000 km, and -2000 km, respectively. For the light wind, it appears that the flow would not even be able to cross the ridge if it was greater than 200 km to the crest! For real-world situations, of course, V and $\frac{\partial\theta}{\partial p}$ would change and must permit the air to eventually cross the crest, otherwise mass would build-up indefinitely on the upwind side.

14.3.6.3 Conversion of Cold Fronts into Shear Lines

As a cold front moves over a warmer surface, thickness warming can occur from below. Such warming is particularly pronounced when cold air moves over warmer water with its large thermal inertia. Over land, surface cooling after cold frontal passage can diminish the rate of thickness modification.

After the thickness gradient is removed, however, the low-level wind convergence can remain particularly if deep cumulus convection resulting from the low-level convergence reinforces the convergence. This conversion of a front into a convergence zone frequently occurs in the winter in the tropics where the conditionally unstable lower tropospheric air provides the buoyant energy for cumulus convection, and the rate of conversion of the low-level wind to gradient balance is slow as a result of the low latitude.

Figure 14.25 illustrates schematically the conversion of a cold front to a convergence zone. Since the winds are opposing in direction across the zone, this region is also called a *shear line*.

A satellite visualization of this conversion process is shown in Anderson et al. (1974, pg. 3-B-3).

14.3.6.4 Backdoor Cold Fronts

If cold air masses are stably stratified, their forward motion will be impeded by mountain barriers unless they have sufficient kinetic energy to go over the barrier. If the flow is *blocked*, however, the airflow will decrease in speed and move towards lower pressure. even when the airflow above the cold air moves in the opposite direction. Substantial warm advection can occur above the cold air mass so that the atmosphere becomes even more stable.

A measure as to whether blocking will occur is the Froude number (e.g., see Section 13.2.4). Defined as:

$$Fr = \overline{V} \Big/ \left[z_{\mathrm{Gmax}} \left(\frac{g}{\theta} \frac{\partial \theta}{\partial z} \right)^{1/2} \right],$$

flows tend to become blocked when $Fr < 0.75$ to 1.5 or so. Greater stratification (i.e., large $\partial\theta/\partial z$), higher terrain (i.e., z_{Gmax}), and lighter winds (i.e., $\overline{V}$) are associated with blocking.

This blocking is a common occurrence east of the Appalachians and Rocky Mountains particularly in the cooler seasons. East of the Appalachians, such blocking of the westward movement of a cold air mass and resultant southward flow is called a *backdoor cold front*. It is referred to as backdoor because it comes from an unusual direction (i.e., from the northeast or north) and is associated with a synoptic extratropical cyclone pattern which would suggest warm advection.

Figure 14.26 schematically illustrates the development and movement of a backdoor cold front due to the presence of a mountain barrier. Since the cold air frequently undercuts a region of ascending warm advection aloft which has a layer of above freezing temperatures, precipitation often falls as sleet or freezing rain after passing through

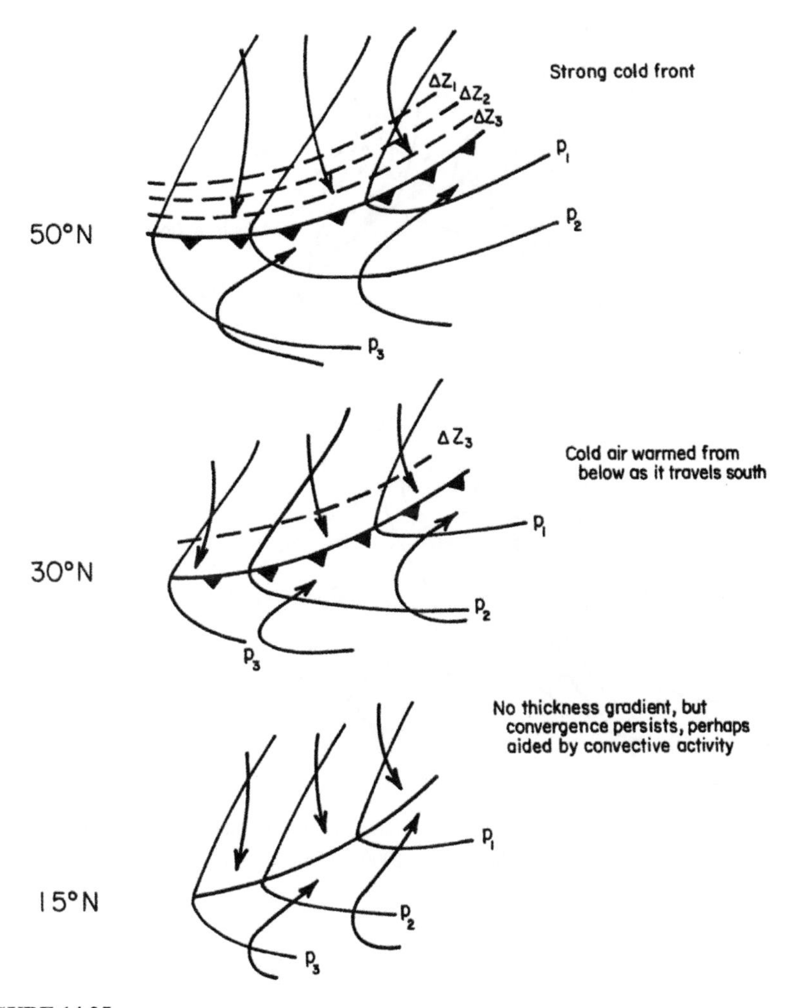

FIGURE 14.25

Schematic of the conversion of a cold front to a shear line as a front moves equatorward. Parcel trajectories are indicated by arrows.

below 0°C air near the surface. Fog and low visibilities are also often associated with the cold, stable, near surface air mass which is capped by warm air aloft.

Anderson et al. (1974, pp. 3-B-15 to 3-B-18) illustrate the cloud pattern associated with a backdoor cold front. A study of the effect of blocking by elevated terrain on a cold front (the Front Range of the Colorado Rockies) is given in Wesley and Pielke (1990).

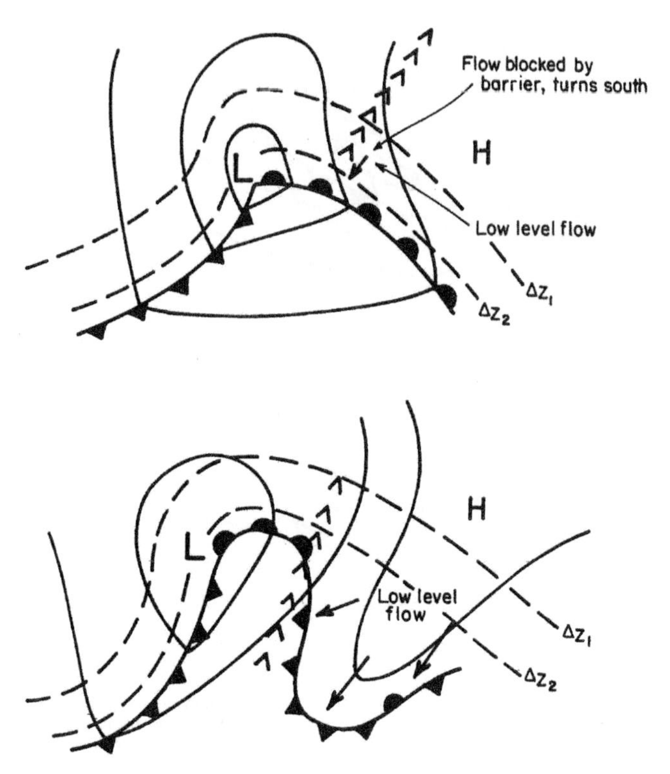

FIGURE 14.26

Generation of a backdoor cold front as a result of blocking. 1000-500 mb thickness, ∇z, continue to move northward since most of the flow in the 1000-500 mb layer is above the cold domes.

14.3.6.5 *Depiction of Fronts in the Southern Hemisphere*

While the dynamics of meteorological flows must be identical in the Southern Hemisphere, their visualization is different since $f < 0$ south of the equator. One can look at a mirror image turned upside down of a Northern Hemisphere weather map to see how the identical features would appear in a Southern Hemisphere representation. Alternatively, turn a Northern Hemisphere analysis over, invert it, and look at it through a light table.

Figure 14.27 illustrates the form that the same extratropical cyclone would have in both hemispheres.

In the context of geostrophic and gradient balance, winds in the Southern Hemisphere blow clockwise around cyclones, counterclockwise around anticyclones. Cold advection is associated with veering winds, warm advection with backing winds in the southern mid and high latitudes.

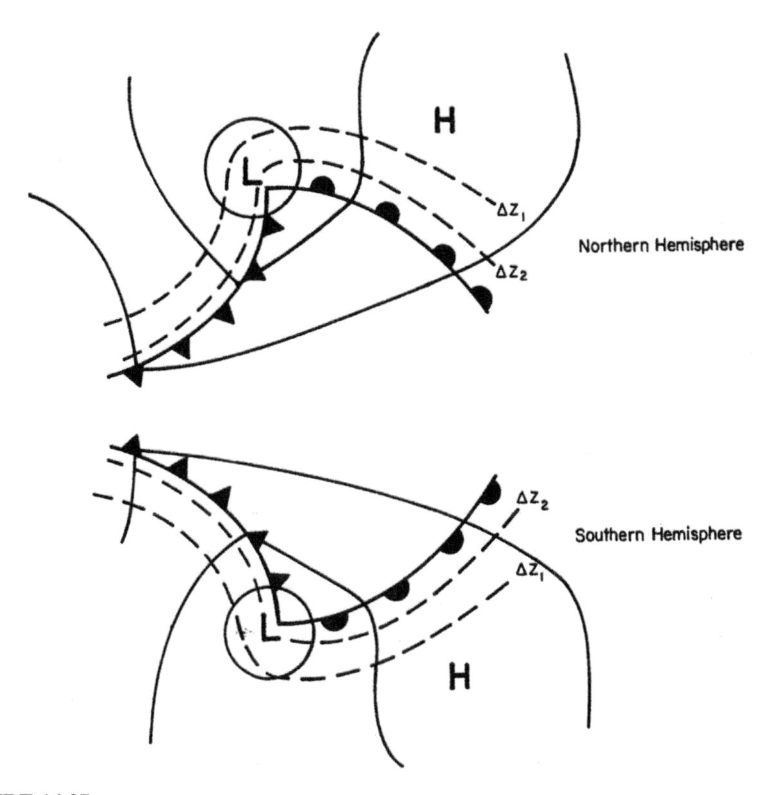

FIGURE 14.27

Representation of the same extratropical cyclone as would appear in Northern and Southern Hemispheric analyses.

14.3.6.6 *Circulation Around an Anticyclone*

Equation 14.71 can be used to describe expected vertical motion around a large lower tropospheric anticyclone. As illustrated in Fig. 14.28, equatorward moving air east of the high pressure region requires $\beta v < 0$. Thus, if the anticyclone is circular such that $d\xi_z/dt = 0$, $d\left(\partial\theta/\partial p\right)/dt > 0$ results. In contrast, west of the anticyclone, $\beta v > 0$, so that $d\left(\partial\theta/\partial p\right)/dt < 0$ results in ascent west of the high.

Therefore, one should expect general subsidence east of a symmetric ridge and ascent on the west side. Precipitation would be expected to be more likely on the west.

In the subtropics, subtropical anticyclones are often configured as more-or-less symmetric high pressure regions. These synoptic realizations of the Hadley cell develop in this form over the ocean areas, as a result of lower pressure caused by surface heating over the land surfaces. The eastern subtropical oceans are characterized by net subsidence and low boundary layer inversions which, to a large extent, can be related to the requirement for a conservation of potential vorticity. In the

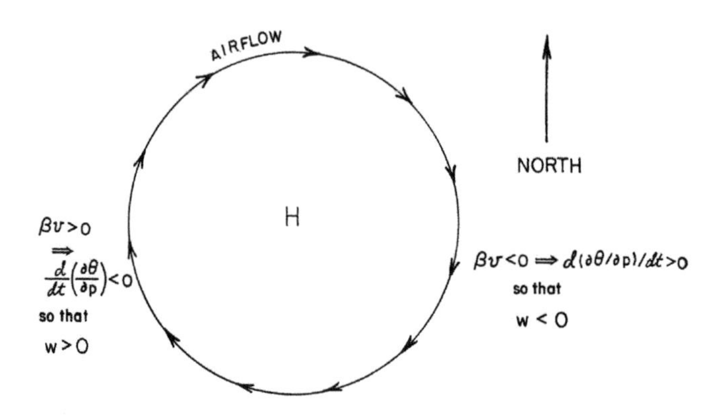

FIGURE 14.28

Schematic flow around an anticyclone in which $\frac{d\xi}{dt} = 0$.

western subtropical oceanic areas, average ascent tends to eliminate the inversion and to provide an atmosphere which is more conducive to precipitation.

14.3.6.7 *Upper-Level Fronts*

Figure 14.29 illustrates the influence of horizontal convergence in the upper troposphere on the generation of an upper-level strong air flow. Referred to as the *subtropical jet*, the convergence at these heights is a direct result of poleward moving air associated with the Hadley cell which is the prime planetary circulation associated with poleward transport of warm air from equatorial regions. Palmén and Newton (1969, pp. 112-113) discuss this jet in more detail. Since the horizontal gradient of thickness only occurs in the layers where the horizontal convergence occurs, the subtropical jet is an upper tropospheric feature exclusively. It directly influences low-level flows only when it becomes coupled through deep cumulus convection and/or interactions with the polar jet. Since cold air is on the poleward side of the gradient, the jet is westerly, as explained by the thermal wind relation (i.e., see Fig. 14.13).

14.3.6.8 *Orientation of Cumulus Convection with Respect to the Wind Shear*

The patterning of shallow to moderate depth cumulus, as viewed from satellite imagery, can often be used to estimate the wind shear, and therefore, the thickness gradient through the layer of the cloud. Two schematic illustrations of cloud patterning associated with cold and warm advection are shown in Fig. 14.30.

Shallow and moderate depth cumulus clouds are oriented in the direction of the shear because at low levels they tend to move in the direction of the wind at that level. As they develop vertically, to the extent that ambient air is entrained within the cumulus clouds, the movement tends to be in the direction of the wind at that level. The appearance of the clouds as viewed from aloft is in the direction of the shear

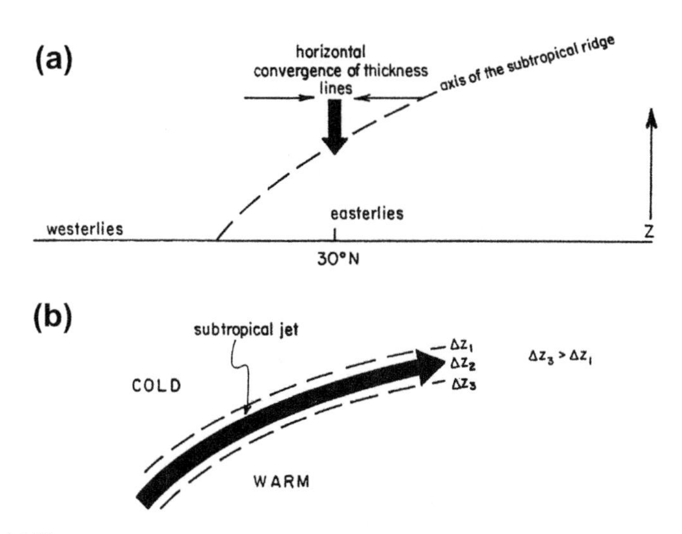

FIGURE 14.29

Illustration of the mechanisms of generation of the subtropical jet for (a) a vertical cross section, and (b) a horizontal cross section.

because of the turning of the cloud with height. With cold advection, cumulus clouds back with height (Fig. 14.29a), while cumulus clouds tend to veer with height under warm advection (Fig. 14.29b).

14.3.6.9 Inertial Instability

Equation 4.21 with $i = 1$ and 2 can be written as:

$$\begin{aligned} \frac{\partial u}{\partial t} &= fv - v\frac{\partial u}{\partial y} = v\left(f - \frac{\partial u}{\partial y}\right) \\ \frac{\partial v}{\partial t} &= -fu \end{aligned} \tag{14.73}$$

if variations in x and z are ignored, and the pressure gradient and frictional forces are neglected. The top equation in Eq. 14.73 can be rewritten as:

$$\begin{aligned} & f\frac{\partial u}{\partial t} = fv\left(f - \frac{\partial u}{\partial y}\right) \\ \text{or} \quad & \frac{\partial^2 v}{\partial t^2} + fv\left(f - \frac{\partial u}{\partial y}\right) = 0 \\ \text{since} \quad & -f\frac{\partial u}{\partial t} = \frac{\partial}{\partial t}(fu) = \frac{\partial^2 v}{\partial t^2}. \end{aligned} \tag{14.74}$$

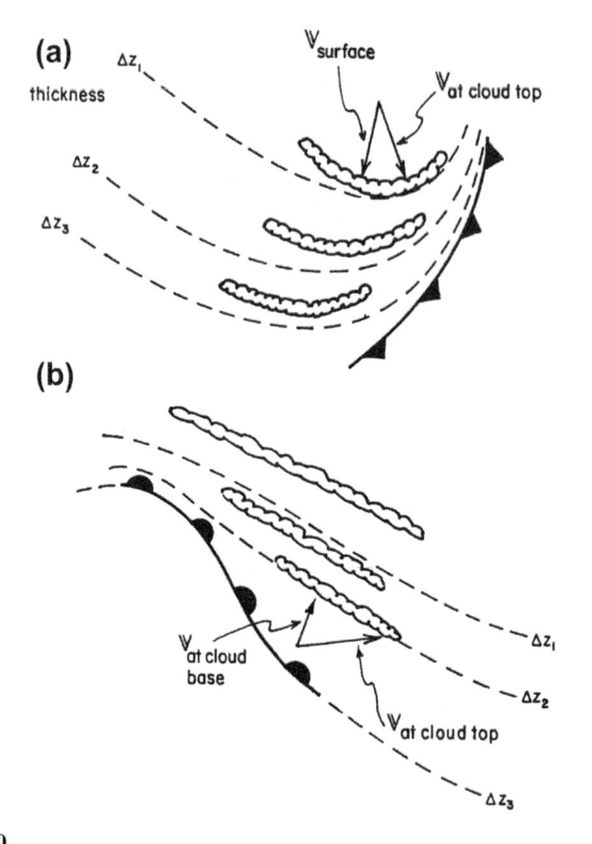

FIGURE 14.30

Schematic illustration of the relation between cumulus cloud orientation, wind shear, and thickness for (a) cold advection, and (b) warm advection.

For a fixed value of $\frac{\partial u}{\partial y}$, Eq. 14.74 can be treated as an ordinary differential equation of the form:

$$\left[\frac{d^2}{dt^2} + f\left(f - \frac{\partial u}{\partial y}\right)\right] v = 0$$

which has a periodic solution if $f > \partial u / \partial y$, but an exponential solution if $f < \frac{\partial u}{\partial y}$ (Spiegel 1967, Chapter 4).

In the synoptic atmosphere, values of $\partial u / \partial y$ are largest south of the polar jet (i.e., anticyclone shear); hence, this region would correspond to a location where we would expect preferred amplification of waves along the jet due to this mechanism of *inertial instability*.

14.3.6.10 *Jet Streaks*

Jet streaks represent localized regions of stronger winds embedded within the jet stream. The jet stream itself results from the horizontal gradient of thickness, as

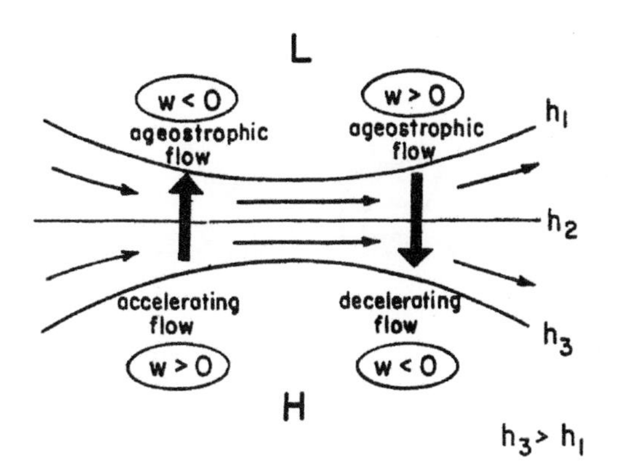

FIGURE 14.31

An idealized jet streak with associated vertical motion pattern.

discussed in Section 14.3.3 (i.e., see Eq. 14.51). Jet streaks on the synoptic scale would, therefore, be expected to be associated with locations of particularly strong horizontal temperature gradients.

Figure 14.31 illustrates an idealized jet streak, and the associated vertical motion pattern, as the wind flow geostrophically adjusts to the different horizontal height field. In the entrance region of the jet streak, the ageostrophic flow is towards lower heights as the air accelerates, but requires a period of time to come into balance with the tighter horizontal gradient of height. In this portion of the jet streak, the actual wind speed is less than the gradient wind. The result is descent in the troposphere on the lower height side of the jet with ascent on the higher height side of the jet. In the exit region of the jet, the opposite pattern occurs as the flow decelerates in response to the weaker horizontal gradient of the height field, but requires time to again achieve gradient balance. In this region, the actual wind speed is greater than the gradient wind.

This pattern of ascent and descent would occur on the synoptic scale in the absence of tropospheric temperature advection. Carlson (1991) discusses jet dynamics in detail, including how temperature advection modifies the upward/downward motion couplet shown in Fig. 14.31. Uccellini et al. (1987) describes how jet streaks are intimately involved in cyclogenesis along the east coast of the United States.

14.3.6.11 *Use of Thermodynamic Cross Sections*

As discussed in Table 14.2, since potential temperature is conserved with respect to dry motions,[10] when $\frac{\partial \theta}{\partial z} > 0, \theta$ can be used as a parcel tracer in an atmosphere without saturation. An additional requirement is that the parcel not entrain adjacent air with different values of θ.

[10]Heat changes due to radiative flux divergence must also be small over the period of interest.

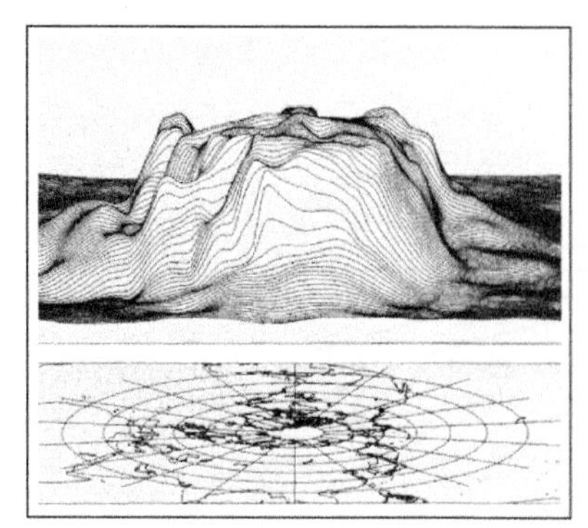

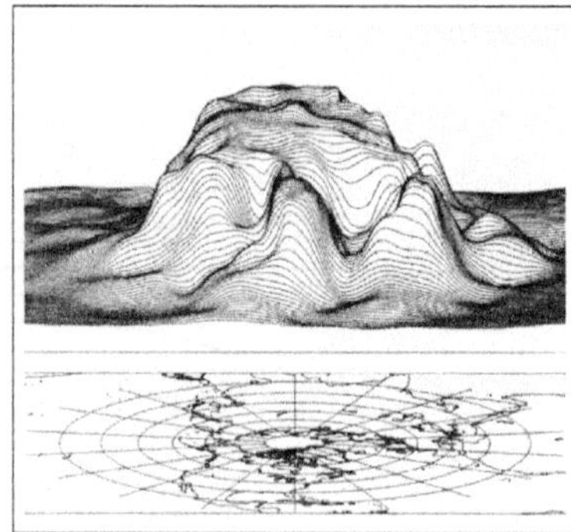

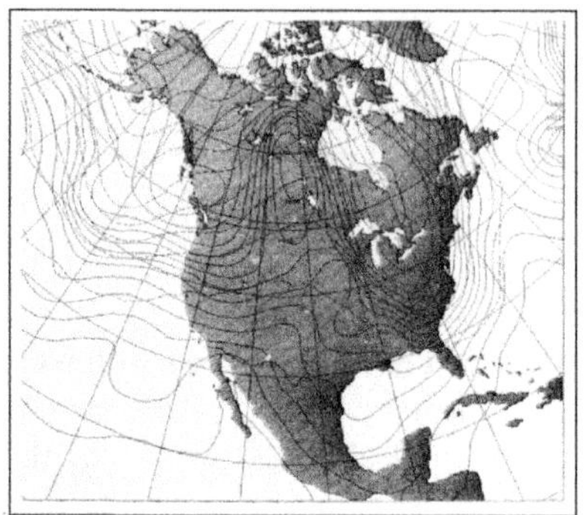

FIGURE 14.32

Illustration of Northern Hemispheric θ surfaces as viewed from two different perspectives (from Nagle 1979). The Montgomery stream function is derived from the generalized vertical coordinate σ from Chapter 6 set equal to θ so that $\frac{1}{\rho}\nabla_z p = \nabla_\theta \left[C_p T + gz\right]$. In this derivation, the small variation of C_p with water vapor content is ignored.

Similarly, the equivalent potential temperature and wet bulb potential temperature can be used as a parcel tracer as long as $\partial\theta_E/\partial z = \partial\theta_W/\partial z > 0$. These temperatures, unlike θ, are also conserved under saturated motion.

It therefore appears useful to use these quantities to estimate atmospheric trajectories. In a dry atmosphere, for example, air parcels would tend to move along θ surfaces, such as illustrated in Fig. 14.32 reproduced from Nagle (1979). In a moist atmosphere, parcels would tend to move along θ_W surfaces.

The application of these surfaces to estimate trajectories in a dry environment are most useful when the atmosphere is absolutely stable (i.e., $\partial\theta/\partial z > 0$) and radiative flux divergence effects are small as mentioned previously. Such thermodynamic structures generally occur above the planetary boundary layer in the remaining portion of the troposphere and in the stratosphere. Near the surface, of course, particularly during sunny days, superadiabatic layers often occur resulting in values of θ which are multi-valued with height. Therefore, a determination of the appropriate layer near the ground surface to follow a trajectory is unclear. In addition, strong vertical mixing as a result of the superadiabatic layer can produce substantial parcel entrainment of air with different values of θ.

In a saturated environment, in the absence of significant entrainment of air with different thermodynamic properties, trajectories will follow surfaces of constant θ_W and θ_E. If the saturated air has a structure such that $\partial\theta_W/\partial z = \partial\theta_E/\partial z < 0$, however, the same problem occurs as when $\partial\theta/\partial z < 0$, namely θ_W and θ_E are multi-valued in height. In such a saturated atmosphere in which conditional instability is realized, cumulus convection will occur with the resultant vertical mixing making trajectory assessments ambiguous. In contrast to profiles of θ, therefore, θ_W and θ_E are often multi-valued even above the planetary boundary layer. Such multi-valued profiles most often occur in air of tropical origin.

In summary, the use of dry and moist isentropic surfaces to estimate trajectory motion are most useful when:

dry motion:
- $\partial\theta/\partial z > 0$
- heating or cooling due to radiative flux divergence is small.

moist motion:
- $\partial\theta_E/\partial z = \partial\theta_W/\partial z > 0$
- heating or cooling due to radiative flux divergence is small.

(4.21)

Since regions of large horizontal thickness gradient also have a large horizontal gradient of θ, θ_E, and θ_W, significant parcel ascent or descent occurs in such regions. The application of θ surfaces in dry air or θ_W surfaces (θ_E surfaces) in saturated air, therefore, provides excellent resolution of parcel motion in such an atmosphere, providing the criteria listed in Eq. 4.21 are satisfied.

In applying these surfaces to estimate parcel motion:

- Use θ surfaces when the relative humidity of the parcel is less than 100%. The simultaneous occurrence of regions with $\partial\theta_E/\partial z = \partial\theta_W/\partial z < 0$ is of no relevance *as long as the parcel remains unsaturated.*
- Use θ_W (θ_E) surfaces when the relative humidity of the parcel reaches 100%. Precipitation can be estimated from the condensation that must occur with continued ascent after a relative humidity of 100% is reached for the first time.
  If $\partial\theta_E/\partial z = \partial\theta_W/\partial z \leq 0$, however, the use of θ_E (θ_W) surfaces is of little value, unless the region of conditional instability is very narrow.

An early, effective discussion of the application of isentropic trajectories is presented in Rossby (1941, pp. 637-641). Artz et al. (1985) provide an evaluation of differences obtained when an isentropic trajectory analysis is used instead of a constant height evaluation.

Problems for Chapter 14

The reader is recommended to review a set of tests and quizzes that are available in:

Pielke Sr., R.A. 2002c: *Synoptic Weather Lab Notes*. Colorado State University, Department of Atmospheric Science Class Report No. 1, Final Version, August 20, 2002. http://pielkeclimatesci.files.wordpress.com/2010/01/nr-77.pdf.

Chapter 14 Additional Readings

Bluestein, H.B., 1992: *Synoptic-Dynamic Meteorology in Midlatitudes: Volume I: Principles of Kinematics and Dynamics*. Oxford University Press, 448 pp.

Bluestein, H.B., 1993: *Synoptic-Dynamic Meteorology in Midlatitudes: Volume II: Observations and Theory of Weather Systems*. Oxford University Press, 608 pp.

Carlson, T.N., 2012: *Mid-Latitude Weather Systems*. Pennsylvania State University Press, 507 pp.

Lackmann, G., 2012: *Midlatitude Synoptic Meteorology: Dynamics, Analysis, and Forecasting*. American Meteorology Society, 345 pp.

Pielke Sr., R.A. 2002c: *Synoptic Weather Lab Notes*. Colorado State University, Department of Atmospheric Science Class Report No. 1, Final Version, August 20, 2002. http://pielkeclimatesci.files.wordpress.com/2010/01/nr-77.pdf.

Rauber, R.M., J.E. Walsh, and D.J. Charlevoix, 2012: *Severe and Hazardous Weather: An Introduction to High Impact Meteorology*. Kendall/Hunt Publishing Company, 4th edition, 550 pp.

APPENDIX

The Solution of Eqs. 9.26 and 9.45

From Ahlberg et al. (1967:15), equations of the form 9.26 and 9.45 for a periodic condition can be written as

$$
\begin{aligned}
b_1x_1 + c_1x_2 + a_1x_D &= d_1 \\
a_2x_1 + b_2x_2 + c_2x_3 &= d_2 \\
&\vdots \\
a_ix_{i-1} + b_ix_i + c_ix_{i+a} &= d_i \\
&\vdots \\
a_{D-1}x_{D-2} + b_{D-1}x_{D-1} + c_{D-1}x_D &= d_{D-1} \\
c_Dx_1 + a_Dx_{D-1} + b_Dx_D &= d_D
\end{aligned}
\tag{A-1}
$$

for $i = 1$ to D.

For Eq. 9.26 $\left(x_i = \phi_j^{\tau+1}\right)$

$$
a_i = \frac{-\Delta t}{\Delta z_j}\frac{K_{j-\frac{1}{2}}}{\Delta z_{j-\frac{1}{2}}}\beta_{\tau+1}; \quad c_i = \frac{-\Delta t}{\Delta z_j}\frac{K_{j+\frac{1}{2}}}{\Delta z_{j+\frac{1}{2}}}\beta_{\tau+1};
$$

$$
b_i = \left[1 + \frac{\Delta t}{\Delta z_j}\frac{K_{j+\frac{1}{2}}}{\Delta z_{j+\frac{1}{2}}}\beta_{\tau+1} + \frac{\Delta t}{\Delta z_j}\frac{K_{j-\frac{1}{2}}}{\Delta z_{j-\frac{1}{2}}}\beta_{\tau+1}\right];
$$

$$
d_i = \phi_j^\tau + \frac{\Delta t}{\Delta z_j}\left\{\frac{K_{j+\frac{1}{2}}\,\beta_\tau\left(\phi_{j+1}^\tau - \phi_j^\tau\right)}{\Delta z_{j+\frac{1}{2}}} - \frac{K_{j-\frac{1}{2}}}{\Delta z_{j-\frac{1}{2}}}\beta_\tau\left(\phi_j^\tau - \phi_{j-1}^\tau\right)\right\};
$$

for $i = 2$ to $D - 1$.

$$
a_i = \frac{-\Delta t}{\Delta z_1}\frac{K_{\frac{1}{2}}}{\Delta z_{\frac{1}{2}}}\beta_{\tau+1}; \quad c_i = \frac{-\Delta t}{\Delta z_1}\frac{K_{1\frac{1}{2}}}{\Delta z_{1\frac{1}{2}}}\beta_{\tau+1};
$$

$$
b_i = \left[1 + \frac{\Delta t}{\Delta z_1}\frac{K_{1\frac{1}{2}}}{\Delta z_{1\frac{1}{2}}}\beta_{\tau+1} + \frac{\Delta t}{\Delta z_1}\frac{K_{\frac{1}{2}}}{\Delta z_{\frac{1}{2}}}\beta_{\tau+1}\right];
$$

Mesoscale Meteorological Modeling, Third Edition, Volume 98. http://dx.doi.org/10.1016/B978-0-12-385237-3.00024-4

$$d_i = \phi_1^\tau + \frac{\Delta t}{\Delta z_1}\left\{\frac{K_{1\frac{1}{2}}\beta_\tau}{\Delta z_{1\frac{1}{2}}}\left(\phi_2^\tau - \phi_1^\tau\right) - \frac{K_{\frac{1}{2}}\beta_\tau}{\Delta z_{\frac{1}{2}}}\left(\phi_1^\tau - \phi_D^\tau\right)\right\};$$

$$a_D = \frac{-\Delta t}{\Delta z_D}\frac{K_{D-\frac{1}{2}}}{\Delta z_{D-\frac{1}{2}}}\beta_{\tau+1}; \quad c_D = \frac{-\Delta t}{\Delta z_D}\frac{K_{D+\frac{1}{2}}}{\Delta z_{D+\frac{1}{2}}}\beta_{\tau+1};$$

$$b_D = \left[1 + \frac{\Delta t}{\Delta z_D}\frac{K_{\frac{1}{2}}}{\Delta z_{\frac{1}{2}}}\beta_{\tau+1} + \frac{\Delta t}{\Delta z_D}\frac{K_{D-\frac{1}{2}}}{\Delta z_{D-\frac{1}{2}}}\beta_{\tau+1}\right];$$

$$d_D = \phi_D^\tau + \frac{\Delta t}{\Delta z_D}\left\{\frac{K_{\frac{1}{2}}}{\Delta z_{\frac{1}{2}}}\beta_\tau\left(\phi_1^\tau - \phi_D^\tau\right) - \frac{K_{D-\frac{1}{2}}}{\Delta z_{D-\frac{1}{2}}}\beta_\tau\left(\phi_D^\tau - \phi_{D-1}^\tau\right)\right\},$$

where $\Delta z_{1/2} = \Delta z_D = \Delta z_{D+1/2} = \Delta z_1$ should be assumed.

For Eq. 9.45 ($x_i = N_i$)

$$\left.\begin{aligned} &a_i = \alpha_i; \quad b_i = 2; \quad c_i = \mu_i; \\ &d_i = 3\frac{\mu_i}{h_{i+1}}\left(\phi_{i+1} - \phi_i\right) + 3\frac{\alpha_i}{h_i}\left(\phi_i - \phi_{i-1}\right) \end{aligned}\right\} \quad \text{for } i = 2 \text{ to } D-1$$

$$a_1 = \alpha_1; \quad b_1 = 2; \quad c_1 = \mu_1;$$

$$d_1 = 3\frac{\mu_1}{h_2}\left(\phi_2 - \phi_1\right) + 3\frac{\alpha_1}{h_1}\left(\phi_1 - \phi_D\right);$$

$$a_D = h_1/\left(h_D + h_1\right); \quad b_D = 2; \quad c_D = 1 - \alpha_D;$$

$$d_D = 3\frac{\mu_D}{h_1}\left(\phi_1 - \phi_D\right) + 3\frac{\alpha_D}{h_D}\left(\phi_D - \phi_{D-1}\right),$$

where $h_1 = h_D$ should be assumed.

The procedure to solve (A-1) involves letting the righthand side of the top and second from the bottom equations in (A-1) be written as $d_1 - a_1 x_D$ and $d_{D-1} - c_{D-1}x_D$, then solving the first $D-1$ equations, as performed in Section 9.2. The result is a solution in terms of x_D, whose value is determined algebraically from the last equation in (A-1).

The algorithm to solve (A-1) is similar in form to that given by Eq. 9.47 and, following Ahlberg et al. (1967), can be written in general form in terms of the coefficients of (A-1) as

$$x_i = u_i + q_i x_{i+1} + s_i x_D, \quad i = 1, \ldots, D-1, \tag{A-2}$$

where

$$\begin{aligned} s_i &= -a_i s_{i-1}/p_i, & s_0 &= 1, \\ p_i &= b_i + a_i q_{i-1}, & q_0 &= 0, \\ q_i &= -c_i/p_i, & & \\ u_i &= \left(d_i - a_i u_{i-1}\right)/p_i, & u_o &= 0. \end{aligned}$$

If the equation

$$x_i = t_i x_D + v_i, \quad i = 1, \ldots, D-1 \tag{A-3}$$

is defined, then substituting into (A-2) for x_{i+1} yields

$$x_i = u_i + q_i v_{i+1} + \left(q_i t_{i+1} + s_i\right) x_D, \quad i = 1, \ldots, D-1. \tag{A-4}$$

By equating like terms between (A-3) and (A-4) produces

$$\begin{aligned} t_i &= q_i t_{i+1} + s_i, \quad i = 1, \ldots, D-1, \\ v_i &= u_i + q_i v_{i+1}, \end{aligned} \tag{A-5}$$

where $v_D = 0$ and $t_D = 1$, as required by (A-2) for $i = D - 1$.

After obtaining values of t_i and v_i from (A-5), x_D is obtained algebraically from the equation

$$c_d \left(t_1 x_D + v_1\right) + a_D \left(t_{D-1} x_D + v_{D-1}\right) + b_D x_D = d_D,$$

where (A-3) is used to substitute into the last line of (A-1). The remaining values of x_i are then determined from A-3.

Eliminating [illegible] between (A-7) and (A-1) produces

[illegible]

where [illegible] (A-2) [illegible]

After [illegible] is obtained [illegible]

the equation

[illegible]

[illegible]

APPENDIX

Model Summaries

B

This Appendix overviews the current status of mesoscale models. Since the first and second editions of the book, the advancement of computer power has facilitated much higher spatial resolution, longer time integrations, and the use of ensemble (multiple realization) model runs. As summarized in the article by Hong and Dudhia (2012), current real-time model runs with grid sizes of 1.5 km provide reliable information for weather forecasts including precipitation. At this resolution, some presentations from operational centers showed the importance of turbulence mixing length and horizontal diffusion in improving the simulation of boundary layer clouds (stratocumulus, shallow cumulus, and fog).

As in the first and second editions, summaries of several mesoscale models are included in this Appendix to illustrate how the models can be described. These models are by no means the only state-of-the-art codes. They are presented to demonstrate how models can be decomposed into their component parts, following the individual chapters in this book. To avoid misrepresentation, the text provided by each modeling group was retained verbatim as much as possible with only minor editorial revision to conform more with the notation in the text. Each modeling group reviewed their model description in the preparation of the Appendix. Clearly the suite of mesoscale models has expanded since the first and second edition of this book appeared!

Date Information Provided: May 2012
Model Name: Coupled Ocean/Atmosphere Mesoscale Prediction System (COAMPS) Coupled Ocean/Atmosphere Mesoscale Prediction System – Tropical Cyclone (COAMPS-TC)
Is the model still under active development? Yes
Name(s) of Developers: James D. Doyle, Richard M. Hodur, Rick Allard, Clark Amerault, Nancy Baker, Ed Barker, Brian Billings, Steve Burk, Tim Campbell, Sue Chen, John Cook, James Cummings, Roger Daley, Mike Frost, Sasa Gabersek, Dan Geiszsler, Chris Golaz, Tracy Haack, Eric Hendricks, Teddy Holt, Yi Jin, Qingfang Jiang, Hao Jin, Hung Chi Kuo, Chi-Sann Liou, Ming Liu, Paul May, Art Mirin, Jason Nachamkin, Julie Pullen, Pat Pauley, Alex Reinecke, Keith Sashegyi, Jerome Schmidt, William Thompson, Pedro Tsai, Shouping Wang.
Organization: Naval Research Laboratory, Address: 7 Grace Hopper Ave

Mesoscale Meteorological Modeling, Third Edition, Volume 98. http://dx.doi.org/10.1016/B978-0-12-385237-3.00025-6

Telephone: (831) 656-4716
Fax: (831) 656-4769
E-mail: james.doyle@nrlmry.navy.mil
Website: http://www.nrlmry.navy.mil/coamps-web/web/home
http://www.nrlmry.navy.mil/coamps-web/web/tc

The complete COAMPS description can be found at: http://pielkeclimatesci.files.wordpress.com/2013/04/coamps-summary-v3.pdf

1. Group: Marine Meteorology Division, Naval Research Laboratory, Monterey, CA. The coupled model development with the ocean and wave models performed in collaboration with Oceanography Division, Naval Research Laboratory, Stennis, MS.

   COAMPS has been operational since 1998 with the primary operational customer being Fleet Numerical Meteorology and Oceanography Center (FNMOC). COAMPS is currently being used for real-time and research applications by many universities and academic partners, U.S. government agencies including the DoD, and other international partners.
2. Equations used for:

   (a) motion (or momentum) (Following Klemp and Wilhelmson 1978a, Hodur 1997)

   i. Horizontal

$$\frac{\partial u}{\partial t} + C_p\theta_v\left(\frac{\partial \pi'}{\partial x} + G_x\frac{\partial \pi'}{\partial \sigma}\right) + K_D\left(\frac{\partial D_3}{\partial x} + G_x\frac{\partial D_3}{\partial \sigma}\right)$$
$$= -u\left(\frac{\partial u}{\partial x}\right)_\sigma - v\left(\frac{\partial u}{\partial y}\right)_\sigma - \dot{\sigma}\left(\frac{\partial u}{\partial \sigma}\right)$$
$$+ fv + D_u + K_H\nabla^4 u$$
$$\frac{\partial v}{\partial t} + C_p\theta_v\left(\frac{\partial \pi'}{\partial y} + G_y\frac{\partial \pi'}{\partial \sigma}\right) + K_D\left(\frac{\partial D_3}{\partial y} + G_y\frac{\partial D_3}{\partial \sigma}\right)$$
$$= -u\left(\frac{\partial v}{\partial x}\right)_\sigma - v\left(\frac{\partial v}{\partial y}\right)_\sigma - \dot{\sigma}\left(\frac{\partial v}{\partial \sigma}\right)$$
$$- fu + D_v + K_H\nabla^4 v$$

   ii. Vertical

$$\frac{\partial w}{\partial t} + C_p\theta_v G_z\frac{\partial \pi'}{\partial \sigma} + K_D G_z\frac{\partial D_3}{\partial \sigma}$$
$$= g\left(\frac{\theta'}{\overline{\theta}} + 0.608q_v' - q_c - q_r - q_s - q_i - q_g\right)$$
$$- u\left(\frac{\partial w}{\partial x}\right)_\sigma - v\left(\frac{\partial w}{\partial y}\right)_\sigma - \dot{\sigma}\left(\frac{\partial w}{\partial \sigma}\right) + D_w + K_H\nabla^4 w$$

(b) Heat

$$\frac{\partial \theta}{\partial t} = -u\left(\frac{\partial \theta}{\partial x}\right)_{\sigma} - v\left(\frac{\partial \theta}{\partial y}\right)_{\sigma} - \dot{\sigma}\left(\frac{\partial \theta}{\partial \sigma}\right) + \frac{Q_{\theta}}{\overline{\rho}} + D_{\theta} + K_H \nabla^4(\theta - \overline{\theta}),$$

(c) Mass

(d) Pressure

$$\frac{\partial \pi'}{\partial t} + \frac{\bar{c}^2}{C_p \bar{\rho} \overline{\theta}_v^2}(D_3) = -u\left(\frac{\partial \pi'}{\partial x}\right)_{\sigma} - v\left(\frac{\partial \pi'}{\partial y}\right)_{\sigma} - \dot{\sigma}\left(\frac{\partial \pi'}{\partial \sigma}\right)$$
$$- \frac{R_d \overline{\pi}}{C_v} \nabla_3 \bullet V + \frac{c^2}{C_p \theta_v^2} \frac{d\theta_v}{dt}$$

(e) Moisture

i. Ice

$$\frac{\partial q_i}{\partial t} = -u\left(\frac{\partial q_i}{\partial x}\right)_{\sigma} - v\left(\frac{\partial q_i}{\partial y}\right)_{\sigma} - \dot{\sigma}\left(\frac{\partial q_i}{\partial \sigma}\right)$$
$$+ D_{q_i} + K_H \nabla^4 q_i + \frac{S_i}{\rho}$$

$$\frac{\partial q_s}{\partial t} = -u\left(\frac{\partial q_s}{\partial x}\right)_{\sigma} - v\left(\frac{\partial q_s}{\partial y}\right)_{\sigma} - \dot{\sigma}\left(\frac{\partial q_s}{\partial \sigma}\right)$$
$$+ \frac{G_z}{\overline{\rho}} \frac{\partial}{\partial \sigma}\left(\overline{\rho} V_s q_s\right) + D_{q_s} + K_H \nabla^4 q_s + \frac{S_s}{\rho}$$

$$\frac{\partial q_g}{\partial t} = -u\left(\frac{\partial q_g}{\partial x}\right)_{\sigma} - v\left(\frac{\partial q_g}{\partial y}\right)_{\sigma} - \dot{\sigma}\left(\frac{\partial q_g}{\partial \sigma}\right)$$
$$+ D_{q_g} + K_H \nabla^4 q_g + \frac{S_g}{\rho}$$

ii. Liquid water

$$\frac{\partial q_c}{\partial t} = -u\left(\frac{\partial q_c}{\partial x}\right)_{\sigma} - v\left(\frac{\partial q_c}{\partial y}\right)_{\sigma} - \dot{\sigma}\left(\frac{\partial q_c}{\partial \sigma}\right)$$
$$+ D_{q_c} + K_H \nabla^4 q_c + \frac{S_c}{\rho}$$

$$\frac{\partial q_r}{\partial t} = -u\left(\frac{\partial q_r}{\partial x}\right)_{\sigma} - v\left(\frac{\partial q_r}{\partial y}\right)_{\sigma} - \dot{\sigma}\left(\frac{\partial q_r}{\partial \sigma}\right)$$
$$+ \frac{G_z}{\overline{\rho}} \frac{\partial}{\partial \sigma}\left(\overline{\rho} V_r q_r\right) + D_{q_r} + K_H \nabla^4 q_r + \frac{S_r}{\rho}$$

iii. Water vapor

$$\frac{\partial q_v}{\partial t} = -u\left(\frac{\partial q_v}{\partial x}\right)_\sigma - v\left(\frac{\partial q_v}{\partial y}\right)_\sigma - \dot{\sigma}\left(\frac{\partial q_v}{\partial \sigma}\right) + D_{q_v} + K_H \nabla^4 \left(q_v - \bar{q}_v\right) + \frac{S_v}{\rho}$$

where

$$\dot{\sigma} = G_x u + G_y v + G_z w$$

$$G_x = \frac{\partial \sigma}{\partial x} = \left(\frac{\sigma - z_{top}}{z_{top} - z_{sfc}}\right)\frac{\partial z_{sfc}}{\partial x}$$

$$G_y = \frac{\partial \sigma}{\partial y} = \left(\frac{\sigma - z_{top}}{z_{top} - z_{sfc}}\right)\frac{\partial z_{sfc}}{\partial y}$$

$$G_z = \frac{\partial \sigma}{\partial z} = \frac{z_{top}}{z_{top} - z_{sfc}}$$

In the above equations, p is the pressure; ρ the density; R_d the gas constant for dry air; T the temperature; q_v, q_c, q_r, q_i, q_g, and q_s the mixing ratios of water vapor, cloud droplets, raindrops, ice crystals, graupel, and snowflakes, respectively; p_{00} a constant reference pressure; C_p the specific heat at constant pressure for the atmosphere; C_v is the specific heat at constant volume; u, v, and w the wind components in the x, y, and z directions, respectively; f the Coriolis force; g the acceleration due to gravity; S_v, S_c, S_r, S_i, S_g, and S_s sources and sinks of q_v, q_c, q_r, q_i, q_g, and q_s, respectively; Q_θ sources and sinks of heat; V_r and V_s the terminal velocities of raindrops and snowflakes, respectively; θ_{std} the standard atmospheric temperature; $q*_v$ the saturation mixing ratio corresponding to the standard atmospheric temperature; and D_3 the density and potential temperature–weighted three-dimensional divergence.

Additional prognostic equations associated with the Khairoutdinov and Kogan (2000) include a drizzle parameterization (not shown here).

(f) Other chemical species

(g) Aerosols

A mineral-dust aerosol model is fully embedded in COAMPS as an inline module of the prediction system, using COAMPS meteorological fields at each time step and each grid point. With the same grid structure as its dynamics model, it has multiple nested grids and interactions of grid nests. The mass conservation equation for a dust particle of radius r in generalized form is

$$\frac{\partial C_r}{\partial t} = -\frac{\partial u C_r}{\partial x} - \frac{\partial v C_r}{\partial y} - \frac{\partial (w + v_f) C_r}{\partial z} + D_x + D_y + D_z + C_{src} - C_{snk}$$

where u, v, and w are the components of the wind vector in the x, y, and z directions; v_f is the particle settling velocity; D_x, D_y, and D_z are turbulent mixing in x, y, and z ; C_{src} is dust source term, i.e., dust mobilization from erodible dessert lands; C_{snk} is the dust sink term which includes dry deposition at the surface and wet removal by precipitation. The dust emission is mainly a function of surface wind stress in a formula which describes the vertical dust flux F being proportional to friction velocity ($u*$) raised to the forth power or the square of wind stress:

$$F = A \quad 1.42 \times 10^{-5} \times u*^4 \quad \text{when} \quad u* \geq u*_t$$

Coefficient A is the fraction of the model grid box that is dust erodible, ranging from 0.0 to 1.0, and $u*_t$ is the threshold friction velocity and land-type dependent. The fractional erodibility is derived using a 1 km resolution land cover dataset developed by Naval Research Laboratory (Walker et al. 2009) for southwest and east Asia, and a global land-use database by the U. S. Geological Survey.

Dust aerosol is modeled with a bimodal lognormal size distribution in mass. The mass median radius and geometric standard deviation in the size distribution are predicted by a sandblasting scheme at each time step, based on the particle's saltating kinetic energy in a dynamical environment. The number of size bins range from 5 to 50 depending on computational efficiency and modeling accuracy requirement.

Dust advection in both horizontal and vertical directions uses a 3rd, or 5th, or 7th order-accurate flux-form, positive-definite, mass-conserved scheme (Bott 1989a,b). The subgrid scale turbulent mixing is solved semi-implicitly with the eddy diffusivities for scalar by the TKE closure. The details of the dust aerosol microphysics and aerosol optics can be found in Liu et al. (2003, 2007).

(h) Ocean circulation
The COAMPS system has the capability to operate in a fully coupled (two-way) air-sea interaction mode (Chen et al. 2010). The atmospheric module within COAMPS-TC is coupled to the NRL-developed Navy Coastal Ocean Model (NCOM) (Martin 2000; Martin et al. 2006) to represent air-sea interaction processes.

(i) Ocean wave models
The COAMPS system has two interchangeable wave models Simulating WAves Nearshore (SWAN) and Wave Watch III (WWIII) to predict ocean surface waves. When operating in the fully coupled air-sea-wave coupled mode, forcings and feedbacks are exchanged between the wave and the atmosphere and ocean circulation models. COAMPS has also been coupled to the Wave Model (WAM) for extratropical and tropical cyclones (Doyle 1995, 2002).

(j) Coupled interface
COAMPS makes use of the Earth System Model Framework (ESMF) to facilitate the air-sea-wave coupling. This is done through a generalized exchange of information between models that allow each model to operate using

independent resolutions and projections, with the feedbacks through the various components taking place on an exchange grid.

(k) Tropical Cyclone (TC) capability, COAMPS-TC
The COAMPS-TC is comprised of data quality control, analysis, initialization, and forecast model sub-components (Doyle et al. 2012b). The NRL Atmospheric Variational Data Assimilation System (NAVDAS) is used to blend conventional and remote sensing observations. Synthetic observations following (Liou and Sashegyi 2012) are used to specify the tropical cyclone intensity and structure and a relocation procedure is used to initialize the position.

(l) Large Eddy Simulation (LES) application
COAMPS options have been formulated to allow for high-resolution applications within the atmospheric boundary layer. Several new closures were developed along with an anelastic dynamical core for efficiency purposes (Doyle et al. 2009; Golaz et al. 2009).

3. Horizontal domain sizes used:
Horizontal domain sizes varying both operationally and in a research mode. Typical domain sizes for a 45 km resolution coarse mesh might be 9000 km × 6750 km, with a fine mesh of 5 km covering an area of 1000 km × 1000 km.

4. Horizontal grid increment(s):
Grid increments vary from 54 km to 10 m used in large-eddy simulation mode.

5. Vertical domain sizes used:
The vertical domain depth is typically 30 km, with recent advancements that allow the model top up to 60 km. Idealized applications have used shallow to very deep model domains.

6. Vertical grid increment(s):
For real data cases, vertically stretched from 10 m at the lowest model level. Idealized applications have used uniform and stretched vertical grid increments.

7. Initialization procedure:
NAVDAS (Daley and Barker 2000) is used to blend observations of winds, temperature, moisture, and pressure from a plethora of sources such as radiosondes, pilot balloons, satellites, surface measurements, ships, buoys, and aircraft. The sea-surface temperature is analyzed directly on the model computational grid using the Navy Coupled Ocean Data Assimilation (NCODA) system, which makes use of all available satellite, ship, float, and buoy observations. Both the NCODA and NAVDAS systems are applied using a data assimilation cycle in which the first guess for the analysis is derived from the previous short-term forecast.

8. Solution techniques for dynamic core equations:

(a) advection
(b) vertical flux divergence
(c) horizontal pressure gradient force
(d) Coriolis term

The model equations are solved time-splitting method, which allows for large time steps to be taken for the slow modes and shorter time steps for the faster sound modes (Klemp and Wilhelmson 1978a; Skamarock and Klemp 1992).

9. Coordinate system used:
COAMPS makes use of a terrain following coordinate (Gal-Chen and Sommerville 1975a).

10. Nested domains:
Generalized nested domains, with the model supporting an arbitrary number of grid meshes. The COAMPS-TC option makes use of moving nested grids that automatically follow the tropical cyclone center.

11. Lateral boundary condition for outermost grid:
Lateral boundaries for real data applications follow Davies (1976). Global model boundary conditions are typically derived from NOGAPS, NAVGEM, or GFS. Other options are available as well. Idealized application options include i) wall, ii) periodic, and iii) open (or radiation).

12. Top boundary condition:
Rigid lid, sponge (explicit and implicit Rayleigh damping), and radiation (linear gravity wave condition) options are available.

13. Surface boundary treatment:
The surface fluxes are represented by a hybrid scheme as described in Wang et al. (2002). The scheme uses Richardson number as a stability parameter as in Louis et al. (1982); it is further tuned to match the stability functional dependence as described in Fairall et al. (1996). The surface momentum roughness length over water follows the Charnock parameterization for wind speed less than 35 m s^{-1}, above which the roughness is specified as discussed in Donelan et al. (2004). The temperature and moisture roughness lengths are set to be one tenth of the momentum roughness over land. They are functions of both the momentum roughness and the roughness Reynolds number as described in Fairall et al. (2003) over water conditions.

The single layer vegetation canopy approach from the WRF NOAH land surface model (LSM) based on Chen and Dudhia (2001) and Ek et al. (2003) is used to compute latent and sensible heat fluxes over land. Volumetric soil moisture and temperature is predicted at four soil layers. Urban effects on surface fluxes can also be activated using the WRF single–layer urban canopy model (Kusaka et al. 2001) or a multi-layer urban model (Holt and Pullen 2007).

14. Parameterization of subgrid-scale flux divergence for:

(a) motion (or momentum)
(b) heat
(c) moisture in its three phases, and
(d) other chemical species

For each, separate those for turbulence and for other subgrid-scale processes. For turbulence parameterization, the Mellor-Yamada's 1.5 order turbulence closure model that predicts turbulence kinetic energy and determines the turbulent fluxes in terms of the down-gradient transport approach (Mellor and Yamada 1982). To account for latent heat effect of clouds on the buoyancy in turbulence mixing, the subgrid-scale cloud fraction is calculated using a Gaussian distribution of turbulent fluctuations of conserved variables (Sommeria and Deardorff 1977). For subgrid-scale parameterization in the LES version (Golaz et al. 2002), both the TKE model (Deardorff 1980) and the local equilibrium model (Stevens et al. 1999) are included.

15. Flux divergence from cumulus cloud parameterization:

(a) motion (or momentum)
(b) heat
(c) moisture in its three phases

The Kain-Fritsch convective parameterization scheme (Kain and Fritsch 1990) is used for most mid-latitude convective forecasts in COAMPS. An updated version of the KF scheme (Kain 2004) is also available as a separate option. Options also include a simplified Arakawa Schubert cumulus parameterization (Arakawa and Schubert 1974) parameterization of shallow cumulus convection (Tiedtke et al. 1988).

16. Radiational flux divergence:

(a) longwave flux divergence
(b) shortwave flux divergence

COAMPS has two radiative transfer (RT) models being implemented: one is the Fu-Liou (1992, 1993) four-stream scattering-absorption algorithm (with a two-stream option), integrating over six shortwave bands and 12 longwave bands; the other is the Harshvardhan et al. (1987) RT model fast-speed two-stream algorithm of three shortwave bands and six longwave bands.

17. Stable clouds and precipitation:
COAMPS moist physics provide estimates of source/sink terms for the scalar equations involving potential temperature, water vapor, the number concentration of aerosol, cloud droplets and rain droplets, and each of the five condensate species (q_c, q_r, q_i, q_s, and q_g) currently carried by the model. Numerical treatment of the advective terms is performed with either a standard leapfrog approach or an optional hybrid time-differencing scheme that allows for integration of the scalar fields on a forward time step. The choice of advection schemes currently includes the 3rd and 5th order positive-definite and monotonic schemes of Bott (1989a,b; 1992), and the selective advection method described by Blossey and Durran (2008).

The treatment of the microphysical source terms for the operational bulk cloud microphysics scheme is a Lin et al. (1983) variant described by Rutledge and Hobbs (1983, 1984). Another microphysical scheme implemented in COAMPS is the Thompson scheme (Thompson et al. 2008).

18. Other parameterizations:
A physically consistent method is developed in COAMPS to include dissipative heating based on turbulent kinetic energy (TKE) dissipation to ensure energy conservation (Jin et al. 2007). The dissipative heating rate is considered at all levels of the model, including both the surface layer and the layers above, in contrast to other studies where dissipative heating is considered at the surface only due to lack of TKE representation in the models. Furthermore, the TKE dissipation in this method is computed semi-implicitly to maintain computational stability. Additionally, for tropical cyclones, a sea spray parameterization option is available (Fairall et al. 2009).
19. Summary of how these physical processes are modeled each time step at individual grid points:
    (a) the total diabatic heating per time step
        i. from resolvable scales
        ii. from subgrid scales
    (b) the total moistening/drying per time step
        i. from resolvable scales
        ii. from subgrid scales
    (c) the total acceleration/deceleration per time step
        i. from resolvable scales
        ii. from subgrid scales
20. Phenomena studied:
    (a) for operational weather forecasting
    Over 70 areas are being run at FNMOC at least twice daily around the world for many different applications. A GUI based system, the COAMPS-On Demand System (COAMPS-OS) is used in operations (and in research applications) to the tailor domain set up, model options, graphics, and linked models and products based on the application and requirements.
    (b) for research applications
    COAMPS is being run in real-time for all hurricanes in the western Atlantic and the eastern and central Pacific in support of the Hurricane Forecast Improvement Project (HFIP); and for all tropical cyclones in the western Pacific, Indian Ocean, and southern Hemisphere for evaluation by the Joint Typhoon Warning Center (JTWC).

    Research has focused on a variety of mesoscale and synoptic-scale topics from the tropics to the Arctic, and from the ocean bottom to the middle atmosphere. In the past, COAMPS has been used in support of many research and field programs including ITOP, LABSEA, Sierra Rotor Experiment, AOSN, AOSN-II, AESOP, Indonesian Straits, T-REX, CBLAST, MAP, America's Cup, ACE, VOCALS, CALJET, PACJET, COAST, COMPARE, FASTEX, GUFMEX, LEADEX, SHAREM, SWADE, PHILEX, and VOCAR.

21. Computer(s) used: examples of time of integration for specific model runs Cray-XT5: 75 minutes wall time for a 120 hr forecast on a triple-nested (45/15/5 km; 281 × 151, 121 × 121, 181 × 181 grid; 40 levels) using 160 processors.

Date Information Provided: March 2012
Model Name: HARMONIE (Hirlam Aladin Regional/Mesoscale Operational NWP in Europe).
Is the model still under active development? Yes, HARMONIE is under active development.
Name(s) of Developers: National Meteorological Services (see below) participating with both the HIRLAM-B and ALADIN consortia.
Organization:
NMSs from the following countries: HIRLAM-B: Denmark, Estonia, Finland, Iceland, Ireland, Lithuania, Netherlands, Norway, Spain, and Sweden.
ALADIN: Algeria, Austria, Belgium, Bulgaria, Croatia, Czech Republic, France, Hungary, Morocco, Poland, Portugal, Romania, Slovakia, Slovenia, Tunisia, and Turkey.
Websites:
HIRLAM-B: http://hirlam.org/
ALADIN: http://www.cnrm.meteo.fr/aladin/

1. Group: Modeling groups within NMSs participating with both the HIRLAM-B and ALADIN consortia.
2. Equations used for:
   - (a) motion (or momentum)
     - i. horizontal
     - ii. vertical
   - (b) heat
   - (c) mass
   - (d) pressure
   - (e) moisture
     - i. ice
     - ii. liquid water
     - iii. water vapor
   - (f) other chemical species

   HARMONIE is based on fully elastic, non-hydrostatic Euler equations (Bubnova et al. 1995), with prognostic equations for 12 3D-variables (Benard et al. 2010):

   - horizontal momentum equations (u- and v-component)
   - equation for vertical divergence
   - equation for temperature
   - equation for non-hydrostatic pressure term
   - 5 equations for specific contents of humidity, cloud water, ice crystals, rain, snow and graupel.
   - equation for turbulent kinetic energy (TKE)

In order to close the system one prognostic 2D-equation for hydrostatic surface pressure is required.

3. Horizontal domain sizes used:

4. Horizontal grid increment(s):

Horizontal domain and grid size is freely adjustable. However, the default grid-size is 2.5 km. Lambert conformal projection is used by default. Horizontal grid is non-staggered Arakawa-A grid.

5. Vertical domain sizes used:

6. Vertical grid increment(s):

Vertical discretization is based on finite differences in a non-staggered grid. A mass-based terrain-following hybrid-pressure coordinate (Laprise 1992a) is used. The default number of levels is 65 with the model top at 10 hPa. In addition, a vertical finite element method is under development.

7. Initialization procedure:

For upper air variables (two wind components, temperature, humidity, and surface pressure) 3D variational assimilation is used. The observation data types which are assimilated by default presently are conventional observations (TEMP, SYNOP, AIREP, PILOT, SATOB, SHIP, DRIBU) and AMSU-A/ATOVS radiances over sea. Additionally, it is possible to assimilate AMSU-A over land and sea ice, AMSU-B, geostationary and MODIS atmospheric motion vectors, SEVIRI cloud-cleared radiances, GPS zenith total delay, wind profilers, radar radial winds, and reflectivity. Assimilation of surface variables (surface temperature, soil moisture, SST, snow depth) are done with optimum interpolation (simplified extended Kalman Filter method under development). To reduce noise and spin-up, analyses can be initialized out by incremental digital filter initialization (Lynch et al. 1997).

8. Solution techniques for dynamic core equations:

(a) advection
(b) vertical flux divergence
(c) horizontal pressure gradient force
(d) Coriolis term

HARMONIE is based on two-time level semi-implicit Semi-Lagrangian discretization scheme on Arakawa A-grid. The double Fourier spectral decomposition (Haugen and Machenhauer 1993) is used for most parameters. A fourth-order spectral horizontal diffusion is applied for each spectral variable. For non-spectral variables (i.e., water condensates) semi-Lagrangian horizontal diffusion is used (Vána et al. 2008). Specific humidity and TKE are not diffused.

9. Nested domains:

One-way nesting procedure directly with host model is the default option. Multiple one-way nesting procedure is also possible.

10. Lateral boundary condition for outermost grid:

Davies relaxation method is applied. HARMONIE is able to use boundary data from ECMWF (default), HIRLAM, and ALADIN models in the outermost domain.

11. Top boundary condition:
At the upper boundary a condition of zero vertical velocity is imposed.
MODEL PHYSICS: The HARMONIE system contains several complete parameterization packages. The default physics package (AROME physics) is described below (Seity et al. 2011). References to the components of other packages (ALARO-, HIRLAM- and ECMWF-physics) are given under the title "optional."

12. Surface boundary treatment:

(a) surface momentum fluxes
(b) surface heat fluxes
(c) surface moisture fluxes
(d) fluxes of other chemical species

The externalized surface scheme SURFEX is used for the soil description. Surface physiographies are prescribed using the 1 km resolution ECOCLIMAP database (Masson et al. 2003). SURFEX in HARMONIE assumes a tile approach, distinguishing 4 surface types:

- Soil: a three-layer force-restore ISBA scheme (Noilhan and Planton 1989)
- Sea surface fluxes (ECUME empirical formula, Belamari and Pirani 2007)
- Urban areas: the Town Energy Budget (TEB) urban canyon model (Masson 2000)
- Lakes: the FLake model (Kourzeneva 2010; Mironov et al. 2010)

Optional: ISBA scheme alone is an alternative to SURFEX.

13. Parameterization of subgrid-scale flux divergence for:

(a) motion (or momentum)
(b) heat
(c) moisture in its three phases, and
(d) other chemical species.

TURBULENCE: Turbulence parameterization is based on 1D prognostic TKE with a diagnostic mixing length (Bougeault and Lacarrere 1989; Cuxart et al. 2000).

14. Flux divergence from cumulus cloud parameterization:

(a) motion (or momentum)
(b) heat
(c) moisture in its three phases

DEEP CONVECTION: At the default grid size (2.5 km) of HARMONIE, the deep convection is assumed to be explicitly resolved. Therefore, the mass-flux convection scheme with a moist convergence closure (Bechtold et al. 2001) is inactivated.
Optional: two other physics packages with the following convection parameterizations; Gerard (2007), and Kain (2004).

SHALLOW CONVECTION: A combined eddy diffusivity- mass-flux scheme for parameterization of dry thermals and non-precipitating shallow cumuli (EDMF-m, Siebesma et al. 2007; de Rooy and Siebesma 2008, 2010; Neggers et al. 2009). Optional: Mass-flux based shallow convection scheme by Pergaud et al. (2009).

15. Radiational flux divergence:

- **(a)** longwave flux divergence
- **(b)** shortwave flux divergence

RADIATION: Longwave radiation scheme follows the RRTM method of Mlawer et al. (1997) with 16 spectral bands associated with water vapor, carbon dioxide, methane and nitrous oxide. Shortwave radiation is based on Fouquart and Bonnel (1980) with 6 spectral bands. Liquid and ice cloud optical properties are following Morcrette and Fouquart (1986) and Ebert and Curry (1992), respectively.

Optional schemes: ECMWF's SW part of the RRTM scheme (Mlawer and Clough 1997). HIRLAM's fast radiation scheme with one spectral band for both LW and SW (Savijärvi 1990) and ALARO-scheme (Ritter and Geleyn 1992; Geleyn et al. 2005)

16. Stable clouds and precipitation:
MICROPHYSICS: A single-moment 6-class microphysics scheme, which have prognostic treatment for specific contents of water vapor, cloud water, ice crystals, rain, snow, and graupel (Pinty and Jabouille 1998; Lascaux et al. 2006). Sedimentation of the falling hydrometeors is treated with a probability density function-based scheme (Bouteloup et al. 2011), which allows for a numerically stable solution with relatively large time steps. Statistical subgrid condensation scheme follows Bougeault (1982) and Bechtold et al. (1995).

Optional: Single moment 3-class microphysics scheme (in HIRLAM package; Rasch and Kristjáansson 1998).

17. Other parameterizations:

18. Summary of how these physical processes are modeled each time step at individual grid points:

- **(a)** the total diabatic heating per time step
  - **i.** from resolvable scales
  - **ii.** from subgrid scales
- **(b)** the total moistening/drying per time step
  - **i.** from resolvable scales
  - **ii.** from subgrid scales
- **(c)** the total acceleration/deceleration per time step
  - **i.** from resolvable scales
  - **ii.** from subgrid scales
  - **iii.** Calculations within a time step is organized as described in Seity et al. (2011):

- Inverse bi-Fourier transform to grid point space
- Calculation of physical parameterization tendencies (adjustment, radiation, surface, shallow convection, turbulence, and microphysics)
- Calculation of the semi-Lagrangian advection scheme
- Direct bi-Fourier transforms to spectral space
- Calculation of the spectral part of horizontal diffusion
- Calculation of a semi-implicit solver

19. Phenomena studied:

(a) for operational weather forecasting
(b) for research applications

HARMONIE is primarily designed to be used for operational weather forecasting. However, the model is also applicable for research applications. For instance, a version suitable for climate studies has been set up.

20. Computer(s) used: examples of time of integration for specific model runs: HARMONIE has been used in variety of computing platforms; from Linux-based laptops to several parallel High Performance Computing environments (both scalar and vector machines).

Date Information Provided: February 2012
Model: The Operational High-Resolution Hurricane Weather Research and Forecasting Modeling System (HWRF)
Name(s): Vijay Tallapragada and Sundararaman G. Gopalakrishnan
Is the model still under active development? Yes, HWRF is still under active development
Original HWRF developers: Sundararaman G. Gopalakrishnan, Robert Tuleya, Zavisa Janjic, and Thomas Black
Advanced HWRF System developers: Xuejin Zhang, Samuel Traham, Qingfu Liu, Jian-Wen Bao, Sundararaman G. Gopalakrishnan, Young Kwon, Dmitry Sheinin, Vijay Tallapragada, Brad Ferrier, Richard Yablonsky, and Mingjing Tong
Organizations: AOML/HRD, 4301 Rickenbacker CSWY, Miami, FL 33149
Telephone: (305) 361-4357
Fax: (305) 361-4402
Email: Gopal@noaa.gov
NCEP/EMC, 5200 Auth Road, Camp Springs, MD 20746-4304
Telephone: (301)763-8000 ext. 7232 Fax: (301)763-8545
Email: Vijay.Tallapragada@noaa.gov

1. Groups: Modeling group at AOML/HRD and Hurricane Modeling group at NCEP/EMC.
2. Equations and numerical solution: HWRF is an ocean-atmosphere coupled system. It consists of three components: the atmospheric model, POM-TC ocean component, and the coupler. The atmospheric model is based on NCEP's fully compressible Eulerian WRF NMM dynamic core (Janjic 1979, 1984a,b, 2003, 2011; Janjic et al. 2001, 2010), The basic features of the model dynamics are:

(a) Vertical coordinate: Terrain-following hybrid sigma-pressure coordinate
(b) Horizontal projection: Rotated latitude-longitude
(c) Vertical grid: Lorenz staggering
(d) Horizontal grid: Arakawa E-grid staggering
(e) Solution techniques for dynamic core equations

Spatial discretization principles

- Conservation of important properties of the continuous system (e.g., Arakawa 1966, 1972; Janjic 1977, 1984a,b)
- Nonlinear energy cascade controlled through energy and enstrophy conservation (Janjic 1984a,b)
- A number of first order and quadratic quantities conserved
- A number of properties of differential operators preserved
- Consistent transformations between kinetic and potential energies
- Mass-conserving positive definite monotone Eulerian tracer advection (replaced older Lagrangian one) Time stepping
- Non-split time integration
- Modified Adams-Bashforth scheme for horizontal advection of basic dynamic variables and Coriolis terms
- Crank Nicholson for vertical advection of basic dynamic variables
- Forward-backward scheme for horizontally propagating fast waves
- Implicit scheme for vertically propagating sound waves

3. Movable nesting techniques: A Hurricane is an intense atmospheric circulation characterized by strong multi-scale interactions between convective clouds, typically on the order of few kilometers, and a larger-scale environment, typically on the order of several hundred to thousands of kilometers. Forecasting such a system requires a large-size domain with high resolution. However, at this time, it is not practical to run regional models with uniform resolution on the order of 1-10 km over such huge domains. In order to forecast tropical cyclones in a multi-scale environment, the vortex-following nesting technique is developed within the WRF-NMM dynamic core and solver (Zhang et al. 2012). The modified dynamic core and solver is adopted by HWRF. The main features are:

   (a) Unlimited telescopic grid refinement
   (b) Vortex-following nest movement algorithm
   (c) Two-way interaction
   (d) Mass and terrestrial information adjusted on terrain

4. Typical domain size for operations and outer boundary conditions:

   (a) Static domain: size, 77.4° × 77.58°; grid space, 0.18°, boundary conditions derived from GFS forecasts
   (b) Intermediate domain: size, 10.44° × 10.14°; grid space, 0.06°
   (c) Innermost domain: size, 6.12° × 5.42°; grid space, 0.02°.

5. Vertical size: 42 levels with model top at 50 hPa and hybrid transitional level from sigma to pressure at 420 hPa.
6. Initialization procedure: vortex cycling and intensity correction with relocation and Grid point Statistical Interpolation (GSI) data assimilation technique.
7. Surface boundary treatment: GFDL surface layer parameterization (Kurihara and Tuleya 1974; Moon et al. 2007; Gopalakrishnan et al. 2011) with Cd and Ck fitted by using oceanic observation data is discussed in Gopalakrishnan et al. (2013). Over land, a slab model is used to predict surface temperature (Tuleya 1994) and Cd and Ck are based on land use.
8. Parameterization of subgrid-scale flux divergence: In the vertical, HWRF uses a non-local PBL scheme (Troen and Mahrt 1986; Hong and Pan 1996) based on the current operational GFS global and GFDL hurricane models. The vertical diffusivity is fitted by observation data in the hurricane wind regime (Zhang et al. 2011; Gopalakrishnan et al. 2013). In the horizontal, subgrid-scale fluxes are estimated by a 2nd order Smagorinsky-type lateral diffusion scheme dependent on flow deformation and local TKE (Janjic 1990; Janjic et al. 2010).
9. Microphysics parameterization: A slightly modified version of the Ferrier microphysics scheme (Ferrier 2005) is employed in HWRF. In the Tropics, the concentration of droplets is set to 60 cm^{-3}, and the onset of grid-resolved condensation above the planetary boundary layer in the outermost 27 km domain is set to 97.5%. It is a single-moment 5-class scheme that calculates changes in mixing ratios of water vapor, cloud water, cloud ice, rain, and snow, but behaves like a 6-class scheme (with graupel) because it includes a diagnostic array to track the growth of snow by accretion and freezing of liquid water drops. For purposes of computational speed and efficiency, all forms of hydrometeors are combined into a total condensate array, which is advected by the model dynamics.
10. Cumulus cloud parameterization: HWRF uses the SAS cumulus parameterization also employed in the GFS (Pan and Wu 1995; Hong and Pan 1998; Pan 2003). This scheme, which is based on Arakawa and Schubert (1974) and simplified by Grell (1993), was made operational in NCEP's global model in 1995 and in the GFDL hurricane model in 2003. A major simplification to the original Arakawa-Shubert scheme was made by considering a random cloud top at a specified time and not the spectrum of cloud sizes as in the original, computationally expensive Arakawa and Schubert scheme. A new shallow convection scheme developed in the GFS model (Han and Pan 2006) is adopted in HWRF. The scheme employs the mass-flux approach that produces heating throughout the convection layers due to the dominant environmental subsidence warming.
11. Radiation flux scheme: The longwave radiation scheme follows the simplified exchange method of Fels and Schwarzkopf (1975) and Schwarzkopf and Fels (1991), with calculation over spectral bands associated with carbon dioxide, water vapor, and ozone. The shortwave radiation follows the GFDL version of the Lacis and Hansen (1974) parameterization employing a broadband approach that accounts for effects of atmospheric water vapor, ozone, and carbon dioxide (Sasamori et al. 1972). Radiative effects from liquid and ice are accounted for

from grid-scale (Harshvardhan et al. 1989; Xu and Randall 1996) and convective clouds (Slingo 1987), in which maximum overlap is assumed within cloud layers and random overlap is assumed between cloud layers.

12. Coupling: A coupler was developed to act as an interface between the HWRF atmospheric component and the POM-TC ocean component. During forecast integration of the HWRF atmospheric component, the east-west and north-south momentum stresses at the air-sea interface are passed from the atmosphere to the ocean, along with temperature flux and the shortwave radiation incident on the ocean surface. During forecast integration of the POM-TC ocean component, the SST is passed from the ocean to the atmosphere (Gopalakrishnan et al. 2011).
13. POM-TC ocean component: A version of the Princeton Ocean Model (Mellor 2004) adapted at the University of Rhode Island for HWRF coupled hurricane-ocean model prediction (Gopalakrishnan et al. 2012). The ocean is initialized in the Atlantic using the feature-based modeling procedure of Yablonsky and Ginis (2008). In addition to vertical mixing, POM-TC captures the fully three-dimensional processes of upwelling/downwelling and horizontal advection, both of which are important for hurricane-induced sea surface temperature cooling and subsequent hurricane intensity change (Yablonsky and Ginis 2009, 2013).

Date Information Provided: May 2012
Model Name: MM5 (PSU research and military-defense versions within US Army MMS-Profiler, US Marines METMFR NEXGEN and DTRA ROFS (Stauffer et al. 2007a)
Is the model still under active development? No, but we are further developing our MM5 physics and data assimilation methods in WRF-ARW (e.g., Lei et al. 2012; Stauffer 2012)
Name(s) of Developers: David Stauffer, Aijun Deng, Annette Gibbs, Glenn Hunter
Organization: Penn State University (PSU), Department of Meteorology **Address:** 503 Walker Building, University Park, PA 16802
Telephone: 814-863-3932
Fax: 814-865-9429
E-mail: stauffer@meteo.psu.edu, deng@meteo.psu.edu, gibbs@meteo.psu.edu, hunter@neteo.psu.edu
Website: http://ploneprod.met.psu.edu/people/drs8

1. Group: Penn State University Numerical Weather Prediction (NWP) Group, led by Dr. David Stauffer
2. Equations used for: nonhydrostatic, fully compressible, primitive equations with variables in "flux-form" (e.g., p^*u, p^*v, p^*T) but p^* varies only in x and y and is defined as the reference state surface pressure minus the pressure at the top of the model (see Dudhia 1993; and Grell et al. 1995 for detailed descriptions of the variables in all of the equations provided below)

(a) motion (or momentum)

i. horizontal

$$
\begin{aligned}
\frac{\partial p^* u}{\partial t} &= -m^2 \left[\frac{\partial p^* uu/m}{\partial x} + \frac{\partial p^* vu/m}{\partial y} \right] - \frac{\partial p^* u \dot{\sigma}}{\partial \sigma} + uDIV \\
&\quad - \frac{mp^*}{\rho} \left[\frac{\partial p'}{\partial x} - \frac{\sigma}{p^*} \frac{\partial p^*}{\partial x} \frac{\partial p'}{\partial \sigma} \right] + p^* fv - p^* ew \cos\theta + D_u \\
\frac{\partial p^* v}{\partial t} &= -m^2 \left[\frac{\partial p^* uv/m}{\partial x} + \frac{\partial p^* vv/m}{\partial y} \right] - \frac{\partial p^* v \dot{\sigma}}{\partial \sigma} + vDIV \\
&\quad - \frac{mp^*}{\rho} \left[\frac{\partial p'}{\partial y} - \frac{\sigma}{p^*} - \frac{\partial p^*}{\partial y} \frac{\partial p'}{\partial \sigma} \right] - p^* fu \\
&\quad + p^* ew \sin\theta + D_v
\end{aligned}
$$

$$
DIV = m^2 \left[\frac{\partial p^* u/m}{\partial x} + \frac{\partial p^* v/m}{\partial y} \right] + \frac{\partial p^* \dot{\sigma}}{\partial \sigma},
$$

$$
\dot{\sigma} = -\frac{\rho_0 g}{p^*} w - \frac{m\sigma}{p^*} \frac{\partial p^*}{\partial x} u - \frac{m\sigma}{p^*} \frac{\partial p^*}{\partial y} v.
$$

ii. vertical

$$
\begin{aligned}
\frac{\partial p^* w}{\partial t} &= -m^2 \left[\frac{\partial p^* uw/m}{\partial x} + \frac{\partial p^* vw/m}{\partial y} \right] - \frac{\partial p^* w \dot{\sigma}}{\partial \sigma} + wDIV \\
&\quad + p^* g \frac{\rho_0}{\rho} \left[\frac{1}{p^*} \frac{\partial p'}{\partial \sigma} + \frac{T_v'}{T} - \frac{T_0 p'}{T p_0} \right] \\
&\quad - p^* g[(q_c + q_r)] + p^* e(u \cos\theta - v \sin\theta) + D_w.
\end{aligned}
$$

(b) heat

(c) temperature

$$
\begin{aligned}
\frac{\partial p^* T}{\partial t} &= -m^2 \left[\frac{\partial p^* uT/m}{\partial x} + \frac{\partial p^* vT/m}{\partial y} \right] - \frac{\partial p^* T \dot{\sigma}}{\partial \sigma} + TDIV \\
&\quad + \frac{1}{\rho C_p} \left[p^* \frac{Dp'}{Dt} - \rho_0 g p^* w - D_{p'} \right] + p^* \frac{\dot{Q}}{C_p} + D_T,
\end{aligned}
$$

(d) perturbation pressure

$$
\begin{aligned}
\frac{\partial p^* p'}{\partial t} &= -m^2 \left[\frac{\partial p^* up'/m}{\partial x} + \frac{\partial p^* vp'/m}{\partial y} \right] - \frac{\partial p^* p' \dot{\sigma}}{\partial \sigma} + p' DIV \\
&\quad - m^2 p^* \gamma p \left[\frac{\partial u/m}{\partial x} - \frac{\sigma}{mp^*} \frac{\partial p^*}{\partial x} \frac{\partial u}{\partial \sigma} + \frac{\partial v/m}{\partial y} - \frac{\sigma}{mp^*} \frac{\partial p^*}{\partial y} \frac{\partial v}{\partial \sigma} \right] \\
&\quad + \rho_0 g \gamma p \frac{\partial w}{\partial \sigma} + p^* \rho_0 g w
\end{aligned}
$$

(e) moisture mixing ratios

i. cloud water (ice) mixing ratio

$$\frac{\partial p^* q_c}{\partial t} = -m^2 \left[\frac{\partial p^* u q_c / m}{\partial x} + \frac{\partial p^* v q_c / m}{\partial y} \right] - \frac{\partial p^* q_c \dot{\sigma}}{\partial \sigma} + \delta_{nh} q_c DIV + p^* \left(P_{ID} + P_{II} - P_{RC} - P_{RA} + P_{CON} \right) + D_{q_c}$$

ii. rain water (snow) mixing ratio

$$\frac{\partial p^* q_r}{\partial t} = -m^2 \left[\frac{\partial p^* u q_r / m}{\partial x} + \frac{\partial p^* v q_r / m}{\partial y} \right] - \frac{\partial p^* q_r \dot{\sigma}}{\partial \sigma} + \delta_{nh} q_r DIV - \frac{\partial V_f \rho g q_r}{\partial \sigma} + p^* \left(P_{RE} + P_{RC} + P_{RA} \right) + D_{q_r}$$

iii. water vapor mixing ratio

$$\frac{\partial p^* q_v}{\partial t} = -m^2 \left[\frac{\partial p^* u q_v / m}{\partial x} + \frac{\partial p^* v q_v / m}{\partial y} \right] - \frac{\partial p^* q_v \dot{\sigma}}{\partial \sigma} + \delta_{nh} q_v DIV + p^* \left(-P_{RE} - P_{CON} - P_{II} - P_{ID} \right) + D_{q_v}$$

(f) other chemical species

3. Horizontal domain sizes used: In mobile, Humvee-based Nowcast systems, Army MMS-Profiler (Profiler) and Marines METMFR NEXGEN (NEXGEN): 36 km: 101 × 101 × 30; 12 and 4 km: 127 × 127 × 30; in DTRA Reachback prediction system, ROFS: variable horizontal and vertical resolutions and domain sizes, typical innermost domain at ~1.3 km grid spacing.
4. Horizontal grid increment(s): Profiler and NEXGEN: 36 km, 12 km, and 4 km; ROFS: variable to ~1.3 km in innermost nest.
5. Vertical domain sizes used: Profiler and NEXGEN: 30 L; ROFS: 30, 45, 62 L.
6. Vertical grid increment(s): Variable normalized reference pressure sigma distribution, stretched in vertical to ptop = 50 hPa or 100 hPa, with surface layer typically ~30 m AGL, and the thickness of the layers increasing gradually with height to the model ptop.
7. Initialization procedure: Multi-scale Four Dimensional Data Assimilation (FDDA), combination of analysis nudging and observation nudging strategies across a nested-grid framework (e.g., Stauffer and Seaman 1994; Schroeder et al. 2006a,b).
8. Solution techniques for dynamic core equations: Arakawa B grid staggering in horizontal, usually on Lambert Conformal projection (Polar Stereographic and Mercator also available), vertical velocity and turbulent kinetic energy (TKE) staggered with other prognostic variables in the vertical, leapfrog with Asselin filter in time, time-splitting (split-implicit), second-order in space, fourth-order explicit numerical diffusion on interior and second-order diffusion near lateral boundaries (Grell et al. 1995)

- **(a)** Advection flux-form (Grell et al. 1995)
- **(b)** Vertical flux divergence: Explicit with dynamic time step in surface layer; implicit above surface layer (Stauffer et al. 1999; Shafran et al. 2000)
- **(c)** Horizontal pressure gradient force: Computed based on perturbation pressure (Grell et al. 1995)
- **(d)** Coriolis term: 3D Coriolis (Grell et al. 1995).

9. Coordinate system used: Normalized pressure sigma using constant dry, hydrostatic reference state $\sigma = (p_{sr} - p_t)/(p_{sr} - p_t) = (p_{sr} - p_t)/p^*$; using hydrostatic reference state pressure p_r and surface pressure p_{sr}, $\partial\sigma = \frac{-\rho_0 g}{p^*}\partial z$.

10. Nested domains: typically three nests, one-way nesting, either as separate jobs with 30-60 minute update of lateral boundary condition tendencies, or concurrent in same job with lateral boundary tendencies updated every coarser-grid advection step.

11. Lateral boundary condition for outermost grid: Relaxation; global models: 1-deg NOGAPS or 1/2 degree GFS, other sources for research including the 32 km NARR and 13 km NDAS.

12. Top boundary condition: Rigid or radiative.

13. Surface boundary treatment: Friction, heat and moisture fluxes using MO similarity theory (Grell et al. 1995), variable land-use categories. Surface layer model in our PSU G-S PBL (Stauffer et al. 1999, Shafran et al. 2000) is based on four stability regimes as in Blackadar scheme (Zhang and Anthes 1982).

(Note that when using the Noah land-surface model (described in 14) below), surface fluxes are computed within Noah based on model time-varying soil temperature and moisture, responding to model-predicted surface-layer fields and precipitation, using standard resistance formulations).

- **(a)** Surface momentum fluxes: Similarity theory as in Blackadar PBL (Grell et al. 1995)
- **(b)** Surface heat fluxes: Similarity theory as in Blackadar PBL (Grell et al. 1995)
- **(c)** Surface moisture fluxes: Similarity theory as in Blackadar PBL (Grell et al. 1995)
- **(d)** Fluxes of other chemical species:

14. Parameterization of subgrid-scale flux divergence: Using 1.5-order (level 2.5) TKE local closure PSU G-S PBL for vertical fluxes, including moist effects in stability parameter and length scales in saturated layers (Stauffer et al. 1999; Shafran et al. 2000), and eddy diffusivity/length scale adjustments for reduction of unresolved horizontal convective roles (Deng and Stauffer 2006).

- **(a)** Motion (or momentum): 1.5-order (level 2.5) TKE-based diffusivities for momentum flux divergences in u and v
- **(b)** Heat: 1.5-order (level 2.5) TKE-based diffusivities for vertical flux divergence in θ_{il} (ice-liquid water potential temperature, conserved for dry or saturated adiabatic processes), and then converted into tendency for temperature

(c) Moisture in its three phases: 1.5-order (level 2.5) TKE-based diffusivities for vertical flux divergence in q_t (total water mixing ratio, conserved for dry or saturated adiabatic processes), and then converted into tendencies for q_l (cloud liquid) or q_i (cloud ice), and q_v (vapor) and

(d) Other chemical species: For each, separate those for turbulence and for other subgrid-scale processes.

15. Flux divergence from cumulus cloud parameterization: Convective available potential energy (CAPE) closure for CPS convective heating and convective precipitation, using different versions of Kain-Fritsch CPS (Kain and Fritsch 1990, Kain and Fritsch 1993, Kain 2004).

(a) Motion (or momentum): None

(b) Heat: CAPE removal over a convective time scale is based on model grid length for temperature tendency, and determined by the temperature difference between the updraft and the environment scaled by the convective time scale.

(c) Moisture in its three phases: CAPE removal over a convective time scale is based on model grid length for moisture tendencies, and determined by the moisture difference between the updraft and the environment scaled by the convective time scale. For cloud water, in the PSU shallow/deep convective parameterization scheme (Deng et al. 2003a,b), the subgrid-scale cloud water and ice are detrained into the environment and provide source terms for the predictive equations for cloud water/ice and cloud fraction. When the grid box becomes saturated, the PSU convective scheme converts the subgrid-scale water onto the resolved scale where the explicit microphysics takes over the job to further process the cloud water/ice.

16. Radiational flux divergence: Longwave and shortwave schemes that interact with the atmosphere, cloud and precipitation fields, and with the surface.

(a) Longwave flux divergence: RRTM (Mlawer et al. 1997) Two-stream, correlated-k method for multiple (16) bands, including effects of water vapor, carbon dioxide, and ozone.

(b) Shortwave flux divergence: Dudhia scheme (Dudhia 1989) Broadband downward integration of solar flux accounting for clear-air scattering, water vapor absorption, and cloud albedo and absorption, based on theoretical results from Stephens (1978b).

17. Stable clouds and precipitation: Simple-ice microphysics (no mixed-phase processes), either liquid or ice cloud, or rain or snow mixing depending on temperature; supersaturation removal, warm rain and ice microphysics, prognostic equations for cloud water/cloud ice and rain/snow mixing ratios (Dudhia 1993, Grell et al. 1995). PSU convective scheme (Deng et al. 2003a,b) predicts the cloud water/ice content and cloud fraction for shallow convection. It also allows the shallow clouds to produce precipitation/drizzle that can evaporate in the sub-cloud layer and affect the stability of the layer.

18. Other parameterizations: Land-surface model (LSM):
Force-restore (single-layer) LSM for ground temperature based on surface energy budget equation, sea-surface temperature specified from observations and held constant in time (Grell et al. 1995, Schroeder et al. 2006a,b).

Noah LSM, prognostic equations for canopy water and four-layer soil temperature and soil moisture (Ek et al. 2003, Chen and Dudhia 2001), coupled to PSU G-S PBL as in Reen and Stauffer (2010).

PSU shallow – deep convection scheme (Deng et al. 2003a,b): In the PSU convective scheme, convection is triggered by the vertical velocities determined by various factors including the TKE within the PBL. A plume model is used to calculate the parcel properties based on the CAPE, buoyancy, entrainment and detrainment. The convective parameterization closure is determined using a hybrid approach combining the boundary layer TKE and CAPE removal, depending on the depth of the convective updraft. In addition, there are two predictive equations for cloud water and cloud fraction, with comprehensive cloud production and dissipation processes that take into account the cloud microphysical processes. The PSU convective scheme shallow convection can also transition to a deep convection regime when the cloud layer becomes sufficiently deep. The subgrid clouds produced by the convection scheme can smoothly transition to explicit clouds when the grid cell becomes saturated. The subgrid-scale partial clouds in the PSU cloud scheme can interact with the atmospheric radiation calculation to produce more realistic heating profiles and surface radiative fluxes.

19. Summary of how these physical processes are modeled each time step at individual grid points: Forward-in-time differencing from back time step over the advection time step, substeps used for numerical stability for sound waves, precipitation fallout, and PBL tendencies; radiation and convective tendencies are recalculated after a certain number of model time steps; second-order centered finite differences in space used for physical quantities except first-order upstream is used for precipitation fallout; implicit temporal methods are used for 1D column calculations of vertical sound waves and vertical diffusion.

(a) The total diabatic heating per time step

i. From resolvable scales: Simple-ice microphysics (Dudhia 1993, Grell et al. 1995) Longwave and shortwave radiation (Mlawer et al. 1997, Dudhia 1989)

ii. From subgrid scales: Kain-Fritsch CPS (Kain and Fritsch 1990, Kain and Fritsch 1993, Kain 2004) 1.5-order (level 2.5) PSU G-S TKE PBL (Stauffer et al. 1999, Shafran et al. 2000).

20. The total moistening/drying per time step:

(a) From resolvable scales: Simple-ice microphysics (Dudhia 1993, Grell et al. 1995)

(b) From subgrid scales: Kain-Fritsch CPS (Kain and Fritsch 1990, Kain and Fritsch 1993, Kain 2004) 1.5-order (level 2.5) PSU G-S TKE PBL (Stauffer et al. 1999, Shafran et al. 2000)

21. The total acceleration/deceleration per time step:

(a) From resolvable scales: Simple-ice microphysics includes water/ice-loading terms in vertical equation of motion to reduce upward vertical motions due to precipitation drag (Grell et al. 1995, Dudhia 1993)
(b) From subgrid scales: 1.5-order (level 2.5) PSU G-S TKE PBL (Stauffer et al. 1999, Shafran et al. 2000)

22. Phenomena studied:

(a) For operational weather forecasting: Profiler, NEXGEN: General for all-terrain, all-climate, mesoscale phenomena worldwide, nowcast on the battlefield for artillery, battlespace awareness (Schroeder et al. 2006a,b, Stauffer et al. 2007a) DTRA ROFS: General for all-terrain, all-climate, mesoscale phenomena worldwide, for hazard prediction and consequence management for worldwide events (Stauffer et al. 2007a,b, 2009).
(b) For research applications: Many diverse research topics ranging from atmospheric process studies (e.g., Reen et al. 2006, 2013; Hanna et al. 2010), data assimilation methods and applications (e.g., Reen and Stauffer 2010, Schroeder et al. 2006a,b, Stauffer et al. 2007a, Otte et al. 2001, Leidner et al. 2001), ensemble modeling (e.g., Kolczynski et al. 2009, Lee et al. 2009), support of air-quality modeling and atmospheric transport and dispersion modeling (e.g., Tanrikulu et al. 2000, Deng et al. 2004, Deng and Stauffer 2006) boundary-layer, land-surface physics (e.g., Reen et al. 2006, 2013; Hanna et al. 2010), shallow and deep convection parameterization (e.g., Deng et al. 2003a,b).

23. Computer(s) used: examples of time of integration for specific model runs: http://www.mmm.ucar.edu/mm5/mpp/helpdesk/20040304a.html

Date Information Provided: August 2012
Model Name: Brazilian developments on the Regional Atmospheric Modeling System (BRAMS)
Name(s) of Developers: Saulo R. Freitas, Karla M. Longo, Enio Souza, Edmilson Freitas, Maria Assucão F. da Silva Dias, and Pedro Silva Dias.
Advanced BRAMS Developers: Marcos Longo, Demerval Moreira, Rodrigo Gevaerd, Jairo Panetta, Alvaro Fazenda, Luiz F. Rodrigues, and Gustavo G. Carrió (BRAMS-MLEF).
Organizations: Brazilian Institute for Space Research (INPE), the University of São Paulo (USP), Federal University of Campina Grande.
Website: http://brams.cptec.inpe.br/

BRAMS is derived from the Regional Atmospheric Modeling System (RAMS) and shares with the latter the same general structure, dynamic core, microphysical and radiative modules; however, BRAMS contains a set of new features:

1. Advection and transport schemes: BRAMS includes coupled "online" Eulerian transport modules (CATT) designed to study emission, deposition and transport of gases and aerosols associated with biomass burning. It uses a monotonic advection scheme (Walcek 2000; Freitas et al. 2011). CATT contains a plume rise model (Freitas et al. 2006, 2007a,b) and it is coupled to the Brazilian Biomass Burning Emission Model (3BEM, Longo et al. 2010, 2013), which provides on a daily basis, the total amount of trace gases and aerosol particles emitted by vegetation fires.
2. Convection parameterization: BRAMS includes an option to use an ensemble version of a deep and shallow cumulus scheme based on the mass flux approach (Grell and Devenyi 2002). One of the advantages of this formulation is that it can adequately describe the vertical transport of tracers when convection is parameterized.
3. Initialization of soil moisture and surface scheme: An operational offline estimation based on a remote-sensing rainfall product daily provides the initial conditions for soil moisture (Gevaerd and Freitas 2006). BRAMS uses improved land use, soil type, and normalized difference vegetative index data sets from the PROVEG project (Sestini et al. 2003), the RADAMBRASIL project (Rossato et al. 2002), and the MODIS (Moderate Resolution Imaging Spectroradiometer) data, respectively. In addition, the various biophysical parameters used for the vegetation soil scheme (common with RAMS) have been adapted for tropical and subtropical biomes and soils, using observations and/or estimations obtained during recent field campaigns.
4. Chemistry modules: CCATT-BRAMS takes into account kinetic and photochemical reactions. Photolysis rates can either be used from LUT or calculated online using the FAST-TUV (Tropospheric Ultraviolet and Visible Radiation Model) which is fully coupled with the aerosol and microphysics modules. Gas-phase chemistry uses SPACK pre-processor (RACM, RELACS, and CB07) (Freitas E.D. et al. 2005, 2007; Freitas S.R. et al. 2000, 2005, 2006, 2007, 2009, 2010, 2011, 2012).
5. Scaling of BRAMS in massive parallel machines: Continuous changes have been made in the model parallelism to benefit from the ever-increasing number of processor cores on computing clusters (among the most recent, Fazenda et al. 2011; Rodrigues et al. 2009, 2010a,b). The execution time of BRAMS version 4 originally increased from 100 to 1000 cores, due to domain decomposition (and composition) algorithms at the master process. New algorithms at BRAMS version 5 eliminated these inefficiencies. Scaling to 10,000 cores revealed new bottlenecks: I/O, dynamic load imbalance, and post-processing. These bottlenecks were removed by parallel I/O, process virtualization, and a parallel post-processor incorporated into the model's code. Currently, BRAMS version 5 scales to 10,000 cores on 5 km resolution runs over all of South America.

6. BRAMS-MLEF: The maximum likelihood ensemble filter has been recently implemented into BRAMS. The coupled system allows the optimal estimation of model parameters in addition to state variables and can be easily tailored to assimilate the statistics of a wide variety of observations.

Date Information Provided: August 2012
Model Name: The Japan Meteorological Agency nonhydrostatic model (NHM)
Is the model still under active development? Yes
Name(s) of Developers: Kazuo Saito, Teruyuki Kato, Hisaki Eito, Chiashi Muroi, Jun-ichi Ishida, Kohei Aranami, Tsukasa Fujita, Yoshinori Yamada, Tabito Hara, and Masami Narita
Organization: Japan Meteorological Agency (JMA), 1-3, Otemachi Chiyoda-ku, Tokyo, Japan and Meteorological Research Institute (MRI), 1-1, Nagamine, Ibaraki, Japan
E-mail: heito@met.kishou.go.jp, ksaito@mri-jma.go.jp

1. Group: JMA and MRI
2. Equations used for:
   (a) Motion (or momentum): fully compressible with map factors (Saito et al. 2006, 2007, Saito 2012)
      i. horizontal 4th order flux form
      ii. vertical 2nd order flux form
   (b) Heat: prognostic potential temperature
   (c) Mass: flux form
   (d) Pressure: horizontally and vertically implicit
   (e) Moisture:
      i. Ice: cloud ice, snow and graupel, double moment
      ii. Liquid water, cloud, rain
      iii. Water vapor, mixing ratio
   (f) Other chemical species:
3. Horizontal domain sizes used: 3600 km*2900 km for operational NWP
4. Horizontal grid increment(s): 5 km for operational NWP
5. Vertical domain sizes used: 22 km for operational NWP
6. Vertical grid increment(s): stretched 40 m-1080 m
7. Initialization procedure: Nonhydrostatic 4DVAR (Honda and Sawada 2008)
8. Solution techniques for dynamic core equations:
   (a) Advection: Flux form 4th order with advection correction
   (b) Vertical flux divergence
   (c) Horizontal pressure gradient force: Split-explicit
   (d) Coriolis term: considered
9. Coordinate system used: terrain-following, generalized hybrid coordinates

10. Nested domains: one-way
11. Lateral boundary condition for outermost grid: JMA global model
12. Top boundary condition: Ridged, with Rayleigh damping
13. Surface boundary treatment:
    - (a) Surface momentum fluxes: Monin-Obukov Similarity law
    - (b) Surface heat fluxes: Similarity law: Louis et al. (1982)
    - (c) Surface moisture fluxes: Similarity law: Louis et al. (1982)
    - (d) Fluxes of other chemical species:
14. Parameterization of subgrid-scale flux divergence for Mellor-Yamada-Nakanishi-Niino level 3 (Nakanishi and Niino 2006)
15. Flux divergence from cumulus cloud parameterization: Modified Kain-Fritsch (Narita 2006)
16. Radiational flux divergence: Nagasawa and Kitagawa (2005)
17. Stable clouds and precipitation
18. Other parameterizations: Mellor Yamada-Nakanishi-Niino level 3 turbulent closure model
19. Summary of how these physical processes are modeled each time step at individual grid points:
    - (a) The total diabatic heating per time step
        - i. From resolvable scales: bulk cloud microphysics
        - ii. From subgrid scales: level 3 closure model
    - (b) The total moistening/drying per time step
        - i. From resolvable scales: bulk cloud microphysics
        - ii. From subgrid scales: level 3 closure model
    - (c) The total acceleration/deceleration per time step
        - i. From resolvable scales
        - ii. From subgrid scales
20. Phenomena studied:
    - (a) For operational weather forecasting: mesoscale NWP at JMA
    - (b) For research applications: several applications
21. Computer(s) used: examples of time of integration for specific model runs HITAC SR16000, 33 hours forecast, 8 times per day for operation Earth simulator, more than one-year for regional climate study

APPENDIX

Geolocation of the Satellite Field of View[1]

This appendix address a simple geometry to determine the geolocation of the satellite field of view (FOV) through an example of a conical-scanning sensor. A conical-scanning sensor mounts a parabolic antenna on the rotation disk. The disk is rotated counter-clockwise at a constant rotation speed (rotations per minute), and an antenna is tilted at a certain angle, represented by the off-nadir viewing angle. The Earth Viewing Sector is defined as a portion of the angle where the antenna samples emerging radiance during a rotation of the antenna. Within the viewing sector, a sensor samples microwave signals for every short-time period, dt_s. Thus, the sensor rotation angle (δ) is predicted by

$$\delta = \delta_0 + \dot{\delta} dt$$

where δ_0 is the rotation angle (Unit = radiance) at epoch time, $\dot{\delta}$ (Unit = radiance per second) is rotation speed, dt is the time progress since epoch time (Fig. C.1). When predicting scanning positions, dt can be the nth step of the sampling interval.

$$dt = n \cdot dt_s$$

For a given set of δ, the sensor view angle (α), and semi-major axis (a = the distance between Earth's center and the satellite), and the Earth's radius (r_e = the distance between Earth's center and the local ellipsoid), the scanning position in the Cartesian coordinate can be predicted as follows. In this solution, the Earth's shape is treated as a spherical oblate in terms of the distance between the Earth's center and the Earth's ellipsoidal surface as a function of latitude. However, curvature assumes a local sphere of influence when estimating the scanning position vector for simplicity (Fig. C.1). This simplicity is valid, because the Earth's radius only changes ~1% from equator to pole and the typical swath width of the Earth observing satellite sensor is up to ~2000 km.

A conical scanning sensor has a constant off-nadir angle (α). The Earth incidence angle (β) can be derived via the law of sines (Fig. C.1).

$$\beta = \sin^{-1}\left(\frac{a}{r_e}\sin(\alpha)\right)$$

[1]This appendix was written by Dr. Toshi Matsui, NASA Goddard Space Flight Center, Greenbelt, MD.

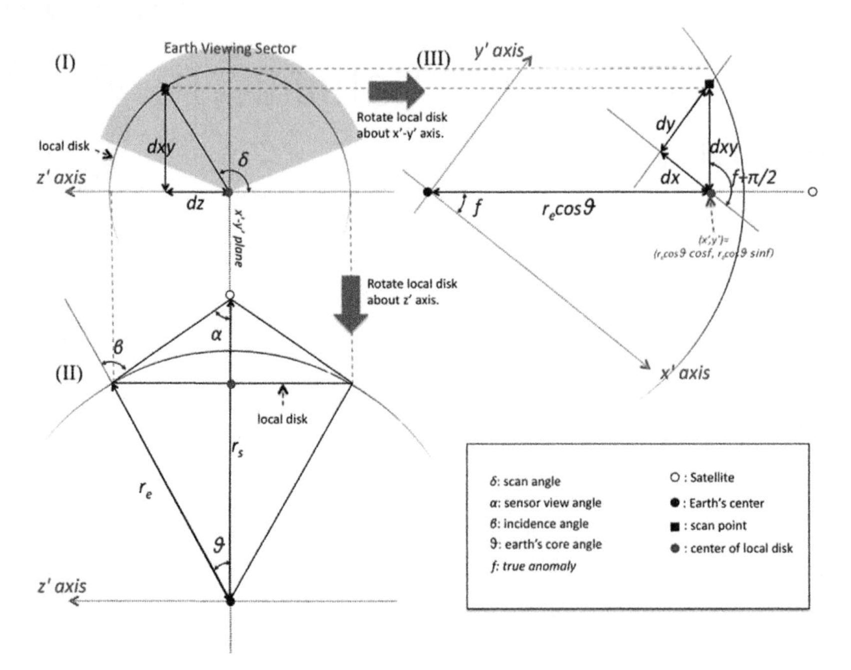

FIGURE C.1

Geometry of scanning positions on the Earth's ellipsoid in a pseudo three-dimensional Cartesian coordinate (x', y', z'). (The color version of this figure is presented in the plate section in the back of the book and the online web version.)

Thus, Earth's core angle (ϑ) can be estimated by (Fig. C.1)

$$\vartheta = \pi - \alpha - (\pi - \beta) = \beta - \alpha$$

Therefore the distance between Earth's center and the local disk center is $r_e \cos \vartheta$ and the radius of the local disk is $r_e \sin \vartheta$.

If one assumes that the orbit eclipse is placed in the pseudo $x' - y' - z'$ coordinate, the *pseudo* Cartesian position of the center of the local disk is represented by the true anomaly (f) (Figs. 12.1 and 12.2),

$$\begin{pmatrix} x' \\ y' \\ z' \end{pmatrix} = \begin{pmatrix} r_e \cos \vartheta \cos f \\ r_e \cos \vartheta \sin f \\ 0 \end{pmatrix}$$

Scanning perturbation distances are represented by dx, dy, and dz.

$$dx = dxy \cos\left(f + \frac{\pi}{2}\right)$$
$$dy = dxy \sin\left(f + \frac{\pi}{2}\right)$$
$$dz = r_e \sin \vartheta \cos \delta$$

where

$$dxy = r_e \sin\vartheta \sin\delta.$$

Thus, a scan position vector lying in the *pseudo* $x' - y' - z'$ coordinate can be represented by

$$\vec{P}_{x'-y'-z'} = \begin{pmatrix} x' \\ y' \\ z' \end{pmatrix} = \begin{pmatrix} r_e \cos\vartheta \cos f + dx \\ r_e \cos\vartheta \sin f + dy \\ 0 + dz \end{pmatrix}$$

where $\vec{P}_{x'-y'-z'}$ is rotated for the argument of perigee (ω), inclination angle (i), and the right ascension of the ascending node (Ω) by the matrix rotation method in order to estimate the position in the ECEF Cartesian coordinate.

References

Abbs, D.J., and R.A. Pielke, 1987: Numerical simulations of orographic effects on NE Colorado snowstorms. *Meteor. Atmos. Phys.,* **37**, 1-10.

Abdullah, A.J., 1966: The musical sound of tornadoes. *Mon. Wea. Rev.,* **94(4)**, 213-220.

Abe, S., and T. Yoshida, 1982: The effect of the width of a peninsula to the sea breeze. *J. Meteor. Soc. Japan,* **60**, 1074-1084.

Abele, J., and D. Clement, 1980: Chebyshev approach to fit atmospheric aerosol size distributions. *Contrib. Atmos. Phys.,* **53**, 469-485.

Achtemeier, G.L., 1983: The relationship between the surface wind field and convective precipitation over the St. Louis area. *J. Climate Appl. Meteor.,* **22**, 982-999.

Ackerman, S.A., and S.K. Cox, 1982: The Saudi Arabian heat low: Aerosol distributions and thermodynamic structure. *J. Geophys. Res.,* **87**, 8991-9002.

Ackerman, T.P., K.-N. Lion, and C.B. Levoy, 1976: Infrared radiative transfer in polluted atmospheres. *J. Appl. Meteor.,* **15**, 28-35.

Adams, J.C., A.K. Cline, M.A. Drake, and R.A. Sweet, 1975: NCAR Software Support Library, Volume I. NCAR Technical Note/1A-105, Boulder, Colorado.

Adams-Selin, R.D., S.C. van den Heever, and R.H. Johnson, 2013a: Sensitivity of bow echo simulation to microphysical parameterizations. *Wea. Forecasting,* in press, http://dx.doi.org/10.1175/WAF-D-12-00108.1.

Adams-Selin, R.D., S.C. van den Heever, and R.H. Johnson, 2013b: Impact of graupel parameterization schemes on idealized bow echo simulations. Mon. Wea. Rev., 141, 1241–1262. http://dx.doi.org/10.1175/MWR-D-12-00064.1.

Adcroft, A., C. Hill, and J., Marshall, 1997: Representation of topography by shaved cells in a height coordinate ocean model. *Mon. Wea. Rev.,* **25**, 2293-2315.

Adler, R.F., H.-Y.M. Yeh, N. Prasad, W.-K. Tao, and J. Simpson, 1991: Microwave simulations of a tropical rainfall system with a three-dimensional cloud model. *J. Appl. Meteor.,* **30**, 924-953.

Agarwal, R.K., 1981: A third-order-accurate upwind scheme for Navier-Stokes solutions at high Reynolds numbers. *AIAA 19th Aerospace Sciences Meeting*, January 12-15, 1981, St. Louis, Missouri, Paper No. AIAA-81-0112, 14 pp.

Ahlberg, J.H., E.N. Nilson, and J.L. Walsh, 1967: *Theory of Splines and Their Applications.* Academic Press, New York, 284 pp.

Mesoscale Meteorological Modeling, Third Edition, Volume 98. http://dx.doi.org/10.1016/B978-0-12-385237-3.00027-X

Akhalkatsi, M., and G. Gogoberidze, 2009: Infrasound generation by tornadic supercell storms. *Quart. J. Roy. Meteor. Soc.,* **135:641**, 935-940.

Akhalkatsi, M., and G. Gogoberidze, 2011: Spectrum of infrasound radiation from supercell storms. *Quart. J. Roy. Meteor. Soc.,* **137:654**, 229-235.

Akiyama, T., 1978: Mesoscale pulsation of convective rain in medium-scale disturbances developed in Bain front. *J. Meteor. Soc. Japan,* **56**, 267-283.

Al Nakshabandi, G., and H. Kohnke, 1965: Thermal conductivity and diffusivity of soils as related to moisture tension and other physical properties. *Agric. Meteor.,* **2**, 271-279.

Alaka, M.A., 1960: The airflow over mountains. WMO Technical Report 34, 135 pp.

Alam, J.M., and J.C. Lin, 2008: Toward a fully Lagrangian atmospheric modeling system, *Mon. Wea. Rev.,* **136(12)**, 4653-4667, http://dx.doi.org/10.1175/2008MWR2515.1.

Alapaty, K., N. Seaman, D. Niyogi, and A. Hanna, 2001: Assimilating surface data to improve the accuracy of atmospheric boundary layer simulations. *J. Appl. Meteor.,* **40**, 2068-2082.

Alberty, R.L., and S.L. Barnes, 1979: *SESAME 1979 plans for operations and data archival.* NOAA/ERL, Boulder, Colorado.

Alberty, R.L., D.W. Burgess, C.E. Hand, and J.F. Weaver, 1979: *SESAME 1979 operations summary.* NOAA/ERL, Boulder, Colorado.

Allender, J.H., 1979: Model and observed circulation throughout the annual temperature cycle of Lake Michigan. *J. Phys. Ocean.,* **9**, 573-579.

Alonso, L., G. Gangoiti, M. Navazo, M.M. Millán, and E. Mantilla, 2000: Transport of tropospheric ozone over the bay of Biscay and the eastern Cantabrian coast of Spain. *J. Appl. Meteor.,* **39**, 475-486.

Alpert, P., and J. Neumann, 1983: A simulation of a Lake Michigan winter land breeze on the 7th of November 1978. *Mon. Wea. Rev.,* **111**, 1873-1881.

Alpert, P., and T. Sholokhman, Eds., 2011: *Factor Separation in the Atmosphere, Applications and Future Prospects.* Cambridge University Press, 292 pp.

Alpert, P., A. Cohen, E. Doron, and J. Neumann, 1982: A model simulation of the summer circulation from the eastern Mediterranean past Lake Kinneret in the Jordan Valley. *Mon. Wea. Rev.,* **110**, 994-1006.

Alpert, P., M. Tsidulko, and U. Stein, 1995: Can sensitivity studies yield absolute comparisons for the effects of several processes? *J. Atmos. Sci.,* **52**, 597-601.

Alpert, P., M. Tsidulko, S. Krichak, and U. Stein, 1996a: A multi-stage evolution of an ALPEX cyclone. *Tellus,* **48A**, 209-220.

Alpert, P., S.O. Krichak, T.N. Krishnamurti, U. Stein, and M. Tsidulko, 1996b: The relative roles of lateral boundaries, initial conditions, and topography in mesoscale simulations of lee cyclogenesis. *J. Atmos. Sci.,* **35**, 1091-1099.

Alpert, P., Y.J. Kaufman, Y. Shay-El, D. Tanre, A. da Silva, S. Schubert, and J.H. Joseph, 1998: Quantification of dust-forced heating of the lower troposphere. *Nature,* **395**, 367-370.

Amiro, B.D., J.I. MacPherson, and R.L. Desjardins, 1999: BOREAS flight measurements of forest-fire effects on carbon dioxide and energy fluxes. *Agric. Forest Meteor.,* **96**, 199-208.

Anagnostopoulos, G.G., D. Koutsoyiannis, A. Christofides, A. Efstratiadis, and N. Mamassis, 2010: A comparison of local and aggregated climate model outputs with observed data. *Hydrol. Sci. J.,* **55(7)**, 1094-1110.

Anderson, G.E., 1971: Mesoscale influences on wind fields. *J. Appl. Meteor.,* **10**, 377-386.

Andersen, O.J., 1981: The katabatic wind field and nocturnal inversions in valleys: A 2-dimensional model. Report No. 1, Department of Meteorology, Univ. of Bergen, Norway.

Andersen, T., and J.M. Shepherd, 2013: A global spatio-temporal analysis of inland tropical cyclone maintenance or intensification. *Int. J. Climatol.,* http://dx.doi.org/10.1002/joc.3693.

Anderson, R.K., G.R. Farr, J.P. Ashman, A.H. Smith, and L.F. Ritter, 1974: Application of meteorological satellite data in analysis and forecasting. Volume 51 of ESSA technical report NESC, U.S. Dept. of Commerce, National Oceanic and Atmospheric Administration, National Environmental Satellite Center, Washington DC, 350 pp.

Anderson, S.F., 1979: *Effects of New York City on the horizontal and vertical structure of sea breeze fronts*. Volume II, Observations of Sea Breeze Frontal Slopes and Vertical Velocities over an Urban Area. R.D. Bornstei, P.I., Report from the Dept. of Meteorology, San Jose State University, San Jose, California.

André, J.C., G. DeMoor, P. Lacarrere, G. Therry, and R. du Vachat, 1978: Modeling the 24-hour evolution of the mean and turbulent structures of the planetary boundary layer. *J. Atmos. Sci.,* **35**, 1862-1883.

Andreas, E.L., and G. Treviño, 2000: Comments on "A physical interpretation of von Karman's constant based on asymptotic considerations–a new value". *J. Atmos. Sci.,* **57**, 1189-1192.

Andreyev, S.D., and L.S. Ivlev, 1980: Infrared radiation absorption by various atmospheric aerosol fractions. *Izv. Atmos. Oceanic Phys.,* **16**, 663-669.

Annamalai, H., K. Hamilton, and K.R. Sperber, 2007: The South Asian summer monsoon and its relationship with ENSO in the IPCC AR4 simulations. *J. Climate,* **20(6)**, 1071-1092.

Anthes, R.A., 1970: Numerical experiments with a two-dimensional horizontal variable grid. *Mon. Wea. Rev.,* **98**, 810-822.

Anthes, R.A., 1972: The development of asymmetries in a three-dimensional numerical model of the tropical cyclone. *Mon. Wea. Rev.,* **100**, 461-476.

Anthes, R.A., 1974a: Data assimilation and initialization of hurricane prediction models. *J. Atmos. Sci.,* **31**, 702-719.

Anthes, R.A., 1974b: The dynamics and energetics of mature tropical cyclones. *Rev. Geophys. Space Phys.,* **12**, 495-522.

Anthes, R.A., 1977: A cumulus parameterization scheme utilizing a one-dimensional cloud model. *Mon. Wea. Rev.,* **105**, 270-286.

Anthes, R.A., 1978: The height of the planetary boundary layer and the production of circulation in a sea breeze model. *J. Atmos. Sci.,* **35**, 1231-1239.

Anthes, R.A., 1982: *Tropical Cyclones, Their Evolution, Structure and Effect.* AMS monograph, Boston, Massachusetts.

Anthes, R.A., 1983: A review of regional models of the atmosphere in middle latitudes. *Mon. Wea. Rev.,* **111**, 1306-1335.

Anthes, R.A., and T.T. Warner, 1974: Prediction of mesoscale flows over complex terrain. U.S. Army Research Development Technical Report ECOM-5532. U.S. Army Electronics Command, White Sands Missile Range, New Mexico.

Anthes, R.A., and T.T. Warner, 1978: Development of hydrodynamic models suitable for air pollution and other mesometeorological studies. *Mon. Wea. Rev.,* **106**, 1045-1078.

Anthes, R.A., J.W. Trout, and S.L. Rosenthal, 1971: Comparisons of tropical cyclone simulations with and without the assumption of circular symmetry. *Mon. Wea. Rev.,* **99**, 759-766.

Appel, B.R., E.L. Kothny, E.M. Hoffer, G.M. Hidy, and J.J. Wesolowski, 1978: Sulfate and nitrate data from the California aerosol characterization experiment (ACHEX). *Environ. Sci. Tech.,* **12**, 418-425.

Arakawa, A., 1966: Computational design for long-term numerical integration of the equations of fluid motion: Two-dimensional incompressible flow. Part 1. *J. Comput. Phys.,* **1**, 119-143.

Arakawa, A., 1972: Design of the UCLA general circulation model. Tech. Report No. 7, Department of Meteorology, University of California, Los Angeles, 116 pp.

Arakawa, A., and V.R. Lamb, 1977: Computational design of the basic dynamical processes of the UCLA general circulation model. *Meth. Comput. Phys.,* **17**, 173-265.

Arakawa, A., and W.H. Schubert, 1974: Interaction of a cumulus cloud ensemble with the large-scale environment. Part I. *J. Atmos. Sci.,* **31**, 674-701.

Arakawa, S., K. Yamada, and T. Toya, 1982: A study of foehn in the Hokuriku district using the AMe DAS data. *Papers Meteor. Geophys.,* **33**, 149-163.

Arola, A., 1999: Parameterization of turbulent and mesoscale fluxes for heterogeneous surfaces. *J. Atmos. Sci.,* **56**, 584-598.

Arritt, R.W., and R.A. Pielke, 1986: Interactions of nocturnal slope flows with ambient winds. *Bound.-Layer Meteor.,* **37**, 183-195.

Arritt, R.W., R.T. McNider, and R.A. Pielke, 1987: Numerical model evaluation of the extension of the critical dividing streamline hypothesis to mesoscale two-dimensional terrain. *Atmos. Environ.,* **21**, 1905-1913.

Arritt, R.W., R.A. Pielke, and M. Segal, 1988: Variations of sulfur dioxide deposition velocity resulting from terrain-forced mesoscale circulations. *Atmos. Environ.,* **22**, 715-723.

Artz, R., R.A. Pielke, and J. Galloway, 1985: Comparison of the ARL/ATAD constant level and the NCAR isentropic trajectory analyses for selected case studies., *Atmos. Environ.,* **19**, 47-63.

Arya, S.P., 1988: *Introduction to Micrometeorology.* Academic Press, Harcourt Brace Jovanovich Publishers, San Diego, CA, 307 pp.

Ashby, M., 1999: Modelling the water and energy balances of Amazonian rainforest and pasture using Anglo-Brazilian Amazonian climate observation study data. *Agric. Forest Meteor.,* **94**, 79-101.

Asai, T., 1965: A numerical study of the air-mass transformation over the Japan Sea in winter. *J. Meteor. Soc. Japan,* **43**, 1-15.

Asai, T., and S. Mitsumoto, 1978: Effects of an inclined land surface on the land and sea breeze circulation: A numerical experiment. *J. Meteor. Soc. Japan,* **56**, 559-570.

Astitha, M., and G. Kallos, 2009: Gas-phase and aerosol chemistry interactions in South Europe and the Mediterranean Region. *Env. Fluid Mech.,* **9**, 3-22.

Atkinson, B.W., 1981: *Mesoscale Atmospheric Circulations.* Academic Press, New York, 495 pp.

Atlas, D., and S.-H. Chou, 1982: Coast-ocean-atmosphere-ocean mesoscale interaction. NASA Technical Memorandum 83903.

Atlas, D., S.-H. Chou, and W.P. Byerly, 1983: The influence of coastal shape on winter mesoscale air-sea interaction. *Mon. Wea. Rev.,* **111**, 245-252.

Atwater, M.A., 1971a: The radiation budget for polluted layers of the urban environment. *J. Appl. Meteor.,* **10**, 205-214.

Atwater, M.A., 1971b: Radiation effects of pollutants in the atmospheric boundary layer. *J. Atmos. Sci.,* **28**, 1367-1373.

Atwater, M.A., 1974: The radiation model. Section 4, Volume I, CEM Report No. 4131-4099. A description of a general three-dimensional numerical simulation model of a coupled air-water and/or air-land boundary layer. Center for the Environment and Man, Hartford, CT, 67-82.

Atwater, M.A., 1977: Urbanization and pollutant effects on the thermal structure in four climatic regions. *J. Appl. Meteor.,* **16**, 888-895.

Atwater, M.A., and J.T. Ball, 1981: A surface solar radiation model for cloudy atmospheres. *Mon. Wea. Rev.,* **109**, 878-895.

Atwater, M.A., and P. Brown Jr., 1974: Numerical calculation of the latitudinal variation of solar radiation for an atmosphere of varying opacity. *J. Appl. Meteor.,* **13**, 289-297.

Auer, A.H., and J.D. Marwitz, 1968: Estimates of air and moisture flux into hailstorms on the high plains. *J. Appl. Meteor.,* **7**, 196-198.

Avissar, R., 1991: A statistical-dynamical approach to parameterize subgrid-scale land-surface heterogeneity in climate models. *Surv. Geophys.,* **12**, 155-178.

Avissar, R., 1992: Conceptual aspects of a statistical-dynamical approach to represent landscape subgrid-scale heterogeneities in atmospheric models. *J. Geophys. Res.,* **97**, 2729-2742.

Avissar, R., 1995: Recent advances in the representation of land-atmosphere interactions in general circulation models. *Rev. Geophys.,* Supplement **33**, 1005-1010.

Avissar, R., and F. Chen, 1995: An approach to represent mesoscale (subgrid-scale) fluxes on GCMs demonstrated with simulations of local deforestation in Amazonia. *Space and time scale variability and interdependencies in hydrological processes,* R.A. Feddes, Ed., Press Syndicate of the University of Cambridge, UK, 89-109.

Avissar, R., and R.D. Lawford, Eds. 1999: GCIP-GEWEX Continental-Scale International Project Part II. *J. Geophys. Res.,* **104, D16**, 19275-19757.

Avissar, R., and Y. Liu, 1996: Three-dimensional numerical study of shallow convective clouds and precipitation induced by land surface forcing. *J. Geophys. Res.,* **101**, 7499-7518.

Avissar, R., and H. Pan, 2000: Simulations of the summer hydrometeorological processes of Lake Kinneret. *J. Hydrometeor.,* **1**, 95-109.

Avissar, R., and R.A. Pielke, 1989: A parameterization of heterogeneous land surfaces for atmospheric numerical models and its impact on regional meteorology. *Mon. Wea. Rev.,* **117**, 2113-2136. http://dx.doi.org/10.1175/1520-0493(1989)117<2113:APOHLS>2.0.CO;2.

Avissar, R., and R.A. Pielke, 1991: The impact of plant stomatal control on mesoscale atmospheric circulations. *Agric. Forest Meteor.,* **54**, 353-372.

Avissar, R., M.D. Moran, R.A. Pielke, G. Wu, and R.N. Meroney, 1990: Operating ranges of mesoscale numerical models and meteorological wind tunnels for the simulation of sea and land breezes. *Bound.-Layer Meteor.,* Special Anniversary Issue, Golden Jubilee, **50**, 227-275.

AWS, 1979: The use of the skew T, Log P diagram in analysis and forecasting. Air Weather Service, Scott AFB, Illinois, 62225.

Bader, D.C., 1981: Simulation of the daytime boundary layer evolution in deep mountain valleys. M.S. Thesis, Department of Atmospheric Science, Colorado State University, 100 pp.

Baer, F., and B. Katz, 1980: Normal mode analysis. Chapter 3 in: *Multidisciplinary Research Program in Atmospheric Science – A Quasi-Biennial Report,* O.E. Thompson, P.I., Ed., Department of Meteorology, University of Maryland, College Park, Maryland, 45-62.

Baer, F., and T.J. Simons, 1970: Computational stability and time truncation of coupled on linear equations with exact solutions. *Mon. Wea. Rev.,* **98**, 665-679.

Bailey, M.J., K.M. Carpenter, L.R. Lowther, and C.W. Passant, 1981: A mesoscale forecast for 14 August 1975 – the Hampstead storm. *Meteor. Mag.,* **110**, 147-161.

Baines, P.G., 1977: Upstream influence and Long's model in stratified flows. *J. Fluid Mech.,* **82**, 147-159.

Baines, P.G., 1979: Observations of stratified flows over two-dimensional obstacles in fluid of finite depth. *Tellus,* **31**, 351-371.

Baines, P.G., 1995: *Topographic Effects in Stratified Flows.* Cambridge University Press, New York, 482 pp.

Baines, P.G., and P.A. Davies, 1980: Laboratory studies of topographic effects in rotating and/or stratified fluids. GARP Series No. 23, WMO, Geneva, Switzerland, 233-299.

Baker, D.G., 1971: A study of high pressure ridges to the east of the Appalachian mountains. Ph.D. Dissertation, Massachusetts Institute of Technology, Cambridge, Massachusetts.

Baker, R.D., G. Schubert, and P.W. Jones, 1998: Cloud-level penetrative compressible convection in the Venus atmosphere. *J. Atmos. Sci.,* **55(1)**, 3-18.

Baker, R.D., B.H. Lynn, A. Boone, W.-K. Tao, and J. Simpson, 2001: The influence of soil moisture, coastline curvature, and land-breeze circulations on sea-breeze initiated precipitation. *J. Hydrometeor.,* **2**, 193-211.

Baldi, M., G.A. Dalu, R.A. Pielke Sr., and F. Meneguzzo, 2005: Analytical evaluation of mesoscale fluxes and pressure field. *Environ. Fluid Mech.,* **5**, 1-2, http://dx.doi.org/10.1007/s10652-005-8089-6, 3-33.

Baldi, M., G.A. Dalu, and R.A. Pielke Sr., 2008: Vertical velocities and available potential energy generated by landscape variability - theory. *J. Appl. Meteor. Climatol.,* **47**, 397-410, http://dx.doi.org/10.1175/2007JAMC1539.1.

Ballentine, R.J., 1980: A numerical investigation of New England coastal frontogenesis. *Mon. Wea. Rev.,* **108**, 1479-1497.

Banfield, D., B. Conrath, J.C. Pearl, M.D. Smith, and P. Christensen, 2000: Thermal tides and stationary waves on Mars as revealed by Mars Global Surveyor thermal emission spectrometer. *J. Geophys. Res.,* **105(E4)**, 9521-9537.

Bankes, S., 1993: Exploratory modeling for policy analysis. *Oper. Resh.,* **41**, 435-449.

Bannon, P.R., 1995a: Hydrostatic adjustment: Lamb's problem. *J. Atmos. Sci.,* **52**, 1743-1752.

Bannon, P.R., 1995b: Potential vorticity conservation, hydrostatic adjustment, and the anelastic approximation. *J. Atmos. Sci.,* **52**, 2302-2312.

Bannon, P.R., 1996: Nonlinear hydrostatic adjustment. *J. Atmos. Sci.,* **53**, 3606-3617.

Bannon, P.R., C.H. Bishop, and J.B. Kerr, 1997: Does the surface pressure equal the weight per unit area of a hydrostatic atmosphere? *Bull. Amer. Meteor. Soc.,* **78**, 2637-2642.

Banta, R.M., 1982: An observational and numerical study of mountain boundary layer flow. Department of Atmospheric Science Paper No. 350, Colorado State University, Fort Collins.

Banta, R.M., 1984: Daytime boundary-layer evolution over mountainous terrain, Pt. 1, Observations of the dry circulations. *Mon. Wea. Rev.,* **112**, 340-356.

Banta, R., and W. Blumen, 1990: *Atmospheric Processes Over Complex Terrain.* Meteorological Monographs (American Meteorological Society), W. Blumen, Ed., No. 45. 323 pp.

Banta, R., and W.R. Cotton, 1981: An analysis of the structure of local wind systems in a broad mountain basin. *J. Appl. Meteor.,* **20**, 1255-1266.

Banta, R.M., and P.T. Gannon Sr., 1995: Influence of soil moisture on simulations of katabatic flow. *Theor. Appl. Climatol.,* **52**, 85-94.

Banta, R.M., C.J. Senff, A.B. White, M. Trainer, R.T. McNider, R.J. Valente, S.D. Mayor, R.J. Alvarez, R.M. Hardesty, D. Parrish, and F.C. Fehsenfeld, 1998: Daytime buildup and nighttime transport of urban ozone in the boundary layer during a stagnation episode. *J. Geophys. Res.,* **103**, 22,519-22,544.

Barlage, M., X. Zeng, K. Mitchell, and H. Wei, 2005: A global 0.05-deg maximum albedo dataset of snow-covered land based on MODIS observations. *Geophys. Res. Lett.,* **32**, http://dx.doi.org/10.1029/2005GL022881.

Barnes, J.R., J.B. Pollack, R.M. Haberle, C.B. Leovy, R.W. Zurek, H. Lee, and J. Schaeffer, 1993: Mars atmospheric dynamics as simulated by the NASA Ames General Circulation Model 2. Transient baroclinic eddies. *J. Geophys. Res.,* **98(E2)**, 3125-3148.

Barnett, K.M., and R.D. Reynolds, 1981: Assessing the local wind field at Sierra Grande Mountains in New Mexico with instrumentation. Battelle Pacific Northwest Laboratory Report PNL-3623.

Baron, J.E., R.A. Simpson, G.L. Tyler, H.J. Moore, and J.K. Harmon, 1998: Estimation of Mars radar backscatter from measured surface rock populations, *J. Geophys. Res.-Planets*, **103(E10)**, 22695-22712.

Barth, E.L., 2010: Cloud formation along mountain ridges on Titan. *Planet Space Sci., 58(13)*, 1740-1747.

Barth, E.L., and S.C.R. Rafkin, 2007: TRAMS: A new dynamic cloud model for Titan's methane clouds, *Geophys. Res. Lett.,* **34(3)**, L03203.

Barth, E.L., and S.C.R. Rafkin, 2010: Convective cloud heights as a diagnostic for methane environment on Titan. *Icarus*, **206(2)**, 467-484.

Bastidas, L.A., H.V. Gupta, S. Sorooshian, W.J. Shuttleworth, and Z.L. Yang, 1999: Sensitivity analysis of a land surface scheme using multicriteria methods. *J. Geophys. Res.,* **104**, 19,481-19,490.

Basu, S., J. Wilson, M. Richardson, and A. Ingersoll, 2006: Simulation of spontaneous and variable global dust storms with the GFDL Mars GCM. *J. Geophys. Res.,* **111(E9)**, E09004.

Bates, J.R., and A. McDonald, 1982: Multiply-upstream, semi-Lagrangian advection schemes: Analysis and application to a multi-level primitive equation model. *Mon. Wea. Rev.,* **110**, 1831-1842.

Battaglia A., and S. Tanelli, 2011: Doppler Multiple Scattering Simulator (DOMUS), *IEEE Trans. Geosci. and Remote Sens.,* **49:1**, 442-450.

Battaglia, A., F. Prodi, F. Porcu, and D.-B. Shin, 2007: 3D effects in MW radiative transport inside precipitating clouds: Modeling and applications. Chapter in: *Measuring Precipitation from Space: EURAINSAT and the Future*, V. Levizzani, P. Bauer, and F.J. Turk, Eds., Springer, 113-126.

Battaglia, A., S. Tanelli, S. Kobayashi, D. Zrnic, R.J. Hogan, and C. Simmer, 2010: Multiple-scattering in radar systems: A review. *J. Quant. Spec. Rad. Transf.,* **111:6**, 917-947.

Batteen, M.L., and Y.-J. Han, 1981: On the computational noise of finite difference schemes used in ocean models. *Tellus,* **33**, 387-396.

Bauer, P., T. Auligné, W. Bell, A. Geer, V. Guidard, S. Heilliette, M. Kazumori, M.-J., Kim, E.H.-C. Liu, A.P. McNally, B. Macpherson, K. Okamoto, R. Renshaw, and L.-P. Riishøjgaard, 2011: Satellite cloud and precipitation assimilation at operational NWP centres. *Quart. J. Roy. Meteor. Soc.,* **137**, 1934-1951, http://dx.doi.org/10.1002/qj.905.

Baumhefner, D.P., and D.J. Perkey, 1982: Evaluation of lateral boundary errors in a limited-domain model. *Tellus,* **34**, 409-428.

Beard, K.V., and H.R. Pruppacher, 1971: A wind tunnel investigation of the rate of evaporation of small water droplets falling at terminal velocity in air. *J. Atmos. Sci.,* **28**, 1455-1464.

Bechtold, P., J. Cuijpers, P. Mascart, and P. Trouilhet, 1995: Modelling of trade-wind cumuli with a low-order turbulence model-toward a unified description of Cu and Sc clouds in meteorological models. *J. Atmos. Sci.,* **52**, 455-463.

Bechtold, P., E. Bazile, F. Guichard, P. Mascart, and E. Richard, 2001: A mass-flux convection scheme for regional and global models. *Quart. J. Roy. Meteor. Soc.,* **127**, 869-886.

Bedard Jr., A.J., 2005: Low-frequency atmospheric acoustic energy associated with vortices produced by thunderstorms. *Mon. Wea. Rev.,* **133**, 241-263.

Bélair, S., J. Mailhot, J.W. Strapp, and J.I. MacPherson, 1999: An examination of local versus nonlocal aspects of a TKE-based boundary layer scheme in clear convective conditions. *J. Appl. Meteor.,* **38**, 1499-1518.

Belamari, S., and A. Pirani, 2007: Validation of the optimal heat and momentum fluxes using the ORCA2-LIM global ocean-ice model. Marine EnviRonment and Security for the European Area - Integrated Project (MERSEA IP), Deliverable D4.1.3, 88 pp.

Beljaars, A.C.M., and P. Viterbo, 1998: Role of the boundary layer in a numerical weather prediction model. In: *Clear and Cloudy Boundary Layers*, A.M. Holstlag, and P.G. Duynkerke, Eds., Royal Netherlands Academy of Arts and Sciences, Amsterdam, 287-304.

Beljaars, A.C.M., P. Viterbo, M.J. Miller, and A.K. Betts, 1996: The anomalous rainfall over the United States during July 1993: Sensitivity to land surface parameterization and soil moisture anomalies. *Mon. Wea. Rev.,* **124**, 362-383.

Bell, R.S., 1978: The forecasting of orographically enhanced rainfall accumulations using 10-level model data. *Meteor. Mag.,* **107**, 113-124.

Bénard, P., J. Vivoda, J. Masek, P. Smolíková, K. Yessad, C. Smith, R. Brozková, and J.-F. Geleyn, 2010: Dynamical kernel of the Aladin-NH spectral limited-area model: Revised formulation and sensitivity experiments. *Quart. J. Roy. Meteor. Soc.,* **136**, 155-169.

Bender, A., and E.D. Freitas, 2013: Evaluation of BRAMS' Turbulence schemes during a squall line occurrence in São Paulo, Brazil. *Amer. J. Environ. Eng.,* **3:1**, 1-7. http://dx.doi.org/10.5923/j.ajee.20130301.01.

Bender, M.A., and Y. Kurihara, 1983: The energy budgets for the eye and eyewall of a numerically simulated tropical cyclone. *J. Meteor. Soc. Japan,* **61**, 239-243.

Beniston, M.G., and G. Sommeria, 1981: Use of a detailed planetary boundary model for parameterization purposes. *J. Atmos. Sci.,* **38**, 780-797.

Bennett, M., and A.E. Saab, 1982: Modelling of the urban heat island and of its interaction with pollutant dispersion. *Atmos. Environ.,* **16**, 1797-1822.

Bennetts, D.A., and J.C. Sharp, 1982: The relevance of conditional symmetric instability to the prediction of mesoscale frontal rainbands. *Quart. J. Roy. Meteor. Soc.,* **108**, 595-602.

Benoit, R., M. Desgagné, P. Pellerin, S. Pellerin, Y. Chartier, and S. Desjardins, 1997a: The Canadian MC2: A semi-Lagrangian, semi-implicit wideband atmospheric model suited for finescale process studies and simulation. *Mon. Wea. Rev.,* **125**, 2382-2415.

Benoit, R., S. Pellerin, and W. Yu. 1997b: MC2 model performance during the Beaufort and Arctic Storm Experiment. In: Numerical Methods in Atmospheric and Oceanic Modelling, The Andre J. Robert Memorial Volume, C.A. Lin, R. Laprise, and H. Ritchie, Eds., 221-244.

Benoit, R., M. Desgagné, P. Pellerin, S. Pellerin, S. Desjardins, and Y. Chartier, 1997c: The Canadian MC2: A semi-Lagrangian, semi-implicit wide-band atmospheric model suited for fine-scale process studies and simulation. *Mon. Wea. Rev.,* **125**, 2382-2415.

Benoit, R., P. Pellerin, N. Kouwen, H. Ritchie, N. Donaldson, P. Joe, and E.D. Soulis, 2000: Toward the use of coupled atmospheric and hydrologic models at regional scale. *Mon. Wea. Rev.,* **128**, 1681-1706.

Benson, J.L., P.B. James, B.A. Cantor, and R. Remigio, 2006: Interannual variability of water ice clouds over major martian volcanoes observed by MOC. *Icarus,* **184(2)**, 365-371.

Berg, W., T. L'Ecuyer, and S. van den Heever, 2008: Evidence for the impact of aerosols on the onset and microphysical properties of rainfall from a combination of satellite observations and cloud-resolving model simulations. *J. Geophys. Res.,* **113**, D14S23, http://dx.doi.org/10.1029/2007JD009649.

Bergmann, J.C., 1998: A physical interpretation of von Karman's constant based on asymptotic considerations - a new value. *J. Atmos. Sci.,* **55**, 3403-3405.

Bergmann, J.C., 2000: Reply. *J. Atmos. Sci.,* **57**, 1193-1195.

Bergstrom, R.W., 1972: Predictions of the spectral absorption and extinction coefficients of an urban air pollution aerosol model. *Atmos. Environ.,* **6**, 247-258.

Bergstrom, R.W., and R. Viskanta, 1973a: Prediction of the solar radiant flux and heating rates in a polluted atmosphere. *Tellus,* **25**, 486-498.

Bergstrom, R.W., and R. Viskanta, 1973b: Modeling of the effects of gaseous and particulate pollutants in the urban atmosphere. Part II. Pollutant dispersion. *J. Appl. Meteor.,* **12**, 913-918.

Berkofsky, L., 1977: The relation between surface albedo and vertical velocity in a desert. *Contrib. Atmos. Phys.,* **50**, 312-320.

Berkofsky, L., 1993: Comments on "Derivation of slope flow equations using two different coordinate representations". *J. Atmos. Sci.,* **50**, 1444-1445.

Bernardet, L.R., L.D. Grasso, J.E. Nachamkin, C.A. Finley, and W.R. Cotton, 2000: Simulating convective events using a high-resolution mesoscale model. *J. Geophys. Res.,* **105**, 14,963-14,982.

Berry, E.X., 1967: Cloud droplet growth by collection. *J. Atmos. Sci.,* **24**, 688-701.

Betts, A.K., 1974: Thermodynamic classification of tropical convective soundings. *Mon. Wea. Rev.,* **102**, 760-764.

Betts, A.K., 1975: Parametric interpretation of trade-wind cumulus budget studies. *J. Atmos. Sci.,* **32**, 1934-1945.

Betts, A.K., 1976: Modelling subcloud layer structure and interaction with a shallow cumulus layer. *J. Atmos. Sci.,* **33**, 2363-2382.

Betts, A.K., 1982: Saturation point analysis of moist convective overturning. *J. Atmos. Sci.,* **39**, 1484-1505.

Betts, A.K., 1986: A new convective adjustment scheme. Part I: Observational and theoretical basis. *Quart. J. Roy. Meteor. Soc.,* **112**, 677-692.

Betts, A.K., and M.J. Miller, 1986: A new convective adjustment scheme. Part II: Single column tests using GATE wave, BOMEX, ATEX, and arctic air-mass data sets. *Quart. J. Roy. Meteor. Soc.,* **112**, 693-709.

Betts, A.K., R.W. Grover, and M.W. Moncrieff, 1976: Structure and motion of tropical squall lines over Venezuela. *Quart. J. Roy. Meteor. Soc.,* **102**, 395-404.

Betts, A.K., J.H. Ball, A.C.M. Beljaars, M.J. Miller, and P. Viterbo, 1996: The land-surface-atmosphere interaction: a review based on observational and global modelling perspectives. *J. Geophys. Res.,* **101**, 7209-7225.

Betts, A.K., P. Viterbo, and A.C.M. Beljaars, 1998: Comparison of the land-surface interaction in the ECMWF reanalysis model with the 1987 FIFE Data. *Mon. Wea. Rev.,* **126**, 186-198.

Beven, K.J., 1982: On subsurface stormflow, an analysis of response times. *Hydrol. Sci. J.,* **27**, 505-521.

Beven, K.J., 1984: Infiltration into a class of vertically non-uniform soils. *Hydrol. Sci. J.,* **29**, 424-434.

Bhumralkar, C.M., 1972: An observational and theoretical study of atmospheric flow over a heated island. NSF Final Report, Grant No. GA-14156. Prepared by the Rosenstiel School of Marine and Atmospheric Science, University of Miami, Coral Gables, Florida.

Bhumralkar, C.M., 1973: An observational and theoretical study of atmospheric flow over a heated island. Parts I and II. *Mon. Wea. Rev.,* **101**, 719-745.

Biggs, W.G., and M.E. Graves, 1962: A lake breeze index. *J. Appl. Meteor.,* **1**, 474-480.

Bitan, A., 1981: Lake Kinneret (Sea of Galilee) and its exceptional wind system. *Bound.-Layer Meteor.,* **21**, 477-487.

Bjerknes, J., and H. Solberg, 1921: *Meteorological conditions for the formation of rain.* Geofysiske Publikationer, Volume 2, No. 3, Oslo, Norway, I Kommission hos Cammermeyers Boghandel, 60 pp.

Black, P.G., and G.J. Holland, 1995: The boundary layer of Tropical Cyclone Kerry (1979). *Mon. Wea. Rev.,* **123**, 2007-2008.

Black, P.G., E.A. D'Asaro, W.M. Drennan, J R. French, P.P. Niiler, T.B. Sanford, E.J. Terrill, E.J. Walsh, and J.A. Zhang, 2007: Air-sea exchange in hurricanes: Synthesis of observations from the Coupled Boundary Layer Air-Sea Transfer experiment. *Bull. Amer. Meteor. Soc.,* **88**, 357-374.

Black, T.J., 1994: The new NMC mesoscale Eta model: Description and forecast examples. *Wea. Forecasting,* **9**, 265-278.

Blackadar, A.K., 1979: High resolution models of the planetary boundary layer. *Adv. Environ. Sci. Eng.,* **I**, J.R. Pfafflin, and E.N. Ziegler, Eds., Gordon and Breach Science Publishers, 50-85.

Blackford, B.L., 1978: Wind-driven inertial currents in the Magdalen Shallows, Gulf of St. Lawrence. *J. Phys. Oceanogr.,* **8**, 655-665.

Blanchard, D.O., 1983: Variability of the convective field pattern in south Florida and its relationship to the synoptic flow. M.S. Thesis, Department of Atmospheric Science, Colorado State University, Fort Collins, Colorado.

Bleck, R., 1978: On the use of hybrid vertical coordinates in numerical weather prediction models. *Mon. Wea. Rev.,* **106**, 1233-1244.

Bleck, R., and D.B. Boudra, 1981: Initial testing of a numerical ocean circulation model using a hybrid (quasi-isopycnin) vertical coordinate. *J. Phys. Oceanogr.,* **11**, 755-770.

Bleck, R., and C. Mattocks, 1984: A preliminary analysis of the role of potential vorticity in Alpine lee cyclogenesis. *Beitr. Phys. Atmos.,* **57**, 357-368.

Blondin, C., 1978: Un modele de meso-echelle-conception-utilisation-developement. Note Technique de L'Etablissement d'Etudes et de Recherches Météorologiques, Direction de la Météorologie, Ministére des Transports, Paris.

Blossey P.N., and D.R. Durran 2008: Selective monotonicity preservation in scalar advection. *J. Comput. Phys.,* **227**, 5160-5183.

Bluestein, H.B., 1992: *Synoptic-Dynamic Meteorology in Midlatitudes: Volume I: Principles of Kinematics and Dynamics*. Oxford University Press, 448 pp.

Bluestein, H.B., 1993: *Synoptic-Dynamic Meteorology in Midlatitudes: Volume II: Observations and Theory of Weather Systems*. Oxford University Press, 608 pp.

Blumen, W., and C.D. McGregor, 1976: Wave drag by three-dimensional mountain lee-waves in nonplanar shear flow. *Tellus,* **28**, 287-298.

Blumenthal, D.L., W.H. White, and T.B. Smith, 1978: Anatomy of a Los Angeles smog episode: Pollutant transport in the daytime sea breeze regime. *Atmos. Environ.,* **15**, 893-907.

Bolton, D., 1980: The computation of equivalent potential temperature. *Mon. Wea. Rev.,* **108**, 1046-1053.

Bonan, G.B., F.S. Chapin III, and W. Wu, 1999: The impact of tundra ecosystems on the surface energy budget and climate of Alaska. *J. Geophys. Res.,* **104**, 6647-6660.

Bond, T.C., S.J. Doherty, D.W. Fahey, P.M. Forster, T. Berntsen, B.J. DeAngelo, M.G. Flanner, S. Ghan, B. Kärcher, D. Koch, S. Kinne, Y. Kondo, P.K. Quinn, M.C. Sarofim, M.G. Schultz, M. Schulz, C. Venkataraman, H. Zhang, S. Zhang, N. Bellouin, S.K. Guttikunda, P.K. Hopke, M.Z. Jacobson, J.W. Kaiser, Z. Klimont, U. Lohmann, J.P. Schwarz, D. Shindell, T. Storelvmo, S.G. Warren, and C.S. Zender, 2013: Bounding the role of black carbon in the climate system: A scientific assessment. *J. Geophys. Res.,* in press, http://dx.doi.org/10.1002/jgrd.50171.

Boone, A., J.C. Calvet, and J. Noilhan, 1999: Inclusion of a third soil layer in a land surface scheme using the force-restore method. *J. Appl. Meteor.,* **38**, 1611-1630.

Boone, A., V. Masson, T. Meyers, and J. Noilhan, 2000: The influence of the inclusion of soil freezing on simulations by a soil-vegetation-atmosphere transfer scheme. *J. Appl. Meteor.,* **39**, 1544-1569.

Bornstein, R., and Q. Lin, 2000: Urban heat islands and summertime convective thunderstorms in Atlanta: three case studies. *Atmos. Environ.,* **34**, 507-516.

Bornstein, R.D., and T. Oke, 1980: Influence of pollution and urbanization on urban climates. *Adv. Environ. Sci. Eng.,* **3**, 171-202.

Bornstein, R.D., and W.T. Thompson, 1981: Effects of frictionally retarded sea breeze and synoptic frontal passages on sulfur dioxide concentrations in New York City. *J. Appl. Meteor.,* **20**, 843-858.

Bosart, L.F., 1975: New England coastal frontogenesis. *Quart. J. Roy. Meteor. Soc.,* **101**, 957-978.

Bosart, L.F., 1981: The President's Day snowstorm of 18-19 February 1979: A subsynoptic-scale event. *Mon. Wea. Rev.,* **109**, 1542-1566.

Bosart, L.F., and F. Sanders, 1981: The Johnstown flood of July 1977: A long-lived convective system. *J. Atmos. Sci.,* **38**, 1616-1642.

Bosart, L.F., C.J. Vaudo, and J.H. Helsdon Jr., 1972: Coastal frontogenesis. *J. Appl. Meteor.,* **11**, 1236-1258.

Bosilovich, M.G., and W.-Y. Sun, 1998: Monthly simulation of surface layer fluxes and soil properties during FIFE. *J. Atmos. Sci.,* **55**, 1170-1183.

Bossert, J.E., and W.R. Cotton, 1994a: Regional-scale flows in mountainous terrain. Part I: A numerical and observational comparison. *Mon. Wea. Rev.,* **122**, 1449-1471.

Bossert, J.E., and W.R. Cotton, 1994b: Regional-scale flows in mountainous terrain. Part II: Simplified numerical experiments: *Mon. Wea. Rev.,* **122**, 1472-1489.

Bott, A., 1989a: A positive definite advection scheme obtained by nonlinear renormalization of the advective fluxes. *Mon. Wea. Rev.,* **117**, 1006-1015.

Bott, A., 1989b: Reply. *Mon. Wea. Rev.,* **117**, 2633-2636.

Bott, A., 1992: Monotone flux limitation in the area-preserving flux-form advection algorithm. *Mon. Wea. Rev.,* **120**, 2592-2602.

Böttcher, M., 1996: A semi-Lagrangian advection scheme with modified exponential splines. *Mon. Wea. Rev.,* **124**, 716-729.

Boudra, D.B., 1977: A numerical study describing regional modification of the atmosphere by the Great Lakes. University of Michigan, Ph.D. Dissertation, Ann Arbor.

Boudra, D.B., 1981: A study of the early winter effects of the Great Lakes. I: Comparison of very fine scale numerical simulations with observed data. *Mon. Wea. Rev.,* **109**, 2507-2526.

Bougeault, P., 1982: Cloud-ensemble relations based on the gamma probability distribution for the higher-order models of the planetary boundary layer. *J. Atmos. Sci.,* **39**, 2691-2700.

Bougeault, P., and P. Lacarrere, 1989: Parameterization of orography-induced turbulence in a meso-beta-scale model. *Mon. Wea. Rev.,* **117**, 1870-1888.

Bougeault, P., A. Jansa Clar, B. Benech, B. Carissimo, J. Pelon, and E. Richard, 1990: Momentum budget over the Pyrenees: The PYREX Experiment. *Bull. Amer. Meteor. Soc.,* **71**, 806-818.

Bougeault, P., B. Benech, P. Bessemoulin, B. Carissimo, A. Jansa Clar, J. Pelon, M. Petitdidier, and E. Richard, 1997: Pyrex: a summary of findings. *Bull. Amer. Meteor. Soc.,* **78**, 637-650.

Bouteloup, Y., Y. Seity, and E. Bazile, 2011: Description of the sedimentation scheme used operationally in all Météo-France NWP models. *Tellus A*, **63**, 300-311.

Bowman, H.S., and A.J. Bedard Jr., 1971: Observations of infrasound and subsonic disturbances related to severe weather. *Geophys. J. Royal Astr. Soc.*, **26**, 215-242.

Bradberry, J.S., 1981: Mesoscale structure of an Oklahoma squall line. *Mon. Wea. Rev.,* **109**, 1110-1117.

Brand, S., R.P. Chambers, W.J. Woo, J.C. Cermak, J.E. Lou, and M. Denard, 1979: A preliminary analysis of mesoscale effects of topography on tropical cyclone-associated surface winds. NAVENVPREDRESCHFAC Technical Report TR79-04. Naval Environmental Prediction Research Facility, Monterey.

Brankovíc, C., 1981: A transformed isentropic coordinate and its use in an atmospheric model. *Mon. Wea. Rev.,* **109**, 2029-2039.

Bremer, D.J., and J.M. Ham, 1999: Effect of spring burning on the surface energy balance in a tallgrass prairie. *Agric. Forest Meteor.,* **97**, 43-54.

Brewer, D.A., E.E. Remsberg, and G.E. Woodbury, 1981: A diagnostic model for studying daytime urban air-quality trends. NASA Technical Paper 1843.

Briggs, G., K. Klaasen, T. Thorpe, J. Wellman, and W. Baum, 1977: Martian dynamical phenomena during June-November 1976: Viking Orbiter imaging results. *J. Geophys. Res.,* **82(28)**, 4121-4149.

Bright, D.A., and S.L. Mullen, 2002: Short-term ensemble forecasts of precipitation during the Southwest Monsoon. *Wea. Forecasting*, **17**, 1080-1100.

Brimelow, J., J. Hanesiak, and R.L. Raddatz, 2010a: Validation of soil moisture simulations from the PAMII model, and an assessment of their sensitivity to uncertainties in soil hydraulic parameters. *Agric. Forest Meteor.,* **150**, 100-114.

Brimelow, J., J. Hanesiak, and R.L. Raddatz, and M. Hayashi, 2010b: Validation of ET estimates from the Canadian prairie agrometeorological model for contrasting vegetation types and growing seasons. *Can. Water Res. J.,* **35**, 209-230.

Brinkmann, W.A.R., 1974: Strong downslope winds at Boulder, CO. *Mon. Wea. Rev.,* **102**, 592-602.

Broder, B., H.U. Dütsch, and W. Graber, 1981: Ozone fluxes in the nocturnal planetary boundary layer over hilly terrain. *Atmos. Environ.,* **15**, 1195-1199.

Brown, J.M., 1979: Mesoscale unsaturated downdrafts driven by rainfall evaporation: A numerical study. *J. Atmos. Sci.,* **36**, 313-338.

Brown Jr., P.S., and J.P. Pandolfo, 1980: A gravity-wave problem with the upstream difference method. *J. Comput. Phys.,* **37**, 141-150.

Brown, M.E., A.H. Bouchez, and C.A. Griffith, 2002: Direct detection of variable tropospheric clouds near Titan's south pole. *Nature,* **420(6917)**, 795-797.

Browning, K.A., 1980: Local weather forecasting. *Proc. Roy. Soc. London A*, **371**, 179-211.

Browning, K.A., Ed., 1982: *Nowcasting*. Academic Press, London and New York, 256 pp.

Browning, K.A., 1986: Conceptual models of precipitation systems. *Wea. Forecasting,* **1**, 23-41.

Browning, K.A., and T.W. Harrold, 1970: Air motion and precipitation growth at a cold front. *Quart. J. Roy. Meteor. Soc.,* **96**, 369-389.

Brunke, M.A., C.W. Fairall, X. Zeng, L. Eymard, and J. Curry, 2003: Which bulk aerodynamic algorithms are least problematic in computing ocean surface turbulent fluxes? *J. Climate*, **16**, 619-635.

Brunke, M.A., M. Zhou, X. Zeng, and E.L. Andreas, 2006: An intercomparison of model bulk aerodynamic algorithms used over sea ice with data from the SHEBA experiment. *J. Geophys. Res.,* **111**, C09001, http://dx.doi.org/10.1029/2005JC002907.

Brutsaert, W., 1982a: *Evaporation into the Atmosphere: Theory, History and Applications,* D. Reidel, Norwell, MA, 299 pp.

Brutsaert, W.H., 1982b: Exchange processes at the earth-atmosphere interface. *Engineering Meteorology*, E. Plate, Ed., Elsevier, New York, 319-369.

Brutsaert, W.H., 1998: Land-surface water vapor and sensible heat flux: spatial variability, homogeneity, and measurement scales. *Water Resour. Res.,* **34**, 2433-2442.

Bryant, N.A., L.F. Johnson, A.J. Brazel, R.C. Balling, C.F. Hutchinson, and L.R. Beck, 1990: Measuring the effect of overgrazing in the Sonoran desert. *Climatic Change,* **17**, 243-264.

Bubnová, R., G. Hello, P. Bénard, and J.-F. Geleyn, 1995: Integration of the fully elastic equations cast in the hydrostatic pressure terrain-following in the framework of the ARPEGE/ALADIN NWP system. *Mon. Wea. Rev.,* **123**, 515-535.

Buckingham, E., 1914: On physically similar systems; illustrations of the use of dimensional equations. *Phys. Rev. IV*, **4**, 345.

Bugnion, V., C. Hill, and P.H. Stone, 2006: An adjoint analysis of the meridional overturning circulation in a hybrid coupled model. *J. Climate,* **19**, 3751-3767.

Burpee, R.W., 1979: Peninsula-scale convergence in the south Florida sea breeze. *Mon. Wea. Rev.,* **107**, 852-860.

Businger, J.A., 1973: Turbulent transfer in the atmosphere surface layer. Chapter 2 in *Workshop in Micrometeorology*, American Meteorological Society, Boston, Massachusetts.

Businger, J.A., J.C. Wyngaard, Y. Izumi, and E.F. Bradley, 1971: Flux-profile relationships in the atmospheric surface layer. *J. Atmos. Sci.,* **28**, 181-189.

Buzzi, A., A. Trevisan, and E. Tossi, 1985: Isentropic analysis of a case of Alpine cyclogenesis. *Beitr Phys. Atmos.,* **58**, 273-284.

Buzzi, A., A. Trevisan, S. Tibaldi, and E. Tossi, 1987: A unified theory of orographic influences upon cyclogenesis. *Meteor. Atmos. Phys.,* **36**, 1-107.

Byers, R.B., 1959: *General Meteorology*. McGraw-Hill Book Co., New York, NY, 540 pp.

Byers, H.R., 1965: *Elements of Cloud Physics.* The University of Chicago Press, Chicago, Illinois, 191 pp.

Byers, H.R., and H.R. Rodebush, 1948: Causes of thunderstorms of the Florida peninsula. *J. Meteor.,* **5**, 275-280.

Byun, D.W., 1999a: Dynamically consistent formulations in meteorological and air quality models for multiscale atmospheric studies. Part I: Governing equations in a generalized coordinate system. *J. Atmos. Sci.,* **56**, 3789-3807.

Byun, D.W., 1999b: Dynamically consistent formulations in meteorological and air quality models for multiscale atmospheric studies. Part II: Mass conservation issues. *J. Atmos. Sci.,* **56**, 3808-3820.

Cai, X.-M., and D.G. Steyn, 2000: Modelling study of sea breezes in a complex coastal environment. *Atmos. Environ.,* **34**, 2873-2885.

Calder, K.L., 1977: Multiple-source plume models of urban air pollution – their general structure. *Atmos. Environ.,* **11**, 403-414.

Camillo, P., and T.J. Schmugge, 1981: A computer program for the simulation of heat and moisture flow in soils. NASA Technical Memo 82121, NAS 5-24350. Available from NTIS.

Cantor, B.A., P.B. James, M. Caplinger, and M.J. Wolff, 2001: Martian dust storms: 1999 Mars Orbiter Camera observations, *J. Geophys. Res.,* **106(E10)**, 23653-23687.

Caracena, F., R.A. Maddox, L.R. Hoxit, and C.F. Chappell, 1979: Mesoanalysis of the Big Thompson storm. *Mon. Wea. Rev.,* **107**, 1-17.

Carbone, R.E., 1982: A severe frontal rainband. Part I: Stormwide hydrodynamics structure. *J. Atmos. Sci.,* **39**, 258-279.

Carbone, R.E., J.W. Wilson, T.D. Keenan, and J.M. Hacker, 2000: Tropical island convection in the absence of significant topography. Part I: Lifecycle of diurnally forced convection. *Mon. Wea. Rev.,* **128**, 3459-3480.

Carl, D.M., T.C. Tarbell, and H.A. Panofsky, 1973: Profiles of wind and temperature from towers over homogeneous terrain. *J. Atmos. Sci.,* **30**, 788-794.

Carlson, T.N., 1991: *Midlatitude Weather Systems.* Academic Press, New York, NY, 507 pp.

Carlson, T.N., 2012: *Mid-latitude Weather Systems.* Pennsylvania State University Press, 507 pp.

Carlson, T.N., and J.A. Augustine, 1978: Temperature mapping of land use in urban areas using satellite data. *Earth Mineral Sci.,* **47**, 41-45.

Carlson, T.N., and S.G. Benjamin, 1980: Radiative heating rates for Saharan dust. *J. Atmos. Sci.,* **37**, 193-213.

Carpenter, K.M., 1979: An experimental forecast using a nonhydrostatic mesoscale model. *Quart. J. Roy. Meteor. Soc.,* **105**, 629-655.

Carpenter, K.M., 1982a: Model forecasts for locally forced mesoscale systems. *Nowcasting*, K. Browning, Ed., Academic Press, London, 256 pp.

Carpenter, K.M., 1982b: Radiation conditions for the lateral boundaries of limited-area numerical models. *Quart. J. Roy. Meteor. Soc.,* **108**, 717-719.

Carpenter, K.M., and L.R. Lowther, 1981: Effect of varying the levels used in Florida sea-breeze simulation. British Meteorological Office (Met 0 11) Technical Note 150.

Carpenter, K.M., and L.R. Lowther, 1982: An experiment on the initial conditions for a mesoscale forecast. *Quart. J. Roy. Meteor. Soc.,* **108**, 643-660.

Carrió, G.G., and W.R. Cotton, 2010: Investigations of aerosol impacts on hurricanes: Virtual seeding flights. *Atmos. Chem. Phys.,* **11**, 2557-2567.

Carrió, G.G., and W.R. Cotton, 2011: Urban growth and aerosol effects on convection over Houston. Part II: Dependence of aerosol effects on instability. *Atmos. Res.,* http://dx.doi.org/10.1016/j.atmosres.2011.06.022.

Carrió, G.G., W.R. Cotton, and W.Y.Y. Cheng, 2010: Effects of the Urban growth of Houston on convection and precipitation. Part I: the August 2000 case. *Atmos. Res.,* **96**, 560-574.

Carroll, J.J., and R.L. Baskett, 1979: Dependence of air quality in a remote location on local and mesoscale transports: A case study. *J. Appl. Meteor.,* **18**, 474-486.

Carsel, R.F., and R.S. Parrish, 1988: Developing joint probability distributions of soil water retention characteristics. *Water Resource Res.,* **24**, 755-769.

Carson, R.B., 1954: Some objective quantitative criteria for summer showers at Miami, Florida. *Mon. Wea. Rev.,* **82**, 9-28.

Carter, W.M., J.M. Shepherd, S. Burian, and I. Jeyachandran, 2012: Integration of lidar data into a coupled mesoscale-land surface model: A theoretical assessment of sensitivity of urban-coastal mesoscale circulations to urban canopy. *J. Atmos. Oceanic Tech.,* http://dx.doi.org/10.1175/2011JTECHA1524.1.

Case, J.L., S.V. Kumar, J. Srikishen, and G.J. Jedlovec, 2011: Improving numerical weather predictions of summertime precipitation over the Southeastern United States through a high-resolution initialization of the surface state. *Wea. Forecasting*, **26**, 785-807.

Cassano, J.J., and T.R. Parish, 2000: An analysis of the nonhydrostatic dynamics in numerically simulated Antarctic katabatic flows. *J. Atmos. Sci.,* **57**, 891-898.

Castro, C.L., R.A. Pielke Sr., and G. Leoncini, 2005: Dynamical downscaling: Assessment of value retained and added using the Regional Atmospheric Modeling System (RAMS). *J. Geophys. Res., Atmospheres,* **110, No. D5,** D05108, http://dx.doi.org/10.1029/2004JD004721.

Castro, C.L., R.A. Pielke Sr., J. Adegoke, S.D. Schubert, and P.J. Pegion, 2007: Investigation of the summer climate of the contiguous U.S. and Mexico using the Regional Atmospheric Modeling System (RAMS). Part II: Model climate variability. *J. Climate*, **20**, 3866-3887.

Caughey, S.J., B.A. Crease, and W.T. Roach, 1982: A field study of nocturnal stratocumulus – II turbulence structure and entrainment. *Quart. J. Roy. Meteor. Soc.,* **108**, 125-144.

Cautenet S., D. Poulet, C. Delon, R. Delmas, J.M. Grégoire, J.M. Pereiras, S. Cherchali, O. Amram, and G. Flouzat, 1999: Simulation of carbon monoxide redistribution over Central Africa during biomass burning events (Experiment for regional Sources and Sinks of Oxidant (EXPRESSO)). *J. Geophys. Res.,* **104**, 30641-30657.

Cautenet, G., F. Guillard, B. Marticorena, G. Bergametti, F. Dulac, and J. Edy, 2000: Modeling of a Saharan dust event. *Meteor. Z.,* **9**, 221-230.

Caya, A., R. Laprise, and P. Zwack, 1998: Consequences of using the splitting method for implementing physical forcings in a semi-implicit semi-Lagrangian model. *Mon. Wea. Rev.,* **126**, 1707-1713.

Centre for Advanced Studies in Atmospheric and Fluids Science, 1983: A brief review of research and developments in atmospheric sciences. Report of the Indian Institute of Technology, New Delhi, India.

Cermak, J.E., 1970: Air motion in and near cities – determination by laboratory simulation. Report available from Fluid Dynamics and Diffusion Lab. Colorado State University, Fort Collins.

Cermak, J.E., 1971: Laboratory simulation of the atmospheric boundary layer. *AIAAJ*, **9**, 1746-1754.

Cermak, J.E., 1975: Applications of fluid mechanics to wind engineering – A Freeman Scholar lecture. *J. Fluids Eng.,* **97**, 9-38.

Cermak, J.E. 1996: Thermal effects on flow and dispersion over urban areas: Capabilities for prediction by physical modeling. *Atmos. Environ.,* **30**, 393-401.

Cerni, T.A., 1982: Comments on the ratio of diffuse to direct solar irradiance (perpendicular to the sun's rays) with clear skies – a conserved quantity throughout the day. *J. Appl. Meteor.,* **21**, 886-887.

Ceselski, B.F., 1974: Cumulus convection in weak and strong tropical disturbance. *J. Appl. Sci.,* **31**, 1241-1255.

Chaboureau, J.-P., J.-P. Cammas, P.J. Mascart, J.-P. Pinty, and J.-P. Lafore, 2002: Mesoscale model cloud scheme assessment using satellite observations. *J. Geophys. Res.,* **107**, 4301.

Chang, C., 1970: A mesoscale numerical model of airflow over the Black Hills. M.S. Thesis, South Dakota School of Mines and Technology, Rapid City.

Chang, C.B., D.J. Perkey, and C.W. Kreitzberg, 1981: A numerical case study of the squall line of 6 May 1975. *J. Atmos. Sci.,* **38**, 1601-1615.

Chang, C.H., 1977: Ice generation in clouds. South Dakota School of Mines and Technology, Department of Meteorology, M.S. Thesis, Rapid City.

Chang, J.T., and P.J. Wetzel, 1991: Effects of spatial variations of soil moisture and vegetation on the evolution of a prestorm environments: a numerical case study. *Mon. Wea. Rev.,* **119**, 1368-1390.

Chang, S.W., 1982: The orographic effects induced by an island mountain range on propagating tropical cyclones. Science Applications, Inc., McLean, Virginia.

Chang, S.W., and R.A. Anthes, 1978: Numerical simulations of the ocean's nonlinear, baroclinic response to translating hurricanes. *J. Phys. Ocean.,* **8**, 468-480.

Chang, S.W., and H.D. Orville, 1973: Large-scale convergence in a numerical cloud model. *J. Atmos. Sci.,* **30**, 947-950.

Chang, S., D. Hahn, C.-H. Yang, and D. Norquist, 1999: Validation study of the CAPS model land surface scheme using the 1987 Cabauw/PILPS dataset. *J. Appl. Meteor.,* **38**, 405-422.

Changnon Jr., S.A., 1980: Evidence of urban and lake differences on precipitation in the Chicago area. *J. Appl. Meteor.,* **19**, 1137-1159.

Changnon Jr., S.A., 1982: Visibility changes caused by St. Louis. *Atmos. Environ.,* **16**, 595-598.

Chappell, C.F., D.R. Smith, and E.C. Nickerson, 1978: Numerical simulation of clouds and snowfall over mountainous terrain. *Preprints, Conferences on Cloud Physics, Atmos. Electr.,* American Meteorological Society, Boston, 259-265.

Chapin, F.S., III, L.D. Hinzman, W. Wu, E. Lilly, G. Vourlitis, and E. Kim, 1999: Surface energy balance on the Arctic tundra: Measurements and models. *J. Climate,* **12**, 2583-2604.

Charlock, T.P., 1982: Cloud optical feedback and climate stability in a radiative-convective model. *Tellus,* **34**, 245-254.

Chase, T.N., R.A. Pielke Sr., T.G.F. Kittel, J.S. Baron, and T.J. Stohlgren, 1999: Potential impacts on Colorado Rocky Mountain weather due to land use changes on the adjacent Great Plains. *J. Geophys. Res.,* **104**, 16673-16690.

Chaudhry, F.H., and J.E. Cermak, 1971: Wind-tunnel modeling of flow and diffusion over an urban complex. Project Themis Technical Report No. 17, Fluid Dynamics and Diffusion Lab, Colorado State University, Fort Collins.

Chen, C., and H.D. Orville, 1980: Effects of mesoscale convergence on cloud convection. *J. Appl. Meteor.,* **19**, 256-274.

Chen, C.-C., D.R. Durran and G.J. Hakim, 2005: Mountain-wave momentum flux in an evolving synoptic-scale flow. *J. Atmos. Sci.,* **62**, 3213-3231.

Chen, C.-C., G.J. Hakim, and D.R. Durran 2007: Transient mountain waves and their interaction with large scales. *J. Atmos. Sci.,* **64**, 2378-2400.

Chen, D.-X., and M.B. Coughenour, 1994: GEMTM: A general model for energy and mass transfer of land surfaces and its application at the FIFE sites. *Agric. Forest Meteor.,* **68**, 145-171.

Chen, F., and R. Avissar, 1994: Impact of land-surface moisture variability on local shallow convective cumulus and precipitation in large-scale models. *J. Appl. Meteor.,* **33**, 1382-1401.

Chen, F., and J. Dudhia, 2001: Coupling an advanced land-surface/hydrology model with the Penn State/NCAR MM5 modeling system. Part I: Model implementation and sensitivity. *Mon. Wea. Rev.,* **129**, 569-585.

Chen, F., K. Mitchell, J. Schaake, Y. Xue, H.L. Pan, V. Koren, Q.Y. Duan, K. Ek, and A. Betts, 1996: Modeling of land-surface evaporation by four schemes and comparison with FIFE observations. *J. Geophys. Res.,* **101**, 7251-7268.

Chen, F., Z. Janjić, and K. Mitchell, 1997: Impact of atmospheric surface layer parameterization in the new land-surface scheme of the NCEP mesoscale Eta numerical model. *Bound.-Layer Meteor.,* **85**, 391-421.

Chen, F., R.A. Pielke Sr., and K. Mitchell, 2001: Development and application of land-surface models for mesoscale atmospheric models: Problems and promises. In *Observations and Modeling of the Land-Surface Hydrological Processes*, Water Science and Application, Vol. 3, V. Lakshmi, J. Alberston, and J. Schaake, Eds., American Geophysical Union, 107-135, http://dx.doi.org/10.1029/WS003p0107.

Chen, J.H., 1973: Numerical boundary conditions and computational modes. *J. Comput. Phys.,* **13**, 522-535.

Chen, R.-R., N.S. Berman, D.L. Boyer, and H.J.S. Fernando, 1999a: Physical model of nocturnal drainage flow in complex terrain. *Contrib. Atmos. Phys.,* **72**, 219-242.

Chen, S.-J., D.-K. Lee, Z.-Y. Tao, and Y.-H. Kuo, 1999b: Mesoscale convective system over the Yellow Sea – A numerical case study. *Meteor. Atmos. Phys.,* **70**, 185-199.

Chen, S., T.J. Campbell, H. Jin, S. Gaberšek, R.M. Hodur, and P. Martin, 2010: Effect of two-way air-sea coupling in high and low wind speed regimes. *Mon. Wea. Rev.,* **138**, 3579-3602.

Chevallier, F., F. Chéruy, N.A. Scott, and A. Chédin, 1998: A neural network approach for a fast and accurate computation of a long wave radiative budget. *J. Appl. Meteor.,* **37**, 1385-1397.

Ching, J.K.S., J.F. Clarke, and J.M. Godowitch, 1983: Modulation of heat flux by different scales of advection in an urban environment. *Bound.-Layer Meteor.,* **25**, 171-191.

Chisholm, A.J., 1970: Alberta hailstorms: A radar study and model. McGill University, Ph.D. Thesis, Montreal, Canada.

Chopra, K.P., 1973: Atmospheric and oceanic flow problems introduced by islands. *Adv. Geophys.,* **16**, 297-421.

Chou, S.-H., and D. Atlas, 1981: Estimating ocean-air heat fluxes during cold air outbreaks by satellite. NASA Technical Memorandum 83854, NASA Goddard.

Chou, S.-H., and D. Atlas, 1982: Satellite estimates of ocean-air heat fluxes during cold air outbreaks. *Mon. Wea. Rev.,* **110**, 1434-1450.

Chow, W.T.L., D. Brennan, and A.J. Brazel, 2012: Urban heat island research in Phoenix, Arizona: Theoretical contributions and policy applications. *Bull. Amer. Meteor. Soc.,* **93**, 517-530. http://dx.doi.org/10.1175/BAMS-D-11-00011.1.

Christensen, O., and L.P. Prahm, 1976: A pseudospectral model for dispersion of atmospheric pollutants. *J. Appl. Meteor.,* **15**, 1284-1294.

Chrust, M.F., C.D. Whiteman, and S.W. Hoch, 2013: Observations of thermally driven wind jets at the exit of Weber Canyon, Utah. *J. Appl. Meteor. Climatol.,* 52, 1187–1200. http://dx.doi.org/10.1175/JAMC-D-12-0221.1.

Chýlek, P., and V. Ramaswamy, 1982: Simple approximation for infrared emissivity of water clouds. *J. Atmos. Sci.,* **39**, 171-177.

Cionco, R.M., 1994: Overview of the project WIND data. In: *Mesoscale Modeling of the Atmosphere.* R.A. Pielke, and R.P. Pearce, Eds., Boston, MA, AMS, 63-71.

Cione, J.J., P.G. Black, and S.H. Houston, 2000: Surface observations in the hurricane environment. *Mon. Wea. Rev.,* **128**, 1550-1561.

Clapp, R., and G. Hornberger, 1978: Empirical equations for some soil hydraulic properties. *Water Resource Res.,* **14**, 601-604.

Clancy, R.M., J.D. Thompson, J.D. Lee, and H.E. Hurlburt, 1979: A model of mesoscale air-sea interaction in a sea breeze-coastal upwelling regime. *Mon. Wea. Rev.,* **107**, 1476-1505.

Clancy, R.T., M.J. Wolff, B.A. Whitney, B.A. Cantor, M.D. Smith, and T.H. McConnochie, 2010: Extension of atmospheric dust loading to high altitudes during the 2001 Mars dust storm: MGS TES limb observations, *Icarus,* **207(1)**, 98-109.

Clark, A.J., S.J. Weiss, J.S. Kain, I.L. Jirak, M. Coniglio, C.J. Melick, C. Siewert, R.A. Sobash, P.T. Marsh, A.R. Dean, M. Xue, F. Kong, K.W. Thomas, Y. Wang, K. Brewster, J. Gao, X. Wang, J. Du, D.R. Novak, F.E. Barthold, M.J. Bodner, J.J. Levit, C.B. Entwistle, T.L. Jensen, and J. Correia Jr., 2012: An Overview of the 2010 Hazardous Weather Testbed Experimental Forecast Program Spring Experiment. *Bull. Amer. Meteor. Soc.,* **93, 1**, 55-74.

Clark, C.A., and R.W. Arritt, 1995: Numerical simulations of the effect of soil moisture and vegetation cover on the development of deep convection. *J. Appl. Meteor.,* **34**, 2029-2045.

Clark, T.L., 1973: Numerical modeling of the dynamics and microphysics of warm cumulus convection. *J. Atmos. Sci.,* **30**, 857-878.

Clark, T.L., 1977: A small-scale dynamic model using a terrain-following coordinate transformation. *J. Comput. Phys.,* **24**, 186-215.

Clark, T.L., 1979: Numerical simulations with a three-dimensional cloud model: Lateral boundary condition experiments and multicellular severe storm simulations. *J. Atmos. Sci.,* **36**, 2191-2215.

Clark, T.L., and R.D. Farley, 1984: Severe downslope windstorm calculations in two and three spatial dimensions using anelastic interactive grid nesting: A possible mechanism for gustiness. *J. Atmos. Sci.,* **41**, 329-350.

Clark, T.L., and R. Gall, 1982: Three-dimensional numerical model simulations of airflow over mountainous terrain: A comparison with observations. *Mon. Wea. Rev.,* **110**, 766-791.

Clark, T.L., and W.R. Peltier, 1977: On the evolution and stability of finite-amplitude mountain waves. *J. Atmos. Sci.,* **34**, 1714-1730.

Clarke, R.H., R.K. Smith, and D.G. Reid, 1981: The morning glory of the Gulf of Carpentaria: An atmospheric undular bore. *Mon. Wea. Rev.,* **109**, 1726-1750.

Clements, W.E., and C.J. Nappo, 1983: Observation of a drainage flow event on a high-altitude simple slope. *J. Climate Appl. Meteor.,* **22**, 331-335.

Climate and Man, 1941: U.S. Dept. of Agriculture Handbook. US Govt. Printing Office, Washington, 1248 pp.

Colaprete, A., J.R. Barnes, R.M. Haberle, and F. Montmessin, 2008: CO_2 clouds, CAPE and convection on Mars: Observations and general circulation modeling. *Planet Space Sci.,* **56(2)**, 150-180.

Colby, F.P. Jr., 1980: The role of convective instability in an Oklahoma squall line. *J. Atmos. Sci.,* **37**, 2113-2119.

Colle, B.A., K.J. Westrick, and C.F. Mass, 1999: Evaluation of MM5 and Eta-10 precipitation forecasts over the Pacific Northwest during the cool season. *Wea. Forecasting,* **14**, 137-154.

Colle, B.A., R. Smith, and D. Wesley, 2012: Theory, Observations, and Predictions of Orographic Precipitation. Chapter 6 in *Mountain Weather Research and Forecasting: Recent Progress and Current Challenges*, Springer, 291-344.

Collier, C.G., 1975: A representation of the effects of topography on surface rainfall within moving baroclinic disturbances. *Quart. J. Roy. Meteor. Soc.,* **101**, 407-422.

Collier, C.G., 1977: The effect of model grid length and orographic rainfall efficient on computed surface rainfall. *Quart. J. Roy. Meteor. Soc.,* **103**, 247-253.

Collins, M., S.R. Lewis, and P.L. Read, 1995: Regular and irregular baroclinic waves in a martian general circulation model: A role for diurnal forcing? *Adv. Space Res.,* **16(6)**, 3-7.

Colton, D.E., 1976: Numerical simulation of the orographically induced precipitation distribution for use in hydrologic analysis. *J. Appl. Meteor.,* **15**, 1241-1251.

Comarazamy, D.E., J.E. González, J.C. Luvall, D.L. Rickman, and R.D. Bornstein, 2013: Climate impacts of land-cover and land-use changes in tropical islands under conditions of global climate change. *J. Climate*, **26**, 1535-1550. http://dx.doi.org/10.1175/JCLI-D-12-00087.1.

Committee on Urban Meteorology: Scoping the Problem, Defining the Needs, Board on Atmospheric Sciences and Climate, Division on Earth and Life Sciences, National Research Council, 2012: *Urban Meteorology: Forecasting, Monitoring, and Meeting Users' Needs*, National Academies Press, 176 pp.

Compo, G.P., J.S. Whitaker, P.D. Sardeshmukh, N. Matsui, R.J. Allan, X. Yin, B.E. Gleason, R.S. Vose, G. Rutledge, P. Bessemoulin, S. Brónnimann, M. Brunet, R.I. Crouthamel, A.N.

Grant, P.Y. Groisman, P.D. Jones, M.C. Kruk, A.C. Kruger, G.J. Marshall, M. Maugeri, H.Y. Mok, Ø. Nordli, T.F. Ross, R.M. Trigo, X.L. Wang, S.D. Woodruff, and S.J. Worley, 2011: The Twentieth Century Reanalysis Project. *Quart. J. Roy. Meteor. Soc.,* **137**, 1-28, http://dx.doi.org/10.1002/qj.776.

Conrath, B.J., 1975: Thermal structure of the Martian atmosphere during the dissipation of the dust storm of 1971. *Icarus*, 24(1), 36-46.

Cooper, H.J., M. Garstang, and J. Simpson, 1982: The diurnal interaction between convection and peninsular-scale forcing over south Florida. *Mon. Wea. Rev.,* **110**, 486-503.

Copeland, J.H., 1995, Ph.D. Dissertation: Impact of soil moisture and vegetation distribution on July 1989 climate using a regional climate model. Department of Atmospheric Science, Colorado State University, 124 pp.

Copeland, J.H., R.A. Pielke, and T.G.F. Kittel, 1996: Potential climatic impacts of vegetation change: A regional modeling study. *J. Geophys. Res.,* **101**, 7409-7418.

Costa, A.A., W.R. Cotton, R.L. Walko, and R.A. Pielke Sr., 2001: Coupled ocean-cloud-resolving simulations of the air-sea interaction over the equatorial western Pacific. *J. Atmos. Sci.,* **58**, 3357-3375.

Cotton, W.R., 1975: Theoretical cumulus dynamics. *Rev. Geophys. Space Phys.,* **13**, 419-448.

Cotton, W.R., and R.A. Anthes, 1989: *Storm and Cloud Dynamics*. Academic Press, San Diego, CA, 883 pp.

Cotton, W.R., and R.A. Pielke, 1995: *Human Impacts on Weather and Climate.* Cambridge University Press, New York, 288 pp.

Cotton, W.R., and R.A. Pielke Sr., 2007: *Human Impacts on Weather and Climate.* Cambridge University Press, 330 pp.

Cotton, W.R., and G.J. Tripoli, 1978: Cumulus convection in shear flow three-dimensional numerical experiments. *J. Atmos. Sci.,* **35**, 1503-1521.

Cotton, W.R., R.A. Pielke, and P.T. Gannon, 1976: Numerical experiments on the influence of the mesoscale circulation on the cumulus scale. *J. Atmos. Sci.,* **33**, 252-261.

Cotton, W.R., R.L. George, and K.R. Knupp, 1982a: An intense, quasi-steady thunderstorm over mountainous terrain. Part I: Evolution of the storm-initiating mesoscale circulation. *J. Atmos. Sci.,* **39**, 328-342.

Cotton, W.R., M.A. Stephens, T. Nehrkorn, and G.J. Tripoli, 1982b: The Colorado State University three-dimensional cloud/mesoscale model – 1981. Part II: An ice phase parameterization. *J. Rech. Atmos.,* **16**, 295-320.

Cotton, W.R., R.L. George, P.J. Wetzel, and R.L. McAnelly, 1983: A long-lived mesoscale convective complex. Part I—The mountain generated component. *Mon. Wea. Rev.,* **111**, 1893-1918.

Cotton, W.R., G. Thompson, and P.W. Mielke, 1994: Realtime mesoscale prediction on workstations. *Bull. Amer. Meteor. Soc.,* **75**, 349-362.

Cotton, W.R., R.A. Pielke Sr., R.L. Walko, G.E. Liston, C. Tremback, H. Jiang, R.L. McAnelly, J.Y. Harrington, M.E. Nicholls, G.G. Carrió, and J.P. McFadden, 2003: RAMS 2001: Current status and future directions. *Meteor. Atmos. Phys.,* **82**, 5-29.

Cotton, W.R., G. Bryan, and S. van den Heever, 2011: *Storm and Cloud Dynamics*, 2nd edition. Academic Press, Elsevier Publishers, The Netherlands, 820 pp.

Cotton, W.R., G.M. Krall, and G.G. Carrió, 2012: Potential indirect effects of aerosol on tropical cyclone intensity: convective fluxes and cold-pool activity. *Tropical Cyclone Res. Rev.,* **1(3)**, 293-306.

Coulson, K.L., 1975: *Solar and Terrestrial Radiation*. Academic Press, New York, 322 pp.

Coustenis, A., and B. Bézard, 1995: Titan's atmosphere from Voyager infrared observations: IV. Latitudinal variations of temperature and composition. *Icarus*, **115(1)**, 126-140.

Covez, L., 1971: Mountain waves in a turbulent atmosphere. *Tellus,* **23**, 104-109.
Cox, R., B.L. Bauer, and T. Smith, 1998: A mesoscale model intercomparison. *Bull. Amer. Meteor. Soc.,* **79**, 265-283.
Cram, J.M., R.A. Pielke Sr., and W.R. Cotton, 1992a: Numerical simulation and analysis of a prefrontal squall line. Part I: Observations and basic simulation results. *J. Atmos. Sci.,* **49**, 189-208.
Cram, J.M., R.A. Pielke Sr., and W.R. Cotton, 1992b: Numerical simulation and analysis of a prefrontal squall line. Part II: Propagation of the squall line as an internal gravity wave. *J. Atmos. Sci.,* **49**, 209-225.
Crook, N.A., 1996: Sensitivity of moist convection forced by boundary layer processes to low-level thermodynamic fields. *Mon. Wea. Rev.,* **124**, 1767-1785.
Crook, N.A., and J.B. Klemp, 2000: Lifting by convergence lines. *J. Atmos. Sci.,* **57**, 873-890.
Csanady, G.T., 1975: Lateral momentum flux in boundary currents. *J. Phys. Oceanogr.,* **5**, 705-717.
Cuijpers, J.W.M., and A.A.M. Holtslag, 1998: Impact of skewness and nonlocal effects on scalar and buoyancy fluxes in convective boundary layers. *J. Atmos. Sci.,* **55**, 151-162.
Cullen, M.J.P., 1976: On the use of artificial smoothing in Galerkin and finite difference solutions of the primitive equations. *Quart. J. Roy. Meteor. Soc.,* **102**, 77-93.
Cunning, J.B., R.L. Holle, P.T. Gannon, and A.I. Watson, 1982: Convective evolution and merger in the FACE experimental area: Mesoscale convergence and boundary layer interactions. *J. Appl. Meteor.,* **21**, 953-977.
Curry, J., and P. Webster, 1999: *Thermodynamics of Atmospheres and Oceans. Int. Geophys.,* **65**, Academic Press, 471 pp.
Cutrim, E., D.W. Martin, and R. Rabin, 1995: Enhancement of cumulus clouds over deforested lands in Amazonia. *Bull. Amer. Meteor. Soc.,* **76**, 1801-1805.
Cuxart, J., P. Bougeault, and J.-L. Redelsberger, 2000: A turbulence scheme allowing for mesoscale and large-eddy simulations. *Quart. J. Roy. Meteor. Soc.,* **126**, 1-30.
Dai, Y.J., and Q. Zeng, 1997: A land surface model (IAP94) for climate studies, Part I: formulation and validation in off-line experiments. *Adv. Atmos. Sci.,* **14**, 433-459.
Dai, Y.J., F. Xue, and Q. Zeng, 1998: A land surface model (IAP94) for climate studies, Part II: implementation and preliminary results of coupled model with IAP GCM. *Adv. Atmos. Sci.,* **15**, 47-62.
Daley, R., 1979: The application of non-linear normal mode initialization to an operational forecast model. *Atmos.-Ocean,* **17**, 97-124.
Daley, R., 1980: The development of efficient time integration schemes using model normal modes. *Mon. Wea. Rev.,* **108**, 100-110.
Daley, R., 1981: Normal mode initialization. *Rev. Geophys. Space Phys.,* **19**, 450-468.
Daley, R., and E. Barker, 2000: NAVDAS – formulation and diagnostics. *Mon. Wea. Rev.,* **129**, 869-883.
Dalu, G.A., 1978: A parameterization of heat convection for a numerical sea breeze model. *Quart. J. Roy. Meteor. Soc.,* **104**, 797-807.
Dalu, G.N., and J.S.A. Green, 1980: Energetics of diabatic mesoscale circulation: A numerical study. *Quart. J. Roy. Meteor. Soc.,* **106**, 727-734.
Dalu, G.A., and R.A. Pielke, 1989: An analytical study of the sea breeze. *J. Atmos. Sci.,* **46**, 1815-1825.
Dalu, G.A., and R.A. Pielke, 1993: Vertical heat fluxes generated by mesoscale atmospheric flow induced by thermal inhomogeneities in the PBL. *J. Atmos. Sci.,* **50**, 919-926.

Dalu, G.A., R.A. Pielke, R. Avissar, G. Kallos, M. Baldi, and A. Guerrini, 1991: Linear impact of thermal inhomogeneities on mesoscale atmospheric flow with zero synoptic wind. *Ann. Geophys.,* **9**, 641-647.

Dalu, G.A., R.A. Pielke, M. Baldi, and X. Zeng 1996: Heat and momentum fluxes induced by thermal inhomogeneities with and without large-scale flow. *J. Atmos. Sci.,* **53**, 3286-3302.

Dalu, G.A., R.A. Pielke, P.L. Vidale, and M. Baldi, 2000: Heat transport and weakening of the atmospheric stability induced by mesoscale flows. *J. Geophys. Res.,* **105**, 9349-9363.

Dalu, G.A., M. Baldi, R.A. Pielke Sr., and G. Leoncini, 2003: Mesoscale nonhydrostatic and hydrostatic pressure gradient forces: Theory and parameterization. *J. Atmos. Sci.,* **60**, 2249-2266.

Danard, M., 1977: A simple model for mesoscale effects of topography on surface winds. *Mon. Wea. Rev.,* **105**, 572-581.

Danielson, K.S., and W.R. Cotton, Eds., 1977: Space Log 1977. Colorado State University, Department of Atmospheric Science, Fort Collins, Colorado.

Darby, L.S., 2001, The sea breeze at Monterey Bay: Comparisons between lidar measurements and two-dimensional numerical simulations. Colorado State University, Department of Atmospheric Science, M.S. Thesis, 106 pp.

Das, S., D. Johnson, and W.-K. Tao, 1999: Single-column and cloud ensemble model simulations of TOGA-COARE convective systems. *J. Meteor. Soc. Japan,* **77**, 803-826.

Davey, C.A., R.A. Pielke Sr., and K.P. Gallo, 2006: Differences between near-surface equivalent temperature and temperature trends for the eastern United States - Equivalent temperature as an alternative measure of heat content. *Global and Planetary Change*, **54**, 19-32.

Davies, H.C., 1976: A lateral boundary formulation for multi-level prediction models. *Quart. J. Roy. Meteor. Soc.,* **102**, 405-418. http://dx.doi.org/10.1002/qj.49710243210.

Davis, H.C., 1983: Limitations of some common lateral boundary schemes used in regional NWP models. *Mon. Wea. Rev.,* **111**, 1002-1012.

Day, S., 1953: Horizontal convergence and the occurrence of summer precipitation at Miami, Florida. *Mon. Wea. Rev.,* **81**, 155-161.

de Jong, B., 1973: *Net Radiation Received by a Horizontal Surface at the Earth.* Delft University Press, Nijgh-Wolters-Noordhoff University Publishers, Rotterdam, The Netherlands, 99 pp.

DeRidder, K., and G. Schayes, 1997: The IAGL land surface model. *J. Appl. Meteor.,* **36**, 167-182.

de Rooy, W.C., and A.P. Siebesma, 2008: A Simple Parameterization for Detrainment in Shallow Cumulus. *Mon. Wea. Rev.,* **136**, 560-576.

de Rooy, W.C., and P.A. Siebesma, 2010: Analytical expressions for entrainment and detrainment in cumulus convection. *Quart. J. Roy. Meteor. Soc.,* **136**, 1216-1227.

de Wispelaere, C. (Ed.), 1981: *Air Pollution Modeling and Its Application.* Plenum, New York, 873 pp.

DeWekker, S.F.J., S. Zhong, J.D. Fast, and C.D. Whiteman, 1998: A numerical study of the thermally driven plain-to-basin wind over idealized basin topographies. *J. Appl. Meteor.,* **37**, 606-622.

Deardorff, J.W., 1966: The contragradient heat flux in the lower atmosphere and in the laboratory. *J. Atmos. Sci.,* **23**, 503-506.

Deardorff, J.W., 1973: Three-dimensional modeling of the planetary boundary layer. *Workshop on Micrometeorology*, A. Haugen, Ed., American Meteorological Society, Boston, 392 pp.

Deardorff, J.W., 1974: Three-dimensional numerical study of the height and mean structure of a heated planetary boundary layer. *Bound.-Layer Meteor.,* **7**, 81-106.

Deardorff, J.W., 1978: Efficient prediction of ground surface temperature and moisture, with inclusion of a layer of vegetation. *J. Geophys. Res.,* **83**, 1889-1903.

Deardorff, J.W., 1980: Stratocumulus-capped mixed layers derived from a three-dimensional model. *Bound.-Layer Meteor.,* **18**, 495-527.

Deardorff, J.W., 1981: On the distribution of mean radiative cooling at the top of a stratocumulus-capped mixed layer. *Quart. J. Roy. Meteor. Soc.,* **107**, 191-202.

Deaven, D.G., 1974: A solution for boundary problems in isentropic coordinate models. Ph.D. Dissertation, Pennsylvania State University, University Park.

Deaven, D.G., 1976: Solution for boundary problems in isentropic coordinate models. *J. Atmos. Sci.,* **33**, 1702-1713.

Decker, M., and X. Zeng, 2009: Impact of modified Richards equation on global soil moisture simulation in the Community Land Model (CLM3.5). *J. Adv. Modeling Earth Systems,* **1**, Art. #5, 22 pp., http://dx.doi.org/10.3894/JAMES.2009.1.5.

Defant, F., 1950: Theorie der land- und seewind. *Arch. Meteor. Geophys. Bioklimatol. Ser. A,* **2**, 404-425.

Defant, F., 1951: Local winds. *Compendium of Meteorology*. American Meteorological Society, Boston, MA, 655-672.

Delage, Y., and P.A. Taylor, 1970: A numerical study of heat island circulations. *Bound.-Layer Meteor.,* **1**, 201-226.

Dell'Osso, L., 1984: High-resolution experiments with the ECMWF-model: A case study. *Mon. Wea. Rev.,* **112**, 1853-1883.

Deng, A., 1999: A shallow convection parameterization scheme for mesoscale models. The Pennsylvania State Univ., Ph.D. Thesis, Department of Meteorology, 151 pp.

Deng, A., and D.R. Stauffer, 2006: On improving 4-km mesoscale model simulations. *J. Appl. Meteor.,* **45**, 361-381.

Deng, A., N.L. Seaman, and J.S. Kain, 1999: Evaluation of the Penn State shallow convection scheme in the marine atmosphere using the ASTEX first Lagrangian experiment. *Preprints, 9th PSU/NCAR MM5 Users' Workshop*, 23-24 June, Boulder, CO, 4 pp.

Deng, A., N.L. Seaman, and A.M. Lario-Gibbs, 2000: A shallow convection scheme of 3-D regional scale air quality applications. *11th AMS Conference on Applications of Air Pollution Meteorology with A&WMA*, Long Beach, CA, 9-14 January, 102-106.

Deng, A., N.L. Seaman, and J.S. Kain, 2003a: A shallow-convection parameterization for mesoscale models Part I: Sub-model description and preliminary applications. *J. Atmos. Sci.,* **60**, 34-56.

Deng, A., N.L. Seaman, and J.S. Kain, 2003b: A shallow-convection parameterization for mesoscale models Part II: Verification and sensitivity studies. *J. Atmos. Sci.,* **60**, 57-78.

Deng, A., N.L. Seaman, G.K. Hunter and D.R. Stauffer, 2004: Evaluation of interregional transport using the MM5-SCIPUFF system. *J. Appl. Meteor.,* **43**, 1864-1886.

Derickson, R.G., 1974: Three dimensional modeling of cold orographic cloud systems. *Preprints, AMS Conference on Cloud Physics*, Tucson, AZ, October 21-24, 227-232.

Derickson, R.G., 1992: Finite difference methods in geophysical flow simulations. Colorado State University, Ph.D. Dissertation, Department of Civil Engineering, 320 pp.

Derickson, R.G., and R.A. Pielke, 2000: A preliminary study of Burgers Equation with symbolic computation. *J. Comput. Phys.,* **162**, 219-244.

Dickerson, M.H., 1978: MASCON - A mass-consistent atmospheric flux model for regions with complex terrain. *J. Appl. Meteor.,* **17**, 241-253.

Dickerson, M.H., and P.H. Gudiksen, 1980: ASCOT-FY-1979 progress report. UCRL-42899/ascot/ 1.Lawrence Livermore National Laboratory, Livermore, California. Available from NTIS.

Dickerson, M.H., and P.H. Gudiksen, 1981: ASCOT-FY-1981 progress report. Lawrence Livermore National Laboratory Report UCID-18878-81, ASCOT-2, Livermore, California.

Dickinson, R.E., 1984: Modeling evapotranspiration for three-dimensional global climate models. *AGU Geophysical Monograph, Maurice Ewing Vol. 5.*, **29**, 58-72.

Dickison, R.B.B., M.J. Haggis, and R.C. Rainey, 1983: Spruce budworm moth flight and storms: Case study of a cold front system. *J. Climate Appl. Meteor.,* **22**, 278-286.

Ding, Y., J. Zhang, and Z. Zhao, 2000: An improved land-surface process model and its simulation experiment - Part II: coupling simulation experiment of land-surface process model with regional climate model. *Acta Meteorologica Sinica,* **14**, 30-45.

Dirmeyer, P.A., 1994: Vegetation stress as a feedback mechanism in midlatitude drought. *J. Climate,* **7**, 1463-1483.

Dirmeyer, P.A., F.J. Zeng, A. Ducharne, J.C. Morrill, and R.D. Koster, 2000: The sensitivity of surface fluxes to soil water content in three land surface schemes. *J. Hydrometeorol.,* **1**, 121-134.

Dolman, A.J., A. Verhagen, and C.A. Rovers, Eds., 2003: *Global Environmental Change and Land Use*. Kluwer Academic Publishers, 210 pp.

Doneaud, A.A., J.R. Miller Jr., D.L. Priegnitz, and L. Viswanath, 1983: Surface mesoscale features as potential storm predictors in the northern Great Plains – two case studies. *Mon. Wea. Rev.,* **111**, 273-293.

Donelan, M.A., B.K. Haus, N. Reul, W.J. Plant, M. Stiassnie, H.C. Graber, O.B. Brown, and E.S. Saltzman, 2004: On the limiting aerodynamic roughness of the ocean in very strong winds. *Geophys. Res. Lett.,* **31**, L18306, http://dx.doi.org/10.1029/2004GL019460.

Donovan, David, and Coauthors, 2004: The EarthCARE Simulator: Users guide and final report. European Space Agency Contract 15346/01/NL/MM, 198 pp.

Doran, E., 1979: Objective analysis of mesoscale flow fields in Israel and trajectory calculations. *Israel J. Earth Sci.,* **28**, 33-41.

Doran, E., and J. Neumann, 1977: Land and mountain breezes with special attention to Israel's Mediterranean coastal plain. G. Steinitz Memorial Volume, *Israel Meteor. Res. Papers*, **1** 109-122.

Doran, J.C., and T.W. Horst, 1983: Observations and models of simple nocturnal slope flows. *J. Atmos. Sci.,* **40**, 708-717.

Doran, J.C., and S. Zhong, 2000: Thermally driven gap winds into the Mexico City Basin. *J. Appl. Meteor.,* **39**, 1330-1340.

Douville, H., P. Viterbo, J.-F. Mahfouf, and A.C.M. Beljaars, 1999: Evaluation of the optimum interpolation and nudging techniques for soil moisture analysis using FIFE data. *Mon. Wea. Rev.,* **128**, 1733-1756.

Doyle, J.D., 1995: Coupled ocean wave/atmosphere mesoscale model simulations of cyclogenesis. *Tellus,* **47A**, 766-788.

Doyle, J.D., 2002: Coupled atmosphere-ocean wave simulations under high wind conditions. *Mon. Wea. Rev.,* **130**, 3087-3099.

Doyle, J.D., and D.R. Durran, 2002: The dynamics of mountain-wave induced rotors. *J. Atmos. Sci.,* **59**, 186-201.

Doyle, J.D., and D.R. Durran, 2007: Rotor and sub-rotor dynamics in the lee of three-dimensional terrain. *J. Atmos. Sci.,* **64**, 4202-4221.

Doyle, J.D., D.R. Durran, C. Chen, B.A. Colle, M. Georgelin, V. Grubisic, W.R. Hsu, C.Y. Huang, D. Landau, Y.L. Lin, G.S. Poulos, W.Y. Sun, D.B. Weber, M.G. Wurtele, and M. Xue,

2000: An intercomparison of model-predicted wave breaking for the 11 January 1972 Boulder windstorm. *Mon. Wea. Rev.,* **128**, 901-914.

Doyle, J.D., V. Grubišić, W.O.J. Brown, S.F.J. De Wekker, A. Dörnbrack, Q. Jiang, S.D. Mayor, M. Weissmann, 2009: Observations and numerical simulations of subrotor vortices during T-REX. *J. Atmos. Sci.,* **66**, 1229-1249.

Doyle, J.D., Carolyn A. Reynolds, Clark Amerault, Jonathan Moskaitis, 2012a: Adjoint sensitivity and predictability of tropical cyclogenesis. *J. Atmos. Sci.,* **69**, 3535-3557.

Doyle, J.D., Y. Jin, R. Hodur, S. Chen. H. Jin, J. Moskaitis, A. Reinecke, P. Black, J. Cummings, E. Hendricks, T. Holt, C. Liou, M. Peng, C. Reynolds, K. Sashegyi, J. Schmidt, and S. Wang, 2012b: Real time tropical cyclone prediction using COAMPS-TC. *Adv. Geophys.,* **28**, 15-28.

Draine, B.T., and P.J. Flatau, 1994: Discrete-dipole approximation for scattering calculations. *J. Opt. Soc. Amer.,* **11**, 1491-1499.

Draxler, R.R., and G.D. Hess, 1998: An overview of the HYSPLIT_4 modelling system for trajectories, dispersion, and deposition. *Aust. Meteorol. Mag.,* **47**, 295-308.

Driscoll, S., A. Bozzo, L.J. Gray, A. Robock, and G. Stenchikov, 2012: Coupled Model Intercomparison Project 5 (CMIP5) simulations of climate following volcanic eruptions. *J. Geophys. Res.,* **117**, D17105, http://dx.doi.org/10.1029/2012JD017607.

Droegemeier, K.K., J.D. Smith, S. Businger, C. Doswell III, J. Doyle, C. Duffy, E. Foufoula-Georgiou, V. Krajewski, M. Lemone, C. Mass, R. Pielke Sr., P. Ray, S. Rutledge, and E. Zipser, 2000: Hydrological aspects of weather prediction and flood warnings: Report of the Ninth Prospectus Development Team of the U.S. Weather Research Program. *Bull. Amer. Meteor. Sci.,* **81**, 2665-2680.

Druilhet, A., A. Herrada, J.-P. Pages, J. Saissai, C. Allet, and M. Ravaut, 1982: Etude expérimentale de la couche lmite interne associée à la brise de mer. *Bound.-Layer Meteor.,* **22**, 511-524.

Dudhia, J., 1989: Numerical study of convection observed during the Winter Monsoon Experiment using a mesoscale two-dimensional model. *J. Atmos. Sci.,* **46**, 3077-3107.

Dudhia, J., 1993: A nonhydrostatic version of the Penn State/NCAR mesoscale model: Validation tests and simulations of an Atlantic cyclone and cold front. *Mon. Wea. Rev.,* **121**, 1493-1513.

Dudhia, J., and M.W. Moncrieff, 1989: A three-dimensional numerical study of an Oklahoma squall line containing right-flank supercells. *J. Atmos. Sci.,* **46**, 3363-3391.

Durran, D.R., 1981: The effects of moisture on mountain lee waves. NCAR Cooperative Ph.D. Thesis No. 65 with the Massachusetts Institute of Technology.

Durran, D.R., 1986: Another look at downslope wind storms, Pt. 1: Development of analogs to supercritical flow in an infinitely deep continuously stratified fluid. *J. Atmos. Sci.,* **43**, 2527-2543.

Durran, D.R., 1991: The third-order Adams-Bashford method: An attractive alternative to leapfrog time differencing. *Mon. Wea. Rev.,* **119**, 702-720.

Durran, D.R., 2000: Comments on 'The differentiation between grid spacing and resolution and their application to numerical modeling.' *Bull. Amer. Meteor. Soc.,* **81**, 2478.

Durran, D.R., and J.B. Klemp, 1982a: On the effects of moisture on the Brunt-Väisälä frequency. *J. Atmos. Sci.,* **39**, 2152-2158.

Durran, D.R., and J.B. Klemp, 1982b: The effects of moisture on trapped mountain lee waves. *J. Atmos. Sci.,* **26**, 241-254.

Dutton, J.A., 1976: *The Ceaseless Wind: An Introduction to the Theory of Atmospheric Motion.* McGraw Hill, New York, 579 pp.

Dutton, J.A., 2002: *The Ceaseless Wind: An Introduction to the Theory of Atmospheric Motion.* Courier Dover Publications, 640 pp.

Dutton, J.A., and G.H. Fichtl, 1969: Approximate equations of motion for gases and liquids. *J. Atmos. Sci.,* **26**, 241-254.

Dyer, A.J., 1974: A review of flux-profile relationships. *Bound.-Layer Meteor.,* **42**, 9-17.

Dyer, A.J., and E.F. Bradley, 1982: An alternative analysis of flux-gradient relationships at the 1976 ITCE. *Bound.-Layer Meteor.,* **22**, 3-19.

Eastman, J.L., 1995: Numerical simulation of Hurricane Andrew - Rapid intensification. *21st Conference on Hurricanes and Tropical Meteorology*, 24-28 April 1995, Miami, Florida, AMS, Boston, 111-113.

Eastman, J.L. R.A. Pielke, and W.A. Lyons, 1995: Comparison of lake-breeze model simulations with tracer data. *J. Appl. Meteor.*, **34**, 1398-1418.

Eastman, J.L., M.E. Nicholls, and R.A. Pielke, 1996: A numerical simulation of Hurricane Andrew. *Second International Symposium on Computational Wind Engineering*, 4-8 August 1996, Fort Collins, CO.

Eastman, J.L., M.B. Coughenour, and R.A. Pielke, 2001a: The effects of CO_2 and landscape change using a coupled plant and meteorological model. *Global Change Biology*, 7, 797-815.

Eastman, J.L., M.B. Coughenour, and R.A. Pielke, 2001b: Does grazing affect regional climate? *J. Hydrometeor.,* **2**, 243-253.

Ebert, E., and J.A. Curry, 1992: A parameterization of ice cloud optical properties for climate models. *J. Geophys. Res.,* **97**, 3831-3835.

Eidenshink, J.C., and R.H. Haas, 1992: Analyzing vegetation dynamics of land systems with satellite data. *Geocarto Intl.,* **1**, 53-61.

Egan, B.A., 1975: Turbulent diffusion in complex terrain. In: *Lectures on Air Pollution and Environmental Impact Analysis*, D. Haugen, Ed., American Meteorological Society, Boston, Massachusetts, 112-135.

Egger, J., 1981: On the linear two-dimensional theory of thermally induced slope winds. *Contrib. Atmos. Phys.,* **54**, 465-481.

Egger, J., 1999: Inertial oscillations revisited. *J. Atmos. Sci.,* **56**, 2951-2954.

Ek, M.B., K.E. Mitchell, Y. Lin, E. Rogers, P. Grummann, V. Koren, G. Gayno, and J.D. Tarpley, 2003: Implementation of Noah land surface model advances in the National Center for Environmental Prediction operational mesoscale Eta model. *J. Geophys. Res.,* **108**, 8851, http://dx.doi.org/10.1029/2002JD003296.

Eliassen, A., 1980: A review of long-range transport modeling *J. Appl. Meteor.,* **19**, 231-240.

Eliassen, A., and E. Palm, 1960: On the transfer of energy in stationary mountain waves. *Geophys. Norv.*, **22**, 1-23.

Eliasson, S., S.A. Buehler, M. Milz, P. Eriksson, and V.O. John, 2010: Assessing modelled spatial distributions of ice water path using satellite data. *Atmos. Chem. Phys. Discuss.*, **10**, 12185-12224.

Ellehoj, M.D., H.P. Gunnlaugsson, K.M. Bean, B.A. Cantor, L. Drube, D. Fisher, B.T. Gheynani, A-M. Harri, C. Holstein-Rathlou, H. Kahanpää, M.T. Lemmon, M.B. Madsen, M.C. Malin, J. Polkko, P. Smith, L.K. Tamppari, P.A. Taylor, W. Weng, and J. Whiteway, 2010: Convective vortices and dust devils at the Phoenix Mars mission landing site. *J. Geophys. Res.*, **115**, E00E16, http://dx.doi.org/10.1029/2009JE003413.

Ellenton, G.E., and M.B. Danard, 1979: Inclusion of sensible heating in convective a parameterization applied to lake-effect snow. *Mon. Wea. Rev.,* **107**, 551-565.

Elsberry, R.L., 1978: Prediction of atmospheric flows on nested grids. In: *Computational Techniques for Interface Problems*, K.C. Park, and D.K. Gartling, Eds., ADM-Volume 30, American Society of Mechanical Engineering, New York, 67-88.

Eltahir, E.A.B., 1996: Role of vegetation in sustaining large-scale atmospheric circulations in the tropics. *J. Geophys. Res.,* **101**, 4255-4268.

Emanuel, K.A., 1982a: Forced and free mesoscale motions in the atmosphere. *Proceedings of the CIMMS Symposium*, Norman, Oklahoma, May 12-16, 1980.

Emanuel, K.A., 1982b: Inertial instability and mesoscale convective systems. Part II: Symmetric CISK in a baroclinic flow. *J. Atmos. Sci.,* **39**, 1080-1097.

Emanuel, K.A., 1991: The theory of hurricanes. *Ann. Rev. Fluid Mech.*, **23(1)**, 179-196, http://dx.doi.org/10.1146/annurev.fl.23.010191.001143.

Emanuel, K.A., 2005: *Divine Wind: The History and Science of Hurricanes*. Oxford University Press, 296 pp.

Emery, K.O, and G.T. Csanady, 1973: Surface circulation of lakes and nearly land-locked seas. *Proceedings of the National Academy of Science USA*, **70**, 93-97.

Emori, S., 1998: The interaction of cumulus convection with soil moisture distribution: An idealized simulation. *J. Geophys. Res.,* **103**, 8873-8884.

Entekhabi, D., 1995: Recent advances in land-atmosphere interaction research 1995. *Rev. Geophys.,* Supplement **33**, 995-1003.

Entekhabi, D., I. Rodriguez-Iturbe, and R.L. Bras, 1992: Variability in large-scale water balance with land surface-atmosphere interaction. *J. Climate,* **5**, 798-813.

Entin, J.K., A. Robock, K.Y. Vinnikov, V. Zabelin, S. Liu, A. Namkhai, and Ts. Adyasuren, 1999: Evaluation of Global Soil Wetness Project soil moisture simulations. *J. Meteor. Soc. Japan,* **77**, 183-198.

EPA, 1980: Guideline for applying the Airshed Model to urban areas. EPA Report 450/4-80-020, October 1980.

Epifanio, C.C., and D.R. Durran, 2002: Lee vortex formation in free-slip stratified flow over ridges. Part II: Mechanisms of vorticity and PV production in nonlinear viscous wakes. *J. Atmos. Sci.,* **59**, 1166-1181.

Errico, R.M., P. Bauer, and J.-F. Mahfouf, 2007: Issues regarding the assimilation of cloud and precipitation data. *J. Atmos. Sci.,* **64**, 3785-3798.

Estoque, M.A., 1961: A theoretical investigation of the sea breeze. *Quart. J. Roy. Meteor. Soc.,* **87**, 136-146.

Estoque, M.A., 1962: The sea breeze as a function of prevailing synoptic situation. *J. Atmos. Sci.,* **19**, 244-250.

Estoque, M.A., 1973: Numerical modeling of the planetary boundary layer. *Workshop in Micrometeorology*, Chapter VI, American Meteorological Society, Boston, Massachusetts.

Estoque, M.A., and C.M. Bhumralkar, 1969: Flow over a localized heat source. *Mon. Wea. Rev.,* **97**, 850-859.

Estoque, M.A., and J.M. Gross, 1981: Further studies of a lake breeze, Part II: Theoretical study. *Mon. Wea. Rev.,* **109**, 619-634.

Estoque, M.A., and K. Ninomiya, 1976: Numerical simulation of Japan Sea effect snowfall. *Tellus,* **28**, 243-253.

Estoque, M.A., J. Gross, and H.W. Lai, 1976: A lake breeze over southern Lake Ontario. *Mon. Wea. Rev.,* **104**, 386-396.

Estoque, M.A., S.V. Almazan, and J.C. Mondares, 1994: A sea breeze rainfall model. *Atmosfera,* **7**, 221-240.

Evans, K.F., and G.L. Stephens, 1991: A new polarized atmospheric radiative transfer model. *J. Quant. Spectrosc. Radiat. Transfer*, **46**, 413-423.

Fairall, C.W., E.F. Bradley, D.P. Rogers, J.B. Edson, and G.S. Young, 1996: Bulk parameterization of air - sea fluxes in TOGA COARE. *J. Geophys. Res.,* **101(C2)**, 3747- 3767.

Fairall, C.W., E.F. Bradley, J.E. Hare, A.A. Grachev, and J.B. Edson, 2003: Bulk parameterization of air-sea fluxes: Updates and verification for the COARE algorithm. *J. Climate,* **16(4)**, 571- 591.

Fairall, C.W., M.L. Banner, W.L. Peirson, W. Asher, and R.P. Morison, 2009: Investigation of the physical scaling of sea spray spume droplet production. *J. Geophys. Res.,* **114**, C10001, http://dx.doi.org/10.1029/2008JC004918.

Fall, S., N. Diffenbaugh, D. Niyogi, R.A. Pielke Sr., and G. Rochon, 2010: Temperature and equivalent temperature over the United States (1979-2005). *Int. J. Climatol.,* http://dx.doi.org/10.1002/joc.2094.

Famiglietti, J.S., and E.F. Wood, 1991: Evapotranspiration and runoff from large land areas: Land surface hydrology for atmospheric General Circulation Models. *Surv. Geophys.*, **12**, 179-204.

Fanelli, P.F., and P.R. Bannon, 2005: Nonlinear adjustment to thermal forcing. *J. Atmos. Sci.,* **62**, 4253-4272.

Fankhauser, J.C., 1971: Thunderstorm-environment interactions determined from the aircraft and radar observations. *Mon. Wea. Rev.,* **99**, 171-192.

Fazenda, A.L., J. Panetta, D.M. Katsurayama, L.F. Rodrigues, and L.F.G. Motta, 2011: Challenges and solutions to improve the scalability of an operational regional meteorological forecasting model. *Int. J. High Perf. Sys. Architecture*, **3**, Nbr 2/3.

Feingold, G., and S.M. Kreidenweis, 2000: Does heterogeneous processing of aerosol increase the number of cloud droplets? *J. Geophys. Res.,* **105**, 24351-24361.

Feliks, Y., and A. Huss, 1982: Spurious mass loss in some mesometeorological models. *Bound.-Layer Meteor.,* **24**, 387-391.

Fels, S.B., and M.D. Schwarzkopf, 1975: The simplified exchange approximation: A new method for radiative transfer calculation. *J. Atmos. Sci.,* **32**, 1475-1488.

Fenton, L.K., A.D. Toigo, and M.I. Richardson, 2005: Aeolian processes in Proctor Crater on Mars: Mesoscale modeling of dune-forming winds. *J. Geophys. Res.*, **110(E6)**, E06005.

Fenton, L.K., P.E. Geissler, and R.M. Haberle, 2007: Global warming and climate forcing by recent albedo changes on Mars. *Nature*, **446(7136)**, 646-649.

Fernandez, W., 1982: Environmental conditions and structure of the West African and eastern tropical Atlantic squall lines. *Arch. Meteor. Geophys. Bioklimatol. Ser. A*, **31**, 71-89.

Ferri, F., P.H. Smith, M. Lemmon, and N.O. Rennó, 2003: Dust devils as observed by Mars Pathfinder. *J. Geophys. Res.*, **108(E12)**, 5133.

Ferrier, B.S., 2005: An efficient mixed-phase cloud and precipitation scheme for use in operational NWP models. *Eos, Trans. Amer. Geophys. Union,* **86(18)**, Jt. Assem. Suppl., A42A-02.

Feser, F., B. Rockel, H. von Storch, J. Winterfeldt, and M. Zahn, 2011: Regional climate models add value to global model data—A review and selected examples. *Bull. Amer. Meteor. Soc.*, **92**, 1181-1192, http://dx.doi.org/10.1175/2011BAMS3061.1.

Finkele, K., 1998: Inland and offshore propagation speeds of a sea breeze from simulations and measurements. *Bound.-Layer Meteor.,* **87**, 307-329.

Finley, C.A., W.R. Cotton, and R.A. Pielke, 2001: Numerical simulation of tornadogenesis in a high-precipitation supercell: Part I: Storm evolution and transition into a bow echo. *J. Atmos. Sci.,* **58**, 1597-1629.

Fisher, E.L., 1961: A theoretical study of the sea breeze. *J. Meteor.,* **18**, 215-233.

Fisher, J.A., M.I. Richardson, C.E. Newman, M.A. Szwast, C. Graf, S. Basu, S P. Ewald, A.D. Toigo, and R.J. Wilson, 2005: A survey of Martian dust devil activity using Mars Global Surveyor Mars Orbiter Camera images. *J. Geophys. Res.*, **110(E3)**, E03004.

Foken, T., and Skeib, G., 1983: Profile measurement in the atmospheric near-surface layer and the use of suitable universal functions for the determination of the turbulent energy exchange. *Boundary-Layer Meteorol.*, **25**, 55-62.

Fontana, P.H., 1979: Effects of New York City on the horizontal and vertical structure of sea breeze fronts. Volume 1: Observations of frictional retardation of sea breeze fronts. R.D. Bornstein, P.I., Report from Department of Meteorology, San Jose State University, San Jose, CA.

Foote, G.B., and J.C. Fankhauser, 1973: Airflow and moisture budge beneath a northeast Colorado hailstorm. *J. Appl. Meteor.,* **12**, 1330-1353.

Fortunato, A.B., and A.M. Baptista, 1996: Evaluation of horizontal gradients in sigma-coordinate shallow water models. *Atmos.-Ocean,* **34**, 489-514.

Fortune, M., 1980: Properties of African squall lines inferred from time-lapse satellite imagery. *Mon. Wea. Rev.,* **108**, 153-168.

Fortune, M., 1989: The evolution of the vortical patterns and vorticals in mesoscale convective complexes. Colorado State University, Ph.D. Dissertation, Atmospheric Science Department, Fort Collins, CO, 183 pp.

Fosberg, M.A., W.E. Marlatt, and L. Krupnak, 1976: Estimating air-flow patterns over complex terrain. U.S. Forest Service, Rocky Mountain Forest Range Experiment Station Research Paper RM-162, 1-16.

Foufoula-Georgiou, E., and P. Kumar, Eds., 1994: *Wavelets in Geophysics.* Academic Press, New York, 373 pp.

Fouquart, Y., and B. Bonnel, 1980: Computations of solar heating of the earth's atmosphere: A new parameterization. *Beitr. Phys. Atmos.,* **53**, 35-62.

Fox, D.G., and J.W. Deardorff, 1972: Computer methods for simulation of multidimensional, nonlinear, subsonic, incompressible flow. *J. Heat Transfer, Ser. C*, **94**, 337-346.

Fox, D.G., and S.A. Orszag, 1973: Pseudospectral approximation to two-dimensional turbulence. *J. Comp. Physiol.*, **11**, 612-619.

Fox-Rabinovitz, M.S., G.L. Stenchikov, M.J. Suarez, and L.L. Takas, 1997: A finite-difference GCM dynamical core with a variation-resolution stretched grid. *Mon. Wea. Rev.,* **125**, 2943-2968.

Fraedrich, K., 1976: A mass budget of an ensemble of transient cumulus clouds determined from direct cloud observations. *J. Atmos. Sci.,* **33**, 262-268.

Frank, N.L., P.L. Moore, and G.E. Fisher, 1967: Summer shower distribution over the Florida Peninsula as deduced from digitized radar data. *J. Appl. Meteor.,* **6**, 309-316.

Frank, W.M., 1978: The life cycles of GATE convective systems. *J. Atmos. Sci.,* **35**, 1256-1264.

Frank, W.M., 1980: Modulations of the net tropospheric temperature during GATE. *J. Atmos. Sci.,* **37**, 1056-1064.

Fraser, A.B., R.C. Easter, and P.V. Hobbs, 1973: A theoretical study of the flow of air and fallout of solid precipitation over mountainous terrain. Part I. Airflow model. *J. Atmos. Sci.,* **30**, 801-812.

Freedman, J.M., D.R. Fitzjarrald, K.E. Moore, and R.K. Sakai, 2001: Boundary layer clouds and vegetation-atmosphere feedbacks. *J. Climate,* **14**, 180-197. http://dx.doi.org/10.1175/1520-0442(2001)013<0180:BLCAVA>2.0.CO;2.

Freitas, E.D., L.D. Martins, P.L. Silva Dias, and M.F. Andrade, 2005: A simple photochemical module implemented in RAMS for tropospheric ozone concentration forecast in the Metropolitan Area of São Paulo - Brazil: Coupling and validation. *Atmos. Environ.,* **39:34**, 6352-6361.

Freitas, E.D., C. Rozoff, W.R. Cotton, and P.L. Silva Dias, 2007: Interactions of an urban heat island and sea breeze circulations during winter over the Metropolitan Area of São Paulo - Brazil. *Bound.-Layer Meteor.,* **122**, 43-65.

Freitas, S.R., M.A.F. Silva Dias, and P.L. Silva Dias, 2000: Modeling the convective transport of trace gases by deep and moist convection. *Hybrid Methods in Engineering*, **3**, 317-330.

Freitas, S.R., K.M. Longo, M.A.F. Silva Dias, P.L. Silva Dias, R. Chatfield, E. Prins, P. Artaxo, G. Grell, and F.S. Recuero, 2005: Monitoring the transport of biomass burning emissions in South America. *Environ. Fluid Mech.*, **5**, 135-167.

Freitas, S.R., K.M. Longo, and M.O. Andreae, 2006: Impact of including the plume rise of vegetation fires in numerical simulations of associated atmospheric pollutants. *Geophys. Res. Lett.,* **33**, L17808.

Freitas, S.R., K.M. Longo, R. Chatfield, D. Latham, M.A.F. Silva Dias, M.O. Andreae, E. Prins, J.C. Santos, R. Gielow, and J.A. Carvalho Jr., 2007: Including the sub-grid scale plume rise of vegetation fires in low resolution atmospheric transport models. *Atmos. Chem. Phys.*, **7**, 3385-3398.

Freitas, S.R., K.M. Longo, M.A.F. Silva Dias, R. Chatfield, P. Silva Dias, P. Artaxo, M.O. Andreae, G. Grell, L.F. Rodrigues, and A. Fazenda, 2009: The Coupled Aerosol and Tracer Transport model to the Brazilian developments on the Regional Atmospheric Modeling System (CATT-BRAMS) Part 1: Model description and evaluation. *Atmos. Chem. Phys.*, **9**, 2843-2861.

Freitas, S.R., K.M. Longo, J. Trentmann, and D. Latham, 2010: Sensitivity of 1-D smoke plume rise models to the inclusion of environmental wind drag. *Atmos. Chem. Phys.*, **10**, 585-594.

Freitas, S.R., K.M. Longo, M.F. Alonso, M. Pirre, V. Marecal, G. Grell, R. Stockler, R.F. Mello, and M. Sánchez Gácita, 2011: PREP-CHEM-SRC 1.0: a preprocessor of trace gas and aerosol emission fields for regional and global atmospheric chemistry models. *Geoscientific Model Development*,**4**, 419-433.

Freitas, S.R., L.F. Rodrigues, K.M. Longo, and J. Panetta, 2012: Impact of a monotonic advection scheme with low numerical diffusion on transport modeling of emissions from biomass burning. *J. Adv. Modeling Earth Sys.*, **4**, p. M01001, http://dx.doi.org/10.1029/2011MS000084.

French, J.R., G. Vali, and R.D. Kelly, 1999: Evolution of small cumulus clouds in Florida: observations of pulsating growth. *Atmos. Res.,* **52**, 143-165.

Frenzen, P., and C.A. Vogel, 1992: The turbulent kinetic energy budget in the atmospheric surface layer: A review and an experimental reexamination in the field. *Bound.-Layer Meteor.,* **60**, 49-76.

Friedrich, K., N. Mölders, and G. Tetzlaff, 2000: On the influence of surface heterogeneity on the Bowen ratio: A theoretical case study. *Theor. Appl. Climatol.,* **65**, 181-196.

Friend, A.L., D. Djurié, and K.C. Brundidge, 1977: A combination of isentropic and sigma coordinates in numerical weather prediction. *Contrib. Atmos. Phys.,* **50**, 290-295.

Fritsch, J.M., 1975: Synoptic-mesoscale budget relationships for a tornado producing squall line. *Proceedings, 9th AMS Conference on Severe Local Storms*, October 21-23 1975, Norman Oklahoma, 165-172.

Fritsch, J.M., and C.F. Chappell, 1980a: Numerical prediction of convectively driven mesoscale pressure systems. Part I: Convective parameterization. *J. Atmos. Sci.,* **37**, 1722-1733.

Fritsch, J.M., and C.F. Chappell, 1980b: Numerical prediction of convectively driven mesoscale pressure systems. Part II: Mesoscale model. *J. Atmos. Sci.,* **37**, 1734-1762.

Fritsch, J.M., and J.S. Kain, 1993: Convective parameterization for mesoscale models: The Fritsch-Chappell Scheme. *The Representation of Cumulus Convection in Numerical Models, Meteor. Monogr.,* No. 46, Amer. Meteor. Soc., 159-164.

Fritsch, J.M., and R.A. Maddox, 1980: Analyses of upper tropospheric wind perturbations associated with midlatitude mesoscale convective complexes. *Preprints, Conference on Weather Forecasting and Analysis*, American Meteorological Society, Boston, 339-345.

Fritsch, J.M., and R.A. Maddox, 1981a: Convectively driven mesoscale systems aloft. Part I: Observations. *J. Appl. Meteor.,* **20**, 9-19.

Fritsch, J.M., and R.A. Maddox, 1981b: Convectively driven mesoscale systems aloft. Part II: Numerical simulation. *J. Appl. Meteor.,* **20**, 20-26.

Fritsch, J.M., C.F. Chappell, and L.K. Hoxit, 1976: The use of large scale budgets for convective parameterizations. *Mon. Wea. Rev.,* **104**, 1408-1418.

Fritsch, J.M., E.L. Magaziner, and C.F. Chappell, 1980: Analytical initialization for three-dimensional numerical models. *J. Appl. Meteor.,* **19**, 809-818.

Fu, Q., and K.-N. Liou, 1992: On the correlated k-distribution method for radiative transfer in nonhomogenous atmospheres. *J. Atmos. Sci.,* **49**, 2139-2156.

Fu, Q., and K.-N. Liu, 1993: Parameterization of the radiative properties of cirrus clouds. *J. Atmos. Sci.,* **50**, 2008-2025.

Fuggle, R.F., and T.R. Oke, 1976: Long-wave radiative flux divergence and nocturnal cooling of the urban atmosphere. I: Above roof-level. *Bound.-Layer Meteor.,* **10**, 113-120.

Fujita, T.T., 1980: In search of mesoscale wind fields in landfalling hurricanes. *13th Conference on Hurricanes and Tropical Meteorology*, American Meteorological Society, Miami Beach, December 1-5.

Fujita, T.T., and R.M. Wakimoto, 1982: Effects of miso- and mesoscale observations on PAM winds obtained during project NIMROD. *J. Appl. Meteor.,* **21**, 840-858.

Fukumori, I., T. Lee, B. Cheng, and D. Menemenlis, 2004: The origin, pathway, and destination of Niño-3 water estimated by a simulated passive tracer and its adjoint. *J. Phys. Oceanogr.,* **34**, 582-604.

Fukutome, S., C. Frei, D. Luthi, and C. Schar, 1999: The interannual variability as a test ground for regional climate simulations over Japan. *J. Meteor. Soc. Japan,* **77**, 649-672.

Fuller, J.G., 1978: *The poison that fell from the sky*, Random House, New York (shortened version published in *Reader's Digest*, August 1977, 191-236).

Furukawa, T., 1973: numerical experiments of the airflow over mountains. I. Uniform current with constant static stability. *J. Meteor. Soc. Japan,* **51**, 400-419.

Fyfe, J.C., W.J. Merryfield, V. Kharin, G.J. Boer, W.-S. Lee, and K. von Salzen, 2011: Skillful predictions of decadal trends in global mean surface temperature. *Geophys. Res. Lett.,* **38**, L22801, http://dx.doi.org/10.1029/2011GL049508.

Gage, K.S., 1979: Evidence of a $k^{-5/3}$ low inertial range in mesoscale two-dimensional turbulence. *J. Atmos. Sci.,* **36**, 1950-1954.

Gal-Chen, T., and R.C.J. Somerville, 1975a: On the use of a coordinate transformation for the solution of the Navier-Stokes equations. *J. Comput. Phys.,* **17**, 209-228.

Gal-Chen, T., and R.C.J. Somerville, 1975b: Numerical solution of the Navier-Stokes equations with topography. *J. Comput. Phys.,* **17**, 276-309.

Gall, R., J. Franklin, F. Marks, E.N. Rappaport, and F. Toepfer, 2013: The Hurricane Forecast Improvement Project. *Bull. Amer. Meteor. Soc.,* 94, 329-343. http://dx.doi.org/10.1175/BAMS-D-12-00071.1.

Gallus Jr., W.A., 1999: Eta simulations of three extreme precipitation events: sensitivity to resolution and convective parameterization. *Wea. Forecasting,* **14**, 405-426.

Gallus Jr., W.A., and J.B. Klemp, 2000: Behavior of flow over step orography. *Mon. Wea. Rev.,* **128**, 1153-1164.

Galmarini, S., and P. Thunis, 1999: On the validity of Reynolds assumptions for running-mean filters in the absence of a spectral gap. *J. Atmos. Sci.,* **56**, 1785-1796.

Galmarini, S., F. Michelutti, and P. Thunis, 2000: Estimating the contribution of leonard and cross terms to the subfilter scale from atmospheric measurements. *J. Atmos. Sci.,* **57**, 2968-2976.

Gamache, J.F., and R.A. Houze Jr., 1982: Mesoscale air motions associated with a tropical squall line. *Mon. Wea. Rev.,* **110**, 118-135.

Gangoiti, G., M.M. Millán, R. Salvador, E. Mantilla, 2001: Long-Range transport and re-circulation of pollutants in the Western Mediterranean during the RECAPMA Project. *Atmos. Environ.,* **35**, 6267-6276.

Gangoiti, G., L. Alonso, M. Navazo, A. Albizuri, G. Perez-Landa, M. Matabuena, V. Valdenebro, M. Maruri, J.A. García, M.M. Millán, 2002: Regional transport of pollutants over the Bay of Biscay: analysis of an ozone episode under a blocking anticyclone in west-central Europe. *Atmos. Environ.,* **36**, 1349-1361.

Gannon Sr., P.T., 1978: Influence of earth surface and cloud properties on the south Florida sea breeze. NOAA Technical Report ERL 402-NHEML-2.

Gao, X., S. Sorooshian, and H.V. Gupta, 1996: Sensitivity analysis of the biosphere-atmosphere transfer scheme. *J. Geophys. Res.,* **101**, 7279-7289.

Garratt, J.R., 1992: *The Atmospheric Boundary Layer.* Cambridge University Press, Cambridge, 316 pp.

Garratt, J.R., and G.D. Hess, 2003. Neutrally stratified boundary layer. In: *Encyclopedia of Atmospheric Sciences*, J.R. Holton, J.A. Curry, and J.A. Pyle, Eds., Academic Press, UK, 262-271.

Garratt, J.R., and R.A. Pielke, 1989: On the sensitivity of mesoscale models to surface-layer parameterization constants. *Bound.-Layer Meteor.,* **48**, 377-387.

Garratt, J.R., R.A. Pielke, W. Miller, and T.J. Lee, 1990: Mesoscale model response to random, surface-based perturbations – A sea-breeze experiment. *Bound.-Layer Meteor.,* **52**, 313-334.

Garrett, A.J., 1978: Numerical simulations of atmospheric convection over the southeastern U.S. in undisturbed conditions. Report No. 47, Atmospheric Science Group, University of Texas, College of Engineering, Austin, Texas.

Garrett, A.J., 1980: Orographic cloud over the eastern slopes of Mauna Loa Volcano, Hawaii, related to insolation and wind. *Mon. Wea. Rev.,* **108**, 931-941.

Garrett, A.J., 1983a: Drainage flow prediction with a one-dimensional model including canopy, soil and radiation parameterization. *J. Appl. Meteor.,* **22**, 79-91.

Garrett, A.J., 1983b: Treatment of ground-air and water-air exchange. *AMS/EPA Specialty Conference on Air Quality Modeling of the Nonstationary, Nonhomogeneous Urban Boundary Layers*, October 31-November 4, Baltimore, MD.

Garstang, M., and D. Fitzjarrald, 1999: *Observations of Surface to Atmospheric Interactions in the Tropics.* Oxford University Press, New York, 405 pp.

Garstang, M., P.D. Tyson, and G.D. Emmitt, 1975: The structure of heat islands. *Rev. Geophys. Space Phys.,* **13**, 139-165.

Garstang, M., R.A. Pielke, and W.J. Snow, 1980: A comparison of model predicted to observed winds in the coastal zone. Final report on Contract B-93492-A-Q for the period 3/80-10/80.

Gedzelman, S.D., and W.L. Donn, 1979: Atmospheric gravity waves and coastal cyclones. *Mon. Wea. Rev.,* **107**, 667-681.

Geisler, J.E., and F.P. Bretherton, 1969: The sea-breeze forerunner. *J. Atmos. Sci.,* **26**, 82-95.

Geleyn, J.-F., R. Fournier, G. Hello, and N. Pristov, 2005: A new "bracketing" technique for a flexible and economical computation of thermal radiative fluxes, on the basis of the Net Exchange Rate (NER) formalism. WGNE Blue Book 4-07.

Gentry, R.C., and P.L. Moore, 1954: Relation of local and general wind interaction near the sea coast to time and location of air-mass showers. *J. Meteor.,* **11**, 507-511.

George, R.L., 1979: Evolution of mesoscale convective systems over mountainous terrain. Colorado State University, Department of Atmospheric Science Paper No. 318, Fort Collins, Colorado.

Georges, T.M., and G.E. Greene, 1975: Infrasound from convective storms. Part IV. Is it useful for warning? *J. Appl. Meteor.,* **14**, 1303-1316.

Georgescu, M., D.B. Lobell, and C.B. Field, 2009a: Potential impact of U.S. biofuels on regional climate, *Geophys. Res. Lett.,* **36**, L21806, http://dx.doi.org/10.1029/2009GL040477.

Georgescu, M., G. Miguez-Macho, L.T. Steyaert, and C.P. Weaver, 2009b: Climatic effects of 30 years of landscape change over the Greater Phoenix, AZ, region: 1. Surface energy budget changes, *J. Geophys. Res.,* **114**, D05110, http://dx.doi.org/10.1029/2008JD010745.

Georgescu, M., G. Miguez-Macho, L.T. Steyaert, and C.P. Weaver, 2009c: Climatic effects of 30 years of landscape change over the Greater Phoenix, AZ, region: 2. Dynamical and thermodynamical response. *J. Geophys. Res.,* **114**, D05111, http://dx.doi.org/10.1029/2008JD010762.

Georgescu, M., D.B. Lobell, and C.B. Field, 2011a: Direct climate effects of perennial bioenergy crops in the United States. *Proc. Natl. Acad. Sci.*, (USA), **110 (10)**, 4134-4139, http://dx.doi.org/10.1073/pnas.1008779108.

Georgescu, M., M. Moustaoui, A. Mahalov, and J. Dudhia, 2011b: An alternative explanation of the semi-arid urban area "oasis effect". *J. Geophys. Res.,* **116**, D24113, http://dx.doi.org/10.1029/2011JD016720.

Georgescu, M., M. Moustaoui, A. Mahalov, and J. Dudhia, 2012a: Summer-time climate impacts of projected megapolitan expansion in Arizona. *Nature - Climate Change*, http://dx.doi.org/10.1038/nclimate1656.

Georgescu, M., A. Mahalov, and M. Moustaoui, 2012b: Seasonal hydro-climatic impacts of Sun corridor expansion. *Environ. Res. Lett.*, **7**, 034026, http://dx.doi.org/10.1088/1748-9326/7/3/034026.

Georgescu, M., D.B. Lobell, C. B. Field, and A. Mahalov, 2013: Simulated hydro-climatic impacts of projected Brazilian sugarcane expansion. *Geophys. Res. Lett.,* 40, 972–977, http://dx.doi.org/10.1002/grl.50206.

Gerard, L., 2007: An integrated package for subgrid convection, clouds and precipitation compatible with meso-gamma scales. *Quart. J. Roy. Meteor. Soc.,* **133**, 711-730.

Gevaerd, R., and S.R. Freitas, 2006.: Estimativa operacional da umidade do solo para iniciacao de modelos de previsao numerica da atmosfera. Parte I: Descricao da metodologia e validacao. *Revista Brasileira de Meteorologia*, **21, 3**, 1-15.

Gierasch, P.J., 1974: Martian dust storms. *Rev. Geophys.*, **12(4)**, 730-734.

Gierasch, P., and R. Goody, 1968: A study of the thermal and dynamical structure of the martian lower atmosphere. *Planet Space Sci.*, **16(5)**, 615-646.

Gierasch, P., and C. Sagan, 1971: A preliminary assessment of Martian wind regimes. *Icarus*, **14(3)**, 312-318.

Gillies, R.R., T.N. Carlson., J. Cui, W.P. Kustas, and K.S. Humes, 1997: A verification of the "triangle" method for obtaining surface soil water content and energy fluxes from remote measurements of the Normalized Difference Vegetation Index (NDVI) and surface radiant temperature. *Int. J. Remote Sens.*, **18**, 3145-3166.

Giorgi, F., M.R. Marinucci, and G.T. Bates, 1993a: Development of a second-generation regional climate model (RegCM2). Part I: Boundary-layer and radiative transfer processes. *Mon. Wea. Rev.,* **121**, 2794-2813.

Giorgi, F., M.R. Marinucci, and G.T. Bates, 1993b: Development of a second-generation regional climate model (RegCM2). Part II: Convective processes and assimilation of lateral boundary conditions. *Mon. Wea. Rev.,* **121**, 2814-2832.

Giovannini, G., S. Lucchesi, and M. Giachetti, 1988: Effect of heating on some physical and chemical parameters related to soil aggregation and erodibility. *Soil Science*, **146**, 255-261.

Givati, A., and D. Rosenfeld, 2004: Quantifying precipitation suppression due to air pollution. *J. Appl. Meteor.,* **43**, 1038-1056.

Glendening, J.W., 2000: Budgets of lineal and nonlineal turbulent kinetic energy under strong shear conditions. *J. Atmos. Sci.,* **57**, 2297-2318.

Gocho, Y., 1978: Numerical experiment of orographic heavy rainfall due to a stratiform cloud. *J. Meteor. Soc. Japan,* **56**, 405-423.

Gocho, Y., 1982: Statistical study on the relations among characteristics of rainfall around the Suzuka Mountains and meteorological conditions during warm season. *J. Meteor. Soc. Japan,* **60**, 739-757.

Goddard, L., A. Kumar, A. Solomon, D. Smith, G. Boer, P. Gonzalez, V. Kharin, W. Merryfield, C. Deser, S.J. Mason, B.P. Kirtman, R. Msadek, R. Sutton, E. Hawkins, T. Fricker, G. Hegerl, C.A.T. Ferro, D.B. Stephenson, G.A. Meehl, T. Stockdale, R. Burgman, A.M. Greene, Y. Kushnir, M. Newman, J. Carton, I. Fukumori, T. Delworth, 2012: A verification framework for interannual-to-decadal predictions experiments. *Climate Dyn.,* **40**, Issue 1-2, 245-272.

Golaz, J.-C., V.E. Larson, and W.R. Cotton, 2002: A PDF-based model for boundary layer clouds. Part I: Method and model description. *J. Atmos. Sci.,* **59**, 3540-3551. http://dx.doi.org/10.1175/1520-0469(2002)059<3540:APBMFB>2.0.CO;2.

Golaz, J.-C., J.D. Doyle, and S. Wang, 2009: One-way nested large-eddy simulation over the Askervein Hill. *J. Adv. Model. Earth Sys.*, **1, 6**, http://dx.doi.org/10.3894/JAMES.2009.1.6.

Golden, J.H., and J.D. Sartor, 1978: AMS Workshop on mesoscale interactions with cloud processes. October 24-25, 1977, Boulder, CO. *Bull. Amer. Meteor. Soc.,* **59**, 720-730.

Goldreich, Y., P.D. Tyson, R.G. Van Gogh, and G.P.N. Venter, 1981: Enhancement and suppression of urban heat plumes over Hojannesburg. *Bound.-Layer Meteor.,* **21**, 115-126.

Gollvik, S., 1999: On the effects of horizontal diffusion, resolution and orography on precipitation forecasting in a limited area model. *Meteorol. Appl.,* **6**, 49-58.

Gómez-Tejedor, J.A., M.J. Estrela, and M.M. Millan, 1999: A mesoscale model application to fire weather winds. *Int. J. Wildland Fire*, **9**, 255-263.

Goodin, W.R., G.J. McRae, and J.H. Seinfeld, 1980: An objective analysis technique for constructing three-dimensional urban scale wind fields. *J. Appl. Meteor.,* **19**, 98-108.

Goodin, W.R., G.J. McRae, and J.H. Seinfeld, 1981. Reply. *J. Appl. Meteor.,* **20**, 92-94.

Gopalakrishnan, S.G., M. Sharan, R.T. McNider, and M.P. Singh, 1998: Study of radiative and turbulent processes in the stable boundary layer under weak wind conditions. *J. Atmos. Sci.,* **55**, 954-960.

Gopalakrishnan, S.G., Q. Liu, T. Marchok, D. Sheinin, N. Surgi, R. Tuleya, R. Yablonsky, and X. Zhang, 2011: Hurricane Weather and Research and Forecasting (HWRF) model scientific documentation. NOAA/NCAR/Development Tech Center, 81 pp. Available online at: http://www.dtcenter.org/HurrWRF/users/docs/scientific_documents/HWRFScientificDocumentation_August2011.pdf.

Gopalakrishnan, S., Q. Liu, T. Marchok, D. Sheinin, V. Tallapragada, M. Tong, R. Tuleya, R. Yablonsky, and X. Zhang, 2012: Hurricane Weather Research and Forecasting (HWRF) Model: 2012 scientific documentation. L. Bernardet, Ed., 96 pp. Available from DTCenter.org.

Gopalakrishnan, S.G., F.D. Marks Jr., J.A. Zhang, X. Zhang, J.-W. Bao, and V. Tallapragada, 2013: A study of the impacts of vertical diffusion on the structure and intensity of the tropical cyclones using the high resolution HWRF system. *J. Atmos. Sci.,* **70**, 524-541. http://dx.doi.org/10.1175/JAS-D-11-0340.1.

Grabowski, W.W., 1998: Toward cloud resolving modeling of large-scale tropical circulations: A simple cloud microphysics parameterization. *J. Atmos. Sci.,* **21**, 3283-3298.

Grabowski, W.W., 2000: Cloud microphysics and the tropical climate: cloud-resolving model perspective. *J. Climate,* **13**, 2306-2322.

Grabowski, W.W., 2001: Coupling cloud processes with the large-scale dynamics using the cloud-resolving convection parameterization (CRCP). *J. Atmos. Sci.,* **58**, 978-997.

Grams, G.W., I.H. Blifford Jr., B.G. Schuster, and J.J. DeLuisi, 1972: Complex index of refraction of airborne fly ash determined by laser radar and collection of particles at 13 km. *J. Atmos. Sci.,* **29**, 900-905.

Grant, L.D., S.C. van den Heever, and L. Lu, 2013: Aerosol-cloud-land surface interactions within tropical convection simulations. Extended Abstract, *19th International Conference on Nucleation and Atmospheric Aerosols*, 24-28 June 2013, Fort Collins, CO.

Grasso, L.D., 2000: The differentiation between grid spacing and resolution and their application to numerical modeling. *Bull. Amer. Meteor. Soc.*, **81**, 579-580.

Grasso, L.D., and T.J. Greenwald, 2004: Analysis of 10.7-μm brightness temperatures of a simulated thunderstorm with two-moment microphysics. *Mon. Wea. Rev.,* **132**, 815-825.

Grasso, L., and D. Lindsey, 2011: An example of the use of synthetic 3.9 m GOES-12 imagery for two-moment microphysical evaluation. *Int. J. Remote Sens.*, **32:8**, 2337-2350.

Grasso, L., M. Sengupta, J. Dostalek, R. Brummer, and M. Demaria, 2008: Synthetic satellite imagery for current and future environmental satellites. *Int. J. Remote Sens.*, **29:15**, 4373-4384.

Grasso, L., M. Sengupta, and M. Demaria, 2010: Comparison between observed and synthetic 6.5 and 10.7 μm GOES-12 imagery of thunderstorms that occurred on 8 May 2003. *Int. J. Remote Sens.*, **31:3**, 647-663.

Greeley, R., R.O. Kuzmin, S.C.R. Rafkin, T.I. Michaels, and R. Haberle, 2003: Wind-related features in Gusev crater, Mars. *J. Geophys. Res.*, **108(E12)**, 8077.

Green, J.S.A., and G.A. Dalu, 1980: Mesoscale energy-generated in the boundary layer. *Quart. J. Roy. Meteor. Soc.,* **106**, 721-726.

Greene, E.M., G.E. Liston, and R.A. Pielke, 1999: Simulation of above treeline snowdrift formation using a numerical snow-transport model. *Cold Regions Sci. Tech.*, **30**, 135-144.

Greenwald, T.J., R. Hertenstein, and T. Vukićević, 2002: An all-weather observational operator for radiance data assimilation with mesoscale forecast models. *Mon. Wea. Rev.,* **130**, 1882-1897.

Grell, G.A., 1993: Prognostic evaluation of assumptions used by cumulus parameterization. *Mon. Wea. Rev.,* **121**, 764-787.

Grell, G., and A. Baklanov, 2011: Integrated modeling for forecasting weather and air quality: A call for fully coupled approaches. *Atmos. Environ.,* 45, 6845–6851, http://dx.doi.org/10.1016/j.atmosenv.2011.01.017.

Grell, G., and D. Devenyi, 2002: A generalized approach to parameterizing convection combining ensemble and data assimilation techniques. *Geophys. Res. Lett.,* **29(14)**, http://dx.doi.org/10.1029/2002GL015311.

Grell, G.A., Y.-H. Kuo, and R. Pasch, 1991: Semi-prognostic test of cumulus parameterization schemes in the middle latitudes. *Mon. Wea. Rev.,* **119**, 5-31.

Grell, G.A., J. Dudhia, and D.R. Stauffer, 1994: A description of the fifth-generation Penn State/NCAR mesoscale modeling system (MM5). NCAR Technical Note, NCAR/TN-397+STR, 117 pp.

Grell, G.A., J. Dudhia and D.R. Stauffer, 1995: A description of the fifth-generation Penn State/NCAR Mesoscale Model (MM5). NCAR Technical Note, NCAR/TN-398+STR, 138 pp.

Griffith, C.A., T. Owen, G.A. Miller, and T. Geballe, 1998: Transient clouds in Titan's lower atmosphere. *Nature,* **395(6702)**, 575-578.

Griffith, K.T., S.K. Cox, and R.G. Knollenberg, 1980: Infrared radiative properties of tropical cirrus clouds inferred from aircraft measurements. *J. Atmos. Sci.,* **37**, 1077-1087.

Grim, J.A., R.M. Rauber, M.K. Ramamurthy, B.F. Jewett and M. Han, 2007: High-resolution observations of the trowel-warm-frontal region of two continental winter cyclones. *Mon. Wea. Rev.,* **135**, 1629-1646.

Grimmond, C.S.B., and T.R. Oke, 1999: Heat storage in urban areas: local-scale observations and evaluation of a simple model. *J. Appl. Meteor.,* **38**, 922-940.

Gross, M.G., 1977: *Oceanography: A View of the Earth.* 2nd Edition, Prentice-Hall, Englewood Cliffs, New Jersey, 581 pp.

Gryning, S.-E., and E. Lyck, 1983: A tracer investigation of the atmospheric dispersion in the Dyrnaes Valley, Greenland, RisøNational Laboratory, Denmark Report RisøR-481.

Gu, L., J.D. Fuentes, M. Garstang, J. Tota da Silva, R. Heitz, J. Sigler, and H.H. Shugart, 2001: Cloud modulation of surface solar irradiance at a pasture site in southern Brazil. *Agric. Forest Meteor.,* **106(2)**, 117-129, http://dx.doi.org/10.1016/S0168-1923(00)00209-4.

Gube, M., J. Schmetz, and E. Raschke, 1980: Solar radiative transfer in a cloud field. *Contrib. Atmos. Phys.,* **53**, 24-34.

Guiraud, J.P., and R.K.H. Zeytounian, 1982: A note on the adjustment to hydrostatic balance. *Tellus,* **34**, 50-54.

Gunn, R., and J.S. Marshall, 1958: The distribution with size and aggregate snowflakes. *J. Meteor.,* **15**, 452-461.

Guo, X., W.G. Lawson, M.I. Richardson, and A. Toigo, 2009: Fitting the Viking lander surface pressure cycle with a Mars General Circulation Model. *J. Geophys. Res.-Planets,* **114**, E07006, http://dx.doi.org/10.1029/2008JE003302.

Gupta, H.V., L.A. Bastidas, S. Sorooshian, W.J. Shuttleworth, and Z.L. Yang, 1999: Parameter estimation of a land surface scheme using multicriteria methods. *J. Geophys. Res.,* **104**, 19,491-19,503.

Gyakum, J.R., M. Carrera, D.-L. Zhang, S. Miller, J. Caveen, R. Benoit, T. Black, A. Buzzi, C. Chouinard, M. Fantini, C. Folloni, J.J. Katzfey, Y.-H. Kuo, F. Lalaurette, S. Low-Nam, J. Mailhot, P. Malguzzi, J.L. McGregor, M. Nakamura, G. Tripoli, and C. Wilson, 1996: A regional model intercomparison using a case of explosive oceanic cyclogenesis. *Wea. Forecasting,* **11**, 521-543.

Haberle, R.M., 1986: Interannual variability of global dust storms on Mars. *Science*, **234(4775)**, 459-461.

Haberle, R.M., C.B. Leovy, and J.B. Pollack, 1982: Some effects of global dust storms on the atmospheric circulation of Mars. *Icarus*, **50**, 322-367.

Haberle, R.M., H.C. Houben, R. Hertenstein, and T. Herdtle, 1993: A boundary layer model for Mars: Comparison with Viking Entry and Lander data. *J. Atmos. Sci.,* **50**, 1544-1559.

Haberle, R.M., J.B. Pollack, J.R. Barnes, R.W. Zurek, C.B. Leovy, J.R. Murphy, H. Lee, and J. Schaeffer, 1993: Mars atmospheric dynamics as simulated by the NASA Ames General Circulation Model 1. The zonal-mean circulation. *J. Geophys. Res.*, **98(E2)**, 3093-3123.

Hachey, H.B., 1934: Movements resulting from mixing of stratified waters. *J. Biol. Board Can.*, **1**, 133-143.

Hack, J.J., and W.H. Schubert, 1981: Lateral boundary conditions for tropical cyclone models. *Mon. Wea. Rev.,* **109**, 1404-1420.

Hadley, G., 1962: *Linear Programming*. Addison-Wesley, Reading, Pennsylvania, 520 pp.

Haertel, P.T., and D.A. Randall, 2002: Could a pile of slippery sacks behave like an ocean? *Mon. Wea. Rev.,* **130**, 2975-2988.

Haertel, P.T., and K.H. Straub, 2010: Simulating convectively coupled Kelvin waves using Lagrangian overturning for a convective parametrization. *Quart. J. Roy. Meteor. Soc.,* **136**, 1598-1613.

Haertel, P.T., D.A. Randall, and T.G. Jensen, 2004: Simulating upwelling in a large lake using slippery sacks. *Mon. Wea. Rev.,* **132**, 66-77.

Haferman, J.L., W.F. Krajewski, T.F. Smith, and A. Sanchez, 1993: Radiative transfer for a three dimensional raining cloud. *Appl. Opt.*, **32**, 2795-2802.

Hafner, J., and S.Q. Kidder, 1999: Urban heat island modeling in conjunction with satellite-derived surface/soil parameters. *J. Appl. Meteor.,* **38**, 448-465.

Halberstam, I., and J.P. Schieldge, 1981: Anomalous behavior of the atmospheric surface layer over a melting snowpack. *J. Appl. Meteor.,* **20**, 255-265.

Hales, J.M., 1975: Atmospheric transformations of pollutants. In: *Lectures on Air Pollution and Environmental Impact Analysis*, D. Haugen, Ed., American Meteorological Society, Boston, Massachusetts.

Hales, J.M., 1983: Parameterization of removal mechanism. *AMS/EPA Specialty Conference on Air Quality Modeling of the Nonstationary, Nonhomogeneous Urban Boundary Layer*, October 31-November 4, Baltimore, MD.

Hall, F.G., Ed., 1999: BOREAS in 1999: Experiment in Science Overview. *J. Geophys. Res.,* **104, D22**, 27627-27971.

Halldin, S., and S.-E. Gryning, 1999: Boreal forests and climate. *Agric. Forest Meteor.,* **98-99**, 1-4.

Haltiner, G.J., 1971: *Numerical Weather Prediction.* Wiley, New York, 317 pp.

Haltiner, G.J., and F.L. Martin, 1957: *Dynamical and Physical Meteorology*. McGraw-Hill, New York, 470 pp.

Haltiner, G.J., and R.T. Williams, 1980: *Numerical Prediction and Dynamic Meteorology*. 2nd Edition, John Wiley and Sons, New York, 477 pp.

Hamilton, P., and M. Rattray Jr., 1978: A numerical model of the depth-dependent, wind-driven upwelling circulation on a continental shelf. *J. Phys. Oceanogr.,* **8**, 437-457.

Han, J., and H.-L. Pan, 2006: Sensitivity of hurricane intensity forecasts to convective momentum transport parameterization. *Mon. Wea. Rev.,* **134**, 664-674.

Han, M., R.M. Rauber, M.K. Ramamurthy, B.F. Jewett, and J.A. Grim, 2007: Mesoscale dynamics of the trowal and warm frontal regions of two continental winter cyclones. *Mon. Wea. Rev.*, **135**, 1647-1670.

Han, M., S.A. Braun, P.O.G. Persson, and J.-W. Bao, 2009: Alongfront variability of precipitation associated with a midlatitude frontal zone: TRMM observation and MM5 simulation. *Mon. Wea. Rev.*, **137**, 1008-1028.

Han, M., S.A. Braun, W.S. Olson, P. Ola G. Persson, and J.-W. Bao 2010: Application of TRMM PR and TMI measurements to assess cloud microphysical schemes in the MM5 for a winter storm. *J. Appl. Meteor. Climatol.*, **49**, 1129-1148.

Han, Y., P. van Delst, Q. Liu, F. Weng, B. Yan, R. Treadon, and J. Derber, 2006: JCSDA Community Radiative Transfer Model (CRTM)—version 1. NOAA Tech. Rep., NESDIS 122, 40 pp.

Hane, C.E., 1973: The squall line thunderstorm: Numerical experimentation. *J. Atmos. Sci.,* **30**, 1672-1690.

Hane, C.E., 1978: Scavenging of urban pollutants by thunderstorm rainfall: Numerical experimentation. *J. Appl. Meteor.,* **17**, 699-710.

Hänel, G., 1971: New results concerning the dependence of visibility on relative humidity and their significance in a model for visibility forecast. *Contrib. Atmos. Phys.,* **44**, 137-167.

Hänel, G., R. Busen, C. Hillenbrand, and R. Schloss, 1982: Light absorption measurements: New techniques. *App. Opt.*, **21**, 382-386.

Hanesiak, J., A. Tat, and R.L. Raddatz, 2009: Initial soil moisture as a predictor of subsequent summer severe weather in the cropped grassland of the Canadian Prairie Provinces. *Int. J. Climatol.,* **29**, 899-909.

Hanna, S.R., 1979: Some statistics of Lagrangian and Eulerian wind fluctuation. *J. Appl. Meteor.,* **18**, 518-531.

Hanna, S.R., 1994: Mesoscale meteorological model evaluation techniques with emphasis on needs of air quality models. In: *Mesoscale Modeling of the Atmosphere*, R.A. Pielke Sr., and R.P. Pearce (Eds.), Boston, MA, American Meteorological Society, 47-58.

Hanna, S.R., and F.A. Gifford, 1975: Meteorological effects of energy dissipation at large power parks. *Bull. Amer. Meteor. Soc.,* **56**, 1069-1077.

Hanna, S.R., and S.D. Swisher, 1971: Meteorological effects of the heat and moisture produced by man. *Nucl. Saf.*, **12**, 114-122.

Hanna, S.R., and R. Yang, 2001: Evaluations of mesoscale models' predictions of near-surface winds, temperature gradients, and mixing depths. *J. Appl. Meteor.,* **40**, 1095-1104.

Hanna, S.R., B. Reen, E. Hendrick, L. Santos, D.R. Stauffer, A. Deng, J. McQueen, M. Tsidulko, Z. Janjic, D. Jovic, and R.I. Sykes, 2010: Comparison of observed, MM5 and WRF-NMM model-simulated, and HPAC-assumed boundary-layer meteorological variables for 3 days during the IHOP field experiment. *Bound.-Layer Meteor.,* **134**, 285-305.

Hansen, J.E., 1971: Multiple scattering of polarized light in planetary atmospheres. Part I. The doubling method. *J. Atmos. Sci.,* **28**, 120-125.

Hansen, J.E., and L.D. Travis, 1974: Light scattering in planetary atmospheres. *Space Sci. Rev.*, **16**, 527-610, http://dx.doi.org/10.1007/BF00168069.

Harada, A., 1981: Urban and industrial effects on precipitation and thunder days in Hokkaido, Japan. *Pap. Meteor. Geophys.,* **32**, 233-245.

Harding, R.J., E.M. Blyth, O.A. Tuinenburg, and A. Wiltshire, 2013: Land atmosphere feedbacks and their role in the water resources of the Ganges basin. *Sci. Total Environ.* http://dx.doi.org/10.1016/j.scitotenv.2013.03.016.

Harrington, J.Y., G. Feingold, and W.R. Cotton, 2000: Radiative impacts on the growth of a population of drops within simulated summertime Arctic stratus. *J. Atmos. Sci.,* **57**, 766-785.

Harrison, R., and B. McGoldrick, 1981: Mapping artificial heat release in Great Britain. *Atmos. Environ.,* **15**, 667-674.

Harshvardhan, R. Davies, D.A. Randall, and T.G. Corsetti, 1987: A fast radiation parameterization for atmospheric circulation models. *J. Geophys. Res.,* **92**, 1009-1016.

Harshvardhan, D.A. Randall, T.G. Corsetti, and D.A. Dazlich, 1989: Earth radiation budget and cloudiness simulations with a general circulation model. *J. Atmos. Sci.,* **46**, 1922-1942.

Hartman, M.D., J.S. Baron, R.B. Lammers, D.W. Cline, L.E. Band, G.E. Liston, and C. Tague, 1999: Simulations of snow distribution and hydrology in a mountain basin. *Water Resour. Res.,* **35**, 1587-1603.

Hartsell, C., 1970: Case study of a traveling hailstorm. Report No. 70-1. Institute of Atmospheric Science, South Dakota School of Mines and Technology, Rapid City, SD.

Haugen, J., and B. Machenhauer, 1993: A spectral limited-area model formulation with time-dependent boundary conditions applied to shallow-water equations. *Mon. Wea. Rev.,* **121**, 2618-2630.

Haurwitz, B., 1947: Comments on the sea breeze circulation. *J. Meteor.,* **4**, 1-8.

Hawkins, J.D., and D.W. Stuart, 1980: Low-level atmospheric changes over Oregon's coastal upwelling region. *Mon. Wea. Rev.,* **108**, 1029-1040.

Hayden, B.P., 1998: Ecosystem feedbacks on climate at the landscape scale. *Phil. Trans. R. Soc. Lond. B.,* **353**, 5-18.

Haynes, J.M., R.T. Marchand, Z. Luo, A. Bodas-Salcedo, and G.L. Stephens, 2007: A multi-purpose radar simulation package: QuickBeam. *Bull. Amer. Meteor. Soc.,* **88**, 1723-1727.

Hearn, A.C, 1973: Reduce 2 user manual. Report prepared in part from support by the NSF under Grant No. GJ-32181 and the Advanced Research Projects Agency of the Office of the Department of Defense under Contract No. DAHC 15-73-C-0363.

Heavens, N.G., D.J. McCleese, M.I. Richardson, D.M. Kass, A. Kleinböhl, and J.T. Schofield, 2011: Structure and dynamics of the Martian lower and middle atmosphere as observed by the Mars Climate Sounder: 2. Implications of the thermal structure and aerosol distributions for the mean meridional circulation. *J. Geophys. Res.*, **116(E1)**, E01010.

Heicklen, J., 1976: *Atmospheric Chemistry*. Academic Press, New York, 406 pp.

Heidinger, A.K., C. O'Dell, R. Bennartz, and T. Greenwald, 2006: The successive-order-of-interaction radiative transfer model. Part I: Model development. *J. Appl. Meteor. Climatol.,* **45**, 1388-1402.

Henderson-Sellers, A. 1984: A new formula for latent heat of vaporization of water as a function of temperature. *Quart. J. Roy. Meteor. Soc.,* **110:466**, 1186-1190.

Henderson-Sellers, A., Z.-L. Yang, and R.E. Dickinson, 1993: The Project for Intercomparison of Land-surface Parameterization Schemes. *Bull. Amer. Meteor. Soc.,* **74**, 1335-1349.

Henderson-Sellers, A., A.J. Pitman, P.K. Love, P. Irannejad, and T.H. Chen, 1995: The project for intercomparison of land surface parameterization schemes (PILPS): Phases 2 and 3. *Bull. Amer. Meteor. Soc.,* **76**, 489-503.

Héreil, P., and R. Laprise, 1996: Sensitivity of internal gravity waves solutions to the time step of a semi-implicit semi-Lagrangian nonhydrostatic model. *Mon. Wea. Rev.,* **124**, 972-999.

Herzegh, P.H., and P.V. Hobbs, 1980: The mesoscale and microscale structure and organization of clouds and precipitation in midlatitude cyclones. II: Warm-frontal clouds. *J. Atmos. Sci.,* **37**, 597-611.
Herzegh, P.H., and P.V. Hobbs, 1981: The mesoscale and microscale structure and organization of clouds and precipitation in midlatitude cyclones. IV. Vertical air motions and microphysical structures of prefrontal surge clouds and cold-frontal clouds. *J. Atmos. Sci.,* **38**, 1771-1784.
Hess, S.L., R.M. Henry, C.B. Leovy, J.A. Ryan, J.E. Tillman, T.E. Chamberlain, H.L. Cole, R.G. Dutton, G.C. Greene, W.E. Simon, and J.L. Mitchell, 1976: Preliminary meteorological results on Mars from Viking-1 Lander. *Science,* **193(4255)**, 788-791.
Hess, S.L., R.M. Henry, C.B. Leovy, J.A. Ryan, and J.E. Tillman, 1977: Meteorological results from the surface of Mars: Viking 1 and 2. *J. Geophys. Res.*, **82(28)**, 4559-4574.
Hickey, J.R., L.L. Stowe, H. Jocobowitz, P. Pellegrino, R.H. Maschkoff, F. House, and T.H. Vonder Haar, 1980: Initial solar irradiance determinations form Nimbus 7 cavity radiometer measurements. *Science,* **208**, 281-283.
Hicks, B.B., 1978: Comments on the characteristics of turbulent velocity components in the surface layer under convective conditions. *Bound.-Layer Meteor.,* **15**, 255-258.
Hildebrand, F.B., 1962: *Advanced Calculus for Applications.* Prentice-Hall, Englewood Cliffs, NJ, 646 pp.
Hildebrand, F.B., 1976: *Advanced Calculus for Applications.* 2nd Edition, Prentice-Hall, Englewood Cliffs, New Jersey, 733 pp.
Hill, F.F., K.A. Browning, and M.J. Bader, 1981: Radar and rain-gauge observations of orographic rain over south Wales. *Quart. J. Roy. Meteor. Soc.,* **107**, 643-670.
Hill, G.E., 1974: Factors controlling the size and spacing of cumulus clouds as revealed by numerical experiments. *J. Atmos. Sci.,* **31**, 646-673.
Hill, G.E., 1978: Observations of precipitation-forced circulations in winter orographic storms. *J. Atmos. Sci.,* **35**, 1463-1472.
Hill, J.D., 1971: Snow squalls in the lee of Lake Erie and Lake Ontario. NOAA Technical Memorandum NWS ER-43, August 1971.
Hinson, D.P., and R.J. Wilson, 2004: Temperature inversions, thermal tides, and water ice clouds in the atmosphere of Mars. *J. Geophys. Res.*, **109**, E01002.
Hinson, D.P., M. Pätzold, S. Tellmann, B. Häusler, and G.L. Tyler, 2008: The depth of the convective boundary layer on Mars. *Icarus*, **198(1)**, 57-66.
Hjelmfelt, M., 1980: Numerical simulation of the effects of St. Louis on boundary layer airflow and convection. University of Chicago, Ph.D. Dissertation, Chicago, Illinois.
Hjelmfelt, M.R., 1982: Numerical simulation of the effects of St. Louis on mesoscale boundary layer airflow and vertical air motion: Simulations of urban vs. non-urban effects. *J. Appl. Meteor.,* **21**, 1239-1257.
Hjelmfelt, M.R., and R.R. Braham Jr., 1983: Numerical simulation of the airflow over Lake Michigan for a major lake-effect snow event. *Mon. Wea. Rev.,* **111**, 205-219.
Hobbs, P.V., 1978: Organization and structure of clouds and precipitation on the mesoscale and microscale in cyclonic storms. *Rev. Geophys. Space Phys.,* **16**, 741-755.
Hobbs, P.V., and P.O.G. Persson, 1982: The mesoscale and microscale structure and organization of clouds and precipitation in mid-latitude cyclones. 5. The substructure of narrow cold-frontal rainbands. *J. Atmos. Sci.,* **39**, 280-295.
Hobbs, P.V., R.C. Eastern, and A.B. Fraser, 1973: A theoretical study of the flow of air and fallout of solid precipitation over mountainous terrain. Part II. Microphysics. *J. Atmos. Sci.,* **30**, 813-823.

Hobbs, P.V., T.J. Matejka, P.H. Herzegh, J.D. Locatelli, and R.A. Houze Jr., 1980: The mesoscale and microscale structure and organization of clouds and precipitation in mid-latitude cyclone. I. A case study of a cold front. *J. Atmos. Sci.,* **37**, 568-596.

Hodur, R.M., 1997: The Naval Research Laboratory's Coupled Ocean/Atmosphere Mesoscale Prediction System (COAMPS). *Mon. Wea. Rev.,* **125**, 1414-1430.

Hof, H., 1999: Land surface data sets. *IGBP Global Change Newsletter,* **39**, 18-19.

Hoffman, M.J., S.J. Greybush, R.J. Wilson, G. Gyarmati, R.N. Hoffman, E. Kalnay, K. Ide, E.J. Kostelich, T. Miyoshi, and I. Szunyogh, 2010: An ensemble Kalman filter data assimilation system for the martian atmosphere: Implementation and simulation experiments. *Icarus,* **209,2**, 470-481.

Hogan, R.J., 2008: Fast lidar and radar multiple-scattering models: Part 1: Small-angle scattering using the photon variance-covariance method. *J. Atmos. Sci.,* **65**, 3621-3635.

Högström, U., 1988: Non-dimensional wind and temperature profiles in the atmospheric surface layer: A re-evaluation. *Bound.-Layer Meteor.,* **42**, 55-78.

Högström, U., 1996: Review of some basic characteristics of the atmospheric surface layer. *Bound.-Layer Meteor.,* **78**, 215-246.

Hoinka, K.P., 1980: Synoptic-scale atmospheric features and foehn. *Contrib. Atmos. Phys.,* **53**, 486-508.

Hoke, J.E., and R.A. Anthes, 1976: The initialization of numerical models by a dynamic-initialization technique. *Mon. Wea. Rev.,* **104**, 1551-1556.

Hoke, J.E., and R.A. Anthes, 1977: Dynamic initialization of a three-dimensional primitive-equation model of Hurricane Alma of 1962. *Mon. Wea. Rev.,* **111**, 1046-1051.

Holle, R.L., and M.W. Maier, 1980: Tornado formation from downdraft interaction in the FACE mesonetwork. *Mon. Wea. Rev.,* **108**, 1010-1028.

Holle, R.L., and A.I. Watson, 1983: Duration of convective events related to visible cloud, convergence, radar, and rain gauge parameters over south Florida. *Mon. Wea. Rev.,* **111**, 1046-1051.

Holley, R.M., 1972: Surface temperature of a tropical island and surrounding ocean measured with an airborne infrared radiometer. Florida State University, M.S. Thesis.

Holroyd, E.W., 1982: some observations on mountain-generated cumulonimbus rainfall on the northern Great Plains. *J. Appl. Meteor.,* **21**, 560-565.

Holt, T., and J. Pullen, 2007: Urban canopy modeling of the New York City metropolitan area: A comparison and validation of single- and multi-layer parameterizations. *Mon. Wea. Rev.,* **135**, 1906-1930.

Holton, J.R., 1972: *An Introduction to Dynamic Meteorology*. Academic Press, New York, 319 pp.

Holton, J.R., 2004: *An Introduction to Dynamic Meteorology*. Fourth Edition, Academic Press, 535 pp.

Holtslag, A.A.M., 1998: Fluxes and gradients in atmospheric boundary layers. In *Clear and Cloudy Boundary Layers*, A.A.M. Holtslag, and P.G. Duynkerke, Eds., Royal Netherlands Academy of Arts and Sciences, PO Box 19121, 1000 G Amsterdam, The Netherlands.

Holtslag, A.A.M., and B.A. Boville, 1993: Local versus nonlocal boundary-layer diffusion in a global climate model. *J. Climate*, **6**, 1825-1842.

Holtslag, A.A.M., and P.G. Duynkerke, Eds., 1998: *Clear and Cloudy Boundary Layers*. Royal Netherlands Academy of Arts and Sciences, 372 pp.

Holtslag, A.A.M., G. Svensson, S. Basu, B. Beare, F.C. Bosveld, and J. Cuxart, 2012: Overview of the GEWEX Atmospheric Boundary Layer Study (GABLS). *Proceedings of the Workshop on Diurnal Cycles and the Stable Boundary Layer*, 7-10 November 2011, Reading, UK. Available at: http://www.wageningenur.nl/en/Publications.htm?publicationId=publication-way-343232353239.

Homan, J.H., and D.G. Vincent, 1983: Mesoscale analysis of surface variables during the severe storm outbreak of April 10-11, 1979. *Mon. Wea. Rev.,* **111**, 1122-1130.

Honda, Y., and K. Sawada, 2008: A new 4D-Var for mesoscale analysis at the Japan Meteorological Agency. *CAS/JSC WGNE Res. Act. Atmos. Ocean. Model.*, **38**, 01.7-01.8.

Hong, S.-H., and H.-L. Pan, 1996: Nonlocal boundary layer vertical diffusion in a medium-range forecast model. *Mon. Wea. Rev.,* **124**, 2322-2339.

Hong, S.-Y., and J. Dudhia, 2012: Next-generation numerical weather prediction: Bridging parameterization, explicit clouds, and large eddies. *Bull. Amer. Meteor. Soc.*, **93**, ES6-ES9, http://dx.doi.org/10.1175/2011BAMS3224.1.

Hong, S.-Y., and H.-L. Pan, 1998: Convective trigger function for a mass-flux cumulus parameterization scheme. *Mon. Wea. Rev.,* **126**, 2599-2620.

Hong, S.-Y., and H.-L. Pan, 1998: Convective trigger function for a mass flux cumulus parameterization scheme. *Mon. Wea. Rev.,* **126**, 2621-2639.

Hong, X., M.J. Leach, and S. Raman, 1995: A sensitivity study of convective cloud formation by vegetation forcing with different atmospheric conditions. *J. Appl. Meteor.,* **34**, 2008-2028.

Hootman, B.W., and W. Blumen, 1983: Analysis of nighttime drainage winds in Boulder, Colorado during 1980. *Mon. Wea. Rev.,* **111**, 1052-1061.

Horne, F.E., and M.L. Kavvas, 1997: Physics of the spatially averaged snowmelt process. *J. Hydrol.*, **191**, 179-207.

Horst, T.W., and J.C. Doran, 1981: Observations of the structure and development of nocturnal slope winds. *Preprints, 2nd Conference on Mountain Meteorology*, American Meteorological Society, Boston, November 10-13, 1981, Steamboat Springs, Colorado.

Hoskins, B.J., and F.P. Bretherton, 1972: Atmospheric frontogenesis models: Mathematical formulation and solution. *J. Atmos. Sci.,* **29**, 11-37.

Hou, A.Y., R. Kakar, S. Neeck, A. Azarbarzin, C. Kummerow, M. Kojima, R. Oki, K. Nakamura, and T. Iguchi, 2013: The Global Precipitation Measurement (GPM) Mission. *Bull. Amer. Meteor. Soc.,* submitted.

Houghton, J.T., Y. Ding, D.J. Griggs, M. Noguer, P.J. van der Linden, and D. Xiaosu, Eds., 2001: Climate Change 2001: The Scientific Basis—Contribution of Working Group I to the Third Assessment Report of the Intergovernmental Panel on Climate Change (IPCC), Cambridge Univ. Press, New York, 944 pp.

Hourdin, F., P. Le Van, F. Forget, and O. Talagrand, 1993: Meteorological variability and the annual surface pressure cycle on Mars. *J. Atmos. Sci.*, **50(21)**, 3625-3640.

Houze Jr., R.A., 1981: Structures of atmospheric precipitation systems: A global survey. *Radio Sci.*, **16**, 671-689.

Houze Jr., R.A., 1982: Cloud clusters and large-scale vertical motions in the tropics. *J. Meteor. Soc. Japan,* **60**, 396-410.

Houze Jr., R.A., 1989: Observed structure of mesoscale convective systems and implications for large-scale heating. *Quart. J. Roy. Meteor. Soc.,* **115**, 425-461.

Houze Jr., R.A., and A.K. Betts, 1981: Convection in GATE. *Rev. Geophys. Space Phys.,* **19**, 541-576.

Houze Jr., R.A., and C.-P. Cheng, 1981: Inclusion of mesoscale updrafts and downdrafts in computations of vertical fluxes by ensembles of tropical clouds. *J. Atmos. Sci.,* **38**, 1751-1770.

Houze Jr., R.A., and P.V. Hobbs, 1982: Organization and structure of precipitation cloud systems. *Adv. Geophys.,* **24**, 225-315.

Houze Jr., R.A., P.V. Hobbs, P.H. Herzegh, and D.B. Parsons, 1979: Size distributions of precipitation particles in frontal clouds. *J. Atmos. Sci.,* **36**, 156-162.

Houze Jr., R.A., S.A. Rutledge, T.J. Matejka, and P.V. Hobbs, 1981a: The mesoscale and microscale structure and organization of clouds and precipitation in midlatitude cyclones. III: Air motions and precipitation growth in a warm-frontal rainband. *J. Atmos. Sci.,* **38**, 639-649.

Houze Jr., R.A., S.G. Geotis, F.D. Marks Jr., and A.K. West, 1981b: Winter monsoon convection in the vicinity of north Borneo, Part I: Structure and time variation of the clouds and precipitation. *Mon. Wea. Rev.,* **109**, 1595-1614.

Hovermale, J.B., 1965: A non-linear treatment of the problem of airflow over mountains. Ph.D. Thesis, Pennsylvania State University, University Park.

Hoxit, L.R., and C.F. Chappell, 1975a: Tornado outbreak of April 3-4 1974. Synoptic Analysis. NOAA Techincal Report NOAA TR ERL 338-APCL 37.

Hoxit, L.R., and C.F. Chappell, 1975b: An analysis of the mesoscale circulations which produced the April 3, 1974 tornadoes in northern Indiana. *Proceedings, 9th AMS Conference on Severe Local Storms*, October 21-23, Norman, Oklahoma, 256-263.

Hoxit, L.R., R.A. Maddox, C.F. Chappell, F.L. Zurkerberg, H.M. Mogil, I. Jones, D.R. Greene, R.E. Saffle, and R.A. Scofield, 1978: Meteorological aspects of the Johnstown, PA flash flood 19-20 July 1977. NOAA Technical Report ERL 401-APCL 43, 1-71.

Hsu, H.-M., 1979: Numerical simulations of mesoscale precipitation systems. University of Michigan, Ph.D. Dissertation, Department of Atmospheric and Oceanic Science, Ann Arbor.

Hsu, S., 1969: Mesoscale structure of the Texas coast sea breeze. Report No. 16, Atmospheric Science Group, University of Texas, College of Engineering, Austin.

Hsu, S.-A., 1970: Coastal air circulation system. Observations and empirical model. *Mon. Wea. Rev.,* **98**, 487-509.

Hsu, W.-R., and W.-Y. Sun, 1994: A numerical study of a low-level jet and its accompanying secondary circulation in a Mei-Yu System. *Mon. Wea. Rev.,* **122**, 324-340.

Hua, B.-L., and F. Thomasset, 1983: A numerical study of the effects of coastline geometry on wind-induced upwelling in the Gulf of Lions. *J. Phys. Oceanogr.,* **13**, 678-694.

Hueso, R., and A. Sánchez-Lavega, 2006: Methane storms on Saturn's moon Titan. *Nature,* **442(7101)**, 428-431.

Hughes, R.L., 1978: A numerical simulation of mesoscale flow over mountains terrain. Paper No. 303, US ISSN 0067-0340, Department of Atmospheric Science, Colorado State University, Fort Collins.

Hunt, G.E., A.O. Pickersgill, P.B. James, and N. Evans, 1981: Daily and seasonal Viking observations of Martian bore wave systems. *Nature,* **293(5834)**, 630-633.

Hunt, J.C.R., W.H. Snyder, and R.E. Lawson Jr., 1978: Flow structure and turbulent diffusion around a three-dimensional hill. Part I. Flow structure. U.S. Environmental Protection Agency, Office of Research and Development, Report EPA-600/4-78-041, 1-83.

Huschke, R.E., 1959: *Glossary of Meteorology.* American Meteorological Society, Boston, MA, 638 pp.

Hutchinson, M.F., 1995: Stochastic space-time weather models from ground-based date. *Agric. Forest Meteor.,* **73**, 237-264.

Hutchinson, M.F., 1998a: Interpolation of rainfall data with thin plate smoothing splines: I two dimensional smoothing of data with short range correlation. *J. Geog. Info. Dec. Anal.,* **2**, 152-167.

Hutchinson, M.F., 1998b: Interpolation of rainfall data with thin plate smoothing splines: II analysis of topographic dependence. *J. Geographic Info. Decision Anal.,* **2**, 168-185.

Hyun, J.M., and M. Kim, 1979: The effect of nonuniform wind shear on the intensification and reflection of mountain waves. *J. Atmos. Sci.,* **36**, 2379-2384.

Ichinose, T., K. Shimodozono, and K. Hanaki, 1999: Impact of anthropogenic heat on urban climate in Tokyo. *Atmos. Environ.,* **33**, 3897-3909.

Idso, S.B., 1981: Surface energy balance and the genesis of deserts. *Arch. Meteor. Geophys. Bioklimatol.,* **30**, 253-260.

Idso, S.B., R.D. Jackson, B. Kimball, and F. Nakayama, 1975a: The dependence of bare soil albedo on soil water content. *J. Appl. Meteor.,* **14**, 109-113.

Idso, S., R.D. Jackson, and R.J. Reginato, 1975b: Detection of soil moisture by remote surveillance. *Amer. Sci.,* **63**, 549-557.

Idso, S., J.A. Aase, and R.D. Jackson, 1975c: Net radiation-soil heat flux relations as influenced by soil water content variations. *Bound.-Layer Meteor.,* **9**, 113-122.

Igel, A.L., and S.C. van den Heever, 2013: The role of latent heating in warm frontogenesis. *Quart. J. Roy. Meteor. Soc.,* http://dx.doi.org/10.1002/qj.2118.

Igel, A.L., S.C. van den Heever, C.M. Naud, S.M. Saleeby, and D.J. Posselt, 2013: Sensitivity of warm frontal processes to cloud-nucleating aerosol concentrations. *J. Atmos. Sci.*, 70, 1768–1783, http://dx.doi.org/10.1175/JAS-D-12-0170.1.

Iguchi, T., T. Matsui, J.J. Shi, W.-K. Tao, A.P. Khain, A. Hou, R. Cifelli, A. Heymsfield, and A. Tokay, 2012a: Numerical analysis using WRF-SBM for the cloud microphysical structures in the C3VP field campaign: Impacts of supercooled droplets and resultant riming on snow microphysics. *J. Geophys. Res.,* **117**, D23206, http://dx.doi.org/10.1029/2012JD018101.

Iguchi, T., T. Matsui, A. Tokay, and P. Kollias, and W.-K. Tao, 2012b: Two distinct modes seen in one-day rainfall event on the MC3E field campaign: Analyses of disdrometric data and WRF-SBM simulation. *Geophys. Res. Lett.,* **39**, L24805, http://dx.doi.org/10.1029/2012GL053329.

Ingersoll, A.P., D. Crisp, and A.W. Grossman, 1987: Estimates of convective heat fluxes and gravity wave amplitudes in the Venus middle cloud layer from VEGA balloon measurements. *Adv. Space Res.*, **7(12)**, 343-349.

IPCC, 2007: Changes in atmospheric constituents and radiative forcing: Climate change: The physical science basis, Cambridge University Press, New York, USA, and Cambridge, UK.

Iribarne, J.V., and W.L. Godson, 1973: *Atmospheric Thermodynamics*. D. Reidel Publishing, Boston, MA, 222 pp.

Ishihara, M., and Z. Yanagisawa, 1982: Structure of a tropical squall line observed in the western tropical Pacific during MONEX. *Pap. Meteor. Geophys.,* **33**, 117-135.

Jacobs, C.A., and P.S. Brown Jr., 1974: IFYCL Final Report, Vol. IV. Three-dimensional results. CEM Report No. 4131-509d. The Center for the Environment and Man, Hartford, CT.

Jakosky, B.M., 1983: The role of seasonal reservoirs in the Mars water cycle: II. Coupled models of the regolith, the polar caps, and atmospheric transport. *Icarus*, **55(1)**, 19-39.

James, E.P., O.B. Toon, and G. Schubert, 1997: A numerical microphysical model of the condensational Venus cloud. *Icarus*, **129(1)**, 147-171.

Janjić, Z.I., 1977: Pressure gradient force and advection scheme used for forecasting with steep and small scale topography. *Contrib. Atmos. Phys.,* **50**, 186-199.

Janjić, Z.I., 1979: Forward-backward scheme modified to prevent two-grid-interval noise and its application in sigma coordinate models. *Contrib. Atmos. Phys.,* **52**, 69-84.

Janjić, Z.I., 1984a: Non-linear advection schemes and energy cascade on semi-staggered grids. *Mon. Wea. Rev.,* **112**, 1234-1245.

Janjić, Z.I., 1984b: Non-linear advection Eulerian schemes. Workshop on Limited Area Numerical Weather Prediction Models for Computers of Limited Power, Part I. WMO, Programme on Short- and Medium-range Weather Prediction Research, PSMP Rept. Series No. 8, 117-156, Case Postale 2300, CH-1211 Geneve 2.

Janjić, Z.I., 1990: A step-mountain coordinate: physical package. *Mon. Wea. Rev.,* **118**, 1429-1443.

Janjić, Z.I., 1995: A note on the performance of the multiply-upstream semi-Lagrangian advection schemes for one-dimensional nonlinear momentum conservation equation. *Meteor. Atmos. Phys.*, **55**, 1-16.

Janjić, Z.I., 2003: A nonhydrostatic model based on a new approach. *Meteor. Atmos. Phys.,* **82**, 271-285.

Janjić, Z.I., 2011: Recent advances in global nonhydrostatic modeling at NCEP. Invited lecture at the ECMWF Workshop on Non-hydrostatic Modeling. In: Proceedings of the ECMWF Workshop on Non-hydrostatic Modelling, 8-10 November 2010, European Centre for Medium-Range Weather Forecasts, Shinfield Park, Reading, Berkshire, UK., 186 pp. Presentation: http://www.ecmwf.int/newsevents/meetings/workshops/2010/Non_hydrostatic_Modelling/presentations/Janjic.pdf.

Janjić, Z.I., J.P. Gerrity, and S. Nickovic, 2001: An alternative approach to nonhydrostatic modeling. *Mon. Wea. Rev.,* **129**, 129, 1164-1178.

Janjić, Z.I., R. Gall, and M.E. Pyle, 2010: Scientific documentation for the NMM Solver. NCAR Technical Note NCAR/TN-477+STR, 54 pp.

Jankov, I., D.L. Birkenheuer, R.L. Brummer, L.D. Grasso, D. Hillger, D.T. Lindsey, P.J. Neiman, M. Sengupta, H. Yuan, D. Zupanski, and M. Zupanski, 2011: An evaluation of five ARW-WRF microphysics schemes using synthetic GOES imagery for an atmospheric river event affecting the California coast. *J. Hydrometeor.,* **12**, 618-633, http://dx.doi.org/doi:10.1175/2010JHM1282.1.

Jazcilevich, A., V. Fuentes, E. Jauregui, and E. Luna, 2000: Simulated urban climate response to historical land use modification in the basin of Mexico. *Climate Change,* **44**, 515-536.

Jiang, H., W.R. Cotton, J.O. Pinto, J.A. Curry, and M.J. Weissbluth, 2000: Cloud resolving simulations of mixed-phase arctic stratus observed during BASE: Sensitivity to concentration of ice crystals and large-scale heat and moisture advection. *J. Atmos. Sci.,* **57**, 2105-2117.

Jiang, J.H., H. Su, C. Zhai, V.S. Perun, A. Del Genio, L.S. Nazarenko, L.J. Donner, L. Horowitz, C. Seman, J. Cole, A. Gettelman, M.A. Ringer, L. Rotstayn, S. Jeffrey, T. Wu, F. Brient, J.-L. Dufresne, H. Kawai, T. Koshiro, M. Watanabe, T.S. L'Ecuyer, E.M. Volodin, T. Iversen, H. Drange, M.D.S. Mesquita, W.G. Read, J.W. Waters, B. Tian, J. Teixeira, G.L. Stephens, 2012: Evaluation of cloud and water vapor simulations in CMIP5 climate models using NASA "A-Train" satellite observations. *J. Geophys. Res.*, **117**, D14105, http://dx.doi.org/10.1029/2011JD017237.

Jiang, L., L. Lu, W.P. Xing, L. Zhang, I. Baker, and J. Zuo, 2011: Evaluating surface energy budgets simulated by SiB3 at three different climate-ecosystem tower sites. *Scientia Meteorologica Sinica*, **31:4**, 140-147.

Jin, M., and J.M. Shepherd, 2008: Aerosol relationships to warm season clouds and rainfall at monthly scales over east China: Urban-land vs. Ocean. *J. Geophys. Res.*, **113**, D24S90, http://dx.doi.org/10:1029/2008JD010276.

Jin, M., J.M. Shepherd, M.D. King, 2005: Urban aerosols and their interaction with clouds and rainfall: A case study for New York and Houston. *J. Geophys. Res.*, **110**, D10S20, http://dx.doi.org/10.1029/2004JD005081.

Jin, M., J.M. Shepherd, and W. Zheng, 2010: Urban surface temperature reduction via the urban aerosol direct effect: A remote sensing and WRF model sensitivity study. *Adv. Meteor.*, Article ID 681587, 14 pp., http://dx.doi.org/10.1155/2010/681587.

Jin, Y., W.T. Thompson, S. Wang, and C-S Liou, 2007: A numerical study of the effect of dissipative heating on tropical cyclone intensity. *Wea. Forecasting,* **22**, 950-966.

Jin, Z., T.P. Charlock, W.L. Smith Jr., and K. Rutledge, 2004: A parameterization of ocean surface albedo. *J. Geophys. Res.*, **31**, L22301, http://dx.doi.org/10.1029/2004GL021180.

Jiusto, J.E., and G. Bosworth, 1971: Fall velocity of snowflakes. *J. Appl. Meteor.,* **10**, 1352-1354.

Jiusto, J.E., and M.L. Kaplan, 1972: Snowfall from lake-effect storms. *Mon. Wea. Rev.,* **100**, 62-66.

Jiusto, J.E., D.A. Paine, and M.L. Kaplan, 1970: Great Lakes snowstorms, Part 2: Synoptic and climatological aspects. U.S. Department of Commerce, ESSA Grant Report E22-13-69(G).

Johnson Jr., A., and J.J. O'Brien, 1973: A study of an Oregon sea breeze event. *J. Appl. Meteor.,* **12**, 1267-1283.

Johnson, D.R., and L.W. Uccellini, 1983: A comparison of methods for computing the sigma-coordinate pressure gradient force for flow over sloped terrain in a hybrid theta-sigma model. *Mon. Wea. Rev.,* **111**, 870-886.

Johnson, R.H., 1977: Effects of cumulus convection on the structure and growth of the mixed layer over south Florida. *Mon. Wea. Rev.,* **105**, 713-724.

Johnson, R.H., 1982: Vertical motion in near-equatorial winter monsoon convection. *J. Meteor. Soc. Japan,* **60**, 682-690.

Johnson, R.H., and D.C. Kriete, 1982: Thermodynamics and circulation characteristic of winter monsoon tropical mesoscale convection. *Mon. Wea. Rev.,* **110**, 1898-1911.

Johnson, R.H., and M.E. Nicholls, 1983: A composite analysis of the boundary layer accompanying a tropical squall line. *Mon. Wea. Rev.,* **111**, 308-319.

Johnson, R.H., and J.J. Toth, 1982a: Topographic effects and weather forecasting in the Colorado PROFS mesonetwork area. *Preprints, AMS Conference on Weather Forecasting and Analysis*, Seattle, Washington.

Johnson, R.H., and J.J. Toth, 1982b: A climatology of the July 1981 surface flow over northeast Colorado. Colorado State University, Department of Atmospheric Science Paper No. 342, Fort Collins.

Johnson, R.H., T.M. Rickenbach, S.A. Rutledge, P.E. Ciesielski, and W.H. Schubert, 1999: Trimodal characteristics of tropical convection. *J. Climate,* **12**, 2397-2418.

Jones, A.S., I.C. Guch, and T.H. Vonder Haar, 1998: Data assimilation of satellite-derived heating rates as proxy surface wetness data into a regional atmospheric mesoscale model: Part I. Methodology, *Mon. Wea. Rev.,* **126**, 634-645.

Jones, R.W., 1973: A numerical experiment on the prediction of the northeast (winter) monsoon in southwest Asia. NOAA Technical Report ERL 272-WMPO-3, 1-56.

Jones, R.W., 1977a: Noise control for a nested grid tropical cyclone model. *Contrib. Atmos. Phys.,* **50**, 393-402.

Jones, R.W., 1977b: A nested grid for a three-dimensional model. *J. Atmos. Sci.,* **34**, 1528-1553.

Jones, R.W., 1980: A three-dimensional tropical cyclone model with release of latent heat by the resolvable scales. *J. Atmos. Sci.,* **37**, 930-938.

Junge, C.E., 1963: *Air Chemistry and Radioactivity*. Academic Press, New York, 382 pp.

Kabat, P., Ed., 1999: Global Change Newsletter- The IGBP-BAHC Special Issue, **39**, September, 32 pp.

Kabat, P., Claussen, M., Dirmeyer, P.A., J.H.C. Gash, L. Bravo de Guenni, M. Meybeck, R.A. Pielke Sr., C.J. Vorosmarty, R.W.A. Hutjes, and S. Lutkemeier, Editors, 2004: *Vegetation, Water, Humans and the Climate: A New Perspective on an Interactive System*. Springer, Berlin, Global Change - The IGBP Series, 566 pp.

Kader, B.A., and Yaglom, A.M., 1990: Mean fields and fluctuation moments in unstably stratified turbulent boundary layers. *J. Fluid Mech.*, **212**, 637-661.

Kahn, R., and P. Gierasch, 1982: Long cloud observations on Mars and implications for boundary layer characteristics over slopes. *J. Geophys. Res.*, **87(A2)**, 867-880.

Kahre, M.A., J.R. Murphy, and R.M. Haberle, 2006: Modelling the Martian dust cycle and surface dust reservoirs with the NASA Ames general circulation model. *J. Geophys. Res. E: Planets*, **111(6)**, E06008, http://dx.doi.org/10.1029/2005JE002588.

Kahre, M.A., R.J. Wilson, J.L. Hollingsworth, and R.M. Haberle, 2010: Using assimilation techniques to model Mars' dust cycle with the NASA Ames and NOAA/GFDL Mars General Circulation Models. *Bull. Amer. Astronomical Soc.*, **42**, 1031.

Kain, J.S., 2004: The Kain-Fritsch convective parameterization. An update. *J. Appl. Meteor.,* **43**, 170-181.

Kain, J.S., and J.M. Fritsch, 1990: A one-dimensional entraining/detraining plume model and its application in convective parameterization. *J. Atmos. Sci.,* **47**, 2784-2802.

Kain, J.S., and J.M. Fritsch, 1993: Convective parameterization for mesoscale models: The Kain-Fritsch Scheme. *The Representation of Cumulus Convection in Numerical Models, Meteor. Monogr.,* No. 46, Amer. Meteor. Soc., 165-170.

Kallo, K.P., and T.W. Owen, 1999: Satellite-based adjustments for urban heat island bias. *J. Appl. Meteor.,* **38**, 806-813.

Kallos, G., P. Kassomenos, and R.A. Pielke, 1993: Synoptic and mesoscale weather conditions during air pollution episodes in Athens, Greece. *Bound.-Layer Meteor.,* **62**, 163-184.

Kallos, G., V. Kotroni, K. Lagouvardos, and A. Papadopoulos, 1998: On the long-range transport of air pollutants from Europe to Africa. *Geophys. Res. Letters,* **25**, 619-622.

Kalnay, E., 2002: *Atmospheric Modeling, Data Assimilation and Predictability*. Cambridge University Press, 364 pp.

Kalnay, E., M. Kanamitsu, R. Kistler, W. Collins, D. Deaven, L. Gandin, M. Iredell, S. Saha, G. White, J. Woollen, Y. Zhu, M. Chelliah, W. Ebisuzaki, W. Higgins, J. Janowiak, K.C. Mo, C. Ropelewski, J. Wang, A. Leetmaa, R. Reynolds, R. Jenne, and D. Joseph, 1996: The NCEP/NCAR 40-Year Reanalysis Project. *Bull. Amer. Meteor. Soc.,* **77**, 437-471.

Kalthoff, N., F. Fiedler, M. Kohler, O. Kolle, H. Mayer, and A. Wenzel, 1999: Analysis of energy balance components as a function of orography and land use and comparison of results with the distribution of variables influencing local climate. *Theor. Appl. Climatol.,* **62**, 65-84.

Kaminski, T., R. Giering, and M. Heimann, 1997: Sensitivity of the seasonal cycle of CO_2 at remote monitoring stations with respect to seasonal surface exchange fluxes determined with the adjoint of an atmospheric transport model. *Phys. Chemistry of the Earth,* **21**, 457-463.

Kanamitsu, M., and H. Kanamaru, 2007: Fifty-seven-year California reanalysis downscaling at 10 km (CaRD10): Part I. System detail and validation with observations. *J. Climate*, **20**, 5553-5571.

Kanda, M., and Y. Inoue, 2001: Numerical study on cloud lines over an urban street induced by heat island in Tokyo metropolitan area. *Bound.-Layer Meteor.,* **98**, 251-273.

Kang, S.-D., F. Kimura, and S. Takahashi, 1998: A numerical study on the Karman vortex generated by divergence of momentum flux in flow past an isolated mountain. *J. Meteor. Soc. Japan,* **76**, 925-935.

Kaplan, W., 1952: *Advanced Calculus*. Addison-Wesley. Reading, MA.

Kaplen, M.L., J.W. Zack, V.C. Wong, and J.J. Tuccillo, 1982: Initial results from a mesoscale atmospheric simulation system and comparisons with the AVE-SESAME I data set. *Mon. Wea. Rev.,* **110**, 1564-1590.

Kasahara, A., 1974: Various vertical coordinate systems used for numerical weather prediction. *Mon. Wea. Rev.,* **102**, 509-522.

Kasahara, A., 1982: Nonlinear normal mode initialization and the bounded derivative method. *Rev. Geophys. Space Phys.,* **20**, 385-397.

Kassomenos, P., V. Kotroni, and G. Kallos, 1995: Analysis of climatological and air quality observations from greater Athens area. *Atmos. Environ.,* **29**, 3671-3688.

Kasten, F., 1969: Visibility forecast in the phase of precondensation. *Tellus,* **21**, 631-635.

Kauhanen, J., T. Siili, S. Järvenoja, and H. Savijärvi, 2008: The Mars limited area model and simulations of atmospheric circulations for the Phoenix landing area and season of operation. *J. Geophys. Res.*, **113(E3)**, E00A14, http://dx.doi.org/10.1029/2007JE003011.

Kawamura, T., H. Takami, and K. Kuwahara, 1986: Computation of high Reynolds number around a circular cylinder with surface roughness. *Fluid Dyn. Res.*, **1**, 145-162.

Keen, C.S., and W.A. Lyons, 1978: Lake/land breeze circulations on the western shore of Lake Michigan. *J. Appl. Meteor.,* **17**, 1843-1855.

Keen, C.S., W.A. Lyons, and J.A. Schuh, 1979: Air pollution transport studies in a coastal zone using kinematic diagnostic analysis. *J. Appl. Meteor.,* **18**, 606-615.

Kerschgens, M., 1983: Parameterization of radiation for the modeling of the urban-boundary layer. *AMS/EPA Specialty Conference on Air Quality Modeling of the Nonstationary, Nonhomogeneous Urban Boundary Layer*, October 31-November 4, Baltimore, MD.

Kessler, E., 1969: On the distribution and continuity of water substance in atmospheric circulations. *Meteor. Mag.,* **32**, 1-84.

Kessler, R.C., and R.A. Pielke, 1982: A numerical study of airflow over adjacent ridges. Unpublished manuscript.

Kessler, R.C., and R.A. Pielke, 1983: A numerical study of the effect of latent heat release on the airflow over a mountain. Unpublished manuscript.

Keyser, D., and R.A. Anthes, 1977: The applicability of a mixed-layer model of the planetary boundary layer to real-data forecasting. *Mon. Wea. Rev.,* **105**, 1351-1371.

Keyser, D., and D.R. Johnson, 1982: Effects of diabatic heating on the ageostrophic circulation of an upper tropospheric jet streak. NASA Contractor Report 3497, NASA Marshall Space Flight Center.

Khairoutdinov, M., and Y. Kogan, 2000: A new cloud physics parameterization in a large-eddy simulation model of marine stratocumulus. *Mon. Wea. Rev.,* **128**, 229-243.

Kidder, S.Q., and T.H. Vonder Haar, 1995: *Satellite Meteorology*. Academic Press, 466 pp.

Kikuchi, Y., S. Arakawa, F. Kimura, K. Shirasaki, and Y. Nagano, 1981: Numerical study on the effects of mountains on the land and sea breeze circulation in the Kanto district. *J. Meteor. Soc. Japan,* **59**, 67-85.

Kimura, F., and T. Kuwagata, 1995: Horizontal heat fluxes over complex terrain computed using a simple mixed-layer model and a numerical model. *J. Appl. Meteor.,* **34**, 549-558.

Kimura, F., and S. Takahashi, 1991: The effects of land-use and anthropogenic heating on the surface temperature in the Tokyo metropolitan area: a numerical experiment. *Atmos. Environ.,* **25B**, 155-164.

Kimura, R., and T. Eguchi, 1978: On dynamical processes of sea and land breeze circulations. *J. Meteor. Soc. Japan,* **59**, 67-85.

Kirshbaum, D.J., and D.R. Durran, 2005: Observations and modeling of banded orographic convection. *J. Atmos. Sci.,* **62**, 1463-1479.

Kirshbaum, D.J., G.H. Bryan, R. Rotunno, and D.R. Durran, 2007: The triggering of orographic rainbands by small-scale topography. *J. Atmos. Sci.,* **64**, 1530-1549.

Kleinböhl, A., et al. 2009: Mars Climate Sounder limb profile retrieval of atmospheric temperature, pressure, and dust and water ice opacity. *J. Geophys. Res. E: Planets,* **114(10)**, http://dx.doi.org/10.1029/2009JE003358.

Klemp, J.B., and D.K. Lilly, 1975: The dynamics of wave-induced downslope winds. *J. Atmos. Sci.,* **32**, 320-339.

Klemp, J.B., and D.K. Lilly, 1978: Numerical simulation of hydrostatic mountain waves. *J. Atmos. Sci.,* **35**, 78-107.

Klemp, J.B., and D.K. Lilly, 1980: Mountain waves and momentum flux. GARP Series No. 23, June 1980, WMO, Geneva, Switzerland, 116-141.

Klemp, J.B., and R.B. Wilhelmson, 1978a: The simulation of three-dimensional convective storm dynamics. *J. Atmos. Sci.,* **35**, 1070-1096.

Klemp, J.B., and R.B. Wilhelmson, 1978b: Simulations of right- and left-moving storms produced through storm splitting. *J. Atmos. Sci.*, **35**, 1097-1110.

Klink, K., 1995: Temporal sensitivity of regional climate to land-surface heterogeneity. *Phys. Geog.,* **16**, 289-314.

Knowles, C.E., and J.J. Singer, 1977: Exchange through a barrier island inlet: Additional evidence of upwelling off the northeast coast of North Carolina. *J. Phys. Oceanogr.,* **7**, 146-152.

Knowles, Captain J.B., 1993: The influence of forest fire induced albedo differences on the generation of mesoscale circulations. Department of Atmospheric Science, M.S. Thesis, Colorado State University, 86 pp.

Knox, J.B., M.C. MacCracken, M.H. Dickerson, P.M. Gresho, F.M. Luther, and R.C. Orphan, 1983: Program report for FY 1982. Atmospheric and Geophysical Sciences Division of the Physics Department. Lawrence Livermore Laboratory Report UCRL-51444-82.

Knupp, K.R., 1987: Downdrafts within High Plains cumulonimbi, Pt. 1: General kinematic structure. *J. Atmos. Sci.,* **44**, 987-1008.

Knupp, K.R., and W.R. Cotton, 1982a: An intense, quasi-steady thunderstorm over mountainous terrain. Part II: Doppler radar observations of the storm morphological structure. *J. Atmos. Sci.,* **39**, 343-358.

Knupp, K.R., and W.R. Cotton, 1982b: An intense quasi-steady thunderstorm over mountainous terrain. Part III: Doppler radar observations of the turbulent structure. *J. Atmos. Sci.,* **39**, 359-368.

Koch, S.E., and J. McCarthy, 1982: The evolution of an Oklahoma dryline, Part II: Boundary-layer forcing of mesoconvective systems. *J. Atmos. Sci.,* **39**, 237-257.

Koch, W., and F. Feser, 2006: Relationship between SAR-derived wind vectors and wind at 10-m height represented by a mesoscale model. *Mon. Wea. Rev.,* **134**, 1505-1517.

Kocin, P.I., and L.W. Uccellini, 2004: *Northeast Snowstorms*. Meteor. Monogr., No. 54, American Meteorological Society, 818 pp.

Kolczynski Jr., W.C., D.R. Stauffer, S.E. Haupt, and A. Deng, 2009: Ensemble variance calibration for representing meteorological uncertainty for atmospheric transport and dispersion modeling. *J. Appl. Meteor. Climatol.,* **48**, 2001-2021.

Kondo, H., 1981: A numerical simulation of the influence of the large scale field on a mesoscale disturbance. *J. Meteor. Soc. Japan,* **59**, 123-132.

Kondo, J., Y. Sasano, and T. Ishii, 1979: On wind-driven current and temperature profiles with diurnal period in the oceanic planetary boundary layer. *J. Phys. Oceanogr.,* **9**, 360-372.

Kondratyev, J., 1969: *Radiation in the Atmosphere*. Academic Press, New York, 912 pp.

Korb, G.J., R.M. Welch, and W.G. Zdunkowski, 1975: An approximate method for the determination of infrared fluxes in scattering and absorbing media. *Contrib. Atmos. Phys.,* **48**, 84-95.

Kossmann, M., U. Corsmeier, S.F.J. DeWekker, F. Fielder, R. Vogtlin, N. Kalthoff, H. Gusten, and B. Neininger, 1999: Observations of handover processes between the atmospheric boundary layer and the free troposphere over mountainous terrain. *Contrib. Atmos. Phys.,* **72**, 329-350.

Kotroni, V., G. Kallos, and K. Lagouvardos, 1997: Convergence zones over the Greek peninsula and associated thunderstorm activity. *Quart. J. Roy. Meteor. Soc.,* **123**, 1961-1984.

Kotroni, V., G. Kallos, K. Lagouvardos, M. Varinou, and R. Walko, 1999a: Numerical simulations of the meteorological and dispersion conditions during an air pollution episode over Athens, Greece. *J. Appl. Meteor.,* **38**, 432-447.

Kotroni, V., K. Lagouvardos, and G. Kallos, 1999b: Model investigation of a cloudband associated with a cold front over Eastern Mediterranean. *Phys. Chem. Earth,* **24**, 633-636.

Kotroni, V., K. Lagouvardos, G. Kallos, and D. Ziakopoulos, 1999c: Severe flooding over central and southern Greece associated with pre-cold frontal orographic lifting. *Quart. J. Roy. Meteor. Soc.,* **125**, 967-991.

Kourzeneva, E., 2010: External data for lake parameterization in numerical weather prediction and climate modeling. *Boreal Env. Res.*, **15**, 165-177.

Kowalczyk, G.S., C.E. Choquette, and G.E. Gordon, 1978: Chemical element balances and identification of air pollution sources in Washington, D.C. *Atmos. Environ.,* **12**, 1143-1153.

Kozo, T.L., 1982a: An observational study of sea breezes along the Alaskan Beaufort sea coast: Part I. *J. Appl. Meteor.,* **21**, 891-905.

Kozo, T.L., 1982b: An mathematical model of sea breezes along the Alaskan Beaufort sea coast: Part II. *J. Appl. Meteor.,* **21**, 906-924.

Kozo, T.L., 1982c: Mesoscale wind phenomena along the Alaskan Beaufort sea coast. In: *The Alaskan Beaufort Sea*, E. Reimnitz, Ed., Pacific Arctic Branch of Marine Geology, U.S. Department of Interior.

Krasnopolsky, V.A., 2006: Chemical composition of Venus atmosphere and clouds: Some unsolved problems. *Planet Space Sci.*, **54(13-14)**, 1352-1359.

Kreitzberg, C.W., and D.J. Perkey, 1976: Release of potential instability. Part I: A sequential plume model within a hydrostatic primitive equation model. *J. Atmos. Sci.,* **33**, 40-63.

Kreitzberg, C.W., and D.J. Perkey, 1977: Release of potential instability. Part II. The mechanism of convective/mesoscale interaction. *J. Atmos. Sci.,* **34**, 1579-1595.

Krishnamurthy, R., R. Calhoun, B. Billings, and J. Doyle, 2013: Mesoscale model evaluation with coherent Doppler Lidar for wind farm assessment. *Remote Sens. Lett.*, 4, 579-588, http://dx.doi.org/10.1080/2150704X.2013.769285.

Krishnamurti, T.N., 1962: Numerical integration of primitive equations by a quasi-Lagrangian advective scheme. *J. Appl. Meteor.,* **1**, 508-521.

Krishnamurti, T.N., and L. Bounoua, 1995: *An Introduction to Numerical Weather Prediction Techniques*. CRC Press, Washington DC, 304 pp.

Krishnamurti, T.N., and M. Kanamitsu, 1973: A study of a coasting easterly wave. *Tellus,* **25**, 568-585.

Krishnamurti, T.N., and W.J. Moxim, 1971: On parameterization of convective and nonconvective latent heat releases. *J. Appl. Meteor.,* **10**, 3-13.

Krishnamurti, T.N., M. Kanamitsu, B. Ceselski, and M.B. Mathur, 1973: Florida State University's Tropical Prediction Model. *Tellus,* **25**, 523-535.

Krishnamurti, T.N., Y. Ramanathan, H.-L. Pan, R.J. Rasch, and J. Molinari, 1980: Cumulus parameterization and rainfall rates. *Mon. Wea. Rev.*, **108**, 465-472.

Krishnamurti, T.N., V. Wong, H.-L. Pan, R. Pasch, J. Molinari, and P. Ardanuy, 1983: A three-dimensional planetary boundary layer model for the Somali jet. *J. Atmos. Sci.,* **40**, 894-908.

Kuettner, J.P., 1986: Aim and conduct of ALPEX. In: *Global Atmospheric Research Programme*, GARP Publications Series No. 27, WMO/TD-No. 108, 3-13.

Kuhn, P., 1963: Radiometeorsonde observations of infrared flux emissivity of water vapor. *J. Appl. Meteor.,* **2**, 368-378.

Kumar, P., M.P. Singh, and A.N. Natarajan, 1998: An analytical model for mountain wave in stratified atmosphere. *MAUSAM,* **49**, 433-438.

Kummerow, C., 1993: On the accuracy of the Eddington approximation for radiative transfer in the microwave frequencies. *J. Geophys. Res.,* **98**, 2757-2765.

Kundzewicz, Z.W., and E.Z. Stakhiv, 2010: Are climate models "ready for prime time" in water resources management applications, or is more research needed? Editorial. *Hydrol. Sci. J.*, **55(7)**, 1085-1089.

Kuo, H.L., 1965: On the formation and intensification of tropical cyclones through latent heat release by cumulus convection. *J. Atmos. Sci.*, **22**, 40-63.

Kuo, H.L., 1968: The thermal interaction between the atmosphere and the earth and propagation of diurnal temperature waves. *J. Atmos. Sci.,* **25**, 682-706.

Kuo, H.L., 1974: Further studies of the parameterization of the influence of cumulus convection on large-scale flow. *J. Atmos. Sci.*, **31**, 1232-1240.

Kuo, H.L., 1979: Infrared cooling rate in a standard atmosphere. *Contrib. Atmos. Phys.,* **52**, 85-94.

Kuo, H.L., and W.H. Raymond, 1980: A quasi-one-dimensional cumulus cloud model and parameterization of cumulus heating and mixing effects. *Mon. Wea. Rev.,* **108**, 991-1009.

Kuo, Y.-H., and Y.-R. Guo, 1989: Dynamic initialization using observations from a network of profilers and its impact on short-range numerical weather prediction. *Mon. Wea. Rev.,* **117**, 1975-1998.

Kuo, Y.-H., R.J. Reed, and S. Low-Nam, 1992: Thermal structure and airflow in a model simulation of an occluded marine cyclone. *Mon. Wea. Rev.,* **120**, 2280-2297.

Kurihara, Y., 1973: A scheme for moist convective adjustment. *Mon. Wea. Rev.,* **101**, 547-553.

Kurihara, Y., 1975: Budget analysis of a tropical cyclone simulated in an axisymmetric numerical model. *J. Atmos. Sci.,* **32**, 25-59.

Kurihara, Y., 1976: On the development of spiral bands in a tropical cyclone. *J. Atmos. Sci.,* **33**, 940-958.

Kurihara, Y., and M.A. Bender, 1979: Supplementary note on a scheme of dynamic initialization of the boundary layer in a primitive equation model. *Mon. Wea. Rev.,* **107**, 1219-1221.

Kurihara, Y., and M.A. Bender, 1982: Structure and analysis of the eye of a numerically simulated tropical cyclone. *J. Meteor. Soc. Japan,* **60**, 381-395.

Kurihara, Y., and R.E. Tuleya, 1974: Structure of a tropical cyclone developed in a three-dimensional numerical simulation model. *J. Atmos. Sci.,* **31**, 893-919.

Kurihara, Y., and R.E. Tuleya, 1978: A scheme for dynamic initialization of the boundary layer in a primitive equation model. *Mon. Wea. Rev.,* **106**, 113-123.

Kusaka, H., H. Kondo, Y. Kikegawa, and F. Kimura, 2001: A simple single-layer urban canopy model for atmospheric models: Comparison with multi-layer and slab models. *Bound.-Layer Meteor.,* **101**, 329-358.

Kuwagata, T., and F. Kimura, 1997: Daytime boundary layer evolution in a deep valley: Part II: Numerical simulation of the cross-valley circulation. *J. Appl. Meteor.,* **36**, 883-895.

Kuzmin, R.O., R. Greeley, S.C.R. Rafkin, and R. Haberle, 2001: Wind-related modification of some small impact craters on Mars. *Icarus*, **153(1)**, 61-70.

Lacis, A.A., and J.E. Hansen, 1974: A parameterization for the absorption of solar radiation in the Earth's atmosphere. *J. Atmos. Sci.,* **31**, 118-133.

Lackmann, G., 2012: *Midlatitude Synoptic Meteorology: Dynamics, Analysis, and Forecasting*. American Meteorology Society, 345 pp.

Lagouvardos, K., V. Kotroni, S. Dobricic, S. Nickovic, and G. Kallos, 1996: The storm of October 21-22, 1994, over Greece: Observations and model results. *J. Geophys. Res.,* **101**, 26217-26226.

Lagouvardos, K., V. Kotroni, and G. Kallos, 1998: An extreme cold surge over the Greek peninsula. *Quart. J. Roy. Meteor. Soc.,* **124**, 2299-2327.

Lagouvardos, K., V. Kotroni, S. Nickovic, D. Jovic, and G. Kallos, 1999: Observations and model simulations of a winter sub-synoptic vortex over the central Mediterranean. *Meteorol. Appl.,* **6**, 371-383.

Lakhtakia, M.N., and T.T. Warner, 1994: A comparison of simple and complex treatments of surface hydrology and thermodynamics suitable for mesoscale atmospheric models. *Mon. Wea. Rev.,* **122**, 880-896.

Lal, M., 1979: Application of Pielke model to air quality studies. *Mausam,* **30**, 69-78.

Lalas, D.P., V.R. Veirs, G. Karras, and G. Kallos, 1982: An analysis of the SO_2 concentration levels in Athens, Greece. *Atmos. Environ.,* **16**, 531-544.

Lalas, D.P., D.N. Asimakopoulos, D.G. Deligiorgi, and C.G. Helmis, 1983: Sea breeze circulation and photochemical pollution in Athens, Greece. *Atmos. Environ.,* **17**, 1621-1632.

Lamb, H., 1908: On the theory of waves propagated vertically in the atmosphere. *Proc. London. Math. Soc.*, **7**, 122-141.

Lamb, H., 1932: *Hydrodynamics*. Dover, 738 pp.

Landsberg, H.E., 1981: *The Urban Climate*. Academic Press, New York, NY, 275 pp.

Landsea, C., and J. Knaff, 2000: How much skill was there in forecasting the very strong 1997-98 El Niño? *Bull. Amer. Meteor. Soc.,* **81**, 2107-2119.

Lange, R., 1981: Modeling a multiple tracer release experiment during nocturnal drainage flow in complex terrain. *Preprints, AMS 2nd Conference on Mountain Meteorology*, Steamboat Springs, Colorado, November 10-13.

Lanicci, J.M., T.N. Carlson, and T.T. Warner, 1987: Sensitivity of the great Plains severe-storm environment to soil-moisture distribution. *Mon. Wea. Rev.,* **115**, 2660-2673.

Lapidus, A., 1967: A detached shock calculation by second-order finite differences. *J. Comput. Phys.,* **2**, 154-177.

Laprise R., 1992a: The Euler equations of motion with hydrostatic pressure as an independent variable. *Mon. Wea. Rev.*, **120**, 197-207.

Laprise, R., 1992b: The resolution of global spectral models. *Bull. Amer. Meteor. Soc.*, **9**, 1453-1454.

Laprise, J.P.R., and A. Plante, 1995: A class of Semi-Lagrangian Integrated-Mass (SLIM) numerical transport algorithms. *Mon. Wea. Rev.*, **123**, 553-565.

Laprise, R., D. Caya, M. Giguére, G. Bergeron, H. Côté, J.-P. Blanchet, G.J. Boer, and N.A. McFarlane, 1998: Climate and climate change in western Canada as simulated by the Canadian Regional Climate Model. *Atmos.-Ocean*, **26**, 119-167.

Lareau, N.P., E. Crosman, C.D. Whiteman, J.D. Horel, S.W. Hoch, W.O.J. Brown, T.W. Horst, 2013: The persistent cold-air pool study. *Bull. Amer. Meteor. Soc.,* **94**, 51-63. http://dx.doi.org/10.1175/BAMS-D-11-00255.1.

Larsen, S.E., H.E. Jørgensen, L. Landberg, and J.E. Tillman, 2002: Aspects of the atmospheric surface layers on Mars and Earth. *Bound.-Layer Meteor.*, **105(3)**, 451-470.

Lascaux, F., E. Richard, and J.-P. Pinty, 2006: Numerical simulations of three map IOPs and the associated microphysical processes. *Quart. J. Roy. Meteor. Soc.*, **132**, 1907-1926.

Lavoie, R.L., 1972: A mesoscale numerical model of lake-effect storms. *J. Atmos. Sci.*, **29**, 1025-1040.

Lavoie, R.L., 1974: A numerical model of trade wind weather over Oahu. *Mon. Wea. Rev.*, **102**, 630-637.

Lazier, J., and H. Sandstrom, 1978: Migrating thermal structure in a freshwater thermocline. *J. Phys. Oceanogr.*, **8**, 1070-1079.

Lebreton, J.-P., and D.L. Matson, 2002: The Huygens probe: science, payload and mission overview. *Space Sci. Rev.*, **104**, 59-100.

Lebreton, J.-P., O. Witasse, C. Sollazzo, T. Blancquaert, P. Couzin, Anne-Marie Schipper, J.B. Jones, D.L. Matson, L.I. Gurvits, D.H. Atkinson, B. Kazeminejad, and M. Pérez-Ayúcar, 2005: An overview of the descent and landing of the Huygens probe on Titan. *Nature*, **438(7069)**, 758-764.

Leduc, R., G. Jacques, M. Ferland, and C. Lelièvre, 1981: Ilot de Chaleur a Quebec: Cas d'hiver. *Bound.-Layer Meteor.*, **21**, 315-324.

Lee, H.N., 1981: An alternate pseudospectral model for pollutant transport, diffusion and deposition in the atmosphere. *Atmos. Environ.*, **15**, 1017-1024.

Lee, J.A., L.J. Peltier, S.E. Haupt, J.C. Wyngaard, D.R. Stauffer, and A. Deng, 2009: Improving SCIPUFF dispersion forecasts with NWP ensembles. *J. Appl. Meteor. Climatol.*, **48**, 2305-2319.

Lee, J.D., 1973: Numerical simulation of the planetary boundary layer over Barbodos, W.I. Florida State University, Ph.D. Thesis, Tallahassee.

Lee, J.T., S. Barr, W.H. Snyder, and R.E. Lawson Jr., 1981: Wind tunnel studies of flow channeling in valleys. *Preprints, AMS 2nd Conference on Mountain Meteorology*, Steamboat Springs, Colorado, November 9-12, 331-338.

Lee, R., 1978: *Forest Micrometeorology*. Columbia University Press, New York, 276 pp.

Lee, R.L., and D.B. Olfe, 1974: Numerical calculations of temperature profiles over an urban heat island. *Bound.-Layer Meteor.*, **7**, 39-52.

Lee, T.J., and R.A. Pielke, 1992: Estimating the soil surface specific humidity. *J. Appl. Meteor.*, **31**, 480-484.

Lee, T.J., R.A. Pielke, R.C. Kessler, and J. Weaver, 1989: Influence of cold pools downstream of mountain barriers on downslope winds and flushing. *Mon. Wea. Rev.*, **117**, 2041-2058.

Lee, T.J., R.A. Pielke, T.G.F. Kittel, and J.F. Weaver, 1993: Atmospheric modeling and its spatial representation of land surface characteristics. *Environmental Modeling with GIS*, M. Goodchild, B. Parks, and L.T. Steyaert, Eds. Oxford University Press, 108-122.

Lee, T.J., R.A. Pielke, and P.W. Mielke Jr., 1995: Modeling the clear-sky surface energy budget during FIFE87. *J. Geophys. Res.*, **100**, 25585-25593.

Lee, T.-Y., Y.-Y. Park, and Y.-L. Lin, 1998: A numerical modeling study of mesoscale cyclogenesis to the east of the Korean Peninsula. *Mon. Wea. Rev.*, **126**, 2305-2329.

Leelasakultum, K., N. Mölders, H.N.Q. Tran, H.N.Q., and G.A. Grell, 2012: Potential impacts of the introduction of low-sulfur fuel on $PM_2.5$ concentrations at breathing level in a subarctic city. *Adv. Meteor.*, http://dx.doi.org/10.1155/2012/427078.

Leffler, R.L., and J.L. Foster, 1974: Snowfall on the Allegheny plateau of Maryland and West Virginia. *Weatherwise*, **27**, 199-201.

Lehner, M., and C.D. Whiteman, 2012: The thermally driven cross-basin circulation in idealized basins under varying wind conditions. *J. Appl. Meteor. Climatol.,* **51**, 1026-1045.

Lehner, M., and C.D. Whiteman, 2013: Physical mechanisms of the thermally driven cross-basin circulation. *Quart. J. Roy. Meteor. Soc.,* http://dx.doi.org/10.1002/qj.2195.

Lei, L., D.R. Stauffer, and A. Deng, 2012: A hybrid nudging - ensemble Kalman filter approach to data assimilation in WRF/DART. *Quart. J. Roy. Meteor. Soc.,* **138**, 2066-2078.

Leidner, S.M., D.R. Stauffer, and N.L. Seaman, 2001: Improving California coastal zone numerical weather prediction by dynamic initialization of the marine layer. *Mon. Wea. Rev.,* **129**, 275-294.

LeMone, M.A., R.L. Grossman, R.L. Coulter, M.L. Wesley, G.E. Klazura, G.S. Poulos, W. Blumen, J.K. Lundquist, R.H. Cuenca, S.F. Kelly, E.A. Brandes, S.P. Oncley, R.T. McMillen, and B.B. Hicks, 2000: Land atmosphere interaction research, early results, and opportunities in the walnut river watershed in southeast Kansas: CASES and ABLE. *Bull. Amer. Meteor. Soc.,* **81**, 757-779.

Lenschow, D.H., B.B. Stankov, and L. Mahrt, 1979: The rapid morning boundary-layer transition. *J. Atmos. Sci.,* **36**, 2108-2124.

Leoncini, G., R.A. Pielke Sr., and P. Gabriel, 2008: From model based parameterizations to Lookup Tables: An EOF approach. *Wea. Forecasting*, **23**, 1127-1145.

Letts, M.G., N.T. Roulet, N.T. Comer, M.R. Skarupa, and D.L. Vereghy, 2000: Parameterization of peatland hydraulic properties for the Canadian Land Surface Scheme. *Atmos.-Ocean,* **38**, 141-160.

Levin, Z., E. Ganor, and V. Gladstein, 1996: The effects of desert particles coated with sulfate on rain formation in the eastern Mediterranean. *J. Appl. Meteor.,* **35**, 1511-1523.

Levin, Z., A. Teller, E. Ganor, and Y. Yin, 2005: On the interactions of mineral dust, sea-salt particles and clouds: A measurement and modeling study from the Mediterranean Israeli Dust Experiment campaign. *J. Geophys. Res.,* **110**, D20202, http://dx.doi.org/10.1029/2005JD005810.

Lewellen, W.S., 1983: Parameterization of subgrid-scale fluxes and estimation of dispersion. *AMS/EPA Specialty Conference on Air Quality Modeling of the Nonstationary, Nonhomogeneous Urban Boundary Layer*, October 31-November 4, Baltimore, MD.

Lewis, S.R., and P.R. Barker, 2005: Atmospheric tides in a Mars general circulation model with data assimilation. *Adv. Space Res.*, **36(11)**, 2162-2168.

Lewis, S.R., P.L. Read, B.J. Conrath, J.C. Pearl, and M.D. Smith, 2007: Assimilation of thermal emission spectrometer atmospheric data during the Mars Global Surveyor aerobraking period. *Icarus*, **192(2)**, 327-347.

Ley, B.E., and W.R. Peltier, 1981: Propagating mesoscale cloud bands. *J. Atmos. Sci.,* **38**, 1206-1219.

Li, J., F. Sun, S. Seemann, E. Weisz, and A.H.-L. Huang, 2004: GIFTS sounding retrieval algorithm development. *20th International Conference on Interactive Information and Processing Systems (IIPS) for Meteorology, Oceanography, and Hydrology*, Amer. Meteor. Soc., Seattle, Wash., 11-15 Jan.

Li, X., W.-K. Tao, T. Matsui, C. Liu, and H. Masunaga, 2010: Improving a spectral bin microphysical scheme using long-term TRMM satellite observations. *Quart. J. Roy. Meteor. Soc.,* **136(647)**, 382-399.

Li, Y., and J.R. Bates, 1996: A study of the behaviour of semi-Lagrangian models in the presence of orography. *Quart. J. Roy. Meteor. Soc.,* **122**, 1675-1700.

Lilly, D.K., 1961: A proposed staggered-grid system for numerical integration of dynamic equations. *Mon. Wea. Rev.,* **89**, 59-65.

Lilly, D.K., Ed., 1975: Open SESAME. *Proceedings of SESAME Open Meeting*. Prepared by NOAA, ERL, Boulder, Colorado, September 4-6, 1974.

Lilly, D.K., 1979: The dynamical structure and evolution of thunderstorms and squall lines. *Ann. Rev. Earth Planet. Sci.,* **7**, 117-161.

Lilly, D.K., 1981: Wave-permeable lateral boundary conditions for convective cloud and storm simulations. *J. Atmos. Sci.,* **38**, 1313-1316.

Lilly, D.K., 1982: Gravity waves and mountain waves. Lecture notes at the NATO Advanced Study Institute on Mesoscale Meteorology – Theory, Observations, and Models, July 13-31, Gascogne, France, Reidel, New York.

Lilly, D.K., 1983: Stratified turbulence and the mesoscale variability of the atmosphere. *J. Atmos. Sci.,* **40**, 749-761.

Lilly, D.K., 1986: The structure, energetics and propagation of rotating convective storms. Part II: Helicity and storm stabilization. *J. Atmos. Sci.,* **43**, 126-140. http://dx.doi.org/10.1175/1520-0469(1986)043<0126:TSEAPO>2.0.CO;2.

Lilly, D.K., 1996: A comparison of incompressible, anelastic and Boussinesq dynamics. *Atmos. Res.,* **40**, 143-151.

Lilly, D.K., and P.J. Kennedy, 1973: Observations of a stationary mountain wave and its associated momentum flux and energy dissipation. *J. Atmos. Sci.,* **30**, 1135-1152.

Lilly, D.K., and J.B. Klemp, 1979: The effects of terrain shape on nonlinear hydrostatic mountain waves. *J. Fluid Mech.,* **95**, 241-261.

Lilly, D.K., and J.B. Klemp, 1980: Comments on the evolution and stability of finite-amplitude mountain waves. Part II: Surface wave drag and severe downslope windstorms. *J. Atmos. Sci.,* **37**, 2119-2121.

Lilly, D.K., and E.J. Zipser, 1972: The Front Range windstorm of 11 January 1972. *Weatherwise,* **25**, 56-63.

Lilly, D.K., J.M. Nickolls, R.M. Chervin, P.J. Kennedy, and J.B. Klemp, 1982: Aircraft measurements of wave momentum flux over the Colorado Rocky Mountains. *Quart. J. Roy. Meteor. Soc.,* **108**, 625-642.

Lin, J.C., C. Gerbig, S.C. Wofsy, A.E. Andrews, B.C. Daube, K.J. Davis, and C.A. Grainger, 2003: A near-field tool for simulating the upstream influence of atmospheric observations: The Stochastic Time-Inverted Lagrangian Transport (STILT) model. *J. Geophys. Res.,* **108**, 4493, http://dx.doi.org/10.1029/2002JD003161.

Lin, J.C., D. Brunner, and C. Gerbig, 2011: Studying atmospheric transport through Lagrangian models. *Eos, Trans. Amer. Geophys. Union,* **92(21)**, 177-178.

Lin, S.-J., and R.B. Rood, 1997: An explicit flux-form semi-Lagrangian shallow-water model on the sphere. *Quart. J. Roy. Meteor. Soc.,* **123**, 2477-2498.

Lin, Y.-L., 2007: *Mesoscale Dynamics*. Cambridge University Press, 630 pp.

Lin, Y.-L., R.D. Farley, and H.D. Orville, 1983: Bulk parameterization of the snow field in a cloud model. *J. Climate Appl. Meteor.,* **22**, 1065-1092.

Lin, Y.-L., J. Han, D.W. Hamilton, and C.-Y. Huang, 1999: Orographic influence on a drifting cyclone. *J. Atmos. Sci.,* **56**, 534-562.

Lindzen, R.S., 2005: *Dynamics in Atmospheric Physics*. Cambridge University Press, 324 pp.

Linkin, V.M., V.V. Kerzhanovich, A.N. Lipatov, K.M. Pichkadze, A.A. Shurupov, A.V. Terterashvili, A.P. Ingersoli, D. Crisp, A.W. Grossman, R.E. Young, A. Seiff, B. Ragen, J.E. Blamont, L.S. Elson, and R.A. Preston, 1986: VEGA balloon dynamics and vertical winds in the Venus middle cloud region. *Science,* **231(4744)**, 1417-1419.

Liou, C.-S., and K.D. Sashegyi, 2012: On the initialization of tropical cyclones with a three dimensional variational analysis. *Natural Hazards,* **63**, 1375-1391.

Liou, K.-N., 1980: *An Introduction to Atmospheric Radiation.* Academic Press, New York, 392 pp.

Liou, K.-N., 2002: *An Introduction to Atmospheric Radiation.* 2nd Ed., Academic Press, 583 pp.

Liou, K.-N., and S.C.S. Ou, 1981: Parameterization of infrared radiative transfer in cloudy atmospheres. *J. Atmos. Sci.,* **38**, 2707-2716.

Liou, K.-N., and G.D. Wittman, 1979: Parameterization of the radiation balance of clouds. *J. Atmos. Sci.,* **38**, 2707-2716.

Lipps, F.B., and R.S. Hemler, 1982: A scale analysis of deep moist convection and some related numerical calculations. *J. Atmos. Sci.,* **39**, 2192-2210.

Lipton, A.E., and R.A. Pielke, 1986: Vertical normal modes of a mesoscale model using a scaled height coordinate. *J. Atmos. Sci.*, **43**, 1650-1655.

List, R.J., 1971: *Smithsonian Meteorological Tables.* 6th Revised Edition. Smithsonian Institution Press, Washington, DC.

Liston, G.E., 1995: Local advection of momentum, heat, and moisture during the melt of patchy snow covers. *J. Appl. Meteor.,* **34**, 1705-1715.

Liston, G.E., 1999: Interrelationships among snow distribution, snowmelt, and snow cover depletion: Implications for atmospheric, hydrologic, and ecologic modeling. *J. Appl. Meteor.,* **38**, 1474-1487.

Liston, G.E., and R.A. Pielke, 2000: A climate version of the Regional Atmospheric Modeling System. *Theor. Appl. Climatol.*, **66**, 29-47.

Liston, G.E., and M. Sturm, 1998: A snow-transport model for complex terrain. *J. Glaciol.*, **44**, 498-516.

Liston, G.E., R.A. Pielke, and E.M. Greene, 1999: Improving first-order snow-related deficiencies in a regional climate model. *J. Geophys. Res.,* **104**, 19559-19567.

Liu, H., S. Liu, and J. Sang, 1999: A modified SiB to simulate momentum, heat and water transfer over various underlying surfaces. *Chi. J. Atmos. Sci.,* **23**, 189-201.

Liu, M., D. Westphal, S. Wang, A. Shimizu, N. Sugimoto, J. Zhou, and Y. Chen, 2003: A high-resolution numerical study of Asian dust storms of April 2001. *J. Geophys. Res.-Atmos.,* **108:D23**, 8653, http://dx.doi.org/10.1029/2002JD003178.

Liu, M., D.L. Westphal, A.L. Walker, T.R. Holt, K.A. Richardson, and S.D. Miller, 2007: COAMPS real-time dust storm forecasting during Operation Iraqi Freedom. *Wea. Forecasting,* **22**, 192-206.

Liu, Q., and F. Weng, 2006: Advanced doubling-adding method for radiative transfer in planetary atmospheres. *J. Atmos. Sci.,* **63**, 3459-3465.

Liu, Y., D.-L. Zhang, and M.K. Yau, 1997: A multiscale numerical study of Hurricane Andrew (1992). Part I: Explicit simulation and verification. *Mon. Wea. Rev.,* **125**, 3073-3093.

Liu, Y., C.P. Weaver, and R. Avissar, 1999a: Toward a parameterization of mesoscale fluxes and moist convection induced by landscape heterogeneity. *J. Geophys. Res.,* **104**, 19,515-19,533.

Liu, Y., D.-L. Zhang, and M.K. Yau, 1999b: A multiscale numerical study of Hurricane Andrew, 1992: Part III: Dynamically-induced vertical motion. *Mon. Wea. Rev.,* **127**, 2597-2616.

Lo, J.C.-F., Z.-L. Yang, and R.A. Pielke Sr., 2008: Assessment of three dynamical climate downscaling methods using the Weather Research and Forecasting (WRF) Model. *J. Geophys. Res.*, **113**, D09112, http://dx.doi.org/10.1029/2007JD009216.

Lo, M.-H., and J.S. Famiglietti, 2013: Irrigation in California's Central Valley strengthens the southwestern U.S. water cycle. *Geophys. Res. Lett.*, **40**, http://dx.doi.org/10.1002/GRL.50108.

Locatelli, J.D., P.V. Hobbs, and J.A. Werth, 1982: Mesoscale structures of vortices in polar air streams. *Mon. Wea. Rev.,* **110**, 1417-1433.

Long Jr., P.E., and F.J. Hicks, 1975: Simple properties of Chapeau functions and their application to the solution of the advection equation. NOAA NWS TDL Office Note 75-8. Techniques Development Lab, Gramex Bldg., Silver Spring, Maryland.

Long, R.L., 1977: Three layer circulations in estuaries and harbors. *J. Phys. Oceanogr.,* **7**, 415-421.

Long, R.R., 1953: Some aspects of the flow of stratified fluids. I. A theoretical investigation. *Tellus,* **5**, 42-58.

Long, R.R., 1955: Some aspects of the flow of stratified fluids. II. Continuous density gradients. *Tellus,* **7**, 341-357.

Longo, K.M., S.R. Freitas, M.O. Andreae, A. Setzer, E. Prins, and P. Artaxo, 2010: The Coupled Aerosol and Tracer Transport model to the Brazilian developments on the Regional Atmospheric Modeling System (CATT-BRAMS) Part 2: Model sensitivity to the biomass burning inventories. *Atmos. Chem. Phys.*, **10**, 5785-5795.

Longo, K.M., S.R. Freitas, M. Pirre, V. Marécal, L.F. Rodrigues, J. Panetta, M.F. Alonso, N.E. Rosário, D.S. Moreira, M.S. Gácita, J. Arteta, R. Fonseca, R. Stockler, D.M. Katsurayama, A. Fazenda, and M. Bela, 2013: The Chemistry CATT-BRAMS Model (CCATT-BRAMS 4.5): A regional atmospheric model system for integrated air quality and weather forecasting and research. *Geosci. Model Dev. Discuss.*, **6**, 1173-1222, http://dx.doi.org/10.5194/gmdd-6-1173-2013.

Loose, T., and R.D. Bornstein, 1977: Observations of mesoscale effects on frontal movement through an urban area. *Mon. Wea. Rev.,* **105**, 562-571.

Louis, J.-F., 1979: Parametric model of vertical eddy fluxes in the atmosphere. *Bound.-Layer Meteor.,* **17**, 187-202.

Louis, J.-F., M. Tiedtke, and J.F. Geleyn, 1982: A short history of the operational PBL-parameterization of ECMWF. *Proc. of the 1981 ECMWF workshop on planetary boundary layer parameterization*, Shinfield Park, Reading, Berkshire, UK, European Centre for Medium Range Weather Forecasts, 59-79.

Lovejoy, S., and D. Schertzer, 2013: *The Weather And The Climate: Emergent Laws And Multifractal Cascades.* Cambridge University Press, 496 pp.

Lu, E., 2004: Louis surface flux scheme and look-up table approach for parameterization. Available online at: http://cires.colorado.edu/science/groups/pielke/classes/atmo595e/ErLu_2.pdf.

Lu, L., 2011: The influence of realistic vegetation phenology on regional climate modeling. In: *Remote Sensing of Protected Lands*, Taylor and Francis, New York, USA, 407-435.

Lu, L., and J. Shuttleworth, 2002: Incorporating NDVI-derived LAI into the climate version of RAMS and its impact on regional climate. *J. Hydrometeor.,* **6**, 347-362.

Lu, L., R.A. Pielke, G.E. Liston, W.J. Parton, D. Ojima, and M. Hartman, 2001: Implementation of a two-way interactive atmospheric and ecological model and its application to the central United States. *J. Climate*, **14**, 900-919.

Lu, L., A.S. Denning, M.A. da Silva Dias, P. Silva-Dias, M. Longo, S.R. Freitas, and S. Saatchi, 2005: Mesoscale circulation and atmospheric CO_2 variation in the Tapajos Region, Para, Brazil. *J. Geophys. Res.,* **110**, D21102, http://dx.doi.org/10.1029/2004JD005757.

Ludwig, F.L., and G. Byrd, 1980: An efficient method for deriving mass-consistent flow fields from wind observations in rough terrain. *Atmos. Environ.,* **14**, 585-587.

Lumley, J.L., and H.A. Panofsky, 1964: *The Structure of Atmospheric Turbulence.* Interscience Monographs and Texts in Physics and Astronomy, Vol. 12, Interscience, New York, NY, 239 pp.

Lunine, J.I., and S.K. Atreya, 2008: The methane cycle on Titan. *Nature Geoscience*, **1(3)**, 159-164.

Lynch, A.H., F.S. Chapin III, L.D. Hinzman, W. Wu, E. Lilly, G. Vourlitis, and E. Kim, 1999a: Surface energy balance on the Arctic tundra: Measurements and models. *J. Climate,* **12**, 2585-2606.

Lynch, P., D. Giard, and V. Ivanovici, 1997: Improving the efficiency of a digital filtering scheme for diabatic initialization. *Mon. Wea. Rev.,* **125**, 1976-1982.

Lynn, B.H., F. Abramopoulos, and R. Avissar, 1995a: Using similarity theory to parameterize mesoscale heat fluxes generated by subgrid-scale landscape discontinuities in GCMs. *J. Climate,* **8**, 932-951.

Lynn, B.H., D. Rind, and R. Avissar, 1995b: The importance of mesoscale circulations generated by subgrid-scale landscape-heterogeneities in general circulation models. *J. Climate,* **8**, 191-205.

Lyons, J.J., and R.K. Steedman, 1981: Stagnation and nocturnal temperature jumps in a desert region of low relief. *Bound.-Layer Meteor.,* **21**, 369-387.

Lyons, T.J., R.C.G. Smith, and H. Xinmei, 1996: The impact of clearing for agriculture on the surface energy budget. *Int. J. Climatol.,* **16**, 551-558.

Lyons, W.A., 1972: The climatology and prediction of the Chicago lake breeze. *J. Appl. Meteor.,* **11**, 1259-1270.

Lyons, W.A., 1975: *Lectures on Air Pollution and Environmental Impact Analysis.* D. Haugen, Ed., American Meteorological Society, Boston, MA, 136-208.

Lyons, W.A., and H.S. Cole, 1976: Photochemical oxidant transport: Mesoscale lake breeze and synoptic-scale aspects. *J. Appl. Meteor.,* **15**, 733-743.

Lyons, W.A., and C.S. Keen, 1976: Particulate transport in a Great Lakes coastal environment. *Proceedings, Second Federal Conference of the Great Lakes*, 222-237.

Lyons, W.A., R.A. Pielke, W.R. Cotton, C.S. Keen, D.A. Moon, and N.R. Lincoln, 1992: Some considerations of the role of the land/lake breeze in causing elevated ozone levels in the southern Lake Michigan region. Chapter 9 in *Environmental Modeling*, P. Melli, and P. Zannetti, Eds. Computational Mechanics Publications, Southampton, UK, 151-171.

Lyons, W.A., C.J. Tremback, and R.A. Pielke, 1995: Applications of the Regional Atmospheric Modeling System (RAMS) to provide input to photochemical grid models for the Lake Michigan Ozone Study (LMOS). *J. Appl. Meteor.,* **34**, 1762-1786.

MacDonald, A.E., J.L. Lee, and Y. Xie, 2000: The use of quasi-nonhydrostatic models for mesoscale weather prediction. *J. Atmos. Sci.,* **57**, 2493-2517.

MacHattie, L.B., 1968: Kananaskis Valley winds in summer. *J. Appl. Meteor.,* **7**, 348-352.

Machenhauer, B., 1979: The spectral model: Numerical methods used in atmospheric models. GARP Report 17, Vol. II, 121-275.

Maddox, R.A., 1980a: Mesoscale convective clusters. *Bull. Amer. Meteor. Soc.,* **61**, 1374-1387.

Maddox, R.A., 1980b: An objective technique for separating macroscale and mesoscale features in meteorological data. *Mon. Wea. Rev.,* **108**, 1108-1121.

Maddox, R.A., 1980c: A satellite based study of midlatitude, mesoscale convective complexes. *Preprints, 8th AMS Conference on Weather Forecasting and Analysis*, Denver, Colorado, June 10-13, 329-338.

Maddox, R.A., D.J. Perkey, and J.M. Fritsch, 1981: Evolution of upper tropospheric features during the development of a mesoscale convective complex. *J. Atmos. Sci.,* **38**, 1664-1674.

Maddukuri, C.S., 1982: A numerical simulation of an observed lake breeze over southern Lake Ontario. *Bound.-Layer Meteor.,* **23**, 369-387.

Magono, C., C. Nakamura, and Y. Yoshida, 1982: Nocturnal cooling of the Moshiri Basin, Hokkaido in midwinter. *J. Meteor. Soc. Japan,* **60**, 1106-1116.

Mahfouf, J.-F., and B. Bilodeau, 2007: Adjoint sensitivity of surface precipitation to initial conditions. *Mon. Wea. Rev.,* **135**, 2879-2896.

Mahmood, R., R.A. Pielke Sr., K.G. Hubbard, D. Niyogi, G. Bonan, P. Lawrence, B. Baker, R. McNider, C. McAlpine, A. Etter, S. Gameda, B. Qian, A. Carleton, A. Beltran-Przekurat, T. Chase, A.I. Quintanar, J.O. Adegoke, S. Vezhapparambu, G. Conner, S. Asefi, E. Sertel, D.R. Legates, Y. Wu, R. Hale, O.W. Frauenfeld, A. Watts, M. Shepherd, C. Mitra, V.G. Anantharaj, S. Fall, R. Lund, A. Nordfelt, P. Blanken, J. Du, H.-I. Chang, R. Leeper, U.S. Nair, S. Dobler, R. Deo, and J. Syktus, 2010: Impacts of land use land cover change on climate and future research priorities. *Bull. Amer. Meteor. Soc.*, **91**, 37-46, http://dx.doi.org/10.1175/2009BAMS2769.1.

Mahmood, R., T. Keeling, S.A. Foster, and K.G. Hubbard, 2013a: Did irrigation impact 20th century air temperature in the High Plains aquifer region? *Appl. Geog.*, **38**, 11-21, http://dx.doi.org/10.1016/j.apgeog.2012.11.002.

Mahmood, R., R.A. Pielke Sr., K. Hubbard, D. Niyogi, P. Dirmeyer, C. McAlpine, A.M. Carleton, R. Hale, S. Gameda, A. Beltran-Przekurat, B. Baker, R. McNider, D.R. Legates, M. Shepherd, J. Du, P.D. Blanken, O.W. Frauenfeld, U.S. Nair, and S. Fall, 2013b: Land cover changes and their biogeophysical effects on climate. *Int. J. Climatol.,* http://dx.doi.org/10.1002/joc.3736.

Mahrer, Y., 1984: An improved numerical approximation of the horizontal gradient in a terrain-following coordinate system. *Mon. Wea. Rev.,* **112**, 918-922.

Mahrer, Y., and R.A. Pielke, 1975: The numerical study of the air flow over mountains using the University of Virginia mesoscale model. *J. Atmos. Sci.,* **32**, 2144-2155.

Mahrer, Y., and R.A. Pielke, 1976: The numerical simulation of the airflow over Barbados. *Mon. Wea. Rev.,* **104**, 1392-1402.

Mahrer, Y., and R.A. Pielke, 1977a: The effects of topography on sea and land breezes in a two-dimensional numerical model. *Mon. Wea. Rev.*, **105**, 1151-1162.

Mahrer, Y., and R.A. Pielke, 1977b: A numerical study of the airflow over irregular terrain. *Beitrage zur Physik der Atmosphare*, **50**, 98-113.

Mahrer, Y., and R.A. Pielke, 1978a: A test of an upstream spline interpolation technique for the advective terms in a numerical mesoscale model. *Mon. Wea. Rev.,* **106**, 818-830.

Mahrer, Y., and R.A. Pielke, 1978b: The meteorological effect of the changes in surface albedo and moisture. *Israel Meteorological Research Papers, (IMS)*, **2**, 55-70.

Mahrt, L.J., 1976: Mixed layer moisture structure. *Mon. Wea. Rev.,* **104**, 1403-1407.

Mahrt, L., 1982: Momentum balance of gravity flows. *J. Atmos. Sci.*, **39**, 2701-2711.

Mahrt, L.J., and R.C. Heald, 1981: Nocturnal surface temperatures over undulating terrain. *Preprints, 15th AMS Conference on Agriculture and Forest Meteorology and 5th AMS Conference on Biometeorology*, Anaheim, CA, April 1-3, 197-199.

Mahrt, L.J., and S. Larsen, 1982: Small scale drainage flow. *Tellus,* **34**, 579-587.

Mailhot, J., and C. Chouinard, 1989: Numerical forecasts of explosive winter storms: Sensitivity experiments with a mesα scale model. *Mon. Wea. Rev.,* **117**, 1311-1343. http://dx.doi.org/10.1175/1520-0493(1989)117<1311:NFOEWS>2.0.CO;2.

Majewski, D., D. Liermann, P. Prohl, B. Ritter, M. Buchhold, T. Hanisch, G. Paul, W. Wergen, and J. Baumgardner, 2002: The Operational Global Icosahedral-Hexagonal Gridpoint Model GME: Description and high-resolution tests. *Mon. Wea. Rev.*, **130(2)3**, 19-338.

Mak, M., 1981: An inquiry on the nature of CISK. Part 1. *Tellus,* **33**, 531-537.

Mak, M., 2011: *Atmospheric Dynamics*. Cambridge University Press, 536 pp.

Makar, P.A., and S.R. Karpik, 1996: Basis-spline interpolation on the sphere: Applications to semi-Lagrangian advection. *Mon. Wea. Rev.,* **124**, 182-199.

Malkus, J.S., and M.E. Stern, 1953: The flow of a stable atmosphere over a heated island. Part I. *J. Meteor.,* **10**, 30-41.

Manins, P.C., and B.L. Sawford, 1979a: A model of katabatic winds. *J. Atmos. Sci.,* **36**, 619-630.

Manins, P.C., and B.L. Sawford, 1979b: Katabatic winds: A field case study. *Quart. J. Roy. Meteor. Soc.,* **105**, 1105-1025.

Manins, P.C., and B.L. Sawford, 1982: Mesoscale observations of upstream blocking. *Quart. J. Roy. Meteor. Soc.,* **108**, 427-434.

Mannouji, N., 1982: A numerical experiment on the mountain and valley winds. *J. Meteor. Soc. Japan,* **60**, 1085-1105.

Manton, M.J., and G.P. Ayers, 1982: On the number concentration of aerosols in towns. *Bound.-Layer Meteor.,* **22**, 171-181.

Marchuk, G.I., 1995: *Adjoint equations and analysis of complex systems*. Kluwer Academic Publishers, Dordrecht, The Netherlands, 277 pp.

Marchuk, G.I., V.P. Kochergin, V.I. Klimok, and V.A. Sukhorukov, 1977: On the dynamics of the ocean surface mixed layer. *J. Phys. Oceanogr.,* **7**, 865-875.

Marks Jr., F.D., and P.M. Austin, 1979: Effects of the New England coastal front on the distribution of precipitation. *Mon. Wea. Rev.,* **107**, 53-67.

Marshall Jr., C.H., K.C. Crawford, K.E. Mitchell, and D.J. Stensrud, 2003: The impact of the land-surface physics in the operational NCEP Eta Model on simulating the diurnal cycle: evaluation and testing using Oklahoma Mesonet data. *Wea. Forecasting*, **18**, 748-768.

Marshall Jr., C.H., R.A. Pielke Sr., L.T. Steyaert, and D.A. Willard, 2004: The impact of anthropogenic land-cover change on the Florida peninsula sea breezes and warm season sensible weather. *Mon. Wea. Rev.*, **132**, 28-52.

Marshall, J.S., and W.M.K. Palmer, 1948: The distribution of raindrops with size. *J. Meteor.,* **5**, 165-166.

Marshall, J., A. Adcroft, C. Hill, L. Perelman, and C. Heisey, 1997: A finite-volume, incompressible Navier Stokes model for studies of the ocean on parallel computers. *J. Geophys. Res.,* **102**, 5753-5766.

Marshall, S., and R.J. Oglesby, 1994: An improved snow hydrology for GCMs. Part 1: snow cover fraction, albedo, grain size, and age. *Climate Dyn.,* **10**, 21-37.

Marshall, S., J.O. Roads, and G. Glatzmaier, 1994: Snow hydrology in a general circulation model. *J. Climate,* **7**, 1251-1269.

Marticorena, B., and G. Bergametti, 1995: Modeling the atmospheric dust cycle: 1. Design of a soil derived dust emission scheme. *J. Geophys. Res.,* **100(D8)**, 16415-16430.

Martin, C.L., 1981: Numerical accuracy in a mesoscale meteorological model. Department Environmental Science, M.S. Thesis, University of Virginia.

Martin, C.L., and R.A. Pielke, 1983: The adequacy of the hydrostatic assumption in sea breeze modeling over flat terrain. *J. Atmos. Sci.,* **40**, 1472-1481.

Martin, P.J., 2000: Description of the NAVY coastal ocean model version 1.0. NRL Rep. NRL/FR/7322-00-9962, 42 pp.

Martin, P.J., J.W. Book, and J.D. Doyle, 2006: Simulation of the northern Adriatic circulation during winter 2003. *J. Geophys. Res.,* **111**, C03S12. http://dx.doi.org/10.1029/2006JC003511.

Marwitz, J.D., 1972: Precipitation efficiency of thunderstorms on the High Plains. *J. Rech. Atmos.,* **6**, 367-370.

Marwitz, J.D., 1983: The kinematics of orographic airflow during Sierra storms. *J. Atmos. Sci.,* **40**, 1218-1227.

Mason, B.J., 1956: On the melting of hailstones. *Quart. J. Roy. Meteor. Soc.,* **82**, 209-216.

Mason, P.J., and R.I. Sykes, 1978: A simple Cartesian model of boundary layer flow over topography. *J. Comput. Phys.,* **28**, 198-210.

Mason, P.J., and R.I. Sykes, 1979: Three-dimensional numerical integrations of the Navier-Stokes equations for flow over surface-mounted obstacles. *J. Fluid Mech.,* **91**, 433-450.

Mass, C., 1981: Topographically forced convergence in western Washington State. *Mon. Wea. Rev.,* **109**, 1335-1347.

Mass, C., 1982: The topographically forced diurnal circulations of western Washington State and their influence on precipitation. *Mon. Wea. Rev.,* **110**, 170-183.

Mass, C.F., and Y.-H. Kuo, 1998: Regional real-time numerical weather prediction: current status and future potential. *Bull. Amer. Meteor. Soc.,* **79**, 253-263.

Masson, V., 2000: A physically-based scheme for the urban energy budget in atmospheric models. *Bound.-Layer Meteor.,* **94**, 357-397.

Masson, V., J.-L. Champeaux, F. Chauvin, C. Meriguet, and R. Lacaze, 2003: A global database of land surface parameters at 1-km resolution in meteorological and climate models. *J. Climate,* **16**, 1261-1282.

Masunaga, H., M. Satoh, and H. Miura, 2008: A joint satellite and global cloud-resolving model analysis of a Madden-Julian Oscillation event: Model diagnosis. *J. Geophys. Res.,* **113**, D17210, http://dx.doi.org/10.1029/2008JD009986.

Masunaga, H., T. Matsui, W.-K. Tao, A.Y. Hou, C. Kummerow, T. Nakajima, P. Bauer, W. Olson, M. Sekiguchi, and T.Y. Nakajima, 2011: Satellite Data Simulator Unit: Multi-Sensor and Multi-Frequency Satellite Simulator package. *Bull. Amer. Meteor. Soc.,* **91**, 1625-1632. http://dx.doi.org/10.1175/2010BAMS2809.1.

Mathur, M.B., 1970: A note on an improved quasi-Lagrangian advection scheme for primitive equations. *Mon. Wea. Rev.,* **98**, 214-219.

Mathur, M.B., 1974: A multiple grid primitive equation model to simulate the development of an asymmetric hurricane (Isbell 1964). *J. Atmos. Sci.,* **31**, 371-393.

Mathur, M.B., 1975: Development of banded structure in a numerically simulated hurricane. *J. Atmos. Sci.,* **32**, 512-522.

Mathur, M.B., 1997: Development of an eye-wall like structure in a tropical cyclone model simulation. *Dyn. Atmospheres Oceans*, **27**, 527-547.

Matson, D.L., L.J. Spilker, and J.P. Lebreton, 2002: The Cassini-Huygens mission to the saturnian system. *Space Sci. Rev.*, **104**, 1-58.

Matson, M., and R.V. Legeckis, 1980: Urban heat islands detected by satellite. *Bull. Amer. Meteor. Soc.,* **61**, 212.

Matson, M., E.P. McClain, D.F. McGinnis Jr., and J.A. Pritchard, 1978: Satellite detection of urban heat islands. *Mon. Wea. Rev.,* **106**, 1725-1734.

Matsui, T., G. Leoncini, R.A. Pielke Sr., and U.S. Nair, 2004: A new paradigm for parameterization in atmospheric models: Application to the new Fu-Liou radiation code. Atmospheric Science Paper No. 747, Colorado State University, Fort Collins, CO 80523, 32 pp.

Matsui, T., X. Zeng, W.-K. Tao, H. Masunaga, W. Olson, and S. Lang, 2009: Evaluation of long-term cloud-resolving model simulations using satellite radiance observations and multifrequency satellite simulators. *J. Atmos. Oceanic Tech.,* **26**, 1261-1274.

Matsui, T., T. Iguchi, X. Li, M. Han, W.-K. Tao, W. Petersen, T. L'Ecuyer, R. Meneghini, W. Olson, C.D. Kummerow, A.Y. Hou, M.R. Schwaller, E.F. Stocker, and J. Kwiatkowski,

2013: GPM satellite simulator over ground validation sites. *Bull. Amer. Meteor. Soc.*, http://dx.doi.org/10.1175/BAMS-D-12-00160.1.

Mayr, G.J., and A. Gohm, 2000: 2D airflow over a double bell-shaped mountain. *Meteor. Atmos. Phys.*, **72**, 13-27.

McAnelly, R.L, and W.R. Cotton, 1986: Meso-beta-scale characteristics of an episode of meso-alpha-scale convective complexes. *Mon. Wea. Rev.*, **114**, 1740-1770.

McCarthy, J., and S.E. Koch, 1982: The evolution of an Oklahoma dryline. Part I: A meso- and subsynoptic-scale analysis. *J. Atmos. Sci.*, **39**, 225-236.

McClatckey, R.A., and J.E.A. Selby, 1972: Atmospheric transmittance, 7-30 μm: Attenuation of CO_2 laser radiation. Environmental Research Paper No. 419, AFCRL-72-0611.

McCleese, D.J., J.T. Schofield, F.W. Taylor, S.B. Calcutt, M.C. Foote, D.M. Kass, C.B. Leovy, D.A. Paige, P.L. Read, and R.W. Zurek, 2007: Mars Climate Sounder: An investigation of thermal and water vapor structure, dust and condensate distributions in the atmosphere, and energy balance of the polar regions. *J. Geophys. Res. E: Planets*, **112(5)**, E05S06, http://dx.doi.org/10.1029/2006JE002790.

McCleese, D.J., G. Heavens, J.T. Schofield, W.A. Abdou, J.L. Bandfield, S.B. Calcutt, P.G.J. Irwin, D.M. Kass, A. Kleinböhl, S.R. Lewis, D.A. Paige, P.L. Read, M.I. Richardson, J.H. Shirley, F.W. Taylor, N. Teanby, and R.W. Zurek, 2010: The structure and dynamics of the Martian lower and middle atmosphere as observed by the Mars Climate Sounder: Seasonal variations in zonal mean temperature, dust, and water ice aerosols. *J. Geophys. Res.*, **115**, E12016, http://dx.doi.org/10.1029/2010JE003677.

McCumber, M.D., 1980: A numerical simulation of the influence of heat and moisture fluxes upon mesoscale circulation. University of Virginia, Ph.D. dissertation, Dept. of Environmental Science.

McCumber, M.D., and R.A. Pielke, 1981: Simulation of the effects of surface fluxes of heat and moisture in a mesoscale numerical model. Part I: Soil layer. *J. Geophys. Res.*, **86**, 9929-9938.

McDonald, A., 1999: An examination of alternative extrapolations to find the departure point position in a "two-time-level" semi-Lagrangian integration. *Mon. Wea. Rev.*, **127**, 1985-1993.

McDonald, J., 1960: Direct absorption of solar radiation by atmospheric water vapor. *J. Meteor.*, **17**, 319-328.

McEwan, M.J., and L.F. Phillips, 1975: *Chemistry of the Atmosphere.* Wiley, New York

McGouldrick, K., and O.B. Toon, 2007: An investigation of possible causes of the holes in the condensational Venus cloud using a microphysical cloud model with a radiative-dynamical feedback. *Icarus*, **191(1)**, 1-24.

McKee, T.B., D.C. Bader, and K. Hansen, 1983: Synoptic influence of urban circulation. *AMS/EPA Specialty Conference on Air Quality Modeling of the Nonstationary, Nonhomogeneous Urban Boundary Layer*, October 31-November 4, Baltimore, MD.

McKenna, D.S., P. Konopka, J.-U. Grooß, G. Gunther, and R. Muller, 2002: A new Chemical Lagrangian Model of the Stratosphere (CLaMS) 1. Formulation of advection and mixing. *J. Geophys. Res.*, **107(D16)**, http://dx.doi.org/10.1029/2000JD000114.

McNider, R.T., 1981: Investigation of the impact of topographic circulations on the transport and dispersion of air pollutants. Ph.D. Dissertation, University of Virginia, Department of Environmental Sciences, Charlottesville, 210 pp.

McNider, R.T., 1982: A note on velocity fluctuations in drainage flows. *J. Atmos. Sci.*, **39**, 1658-1660.

McNider, R.T., 1983: Transport and diffusion models for the urban boundary layer. *AMS/EPA Specialty Conference on Air Quality Modeling of the Nonstationary, Nonhomogeneous Urban Boundary Layer*, October 31-November 4, Baltimore, MD.

McNider, R.T., and F.J. Kopp, 1990: Specification of the scale and magnitude of thermals used to initiate convection in cloud models. *J. Appl. Meteor.,* **29**, 99-104.

McNider, R.T., S.R. Hanna, and R.A. Pielke, 1980: Sub-grid scale plume dispersion in coarse resolution mesoscale models. *Proceedings of the Second AMS Joint Conference on Applications of Air Pollution Meteorology*, New Orleans, Louisiana, 424-429.

McNider, R.T., A.P. Mizzi, and R.A. Pielke, 1982: Numerical investigation of low level jets in coastal zones. *Proceedings of the First International Conference on Meteorology and Air/Sea Interactions of the Coastal Zone*, May 10-14, The Hague, Netherlands, 190-195.

McNider, R.T., A.J. Song, D.M. Casey, P.J. Wetzel, W.L. Crosson, and R.M. Rabin, 1994: Toward a dynamic- thermodynamic assimilation of satellite surface temperature in numerical atmospheric models. *Mon. Wea. Rev.,* **122**, 2784-2803.

McNider, R.T., W.M. Lapenta, A.P. Biazar, G.J. Jedlovec, R.J. Suggs, and J. Pleim, 2005: Retrieval of model grid-scale heat capacity using geostationary satellite products, Part I: First-case-study application. *J. Appl. Meteor.,* **9**, 1346-1360, http://dx.doi.org/10.1175/JAM2270.1.

McNider, R.T., G.J. Steeneveld, A.A.M. Holtslag, R.A. Pielke Sr., S. Mackaro, A. Pour-Biazar, J. Walters, U. Nair, and J. Christy, 2012: Response and sensitivity of the nocturnal boundary layer over land to added longwave radiative forcing. *J. Geophys. Res.*, **117**, D14106, http://dx.doi.org/10.1029/2012JD017578.

McPhee, M.G., 1979: The effect of the oceanic boundary layer on the mean drift of pack ice: Application of a simple mode. *J. Phys. Oceanogr.,* **9**, 388-400.

McPherson, R.D., 1970: A numerical study of the effect of a coastal irregularity on the sea breeze. *J. Appl. Meteor.,* **9**, 767-777.

McQueen, J.T., R.R. Draxler, and G.D. Rolph, 1995: Influence of grid size and terrain resolution on wind-field predictions from an operational mesoscale model. *J. Appl. Meteor.,* **34**, 2166-2181.

McRae, G.J., and J.H. Seinfeld, 1983: Development of a second-generation mathematical model for urban air pollution – II: Evaluation of model performance. *Atmos. Environ.,* **17**, 491-499.

McRae, G.J., F.H. Shair, and J.H. Seinfeld, 1982a: Convective downmixing of plumes in a coastal environment. *J. Appl. Meteor.,* **20**, 1312-1324.

McRae, G.J., W.R. Goodin, and J.H. Seinfeld, 1982b: Development of a second-generation mathematical model for urban air pollution – I. Model formulation. *Atmos. Environ.,* **16**, 679-696.

Mechem, D.B., and Y.L. Kogan, 2003: Simulating the transition from drizzling marine stratocumulus to boundary layer cumulus with a mesoscale model. *Mon. Wea. Rev.,* **131**, 2342-2360.

Meirold-Mautner, I., C. Prigent, E. Defer, J.R. Pardo, J.-P. Chaboureau, J.-P. Pinty, M. Mech, and S. Crewell, 2007: Radiative transfer simulations using mesoscale cloud model outputs: Comparisons with passive microwave and infrared satellite observations for midlatitudes. *J. Atmos. Sci.,* **64**, 1550-1568.

Melgarejo, J., 1980: A numerical simulation of wind over Gotland with a two-dimensional mesometeorological boundary layer model. Report of the Swedish Meteorological and Hydrological Institute, No. 11.

Mellor, G.L., 2004: Users guide for a three-dimensional, primitive equation, numerical ocean model (June 2004 version). *Prog. in Atmos. and Oceanic Sci.*, Princeton University, 56 pp.

Mellor, G.L., and T. Yamada, 1974: A hierarchy of turbulence closure models for the planetary boundary Lares. *J. Atmos. Sci.,* **31**, 1791-1806.

Mellor, G.L., and T. Yamada, 1982: Development of a turbulence closure model for geophysical fluid problems. *Rev. Geophys. Space Phys.*, **20**, 851-875.

Melville, W.K., 1977: Wind stress and roughness length over breaking waves. *J. Phys. Oceanogr.,* **7**, 702-710.

Meneghini, R., 1996: Analysis of radar and radar-radiometer methods for spaceborne measurements of precipitation. Kyoto University, Ph.D. Dissertation.

Menenti, M., and J.C. Ritchie, 1994: Estimation of effective aerodynamic roughness of Walnut Gulch watershed with laser altimeter measurements. *Water Res. Research*, **30:5**, 1329–1337, http://dx.doi.org/10.1029/93WR03055.

Merceret, F.J., 1975: Relating rainfall rate to the slope of raindrop size spectra. *J. Appl. Meteor.,* **14**, 259-260.

Merilees, P.E., and S.A. Orszag, 1979: The pseudospectral method. Numerical methods used in atmospheric models. GARP Report 17, Volume II, 276-299.

Meroney, R.N., 1998: Spurious or virtual correlation errors commonly encountered in reduction of scientific data. *J. Wind Eng.,* **77&78**, 543-553.

Meroney, R.N., A.J. Bowen, B. Lindley, and J.R. Pearse, 1978: Wind characteristics over complex terrain: Laboratory simulation and field measurements at Rakaia Gorge. New Zealand. Final Report: Part II. Fluid Mechanics and Wind Engineering Program, Colorado State University, Fort Collins, Colorado.

Mesinger, F., 1996: Improvements in quantitative precipitation forecasts with the Eta Regional Model at the National Centers for Environmental Prediction: The 48-km upgrade. *Bull. Amer. Meteor. Soc.,* **77**, 2637-2649; Corrigendum, **78**, 506.

Mesinger, F., 1997: Dynamics of limited-area models: Formulation and numerical methods. *Meteor. Atmos. Phys.,* **63**, 3-14.

Mesinger, F., 1998: Comparison of quantitative precipitation forecasts by the 48- and by the 29-km Eta model: An update and possible implications. Preprints, *12th Conf. on Numerical Weather Prediction,* Phoenix, AZ, Amer. Meteor. Soc., J22-J23.

Mesinger, F., and A. Arakawa, 1976: Numerical methods used in atmospheric models. *GARP Publ. Ser.*, **17**, 1-64.

Mesinger, F., and Black, 1992: On the impact on forecast accuracy of the step-mountain (Eta) vs. sigma coordinate. *Meteor. Atmos. Phys.,* **50**, 47-60.

Mesinger, F., Z.I. Janjić, S. Nickovic, D. Gavrilov and D.G. Deaven, 1988: The step-mountain coordinate: Model description, and performance for cases of Alpine lee cyclogenesis and for a case of an Appalachian redevelopment. *Mon. Wea. Rev.*, **116**, 1493-1518.

Mesinger, F., T.L. Black, and M.E. Baldwin, 1997: Impact of resolution and of the Eta coordinate on-skill of the Eta Model precipitation forecasts. *Numerical Methods in Atmospheric- and Oceanic Modelling*. C. Lin, R. Laprise, and H. Ritchie, Eds., The Andre J. Robert Memorial Monograph, Soc./NRC Research-Press, 399-423.

Method, T.J., and T.N. Carlson, 1982: Radiative heating rates and some optical properties of the St. Louis aerosol, as inferred from aircraft measurements. *Atmos. Environ.,* **16**, 53-66.

Meyers, M.P., and W.R. Cotton, 1992: Evaluation of the potential for wintertime quantitative precipitation forecasting over mountainous terrain with an explicit cloud model. Part I: Two-dimensional sensitivity experiments. *J. Appl. Meteor.,* **31**, 26-50.

Meyers, M.P., R.L. Walko, J.Y. Harrington, and W.R. Cotton, 1997: New RAMS cloud microphysics parameterization. Part II: The two-moment scheme. *Atmos. Res.,* **45**, 3-39.

Michaels, T.I., 2006: Numerical modeling of Mars dust devils: Albedo track generation. *Geophys. Res. Lett.*, **33(19)**, L19S08.

Michaels, T.I., and S.C.R. Rafkin, 2004: Large-eddy simulation of atmospheric convection on Mars. *Quart. J. Royal Meteorol. Soc.*, **130(599)**, 1251-1274.

Michaels, T.I., and S.C.R. Rafkin, 2008: Meteorological predictions for candidate 2007 Phoenix Mars Lander sites using the Mars Regional Atmospheric Modeling System (MRAMS). *J. Geophys. Res.-Planets*, **113**, E00A07, http://dx.doi.org/10.1029/2007JE003013.

Michaels, T.I., A. Colaprete, and S.C.R. Rafkin, 2006: Significant vertical water transport by mountain-induced circulations on Mars. *Geophys. Res. Lett.*, **33**, L16201, http://dx.doi.org/10.1029/2006GL026562.

Mielke, P.W., 1984: Meteorological applications of permutation techniques based on distance functions. In *Handbook of Statistics*, Volume 4, P.R. Krishnaiah, and P.K. Sen, Eds., North-Holland, New York, 813-830.

Mielke, P.W., 1991: The application of multivariate permutation methods based on distance functions in the earth sciences. *Earth Sci. Rev.*, **31**, 55-71.

Mielke, P.W., and K.J. Berry, 2000: Euclidean distance based permutation methods in atmospheric science. *Data Mining and Knowledge Disc.,* **7**, 7-27.

Mihailovic, D.T., and G. Kallos, 1997: A sensitivity study of a coupled soil-vegetation boundary-layer scheme for use in atmospheric modeling. *Bound.-Layer Meteor.,* **82**, 283-315.

Mihailovic, D.T., and M. Ruml, 1996: Design of land-air parameterization scheme (LAPS) for modeling boundary layer surface processes. *Meteor. Atmos. Phys.,* **6**, 65-81.

Mihailovic, D.T., R.A. Pielke, B. Rajkovic, T.J. Lee, and M. Jeftic, 1993: A resistance representation of schemes for evaporation from bare and partly plant-covered surfaces for use in atmospheric models. *J. Appl. Meteor.,* **32**, 1038-1054.

Mihailovic, D.T., B. Rajkovic, L. Dekic, R.A. Pielke, T.J. Lee, and Z. Ye, 1995: The validation of various schemes for parameterizing evaporation from bare soil for use in meteorological models: A numerical study using in situ data. *Bound.-Layer Meteor.,* **76**, 259-289.

Mihailovic, D.T., B. Rajkovic, B. Lalic, D. Jovic, and L. Dekic, 1998: Partitioning the land surface water simulated by a land-air surface scheme. *J. Hydro.,* **211**, 17-33.

Mihailovic, D.T., G. Kallos, I.D. Arsenic, B. Lalic, B. Rajkovic, and A. Papadopoulos, 1999a: Sensitivity of soil surface temperature in a force-restore equation to heat fluxes and deep soil temperature. *Int. J. Climatol.,* **19**, 1617-1632.

Mihailovic, D.T., G. Kallos, B. Lalic, A. Papadopoulos, and I. Arsenic, 1999b: Parameterization of hydrological processes for application to regional and mesoscale modeling. *Global Atmos. Ocean Sys.,* **7**, 73-89.

Mihailovic, D.T., T.J. Lee, R.A. Pielke Sr., B. Lalic, I. Arsenic, B. Rajkovic, and P.L. Vidale, 2000: Comparison of different boundary layer schemes using single point micrometeorological field data. *Theor. Appl. Climatol.,* **67**, 135-151.

Millán, M.M., 2002: Ozone Dynamics in the Mediterranean Basin: A collection of scientific papers resulting from the MECAPIP, RECAPMA and SECAP Projects. Air Pollution Research Report 78, European Commission, DG RTD I.2, LX 46 2/82, Rue de la Loi, 200, B-1040, Brussels, 287 pp.

Millán, M.M., M. Navazo, and A. Ezcurra, 1987: Meso-meteorological analysis of air pollution cycles in Spain. *Proc. 4th European Symposium on Physico-Chemical Behavior of Atmospheric Pollutants*, Stressa, Sep.1986, (p. 614-619). G. Angeletti and G. Restelli, Eds.

Published for the Commission of the European Communities, D. Reidel Pub. Co., Dordrecht, Holland, 809 pp.

Millán, M.M., R. Salvador, E. Mantilla, and B. Artíñano, 1996: Meteorology and photochemical air pollution in southern europe: Experimental results from EC research project. *Atmos. Environ.,* **30**, 1909-1924.

Millán, M.M., R. Salvador, E. Mantilla, and G. Kallos, 1997: Photo-oxidant dynamics in the Western Mediterranean in Summer: Results from European Research Projects. *J. Geophys. Res.,* **102**, D7, 8811-8823.

Millán, M.M., M.J. Estrela, and C. Badenas, 1998: Meteorological processes relevant to forest fire dynamics on the Spanish Mediterranean Coast. *J. Appl. Meteor.,* **37**, 83-100, http://dx.doi.org/10.1175/1520-0450(1998)037<0083:MPRTFF>2.0.CO;2.

Millán, M.M., E. Mantilla, R. Salvador, A. Carratalá, M.J., Sanz, L. Alonso, G. Gangoiti, and M. Navazo, 2000: Ozone cycles in the Western Mediterranean Basin: Interpretation of monitoring data in complex coastal terrain. *J. Appl. Meteor.,* **39**, 487-508.

Millán, M.M., M.J. Sanz, R. Salvador, and E. Mantilla, 2002: Atmospheric dynamics and ozone cycles related to nitrogen deposition in the western Mediterranean. *Environ. Poll.*, **118(2)**, 167-186.

Miller, M.J., and R.P. Pearce, 1974: A three-dimensional primitive equation model of cumulonimbus convection. *Quart. J. Roy. Meteor. Soc.,* **100**, 133-154.

Miller, M.J., and A.J. Thorpe, 1981: Radiation conditions for the lateral boundaries of limited-area numerical models. *Quart. J. Roy. Meteor. Soc.,* **107**, 615-628.

Mironov, D., E. Heise, E. Kourzeneva, B. Ritter, N. Schneider, and A. Terzhevik, 2010: Implementation of the lake parameterisation scheme FLake into numerical weather prediction model COSMO. *Boreal. Environ. Res.*, **15**, 218-230.

Mishchenko, M.I., L.D. Travis, and D.W. Mackowski, 1996: T-matrix computations of light scattering by nonspherical particles: A review. *J. Quant. Spectrosc. Radiat. Transfer*, **55(5)**, 535-575.

Mishra, V., K.A. Cherkauer, D. Niyogi, M. Lei, B.C. Pijanowski, D.K. Ray, L.C. Bowling, and G. Yang, 2010: A regional scale assessment of land use/land cover and climatic changes on water and energy cycle in the upper Midwest United States. *Int. J. Climatol.,* **30**, 2025-2044, http://dx.doi.org/10.1002/joc.2095.

Mitchell, J.L., R.T. Pierrehumbert, D.M.W. Frierson, and R. Caballero, 2006: The dynamics behind Titan's methane clouds. *Proc. Natl. Acad. Sci.*, **103(49)**, 18421-18426.

Mitra, C., M. Shepherd, and T. Jordan, 2011: On the relationship between the pre-monsoonal rainfall climatology and urban land cover dynamics in Kolkata city, India. *Int. J. Climatol.,* 32, 1443–1454, http://dx.doi.org/10.1002/joc.2366.

Mitsumoto, S., H. Ueda, and H. Ozoe, 1983: A laboratory experiment on the dynamics of the land and sea breeze. *J. Atmos. Sci.,* **40**, 1228-1240.

Mizzi, A.P., 1982: A numerical investigation of the mesoscale atmospheric circulation in the Oregon coastal zone with a coupled atmosphere-ocean model. M.S. Thesis, University of Virginia, 114 pp.

Mizzi, A.P., and R.A. Pielke, 1984: A numerical study of the mesoscale atmospheric circulation observed during a coastal upwelling event on August 23, 1972. Part I: Sensitivity studies. *Mon. Wea. Rev.,* **112**, 76-90.

Mlawer, E.J., and S. Clough, 1997: Shortwave and longwave enhancements in the rapid radiative transfer model. Proc., 7th Atmospheric Radiation Measurement (ARM) Science Team Meeting, U.S. Department of Energy, CONF-9603149, 223-226.

Mlawer, E.J., S.J. Taubman, P.D. Brown, M.J. Iacono and S.A. Clough, 1997: Radiative transfer for inhomogeneous atmospheres: RRTM, a validated correlated-k model for the longwave. *J. Geophys. Res.,* **102(D14)**, 16663-16682.

Mocko, D.M., and W.R. Cotton, 1995: Evaluation of fractional cloudiness parameterizations for use in a mesoscale model. *J. Atmos. Sci.,* **52**, 2884-2901.

Mohr, P.J., and B.N. Taylor, 2000: The fundamental physical constants: a recent least-squares adjustment has produced a new set of recommended values of the basic constants and conversion factors of physics and chemistry. *Phys. Today,* **53**, BG6-BG13.

Mohr, K.I., J.S. Famiglietti, A. Boone, and P.J. Starks, 2000: Modeling soil moisture and surface flux variability with an untuned land surface scheme: a case study from the Southern Great Plains 1997 hydrology experiment. *J. Hydrometeor.,* **1**, 154-169.

Molinari, J., 1982: A method for calculating the effects of deep cumulus convection in numerical models. *Mon. Wea. Rev.,* **110**, 1527-1534.

Mölders, N., 1999a: On the atmospheric response to urbanization and open-pit mining under various geostrophic wind conditions. *Meteor. Atmos. Phys.,* **71**, 205-228.

Mölders, N., 1999b: On the effects of different flooding stages of the oder and different land-use types on the distributions of evapotranspiration, cloudiness and rainfall in the Brandenburg-Polish border area. *Contrib. Atmos. Phys.,* **72**, 1-25.

Mölders, N., 2013: Investigations on the impact of single direct and indirect, and multiple emission-control measures on cold-season near-surface PM_2.5-concentrations in Fairbanks, Alaska. *Atmos. Pollution Res.*, **6**, 87-100.

Mölders, N., and G. Kramm, 2007: Influence of wildfire induced land-cover changes on clouds and precipitation in Interior Alaska – A case study. *Atmos. Res.*, **84**, 142-168, http://dx.doi.org/10.1016/j.atmosres.2006.06004.

Mölders, N., and G. Kramm, 2010: A case study on wintertime inversions in interior Alaska with WRF. *Atmos. Res.*, **95,** 314-332.

Mölders, N., and A. Raabe, 1997: Testing the effect of a two-way-coupling of a meteorological and a hydrologic model on the predicted local weather. *Atmos. Res.,* **45**, 81-107.

Mölders, N., S.E. Porter, C.F. Cahill, and G.A. Grell, 2010: Influence of ship emissions on air quality and input of contaminants in southern Alaska National Parks and Wilderness Areas during the 2006 tourist season. *Atmos. Environ.*, **44**, 1400-1413.

Mölders, N., H.N.Q. Tran, P. Quinn, K. Sassen, G.E. Shaw, and G. Kramm, 2011: Assessment of WRF/Chem to simulate sub-arctic boundary layer characteristics during low solar irradiation using radiosonde, sodar, and surface data. *Atmos. Pollution Res.*, **2**, 283-299.

Mölders, N., H.N.Q. Tran, C.F. Cahill, K. Leelasakultum, and T.T. Tran, 2012: Assessment of WRF/Chem PM_2.5 forecasts using mobile and fixed location data from the Fairbanks, Alaska winter 2008/09 field campaign. *Atmos. Pollution Res.*, **3**, 180-191.

Molinari, J., 1985: A general form of Kuo's cumulus parameterization. *Mon. Wea. Rev.,* **113**, 1411-1416.

Molinari, J., and T. Corsetti, 1985: Incorporation of cloud-scale and mesoscale down-drafts into a cumulus parameterization: Results of one- and three-dimensional integrations. *Mon. Wea. Rev.,* **113**, 485-501.

Molinari, J., and M. Dudek, 1992: Parameterization of convective precipitation in mesoscale numerical models: A critical review. *Mon. Wea. Rev.,* **120**, 326-344.

Monaghan, A.J., K. MacMillan, S.M. Moore, P.S. Mead, M.H. Hayden, and R.J. Eisen, 2012: A regional climatography of West Nile, Uganda, to support human plague modeling. *J. Appl. Meteor. Climatol.,* **51**, 1201-1221.

Montabone, L., S.R. Lewis, and P.L. Read, 2005: Interannual variability of Martian dust storms in assimilation of several years of Mars global surveyor observations. *Adv. Space Res.*, **36(11)**, 2146-2155.

Monteith, J.L., Ed. 1975a: *Vegetation and the Atmosphere. Volume 1, Principles*. Academic Press, New York, 278 pp.

Monteith, J.L., Ed. 1975b: *Vegetation and the Atmosphere. Volume 2, Case Studies*. Academic Press, New York, 439 pp.

Monteith, J.L., 1981: Evaporation and surface temperature. *Quart. J. Roy. Meteor. Soc.,* **107**, 1-27.

Moon, I., I. Ginis, T. Hara, and B. Thomas 2007: Physics-based parameterization of air-sea momentum flux at high wind speeds and its impact on hurricane intensity predictions. *Mon. Wea. Rev.,* **135**, 2869-2878.

Moore, A.M., H.G. Arango, E. Di Lorenzo, A.J. Miller, and B.D. Cornuelle, 2009: An adjoint sensitivity analysis of the Southern California current circulation and ecosystem. *J. Phys. Oceanogr.,* **39**, 702-720.

Moores, W.H., 1982: Direct measurements of radiative and turbulence flux convergencies in the lowest 1000 m of the convective boundary layer. *Bound.-Layer Meteor.,* **22**, 283-294.

Moran, M.D., 1992: Numerical modelling of mesoscale atmospheric dispersion. Colorado State University, Department of Atmospheric Science, Ph.D. Dissertation, 758 pp.

Moran, M.D., and R.A. Pielke, 1996a: Evaluation of a mesoscale atmospheric dispersion modeling system with observations from the 1980 Great Plains mesoscale tracer field experiment. Part I: Data sets and meteorological simulations. *J. Appl. Meteor.,* **35**, 281-307.

Moran, M.D., and R.A. Pielke, 1996b: Evaluation of a mesoscale atmospheric dispersion modeling system with observations from the 1980 Great Plains mesoscale tracer field experiment. Part II: Dispersion simulations. *J. Appl. Meteor.,* **35**, 308-329.

Morcrette, J.-J., and Y. Fouquart, 1986: The overlapping of cloud layers in shortwave radiation parameterizations. *J. Atmos. Sci.,* **43**, 321-328.

Moroz, W.J., 1967: A lake breeze on the eastern shore of Lake Michigan: Observations and model. *J. Atmos. Sci.,* **24**, 337-355.

Morse, B.J., 1973: An analytical study of mesh refinement applied to the wave equation. NOAA Technical Memorandum WMPO-5, August, 1973.

Moss, M.S., and R.W. Jones, 1978: A numerical simulation of hurricane landfall. NOAA Technical Memo ERL-NHEML-3, 1-15.

Moudden, Y., and J.C. McConnell, 2005: A new model for multiscale modeling of the Martian atmosphere, GM3. *J. Geophys. Res.*, **110(E4)**, E04001.

Mullen, S.L., 1979: An investigation of small synoptic-scale cyclones in polar air stream. *Mon. Wea. Rev.,* **107**, 1636-1647.

Mullen, S.L., and R. Buizza, 2002: The impact of horizontal resolution and ensemble size on probabilistic precipitation forecasts by the ECMWF ensemble prediction system. *Wea. Forecasting*, **17**, 173-191.

Muller, R.A., 1966: Snowbelts of the Great Lakes. *Weatherwise,* **19**, 248-255.

Mumpower, J., P. Middleton, R.L. Dennis, T.R. Stewart, and V. Viers, 1981: Visual air quality assessment: Denver case study. *Atmos. Environ.,* **15**, 2433-2441.

Munn, R.E., 1966: *Descriptive Micrometeorology*. Academic Press, New York, 245 pp.

Murdoch, D.C., 1957: *Linear Algebra for Undergraduates*. Wiley, New York, 239 pp.

Murphy, R.E., Ed., 1992: FIFE Special Issue, *J. Geophys. Res.,* **97, No. D17**, 18343-19109.

Murray, F.W., 1967: On the computation of saturation vapor pressure. *J. Appl. Meteor.*, **6**, 203-204. http://dx.doi.org/10.1175/1520-0450(1967)006<0203:OTCOSV>2.0.CO;2.

Murray, F.W., 1970: Numerical models of a tropical cumulus cloud with bilateral and axial symmetry. *Mon. Wea. Rev.,* **98**, 14-28.

Myhre, Gunnar, F. Stordal, M. Johnsrud, A. Ignatov, M.I. Mishchenko, I.V. Geogdzhayev, D. Tanré, J.-L. Deuzé, P. Goloub, T. Nakajima, A. Higurashi, O. Torres, and B. Holben 2004: Intercomparison of satellite retrieved aerosol optical depth over the ocean. *J. Atmos. Sci.*, **61**, 499-513.

Myhre, G., A. Grini, J. Haywood, F. Stordal, B. Chatenet, D. Tanre, J. Sundet, I. Isaksen, 2003: Modelling the radiative impact of mineral dust aerosol during the Saharan Dust Experiment (SHADE) campaign. *J. Geophys. Res.*, **108(D18)**, 8579, http://dx.doi.org/10.1029/2002JD002566, 2003.

Nachamkin, J.E., and W.R. Cotton, 2000: Interactions between a developing mesoscale convective system and its environment. Part II: Numerical simulation. *Mon. Wea. Rev.*, **128**, 1225-1244.

Nachamkin, J., R.L. McAnelly, and W.R. Cotton, 2000: Interactions between a developing mesoscale convective system and its environment. Part I. Observational analysis. *Mon. Wea. Rev.*, **128**, 1205-1224.

Nadezhina, E.D., and O.B. Shklyarevich, 1996: A numerical simulation of the atmospheric boundary layer in a coastal area with allowance for vegetation. *Russ. Meteorol. Hydrol.*, **11**, 21-28.

Nagasawa, R., and H. Kitagawa, 2005: Modification of radiation scheme of JMA-NHM. Proc. Autumn Conf. JMSJ. 87, B403 (in Japanese).

Nagle, F.W., 1979: Visualizing a cold dome. *Bull. Amer. Meteor. Soc.*, **60**, 128.

Nair, U.S., M.R. Hjelmfelt, and R.A. Pielke, 1997: Numerical simulation of the June 9-10 1972 Black Hills storm using CSU RAMS. *Mon. Wea. Rev.*, **125**, 1753-1766.

Nair, U.S., R.C. Weger, K.S. Kuo, and R.M. Welch, 1998: Clustering, randomness, and regularity in cloud fields 5. The nature of regular cumulus cloud fields. *J. Geophys. Res.*, **103**, 11,363-11,380.

Nair, U.S., Y. Wu, J. Kala, T.J. Lyons, R.A. Pielke Sr., and J.M. Hacker, 2011: The role of land use change on the development and evolution of the West Coast trough, convective clouds, and precipitation in Southwest Australia. *J. Geophys. Res.-Atmos.*, **116**, D07103, http://dx.doi.org/10.1029/2010JD014950.

Nakajima, T., and M.D. King, 1990: Determination of the optical thickness and effective particle radius of clouds from reflected solar radiation measurements. Part I: Theory. *J. Atmos. Sci.*, **47**, 1878-1893.

Nakajima, T.Y., K. Suzuki, and G.L. Stephens, 2010: Droplet growth in warm water clouds observed by the A-Train. Part I: Sensitivity analysis of the MODIS-derived cloud droplet sizes. *J. Atmos. Sci.*, **67**, 1884-1896.

Nakanishi, M., and H. Niino, 2006: An improved Mellor-Yamada level-3 model: Its numerical stability and application to a regional prediction of advection fog. *Bound.-Layer Meteor.*, **119**, 397-407.

Namias, J., 1983: The history of polar front and air mass concepts in the United States - An eyewitness account. *Bull. Amer. Meteor. Soc.*, **1**, 734-755.

Nappo, C.J., 2002: *An Introduction to Atmospheric Gravity Waves.* Academic Press, 276 pp.

Narita, M., 2006: On the modification of Kain-Fritsch convective parameterization scheme in JMA nonhydrostatic model. *Proc., 8th Workshop on Nonhydrostatic Models*, 31-32 (in Japanese).

Naud, C.M., D.J. Posselt, and S.C. van den Heever, 2012: Observational analysis of cloud and precipitation in midlatitude cyclones: Northern versus Southern Hemisphere warm fronts. *J. Climate*, 25, 5135-5151, http://dx.doi.org/10.1175/JCLI-D-11-00569.1.

NBS, 1974: Fundamental physical constants. National Bureau of Standards Special Publication 398, U.S. Government Printing Office, SD Catalog No. C13.10:398, Stock Number 0303-01331, 1 card.

Neggers, R., M. Kohler, and A. Beljaars, 2009: A dual mass flux framework for boundary layer convection. Part I: Transport. *J. Atmos. Sci.,* **66**, 1465-1487.

Negri, A.J., and T.H. Vonder Haar, 1980: Moisture convergence using satellite-derived wind fields: A severe local storm case study. *Mon. Wea. Rev.,* **108**, 1170-1182.

Neumann, J., 1951: Land breezes and nocturnal thunderstorms. *J. Meteor.,* **8**, 60-67.

Neumann, J., 1977: On the rotation rate of the direction of sea and land breezes. *J. Atmos. Sci.,* **34**, 1913-1917.

Neumann, J., and Y. Mahrer, 1971: A theoretical study of the land and sea breeze circulations. *J. Atmos. Sci.,* **28**, 532-542.

Neumann, J., and Y. Mahrer, 1974: A theoretical study of the sea and land breezes of circular islands. *J. Atmos. Sci.,* **31**, 2027-2039.

Neumann, J., and Y. Mahrer, 1975: A theoretical study of the lake and land breezes of circular lakes. *Mon. Wea. Rev.,* **103**, 474-485.

Newiger, M., and K. Bähnke, 1981: Influence of cloud composition and cloud geometry on the absorption of solar radiation. *Contrib. Atmos. Phys.,* **54**, 370-382.

Newman, C.E., S.R. Lewis, P.L. Read, and F. Forget, 2002: Modeling the Martian dust cycle 2. Multiannual radiatively active dust transport simulations. *J. Geophys. Res.*, **107(E12)**, 5124.

Newton, C.W., 1966: Circulations in large sheared cumulonimbus. *Tellus,* **18**, 699-712.

Nicholls, M.E., and R.A. Pielke, 1994a: Thermal compression waves. I: Total energy transfer. *Quart. J. Roy. Meteor. Soc.*, **120**, 305-332.

Nicholls, M.E., and R.A. Pielke, 1994b: Thermal compression waves. II: Mass adjustment and vertical transfer of total energy. *Quart. J. Roy. Meteor. Soc.*, **120**, 333-359.

Nicholls, M.E., and R.A. Pielke Sr., 2000: Thermally-induced compression waves and gravity waves generated by convective storms. *J. Atmos. Sci.*, **57**, 3251-3271.

Nicholls, M.E., R.A. Pielke, and W.R. Cotton, 1991: A two-dimensional numerical investigation of the interaction between sea-breezes and deep convection over the Florida peninsula. *Mon. Wea. Rev.,* **119**, 298-323.

Nicholls, M.E., R.A. Pielke, and R. Meroney, 1993: Large eddy simulation of microburst winds flowing around a building. *J. Wind Eng. Indus. Aerodyn.*, **46 & 47**, 229-237.

Nicholls, M.E., R.A. Pielke, J.L. Eastman, C.A. Finley, W.A. Lyons, C.J. Tremback, R.L. Walko, and W.R. Cotton, 1995: Applications of the RAMS numerical model to dispersion over urban areas. In: *Wind Climate in Cities*, J.E. Cermak, Editor. Kluwer Academic Publishers, The Netherlands, 703-732.

Nickerson, E.C., 1979: On the numerical simulation of airflow and clouds over mountainous terrain. *Contrib. Atmos. Phys.,* **52**, 161-177.

Nickerson, E.C., and E.L. Magaziner, 1976: A three-dimensional simulation of winds and non-precipitating orographic clouds over Hawaii. NOAA Technical Report ERL 377-APCL 39, 1-35.

Nicosia, D.J., and R.H. Grumm, 1999: Mesoscale band formation in three major northeastern United States snowstorms. *Wea. Forecasting,* **14**, 346-368.

Nie, D., T. Demetriades-Shah, and E.T. Kanemasu, 1992: Surface energy fluxes on four slope sites during FIFE 1988. *J. Geophys. Res.,* **97**, 18641-18649.

Nielsen, K.L., 1964: *Methods in Numerical Analysis*, 2nd Edition. MacMillan, New York, 408 pp.

Niemann, H.B., S.K. Atreya, S.J. Bauer, G.R. Carignan, J.E. Demick, R L. Frost, D. Gautier, J.A. Haberman, D.N. Harpold, D.M. Hunten, G. Israel, J.I. Lunine, W.T. Kasprzak, T.C. Owen, M. Paulkovich, F. Raulin, E. Raaen, and S.H. Way, 2005: The abundances of constituents of Titan's atmosphere from the GCMS instrument on the Huygens probe. *Nature*, **7069**, 779-784.

Nikuradse, J., 1933: Stroömungesetze, in rauhen Rohren, Forschungsheft 361. Referred to by Businger 1973.

Nilsson, B., 1979: Meteorological influence on aerosol extinction in the 0.2-40 μm wavelength range. *Appl. Opt.*, **18**, 3457-3473.

Ninomiya, K., 1992: Multi-scale features of Baiu, the summer monsoon over Japan and the East Asia. *J. Meteor. Soc. Japan,* **70**, 467-495.

Ninomiya, K., 2000: Large- and meso-alpha-scale characteristics of Meiyu/Baiu front associated with intense rainfalls in 1-10 July 1991. *J. Meteor. Soc. Japan,* **78**, 141-157.

Ninomiya, K., M. Ikawa, and T. Akiyama, 1981: Long-level medium-scale cumulonimbus cluster in Asian subtropical humid region. *J. Meteor. Soc. Japan,* **59**, 564-577.

Nitta, T., 1976: Large-scale heat and moisture budgets during the air mass transformation experiment. *J. Meteor. Soc. Japan,* **54**, 1-14.

Niyogi, D.S., and S. SethuRaman, 1997: Comparison of four different stomatal resistance schemes using FIFE observations. *J. Appl. Meteor.,* **36**, 903-917.

Niyogi, D.S., S. SethuRaman, and K. Alapaty, 1999: Uncertainty in the specification of surface characteristics, Part II: Hierarchy of interaction - explicit statistical analysis. *Bound.-Layer Meteor.,* **91**, 341-366.

Niyogi, D.S., P. Pyle, M. Lei, S. Arya, C. Kishtawal, M. Shepherd, F. Chen, and B. Wolfe, 2011: Urban modification of thunderstorms: An observational storm climatology and model case study for the Indianapolis urban region. *J. Appl. Meteor. Climatol.,* **50**, 1129-1144.

Nkemdirim, L.C., 1980: A test of a lapse rate/wind speed model for estimating heat island magnitude in an urban airshed. *J. Appl. Meteor.,* **19**, 749-756.

Nobis, T.E., 2010: Coupling an urban parameterization to an atmospheric model using an operational configuration. Department of Atmospheric Sciences, Ph.D. Dissertation, Colorado State University, 182 pp.

Noilhan, J., and S. Planton, 1989: A simple parameterization of land surface processes for meteorological models. *Mon. Wea. Rev.,* **117**, 536-549.

Noonkester, V.R., 1979: Coastal marine fog in southern California. *Mon. Wea. Rev.,* **107**, 830-851.

North, G.R., and T.L. Erukhimova, 2009: *Atmospheric Thermodynamics: Elementary Physics and Chemistry*. Cambridge University Press, 280 pp.

Noto, K., 1996: Dependence of heat island phenomena on stable stratification and heat quantity in a calm environment. *Atmos. Environ.,* **30**, 475-485.

Novak, D.R., B.A. Colle, and R. McTaggart-Cowan, 2009: The role of moist processes in the formation and evolution of mesoscale snowbands within the Comma Head of northeast US cyclones. *Mon. Wea. Rev.*, **137**, 2662-2686.

Novak, D.R., B.A. Colle, and A.R. Aiyyer, 2010: Evolution of mesoscale precipitation band environments within the comma head of northeast US cyclones. *Mon. Wea. Rev.*, **138**, 2354-2374.

NRC (National Research Council), 2005: Radiative forcing of climate change: Expanding the concept and addressing uncertainties. Committee on Radiative Forcing Effects on Climate Change, Climate Research Committee, Board on Atmospheric Sciences and Climate, Division on Earth and Life Studies, The National Academies Press, Washington, D.C., 208 pp.

NRC (National Research Council), 2012: Urban meteorology: forecasting, monitoring, and meeting users' needs. Washington, DC: The National Academies Press, 176 pp.

Nunez, M., and T.R. Oke, 1976: Long-wave radiative flux divergence and nocturnal cooling of the urban atmosphere. II. Within an urban canyon. *Bound.-Layer Meteor.,* **10**, 121-135.

Nunez, M., and T.R. Oke, 1977: The energy balance of an urban canyon. *J. Appl. Meteor.,* **16**, 11-19.

Nykanen, D.K., E. Foufoula-Georgiou, and W.M. Lapenta, 2001: Impact of small-scale precipitation variability on larger-scale spatial organization of land-atmosphere fluxes. *J. Hydrometeor.,* **2**, 105-121. http://dx.doi.org/10.1175/1525-7541(2001)002<0105:IOSSRV>2.0.CO;2.

O'Brien, J.J., 1970: Alternative solutions to the classical vertical velocity problem. *J. Appl. Meteor.,* **9**, 197-203.

O'Brien, J.J., and H.E. Hurlburt, 1972: A numerical model of coastal upwelling. *J. Phys. Oceanogr.,* **2**, 14-26.

Oerlemans, J., 1980: A case study of a subsynoptic disturbance in a polar outbreak. *Quart. J. Roy. Meteor. Soc.,* **106**, 313-325.

Ogura, Y., 1972: Clouds and convection. GARP Publication Series No. 8, WMO, Geneva, Switzerland, 20-29.

Ogura, Y., and J.G. Charney, 1961: A numerical model of thermal convection in the atmosphere. *Proceedings, International Symposium on Numerical Weather Prediction*, 431-450.

Ogura, Y., and M.-T. Liou, 1980: The structure of a midlatitude squall line: A case study. *J. Atmos. Sci.,* **37**, 553-567.

Ogura, Y., and N.A. Phillips, 1962: Scale analysis of deep and shallow convection in the atmosphere.
J. Atmos. Sci., **19**, 173-179.

Ogura, Y., and T. Takahashi, 1971: Numerical simulation of the life cycle of a thunderstorm cell. *Mon. Wea. Rev.,* **99**, 895-911.

Ogura, Y., and T. Takahashi, 1973: The development of warm rain in a cumulus model. *J. Atmos. Sci.,* **30**, 262-277.

Ogura, Y., Y.-L. Chen, J. Russell, and S.-T. Soong, 1979: On the formation of organized convective systems observed over the eastern Atlantic. *Mon. Wea. Rev.,* **107**, 426-441.

Ohata, T., K. Higuchi, and K. Ikegami, 1981: Mountain-valley wind system in the Khumbu Himal, East Nepal. *J. Meteor. Soc. Japan,* **59**, 753-762.

O'Hirok, W., and C. Gautier, 1998a: A three-dimensional radiative transfer model to investigate the solar radiation within a cloudy atmosphere. Part I: Spatial effects. *J. Atmos. Sci.,* **55**, 2162-2179.

O'Hirok, W., and C. Gautier, 1998b: A three-dimensional radiative transfer model to investigate the solar radiation within a cloudy atmosphere. Part II: Spectral effects. *J. Atmos. Sci.,* **55**, 3065-3076.

Oke, T.R., 1973: City size and the urban heat island. *Atmos. Environ.,* **7**, 769-779.

Oke, T.R., 1976: The distinction between canopy and boundary-layer urban heat islands. *Atmosfera,* **14**, 268-277.

Oke, T.R., 1978: *Boundary Layer Climates*. Methuen, London, 372 pp.

Oke, T.R., 1982: The energetic basis of the urban heat island. *Quart. J. Roy. Meteor. Soc.,* **108**, 1-24.

Oke, T.R., and R.F. Fuggle, 1972: Comparison of urban/rural counter and net radiation at night. *Bound.-Layer Meteor.,* **2**, 290-308.

Okeyo, A.E., 1982: A two-dimensional numerical model of the lake-land and sea-land breezes over Kenya. Department of Meteorology, M.S. thesis, University of Nairobi, Kenya.

Okoola, R.E., 2000: The characteristics of cold air outbreaks over the eastern highlands of Kenya. *Meteor. Atmos. Phys.,* **73**, 177-187.

Oleson, K.W., and coauthors, 2010: Technical Description of version 4.0 of the Community Land Model (CLM), NCAR Technical Note (NCAR/TN-478+STR), 257 pp.

Olfe, D.B., and R.L. Lee, 1971: Linearized calculations of urban heat island convection effects. *J. Atmos. Sci.,* **28**, 1374-1388.

Oki, T., T. Nishimura, and P. Dirmeyer, 1999: Assessment of annual runoff from land surface models using Total Runoff Integrating Pathways (TRIP). *J. Meteor. Soc. Japan,* **77**, 235-255.

Oliger, J., and A. Sundstroöm, 1976: Theoretical and practical aspects of some initial-boundary value problems in fluid dynamics. Report STAN-CS-76-578, Stanford University, Computer Science Department (available from the Publications Office).

Oochs, H.J. III, and D.B. Johnson, 1980: Urban effects on the properties of first echoes. *J. Appl. Meteor.,* **19**, 1160-1166.

Ookouchi, Y., M. Uryu, and R. Sawada, 1978: A numerical study of the effects of a mountain on the land and sea breezes. *J. Meteor. Soc. Japan,* **56**, 368-385.

Ooyama, K.V., 1971: A theory of parameterization of cumulus convection. *J. Meteor. Soc. Japan,* **49**, 744-756.

Ooyama, K.V., 1982: Conceptual evolution of the theory and modeling of the tropical cyclone. *J. Meteor. Soc. Japan,* **60**, 369-380.

Orgill, M.M., 1981: A planning guide for future studies. DOE Contract Report prepared by Battelle Pacific North west Laboratory, Report No. PNL-3656, ASCOT/80/4.

Orlanski, I., 1975: A rational subdivision of scales for atmospheric process. *Bull. Amer. Meteor. Soc.,* **56**, 527-530.

Orlanski, I., 1976: A simple boundary condition for unbounded hyperbolic flows. *J. Comput. Phys.,* **21**, 251-269.

Orlanski, I., 1981: The quasi-hydrostatic approximation. *J. Atmos. Sci.,* **38**, 572-582.

Orlanski, I., B. Ross, and L. Polinsky, 1974: Diurnal variation of the planetary boundary layer in a mesoscale model. *J. Atmos. Sci.,* **31**, 965-989.

Orszag, S.A., 1971: Numerical simulation of incompressible flows within simple boundaries: Accuracy. *J. Fluid Mech.,* **49**, 76-112.

Orville, H.D., 1980: Numerical modeling of clouds. In *Lecture Notes*. IFAORS Short Course 450 on Clouds: Their Formation, Properties, and Effects. Williamsburg, Virginia, 1-5 December 1980. Held at the Institute for Atmospheric Optics and Remote Sensing. Post Office Box P, Hampton, VA 23666.

Orville, H.D., P.A. Eckhoff, J.E. Peak, J.H. Hirsch, and F.J. Kopp, 1981: Numerical simulation of the effects of cooling tower complexes on clouds and severe weather. *Atmos. Environ.,* **15**, 823-836.

Otkin, J.A., T.J. Greenwald, J. Sieglaff, and H.-L. Huang, 2009: Validation of a large-scale simulated brightness temperature dataset using SEVIRI Satellite observations. *J. Appl. Meteor. Climatol.,* **48**, 1613-1626.

Otte, T.L., N.L. Seaman, and D.R. Stauffer, 2001: A heuristic study on the importance of anisotropic error distributions in data assimilation. *Mon. Wea. Rev.,* **129**, 766-783.

Otterman, J., 1974: Baring high albedo soils by desertification – a hypothesized desertification mechanism. *Science,* **184**, 531-533.

Otterman, J., 1975: Possible rainfall reduction through reduced surface temperature due to overgrazing. NASA Report, Technical Inf. Division, Code 250, Goddard Space Flight Center, Greenbelt, Maryland.

Otterman, J., 1981a: Plane with protrusions as an atmospheric boundary. *J. Geophys. Res.,* **86**, 6627-6630.

Otterman, J., 1981b: Satellite and field studies of man's impact on the surface in arid regions. *Tellus,* **33**, 68-77.

Ozoe, H., T. Shibata, H. Sayama, and H. Ueda, 1983: Characteristics of air pollution in the presence of land and sea breeze – a numerical simulation. *Atmos. Environ.,* **17**, 35-42.

Paegle, J., W.G. Zdunkowski, and R.M. Welch, 1976: Implicit differencing of predictive equations of the boundary layer. *Mon. Wea. Rev.,* **104**, 1321-1324.

Paeth, H., A. Scholten, P. Friederichs, and A. Hense, 2008: Uncertainties in climate change prediction: El Niño-Southern Oscillation and monsoons. *Global Planetary Change*, **60(3-4)**, 265-288, http://dx.doi.org/10.1016/j.gloplacha.2007.03.002.

Palau, J.L., G. Perez-Landa, J. Meliá, D. Segarra, M.M. Millán, 2005a: A study of dispersion in complex terrain under winter conditions using high-resolution mesoscale and Lagrangian particle models. *Atmos. Chem. Phys. (ACP),* **5**, 11965-12030 (European Geosciences Union).

Palau, J.L., G. Perez-Landa, J.J. Dieguez, C. Monter, and M.M. Millán, 2005b: The importance of meteorological scales to forecast air pollution scenarios on coastal complex terrain. *Atmos. Chem. Phys. (ACP),* **5**, 2771-2785.

Palmén, E., and C.W. Newton, 1969: *Atmospheric Circulation Systems.* Academic Press, New York, NY, 603 pp.

Palmer, T.N., F.J. Doblas-Reyes, A. Weisheimer, and M.J. Rodwell, 2008: Toward seamless prediction: Calibration of climate change projections using seasonal forecasts. *Bull. Amer. Meteor. Soc.*, **89**, 459-470, http://dx.doi.org/10.1175/BAMS-89-4-459.

Paltridge, G.W., and C.M.R. Platt, 1976: *Radiative Processes in Meteorology and Climatology.* Elsevier, New York, 318 pp.

Paltridge, G.W., and C.M.R. Platt, 1981: Aircraft measurements of solar and infrared radiation and the microphysics of cirrus clouds. *Quart. J. Roy. Meteor. Soc.,* **107**, 367-380.

Palumbo, A., and A. Mazzarella, 1980: Rainfall statistical properties in Naples. *Mon. Wea. Rev.,* **108**, 1041-1045.

Pan, H.-L, 2003: The GFS Atmospheric Model. NCEP Office Note, No. 442, 14 pp. [Available from NCEP, 5200 Auth Road, Washington, DC 20233].

Pan, H.-L., and J. Wu, 1995: Implementing a Mass Flux Convection Parameterization Package for the NMC Medium-Range Forecast Model. NMC Office Note, No. 409, 40 pp., Available from NCEP, 5200 Auth Road, Washington, DC 20233.

Pandolfo, J.P., C.A. Jacobs, R.J. Ball, M.A. Atwater, and J.A. Sekorski, 1976: Refinement and validation of an urban meteorological-pollutant model. U.S. Environmental Protection Agency, Office of Research Development, Report EPA-600/4-76-037, 1-21.

Panin, B.D., R.P. Repinskaya, K. Buzian, and W. Feng-Lei, 1999: A diabatic regional model with a nested grid. *Russ. Meteorol. Hydrol.,* **3**, 25-33.

Panofsky, H.A., and G.W. Brier, 1968: *Some Applications of Statistics to Meteorology*. Pennsylvania State University, 224 pp.

Papineau, J.M., 1998: Orographic precipitation: A mesoscale modeling perspective. Department of Atmospheric Science, Ph.D. Dissertation, Colorado State University, 131 pp.

Paredes-Miranda, G., W.P. Arnott, H. Moosmüller, M.C. Green, and M. Gyawali, 2013: Black carbon aerosol concentration in five cities and its scaling with city population. *Bull. Amer. Meteor. Soc.*, **94**, 41-50. http://dx.doi.org/10.1175/BAMS-D-11-00225.1.

Park, S.K., and K.K. Droegemeier, 2000: Sensitivity analysis of a 3D convective storm: Implications for variational data assimilation and forecast error. *Mon. Wea. Rev.,* **128**, 140-159.

Park, S.-U., and D.N. Sikdar, 1982: Evolution of the dynamics and thermodynamics of a mesoscale convective system: A case study. *Mon. Wea. Rev.,* **110**, 1024-1040.

Parlange, M.B., A.T. Cahill, D.R. Nielsen, J.W. Hopmans, and O. Wendroth, 1998: Review of heat and water movement in field soils. *Soil and Tillage Research,* **47**, 5-10.

Parrish, D.F., and J.C. Derber, 1992: The National Meteorological Center's spectral statistical-interpolation analysis system. *Mon. Wea. Rev.,* **120**, 1747-1763.

Parrish, T.R., 1982: Barrier winds along the Sierra Nevada Mountains. *J. Appl. Meteor.,* **21**, 925-930.

Pasquill, F., 1961: The estimation of the dispersion of windborne material. *Meteor. Mag.,* **90**, 33-49.

Passarelli Jr., R.E., and H. Boehme, 1983: The orographic modulation of pre-warm-front precipitation in southern New England. *Mon. Wea. Rev.,* **111**, 1062-1070.

Passner, J.E., and J.M. Noble, 2006: Acoustic energy measured in severe storms during a field study in June 2003. ARL-TR-3749, Army Research Laboratory.

Pastushkov, R.S., 1975: The effects of vertical wind shear on the evolution of convective clouds. *Quart. J. Roy. Meteor. Soc.,* **101**, 281-291.

Patnack, P.C., B.E. Freeman, R.M. Traci, and G.T. Phillips, 1983: Improved simulations of mesoscale meteorology. Report Atmospheric Science Laboratory, White Sands Missile Range, NM ASL CR-83-0127-1.

Patrinos, A.N.A., and A.L. Kistler, 1977: A numerical study of the Chicago lake breeze. *Bound.-Layer Meteor.,* **12**, 93-123.

Patrinos, A.N.A., and M.J. Leach, 1982: On the use of the pseudo-spectral technique in air-pollution modelling. *Preprints, AMS 3rd Joint Conference on Applications of Air Pollution Meteorology*, San Antonio, Texas, January 11-15.

Pauwels, V.R.N., and E.F. Wood, 1999: A soil-vegetation-atmosphere transfer scheme for the modeling of water and energy balance processes in high latitudes. 2. Application and validation. *J. Geophys. Res.,* **104**, 27,823-27,839.

Pearson, R.A., 1973: Properties of the sea breeze front as shown by a numerical model. *J. Atmos. Sci.,* **30**, 1050-1060.

Pearson, R.A., 1975: On the asymmetry of the land-breeze sea breeze circulation. *Quart. J. Roy. Meteor. Soc.,* **101**, 529-536.

Pedlosky, J., 2003: *Waves in the Ocean and Atmosphere: Introduction to Wave Dynamics.* Springer, 260 pp.

Pellerin, P., R. Laprise, and I. Zawadzki, 1995: The performance of a semi-Lagrangian transport scheme for the advection-condensation problem. *Mon. Wea. Rev.,* **123**, 3318-3330.

Peltier, W.R., and T.L. Clark, 1979: The evolution and stability of finite-amplitude mountain waves. Part II. Surface wave drag and severe downslope windstorms. *J. Atmos. Sci.,* **36**, 1498-1529.

Peltier, W.R., and T.L. Clark, 1983: Nonlinear mountain waves in two and three spatial dimensions. *Quart. J. Roy. Meteor. Soc.,* **109**, 527-548.

Pepper, D.W., C.D. Keen, and P.E. Long Jr., 1979: Modeling the dispersion of atmospheric pollution using cubic splines and Chapeau functions. *Atmos. Environ.,* **13**, 223-237.

Pergaud, J., V. Masson, S. Malardel, and F. Couvreux, 2009: A parameterization of dry thermals and shallow cumuli for mesoscale numerical weather prediction. *Bound.-Layer Meteor.,* **132**, 83-106.

Perkey, D.J., 1976: A description and preliminary results from a fine-mesh model for forecasting quantitative precipitation. *Mon. Wea. Rev.,* **104**, 1513-1526.

Perkey, D.J., and C.W. Kreitzberg, 1976: A time-dependent lateral boundary scheme for limited-area primitive equation models. *Mon. Wea. Rev.,* **104**, 744-755.

Petersen, A.C., and A.A.M. Holtslag, 1999: A first-order closure for covariances and fluxes of reactive species in the convective boundary layer. *J. Appl. Meteor.,* **38**, 1758-1776.

Petersen, W.A., L.D. Carey, S.A. Rutledge, J.C. Knievel, N.J. Doesken, R.H. Johnson, T.B. McKee, T. Vonder Haar, and J.F. Weaver, 1999: Mesoscale and radar observations of the Fort Collins flash flood of 28 July 1997. *Bull. Amer. Meteor. Soc.,* **80**, 191-216.

Peterson, J.T., 1970: Distribution of sulfur dioxide over metropolitan St. Louis, as described by empirical eigen vectors, and its relation to meteorological parameters. *Atmos. Environ.,* **4**, 501-518.

Peterson, T.C., K.M. Willett, and P.W. Thorne, 2011: Observed changes in surface atmospheric energy over land. *Geophys. Res. Lett.*, **38**, L16707, http://dx.doi.org/10.1029/2011GL048442.

Petterssen, S., 1956: *Weather Analysis and Forecasting*. McGraw-Hill, New York.

Pettré, P., 1982: On the problem of violent valley winds. *J. Atmos. Sci.,* **39**, 542-554.

Petty, G.W., 2006: *A First Course in Atmospheric Radiation.* 2nd Ed., Sundog Publishing, 460 pp.

Philandras, C.M., D.A. Metaxas, and P.Th. Nastos, 1999: Climate variability and urbanization in Athens. *Theor. Appl. Climatol.,* **63**, 65-72.

Phillips, N.A., 1957: A coordinate system having some special advantages for numerical forecasting. *J. Meteor.,* **14**, 184-185.

Physick, W., 1976: A numerical model of the sea-breeze phenomenon over a lake or gulf. *J. Atmos. Sci.,* **33**, 2107-2135.

Physick, W.L., 1980: Numerical experiments on the inland penetration of the sea breeze. *Quart. J. Roy. Meteor. Soc.,* **106**, 735-746.

Physick, W.L., 1986: Application of a mesoscale flow model in the irregular terrain of the Grand Canyon National Park. Prepared for Donald Henderson, National Park Service, Department of the Interior, Denver, Co 80225, Contract NA-85-RAH05045, Amendment 17, Item 15, R.A. Pielke and M. Segal, P.I., 98 pp.

Pichler, H., R. Steinacker, E. Hagenauer, and A. Jager, 1995: ALPEX-simulation. *Meteor. Atmos. Phys.,* **56**, 197-208.

Pickett, R.L., and D.A. Dossett, 1979: Mirex and the circulation of Lake Ontario. *J. Phys. Oceanogr.*, **9**, 441-445.

Pielke Jr., R.A., 2010: *The Climate Fix: What Scientists and Politicians Won't Tell You About Global Warming*. Basic Books, 276 pp.

Pielke Jr., R.A., and R.A. Pielke Sr., 1997: *Hurricanes: Their Nature and Impacts on Society.* John Wiley and Sons, England, 279 pp.

Pielke Jr., R.A., and R.A. Pielke Sr., Eds., 2000: *Storms.* Volumes I and II. Routledge Press, London.

Pielke Jr., R.A., D. Sarewitz, R. Byerly Jr., and D. Jamieson, 1999: Prediction in the Earth sciences and environmental policy making. *Eos, Trans. Amer. Geophys. Union,* **80**, 1, 312-313.

Pielke, R.A., 1972: Comparison of a hydrostatic and anelastic dry shallow primitive equation model. NOAA Technical Memo., ERL OD-13, February, 42 pp.

Pielke, R.A., 1974a: A three-dimensional numerical model of the sea breezes over south Florida. *Mon. Wea. Rev.,* **102**, 115-139.

Pielke, R.A., 1974b: A comparison of three-dimensional and two-dimensional numerical predictions of sea breezes. *J. Atmos. Sci.,* **31**, 1577-1585.

Pielke, R.A., 1976: Inadvertent weather modification potentials due to microwave transmissions and the thermal heating at SPS rectenna sites. In consultation with M. Garstang, J. Simpson and R.H. Simpson for Lockheed Electronics Company, Inc., 26 pp.

Pielke, R.A., 1978: The role of man and machine in the weather service of the future. *Preprint Volume of the American Meteorological Society Conference on Weather Forecasting and Analysis and Aviation Meteorology*, October 16-19, Silver Spring, Maryland, 271-272.

Pielke, R.A., 1981: An overview of our current understanding of the physical interactions between the sea- and land-breeze and the coastal waters. *Ocean Management*, **6**, 87-100.

Pielke, R.A. 1984: *Mesoscale Meteorological Modeling*. 1st Edition, Academic Press, New York, NY, 612 pp.

Pielke, R.A., 1990: *The Hurricane*. Routledge Press, London, England, 228 pp.

Pielke, R.A., 1991: A recommended specific definition of 'resolution.' *Bull. Amer. Meteor. Soc.*, **72**, 1914.

Pielke, R.A., 1994: The status of mesoscale meteorological models. *Planning and Managing Regional Air Quality: Modeling and Measurement Studies,* P.A. Solomon, and T.A. Silver, Eds., Lewis Publishers, 435-458.

Pielke, R.A., 1995: Synoptic Weather Lab Notes. Dept. of Atmospheric Science Class Report No. 1, Colorado State University, Fort Collins, CO.

Pielke, R.A., 1998: Climate prediction as an initial value problem. *Bull. Amer. Meteor. Soc.*, **79**, 2743-2746.

Pielke, R.A., 2001a: Influence of the spatial distribution of vegetation and soils on the prediction of cumulus convective rainfall. *Rev. Geophys.,* **39**, 151-177.

Pielke, R.A., 2001b: Further comments on "The differentiation between grid spacing and resolution and their application to numerical modeling". *Bull. Amer. Meteor. Soc.*, **82**, 699.

Pielke, R.A., 2001c: Earth system modeling – An integrated assessment tool for environmental studies. In: Present and Future of Modeling Global Environmental Change: Toward Integrated Modeling, T. Matsuno, and H. Kida, Eds., Terra Scientific Publishing Company, Tokyo, Japan, 311-337.

Pielke Sr., R.A., 2002a: Overlooked issues in the U.S. National Climate and IPCC assessments. *Climatic Change*, **52**, 1-11.

Pielke Sr., R.A., 2002b: *Mesoscale Meteorological Modeling*. 2nd Edition, Academic Press, San Diego, CA, 676 pp.

Pielke Sr., R.A., 2002c: *Synoptic Weather Lab Notes.* Colorado State University, Department of Atmospheric Science Class Report No. 1, Final Version, August 20, 2002. http://pielkeclimatesci.files.wordpress.com/2010/01/nr-77.pdf.

Pielke Sr., R.A., Editor in Chief., 2013: *Climate Vulnerability, Understanding and Addressing Threats to Essential Resources*, 1st Edition. J. Adegoke, F. Hossain, G. Kallos, D. Niyogi, T. Seastedt, K. Suding, and C. Wright, Eds., Academic Press, 1570 pp.

Pielke, R.A., and R.W. Arritt, 1984: A proposal to standardize models. *Bull. Amer. Meteor. Soc.*, **65**, 10, 1082.

Pielke, R.A., and R. Avissar, 1990: Influence of landscape structure on local and regional climate. *Landscape Ecology*, **4**, 133-155.

Pielke, R.A., and W.R. Cotton, 1977: A mesoscale analysis over south Florida for a high rainfall event. *Mon. Wea. Rev.,* **105**, 343-362.

Pielke, R.A., and J.M. Cram, 1989: A terrain-following coordinate system – Derivation of diagnostic relationships. *Meteor. Atmos. Phys.,* **40**, 189-193.

Pielke, R.A., and E. Kennedy, 1980: Mesoscale terrain features, January 1980. Report #UVA-ENV SCI-MESO-1980-1, 29 pp.

Pielke, R.A., and T.J. Lee, 1991: Influence of sea spray and rainfall on the surface wind profile during conditions of strong winds. *Bound.-Layer Meteor.,* **55**, 305-308.

Pielke, R.A., and Y. Mahrer, 1978: Verification analysis of the University of Virginia three-dimensional mesoscale model prediction over south Florida for July 1, 1973. *Mon. Wea. Rev.,* **106**, 1568-1589.

Pielke, R.A., and P. Mehring, 1977: Mesoscale climatology in mountainous terrain - mean monthly temperatures. *Mon. Wea. Rev.*, **105**, 108-112.

Pielke, R.A., and R.P. Pearce, Editors, 1994: *Mesoscale Modeling of the Atmosphere*. American Meteorological Society Monograph, Volume 25, 167 pp.

Pielke Sr., R.A., and R.L. Wilby, 2012: Regional climate downscaling - what's the point? *Eos Forum*, **93**, No. 5, 52-53, http://dx.doi.org/10.1029/2012EO050008.

Pielke, R.A., and X. Zeng, 1989: Influence on severe storm development of irrigated land. *Natl. Wea. Dig.*, **14**, 16-17.

Pielke, R.A., M. Segal, R.T. McNider, and Y. Mahrer, 1985: Derivation of slope flow equations using two different coordinate representations. *J. Atmos. Sci.,* **42**, 1102-1106.

Pielke, R.A., G. Kallos, and M. Segal, 1989: Horizontal resolution needs for adequate lower tropospheric profiling involved with thermally-forced atmospheric systems. *J. Atmos. Oceanic Tech.*, **6**, 741-758.

Pielke, R.A., A. Song, P.J. Michaels, W.A. Lyons, and R.W. Arritt, 1991: The predictability of sea-breeze generated thunderstorms. *Atmosfera*, **4**, 65-78.

Pielke, R.A., W.R. Cotton, R.L. Walko, C.J. Tremback, W.A. Lyons, L.D. Grasso, M.E. Nicholls, M.D. Moran, D.A. Wesley, T.J. Lee, and J.H. Copeland, 1992: A comprehensive meteorological modeling system – RAMS. *Meteor. Atmos. Phys.,* **49**, 69-91.

Pielke, R.A., M.E. Nicholls, and A.J. Bedard, 1993a: Using thermal compression waves to assess latent heating from clouds. *EOS*, **74**, 493.

Pielke, R.A., M. Segal, R.T. McNider, and Y. Mahrer, 1993b: Reply. *J. Atmos. Sci.,* **50**, 1446-1447.

Pielke Sr., R.A., J. Eastman, L.D. Grasso, J. Knowles, M. Nicholls, R.L. Walko, and X. Zeng, 1995a: Atmospheric vortices. In: *Fluid Vortices*, S. Green, Editor, Kluwer Academic Publishers, The Netherlands, 617-650.

Pielke Sr., R.A., L.R. Bernardet, P.J. Fitzpatrick. S.C. Gillies, R.F. Hertenstein, A.S. Jones, X. Lin, J.E. Nachamkin, U.S. Nair, J.M. Papineau, G.S. Poulos, M.H. Savoie, and P.L. Vidale, 1995b: Standardized test to evaluate numerical weather prediction algorithms. *Bull. Amer. Meteor. Soc.*, **76**, 46-48.

Pielke Sr., R.A., X. Zeng, T.J. Lee, and G.A. Dalu, 1997: Mesoscale fluxes over heterogeneous flat landscapes for use in larger scale models. *J. Hydrology*, **190**, 317-336.

Pielke Sr., G.E. Liston, J.L. Eastman, L. Lu, and M. Coughenour, 1999a: Seasonal weather prediction as an initial value problem. *J. Geophys. Res.,* **104**, 19463-19479.

Pielke Sr., R.A., R.L. Walko, L.T. Steyaert, P.L. Vidale, G.E. Liston, W.A. Lyons, and T.N. Chase, 1999b: The influence of anthropogenic landscape changes on weather in south Florida. *Mon. Wea. Rev.,* **127**, 1663-1673.

Pielke Sr., R.A., C. Davey, and J. Morgan, 2004: Assessing "global warming" with surface heat content. *Eos, Trans. Amer. Geophys. Union,* **85**, No. 21, 210-211.

Pielke Sr., R.A., K. Wolter, O. Bliss, N. Doesken, and B. McNoldy, 2006a: The July 2005 Denver heat wave: How unusual was it? *Nat. Wea. Dig.*, **31**, 24-35.

Pielke Sr., R.A., T. Matsui, G. Leoncini, T. Nobis, U. Nair, E. Lu, J. Eastman, S. Kumar, C. Peters-Lidard, Y. Tian, and R. Walko, 2006b: A new paradigm for parameterizations in numerical weather prediction and other atmospheric models. *National Wea. Digest*, **30**, 93-99.

Pielke Sr., R.A., D. Stokowski, J.-W. Wang, T. Vukicevic, G. Leoncini, T. Matsui, C. Castro, D. Niyogi, C.M. Kishtawal, A. Biazar, K. Doty, R.T. McNider, U. Nair, and W.K.

Tao, 2007a: Satellite-based model parameterization of diabatic heating. *Eos, Trans. Amer. Geophys. Union,* **88:8**, 20 February, 96-97.

Pielke, R.A. Sr., J. Adegoke, A. Beltran-Przekurat, C.A. Hiemstra, J. Lin, U.S. Nair, D. Niyogi, and T.E. Nobis, 2007b: An overview of regional land use and land cover impacts on rainfall. *Tellus B*, **59**, 587-601.

Pielke Sr., R., K. Beven, G. Brasseur, J. Calvert, M. Chahine, R. Dickerson, D. Entekhabi, E. Foufoula-Georgiou, H. Gupta, V. Gupta, W. Krajewski, E. Philip Krider, W.K.M. Lau, J. McDonnell, W. Rossow, J. Schaake, J. Smith, S. Sorooshian, and E. Wood, 2009: Climate change: The need to consider human forcings besides greenhouse gases. *Eos, Trans. Amer. Geophys. Union,* **90:45**, 413.

Pielke Sr., R.A., A. Pitman, D. Niyogi, R. Mahmood, C. McAlpine, F. Hossain, K. Goldewijk, U. Nair, R. Betts, S. Fall, M. Reichstein, P. Kabat, and N. de Noblet-Ducoudre, 2011: Land use/land cover changes and climate: Modeling analysis and observational evidence. *WIREs Clim Change 2011*, **2**, 828-850. http://dx.doi.org/10.1002/wcc.144.

Pielke Sr., R.A., R. Wilby, D. Niyogi, F. Hossain, K. Dairuku, J. Adegoke, G. Kallos, T. Seastedt, and K. Suding, 2012: Dealing with complexity and extreme events using a bottom-up, resource-based vulnerability perspective. Extreme Events and Natural Hazards: The Complexity Perspective Geophysical Monograph Series 196, 10.1029/2011GM001086.

Pierrehumbert, R.T., 1985: A theoretical model of orographically modified cyclogenesis. *J. Atmos. Sci.,* **42**, 1244-1258.

Pilié, R.J., E.J. Mark, C.W. Rogers, U. Katz, and W.C. Kocmond, 1979: The formation of marine fog and the development of fog-stratus systems along the California coast. *J. Appl. Meteor.,* **18**, 175-1286.

Pilinis, C., P. Kassomenos, and G. Kallos, 1993: Modeling of photochemical pollution in Athens, Greece. Application of the RAMS-CALGRID modeling system. *Atmos. Environ.,* **27B**, 353-370.

Pinel, J., and S. Lovejoy, 2013: Atmospheric waves as scaling, turbulent phenomena. Atmos. Chem. Phys. Discuss., 13, 14797-14822, http://dx.doi.org/10.5194/acpd-13-14797-2013.

Pinty, J.-P., and P. Jabouille, 1998: A mixed-phased cloud parameterization for use in a mesoscale non-hydrostatic model: Simulations of a squall line and of orographic precipitation. *Preprints, Conf. on Cloud Physics*, Everett, WA, Amer. Meteor. Soc., 217-220.

Pinty, J.-P., P. Mascart, E. Richard, and R. Rosset, 1989: An investigation of mesoscale flows included by vegetation inhomogeneities using an evapotranspiration model calibrated against HAPEX-MOBILHY data. *J. Appl. Meteor.,* **9**, 976-992.

Pinty, J.-P., R. Benoit, E. Richard, and R. Laprise, 1995: Simple tests of a semi-implicit semi-Lagrangian model on 2D mountain wave problems. *Mon. Wea. Rev.,* **123**, 3042-3058.

Pitman, A.J., 2003: The evolution of, and revolution in, land surface schemes designed for climate models. *Int. J. Climatol.*, **23**, 479-510, http://dx.doi.org/10.1002/joc.893.

Pitman, A.J., M. Zhao, and C.E. Desborough, 1999: Investigating the sensitivity of a land surface scheme's simulation of soil wetness and evaporation to spatial and temporal leaf area index variability within the global soil wetness project. *J. Meteor. Soc. Japan,* **77**, 281-290.

Pitman, A.J., N. de Noblet-Ducoudré, F.B. Avila, L.V. Alexander, J.P. Boisier, V. Brovkin, C. Delire, F. Cruz, M.G. Donat, V. Gayler, B. van den Hurk, C. Reick, and A. Voldoire, 2012: Effects of land cover change on temperature and rainfall extremes in multi-model ensemble simulations. *Earth System Dynamics*, **3:2**, 213-231, http://dx.doi.org/10.5194/esd-3-213-2012.

Plank, V.G., 1966: Wind conditions in situations of pattern form and non-pattern form cumulus convection. *Tellus,* **18**, 1-12.

Platt, C.M.R., 1973: Lidar and radioinetric observations of cirrus clouds. *J. Atmos. Sci.,* **30**, 1191-1204.

Platt, C.M.R., 1981: The effect of cirrus of varying optical depth on the extraterrestrial net radiative flux. *Quart. J. Roy. Meteor. Soc.,* **107**, 671-678.

Pleim, J.E., and A. Xiu, 1995: Development and testing of a surface flux and planetary boundary layer model for application in mesoscale models. *J. Appl. Meteor.,* **34**, 16-32.

Polkinghorne, R., and T. Vukićević, 2011: Data assimilation of cloud-affected radiances in a cloud-resolving model. *Mon. Wea. Rev.,* **139**, 755-773.

Pollack, J.B., 1982: Properties of dust in the martian atmosphere and its effect on temperature structure. *Adv. Space Res.*, **2:2**, 45-56.

Pollack, J.B., D.S. Colburn, F.M. Flasar, R. Kahn, C.E. Carlston, and D. Pidek, 1979: Properties and effects of dust particles suspended in the Martian atmosphere. *J. Geophys. Res.,* **84:B6**, 2929-2945.

Porch, W.M., 1982: Implication of spatial averaging in complex-terrain wind studies. *J. Appl. Meteor.,* **21**, 1258-1265.

Porco, C.C., 2005: Imaging of Titan from the Cassini spacecraft. *Nature*, **434**, 156-165.

Poreh, M., 1996: Investigation of heat islands using small scale models. *Atmos. Environ.,* **30**, 467-474.

Porte-Agel, F., C. Meneveau, and M.B. Parlange, 2000: A scale-dependent dynamic model for large-eddy simulation: Application to a neutral atmospheric boundary layer. *J. Fluid Mech.,* **415**, 261-284.

Potty, K.V.J., and S. Sethu Raman, 2000: Numerical simulation of monsoon depressions over India with a high-resolution nested regional model. *Meteorol. Appl.,* **7**, 45-60.

Poulos, G.S., and J.E. Bossert, 1995: An observational and prognostic numerical investigation of complex terrain dispersion. *J. Appl. Meteor.,* **34**, 650-660.

Poulos, G.S., and R.A. Pielke, 1994: A numerical analysis of Los Angeles Basin pollution transport to the Grand Canyon under stably stratified, southwest flow conditions. *Atmos. Environ.,* **28**, 3329-3357.

Poulos, G.S., J.E. Bossert, T.B. McKee, and R.A. Pielke, 2000: The interaction of katabatic flow and mountain waves. Part I: Observations and idealized simulations. *J. Atmos. Sci.,* **57**, 1919-1936.

Powell, M.D., 1980: Evaluations of diagnostic marine boundary-layer models applied to hurricanes. *Mon. Wea. Rev.,* **108**, 757-766.

Powell, M.D., S.H. Houston, and T.A. Reinhold, 1996: Hurricane Andrew's landfall in south Florida. Part I: Standardizing measurements for documentation of surface wind fields. *Wea. Forecasting,* **11**, 304-328.

Price, G.V., and A.K. MacPherson, 1973: A numerical weather forecasting method using cubic splines on a variable mesh. *J. Appl. Meteor.,* **12**, 1102-1113.

Price, J.C., 1979: Assessment of the urban heat island effect through the use of satellite data. *Mon. Wea. Rev.,* **107**, 1554-1557.

Priestly, C.H.B., 1959: *Turbulent Transfer in the Lower Atmosphere.* The University of Chicago Press, 130 pp.

Project METROMEX, 1976: METROMEX update. *Bull. Amer. Meteor. Soc.,* **57**, 304-308.

Prudhomme, C., R.L. Wilby, S. Crooks, A.L. Kay, and N.S. Reynard, 2010: Scenario-neutral approach to climate change impact studies: Application to flood risk. *J. Hydrol.*, **390**, 198-209, http://dx.doi.org/10.1016/j.jhydrol.2010.06.043.

Pruppacher, H.R., 1982: Cloud and precipitation physics and the water budget of the atmosphere. In: *Engineering Meteorology*, E. Plate, Ed., Elsevier, New York, 71-124.

Pruppacher, H.R., and J.D. Klett, 1978: *Microphysics of Clouds and Precipitation.* Reidel, Holland, 714 pp.

Purdom, J.F.W., and K. Marcus, 1982: Thunderstorm trigger mechanisms over the southeast United States. *Preprints, 12th AMS Conference on Severe Local Storms*, January 11-15, San Antonio, TX, 487-488.

Putzig, N.E., M.T. Mellon, K.A. Kretke, and R.E. Arvidson, 2005: Global thermal inertia and surface properties of Mars from the MGS mapping mission. *Icarus*, **173(2)**, 325-341.

Qian, J.H., F.H.M. Semazzi, and J.S. Scroogs, 1998: A global nonhydrostatic semi-Lagrangian atmospheric model with orography. *Mon. Wea. Rev.*, **126**, 747-771.

Qingcun, Z., Y.J. Dai, and F. Xue, 1998: Simulation of the Asian monsoon by IAP AGCM coupled with an advanced land surface model (IAP94). *Adv. Atmos. Sci.*, **15**, 1-16.

Qu, W., A. Henderson-Sellers, A.J. Pitman, T.H. Chen, F. Abramopoulos, A. Boone, S. Chang, F. Chen, Y. Dai, R.E. Dickinson, L. Dumenil, M. Ek, N. Gedney, Y.M. Gusev, J. Kim, R. Koster, E.A. Kowalczyk, J. Lean, D. Lettenmaier, X. Liang, J.-F. Mahfouf, H.-T. Mengelkamp, K. Mitchell, O.N. Nasonova, J. Noilhan, A. Robock, C. Rosenzweig, J. Schaake, C.A. Schlosser, J.-P. Schultz, A.B. Shmakin, D.L. Verseghy, P. Wetzel, E.F. Wood, Z.-L. Yang, and Q. Zeng, 1998: Sensitivity of latent heat flux from PILPS land-surface schemes to perturbations of surface air temperature. *J. Atmos. Sci.*, **55**, 1909-1927.

Queney, P., 1947: Theory of perturbations in stratified currents with applications to air flow over mountain barriers. Department of Meteorology, University of Chicago, Misc. Report No. 23, University of Chicago Press, Illinois.

Queney, P., 1948: The problem of air flow over mountains: A summary of theoretical studies. *Bull. Amer. Meteor. Soc.*, **29**, 16-26.

Rabier, F., P. Courtier, and O. Talagrand, 1992: An application of adjoint models to sensitivity analysis. *Contrib. Atmos. Phys.*, **65**, 177-192.

Raddatz, R.L., 2005: Moisture recycling on the Canadian Prairies for summer droughts and pluvials from 1997 to 2003. *Agric. Forest Meteor.*, **131**, 13-26.

Raddatz, R.L., 2007: Evidence for the influence of agriculture on weather and climate through the transformation and management of vegetation: illustrated by examples from the Canadian Prairies. *Agric. Forest Meteor.*, **142**, 186-202.

Raddatz, R.L., and M.L. Khandekar, 1979: Upslope enhanced extreme rainfall events over the Canadian western plains: A mesoscale numerical simulation. *Mon. Wea. Rev.*, **107**, 650-661.

Raddatz, R.L., T.N. Papakyriakou, K.A. Swystun, and M. Tenuta, 2009: Evaporation from a wetland tundra sedge fen: Surface resistance of peat for land-surface schemes. *Agric. Forest Meteor.*, **149**, 851-861.

Raddatz, R.L., R.J. Galley, L.M. Candlish, M.G. Asplin, and D.G. Barber, 2013: Integral profile estimates of sensible heat flux from an unconsolidated sea-ice surface. *Atmosphere-Ocean*, **51**, 135-144, http://dx.doi.org/10.1080/07055900.2012.759900.

Rafkin, S.C.R., 2009: A positive radiative-dynamic feedback mechanism for the maintenance and growth of Martian dust storms. *J. Geophys. Res.*, **114(E1)**, E01009.

Rafkin, S.C.R., 2012: The potential importance of non-local, deep transport on the energetics, momentum, chemistry, and aerosol distributions in the atmospheres of Earth, Mars, and Titan. *Planet. Space Sci.*, **60**, 147-154.

Rafkin, S.C.R., and T.I. Michaels, 2003: Meteorological predictions for 2003 Mars Exploration Rover high-priority landing sites. *J. Geophys. Res.*, **108(E12)**, 8091, http://dx.doi.org/10.1029/2002JE002027.

Rafkin, S.C.R., R.M. Haberle, and T.I. Michaels, 2001: The Mars Regional Atmospheric Modeling System: Model description and selected simulations. *Icarus*, **151(2)**, 228-256.

Rafkin, S.C.R., M.R.V. Sta. Maria, and T.I. Michaels, 2002: Simulation of the atmospheric thermal circulation of a Martian volcano using a mesoscale numerical model. *Nature*, **419(6908)**, 697-699.

Ramanathan, V., M.V. Ramana, G. Roberts, D. Kim, C. Corrigan, C. Chung, and D. Winker, 2007: Warming trends in Asia amplified by brown cloud solar absorption. *Nature*, **448**, 575-578.

Randall, D.A., M. Khairoutdinov, A. Arakawa, and W. Grabowski, 2003: Breaking the cloud-parameterization deadlock. *Bull. Amer. Meteor. Soc.,* **84**, 1547-1564.

Randerson, D., and A.H. Thompson, 1964: Investigation of a Tiros III photograph of the Florida peninsula taken on 14 July 1961. Science Report No. 6 Prepared under Contract AF 19(604)-8450 for Air Force Cambridge Research Labs, Office of Aerospace Research, Bedford, Massachusetts by Texas A&M.

Rannou, P., F. Montmessin, F. Hourdin, and S. Lebonnois, 2006: The latitudinal distribution of clouds on Titan. *Science*, **311(5758)**, 201-205.

Rao, G.V., 1971: A numerical study of the frontal circulation in the atmospheric boundary layer. *J. Appl. Meteor.,* **10**, 26-35.

Rasch, P.J., and J.E. Kristjáansson, 1998: A comparison of the CCM3 model climate using diagnosed and predicted condensate parameterizations. *J. Climate,* **11**, 1587-1614.

Rasmussen, E., 1979: The polar low as an extratropical CISK disturbance. *Quart. J. Roy. Meteor. Soc.,* **105**, 531-549.

Rasmussen, E., 1981: An investigation of a polar low with a spiral cloud structure. *J. Atmos. Sci.,* **38**, 1785-1792.

Rasmussen, E., 1983: A review of mesoscale disturbances in cold air masses. In: Mesoscale Meteorology — Theories, Observations and Models NATO ASI Series Volume 114, Lilly, D.K. and T, Gal-Chen, Eds., 247-283, Reidel, Holland, http://dx.doi.org/10.1007/978-94-017-2241-4_13.

Rasmussen, E.A., 2000: Polar lows for storms. In: *Storms*, Volume II. R.A. Pielke Jr., and R.A. Pielke Sr. Eds., Routledge Press, London, 255-269.

Rauber, R.M., J.E. Walsh, B. Rauber, and D.J. Charlevoix, 2009: *Severe and Hazardous Weather: An Introduction to High Impact Meteorology*. Kendall/Hunt Publishing Company, 642 pp.

Rauber, R.M., J.E. Walsh, and D.J. Charlevoix, 2012: *Severe and Hazardous Weather: An Introduction to High Impact Meteorology*. Kendall/Hunt Publishing Company, 4th edition, 550 pp.

Raupach, M.R., and R.H. Shaw, 1982: Averaging procedures for flow within vegetation canopies. *Bound.-Layer Meteor.,* **22**, 79-90.

Ray, D., 2013: Dry season precipitation over the Mesoamerican Biological Corridor is more sensitive to deforestation than to greenhouse gas driven climate change. *Climatic Change,* http://dx.doi.org/10.1007/s10584-013-0753-0.

Ray, D., R.A. Pielke Sr., U.S. Nair, and D. Niyogi, 2010: The roles of atmospheric and land surface data in dynamic regional downscaling. *J. Geophys. Res.,* **115**, D05102, http://dx.doi.org/10.1029/2009JD012218.

Raymond, D.J., 1975: A model for predicting the movement of continuously propagating convective storms. *J. Atmos. Sci.,* **32**, 1308-1317.

Raymond, D.J., 1983: Wave-CISK in mass flux form. *J. Atmos. Sci.,* **40**, 2561-2572.

Reason, C.J.C., K.J. Tory, and P.L. Jackson, 1999: Evolution of a southeast Australian coastally trapped disturbance. *Meteor. Atmos. Phys.,* **70**, 141-165.

Reed, R.J., 1979: Cyclogenesis in polar air streams. *Mon. Wea. Rev.,* **107**, 38-52.

Reed, R.J., 1981: A case study of a bora-like windstorm in western Washington. *Mon. Wea. Rev.,* **109**, 2382-2393.

Reen, B.P., and D.R. Stauffer, 2010: Data assimilation strategies in the planetary boundary layer. *Bound.-Layer Meteor.,* **137**, 237-269. http://dx.doi.org/10.1007/s10546-010-9528-6.

Reen, B.P., D.R. Stauffer, K.J. Davis and A.R. Desai, 2006: A case study on the effects of heterogeneous soil moisture on mesoscale boundary layer structure in the southern Great Plains, USA. Part II: Mesoscale modeling. *Bound.-Layer Meteor.,* **120**, 275-314.

Reen, B.P., D.R. Stauffer, and K.J. Davis, 2013: Land-surface heterogeneity effects in the planetary boundary layer. *Bound.-Layer Meteor.,* submitted.

Reible, D.D., F.H. Shair, and R. Aris, 1983: A two-layer model of the atmosphere indicating the effects of mixing between the surface layer and the air aloft. *Atmos. Environ.,* **17**, 25-33.

Reid, J.D., L.O. Grant, R.A. Pielke, and Y. Mahrer, 1976: Observations and numerical modeling of seeding agent delivery from ground based generations to orographic cloud base. *Proceedings, International Weather Modification Conference*, August 26, Boulder, Colorado, 521-527.

Reynolds, R.W., and T.M. Smith, 1994: Improved global sea surface temperature analyses using optimum interpolation. *J. Climate*, **7**, 929-948.

Rhea, J.O., 1977: Orographic precipitation model for hydrometeorological use. Ph.D. Dissertation, Department of Atmospheric Science, Colorado State University, Fort Collins, Colorado.

Richards Jr., J.A., F.W. Sears, M.R. Wehr, and M.W. Zemansky, 1962: *Modern College Physics*. Addison-Wesley, Reading, MA, 755 pp.

Richtmyer, R.D., and K.W. Morton, 1967: *Difference Methods for Initial-Value Problems*. Interscience Publishers, New York, NY, 405 pp.

Richwien, B.A., 1978: The damming effect of the southern Appalachians. *Proceedings, AMS Conference on Weather Forecasting Analysis and Aviation Meteorology*, 94-101.

Richwien, B.A., 1980: The damming effect of the southern Appalachians. *Natl. Wea. Dig.*, **5**, 2-12.

Rieck, R.W., Ed. 1979: Facsimile products. NWS Forecasting Handbook No. 1, Facsimile Products, NOAA.

Ritchie, H., and M. Tanguay, 1996: A comparison of spatially averaged Eulerian and semi-Lagrangian treatment of mountains. *Mon. Wea. Rev.,* **124**, 167-181.

Ritter, B., and J.-F. Geleyn, 1992: A comprehensive radiation scheme for numerical weather prediction models with potential applications in climate simulations. *Mon. Wea. Rev.,* **120**, 303-325.

Roach, G.F., 1970: *Green's Functions: Introductory Theory with Applications*. Van Nostrand, London, 279 pp.

Roach, W.T., and A. Slingo, 1979: A high resolution infrared radiative transfer scheme to study the interaction of radiation with cloud. *Quart. J. Roy. Meteor. Soc.,* **105**, 603-614.

Robinson, G.D., Ed., 1977: Inadvertent weather modification workshop. Final Report to NSF under Grant No. ENV-77-10186. Prepared at the Center for the Environment and Man, Hartford, CT.

Rockel, B., C.L. Castro, R.A. Pielke Sr., H. von Storch, and G. Leoncini, 2008: Dynamical downscaling: Assessment of model system dependent retained and added variability for two different regional climate models. *J. Geophys. Res.*, **113**, D21107, http://dx.doi.org/10.1029/2007JD009461.

Rodrigues, E.R., F.L. Madruga, P.O.A. Navaux, and J. Panetta, 2009: Multi-core aware mapping of parallel applications. *Computers and Communications, 2009.* ISCC 2009. pp.811,817, 5-8 July 2009. http://dx.doi.org/10.1109/ISCC.2009.5202271.

Rodrigues, E.R., P.O.A. Navaux, J. Panetta, A.L. Fazenda, C.L. Mendes, and L.V. Kale, 2010a: A comparative analysis of load balancing algorithms applied to a weather forecast model. *22nd International Symposium on Computer Architecture and High Performance Computing*, SBAC-PAD.

Rodrigues, E.R., P.O.A. Navaux, J. Panetta, C.L. Mendes, and L.V. Kale, 2010b: Optimizing an MPI Weather Forecasting Model via Processor Virtualization. *17th. International Conference on High Performance Computing*, HiPC.

Rogers, D.P., 1995: Air-sea interaction: Connecting the ocean and atmosphere. *Rev. Geophys.,* Supplement **33**, 1377-1383.

Rõõm, R., and A. Männik, 1999: Responses of different nonhydrostatic, pressure-coordinate models to orographic forcing. *J. Atmos. Sci.,* **56**, 2553-2570.

Rosenberg, N., 1974: *Microclimate: The Biological Environment.* Wiley, New York, NY, 315 pp.

Rosenthal, S.L., 1970: A circularly symmetric primitive equation model of tropical cyclone development containing an explicit water vapor cycle. *Mon. Wea. Rev.,* **98**, 643-663.

Rosenthal, S.L., 1971: The response of a tropical cyclone model to variations in boundary layer parameters, initial conditions, lateral boundary conditions and domain size. *Mon. Wea. Rev.,* **99**, 767-777.

Rosenthal, S.L., 1978: Numerical simulation of tropical cyclone development with latent heat release by the resolvable scales. I. Model description and preliminary results. *J. Atmos. Sci.,* **35**, 258-271.

Rosenthal, S.L., 1979a: Cumulus effects in hurricane models - to parameterize or not to parameterize. *Proceedings of the Seminar on the Impact of GATE on Large-Scale Numerical Modeling of the Atmosphere and Ocean*, Woods Hole, MA, August 20-29, US GARP Program, 242-248.

Rosenthal, S.L., 1979b: The sensitivity of simulated hurricane development to cumulus parameterization details. *Mon. Wea. Rev.,* **107**, 193-197.

Rosenthal, S.L., 1980: Numerical simulation of tropical cyclone development with latent heat release by the resolvable scales, II: Propagating small-scale features observed in the prehurricane phase. NOAA Technical Report ERL413-AOML29.

Ross, B.B., and I. Orlanski, 1978: The circulation associated with a cold front. Part II. Moist case. *J. Atmos. Sci.,* **35**, 445-465.

Ross, B.B., and I. Orlanski, 1982: The evolution of an observed cold front. Part I: Numerical simulation. *J. Atmos. Sci.,* **39**, 296-327.

Rossato, L., R.S. Alvalá, and J. Tomasella, 2002: Climatologia da umidade do solo no Brasil. *Anais do XII Congresso Brasileiro de Meteorologia* 1910-1915.

Rossby, C.G., 1941: The scientific basis of modern meteorology. In: *Climate and Man, Yearbook of Agriculture 1941*, U.S. Dept of Agriculture Handbook, 599-655.

Rothman, L., C. Rinsland, A. Goldman, S.T. Massie, and D.P. Edwards, 1998: The HITRAN molecular spectroscopic database and HAWKS (HITRAN atmospheric workstation): 1996 edition. *J. Quantitative Spectroscopy and Radiative Transfer*, **60**, 665-710, http://dx.doi.org/10.1016/S0022-4073(98)00078-8.

Rotstayn, L.D., 1999: Climate sensitivity of the CSIRO GCM: Effect of cloud modeling assumptions. *J. Climate,* **12**, 334-356.

Rotstayn, L.D., B.F. Ryan, and J.J. Katzfey, 2000: A scheme for calculation of the liquid fraction in mixed-phase stratiform clouds in large-scale models. *Mon. Wea. Rev.,* **128**, 1070-1088.

Rotunno, R., 1983: On the linear theory of the land- and sea-breeze. *J. Atmos. Sci.,* **40**, 1999-2005.

Rotunno, R., J.A. Curry, C.W. Fairall, C.A. Friehe, W.A. Lyons, J.E. Overland, R.A. Pielke, D.P. Rogers, S.A. Stage, G.L. Geernaert, J.W. Nielsen, and W.A. Sprigg, 1992: *Coastal Meteorology - A Review of the State of the Science.* Panel on Coastal Meteorology, Committee on Meteorological Analysis, Prediction, and Research, Board on Atmospheric Sciences and Climate, Commission on Geosciences, Environment, and Resources, National Research Council, National Academy Press, Washington, D.C., 99 pp.

Rüshøjgaard, L.P., S.E. Cohn, Y. Li, and R. Ménard, 1998: The use of spline interpolation in semi-Lagrangian transport models. *Mon. Wea. Rev.,* **126**, 2008-2016.

Rutledge, S.A., and P.V. Hobbs, 1983: The mesoscale and microscale structure and organization of clouds and precipitation in midlatitude cyclones, VIII: A model for the "seeder-feeder" process in warm-frontal rainbands. *J. Atmos. Sci.,* **40**, 1185-1206.

Ryall, D.B., and R.H. Maryon, 1998: Validation of the UK Met. Office's NAME model against the ETEX dataset. *Atmos. Environ.,* **32**, 4265-4276.

Ryznar, E., and J.S Touma, 1981: Characteristics of true lake breezes along the eastern shore of Lake Michigan. *Atmos. Environ.,* **15**, 1201-1205.

Rutledge, S.A., and P.V. Hobbs, 1984: The mesoscale and microscale structure and organization of clouds and precipitation in midlatitude cyclones. XII: A diagnostic modeling study of precipitation development in narrow cold-frontal rainbands. *J. Atmos. Sci.,* **41**, 2949-2972.

Sackinger, P.A., D.D. Reible, and F.H. Shair, 1982: Uncertainties associated with the estimation of mass balances and Gaussian parameters from atmospheric tracer studies. *J. Air Pollut. Control Assoc.*, **32**, 720-724.

Sahashi, K., 1981: Numerical experiment of land and sea breeze circulation with undulating orography. Part I: Model. *J. Meteor. Soc. Japan,* **59**, 254-261.

Sailor, D.J., 1995: Simulated urban climate response to modifications in surface albedo and vegetative cover. *J. Appl. Meteor.,* **34**, 1694-1704.

Saito, K., 1997: Semi-implicit fully compressible version of the MRI Mesoscale Nonhydrostatic Model – Forecast experiment of the 6 August 1993 Kagashima torrential rain. *Geophys. Magazine*, **2**, Series 2, 109-137.

Saito, K., 2012: The JMA Nonhydrostatic Model and Its Applications to Operation and Research, Atmospheric Model Applications, Dr. Ismail Yucel, Ed., ISBN: 978-953-51-0488-9, InTech, Available from: http://www.intechopen.com/books/atmospheric-model-applications/the-jma-nonhydrostatic-model-and-its- applications-to-operation-and-research.

Saito, K., T. Fujita, Y. Yamada, J. Ishida, Y. Kumagai, K. Aranami, S. Ohmori, R. Nagasawa, S. Kumagai, C. Muroi, T. Kato, H. Eito, and Y. Yamazaki, 2006: The operational JMA Nonhydrostatic Mesoscale Model. *Mon. Wea. Rev.,* **134**, 1266-1298.

Saito, K., J. Ishida, K. Aranami, T. Hara, T. Segawa, M. Narita and Y. Honda, 2007: Nonhydrostatic atmospheric models and operational development at JMA. *J. Meteor. Soc. Japan.*, **85B**, 271-304.

Saito, T., 1981: The relationship between the increased rate of downward long-wave radiation by atmospheric pollution and the visibility. *J. Meteor. Soc. Japan,* **59**, 254-261.

Sakaguchi, K., X. Zeng, and M.A. Brunke, 2012: The hindcast skill of the CMIP ensembles for the surface air temperature trend. *J. Geophys. Res.,* **117**, D16113, http://dx.doi.org/10.1029/2012JD017765.

Sakakibara, H., 1981: Heavy rainfall from very shallow convective clouds. *J. Meteor. Soc. Japan,* **59**, 387-394.

Saleeby, S.M., and S.C. van den Heever, 2013: Developments in the CSU-RAMS aerosol model: Emissions, nucleation, regeneration, deposition, and radiation. *J. Appl. Meteor. Climatol.*, http://dx.doi.org/10.1175/JAMC-D-12-0312.1.

Salmon, J.R., J.L. Walmsley, and P.A. Taylor, 1981: Development of a model of neutrally stratified boundary layer flow over real terrain. Internal report. Boundary Layer Division, Atmospheric Environment Service, Downsview, Ontario, Canada.

Salvador, R., M.M. Millán, E. Mantilla, and J.M. Baldasano, 1997: Mesoscale modelling of atmospheric processes over the western Mediterranean during summer. *Int. J. Environ. Poll.* **8**, Nos. 3-6, 513-529.

Salvador, R., J. Calbó, and M.M. Millán, 1999: Horizontal grid size selection and its influence on mesoscale model simulations. *J. Appl. Meteor.,* **38**, 1311-1329.

Sangster, W.E., 1977: An updated objective forecast technique for Colorado downslope winds. NOAA Technical Memo. NWS CR-61, 24 pp.

Sardie, J.M., and T.T. Warner, 1983: On the mechanism for the development of polar lows. *J. Atmos. Sci.*, **40**, 869-881.

Sarewitz, D., R.A. Pielke Jr., and R. Byerly, Eds., 2000: *Prediction: Science Decision Making and the Future of Nature.* Island Press, Covelo, CA, 405 pp.

Sasaki, Y., 1970a: Some basic formalisms in numerical variational analysis. *Mon. Wea. Rev.,* **98**, 875-883.

Sasaki, Y., 1970b: Numerical variational analysis formulated under the constraints as determined by longwave equations and low-pass filter. *Mon. Wea. Rev.,* **98**, 884-898.

Sasaki, Y., 1970c: Numerical variational analysis with weak constraint and application to surface analysis of a severe gust front. *Mon. Wea. Rev.,* **98**, 899-910.

Sasaki, Y., and J.M. Lewis, 1970: Numerical variational objective analysis of the planetary boundary layer in conjunction with squall line formation. *J. Meteor. Soc. Japan,* **48**, 381-393.

Sasamori, T., 1968: Radiative cooling calculation for application to general circulation experiments. *J. Appl. Meteor.,* **7**, 721-729.

Sasamori, T., 1970: A numerical study of atmospheric and soil boundary layers. *J. Atmos. Sci.,* **27**, 1122-1137.

Sasamori, T., 1972: A linear harmonic analysis of atmospheric motion with radiative dissipation. *J. Meteor. Soc. Japan,* **50**, 505-517.

Sasamori, T., J. London, and D.V. Hoyt, 1972: Radiation budget of the Southern Hemisphere. *Meteor. Mag.,* **35**, 9-23.

Sato, Y., K. Suzuki, T. Iguchi, I.-J. Choi, H. Kadowaki, and T. Nakajima, 2012: Characteristics of correlation statistics between droplet radius and optical thickness of warm clouds simulated by a three-dimensional regional-scale spectral bin microphysics cloud model. *J. Atmos. Sci.,* **69:2**, 484-503.

Satomura, T., 2000: Diurnal variation of precipitation over the Indo-China peninsula: two-dimensional numerical simulation. *J. Meteor. Soc. Japan,* **78**, 461-474.

Savijärvi, H., 1990: Fast radiation parameterization schemes for mesoscale and short-range forecast models. *J. Appl. Meteor. Climatol.,* **29**, 437-447.

Savijärvi, H., and S. Järvenoja, 2000: Aspects of the fine-scale climatology over Lake Tanganyika as resolved by a mesoscale model. *Meteor. Atmos. Phys.,* **73**, 77-88.

Savijärvi, H., and J. Kauhanen, 2008: Surface and boundary-layer modelling for the Mars Exploration Rover sites, *Quart. J. Roy. Meteor. Soc.,* **134:632**, 635-641.

Sawai, T., 1978: Formation of the urban air mass and the associated local circulations. *J. Meteor. Soc. Japan,* **56**, 159-173.

Schaeffer, J.T., 1974: A simulative model of dryline motion. *J. Atmos. Sci.,* **31**, 956-964.

Schaller, E.L., M.E. Brown, H.G. Roe, and A.H. Bouchez, 2006: A large cloud outburst at Titan's south pole. *Icarus,* **182,1**, 224-229.

Schecter, D.A., and M.E. Nicholls, 2010: Generation of infrasound by evaporating hydrometeors in a cloud model. *J. Appl. Meteorol. Climatol.*, **49:4**, 664-675, http://dx.doi.org/10.1175/2009JAMC2226.1.

Schecter, D.A., M.E. Nicholls, J. Persing, A.J. Bedard Jr., and R.A. Pielke Sr., 2008: Infrasound emitted by tornado-like vortices: Basic theory and a numerical comparison to the acoustic radiation of a single-cell thunderstorm. *J. Atmos. Sci.*, **65**, 685-713.

Schere, K.L., and K.L. Demerjian, 1978: A photochemical box model for urban air quality simulation. *4th Joint Conference on Sens. Environmental Pollution*, 427-433.

Schlesinger, R.E., 1973: A numerical model of deep moist convection. I. Comparative experiments for variable ambient moisture and wind shear. *J. Atmos. Sci.,* **30**, 835-856.

Schlesinger, R.E., 1980: A three-dimensional numerical model of an isolated thunderstorm. II. Dynamics of updraft splitting and mesovortex couplet evolution. *J. Atmos. Sci.,* **37**, 395-420.

Schlesinger, R.E., 1982a: Three-dimensional numerical modeling of convective storms: A review of milestones and challenges. *12th AMS Conference on Severe Local Storms*, San Antonio, TX, January 11-15, 506-515.

Schlesinger, R.E., 1982b: Effects of mesoscale lifting, precipitation and boundary layer shear on severe storm dynamics in a three-dimensional numerical modeling study. *12th AMS Conference on Severe Local Storms*, San Antonio, TX, January 11-15, 536-541.

Schlosser, C.A., A.G. Slater, A. Robock, A.J. Pitman, K.Y. Vinnikov, A. Henderson-Sellers, N.A. Speranskaya, K. Mitchell, and the PILPS 2(D) Contributors, 2000: Simulations of a Boreal Grassland hydrology at Valdi, Russia: PILPS Phase 2(D). *Mon. Wea. Rev.,* **128**, 301-321.

Schmid, H.P., and B. Bünzli, 2007: The influence of surface texture on the effective roughness length. *Quart. J. Royal Meteor. Soc.*, 121, 1–21, http://dx.doi.org/10.1002/qj.49712152102.

Schmidt, F.H., 1947: An elementary theory of the land- and sea-breeze circulation. *J. Meteor.*, **4**, 9-15.

Schmidt, R., and K. Housen, 1995: Problem solving with dimensional analysis. *Indust. Phys.,* **1**, 21-24.

Scofield, R.A., and C.E. Weiss, 1977: A report on the Chesapeake Bay Region Nowcasting Experiment. NOAA Tech. Memo NESS 94.

Schofield, J.T., J.R. Barnes, D. Crisp, R.M. Haberle, S. Larsen, J.A. Magalhães, J.R. Murphy, A. Seiff, and G. Wilson, 1997: The Mars Pathfinder Atmospheric Structure Investigation/Meteorology (ASI/MET) Experiment. *Science,* **278:5344**, 1752-1758.

Schrieber, K., R. Stull, and Q. Zhang, 1996: Distributions of surface-layer buoyancy versus lifting condensation level over a heterogeneous land surface. *J. Atmos. Sci.,* **53**, 1086-1107.

Schroeder, A.J., D.R. Stauffer, N.L. Seaman, A. Deng, A.M. Gibbs, G.K. Hunter and G.S. Young, 2006a: Evaluation of a high-resolution, rapidly relocatable meteorological nowcasting and prediction system. *Mon. Wea. Rev.,* **134**, 1237-1265.

Schroeder, A.J., D.R. Stauffer, N.L. Seaman, A. Deng, A.M. Gibbs, G.K. Hunter and G.S. Young, 2006b: Unleashing an automated mobile weather prediction system, summary of Schroeder et al. 2006 from *Mon. Wea. Rev.* and sidebar, written by Prof. Stauffer and appearing in Papers of Note, Bull. Amer. Meteor. Soc., July 2006, 878-880.

Schulman, E.E., 1970: The Antarctic circumpolar current. *Proc., 1970 Summer Computer Simulation Conference*, Denver, CO, 955-961.

Schultz, D.M., 1999: Lake-effect snowstorms in northern Utah and western New York with and without lightning. *Wea. Forecasting,* **14**, 1023-1031.

Schultz, D.M., and G. Vaughan, 2011: Occluded fronts and the occlusion process: A fresh look at conventional wisdom. *Bull. Amer. Meteor. Soc.*, **92**, 443-466.

Schultz, J.-P., L. Dumenil, J. Polcher, C.A. Schlosser, and Y. Xue, 1998: Land surface energy and moisture fluxes: comparing three models. *J. Appl. Meteor.,* **37**, 288-307.

Schultz, P., 1995: An explicit cloud physics parameterization for operational numerical weather prediction. *Mon. Wea. Rev.*, **123**, 3331-3343.

Schultz, P., and T.T. Warner, 1982: Characteristics of summer-time circulations and pollutant ventilation in the Los Angeles Basin. *J. Appl. Meteor.,* **21**, 672-682.

Schuur, T.J., and S.A. Rutledge, 2000a: Electrification of stratiform regions in mesoscale convective systems. Part I: An observational comparison of symmetric and asymmetric MCSs. *J. Atmos. Sci.,* **57**, 1961-1982.

Schuur, T.J., and S.A. Rutledge, 2000b: Electrification of stratiform regions in mesoscale convective systems. Part II: Two-dimensional numerical model simulations of a symmetric MCS. *J. Atmos. Sci.,* **57**, 1983-2006.

Schwartz, B.E., and L.F. Bosart, 1979: The diurnal variability of Florida rainfall. *Mon. Wea. Rev.,* **107**, 1535-1545.

Schwarzkopf, M.D., and S. Fels, 1991: The simplified exchange method revisited: An accurate, rapid method for computation of infrared cooling rates and fluxes. *J. Geophys. Res.,* **96(D5)**, 9075-9096.

Schwerdtfeger, W., 1974: Mountain barrier effect on the flow of stable air north of the Brooks Range. *Proc., 24th Alaskan Science Conference*, Geophysical Institute University of Alaska, Fairbanks, 204-208.

Schwerdtfeger, W., 1975: The effect of the Antarctic peninsula on the temperature regime of the Weddell Sea. *Mon. Wea. Rev.,* **103**, 45-51.

Science News, 1976: Dioxin toxicity data sent to aid Italy. **110**, 359.

Scorer, R.S., 1949: Theory of waves in the lee of mountains. *Quart. J. Roy. Meteor. Soc.,* **75**, 41-56.

Scott, B.C., 1982: Theoretical estimates of the scavenging coefficient for soluble aerosol particles as a function of precipitation type, rate and altitude. *Atmos. Environ.,* **16**, 1753-1762.

Seaman, C.J., M. Sengupta, and T.H. Vonder Haar, 2010: Mesoscale satellite data assimilation: Impact of cloud-affected infrared observations on a cloud-free initial model state. *Tellus A,* **62:3**, 298-318.

Seaman, N.L., 1982: A numerical simulation of three-dimensional mesoscale flows over mountainous terrain. University of Wyoming, Dept. of Atmospheric Science Report No. AS 135 under NSF Grant No. ATM-77-17540.

Seaman, N.L., 2000: Meteorological modeling for air-quality assessments. *Atmos. Environ.,* **34**, 2231-2259.

Seaman, N.L., and R.A. Anthes, 1981: A mesoscale semi-implicit numerical model. *Quart. J. Roy. Meteor. Soc.,* **107**, 167-190.

Seaman, N.L., and S.A. Michelson, 2000: Mesoscale meteorological structure of a high-ozone episode during the 1995 NARSTO-Northeast study. *J. Appl. Meteor.,* **39**, 109-123.

Seaman, N.L., J.S. Kain, and A. Deng, 1996: Development of a shallow convection parameterization for mesoscale models. *Preprints, 11th Conf. on Numerical Weather Prediction,* Amer. Meteor. Soc., Aug. 19-23, Norfolk, VA, 340-342.

Segal, M., and R.A. Pielke, 1981: Numerical model simulation of human biometeorological heat load conditions - summer day case study for the Chesapeake Bay area. *J. Appl. Meteor.,* **20**, 735-749.

Segal, M., R.T. McNider, R.A. Pielke, and D.S. McDougal, 1982a: A numerical model simulation of the regional air pollution meteorology of the greater Chesapeake Bay area - summer day case study. *Atmos. Environ.*, **16**, 1381-1397.

Segal, M., Y. Mahrer, and R.A. Pielke, 1982b: Application of a numerical mesoscale model for the evaluation of seasonal persistent regional climatological patterns. *J. Appl. Meteor.,* **21**, 1754-1762.

Segal, M., R.A. Pielke, and Y. Mahrer, 1983a: On climatic changes due to a deliberate flooding of the Qattara depression (Egypt). *Climatic Change,* **5**, 73-83.

Segal, M., Y. Mahrer, and R.A. Pielke, 1983b: A study of meteorological patterns associated with a lake confined by mountains - the Dead Sea case. *Quart. J. Roy. Meteor. Soc.,* **109**, 549-564.

Segal, M., R. Avissar, M.C. McCumber, and R.A. Pielke, 1988: Evaluation of vegetation effects on the generation and modification of mesoscale circulations. *J. Atmos. Sci.,* **45**, 2268-2292.

Segal, M., J.R. Garratt, G. Kallos, and R.A. Pielke, 1989a: The impact of wet soil and canopy temperatures on daytime boundary-layer growth. *J. Atmos. Sci.,* **46**, 3673-3684.

Segal, M., W. Schreiber, G. Kallos, R.A. Pielke, J.R. Garratt, J. Weaver, A. Rodi, and J. Wilson, 1989b: The impact of crop areas in northeast Colorado on midsummer mesoscale thermal circulations. *Mon. Wea. Rev.,* **117**, 809-825.

Segal, M., J.R. Garratt, R.A. Pielke, and Z. Ye., 1991a: Scaling and numerical model evaluation of snow-cover effects on the generation and modification of daytime mesoscale circulations. *J. Atmos. Sci.*, **48**, 1024-1042.

Segal, M., J.H. Cramer, R.A. Pielke, J.R. Garratt, and P. Hildebrand, 1991b: Observational evaluation of the snow-breeze. *Mon. Wea. Rev.*, **119**, 412-424.

Segal, M., J.R. Garratt, R.A. Pielke, P. Hildebrand, F.A. Rogers, and J. Cramer, 1991c: On the impact of snow cover on daytime pollution dispersion. *Atmos. Environ.,* **25B**, 177-192.

Seiff, A., and D.B. Kirk, 1982: Structure of the Venus mesosphere and lower thermosphere from measurements during entry of the pioneer Venus probes, *Icarus,* **49,1**: 49-70.

Seigel, R.B., and S.C. van den Heever, 2012a: Mineral dust pathways into supercell storms. *J. Atmos. Sci.*, **69**, 1453-1473.

Seigel, R.B., and S.C. van den Heever, 2012b: Simulated density currents beneath embedded stratified layers. *J. Atmos. Sci.*, **69**, 2192-2200.

Seigel, R.B., and S.C. van den Heever, 2013: Squall line intensification vis hydrometeor recirculation. *J. Atmos. Sci.*, in press.

Seigel, R.B., S.C. van den Heever, and S.M. Saleeby, 2012: Assessing mineral dust indirect effects and radiation impacts on a simulated idealized nocturnal squall line. Atmos. Chem. Phys. Discuss., 12, 29607-29655, http://dx.doi.org/10.5194/acpd-12-29607-2012.

Seigneur, C., 1994: The status of mesoscale air quality models. In *Planning and Managing Regional Air Quality Modeling and Measurement Studies. Part III: Studies Supporting Project Planning*, P.A. Solomon, Ed., CRC Press, Inc. Boca Raton, FL, 403-434.

Seinfeld, J.H., 1975: *Air Pollution: Physical and Chemical Fundamentals.* McGraw-Hill, New York, NY, 523 pp.

Seinfeld, J.H., and S. Pandis, 1997: *Atmospheric Chemistry and Physics: Air Pollution to Climate.* John Wiley and Sons, 1326 pp.

Seinfeld, J.H., G.R. Carmichael, R. Arimoto, W.C. Conant, F.J. Brechtel, T.S. Bates, T.A. Cahill, A.D. Clarke, S.J. Doherty, P.J. Flatau, B.J. Huebert, J. Kim, K.M. Markowicz, P.K. Quinn, L.M. Russell, P.B. Russell, A. Shimizu, Y. Shinozuka, C.H. Song, Y. Tang, I. Uno, A.M. Vogelmann, R.J. Weber, J.-H. Woo, and X.Y. Zhang, 2004: ACE-ASIA regional climate and atmospheric chemical effects of Asian dust and pollution, *Bull. Amer. Meteor. Soc.,* **85**, 367-380.

Seity, Y., P. Brousseau, S. Malardel, G. Hello, P. Bénard, F. Bouttier, C. Lac, and V. Masson, 2011: The AROME-France Convective-Scale Operational Model. *Mon. Wea. Rev.,* **139**, 976-991.

Selby, S.M., Ed., 1967: *CRC Standard Mathematical Tables*. 15th Edition. The Chemical Rubber Co., Cleveland, OH.

Sellers, W.D., 1965: *Physical climatology*. University of Chicago Press, IL, 272 pp.

Sellers, P.J., Y. Mintz, Y.C. Sud, and A. Dalcher, 1986: A simple biosphere model (SiB) for use within general circulation models. *J. Atmos. Sci.,* **43**, 505-531.

Sellers, P.J., F.G. Hall, R.D. Kelly, A. Black., D. Baldocchi, J. Berry, M. Ryan, K.J. Ranson, P.M. Crill, D.P. Lettenmaier, H. Margolis, J. Cihlar, J. Newcomer, D. Fitzjarrald, P.G. Jarvis, S.T. Gower, D. Halliwell, D. Williams, B. Goodison, D.W. Wickland, and F.E. Guertin, Eds., 1997: BOREAS Special Issue. *J. Geophys. Res.,* **102, D24**, 28731-29745.

Sen, O.L., W.J. Shuttleworth, and Z.-L. Yang, 2000: Comparative evaluation of BATS2, BATS, and SiB2 with Amazon data. *J. Hydrometeor.,* **1**, 135-153.

Sergeev, B.N., 1983: Numerical simulation of an atmospheric front with a closed system and precipitation. *Sov. Meteor. Hydro.*, **4**, 16-23.

Sestini, M., E. Reimer, D. Valeriano, R. Alvalá, E. Mello, C. Chan, and C. Nobre, 2003: Mapa de cobertura da terra da Amazônia legal para uso em modelos meteorológicos. *Anais XI Simpósio Brasileiro de Sensoriamento Remoto*, 2901-2906.

Seth, A., and F. Giorgi, 1998: The effects of domain choice on summer precipitation simulation and sensitivity in a regional climate model. *J. Climate,* **11**, 2698-2712.

SethuRaman, S., and J.E. Cermak, 1973: Stratified shear flows over a simulated three-dimensional urban heat island. Project THEMIS Technical Report No. 23. Fluid Dynamics and Diffusion Lab., Colorado State University, Fort Collins.

Shafran, P.C., N.L. Seaman, and G.A. Gayno, 2000: Evaluation of numerical predictions of boundary layer structure during the Lake Michigan ozone study. *J. Appl. Meteor.,* **39**, 55-69.

Shao, Y., and A. Henderson-Sellers, 1996: Modeling soil moisture: A project for intercomparison of land surface parameterization schemes Phase 2(b). *J. Geophys. Res.,* **101**, 7227-7250.

Shao, Y., and P. Irannejad, 1999: On the choice of soil hydraulic models in land-surface schemes. *Bound.-Layer Meteor.,* **90**, 83-115.

Shapiro, R., 1970: Smoothing, filtering and boundary effects. *Rev. Geophys. Space Phys.,* **8**, 359-387.

Shapiro, M.A., 1981: Frontogenesis and geostrophically forced secondary circulations in the vicinity of jet stream-frontal zone systems. *J. Atmos. Sci.,* **38**, 954-973.

Shapiro, M.A., 1982: Mesoscale weather systems of the Central United States. Report, CIRES and NOAA. Boulder, CO 80309.

Sharan, M., S.G. Golpalakrishnan, and R.T. McNider, 1999: A local parameterization scheme for sigma (w) under stable conditions. *J. Appl. Meteor.,* **38**, 617-622.

Sharan, M., S.G. Golpalakrishnan, R.T. McNider, and M.P. Singh, 2000: A numerical investigation of urban influences on local meteorological conditions during the Bhopal gas accident. *Atmos. Environ.,* **34**, 539-552.

Shaw, B.L., R.A. Pielke, and C.L. Ziegler, 1997: A three-dimensional numerical simulation of a Great Plains dryline. *Mon. Wea. Rev.,* **125**, 1489-1506.

Sheffield, A.M., S.C. van den Heever, and S.M. Saleeby, 2013: Growth of cumulus congestus clouds when impacted byaerosols. *J. Atmos. Sci.*, accepted.

Sheih, C.M., 1977: Mathematical modeling of particulate thermal coagulation and transport downstream of an urban source. *Atmos. Environ.,* **11**, 1185-1190.

Sheih, C.M., 1978: A puff-on-cell model for computing pollutant transport and diffusion. *J. Appl. Meteor.,* **17**, 140-147.

Sheng, P.Y., W. Lick, R.T. Gedney, and F.B. Molls, 1978: Numerical computations of three-dimensional circulations in Lake Erie: A comparison of a free-surface model and a rigid-lid model. *J. Phys. Oceanogr.,* **8**, 713-727.

Sheng, L., K.H. Schlunzen, and Z. Wu, 2000: Three-dimensional numerical simulation of the mesoscale wind structure over Shandong Peninsula. *Acta Meteorol. Sinica,* **14**, 96-107.

Shepherd, J.M., 2013: Impacts of urbanization on precipitation and storms: Physical insights and vulnerabilities. In: *Climate Vulnerability, Understanding and Addressing Threats to Essential Resources*, Roger Pielke, Chief Ed., 5, 109-125.

Shepherd, J.M., and T.L. Mote, 2011: Can cities create their own snowfall?: What observations are required to find out? *Earthzine, Special Urban Monitoring Theme Issue.* http://www.earthzine.org/2011/09/06/can-cities-create-their-own-snowfall-what-observations-are-
required-to-find-out/.

Shepherd, J.M., W. Shem, M. Manyin, L. Hand, and D. Messen, 2010a: Modeling urban effects on the precipitation component of the water cycle. Invited chapter for the book: Geospatial Analysis and Modeling of Urban Environments. X. Yao, and H. Jaing, Eds., Springer Book Series and GIScience, 445 pp.

Shepherd, J.M., J.A. Stallins, M. Jin, and T.L. Mote, 2010b: Urban effects on precipitation and associated convective processes. Chapter 12 in *The Routledge Handbook of Urban Ecology.* Ian Douglas, D. Goode, M. Hourck, and R. Wang, Eds., Taylor and Francis US, 688 pp.

Shepherd, J.M., W.M. Carter, M. Manyin, D. Messen, and S. Burian, 2010c: The impact of urbanization on current and future coastal convection: A case study for Houston. *Environ. Planning*, **37**, 284-304.

Shepherd, J.M., T.L. Mote, S. Nelson, S. McCutcheon, P. Knox, M. Roden, and J. Dowd, 2011: An overview of synoptic and mesoscale factors contributing to the disastrous Atlanta flood of 2009. *Bull. Amer. Meteor. Soc.,* **92**, 861-870. http://dx.doi.org/10.1175/2010BAMS3003.1.

Sherman, C.E., 1978: A mass-consistent model for wind fields over complex terrain. *J. Appl. Meteor.,* **17**, 312-319.

Sheynin, O.B., 1973: R.J. Boscovich's work on probability. *Arch. Hist. Exact Sci.*, **9**, 306-324.

Shi, J.J., W.-K. Tao, T. Matsui, A. Hou, S. Lang, C. Peters-Lidard, G. Jackson, R. Cifelli, S. Rutledge, and W. Petersen, 2010: Microphysical properties of the January 20-22 2007 snow events over Canada: Comparison with in-situ and satellite observations. *J. Appl. Meteor. Climatol.,* **49(11)**, 2246-2266.

Shi, Y., X.S. Feng, F.S. Wei, and W. Jiang, 2000: Three-dimensional nonhydrostatic numerical simulation for the PBL of an open-pit mine. *Bound.-Layer Meteor.,* **94**, 197-224.

Shin, S., and S. Reich, 2009: Hamiltonian particle-mesh simulations for a non-hydrostatic vertical slice model. *Atmos. Sci. Lett.,* **10**, 233-240.

Shreffler, J.H., 1979: Heat island convergence in St. Louis during calm periods. *J. Appl. Meteor.,* **18**, 1512-1520.

Shreffler, J.H., 1982: Intercomparisons of upper air and surface winds in an urban region. *Bound.-Layer Meteor.,* **22**, 345-356.

Shrestha, B.M., R.L. Raddatz, R.L. Desjardins, and D.E. Worth, 2012: Continuous cropping and moist deep convection on the Canadian prairies. *Atmosphere*, **3**, 573-590, http://dx.doi.org/10.3390/atmos3040573.

Siebesma, A.P., P.M.M. Soares, and J. Teixeira, 2007: A combined eddy-diffusivity mass-flux approach for the convective boundary layer. *J. Atmos. Sci.,* **64**, 1230-1248.

Sievers, V., R. Forkel, and W. Zdunkowski, 1983: Transport equations for heat and moisture in the soil and their application to boundary layer problems. *Contrib. Atmos. Phys.,* **56**, 58-83.

Siili, T., J. Kauhanen, H. Savijärvi, A.-M. Harri, W. Schmidt, S. Järvenoja, P.L. Read, L. Montabone, and S.R. Lewis, 2006: Simulations of atmospheric circulations for the Phoenix landing area and season-of-operation with the Mars Limited Area Model (MLAM). *Fourth International Conference on Mars Polar Science and Exploration*, Lunar and Planetary Institute, Davos, Switzerland.

Silva Dias, M.R., 1979: Linear spectral model of tropical mesoscale systems. Ph.D. Dissertation, Atmospheric Science Paper No. 311, Colorado State University, Fort Collins, CO 80523.

Simpson, J., 1976: Precipitation augmentation from cumulus clouds and systems: Scientific and technological foundation, 1975. *Adv. Geophys.,* **19**, 1-72.

Simpson, J., and V. Wiggert, 1969: Models of precipitating cumulus towers. *Mon. Wea. Rev.,* **97**, 471-489.

Simpson, J., N.E. Westcott, R.J. Clerman, and R.A. Pielke, 1980: On cumulus mergers. *Arch. Meteor. Geophys. Bioklimatol.,* **29**, 1-40.

Simpson, J., G.V. Helvoirt, and M. McCumber, 1982: Three-dimensional simulations of cumulus congestus clouds on GATE Day 261. *J. Atmos. Sci.,* **39**, 126-145.

Simpson, J.E., 1982: Gravity currents in the laboratory, atmosphere and ocean. *Ann. Rev. Fluid Mech.,* **14**, 213-234.

Simpson, J.E., 1983: Cumulus clouds: numerical models, observations and entrainment. In: *Mesoscale Meteorology-Theories, Observations and Models*. D. Reidel Publishing Company. 413-445.

Simpson, J.E., 1994: *Sea Breeze and Local Winds*. Cambridge University Press, 248 pp.

Simpson, J.E., 1996: Diurnal changes in sea-breeze direction. *J. Appl. Meteor.,* **35**, 1166-1169.

Simpson, J.E., 2007: *Sea Breeze and Local Winds*. Cambridge University Press, 252 pp.

Simpson, J.E., D.A. Mansfield, and J.R. Milford, 2006: Inland penetration of sea-breeze fronts. *Quart. J. Royal Meteor. Soc.,* http://dx.doi.org/10.1002/qj.49710343504.

Simpson, R.H., 1978: On the computation of equivalent potential temperature. *Mon. Wea. Rev.,* **106**, 124-130.

Simpson, R.H., and R.A. Pielke, 1976: Hurricane development and movement. *Appl. Mech. Rev.,* **29**, 601-609.

Sinha, S.K., S.G. Narkhedkar, and S. Rajamani, 1998: Application of Sasaki's numerical variational technique to the analysis of height and wind fields over Indian region. *Mausam,* **49**, 1-10.

Sisterson, D.L., and R.A. Dirks, 1978: Structure of the daytime urban moisture field. *Atmos. Environ.,* **12**, 1943-1949.

Skamarock, W.C., and J.B. Klemp, 1992: The stability of time-split numerical methods for the hydrostatic and the nonhydrostatic elastic equations. *Mon. Wea. Rev.,* **120**, 2109-2127.

Skamarock, W.C., J.B. Klemp, J. Dudhia, D.O. Gill, D.M. Barker, W. Wang, and J.G. Powers, 2008: A description of the Advanced Research WRF Version 3. NCAR Technical Note TN-468+STR. 113 pp.

Skibin, D., and A. Hod, 1979: Subjective analysis of mesoscale flow patterns in northern Israel. *J. Appl. Meteor.,* **18**, 329-337.

Skinner, T., and N. Tapper, 1994: Preliminary sea breeze studies over Bathurst and Melville Islands, Northern Australia, as part of the Island Thunderstorm Experiment (ITEX). *Meteor. Atmos. Phys.,* **53**, 77-94.

Slingo, A., R. Brown, and C.L. Wrench, 1982: A field study of nocturnal stratocumulus-III high resolution radiative and microphysical observations. *Quart. J. Roy. Meteor. Soc.,* **108**, 145-165.

Slingo, J.M., 1987: The development and verification of a cloud prediction model for the ECMWF model. *Quart. J. Roy. Meteor. Soc.,* **113**, 899-927.

Smith, D.E., M.T. Zuber, H.V. Frey, J.B. Garvin, J.W. Head, D.O. Muhleman, G.H. Pettengill, R.J. Phillips, S.C. Solomon, H.J. Zwally, W.B. Banerdt, T.C. Duxbury, M.P. Golombek, F.G. Lemoine, G.A. Neumann, D.D. Rowlands, O. Aharonson, P.G. Ford, A.B. Ivanov, C.L. Johnson, P.J. McGovern, J.B. Abshire, R.S. Afzal, X. Sun 2001: Mars Orbiter Laser Altimeter: Experiment summary after the first year of global mapping of Mars. *J. Geophys. Res.,* **106(E10)**, 23689-23722.

Smith, J.K., and T.B. McKee, 1983: Undisturbed clear day diurnal wind and temperature pattern in northeastern Colorado. CSU Atmospheric Science Paper No. 365, Fort Collins, CO.

Smith, M.D., J.C. Pearl, B.J. Conrath, and P.R. Christensen, 2001: Thermal Emission spectrometer results: Mars atmospheric thermal structure and aerosol distribution. *J. Geophys. Res.,* **106**, 23929-23945.

Smith, M.D., B.J. Conrath, J.C. Pearl, and P.R. Christensen, 2002: Thermal emission spectrometer observations of Martian planet-encircling dust storm 2001A. *Icarus,* **157(1)**, 259-263.

Smith, M.D., M.J. Wolff, M.T. Lemmon, N. Spanovich, D. Banfield, C.J. Budney, R.T. Clancy, A. Ghosh, G.A. Landis, P. Smith, B. Whitney, P.R. Christensen, and S.W. Squyres, 2004: First Atmospheric Science results from the Mars Exploration Rovers Mini-TES. *Science,* **306(5702)**, 1750-1753.

Smith, R.B., 1979: Influence of mountains on the atmosphere. *Adv. Geophys.,* **21**, 217-230.

Smith, R.B., 1982a: A differential advection model of orographic rain. *Mon. Wea. Rev.,* **110**, 306-309.

Smith, R.B., 1982b: Synoptic observations and theory of orographically disturbed wind and pressure. *J. Atmos. Sci.,* **39**, 60-70.

Smith, R.B. 1984: A theory of lee cyclogenesis. *J. Atmos. Sci.,* **41**, 1159-1168.

Smith, R.B., and Y.-L. Lin, 1982: The addition of heat to a stratified airstream with application to the dynamics of orographic rain. *Quart. J. Roy. Meteor. Soc.,* **108**, 353-378.

Smith, R.C., 1955: Theory of air flow over a heated land mass. *Quart. J. Roy. Meteor. Soc.,* **81**, 382-395.

Smith, R.C., 1957: Air motion over a heated land mass: II. *Quart. J. Roy. Meteor. Soc.,* **83**, 248-256.

Smolarkiewicz, P.K., 1983: A simple positive definite advection scheme with small implicit diffusion. *Mon. Wea. Rev.,* **111**, 479-486.

Smolarkiewicz, P.K., and L.G. Margolin, 1997: On forward-in-time differencing for fluids: An Eulerian/semi-Lagrangian nonhydrostatic model for stratified flows. *Atmos.-Ocean Special,* **35**, 127-152.

Smolarkiewicz, P.K., and L.G. Margolin, 1998: MPDATA: A finite-difference solver for geophysical flows. *J. Comput. Phys.,* **140**, 459-480.

Smolarkiewicz, P.K., and J.A. Pudykiewicz, 1992: A class of semi-Lagrangian approximations for fluids. *J. Atmos. Sci.,* **49**, 2082-2096.

Smolarkiewicz, P.K., and R. Rotunno, 1989: Low Froude number flow past three-dimensional obstacles. Part I: Baroclinically generated lee vortices. *J. Atmos. Sci.,* **46**, 1154-1164.

Smull, B.F., 1995: Convectively-induced mesoscale weather systems in the tropical and warm-season midlatitude atmosphere. *Rev. Geophys.,* Supplement **33**, 897-906.

Snook, J.S., and R.A. Pielke, 1995: Diagnosing a Colorado heavy snow event with a nonhydrostatic mesoscale numerical model structured for operational use. *Wea. Forecasting*, **10**, 261-285.

Snook, J.S., P.A. Stamus, J. Edwards, Z. Christidis, and J.A. McGinley, 1998: Local-domain mesoscale analysis and forecast model support for the 1996 Centennial Olympic Games. *Wea. Forecasting,* **13**, 138-150.

Snow, J.W., 1981: Wind power assessment along the Atlantic and Gulf Coasts of the U.S. University of Virginia, Dept. of Environmental Sciences, Ph.D. Dissertation, Charlottesville, 183 pp.

Solomon, S., D. Qin, M. Manning, Z. Chen, M. Marquis, K.B. Averyt, M. Tignor, and H.L. Miller, Eds., 2007: Contribution of Working Group I to the Fourth Assessment Report of the Intergovernmental Panel on Climate Change, 2007, Cambridge Univ. Press, Cambridge, U.K.

Solomos, S., G. Kallos, J. Kushta, M. Astitha, C. Tremback, A. Nenes, and Z. Levin, 2011: An integrated modeling study on the effects of mineral dust and sea salt particles on clouds and precipitation. *Atmos. Chem. Phys.*, **11**, 873-892, http://dx.doi.org/10.5194/acp11-873-2011.

Solomos, S., G. Kallos, E. Mavromatidis, and J. Kushta, 2012: Density currents as a desert dust mobilization mechanism. *Atmos. Chem. Phys. Discuss.*, **12**, 21579-21614, http://dx.doi.org/10.5194/acpd-12-21579-2012.

Somieski, F., 1981: Linear theory of three-dimensional flow over mesoscale mountains. *Contrib. Atmos. Phys.,* **54**, 315-334.

Sommeria, G., 1976: Three-dimensional simulation of turbulent processes in an undisturbed tradewind boundary layer. *J. Atmos. Sci.,* **33**, 216-241.

Sommeria, G., and J.W. Deardorff, 1977: Subgrid-scale condensation in models of nonprecipitating clouds. *J. Atmos. Sci.,* **33**, 216-241.

Song, J.L., R.A. Pielke, M. Segal, R.W. Arritt, and R. Kessler, 1985: A method to determine non-hydrostatic effects within subdomains in a mesoscale model. *J. Atmos. Sci.*, **42**, 2110-2120.

Song, Y., and D. Haidvogel, 1994: A semi-implicit ocean circulation model using a generalized topography-following coordinate system. *J. Comput. Phys.,* **115**, 228-244.

Soong, S.-T., and W.-K. Tao, 1980: Response of deep tropical cumulus clouds to mesoscale processes. *J. Atmos. Sci.,* **37**, 2016-2034.

Sousounis, P.J., G.E. Mann, G.S. Young, R.B. Wagenmaker, B.D. Hoggatt, and W.J. Badini, 1999: Forecasting during the Lake-ICE/SNOWBANDS field experiments. *Wea. Forecasting,* **14**, 955-975.

Sorbjan, Z., 1989: *Structure of the Atmospheric Boundary Layer*. Prentice Hall, Englewood Cliffs, NJ, 317 pp.

Sorbjan, Z., 2007: Statistics of shallow convection on Mars based on large-eddy simulations. Part 1: shearless conditions. *Bound.-Layer Meteor.,* **123(1)**, 121-142.

Sorbjan, Z., and M. Uliasz, 1982: Some numerical urban boundary-layer studies. *Bound.-Layer Meteor.,* **22**, 481-502.

Spiegel, M.R., 1967: *Applied Differential Equations.* Prentice-Hall, Englewood Cliff, NJ, 654 pp.

Spiga, A., 2011: Elements of comparison between Martian and terrestrial mesoscale meteorological phenomena: Katabatic winds and boundary layer convection, *Planetary Space Sci.*, **59(10)**, 915-922.

Spiga, A., and F. Forget, 2009: A new model to simulate the Martian mesoscale and microscale atmospheric circulation: Validation and first results. *J. Geophys. Res.*, **114**, E02009, http://dx.doi.org/10.1029/2008JE003242.

Spiga, A., F. Forget, J.-B. Madeleine, L. Montabone, S.R. Lewis, and E. Millour, 2011: The impact of martian mesoscale winds on surface temperature and on the determination of thermal inertia. *Icarus*, **212(2)**, 504-519.

Spiegel, M.R., 1987: *Applied Differential Equations*, 2nd Edition. Prentice-Hall, Inc. Englewood Cliffs, N.J., 412 pp.

Spurr, R.J.D., 2001: Linearized radiative transfer theory: A general discrete ordinate approach to the calculation of radiances and analytic weighting functions, with application to atmospheric remote sensing. Ph.D. Thesis, Technical University of Eindhoven, Eindhoven, The Netherlands, 230 pp.

Spurr, R.J.D., T.P. Kurosu, and K.V. Chance, 2001: A linearized discrete ordinate radiative transfer model for atmospheric remote-sensing retrieval. *J. Quant. Spectrosc. Radiat. Transfer*, **68**, 689-735.

Spyrou, C., C. Mitsakou, G. Kallos, P. Louka, and G. Vlastou, 2010: An improved limited-area model for describing the dust cycle in the atmosphere. *J. Geophys. Res.*, **115**, D17211, http://dx.doi.org/10.1029/2009JD013682.

Srivastava, R.C., 1971: Size distribution of raindrops generated by their breakup and coalescence. *J. Atmos. Sci.*, **28**, 410-415.

Sta. Maria, M.R.V., S.C.R. Rafkin, and T.I. Michaels, 2006: Numerical simulation of atmospheric bore waves on Mars. *Icarus*, **185(2)**, 383-394.

Stage, S.A., 1983: Boundary layer evolution in the region between shore and cloud edge during cold air outbreaks. *J. Atmos. Sci.*, **40**, 1453-1471.

Staley, D.O., and G.M. Jurica, 1970: Flux emissivity tables for water vapor, carbon dioxide and ozone. *J. Appl. Meteor.*, **9**, 365-372.

Stamnes, K.H., S.-C. Tsay, W. Wisecombe, and K. Jayaweera, 1988: Numerical stable algorithm for discrete-ordinate-method radiative transfer in multiple scattering and emitting layered media. *Appl. Opt.*, **27**, 2502-2509.

Staniforth, A., and J. Cote, 1991: Semi-Lagrangian integration schemes for atmospheric models–a review. *Mon. Wea. Rev.*, **119**, 2206-2223.

Stauffer, D.F., 2012: Uncertainty in environmental NWP modeling. In: *Handbook of Environmental Fluid Dynamics, Volume Two: Systems, Pollution, Modeling, and Measurements*, H. Fernado, Ed., CRC Press, 411-424.

Stauffer, D.R., and N.L. Seaman, 1994: Multiscale four-dimensional data assimilation. *J. Appl. Meteor.*, **33**, 416-434.

Stauffer, D.R., R.C. Munoz, and N.L. Seaman, 1999: In-cloud turbulence and explicit microphysics in the MM5. *Preprints, Ninth PSU/NCAR MM5 Modeling Systems Users' Workshop*, Boulder, Colorado, June 23-24, 177-180.

Stauffer, D.R., A. Deng, G.K. Hunter, A.M. Gibbs, J.R. Zielonka, K. Tinklepaugh, and J. Dobek, 2007a: NWP goes to war. *Preprints, 22nd Conference on Weather Analysis and Forecasting/18th Conference on Numerical Weather Prediction*, Park City, UT, June 25-29, 13 pp.

Stauffer, D.R., G.K. Hunter, A. Deng, J.R. Zielonka, K. Tinklepaugh, P. Hayes and C. Kiley, 2007b: On the role of atmospheric data assimilation and model resolution on model forecast accuracy for the Torino Winter Olympics. *Preprints, 22nd Conference on Weather Analysis and Forecasting/18th Conference on Numerical Weather Prediction*, Park City, UT, June 25-29, 7 pp.

Stauffer, D.R., G.K. Hunter, A. Deng, J.R. Zielonka, K. Dedrick, C. Broadwater, A. Grose, C. Pavloski, and J. Toffler, 2009: Realtime high-resolution mesoscale modeling for the Defense Threat Reduction Agency. *23rd Conference on Weather Analysis and Forecasting/19th Conference on Numerical Weather Prediction*, Omaha, NE, Jun 1-5, 10 pp.

Steeneveld, G.J., A.A.M. Holtslag, R.T. McNider, and R.A Pielke Sr., 2011: Screen level temperature increase due to higher atmospheric carbon dioxide in calm and windy nights revisited. *J. Geophys. Res.*, **116**, D02122, http://dx.doi.org/10.1029/2010JD014612.

Stein, U., and P. Alpert, 1993: Factor separation in numerical simulations. *J. Atmos. Sci.,* **50**, 2107-2115.

Stensrud, D.J., 2007: *Parameterization Schemes: Keys to Understanding Numerical Weather Prediction Models*. Cambridge University Press, 478 pp.

Stephens, G.L., 1978a: Radiation profiles in extended water clouds. II: Parameterization schemes. *J. Atmos. Sci.,* **35**, 2123-2132.

Stephens, G.L., 1978b: Radiation profiles in extended water clouds. I Theory. *J. Atmos. Sci.,* **35**, 2111-2122.

Stephens, G.L., 1984: Parameterization of radiation for numerical weather prediction models. *Mon. Wea. Rev.*, **112**, 826-867.

Stephens, G.L., and P.J. Webster, 1981: Clouds and climate: sensitivity of simple systems. *J. Atmos. Sci.,* **38**, 235-247.

Stephens, G.L., D.G. Vane, R.J. Boain, G.G. Mace, K. Sassen, Z. Wang, A.J. Illingworth, E.J. O'Connor, W.B. Rossow, S.L. Durden, S.D. Miller, R.T. Austin, A. Benedetti, C. Mitrescu, 2002: The cloudsat mission and the A-train. *Bull. Amer. Meteor. Soc.,* **83**, 1771-1790.

Stephens, G.L., N.B. Wood, and L.A. Pakula, 2004: On the radiative effects of dust on tropical convection. *Geophys. Res. Lett.,* **31**, L23112, http://dx.doi.org/10.1029/2004GL021342.

Stephens, G.L., T. L'Ecuyer, R. Forbes, A. Gettlemen, J.-C. Golaz, A. Bodas-Salcedo, K. Suzuki, P. Gabriel, and J. Haynes, 2010: Dreary state of precipitation in global models. *J. Geophys. Res.*, **115**, D24211, http://dx.doi.org/10.1029/2010JD014532.

Stern, M.E., and J.S. Malkus, 1953: The flow of a stable atmosphere over a heated island. Part II. *J. Meteor.,* **10**, 105-120.

Stevens, B., C.-H. Moeng, and P.P. Sullivan, 1999: Large-eddy simulations of radiatively driven convection: Sensitivities to the representation of small scales. *J. Atmos. Sci.,* **56**, 3963-3984.

Steyn, D.G., and K.W. Ayotte, 1985: Application of two-dimensional terrain height spectra to mesoscale modeling. *J. Atmos. Sci.,* **42**, 2884-2887.

Stohl, A., 1998: Computation, accuracy and applications of trajectories–a review and bibliography. *Atmos. Environ.,* **32**, 947-966.

Stohl, A., C. Forster, A. Frank, P. Seibert, and G. Wotawa, 2005: Technical note: The Lagrangian particle dispersion model FLEXPART version 6.2. *Atmos. Chem. Phys.,* **5**, 2461-2474.

Stohlgren, T.J., T.N. Chase, R.A. Pielke, T.G.F. Kittel, and J. Baron, 1998: Evidence that local land use practices influence regional climate and vegetation patterns in adjacent natural areas. *Global Change Biology*, **4**, 495-504.

Stone Jr., B., 2012: *The City and the Coming Climate: Climate Change in the Places We Live*. Cambridge University Press, 198 pp.

Storer, R.L., and S.C. van den Heever, 2012: Microphyiscal processes evident in aerosol forcing of tropical deep convection. *J. Atmos. Sci.*, http://dx.doi.org/10.1175/JAS-D-12-076.1.

Strommen, N.D., and J.R. Harman, 1978: Seasonally changing patterns of lake-effect snowfall in western lower Michigan. *Mon. Wea. Rev.*, **106**, 503-509.

Stull, R.B., 1988: *An Introduction to Boundary Layer Meteorology.* Kluwer Academic Publishers, The Netherlands, 666 pp.

Stull, R.B., 2000: *Meteorology for Scientists and Engineers.* 2nd Edition, Brooks/Cole Thomson Learning, 502 pp.

Sturm, M., J. Holmgren, and G.E. Liston, 1995: A seasonal snow cover classification system for local to global applications. *J. Climate,* **8**, 1261-1283.

Sud, Y.C., W.C., Chao, and G.K. Walker, 1993: Dependence of rainfall on vegetation: theoretical considerations, simulation experiments, observations, and inferences from simulated atmospheric soundings. *J. Arid Environ.,* **25**, 5-18.

Sud, Y.C., K.M. Lau, G.K. Walker, and J.H. Kim, 1995: Understanding biosphere – precipitation relationships: Theory, model simulations and logical inferences. *MAUSAM,* **46**, 1-14.

Sun, W.-Y., 1980: A forward-backward time integration scheme to treat internal gravity waves. *Mon. Wea. Rev.,* **108**, 402-407.

Sun, W.-Y., 1984a: Numerical-analysis for hydrostatic and nonhydrostatic equations of inertial internal gravity waves. *Mon. Wea. Rev.,* **112**, 259-268.

Sun, W.-Y., 1984b: Rainbands and symmetric instability. *J. Atmos. Sci.,* **41**, 3412-3426.

Sun, W.-Y., 1987: Mesoscale convection along the dryline. *J. Atmos. Sci.,* **44**, 1394-1403.

Sun, W.-Y., 1993: Numerical experiments for advection equation. *J. Comput. Phys.,* **108**, 264-271.

Sun, W.-Y., 1995: Pressure gradient in a sigma coordinate. *Terr. Atmos. Ocean.,* **6**, 579-590.

Sun, W.-Y., and J.-D. Chern, 1994: Numerical experiments of vortices in the wakes of large idealized mountains. *J. Atmos. Sci.,* **51**, 191-209.

Sun, W.-Y., and Y. Ogura, 1979: Boundary-layer forcing as a possible trigger to a squall line formation. *J. Atmos. Sci.,* **36**, 235-254.

Sun, W.-Y., and Y. Ogura, 1980: Modeling the evolution of the convection planetary boundary layer. *J. Atmos. Sci.,* **37**, 1558-1572.

Sun, W.-Y., and I. Orlanski, 1981a: Large mesoscale convection and sea breeze circulation. Part I: linear stability analysis. *J. Atmos. Sci.,* **38**, 1675-1693.

Sun, W.-Y., and I. Orlanski, 1981b: Large mesoscale convection and sea breeze circulation. Part II: nonlinear numerical model. *J. Atmos. Sci.,* **38**, 1694-1709.

Sun, W.-Y., and C.-C. Wu, 1992: Formation and diurnal variation of the dryline. *J. Atmos. Sci.,* **49**, 1606-1619.

Sun, W.-Y., and K.-S. Yeh, 1997: A general semi-Lagrangian advection scheme employing forward trajectories. *Quart. J. Roy. Meteor. Soc.,* **123**, 2463-2476.

Sun, W.-Y., K.-S. Yeh, and R.-Y. Sun, 1996: A simple-Lagrangian scheme for advection equations. *Quart. J. Roy. Meteor. Soc.,* **122**, 1211-1226.

Sun, Z., J. Liu, X. Zeng, and H. Liang, 2013: Parameterization of instantaneous global horizontal irradiance at the surface. Part II: Cloudy-sky component, *J. Geophys. Res.,* http://dx.doi.org/10.1029/2012JD017557, in press.

Sundquist, H., 1979: Vertical coordinates and related discretization. "Numerical Methods Used in Atmospheric Models," GARP Report, 17, Volume II, 1-150.

Sutherland, B., 2010: *Internal Gravity Waves.* Cambridge University Press, 394 pp.

Svendsen, H., and R.O.R.Y. Thompson, 1978: Wind-driven circulation in fjord. *J. Phys. Oceanogr.,* **8**, 703-712.

Swan, P.R., and I.Y. Lee, 1980: Meteorological and air pollution modeling for an urban airport. *J. Appl. Meteor.,* **19**, 534-544.

Sweet, W., R. Fett, J. Kerling, and P. LaViolette, 1981: Air-sea interaction effects in the lower troposphere across the north wall of the Gulf stream. *Mon. Wea. Rev.,* **109**, 1042-4052.

Tafferner, A., and J. Egger, 1990: Test of theories of lee cyclogenesis. *J. Atmos. Sci.,* **47**, 2417-2428.

Tag, P.M., and T.E. Rosmond, 1980: Accuracy and energy conservation in a three-dimensional anelastic model. *J. Atmos. Sci.,* **37**, 2150-2168.

Tag, P.M., F.W. Murray, and L.R. Kvenig, 1979: A comparison of several forms of eddy viscosity parameterization in a two-dimensional cloud model. *J. Appl. Meteor.,* **18**, 1429-1441.

Taha, H., 1999: Modifying a mesoscale meteorological model to better incorporate urban heat storage: A bulk-parameterization approach. *J. Appl. Meteor.,* **38**, 466-473.

Taha, H., St. Konopacki, and S. Gabersek, 1999: Impacts of large-scale surface modifications on meteorological conditions and energy use: A 10-region modeling study. *Theor. Appl. Climatol.,* **62**, 175-185.

Takahashi, N., H. Uyeda, K. Kikuchi, and K. Iwanami, 1996: Mesoscale and convective scale features of heavy rainfall events in late period of the Baiu season in July 1988, Nagasaki prefecture. *J. Meteor. Soc. Japan,* **74**, 539-561.

Takeda, T., K. Isono, M. Wada, Y. Ischizaka, K. Okada, Y. Fujiyoshi, M. Maruyama, Y. Igawa, and K. Nagaya, 1982: Modification of convective snow-clouds in landing the Japan Sea coastal region. *J. Meteor. Soc. Japan,* **60**, 967-977.

Takle, E.S., W.J. Gutowski Jr., R.W. Arritt, Z. Pan, C.J. Anderson, R. Ramos da Silva, D. Caya, S.-C. Chen, F. Giorgi, J.H. Christensen, S.-Y. Hong, H.-M. Juang, J. Katzfey, W.M. Lapenta, R. Laprise, G.E. Liston, P. Lopez, J. McGregor, R.A. Pielke Sr., and J.O. Roads, 1999: Project to Intercompare Regional Climate Simulations (PIRCS): Description and initial results. *J. Geophys. Res.,* **104**, 19443-19461.

Tamppari, L.K., J. Barnes, E. Bonfiglio, B. Cantor, A.J. Friedson, A. Ghosh, M.R. Grover, D. Kass, T.Z. Martin, M. Mellon, T. Michaels, J. Murphy, S.C.R. Rafkin, M.D. Smith, G. Tsuyuki, D. Tyler, and M. Wolff, 2008: Expected atmospheric environment for the Phoenix landing season and location. *J. Geophys. Res.-Planets*, **113**, E00A20, http://dx.doi.org/10.1029/2007JE003034.

Tamppari, L.K.,D. Bass, B. Cantor, I. Daubar, C. Dickinson, D. Fisher, K. Fujii, H.P. Gunnlauggson, T.L. Hudson, D. Kass, A. Kleinböhl, L. Komguem, M.T. Lemmon, M. Mellon, J. Moores, A. Pankine, J. Pathak, M. Searls, F. Seelos, M.D. Smith, S. Smrekar, P. Taylor, C. Holstein-Rathlou, W. Weng, J. Whiteway, and M. Wolff, 2010: Phoenix and MRO coordinated atmospheric observations, *J. Geophys. Res.-Planets*, **115**, E00E17, http://dx.doi.org/10.1029/2009JE003415.

Tanelli, S., E. Im, R. Mascelloni, and L. Facheris, 2005: Spaceborne Doppler radar measurements of rainfall: correction of errors induced by pointing uncertainties. *J. Atmos. Oceanic Tech.,* **22**, 1676-1690.

Tanrikulu, S., D.R. Stauffer, N.L. Seaman, and A.J. Ranzieri, 2000: A field-coherence technique for meteorological field-program design for air-quality studies. Part II: Evaluation in the San Joaquin Valley. *J. Appl. Meteor.,* **39**, 317-334.

Tao, W.-K., and J. Simpson, 1989: Modeling study of a tropical squall-type convective line. *J. Atmos. Sci.,* 46177-46202.

Tao, W.-K., and J. Simpson, 1993: Goddard cumulus ensemble model. Part I: Model description. *Terrestrial, Atmos. Oceanic Sci.*, **4**, 35-72.

Tao, W.-K., J. Simpson, C.-H. Sui, C.-L. Shie, B. Zhou, K.M. Lau, and M. Moncrieff, 1999: Equilibrium states simulated by cloud-resolving models. *J. Atmos. Sci.,* **56**, 3128-3139.

Tao, W.-K., C.-L. Shie, and J. Simpson, 2001: Comments on "The sensitivity study of radiative-convective equilibrium in the tropics with a convective resolving model". *J. Atmos. Sci.,* 58, 1328-1333.

Tao, W.-K., D. Starr, A. Hou, P. Newman, and Y. Sud, 2003: Summary of cumulus parameterization workshop. *Bull. Amer. Meteor. Soc.,* **84**, 1055-1062.

Tao W.-K., J. Chern, R. Atlas, D. Randall, X. Lin, M. Khairoutdinov, J. Li, D. Waliser, A. Hou, C. Peters-Lidard, W. Lau, and J. Simpson, 2009: A Multi-scale modeling system: Development, applications and critical issues. *Bull. Amer. Meteor. Soc.*, **90(4)**, 515-534.

Tapp, M.C., and P.W. White, 1976: A non-hydrostatic mesoscale model. *Quart. J. Roy. Meteor. Soc.,* **102**, 277-296.

Tapper, N.J., P.D. Tyson, I.F. Owens, and W.J. Hastie, 1981: Modeling the winter urban heat island over Christchurch, New Zealand. *J. Appl. Meteor.,* **20**, 365-376.

Taylor, C.M., F. Said, and T. Lebel, 1997: Interactions between the land surface and mesoscale rainfall variability during HAPEX-Sahel. *Mon. Wea. Rev.,* **125**, 2211-2227.

Taylor, C.M., R.J. Harding, R.A. Pielke Sr., P.L. Vidale, R.L. Walko, and J.W. Pomeroy, 1998: Snow breezes in the boreal forest. *J. Geophys. Res.*, **103**, 23087-23101.

Taylor, J.P., and A.S. Ackerman, 1999: A case-study of pronounced perturbations to cloud properties and boundary-layer dynamics due to aerosol emissions. *Quart. J. Roy. Meteor. Soc.,* **125**, 2543-2661.

Taylor, P.A., 1974: Urban meteorological modelling – some relevant studies. *Adv. Geophys.,* **18B**, 173-185.

Temperton, C., 1973: Some experiments in dynamic initialization for a simple primitive equation model. *Quart. J. Roy. Meteor. Soc.,* **99**, 303-319.

Tenhunen, J.D., and P. Kabat, Editors, 1999: Integrating hydrology, ecosystem dynamics, and biogeochemistry in complex landscapes. Report of the Dahlem Workshop on Integrating Hydrology, Ecosystem Dynamics, and Biogeochemistry in Complex Landscapes, January 18-23, 1998, Wiley, New York, 367 pp.

Tennekes, H., 1978: Turbulent flow in two and three dimensions. *Bull. Amer. Meteor. Soc.,* **59**, 22-28.

Tennekes, H., and J.L. Lumley, 1972: *A First Course in Turbulence.* MIT Press, Cambridge, MA, 300 pp.

Terjung, W.H., and P.A. O'Rourke, 1981: Relative influence of vegetation on urban energy budgets and surface temperatures. *Bound.-Layer Meteor.,* **25**, 255-263.

Thompson, G. 1993: Prototype real-time mesoscale prediction during 1991-92 winter season and statistical verification of model data. Colorado State University, M.S. Thesis, Dept. of Atmospheric Science Paper No. 521, Fort Collins, CO 80523, 105 pp.

Thompson, G., P.R. Field, R.M. Rasmussen, and W.D. Hall, 2008: Explicit forecasts of winter precipitation using an improved bulk microphysics scheme. Part II: Implementation of a new snow parameterization. *Mon. Wea. Rev.,* **136**, 5095-5115.

Thompson, W.T., 1979: Effects of New York City on the horizontal and vertical structure of sea breeze fronts, Vol. III. Effects of frictionally retarded sea breeze and synoptic frontal passages on sulfur dioxide concentrations. R.D. Borstein, P.I., Report from the Dept. of Meteorology, San Jose State University, San Jose, California.

Thomson, D.J., 1987: Criteria for the selection of stochastic models of particle trajectories in turbulent flows. *J. Fluid Mech.*, **180**, 529-556.

Thorpe, A.J., M.J. Miller, and M.W. Moncrieff, 1982: Two-dimensional convection in non-constant shear: a model of mid-latitude squall lines. *Quart. J. Roy. Meteor. Soc.,* **108**, 739-762.

Thuburn, J., 2006: Some conservation issues for the dynamical cores of NWP and climate models. *J. Comput. Phys.,* **227**, 3715-3730.

Thunis, P., and R. Bornstein, 1996: Hierarchy of mesoscale flow assumptions and equations. *J. Atmos. Sci.*, **53**, 380-397.

Thunis, P., and A. Clappier, 2000: Formulation and evaluation of a nonhydrostatic Mesoscale Vorticity Model (TVM). *Mon. Wea. Rev.,* **128**, 3236-3251.

Tibaldi, S., and A. Buzzi, 1983: Effects of orography on Mediterranean lee cyclogenesis and its relationships to European blocking. *Tellus,* **35A**, 269-286.

Tibaldi, S., A. Buzzi, and A. Speranza, 1990: Orographic cyclogenesis. Extratropical cyclones: The Eric Palmen Memorial Volume, C.W. Newton, and E.O. Holopainen, Eds., Amer. Meteor. Soc., 107-127.

Tiedtke, M., 1996: An extension of cloud-radiation parameterization in the ECMWF model: The representation of subgrid-scale variations of optical depth. *Mon. Wea. Rev.,* **124**, 745-750.

Tiedtke, M., W.A. Heckley, and J. Slingo, 1988: Tropical forecasting at ECMWF: The influence of physical parameterization on the mean structure of forecasts and analyses. *Quart. J. Roy. Meteor. Soc.,* **114**, 639-644.

Tijm, A.B.C., and A.J. Van Delden, 1998: The role of sound waves in sea-breeze initiation. *Quart. J. Roy. Meteor. Soc.,* **125**, 1997-2018.

Tijm, A.B.C., A.A.M. Holtslag, and A.J. van Delden, 1999a: Observations and modeling of the sea breeze with the return current. *Mon. Wea. Rev.,* **27**, 625-640.

Tijm, A.B.C., A.J. Van Delden, and A.A.M. Holtslag, 1999b: The inland penetration of sea breezes. *Contrib. Atmos. Phys.,* **72**, 317-328.

Tillman, J.E., L. Landberg, and S.E. Larsen, 1994: The boundary layer of Mars: Fluxes, stability, turbulent spectra, and growth of the mixed layer. *J. Atmos. Sci.*, **51(12)**, 1709-1727.

Toigo, A.D., and M.I. Richardson, 2003: Meteorology of proposed Mars Exploration Rover landing sites. *J. Geophys. Res.*, **108(E12)**, 8092.

Toigo, A.D., M.I. Richardson, R.J. Wilson, H. Wang, and A.P. Ingersoll, 2002: A first look at dust lifting and dust storms near the south pole of Mars with a mesoscale model. *J. Geophys. Res.*, **107(E7)**, 5050.

Tompkins, A.M., 2000: The impact of dimensionality on long-term cloud-resolving model simulations. *Mon. Wea. Rev.,* **128**, 1521-1535.

TPBS, 1974: Performance characteristics of the operational models at the National Meteorological Center (NMC). Technical Procedures Bulletin No. 107. National Weather Service.

Tran, H.N.Q., and N. Mölders, 2012a: Investigations on meteorological conditions for elevated PM_2.5 in Fairbanks, Alaska. *Atmos. Res.* **99 (1)**, 39-49.

Tran, H.N.Q., and N. Mölders, 2012b. Numerical investigations on the contribution of point source emissions to the PM_2.5 concentrations in Fairbanks, Alaska. *Atmos. Pollution Res.*, **3**, 199-210.

Trapp, R.J., 2013: *Mesoscale-Convective Processes in the Atmosphere*, Cambridge University Press, 400 pp.

Tremback, C.J., 1990: Numerical simulation of a mesoscale convective complex: model development and numerical results. Ph.D. dissertation, Atmos. Sci. Paper 465, Colorado State University, Department of Atmospheric Science, Fort Collins, Colorado, 247 pp.

Tremback, C.J., J. Powell, W.R. Cotton, and R.A. Pielke, 1987: The forward-in-time upstream advection scheme-extension to higher orders. *Mon. Wea. Rev.,* **115**, 540-555.

Trenberth, K.E., 1978: On the interpretation of the diagnostic quasi-geostrophic Omega equation. *Mon. Wea. Rev.*, **106**, 131-137.

Tripoli, G.J., and W.R. Cotton, 1980: A numerical investigation of several factors contributing to the observed variable intensity of deep convection over south Florida. *J. Appl. Meteor.,* **19**, 1037-1063.

Tripoli, G.J., and W.R. Cotton, 1981: The use of ice-liquid water potential temperature as a thermodynamic variable in deep atmospheric models. *Mon. Wea. Rev.,* **109**, 1094-1102.

Tripoli, G.J., and W.R. Cotton, 1982: The Colorado State University three-dimensional cloud/mesoscale model – 1982. Part I: General theoretical framework and sensitivity experiments. *J. Rech. Atmos.,* **16**, 185-219.

Tripoli, G.J., and W.R. Cotton, 1989a: A numerical study of an observed orogenic mesoscale convective system. Part 1. Simulated genesis and comparison with observations. *Mon. Wea. Rev.,* **117**, 273-304.

Tripoli, G.J., and W.R. Cotton, 1989b: A numerical study of an observed orogenic mesoscale convective system. Part 2. Analysis of governing dynamics. *Mon. Wea. Rev.,* **117**, 305-328.

Troen, I., and L. Mahrt, 1986: A simple model of the atmospheric boundary layer: Sensitivity to surface evaporation. *Bound.-Layer Meteor.,* **37**, 129-148.

Truong, N.M., T.T. Tien, R.A. Pielke Sr., C.L. Castro, and G. Leoncini, 2009: A modified Kain-Fritsch scheme and its application for simulation of an extreme precipitation event in Vietnam. *Mon. Wea. Rev.*, **137**, 766-789, http://dx.doi.org/10.1175/2008MWR2434.1.

Truong, N.M., V.T. Hang, R.A. Pielke Sr., C.L. Castro, and K. Dairaku, 2012: Synoptic-scale physical mechanisms associated with Mei-yu front: A numerical case study in 1999. *Asia-Pacific J. Atmos. Sci.*, **48 (4)**, 433-448, http://dx.doi.org/10.1007/s13143-012-0039-x.

Tso, C.P., 1995: A survey of urban heat island studies in low tropical cities. *Atmos. Environ.,* **30**, 507-519.

Tsonis, A.A., 2007: *An Introduction to Atmospheric Thermodynamics*. Second Edition, Cambridge University Press, 198 pp.

Tucker, D.F., and K.S. Zentmire, 1999: On the forecasting of orogenic mesoscale convective complexes. *Wea. Forecasting,* **14**, 1017-1022.

Tuinenburg, O.A., R.W.A. Hutjes, C.M.J. Jacobs, and P. Kabat, 2011: Diagnosis of local land-atmosphere feedbacks in India. *J. Climate*, **24**, 251-266. http://dx.doi.org/10.1175/2010JCLI3779.1.

Tuleya, R.E., 1994: Tropical storm development and decay: Sensitivity to surface boundary conditions. *Mon. Wea. Rev.,* **122**, 291-304.

Tuleya, R.E., and Y. Kurihara, 1978: A numerical simulation of the landfall of tropical cyclones. *J. Atmos. Sci.,* **35**, 242-257.

Tuleya, R.E., M.A. Bender, and Y. Kurihara, 1984: A simulation study of the landfall of tropical cyclones using a movable nested-mesh model. *Mon. Wea. Rev.,* **12**, 124-136.

Tunick, A., 1999: A review of previous works on observing the atmospheric boundary layer through meteorological measurements. Army Res. Lab., MR-448, 1-41.

Turner, B.D., 1969: Workbook of atmospheric dispersion estimates. *U.S. Public Health Serv. Publ. 999-AP-26*, 84 pp.

Twitty, J.T., and J.A. Weinman, 1971: Radiative properties of carbonaceous aerosols. *J. Appl. Meteor.,* **10**, 725-731.

Twomey, S., 1978: The influence of aerosols on radiative properties of clouds. In: *Climatic Change and Variability: A Southern Perspective*, Pittock et al., Ed., Cambridge University Press, New York, 455 pp.

Tyler, D., and J.R. Barnes, 2005: A mesoscale model study of summertime atmospheric circulations in the north polar region of Mars. *J. Geophys. Res.-Planets*, **110(E6)**, 110, E06007, http://dx.doi.org/10.1029/2004JE002356.

Tyler, D., J.R. Barnes, and R.M. Haberle, 2002: Simulation of surface meteorology at the Pathfinder and VL1 sites using a Mars mesoscale model. *J. Geophys. Res.-Planets*, **107(E4)**, http://dx.doi.org/10.1029/2001JE001618.

Tyler, D., J.R. Barnes, and E.D. Skyllingstad, 2008: Mesoscale and large-eddy simulation model studies of the Martian atmosphere in support of Phoenix. *J. Geophys. Res.-Planets*, 113(E8), E00A12, http://dx.doi.org/10.1029/2007JE003012.

Uccellini, L.W., 1980: On the role of upper tropospheric jet streaks and leeside cyclogenesis in the development of low-level jets in the Great Plains. *Mon. Wea. Rev.,* **108**, 1689-1696.

Uccellini, L.W., and D.R. Johnson, 1979: The coupling of upper and lower tropospheric jet streaks and implications for the development of sever convective storms. *Mon. Wea. Rev.,* **107**, 682-703.

Uccellini, L.W., D.R. Johnson, and R.E. Schlesinger, 1979: An isentropic and sigma coordinate hybrid numerical model: Model development and some initial tests. *J. Atmos. Sci.,* **36**, 390-414.

Uccellini, L.W., P.J. Korin, and C.H. Wash, 1981: *The President's Day cyclone, 17-19 February 1979: An analysis of jet streak interaction prior to cyclogenesis.* NASA Technical Memorandum 82077, NASA Goddard.

Uccellini, L.W., R.A. Petersen, K.F. Brill, P.J. Kocin, and J.J. Tuccillo, 1987: Synergistic interactions between an upper-level jet streak and diabatic processes that influence the development of a low-level jet and a secondary coastal cyclone. *Mon. Wea. Rev.*, **115**, 2227-2261.

Ueda, H., 1983: Effects of external parameters on the flow field in the coastal region – A linear model. *J. Climate Appl. Meteor.,* **22**, 312-321.

Ulanski, S.L., and M. Garstang, 1978: The role of surface divergence and vorticity in the life cycle of convective rainfall. Part I. Observations and analysis. *J. Atmos. Sci.,* **35**, 1047-1062.

Uliasz, M., 1994: Subgrid scale parameterizations. In: *Mesoscale modeling of the atmosphere.* R. Pearce, and R.A. Pielke, Eds., American Meteorological Society, 13-19.

Uliasz, M., R.A. Stocker, and R.A. Pielke, 1996: Regional modeling of air pollution transport in the southwestern United States. Chapter 5 in *Environmental Modeling, Volume III,* P. Zannetti, Ed., Computational Mechanics Publications, 145-181.

Ulrickson, B.L., and C.F. Mass, 1990: Numerical investigation of mesoscale circulations over the Los Angels basin. Part 1, A verification study. *Mon. Wea. Rev.,* **118**, 2138-2161.

USDA, 1941: *Climate and Man, Yearbook of Agriculture 1941*, U.S. Government Printing Office, 1248 pp.

USDA, 1951: *Soil Survey Manual.* USDA Agricultural Handbook 18. U.S. Government Printing Office, Washington, DC.

Vali, G., R.D. Kelly, J. French, S. Haimov, D. Leon, R.E. McIntosh, and A. Pazmany, 1998: Finescale structure and microphysics of coastal stratus. *J. Atmos. Sci.,* **55**, 3540-3564.

van den Hurk, B.J.J.M., P. Viterbo, A.C.M. Beljaars, and A.K. Betts, 2000: Offline validation of the ERA40 surface scheme. *Tech. Memor.,* **295**, 1-42.

van der Ent, R.J., H.H.G. Savenije, B. Scheafli, and S.C. Steele-Dunne, 2010: Origin and fate of atmospheric moisture over continents. *Water Resour. Res.*, **46**, W09525, http://dx.doi.org/10.1029/2010WR009127.

Van der Hoven, I., 1967: Atmospheric transport and diffusion at coastal sites. *Nucl. Safety,* **8**, 490-499.

van Egmond, N.D., and D. Onderdelinden, 1981: Objective analysis of air pollution monitoring network data; spatial interpolation and network density. *Atmos. Environ.,* **15**, 1035-1046.

van Genuchten, M.Th., 1980: A closed-form equation for predicting the hydraulic conductivity of unsaturated soils. *Soil Sci. Soc. Amer. J.*, **44**, 892-898.

van Haren, R., G.J. van Oldenborgh, G. Lenderink, M. Collins, and W. Hazeleger, 2012: SST and circulation trend biases cause an underestimation of European precipitation trends. *Climate Dyn.,* http://dx.doi.org/10.1007/s00382-012-1401-5.

van Oldenborgh, G.J., F.J. Doblas-Reyes, B. Wouters, and W. Hazeleger, 2012: Decadal prediction skill in a multi-model ensemble. *Climate Dyn.,* http://dx.doi.org/10.1007/s00382-012-1313-4.

Vána, F., P. Bénard, J.-F. Geleyn, A. Simon, and Y. Seity, 2008: Semi-Lagrangian advection scheme with controlled damping: An alternative to nonlinear horizontal diffusion in a numerical weather prediction model. *Quart. J. Roy. Meteor. Soc.,* **134**, 523-537.

Varinou, M., G. Kallos, G. Tsiligiridis, and G. Sistla, 1999: The role of anthropogenic and biogenic emissions on tropospheric ozone formation over Greece. *Phys. Chem. Earth,* **24**, 507-513.

Veljovic, K., B. Rajkovic, M.J. Fennessy, E.L. Altshuler, and F. Mesinger, 2010: Regional climate modeling: Should one attempt improving on the large scales? Lateral boundary condition scheme: Any impact? *Meteorologische Zeits.*, **19:3**, 237-246. http://dx.doi.org/10.1127/0941-2948/2010/0460.

Vergeiner, I., 1971: An operational linear lee wave model for arbitrary basic flow and two-dimensional topography. *Quart. J. Roy. Meteor. Soc.,* **97**, 30-60.

Verseghy, D.L., 2000: The Canadian Land Surface Scheme (CLASS): Its history and future. *Atmos.-Ocean,* **38**, 1-13.

Vidale, P.L., R.A. Pielke, A. Barr, L.T. Steyaert, 1997: Case study modeling of turbulent and mesoscale fluxes over the BOREAS region. *J. Geophys. Res.,* **102**, 29167-29188.

Viskanta, R., and R.A. Daniel, 1980: Radiative effects of elevated pollutant layers on temperature structure and dispersion in an urban atmosphere. *J. Appl. Meteor.,* **19**, 53-70.

Viskanta, R., and T.L. Weirich, 1978: Feedback between radiatively interacting pollutants and their dispersion in the urban boundary layer. WMO Symposium on Boundary Layer Physics and Applied Specific Problems of Air Pollution, WMO No. 510, 31-38.

Viskanta, R., R.W. Bergstrom, and R.O. Johnson, 1977a: Effects of air pollution on thermal structure and dispersion in an urban planetary boundary layer. *Contrib. Atmos. Phys.,* **50**, 419-440.

Viskanta, R., R.W. Bergstrom, and R.O. Johnson, 1977b: Radiative transfer in a polluted urban planetary boundary layer. *J. Atmos. Sci.,* **34**, 1091-1103.

Viterbo, P., and A.C.M. Beljaars, 1995: An improved surface parameterization scheme in the ECMWF model and its validation. *J. Climate,* **8**, 2716-2748.

Viterbo, P., and A.K. Betts, 1999: Impact of ECMWF reanalysis soil water on forecasts of the July 1993 Mississippi flood. *J. Geophys. Res.,* **104**, 19,361-19,366.

Viterbo, P., A. Beljaars, J.-F. Mahfouf, and J. Teixeira, 1999: The representation of soil moisture freezing and its impact on the stable boundary layer. *Quart. J. Roy. Meteor. Soc.,* **125**, 2401-2426.

Vogel, S., 1998: Exposing life's limits with dimensionless numbers. *Phys. Today,* **51**, 22-27.

von Storch, H., 1978: Construction of optimal numerical filters fitted for noise damping in numerical simulation models. *Contrib. Atmos. Phys.,* **51**, 189-197.

Vukićević, T., 1998: Optimal initial perturbations for two cases of extratropical cyclogenesis. *Tellus,* **50**, 143-166.
Vukićević, T., and P. Hess, 2000: Analysis of tropospheric transport in the Pacific basin using the adjoint technique. *J. Geophys. Res.,* **105**, 7213-7230.
Vukićević, T., and K. Raeder, 1995: Use of an adjoint model for finding triggers for Alpine lee cyclogenesis. *Mon. Wea. Rev.,* **123**, 800-816.
Vukićević, T., B.H. Braswell, and D. Schimel, 2001: A diagnostic study of temperature controls on global terrestrial carbon exchange. *Tellus B*, **53**, 150-170.
Vukićević, T., M. Sengupta, A.S. Jones, and T. Vonder Haar, 2006: Cloud-resolving satellite data assimilation: Information content of IR window observations and uncertainties in estimation. *J. Atmos. Sci.,* **63:3**, 901-919.
Vukovich, F.M., and W.J. King, 1980: A theoretical study of the St. Louis Heat Island: Comparisons between observed data and simulation results on the urban heat island circulation. *J. Appl. Meteor.,* **19**, 761-770.
Vukovich, F.M., J.W. Dunn, III, and B.W. Crissman, 1976: A theoretical study of the St. Louis heat island: The wind and temperature distribution. *J. Appl. Meteor.,* **15**, 417-440.
Vukovich, F.M., W.J. King, J.W. Dunn, III, and J.J.B. Worth, 1979: Observations and simulations of the diurnal variation of the urban heat island circulation and associated variations of the ozone distribution: A case study. *J. Appl. Meteor.,* **18**, 836-854.
Walcek, C.J., 2000: Minor flux Adjustment near mixing ratio extremes for simplified yet highly accurate monotonic calculation of tracer advection. *J. Geophys. Res.,* **105**, 9335-9348.
Waliser, D.E., J.-. F. Li, C.P. Woods, R.T. Austin, J. Bacmeister, J. Chern, A. Del Genio, J.H. Jiang, Z. Kuang, H. Meng, P. Minnis, S. Platnick, W.B. Rossow, G.L. Stephens, S. Sun-Mack, W.-K. Tao, A.M. Tompkins, D.G. Vane, C. Walker, and D. Wu, 2009: Cloud ice: A climate model challenge with signs and expectations of progress, *J. Geophys. Res.,* **114**, D00A21, http://dx.doi.org/10.1029/2008JD010015.
Walker, A.L., M. Liu, S.D. Miller, K.A. Richardson, and D.L. Westphal, 2009: Development of a dust source database for mesoscale forecasting in southwest Asia. *J. Geophys. Res.,* **114**, D18207, http://dx.doi.org/10.1029/2008JD011541.
Walko, R.L., and R. Avissar, 2008a: The Ocean-Land-Atmosphere Model (OLAM). Part I: Shallow water tests. *Mon. Wea. Rev.*, **136**, 4033-4044.
Walko, R.L., and R. Avissar, 2008b: The Ocean-Land-Atmosphere Model (OLAM). Part II: Formulation and tests of the nonhydrostatic dynamic core. *Mon. Wea. Rev.*, **136**, 4045-4062.
Walko, R.L., and R. Avissar, 2011: A direct method for constructing refined regions in unstructured conforming triangular-hexagonal computational grids: Application to OLAM. *Mon. Wea. Rev.*, **139**, 3923-3937.
Walko, R.L., C.J. Tremback, R.A. Pielke, and W.R. Cotton, 1995: An interactive nesting algorithm for stretched grids and variable nesting ratios. *J. Appl. Meteor.,* **34**, 994-999.
Walko, R.L., L.E. Band, J. Baron, T.G.F. Kittel, R. Lammers, T.J. Lee, R.A. Pielke Sr., C. Taylor, C. Tague, C.J. Tremback, and P.L. Vidale, 1998: Coupled atmosphere-terrestrial ecosystem-hydrology models for environmental modeling. Department of Atmospheric Science Class Report #9, Colorado State University, Fort Collins, CO, 53 pp.
Walko, R.L., W.R. Cotton, G. Feingold, and B. Stevens, 2000a: Efficient computation of vapor and heat diffusion between hydrometeors in a numerical model. *Atmos. Res.*, **53**, 171-183.
Walko, R.L., L.E. Band, J. Baron, T.G.F. Kittel, R. Lammers, T.J. Lee, D.S. Ojima, R.A. Pielke, C. Taylor, C. Tague, C.J. Tremback, and P.L. Vidale, 2000b: Coupled atmosphere-biophysics-hydrology models for environmental modeling. *J. Appl. Meteor.,* **39**, 931-944.

Walko, R.L., Tremback, C.J., Bell, M.J., 2001. HYPACT Hybrid Particle and Concentration Transport Model, User's Guide. Mission Research Corporation, Fort Collins, CO, 113 pp, available at: http://gate1.baaqmd.gov/pdf/1211_HYPACT_HYbrid_Particle_Concentration_Transport_Model_Version_1.2.0_Users_Guide_2001.pdf.

Wallace, J.M., and P.V. Hobbs, 1977: *Atmospheric Science: An Introductory Survey.* Academic Press, New York, 467 pp.

Wallace, J.M., S. Tibaldi, and A.J. Simmons, 1983: Reduction of systematic forecast errors in the ECMWF model through the introduction of an envelope orography. *Quart. J. Roy. Meteor. Soc.,* **109**, 683-717.

Walmsley, J.L., J.R. Salmon, and P.A. Taylor, 1982: On the application of a model of boundary-layer flow over low hills to real terrain. *Bound.-Layer Meteor.,* **23**, 17-46.

Walsh, J.E., 1974: Sea breeze theory and applications. *J. Atmos. Sci.,* **31**, 2012-2026.

Walter, B.A., and J.E. Overland, 1982: Response of stratified flow in the lee of the Olympic Mountains. *Mon. Wea. Rev.,* **110**, 1458-1473.

Walters, M.K., 2000: Comments on 'The differentiation between grid spacing and resolution and their application to numerical modeling.' *Bull. Amer. Meteor. Soc.*, **81**, 2475-2477.

Wang, D.P., 1979: Wing-driven circulation in the Chesapeake Bay, Winter 1975. *J. Phys. Oceanogr.,* **9**, 564-572.

Wang, H., 2007: Dust storms originating in the northern hemisphere during the third mapping year of Mars Global Surveyor. *Icarus*, **189(2)**, 325-343.

Wang, H., M.I. Richardson, R.J. Wilson, A.P. Ingersoll, A.D. Toigo, and R.W. Zurek, 2003: Cyclones, tides, and the origin of a cross-equatorial dust storm on Mars. *Geophys. Res. Lett.*, **30(9)**, 1488.

Wang, J.-J., H.-M.H. Juang, K. Kodama, S. Businger, Y.-L. Chen, and J. Partain, 1998: Application of the NCEP Regional Spectral Model to improve mesoscale weather forecasts in Hawaii. *Wea. Forecasting,* **13**, 560-575.

Wang, S., Q. Wang, and J. Doyle, 2002: Some improvement of Louis surface flux parameterization. *Preprints, 15th Symp. on Boundary Layers and Turbulence*, Wageningen, Netherlands, Amer. Meteor. Soc., 547-550.

Wang, T.-A., and Y.-L. Lin, 1999: Wave ducting in a stratified shear flow over a two-dimensional mountain. Part II: Implications for the development of high-drag states for severe downslope windstorms. *J. Atmos. Sci.,* **56**, 437-452.

Wang, Y., 1996: On the forward-in-time upstream advection scheme for non-uniform and time-dependent flow. *Meteor. Atmos. Phys.*, **61**, 27-38.

Wang, Z., M. Barlage, X. Zeng, R.E. Dickinson, and C.B. Schaaf, 2005: The solar zenith angle dependence of desert albedo. *Geophys. Res. Lett.*, **32**, L05403, http://dx.doi.org/10.1029/2004GL021835.

Wang, Z., X. Zeng, and M. Barlage, 2007: MODIS BRDF-based land surface albedo parameterization for weather and climate models. *J. Geophys. Res.-Atmospheres*, **112**, D02103, http://dx.doi.org/10.1029/2005JD006736.

Wang, Z., X. Zeng, and M. Decker, 2010: Improving the snow processes in the Noah land model. *J. Geophys. Res.*, **115**, D20108, http://dx.doi.org/10.1029/2009JD013761.

Warner, C., 1982: Mesoscale features and cloud organization on 10-12 December 1978 over the south China Sea. *J. Atmos. Sci.,* **39**, 1619-1641.

Warner, C., J. Simpson, D.W. Martin, D. Suchman, F.R. Mosher, and R.F. Reinking, 1979: Shallow convection on Day 261 of GATE: mesoscale arcs. *Mon. Wea. Rev.,* **107**, 1617-1635.

Warner, T.T., 1983: Fundamental equations and numerical techniques of urban boundary layer modeling. *AMS/EPA Specialty Conference on Air Quality Modeling of the Nonstationary, Nonhomogeneous Urban Boundary Layer*, October 31-November 4, Baltimore, MD.

Warner, T.T., 2011: *Numerical Weather and Climate Prediction.* Cambridge University Press, 548 pp.

Warner, T.T., and H.-M. Hsu, 2000: Nested-model simulation of moist convection: The impact of coarse-grid parameterized convection on fine-grid resolved convection. *Mon. Wea. Rev.,* **128**, 2211-2231.

Warner, T.T., and R.-S. Sheu, 2000: Multiscale local forcing of the Arabian Desert daytime boundary layer, and implications for the dispersion of surface-released contaminants. *J. Appl. Meteor.,* **39**, 686-707.

Warner, T.T., R.A. Anthes, and A.L. McNab, 1978: Numerical simulations with a three-dimensional mesoscale model. *Mon. Wea. Rev.,* **106**, 1079-1099.

Warner, T.T., R.A. Peterson, and R.E. Treadon, 1997: A tutorial on lateral boundary conditions as a basic and potentially serious limitation to regional numerical weather prediction. *Bull. Amer. Meteor. Soc.,* **78**, 2599-2617.

Weaver, C., A. da Silva, M. Chin, P. Ginoux, O. Dubovik, D. Flittner, A. Zia, L. Remer, B. Holben, and W. Gregg, 2007: Direct insertion of MODIS radiances in a global aerosol transport model. *J. Atmos. Sci.,* **64**, 808-827.

Weaver, J.F., and F.P. Kelly, 1982: A mesoscale, climatologically-based forecast technique for Colorado. *Preprints, AMS Conf. on Weather Forecasting and Analysis*, Seattle, Washington, June 28-July 1.

Weber, M.R., and C.B. Baker, 1982: Comments on "The ratio of diffuse to direct solar irradiance (perpendicular to the sun's rays) with clear skies – a conserved quantity throughout the day". *J. Appl. Meteor.,* **21**, 883-886.

Webster's New World Dictionary of American English, 1988: Third College Edition, Victoria Neufeldt, and David Bernard Guralnik, Eds., Distributed by Prentice Hall Trade, 1574 pp.

Weisberg, R.H., 1976: The nontidal flow in the Providence River of Narragansett Bay: A stochastic approach to estuarine circulation. *J. Phys. Oceanogr.,* **6**, 721-734.

Weisman, M.L., J.B. Klemp, and W.C. Skamarock, 1997: The resolution-dependence of explicitly-modeled convective systems. *Mon. Wea. Rev.,* **125**, 527-548.

Weiss, C.E., and J.F.W. Purdom, 1974: The effect of early-morning cloudiness on squall-line activity. *Mon. Wea. Rev.,* **102**, 400-402.

Welander, P., 1955: Studies on the general development of motion in a two-dimensional, ideal fluid. *Tellus,* **7**, 141-156.

Welch, R.M., and W.G. Zdunkowski, 1976: A radiation model for the polluted atmospheric boundary layer. *J. Atmos. Sci.,* **33**, 2170-2184.

Welch, R.M., and W.G. Zdunkowski, 1981a: The effect of cloud shape on radiative characteristics. *Contrib. Atmos. Phys.,* **54**, 482-491.

Welch, R.M., and W.G. Zdunkowski, 1981b: Improved approximation for diffuse solar radiation on oriented sloping terrain. *Contrib. Atmos. Phys.,* **54**, 362-369.

Welch, R.M., J. Paegle, and W.G. Zdunkowski, 1978: Two-dimensional numerical simulation of the effects of air pollution upon the urban-rural complex. *Tellus,* **30**, 136-150.

Welch, R.M., S.K. Cox, and J.M. Davis, 1980: *Solar radiation and clouds.* Meteorological Monographs, 17, AMS, Boston, 96 pp.

Wen, X., S. Lu, and J. Jin, 2012: Integrating remote sensing data with WRF for improved simulations of oasis effects on local weather processes over an arid region in northwestern China. *J. Hydrometeor.,* **13**, 573-587.

Weng, F., 1992: A multi-layer discrete-ordinate method for vector radiative transfer in a vertically-inhomogeneous, emitting and scattering atmosphere I: Theory. *J. Quant. Spectrosc. Radiat. Transfer*, **47**, 19-33.

Wergen, W., 1981: Nonlinear normal model initialization of a multi-level fine-mesh model with steep orography. *Contrib. Atmos. Phys.*, **53**, 389-402.

Wesley, D.A., and R.A. Pielke, 1990: Observations of blocking-induced convergence zones and effects on precipitation in complex terrain. *Atmos. Res.*, **25**, 235-276.

Wesley, D.A., G. Poulos, J. Snook, P. Kennedy, M. Meyers, and G. Byrd, 2013: Extreme snowfall variations and cold air damming in the Front Range heavy snowstorm of 17-20 March 2003. *Nat. Wea. Dig.*, **36**, No. 1, in press.

Wetzel, P.J., 1978: A detailed parameterization of the atmospheric boundary layer. Colorado State University, Ph.D. Dissertation, Dept. of Atmospheric Science, Paper No. 302, Fort Collins, CO, 165 pp.

Wetzel, P.J., W.R. Cotton, and R.L. McAnelly, 1983: A long-lived mesoscale convective complex. Part II: Evolution and structure of the mature complex. *Mon. Wea. Rev.*, **111**, 1919-1937.

Whiteman, C.D., 1980: Breakup of temperature inversions in Colorado mountain valleys. Dept. of Atmospheric Science, Paper No. 328, Colorado State University, Fort Collins, CO, 270 pp.

Whiteman, C.D., 1982: Breakup of temperature inversions in deep mountain valleys: Part I. Observations. *J. Appl. Meteor.*, **21**, 270-289.

Whiteman, C.D., 2000: *Mountain Meteorology: Fundamentals and Applications*, Oxford University Press, 376 pp.

Whiteman, C.D., and J.C. Doran, 1993: The relationship between overlying synoptic-scale flows and winds within a valley. *J. Appl. Meteor.*, **32**, 1669-1682.

Whiteman, C.D., and T.B. McKee, 1978: Air pollution implications of descent in mountain valleys. *Atmos. Environ.*, **12**, 2151-2158.

Whiteman, C.D., and T.B. McKee, 1982: Breakup of temperature inversions in deep mountain valleys: Part II. Thermodynamic model. *J. Appl. Meteor.*, **21**, 290-302.

Whiteman, C.D., X. Bian, and S. Zhong, 1999: Wintertime evolution of the temperature inversion in the Colorado plateau basin. *J. Appl. Meteor.*, **38**, 1103-1117.

Whiteman, C.D., S.W. Hoch, M. Lehner, and T. Haiden, 2010: Nocturnal cold air intrusions into Arizona's Meteor Crater: Observational evidence and conceptual model. *J. Appl. Meteor. Climatol.*, **49**, 1894-1905.

Wicker, L.J., and W.C. Skamarock, 1998: A time-splitting scheme for the elastic equations incorporating second-order Runge-Kutter time differencings. *Mon. Wea. Rev.*, **126**, 1992-1999.

Williams, R.T., 1967: Atmospheric frontogenesis: A numerical experiment. *J. Atmos. Sci.*, **24**, 627-641.

Williams, R.T., 1972: Quasi-geostrophic versus nongeostrophic frontogenesis. *J. Atmos. Sci.*, **29**, 3-10.

Wilson, R.J., 1997: A general circulation model simulation of the Martian polar warming. *Geophys. Res. Lett.*, **24**, 123-126.

Wilson, R.J., and K.P. Hamilton, 1996: Comprehensive model simulation of thermal tides in the Martian atmosphere. *J. Atmos. Sci.*, **53**, 1290-1326.

Wing, D.R., and G.L. Austin, 2006: Global Mars mesoscale meteorological model. *Icarus*, **185**, 370-382, http://dx.doi.org/10.1016/j.icarus.2006.07.016.

Winkler, J.A., R.H. Skaggs, and D.G. Baker, 1981: Effect of temperature adjustments on the Minneapolis-St. Paul urban heat island. *J. Appl. Meteor.,* **20**, 1295-1300.

Winninghoff, F.J., 1968: On the adjustment toward a geostrophic balance in a simple primitive equation model with application to the problems of initialization and objective analysis. University of California, Ph.D. Thesis, Los Angeles.

Winstead, N.S., and G.S. Young, 2000: An analysis of exit-flow drainage jets over the Chesapeake Bay. *J. Appl. Meteor.,* **39**, 1269-1281.

Wipperman, F., 1981: The applicability of several approximations in mesoscale modelling – a linear approach. *Contrib. Atmos. Phys.,* **54**, 298-308.

Wipperman, F., and G. Gross, 1981: On the construction of orographically influenced wind roses for given distributions of the large-scale wind. *Beitr. Phys. Atmos.,* **54**, 492-501.

Wolfram, S., 1999: *Mathematica Book, Version 4.* 4th Edition, Cambridge University Press. 1496 pp.

Wolkenberg, P., D. Grassi, and V. Formisano, 2009: The impact of Martian aerosols on the retrieval of temperature profiles from PFS measurements. *Icarus*, **110(17)**, 1908-1925.

Wong, V.C., J.W. Zach, M.L. Kaplan, and S.L. Chuang, 1983a: A numerical investigation of the effects of cloudiness on mesoscale atmosphere circulation. *Preprints, AMS 5th Conference on Numerical Weather Atmospheric Radiation*, Oct. 31- Nov. 4.

Wong, V.C., J.W. Zach, M.L. Kaplan, and G.D. Coats, 1983b: A nested-grid limited-area model for short term weather forecasting. *Preprints, AMS 6th Conference on Numerical Weather Prediction*, June 6-9, Omaha, Nebraska, 9-15.

Wooldridge, G.L., and M.M. Orgill, 1978: Airflow, diffusion, and momentum flux patterns in a high mountain valley. *Atmos. Environ.,* **12**, 803-808.

Wortman-Vierthaler, M., and N. Moussiopoulos, 1995: Numerical tests of a refined flux corrected transport scheme. *Environ. Software,* **10**, 157-175.

Wu, X., M.W. Moncrieff, and K.A. Emanuel, 2000a: Evaluation of large-scale forcing during TOGA COARE for cloud-resolving models and single-column models. *J. Atmos. Sci.,* **57**, 2977-2985.

Wu, A., T.A. Black, D.L. Verseghy, M.D. Novak, and W.G. Bailey, 2000b: Testing the alpha and beta methods of estimating evaporation from bare and vegetated surfaces in CLASS. *Atmos.-Ocean,* **38**, 15-35.

Wu, L., and G.W. Petty, 2010: Intercomparison of bulk microphysics schemes in model simulations of polar lows. *Mon. Wea. Rev.,* **138**, 2211-2228. http://dx.doi.org/10.1175/2010MWR3122.1.

Wu, W., A.H. Lynch, and A. Rivers, 2005: Estimating the uncertainty in a regional climate model related to initial and lateral boundary conditions. *J. Climate*, **18**, 917-933.

Wu., Y., U.S. Nair, R.A. Pielke Sr., R.T. McNider, S.A. Christopher, and V. Anantharaj, 2009: Impact of land surface heterogeneity on mesoscale atmospheric dispersion. *Bound.-Layer Meteor.*, **133**, No 3, http://dx.doi.org/10.1007/s10546-009-9415-1, 367-389.

Wyatt, M.G., 2012: A multidecadal climate signal propagating across the Northern Hemisphere through indices of a synchronized network. Department of Geology, University of Colorado, Ph.D. Dissertation, Boulder, CO, 201 pp.

Wyngaard, J.C., 1982: Boundary layer modeling. In *Atmospheric Turbulence and Air Pollution Modelling*, F.T.M. Nieuwstadt, and H. Van Dop, Eds., Reidel, Holland, 69-106.

Wyngaard, J.D., 1983: *Lectures on the Planetary Boundary Layer. Mesoscale Meteorology - Theories, Observations, and Models*, T. Gal-Chen and D.K. Lilly, Eds, Reidel. Dordrecht, Holland, 781 pp.

Wyser, K., L. Rontu, and H. Savijarvi, 1999: Introducing the effective radius into a fast radiation scheme of a mesoscale model. *Contrib. Atmos. Phys.,* **72**, 205-218.

Xian, Z., and R.A. Pielke, 1991: The effects of width of land masses on the development of sea breezes. *J. Appl. Meteor.,* **30**, 1280-1304.

Xiao, F., 2000: A class of single-cell high-order semi-Lagrangian advection schemes. *Mon. Wea. Rev.,* **28**, 1165-1176.

Xu, K.-M., and D.A. Randall, 1996: A semiempirical cloudiness parameterization for use in climate models. *J. Atmos. Sci.,* **53**, 3084-3102.

Xu, L., S. Raman, R.V. Madala, and R. Hodur, 1996: A non-hydrostatic modeling study of surface moisture effects on mesoscale convection induced by sea breeze circulation. *Meteor. Atmos. Phys.,* **58**, 103-122.

Xu, Q., B. Zhou, S.D. Burk, and E.H. Barker, 1999: An air-soil layer coupled scheme for computing surface heat fluxes. *J. Appl. Meteor.,* **38**, 211-223.

Xu, Z., and Z.-L. Yang, 2012: An improved dynamical downscaling method with GCM bias corrections and its validation with 30 years of climate simulations. *J. Climate,* http://dx.doi.org/10.1175/JCLI-D-12-00005.1.

Xue, H., Z. Pan, and J.M. Bane Jr., 2000a: A 2D coupled atmosphere-ocean model study of air-sea interactions during a cold air outbreak over the Gulf Stream. *Mon. Wea. Rev.,* **128**, 973-996.

Xue, M., K.K. Droegemeier, and V. Wong, 2000b: The Advanced Regional Prediction System (ARPS)—A multi-scale nonhydrostatic atmospheric simulation and prediction model. Part I: Model dynamics and verification, *Meteor. Atmos. Phys.,* **75**, 161-193.

Xue, M., K.K. Droegemeier, V. Wong, A. Shapiro, K. Brewster, F. Carr, D. Weber, Y. Liu, and D. Wang, 2001: The Advanced Regional Prediction System (ARPS)—A multi-scale nonhydrostatic atmospheric simulation and prediction tool. Part II: Model physics and applications, *Meteor. Atmos. Phys.,* **76**, 143-146.

Yablonsky, R.M., and I. Ginis, 2008: Improving the ocean initialization of coupled hurricane-ocean models using feature-based data assimilation. *Mon. Wea. Rev.,* **136**, 2592-2607.

Yablonsky, R.M., and I. Ginis, 2009: Limitation of one-dimensional ocean models for coupled hurricane-ocean model forecasts. *Mon. Wea. Rev.,* **137**, 4410-4419.

Yablonsky, R.M., and I. Ginis, 2013: Impact of a warm ocean eddy's circulation on hurricane-induced sea surface cooling with implications for hurricane intensity. *Mon. Wea. Rev.,* **141**, 997-1021.

Yamada, T., 1977: A numerical experiment on pollutant dispersion in a horizontally-homogeneous atmospheric boundary layer. *Atmos. Environ.,* **11**, 1015-1024.

Yamada, T., 1978: A three-dimensional numerical study of complex atmospheric circulations produced by terrain. *Proc., AMS Conference Sierra Nevada Meteorology*, 61-67.

Yamada, T., 1981: A numerical simulation of nocturnal drainage flow. *J. Meteor. Soc. Japan,* **59**, 108-122.

Yamada, T., 1982: A numerical model study of turbulent airflow in and above a forest canopy. *J. Meteor. Soc. Japan,* **60**, 439-454.

Yamada, T., 1999: Numerical simulations of airflows and tracer transport in the southwestern United States. *J. Appl. Meteor.,* **39**, 399-411.

Yamada, T., and R.N. Meroney, 1971: *Numerical and wind tunnel simulation of response of stratified shear layers to nonhomogeneous surface features* Project THEMIS Tech. Rep. No. 9. Prepared by the Fluid Dynamics and Diffusion Laboratory, Colorado State University, Fort Collins.

Yamamoto, G., 1962: Direct absorption of solar radiation by atmospheric water vapor, carbon dioxide and molecular oxygen. *J. Atmos. Sci.,* **19**, 182-188.

Yamasaki, M., 1977: A preliminary experiment of the tropical cyclone without parameterizing the effects of cumulus convection. *J. Meteor. Soc. Japan,* **55**, 11-31.
Yamazaki, K., and K. Ninomiya, 1981: Response of Arakawa-Schubert cumulus parameterization model to real data in heavy rainfall areas. *J. Meteor. Soc. Japan,* **59**, 547-563.
Yamazaki, N., and T.-C. Chen, 1993: Analysis of the East Asian monsoon during early summer of 1979: Structure of the Baiu front and its relationship to large-scale fields. *J. Meteor. Soc. Japan,* **71**, 339-355.
Yang, A., G. Sun, L. Lu, Z. Guo, and Y. Liu, 2011: Deriving aerodynamic roughness length and zero-plane displacement height from MODIS product for Eastern China. *Scientia Meteorologica Sinica*, **31, 4**, 81-92.
Yang, F., K. Mitchell, Y.-T. Hou, Y. Dai, X. Zeng, Z. Wang, and X. Liang, 2008: Dependence of land surface albedo on solar zenith angle: observations and model parameterizations. *J. Appl. Meteor. Climatol.,* **47**, 2963-2982, http://dx.doi.org/10.1175/2008JAMC1843.1.
Yang, P., H. Wei, H-L. Huang, B.A. Baum, Y.X. Hu, G.W. Kattawar, M.I. Mischenko, and Q. Fu, 2005: Scattering and absorption property database for nonspherical ice particles in the near- through far-infrared spectral region. *Appl. Opt.*, **44**, 5512-5523.
Yang,Y.J., B.W Wu, C. Shi, J.H. Zhang, Y.B. Li, W.A. Tang, H.-Y. Wen, H.-Q. Zhang, and T. Shi, 2012: Impacts of urbanization and station-relocation on surface air temperature series in Anhui Province, China. *Pure Appl. Geophys.*, http://dx.doi.org/10.1007/s00024-012-0619-9.
Yang, Z.-L., R.E. Dickinson, W.J. Shuttleworth, and M. Shaikh, 1998: Treatment of soil, vegetation and snow in land surface models: A test of the biosphere-atmosphere transfer scheme with the HAPEX-MOBILHY, ABRACOS and Russian data. *J. Hydrometeor.,* **212-213**, 109-127.
Yang, Z.-L., Y. Dai, R.E. Dickerson, and W.J. Shuttleworth, 1999a: Sensitivity of ground heat flux to vegetation cover fraction and leaf area index. *J. Geophys. Res.,* **104**, 19,505-19,514.
Yang, Z.-L., G.-Y. Niu, and R.E. Dickinson, 1999b: Comparing snow simulations from NCAR LSM and BATS using PILPS 2D data. *Preprints, 14th AMS Conference on Hydrology*, Dallas, TX, 316-319.
Yap, D., and T.R. Oke, 1974: Sensible heat fluxes over an urban area – Vancouver, B.C. *J. Appl. Meteor.,* **13**, 880-890.
Yarker, M.B., D. PaiMazumder, C.F. Cahill, J. Dehn, A. Prakash, and N. Mölders, 2010: Theoretical investigations on potential impacts of high-latitude volcanic emissions of heat, aerosols and water vapor and their interactions on clouds and precipitation. *The Open Atmos. Sci. J.*, **4**, 12-23.
Yau, M.K., and R. Michaud, 1982: Numerical simulation of a cumulus ensemble in three dimensions. *J. Atmos. Sci.,* **39**, 1062-1079.
Ye, Z., and R.A. Pielke, 1993: Atmospheric parameterization of evaporation from non-plant-covered surfaces. *J. Appl. Meteor.*, **32**, 1248-1258.
Ye, Z.J., M. Segal, and R.A. Pielke, 1987: Effects of atmospheric thermal stability and slope steepness on the development of daytime thermally-induced upslope flow. *J. Atmos. Sci.*, **44**, 3341-3354.
Ye, Z., M. Segal, J.R. Garratt, and R.A. Pielke, 1989: On the impact of cloudiness on the characteristics of nocturnal downslope flows. *Bound.-Layer Meteor.*, **49**, 23-51.
Ye, Z.J., J.R. Garratt, M. Segal, and R.A. Pielke, 1990a: On the impact of atmospheric thermal stability on the characteristics of nocturnal downslope flows. *Bound.-Layer Meteor.*, **51**, 77-97.
Ye, Z.J., M. Segal, and R.A. Pielke, 1990b: A comparative study of daytime thermally induced upslope flow on Mars and Earth, *J. Atmos. Sci.*, **47(5)**, 612-628.
Yenai, M., 1975: Tropical meteorology. *Rev. Geophys. Space Phys.,* **13**, 685-808.

Yih, C.-S., 1980: *Stratified Flows*. Academic Press, New York, 418 pp.

Yimin, M., and T.J. Lyons, 2000: Numerical simulation of a sea breeze under dominant synoptic conditions at Perth. *Meteor. Atmos. Phys.*, **73**, 89-103.

Yonetani, T., 1982: Increase in number of days with heavy precipitation in Tokyo urban area. *J. Appl. Meteor.*, **21**, 1466-1471.

Yonetani, T., 1983: Enhancement and initiation of a cumulus by a heat island. *J. Meteor. Soc. Japan*, **61**, 244-253.

Yoshikado, H., 1981: Statistical analyses of the sea breeze pattern in relation to general weather patterns. *J. Meteor. Soc. Japan*, **59**, 98-107.

Young, G.S., and R.A. Pielke, 1983: Application of terrain height variance spectra to mesoscale modeling. *J. Atmos. Sci.*, **40**, 2555-2560.

Young, G.S., R.A. Pielke, and R.C. Kessler, 1984: A comparison of the terrain height variance spectra of the front range with that of a hypothetical mountain. *J. Atmos. Sci.*, **41**, 1249-1250.

Young, R.E., R.L. Walterscheid, G. Schubert, A. Seiff, V.M. Linkin, and A.N. Lipatov, 1987: Characteristics of gravity waves generated by surface topography on Venus: Comparison with the VEGA balloon results. *J. Atmos. Sci.*, **44(18)**, 2628-2639.

Young, W.R., P.B. Rhines, and C.J.R. Garrett, 1982: Shear-flow dispersion, internal waves and horizontal mixing in the ocean. *J. Phys. Oceanogr.*, **12**, 515-527.

Zalucha, A.M., R.A. Plumb, and R.J. Wilson, 2010: An analysis of the effect of topography on the Martian Hadley Cells. *J. Atmos. Sci.*, **67(3)**, 673-693.

Zannetti, P., 1990: *Air Pollution Modeling: Theories, Computational Methods, and Available Software*. Computational Mechanics Publications, Boston, 444 pp.

Zardi, D., and C.D. Whiteman, 2013: Diurnal Mountain Wind Systems. Chapter 2 in: *Mountain Weather Research and Forecasting*. Chow et al. Eds., Springer, Berlin, 35-119.

Zarnecki, J.C., 2002: Huygens' surface science package. *Space Sci. Rev.*, **104**, 593-611.

Zdunkowsk, W., and A. Bott, 2004: *Thermodynamics of the Atmosphere: A Course in Theoretical Meteorology*. Cambridge University Press, 266 pp.

Zdunkowski, W.G., and G. Korb, 1974: An approximate method for the determination of longwave radiation fluxes in scattering and absorbing media. *Beitr. Phys. Atmos.*, **47**, 129-144.

Zdunkowski, W.G., and K.-N. Liou, 1976: Humidity effects on the radiative properties of a hazy atmosphere in the visible spectrum. *Tellus*, **28**, 31-36.

Zdunkowski, W.G., R.M. Welch, and J. Paegle, 1976: One-dimensional numerical simulation of the effects of air pollution on the planetary boundary layer. *J. Atmos. Sci.*, **33**, 2399-2414.

Zdunkowski, W.G., R.M. Welch, and R.C. Hanson, 1980: Direct and diffuse solar radiation on oriented sloping surfaces. *Contrib. Atmos. Phys.*, **53**, 449-468.

Zeller, K.F., and N.T. Nikolov, 2000: Quantifying simultaneous fluxes of ozone, carbon dioxide and water vapor above a subalpine forest ecosystem. *Environ. Pollution*, **107**, 1-20.

Zeng, N., and J.D. Neelin, 1999: A land-atmosphere interaction theory for the tropical deforestation problem. *J. Climate*, **12**, 857-872.

Zeng, X., and A. Beljaars, 2005: A prognostic scheme of sea surface skin temperature for modeling and data assimilation. *Geophys. Res. Lett.*, **32**, L14605, 10.1029/2005GL023030.

Zeng, X., and M. Decker, 2009: Improving the numerical solution of soil moisture-based Richards equation for land models with a deep or shallow water table. *J. Hydrometeor.*, **10**, 308-319. http://dx.doi.org/10.1175/2008JHM1011.1.

Zeng, X., and R.E. Dickinson, 1998: Impact of diurnally-varying skin temperature on surface fluxes over the tropical Pacific. *Geophys. Res. Lett.*, **25**, 1411-1414.

Zeng, X., and R.A. Pielke, 1993: Error-growth dynamics and predictability of surface thermally-induced atmospheric flow. *J. Atmos. Sci.*, **50**, 2817-2844.

Zeng, X., and R.A. Pielke, 1995a: Further study on the predictability of landscape-induced atmospheric flow. *J. Atmos. Sci.,* **52**, 1680-1698.

Zeng, X., and R.A. Pielke, 1995b: Landscape-induced atmospheric flow and its parameterization in large-scale numerical models. *J. Climate,* **8**, 1156-1177.

Zeng, X., M. Zhao, and R.E. Dickinson, 1998: Intercomparison of bulk aerodynamic algorithms for the computation of sea surface fluxes using TOGA COARE and TAO data. *J. Climate,* **11**, 2628-2644.

Zeng, X., M. Zhao, R.E. Dickinson, and Y. He, 1999a: A multiyear hourly sea surface skin temperature data set derived from the TOGA TAO bulk temperature and wind speed over the tropical Pacific. *J. Geophys. Res.,* **104**, 1525-1536.

Zeng, X., M. Xhao, B. Su, and H. Wang, 1999b: Study on a boundary-layer numerical model with inclusion of heterogeneous multi-layer vegetation. *Adv. Atmos. Sci.*, **16**, 431-442.

Zeng, X., R.E. Dickinson, A. Walker, M. Shaikh, R.S. DeFries, and J. Qi, 2000a: Derivation and evaluation of global 1-km fractional vegetation cover data for land modeling. *J. Appl. Meteor.,* **39**, 826-839.

Zeng, X., M. Zhao, and B. Su, 2000b: A numerical study of land-surface heterogeneity from "combined approach" on atmospheric process, Part I: Principle and method. *Adv. Atmos. Sci.,* **17**, 103-120.

Zeng, X., W.-K. Tao, T. Matsui, S. Xie, S. Lang, M. Zhang, D. Starr, and X. Li, 2011: Estimating the ice crystal enhancement factor in the Tropics. *J. Atmos. Sci.,* **68**, 1424-1434. http://dx.doi.org/10.1175/2011JAS3550.1.

Zeng, X., Z. Wang, and A. Wang, 2012: Surface skin temperature and the interplay between sensible and ground heat fluxes over arid regions. *J. Hydrometeor.,* **13**, 1359-1370. http://dx.doi.org/10.1175/JHM-D-11-0117.1.

Zhang, D., and R.A. Anthes, 1982: A high-resolution model of the planetary boundary layer-sensitivity tests and comparisons with SESAME-79 data. *J. Appl. Meteor.,* **21**, 1594-1609.

Zhang, D.-L., Y. Liu, and M.K. Yau, 2000: A multiscale numerical study of Hurricane Andrew (1992). Part III: Dynamically-induced vertical motion. *Mon. Wea. Rev.,* **128**, 3772-3788.

Zhang, H., G.M. McFarquhar, S.M. Saleeby, and W.R. Cotton, 2007: Impacts of Saharan dust as CCN on the evolution of an idealized tropical cyclone. *Geophys. Res. Lett.,* **34**, L14812, http://dx.doi.org/10.2029/2007GL029876.

Zhang, H., G.M. McFarquhar, W.R. Cotton, and Y. Deng, 2009: Direct and indirect impacts of Saharan dust acting as cloud condensation nuclei on tropical cyclone eyewall development. *Geophys. Res. Lett.,* **36**, L06802, http://dx.doi.org/10.1029/2009GL037276.

Zhang, J.A., R.F. Rogers, D.S. Nolan, and F.D. Marks, 2011: On the characteristic height scales of the hurricane boundary layer. *Mon. Wea. Rev.,* **139**, 2523-2535.

Zhang, X., S.G. Gopalakrishnan, and S. Traham, 2012: Design of multiple moving nests in Hurricane Weather Research and Forecasting Modeling System. *The 30th Conference on Hurricanes and Tropical Meteorology.* Ponte Vedra Beach, FL, 15-20 April 2012.

Zhao, Q., T.L. Black, and M.E. Baldwin, 1997: Implementation of the cloud prediction scheme in the eta model at NCEP. *Wea. Forecasting,* **12**, 697-712.

Zheng, W., H. Wei, Z. Wang, X. Zeng, J. Meng, M. Ek, K. Mitchell, and J. Derber, 2012: Improvement of land surface skin temperature in the NCEP GFS Model and its impact on satellite data assimilation. *J. Geophys. Res.-Atmospheres*, **117**, D06117, http://dx.doi.org/10.1029/2011JD015901.

Zhong, S., J.D. Fast, and X. Bian, 1996: A case study of the Great Plains low-level jet using wind profiler network data and a high-resolution mesoscale model. *Mon. Wea. Rev.,* **124**, 785-806.

Zhou, X.P., 1980: Severe storms research in China. *Bull. Amer. Meteor. Soc.,* **61**, 12-21.

Zhou, Y.P., W.-K. Tao, A.Y. Hou, W.S. Olson, C.-L. Shie, K.-M. Lau, M.-D. Chou, X. Lin, and M. Grecu 2007: Use of high-resolution satellite observations to evaluate cloud and precipitation statistics from cloud-resolving model simulations. Part I: South China Sea Monsoon experiment. *J. Atmos. Sci.,* **64**, 4309-4329.

Ziegler, C.L., and E.N. Rasmussen, 1998: The initiation of moist convection at the dryline: Forecasting issues from a case study perspective. *Wea. Forecasting,* **13**, 1106-1131.

Ziegler, C.L., W.J. Martin, R.A. Pielke, and R.L. Walko, 1995: A modeling study of the dryline. *J. Atmos. Sci.,* **52**, 263-285.

Ziegler, C.L., T.J. Lee, and R.A. Pielke, 1997: Convective initiation at the dryline: A modeling study. *Mon. Wea. Rev.,* **125**, 1001-1026.

Zilitinkevich, S.S., 1970: *Dynamics of the Atmospheric Boundary Layer.* Gidrometeoizdat, Leningrad, 292 pp (in Russian).

Zilitinkevich, S.S., and D.V. Chailikov, 1968: Determining the universal wind-velocity and temperature profiles in the atmospheric boundary layer. *Izvestiya, Atmospheric and Oceanic Physics*, **4**, 165-170 (English translation).

Zipser, E.J., 1971: Internal structure of cloud clusters. *GATE Experimental Design Proposal,* Vol. 2. Annex VII. Interim Scientific Management Group, WMO-ICSU, Geneva.

Zipser, E.J., 1977: Mesoscale and convective-scale downdrafts as distinct components of a squall line structure. *Mon. Wea. Rev.,* **105**, 1568-1589.

Zipser, E.J., 1982: Use of a conceptual model of the life-cycle of mesoscale convective systems to improve very-short-range forecasts. In: *Nowcasting*, K. Browning, Ed., Academic Press, London, 191-204.

Zipser, E.J., and A.J. Bedard Jr., 1982: Front range windstorms revisited. *Weatherwise*, **35**, 82-85.

Zipser, E.J., and C. Gautier, 1978: Mesoscale events within a GATE tropical depression. *Mon. Wea. Rev.,* **106**, 789-805.

Zipser, E.J., R.J. Meitin, and M.A. LeMone, 1981: Mesoscale motion fields associated with a slowly moving GATE convective band. *J. Atmos. Sci.,* **38**, 1725-1750.

Zou, X., A. Barcilon, I.M. Navon, J. Whitaker, and D.G. Caccuci, 1993: An adjoint sensitivity study of blocking in a two-layer isentropic model. *Mon. Wea. Rev.,* **121**, 2833-2857.

Zupanski, D., S.Q. Zhang, M. Zupanski, A.Y. Hou, and S.H. Cheung, 2011a: A prototype WRF-based ensemble data assimilation system for dynamically downscaling satellite precipitation observations. *J. Hydrometeor.,* **12:1**, 118-134.

Zupanski, D., M. Zupanski, L.D. Grasso, R. Brummer, I. Jankov, D. Lindsey, M. Sengupta, and M. Demaria, 2011b: Assimilating synthetic GOES-R radiances in cloudy conditions using an ensemble-based method. *Int. J. Remote Sens.*, **32:24**, 9637-9659.

Zweers, N.C., V.K. Makin, J.W. de Vries, and G. Burgers, 2010: A sea drag relation for hurricane wind speeds. *Geophys. Res. Lett.,* **37**, L21811, http://dx.doi.org/10.1029/2010GL045002.

Index

A

B

C

D

E

F

G

H

I

J

K

L

M

N

O

P

Q

R

S

T

U

V

W

Z

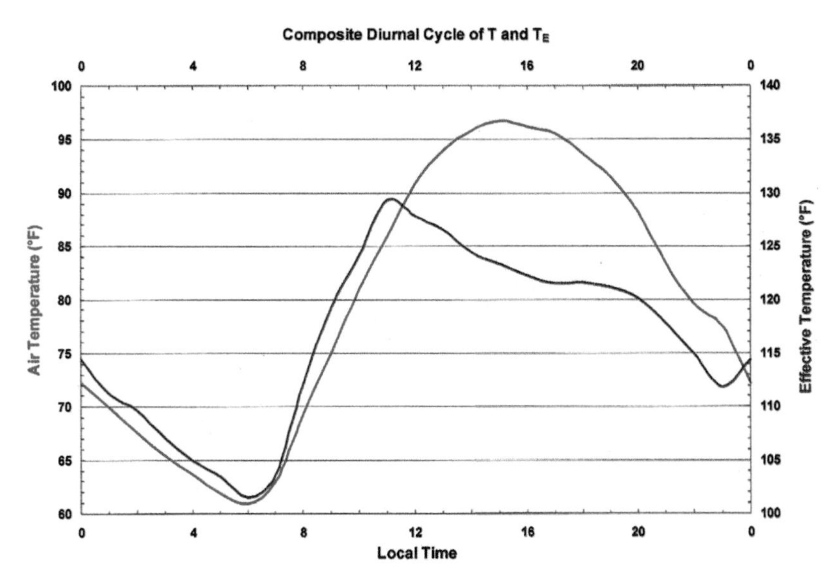

FIGURE 2.3

A daily composite of air temperature (red line) and effective temperature (blue line) for Fort Collins, Colorado, USA. The composite is created by averaging hourly data during the five days with highest air temperature in each of the three years considered in this section – fifteen days total. This shows the pattern of heating and cooling on the station's extreme hottest days. Note how the effective temperature peaks approximately four hours before the air temperature peaks. Typically, the hottest days are characterized by exceptionally low relative humidity in the late afternoon, which explains the premature drop in effective temperature (from Pielke et al. 2006a).

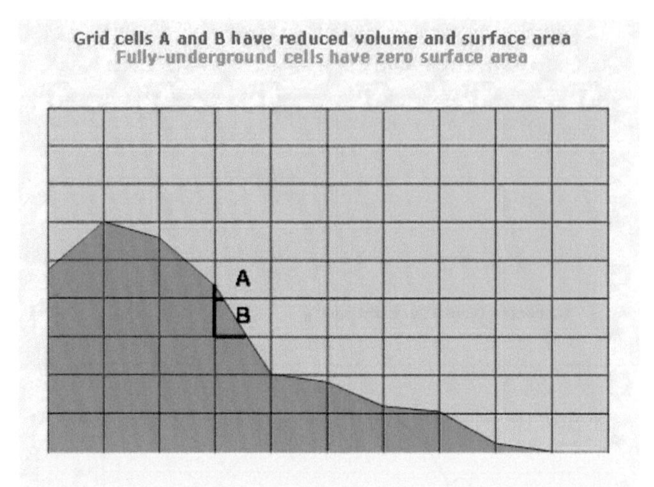
Grid cells A and B have reduced volume and surface area
Fully-underground cells have zero surface area
A
B

FIGURE 9.9

Schematic of cut-cell representation of topography. Bold lines along the faces of cells A and B represent the fractional closures of those face areas, while the green areas represent the fractional reduction in cell volumes. Courtesy of Robert Walko.

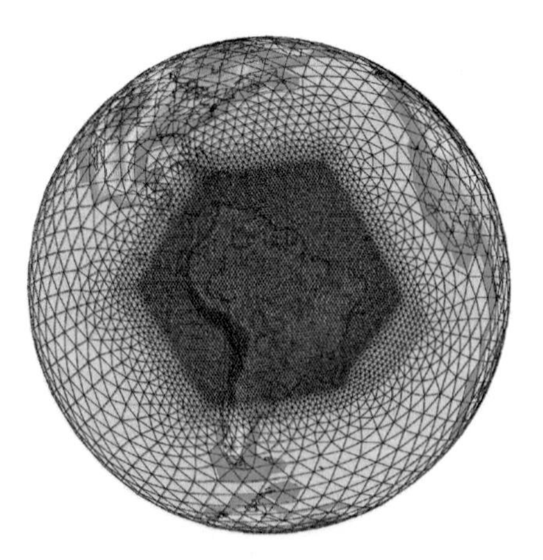

FIGURE 10.6

A global grid coverage with mesoscale grid spacing in selected regions using triangulation based on an icosahedron in the OLAM model. Figure provided courtesy of Robert Walko.

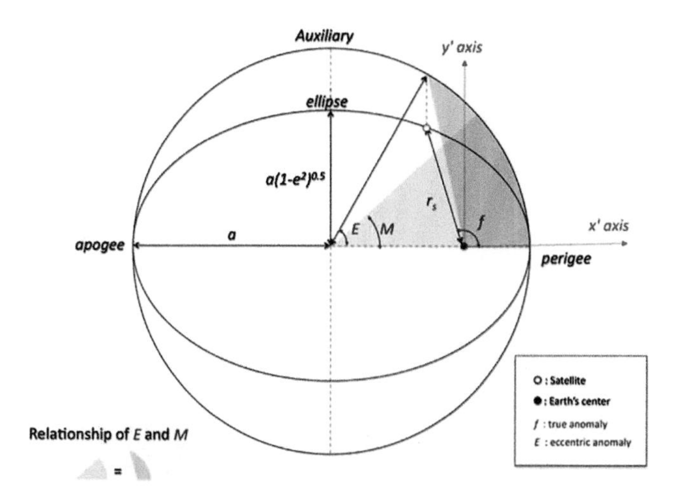

FIGURE 12.2

Relationship between eccentric anomalies (E) and mean anomalies (M).

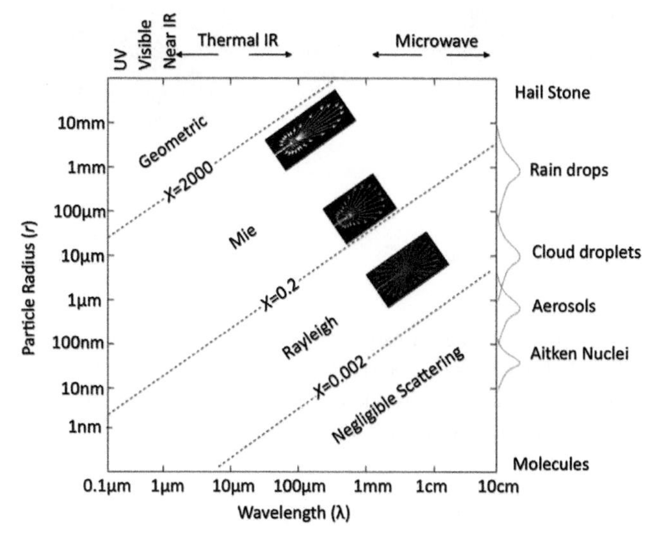

FIGURE 12.3

Non-dimensional shape parameter (X) as a function of particle radius (r) and wavelength (λ) adapted from Petty (2006). The Rayleigh regime corresponds to X ranging from 0.002 to 0.2, where scattering is relatively small and isotropic. The Mie regime corresponds to X ranging from 0.2 to 2000, where scattering is stronger and peaked more in the forward direction. The geometric regime corresponds to X greater than 2000, where most of the scattering is directed forward.

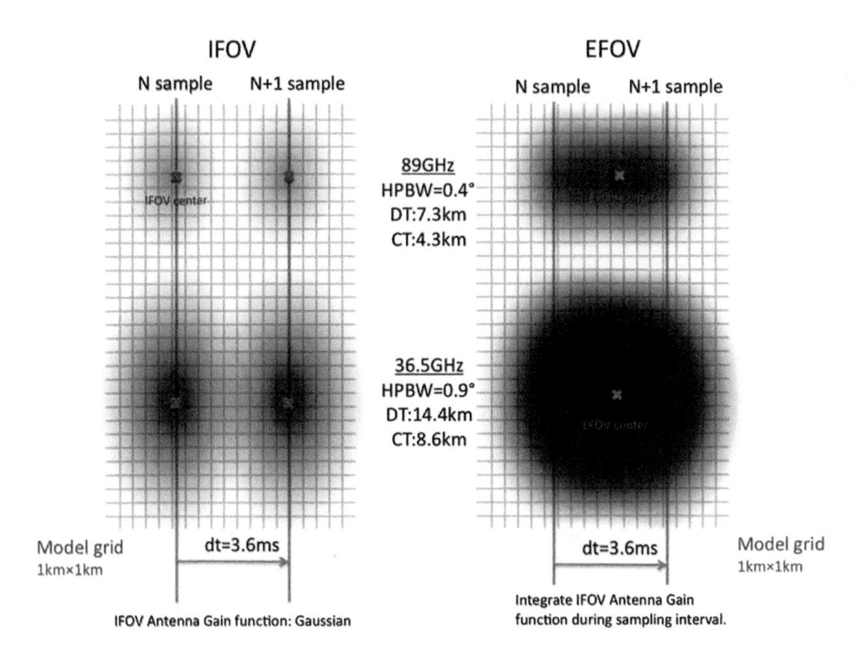

FIGURE 12.4

Scheme of satellite instrument antenna gain patterns over the mesoscale model grid (1 km of grid spacing). Example I is set for half-power beam width (HPBW) = 0.4° and 0.9°, which is approximately 7.3×4.3 km^2 and 14.4×4.3 km^2 of the footprint size for a given satellite altitude (6776.14 km) and off-nadir angles (48.5°) in a conically rotating sensor.

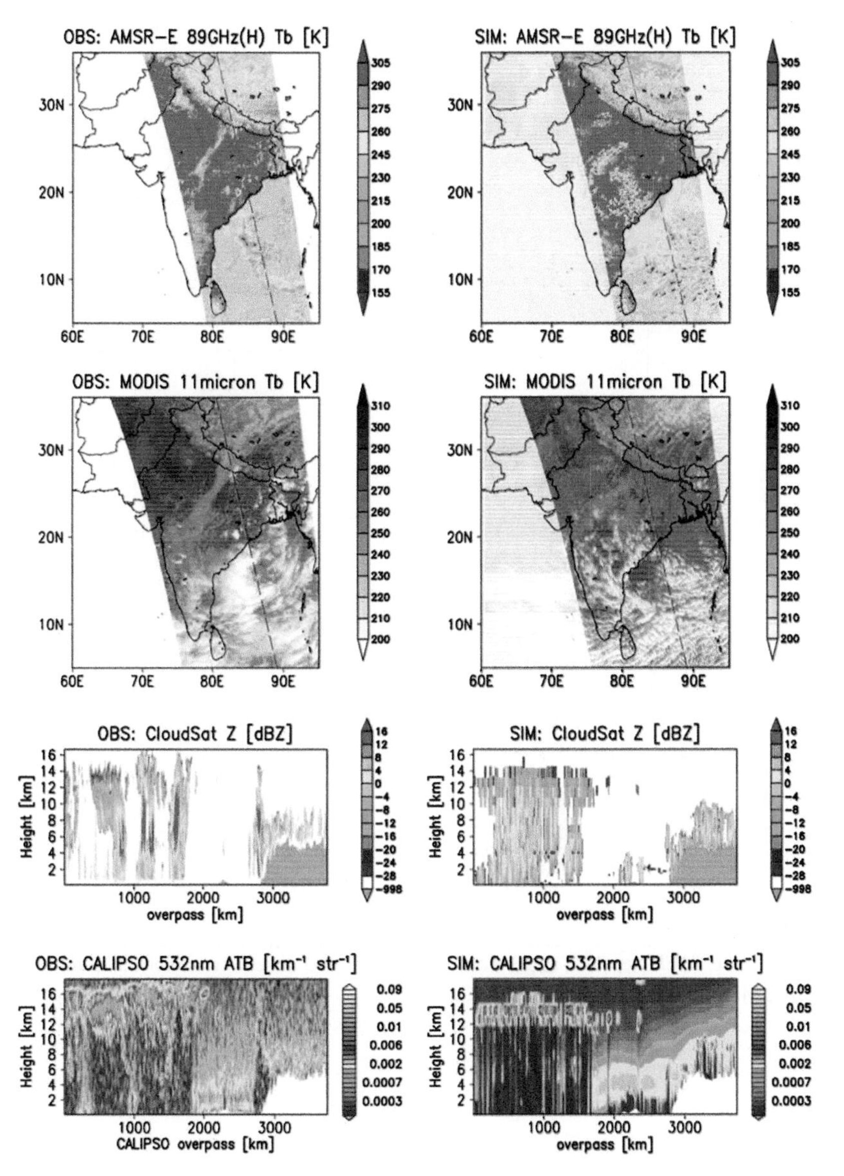

FIGURE 12.5

Observed (left panel) and simulated (right panel) satellite Level1B signals over India.

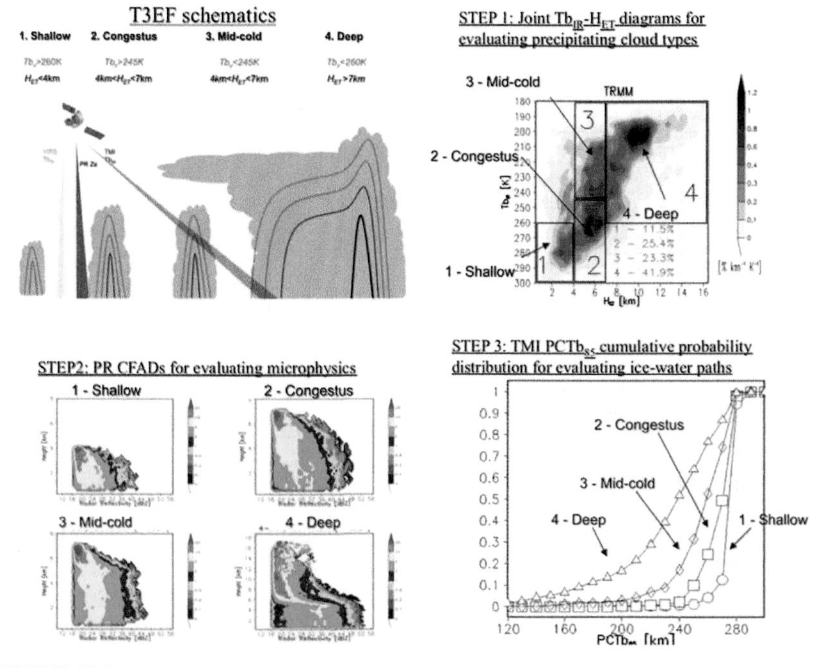

FIGURE 12.6

Schematic of the T3EF step 1, 2, and 3 from the TRMM L1B observation over the South China Sea during May–June in 1998.

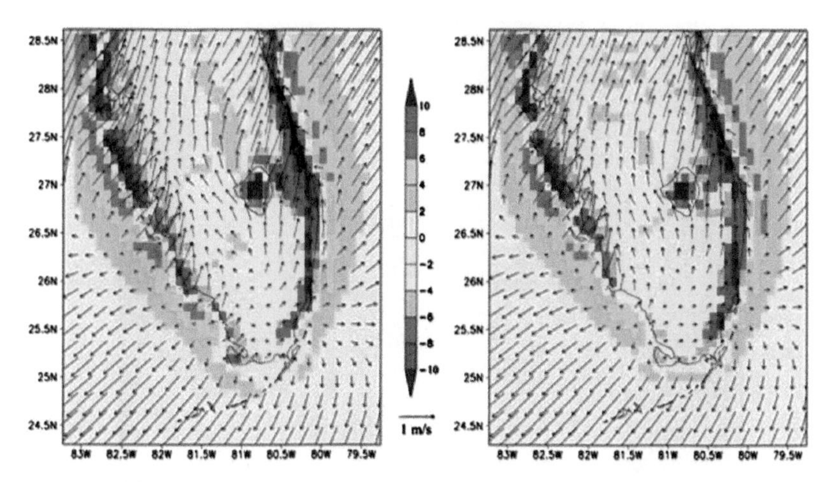

FIGURE 13.5

Two-month average of the 1600 UTC 10 m horizontal wind (vectors) and derived divergence field (color shaded; $10^5\ s^{-1}$) from the simulations of Jul–Aug 1989 with (left) pre-1900 land cover and (right) 1993 land use (from Marshall et al. 2004).

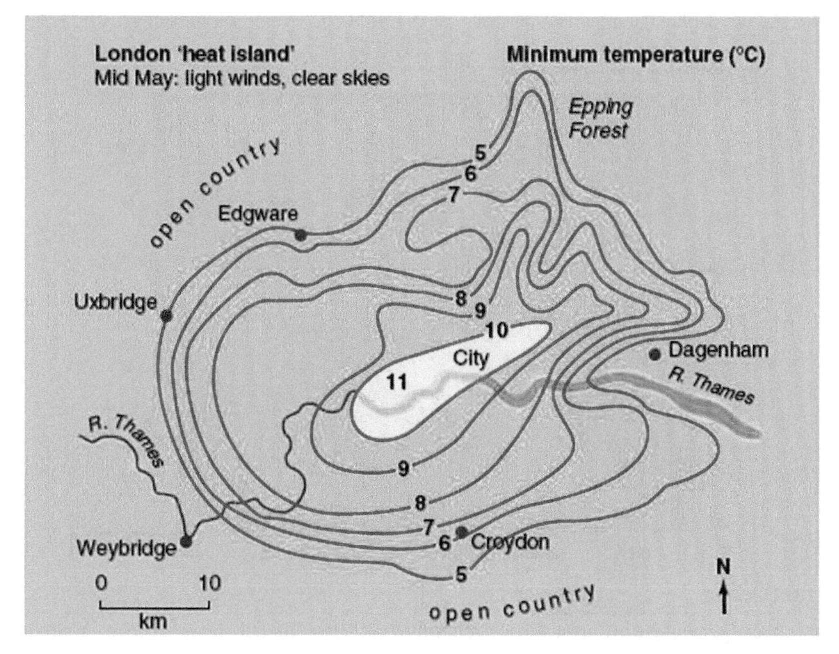

FIGURE 13.15

Illustrates the higher surface temperatures of a metropolitan area (in this case, London, U.K.) relative to the rural area for a calm night in mid-May (from the University of Wisconsin, CIMSS, 2012; available online at: http://cimss.ssec.wisc.edu/climatechange/globalCC/lesson7/UHI2.html).

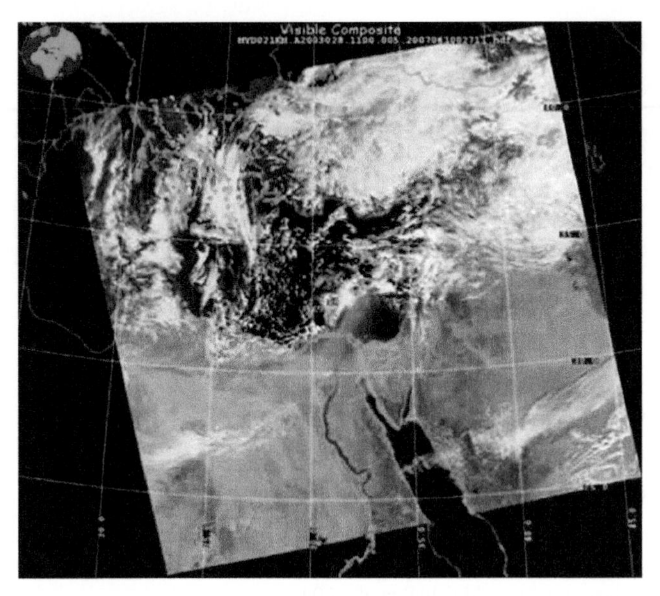

FIGURE 13.23

MODIS-Aqua visible channel on 28 January 2003 1100 UTC (from Solomos et al. 2011).

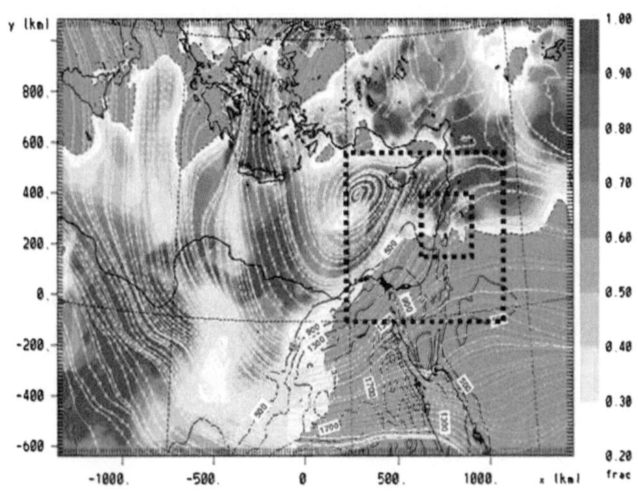

FIGURE 13.24

Cloud cover percentage (grayscale), streamlines at first model layer (green contours), dust – load (red contours in mg m^{-2}). The dashed rectangulars indicate the location of the nested grids (from Solomos et al. 2011).

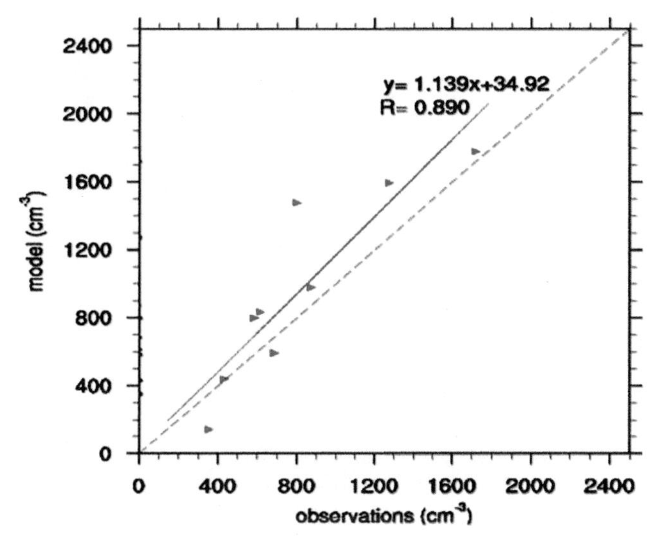

FIGURE 13.25

Comparison of aircraft measurements of natural particles with modeled dust and salt concentrations inside the dust layer (below 2 km). The red line indicates the linear regression line while the dotted line indicates the $y = x$ line (from Solomos et al. 2011).

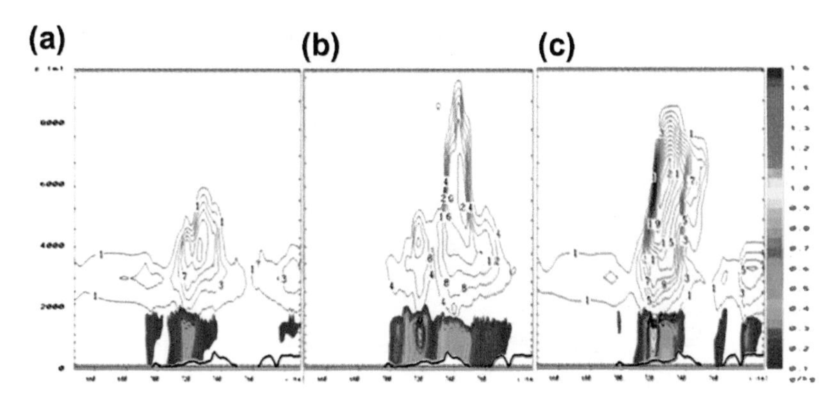

FIGURE 13.26

West to east cross-section of rain mixing ratio (color palette in g kg^{-1}) and ice mixing ratio (black line contours in g kg^{-1}) at the time of highest cloud top over Haifa. a) 0900 UTC 29 January 2003 assuming 5% hygroscopic dust (EXP1). b) 1000 UTC 29 January 2003 assuming 20% hygroscopic dust (EXP2). c) 0900 UTC 29 January 2003 assuming 5% hygroscopic dust and INx10 (EXP3).

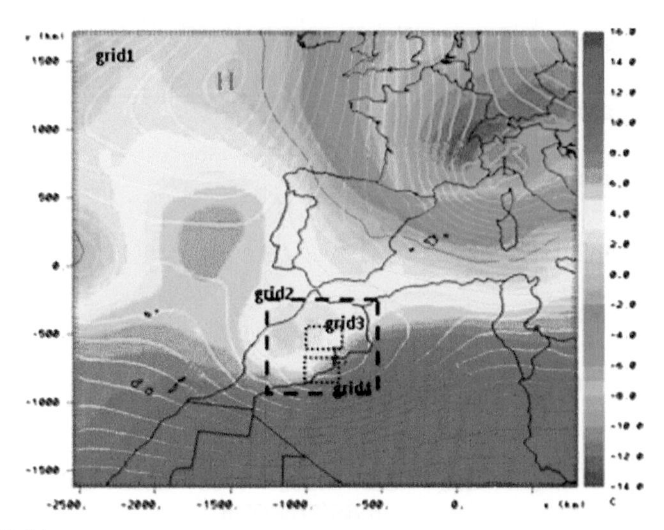

FIGURE 13.27

Geopotential height (white contour lines every 10 gpm) and temperature at 700 hPa (color palette in °C) – 1100 UTC on 31 May 2006. The dashed rectangulars indicate the locations of the nested grids (from Solomos et al. 2012).

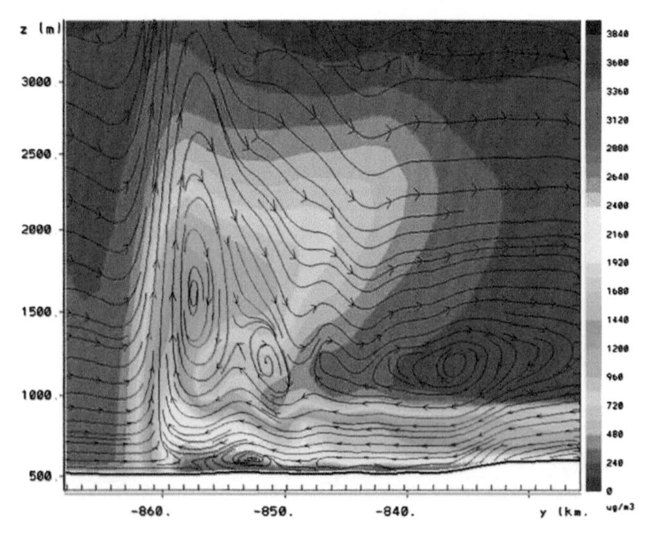

FIGURE 13.28

Dust concentration (color scale in μg m^{-3}) and streamlines at 1850 UTC for a reference frame relative to the propagating speed. The location of the cross section is at 4.4°E and from 31.50°N to 29.80°N. The direction of the motion is from north to south as illustrated with the red arrow on top of the figure (from Solomos et al. 2012).

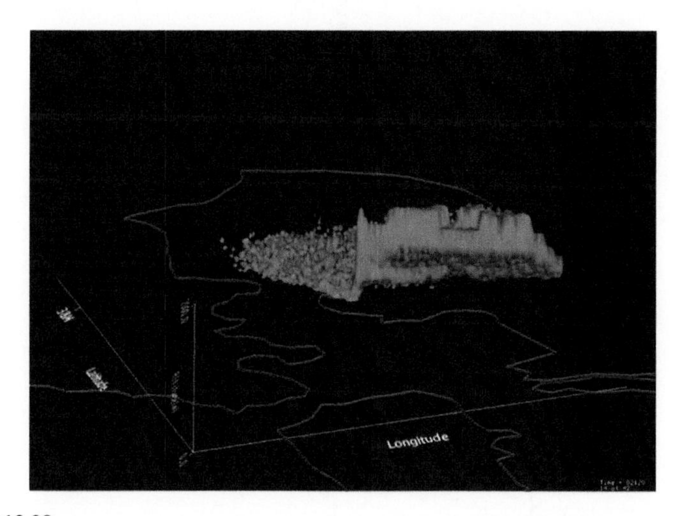

FIGURE 13.29

Simulation of the plume rise due to an industrial accident at the area of San Francisco Bay (26 August 2012).

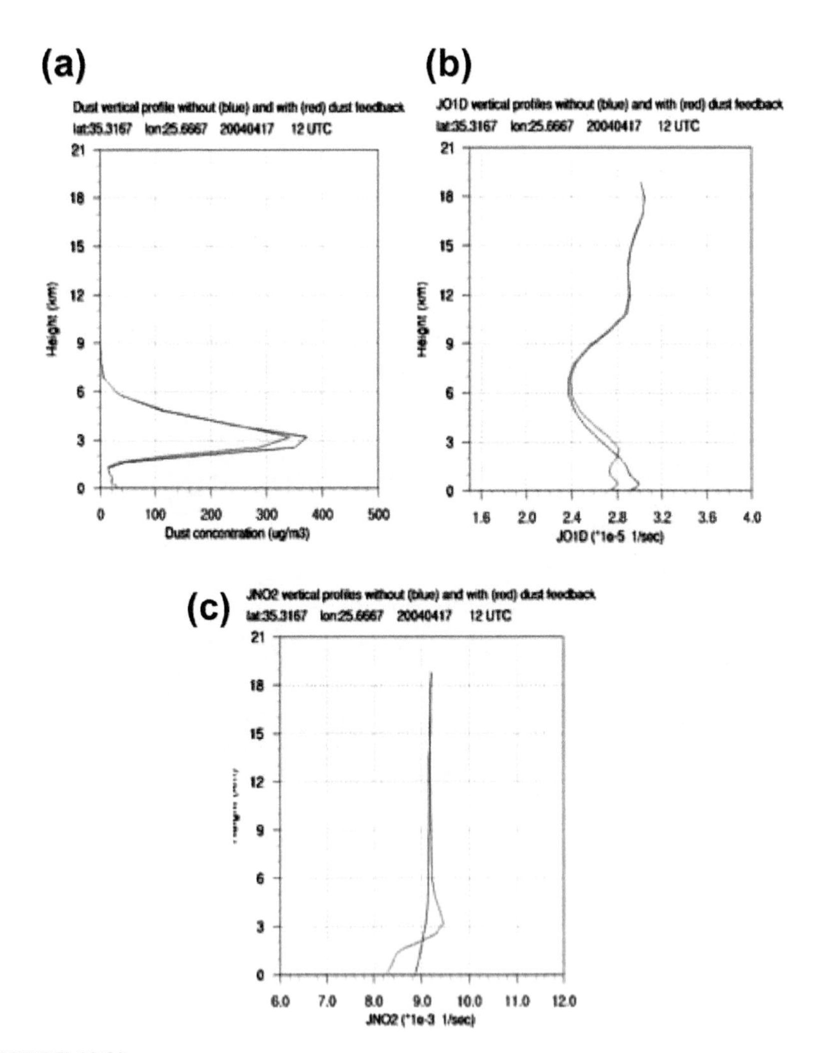

FIGURE 13.30

Vertical profiles of a) dust concentration (g m^{-3}) and photolysis rates of b) JO1D (10^{-5} sec^{-1}), and c) JNO2 (10^{-3} sec^{-1}), with dust impact on radiation (red) and without dust impact on radiation (blue) on 17 July 1200 UTC at Finokalia station, Crete (lat = 35.3°, lon = 25.7°).

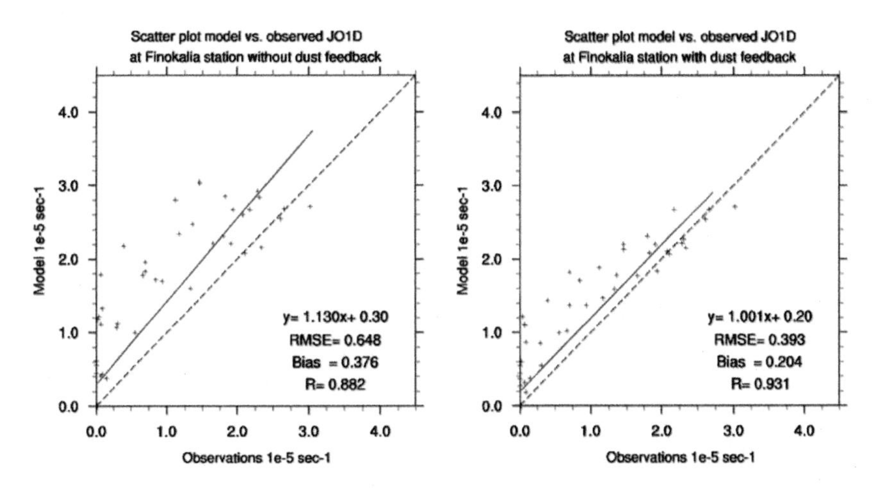

FIGURE 13.31

Scatter plots of modeled versus observed photolysis rates of a) JO1D ($10^{-5}sec^{-1}$) without dust impact on solar radiation, b) JO1D ($10^{-5}sec^{-1}$) with dust impact on solar radiation, at Finokalia station, Crete, on 17 and 18 April 2004. The black dashed line represents the $x = y$ line while the red line represents the regression line of each plot.

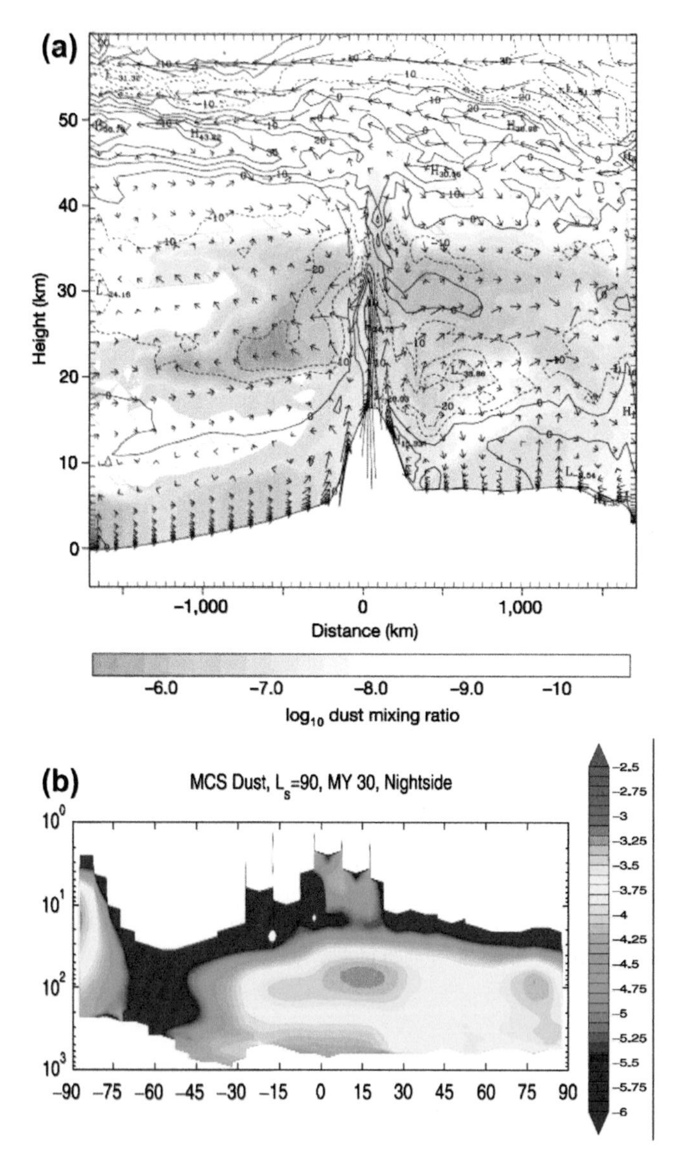

FIGURE 13.35

(a) A vertical, east–west cross-section from grid two simulated by the Mars Regional Atmospheric Modelling System (MRAMS), cutting through the centre of Arsia Mons. (b) MCS image of the dust profile. From Rafkin et al. (2002).

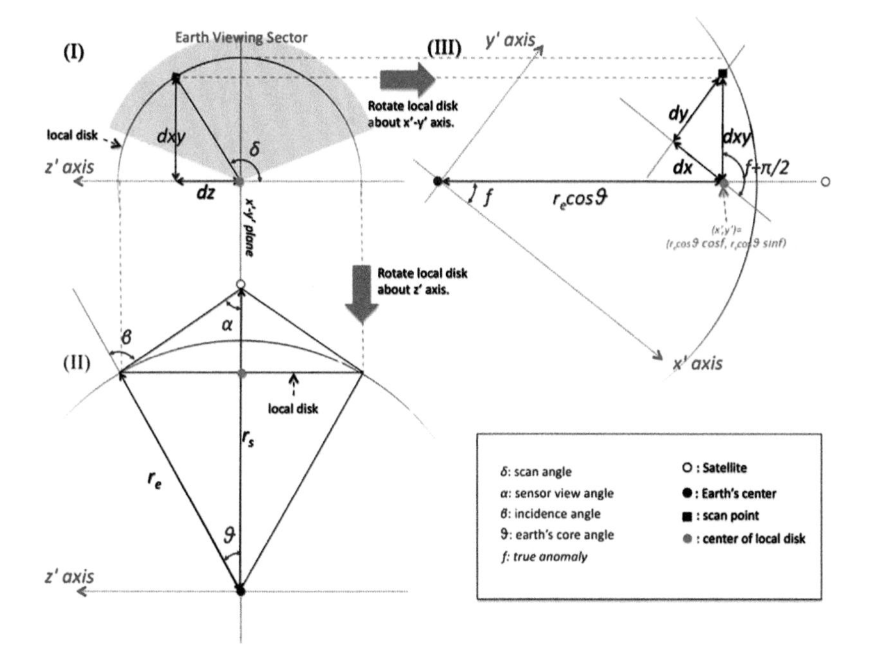

FIGURE C.1

Geometry of scanning positions on the Earth's ellipsoid in a pseudo three-dimensional Cartesian coordinate (x', y', z').

Printed and bound by CPI Group (UK) Ltd, Croydon, CR0 4YY

08/05/2025

01864836-0001